DIAGNOSIS AND MANAGEMENT IN DEMENTIA

DIAGNOSIS AND MANAGEMENT IN DEMENTIA

The Neuroscience of Dementia

VOLUME 1

Edited by

COLIN R. MARTIN
Institute for Clinical and Applied Health Research (ICAHR)
University of Hull
Hull, United Kingdom

VICTOR R. PREEDY
King's College London
London
United Kingdom

ELSEVIER

ACADEMIC PRESS
An imprint of Elsevier

Academic Press is an imprint of Elsevier
125 London Wall, London EC2Y 5AS, United Kingdom
525 B Street, Suite 1650, San Diego, CA 92101, United States
50 Hampshire Street, 5th Floor, Cambridge, MA 02139, United States
The Boulevard, Langford Lane, Kidlington, Oxford OX5 1GB, United Kingdom

Notices
Knowledge and best practice in this field are constantly changing. As new research and experience
broaden our understanding, changes in research methods, professional practices, or medical
treatment may become necessary.

Practitioners and researchers must always rely on their own experience and knowledge in evaluating
and using any information, methods, compounds, or experiments described herein. In using such
information or methods they should be mindful of their own safety and the safety of others, including
parties for whom they have a professional responsibility.

To the fullest extent of the law, neither the Publisher nor the authors, contributors, or editors, assume
any liability for any injury and/or damage to persons or property as a matter of products liability,
negligence or otherwise, or from any use or operation of any methods, products, instructions, or ideas
contained in the material herein.

Library of Congress Cataloging-in-Publication Data
A catalog record for this book is available from the Library of Congress

British Library Cataloguing-in-Publication Data
A catalogue record for this book is available from the British Library

ISBN: 978-0-12-816043-5 (Set)
ISBN: 978-0-12-815854-8 (Volume 1)
ISBN: 978-0-12-815868-5 (Volume 2)

For information on all Academic Press publications visit our
website at https://www.elsevier.com/books-and-journals

Publisher: Nikki Levy
Acquisitions Editor: Natalie Farra
Editorial Project Manager: Timothy Bennett
Production Project Manager: Paul Prasad Chandramohan
Cover Designer: Matthew Limbert

Typeset by TNQ Technologies

Colin R. Martin—*I would like to dedicate this book to my beautiful daughter Dr. Caragh Brien, of whom I am so proud.*

Contents

Part I: Dementia: introductory chapters and setting the scene

Part II: Biomarkers, psychometric instruments and diagnosis

Part III: Pharmacological treatments for dementia

Part IV: Non-pharmacological treatments and procedures

Contributors

Athanasios Alexiou
Novel Global Community Educational Foundation, Hebersham, NSW, Australia; AFNP Med Austria, Wien, Austria

Francesco Amenta
Clinical Research, Telemedicine and Telepharmacy Centre, School of Medicinal and Health Products Sciences, University of Camerino, Camerino, Italy

Nicola Amoroso
Dipartimento Interateneo di Fisica "M. Merlin", Università degli studi di Bari "A. Moro", Istituto Nazionale di Fisica Nucleare - Sez. di Bari, Bari, Italy

Jessica L. Andrews
Office of the DVC Research, The University of Sydney, Sydney, NSW, Australia

Francesco Arba
Stroke Unit, Azienda Ospedaliero-Universitaria Careggi, Florence, Italy

Ubaldo Armato
Histology & Embryology Unit, School of Medicine, University of Verona, Verona, Italy

Ghulam Md Ashraf
King Fahd Medical Research Center, King Abdulaziz University, Jeddah, Saudi Arabia

Lapo Attardo
Music Therapist at ASP Istituti Milanesi Martinitt e Stelline e Pio Albergo Trivulzio, Milan, Italy

Thiago Junqueira Avelino-Silva
Division of Geriatrics, Department of Internal Medicine, University of Sao Paulo Medical School, Sao Paulo, Brazil

Annelise Ayres
Postgraduate Program in Health Sciences, Universidade Federal de Ciências da Saúde de Porto Alegre, Porto Alegre, Rio Grande do Sul, Brazil

Giacinto Bagetta
Section of Preclinical and Translational Pharmacology, Pharmacotechnological Documentation and Transfer Unit (PDTU), Department of Pharmacy, Health Science and Nutrition, University of Calabria, Rende, Italy

Marta Balietti
Center of Neurobiology of Aging, IRCCS INRCA, Ancona, Italy

Gopi Battineni
Telemedicine and Telepharmacy Center, School of Pharmacological Sciences and Health Products, University of Camerino, Camerino, Italy

Siamak Beheshti
Department of Plant and Animal Biology, Faculty of Biological Science and Technology, University of Isfahan, Isfahan, Iran

Lazaros Belbasis
Department of Hygiene and Epidemiology, University of Ioannina Medical School, Ioannina, Greece

Vanesa Bellou
Department of Hygiene and Epidemiology, University of Ioannina Medical School, Ioannina, Greece

Leandro Bueno Bergantin
Department of Pharmacology, Universidade Federal de São Paulo (UNIFESP), São Paulo, SP, Brazil

Waleska Berríos
Department of Neurology, Italian Hospital of Buenos Aires, Buenos Aires, Argentina

Virginia Boccardi
Department of Medicine, Institute of Gerontology and Geriatrics, University of Perugia, Perugia, Italy

Andrea Bosco
Department of Education Science, Psychology, Communication, Università degli Studi di Bari "Aldo Moro", Bari, Italy

Robert Briggs
Centre for Ageing, Neurosciences and the Humanities, Tallaght Hospital, Dublin, Ireland

Johannes Burtscher
Laboratory of Molecular and Chemical Biology of Neurodegeneration, École Polytechnique Fédérale de Lausanne (EPFL), Lausanne, Switzerland

Martin Burtscher
Department of Sport Science, University of Innsbruck, Innsbruck, Austria; Austrian Society for Alpine and High-Altitude Medicine, Innsbruck, Austria

Alessandro O. Caffò
Department of Education Science, Psychology, Communication, Università degli Studi di Bari "Aldo Moro", Bari, Italy

Nohelia Cajas-Salazar
Department of Biology, Research Group Genetic Toxicology and Cytogenetics, Faculty of Natural Sciences and Education, University of Cauca, Popayán, Cauca, Colombia

Michele L. Callisaya
Menzies Institute for Medical Research, University of Tasmania, Hobart, TAS, Australia; Peninsula Clinical School, Central Clinical School, Monash University, Melbourne, Victoria, Australia

Afonso Caricati-Neto
Head of Laboratory of Autonomic and Cardiovascular Pharmacology, Department of Pharmacology, Universidade Federal de São Paulo (UNIFESP), São Paulo, SP, Brazil

Cecilia Carlesi
Neurology Unit, Versilia Hospital, Camaiore (Lucca), Italy

Willian Orlando Castillo-Ordoñez
Department of Biology, Research Group Genetic Toxicology and Cytogenetics, Faculty of Natural Sciences and Education, University of Cauca, Popayán, Cauca, Colombia

Victor T.T. Chan
Department of Ophthalmology and Visual Sciences, The Chinese University of Hong Kong, Shatin, Hong Kong

Stylianos Chatzichronis
AFNP Med Austria, Wien, Austria; National and Kapodistrian University of Athens, Department of Informatics and Telecommunications, Bioinformatics Program, Zografou, Greece

Carol Y. Cheung
Department of Ophthalmology and Visual Sciences, The Chinese University of Hong Kong, Shatin, Hong Kong

Anna M. Chiarini
Histology & Embryology Unit, School of Medicine, University of Verona, Verona, Italy

Virginia Cipollini
NESMOS Department, Faculty of Medicine and Psychology, Sapienza University of Rome, Rome, Italy

Gabriele Cipriani
Neurology Unit, Versilia Hospital, Camaiore (Lucca), Italy

Sylvie Claeysen
IGF, Univ Montpellier, CNRS, INSERM, Montpellier, France

Paul Claffey
Centre for Ageing, Neurosciences and the Humanities, Tallaght Hospital, Dublin, Ireland

Roger Clarnette
Medical School, University of Western Australia, Crawley, WA, Australia

Maria Tiziana Corasaniti
Department of Health Sciences, University "Magna Græcia" of Catanzaro, Catanzaro, Italy

Elise Cornelis
Department Gerontology, Vrije Universiteit Brussel, Brussels, Belgium

Ilaria Dal Prà
Histology & Embryology Unit, School of Medicine, University of Verona, Verona, Italy

Sultan Darvesh
Department of Medical Neuroscience, Dalhousie University, Halifax, NS, Canada; Department of Medicine (Neurology), Dalhousie University, Halifax, NS, Canada

Drew R. DeBay
Department of Medical Neuroscience, Dalhousie University, Halifax, NS, Canada; Department of Medicine (Neurology), Dalhousie University, Halifax, NS, Canada

Paolo Del Dotto
Neurology Unit, Versilia Hospital, Camaiore (Lucca), Italy

Jacques De Reuck
Degenerative & vascular cognitive disorders. Lille, France

Patricia De Vriendt
Department Gerontology, Vrije Universiteit Brussel, Brussels, Belgium

Thanuja Dharmadasa
Brain and Mind Centre, The University of Sydney, Sydney, NSW, Australia

Kathryn Dovey
Faculty of Health Sciences, University of Sydney, Lidcombe, NSW, Australia

H. Fred Downey
Department of Physiology and Anatomy, University of North Texas Health Science Center, Fort Worth, TX, United States; Scientific Educational Center for Biomedical Technology, South Ural State University, Chelyabinsk, Russia

Adam H. Dyer
Department of Medical Gerontology, Trinity College Dublin, Dublin, Ireland

Claudio Eccher
Villa Bianca Hospital, Surgery Unit, Trento, Italy

Kristina Endres
Department of Psychiatry and Psychotherapy, Medical Center, Johannes Gutenberg-University of Mainz, Mainz, Germany

Evangelos Evangelou
Department of Hygiene and Epidemiology, University of Ioannina Medical School, Ioannina, Greece; Department of Epidemiology and Biostatistics, School of Public Health, Imperial College London, London, United Kingdom

Francesca Fernandez
Faculty of Health Sciences, School of Behavioural and Health Sciences, Australian Catholic University, Nudgee, QLD, Australia

Alycia Fong Yan
Faculty of Health Sciences, University of Sydney, Lidcombe, NSW, Australia

Emily Frith
Physical Activity Epidemiology Laboratory, Exercise & Memory Laboratory, Department of Health, Exercise Science and Recreation Management, The University of Mississippi, University, MS, United States

Flavia Barreto Garcez
Division of Geriatrics, Department of Internal Medicine, University of Sao Paulo Medical School, Sao Paulo, Brazil

Patrizia Giannoni
EA7352 CHROME, University of Nîmes, Nîmes, France

Franco Giubilei
NESMOS Department, Faculty of Medicine and Psychology, Sapienza University of Rome, Rome, Italy

Oleg S. Glazachev
Department of Normal Physiology, I.M.Sechenov First Moscow State Medical University (Sechenov University), Moscow, Russia

B.E. Glynn-Servedio
Clinical Pharmacy Specialist—Ambulatory Care, Durham VA Health Care System, Raleigh 1 Community-Based Outpatient Clinic, Raleigh, NC, United States

Angel Golimstok
Department of Neurology, Italian Hospital of Buenos Aires, Buenos Aires, Argentina

Ellen Gorus
Department Gerontology, Vrije Universiteit Brussel, Brussels, Belgium

Rebecca F. Gottesman
Departments of Neurology and Epidemiology, Johns Hopkins School of Medicine, Baltimore, MD, United States

Shizuo Hatashita
Department of Neurology, Shonan Atsugi Hospital, Atsugi, Japan

Bernhard Holle
German Center for Neurodegenerative Diseases e.V. (DZNE), DZNE site Witten, Witten, Germany

Mahboobeh Housseini
Centre for Healthy Brain Ageing (CHeBA), School of Psychiatry, University of New South Wales, Sydney, New South Wales, Australia

William Huynh
Brain and Mind Centre, The University of Sydney, Sydney, NSW, Australia; Prince of Wales Hospital, Randwick, NSW, Australia

Elena Caldarazzo Ienco
Neurology Unit, Versilia Hospital, Camaiore (Lucca), Italy

Caroline Ismeurt
IGF, Univ Montpellier, CNRS, INSERM, Montpellier, France

Oshadi Jayakody
Menzies Institute for Medical Research, University of Tasmania, Hobart, TAS, Australia

Pabiththa Kamalraj
Health and Rehabilitation Sciences, Western University, London, ON, Canada

Karin Wolf-Ostermann
University of Bremen, Department 11, Human and Health Sciences, Bremen, Germany

Kazunori Kawaguchi
School of Health Sciences, Fujita Health University, Toyoake, Aichi, Japan

Sean P. Kennelly
Centre for Ageing, Neurosciences and the Humanities, Tallaght Hospital, Dublin, Ireland

Matthew C. Kiernan
Bushell Chair of Neurology Department of Neurology Royal Prince Alfred Hospital, Sydney, NSW, Australia

Anna E. King
Wicking Dementia Research and Education Centre, University of Tasmania, Hobart, TAS, Australia

Nobuya Kitaguchi
School of Health Sciences, Fujita Health University, Toyoake, Aichi, Japan

Shinsuke Kito
Department of Psychiatry, Jikei University School of Medicine, Minato-ku, Tokyo, Japan

Franziska Laporte Uribe
German Center for Neurodegenerative Diseases e.V. (DZNE), DZNE site Witten, Witten, Germany

Yue Liu
Centre for Healthy Brain Ageing (CHeBA), School of Psychiatry, University of New South Wales, Sydney, New South Wales, Australia

Antonella Lopez
Department of Education Science, Psychology, Communication, Università degli Studi di Bari "Aldo Moro", Bari, Italy

Paul D. Loprinzi
The University of Mississippi, Physical Activity Epidemiology Laboratory, Exercise & Memory Laboratory, Department of Health, Exercise Science, and Recreation Management, 229 Turner Center, University, MS, United States

Lee-Fay Low
Faculty of Health Sciences, University of Sydney, Lidcombe, NSW, Australia

Robert T. Mallet
Department of Physiology and Anatomy, University of North Texas Health Science Center, Fort Worth, TX, United States

Eugenia B. Manukhina
Laboratory for Regulatory Mechanisms of Stress and Adaptation, Institute of General Pathology and Pathophysiology, Moscow, Russia; Laboratory for Molecular Mechanisms of Stress, South Ural State University, Chelyabinsk, Russia; Department of Physiology and Anatomy, University of North Texas Health Science Center, Fort Worth, TX, United States

Gabriella Marucci
School of Medicinal and Health Sciences Products, University of Camerino, Camerino, Italy

Jordi A. Matias-Guiu
Department of Neurology, Hospital Clínico San Carlos, San Carlos Institute for Health Research (IdISSC), Universidad Complutense, Madrid, Spain

Wong Matthew Wai Kin
Centre for Healthy Brain Ageing (CHeBA), School of Psychiatry, University of New South Wales, Sydney, New South Wales, Australia

Patrizia Mecocci
Department of Medicine, Institute of Gerontology and Geriatrics, University of Perugia, Perugia, Italy

D. William Molloy
Centre for Gerontology and Rehabilitation, St. Finbarr's Hospital, University College Cork, Cork City, Ireland

Domenico Monteleone
DG Animal Health and Veterinary Drugs, Ministry of Health, Rome, Italy

Luigi Antonio Morrone
Section of Preclinical and Translational Pharmacology, Pharmacotechnological Documentation and Transfer Unit (PDTU), Department of Pharmacy, Health Science and Nutrition, University of Calabria, Rende, Italy

Michele Moruzzi
School of Medicinal and Health Sciences Products, University of Camerino, Camerino, Italy

Thomas Müller
Department of Neurology, St. Joseph Hospital Berlin-Weissensee, Berlin, Germany

Braidy Nady
Centre for Healthy Brain Ageing (CHeBA), School of Psychiatry, University of New South Wales, Sydney, New South Wales, Australia

Akihiko Nunomura
Department of Psychiatry, Jikei University School of Medicine, Minato-ku, Tokyo, Japan

Angelo Nuti
Neurology Unit, Versilia Hospital, Camaiore (Lucca), Italy

Rónán O'Caoimh
Clinical Sciences Institute, National University of Ireland Galway, Galway City, Ireland; Department of Geriatric Medicine, Mercy University Hospital, Cork City, Ireland

Paul O'Halloran
School of Psychology and Public Health, La Trobe University, Melbourne, VIC, Australia

Marina Padovani
School of Speech-Language Pathology and Audiology, Santa Casa de São Paulo, School of Medical Sciences, São Paulo, Brazil

Graziano Pallotta
Clinical Research, Telemedicine and Telepharmacy Centre, School of Medicinal and Health Products Sciences, University of Camerino, Camerino, Italy

Lucia Paolacci
Department of Medicine, Institute of Gerontology and Geriatrics, University of Perugia, Perugia, Italy

Helen Parker
Faculty of Health Sciences, University of Sydney, Lidcombe, NSW, Australia

Sachdev Perminder Singh
Centre for Healthy Brain Ageing (CHeBA), School of Psychiatry, University of New South Wales, Sydney, New South Wales, Australia; Neuropsychiatric Institute, Euroa Centre, Prince of Wales Hospital, Sydney, New South Wales, Australia

Couratier Philippe
Service Neurologie, Centre de référence SLA et autres maladies du neurone moteur, CHU Limoges, Limoges, France

Anne Poljak
Centre for Healthy Brain Ageing (CHeBA), School of Psychiatry, University of New South Wales, Sydney, New South Wales, Australia; Bioanalytical Mass Spectrometry Facility (BMSF), University of New South Wales, Sydney, New South Wales, Australia; School of Medical Sciences, University of New South Wales, Sydney, New South Wales, Australia

Alfredo Raglio
Music Therapy Research Laboratory, Istituti Clinici Scientifici Maugeri, Pavia, Italy

Innocenzo Rainero
Aging Brain and Memory Clinic, Department of Neuroscience "Rita Levi Montalcini", University of Torino, Torino, Italy

Bridget Regan
Lincoln Centre for Research on Ageing, Australian Institute for Primary Care & Ageing, School of Nursing and Midwifery, La Trobe University, Melbourne, VIC, Australia

Larry D. Reid
Department of Cognitive Science, Rensselaer Polytechnic Institute, Troy, NY, United States

Sven Reinhardt
Department of Psychiatry and Psychotherapy, Medical Center, Johannes Gutenberg-University of Mainz, Mainz, Germany

Jochen René Thyrian
German Center for Neurodegenerative Diseases e.V. (DZNE), DZNE site Rostock/Greifswald, Greifswald, Germany

Valentina Rinnoci
Stroke Unit, Azienda Ospedaliero-Universitaria Careggi, Florence, Italy; IRCCS Don Gnocchi Fundation, Florence, Italy

Sergio del Río-Sancho
Instituto de Ciencias Biomédicas, Departamento de Farmacia, Facultad de Ciencias de la Salud, Universidad CEU Cardenal Herrera, CEU Universities, Valencia, Spain

Laura Rombolà
Section of Preclinical and Translational Pharmacology, Pharmacotechnological Documentation and Transfer Unit (PDTU), Department of Pharmacy, Health Science and Nutrition, University of Calabria, Rende, Italy

Maira Rozenfeld Olchik
Department of Surgery and Orthopedics, Speech Language Pathology Course, Universidade Federal do Rio Grande do Sul, Porto Alegre, Rio Grande do Sul, Brazil

Elisa Rubino
Aging Brain and Memory Clinic, Department of Neuroscience "Rita Levi Montalcini", University of Torino, Torino, Italy

Kazuyoshi Sakai
School of Health Sciences, Fujita Health University, Toyoake, Aichi, Japan

Tsukasa Sakurada
First Department of Pharmacology, Daiichi College of Pharmaceutical Sciences, Fukuoka, Japan

Shinobu Sakurada
Department of Physiology and Anatomy, Tohoku Pharmaceutical University, Sendai, Japan

Marie Y. Savundranayagam
School of Health Studies, Western University, London, ON, Canada

Fúlvio Alexandre Scorza
Department of Neurology/Neurosurgery, Universidade Federal de São Paulo (UNIFESP), São Paulo, SP, Brazil

Damiana Scuteri
Section of Preclinical and Translational Pharmacology, Pharmacotechnological Documentation and Transfer Unit (PDTU), Department of Pharmacy, Health Science and Nutrition, University of Calabria, Rende, Italy

Tatiana V. Serebrovskaya
Department of Hypoxic States, Bogomoletz Institute of Physiology, Kiev, Ukraine

Masahiro Shigeta
Department of Psychiatry, Jikei University School of Medicine, Minato-ku, Tokyo, Japan

Shunichiro Shinagawa
Department of Psychiatry, Jikei University School of Medicine, Minato-ku, Tokyo, Japan

Giuseppina Spano
Department of Education Science, Psychology, Communication, Università degli Studi di Bari "Aldo Moro", Bari, Italy; Department of Agro-Environmental and Territorial Sciences, Università degli Studi di Bari "Aldo Moro", Bari, Italy

Kimberley E. Stuart
Wicking Dementia Research and Education Centre, University of Tasmania, Hobart, TAS, Australia

Kenji Tagai
Department of Psychiatry, Jikei University School of Medicine, Minato-ku, Tokyo, Japan

Toshio Tamaoki
Department of Neuropsychiatry, Graduate School of Medical Science, University of Yamanashi, Chuo, Yamanashi, Japan

Dylan Z. Taylor
Department of Cognitive Science, Rensselaer Polytechnic Institute, Troy, NY, United States

Enea Traini
Telemedicine and Telepharmacy Center, School of Pharmacological Sciences and Health Products, University of Camerino, Camerino, Italy

Fernanda Troili
NESMOS Department, Faculty of Medicine and Psychology, Sapienza University of Rome, Rome, Italy

Alessandro Vacca
Aging Brain and Memory Clinic, Department of Neuroscience "Rita Levi Montalcini", University of Torino, Torino, Italy

James C. Vickers
Wicking Dementia Research and Education Centre, University of Tasmania, Hobart, TAS, Australia

Alicia A. Walf
Department of Cognitive Science, Rensselaer Polytechnic Institute, Troy, NY, United States

Keenan A. Walker
Department of Neurology, Johns Hopkins School of Medicine. Baltimore, MD, United States

Yvonne Wells
Lincoln Centre for Research on Ageing, Australian Institute for Primary Care & Ageing, School of Nursing and Midwifery, La Trobe University, Melbourne, VIC, Australia

Randall J. Woltjer
Department of Neurology, Oregon Health Science University and Portland VA Medical Center, Portland, OR, United States

Paul L. Wood
Metabolomics Unit, Associate Dean of Research, College of Veterinary Medicine, Lincoln Memorial University, Harrogate, TN, United States

Foreword

I am gratified to write the foreword to this comprehensive book on dementia. Professors Preedy and Martin's purpose is to improve and enhance the care of individuals who have been diagnosed with dementia.

Given the scope of what dementia is—a catch-all term for a constellation pathologies that impact deleteriously and chronically on brain function—understanding the spectrum of what we consider the disease and process of dementia is inevitably a complex task. Yet, as this inspiring new book reveals, we are making significant headway.

The two professors, one from the University of Hull and the other from King's College London, leading authorities themselves in applied health research, have brought together other leading authorities from around the world who specialize in dementia clinical and applied research. Their focus is on contemporary treatment, management, and research innovation from a primarily physiological perspective, whilst implicitly emphasizing that the fundamental purpose is enhancing and improving the understanding and care of those with a diagnosis of dementia.

They have not forgotten the vital role that family, friends, and other carers play in supporting the patient with dementia and that these carers are often long-term partners who are themselves very elderly, maybe with multiple pathology themselves, or offspring, no longer in the first flush of youth. Their needs are vital. Improving the outcomes for each individual patient with dementia integrates the patient, family, friends, and other carers with health and social care practitioners in a unique partnership aimed at improving care and quality of life for both patient and those close to the patient.

The impact of the dementia diagnosis is likely to be met by fear, anxiety, and trepidation and perhaps, shame. I am old enough to remember when a diagnosis of cancer was met with the same emotions and, to some extent, stigma. We have learned so much more about cancer, the care of patients with a diagnosis of cancer, and the needs of carers. This book helps move us along the road to an increasingly evidence-based treatment of dementia, a recognition of the trauma of the diagnosis. It enables improved education and support of family, carers, and the public, and, not least, practitioners and researchers.

Surprisingly, though we know much regarding the psychosocial aspects of dementia care from an integrated perspective, the underlying biological substrates and layers of dementia are less clearly understood, particularly from an integrated perspective.

This book has special resonance for me at many levels. It balances the biological aspects of disease, evidence-based treatment, the care of patients and carers, and acknowledges the complex web that enables effective care.

My own journey through nursing and health visiting led me to research the role of the health visitor with older people. It makes me smile, even now, as I remember the lady who opened the door and I asked to see her mother. "She's dead," the lady replied. I quickly discovered the lady to whom I was speaking was 90. Hale and hearty. But there were others, very ill in their fifties and sixties. Others too, desperately caring for their partners. Feeling frightened and alone. Patients with dementia who had other illnesses not diagnosed. So much to do.

I understand well the complex web that produces integrated care, as I went on to research interprofessional, interorganizational relations, as with this book, with the sole purpose of assisting with the development of care for patients and their carers. Influencing policy and advocating are as important as the "hard" sciences, social sciences, in improving care. It is exciting. Fundamental is to be caring, this is my interest in ethics, moral behavior—why are people ill-treated?

It has been a pleasure to write the foreword for this stimulating new book. Victor Preedy and Colin Martin have produced a work of considerable value to both clinicians and researchers—often the same people. The discovery of new knowledge is always exciting, and in this instance, it can help also to prevent ill treatment.

Carolyn Roberts
Lady Roberts is Pro Chancellor at the University of Hull.

Her career has straddled clinical practice, research, consultancy, and management. She has a keen interest in clinical ethics and bioethical issues. As well as experiencing the ups and downs of life, she is ever seeking to overcome disadvantage, to be holistic—as this book does, integrating all aspects of life—plus of course, always interested in the exciting search for new knowledge, new skills, and continuous improvement.

Preface

There are many different types of dementia, and the most common of these include Alzheimer disease and Lewy body, mixed and vascular dementias. Together they account for about 90% of all dementias, though there are others, such as those associated with Parkinson's, Creutzfeldt-Jakob, and Huntingdon diseases. Globally there are 50 million people living with dementia. In the United States there are 5 million people with dementia costing an annual 250 billion dollars, or more. The present trajectory suggests that by 2050 the number of people in the United States with dementia will reach 16 million. Connected with this are the unpaid carers, who currently number 15 million in the United States alone.

Whilst the day-to-day impact of dementia on the individual and family unit is known, the neuroscientific basis and different methods of diagnosis and treatment are diffuse, appearing in different scientific domains. This is addressed in *Diagnosis and Management in Dementia: The Neuroscience of Dementia*, which brings together different fields of dementia into a single-source material. The book covers a wide range of subjects that encompass and interlinks general aspects, methods of diagnosis, and treatment protocols. The book has wide coverage and includes descriptions of the different types of dementias, mortality, gait, environmental factors, biomarkers, imaging, questionnaires, cholinesterase inhibitors, calcium channel blockers, receptor antagonists, group therapies, exercise, cognitive behavioral therapy, and other areas too numerous to list here. The book has over 50 chapters and is divided into the following subsections:

[1] Dementia: Introductory Chapters and Setting the Scene
[2] Biomarkers, Psychometric Instruments, and Diagnosis
[3] Pharmacological Treatments, Other Agents and Strategies
[4] Nonpharmacological Treatments and Procedures

There are of course always difficulties in ascribing chapters to different sections and placing them in order. Some chapters are equally at home in more than one section. However, the excellent indexing system allows material to be rapidly located.

Diagnosis and Management in Dementia: The Neuroscience of Dementia bridges the multiple disciplinary and intellectual divides as each chapter has:

- *Key facts*
- *Mini-dictionary of terms*
- *Summary points*

Diagnosis and Management in Dementia: The Neuroscience of Dementia is designed for research and teaching purposes. It is suitable for neurologists, psychologists, health scientists, public health workers, doctors, pharmacologists, and research scientists.

Those working in the fields of biomarkers, psychometric instruments, imaging, diagnostics, and nondrug treatments will also find the book of interest. It is valuable as a personal reference book and also for academic libraries, as it covers the domains of neurology and health sciences. Contributions are from leading national and international experts including those from world-renowned institutions. It is suitable for undergraduates, postgraduates, lecturers, and academic professors.

The Editors

PART I

Dementia: introductory chapters and setting the scene

CHAPTER 1

Mixed dementia: a neuropathological overview

Jacques De Reuck

Degenerative & vascular cognitive disorders. Lille, France

List of abbreviations

AD Alzheimer disease
ALS amyotrophic lateral sclerosis
CAA cerebral amyloid angiopathy
CBD corticobasal degeneration
CoMBs cortical microbleeds
CoMIs cortical microinfarcts
FTLD frontotemporal lobar degeneration
FUS type fused sarcoma type
LBD Lewy body disease
MRI magnetic resonance imaging
PET positron emission tomography
PSP progressive supranuclear paralysis
TDP type TAR DNA-binding protein type
VaD vascular dementia
WMCs white matter changes

Mini-dictionary of terms

Mixed dementia severe cognitive deficiency due to a combination of different neurodegenerative and/or cerebrovascular diseases.
Neuropathological examination postmortem evaluation of the brain to confirm the clinical suspected diagnosis.
Postmortem MRI this allows the detection of additional postmortem brain lesions, in particular the distribution of small cerebrovascular lesions, and the iron content.
Neurodegenerative diseases progressive diffuse diseases affecting mainly neurons, leading to a global cerebral dysfunction.
Vascular dementia severe cognitive disturbances due to the progressive accumulation of small and large cerebrovascular lesions.

Introduction

The etiological diagnosis of mixed dementia diseases can be made with certainty only after death by an extensive examination of the brain (Jellinger & Attems, 2007). According to several clinicopathological surveys, the main clinical diagnosis is confirmed in 43%

Diagnosis and Management in Dementia
ISBN 978-0-12-815854-8, https://doi.org/10.1016/B978-0-12-815854-8.00001-X

up to 86% of cases by the neuropathological evaluation (Suemoto et al., 2017). This correlation has increased significantly from 65% up to 96% since the development of new biomarkers such as magnetic resonance imaging (MRI), positron emission tomography (PET), cerebrospinal fluid analysis, and genetic markers (Jellinger, 2009).

In the Lille Memory Clinic the clinical diagnosis of Alzheimer disease (AD) is confirmed in 94% of the cases, Lewy body disease (LBD) in 80%, vascular dementia (VaD) in 100%, and mixed AD—VaD in 93% by the neuropathological examination (Bombois et al., 2008). Early previously obtained informed consent from the patients themselves or later from the nearest family allows the autopsy for diagnostic and scientific purposes. The brain tissue samples are acquired from the Neuro-Bank of Lille University, federated to the Centre des Resources Biologiques, which acts as an institutional review board.

The standard procedure for the neuropathological diagnosis of the dementia types consists of examining samples from the primary motor cortex; the associated frontal, temporal, and parietal cortices; the primary and secondary visual cortex; the cingulate gyrus; the basal nucleus of Meynert; the amygdaloid body; the hippocampus; the basal ganglia; the mesencephalon; the pons; the medulla; and the cerebellum. The slides from paraffin-embedded sections are stained with hematoxylin—eosin, Luxol fast blue, and Prussian Perl. In addition, immunostaining for protein tau, β-amyloid, α-synuclein, prion protein, TDP-43, and ubiquitin is performed (De Reuck, 2012a). Fused sarcoma (FUS) histochemistry is carried out in the cases of frontotemporal lobar degeneration (FTLD) that are tau and TDP-43 negative (De Reuck et al., 2016d). In addition, small cerebrovascular lesions can be quantified on microscopical examination of a large complete coronal section of a cerebral hemisphere at the level of the mammillary body.

Postmortem 7.0-tesla MRI is an additional useful tool. Three to six coronal sections of a cerebral hemisphere, a sagittal section of the brain stem, and a horizontal section of the cerebellum allow an extensive evaluation of the whole brain. The brain sections, previously cleaned with formalin, are placed in a plastic box filled with salt-free water. The boxes are inserted in an issuer—receiver cylinder coil, with a 72-mm inner diameter, of a 7.0-tesla MRI Bruker BioSpin SA (Ettlingen, Germany). Three MRI sequences are used: a positioning sequence, a spin-echo T2 sequence, and a T2* sequence. The positioning sequence allows determination of the three-directional position of the brain section inside the magnet. The spin-echo T2 is used to demonstrate hyperintensities corresponding to cortical microinfarcts (CoMIs) and white matter changes (WMCs). The T2* detects cortical microbleeds (CoMBs) and iron deposition (De Reuck et al., 2011a).

MRI improves the evaluation of the degree and distribution of the cerebral atrophy and the WMCs, compared with the neuropathological examination. It also allows the detection of lesions that can be selected for histological examination. In addition, small cerebrovascular lesions, such CoMBs, CoMIs, and lacunes, can be quantified according to their location. The degree of iron load can be evaluated in the basal ganglia and the brain stem structures, but not in the cerebral cortex (De Reuck, 2016a).

Incidence of mixed dementia

The prevalence of mixed brain pathologies compared with unmixed ones, reported in community-based studies, shows an incidence estimated between 19% and 67% for AD-related pathology, between 6% and 39% for LBD pathology, between 28% and 70% for VaD pathologies, and between 13% and 46% for FTLD pathology (Rahimi & Kovacs, 2014). Our 2017 study shows that mixed dementia cases are overall older than unmixed ones and that they are mainly due to the combination of the severity of AD, LBD, and cerebrovascular pathologies related to cerebral amyloid angiopathy (CAA) or arteriosclerotic disease. Together they are responsible for more than 85% of the mixed dementia cases (De Reuck et al., 2017a). CAA severity is determined according to the CERAD criteria (Ellis et al., 1996). There are differences in incidence and age distribution of some types of mixed compared with unmixed forms of neurodegenerative diseases: LBD appears predominately as a mixed form in elderly patients, while FTLD in the adult and corticobasal degeneration (CBD) are mainly single diseases without additional pathology (Table 1.1).

Unmixed and mixed Alzheimer disease

The neuropathological diagnosis of AD is made according to the Braak & Braak (1991). Mixed AD dementia syndromes are mainly due to concomitant cerebrovascular and other neurodegenerative pathologies (Kapasi, DeCarli & Schneider, 2017). AD patients with cerebrovascular disease are older than those without this additional pathology (Toledo et al., 2013). They represent 36% of all cases of mixed dementia diseases (De Reuck et al., 2018) (Fig. 1.1). The most associated cerebrovascular lesions are those

Table 1.1 Comparison of the incidences of unmixed and mixed dementia syndromes.

Main diagnosis	Number of patients	Unmixed dementia	Mixed dementia	P value
Alzheimer disease	107	45%	55%	N.S.
Frontotemporal lobar degeneration	27	79%	21%	≤0.05
Vascular dementia	27	55%	45%	N.S.
Lewy body disease	24	23%	77%	≤0.01
Progressive supranuclear palsy	17	60%	40%	N.S.
Amyotrophic lateral sclerosis	17	60%	40%	N.S.
Corticobasal degeneration	3	100%	0%	N.S.

Frontotemporal lobar degeneration is significantly more often an unmixed entity, while Lewy body disease is mainly mixed and associated with other pathologies (percentage distribution of the different postmortem confirmed types of dementia of the memory clinic of Lille University Hospital). *N.S.*, nonsignificant.

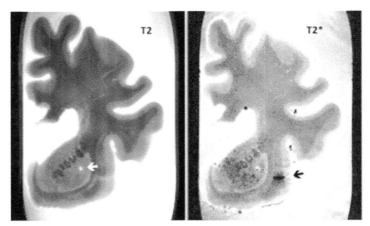

Figure 1.1 A 7.0-tesla MRI of a frontal section of a cerebral hemisphere in a brain with mixed Alzheimer disease and associated cerebrovascular pathology. A lacune in the putamen (white arrow) on the T2 sequence and a microbleed in the insular cortex (black arrow) on the T2* sequence are seen. Mild confluent white matter changes are also observed.

due to CAA (Gorelick et al., 2011). This is predominantly observed in elderly patients compared with the unmixed form occurring mainly in adults. AD−CAA has to be considered as the end stage of AD (De Reuck et al., 2016b). CoMBs are the main hallmarks of the AD−CAA diseases (De Reuck et al., 2015a). However, CoMIs are also increased in AD−CAA brains (De Reuck et al., 2014a). Also, more cortical territorial infarcts and WMCs are observed in the severe AD−CAA cases (De Reuck et al., 2011b). Lacunes due to arteriosclerotic small-vessel disease are another cause, contributive to the development of mixed dementia (Raz, Knoefel & Bhaskar, 2016).

The second most frequent mixed AD dementia is the association with LBD pathology (Jellinger & Attems, 2015). In our series they represent 24% of all mixed dementias (De Reuck et al., 2018). The prevalence of AD combined with LBD pathology remains fairly constant with increasing age (Jellinger & Attems, 2010). Additional cerebrovascular lesions are also frequently observed in this type of mixed dementia (De Reuck et al., 2016b) (Fig. 1.2). Iron deposition is only moderately increased in the caudate nucleus of both unmixed and mixed cases (De Reuck et al., 2014b).

Unmixed and mixed lewy body disease

The neuropathological diagnosis of LBD is made according to the report of the Consortium on DLB International Workshop (McKeith et al., 1996). Unmixed and mixed forms occur mainly in the oldest elderly patients, increasing still further with age (Wakisaka et al., 2003). Mixed LBD represents 16% of all mixed cases (De Reuck et al., 2018). AD pathology is the main association (Nedelska et al., 2015). A 2017 study

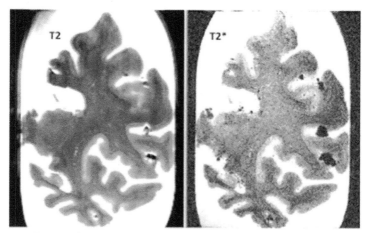

Figure 1.2 A 7.0-tesla MRI of a central hemispheric section of a brain with mixed Alzheimer disease and associated Lewy body pathology. Global cerebral atrophy and moderate diffuse white matter changes on the T2 and T2* sequences are seen.

shows that LBD concomitant with low–level AD pathology occurs in 26%, with intermediate level in 21%, and with high level in 30% (Irwin et al., 2017). Although a high incidence of very small CoMIs is observed in LBD (De Reuck et al., 2014a), other cerebrovascular lesions are overall considered rare in LBD, except for those associated with CAA (De Reuck et al., 2016b). The prevalence of CoMBs appears independent of the coexistence of AD pathology and CAA (De Reuck et al., 2015b) (Fig. 1.3).

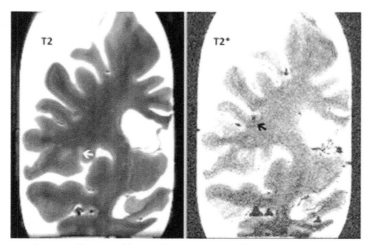

Figure 1.3 A 7.0-tesla MRI of a central hemispheric section of a brain with mixed Lewy body disease and associated Alzheimer pathology. Moderate global cortical atrophy with a small insular infarct (*white arrow*) on the T2 sequence and a cortical microbleed in the parietal cortex (*black arrow*) on the T2* sequence are seen.

CoMBs predominate in the frontal areas and are associated with severe WMCs (Fukui, Oowan, Yamazaki & Kinno, 2013). The WMCs appear to be primarily due to an associated history of vascular disease (Sarro et al., 2016). There is no difference in severity of the WMCs between unmixed and mixed forms of LBD (De Reuck et al., 2018).

Unmixed and mixed frontotemporal lobar degeneration

The neuropathological diagnosis is made according to the consensus criteria of the FTLD consortium (Cairns et al., 2007). FTLD comprises a heterogeneous spectrum of clinical syndromes and is pathologically and genetically heterogeneous (Sieben et al., 2012). Mixed FTLD represents 5% of the overall cases (De Reuck et al., 2018). Amyotrophic lateral sclerosis (ALS) is associated with FTLD in 15% of the cases (Liscic et al., 2008). This association represents 3% of all mixed dementia cases. Concomitant AD and cerebrovascular pathology appear respectively in 3% and 2%. Unmixed FTLD occurs mainly in adults and younger elderly patients, while the mixed form predominates in the middle elderly and oldest patients (De Reuck et al., 2018). CoMBs are mainly observed in the frontal cortex, where the most severe neurodegenerative changes are present. No differences in severity are found between unmixed and mixed cases (De Reuck et al., 2016d). The extension and the severity of the frontal WMCs are related to the severity of the neurodegeneration and the degree of atrophy of the covering cerebral cortex (De Reuck et al., 2012b) (Fig. 1.4). These changes are increased in the mixed cases (Michielse et al., 2010). CoMIs are extremely rare in the pure FTLD brains, while moderately increased in

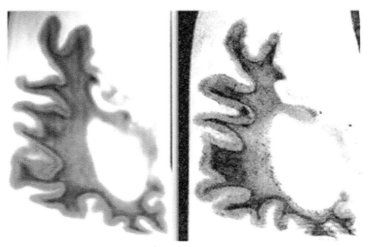

Figure 1.4 A 7.0-tesla MRI of a frontal hemispheric section of a brain with frontotemporal lobar degeneration. Extensive frontal cortical atrophy and severe confluent hyperintensity of the underlying white matter are seen.

the mixed cases compared with normal brains, probably due to associated cerebrovascular disease (De Reuck et al., 2018).

Iron deposition is increased in the basal ganglia of FTLD brains, mainly those of the FUS and TAR DNA-binding protein (TDP) types, compared with other neurodegenerative diseases and normal controls. Although iron deposition increases in normal aging brains, the accumulation is less in the mixed older patients with FTLD (De Reuck et al., 2014b). It is widely accepted that excessive accumulation of iron contributes to the neurodegenerative process, but it is still an open question whether this is an initial event or a consequence of the disease process (Batista-Nascimento, Pimentel, Menezes, Rodrigues-Pousada, 2012). The fact that iron accumulation is less severe in the middle and oldest elderly patients with mixed FTLD can be explained by a more important neuronal loss as the neurodegenerative disease progresses with aging (Grolez et al., 2016).

Progressive supranuclear palsy

The neuropathological criteria for the Steele—Richardson—Olszewski syndrome (progressive supranuclear paralysis, PSP) are those proposed by the US National Institutes of Health in 1993 (Hauw et al., 1994). The patients with a mixed form represent only 1% of the overall group. They are much older than those with the unmixed type. The iron content is decreased in the subthalamic nucleus, the red nucleus, and the substantia nigra (De Reuck et al., 2014b). Microbleeds prevail around the dentate nucleus of the cerebellum and in the tegmentum pontis of the mixed as well as the unmixed PSP brains. CoMIs and CoMBs are rare in both unmixed and mixed types (De Reuck, 2014c). WMCs are increased in the frontal lobes in the unmixed as well as in the mixed form (De Reuck, 2017b).

Amyotrophic lateral sclerosis

The neuropathological criteria proposed by Cruz-Sanchez, Moral, de Belleroche & Rossi (1993) are used for the diagnosis of ALS. There is a link between ALS and FTLD, as both have increased iron deposition in the basal ganglia, although to a lesser degree in the former (De Reuck et al., 2017c). The patients with the mixed form are older and represent 3% of the overall group, mainly due to additional AD pathology (De Reuck, 2018). They have moderate WMCs that are more severe in those patients with a history of arterial hypertension (Moreau et al., 2012). Unmixed and mixed forms both have some increase in CoMBs in the frontal lobes. They both have fewer CoMIs compared with normal brains of the same age, illustrating the favorable vascular risk profile in this disease (Kioumourtzoglou et al., 2015).

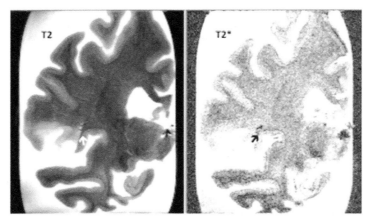

Figure 1.5 A 7.0-tesla MRI of a central hemispheric section of a brain with corticobasal degeneration. Severe atrophy of the insular cortex with a small cortical microbleed (*arrows*) on the T2 and T2* sequences is seen.

Corticobasal degeneration

The diagnostic criteria of CBD are those proposed by the International Consortium of Behavioral Neurology (Armstrong et al., 2013). In our limited series only unmixed CBD cases are observed. As the average age at death is 70 years, it is most probable that CBD remains mainly an unmixed disease. Frontal WMCs are increased in CBD, compared with normal age-controlled brains. CoMBs occur in the regions of most severely affected cerebral cortex. They are also more frequent compared with normal aging brains, but no differences are observed concerning the incidence of CoMIs and other cerebrovascular lesions (De Reuck, 2017c) (Fig. 1.5).

Vascular dementia

The postmortem assessment of VaD is done according to the criteria of McAleese et al. (2016). The mixed form of VaD with AD pathology represents 15% of the overall cases (De Reuck et al., 2018). It occurs more frequently in the old age groups, compared with the unmixed form of VaD, appearing more in the adult group. Different mechanisms are responsible for the vascular lesions in the mixed forms. They can be due to CAA as well as to atherosclerotic vascular disease (Haglund, Kalaria, Slade & Englund, 2006). Vascular cognitive impairment is mainly linked to the presence of lacunar infarcts and diffuse ischemic changes in the white matter (De Reuck et al., 2016e) (Fig. 1.6). CoMIs are mainly increased in the cingulated and inferior frontal gyri in VaD (De Reuck et al., 2016c). In the mixed type multiple larger infarcts are also more frequent (Attems & Jellinger, 2014).

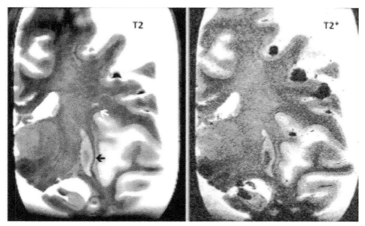

Figure 1.6 A 7.0-tesla MRI of a central hemispheric section of a brain with mixed severe vascular disease associated with Alzheimer pathology. Diffuse hyperintensity of the corona radiata on the T2 sequence is seen. The presence of an infarct in the putamen (*black arrow*) and a lacune in the corona radiata (*white arrow*) is seen on both MRI sequences.

Clinical diagnosis of mixed dementia

Mixed dementia diseases should be suspected mainly in elderly patients, although initially they can appear as a single disease at a younger age (Chui & Ramirez Gomez, 2017). Not all diseases seem to become mixed with aging. FTLD and ALS frequently remain single, due to their favorable vascular profile (De Reuck et al., 2017c). There is evidence of a neuropathological continuity between ALS and the TDP and FUS types of FTLD (Riku et al., 2014). No definite conclusions can be drawn concerning CBD, as the number of cases is low. However, as these patients die at an old age and are part of the complex Pick diseases, a low incidence of associated cerebrovascular pathology can be suspected (De Reuck, 2017c).

Mixed dementia syndromes are, for 85%, due to a combination of AD, VaD, and LBD pathologies (De Reuck et al., 2018). WMCs are not only due to ischemia, but can also be the result of cortical neuronal degeneration, leading to axonal and myelin loss, which can occur to some mild degree even during the normal aging process (De Reuck et al., 2018).

The mixed form of AD—VaD pathology can be identified during life by the combination of MRI and PET using glucose, amyloid, and tau tracers (Heiss, Rosenberg, Thiel, Berlot, De Reuck, 2016). The impact of CAA in mixed dementia cases can be suspected when using the modified Boston criteria: presence of lobar hematomas, superficial siderosis, CoMBs, and WMCs (Greenberg & Charidimou, 2018).

The contribution of LBD pathology in mixed dementias is more difficult to prove with neuroimaging techniques. It can only be suspected on clinical grounds: fluctuation in cognitive function, persistent well-formed visual hallucinations, and spontaneous motor features of parkinsonism are the core features (McKeith et al., 1996).

Key facts of mixed dementia

- The brain is the center of all cognitive functions, including immediate and remote memory, personality, speech and motor actions, and recognition of the surroundings.
- Brain lesions can lead to severe changes in behavior and social integration.
- Neurodegenerative diseases cause progressive impairment mainly involving memory and personality changes.
- Repetitive small and large infarcts and bleeds of the brain can lead to what is called "vascular dementia."
- On clinical grounds only a probable diagnosis of the disease can be made.
- Postmortem examination of the brain is the only way to confirm the clinically suspected disease.
- Postmortem MRI allows additional information concerning the distribution and quantification of mainly small cerebrovascular lesions.

Summary points

- This chapter examines the underlying diseases leading to mixed dementias in post-mortem brains.
- Mixed dementia is due to the combination of different brain diseases and occurs mainly in elderly patients.
- Alzheimer pathology combined with either cerebrovascular disease or Lewy body pathology is the most common cause of mixed dementia.
- Some brain diseases, such as FTLD, ALS, PSP, and CBD, rarely lead to mixed dementia.

References

Armstrong, M. J., Litvan, I., Lang, A. E., Bak, T. H., Bathia, K. P., Borroni, B., et al. (2013). Criteria for the diagnosis of corticobasal degeneration. *Neurology, 80*, 496−503.

Attems, J., & Jellinger, K. A. (2014). The overlap between vascular disease and Alzheimer's disease − lessons from pathology. *BMC Medicine*. https://doi.org/10.1186/s12916-014-0206-2.

Batista-Nascimento, L., Pimentel, C., Menezes, R. A., & Rodrigues-Pousada, C. (2012). Iron and neurodegeneration: From cellular homeostasis to disease. *Oxid Medical Cell Longevity*, 128647.

Bombois, S., Debette, S., Bruandet, A., Delbeuck, X., Delmaire, C., Leys, D., et al. (2008). Vascular subcortical hyperintensities predict conversion to vascular and mixed dementia in MCI patients. *Stroke, 39*, 2046−2051.

Braak, H., & Braak, E. (1991). Neuropathological staging of Alzheimer-related changes. *Acta Neuropathologica, 82*, 239−259.

Cairns, N. J., Bigio, E. H., Mackensie, I. R., Neumann, M., Lee, V. M., Hatanpaa, K. J., et al. (2007). Neuropathologic diagnostic and nosologic criteria for frontotemporal lobar degeneration: Consensus of the consortium for frontotemporal lobar degeneration. *Acta Neuropathologica, 114*, 5–22.

Chui, H. C., & Ramirez Gomez, L. (2017). Vascular contributions to cognitive impairment in late life. *Neurologic Clinics, 35*, 295–323.

Cruz-Sanchez, F. F., Moral, A., de Belleroche, J., & Rossi, M. L. (1993). Amyotrophic lateral sclerosis brain banking: A proposal to standardize protocols and neuropathological diagnostic criteria. *Journal of Neural Transmission, 39*(Suppl. 2), 215–222.

De Reuck, J. (2012a). Histopathological stainings and definition of vascular disruptions in the elderly brain. *Experimental Gerontology, 47*, 834–837.

De Reuck, J. (2016a). Post-mortem magnetic resonance imaging as an additional tool of the neuropathological examination of neurodegenerative and cerebrovascular diseases. *European Neurological Review, 11*, 22–25.

De Reuck, J. (2017b). Cerebrovascular lesions in Pick complex diseases: A neuropathological study with a 7.0-tesla magnetic resonance imaging study. *European Neurological Review, 12*, 84–86.

De Reuck, J., Auger, F., Cordonnier, C., Deramecourt, V., Durieux, N., Pasquier, F., et al. (2011a). Comparison of 7.0-tesla T2*-magnetic resonance imaging of cerebral bleeds in post-mortem brain sections of Alzheimer patients with their neuropathological correlates. *Cerebrovascular Diseases, 31*, 511–517.

De Reuck, J., Auger, F., Durieux, N., Cordonnier, C., Deramecourt, V., Lebert, F., et al. (2015b). Detection of cortical microbleeds in postmortem brains with lewy body dementia: A 7.0-tesla magnetic resonance imaging study with neuropathological correlates. *European Neurology, 74*, 158–161.

De Reuck, J., Auger, F., Durieux, N., Cordonnier, C., Deramecourt, V., Pasquier, F., et al. (2016c). The topography of cortical microinfarcts in neurodegenerative diseases and in vascular dementia: A post-mortem 7.0-tesla magnetic resonance imaging study. *European Neurology, 76*, 57–61.

De Reuck, J., Auger, F., Durieux, N., Cordonnier, C., Deramecourt, V., Pasquier, F., et al. (2016e). Topographic distribution of white matter changes and lacunar infarcts in neurodegenerative and vascular dementia syndromes: A post–mortem 7.0-tesla magnetic resonance imaging study. *European Stroke Journal, 1*, 122–129.

De Reuck, J., Auger, F., Durieux, N., Deramecourt, V., Cordonnier, C., Pasquier, F., et al. (2015a). Topography of cortical microbleeds in Alzheimer's disease with and without cerebral amyloid angiopathy: A post-mortem 7.0 tesla MRI study. *Aging and Disease, 6*, 437–443.

De Reuck, J., Auger, F., Durieux, N., Deramecourt, V., Maurage, C.-A., Lebert, L., et al. (2016d). The topography of cortical microbleeds in frontotemporal lobar degeneration: A post-mortem 7.0-tesla magnetic resonance study. *Folia Neuropathologica, 54*, 1–7.

De Reuck, J., Auger, F., Durieux, N., Deramecourt, V., Maurage, C. A., Pasquier, F., et al. (2018). Cerebrovascular lesions during normal aging: A neuropathological study with 7.0-tesla magnetic resonance imaging. *CR Neurology, 10*, 229–235.

De Reuck, J., Caparros-Lefebvre, D., Deramecourt, V., Defebvre, L., Auger, F., Durieux, N., et al. (2014c). Prevalence of small cerebral bleeds in patients with progressive supranuclear palsy: A neuropathological study with 7.0-tesla magnetic resonance imaging correlates. *Folia Neuropathologica, 52*, 421–427.

De Reuck, J., Deramcourt, V., Auger, F., Durieux, N., Maurage, C.-A., Pasquier, F., et al. (2017a). Cerebrovascular lesions in mixed neurodegenerative dementia: A neuropathological and magnetic resonance study. *European Neurology, 78*, 1–5.

De Reuck, J., Deramecourt, V., Auger, F., Durieux, N., Cordonnier, C., Devos, D., et al. (2014a). Post-mortem 7.0-tesla magnetic resonance study of cortical microinfarcts in neurodegenerative diseases and vascular dementia with neuropathological correlations. *Journal of Neurological Sciences, 346*, 85–89.

De Reuck, J., Deramecourt, V., Auger, F., Durieux, N., Cordonnier, C., Devos, D., et al. (2014b). Iron deposits in post-mortem brains of patients with neurodegenerative and cerebrovascular diseases: A semi-quantitative 7.0 T magnetic resonance study. *European Journal of Neurology, 21*, 1026–1031.

De Reuck, J., Deramecourt, V., Cordonnier, C., Leys, D., Maurage, C. A., & Pasquier, F. (2011b). The impact of cerebral amyloid angiopathy on the occurrence of cerebrovascular lesions in demented patients with Alzheimer features: A neuropathological study. *European Journal of Neurology, 18*, 913–918.

De Reuck, J., Deramecourt, V., Cordonnier, C., Leys, D., Pasquier, F., & Maurage, C.-A. (2012b). Cerebrovascular lesions in patients with frontotemporal lobar degeneration: A neuropathological study. *Neurodegenerative Diseases, 9*, 170–175.

De Reuck, J., Deramecourt, V., Cordonnier, C., Pasquier, F., Leys, D., Maurage, C.-A., et al. (2016b). The incidence of post-mortem neurodegenerative and cerebrovascular pathology in mixed dementia. *Journal of Neurological Sciences, 366*, 164–166.

De Reuck, J., Devos, D., Moreau, C., Auger, F., Durieux, N., Deramecourt, V., et al. (2017c). Topographic distribution of brain iron deposition and small cerebrovascular lesions in amyotrophic lateral,sclerosis and in frontotemporal lobar degeneration: A post-mortem 7.0-tesla magnetic resonance imaging study with neuropathological correlates. *Acta Neurologica Belgica, 117*, 873–878.

De Reuck, J., Maurage, C.-A., Deramecourt, V., Pasquier, F., Cordonnier, C., Leys, D., et al. (2018). Aging and cerebrovascular lesions in pure and in mixed neurodegenerative and vascular dementia brains: A neuropathological study. *Folia Neuropathologica, 56*, 81–87.

Ellis, R., Olichney, J. M., Thal, L. J., Mirra, S. S., Morris, J. C., Beekly, D., et al. (1996). Cerebral amyloid angiopathy in brains of patients with Alzheimer's disease: The CERAD experience, Part XV. *Neurology, 46*, 1592–1596.

Fukui, T., Oowan, Y., Yamazaki, T., & Kinno, R. (2013). Prevalence and clinical implication of microbleeds in dementia with lewy bodies in comparison with microbleeds in Alzheimer's disease. *Dementia and Geriatric Cognitive Disorders Extra, 3*, 148–160.

Gorelick, P. B., Scuteri, A., Black, S. E., Decarli, C., Greenberg, S. M., Iadecla, C., et al. (2011). Vascular contributions to cognitive impairment and dementia: A statement for healthcare professionals from the American heart association/American stroke association. *Stroke, 42*, 2672–2713.

Greenberg, S. M., & Charidimou, A. (2018). Diagnosis of cerebral amyloid angiopathy: Evolution of the Boston criteria. *Stroke, 49*, 491–497.

Grolez, G., Moreau, C., Daniel-Brunaud, V., Delmaire, C., Lopes, R., Pradat, P. F., et al. (2016). The value of magnetic resonance imaging as a biomarker of amyotrophic lateral sclerosis: A systematic review. *BMC Neurology, 16*, 155. https://doi.org/10.1186/s12883-016-0672-6.

Haglund, M., Kalaria, R., Slade, J. Y., & Englund, E. (2006). Differential deposition of amyloid beta peptides in cerebral amyloid angiopathy associated with Alzheimer's disease and vascular dementia. *Acta Neuropathologica, 111*, 430–435.

Hauw, J. J., Daniel, S. E., Dickson, D., Horoupian, D. S., Jellinger, K., Lantos, P. L., et al. (1994). Preliminary NINDS neuropathological criteria for Steele-Richardson-Olszewski syndrome (progressive supranuclear palsy). *Neurology, 44*, 2015–2019.

Heiss, W. D., Rosenberg, G. A., Thiel, A., Berlot, R., & De Reuck, J. (2016). Neuroimaging in vascular cognitive impairment: A state-of-the art review. *BMC Medicine.* https://doi.org/10.1186/s12916-016-0725-0.

Irwin, D. J., Grossman, M., Weintraub, D., Hurtig, H. I., Duda, J. E., Xie, R. X., et al. (2017). Neuropathological and genetic correlates of survival and dementia onset in synucleinopathies: A retrospective study. *The Lancet Neurology*, 55–65.

Jellinger, K. (2009). Criteria for the neuropathological diagnosis of dementing disorders: Routes out of the swamp. *Acta Neuropathologica, 117*, 101–110.

Jellinger, K., & Attems, J. (2007). Neuropathological examination of mixed dementia. *Journal of Neurological Sciences, 257*, 80–87.

Jellinger, K., & Attems, J. (2010). Prevalence of dementia disorders in the oldest-old: An autopsy study. *Acta Neuropathologica, 119*, 421–433.

Jellinger, K. A., & Attems, J. (2015). Challenges of multi-morbidity of the aging brain: A critical update. *Journal of Neural Transmission, 122*, 505–521.

Kapasi, A., DeCarli, C., & Schneider, J. A. (2017). Impact of multiple pathologies on the threshold for clinically overt dementia. *Acta Neuropathologica, 134*, 171–186.

Kioumourtzoglou, M. A., Rotem, R. S., Seals, R. M., Gredal, O., Hansen, J., & Weisskopf, M. G. (2015). Diabetes, obesity, and diagnosis of amyotrophic lateral sclerosis: A population based study. *JAMA Neurology, 72*, 905–911.

Liscic, R. M., Grinberg, L. T., Zidar, J., Gitcho, M. A., & Cairns, N. J. (2008). ALS and FTLD: Two faces of TDP-43 proteinopathy. *European Journal of Neurology, 15*, 772—780.

McAleese, K. E., Alafuzoff, I., Charidimou, A., De Reuck, J., Grinberg, L. T., Hainsworth, A. H., et al. (2016). Post-mortem assessment in vascular dementia: Advances and aspirations. *BMC Medicine, 14*, 129. https://doi.org/10.1186/s12916-016-0676-5.

McKeith, I. G., Galasko, D., Kosaka, K., Perry, E. K., Dickson, D. W., Hansen, L. A., et al. (1996). Consensus guidelines for the clinical and pathological diagnosis of dementia with lewy bodies (DLB): Report of the consortium on DLB international workshop. *Neurology, 47*, 1113—1124.

Michielse, S., Coupland, N., Camicioli, R., Carter, R., Seres, P., Sabino, J., et al. (2010). Selective effects of ageing on brain white matter microstructure: A diffusion tensor imaging tractography study. *NeuroImage, 52*, 1190—1201.

Moreau, C., Brunaud-Danel, V., Dallongeville, J., Duhamel, A., Laurier-Grymonprez, L., De Reuck, J., et al. (2012). Modifying effect of arterial hypertension on amyotrophic lateral sclerosis. *Amyotrophic Lateral Sclerosis, 13*, 194—201.

Nedelska, Z., Ferman, T. J., Boeve, B. F., Przybelski, S. A., Lesnick, T. G., Murray, M. E., et al. (2015). Pattern of brain atrophy rates in autopsy-confirmed dementia with Lewy bodies. *Neurobiology of Aging, 36*, 452—461.

Rahimi, J., & Kovacs, G. G. (2014). Prevalence of mixed pathologies in the aging brain. *Alzheimer's Research and Therapy, 6*, 82. https://doi.org/10.1186/s13195-014-0082-1.

Raz, L., Knoefel, J., & Bhaskar, K. (2016). The neuropathology and cerebrovascular mechanisms of dementia. *Journal of Cerebral Blood Flow and Metabolism, 36*, 172—186.

Riku, Y., Watanabe, H., Yoshida, M., Tatsumi, S., Mimuro, M., Iwasaki, M., et al. (2014). Lower motor neuron involvement in TAR DNA binding protein of 43 kDa-related frontotemporal lobar degeneration and amyotrophic lateral sclerosis. *JAMA Neurology, 71*, 172—179.

Sarro, L., Tosakulwong, N., Schwarz, C. G., Graff-Redford, J., Przybelski, S. A., Lesnick, T. G., et al. (2016). An investigation of cerebrovascular lesions in dementia with Lewy bodies compared to Alzheimer's disease. *Alzheimer's and Dementia.* https://doi.org/10.1016/j.jalz.2016.07.003.

Sieben, A., Van Langenhove, T., Engelborghs, S., Martin, J. J., Boon, P., Cras, P., et al. (2012). The genetic and neuropathology of frontotemporal lobar degeneration. *Acta Neuropathologica, 124*, 353—372.

Suemoto, C. K., Ferretti-Rebustini, R. E., Rodrigues, D. R., Leite, R. E., Soterio, L., Brucki, S. M., et al. (2017). Neuropathological diagnoses and clinical correlates in older adults in Brasil: A cross-sectional study. *PLoS Medicine, 14*, e1002267. https://doi.org/10.1371/journal.pmed.1002267.

Toledo, J. B., Arnold, S. E., Raible, K., Brettschneider, J., Xie, S. X., Grossman, M., et al. (2013). Contribution of cerebrovascular disease in autopsy confirmed neurodegenerative disease cases in the National Alzheimer's Coordinating Centre. *Brain, 136*, 2697—2706.

Wakisaka, Y., Furuta, A., Tanizaki, Y., Kiyohara, Y., Lida, M., & Iwaki, T. (2003). Age-associated prevalence of risk factors of lewy body pathology in a general population: The Hisayama study. *Acta Neuropathologica, 106*, 374—382.

CHAPTER 2

Vascular dementia: an overview

Virginia Cipollini, Fernanda Troili, Franco Giubilei
NESMOS Department, Faculty of Medicine and Psychology, Sapienza University of Rome, Rome, Italy

List of abbreviations

AD Alzheimer's disease
BBB blood—brain barrier
CAA cerebral amyloid angiopathy
CADASIL cerebral autosomal dominant arteriopathy with subcortical infarcts and leukoencephalopathy
CSF cerebrospinal fluid
CVD cerebrovascular disease
DSM-5 *Diagnostic and Statistical Manual of Mental Disorders*, fifth edition
EPVS enlarged perivascular space
MID multiinfarct dementia
MOCA Montreal Cognitive Assessment
MRI magnetic resonance imaging
NINDS—AIREN National Institute of Neurological Disorders and Stroke and the Association Internationale pour la Recherche et l'Enseignement en Neurosciences
NVU neurovascular unit
STRIVE Standards for Reporting Vascular Changes on Neuroimaging
SVD small vessels disease
VaD vascular dementia
VCI vascular cognitive impairment
WMH white matter hyperintensity
WML white matter lesion

Mini-dictionary of terms

Blood—brain barrier the blood—brain barrier is a brain-specific capillary barrier separating the central nervous system from the systemic circulation, and it is essential to maintain the optimal microenvironment in the central nervous system. A dysfunction of the blood—brain barrier can be responsible for the progression of several neurological diseases.
CADASIL CADASIL, or cerebral autosomal dominant arteriopathy with subcortical infarcts and leukoencephalopathy, is a hereditary SVD that may lead to VaD. It is caused by mutations in the NOTCH3 gene and its clinical manifestations include migraine with or without aura, transient ischemic attacks or minor strokes, multiple lacunar infarcts, and dementia.
Cerebral amyloid angiopathy CAA is a disorder characterized by the deposition of amyloid peptides in the walls of the small and medium-sized blood vessels of the leptomeninges and central nervous system. It may cause microhemorrhages, macrohemorrhages, small multiple infarctions, transient neurological symptoms, and eventually dementia.

Diagnosis and Management in Dementia
ISBN 978-0-12-815854-8, https://doi.org/10.1016/B978-0-12-815854-8.00002-1
17

Enlarged perivascular spaces enlarged perivascular spaces (also known as Virchow—Robin spaces) are cerebrospinal fluid—filled cavities that surround penetrating vessels entering the brain parenchyma and correspond with extensions of the subarachnoid space. They serve as a draining channel for the brain and can be visualized on T2-weighted brain MRI.

Neurovascular unit the NVU is constituted by endothelial cells, myocytes, neurons and their processes, astrocytes, and perivascular cells. It is involved in many functions such as cerebral blood flow regulation, blood—brain barrier exchange, immune surveillance, trophic support, and homeostatic balance.

White matter hyperintensities WMHs, also referred to as leukoaraiosis, are a very common finding on brain MRI in older subjects and in patients with stroke and dementia. We can identify different patterns and the extent of WMHs, which might manifest a variety of symptoms. The predominant clinical associations are with stroke, cognitive impairment, and dementia; however, some patients with extensive WMLs might remain asymptomatic.

Introduction

Since Hachinski, Lassen and Marshall (1974) proposed the term "multiinfarct dementia" in 1973, several attempts have been made to find a comprehensive term and diagnostic criteria that could thoroughly describe the complexity of the wide range of cognitive deficits caused by heterogeneous underlying cerebrovascular disease (CVD).

Over the years, the term "vascular dementia" (VaD) was adopted as a more inclusive nosographic category, and efforts were made to propose accurate diagnostic criteria (Chui et al., 1992; Erkinjuntti et al., 2000; Román et al., 1993). On one hand, modern diagnostic criteria allowed one to categorize subgroups in the VaD population to carry out rigorous clinical trials, but, on the other hand, the term vascular dementia was progressively criticized. This criticism led to its proposed replacement with the term "vascular cognitive impairment" (VCI) (O'Brien et al., 2003), which refers to the entire spectrum of vascular brain pathologies that contribute to cognitive impairment, ranging from subjective cognitive decline to overt dementia.

In the *Diagnostic and Statistical Manual of Mental Disorders*, fifth edition (DSM-5), another definition was proposed, which is described under the terms "mild and major vascular neurocognitive disorders," which partly overlap the VCI definition (American Psychiatric Association, 2013).

Multiple criteria (Gorelick et al., 2011; Sachdev et al., 2014; Skrobot et al., 2017) and research guidelines have been formulated, but are not easily interchangeable: this may contribute to variability in prevalence measures in the literature.

Epidemiology and risk factors

According to estimates from the World Alzheimer Report 2015, 46.8 million people worldwide have dementia. With population aging, the number of subjects affected by dementia is expected to reach 115 million in 2050, worldwide (Prince et al., 2015), with individuals >85 years of age being more affected (Prince et al., 2013).

In this scenario, VaD is claimed to be the second most common cause of dementia, accounting for ~15%—20% of cases (Goodman et al., 2017; Rizzi, Rosset & Roriz-Cruz, 2014).

To provide an understanding of the full spectrum of dementia in the general population and to identify risk factors across different populations and life courses, we need to analyze population-based epidemiological research. It has to be known that VaD, rather than VCI, is the term most frequently used in epidemiologic literature. Evaluation of the changes in dementia incidence and prevalence over time is challenging, as changes in diagnostic criteria and other methodological variations can cause data to be not easily comparable.

Although the prevalence and incidence of VaD increase with age, the increased risk of dementia due to CVD seems to decline at very old ages (von Strauss et al., 1999; Corraini et al., 2017). A possible explanation for this change might be that other causes of dementia are more common at very old ages and frequently overlap. It has been in fact observed in autopsy series that a diagnosis of "pure" VaD is uncommon, with mixed pathology prevailing in elderly subjects: the interaction between Alzheimer's pathology and vascular pathology represents one of the most challenging aspects of understanding pathophysiological mechanisms in late-life dementia (Pendlebury & Rothwell, 2009; Schneider et al., 2007).

Since 2010, descriptive epidemiological studies have strengthened the evidence that the incidence and prevalence of dementia are in a declining trend in Europe and North America and that the number of people with dementia can remain stable despite population aging (Satizabal et al., 2016; Wu et al., 2017). This might be the result of a better control of the main risk factors and increased awareness of the importance of building up an adequate cognitive reserve. Still, a lot needs to be revealed in terms of risk and protective factors related to dementia.

About VaD, nonmodifiable risk factors are associated with the disease, such as age and female sex (some evidence in poststroke dementia) (Allan et al., 2011; Kalaria et al. 2016; Leys et al., 2005; Pendlebury & Rothwell, 2009). In contrast with the deep genetic characterization of monogenic disorders responsible for VaD (e.g., cerebral autosomal dominant arteriopathy with subcortical infarcts and leukoencephalopathy) that has been made, no robust genetic risk factors have been identified for sporadic VaD (Haffner et al., 2016; Ikram et al., 2017).

Studies of modifiable risk factors, such as low education, smoking, reduced physical activity, overweight, and high total cholesterol levels in midlife, show controversial results in literature. On the other hand, hypertension in midlife, chronic hyperglycemia, and atrial fibrillation increase the risk of dementia, independent of the associated increase in risk of stroke (Dichgans & Zietemann, 2012; Iadecola, 2013; Iadecola et al., 2016; Ngandu et al., 2015).

As previous studies have described, there are several common vascular risk factors between Alzheimer's disease (AD) and VaD, and this is both relevant and important to the known interaction between Alzheimer's and vascular pathology (O'Brien & Thomas, 2015) (Table 2.1).

In a recent metaanalysis, late-life depression seems to increase the risk of both VaD and AD (Diniz et al. 2013), providing additional evidence of the association between vascular pathology and late-life depression. Instead, other studies suggest that depression can be considered a comorbidity, a prodromal factor, or a consequence of VaD rather than a factor that specifically alters vascular physiology (Gorelick et al., 2011).

It is hoped that, with the course of scientific progress, a better identification and understanding of risk factors amenable to prevention and treatment could lead to the implementation of a public health policy tailored for the at-risk population.

Clinical features

Diagnosis

The diagnosis of VaD can be made only after considering medical history, physical examination, neuroimages, and neuropsychological assessment. VaD is diagnostically

Table 2.1 Risk factors for vascular dementia.

	Modifiable/ nonmodifiable	Possible risk factor for Alzheimer's disease
Demographic factors		
Age	Nonmodifiable	Yes
Female sex	Nonmodifiable	Yes
Low education	Modifiable	Yes
Chronic diseases		
Hypertension	Modifiable	Yes
Diabetes mellitus	Modifiable	Yes
Atrial fibrillation	Modifiable	Yes
Myocardial infarction, angina pectoris	Modifiable	Not clear
Stroke	Modifiable	Not clear
Hypercholesterolemia	Modifiable	Yes
Depression	Modifiable	Yes
Lifestyle factors		
Smoking	Modifiable	Not clear
Reduced physical activity	Modifiable	Yes
Overweight	Modifiable	Yes

There are modifiable and nonmodifiable risk factors associated with vascular dementia. Most of them are also associated with Alzheimer's disease, and this is important to the known interaction between the pathologies of these two diseases.

challenging, and the diagnosis may be not precise given the many causes of dementia, including the potential for a mixed dementia syndrome. AD, dementia with Lewy bodies, frontotemporal dementia, and psychiatric conditions such as depression are the most common alternative types of dementia that enter in differential diagnosis with VaD.

In the clinical setting, to make a diagnosis of VaD we can use diagnostic guidelines, differing in their definition of cognitive impairment and involvement of vascular disease as the leading cause. The most commonly used guidelines are the National Institute of Neurological Disorders and Stroke and the Association Internationale pour la Recherche et l'Enseignement en Neurosciences (NINDS—AIREN) criteria (Román et al., 1993) and the DSM-5 (American Psychiatric Association 2013) (Table 2.2).

Table 2.2 Diagnostic criteria for vascular dementia.

NINDS—AIREN criteria for diagnosis of VaD (Román et al., 1993)	The criteria for the clinical diagnosis of probable VaD include all of the following: Dementia definition: impairment of memory plus at least two other cognitive domains (orientation, attention, language, visuospatial functions, executive functions, motor control, and praxis) should be severe enough to interfere with activities of daily living, not due to physical effects of stroke alone. Cerebrovascular disease definition: presence of focal signs on neurologic examination and evidence of relevant cerebrovascular disease by brain imaging (computed tomography or MRI). A relationship between the above two disorders manifested by the presence of one or more of the following: • onset of dementia within 3 months following a recognized stroke • abrupt deterioration in cognitive functions or fluctuating, stepwise progression of cognitive deficits
DSM-5 criteria for diagnosis of VaD (American Psychiatric Association, 2013)	Major neurocognitive disorder definition: Evidence of significant cognitive decline from a previous level of performance in one or more cognitive domains based on: • concern of the individual, a knowledgeable informant, or the clinician that there has been a significant decline in cognitive function • a substantial impairment in cognitive performance, preferably documented by standardized neuropsychological testing or another quantified clinical assessment The cognitive deficits interfere with independence in everyday activities The cognitive deficits do not occur exclusively in the context of a delirium The cognitive deficits are not better explained by another mental disorder (for example, major depressive disorder or schizophrenia)

Continued

Table 2.2 Diagnostic criteria for vascular dementia.—cont'd

	Mild neurocognitive disorder definition:
	Evidence of modest cognitive decline from a previous level of performance in at least one cognitive domain based on: <ul><li>concern of the individual, a knowledgeable informant, or the clinician that there has been a mild decline in cognitive function</li><li>a modest impairment in cognitive performance, preferably documented by standardized neuropsychological testing or another quantified clinical assessment</li></ul> The cognitive deficits do not interfere with capacity for independence in everyday activities The cognitive deficits do not occur exclusively in the context of a delirium The cognitive deficits are not better explained by another mental disorder (for example, major depressive disorder or schizophrenia) Vascular neurocognitive disorder definition: Criteria are met for major or mild neurocognitive disorder The clinical features are consistent with a vascular etiology, as suggested by either of the following: <ul><li>onset of the cognitive deficit is temporally related to one or more cerebrovascular events</li><li>evidence for decline is prominent in complex attention (including processing speed) and frontal executive function</li></ul> Evidence of cerebrovascular disease from history, physical examination, and/or neuroimaging that is considered sufficient to account for the neurocognitive deficits The symptoms are not better explained by another brain disease or systemic disorder Probable vascular neurocognitive disorder is diagnosed if one of the following is present: <ul><li>clinical criteria are supported by neuroimaging evidence of significant parenchymal injury attributed to cerebrovascular disease (which is supported by neuroimaging)</li><li>neurocognitive syndrome is temporally related to one or more documented cerebrovascular events</li><li>both clinical and genetic (for example, cerebral autosomal dominant arteriopathy with subcortical infarcts and leukoencephalopathy) evidence of cerebrovascular disease is present</li></ul>

The most commonly used guidelines for the diagnosis of VaD are the National Institute of Neurological Disorders and Stroke and the Association Internationale pour la Recherche et l'Enseignement en Neurosciences (*NINDS—AIREN*) criteria and the *Diagnostic and Statistical Manual of Mental Disorders*, fifth edition (*DSM-5*). Both core features include a stepwise progression, presence of focal neurological signs and symptoms, neuroimaging evidence of cerebrovascular disease, a history of multiple ischemic strokes, and a temporal relationship between cerebrovascular disease and dementia. *VaD*, vascular dementia.

Both core features include a stepwise progression, focal neurological signs and symptoms, an unequal distribution of cognitive deficits, a history of multiple ischemic strokes, neuroimaging evidence of CVD, and an association with etiology and, finally, a temporal relationship between CVD and dementia (van der Flier et al., 2018).

The NINDS—AIREN criteria, which are still the most widely used, especially in the research setting, emphasize the heterogeneity of VaD syndromes and pathological subtypes such as ischemic or hemorrhagic strokes, white matter changes, and cerebral hypoxic—ischemic events (Kalaria, 2016). The DSM-5 criteria differ from the previous guidelines, giving less importance to memory deficit, as impairment in any cognitive domain (including executive function) is sufficient for diagnosis. Of course, cognitive changes in VaD are much more variable and less homogeneous than in other disorders, such as AD, and depend on the particular neural substrates affected by the vascular pathology (O'Brien & Thomas, 2015).

There are multiple terms to describe the vascular pathology underlying VaD diagnosis, such as multiinfarct dementia, small vessel disease (SVD) or Binswanger disease, strategic infarct dementia, hypoperfusion dementia, hemorrhagic dementia, hereditary VaD, and AD with cardiovascular disease.

We can recognize two main clinical syndromes of VaD: poststroke VaD, in which cognitive impairment is the immediate consequence of a recent stroke, and VaD without recent stroke, in which cognitive impairment arises from vascular brain injury, detectable only on neuroimaging or neuropathology (Smith, 2017). This second form mostly appears as SVD, which represents a diverse range of pathological changes that affect capillaries, small arteries, and small veins in the brain and is commonly related to lacunar infarcts, microbleeds, enlarged perivascular spaces, leukoaraiosis, and cortical atrophy. SVD causes 20% of strokes and is considered a very common cause of cognitive decline, particularly in the elderly (Fu et al., 2018).

The clinical evaluation of patients with VaD should aim to discover neurological dysfunction in a vascular territory related to prior stroke, or signs of motor and sensory dysfunction related to SVD. The clinical manifestations may include subtle symptoms such as increased tone, hyperreflexia, Babinski responses, frontal lobe release signs, gait disorder with instability and recurrent falls, urinary frequency or urgency, delirium, and personality and mood changes (Jaul & Meiron, 2017). A lower body parkinsonism syndrome with increased leg tone and decreased gait speed can also be frequently detected (Smith, 2017).

Cognitive assessment

Cognitive assessment is very important to make a correct diagnosis of VaD. A global cognitive screening tool can be used to objectively document cognitive impairments. The most commonly used instruments are the Folstein Mini-Mental State exam (Folstein et al., 1975) and the Montreal Cognitive Assessment tool (Nasreddine et al., 2005).

Neuropsychological evaluation might help to differentiate AD from VaD in the early stages of these diseases. However, the data in the literature are conflicting about the possible differences in cognitive performance between AD and VaD.

The most consistent findings suggest that AD is characterized by a greater impairment in episodic memory, whereas patients with VaD have greater deficits in executive/attentional abilities (McGuinness et al., 2010). Nevertheless, other studies have found no differences between the two diseases regarding executive functions (Cavalieri et al., 2010). The results may be so discrepant because of the differences in study populations and the lack of accordance in the definition of VaD.

Multiinfarct VaD and strategic infarct VaD express a variety of cognitive dysfunctions, which depend on the site and extent of the damage in the brain, whereas in SVD we usually find a dysexecutive syndrome (Stokholm et al., 2006). However, executive dysfunctions can also be present in AD patients, under a variety of different patterns, depending on the vascular involvement (Clark et al., 2012).

Early identification of small vessel cognitive impairment is crucial to allow intervention to control vascular risk factors before the onset of dementia. The hypothesis of a temporal continuum of dysexecutive syndrome, based on a multidimensional concept of executive function and on pathophysiological aspects of lesion progression in small vessel VaD, might be of great value for this purpose (De Carolis et al., 2017; Sudo et al., 2017).

Alterations in mood and behavior are common and may be very stressful for both patients and caregivers. Particularly, VaD has been associated with late-life onset depression, apathy (Fuh et al., 2005), and psychosis (Fischer et al., 2015); otherwise, other features such as delusions and hallucinations are less frequent (Gupta et al., 2014).

Neuroimaging

Brain imaging provides an excellent tool for identifying the main vascular lesions that may be linked to a probable VCI. They include infarcts, hemorrhages, and white matter hyperintensities (WMHs). Single or multiple territorial, lacunar, or border zone infarcts are the most frequently observed lesions (Table 2.3).

Magnetic resonance imaging (MRI) is the first-choice neuroimaging technique for the evaluation of patients with suspected VaD, although computed tomography can be used to detect atrophy and some vascular lesions.

WMHs are a frequent finding on MRI, but they can indicate other nonischemic causes. Nevertheless, the etiology might more likely be vascular in elderly subjects: they represent strong predictors of cognitive impairment over the subsequent 3 years (Inzitari et al., 2009).

Both hippocampal atrophy and global atrophy, visualized on T1-weighted MRI, may also be related to either AD or vascular pathology (Jagust et al., 2008). This suggests that the contribution of vascular disease to atrophy can often be underestimated.

Table 2.3 Neuroimaging features of vascular dementia.

Large vessel VaD	1. Multiinfarct dementia (multiple large complete infarcts involving cortical and subcortical areas) 2. Strategic single-infarct dementia 3. Watershed infarction 4. Hypoperfusion encephalopathy
Small vessel VaD	1. Subcortical VaD 2. Lacunes 3. Perivascular spaces
Microhemorrhage and dementia	1. Cerebral amyloid angiopathy 2. Cerebral autosomal dominant arteriopathy with subcortical infarcts and leukoencephalopathy 3. Cerebral autosomal recessive arteriopathy with subcortical infarcts and leukoencephalopathy 4. Hereditary endotheliopathy, retinopathy, nephropathy, and stroke

Magnetic resonance images identify the main vascular lesions that may be linked to a probable vascular cognitive impairment. They include infarcts, hemorrhages, and white matter hyperintensities. Different neuroimaging features are more frequent, in particular, vascular dementia (*VaD*) syndromes and pathological subtypes.

There are some limitations in the use of MRI in clinical practice nowadays; for example, the usually used machines do not easily detect small cortical infarcts. The neuroimaging field is rapidly advancing, and more sophisticated MRI techniques could be available for clinical application in the future.

Neuropathological features

VaD is a term used to describe a heterogeneous group of conditions, also from the neuropathological point of view. In fact, VaD is an entity whose heterogeneous clinical manifestations are due to a substrate of multiple pathogenic and structural factors. Nevertheless, histopathologic evidence, obtained by biopsy or by autopsy, is essential in each guideline to make a diagnosis of definite VaD.

A neuropathological diagnosis identifies the type of underlying cerebrovascular lesions, including lesions in the parenchyma (such as infarcts and white matter changes) and alterations in blood vessels (such as arteriolosclerosis and cerebral amyloid angiopathy [CAA]). Amyloid plaques and neurofibrillary tangles may also be present during the pathological examination and may contribute to cognitive dysfunction. In this case, a diagnosis of VaD cannot be sustained, although dual etiology (vascular and degenerative) can still be diagnosed (Sachdev et al., 2014).

Cognitive impairment may be caused by a single "strategic" infarct. However, vascular brain lesions lower the threshold of AD pathology required to induce dementia (Snowdon et al., 1997). Conversely, AD pathology increases the risk of dementia after stroke (Pendlebury & Rothwell, 2009) and contributes to cognitive decline in patients with VCI (Ye et al., 2015). The interplay between vascular and AD-type pathology as regards cognitive decline might show that the effects are additive (Schneider et al., 2004; Vemuri et al., 2015).

The main alterations of cerebral blood vessels in VaD are: (1) atherosclerosis, which can be due to hypertension, aging, and other vascular risk factors; (2) SVD, caused mostly by arteriolosclerosis; and (3) CAA, in which the damage of cerebral vessels happens because of accumulation of Aβ among adventitia and media layers. All these diseases can lead to various cerebrovascular lesions, including ischemic infarcts, hemorrhages and white matter lesions (WMLs) (McAleese et al., 2016).

Ischemia may occur because of several mechanisms, including atherosclerosis, thrombosis, or vasculopathy of large to medium arteries, often as the result of an atherosclerotic plaque rupture.

Hemorrhagic infarcts can occur in infarcted regions in which the remaining vessels have fragile walls as a result of SVD or CAA, or they may be caused by venous obstruction. Dementia is more frequent in lobar hemorrhages and hemosiderosis (Garcia et al., 2013; Moulin et al., 2016), which may be associated with CAA.

The mechanism underlying WMLs is more unclear and still poorly understood: it probably involves a multifactorial process, which includes blood—brain barrier disruption, hypoxia and hypoperfusion, oxidative stress, neuroinflammation, and alterations in neurovascular unit (NVU) coupling, leading to demyelination and gliosis (Yang et al., 2017) (Fig. 2.1).

Pathological data show that both large and small artery disease is associated with AD dementia independent of infarcts (Arvanitakis et al., 2016). SVD in particular may be linked to dysfunction of the glymphatic pathway that is possibly involved in the clearance of misaggregated proteins. Enlarged perivascular spaces have emerged as potential biomarkers of neurovascular dysfunction (Ramirez et al., 2016; Wardlaw et al., 2013) and impaired clearance and can be identifiable on brain MR images, according to the Standards for Reporting Vascular Changes on Neuroimaging (STRIVE) criteria (Potter et al., 2015; Wardlaw et al., 2013), using axial T2-weighted images as reference.

In VaD other pathophysiological processes may also be involved, including α-synucleinopathy, tau pathology, and TAR DNA-binding protein 43 pathology (van der Flier et al., 2018).

Management and treatment options

First, to manage VaD, it is important to make a correct diagnosis. Indeed, the comorbid effects of AD pathology in some cases may be an obstacle to reliable clinical diagnosis and may hinder research into effective management options.

Theoretically, VaD is preventable and treatable, as there are established primary and secondary prevention measures for the causative CVDs.

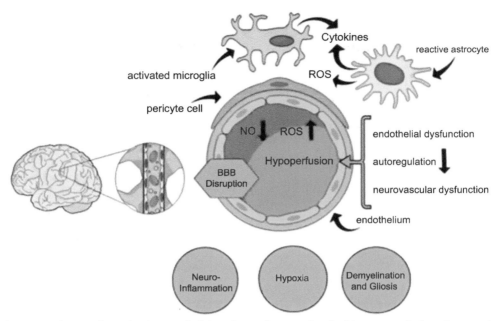

Figure 2.1 Potential mechanism underlying the pathogenesis of white matter lesions in vascular dementia. Endothelial dysfunction, impairment of autoregulation, and alterations in NVU coupling, mediated by NO deficit and oxidative stress, reduce CBF and induce hypoperfusion and BBB disruption. Reactive astrocytes and activated microglia promote neuroinflammation. The resulting hypoxia finally leads to demyelination and gliosis. *BBB*, blood—brain barrier; *CBF*, cerebral blood flow; *NO*, nitric oxide; *NVU*, neurovascular unit; *ROS*, reactive oxygen species.

Primary VaD prevention strategies, in target high-risk groups, should focus on modifying daily lifestyles like smoking, caloric intake, carbohydrate and salt intake in the diet, aerobic and anaerobic physical exercise, optimizing control of diabetes, and control of hypertension (Jaul & Meiron, 2017). In secondary prevention the target is stroke management and prevention of recurrent strokes; the strategies may include monitoring of antidiabetics, antihypertensives, antilipemics, and antiplatelet and anticoagulant medications (McVeigh & Passmore, 2006). Tertiary prevention measures, such as rehabilitation programs after stroke, as well as promoting programs that facilitate social interaction and everyday independent activities are also important (Jaul & Meiron, 2017).

Antiplatelet agents, statins, and drugs that reduce blood pressure represent important treatments for VaD risk factors, even if single-drug strategies do not provide support for these interventions to prevent or treat VaD (O'Brien & Thomas, 2015).

Nowadays, unlike AD, there are no established symptomatic treatments for VaD for cognitive symptoms. Several clinical trials have tried cholinesterase inhibitors and memantine as potential treatments for VaD, based on the evidence of common neuropathological patterns in VaD and AD and in particular on the suggestion of cholinergic deficit in VaD, but they have not led to consistent results. Little cognitive benefit was observed using donepezil in patients with VaD (Wilkinson et al., 2003), although the exact clinical significance of this result is uncertain. Therefore, data are insufficient to support the widespread use of these drugs in VaD, and guideline groups conclude that cholinesterase inhibitors and memantine should not be used in patients with VaD (O'Brien & Thomas, 2015).

Cognitive training and cognitive rehabilitation are specific forms of nonpharmacological intervention to address cognitive and noncognitive outcomes in patients with dementia. In 2013, a Cochrane Review evaluated the effectiveness and impact of cognitive training and cognitive rehabilitation for people with mild AD or VaD, concluding that, even if trial reports indicate some gains resulting from intervention, there is still no indication of any significant benefit derived from cognitive training, maybe because of the absence of standardized outcome measures (Bahar-Fuchs et al., 2013).

Other important points in the management of VaD are identifying and managing comorbidities, with special attention on noncognitive symptoms, and appropriate psychosocial and other support to optimize quality of life for patients and caregivers (O'Brien & Thomas, 2015).

In conclusion, even if there are no proven treatments to reduce the risk of progressive cognitive and functional decline in VaD patients, advances in diagnosis, neuroimaging, trial methods, and harmonization standards for VaD research will help a new generation of trials to improve outcomes in VaD (Smith et al., 2017). Furthermore, advances in the understanding of the pathobiology of the NVU can provide new potential molecular targets for focused and structured interventions in future trials.

Conclusions

VaD is one of the major causes of dementia in elderly, often in conjunction with neurodegenerative diseases such as AD. Although there has been much progress in defining and understanding the relation between CVD and cognitive impairment and dementia, some uncertainties remain. Clinical diagnostic criteria are useful for clinical trials; however, there is still a lack of consensus regarding both the clinical and the pathological definitions (O'Brien & Thomas et al., 2015).

Challenges continue to exist in the differentiation of "pure" VaD from mixed AD and CVD. Nowadays, we have the possibility of detecting AD pathology using in vivo markers such as amyloid PET imaging, tau and amyloid cerebrospinal fluid markers, and in vivo tau imaging, but a combination of biomarkers to define the independent severity of CVD and AD pathology contributing to brain injury in dementia remains an important priority.

Preventing vascular diseases remains the most promising strategy to prevent VaD and possibly dementia in general, although the level of evidence remains low for most interventions. We still need large, properly designed trials for VaD. In the meantime, the control of vascular factors, including treatment of risk factors and secondary stroke prevention, should be considered reasonable strategies for preventing or slowing the progression of cognitive impairment and dementia.

Key facts of vascular dementia

- VaD is one of the major cause of dementia in the elderly, often in conjunction with neurodegenerative diseases such as AD.
- Age, female sex, low education, and vascular risk factors are considered risk factors for developing VaD.
- VCI is the most comprehensive term, and it refers to the entire spectrum of vascular brain pathologies that contribute to the cognitive impairment, ranging from subjective cognitive decline to overt dementia.
- Clinical diagnosis of VaD is challenging and can be made only after considering medical history, physical examination, neuroimages, and neuropsychological assessment.
- The core features for the diagnosis of VaD include a stepwise progression, focal neurological signs and symptoms, an unequal distribution of cognitive deficits, a history of multiple ischemic strokes, neuroimaging evidence of CVD, and a temporal relationship between CVD and dementia.

Summary points

- VaD is the second most common cause of dementia after AD, accounting for ~15%—20% of cases. There are modifiable and nonmodifiable risk factors associated with VaD and most of them are also associated with AD.
- NINDS—AIREN and DSM-5 criteria are the most commonly used guidelines for the diagnosis of VaD. The two main clinical syndromes of VaD are poststroke VaD and VaD without recent stroke.
- Cognitive assessment is very important to make a correct diagnosis of VaD, and a dysexecutive syndrome is the most frequent finding.
- Brain imaging provides an excellent tool for identifying the main vascular lesions that may be linked to a probable VCI, and MRI is the first-choice neuroimaging technique to recognize them. WMH, hippocampal or global atrophy, and cerebral hemorrhages are frequent findings on MRI.
- The pathological examination in VaD usually reveals lesions in the parenchyma, such as infarcts and white matter changes, and alterations in blood vessels, such as arteriosclerosis and CAA.
- Nowadays there are no established symptomatic treatments for VaD cognitive symptoms. The control of vascular factors, including treatment of risk factors, and secondary stroke prevention should be considered reasonable strategies for preventing or slowing the progression of cognitive impairment in VaD.

References

Allan, L. M., Rowan, E. N., Firbank, M. J., Thomas, A. J., Parry, S. W., Polvikoski, T. M., et al. (2011). Long term incidence of dementia, predictors of mortality and pathological diagnosis in older stroke survivors. *Brain, 134*, 3716−3727.

American Psychiatric Association. (2013). *Diagnostic and statistical manual of mental disorders* (5th ed.). American Psychiatric Association.

Arvanitakis, Z., Capuano, A. W., Leurgans, S. E., Bennett, D. A., & Schneider, J. A. (2016). Relation of cerebral vessel disease to Alzheimer's disease dementia and cognitive function in elderly people: A cross-sectional study. *The Lancet Neurology, 15*, 934−943.

Bahar-Fuchs, A., Clare, L., & Woods, B. (2013). Cognitive training and cognitive rehabilitation for mild to moderate Alzheimer's disease and vascular dementia. *Cochrane Database of Systematic Reviews, 5*(6), CD003260.

Cavalieri, M., Enzinger, C., Petrovic, K., Pluta-Fuerst, A., Homayoon, N., Schmidt, H., et al. (2010). Vascular dementia and Alzheimer's disease—are we in a dead-end road? *Neurodegenerative Diseases, 7*, 122−126.

Chui, H. C., Victoroff, J. I., Margolin, D., Jagust, W., Shankle, R., & Katzman, R. (1992). Criteria for the diagnosis of ischemic vascular dementia proposed by the state of California Alzheimer's disease diagnostic and treatment centers. *Neurology, 42*, 473−480.

Clark, R. L., Schiehser, D. M., Weissberger, G. H., Salmon, D. P., Delis, D. C., & Bondi, M. W. (2012). Specific measures of executive function predict cognitive decline in older adults. *Journal of the International Neuropsychological Society, 18*(1), 118−127.

Corraini, P., Henderson, V. W., Ording, A. G., Pedersen, L., Horváth-Puhó, E., & Toft Sørensen, H. (2017). Long-term risk of dementia among survivors of ischemic or hemorrhagic stroke. *Stroke, 48*(1), 180−186.

De Carolis, A., Cipollini, V., Donato, N., Sepe-Monti, M., Orzi, F., & Giubilei, F. (2017). Cognitive profiles in degenerative dementia without evidence of small vessel pathology and small vessel vascular dementia. *Neurological Sciences, 38*(1), 101−107.

Dichgans, M., & Zietemann, V. (2012). Prevention of vascular cognitive impairment. *Stroke, 43*, 3137−3146.

Diniz, B. S., Butters, M. A., Albert, S. M., Dew, M. A., & Reynolds, C. F., 3rd (2013). Late-life depression and risk of vascular dementia and Alzheimer's disease: Systematic review and meta-analysis of community-based cohort studies. *British Journal of Psychiatry, 202*(5), 329−335.

Erkinjuntti, T., Inzitari, D., Pantoni, L., et al. (2000). Research criteria for subcortical vascular dementia in clinical trials. *Journal of Neural Transmission Supplementum, 59*, 23−30.

Fischer, C. E., Qian, W., Schweizer, T. A., Millikin, C. P., Ismail, Z., Smith, E. E., et al. (2015). Lewy bodies, vascular risk factors, and subcortical arteriosclerotic leukoencephalopathy, but not Alzheimer pathology, are associated with development of psychosis in Alzheimer's disease. *Journal of Alzheimer's Disease, 50*, 283−295.

van der Flier, W. M., Skoog, I., Schneider, J. A., Pantoni, L., Mok, V., Chen, C. L. H., et al. (2018). Vascular cognitive impairment. *Nature Reviews Disease Primers, 15*, 4, 18003.

Folstein, M. F., Folstein, S. E., & McHugh, P. R. (1975). "Mini-mental state". A practical method for grading the cognitive state of patients for the clinician. *Journal of Psychiatric Research, 12*, 189−198.

Fuh, J. L., Wang, S. J., & Cummings, J. L. (2005). Neuropsychiatric profiles in patients with Alzheimer's disease and vascular dementia. *Journal of Neurology Neurosurgery and Psychiatry, 76*, 1337−1341.

Fu, Y., & Yan, Y. (2018). Emerging role of immunity in cerebral small vessel disease. *Frontiers in Immunology, 9*, 67.

Garcia, P. Y., Roussel, M., Lamy, C., Canaple, S., Bugnicourt, J. M., Peltier, J., et al. (2013). Cognitive impairment and dementia following intracerebral hemorrhage: A cross-sectional study of a university hospital-based series. *Journal of Stroke and Cerebrovascular Diseases, 22*, 80−86.

Goodman, R. A., Lochner, K. A., Thambisetty, M., Wingo, T. S., Posner, S. F., & Ling, S. M. (2017). Prevalence of dementia subtypes in United States Medicare fee-for-service beneficiaries, 2011-2013. *Alzheimers Dementia, 13*(1), 28−37.

Gorelick, P. B., Scuteri, A., Black, S. E., et al. (2011). Vascular contributions to cognitive impairment and dementia: A statement for healthcare professionals from the American heart association/American stroke association. *Stroke 2011, 42*, 2672–2713.

Gupta, M., Dasgupta, A., et al. (2014). Behavioural and psychological symptoms in poststroke vascular cognitive impairment. *Behavioural Neurology*, 430128.

Hachinski, V. C., Lassen, N. A., & Marshall, J. (1974). Multi-infarct dementia: A cause of mental deterioration in the elderly. *Lancet, 2*, 207–210.

Haffner, C., Malik, R., & Dichgans, M. (2016). Genetic factors in cerebral small vessel disease and their impact on stroke and dementia. *Journal of Cerebral Blood Flow and Metabolism, 36*, 158–171.

Iadecola, C. (2013). The pathobiology of vascular dementia. *Neuron, 80*(4), 844–866.

Iadecola, C., Yaffe, K., Biller, J., Bratzke, L. C., Faraci, F. M., Gorelick, P. B., et al. (2016). Impact of hypertension on cognitive function: A scientific statement from the American heart association. *Hypertension, 68*, e67–e94.

Ikram, M. A., Bersano, A., Manso-Calderón, R., et al. (2017). Genetics of vascular dementia — review from the ICVD working group. *BMC Medicine, 15*, 48.

Inzitari, D., Pracucci, G., Poggesi, A., the LADIS Study Group, et al. (2009). Changes in white matter as determinant of global functional decline in older independent outpatients: Three-year follow-up of LADIS (leukoaraiosis and disability) study cohort. *BMJ, 339*, b2477.

Jagust, W. J., Zheng, L., Harvey, D. J., Mack, W. J., Vinters, H. V., Weiner, M. W., et al. (2008). Neuropathological basis of magnetic resonance images in aging and dementia. *Annals of Neurology, 63*, 72–80.

Jaul, E., & Meiron, O. (2017). Systemic and disease-specific risk factors in vascular dementia: Diagnosis and prevention. *Frontiers in Aging Neuroscience, 9*, 333.

Kalaria, R. N. (2016). Neuropathological diagnosis of vascular cognitive impairment and vascular dementia with implications for Alzheimer's disease. *Acta Neuropathologica, 131*(5), 659–685.

Kalaria, R. N., Akinyemi, R., & Ihara, M. (2016). Stroke injury, cognitive impairment and vascular dementia. *Biochimica et Biophysica Acta, 1862*, 915–925.

Leys, D., Hénon, H., Mackowiak-Cordoliani, M. A., & Pasquier, F. (2005). Poststroke dementia. *The Lancet Neurology, 4*, 752–759.

McAleese, K. E., Alafuzoff, I., Charidimou, A., et al. (2016). Post-mortem assessment in vascular dementia: Advances and aspirations. *BMC Medicine, 14*(1), 129.

McGuinness, B., Barrett, S. L., Craig, D., Lawson, J., & Passmore, A. P. (2010). Executive functioning in Alzheimer's disease and vascular dementia. *International Journal of Geriatric Psychiatry, 25*(6), 562–568.

McVeigh, C., & Passmore, P. (2006). Vascular dementia: Prevention and treatment. *Clinical Interventions in Aging, 1*(3), 229–235.

Moulin, S., Labreuche, J., Bombois, S., Rossi, C., Boulouis, G., Hénon, H., et al. (2016). Dementia risk after spontaneous intracerebral haemorrhage: A prospective cohort study. *The Lancet Neurology, 15*(8), 820–829.

Nasreddine, Z. S., Phillips, N. A., Bedirian, V., Charbonneau, S., Whitehead, V., Collin, I., et al. (2005). The montreal cognitive assessment, MoCA: A brief screening tool for mild cognitive impairment. *Journal of the American Geriatrics Society, 53*, 695–699.

Ngandu, T., Lehtisalo, J., Solomon, A., et al. (2015). A 2 year multidomain intervention of diet, exercise, cognitive training, and vascular risk monitoring versus control to prevent cognitive decline in at-risk elderly people (FINGER): A randomised controlled trial. *Lancet, 385*, 2255–2263.

O'Brien, J. T., & Thomas, A. (2015). Vascular dementia. *Lancet, 386*(10004), 1698–1706.

O'Brien, J. T., Erkinjuntti, T., Reisberg, B., Roman, G., Sawada, T., Pantoni, L., et al. (2003). Vascular cognitive impairment. *The Lancet Neurology, 2*, 89–98.

Pendlebury, S. T., & Rothwell, P. M. (2009). Prevalence, incidence, and factors associated with pre-stroke and post-stroke dementia: A systematic review and meta-analysis. *The Lancet Neurology, 8*, 1006–1018.

Potter, G. M., Chappell, F. M., Morris, Z., & Wardlaw, J. M. (2015). Cerebral perivascular spaces visible on magnetic resonance imaging: Development of a qualitative rating scale and its observer reliability. *Cerebrovascular Diseases, 39*(3–4), 224–231.

Prince, M., Bryce, R., Albanese, E., et al. (2013). The global prevalence of dementia: A systematic review and metaanalysis. *Alzheimers Dement, 9*, 63–75.

Prince, M., Wimo, A., World Alzheimer Report 2015, et al. (2015). The global impact of dementia: An analysis of prevalence, incidence, cost and trends. *Alzheimer's Disease International*. https://www.alz.co.uk/research/WorldAlzheimerReport2015.pdf.

Ramirez, J., Berezuk, C., McNeely, A. A., Gao, F., McLaurin, J., & Black, S. E. (2016). Imaging the perivascular space as a potential biomarker of neurovascular and neurodegenerative diseases. *Cellular and Molecular Neurobiology, 36*(2), 289–299.

Rizzi, L., Rosset, I., & Roriz-Cruz, M. (2014). Global epidemiology of dementia: Alzheimer's and vascular types. *BioMed Research International, 2014*, 908915.

Román, G. C., Tatemichi, T. K., Erkinjuntti, T., et al. (1993). Vascular dementia: Diagnostic criteria for research studies. Report of the NINDS-AIREN international workshop. *Neurology, 43*, 250–260.

Sachdev, P., Kalaria, R., O'Brien, J., et al. (2014). Diagnostic criteria for vascular cognitive disorders: A VASCOG statement. *Alzheimer Disease and Associated Disorders, 28*(3), 206–218.

Satizabal, C. L., Beiser, A. S., Chouraki, V., Chêne, G., Dufouil, C., & Seshadri, S. (2016). Incidence of dementia over three decades in the framingham heart study. *New England Journal of Medicine, 374*, 523–532.

Schneider, J. A., Arvanitakis, Z., Bang, W., & Bennett, D. A. (2007). Mixed brain pathologies account for most dementia cases in community-dwelling older persons. *Neurology, 69*, 2197–2204.

Schneider, J. A., Wilson, R. S., Bienias, J. L., Evans, D. A., & Bennett, D. A. (2004). Cerebral infarctions and the likelihood of dementia from Alzheimer disease pathology. *Neurology, 62*, 1148–1155.

Skrobot, O. A., O'Brien, J., Black, S., et al. (2017). The vascular impairment of cognition classification consensus study. *Alzheimers Dementia, 13*, 624–633.

Smith, E. E. (2017). Clinical presentations and epidemiology of vascular dementia. *Clinical Science, 131*(11), 1059–1068.

Smith, E. E., Cieslak, A., Barber, P., et al. (2017). Therapeutic strategies and drug development for vascular cognitive impairment. *Journal of the American Heart Association: Cardiovascular and Cerebrovascular Disease, 6*(5), e005568.

Snowdon, D. A., Grainer, L. H., Mortimer, J. A., Riley, K. P., Grainer, P. A., & Markesbery, W. R. (1997). Brain infarction and the clinical expression of Alzheimer disease: The nun study. *Journal of the American Medical Association, 277*, 813–817.

Stokholm, J., Vogel, A., Gade, A., & Waldemar, G. (2006). Heterogeneity in executive impairment in patients with very mild Alzheimer's disease. *Dementia and Geriatric Cognitive Disorders, 22*(1), 54–59.

von Strauss, E., Viitanen, M., De Ronchi, D., Winblad, B., & Fratiglioni, L. (1999). Aging and the occurrence of dementia: Findings from a population-based cohort with a large sample of nonagenarians. *Archives of Neurology, 56*, 587–592.

Sudo, F. K., Amado, P., Alves, G. S., Laks, J., & Engelhardt, E. (2017). A continuum of executive function deficits in early subcortical vascular cognitive impairment: A systematic review and meta-analysis. *Dement Neuropsychol, 11*(4), 371–380.

Vemuri, P., Lesnick, T. G., Przybelski, S. A., et al. (2015). Vascular and amyloid pathologies are independent predictors of cognitive decline in normal elderly. *Brain, 138*, 761–771.

Wardlaw, J. M., Smith, E. E., Biessels, G. J., et al. (2013). Neuroimaging standards for research into small vessel disease and its contribution to ageing and neurodegeneration. *The Lancet Neurology, 12*(8), 822–838.

Wardlaw, J. M., Smith, C., & Dichgans, M. (2013). Mechanisms of sporadic cerebral small vessel disease: Insights from neuroimaging. *Lancet Neurology, 12*(5), 483–497. Elsevier Ltd.

Wilkinson, D., Doody, R., Helme, R., the Donepezil 308 Study Group, et al. (2003). Donepezil in vascular dementia: A randomized, placebo-controlled study. *Neurology, 61*, 479–486.

Wu, Y. T., Beiser, A. S., Breteler, M. M. B., Fratiglioni, L., Helmer, C., Hendrie, H. C., et al. (2017). The changing prevalence and incidence of dementia over time — current evidence. *Nature Reviews Neurology, 13*(6), 327–339.

Yang, T., Sun, Y., Lu, Z., Leak, R. K., & Zhang, F. (2017). The impact of cerebrovascular aging on vascular cognitive impairment and dementia. *Ageing Research Reviews, 34*, 15–29.

Ye, B. S., Seo, S. W., Kim, J. H., et al. (2015). Effects of amyloid and vascular markers on cognitive decline in subcortical vascular dementia. *Neurology, 85*, 1687–1693.

CHAPTER 3

Small vessel disease and dementia

Francesco Arba[1], Valentina Rinnoci[1,2]
[1]Stroke Unit, Azienda Ospedaliero-Universitaria Careggi, Florence, Italy; [2]IRCCS Don Gnocchi Fundation, Florence, Italy

List of abbreviations

CMB cerebral microbleed
CT computed tomography
EPVS enlarged perivascular space
MR magnetic resonance
SVD small vessel disease
VaD vascular dementia
WMC white matter changes

Small vessel disease: the problem

Small vessel disease (SVD) is a common pathology of small arteries, capillaries, and venules of the brain (Pantoni, 2010). SVD causes cognitive, psychiatric, and physical disabilities and has been recognized as a major problem for health systems, with relevant consequences for health and social costs (Wardlaw et al., 2013). Nonetheless, the pathology still does not have a recognized cause, and evidence about how to prevent and treat this serious issue is lacking. SVD is responsible for about one-fifth of all strokes worldwide, is an important independent risk factor for future stroke, and is the major contributor to vascular dementia (VaD), which represents almost half of all subtypes of dementia. Life expectancy is now longer than before, and since SVD increases its prevalence with age, understanding the mechanisms of pathology, prevention, and treatment is urgently needed.

Small vessel disease: concepts and imaging features

Small vessels are difficult to visualize in vivo; however, diseased microvasculature leaves hallmarks on the brain that can be detected with neuroimaging techniques. It should be kept in mind, though, that what we see in vivo in the human brain is a neuroimaging surrogate of SVD. In other words, with current diagnostic techniques we are able to detect, evaluate, and quantify only the effects of the affected microvasculature on the brain parenchyma, not the diseased microvasculature itself. The definition of SVD is therefore based on imaging findings rather than on histopathological or clinical findings, and the spectrum of SVD evolves with advances in neuroimaging.

Diagnosis and Management in Dementia
ISBN 978-0-12-815854-8, https://doi.org/10.1016/B978-0-12-815854-8.00003-3

Some years ago an international collaborative group called STRIVE-v1 (Standards for Reporting Vascular Changes on Neuroimaging) (Wardlaw et al., 2013) agreed on standards for reporting neuroimaging markers of SVD and revised the terminology to describe typical lesions of SVD for research and clinical practice (Table 3.1). One of the conclusions of the group was that SVD recognizes various radiological phenotypes, with magnetic resonance (MR)- and computed tomography (CT)-detectable SVD features listed as follows: (1) recent small subcortical infarcts, (2) lacunes of presumed vascular origin, (3) white matter changes (WMCs) of presumed vascular origin, (4) enlarged perivascular spaces (EPVSs), (5) cerebral microbleeds (CMBs) and other hemorrhagic lesions, (6) brain atrophy (Fig. 3.1). Cortical microinfarcts can be added to this list (ter Telgte et al., 2018), although high-field MR is needed to visualize them. Some lesions are more frequent than others, but efforts in both research and clinical practice should be focused on assessing each radiological feature of SVD and how they interact with one another. In addition to these detectable SVD features, "invisible" modifications of the brain parenchyma have been identified, so-called normal-appearing WMCs (Gouw et al. 2011), detectable with diffusion tensor MR imaging, reflecting disorganization into axonal myelin. Such modifications have been named "SVD penumbra" and predate WMCs (Huisa et al., 2015; Rosenberg et al., 2016) and surround lacunar infarcts up to 1.5 times the diameter of the lacune. Also, normal-appearing WMCs have been found around perivascular spaces and microbleeds (ter Telgte et al., 2018).

Small subcortical infarcts and lacunar strokes

Lacunar strokes cause about 25% of all ischemic strokes and are thought to be the consequence of occlusion of small penetrating arteries. Lacune is a Latin term that indicates a hole, as stated in the first description by Miller Fisher with pathologic specimens. Lacunes have been historically considered the first SVD feature (Fisher, 1991) and are one of the etiology subtypes of ischemic stroke. Actually, the radiological appearance of the lacune is a cavity with the same cerebrospinal fluid density (if CT scan) or intensity (if MR) signal. Radiologically evident lacunes can be either symptomatic and cause a stroke or asymptomatic and detected by chance on MR or CT. STRIVE-v1 defined lacunes as round or ovoid fluid-filled cavities ranging from 3 to 15 mm, although the higher cutoff of 20 mm is usually accepted.

The concept of clinically a silent small subcortical infarct (i.e., silent brain infarcts) is slightly different from lacune. Similar to lacunes, silent small subcortical infarcts are sometimes found by chance with MR imaging without a clinical correlate. Conversely, up to 30% of patients with a clinically lacunar syndrome may not present with an MR-detectable small subcortical infarct (Doubal, Dennis & Wardlaw, 2011). In addition, small subcortical infarcts may have a diverse radiological evolution, ranging from a lacunar

Table 3.1 Imaging features and clinical correlates of cerebral small vessel disease.

	SVD feature	Imaging	Definition	Clinical correlate
Ischemic manifestations	Lacunes of presumed vascular origin[a]	MR/CT	Small round or ovoid lesions from 3 to 15 mm (20 mm according to some studies), with the same signal intensity of cerebrospinal fluid	Asymptomatic/ischemic stroke/cognitive impairment
	Recent small subcortical infarct	MR	Small round or ovoid hyperintense lesion ≤ 20 mm detected on DWI in the subcortical structures; possible evolution in lacunes, confluent white matter changes, or disappearance	Asymptomatic/ischemic stroke/cognitive impairment
	White matter changes	MR/CT	Bilateral, symmetric hypodensity (CT) or hyperintensity (MR, T2WI), located in the periventricular or deep white matter	Asymptomatic/cognitive impairment/gait disturbances/mood disorders
	Enlarged perivascular spaces	MR, sometimes CT	Fluid-filled spaces that follow the course of a penetrating vessel through the white or gray matter; appear linear when imaged parallel to the course of the vessel (e.g., centrum semiovale), round or ovoid (≤ 3 mm) when imaged perpendicular to the course of the vessel (e.g., basal ganglia); visible on CT when large ("giant perivascular spaces")	Asymptomatic/cognitive impairment?

Continued

Table 3.1 Imaging features and clinical correlates of cerebral small vessel disease.—cont'd

SVD feature	Imaging	Definition	Clinical correlate
Microinfarcts	MR	Small (<5 mm) areas, located in any part of the brain, hyperintense on DWI or T2WI, hypointense on T1WI, visible in at least two planes	Asymptomatic/cognitive impairment?
Atrophy	MR/CT	A lower brain volume that is not related to a specific macroscopic focal injury such as trauma or infarction	Cognitive impairment
Hemorrhagic features			
Hemorrhage	MR/CT	Hyperdense (CT), hyper-/hypointense (in different MR sequences) lesion primarily located in typical (deep, basal ganglia, thalamus) or atypical (lobar) regions of the brain; extravasation into subarachnoid space and/or ventricles is possible	Hemorrhagic stroke, cognitive impairment
Microbleeds	MR	Small rounded areas (2–5 mm in diameter, sometimes up to 10 mm) of hypointense signal with blooming effect on T2*-GE or SWI	Amyloid spells, cognitive impairment
Cortical superficial siderosis	MR	Well defined, homogeneous hypointense curvilinear signal intensity (black) on T2*-GE or SWI in the superficial layers of the cerebral cortex, within the subarachnoid space, or both	Focal neurological deficits (mainly transient)

CT, computed tomography; DWI, diffusion-weighted imaging; MR, magnetic resonance; SVD, small vessel disease; SWI, susceptibility-weighted imaging; T1WI, T1-weighted imaging; T2WI, T2-weighted imaging; T2*-GE, T2* gradient-echo-weighted imaging.
[a]Lacunes sometimes are the healing stage of a small hemorrhage.

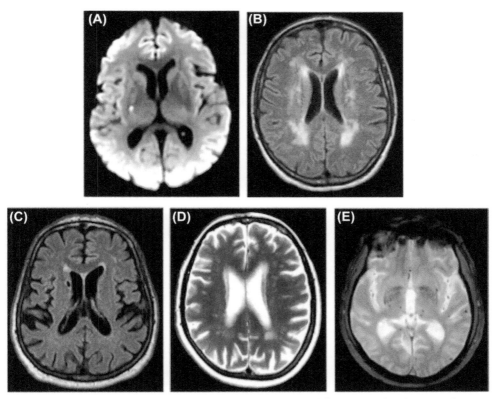

Figure 3.1 Imaging features of small vessel disease. (A) Axial diffusion-weighted image showing a recent small subcortical infarct. (B) Axial fluid attenuated inversion recovery (FLAIR) image showing white matter hyperintensity of presumed vascular origin. (C) Axial FLAIR image showing a lacune of presumed vascular origin. (D) Axial T2-weighted image showing enlarged perivascular spaces. (E) Axial T2* gradient-echo image showing multiple cerebral microbleeds.

cavity to an almost complete disappearance with little visible lesion on conventional MR. This suggests a great heterogeneity in both clinical and radiological presentation of small subcortical infarcts, possibly reflecting the scarce data existing about the pathological correlation of small subcortical infarcts.

White matter changes

WMCs are defined as bilateral, mostly symmetrical changes in density (with CT scan) or signal (with MR imaging) in deep or periventricular white matter. The brain stem could be another localization of such changes, although there is no consensus as to whether brain-stem WMCs should be systematically included in relevant studies. They have been traditionally associated with cardiovascular risk factors, such as age, hypertension, diabetes, and smoking (Debette & Markus, 2010), but their pathogenesis is still not fully understood. The term WMC of presumed vascular origin should be preferred to exclude other white matter lesions with different origins (e.g., multiple sclerosis white matter lesions). White matter lesions are detected with both MR and CT imaging. They appear hyperintense on T2- and hypointense on T1-weighted sequences, depending on the severity of the disease. When using CT, they appear as an X-ray attenuation, mainly in the periventricular white matter. WMCs can be measured with qualitative and quantitative scales, in both MR and CT.

Enlarged perivascular spaces

Perivascular spaces are virtual spaces around cerebral arterioles, veins, and venules. They course from the brain surface through the brain parenchyma and can be followed by leptomeninges. Perivascular spaces, also called Virchow—Robin spaces, are usually not visible with conventional neuroimaging; however, they become increasingly larger with age. EPVSs are associated with morphological signs of SVD such as white matter hyperintensities and lacunes, but not atrophy. EPVSs are detectable as fluid-filled spaces that follow the course of a vessel through gray or white matter, with MR signal intensity similar to that of cerebrospinal fluid, rounded or ovoid appearance with a diameter smaller than 3 mm when imaged perpendicular to the vessel, and linear when imaged parallel to the vessel. EPVSs can be seen on axial, sagittal, or coronal MR sections and can be rated on brain stem, hippocampus (Adams, 2013), basal ganglia, and centrum semiovale (Potter, 2015). With high-resolution MR a small vessel can be seen in the center of the perivascular space, which could differentiate EPVSs from small lacunes. The clinical significance of EPVSs remains controversial: although an association with worse cognitive function has been reported (Arba & Quinn,et al., 2018), perivascular spaces should not be referred to as pathological lesions.

Cerebral microbleeds and other hemorrhagic lesions

CMBs are hemosiderin-laden macrophage deposits. Typically located in the perivascular tissue, consistent with leakage of blood cells, CMBs are associated with SVD and Alzheimer disease. The location of microbleeds (e.g., lobar or cortical) has been included in the Boston criteria for diagnosis of cerebral amyloid angiopathy. The clinical relevance of CMBs has been overlooked in past years; however, emerging data have been showing a meaningful association with cognitive impairment. CMBs are generally not seen on CT, whereas they are detectable with T2*-weighted gradient echo sequences as small (2—5 mm in diameter, up to 10 mm) hypointense lesions, most commonly located in the cortico-subcortical junction, deep gray or white matter, brain stem, or cerebellum.

Other hemorrhagic lesions may occur in SVD: intracerebral hemorrhage and superficial siderosis. Intracerebral hemorrhage is one of the clinical manifestations of SVD (Pantoni, 2010). Nonlobar (i.e., deep or basal ganglia) hemorrhage is mostly due to perforating artery vasculopathy, whereas the presence of lobar hemorrhage should raise suspicion of cerebral amyloid angiopathy. Superficial cortical siderosis is defined as the neuroimaging evidence of blood products (i.e., hemosiderin) in the superficial cortex under the pial meninges. It could be the consequence of subarachnoid hemorrhage, vascular malformations, cerebral amyloid angiopathy, or dural defects or could also be idiopathic. T2*-weighted gradient echo sequences can detect a linear hypointensity over the cortex; however, this can be mimicked by petechial transformation of cortical cerebral infarcts. Superficial siderosis and CMBs may occasionally cause positive or negative focal neurological symptoms, and around 25% of patients with such symptoms (the so-called "amyloid spells") may develop a lobar intracerebral hemorrhage (Charidimou et al., 2012).

Brain atrophy

Brain atrophy occurs with the physiological aging process, but its extent may vary among individuals. It can have different characteristics with regard to localization (global or focal, symmetric or asymmetric) and the tissue predominantly affected (i.e., gray or white matter). As to SVD, many studies report an association between presence and severity of SVD and brain atrophy, including global atrophy, corpus callosum atrophy, central atrophy (i.e., increased ventricular size and atrophy of the basal ganglia), mesencephalic atrophy, hippocampal atrophy, and focal cortical thinning in regions connected to subcortical infarcts. As a consequence, brain atrophy is a useful measure to assess the burden of vascular damage in the brain, and brain atrophy may partly explain the effects of vascular lesions on cognition. Brain atrophy related to SVD is defined as lower brain volume not associated with a macroscopic focal injury such as brain trauma or cortical infarcts.

Cortical microinfarcts

Cortical microinfarcts are small lesions of presumed ischemic origin that can be detected on diffusion-weighted MR, high-resolution MR, or neuropathological examination (van Veluw et al., 2017). Cortical microinfarcts are located in the cortex of cerebral lobes, smaller than 5 mm, hyperintense in diffusion-weighted and T2 sequences, and hypointense in T1 sequences. They are very common in the elderly in both healthy subjects and patients with dementia. Cortical microinfarcts are present in one-fourth of patients without dementia, and almost two-thirds of patients with VaD have cortical microinfarcts, whereas nearly 50% of patients with Alzheimer disease have cortical microinfarcts (Brundel, De Bresser, Van Dillen, Kappelle & Biessels, 2012). Among the causes of cortical microinfarcts we can find SVD, microemboli, and cerebral hypoperfusion.

Frequently associated with lacunar infarcts, WMCs, and cerebral amyloid angiopathy, cortical microinfarcts have been associated with cognitive impairment and cognitive dementia, through perilesional and remote effects on brain function.

Detection of small vessel disease: imaging

Imaging features of SVD are frequently detected in asymptomatic adults, stroke patients, and patients with cognitive impairment, and a number of studies have investigated the presence and frequency of SVD. However, relevant differences in methods and study design make difficult the estimate of the exact prevalence and incidence of SVD. The wide imaging phenotype of SVD has not been systematically evaluated in the majority of the studies; often only one of the imaging hallmarks of SVD has been evaluated, usually WMCs. Also, the spectrum of imaging manifestations of SVD has been enriched through the years: in the late 1980s and early 1990s, SVD and leukoaraiosis (i.e., WMCs) were synonyms; today we assume that even brain atrophy, which was traditionally regarded as a marker of neurodegeneration, could be at least partly referred to as an SVD marker (METACOHORTS Consortium, 2016; ter Telgte et al., 2018).

It is an established fact that SVD prevalence increases with age. WMCs are particularly common in patients with cognitive impairment. It has been found that individuals between 65 and 69 years had a prevalence of 90% of any WMC, whereas the occurrence of WMCs in subjects over 80 years was 98%, with increasing severity of WMC grade (Breteler et al., 1994; Longstreth et al., 1996). In patients with Alzheimer disease, assessment of WMCs with plain CT scan ranges from around 20% to 80%, whereas studies with MR assessment report wider percentages, ranging from 8% to 100% (Breteler et al., 1994; Liao et al., 1996; Longstreth et al., 1996). Around one-fourth of new lesions consistent with WMCs may appear in a time frame of 5 years (Longstreth et al., 2005). Silent brain infarcts, which are very similar to new punctate WMC lesions, have been reported to have an incidence of 3% per year among elderly people, and prevalence ranges from about 6% in patients at 60 years of age to 28% at age 80 years (Bryan et al., 1997). However, SVD is a highly dynamic process, with nonlinear progression of its volume and an annual incidence of both lacunes and microbleeds of about 2% (Van Leijsen, 2017). Due to advances in diagnostic imaging techniques, further longitudinal and population studies are required to get more precise data about the incidence and prevalence of SVD.

Classification and pathogenesis of small vessel disease

As previously stated, SVD is a term that encompasses a wide range of diseases, many of which are systemic. Therefore, SVD could be classified on the basis of the underlying pathology. Every effort should be made to classify SVD when approaching a patient, since this could help to target prevention and eventually treatment. According to the etiopathogenetic classification, we can identify six subtypes of SVD (Pantoni, 2010): (1) arteriolosclerosis, (2) sporadic and hereditary amyloid angiopathy, (3) inherited or genetic SVD distinct from amyloid angiopathy, (4) inflammatory and immunologic SVD, (5) venous collagenosis, (6) other SVD (e.g., postradiation angiopathy).

The pathogenesis of SVD is a debated topic. The main concept is that small vessels are broadly affected, and this may lead to occlusion or rupture of the blood vessel, causing ischemic or hemorrhagic lesion, respectively. The traditional theory advocates that vessel wall modifications cause a reduction in cerebral blood flow with consequent parenchymal damage, particularly in the white matter (Pantoni, 2010). According to this explanation, the ischemic damage can be chronic, causing white matter hyperintensity, or acute, causing a lacunar infarction or a small subcortical infarct. The rupture of the vessel can result in a microbleed or a cerebral hemorrhage, according to the size and location of the vessel. However, this theory has been challenged by a more recent hypothesis that advocates endothelial dysfunction and blood—brain barrier disruption as the first steps toward microstructural damage of the white matter (Rosenberg et al., 2016; Warldaw et al., 2013). Again, neurovascular inflammation and remodeling of the extracellular matrix appear to play a relevant role in pathogenesis of SVD, as reported by an increasing number of studies (Arba & Piccardi,et al., 2018; Shoamanesh et al., 2015). The subsequent vascular oxidative stress may finally result in dysfunction of the neurovascular unit, which increases the susceptibility of the brain to injury by altering regulation of the cerebral blood supply, disrupting blood—brain barrier function and the repair potential of the neurons. As a result, we can detect the anatomical effect of such pathological process on the brain as WMCs and the other hallmarks of SVD.

Although classical vascular risk factors, such as hypertension, diabetes, hypercholesterolemia, and smoke exposure, are likely to be involved in the pathogenesis of SVD, data suggest that other factors may contribute to the development and progression of SVD. In a population study, hypertension (i.e., the more frequent vascular risk factor) accounted for less than 2% of the variance in WMC burden (Wardlaw et al., 2014), suggesting that other factors than hypertension were responsible for variations in WMCs. This is consistent with data from a randomized controlled trial, which showed that in patients with previous lacunar infarction, aggressive reduction in blood pressure did not significantly reduce the incidence of new stroke events (odds ratio [OR] = 0.81; 95% confidence interval [CI] 0.64—1.03) nor the composite outcome of myocardial infarcts and vascular death (OR = 0.84; 95% CI 0.68—1.04) (SPS3 Study Group, 2013). Genetic factors play a major role among emerging risk factors and seem to be particularly important in the development of white matter hyperintensities and dilated perivascular spaces (Duperron et al., 2018). However, it is now recognized that SVD develops throughout the whole life span. The lack of physical exercise is associated with recurrence of stroke and higher progression of white matter hyperintensity (Gow et al., 2012), and studies indicate that high dietary salt intake increases the risk of lacunar stroke and white matter hyperintensity load (Heye et al., 2016; Makin et al., 2017). Similarly, supplementation with B vitamins appears to reduce the progression of white matter hyperintensities in patients with severe SVD at baseline (Cavalieri et al., 2012), suggesting that metabolic factors may also influence WMC burden. There is also evidence that early life exposure to environmental factors such as poor socioeconomic status and educational achievements, as well as lower childhood intelligence, may predate the development of future SVD burden (Backhouse, McHutchison, Cvoro, Shenkin & Wardlaw, 2017). SVD appears, therefore, to be a multifactorial pathology, linked to vascular, genetic, and environmental factors.

Small vessel disease and cognitive impairment: subtypes

SVD largely contributes to vascular mild cognitive impairment and VaD, regarded as the second most common type of dementia after Alzheimer disease (Kalaria, 2018; ter Telgte et al., 2018). The prevalence of VaD is age dependent and doubles approximately every 5 years of age, with estimates ranging from less than 1% in patients at 60 years to around 5% in patients over 85 years of age (Kalaria et al., 2008), with a poorer prognosis compared with Alzheimer disease, due to concurrent comorbidities.

Three main subtypes of vascular lesions that cause cognitive impairment have been classically identified: multiinfarct dementia/cortical VaD, strategic infarct dementia, and subcortical VaD. A further subtype of VaD is mixed dementia, which means dementia with concurrent vascular and neurodegenerative (mainly Alzheimer disease) pathological changes. In addition to those three entities, the term "poststroke dementia" refers to when a cognitive deficit develops soon after a stroke, regardless of the type, size, or number of vascular lesions. However, SVD may contribute to all types of vascular cognitive impairment (Fig. 3.2).

Poststroke dementia occurs in about 10% of patients after stroke, although this figure may vary on the basis of the study type, ranging from 7% in population-based studies to over 40% in hospital-based studies (Pendlebury & Rothwell, 2009). However, it is generally accepted that poststroke dementia occurs in around one-third of stroke survivors. In the aforementioned metaanalysis, pathological changes in SVD such as pre-existing WMCs (OR = 2.5; 95% CI 1.9−3.4) and global cerebral atrophy (OR = 2.6; 95% CI 1.1−6.3) were independently associated with development of subsequent poststroke dementia (Pendlebury & Rothwell, 2009), supporting a possible role for SVD also in poststroke dementia.

Traditionally, poststroke dementia associated with lesions in specific areas (such as thalamus, basal ganglia, limbic system) that are considered important for neurocognitive processes was called strategic infarct dementia. The main feature of strategic infarct dementia is a mismatch between the size of the lesion, which is generally small (i.e.,

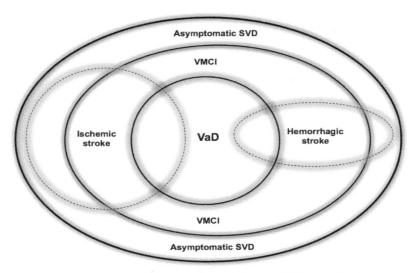

Figure 3.2 Overlap between asymptomatic small vessel disease (*SVD*), cognitive impairment, dementia, and stroke. *VaD*, vascular dementia; *VMCI*, vascular mild cognitive impairment.

lacunar), and the effects on cognition (Desmond et al., 2000). Lacunar strokes are frequent in basal ganglia and thalamus and may contribute to cognitive impairment, and are usually thought to be mild strokes. Nonetheless, lacunar strokes have a similar incidence (around one-fourth) of poststroke cognitive impairment or dementia compared with nonlacunar strokes, although authors have suggested that concurrent associated SVD could contribute to explaining this figure (Makin, Turpin, & Dennis, 2013).

The onset of subcortical ischemic VaD (i.e., dementia purely due to SVD) is more insidious, and temporal relations between cognitive decline, brain imaging features, and evidence of cerebrovascular disease may be not as clear as in other subtypes, where onset is sudden and in accordance to clinically evident stroke (Table 3.2).

Small vessel disease and cognitive impairment: clinical features

As well as the physiopathology of the pathological events, SVD is a disorder with complex clinical manifestations. The anatomical modifications that occur in SVD are the tip of the iceberg of a greater network dysfunction, because the loss of connectivity of subcortical white matter tracts causes a loss of functional connectivity in the brain, which is responsible for the subtle clinical symptoms in vascular cognitive impairment due to SVD.

The clinical presentation is typically subtle, slow, typically with a stepwise progression. Cognitive impairment is frequently due to executive dysfunction, mainly characterized by deficits in abstract problem solving, attention (sustained and separate), set shifting, and verbal fluency and decreased information processing speed. All these characteristics can

Table 3.2 Definition and classification of vascular cognitive impairment.

Vascular cognitive impairment	
VaMCI	**DVaD**
• Includes the four subtypes proposed for the classification of mild CI: amnestic, amnestic plus other domains, nonamnestic single domain, and nonamnestic multiple domain • Diagnosis must be based on the assessment of a minimum of four cognitive domains: executive/attention, memory, language, and visuospatial functions • Functional independence of the subject in activities of daily living is not impaired or is mildly impaired	• Decline in cognitive functions from a prior baseline and a deficit in performance in two or more cognitive domains that affects functional independence of the subject in activities of daily living • Diagnosis must be based on the assessment of a minimum of four cognitive domains: executive/attention, memory, language, and visuospatial functions
Probable VaMCI Presence of CI plus imaging evidence of cerebrovascular disease **AND** – temporal relationship between a vascular event (i.e., clinical stroke) and onset of cognitive deficits **OR** – relationship in the severity and pattern of cognitive decline and the presence of diffuse cerebral subcortical microangiopathy (i.e., evidence of small vessel disease pathology) **AND** No history of progressive cognitive deficits suggesting the presence of a neurodegenerative disorder	**Probable VaD** Presence of CI plus imaging evidence of cerebrovascular disease **AND** – temporal relationship between a vascular event (i.e., clinical stroke) and the onset of cognitive deficits **OR** – relationship in the severity and pattern of cognitive decline and the presence of diffuse cerebral subcortical microangiopathy (i.e., evidence of small vessel disease pathology) **AND** No history of progressive cognitive deficits suggesting the presence of a previous neurodegenerative disorder
Possible VaMCI Presence of CI plus imaging evidence of cerebrovascular disease, **BUT** – no clear relationship between the vascular disease and the onset of CI **OR** – insufficient data for the diagnosis of VaMCI (i.e., no availability of neuroimaging) **OR** – severity of symptoms (e.g., aphasia) precludes proper cognitive assessment **OR** – evidence of neurodegenerative disease or condition that may affect cognition	**Possible VaD** Presence of CI plus imaging evidence of cerebrovascular disease, **BUT** – no clear relationship between the vascular disease and the onset of CI **OR** – insufficient data for the diagnosis of VaD (i.e., no availability of neuroimaging) **OR** – severity of symptoms (e.g., aphasia) precludes proper cognitive assessment **OR** – evidence of neurodegenerative disease or condition that may affect cognition

Unstable VaMCI applies to subjects with the diagnosis of probable or possible VaMCI whose symptoms revert to normal. CI, cognitive impairment; VaMCI, vascular mild cognitive impairment; VaD, vascular dementia.

be detectable in the early stage of the disease. Cognitive impairment due to SVD may also present memory impairment, which is different from that observed in neurodegenerative disorders as Alzheimer disease. Language deficits are not common in cognitive impairment related to SVD, although worse comprehension tasks may occur.

The presence, severity, and progression of WMCs have been associated with cognitive decline in executive functions, attention, speed, and motor control among nondemented independent elderly subjects (Verdelho et al., 2007). Lobar CMBs have been associated with information processing, motor speed, and global cognitive performance (Poels et al., 2012), whereas deep CMBs were associated with motor speed, processing speed, and executive functions. Multiple CMBs are associated with worse cognitive functioning with a dose—response effect (Qiu et al., 2010), supporting a role of CMBs in the occurrence of cognitive impairment.

In addition to cognitive symptoms related to SVD, motor impairment may occur. Dizziness and unsteadiness are early features and could precede gait impairment (parkinsonian gait), with associated axial and arm rigidity and bradykinesia, particularly in lower limbs. Affected patients are at high risk of falls, and the course of pathology can be frequently complicated by bone fractures with consequent hospitalization. Mood disorders, particularly depression and apathy, frequently occur in affected patients. Delirium is also frequent, together with delusions and agitation, and frequently patients require antipsychotic drugs. Clinicians should be aware of the intrinsic higher risk of cerebrovascular events due to antipsychotic drugs; however, treatment is often the trade-off for caregivers to avoid institutionalization of the patient. The course of the pathology could be complicated by dysphagia and dysarthria, which are frequent, as well as hypophonia. Bladder incontinence is a tardive feature, frequently followed by fecal incontinence.

Stroke, ischemic as well as hemorrhagic, may occur at each stage of the pathology. Ischemic stroke is usually a lacunar stroke, caused by occlusion of a penetrating artery and located in the basal ganglia, centrum semiovale, or brain stem. Hemorrhagic stroke is generally located in the basal ganglia (typical hemorrhage location), whereas a lobar or cortical location (atypical hemorrhage) is suggestive of cerebral amyloid angiopathy.

Conclusions

SVD is a heterogeneous pathology of the microcirculation of the brain. Due to the aging of the population, the occurrence of SVD is likely to increase, with important implications for health care systems. Despite advances in detecting imaging features and understanding mechanisms, SVD remains a major challenge for researchers and clinicians. Future efforts in research should focus on prevention and treatment, thus limiting the detrimental effects of the pathology.

Key facts

- SVD affects microcirculation in the brain.
- SVD is detectable with magnetic resonance or computed tomography imaging.
- SVD imaging features may vary from ischemic to hemorrhagic.
- SVD is mainly an age-related disorder, but genetic and inflammatory forms may occur.
- The pathogenesis of SVD is still under debate.
- Dementia related to SVD is the most common form of vascular dementia.
- Clinical manifestations of SVD range from cognitive impairment to stroke, both ischemic and hemorrhagic.

References

Adams, H. H., Cavalieri, M., Verhaaren, B. F., Bos, D., van der Lugt, A., Enzinger, C., et al. (2013). Rating method for dilated Virchow-Robin spaces on magnetic resonance imaging. *Stroke, 44,* 1732–1735.

Arba, F., Piccardi, B., Palumbo, V., Giusti, B., Nencini, P., Gori, A. M., et al., MAGIC Study Group. (2018). Small vessel disease is associated with tissue inhibitor of matrix metalloproteinase-4 after ischaemic stroke. *Translational Stroke Research.* https://doi.org/10.1007/s12975-018-0627-x (Epub ahead of print).

Arba, F., Quinn, T. J., Hankey, G. J., Lees, K. R., Wardlaw, J. M., Ali, M., et al., VISTA Collaboration. (2018). Enlarged perivascular spaces and cognitive impairment after stroke and transient ischemic attack. *International Journal of Stroke, 13,* 47–56.

Backhouse, E. V., McHutchison, C. A., Cvoro, V., Shenkin, S. D., & Wardlaw, J. M. (2017). Early life risk factors for cerebrovascular disease: A systematic review and meta-analysis. *Neurology, 88,* 976–984.

Benavente, O. R., Coffey, C. S., Conwit, R., Hart, R. G., McClure, L. A., Pearce, L. A., et al., SPS3 Study Group. (2013). Blood-pressure targets in patients with recent lacunar stroke: The SPS3 randomised trial. *Lancet, 382*(9891), 507–515. Erratum in: Lancet (2013), 382(9891), 506.

Breteler, M. M., van Swieten, J. C., Bots, M. L., Grobbee, D. E., Claus, J. J., van den Hout, J. H., et al. (1994). Cerebral white matter lesions, vascular risk factors, and cognitive function in a population-based study: The rotterdam study. *Neurology, 44*(7), 1246–1252.

Brundel, M., De Bresser, J., Van Dillen, J. J., Kappelle, L. J., & Biessels, G. J. (2012). Cerebral microinfarcts: A systematic review of neuropathological studies. *Journal of Cerebral Blood Flow and Metabolism, 32,* 425–436.

Bryan, R. N., Wells, S. W., Miller, T. J., Elster, A. D., Jungreis, C. A., Poirier, V. C., et al. (1997). Infarct like lesions in the brain: Prevalence and anatomic characteristics at MR imaging of the elderly— data from the cardiovascular health study. *Radiology, 202,* 47–54.

Cavalieri, M., Schmidt, R., Chen, C., Mok, V., de Freitas, G. R., Song, S., et al., VITATOPS Trial Study Group. (2012). B vitamins and magnetic resonance imaging-detected ischemic brain lesions in patients with recent transient ischemic attack or stroke: The VITAmins TO prevent stroke (VITATOPS) MRI-substudy. *Stroke, 43,* 3266–3270.

Charidimou, A., Peeters, A., Fox, Z., Gregoire, S. M., Vandermeeren, Y., Laloux, P., et al. (2012). Spectrum of transient focal neurological episodes in cerebral amyloid angiopathy: Multicentre magnetic resonance imaging cohort study and meta-analysis. *Stroke, 43*(9), 2324–2330.

Debette, S., & Markus, H. S. (2010). The clinical importance of white matter hyperintensities on brain magnetic resonance imaging: Systematic review and meta-analysis. *BMJ, 341,* c3666.

Desmond, D. W., Moroney, J. T., Paik, M. C., Sano, M., Mohr, J. P., Aboumatar, S., et al. (2000). Frequency and clinical determinants of dementia after ischemic stroke. *Neurology, 54*(5), 1124–1131.

Doubal, F. N., Dennis, M. S., & Wardlaw, J. M. (2011). Characteristics of patients with minor ischaemic strokes and negative MRI: A cross-sectional study. *Journal of Neurology Neurosurgery and Psychiatry, 82,* 540—542.

Duperron, M. G., Tzourio, C., Sargurupremraj, M., Mazoyer, B., Soumaré, A., Schilling, S., et al. (2018). Burden of dilated perivascular spaces, an emerging marker of cerebral small vessel disease, is highly heritable. *Stroke, 49*(2), 282—287.

Fisher, C. M. (1991). Lacunar infarcts-a review. *Cerebrovascular Diseases, 1,* 311—320.

Gouw, A. A., Seewann, A., van der Flier, W. M., Barkhof, F., Rozemuller, A. M., Scheltens, P., et al. (2011). Heterogeneity of SVD: A systematic review of MRI and histopathology correlations. *Journal of Neurology Neurosurgery and Psychiatry, 82,* 126—135.

Gow, A. J., Bastin, M. E., Muñoz Maniega, S., Valdés Hernández, M. C., Morris, Z., Murray, C., et al. (2012). Neuroprotective lifestyles and the aging brain: Activity, atrophy, and white matter integrity. *Neurology, 79*(17), 1802—1808.

Heye, A. K., Thrippleton, M. J., Chappell, F. M., Hernández, M., del, C., Armitage, P. A., et al. (2016). Blood pressure and sodium: Association with MRI markers in cerebral small vessel disease. *Journal of Cerebral Blood Flow and Metabolism, 36*(1), 264—274.

Huisa, B. N., Caprihan, A., Thompson, J., Prestopnik, J., Qualls, C. R., & Rosenberg, G. A. (2015). Long-term blood-brain barrier permeability changes in binswanger disease. *Stroke, 46,* 2413—2418.

Kalaria, R. N. (2018). The pathology and pathophysiology of vascular dementia. *Neuropharmacology, 134,* 226—239.

Kalaria, R. N., Maestre, G. E., Arizaga, R., Friedland, R. P., Galasko, D., Hall, K., et al., World Federation of Neurology Dementia Research Group. (2008). Alzheimer's disease and vascular dementia in developing countries: Prevalence, management, and risk factors. *The Lancet Neurology, 7*(9), 812—826.

Liao, D., Cooper, L., Cai, J., Toole, J. F., Bryan, N. R., Hutchinson, R. G., et al. (1996). Presence and severity of cerebral white matter lesions and hypertension, its treatment, and its control. The ARIC Study. Atherosclerosis Risk in Communities Study. *Stroke, 27,* 2262—2270.

Longstreth, W. T., Jr., Arnold, A. M., Beauchamp, N. J., Jr., Manolio, T. A., Lefkowitz, D., Jungreis, C., et al. (2005). Incidence, manifestations, and predictors of worsening white matter on serial cranial magnetic resonance imaging in the elderly: The cardiovascular health study. *Stroke, 36,* 56—61.

Longstreth, W. T., Jr., Manolio, T. A., Arnold, A., Burke, G. L., Bryan, N., Jungreis, C. A., et al. (1996). Clinical correlates of white matter findings on cranial magnetic resonance imaging of 3301 elderly people. The Cardiovascular Health Study. *Stroke, 27,* 1274—1282.

Makin, S. D. J., Mubki, G. F., Doubal, F. N., Shuler, K., Staals, J., Dennis, M. S., et al. (2017). Small vessel disease and dietary salt intake: Cross-sectional study and systematic review. *Journal of Stroke and Cerebro-vascular Diseases, 26*(12), 3020—3028.

Makin, S. D., Turpin, S., Dennis, M. S., & Wardlaw, J. M. (2013). Cognitive impairment after lacunar stroke: Systematic review and meta-analysis of incidence, prevalence and comparison with other stroke subtypes. *Journal of Neurology Neurosurgery and Psychiatry, 84,* 893—900.

METACOHORTS Consortium. (2016). METACOHORTS for the study of vascular disease and its contribution to cognitive decline and neurodegeneration: An initiative of the Joint Programme for Neurodegenerative Disease Research. *Alzheimers Dement, 12*(12), 1235—1249.

Pantoni, L. (2010). Cerebral small vessel disease: From pathogenesis and clinical characteristics to therapeutic challenges. *The Lancet Neurology, 9*(7), 689—701.

Pendlebury, S. T., & Rothwell, P. M. (2009). Prevalence, incidence, and factors associated with pre-stroke and post-stroke dementia: A systematic review and meta-analysis. *The Lancet Neurology, 8,* 1006—1018.

Poels, M. M., Ikram, M. A., van der Lugt, A., Hofman, A., Niessen, W. J., Krestin, G. P., et al. (2012). Cerebral microbleeds are associated with worse cognitive function: The rotterdam scan study. *Neurology, 78*(5), 326—333.

Potter, G. M., Chappell, F. M., Morris, Z., & Wardlaw, J. M. (2015). Cerebral perivascular spaces visible on magnetic resonance imaging: Development of a qualitative rating scale and its observer reliability. *Cerebrovascular Diseases, 39,* 224—231.

Qiu, C., Cotch, M. F., Sigurdsson, S., Jonsson, P. V., Jonsdottir, M. K., Sveinbjrnsdottir, S., et al. (2010). Cerebral microbleeds, retinopathy, and dementia: The AGES-reykjavik study. *Neurology, 75*(24), 2221−2228.

Rosenberg, G. A., Wallin, A., Wardlaw, J. M., Markus, H. S., Montaner, J., Wolfson, L., et al. (2016). Consensus statement for diagnosis of subcortical small vessel disease. *Journal of Cerebral Blood Flow and Metabolism, 36*(1), 6−25.

Shoamanesh, A., Preis, S. R., Beiser, A. S., Vasan, R. S., Benjamin, E. J., Kase, C. S., et al. (2015). Inflammatory biomarkers, cerebral microbleeds, and small vessel disease: Framingham Heart Study. *Neurology, 84*(8), 825−832.

Ter Telgte, A., van Leijsen, E. M. C., Wiegertjes, K., Klijn, C. J. M., Tuladhar, A. M., & de Leeuw, F. E. (2018). Cerebral small vessel disease: From a focal to a global perspective. *Nature Reviews Neurology, 14*(7), 387−398.

van Veluw, S. J., Shih, A. Y., Smith, E. E., Chen, C., Schneider, J. A., Wardlaw, J. M., et al. (2017). Detection, risk factors, and functional consequences of cerebral microinfarcts. *The Lancet Neurology, 16*, 730−740.

Verdelho, A., Madureira, S., Ferro, J. M., Basile, A. M., Chabriat, H., Erkinjuntti, T., et al., LADIS Study. (2007). Differential impact of cerebral white matter changes, diabetes, hypertension and stroke on cognitive performance among non-disabled elderly. The LADIS study. *Journal of Neurology Neurosurgery and Psychiatry, 78*(12), 1325−1330.

Wardlaw, J. M., Allerhand, M., Doubal, F. N., Valdes Hernandez, M., Morris, Z., Gow, A. J., et al. (2014). Vascular risk factors, large-artery atheroma, and brain white matter hyperintensities. *Neurology, 82*(15), 1331−1338.

Wardlaw, J. M., Smith, E. E., Biessels, G. J., Cordonnier, C., Fazekas, F., Frayne, R., et al. (2013). STandards for ReportIng Vascular changes on nEuroimaging (STRIVE). Neuroimaging standards for research into small vessel disease and its contribution to ageing and neurodegeneration. *The Lancet Neurology, 12*(8), 822−838.

CHAPTER 4

Linking amyotrophic lateral sclerosis and frontotemporal dementia

Couratier Philippe
Service Neurologie, Centre de référence SLA et autres maladies du neurone moteur, CHU Limoges, Limoges, France

Introduction

Amyotrophic lateral sclerosis (ALS) is the most common form of adult-onset motor neuron degeneration that affects the upper and lower motor neurons and the corticospinal tract. ALS is understood as a complex multisystem neurodegenerative disease. Frontotemporal dementia (FTD) is characterized by selective involvement of the frontal and temporal lobes and is associated with changes in behavior and personality, frontal executive deficits, and language dysfunction. Three FTD subtypes are described, including a behavioral variant (bvFTD), which manifests with disinhibition, compulsive or perseverative behavior, overeating, apathy, and emotional blunting associated with a most severe cortical atrophy in the frontal and anterior temporal lobes, and two language variants, semantic variant primary progressive aphasia (svPPA), which exhibits loss of conceptual knowledge for words (left-sided degeneration) or faces and people (right-sided degeneration), due to selective involvement of the anterior temporal lobes, and nonfluent variant primary progressive aphasia (nfvPPA), characterized by agrammatic, nonfluent language output and apraxia of speech.

A possible association between motor neuron disease (MND) and FTD began in 1990 when Mitsuyama reported 71 patients with a presenile dementia and MND (Mitsuyama, 1993). At the same time, Neary et al. (1990) described four patients with rapidly progressive dementia marked by a frontal lobe dysfunction in association with clinical features of MND (). The realization that FTD—MND had a distinctive neuropathology began in the 1980s, with the first reports of ubiquitin-positive immunoreactive inclusions in the cytoplasm of motor neurons (Leigh et al., 1988, 1991). In addition, evidence of ubiquitin-positive inclusions in the extramotor cortex was shown in both pure ALS patients (Okamoto, Hirai, Yamazaki, Nakazato, & Okamoto, 1991) and ALS patients with dementia (Wightman et al., 1992). These ubiquitin-positive inclusions became the pathological hallmark of the FTD—MND syndrome. A key finding determining the link between FTD and MND was the identification of the transactive response DNA-binding protein 43 (TDP-43) in 2006 as the major inclusion protein associated with ubiquitinated inclusions in the vast majority of ALS patients, and in the most common pathological subtype of FTD, referred to as frontotemporal lobar degeneration (FTLD) with TDP-43 pathology (FTLD—TDP)

Diagnosis and Management in Dementia
ISBN 978-0-12-815854-8, https://doi.org/10.1016/B978-0-12-815854-8.00004-5

(Mackenzie et al., 2009; Neumann et al., 2006). The discovery in 2011 that the chromosome 9 open reading frame 72 (*C9orf72*) gene mutation can cause both FTD and ALS has modified the concept that ALS is "purely" a movement disorder and that FTD is "purely" a cognitive or behavioral form of dementia.

The clinical, pathological, and genetic features in favor of a continuum between ALS and FTD will be described.

Clinical features underlying the link between frontotemporal dementias and motor neuron disease

Epidemiological studies estimated the prevalence of FTD at 15–22 per 100,000 inhabitants ages 45–64 years. Two studies reported the incidence of FTD at 3.5–4.1 cases per 100,000 person-years in the age group 45–64 years (Knopman, Petersen, Edland, Cha, & Rocca, 2004; Mercy, Hodges, Dawson, Barker, & Brayne, 2008), with no gender difference (Hou, Yaffe, Pérez-Stable, & Miller, 2006; Rosso et al., 2003). The mean age of onset of FTD is typically in the fifth to the seventh decade of life (Johnson et al., 2005), with approximately 10% having an onset at over 70 years (Seelaar et al., 2008). Median survival in FTD has been estimated at 6–11 years from symptom onset and 3–4 years from diagnosis (Hodges, Davies, Xuereb, Kril, & Halliday, 2003; Rasciovsky, et al., 2005). Hodges et al. (2003) reported the longest survival in nfvPPA (mean 10.6 years from onset) followed by bvFTD (8.2 years) and semantic dementia (SD) (6.9 years). Survival appears to be shorter and decline more rapid when associated with MND (2.4–4.9 years from onset and 1.2–1.4 years from diagnosis) (Rascovsky et al., 2005; Roberson et al., 2005).

The frequency of FTD in ALS patients varies in the literature. A great number of studies (Borroni, et al., 2010; Ratnavalli, Brayne, Dawson, & Hodges, 2002) have suggested that roughly 50% of patients with ALS have some cognitive impairment, marked by an alteration of executive functions, including social cognition and/or language.Some patients may develop a mild behavioral impairment predominantly marked by apathy associated or not with disinhibition, compulsive or perseverative behavior, and psychotic symptoms. Some patients may develop both mild cognitive and behavioral impairment. The detection of these cognitive or behavioral disorders was made difficult because most of the tests used interfered with motor impairment. To determine cognitive and behavioral changes in patients suffering from ALS, the Edinburgh Cognitive and Behavioural ALS Screen (ECAS) was developed by Abrahams and Thomas Bak (Abrahams, Newton, Niven, Foley, & Bak, 2014). With ECAS, ALS-specific and ALS-nonspecificfunctions can be analyzed to enable the distinction from other diseases with cognitive and behavioral impairments. It is crucial to detect these disorders because their presence can affect the care of patients and represents a significant burden for caregivers.

Ten to fifteen percent of ALS patients reach the criteria for diagnosis of FTD according to Rascovsky et al. Bulbar onset of symptoms and lower educational attainment have been associated with cognitive involvement (Knopman et al., 2004), and cognitive impairment in ALS has been associated with shorter survival. Similarly, approximately 15% of FTD patients develop clinical symptoms of motor neuron dysfunction.

Pathological hallmarks underlying the link between frontotemporal dementia and ALS

FTLD can be divided into two major subtypes: FTLD with tau-positive inclusions (FTLD—tau); and FTLD with ubiquitin-positive and TDP-43-positive but tau-negative inclusions (FTLD—TDP) (Mackenzie et al., 2010). FTLD—TDP appears to be the primary pathology underlying the overlap between FTD and MND.

FTLD—TDP

Neumann et al. (2006) identified TDP-43 as the ubiquitinated pathological protein in most cases of FTLD—U as well as the majority of sporadic ALS (sALS) and some familial ALS (fALS) cases. TDP-43 pathology is present in 90% of ubiquitin-positive FTLD cases and non—superoxide dismutase-1 (SOD1) ALS cases, with fused-in-sarcoma (FUS)-positive inclusions accounting for the majority of remaining ubiquitin-positive TDP-43-negative inclusions (Mackenzie & Rademakers, 2008; Neumann, Rademakers et al., 2009; Neumann, Roeber et al., 2009). Another subtype with TDP-43 pathology was described, associated with the familial syndrome of inclusion body myopathy with Paget's disease of bone and FTD caused by mutations in the valosin-containing protein (VCP) gene (Forman et al., 2006). As a result, cases of FTLD with TDP-43 pathology are now designated as FTLD—TDP and the term FTLD—U is no longer recommended.

FTLD—FUS, FTLD—UPS

Approximately 6%—20% of FTLD disorders are characterized by ubiquitin-positive, TDP-43/tau-negative inclusions. Many such cases have shown immunoreactivity with the FUS antibody, but none of the FTLD—FUS cases had FUS gene mutations (Urwin et al., 2010). The FUS protein is involved in DNA repair and regulation of RNA splicing (Seelaar et al., 2010). The morphology of the intranuclear inclusions appears to be pathognomic in this subtype (Vance et al., 2009). FTLD—FUS cases are characterized by a young age at onset, bvFTD, negative family history, and caudate atrophy on MRI (Josephs et al., 2010). FUS-positive inclusions are also found in patients with neuronal filament inclusion disease, who mainly present with bvFTD and a rapid clinical course (Neumann, Rademakers et al., 2009; Neumann, Roeber et al., 2009). Finally, cases of ubiquitin-positive, TDP-43 and FUS-negative inclusions have been termed FTLD—UPS cases. Most of these cases carry a charged multivesicular body protein 2B (CHMP2B) mutation (Holm, Isaacs, & Mackenzie, 2009).

Genetics underlying the link between frontotemporal dementias and motor neuron disease

The majority of families with autosomal dominant FTD have a mutation in one of three genes: the *microtubule-associated protein-tau* gene (MAPT), the *progranulin* gene (GRN), and the *C9orf72* gene. The most important discovery remains that of the *C9orf72* locus on chromosome 9p, responsible for a majority of the hereditary cases of FTD, ALS, and

FTD—ALS. Before the identification of *C9orf72*, only 20%—30% of fALS cases were explained by mutations in the *SOD1* gene and the genes encoding TDP-43 (*TARDBP*) and *FUS*. Mutations in TARDBP are responsible for 4%—6% of fALS and 1% of sALS (Andersen & Al-Chalabi, 2011). Mutations in *TARDBP* cause, rarely, FTD or cortico-basal syndrome (Borroni et al., 2009). Mutations in *FUS* are causative of approximately 1% and 4% of apparent sALS and fALS, respectively. About 40% of patients with bvFTD have a positive family history, but it is more rare in patients with svPPA and nfvPPA (Goldman et al., 2005; Stevens et al., 1998).

C9FTLD/ALS

Two groups identified, in 2011, an expansion of a GGGGCC hexanucleotide repeat in the *C9orf72* gene (DeJesus-Hernandez et al., 2011; Renton et al., 2011). The hexanucleotide repeat is located between two 5′ noncoding exons of *C9orf72*. While a repeat length of greater than 30 has arbitrarily been defined as pathogenic in some studies, it remains unknown how many repeats are needed to cause disease. In most healthy individuals, repeat lengths with a maximum of 20 repeats are found, whereas in FTD and ALS patients, disease-associated expansions of a few hundred to several thousand repeats have been identified using Southern blotting techniques. Most research laboratories now use PCR-based methods to determine the presence of pathogenic repeat expansions. The *C9orf72* expansion is the most frequent mutation associated with familial cases of FTD and ALS—FTD, accounting for roughly 25% of familial FTD (Majounie et al., 2012), about 40% of fALS (Renton et al., 2011), and 6% of sporadic FTD cases. Frequency varies widely across different populations, with the highest frequencies in Finland (29%) (Renton et al., 2011), probably due to a founder effect. The mutation is most frequently found in Caucasian populations, although the expansion has been shown in a few Chinese cases (Jiao et al., 2014). *C9orf72* repeat expansions are even more common in ALS: 61% of fALS and 19% of sALS patients in Finland, 22% of fALS cases in Germany, and only 3.4% in Japan (Cooper-Knock et al., 2012; Konno et al., 2013). Estimates suggest that *C9orf72* expansions are rarely penetrant at <35 years, 50% penetrant by 58 years, and usually >95% penetrant at 80 years. The penetrance may not actually reach 100% at 80 years, as carriers of the expansion without the clinical phenotype have been reported over the age of 80 years. Finally, an Italian study investigated the possibility of anticipation in *C9orf72* carriers and found that *C9orf72*-related FTD families showed possible evidence of anticipation, with a mean difference in age of onset of 7 years earlier in the subsequent generation in ALS cases (Chio et al., 2012).

The age at onset and disease duration appear to be highly variable in *C9orf72* carriers, even within a single family. *C9orf72*-related ALS cases tend to have an earlier age of onset and shorter disease duration compared with ALS patients who do not carry the expansion, although this was not found in all cohorts (Byrne et al., 2012; García-Redondo et al., 2013). Gender may also play a role, as male *C9orf72*-related ALS cases have been reported to have a younger age of onset than non-*C9orf72* ALS cases (Williams et al., 2013).

C9orf72-related FTD most commonly presents with the behavioral variant syndrome, nfvPPA is the second most common FTD variant, while svPPA has rarely been associated with *C9orf72* expansions (Snowden et al., 2012). *C9orf72*-mutated patients appear to have a higher prevalence of psychosis, hallucinations, and delusions, compared with sporadic FTD patients (Kertesz et al., 2013). Therefore, the presence of psychiatric symptoms in the context of FTD—ALS should prompt consideration of a *C9orf72* repeat expansion (Snowden et al., 2013). All ALS phenotypes may be associated with *C9orf72* repeat expansions. Millecamps et al. (2012) compared phenotypic differences between *C9orf72*-related ALS and other ALS-causing genes, and found that *C9orf72*-related cases had a significantly higher incidence of bulbar onset compared with ALS cases with mutations in SOD1, TARDBP, or FUS and other fALS cases. Progressive muscular atrophy and primary lateral sclerosis have been rarely associated with *C9orf72* expansions: 1.6% and 0.9%, respectively (Van Rheenen et al., 2012). Also, *C9orf72* FTLD/ALS phenotypes show higher association with other neurological motor phenotypes, such as parkinsonism (Boeve et al., 2012).

Neuropathological studies in *C9orf72* repeat expansions are characterized by TDP-43 pathology. TDP-43 pathology in *C9orf72* expansion cases has been shown to consist of compact and granular neuronal cytoplasmic inclusions (NCIs), dystrophic neurites, glial cytoplasmic inclusions, and variable presence of neuronal intranuclear inclusions. TDP-43 pathology is found consistently in the frontal and temporal cortices (carriers with clinical FTLD and FTD—ALS syndromes), pyramidal motor system (carriers with clinical ALS), and frequently other limbic (hippocampus and amygdala), brain stem (midbrain/substantia nigra), and subcortical structures (striatum and thalamus) (Murray et al., 2011). Another highly characteristic pathological feature of patients with *C9orf72* expansions is the significant presence of NCIs in the cerebellar granule cell layer, hippocampal pyramidal neurons, and neocortex that stain positive for proteins of the ubiquitin proteasome system, including ubiquitin, ubiquilins, and p62, but are negative for TDP-43 (Pikkarainen, Hartikainen, & Alafuzoff, 2008).

It is interesting to observe a higher than expected coincidence of repeat expansions in individuals carrying other genetic variants of ALS (Van Blitterswijk, DeJesus-Hernandez, & Rademakers, 2012), but also in FTD (Ferrari et al., 2012). While some may represent benign polymorphisms, some of these mutations may act as disease modifiers, contributing to the pleiotropy encountered in patients with *C9orf72* mutations. TMEM106B genotypes were identified as the first genetic factor modifying disease presentation in *C9orf72* expansion carriers (Van Blitterswijk et al., 2014).

Another potential modifier could be CAG repeat expansions in the ATXN2 exon. Lattante et al. screened a large French cohort of ALS, FTD, FTD—ALS, and progressive supranuclear paralysis (PSP) patients for ATXN2 CAG repeat lengths, including over 300 patients who were *C9orf72* expansion carriers (Lattante et al., 2012). They found a significant association with intermediate repeat size (>29 repeats) in patients with fALS, sALS, and familial FTD—ALS. Twenty-three percent of the intermediate-repeat ATXN2 carriers had cooccurrence of pathogenic *C9orf72* expansions, all in the FTD—ALS and fALS subgroups, suggesting that ATXN2 poly(Q) expansions may act as a strong modifier of the FTD phenotype in the presence of a *C9orf72* expansion. In the cohort of *C9orf72* carriers, 3% also carried an intermediate ATXN2 repeat length, whereas PSP and FTD patients had ATXN2 repeat lengths similar to those of controls. Similarly, a multicenter study on *C9orf72* carriers suggested that intermediate ATXN2 expansions possibly act as disease modifiers in *C9orf72* expansion carriers, the effect being most profound in probands with MND or FTD—MND (2.1% vs. 0% in controls, $P = .013$) (Van Blitterswijk et al., 2014). In conclusion, ATXN2 repeat expansions can cause either spinocerebellar ataxia (SCA) or ALS, but also a neuropathological overlap syndrome of SCA2 and FTLD—MND presenting clinically as pure FTLD—ALS without ataxia.

TARDBP mutations, chromosome 1p36.22

TARDBP mutations have a prevalence of 4%—6% in fALS and 1% in sALS. Screening of family members of sALS patients reveals healthy *TARDBP* mutation carriers, implying that the penetrance is incomplete. Cognitive deficits are rarely found in ALS patients with *TARDBP* mutations (Benajiba et al., 2009). Mutations have been found in two French fALS patients with dementia, one Norwegian fALS patient with FTD, and one patient with FTD, PSP, and chorea. So, *TARDBP* mutations can cause sporadic and familial ALS, but rarely FTD—ALS or isolated FTD.

Fused-in-sarcoma mutations, chromosome 16p11.2

FUS mutations have been found in 4% of fALS and <1% of sALS patients. *FUS* mutations have been associated with a juvenile onset of the disease. Cognitive impairment is rarely seen with ALS caused by *FUS* mutations: only three cases had FTD features (Yan et al., 2010) and two others parkinsonism and dementia.

VCP mutations, chromosome 9p13

VCP is a member of the AAA-ATPase gene superfamily acting as a molecular chaperone in endoplasmic reticulum—associated protein degradation. VCP is located on chromosome 9p13.3, and mutations result in an autosomal dominant disorder associated with FTD called inclusion body myopathy (IBM) with Paget's disease of bone and FTD. Affected muscles display hallmarks of IBM. Paget's disease of bone typically shows characteristic radiologic findings or an elevation in serum alkaline phosphatase level and is confirmed by bone biopsy (Nalbandian et al., 2011). Heterogeneity in clinical phenotypes may manifest as IBM in about 90% of patients, Paget's in 50%, and FTD in about one-third of cases. Rarely, *VCP* mutations are associated with sALS, fALS, or familial FTD—ALS.

CHMP2B mutations/frontotemporal dementia-3, chromosome 3p11

Mutations in the *CHMP2B* gene, a rare cause of ALS, have been described to cause FTD only in individuals of Danish and Belgian ancestry. A large Danish family with autosomal dominant FTD caused by mutation in *CHMP2B* on chromosome 3 has been followed for more than 2 decades. Late in the disease course, a motor syndrome typically develops, with features of parkinsonism, dystonia, and myoclonus and pyramidal signs. None of the patients had clinical signs of motor neuron impairment. Neuropathologically, ubiquitin- and p62-positive NCIs are observed in the dentate granule cell layer of the hippocampus and, to a lesser extent, in the frontal cortex. These inclusions are negative for both TDP-43 and FUS. Five *CHMP2B* missense mutations in eight individuals have been identified in a range of FTD–MND spectrum disorders (Isaacs, Johannsen, Holm, & Nielsen, 2011).

SQSTM1 mutations, chromosome 5q35

The sequestosome 1 (SQSTM1) gene is located on 5q35 and encodes p62, a protein that acts as a cargo receptor for the degradation of ubiquitinated proteins through autophagic or proteasomal pathways. *SQSTM1* mutations have been identified in patients with familial (<2%) and sALS (4%). A French cohort identified four heterozygous missense mutations in four unrelated families with FTD, with one family showing clinical symptoms of FTD–ALS and supporting a pathogenic role for the p62 protein in FTD disorders (Le Ber et al., 2013).

UBQLN2 mutations, chromosome Xp11.21

Mutations in *UBQLN2* have been shown to cause dominant X-linked ALS and ALS–dementia and also rarely sALS (Synofzik et al., 2012). *Ubiquilin-2* mutations may disrupt clearance of misfolded or damaged proteins, leading to impaired cellular functioning, especially in motor neurons. UBQLN2-positive inclusions have been observed in non-UBQLN2-linked ALS. Age at disease onset varies widely, from the late teens to the seventh decade, possibly earlier in males than in females, with average disease duration of less than 4 years. Symptoms of FTD may precede motor manifestations. Data on the frequency of UBQLN2 in different disease phenotypes is still very limited, with almost no reports of the mutation causing pure dementia.

Conclusion

FTD and ALS share clinical, genetic, and neuropathological features and neurodegenerative pathways, suggesting that they may be part of a common disease spectrum. The identification of *C9orf72* expansion has ended a long period of research and considerably changed our clinical practice. The understanding of pathological mechanisms in *C9orf72*-related disease is crucial for the development of biomarkers and for the identification of presymptomatic persons at risk. Promising studies have revealed that antisense oligonucleotides could reduce RNA foci and reverse toxicity, which will potentially have important implications for the development of effective and targeted therapies in the future.

Key points

- ALS is not a pure MND. It is common to observe mild cognitive or behavioral impairment.
- ALS may be associated with FTLD, most frequently a behavioral form.
- FTLD may be associated with an MND.
- ALS−FTD and FTLD share common pathological features.
- C9ORF72 mutation is the most common mutation found in familial ALS, in FTLD, and in ALS−FTD.

Summary points

- There are clinical, pathological, and genetic features in favor of a continuum between ALS and FTD.
- The Edinburgh Cognitive and Behavioral ALS Screen allows us to detect cognitive and behavioral impairment in 50% of ALS patients.
- Fifteen percent of FTD patients develop clinical symptoms of motor neuron dysfunction. Ten to fifteen percent of ALS patients reach criteria for diagnosis of FTD.
- FTLD−TDP appears to be the primary pathology underlying the overlap between FTD and MND.
- The most important discovery remains that of the *C9orf72* locus on chromosome 9p, responsible for a majority of the hereditary cases of FTD, ALS, and FTD−ALS.
- The presence of psychiatric symptoms in the context of FTD−ALS should prompt consideration of a *C9orf72* repeat expansion.

References

Abrahams, S., Newton, J., Niven, E., Foley, J., & Bak, T. H. (2014). Screening for cognition and behavior changes in ALS. *Amyotrophic Lateral Sclerosis and Frontotemporal Degeneration, 15*, 9−14.

Andersen, P. M., & Al-Chalabi, A. (November 2011). Clinical genetics of amyotrophic lateral sclerosis: What do we really know? *Nature Reviews Neurology, 7*, 603−615.

Benajiba, L., Le Ber, I., Camuzat, A., Lacoste, M., Thomas-Anterion, C., Couratier, P., et al. (2009). TARDBP mutations in motoneuron disease with frontotemporal lobar degeneration. *Annals of Neurology, 65*, 470−473.

Boeve, B. F., Boylan, K. B., Graff-Radford, N. R., DeJesus-Hernandez, M., Knopman, D. S., Pedraza, O., et al. (2012). Characterization of frontotemporal dementia and/or amyotrophic lateral sclerosis associated with the GGGGCC repeat expansion in C9ORF72. *Brain: A Journal of Neurology, 135*, 765−783.

Borroni, B., Alberici, A., Grassi, M., Turla, M., Zanetti, O., Bianchetti, A., et al. (2010). Is frontotemporal lobar degeneration a rare disorder? Evidence from a preliminary study in Brescia country, Italy. *Journal of Alzheimer's Disease, 19*, 111−116.

Borroni, B., Bonvicini, C., Alberici, A., Buratti, E., Agosti, C., Archetti, S., et al. (November 2009). Mutation within TARDBP leads to frontotemporal dementia without motor neuron disease. *Human Mutation, 30*, E974−E983.

Byrne, S., Elamin, M., Bede, P., Shatunov, A., Walsh, C., Corr, B., et al. (2012). Cognitive and clinical characteristics of patients with amyotrophic lateral sclerosis carrying a C9orf72 repeat expansion: A population-based cohort study. *The Lancet Neurology, 11*, 232–240.

Chiò, A., Borghero, G., Restagno, G., Mora, G., Drepper, C., Traynor, B. J., et al. (2012). Clinical characteristics of patients with familial amyotrophic lateral sclerosis carrying the pathogenic GGGGCC hexanucleotide repeat expansion of C9ORF72. *Brain: A Journal of Neurology, 135*, 784–793.

Cooper-Knock, J., Hewitt, C., Highley, J. R., Brockington, A., Milano, A., Man, S., et al. (March 2012). Clinicopathological features in amyotrophic lateral sclerosis with expansions in C9ORF72. *Brain: A Journal of Neurology, 135*(Pt 3), 751–764.

DeJesus-Hernandez, M., Mackenzie, I. R., Boeve, B. F., Boxer, A. L., Baker, M., Rutherford, N. J., et al. (2011). Expanded GGGGCC hexanucleotide repeat in noncoding region of C9ORF72 causes chromosome 9-linked FTD and ALS. *Neuron, 72*, 245–256.

Ferrari, R., Mok, K., Moreno, J. H., Cosentino, S., Goldman, J., Pietrini, P., et al. (2012). Screening for C9ORF72 repeat expansion in FTLD. *Neurobiology of Aging, 33*(1850), e1–11.

Forman, M. S., Mackenzie, I. R., Cairns, N. J., Swanson, E., Boyer, P. J., Drachman, D. A., et al. (June 2006). Novel ubiquitin neuropathology in frontotemporal dementia with valosin-containing protein gene mutations. *Journal of Neuropathology and Experimental Neurology, 65*, 571–581.

García-Redondo, A., Dols-Icardo, O., Rojas-García, R., Esteban-Pérez, J., Cordero-Vázquez, P., Muñoz-Blanco, J. L., et al. (2013). Analysis of the C9orf72 gene in patients with amyotrophic lateral sclerosis in Spain and different populations worldwide. *Human Mutation, 34*, 79–82.

Goldman, J. S., Farmer, J. M., Wood, E. M., Johnson, J. K., Boxer, A., Neuhaus, J., et al. (2005). Comparison of family histories in FTLD subtypes and related tauopathies. *Neurology, 65*, 1817–1819.

Hodges, J. R., Davies, R., Xuereb, J., Kril, J., & Halliday, G. (2003). Survival in frontotemporal dementia. *Neurology, 61*, 349–354.

Holm, I. E., Isaacs, A. M., & Mackenzie, I. R. A. (2009). Absence of FUS-immunoreactive pathology in frontotemporal dementia linked to chromosome 3 (FTD-3) caused by mutation in the CHMP2B gene. *Acta Neuropathologica, 118*, 719–720.

Hou, C. E., Yaffe, K., Pérez-Stable, E. J., & Miller, B. L. (2006). Frequency of dementia etiologies in four ethnic groups. *Dementia and Geriatric Cognitive Disorders, 22*, 42–47.

Isaacs, A. M., Johannsen, P., Holm, I., & Nielsen, J. E. (2011). FReJA consortium Frontotemporal dementia caused by CHMP2B mutations. *Current Alzheimer Research, 8*, 246–251.

Jiao, B., Tang, B., Liu, X., Yan, X., Zhou, L., Yang, Y., et al. (2014). Identification of C9orf72 repeat expansions in patients with amyotrophic lateral sclerosis and frontotemporal dementia in mainland China. *Neurobiology of Aging, 35*(936), e19–22.

Johnson, J. K., Diehl, J., Mendez, M. F., Neuhaus, J., Shapira, J. S., Forman, M., et al. (June 2005). Frontotemporal lobar degeneration: Demographic characteristics of 353 patients. *Archives of Neurology, 62*(6), 925–930.

Josephs, K. A., Whitwell, J. L., Parisi, J. E., Petersen, R. C., Boeve, B. F., Jack, C. R., Jr., et al. (2010). Caudate atrophy on MRI is a characteristic feature of FTLD-FUS. *European Journal of Neurology, 17*, 969–975.

Kertesz, A., Ang, L. C., Jesso, S., MacKinley, J., Baker, M., Brown, P., et al. (2013). Psychosis and hallucinations in frontotemporal dementia with the C9ORF72 mutation: A detailed clinical cohort. *Cognitive and Behavioral Neurology: Official Journal of the Society for Behavioral and Cognitive Neurology, 26*, 146–154.

Knopman, D. S., Petersen, R. C., Edland, S. D., Cha, R. H., & Rocca, W. A. (2004). The incidence of frontotemporal lobar degeneration in Rochester, Minnesota, 1990 through 1994. *Neurology, 62*(3), 506–508.

Konno, T., Shiga, A., Tsujino, A., Sugai, A., Kato, T., Kanai, K., et al. (April 2013). Japanese amyotrophic lateral sclerosis patients with GGGGCC hexanucleotide repeat expansion in C9ORF72. *Journal of Neurology Neurosurgery and Psychiatry, 84*(4), 398–401.

Lattante, S., Conte, A., Zollino, M., Luigetti, M., Del Grande, A., Marangi, G., et al. (2012). Contribution of major amyotrophic lateral sclerosis genes to the etiology of sporadic disease. *Neurology, 79*, 66–72.

Le Ber, I., Camuzat, A., Guerreiro, R., Bouya-Ahmed, K., Bras, J., Nicolas, G., et al. (2013). SQSTM1 mutations in French patients with frontotemporal dementia or frontotemporal dementia with amyotrophic lateral sclerosis. *JAMA Neurology, 70*, 1403–1410.

Leigh, P. N., Anderton, B. H., Dodson, A., Gallo, J. M., Swash, M., & Power, D. M. (November 11, 1988). Ubiquitin deposits in anterior horn cells in motor neurone disease. *Neuroscience Letters, 93*(2–3), 197–203.

Leigh, P. N., Whitwell, H., Garofalo, O., Buller, J., Swash, M., Martin, J. E., et al. (1991). Ubiquitin-immunoreactive intraneuronal inclusions in amyotrophic lateral sclerosis. Morphology, distribution, and specificity. *Brain: A Journal of Neurology, 114*, 775–788.

Mackenzie, I. R. A., Neumann, M., Bigio, E. H., Cairns, N. J., Alafuzoff, I., Kril, J., et al. (2009). Nomenclature for neuropathologic subtypes of frontotemporal lobar degeneration: Consensus recommendations. *Acta Neuropathologica, 117*, 15–18.

Mackenzie, I. R. A., Neumann, M., Bigio, E. H., Cairns, N. J., Alafuzoff, I., Kril, J., et al. (2010). Nomenclature and nosology for neuropathologic subtypes of frontotemporal lobar degeneration: An update. *Acta Neuropathologica, 119*, 1–4.

Mackenzie, I. R. A., & Rademakers, R. (2008). The role of transactive response DNA-binding protein-43 in amyotrophic lateral sclerosis and frontotemporal dementia. *Current Opinion in Neurology, 21*, 693–700.

Majounie, E., Renton, A. E., Mok, K., Dopper, E. G. P., Waite, A., Rollinson, S., et al. (2012). Frequency of the C9orf72 hexanucleotide repeat expansion in patients with amyotrophic lateral sclerosis and frontotemporal dementia: A cross-sectional study. *The Lancet Neurology, 11*(4), 323–330.

Mercy, L., Hodges, J. R., Dawson, K., Barker, R. A., & Brayne, C. (November 4, 2008). Incidence of early-onset dementias in Cambridgeshire, United Kingdom. *Neurology, 71*(19), 1496–1499.

Millecamps, S., Boillée, S., Le Ber, I., Seilhean, D., Teyssou, E., Giraudeau, M., et al. (2012). Phenotype difference between ALS patients with expanded repeats in C9ORF72 and patients with mutations in other ALS-related genes. *Journal of Medical Genetics, 49*, 258–263.

Mitsuyama, Y. (1993). Presenile dementia with motor neuron disease. *Dementia and Geriatric Cognitive Disorders, 4*, 137–142.

Murray, M. E., DeJesus-Hernandez, M., Rutherford, N. J., Baker, M., Duara, R., Graff-Radford, N. R., et al. (2011). Clinical and neuropathologic heterogeneity of c9FTD/ALS associated with hexanucleotide repeat expansion in C9ORF72. *Acta Neuropathologica, 122*, 673–690.

Nalbandian, A., Donkervoort, S., Dec, E., Badadani, M., Katheria, V., Rana, P., et al. (2011). The multiple faces of valosin-containing protein-associated diseases: Inclusion body myopathy with Paget's disease of bone, frontotemporal dementia, and amyotrophic lateral sclerosis. *Journal of Molecular Neuroscience, 45*, 522–531.

Neary, D., Snowden, J. S., Mann, D. M., Northen, B., Goulding, P. J., & Macdermott, N. (1990). Frontal lobe dementia and motor neuron disease. *Journal of Neurology Neurosurgery and Psychiatry, 53*(1), 23–32.

Neumann, M., Rademakers, R., Roeber, S., Baker, M., Kretzschmar, H. A., & Mackenzie, I. R. A. (2009). A new subtype of frontotemporal lobar degeneration with FUS pathology. *Brain: A Journal of Neurology, 132*, 2922–2931.

Neumann, M., Roeber, S., Kretzschmar, H. A., Rademakers, R., Baker, M., & Mackenzie, I. R. A. (2009). Abundant FUS-immunoreactive pathology in neuronal intermediate filament inclusion disease. *Acta Neuropathologica, 118*, 605–616.

Neumann, M., Sampathu, D. M., Kwong, L. K., Truax, A. C., Micsenyi, M. C., Chou, T. T., et al. (2006). Ubiquitinated TDP-43 in frontotemporal lobar degeneration and amyotrophic lateral sclerosis. *Science, 314*, 130–133.

Okamoto, K., Hirai, S., Yamazaki, T., Sun, X. Y., & Nakazato, Y. (1991). New ubiquitin-positive intraneuronal inclusions in the extra-motor cortices in patients with amyotrophic lateral sclerosis. *Neuroscience Letters, 129*, 233–236.

Pikkarainen, M., Hartikainen, P., & Alafuzoff, I. (2008). Neuropathologic features of frontotemporal lobar degeneration with ubiquitin-positive inclusions visualized with ubiquitin-binding protein p62 immunohistochemistry. *Journal of Neuropathology and Experimental Neurology, 67*, 280–298.

Rasciovsky, K., Salmon, D. P., Lipton, A. M., Leverenz, J. B., DeCarli, C., Jagust, W. J., et al. (2005). Rate of progression differs in frontotemporal dementia and Alzheimer disease. *Neurology, 65*, 397–403.

Ratnavalli, E., Brayne, C., Dawson, K., & Hodges, J. R. (2002). The prevalence of frontotemporal dementia. *Neurology, 58*, 1615–1621.

Renton, A. E., Majounie, E., Waite, A., Simón-Sánchez, J., Rollinson, S., Gibbs, J. R., et al. (2011). A hexanucleotide repeat expansion in C9ORF72 is the cause of chromosome 9p21-linked ALS-FTD. *Neuron, 72*, 257–268.

Roberson, E. D., Hesse, J. H., Rose, K. D., Slama, H., Johnson, J. K., Yaffe, K., et al. (2005). Frontotemporal dementia progresses to death faster than Alzheimer disease. *Neurology, 65*, 719–725.

Rosso, S. M., Donker Kaat, L., Baks, T., Joosse, M., de Koning, I., Pijnenburg, Y., et al. (2003). Frontotemporal dementia in The Netherlands: Patient characteristics and prevalence estimates from a population based study. *Brain: A Journal of Neurology, 126*, 2016–2022.

Seelaar, H., Kamphorst, W., Rosso, S. M., Azmani, A., Masdjedi, R., de Koning, I., et al. (October 14, 2008). Distinct genetic forms of frontotemporal dementia. *Neurology, 71*(16), 1220–1226.

Seelaar, H., Klijnsma, K. Y., de Koning, I., van der Lugt, A., Chiu, W. Z., Azmani, A., et al. (2010). Frequency of ubiquitin and FUS-positive, TDP-43-negative frontotemporal lobar degeneration. *Journal of Neurology, 257*, 747–753.

Snowden, J. S., Harris, J., Richardson, A., Rollinson, S., Thompson, J. C., Neary, D., et al. (2013). Frontotemporal dementia with amyotrophic lateral sclerosis: A clinical comparison of patients with and without repeat expansions in C9orf72. *Amyotroph Lateral Scler Front Degener, 14*, 172–176.

Snowden, J. S., Rollinson, S., Thompson, J. C., Harris, J. M., Stopford, C. L., Richardson, A. M. T., et al. (2012). Distinct clinical and pathological characteristics of frontotemporal dementia associated with C9ORF72 mutations. *Brain: A Journal of Neurology, 135*, 693–708.

Stevens, M., van Duijn, C. M., Kamphorst, W., de Knijff, P., Heutink, P., van Gool, W. A., et al. (1998). Familial aggregation in frontotemporal dementia. *Neurology, 50*, 1541–1545.

Synofzik, M., Maetzler, W., Grehl, T., Prudlo, J., Vom Hagen, J. M., Haack, T., et al. (2012). Screening in ALS and FTD patients reveals 3 novel UBQLN2 mutations outside the PXX domain and a pure FTD phenotype. *Neurobiology of Aging, 33*(2949), e13–17.

Urwin, H., Josephs, K. A., Rohrer, J. D., Mackenzie, I. R., Neumann, M., Authier, A., et al. (2010). FUS pathology defines the majority of tau- and TDP-43-negative frontotemporal lobar degeneration. *Acta Neuropathologica, 120*, 33–41.

Van Blitterswijk, M., DeJesus-Hernandez, M., & Rademakers, R. (2012). How do C9ORF72 repeat expansions cause amyotrophic lateral sclerosis and frontotemporal dementia: Can we learn from other noncoding repeat expansion disorders? *Current Opinion in Neurology, 25*, 689–700.

Van Blitterswijk, M., Mullen, B., Heckman, M. G., Baker, M. C., DeJesus-Hernandez, M., Brown, P. H., et al. (2014). Ataxin-2 as potential disease modifier in C9ORF72 expansion carriers. *Neurobiology of Aging, 35*(2421), e13–17.

Van Blitterswijk, M., Mullen, B., Nicholson, A. M., Bieniek, K. F., Heckman, M. G., Baker, M. C., et al. (2014). TMEM106B protects C9ORF72 expansion carriers against frontotemporal dementia. *Acta Neuropathologica, 127*, 397–406.

Van Rheenen, W., van Blitterswijk, M., Huisman, M. H. B., Vlam, L., van Doormaal, P. T. C., Seelen, M., et al. (2012). Hexanucleotide repeat expansions in C9ORF72 in the spectrum of motor neuron diseases. *Neurology, 79*, 878–882.

Vance, C., Rogelj, B., Hortobágyi, T., De Vos, K. J., Nishimura, A. L., Sreedharan, J., et al. (2009). Mutations in FUS, an RNA processing protein, cause familial amyotrophic lateral sclerosis type 6. *Science, 323*, 1208–1211.

Wightman, G., Anderson, V. E., Martin, J., Swash, M., Anderton, B. H., Neary, D., et al. (1992). Hippocampal and neocortical ubiquitin-immunoreactive inclusions in amyotrophic lateral sclerosis with dementia. *Neuroscience Letters, 139*, 269–274.

Williams, K. L., Fifita, J. A., Vucic, S., Durnall, J. C., Kiernan, M. C., Blair, I. P., et al. (2013). Pathophysiological insights into ALS with C9ORF72 expansions. *Journal of Neurology Neurosurgery and Psychiatry, 84*, 931–935.

Yan, J., Deng, H.-X., Siddique, N., Fecto, F., Chen, W., Yang, Y., et al. (2010). Frameshift and novel mutations in FUS in familial amyotrophic lateral sclerosis and ALS/dementia. *Neurology, 75*, 807–814.

CHAPTER 5

Mortality in dementia: linking in the role of delirium

Flavia Barreto Garcez, Thiago Junqueira Avelino-Silva*

Division of Geriatrics, Department of Internal Medicine, University of Sao Paulo Medical School, Sao Paulo, Brazil

List of abbreviations

AD Alzheimer's disease
DSD delirium superimposed on dementia

Mini-dictionary of terms

Brain stem An anatomical region that connects the deep brain structures with the spinal cord and is responsible for controlling vital functions such as breathing, heart rate, and level of consciousness.
Cognitive domains Parameters of normal cerebral performance in cortical areas (e.g., attention, memory, executive function, language, and visuospatial function) that are tested to diagnose cognitive impairment.
Delirium superimposed on dementia Altered mental status in patients with dementia, characterized by inattention and altered level of consciousness that are not better explained by the preexisting cognitive deficit.
Dementia pathologic burden Pathologic findings associated with clinical dementia syndromes, such as neurofibrillary tangles, cortical amyloid plaques, Lewy bodies, and vascular lesions (infarcts > 10 mm, lacunes, hemorrhages).
Functional decline Loss of independence to perform activities of daily living, which can be basic (e.g., taking a bath, dressing, eating) or instrumental (e.g., handling finances, using a telephone, managing own medications).
Level of arousal Status of vigilance and alertness when a person is awake, which can be measured by the degree of sensory stimuli needed to maintain appropriate responsiveness.

Introduction

Dementia is a global health issue that dramatically burdens patients, caregivers, and society in general (Livingston et al., 2017). Additionally, mortality rates are higher in individuals living with dementia than in nondemented persons, contributing to the sense of hopelessness that often follows the condition (Rao, Suliman, Vuik, Aylin, & Darzi, 2016). Despite an expected reduction in cases of severe cognitive impairment in some countries (Matthews et al., 2013), others will long remain unprepared to manage chronic diseases that emerge with population aging, leaving older adults exposed to their consequences (Prince et al., 2013).

*Senior author.

Diagnosis and Management in Dementia
ISBN 978-0-12-815854-8, https://doi.org/10.1016/B978-0-12-815854-8.00005-7

The disabilities and functional decline associated with dementia are responsible for several other adverse outcomes, such as higher healthcare expenses (Prince, Comas-Herrera, Knapp, Guerchet, & Karagiannidou, 2016), admission to long-term care facilities (Comas-Herrera, Wittenberg, Pickard, & Knapp, 2007), and increased rates of hospitalization (Phelan, Borson, Grothaus, Balch, & Larson, 2012). Unplanned hospital admissions can be particularly problematic, as patients with dementia are more vulnerable to several hospital-associated hazards, including infections, accelerated functional decline (Sands et al., 2003), and delirium (Inouye, Westendorp, & Saczynski, 2014).

Delirium superimposed on dementia

Delirium is more simply defined as an acute confusional state; it is most commonly observed in older adults (Inouye et al., 2014), but it can affect patients in other age ranges to a lesser extent (Dahmani, Delivet, & Hilly, 2014). This disturbance of one's mental status develops rapidly and is characterized by impaired cognition, inattention, and altered levels of consciousness that typically have a fluctuating nature (Inouye et al., 2014). Similar to the case with other models of acute injuries, such as acute heart failure and acute kidney failure, it has been proposed that delirium is the expression of acute brain failure and can result in permanent organic damage (AGS/NIA Delirium Conference Writing Group, 2015). Furthermore, it has been widely established that delirium is associated with several unfavorable outcomes both during hospitalization and after discharge (Witlox et al., 2010).

Cognitive impairment remains one of the strongest predisposing factors of delirium (Inouye, Viscoli, Horwitz, Hurst, & Tinetti, 1993). In an essential prospective study examining hospitalized older adults, delirium was up to four times more common in patients with dementia than in those without (Travers, Byrne, Pachana, Klein, & Gray, 2014). Another large cohort of 2000 patients consecutively admitted to a general hospital showed that 9% of their population had dementia and that 44% of them had delirium on admission (Erkinjuntti, 1986). The prevalence of acute confusional states and dementia can be even higher in other settings, such as geriatric wards and palliative care units (Laurila, Pitkala, Strandberg, & Tilvis, 2004). Moreover, as cognitive impairment is one of the top three predictors of delirium occurrence (Inouye et al., 1993), it is essential to educate healthcare providers on how to detect delirium in patients with dementia (Table 5.1) (Morandi et al., 2017).

On the other hand, delirium has been reported as the fifth cause of acute hospitalizations in patients with Alzheimer's disease (AD) (Rudolph et al., 2010). The clinical overlap between these two conditions is a natural consequence of their physiopathologic mechanisms and of the importance that cognitive reserve plays in their development. When delirium affects a person who has dementia, changing their already compromised baseline mental status, it is usually referred to as delirium superimposed on dementia

Table 5.1 Structure suggested to improve the diagnosis of delirium superimposed on dementia.

DSD feature	Action(s)
Attention	Determine dementia etiology and severity-specific effects on attention
	Determine type of attention to be tested
	Choose which test to use
Level of arousal	Use objective measures to determine level of arousal and fluctuations over time
Motor activity	Use objective measures to determine patterns of motor activity and fluctuations over time
Clinical context	Use information available from family members, medical record review, and evaluation of basic and instrumental activities of daily living to determine changes from baseline
Clinical examination	Perform detailed clinical and neurologic examination
Laboratory testing	Determine a basic set of laboratory evaluation for diagnostic investigation
Neuroimaging	Establish a rationale to indicate neuroimaging to extend diagnostic investigation if necessary

DSD, delirium superimposed on dementia.

(DSD). DSD is often a challenging diagnosis (Fick, Agostini, & Inouye, 2002), and even in specialized units, frequently goes underrecognized or underreported (Laurila et al., 2004). The presence of dementia is one of the most critical risk factors for missing delirium during routine evaluations, and detection rates are significantly increased when a thorough cognitive assessment is performed, especially if it includes the systematic assessment of delirium using tools such as the confusion assessment method (Inouye, 2001). Even so, reaching an accurate diagnosis can be difficult even with a comprehensive assessment of all cognitive domains.

When it comes to distinguishing delirium from dementia, one has to consider the phenomenological intersections that exist between these conditions and reflect on how to escape them (Fick et al., 2002). In that regard, there are two dimensions of the neuropsychiatric exam that allow for the distinction between dementia and delirium: attention and vigilance (Leonard et al., 2016, pp. 1—9). Inattention is a prominent feature of delirium and is usually absent in dementia, even if various degrees of impairment can be observed in moderate-to-severe stages of chronic cognitive disorders (Morandi et al., 2017). Therefore, inattention can be a useful tool to diagnose DSD, especially if the evaluator is meticulous enough to use different strategies to examine their patient's attention (e.g., inattention in persons with dementia will frequently relate to disturbances in working memory and executive function, and the assessment will benefit from tests that account for these peculiarities) (Tieges, Brown, & MacLullich, 2014). Besides attention, level of arousal (or vigilance) might also be a useful target for evaluation, as it is primarily a brain stem function and is generally intact even in advanced stages of dementia (Morandi et al., 2017). Since a minimum state of vigilance is required to complete attention tasks, testing the level of arousal can be a straightforward method to diagnose DSD. Nonetheless, a combined assessment of attention and level of arousal is always preferred and has been demonstrated to be more accurate than other approaches (Quispel-Aggenbach, Oltman, Zwartjes, & Zuidema, 2018).

Cognitive and functional decline in delirium superimposed on dementia

Patients with baseline cognitive impairment are more vulnerable to functional decline when hospitalized (Sager et al., 1996), and this occurrence may apply to cognitive function as well. Although the impact of a hospital admission on cognition has been clearly demonstrated in subjects without preexisting cognitive decline (Ehlenbach et al., 2010), there is increasing concern about the effect a hospitalization might have on someone with dementia, especially if delirium comes into play (Eide et al., 2016; Mathews, Arnold, & Epperson, 2014).

Many of the hospital-related outcomes that have been reported as happening more frequently in persons with AD (e.g., institutionalization, death) have also been associated with delirium. Truthfully, delirium seems to operate as an additive risk factor to hospitalization and is itself a predictor of these outcomes (Fong et al., 2012). In this context, DSD can modify the natural trajectories of cognition and functionality in a person with dementia and over time lead to the occurrence of a wide range of adverse events at a magnitude disproportionate with what would be expected in the presence of dementia or delirium alone (Morandi et al., 2014) (Fig. 5.1).

One previous study screened medical inpatients for dementia and delirium and classified them according to four groups: DSD, delirium alone, dementia alone, and neither delirium nor dementia (McCusker, Cole, Dendukuri, Belzile, & Primeau, 2001). The participants were periodically reevaluated after hospital discharge to measure cognition and function. Patients with DSD experienced the most significant cognitive and functional declines after adjustment for possible confounders such as comorbidity burden and disease severity. Although the authors found long-term deterioration in cognitive function and physical independence caused by delirium alone, this effect was considerably superior when associated with preexisting dementia (McCusker et al., 2001). Further studies examining long-term cognitive outcomes in delirium and DSD have confirmed these results (Davis et al., 2017; Fong et al., 2012, 2009; Gross et al., 2012). Deleterious effects on cognitive performance were observed up to 5 years after these patients had experienced delirium (Gross et al., 2012), indicating its enduring influence on dementia's natural history.

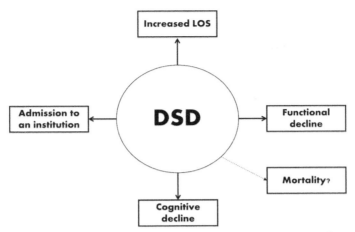

Figure 5.1 Representation of outcomes associated with delirium superimposed on dementia. *DSD,* delirium superimposed on dementia.

Whether this incremental damage to brain functioning shares underlying mechanisms or neuropathological signatures with injuries found in chronic cognitive disorders is still unclear. Notwithstanding the common neurobiological features known to exist between dementia and delirium, uncertainty remains when it comes to developing a satisfactory pathway linking the two conditions (Fong, Davis, Growdon, Albuquerque, & Inouye, 2015). It is unlikely, however, that a single mechanism will be found to justify the neuro-degeneration process triggered by delirium, and it would be naïve to presume that dementia pathologies would provide the only means for long-term cognitive impairment in delirium (Davis et al., 2012). In that light, a few hypotheses have been raised based on animal models and surrogate markers of brain injury in human subjects (Fig. 5.2); a possible pathway proposes that the combined interaction of systemic and neurologic inflammatory processes could lead to the loss of synapses and the accumulation of white matter in the brain (Davis et al., 2015).

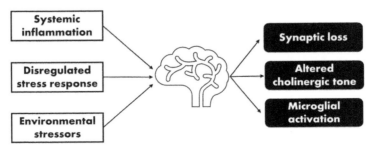

Figure 5.2 Possible neurobiological mechanisms linking delirium to long-term cognitive impairment.

Overall, there is substantial evidence to support delirium as a robust prognostic factor that accelerates cognitive impairment in patients with dementia (Davis et al., 2017; Fong et al., 2009, 2015). As delirium is a potentially preventable condition that can be avoided in up to 40% of cases (Inouye et al., 2014), further investigations should explore effective preventive measures as a means to reduce cognitive decline in vulnerable persons and possibly slow dementia progression in already afflicted patients.

Does delirium play a role in dementia mortality?

Before cognitive outcomes became a primary focus of delirium research, the association between delirium and mortality was widely investigated and consistently verified across several populations (Kennedy et al., 2014; McCusker, Cole, Abrahamowicz, Primeau, & Belzile, 2002; Witlox et al., 2010). Mortality rates have been reported to quadruple in medical and surgical patients who experienced delirium, and twofold increases have been described in intensive care settings (Salluh et al., 2015; Witlox et al., 2010).

A frequent topic of debate is whether the increase in mortality could be directly attributed to delirium or if the underlying disease process that led to delirium would also justify the higher risk of death. Many observational studies have tried to clarify this issue, and delirium has been reported as an independent predictor of death even after statistical adjustments for possible confounders such as such as age, cognitive status, and number of comorbidities (Avelino-Silva, Jerussalmy, Farfel, Curiati, & Jacob-Filho, 2009; Kennedy et al., 2014; Pendlebury et al., 2015). The prognostic impact of delirium occurrence is especially strong on short-term mortality, possibly diminishing after hospital discharge (Pendlebury et al., 2015), but some data suggest that it may influence long-term mortality as well (McCusker et al., 2002; Witlox et al., 2010), particularly if delirium was present on discharge (McAvay et al., 2006).

As patients with dementia have a higher probability of developing delirium (Inouye et al., 1993; Kennedy et al., 2014) and are vulnerable to many of its associated outcomes (Fong et al., 2012), it is imperative to explore delirium as a predictor of death in patients with cognitive disorders. McCusker et al. (2002) prospectively examined a cohort of patients aged 65 or over to investigate whether baseline cognitive function modified the effect of delirium on mortality and found that although delirium was independently associated with a fourfold increase in mortality in patients without dementia, the same could not be said of those who had preexisting dementia. In this group, delirium had a weak and nonsignificant association with mortality after adjusting for possible confounders (McCusker et al., 2002). Nonetheless, recent results from a large cohort investigating mortality in patients with DSD provided a different perspective (Avelino-Silva, Campora, Curiati, & Jacob-Filho, 2017). In this study, the hospital mortality in patients with DSD was three times higher than in those who only had dementia (32% vs. 12% respectively). Caveats from the commentary are that the authors were not

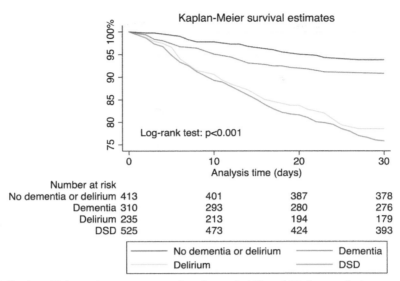

Figure 5.3 Kaplan—Meier estimates representing the probability of 30-day survival among acutely ill hospitalized older adults according to delirium and dementia diagnosis in a tertiary university hospital in Sao Paulo, Brazil. The corresponding log-rank test indicates a statistically significant difference between the groups (N = 1554). *DSD*, delirium superimposed on dementia.

able to demonstrate an association between DSD and long-term mortality and that delirium in patients without preexisting cognitive decline still had the most significant association with mortality (Fig. 5.3) (Avelino-Silva et al., 2017). We were able to obtain updated data from this last cohort exploring the association between DSD and 30-day mortality in acutely ill patients aged +65 years, and the results were consistent with the published data (Fig. 5.3; Table 5.3).

Table 5.2 Characteristics of longitudinal studies that investigated the association between delirium superimposed on dementia and mortality.

Author	Year	Sample size	Delirium assessment	Effect size (OR/HR, 95% CI)	P- value
Avelino-Silva et al.	2017	1409	CAM	2.14 (1.33—3.45)	.002
Bellelli et al.	2007	188	CAM	2.3 (1.1—5.5)	.04
Fick et al.	2013	139	CAM	2.33 (0.82—6.61)	.11
Fong et al.	2012	771	Chart-DEL	5.4 (2.3—12.5)	<.05
McCusker et al.	2002	361	CAM	NS	NS
Morandi et al.	2014	2642	CAM DSM-IV-TR	1.8 (1.1—2.8)	.01

CAM, confusion assessment method; *Chart-DEL*, chart-based delirium identification instrument; *DSM-IV-TR*, Diagnostic and Statistical Manual of Mental Disorders, Fourth Edition; *NS*, non-significant.

Table 5.3 Association between dementia, delirium, and delirium superimposed on dementia and 30-day mortality in acutely ill hospitalized older adults (N = 1554).

Variable	Hospital mortality, n (percent)	Bivariate hazard ratio (95% CI)	Adjusted hazard ratio[a] (95% CI)	Adjusted P- value
No delirium/dementia	46(11)	Referent	Referent	Referent
Dementia	43(13)	1.53 (0.89—2.63)	1.0 (0.58—1.72)	0.987
Delirium	63(25)	3.82 (2.37—6.16)	3.08 (1.91—4.96)	<.001
DSD	148 (27)	4.35 (2.84—6.67)	2.34 (1.48—3.70)	<.001

DSD, delirium superimposed on dementia.
[a]Cox proportional hazards model adjusted for age, sex, nutrition status, and Charlson Comorbidity Index.

The complexity of care inherent to persons with dementia or delirium might be one of the reasons it is so difficult to get a clear picture of how each of these conditions affects prognosis. Delirium, for example, frequently complicates hospitalizations, exposing patients to a myriad of potential iatrogenic events, such as the prescription of psychoactive medications and physical restraints. Delirium patients have also been observed to have a higher risk of developing aspiration pneumonia and pressure ulcers (Inouye, Schlesinger, & Lydon 1999). On the other hand, persons with cognitive disorders share a similar predisposition to hospital-associated hazards and could develop several clinical complications without the "help" of a delirium episode (Rao et al., 2016).

Another relevant factor that could explain the discrepant results for the association between DSD and mortality resides in the challenge that diagnosing delirium, and even more so DSD, represents to most healthcare providers (Fick et al., 2002; Inouye et al., 1999). Misdiagnosis, misclassification, and underreporting could bias research conclusions for many of these issues and understate the importance of DSD as a predictor of mortality (Bellelli et al., 2015). Finally, it is also possible that DSD and delirium in cognitively normal individuals have different phenomenologies. For example, a patient with dementia might have an underlying vulnerability that requires a less noxious insult to trigger the delirious episode; consequently, delirium would have a milder clinical meaning and a reduced effect on prognosis (McCusker et al., 2002).

Despite the discrepancies in existing studies regarding the specific association between DSD and mortality and regarding the long-term survival of these patients, ample evidence points to a significant effect of acute mental disturbances on the short-term survival of hospitalized patients (Table 5.2) (Avelino-Silva et al., 2017; Fick, Steis, Waller, & Inouye, 2013). Further understanding of the mechanisms that lead to this incremental risk is undoubtedly a key to consolidating the role of delirium assessment as a prognostic tool in the management of patients with chronic cognitive disorders.

Final considerations

Delirium is an important prognostic marker for patients with dementia, as it not only reveals the existence of cognitive vulnerability but also accelerates cognitive and functional decline, modifying the long-term trajectories of neurodegenerative disorders (Fong et al., 2012; Gross et al., 2012). Preventive measures and early detection of delirium are critical points to avoid unwanted outcomes, including death (Fick et al., 2002; Inouye et al., 2014).

When there is an additional challenge to the diagnosis of delirium due to preexisting cognitive impairment or inaccessibility, recognition of inattention and impaired level of arousal could be good proxies for delirium diagnosis (Morandi et al., 2017; Richardson et al., 2017, pp. 1–9). Future research should focus on exploring the physiopathology and clinical features that both coincide and diverge in delirium and dementia, providing additional groundwork for epidemiologists to understand the role of DSD as a prognostic factor and predictor of mortality in dementia.

Key facts of delirium superimposed on dementia

- Delirium is an acute and fluctuating disorder of mental state characterized by inattention, altered level of consciousness, and impaired cognition.
- Delirium is a frequent complication of acute illnesses, particularly in patients with baseline vulnerability, such as older adults and persons with dementia.
- Preexisting cognitive impairment denotes a four to eight times higher risk of developing delirium to the affected person.
- When delirium is detected in patients with preexisting cognitive disorders, it is called delirium superimposed on dementia (DSD).
- The two most important features to differentiate delirium from dementia are inattention and impaired level of arousal.

Summary points

- This chapter focuses on describing delirium as an important prognostic factor, including for patients with dementia.
- The diagnosis of DSD is challenging in clinical practice.
- Underdetection of DSD is a frequent phenomenon and can be responsible for underestimating adverse events associated with delirium.
- DSD can accentuate functional decline and accelerate the progression of dementia, modifying the expected trajectory of the disease.
- Evidence exists to suggest that DSD is a predictor of short-term mortality in hospitalized patients with dementia.
- Recognizing the impact of DSD in patients with dementia is an indispensable step toward improving delirium detection and implementing effective preventive measures in this population.

References

AGS/NIA Delirium Conference Writing Group, & Planning Committee and Faculty. (2015). The American Geriatrics Society/National Institute on Aging bedside-to-bench conference: Research agenda on delirium in older adults. *Journal of the American Geriatrics Society, 63*(5), 843–852. https://doi.org/10.1111/jgs.13406.

Avelino-Silva, T. J., Campora, F., Curiati, J. A. E., & Jacob-Filho, W. (2017). Association between delirium superimposed on dementia and mortality in hospitalized older adults: A prospective cohort study. *PLoS Medicine, 14*(3), e1002264. https://doi.org/10.1371/journal.pmed.1002264.

Avelino-Silva, T. J., Jerussalmy, C. S., Farfel, J. M., Curiati, J. A. E., & Jacob-Filho, W. (2009). Predictors of in-hospital mortality among older patients. *Clinics (São Paulo, Brazil), 64*(7), 613–618.

Bellelli, G., Frisoni, G. B., Turco, R., Lucchi, E., Magnifico, F., & Trabucchi, M. (2007). Delirium superimposed on dementia predicts 12-month survival in elderly patients discharged from a postacute rehabilitation facility. *Journals Gerontol - Ser A Biol Sci Med Sci, 62*(11), 1306–1309.

Bellelli, G., Nobili, A., Annoni, G., Morandi, A., Djade, C. D., Meagher, D. J., et al. (2015). Under-detection of delirium and impact of neurocognitive deficits on in-hospital mortality among acute geriatric and medical wards. *European Journal of Internal Medicine*, 1–9. https://doi.org/10.1016/j.ejim.2015.08.006.

Comas-Herrera, A., Wittenberg, R., Pickard, L., & Knapp, M. (2007). Cognitive impairment in older people: Future demand for long-term care services and the associated costs. *International Journal of Geriatric Psychiatry, 22*(10), 1037–1045. https://doi.org/10.1002/gps.1830.

Dahmani, S., Delivet, H., & Hilly, J. (2014). Emergence delirium in children: An update. *Current Opinion in Anaesthesiology, 27*(3), 309–315. https://doi.org/10.1097/ACO.0000000000000076.

Davis, D. H. J., Muniz Terrera, G., Keage, H., Rahkonen, T., Oinas, M., Matthews, F. E., et al. (2012). Delirium is a strong risk factor for dementia in the oldest-old: A population-based cohort study. *Brain: A Journal of Neurology, 135*(Pt 9), 2809–2816. https://doi.org/10.1093/brain/aws190.

Davis, D. H. J., Muniz-Terrera, G., Keage, H. A. D., Stephan, B. C. M., Fleming, J., Ince, P. G., et al. (2017). Association of delirium with cognitive decline in late life: A neuropathologic study of three population-based cohort studies. *Journal of the American Academy of Dermatology Psychiatry, 74*(3), 244–251. https://doi.org/10.1001/jamapsychiatry.2016.3423.

Davis, D. H. J., Skelly, D. T., Murray, C., Hennessy, E., Bowen, J., Norton, S., et al. (2015). Worsening cognitive impairment and neurodegenerative pathology progressively increase risk for delirium. *American Journal of Geriatric Psychiatry: Official Journal of the American Association for Geriatric Psychiatry, 23*(4), 403–415. https://doi.org/10.1016/j.jagp.2014.08.005.

Ehlenbach, W. J., Crane, P. K., Haneuse, S. J. P. A., Carson, S. S., Curtis, J. R., & Larson, E. B. (2010). Association between acute care and critical illness hospitalization. *The Journal of the American Medical Association, 303*(8), 763–770.

Eide, L. S. P., Ranhoff, A. H., Fridlund, B., Haaverstad, R., Hufthammer, K. O., Kuiper, K. K. J., et al. (2016). Delirium as a predictor of physical and cognitive function in individuals aged 80 and older after transcatheter aortic valve implantation or surgical aortic valve replacement. *Journal of the American Geriatrics Society, 64*(6), 1178–1186. https://doi.org/10.1111/jgs.14165.

Erkinjuntti, T. (1986). Dementia among medical inpatients. *Archives of Internal Medicine, 146*(10), 1923. https://doi.org/10.1001/archinte.1986.00360220067013.

Fick, D. M., Agostini, J. V., & Inouye, S. K. (2002). Delirium superimposed on dementia: A systematic review. *Journal of the American Geriatrics Society, 50*(10), 1723–1732. https://doi.org/10.1046/j.1532-5415.2002.50468.x.

Fick, D. M., Steis, M. R., Waller, J. L., & Inouye, S. K. (2013). Delirium superimposed on dementia is associated with prolonged length of stay and poor outcomes in hospitalized older Adults. *Journal of Hospital Medicine, 8*(9), 6–11. https://doi.org/10.1002/jhm.2077.

Fong, T. G., Davis, D., Growdon, M. E., Albuquerque, A., & Inouye, S. K. (2015). The interface between delirium and dementia in elderly adults. *The Lancet Neurology, 14*(8), 823–832.

Fong, T. G., Jones, R. N., Marcantonio, E. R., Tommet, D., Gross, A. L., Habtemariam, D., et al. (2012). Adverse outcomes after hospitalization and delirium in persons with Alzheimer disease. *Annals of Internal Medicine, 156*(12), 848–856. https://doi.org/10.7326/0003-4819-156-12-201206190-00005. W296.

Fong, T. G., Jones, R. N., Shi, P., Marcantonio, E. R., Yap, L., Rudolph, J. L., et al. (2009). Delirium accelerates cognitive decline in Alzheimer's disease. *Neurology, 72*(18), 1570–1575.

Gross, A. L., Jones, R. N., Habtemariam, D. A., Fong, T. G., Tommet, D., Quach, L., et al. (2012). Delirium and long-term cognitive trajectory among persons with dementia. *Archives of Internal Medicine, 172*(17), 1324—1331. https://doi.org/10.1001/archinternmed.2012.3203.

Inouye, S. K. (2001). Nurses' recognition of delirium and its symptoms. *Archives of Internal Medicine, 161*(20), 2467. https://doi.org/10.1001/archinte.161.20.2467.

Inouye, S. K., Schlesinger, S. J., & Lydon, T. (1999). Delirium: A symptom of how hospital care is failing older persons and a window to improve quality of hospital care. *The American Journal of Medicine, 106*(13), 565—573.

Inouye, S. K., Viscoli, C. M., Horwitz, R. I., Hurst, L. D., & Tinetti, M. E. (1993). A predictive model for delirium in hospitalized elderly medical patients based on admission characteristics. *Annals of Internal Medicine, 119*(6), 474—481. https://doi.org/10.7326/0003-4819-119-6-199309150-00005.

Inouye, S. K., Westendorp, R. G. J., & Saczynski, J. S. (2014). Delirium in elderly people. *Lancet, 383*(9920), 911—922.

Kennedy, M., Enander, R. A., Tadiri, S. P., Wolfe, R. E., Shapiro, N. I., & Marcantonio, E. R. (2014). Delirium risk prediction, healthcare use, and mortality of elderly adults in the emergency department. *Journal of the American Geriatrics Society, 62*(3), 462—469. https://doi.org/10.1111/jgs.12692.

Laurila, J. V., Pitkala, K. H., Strandberg, T. E., & Tilvis, R. S. (2004). Detection and documentation of dementia and delirium in acute geriatric wards. *General Hospital Psychiatry, 26*(1), 31—35. https://doi.org/10.1016/j.genhosppsych.2003.08.003.

Leonard, M., Mcinerney, S., Mcfarland, J., Condon, C., Awan, F., Connor, M. O., et al. (2016). *Comparison of cognitive and neuropsychiatric profiles in hospitalised elderly medical patients with delirium, dementia, and comorbid delirium — dementia.* https://doi.org/10.1136/bmjopen-2015-009212.

Livingston, G., Sommerlad, A., Orgeta, V., Costafreda, S. G., Huntley, J., Ames, D., et al. (2017). Dementia prevention, intervention, and care. *Lancet, 390*(10113), 2673—2734. https://doi.org/10.1016/S0140-6736(17)31363-6.

Mathews, S. B., Arnold, S. E., & Epperson, C. N. (2014). Hospitalization and cognitive decline: Can the nature of the relationship be deciphered? *American Journal of Geriatric Psychiatry, 22*(5), 465—480. https://doi.org/10.1016/j.jagp.2012.08.012.

Matthews, F. E., Arthur, A., Barnes, L. E., Bond, J., Jagger, C., Robinson, L., et al. (2013). A two-decade comparison of prevalence of dementia in individuals aged 65 years and older from three geographical areas of England: Results of the cognitive function and ageing study I and II. *Lancet, 382*(9902), 1405—1412. https://doi.org/10.1016/S0140-6736(13)61570-6.

McAvay, G. J., Van Ness, P. H., Bogardus, S. T., Zhang, Y., Leslie, D. L., Leo-Summers, L. S., et al. (2006). Older adults discharged from the hospital with delirium: 1-year outcomes. *Journal of the American Geriatrics Society, 54*(8), 1245—1250. https://doi.org/10.1111/j.1532-5415.2006.00815.x.

McCusker, J., Cole, M., Abrahamowicz, M., Primeau, F., & Belzile, E. (2002). Delirium predicts 12-month mortality. *Archives of Internal Medicine, 162*(4), 457—463.

McCusker, J., Cole, M., Dendukuri, N., Belzile, E., & Primeau, F. (2001). Delirium in older medical inpatients and subsequent cognitive and functional status: A prospective study. *Canadian Medical Association Journal, 165*(5), 575—583.

Morandi, A., Davis, D., Bellelli, G., Arora, R. C., Mbbs, G. A. C., Kamholz, B., et al. (2017). The diagnosis of delirium superimposed on Dementia : An emerging challenge. *Journal of the American Medical Directors Association, 18*(1), 12—18. https://doi.org/10.1016/j.jamda.2016.07.014.

Morandi, A., Davis, D., Fick, D. M., Turco, R., Boustani, M., Lucchi, E., et al. (2014). Delirium superimposed on dementia strongly predicts worse outcomes in older rehabilitation inpatients. *Journal of the American Medical Directors Association, 15*(5), 349—354. https://doi.org/10.1016/j.jamda.2013.12.084.

Pendlebury, S. T., Lovett, N. G., Smith, S. C., Dutta, N., Bendon, C., Lloyd-Lavery, A., et al. (2015). Observational, longitudinal study of delirium in consecutive unselected acute medical admissions: Age-specific rates and associated factors, mortality and re-admission. *British Medical Journal Open, 5*(11), 1—8. https://doi.org/10.1136/bmjopen-2015-007808.

Phelan, E. A., Borson, S., Grothaus, L., Balch, S., & Larson, E. B. (2012). Association between incident dementia and risk of hospitalization. *Journal of the American Medical Association, 307*(2), 165—172. https://doi.org/10.1001/jama.2011.1964.Association.

Prince, M., Bryce, R., Albanese, E., Wimo, A., Ribeiro, W., & Ferri, C. P. (2013). The global prevalence of dementia: A systematic review and metaanalysis. *Alzheimer's and Dementia, 9*(1), 63–75. https://doi.org/10.1016/j.jalz.2012.11.007.

Prince, M., Comas-Herrera, A., Knapp, M., Guerchet, M., & Karagiannidou, M. (2016). World Alzheimer Report 2016: Improving healthcare for people living with dementia. Coverage, quality, and costs now and in the future. *Alzheimer's Disease International*, 1–140.

Quispel-Aggenbach, D. W. P., Oltman, G. A. H., Zwartjes, H. A. H. T., & Zuidema, S. U. (2018). Attention, arousal, and other rapid bedside screening instruments for delirium in older patients: A systematic review of test accuracy studies. *Age and Ageing*, (April), 1–10. https://doi.org/10.1093/ageing/afy058.

Rao, A., Suliman, A., Vuik, S., Aylin, P., & Darzi, A. (2016). Outcomes of dementia: Systematic review and meta-analysis of hospital administrative database studies. *Archives of Gerontology and Geriatrics, 66*, 198–204. https://doi.org/10.1016/j.archger.2016.06.008.

Richardson, S. J., Davis, D. H. J., Bellelli, G., Hasemann, W., Meagher, D., Kreisel, S. H., et al. (2017). *Detecting delirium superimposed on dementia : Diagnostic accuracy of a simple combined arousal and attention testing*. https://doi.org/10.1017/S1041610217000916.

Rudolph, J. L., Zanin, N. M., Jones, R. N., Marcantonio, E. R., Fong, T. G., Yang, F. M., et al. (2010). Hospitalization in community-dwelling persons with Alzheimer's disease: Frequency and causes. *Journal of the American Geriatrics Society, 58*(8), 1542–1548. https://doi.org/10.1111/j.1532-5415.2010.02924.x.

Sager, M. A., Rudberg, M. A., Jalaluddin, M., Franke, T., Inouye, S. K., Seth Landefeld, C., et al. (1996). Hospital Admission Risk Profile (HARP): Identifying older patients at risk for functional decline following acute medical illness and hospitalization. *Journal of the American Geriatrics Society, 44*(3), 251–257. https://doi.org/10.1111/j.1532-5415.1996.tb00910.x.

Salluh, J. I. F., Wang, H., Schneider, E. B., Nagaraja, N., Yenokyan, G., Damluji, A., et al. (2015). Outcome of delirium in critically ill patients: Systematic review and meta-analysis. *British Medical Journal, 350*(May). https://doi.org/10.1136/bmj.h2538. h2538–h2538.

Sands, L. P., Yaffe, K., Covinsky, K., Chren, M.-M., Counsell, S., Palmer, R., et al. (2003). Cognitive screening predicts magnitude of functional recovery from admission to 3 Months after discharge in hospitalized elders. *The Journals of Gerontology Series A: Biological Sciences and Medical Sciences, 58*(1), M37–M45. https://doi.org/10.1093/gerona/58.1.M37.

Tieges, Z., Brown, L. J. E., & MacLullich, A. M. J. (2014). Objective assessment of attention in delirium: A narrative review. *International Journal of Geriatric Psychiatry, 29*(12), 1185–1197. https://doi.org/10.1002/gps.4131.

Travers, C., Byrne, G. J., Pachana, N. A., Klein, K., & Gray, L. C. (2014). Prospective observational study of dementia in older patients admitted to acute hospitals. *Australasian Journal on Ageing, 33*(1), 55–58. https://doi.org/10.1111/ajag.12021.

Witlox, J., Eurelings, L. S. M., de Jonghe, J. F. M., Kalisvaart, K. J., Eikelenboom, P., & van Gool, W. A. (2010). Delirium in elderly patients and the risk of postdischarge mortality, institutionalization, and dementia: A meta-analysis. *Journal of the American Medical Association, 304*(4), 443–451. https://doi.org/10.1001/jama.2010.1013.

CHAPTER 6

Midlife diabetes and the risk of dementia: understanding the link

Adam H. Dyer[1], Sean P. Kennelly[2]

[1]Department of Medical Gerontology, Trinity College Dublin, Dublin, Ireland; [2]Centre for Ageing, Neurosciences and the Humanities, Tallaght Hospital, Dublin, Ireland

Mini-dictionary of terms

Type 2 diabetes mellitus a metabolic disorder characterized by high blood glucose and resistance to the physiological effects of insulin.

Insulin peptide anabolic hormone produced by pancreatic β cells with numerous functions on differing cell types in the body.

Insulin resistance decreased sensitivity to the effect of insulin, characteristic of T2DM.

Hyperglycemia elevated blood sugar levels.

Neuroinflammation inflammation of the CNS, thought to be an important process in various neurodegenerative disorders.

Hypothalamic—pituitary—adrenal axis the main stress response system in the human body, activation of which results in an increase in circulating cortisol levels.

Cortisol a steroid hormone produced mainly in the adrenal gland.

Introduction

Type 2 diabetes mellitus (T2DM) in midlife is an important, yet often underappreciated, risk factor for the development of dementia in later life. T2DM is a common and complex metabolic disorder of altered carbohydrate metabolism characterized by insulin resistance (IR) resulting in numerous microvascular and macrovascular complications. T2DM affects approximately 415 million people worldwide, which is set to increase to almost 600 million by 2035 (Guariguata, Nolan, Beagley, Linnenkamp & Jacqmain, 2014). Further, as a result of our aging demography, the prevalent cases of T2DM are also set to increase, with T2DM disproportionately affecting those age 60 and older (Barbagallo & Dominguez, 2014). Thus, despite the fact that older age is usually a time characterized by health and well-being, there is set to be a large increase in the number of adults growing older with T2DM. The result of these global epidemiological trends, in addition to the better treatment and management of T2DM, is the emergence of novel complications and associations of this common chronic disease. One such (and indeed well-characterized) association, which is often unrecognized, is the increased risk of dementia seen in patients with T2DM.

Diagnosis and Management in Dementia
ISBN 978-0-12-815854-8, https://doi.org/10.1016/B978-0-12-815854-8.00006-9

The disease course for both T2DM and dementia is typified by a long prodromal asymptomatic phase, which becomes clinically apparent many years later when much of the pathophysiological damage has occurred. Teasing out the link between the two illnesses, and being able to predict which patients with T2DM will go on to develop cognitive impairment, is an important public health priority. As a result of recent trials (such as the FINGER trial) that have demonstrated positive effects of multidomain interventions for high-risk dementia groups (Ngandu et al. 2015), targeting populations such as people with T2DM may represent an important preventative strategy. Here, we explore the risk of dementia in patients with diabetes and reflect on potential pathophysiological mechanisms, risk factors, and interventions that may show promise in preventing dementia in patients with T2DM.

Evidence supporting the risk of dementia in diabetes

The increased risk of "all-type" dementia (ATD) conferred by diabetes has been known for some time, with original reports from more than 30 years ago identifying it as an important risk factor (Katzmann et al. 1989). Early studies pointing to the increased risk of dementia in those with diabetes were mainly cross-sectional in nature and it was not until the first well-designed longitudinal studies became of age that the picture became a little clearer. Chief among these are studies such as the Rotterdam and Rochester studies. Just before the turn of the 21st century, the Rotterdam Study, which followed over 6000 dementia-free older adults, associated T2DM with a doubling of dementia risk (Ott et al. 1999). In the Rochester Study, which followed just under 1500 patients with T2DM (age 45 and older) for nearly 10,000 person years, the risk of ATD was shown to be increased in T2DM (Leibson et al. 1997).

Further robust evidence of the increased dementia risk with midlife T2DM comes from a large metaanalysis of 28 prospective studies, which estimated a risk ratio of 1.73 for the later development of ATD (Gudala, Bansai, Schifano & Bhansali, 2013). More recently, studies looking at comparative midlife vascular risk factors for dementia in well-controlled cohorts have demonstrated that the risk conferred by T2DM may even be as big as that conferred by the APOE genotype (one of the most important and well-characterized risk factors for dementia to date) (Gottesman et al. 2017). While many studies have demonstrated the increased risk of dementia in those ages >65 conferred by T2DM in midlife, the risk appears not to be as strong in those ages >85 (Van den Berg, de Craen, Biessels, Gussekloo, Westendorp, 2006), supporting the clinical need for early identification and intervention.

Not only has T2DM emerged as a risk factor for the later development of dementia, but T2DM is also associated with an increase in the risk of mild cognitive impairment (MCI) and indeed accelerates the progression to dementia in those affected by MCI (Luchsinger et al. 2007; Xu et al. 2010). This association has even been seen in middle-age individuals, placing those affected at increased risk as they reach older age (Winkler et al. 2014). Coupled with this evidence and dementia risk, the epidemiology supports that much of the pathophysiological damage done by midlife T2DM may occur well before a clinical manifestation of dementia.

Potential pathoetiological mechanisms

The exact mechanisms by which T2DM may increase dementia risk have attracted a broad and heterogeneous literature. The principle hypotheses linking T2DM and increase dementia risk include glycemic control (hyperglycemia and hypoglycemia), insulin signaling abnormalities and IR, oxidative stress, neuroinflammation, and abnormalities in the function of the hypothalamic—pituitary—adrenal (HPA) axis. It is highly unlikely these prevailing mechanisms act in isolation, and it is best to view the etiology as multi-factorial. In this section, we explore the potential pathophysiological mechanisms by which T2DM increases dementia risk (see Fig. 6.1).

Hyper- and hypoglycemia

T2DM is fundamentally a disorder of impaired glucose tolerance, indicated by hypergly-cemia. There is strong evidence that hyperglycemia impairs cognition in both the acute and the chronic contexts. Interestingly, studies using hyperinsulinemic glucose clamps to maintain hyperglycemia/euglycemia have demonstrated working memory and attentional deficits in patients with T2DM (Sommerfield, Deary & Frier, 2004). An often-cited study supporting the role of hyperglycemia in T2DM-related dementia is the Memory in Diabetes substudy (ACCORD-MIND) in which investigators found that a 1% higher glycated hemoglobin (HbA1c) was associated with significantly lower scores on cognitive tests such as the Digit Symbol Substitution Test (DSST) and the Mini-Mental State Examination (MMSE) (Cukierman-Yaffe et al. 2009). In longitudinal studies of patients with and without a diagnosis of T2DM, higher serum glucose levels have been associated with the incidence of ATD (Crane et al. 2013). Diet-based glycemic load and elevated blood glucose are associated with impairments in perceptual speed and visuospatial as well as general cognitive ability (Seetharaman et al. 2015). Furthermore, there is suggestive evidence that higher glucose levels accelerate the conversion from MCI to dementia and have even been associated with decreased frontal gray matter volumes (Morris et al. 2014).

The role of hypoglycemia on cognition is more complex. When studying patients with T1DM (also characterized by an increased risk of cognitive impairment, micro-/macrovascular complications, and hyperglycemia), cross-sectional studies have demon-strated a link between hypoglycemia and cognitive impairment. While some longitudinal studies have not supported a link, others have demonstrated an increased risk of dementia in those with severe episodes of hypoglycemia, with risk increasing proportionally with the number of episodes of hypoglycemia (Whitmer, Karter, Yaffe, Quesenberry & Selby 2009). Interpretation remains difficult, and cognitive impairment and dementia are risk factors for hypoglycemia due to treatment adherence issues.

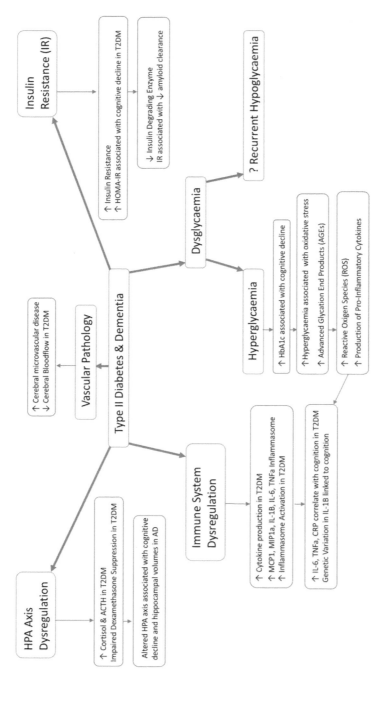

Figure 6.1 Potential mechanisms linking type 2 diabetes mellitus and dementia. The potential mechanisms linking type 2 diabetes mellitus and dementia are detailed, including dysglycemia (increased/decreased blood sugar), activation of the HPA axis, neuroinflammation, and vascular pathology. *ACTH*, adrenocorticotrophic hormone; *AD*, Alzheimer's disease; *CRP*, C-reactive protein; *HbA1c*, hemoglobin A1c; *HOMA-IR*, homeo-static model assessment—insulin resistance; *HPA axis*, hypothalamic—pituitary—adrenal axis; *IL-1B*, interleukin 1B; *IL-6*, interleukin 6; *MCP1*, monocyte chemoattractant protein 1; *MIF1α*, macrophage migration inhibitory factor 1a; *T2DM*, type 2 diabetes mellitus; *TNFα*, tumor necrosis factor α.

Hyperglycemia in the acute context may alter cerebral blood flow and cause osmotic changes and oxidative stress in neurons. High levels of blood glucose are associated with an increase in the formation of advanced glycation end products (AGEs). Levels of AGEs are increased in patients with T2DM and cognitive impairment and are correlated with accelerated cognitive decline in those with and without diabetes (Umegaki, 2014). Further, the receptor for AGEs (esRAGE) has been implicated as a protective factor in the development of MCI (Chen et al. 2011).

AGEs can lead to activation of the innate immune system, acting as damage-associated molecular patterns (DAMPs). DAMPs such as AGEs may activate microglia in the central nervous system (CNS). Such a process may lead to the formation of reactive oxygen species (ROS) and proinflammatory cytokines, with further damage to neurons potentially via mitochondrial dysfunction. Activation of the inflammasome and the innate immune system may be a central mechanism by which T2DM increases dementia risk. Priming of microglia in the brain appears to be important in the pathophysiology of dementia, and AGEs acting as DAMPs may be crucial in predisposing those with T2DM to dementia, setting the stage for further inflammation and pathological damage (Block, Zecca & Hong 2007).

Insulin resistance

IR is a central part of the T2DM syndrome. Insulin receptors are readily located in areas such as the hippocampus, entorhinal cortex, and frontal areas involved in memory and learning. Insulin is a hormone and a growth factor and is rapidly transported over the blood—brain barrier into the CNS. IR has been established as an independent risk factor for cognitive impairment and hence an important mechanism linking T2DM and dementia (Ma, Wang & Li, 2015). IR as measured by the homeostasis model of assessment of IR (HOMA—IR) significantly correlates with cognitive performance in both those with and those without T2DM (Ronnemaa et al. 2008).

The IR that is so characteristic of T2DM is associated with a reduction in synthesis of proteins, such as insulin-degrading enzyme (IDE). Interestingly, IDE has been demonstrated to degrade amyloid-β, and IR may contribute to decreased IDE, leading to the formulation of neuritic plaques (Umegaki, 2014). IR may contribute to tau phosphorylation pathogenesis in Alzheimer's disease (AD) (El Khoury, Gratuze, Papon, Bretteville & Planel, 2014). Such a mechanism, IR leading to impaired clearance of amyloid-β and tau phosphorylation and plaque formation, may be a central part in understanding the link between T2DM and dementia.

In AD neurons, IR has been documented. On postmortem examination of AD patients, it is noted that insulin signaling is greatly reduced (phosphorylation of insulin receptor substrate 1) (Umegaki, 2014). Of note, insulin-like growth factor-1 and one of its binding proteins, IGFBP3, have also been shown to be reduced in AD patients, as have canonical signaling pathways in their action, such as the insulin—PI3K—Akt pathway in AD with diabetes (Duron et al. 2012). Compelling evidence comes from a longitudinal study of over 2000 men, which demonstrated that a low insulin response, assessed by intravenous glucose tolerance test and HOMA-IR, predicted the development of AD during follow-up (median follow-up of 32 years) (Ronnemaa et al. 2008).

Neuroinflammation

Inflammation may have a crucial role to play in understanding the link between T2DM and dementia. A putative hypothesis is that with hyperglycemia and metabolic stress from AGEs acting as DAMPs, immune system activation may result. Following this, the production of proinflammatory cytokines and ROS may lead to a proinflammatory state in the CNS of patients with T2DM. IR is also linked to inflammation, with increased interleukin 6 (IL-6) and C-reactive protein (CRP) reported.

In T2DM, proinflammatory cytokines such as IL-1β are elevated in pancreatic β cells, leading to an increased release of insulin from pancreatic β cells and expression of other cytokines, such as tumor necrosis factor α (TNFα), monocyte chemotactic protein, and macrophage inflammatory protein 1α (Boni-Schnetzler et al. 2008). Importantly, levels of proinflammatory cytokines have been shown to be elevated in patients with diabetes and have an important role to play in the underlying biology, and indeed there may be inflammatory synergy between T2DM and AD (Lue, Andrade, Sabbagh & Walker, 2012). In cognitive decline and dementia, epidemiological studies have demonstrated an increase in proinflammatory mediators such as CRP and IL-6 as risk factors, with TNFα also raised in those with dementia and cognitive impairment (Engelhart et al. 2004; Guerreiro et al. 2007). Genetic variants in IL-1β have also been linked to cognitive function in older adults (Trompet et al. 2008).

Pairing these findings together postulates that via innate immune system activation, particularly with AGEs acting as DAMPs as well as inflammation associated with the diabetic syndrome, CNS damage may result. In the Edinburgh Type 2 Diabetes Study, a correlation between both raised IL-6 and TNFα levels and cognitive deficits in patients with T2DM has been reported (Marioni et al. 2010). A 2017 study examining the relationship between T2DM and dementia has demonstrated that increased CRP and TNFα levels were correlated with cognitive decline, as were levels of malondialdehyde, supporting the role of systemic inflammation as a causative mechanism (Ragy & Kamal, 2017). Interestingly, increased blood—brain barrier permeability has been demonstrated in patients with T2DM, which may add to the ability of systemic inflammation to damage the CNS (Janelidze et al. 2017). While the preceding is a small selection of studies examining the role for inflammation in both T2DM and dementia, further work is needed to fully elucidate the shared inflammatory mechanisms that may result in end organ inflammatory damage to the diabetic brain. At present, it seems that the proinflammatory cytokines shared in both conditions may contribute to central neurodegeneration in T2DM patients.

Glucocorticoid excess

Individuals with T2DM have central dysregulation of the HPA axis, with raised plasma cortisol levels, increased adrenocorticotrophic hormone, and impaired dexamethasone suppression. Elevated levels of glucocorticoids have deleterious effects on hippocampal neurons, and this is another putative mechanism underlying the T2DM—dementia link (Strachan, Reynolds, Frier, Mithcell & Price 2009). In insulin-deficient (diabetic animal models) rats and mice, studies have demonstrated that diabetes impairs hippocampal function through glucocorticoid-mediated effects on both new and mature neurons (Stranahan et al. 2008). Interestingly, in AD, alterations in HPA axis activity and raised plasma cortisol are associated with cognitive decline and altered hippocampal volumes (de Leon et al. 1988, p. p392; O'Brien, Ames, Schweitzer I, Mastwyk & Colman 1996). In the Edinburgh Type 2 Diabetes Study, high fasting cortisol was associated with accelerated cognitive decline in cross-sectional analysis (reductions in letter number sequencing and the DSST) (Reynolds et al. 2010). Taken together, these findings implicate a key role for excess cortisol and HPA dysregulation in mediating the T2DM—dementia risk.

Vascular pathology

Many pathophysiological studies have suggested that the link between T2DM and dementia may be most reflective of an AD-type pathology. However, parsing dementia into vascular and AD separately may not be fully reflective of the real-life situation, and the coexistence and vascular contribution to AD pathology are now better recognized. Neuroimaging studies have shown that cerebral microvascular disease is increased in patients with T2DM, with T2DM patients also having a decreased cerebral blood flow (Roriz-Filho et al. 2009; van Harten et al. 2007). Thus the ischemia and hyperglycemia seen in patients with long-standing T2DM may contribute to the cognitive impairment and vascular disease, meaning that vascular, and in particular small vessel, disease may serve to lower the threshold for the development of dementia (Umegaki, 2014). Thus, the aforementioned mechanisms involved, such as dysglycemia, may contribute to dementia risk in T2DM in numerous ways, contributing to abnormalities of small blood vessels, accelerating vascular as well as AD pathology. An integrative interpretation of the pathobiology underlying the T2DM—dementia link is thus most pragmatic.

Diabetes-related risk factors for dementia

T2DM is a heterogeneous disease, and those affected have varying clinical courses, affected at different times by a myriad of potential complications. Parsing out which adults with T2DM are particularly at risk is a key area of priority for the academic literature, to identify those who may maximally benefit from potential preventative management. Much of this work has focused on correlation with measures of diabetes control, vascular risk, the presence of T2DM-related complications, and serological measurements. In the future, it is perhaps a composite of these measures that may adequately identify those at particular risk of cognitive decline. Interestingly, prediction models have been devised in large cohorts incorporating microvascular disease, diabetic foot, cerebrovascular disease, cardiovascular disease, acute metabolic events, depression, age, and education (Exalto et al. 2013). Such scores help select those with T2DM at highest risk of cognitive decline for screening or participation in clinical trials.

HbA1c is a key clinical measure of diabetes control (and indeed diagnosis in the initial stage) over a long period of time (typically the lifetime of a red blood cell, around 90−120 days). Increased HbA1c has been well validated as a risk factor for T2DM complications and is routinely used in monitoring disease. HbA1c has been investigated in numerous studies examining dementia risk in T2DM and correlates with the presence of cognitive impairment in older adults (Cukierman-Yaffe et al. 2009). Most recently, in the English Longitudinal Study of Ageing following controlling for other risk cofactors for dementia, an increase in HbA1c correlated with accelerated global cognitive decline and executive function, during a mean follow-up of over 8 years (Zheng, Yan, Yang, Zhong & Xie, 2018). This association may reflect the role of hyperglycemia in mediating the T2DM−dementia link as discussed above. Further to diabetes control, the duration of T2DM (Roberts et al. 2008) and presence of T2DM-related complications (Bruce et al. 2008) are both associated with an increased risk of cognitive decline.

Other physiological measures related to the T2DM syndrome have also been found to place individuals with T2DM at higher risk of cognitive decline and dementia. These include measures of central adiposity, total fat mass, systolic blood pressure (indicative of hypertension), cholesterol profiles, and arterial stiffness (Abbatecola et al. 2010a; Mehrabian et al. 2012; Petrova, Prolopenko, Pronina & Mozheyko, 2010). There is also evidence that lifestyle factors such as smoking and lack of exercise may place individuals at increased risk (Feinkohl et al. 2015). Many of these risk factors are well known in predicting the presence of other complications related to T2DM, and again, a composite of simple measures routinely obtained in the clinical setting may be a pragmatic solution to predicting dementia/cognitive decline in patients with T2DM.

Given the association between other T2DM-related complications and cognitive decline/dementia, the presence of these may alert the clinician to screen cognition. Interestingly, metaanalytic studies have shown an association between cognitive impairment and diabetic retinopathy (Crosby-Nwaobi, Sivaprasad & Forbes, 2012). Complications such as diabetic retinopathy, which is already screened for in T2DM, may be a clinically accessible indicator of the presence of T2DM-related complications (such as cognition) in the CNS.

Neuroimaging in patients with type 2 diabetes mellitus

An interesting body of work has emerged examining the structural and functional neuroimaging deficits in patients with T2DM and the risk of cognitive decline. Evidence for cortical and subcortical atrophy, hippocampal and amygdalar atrophy, and decreased gray matter volumes in frontoparietal areas in patients with T2DM has been noted, many of which are similar to the profile of AD seen on neuroimaging (Den Heijer et al. 2003; Garcia-Casares et al. 2014; Kumar et al. 2008; Moran et al. 2013). In those with both MCI and T2DM in comparison with either alone, decreased gray matter volume in the medial temporal gyrus has been noted (Zhang et al. 2014). Interestingly, HbA1c, perturbation of HPA axis feedback, and IR have all been demonstrated as predictive of gray matter atrophy in T2DM, many of which recapitulate the risk factors for cognitive decline and dementia in those with T2DM mentioned earlier (Bruehl et al. 2010; Gold et al. 2007).

Studies using diffusion tensor imaging (DTI) examining white matter changes have demonstrated reduced fractional anisotropy in frontotemporal regions in those with T2DM (Falvey et al. 2013). Further, in cognitively intact patients with T2DM, DTI studies have demonstrated a link between disease duration and an increase in mean diffusivity and axial and transverse diffusivity (Hsu et al. 2012). The presence of neuroimaging abnormalities in individuals without cognitive decline or evidence of cognitive impairment demonstrates that much of the neuropathological damage caused by T2DM may be occurring at a stage when changes in cognition are not clinically apparent. Thus, in line with the aforementioned epidemiological evidence, much of the damage may be occurring before cognitive decline is clinically apparent. Strikingly, a 2017 study has even demonstrated microstructural abnormalities in white and gray matter in adolescents with T2DM, with reduced gray matter volume, left putamen and caudate, and bilateral amygdala (Nouwen et al. 2017).

Functional neuroimaging also supports the increased risk of cognitive decline in T2DM. Reduced connectivity between frontotemporal regions and hippocampus and reduced frontotemporal blood flow and glucose utilization have been noted (Garcia-Caseres et al. 2014; Jones et al. 2014; Niwa et al. 2006). These abnormalities implicate the frontal and temporal cortex in cognitive decline in T2DM, mirroring those seen in AD. Altered attentional network activating in patients with T2DM and abnormalities in the resting state have also been demonstrated (Xia et al. 2015). Many of the neuroimaging findings support the risk of cognitive decline seen in diabetes and are similar in appearance to those seen in AD, supporting the evidence for increased risk of dementia and cognitive decline in those with T2DM.

Interventions to mitigate the risk of dementia in type 2 diabetes

Despite the well-characterized link between T2DM and subsequent dementia risk, there is no good evidence at present that any treatment strategy for T2DM or related complications can mitigate the effect of T2DM on cognition. Part of the reason for this is the lack of acknowledgment of the cognitive complications of T2DM in clinical trials. Well-designed trials specifically examining the effect of interventions for cognition in T2DM are eagerly awaited. Interventions aimed at reducing this risk may consist of intensive glycemic control, the additional use of medications for T2DM or cognition, or lifestyle interventions. Clinical trials examining the effects of interventions for cognition in T2DM are summarized in Table 6.1.

Table 6.1 Potential treatments for cognition in type 2 diabetes.

Study	Participants	Age	Intervention	Duration	Comparison	Outcome	Result
Glycemic control							
ADVANCE study	11,140 with first diagnosis of T2DM ages >30	Mean age 66.6 years	Intensive glucose control regime including gliclazide and other treatments at physician's discretion and also randomized to perindopril—indapamide/other antihypertensives	5 years	Standard control group	Number of patients declining by 3 points on MMSE from baseline	No difference
ACCORD–MIND study	2794 with T2DM and HbA1c of >7.5%	Mean age 62.3 years	40 months of intensive glycemic control with an HbA1c target of 7.0%–7.9%, in addition to intensive blood pressure lowering	40 months	Diabetes education, glucose monitoring equipment, and antidiabetic medications	MMSE, RVALT, Stroop test, DSST	No difference
Araki et al. 2012	1173 with T2DM and other cardiovascular risk factors	Mean age 71.9 years	Intensive regime of treatment for diabetes, hypertension, and dyslipidemia or treatment as usual over 6 years	6 years	Treatment as usual	MMSE	No difference

Diet and exercise

Study	Population	Age	Intervention	Duration	Control	Cognitive assessment	Outcome
Look-AHEAD study	3802 overweight and obese with T2DM	45–76 years	Intensive lifestyle intervention (calorie goals of 1200–1800 and >175 min/ week activity)	10 years	Diabetes support and education	Diagnosis of MCI/ dementia, MMSE, functional assessment questionnaire, TMT-A, RVALT, DSST, Stroop test	No difference
LIFE study	1476 older adults with functional limitations (post hoc analysis of 415 with diabetes)	70–89 years	Physical activity (walking, resistance training, flexibility)	2 years	Health education	MMSE, DSC, HVLT-D	Improvements in global cognitive function in patients with T2DM

Diabetic medications

Study	Population	Age	Intervention	Duration	Control	Cognitive assessment	Outcome
Abbetacola et al. 2006	156 patients	Mean age 74.3 years	Glibenclamide (glyburide)	1 year	Repaglinide	MMSE, global cognitive score (TMT, DSC, DSST, verbal fluency test)	Decline in cognition seen only in the glibenclamide group

Continued

Table 6.1 Potential treatments for cognition in type 2 diabetes.—cont'd

Study	Participants	Age	Intervention	Duration	Comparison	Outcome	Result
Ryan et al. 2006	145 adults with T2DM receiving oral metformin combination therapy	Mean age 60.7 years	Rosiglitazone plus metformin	2 years	Glibenclamide (glyburide)	DSST, RVALT, paired associates learning, pattern recognition memory, spatial working memory test, reaction time, rapid visual information processing	Improvements in working memory correlated with glycemic control
Guo et al. 2014	58 patients with T2DM and depression	Mean age 54.7 years	Treatment with metformin	2 years	Placebo (vitamin C)	WMS, revised	Improved cognition as measured by the WMS
Abbetacola et al. 2010	97 older patients who had recently started treatment for T2DM	Mean age 76 years	Metformin ± rosiglitazone	9 months	Diet	MMSE, RVALT, TMT	Results for metformin/rosiglitazone group remained stable for RVALT

This table details the interventions that have been trialed to target cognition in patients with T2DM. *DSC*, Digit Symbol Coding task; *DSST*, Digit Symbol Substitution Test; *HbA1c*, hemoglobin A1c; *HVLT-D*, Harvard Verbal Learning Test revised; *MCI*, mild cognitive impairment; *MMSE*, Mini-Mental State Examination; *RVALT*, Rey Verbal Auditory Learning Test; *T2DM*, type 2 diabetes mellitus; *TMT*, Trail-Making Test; *WMS*, Wechsler Memory Test.

Intensive Glycemic control

Based on the aforementioned evidence, it may be predicted that tight glycemic control may have a protective cognitive effect in T2DM. In the ADVANCE study (2008), over 11,000 (mean age = 66 years) patients with T2DM were randomized to an intensive glucose control regime (HbA1c of 6.5%) to a control group receiving standard treatment alone for a median of 5 years (Areosa Sastre et al. 2017). Unfortunately, there was no beneficial effect on cognition observed. In another large study, the ACCORD-MIND study, nearly 3000 patients (mean age 62.5 years) with T2DM, high HbA1c, and a high risk of cardiovascular events were assigned to intensive control (HbA1c of <6%) or standard strategy treatment (HbA1c of 7%–7.9%). Treatment was not associated with any effect on DSST score at 40 months (Launer et al. 2011). A third trial of nearly 1200 patients (average age of just under 72) examined the treatment of HbA1c, blood pressure <130/85 mm Hg, and lipid profile using oral hypoglycemics/statins in comparison with baseline treatment. Cognitive function as measured by MMSE showed no improvement (Araki et al. 2012). While the results of trials examining intensive glycemic control seem disappointing, perhaps one reason for the lack of observed effect is the age of the included patients. Throughout this chapter we have seen that much of the cognitive damage caused by T2DM may have its origins in midlife, and perhaps targeting patients earlier in the disease course may prove of benefit. Treating patients of ages >60 may be targeting a group in which much of the CNS damage may have already occurred.

Metformin

In a small trial of 58 patients, treatment with metformin for 24 weeks in patients with both T2DM and depression has been demonstrated to improve cognitive function as measured by the Wechsler Memory Scale—revised (Guo et al. 2014). This is on the background of reports of both increased risk and decreased risk for cognitive decline in T2DM and metformin use. Most recently, in a large cohort of US veterans of >65 years, metformin use was associated with a lower risk of subsequent dementia than sulfonylurea use (Orkaby et al. 2017). Interpretation of these retrospective studies remains difficult and high-quality clinical trials are needed.

Other oral hypoglycemics

Other oral hypoglycemics have also been examined. In a study by Abbatecola et al. (2006), patients were randomized to repaglinide or glibenclamide (77 and 79 patients, respectively). A decline in cognitive function was observed only in the glibenclamide group, which was attributed to the postprandial cognitive glucose excursion. In another study (145 patients), add-on therapy of either rosiglitazone or glyburide was associated with improvement in paired associates learning tests and correlated with the improved glycemic control achieved (Ryan et al. 2006). Rosiglitazone has also been shown to have a protective effect on cognitive decline in a small study of 97 patients (Abbectacola et al. 2010b). Again, future trials are required to examine the cognitive effects of the differing oral hypoglycemics.

Diet and exercise

The effects of lifestyle interventions on cognitive risk in T2DM have also been examined. Two exemplars of this are the Look–AHEAD and the LIFE studies. In the Look–AHEAD study, examining overweight and obese patients with T2DM, a 10-year intensive lifestyle intervention consisting of calorie restriction and physical activity found no difference in the diagnosis of MCI and probable dementia, and no cognitive differences were observed between the groups (however, cognition was not measured at baseline) (Espeland et al. 2017a; Rapp et al. 2017). In secondary analysis of the LIFE study, a 2-year physical activity intervention, cognitive benefit was seen in those assigned to physical activity in comparison with control, but only for those patients with diabetes (Espeland et al. 2017b). Both studies concerned secondary analysis of clinical trials. Interpreting these studies is difficult, and it is currently unclear as to whether diet and exercise have any effect on cognition in patients with T2DM.

Treatment strategies

Thus, a limited number of studies on intensive glycemic control, oral hypoglycemics, and lifestyle interventions demonstrate no clear evidence either for or against a risk reduction for cognition in T2DM. Future trials should assess the efficacy of multidomain interventions in midlife populations of T2DM patients (not just those age 60 and above, as in most of the aforementioned trials). In the FINGER trial, a multidomain intervention was found to be effective in preventing cognitive decline in adults at high risk for developing dementia. T2DM may represent a population that may benefit from a similar intervention; however, the intervention must take place at the right point in the disease course. Therefore, we would propose that future trials should include multidomain interventions and take place at midlife or before, when the epidemiological and pathophysiological evidence tells us that most of the deleterious effects of T2DM on cognition occur.

Conclusion

Thus, it can be seen that there is an abundance of evidence pointing to an increased risk of dementia and cognitive decline in patients with T2DM. Awareness of the risk for dementia attributable to diabetes is low. In a recent study only 35% of T2DM patients were aware of the increased risk of dementia, in comparison with 84% awareness of renal and ophthalmological complications (Dolan et al. 2018). While currently the American Diabetes Association recommends screening populations age 65 and older annually for cognitive decline, issues around competency and time constraints are also highlighted (American Diabetes Association, 2018). Selecting patients who may be at highest risk of cognitive decline is therefore important, not only for identification, but also in the future for participation in potential preventative strategies that may be used. The development of cognitive impairment in patients with T2DM may lead to poorer disease control. In a large cross-sectional study (Health and Retirement Study Diabetes), adults with cognitive impairments were less likely to adhere to treatment (Feil, Zhu & Sultzer, 2012), and in previous studies cognitive impairment was associated with poorer diabetes control (Feil et al. 2009). Thus, developing preventative measures to combat the cognitive decline and dementia risk in T2DM represents a clinically unmet need, which is of significant public health concern at present. Further study on the early-stage pathophysiological mechanisms linking these two common diseases and the creation of multidomain interventions, targeted at patients with T2DM in midlife, are of paramount importance for the field in the coming years.

Key facts of diabetes

- Diabetes is a common and complex disorder of carbohydrate metabolism characterized by IR and hyperglycemia.
- Diabetes is divided into type 1 (insulin dependent) and type 2 (non—insulin dependent) diabetes. Type 1 diabetes is characterized by an endogenous lack of insulin and type 2 diabetes is characterized by a resistance of insulin in target organs.
- It is estimated that type 2 diabetes affects 415 million people worldwide and that this may increase to almost 600 million by 2035.
- Diabetes is characterized by numerous microvascular (ischemia, neuropathy, retinopathy, neuropathy) and macrovascular (myocardial infarction, peripheral vascular disease, stroke) complications.
- Longitudinal studies have demonstrated that type 2 diabetes in midlife is a significant risk factor for the development of dementia (ATD) in later life.
- Management of diabetes is individual and tailored to patients' needs.
- Achieving glycemic control is a key target in therapy and is monitored by HbA1c.
- Management also includes treatment of hyperlipidemia and blood pressure and cardiovascular risk management.
- Smoking cessation, regular exercise, and a healthy diet are key aspects of diabetes management.
- Medications used in the management of T2DM include metformin, sulfonylureas, insulin secretagogues, thiazolidinediones, dipeptidyl peptidase 4 inhibitors, sodium—glucose cotransporter-2 inhibitors, and, in some cases, insulin therapy.

- Despite the fact that cognitive impairment is a recognized complication of T2DM, awareness typically remains low.
- While the American Diabetes Association recommends annual screening of cognition over the age of 65, concerns about competence in the assessment of cognition, time constraints of busy clinics, and lack of an evidence-based intervention at present have also been expressed.

Summary points

- T2DM in midlife is one of the most significant risk factors for the development of dementia in later life.
- Numerous early cross-sectional studies hinted at this relationship, with a significant increase in risk found in subsequent longitudinal studies. More recently, studies examining comparative risks have shown that the risk conferred by T2DM may even be as great as that conferred by the apolipoprotein genotype.
- Hyperglycemia has been demonstrated to correlate with dementia risk in well-designed longitudinal cohort studies and seems to be a central mechanism in mediating the T2DM—dementia risk.
- Hyperglycemia, with a resultant buildup in AGEs, may act as DAMPs in activating microglia in the CNS, the brain's resident immune cells, contributing to neuroinflammation and CNS damage.
- IR and alterations in insulin signaling seen in T2DM are also seen in Alzheimer's dementia and may be important in mediating the shared pathology of both syndromes.
- Systemic inflammation seen in T2DM may serve to accelerate cognitive decline, and there may be an inflammatory synergy between T2DM and AD, with IL-6 and TNFα elevated in patients with T2DM showing evidence of cognitive impairment.
- Glucocorticoid excess and abnormal functioning of the HPA axis may be another shared mechanism, and has been well supported in animal models of diabetes.
- Risk factors for cognitive impairment in T2DM include duration of diabetes, diabetes-associated complications (including diabetic retinopathy), central adiposity, total fat mass, systolic blood pressure (indicative of hypertension), cholesterol profiles, arterial stiffness, and numerous hormone and serological measurements (including vitamin D and leptin).
- There are both structural and functional neuroimaging findings in patients with T2DM and these include cortical and subcortical atrophy, hippocampal and amygdalar atrophy, and decreased gray matter volumes in frontoparietal areas in patients. Many of these are similar to the profile of AD seen on neuroimaging.
- There is no good evidence at present that any single strategy (intensive glycemic control, hypoglycemic medication, diet and exercise) may mitigate the cognitive deficits

seen in T2DM. While the results of trials in these areas have been disappointing, one reason is that the age of the included patients has been typically at least 60 years.

- In the future, in concordance with trials such as the FINGER trial (a multidomain intervention to prevent dementia in at-risk adults), multidomain interventions that target adults at risk in midlife must be designed to evaluate preventative strategies that are of paramount public health importance.

References

Abbatecola, A. M., Lattanzio, F., Molinari, A. M., Cioffi, M., Mansi, L., Rambaldi, P., et al. (2010b). Riso-glitazone and cognitive stability in older individuals with type 2 diabetes and mild cognitive impairment. *Diabetes Care, 33*(8), 1706–1711.

Abbatecola, A. M., Lattanzio, F., Spazzafumo, L., Molinari, A. M., Cloffi, M., Canonico, R., et al. (2010a). Adiposity predits cognitive decline in older persons with diabetes: A 2 Year follow up. *PloS One, 5*(4), e10333.

Abbatecola, A. M., Rizzo, M. R., Barbeieri, M., Grella, R., Arciella, A., Laieta, M. T., et al. (2006). Posprandial plasma glucose excursions and cognitive functioning in aged type 2 diabetics. *Neurology, 67*(2), 235–240.

American Diabetes Association. (January 2018). Medical evaluation and assessment of comorbidities: Standards of medical care in diabetes—2018 American diabetes association. *Diabetes Care, 41*(Suppl. 1). S28-S3.

Araki, A., Limuro, S., Sakurai, T., Umegaki, H., Iijima, K., Nakano, H., et al. (2012). Long-term multiple risk factor interventions in Japenese elderly diabetic patients: The Japanese elderly diabetes intervention trial: Study design, baseline characteristics and effects of intervention. *Geriatrics and Gerontology International,* (Suppl. 1), 7–17.

Areosa Sastre, A., Vernooij, R. W., Gonzalez-Colaco Harmand, M., & Martinez, G. (2017). Effect of the treatment of Type 2 Diabetes mellitus on the development of cognitive impairment and dementia. *Cochrane Database of Systematic Reviews, 15*, 6. CD003804.

Barbagallo, M., & Dominguez, L. J. (2014). Type 2 diabetes mellitus and Alzheimer's disease. *World Journal of Diabetes, 5*(6), 889-9.

Block, M. L., Zecca, L., & Hong, J. S. (2007). Microglia-mediated neurotoxicity: Uncovering the molecular mechanisms. *Nature Reviews Neuroscience, 8*(1), 57–69.

Boni-Schnetzler, M., Thorne, J., Parnaud, G., Marselli, L., Ehses, J. A., Kerr-Conte, J., et al. (2008). Increased interleukin (IL-1beta messenger ribonucleic acid expression in beta-cells of individuals with type 2 diabetes and regulation of IL-1beta in human islets by glucose and autostimulation. *Journal of Clinical Endocrinology & Metabolism, 93*(10), 4065–4074, 2008.

Bruce, D. G., Davis, W. A., Casey, G. P., Starkstein, S. E., Clarnette, R. M., Almeida, O. P., et al. (2008). Predictors of cognitive decline in older Individuals with diabetes. *Diabetes Care, 31*(11), 2103–2107.

Bruehl, H., Sweat, V., Hassenstab, J., Polyakov, C., & Convit, A. (2010). Cognitive impairment in nondiabetic middle aged and older adults is associated with insulin resistance. *Journal of Clinical and Experimental Neuropsychology, 32*(5), 487–493.

Chen, G., Cai, L., Chen, B., Liang, J., Lin, F., Li, L., et al. (2011). Serum level of endogenous secretory receptor for advanced glycation end products and other factors in type 2 diabetic patients with mild cognitive impairment. *Diabetes Care, 32*(12), 2586–2590.

Crane, P. K., Walker, R., Hubbard, R. A., Li, G., Nathan, D. M., Zheng, H., et al. (2013). Glucose levels and risk of dementia. *New England Journal of Medicine, 369*(6), 540–548.

Crosby-Nwaobi, R., Sivaprasad, S., & Forbes, A. (2012). A systematic review of the association of diabetic retinopathy and conitive impairment in people with Type 2 diabetes. *Diabetes Research and Clinical Practice, 96*(2), 101–110.

Cukierman-Yaffe, T., Gerstein, H. C., Williamson, J. G., Lazar, J. D., Lovato, et al. (2009). Relationship between baseline glycemic control and cognitive function in individuals with type 2 diabetes and other cardiovascular risk factors: the action to control cardiovascular risk in diabetes-memory in diabetes (ACCORD-MIND) trial. *Diabetes Care, 32*(2), 221−226, 2009.

De Leon, M. J., McRae, T., Tsai, J. R., George, A. E., Marcus, D. L., Freedman, M., et al. (1988). Abnormal cortisol response in Alzheimer's Disease linked to hippocampal atrophy. *Lancet, 1899*, p392.

Den Heijer, T., Vermer, S. E., van Doojk, E. J., Prins, N. D., Koudstaal, P. J., Hofman, A., et al. (2003). Type 2 diabetes and atrophy of medial temporal lobe structures on brain MRI. *Diabetologia, 46*(12), 1604−1610.

Dolan, C., Glynn, R., Griggin, S., Conroy, C., Loftus, C., Wiehe, P. C., et al. (April 13, 2018). Brain complications of diabetes mellitus: A cross-sectional study of awareness among individuals with diabetes and the general population in Ireland. *Diabetic Medicine.* https://doi.org/10.1111/dme.13639 ([Epud ahead of print]).

Duron, E., Funalot, B., Brunel, N., Coste, J., Qunquis, L., Viollet, C., et al. (2012). Insulin-like growth factor 1 and insulin like growth fActor binding protein 3 in Alzheimers disease. *Journal of Clinical Endocrinology & Metabolism, 97*(12), 4673−4681.

El Khoury, N. B., Gratuze, M., Papon, M. A., Bretteville, A., & Planel, E. (2014). Insulin dysfunction and Tau pathology. *Frontiers in Cellular Neuroscience, 8*, 22.

Engelhart, M. J., Geerlings, M. I., Meijer, J., Kiliaan, A., Ruitenberg, A., et al. (2004). Inflammatory proteins in plasma and the risk of dementia: The Rotterdam study. *Archives of Neurology, 61*(5), 668−672.

Espeland, M. A., Lipska, K., Miller, M. E., Rushing, J., Cohen, R. A., Verghesse, J., et al. (2017b). Effects of physical activity interventino on physical and cognitive function in sedentary adults with and without diabetes. *The Journals of Gerontology Series A Biological Sciences and Medical Sciences, 72*(6), 861−866.

Espeland, M. A., Luchsinger, J. A., Baker, L. D., Neiberg, R., Sahn, S. E., Arnold, S. E., et al. (2017a). Effect of a long-term intensive lifestyle intervention on prevalence of cognitive impairment. *Neurology, 88*(21), 2026−2035.

Exalto, L. G., Biessels, G. J., Karter, A. J., Huang, E. S., Katon, W. J., Minkoff, J. R., et al. (2013). *Risk score for prediction of 10 year dementia risk in individuals with type 2 diabetes: A cohort study.*

Falvey, C. M., Rosano, C., Simonsick, E. M., Harris, T., Strotmeyer, E. S., Satterfield, S., et al. (2013). Macro and microstructural magnetic resonance imaging indices associated with diabetes among community dwelling older adults. *Diabetes Care, 36*(3), 677−682.

Feil, D. G., Pearman, A., Victor, T., Harwood, D., Weinreb, J., Kahle, K., et al. (2009). The role of cognitive impairment and caregiver support in diabetes management of older outpatients. *The International Journal of Psychiatry in Medicine, 39*(2), 199−214.

Feil, D. G., Zhu, C. W., & Sultzer, D. L. (2012). The relationship between cognitive impairment and diabetes self management in a population based community sample of older adults with Type 2 diabetes. *Journal of Behavioral Medicine, 35*(2), 190−199.

Feinkohl, I., Keller, M., Robertson, C. M., Morling, J. R., McLachlan, S., Frier, B. M., et al. (2015). Cardiovascular risk factors and cognitive decline in older people with type 2 diabetes. *Diabetologia, 58*(7), 1637−1645.

Garcia-Casares, N., Berthier, M. L., Jorge, R. E., Gonzalez-Algere, P., Gutierrez-Cardo, A., Rioja Villodres, J., et al. (2014). Structural and functional brain dhanges in middle aged type 2 diabetic patients: A cross sectional study. *Journal of Alzheimer's Disease*, 375−386.

Gold, S. M., Dziobek, I., Sweat, V., Tirsi, A., Rogers, K., Bruehl, H., et al. (2007). Hippocampal damage and memory impairments as possible early brain complications of type 2. *Diabetes, 50*(4), 711−719.

Gottesman, R. F., Albert, M. S., Alonso, A., Coker, L. H., Coresh, J., Davis, S. M., et al. (2017). Associations between midlife vascular risk factors and 25-year. *Incident Dementia in the Atherosclerosis Risk in Communities (AIRC) Cohort, 74*(10), 1246−1254.

Guariguata, L., Nolan, T., Beagley, J., Linnenkamp, U., & Jacqmain, O. (2014). *International diabetes federation diabetes atlas*. Brussels: International Diabetes Federation.

Gudala, K., Bansai, D., Schifano, D., & Bhansali, A. (2013). Diabetes mellitus and risk of dementia: A meta-analysis of prospective observational studies. *Journal of Diabetes Investigation, 4*(6), 640−650, 27.

Guerreiro, R. Y., Santana, I., Bras, J. M., Santiago, B., Paiva, A., et al. (2007). Peripheral inflammatory cytokines as biomarkers in Alzheimer's disease and mild cognitive impairment. *Neurodegenerative Diseases, 4*(6), 406−412.

Guo, M., Mi, J., Jiang, Q. M., Xu, J. M., Tang, Y. Y., Tian, G., et al. (2014). Metformin may produce antidepressant effects through improvement of cognitive function among depressed patients with diabetes mellitus. *Clinical and Experimental Pharmacology and Physiology, 41*(9), 650−656.

Hsu, J. L., Chen, Y. L., Leu, J. G., Jaw, F. S., Lee, C. H., & Tsai, Y. F. (2012). Microstructural white matter abnormalities in type 2 diabetes mellitus. *A Diffusion Tensor Imaging Study, 59*(2), 1098−1105.

Janelidze, S., Hertze, J., Nagga, K., Nilsson, K., Nilsson, C., Swedish BioFINDER Study Group, et al. (2017). Increased blood-brain barrier permeability is associated with dementia and diabetes but not amyloid pathology or APOE genotype. *Neurobiology of Aging, 51*, 104−112.

Jones, N., Riby, L. M., Mitchell, R. L., & Smith, M. A. (2014). Type 2 diabetes and memory: Using neuroimaging to understand the mechanisms. *Current Diabetes Reviews, 10*(2), 118−123.

Katzman, R., Aronson, M., Fuld, P., Kawas, C., Brown, T., Morgenstern, H., et al. (1989). Development of dementing illness in an 80 year-old volunteer cohort. *Annals of Neurology, 25*(4), 317−324.

Kumar, R., Anstey, K. J., Cherbuin, N., Wen, W., & Sachdev, P. S. (2008). Association of type 2 diabetes with depression, brain atrophy and reduced fine motor speed in a 60- to 64- year old community sample. *American Journal of Geriatric Psychiatry, 16*(12), 989−998.

Launer, L. J., Miller, M. E., Williamson, J. D., Lazar, R. M., Gerstein, H. C., Murray, A. M., et al. (2011). Effects of intensive glucose lowering on brain structure and function in people with type 2 diabates (ACCORD-MIND): A randomised open label substudy. *The Lancet Neurology, 10*(11), 969−977.

Leibson, C. L., Rocca, W. A., Hanson, V. A., Cha, R., Kokmen, E., O'Brien, P. C., et al. (1997). Risk of dementia among persons with diabetes mellitus: A population-based cohort study. *American Journal of Epidemiology, 145*(4), 301−308.

Luchsinger, J. A., Reitz, C., Patel, B., Tang, M. X., Manly, J. J., & Mayeux, R. (2007). Relation of diabetes to mild cognitive impairment. *Archives of Neurology, 64*(4), 570−575.

Lue, L. F., Andrade, C., Sabbagh, M., & Walker, D. (2012). Is there inflammatory synergy in type II diabetes mellitus and Alzheimer's disease. *International Journal of Alzheimer's Disease, 2012*, 918680.

Marioni, R. E., Strachan, M. W., Reynolds, R. M., Lowe, G. D., Mitchell, R. J., Fowkes, F. G., et al. (2010). Association between raised inflammatory markers and cognitive decline in elderly people with type 2 diabetes: The Edinburgh type 2 diabetes study. *Diabetes, 59*(3), 710−713.

Ma, L., Wang, J., & Li, Y. (2015). Insulin resistance and cognitive dysfunction. *Clinica Chimica Acta, 444*, 18−23.

Mehrabian, S., Raycheva, M., Gateva, A., Todorova, G., Angelova, P., Traykova, M., et al. (2012). Cognitive dysfunction profile and arterial stiffness in type 2 diabetes. *Journal of Neurological Sciences, 322*(1−2), 152−156.

Moran, C., Phan, T. G., Chen, J., Blizzard, L., Beare, R., Venn, A., et al. (2013). Brain atrophy in type 2 diabetes: Regional distribution and influence on cognition. *Diabetes Care, 36*(12), 4036−4042.

Morris, J. K., Vidoni, E. D., Honea, R. A., Burns, J. M., et al. (2014). Impaired glycaemia increases disease progression in mild cognitive impairment. *Neurobiology of Aging, 35*(3), 585−589.

Ngandu, T., Lethisalo, J., Solomon, A., Levalahti, E., Ahtiluoto, S., Antikanien, R., et al. (2015). A 2 year multidomain intervention of diet, exercise, cognitive training and vascular risk monitoring versus control to prevent cognitive decline in at-risk elderly people (FINGER): A randomised controlled trial. *Lancet, 385*(9984), 2255.

Niwa, H., Koumoto, C., Shiga, T., Takeuchi, J., Mishima, S., Segawa, T., et al. (2006). Clinical analysis of cognitive function in diabetic patients by MMSE and SPECT. *Diabetes Research and Clinical Practice, 72*(2), 142−147.

Nouwen, A., Chambers, A., Chechiacz, M., Higgs, S., Bisset, J., Barret, T. G., et al. (2017). Microstructural abnormalities in white and gray matter in obese adolescents with and without type 2 diabetes. *Neuroimage Clin, 16*, 43−51.

Orkaby, A. R., Cho, K., Cormack, J., Gagnor, D. R., & Driver, J. A. (2017). Metoformin vs sulfonylurea use and risk of dementia in US veterans aged > 65 years with diabetes. *Neurology, 89*(18), 1877−1885.

Ott, A., Stolk, R. P., van Harskamp, F., Pols, H. A., Hofman, A., & Breteler, M. M. (1999). Diabetes mellitus and the risk of dementia. *The Rotterdam Study, 10*(9), 1937−1942, 53.

O'Brien, J. T., Ames, D., Schweitzer, I., Mastwyk, M., & Colman, P. (1996). Enhanced adrenal sensitivity to adrenocorticotrophic hormone (ACTH) is evidence of HPA axis hyperactivity in Alzheimer's disease. *Psychologie Medicale, 26*(1), 7—14.

Petrova, M., Prolopenko, S., Pronina, E., & Mozheyko, E. (2010). Diabetes type 2, hypertension and cognitive dysfunction in middle age women. *Journal of Neurological Sciences, 299*(1—2), 39—41.

Ragy, M. M., & Kamal, N. N. (2017). Linking senile dementia to type 2 diabes: Role of oxidative stress markers, C-reactive protein and tumor necrosis factor-a. *Neurological Research, 38*(7), 587—595.

Rapp, S. R., Luchsinger, J. A., Baker, L. D., Blackburn, G. L., Hazuda, H. P., Demos-McDermott, K. E., et al. (2017). Effect of a long-term intensive lifestyle intervention on cognitive function. *Action for Health in Diabetes Study, 65*(5), 966—972.

Reynolds, R. M., Strachan, M. W., Labad, J., Lee, A. J., Frier, B. M., & Fowkes, F. G. (2010). Morning cortisol levels and cognitive abilities in people with type 2 diabetes. *The Edinburgh Type 2 Diabetes Study, 33*(4), 714—720.

Roberts, R. O., Geda, Y. E., Knopman, D. S., Christianson, T. J., Pankratz, V. S., Boeve, B. P., et al. (2008). Association of duration and severity of diabetes mellitus with mild cognitive impairment. *Archives of Neurology, 65*(8), 1066—1073.

Ronnemaa, E., Zethelius, B., Sundelof, J., Sundstrom, J., Degerman-Gunnarsson, M., Berne, C., et al. (2008). Impaired insulin secretion increases the risk of Alzheimer's Disease. *Neurology, 71*(14), 1065—1071.

Roriz Filho, J. S., Sa-Roriz, T. M., Rosset, I., Camozzato, A. L., Santos, A. C., & Chaves, M. L. (2009). (Pre)diabetes, brain aging, and cognition. *Biochem Biophys Acta, 1792*(5), 432—443.

Ryan, C. M., Freed, M. I., Rood, J. A., Cobitz, A. R., Wterhouse, B. R., & Strachan, M. W. (2006). Improving metabolic control leads to better working memory in adults with type 2. *Diabetes, 29*(2), 345—351.

Seetharaman, S., Andel, R., McEvoy, C., Dahl Aslan, A. K., Finkel, D., & Pedersen, N. L. (2015). Blood glucose, diet-based glycemic load and cognitive aging among dementia-free older adults. *The Journals of Gerontology Series A Biological Sciences and Medical Sciences, 70*(4), 471—479, 2015.

Sommerfield, A. J., Deary, I. J., & Frier, B. M. (2004). Acute hyperglycaemia alters modd state and impairs cognitive performance in people with type 2 diabetes. *Diabetes Care, 27*(10), 2335—2340.

Strachan, M. W., Reynolds, R. M., Frier, B. M., Mitchell, R. J., & Price, J. F. (2009). The Role of metabolic derangements and glucocorticoid excess in the aetiology of cognitive impairment in type 2 diabetes. Implications for future therapeutic strategies. *Diabetes, Obesity and Metabolism, 11*(5), 407—414.

Stranahan, A. M., Arumugam, T. V., Cutler, R. G., Lee, K., Egan, J. M., & Mattson, M. P. (2008). Diabetes impairs hippocampal function through glucocorticoid-mediated effects on new and mature neurons. *Nature Neuroscience, 11*(3), 309—317.

Trompet, S., de Craen, A. J., Slagboom, P., Shepherd, J., Blauw, G. J., Murphy, M. B., et al. (2008). Genetic variation in the interleukin-1 beta-converting enzyme associates with cognitive function. The PROSPER study. *Brain, 131*(4), 1069—1077.

Umegaki, H. (2014). Type 2 diabetes as a risk factor for cognitive impairment: Current insights. *Clinical Interventions in Aging, 9*, 1011—1019.

Van Harten, B., Oosterman, J., Muslimovic, D., van Loon, B. J., Scheltens, P., & Weinstein, H. C. (2007). Cognitive Impairment and MRI correlates in the elderly patients with type 2 diabetes mellitus. *Age and Ageing, 36*(2), 164—170.

Van den Berg, de Craen, A. J., Biessels, Gussekloo, J., & Westendorp, R. G. (2006). The impact diabetes mellitus on cognitive decline in the oldest of the old: A prospective population based study. *Diabetologia, 49*(9), 23.

Whitmer, R. A., Karter, A. J., Yaffe, K., Quesenberry, C. P., Jr., & Selby, J. V. (2009). Hypoglycemic episodes and risk of dementia in older patients with type 2 diabetes mellitus. *Journal of the American Medical Association, 301*(15), 1565—1572.

Winkler, A., Dlugaj, M., Weimer, C., Jockel, K. H., Erbel, R., Dragano, N., et al. (2014). Association of diabetes mellitus and mild cognitive impairment in middle aged men and women. *Journal of Alzheimer's Disease, 42*(4), 1269—1277.

Xia, W., Wang, S., Rao, H., Spaeth, A. M., Wang, P., Yang, Y., et al. (2015). Disrupted resting-state attentional networks in T2DM patients. *Scientific Reports, 5*, 11148.

Xu, W., Caracciolo, B., Wang, H. X., Winblad, B., Backman, L., Qui, C., et al. (2010). Accelerated progression from mild cognitive impairment to dementia in people with diabetes. *Diabetes, 59*(11), 2928—2935.

Zhang, Y., Zhang, X., Zhang, J., Liu, C., Yuan, Q., Yin, X., et al. (2014). Gray matter volume abnormalities in type 2 diabetes mellitus with and without mild cognitive impairment. *Neuroscience Letters, 562*, 1—6.

Zheng, F., Yan, L., Yang, Z., Zhong, B., & Xie, W. (2018). HbA1c, diabetes and cognitive decline: The English longitudinal study of ageing. *Diabetologia, 61*(4), 839—848.

CHAPTER 7

Gait and dementia

Oshadi Jayakody[1], Michele L. Callisaya[1,2]

[1]Menzies Institute for Medical Research, University of Tasmania, Hobart, TAS, Australia; [2]Peninsula Clinical School, Central Clinical School, Monash University, Melbourne, Victoria, Australia

List of abbreviations

AD Alzheimer's disease
ApoE4 apolipoprotein E polymorphism ε4 allele
Aβ β-amyloid
bvFTD behavioral variant of frontotemporal dementia
MCI mild cognitive impairment
MCR motoric cognitive risk syndrome
GOOD initiative Gait, cOgnitiOn & Decline initiative
UPDRS Unified Parkinson's Disease Rating Scale

Mini-dictionary of terms

Absolute gait measures are the mean gait speed, temporal, or spatial measure over a period of time.
Gait variability is how a temporal or spatial gait measure fluctuates from step to step.
Gait speed is calculated as the distance traveled divided by the ambulation time and is hence both a temporal and a spatial measure of gait.
Temporal gait measures are time related and include cadence, step time, stance time, swing time, and single- and double-support time.
Spatial gait measures are distance related and include step length, base of support, or step width.
Dual-task walking is walking while carrying out a cognitive (i.e., talking on the phone, counting backward) or motor activity (i.e., carrying a tray).

Introduction

Dementia is the largest cause of disability in older people and costs approximately US$818 billion globally each year (Livingston et al., 2017). The disease affects not only memory but a wide range of cognitive, behavioral, psychological, and motor skills, resulting in a loss of ability to perform activities of daily living and to maintain relationships with loved ones. Before the clinical onset of dementia there is often a period of cognitive decline. Clinically this has been termed mild cognitive impairment (MCI) and is proposed as a transitional stage between cognitively normal and dementia.

This chapter will (1) outline the evidence for gait as a marker of both MCI and dementia, (2) discuss whether gait might be a useful biomarker in predicting both cognitive

Diagnosis and Management in Dementia
ISBN 978-0-12-815854-8, https://doi.org/10.1016/B978-0-12-815854-8.00007-0

decline and incident dementia, and (3) provide a brief review of the association between gait and dementia biomarkers.

The measurement of gait

Subjective description of gait

Gait evaluations can be performed using direct observation without any equipment. Classifications include nonneurologic or neurologic, and the latter can be subclassified into, for example, ataxic, frontal, and hemiparetic based on the observed gait characteristics (Verghese et al., 2002). Clinical scales such as the Unified Parkinson's Disease Rating Scale (UPDRS) or the Tinetti Gait and Balance scale can also be used to measure gait semiquantitatively.

Quantitative measurement of gait

Absolute gait measures include gait speed and its spatial (i.e., step length) and temporal (i.e., step time) components (see Figs. 7.1 and 7.2). Gait speed can be measured simply in the clinic as the time to walk a fixed distance, or with more expensive equipment such as a computerized walkway (i.e., GAITRite). *Gait variability* is the intraindividual fluctuation of a gait parameter from one step to another (Fig. 7.2) (Callisaya et al., 2011). Gait is multifaceted, but also highly correlated. To address this, factor analysis has been used in some studies to create gait factors (Verghese, Wang, Lipton, Holtzer, & Xue, 2007).

Different conditions under which gait can be assessed

Gait is most commonly assessed under a single-task, in which a person walks at his or her normal walking pace or at a fast pace, which is considered more challenging. Dual-task walking is walking while performing an additional challenging motor or cognitive task (Fig. 7.3).

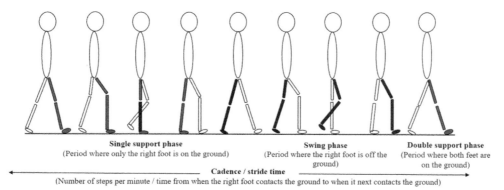

Single support phase
(Period where only the right foot is on the ground)

Swing phase
(Period where the right foot is off the ground)

Double support phase
(Period where both feet are on the ground)

Cadence / stride time
(Number of steps per minute / time from when the right foot contacts the ground to when it next contacts the ground)

Figure 7.1 *Temporal gait measures.* Temporal (timing) gait measures (for the right foot) are depicted.

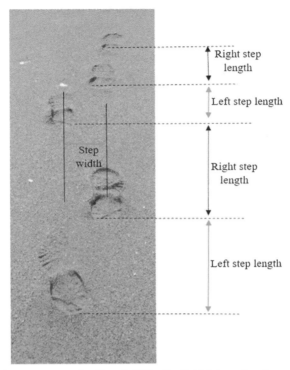

Figure 7.2 *Spatial gait measures and gait variability.* Spatial (distance) gait measures and step length variability (the fluctuation from one step to another) are depicted.

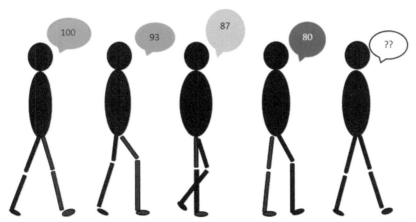

Figure 7.3 *Dual-task walking.* A person performing a cognitive task (subtracting 7's) while walking is shown.

Gait as a marker of mild cognitive impairment and dementia

Even in people without dementia, poorer executive function, processing speed, visuo-spatial function (Martin et al., 2013), and memory (Holtzer, Verghese, Xue, & Lipton, 2006) are associated with slower gait speed. This suggests that gait may be useful in distinguishing between those with normal cognition and those with cognitive impairment.

Gait as a marker of mild cognitive impairment

Studies have generally grouped people with MCI together; however, a small number of studies have subtyped MCI further into memory (amnestic) versus nonmemory (nonamnestic) or single versus multidomain MCI.

People with a diagnosis of MCI were more likely to have gait disturbances rated using the UPDRS than controls (Boyle et al., 2005), and disturbances were more frequently found in those with nonamnestic MCI (Boyle et al., 2005). In a separate study, those with amnestic, but not nonamnestic, MCI were more likely to have neurological gait patterns described as unsteady, ataxic, frontal, parkinsonian, neuropathic, hemiparetic, or spastic compared with controls (Verghese et al., 2008). This suggests that there are visible changes in gait patterns even in the early stages of cognitive impairment.

Gait speed and mild cognitive impairment

A 2017 meta-analysis (n = 11 studies) found that gait speed was slower in people with MCI compared with those who were cognitively normal (effect size d = −0.74; 95% confidence interval [CI] −0.89 to −0.59) (Bahureksa et al., 2017). Few studies have examined MCI subtypes. In the Gait cOgnitiOn & Decline (GOOD) initiative, which combined data from seven countries (n = 1719), gait speed was slower in both the non-amnestic (by −0.09 m/s; 95% CI −0.12 to −0.63) and the amnestic (−0.05 m/s; 95% CI −0.09 to −0.008) groups compared with controls (Allali et al., 2016). In a separate study, under both preferred and fast pace conditions, there was a slowing of gait across stages of cognitive impairment (cognitively healthy, MCI, mild and moderate dementia) (Callisaya et al., 2017). In a subanalysis there was a trend for slower speeds from cognitively healthy to amnestic, nonamnestic, and multidomain MCI. Taken together gait speed appears slower in those with MCI compared with controls, with preliminary evidence that those with nonamnestic MCI or multidomain MCI have the slowest speeds.

Gait speed under dual-task and mild cognitive impairment

Gait speed under dual-task has been examined in fewer studies than under single-task conditions. Differences in the task (i.e., animal naming vs. counting backward) and the methods by which the dual task is assessed make comparing results difficult. One of the first studies to examine dual-task walking in people with MCI found that the speed

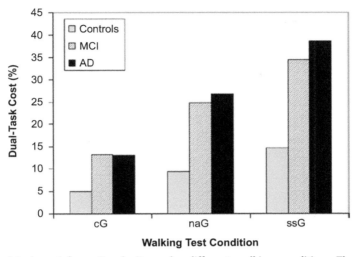

Figure 7.4 *Dual-task cost for gait velocity under different walking conditions.* The differences in dual-task cost during three walking conditions in older adults with normal cognition, mild cognitive impairment, and Alzheimer's disease are shown. *AD*, Alzheimer's disease; *cG*, counting backward from 100 by 1's; *MCI*, mild cognitive impairment; *naG*, naming animals; *ssG*, counting backward from 100 by 7's. *(From Muir, S. W., Speechley, M., Wells, J., Borrie, M., Gopaul, K., & Montero-Odasso, M. (2012). Gait assessment in mild cognitive impairment and Alzheimer's disease: The effect of dual-task challenges across the cognitive spectrum. Gait & Posture, 35(1), 96–100, with permission from the publishers.)*

under the dual-task condition (but not under single-task), and the change in speed from single- to dual-task, was able to distinguish between those who were cognitively normal and those with MCI or dementia (Muir et al., 2012). The percentage change in gait speed for those who were cognitively normal, with MCI, and with dementia is shown in Fig. 7.4. A recent meta-analysis confirmed that dual-task walking speed was slower in people with MCI during verbal tasks such as animal naming (n = 3 studies), counting backward in sevens (n = 2 studies), and also counting backward in ones (n = 4 studies), compared with those who were cognitively normal. The largest effect size was found for serial sevens (d = −1.34; 95% CI −1.74 to −0.93), which is larger than the effect size for gait speed under a single-task described in the previous section (Bahureksa et al., 2017). However, these results should be viewed with caution as the two included studies were from the same research group (Montero-Odasso, Muir, & Speechley, 2012; Muir et al., 2012). In contrast to findings under single-task, one study reported poorer gait in the amnestic versus the nonamnestic group under dual-task (Montero-Odasso et al., 2014). Serial sevens had the largest percentage change between single- and dual-task, with a 14.14% (SD 15.59) difference in the nonamnestic compared with 23.21% (SD 16.31) in the amnestic group. Taken together, dual-task walking may provide a more sensitive marker of MCI than single-task when measuring gait speed. However, it is currently unclear if the dual-task speed or percentage change is a better measure.

Other gait measures and mild cognitive impairment

Under single task, stride length (n = 2 studies), stride time (n = 4 studies), and stride time variability (n = 6) were worse in people with MCI, whereas under dual-task only stride time (n = 2 studies) and stride time variability (n = 3 studies) were worse than in controls in a 2017 meta-analysis (Bahureksa et al., 2017). In the GOOD initiative (n = 1719 participants), there was a linear worsening of all gait measures under single-task from cognitively healthy to amnestic MCI, nonamnestic MCI, Alzheimer's disease (AD), and non-AD dementia, except for swing time, single-support phase, and swing time variability (Allali et al., 2016).

Gait as a marker of dementia

Gait impairments are highly prevalent in dementia (Allan, Ballard, Burn, & Kenny, 2005). Studies have largely examined gait in AD. Where available, this section will also describe changes in gait in non-AD-type dementias (vascular dementia, dementia with Lewy bodies, Parkinson's disease dementia, or frontotemporal dementia).

Disequilibrium, shorter steps, and apraxia are more common in those with AD compared with cognitively healthy controls, with frontal gait disorders (variable base width, disequilibrium, short steps, shuffling, and hesitation with gait initiation and turning) becoming more common with increasing severity (O'keeffe et al., 1996). A further study found that the odds of gait disorders were higher, and Tinetti gait and balance scores worse, in non-AD compared with AD-type dementia. The highest prevalence of gait disorders was in Parkinson's disease with dementia (93%), followed by vascular dementia (79%), dementia with Lewy bodies (75%), AD (25%), and then controls (7%) (Allan et al., 2005).

Gait speed and dementia

In a systematic review of seven studies with quantitative gait analysis, slower gait speeds were found in AD versus cognitively healthy older controls (van Iersel, Hoefsloot, Munneke, Bloem, & Rikkert, 2004). The average range of gait speeds was slower in AD (0.57−0.93 m/s) compared with controls (0.97−1.40 m/s). In the only study in this review that examined vascular dementia, mean gait speed was even slower at 0.52 m/s (Tanaka et al., 1995). A multicounty initiative reported mean gait speeds of 1.05 ± 0.22 m/s in cognitively healthy older people, slowing to 0.74 ± 0.19 m/s in mild AD, 0.72 ± 0.20 m/s in mild non-AD, 0.68 ± 0.21 m/s in moderate AD, and 0.62 ± 0.20 m/s in moderate non-AD-type dementia (Allali et al., 2016). These findings suggest that non-AD-type dementia is associated with slower gait than AD-type dementia.

Gait speed under dual-task and dementia

Gait is slower during dual-task walking in those with AD versus cognitively healthy controls (Camicioli, Howieson, Lehman, & Kaye, 1997; Muir et al., 2012). However, there were no differences between the MCI and the AD groups in one small study (Muir et al., 2012). There is little information regarding gait under dual-task and other types of dementia.

Other gait measures and dementia

Of the determinants of gait speed, step length shows the highest effect size of all gait measures for dementia (Allali et al., 2016). However, other gait measures such as stance and double-support time, as well as gait variability, are worse in AD compared with controls, and most of these measures are worse in non-AD than in AD-type dementia (Allali et al., 2016). Slower speed, shorter stride lengths, and increased gait variability were found in those with Lewy body dementia compared with those with AD during both normal and fast-pace walking (Fritz et al., 2016). In addition, those with a behavioral variant of frontotemporal dementia appear to have greater stride time variability (but not mean stride time) compared with controls under single- and dual-task conditions (Allali et al., 2010).

In summary (see Table 7.1), poor gait performance is common in MCI and dementia compared with cognitively healthy older adults. Poorer gait is found in nonamnestic

Table 7.1 Gait characteristics in mild cognitive impairment and dementia.

	MCI	Dementia
At preferred walking speeds		
Speed	# A < B	# C < D
Step/stride length	# A < B	# C < D
Step/stride time/cadence	# A < B	# C < D
Step/stride width		# C < D
Temporal variability	# A < B	# C < D
Spatial variability	# A < B	# C < D
Neurological gait disorders	#	# C < D
At fast pace		
Speed	# A < B	#
Dual-task walking		
Speed	# B < A	#
Step/stride time	#	
Temporal variability	#	# (bvFTD)

The table shows the differences in gait between cognitive healthy individuals, MCI, and dementia under different walking conditions. # indicates worse gait compared with cognitively healthy older people; < indicates poorer than; A, nonamnestic MCI; B, amnestic MCI; bvFTD, behavioral variant frontotemporal dementia; C, non-Alzheimer's disease; D, Alzheimer's disease; MCI, mild cognitive impairment.

compared with amnestic MCI and in non-AD compared with AD dementia (Allali et al., 2016). Assessing gait under dual-task (rather than single-task) may provide a more sensitive marker of cognitive impairment.

Gait as a predictor of cognitive decline, mild cognitive impairment, and dementia

Gait as a predictor of cognitive decline and mild cognitive impairment

Slower gait speed predicts cognitive decline (Alfaro-Acha, Al Snih, Raji, Markides, & Ottenbacher, 2007; Gale, Allerhand, Sayer, Cooper, & Deary, 2014; Inzitari et al., 2007; Verghese et al., 2007) and risk of cognitive impairment (Deshpande, Metter, Bandinelli, Guralnik, & Ferrucci, 2009). Changes in gait speed may occur many years before a diagnosis of cognitive impairment. In the Oregon Brain Study, 204 older people with a clinical dementia rating scale of zero were followed up for a mean of 9 years, with 46.6% developing MCI. Those who converted to MCI had a faster rate of decline in gait speed than those who did not convert. Decline in gait speed accelerated approximately 14 years for men and 6 years for women before diagnosis (Buracchio, Dodge, Howieson, Wasserman, & Kaye, 2010).

Other studies have examined whether gait is related to specific cognitive domains. Slowing of gait was associated with decline in executive function over 2.5 years (Callisaya et al., 2015). But there were no associations with processing speed, visuospatial function, or memory, suggesting that gait and executive function vary together and may have similar underlying neural pathways. However, slower gait at baseline predicted decline in memory, language, executive function, and visuospatial skills over a median of 4.1 years in those without dementia (Mielke et al., 2013). This topic was the subject of a recent review article, which reported that the average baseline gait speed for those who did not decline was faster (1.1 m/s) compared with those who developed MCI (0.91 m/s) or dementia (0.8 m/s) (Kikkert, Vuillerme, van Campen, Hortobagyi, & Lamoth, 2016). Other studies have examined gait measures other than speed. A rhythm factor (highest loadings on cadence, swing, and stance time) was associated with memory decline, whereas a pace factor (highest loadings on gait speed, step length, and double-support time) was associated with executive decline (Verghese et al., 2007).

Gait as a predictor of dementia

Clinically defined neurological gait abnormalities predicted incident non-AD dementia (hazard ratio [HR] 3.51; 95% CI 1.98 to 6.24), but not AD dementia (HR 1.07; 95% CI 0.57 to 2.02) over a mean follow-up of 6.6 years (Verghese et al., 2002). Hemiparetic, frontal, and unsteady gait specifically increased the risk of vascular dementia. In a 2016 meta-analysis (12 studies; participants n = range 171—3855; follow-up 3—9 years), abnormal gait predicted increased risk of all-cause dementia (HR 1.53; 95% CI 1.42

to 1.65). Findings were lower for AD (HR 1.03; 95% CI 1.01 to 1.05), compared with non-AD dementia (HR 1.89; 95% CI 1.60 to 2.22) or vascular dementia (HR 1.79; 95% CI 1.51 to 2.12) (Beauchet et al., 2016). However, there was considerable heterogeneity in the way that abnormal gait was classified.

Gait speed as a predictor of dementia

In a meta-analysis of 14,140 cognitively healthy older adults, the pooled risk of developing dementia was increased by 66% for those in the lowest compared with the highest category of walking pace (relative risk [RR] 1.66; 95% CI 1.43 to 1.92; n = 1903 dementia cases) (Quan et al., 2017). A 13% increase in risk of dementia was found for every 0.01 m/s slower walking speed (three studies) (Quan et al., 2017). A scoping review identified mean gait speeds of 0.8 m/s in people who developed dementia (n = 2631) compared with 1.11 m/s speed in participants who were dementia free (Kikkert et al., 2016). A further study published in 2017 reported double the risk of developing dementia (HR 2.28; 95% CI 1.76 to 2.98) for those with walking speeds less than 1.0 m/s 7 years earlier (Dumurgier et al., 2017). Associations were stronger for vascular dementia (HR 12.11; 95% CI 4.04 to 36.31) versus AD (HR 2.08; 95% CI 1.55 to 2.80). Greater declines in gait speed over time are also associated with increased risk of dementia (HR per 0.007 m/s/year decrease: 3.39; 95% CI 1.37 to 8.43) (Dumurgier et al., 2017).

Motoric cognitive risk syndrome

Motoric cognitive risk (MCR) is a newly defined syndrome comprising the presence of cognitive complaints and slow gait in older individuals without dementia or mobility disability (Verghese, Wang, Lipton, & Holtzer, 2012). The pooled predictive effect of MCR for risk of dementia is an adjusted HR of 1.93 (95% CI 1.59 to 2.35) (Verghese et al., 2014). MCR appears to be associated with both AD (Verghese et al., 2014) and vascular-type dementia (Verghese et al., 2012). Furthermore, MCR may provide additional diagnostic value over its components—gait speed or a cognitive complaint—alone (Verghese et al., 2012, 2014). MCR is unique in that it does not require psychological tests, and its simplicity (quick and inexpensive) makes it promising as a dementia screening tool at a community level.

Other gait measures

Mean values of swing time and stride length, as well as stride length variability, have individually predicted dementia (median follow-up of 2 years) (Verghese et al., 2007). In the same study, a poor rhythm factor (HR 1.48; 95% CI 1.03 to 2.14) and variability factor (HR 1.37; 95% CI 1.05 to 1.78), but not a pace factor, predicted any dementia. None of the factors predicted AD-type dementia, and only the pace factor (HR 1.60; 95% CI 1.06 to 2.41) predicted vascular dementia (Verghese et al., 2007).

Dual-task walking as a predictor of dementia

As of this writing, only two studies have examined if dual-task walking can assist in the prediction of dementia (Ceïde, Ayers, Lipton, & Verghese, 2018; Montero-Odasso et al., 2017). In the Einstein Aging Study, community-dwelling older people (n = 1156) without cognitive impairment were followed over a median of 1.9 years. Greater intra-individual gait variability (but not gait speed) during dual-task walking was associated with a 50% (HR 1.50; 95% CI 1.06 to 2.12) increase in risk of developing dementia related to vascular disease, but not AD (HR 0.85; 95% CI 0.65 to 1.12) (Ceïde et al., 2018). In contrast, Montero-Odasso et al. found that dual-task gait speed and the percentage change in gait speed (but not single-task gait speed) predicted conversion to any dementia (average follow-up of 2 years) in 112 older people with MCI (Montero-Odasso et al., 2017).

In summary (Table 7.2), slower gait speed predicts cognitive decline and dementia years before diagnosis. Gait speed during dual-task (or the change from single- to dual-task) may be sensitive for predicting dementia once a diagnosis of MCI has been made, but gait variability may be more sensitive before the development of MCI.

Gait and biomarkers of dementia

β-Amyloid

β-Amyloid (Aβ) is considered a major hallmark of AD, but only a few cross-sectional (Del Campo et al., 2016; Nadkarni et al., 2017; Tian et al., 2018; Wennberg et al., 2017) and longitudinal (Tian et al., 2017) studies have examined its association with

Table 7.2 Gait characteristics that predict incident mild cognitive impairment and dementia.

	MCI	Dementia
Preferred speed		
Speed	A	B
Decline in speed	A	B
Step/stride length		B
Swing time		B
Rhythm factor		B
Variability factor		B
Pace factor		B (vascular dementia)
Neurological gait		B (non-Alzheimer's disease dementia)
Dual- task gait		
Speed (from MCI)		B
Variability factor (from cognitively healthy)		B (vascular dementia)

The table shows the gait measures that predict a future diagnosis of MCI or dementia under different walking conditions. *MCI*, mild cognitive impairment; *A*, MCI; *B*, dementia.

gait. Some studies reported that slower gait speed is associated with greater Aβ in widespread areas, including the putamen, precuneus, prefrontal, orbitofrontal and occipital areas, anterior and posterior cingulate, and motor areas (precentral gyrus, postcentral gyrus, Rolandic operculum, and supplementary motor areas) (Del Campo et al., 2016; Wennberg et al., 2017). In contrast, others failed to observe significant associations (Nadkarni et al., 2017; Tian et al., 2017, 2018). In a longitudinal study, higher total cerebral and regional Aβ burden in the putamen, dorsolateral prefrontal cortex, and lateral temporal lobe was associated with gait speed decline in 99 exceptionally healthy older adults (Tian et al., 2017).

Apolipoprotein E polymorphism ε4 allele

Apolipoprotein E polymorphism ε4 allele (ApoE4) is associated with slower gait speed in people with MCI (Shimada, Makizako, Tsutsumimoto, Uemura, & Suzuki, 2015). However, in a large study from eight cohorts (participants n = 23,916; age 44—90 years), there was no association between ApoE4 and physical capability measures, including gait speed (Alfred et al., 2014). The evidence from longitudinal studies is largely negative (Buchman et al., 2009; Melzer, Dik, Van Kamp, Jonker, & Deeg, 2005; Skoog et al., 2016; Verghese et al., 2013), but for one study that found that over 3 years, ApoE4 was associated with gait speed decline only in males (Verghese et al., 2013).

Brain atrophy and vascular disease

Smaller brain volumes in areas associated with dementia are found in people with slower gait, including hippocampal (Callisaya et al., 2013), subcortical (Callisaya, Beare, Phan, Chen, & Srikanth, 2014), and prefrontal regions (Callisaya et al., 2014). White matter hyperintensities alone and in combination with microbleeds and infarcts are also associated with slower gait speeds (Choi et al., 2012; Srikanth et al., 2010).

In summary, both vascular and neurodegenerative pathways appear to contribute to slow gait.

Summary points

- This chapter focuses on the role of gait in the context of dementia.
- Temporal and spatial gait measures progressively worsen from cognitively healthy to MCI to dementia. There is preliminary evidence that (1) those with nonamnestic or multidomain MCI have worse gait than those with amnestic MCI and (2) those with non-AD dementia have worse gait than AD-type dementia.
- Dual-task conditions may be more sensitive in distinguishing MCI from controls than single-task conditions.
- Slow gait under single-task predicts dementia in longitudinal studies.

- Slow gait speed under dual-task may be sensitive for predicting dementia once a diagnosis of MCI has been made, but gait variability may be more sensitive before the development of MCI.
- These relationships may be explained by both neurodegeneration and neurovascular disease.

Key facts of gait

- Gait is the pattern or way someone walks.
- Upright walking evolved more than 4.4 million years ago (Lovejoy, Suwa, Spurlock, Asfaw, & White, 2009).
- The study of gait dates back to the 17th century.
- Until recently gait and cognition were studied separately.
- Gait speed can be measured simply using a fixed distance (e.g., 4 or 10 m) and a stop watch.
- Gait is an important component of a neurological, orthopedic, or geriatric assessment.
- There is a push for gait speed to be included as the sixth vital sign due to its ability to predict a wide range of adverse health outcomes in older age (Fritz & Lusardi, 2009).

References

Alfaro-Acha, A., Al Snih, S., Raji, M. A., Markides, K. S., & Ottenbacher, K. J. (2007). Does 8-foot walk time predict cognitive decline in older Mexicans Americans? *Journal of the American Geriatrics Society, 55*(2), 245−251. doi:JGS1039 [pii]10.1111/j.1532-5415.2007.01039.x.

Alfred, T., Ben-Shlomo, Y., Cooper, R., Hardy, R., Cooper, C., Deary, I. J., et al. (2014). Associations between APOE and low-density lipoprotein cholesterol genotypes and cognitive and physical capability: The HALCyon programme. *Age, 36*(4), 9673.

Allali, G., Annweiler, C., Blumen, H. M., Callisaya, M. L., De Cock, A. M., Kressig, R. W., et al. (2016). Gait phenotype from mild cognitive impairment to moderate dementia: Results from the GOOD initiative. *European Journal of Neurology, 23*(3), 527−541.

Allali, G., Dubois, B., Assal, F., Lallart, E., de Souza, L. C., Bertoux, M., et al. (2010). Frontotemporal dementia: Pathology of gait? *Movement Disorders, 25*(6), 731−737.

Allan, L. M., Ballard, C. G., Burn, D. J., & Kenny, R. A. (2005). Prevalence and severity of gait disorders in Alzheimer's and non-Alzheimer's dementias. *Journal of the American Geriatrics Society, 53*(10), 1681−1687.

Bahureksa, L., Najafi, B., Saleh, A., Sabbagh, M., Coon, D., Mohler, M. J., et al. (2017). The impact of mild cognitive impairment on gait and balance: A systematic review and meta-analysis of studies using instrumented assessment. *Gerontology, 63*(1), 67−83. https://doi.org/10.1159/000445831.

Beauchet, O., Annweiler, C., Callisaya, M. L., De Cock, A.-M., Helbostad, J. L., Kressig, R. W., et al. (2016). Poor gait performance and prediction of dementia: Results from a meta-analysis. *Journal of the American Medical Directors Association, 17*(6), 482−490.

Boyle, P. A., Wilson, R. S., Aggarwal, N. T., Arvanitakis, Z., Kelly, J., Bienias, J. L., et al. (2005). Parkinsonian signs in subjects with mild cognitive impairment. *Neurology, 65*(12), 1901−1906. https://doi.org/10.1212/01.wnl.0000188878.81385.73.

Buchman, A. S., Boyle, P. A., Wilson, R. S., Beck, T. L., Kelly, J. F., & Bennett, D. A. (2009). Apolipoprotein E e4 allele is associated with more rapid motor decline in older persons. *Alzheimer Disease and Associated Disorders, 23*(1), 63.

Buracchio, T., Dodge, H. H., Howieson, D., Wasserman, D., & Kaye, J. (2010). The trajectory of gait speed preceding mild cognitive impairment. *Archives of Neurology, 67*(8), 980—986. doi:67/8/980 [pii] 10.1001/archneurol.2010.159.

Callisaya, M. L., Beare, R., Phan, T. G., Blizzard, L., Thrift, A. G., Chen, J., et al. (2013). Brain structural change and gait decline: A longitudinal population-based study. *Journal of the American Geriatrics Society, 61*(7), 1074—1079.

Callisaya, M. L., Beare, R., Phan, T. G., Chen, J., & Srikanth, V. K. (2014). Global and regional associations of smaller cerebral gray and white matter volumes with gait in older people. *PloS One, 9*(1), e84909. https://doi.org/10.1371/journal.pone.0084909.

Callisaya, M. L., Blizzard, L., Schmidt, M. D., Martin, K. L., McGinley, J. L., Sanders, L. M., et al. (2011). Gait, gait variability and the risk of multiple incident falls in older people: A population-based study. *Age and Ageing, 40*(4), 481—487.

Callisaya, M. L., Blizzard, C. L., Wood, A. G., Thrift, A. G., Wardill, T., & Srikanth, V. K. (2015). Longitudinal relationships between cognitive decline and gait slowing: The tasmanian study of cognition and gait. *The Journals of Gerontology Series A Biological Sciences and Medical Sciences, 70*(10), 1226—1232. https://doi.org/10.1093/gerona/glv066.

Callisaya, M. L., Launay, C. P., Srikanth, V. K., Verghese, J., Allali, G., & Beauchet, O. (2017). Cognitive status, fast walking speed and walking speed reserve—the Gait and Alzheimer Interactions Tracking (GAIT) study. *GeroScience, 39*(2), 231.

Camicioli, R., Howieson, D., Lehman, S., & Kaye, J. (1997). Talking while walking the effect of a dual task in aging and Alzheimer's disease. *Neurology, 48*(4), 955—958.

Ceïde, M. E., Ayers, E., Lipton, R., & Verghese, J. (2018). Walking while talking and risk of incident dementia. *American Journal of Geriatric Psychiatry* (in press).

Choi, P., Ren, M., Phan, T. G., Callisaya, M., Ly, J. V., Beare, R., et al. (2012). Silent infarcts and cerebral microbleeds modify the associations of white matter lesions with gait and postural stability: Population-based study. *Stroke, 43*(6), 1505—1510. doi:STROKEAHA.111.647271 [pii]10.1161/STROKEAHA.111.647271.

Del Campo, N., Payoux, P., Djilali, A., Delrieu, J., Hoogendijk, E. O., Rolland, Y., et al. (2016). Relationship of regional brain β-amyloid to gait speed. *Neurology, 86*(1), 36—43.

Deshpande, N., Metter, E. J., Bandinelli, S., Guralnik, J., & Ferrucci, L. (2009). Gait speed under varied challenges and cognitive decline in older persons: A prospective study. *Age and Ageing, 38*(5), 509—514. doi: afp093 [pii]10.1093/ageing/afp.093.

Dumurgier, J., Artaud, F., Touraine, C., Rouaud, O., Tavernier, B., Dufouil, C., et al. (2017). Gait speed and decline in gait speed as predictors of incident dementia. *Journal of Gerontology: Series A, 72*(5), 655—661.

Fritz, N. E., Kegelmeyer, D. A., Kloos, A. D., Linder, S., Park, A., Kataki, M., et al. (2016). Motor performance differentiates individuals with Lewy body dementia, Parkinson's and Alzheimer's disease. *Gait & Posture, 50*, 1—7.

Fritz, S., & Lusardi, M. (2009). White paper:"walking speed: The sixth vital sign". *Journal of Geriatric Physical Therapy, 32*(2), 2—5.

Gale, C. R., Allerhand, M., Sayer, A. A., Cooper, C., & Deary, I. J. (2014). The dynamic relationship between cognitive function and walking speed: The English longitudinal study of ageing. *Age, 36*(4), 9682. https://doi.org/10.1007/s11357-014-9682-8.

Holtzer, R., Verghese, J., Xue, X., & Lipton, R. B. (2006). Cognitive processes related to gait velocity: Results from the Einstein aging study. *Neuropsychology, 20*(2), 215—223.

van Iersel, M. B., Hoefsloot, W., Munneke, M., Bloem, B. R., & Rikkert, M. M. O. (2004). Systematic review of quantitative clinical gait analysis in patients with dementia. *Zeitschrift für Gerontologie und Geriatrie, 37*(1), 27—32.

Inzitari, M., Newman, A. B., Yaffe, K., Boudreau, R., de Rekeneire, N., Shorr, R., et al. (2007). Gait speed predicts decline in attention and psychomotor speed in older adults: The health aging and body composition study. *Neuroepidemiology, 29*(3—4), 156—162. doi:000111577 [pii]10.1159/000111577.

Kikkert, L. H., Vuillerme, N., van Campen, J. P., Hortobagyi, T., & Lamoth, C. J. (2016). Walking ability to predict future cognitive decline in old adults: A scoping review. *Ageing Research Reviews, 27*, 1—14.

Livingston, G., Sommerlad, A., Orgeta, V., Costafreda, S. G., Huntley, J., Ames, D., et al. (2017). Dementia prevention, intervention, and care. *Lancet, 390*(10113), 2673–2734. https://doi.org/10.1016/S0140-6736(17)31363-6.

Lovejoy, C. O., Suwa, G., Spurlock, L., Asfaw, B., & White, T. D. (2009). The pelvis and femur of Ardipithecus ramidus: The emergence of upright walking. *Science, 326*(5949), 71–71e76.

Martin, K. L., Blizzard, L., Wood, A. G., Srikanth, V., Thomson, R., Sanders, L. M., et al. (2013). Cognitive function, gait, and gait variability in older people: A population-based study. *The Journals of Gerontology Series A Biological Sciences and Medical Sciences, 68*(6), 726–732. doi:10.1093/gerona/gls224gls224 [pii].

Melzer, D., Dik, M., Van Kamp, G. J., Jonker, C., & Deeg, D. J. (2005). The apolipoprotein E e4 polymorphism is strongly associated with poor mobility performance test results but not self-reported limitation in older people. *The Journals of Gerontology Series A: Biological Sciences and Medical Sciences, 60*(10), 1319–1323.

Mielke, M. M., Roberts, R. O., Savica, R., Cha, R., Drubach, D. I., Christianson, T., et al. (2013). Assessing the temporal relationship between cognition and gait: Slow gait predicts cognitive decline in the mayo clinic study of aging. *The Journals of Gerontology Series A Biological Sciences and Medical Sciences, 68*(8), 929–937. doi:10.1093/gerona/gls256gls256 [pii].

Montero-Odasso, M., Muir, S. W., & Speechley, M. (2012). Dual-task complexity affects gait in people with mild cognitive impairment: The interplay between gait variability, dual tasking, and risk of falls. *Archives of Physical Medicine and Rehabilitation, 93*(2), 293–299. doi:10.1016/j.apmr.2011.08.026.

Montero-Odasso, M., Oteng-Amoako, A., Speechley, M., Gopaul, K., Beauchet, O., Annweiler, C., et al. (2014). The motor signature of mild cognitive impairment: Results from the gait and brain study. *The Journals of Gerontology Series A Biological Sciences and Medical Sciences, 69*(11), 1415–1421. https://doi.org/10.1093/gerona/glu155.

Montero-Odasso, M. M., Sarquis-Adamson, Y., Speechley, M., Borrie, M. J., Hachinski, V. C., Wells, J., et al. (2017). Association of dual-task gait with incident dementia in mild cognitive impairment: Results from the gait and brain study. *JAMA Neurology, 74*(7), 857–865. https://doi.org/10.1001/jamaneurol.2017.0643.

Muir, S. W., Speechley, M., Wells, J., Borrie, M., Gopaul, K., & Montero-Odasso, M. (2012). Gait assessment in mild cognitive impairment and Alzheimer's disease: The effect of dual-task challenges across the cognitive spectrum. *Gait & Posture, 35*(1), 96–100.

Nadkarni, N. K., Lopez, O. L., Perera, S., Studenski, S. A., Snitz, B. E., Erickson, K. I., et al. (2017). Cerebral amyloid deposition and dual-tasking in cognitively normal, mobility unimpaired older adults. *Journal of Gerontology: Series A, 72*(3), 431–437.

O'keeffe, S., Kazeem, H., Philpott, R., Playfer, J., Gosney, M., & Lye, M. (1996). Gait disturbance in Alzheimer's disease: A clinical study. *Age and Ageing, 25*(4), 313–316.

Quan, M., Xun, P., Chen, C., Wen, J., Wang, Y., Wang, R., et al. (2017). Walking pace and the risk of cognitive decline and dementia in elderly populations: A meta-analysis of prospective cohort studies. *Journal of Gerontology: Series A, 72*(2), 266–270.

Shimada, H., Makizako, H., Tsutsumimoto, K., Uemura, K., & Suzuki, T. (2015). Apolipoprotein E genotype and physical function among older people with mild cognitive impairment. *Geriatrics and Gerontology International, 15*(4), 422–427.

Skoog, I., Hörder, H., Frändin, K., Johansson, L., Östling, S., Blennow, K., et al. (2016). Association between APOE genotype and change in physical function in a population-based Swedish cohort of older individuals followed over four years. *Frontiers in Aging Neuroscience, 8*, 225.

Srikanth, V., Phan, T. G., Chen, J., Beare, R., Stapleton, J. M., & Reutens, D. C. (2010). The location of white matter lesions and gait–a voxel-based study. *Annals of Neurology, 67*(2), 265–269. https://doi.org/10.1002/ana.21826.

Tanaka, A., Okuzumi, H., Kobayashi, I., Murai, N., Meguro, K., & Nakamura, T. (1995). Gait disturbance of patients with vascular and Alzheimer-type dementias. *Perceptual & Motor Skills, 80*(3), 735–738.

Tian, Q., Bair, W.-N., Resnick, S. M., Bilgel, M., Wong, D. F., & Studenski, S. A. (2018). β-amyloid deposition is associated with gait variability in usual aging. *Gait & Posture, 61*, 346–352.

Tian, Q., Resnick, S. M., Bilgel, M., Wong, D. F., Ferrucci, L., & Studenski, S. A. (2017). β-amyloid burden predicts lower extremity performance decline in cognitively unimpaired older adults. *Journal of Gerontology: Series A, 72*(5), 716—723.

Verghese, J., Annweiler, C., Ayers, E., Barzilai, N., Beauchet, O., Bennett, D. A., et al. (2014). Motoric cognitive risk syndrome Multicountry prevalence and dementia risk. *Neurology, 83*(8), 718—726.

Verghese, J., Holtzer, R., Wang, C., Katz, M. J., Barzilai, N., & Lipton, R. B. (2013). Role of APOE genotype in gait decline and disability in aging. *Journals of Gerontology Series A: Biomedical Sciences and Medical Sciences, 68*(11), 1395—1401.

Verghese, J., Lipton, R. B., Hall, C. B., Kuslansky, G., Katz, M. J., & Buschke, H. (2002). Abnormality of gait as a predictor of non-Alzheimer's dementia. *New England Journal of Medicine, 347*(22), 1761—1768.

Verghese, J., Robbins, M., Holtzer, R., Zimmerman, M., Wang, C., Xue, X., et al. (2008). Gait dysfunction in mild cognitive impairment syndromes. *Journal of the American Geriatrics Society, 56*(7), 1244—1251. doi: JGS1758 [pii]10.1111/j.1532-5415.2008.01758.x.

Verghese, J., Wang, C., Lipton, R. B., & Holtzer, R. (2012). Motoric cognitive risk syndrome and the risk of dementia. *Journals of Gerontology Series A: Biomedical Sciences and Medical Sciences, 68*(4), 412—418.

Verghese, J., Wang, C., Lipton, R. B., Holtzer, R., & Xue, X. (2007). Quantitative gait dysfunction and risk of cognitive decline and dementia. *Journal of Neurology, Neurosurgery & Psychiatry, 78*(9), 929—935.

Wennberg, A., Savica, R., Hagen, C. E., Roberts, R. O., Knopman, D. S., Hollman, J. H., et al. (2017). Cerebral amyloid deposition is associated with gait parameters in the mayo clinic study of aging. *Journal of the American Geriatrics Society, 65*(4), 792—799.

CHAPTER 8

The role of blood pressure and hypertension in dementia

Keenan A. Walker[1], Rebecca F. Gottesman[2]

[1]Department of Neurology, Johns Hopkins School of Medicine. Baltimore, MD, United States; [2]Departments of Neurology and Epidemiology, Johns Hopkins School of Medicine, Baltimore, MD, United States

List of abbreviations

ACE angiotensin-converting enzyme
ARB angiotensin II receptor blocker
Aβ β-amyloid
RCT randomized clinical trial

Mini-dictionary of terms

Systolic blood pressure Maximum arterial pressure during the contraction of the heart's left ventricle.
Diastolic blood pressure The period of low arterial pressure that occurs between the heart's contractions as the heart relaxes.
Hypertension Current ACC/AHA guidelines define hypertension as a systolic blood pressure above 130 mmHg or diastolic blood pressure above 80 mmHg.
Renin-angiotensin system A network of hormones that maintains homeostasis of the body's fluids and controls blood pressure by manipulating the availability of angiotensin II, a vasoconstrictive peptide.
Atherosclerosis A buildup of plaque on the inside of an artery that causes arterial narrowing. Following endothelial damage, macrophage, cholesterol, and calcium can build up over time to form an atherosclerotic plaque.
Arteriolosclerosis A type of age-related small vessel disease within the small arteries (arterioles), which is defined pathologically by fibrohyaline deposits, a thickening of the vessel wall, a narrowing of the lumen, and a loss of vascular smooth muscle cells.
Vascular remodeling A change in vascular structure in response to persistent elevations in blood pressure characterized by an increase in the size of the vascular smooth muscle cells and an accumulation of proteins such as fibronectin and collagen.
Cerebral blood flow The supply of blood to the brain, which is approximately 750 mL per minute in adults, representing 15% of cardiac output.
Cerebral autoregulation The process by which mammals maintain appropriate and consistent cerebral perfusion despite constant changes in systemic blood pressure.
Oligemia A reduction in blood flow to an end organ, such as the brain, which is not severe enough to cause ischemia or other acute tissue damage.

Diagnosis and Management in Dementia
ISBN 978-0-12-815854-8, https://doi.org/10.1016/B978-0-12-815854-8.00008-2

Introduction

The identification of modifiable risk factors to reduce the burden of late-life cognitive decline and dementia has become a priority in the absence of disease-modifying treatments. Hypertension, redefined recently by the American Heart Association as blood pressure above 130/80 mmHg, is one such risk factor and is both highly prevalent and believed to be associated with adverse neurologic outcomes (Whelton et al., 2018). Hypertension has been historically recognized as a strong risk factor for cardiovascular and cerebrovascular disease, but more recent evidence has implicated hypertension as a risk factor for cognitive decline and dementia as well. Over the last several decades there has been a great deal of translational and clinical research on the connection between blood pressure and neurocognitive function. Although this relationship is still not fully understood, this chapter will highlight some of the work that has thus far shed light on how blood pressure may influence late-life cognitive decline and dementia.

Blood pressure and cognitive function

The relationship between blood pressure and cognitive function appears to differ by the age at which blood pressure is measured. High blood pressure during midlife (fourth and fifth decades) has been consistently associated with greater cognitive impairment in older adulthood, especially if blood pressure is untreated (Launer, Masaki, Petrovitch, Foley, & Havlik, 1995). This work has been supported by several longitudinal studies demonstrating steeper subsequent cognitive decline over 20+ years among middle-aged adults with hypertension (e.g., Gottesman et al., 2014). Studies evaluating the effects of late-life blood pressure on cognitive function have been less consistent. Hypertension occurring during the sixth and seventh decades of life has shown a fairly consistent relationship with cognitive impairment and cognitive decline (Tsivgoulis et al., 2009); however, this has not been the case for hypertension occurring for those in their late seventies as well as their eighties and nineties (Pandav, Dodge, DeKosky, & Ganguli, 2003; Reitz, Tang, Manly, Mayeux, & Luchsinger, 2007). In fact, several studies have found hypertension among octogenarians and nonagenarians to be protective against cognitive deficits (Guo, Fratiglioni, Winblad, & Viitanen, 1997).

One explanation for the pattern of stronger and more consistent associations between midlife hypertension and late-life cognitive deficits is that persons with midlife hypertension are more likely to be exposed to the deleterious effects of high blood pressure for a longer period. This hypothesis has been supported by longitudinal studies showing that a longer duration of elevated blood pressure is associated with reduced cognitive abilities in later life (Power, Tchetgen, Sparrow, Schwartz, & Weisskopf, 2013; Swan, Carmelli, & Larue, 1998). It is also possible that midlife represents a critical period prior to the development of neurodegeneration or cerebrovascular disease during which exposure to hypertension might trigger cerebral changes which ultimately impact cognition and lead to dementia. Importantly, midlife blood pressure exposure may also dictate what blood pressure is tolerated later in life: another study that examined the effect of longitudinal blood pressure changes found that adults with midlife hypertension who then develop low diastolic blood pressure in late life may have even greater cognitive deficits and smaller brain volumes as older adults (Muller et al., 2014). Additional longitudinal studies will be needed to determine what blood pressure targets should be used for older adults (both with and without a history of long-term hypertension) to mitigate late-life cognitive deficits.

Hypertension as a risk factor for dementia

Although large- and small-vessel cerebrovascular pathologies have been historically considered a defining feature of vascular dementia, it is now clear that these same neuro-pathological features occur in approximately half of autopsied individuals diagnosed clinically with Alzheimer's disease (Schneider, Arvanitakis, Bang, & Bennett, 2007). These overlapping pathological findings suggest that vascular disease may play a key role in promoting neuronal dysfunction and progressive cognitive impairment in both forms of dementia (Zlokovic, 2011).

Alzheimer's dementia

As highlighted in Table 8.1, hypertension during midlife has shown a fairly consistent association with later Alzheimer's disease and all-cause dementia despite less consistent associations for systolic and diastolic blood pressure levels (Gottesman et al., 2017; Power et al., 2011). Among older adults, low blood pressure, rather than hypertension, appears to be more consistently associated with Alzheimer's disease (Li et al., 2007; Muller et al., 2007; Shah et al., 2005). The combination of these findings suggests that a pattern of hypertension during middle adulthood followed by a late-life drop in blood pressure may make one particularly susceptible to dementia. Although this hypothesis has been supported by several longitudinal studies, few studies to date have been designed to evaluate this hypothesis directly (Glodzik et al., 2014; Muller et al., 2014). It is possible that low blood pressure is simply an epiphenomenon of the pathological brain changes and declining health that accompany dementia.

Vascular dementia

Hypertension has been established as a risk factor for small-vessel disease and stroke, both of which represent common etiological components of vascular dementia. Studies that have examined the association between hypertension and vascular dementia have found midlife hypertension to be most consistently associated with vascular dementia

Table 8.1 Epidemiologic studies examining the association of hypertension and blood pressure with dementia.

Study name	Total N (dementia cases)	Age (SD) at BP assessment	BP exposure variable	Follow-up period (years)	Major findings
Honolulu–asia aging study (Launer et al., 2000)	N = 3547 (118)	53 (5)	SBP, DBP	27	Among men never treated for hypertension, elevated SBP and DBP during midlife was associated with increased risk for dementia. BP was not associated with dementia in treated men.
Rotterdam study (Ruitenberg et al., 2001)	N = 6985 (196)	70 (9)	SBP, DBP	2	Higher SBP and DBP were associated with reduced dementia incidence among participants using antihypertensive medication.
FINMONICA (Kivipelto et al., 2002)	N = 1449 (48)	50 (6)	SBP, DBP	21	Elevated SBP, but not DBP, during midlife was associated with increased risk for dementia.
Canadian study of health and aging (Lindsay et al., 2002)	N = 4088 (157)	68	HTN	5	Hypertension was not associated with risk of dementia.
Xian, China cohort (Qu et al., 2005)	Cross-temporal N = 2197 (37)	65	HTN	3	Hypertension was associated with increased risk of VD but not AD.

Study	N	Age	BP measure		Findings
Religious orders study (Shah et al., 2005)	N = 824 (151)	75 (7)	SBP, DBP	7	Neither SBP nor DBP was associated with incident AD.
Cache County (Hayden et al., 2006)	N = 3308 (185)	74 (6)	HTN	3	Hypertension was not associated with all-cause dementia, AD, or VD.
Adult change in thought (Li et al., 2007)	N = 2356 (204)	75	SBP, DBP	6	Higher SBP was associated with incident dementia among younger (65–74 at baseline) but not older (75+) participants.
Washington heights-Inwood Community Aging project (Muller et al., 2007)	N = 1833 (147)	76 (6)	HTN	4	Hypertension was not associated with all-cause dementia, AD, or dementia associated with stroke.
Three City study (Raffaitin et al., 2009)	N = 7087 (134)	73 (5)	HTN	4	Hypertension was not associated with all-cause dementia, AD, or VD.
Neurological disorder in Central Spain study (Bermejo-Pareja et al., 2010)	N = 3834 (113)	73 (6)	HTN	3	Untreated hypertension was associated with incident all-cause dementia and AD.
Atherosclerosis risk in communities study (Gottesman et al., 2017)	N = 15,744 (1516)	54 (6)	HTN	23	Hypertension and prehypertension during midlife were associated with incident dementia.

AD, Alzheimer's disease; *BP*, blood pressure; *DBP*, diastolic blood pressure; *HTN*, hypertension; *SBP*, systolic blood pressure; *VD*, vascular dementia.

(Rönnemaa, Zethelius, Lannfelt, & Kilander, 2011; Yamada et al., 2003). On the other hand, the associations between late-life hypertension and vascular dementia have been less consistent (e.g., Posner et al., 2002). Although a comparison of the literature might support a stronger and more consistent association between midlife hypertension and vascular dementia, compared with Alzheimer's dementia, it is important to consider that the majority of patients diagnosed clinically as having Alzheimer's dementia have a "mixed" pathology with a high burden of cerebrovascular disease (Schneider et al., 2007).

Blood pressure in relation to brain structural and molecular changes

Neuroimaging studies have offered a window into the underlying neuropathological changes that accompany hypertension. Higher rates of subclinical cerebrovascular pathology have been reliably found among individuals with hypertension. Specifically, hypertension has been associated with the presence of lacunar infarctions, cerebral microbleeds, and greater white matter hyperintensity volume (Gottesman et al., 2010; Shams et al., 2015). Some of the earliest brain abnormalities associated with hypertension may be at the level of the white matter microstructure, as elevated blood pressure, even in younger individuals with normally appearing white matter, has been associated with abnormalities in white matter microstructure (Maillard et al., 2012), which can be detected using MRI diffusion tensor imaging. Hypertension measured during midlife has also been associated with smaller brain volumes and greater rates of atrophy (Jennings et al., 2012); however, elevated blood pressure in older adulthood appears to be protective against brain volume loss (van Velsen et al., 2013).

Several studies have found evidence indicating that high blood pressure may influence Alzheimer's pathology directly. For example, autopsy studies have found that the brain of individuals with a history of hypertension have greater levels of β-amyloid (Aβ) plaques, neurofibrillary tangles, and atrophy compared with those of normotensive individuals (Ashby, Miners, Kehoe, & Love, 2016). In support of these findings, several positron emission tomography studies have found higher levels of Aβ and reduced glucose metabolism among cognitively normal individuals with high blood pressure (Fig. 8.1), particularly in brain regions known to be susceptible to Alzheimer's pathology (Langbaum et al., 2012).

The physiology of hypertension-related brain changes

The physiological mechanisms that underlie the relationship between hypertension and adverse neurological outcomes have not yet been fully elucidated. However, evidence suggests that the hypertension-related alterations to neurovascular pathways described below may pave the way for further pathologic brain changes.

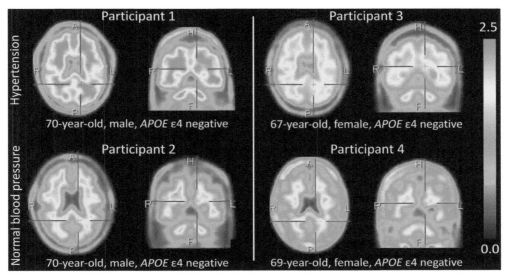

Figure 8.1 Cerebral amyloid measured by florbetapir positron emission tomography in older adults with hypertension and matched participants with normal blood pressure. Red (lighter colors) represents high amyloid levels; blue (darker colors) represents low amyloid levels. Participants are matched on age, sex, race, and *APOE* ε4 status.

Vascular remodeling

Sustained hypertension is associated with a process of hypertrophic remodeling of vascular smooth muscle cells and media, which ultimately results in a reduction in lumen diameter and vessel wall stiffening (Faraco & Iadecola, 2013). Within large cerebral arteries, hypertension can also promote the development of intracranial atherosclerosis (Holmstedt, Turan, & Chimowitz, 2013). Within the smaller arterioles that supply deep gray matter structures and subcortical white matter, hypertension can lead to arteriolosclerosis, a pathological process characterized by a loss of tunica media smooth muscle cells, thickening of the vessel wall, and fibrohyaline deposits (Pantoni, 2010). Together, these vascular changes can set the stage for cerebral infarction and lead to increases in arterial pulse wave velocity (and pulsatile pressure), which itself can have damaging effects on the arterioles and capillaries of end organs such as the brain.

Cerebral perfusion and autoregulation

Through the process of cerebral autoregulation, the brain constricts and dilates small arteries and arterioles to maintain a consistent and continuous high volume of low-pressure blood flow and ensure adequate perfusion in the context of changes to systemic blood pressure (Muller, Van Der Graaf, Visseren, Mali, & Geerlings, 2012). Long-term exposure to high blood pressure can impair cerebrovascular reactivity and disrupt cerebral autoregulation in a way that increases the lower limit of the systemic blood pressure

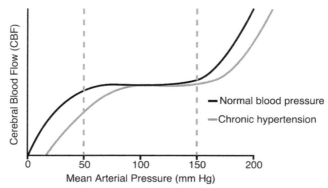

Figure 8.2 The hypothesized shift in cerebral autoregulation curve resulting from chronic hypertension. In healthy adults, cerebral autoregulation is believed to maintain persistent blood flow to the brain between an upper and lower limit of mean arterial pressure. Longstanding hypertension may shift the autoregulatory capacity so that it is less effective at lower blood pressures. *Dotted red (gray in print version) lines* represent the approximate upper and lower limits of blood pressure within which cerebral autoregulation can maintain consistent cerebral blood flow in healthy adults.

needed to adequately perfuse the brain (Fig. 8.2) (Wong et al., 2011). Therefore, among persons with a longstanding history of hypertension who have a shift in cerebral autoregulatory limits, even normal blood pressure levels may not be high enough to maintain adequate cerebral perfusion. Although ischemia can occur in the context of chronic cerebral hypoperfusion, cerebral oligemia, defined as a mild reduction in cerebral blood flow, is believed to be more common. Oligemia has been associated with a number of physiological changes, such as oxidative stress and reduced protein synthesis, that together can promote cerebrovascular dysfunction and neurodegeneration (Zlokovic, 2011).

Blood pressure and Alzheimer's disease neurobiology

There is mounting evidence that perturbations in blood pressure, especially if protracted, may directly promote the development of Alzheimer's pathology (Fig. 8.3). For example, evidence from animal models suggests that reductions in cerebral blood flow can lead to upregulation in neuronal tau phosphorylation, increases in Aβ synthesis, and Aβ oligomerization (Koike, Green, Blurton-Jones, & Laferla, 2010). Additionally, hypertension itself may alter blood—brain barrier transport mechanisms in a way that inhibits the clearance of proteins, such as Aβ. For example, hypertension is associated with upregulation of the receptor for advanced glycation end products, which facilitates the exchange of Aβ from blood across the blood—brain barrier into the brain (Carnevale et al., 2012). A theoretical model illustrating the link between blood pressure and dementia-related brain changes is provided in Fig. 8.4.

Antihypertensive clinical trials for improving cognition

In light of observational and experimental evidence that implicates hypertension as a risk factor for neurodegeneration, cognitive decline, and dementia, a number of randomized

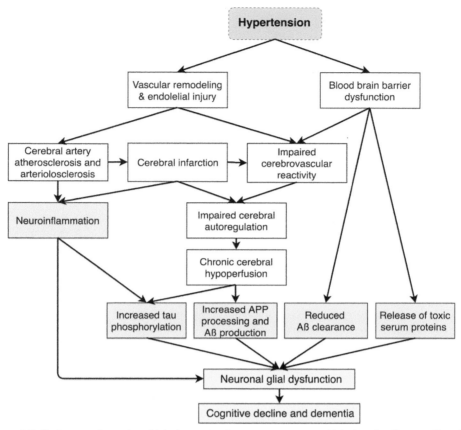

Figure 8.3 Pathways through which hypertension may promote Alzheimer's disease. Sustained hypertension can cause direct structural and functional changes to cerebral vasculature (*white boxes*), promoting cerebrovascular disease, such as lacunar infarctions and microhemorrhages. These same hypertension-induced vascular changes can also promote neuroinflammation and Alzheimer's-specific pathophysiological processes, including tau phosphorylation and Aβ synthesis and oligomerization. Each of these molecular changes is known to cause neuronal and glial dysfunction, which can eventually lead to neurodegeneration and cognitive decline. *Aβ*, beta-amyloid; *APP*, amyloid precursor protein.

clinical trials (RCTs) have been conducted to determine whether a lowering of blood pressure can slow cognitive decline and reduce dementia risk. However, the evidence that has emerged to date from RCTs has been largely inconclusive (McGuinness, Todd, Passmore, & Bullock, 2009). As described in Table 8.2, several large trials have found beneficial effects of antihypertensive agents on dementia-related outcomes; however, several others have failed to replicate these findings. A Cochrane Review published in 2009 of randomized, double-blind, placebo-controlled trials concluded that the evidence for neuroprotective effects from antihypertensive medication use in late life is

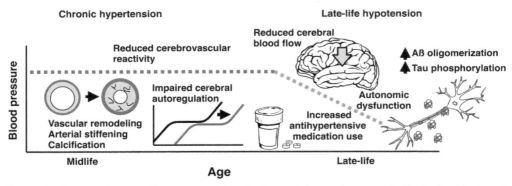

Figure 8.4 Factors that contribute to cognitive decline and dementia among individuals with chronic hypertension and low blood pressure during late life. Chronic hypertension causes changes to the structure and function of arteries and arterioles. These changes include inward vascular remodeling, arterial stiffening, and calcification. These changes can disrupt the function of arteriole endothelial and vascular smooth muscle cells, leading to impaired cerebrovascular reactivity and resulting disruption to cerebral autoregulatory capacity. A subset of individuals with a history of persistent hypertension experience a drop in blood pressure in older adulthood due to factors including increased use of antihypertensive medication, declining health, or autonomic dysfunction resulting from neurodegenerative brain changes. In the context of impaired cerebral autoregulation, a reduction in late-life blood pressure may lead to a reduction in cerebral blood flow that can exacerbate neurodegenerative changes and cognitive decline.

not convincing (McGuinness et al., 2009). Discordant findings between observational studies and RCTs have been attributed to several factors including study design as well as the timing, duration, length, and effectiveness of antihypertensive treatment. As summarized above, the impact of hypertension on adverse cognitive outcomes may take decades to occur, and similarly treatment may need to occur over decades before a reduction in cognitive decline is observed. It is also worth noting that many trials that have examined the neuroprotective effects of antihypertensive treatment assessed cognitive decline or dementia as a secondary endpoint. The SPRINT-MIND trial reported reductions in MCI and in a composite outcome of MCI and dementia in individuals treated to a more aggressive blood pressure threshold (below 120/80 mmHg) compared with results using a standard threshold; no significant difference was found for dementia, however (Williamson et al., 2019).

Preliminary evidence indicates that specific classes of blood-pressure-lowering medication may show stronger neuroprotective effects than others. One meta-analysis of antihypertensive randomized trials found that angiotensin II receptor blockers (ARBs) tended to outperform other types of blood-pressure-lowering medication (Marpillat, Macquin-Mavier, Tropeano, Bachoud-Levi, & Maison, 2013). Consistent with this notion, fairly strong evidence suggests that antihypertensive drugs may have neuroprotective effects that extend beyond the lowering of blood pressure. For example,

Table 8.2 Placebo-controlled randomized clinical trials examining the effect of antihypertensive therapy on cognition and dementia.

Study name	Total N (treatment group)	Age (SD) at baseline	Follow-up period (years)	Antihypertensive agents studied	Outcome of interest	Effect of treatment
HYVET-Cog (Peters et al., 2008)	N = 3336 (1687)	84 (3)	2.2	Diuretic and ACE inhibitor versus placebo	Dementia	No effect
SHEP (Applegate et al., 1994)	N = 4608 (2317)	74	5	Diuretic and β-blocker versus placebo	Cognition	No effect
MRC (Prince, Bird, Blizard, & Mann, 1996)	N = 2584 (1273)	70	4.5	Diuretic or β-blocker versus placebo	Cognition	No effect
Syst-Eur (Forette et al., 1998)	N = 2418 (1238)	70 (7)	2	CCB ± diuretic versus placebo	Dementia	Protective
Syst-Eur follow-up (Forette et al., 2002)	N = 2902 (1327)	68	3.9	CCB ± diuretic versus placebo	Dementia	Protective
PROGRESS (Tzourio et al., 2003)	N = 6105 (3054)	64 (10)	4	ACE inhibitor ± diuretic versus placebo	Cognition, dementia	Protective for participants with recurrent stroke
SCOPE (Lithell et al., 2003)	N = 4964 (2477)	76	4	ARB versus placebo	Cognition, dementia	No effect
PRoFESS (Diener et al., 2008)	N = 20,332 (10,146)	66 (9)	2.4	ARB versus placebo	Cognition, dementia	No effect
TRANSCEND (Anderson et al., 2011)	N = 5231 (2972)	67 (7)	4.7	ARB versus placebo	Cognition, dementia	No effect
SPRINT MIND study (AAIC Press Release, 2018)	N = 9361 (4678)	68	3.3[a]	BP treatment to <120 mmHg versus treatment to <140 mmHg	MCI, dementia	BP treatment <120 mmHg was protective against MCI and MCI/ Dementia combined

ARB, angiotensin II receptor blocker; *ACE,* angiotensin-converting enzyme; *BP,* blood pressure; *CCB,* calcium channel blocker; *HYVET-COG,* Hypertension in the Very Elderly Trial Cognitive Function Assessment; *MCI,* mild cognitive impairment; *MRC,* Medical Research Council; *ProFESS,* Prevention Regimen for Effectively Avoiding Second Strokes; *PROGRESS,* Perindopril Protection against Recurrent Strokes; *SHEP,* Systolic Hypertension in the Elderly Program; *SPRINT MIND,* Systolic Blood Pressure Intervention Trial, Memory and Cognition in Decreased Hypertension; *Syst-Eur,* Systolic Hypertension in Europe; *TRANSCEND,* Telmisartan Randomized Assessment Study in ACE Intolerant Subjects with Cardiovascular Disease.
[a]Follow-up cognitive assessment continued for several additional years.

several preclinical studies indicate that ARBs and angiotensin-converting enzyme inhibitors have direct antiinflammatory effects on microglia and cerebrovascular endothelial cells (Li et al., 2014; Torika, Asraf, Roasso, Danon, & Fleisher-Berkovich, 2016). Despite such evidence for class-specific neuroprotective effects of antihypertensive agents in rodent models, whether these protective pathways can be recapitulated in humans at risk for cognitive decline remains to be seen.

Conclusions

Hypertension is a highly prevalent condition that has been previously associated with cardiovascular disease and stroke. Accumulating evidence from laboratory, neuropsychological, and neuroimaging research indicates that hypertension, especially during middle adulthood, is associated with a number of adverse neurological outcomes that together promote cognitive decline and dementia. Despite evidence from observational and preclinical studies that points to the beneficial effects of lowering blood pressure, evidence from RCTs for neuroprotection resulting from that use of antihypertensive agents has been largely inconsistent. Future work is needed to understand the class-specific effects of antihypertensive agents, as evidence for neuroprotection through multiple pathways is mounting, especially for antihypertensive agents that modulate the renin-angiotensin system.

Key facts of hypertension

- Hypertension is a highly prevalent condition occurring in as much as half the world's adult population and in two out of every three adults over the age of 65.
- High blood pressure can sometimes be managed by lifestyle modifications, such as dietary changes, stress reduction, physical exercise, and weight loss, although many individuals require medications as well.
- Hypertension, which has been consistently associated with stroke and ischemic heart disease, has been identified as one of the most important modifiable risk factors for premature death.
- Hypertension is associated with a number of abnormalities detected on brain imaging, including infarctions, microhemorrhages, and white matter structural changes.
- Persistent hypertension can lead to a number of pathological vascular changes, including vascular stiffness, atherosclerosis, and reduced cerebrovascular reactivity.

Summary points

- Elevated blood pressure during midlife has been consistently associated with late-life cognitive impairment, cognitive decline, and dementia.

- During older adulthood, high blood pressure may be protective, as lower blood pressure during this period has been associated with cognitive impairment and dementia in a number of studies.
- Hypertension has been associated with a more rapid progression of cerebral small vessel disease and brain volume loss.
- Compared with normotensive individuals, individuals with hypertension tend to show higher levels of Alzheimer's disease pathology in their brains as older adults.
- Hypertension is believed to promote dementia-related pathology through multiple neurovascular and Alzheimer's-specific pathways.
- Evidence from randomized clinical trials evaluating the neuroprotective effects of antihypertensive agents has been inconsistent for dementia-related outcomes.

References

AAIC. (2018). *Intensive blood pressure control significantly reduces new cases of MCI, and combined risk of MCI and dementia: SPRINT MIND study.* Retrieved from https://www.alz.org/aaic/releases_2018/AAIC18-Wed-developing-topics.asp.

Anderson, C., Teo, K., Gao, P., Arima, H., Dans, A., Unger, T., et al. (2011). Renin-angiotensin system blockade and cognitive function in patients at high risk of cardiovascular disease: Analysis of data from the ONTARGET and TRANSCEND studies. *The Lancet Neurology, 10*(1), 43−53. https://doi.org/10.1016/S1474-4422(10)70250-7.

Applegate, W. B., Pressel, S., Wittes, J., Luhr, J., Shekelle, R. B., Camel, G. H., et al. (1994). Impact of the treatment of isolated systolic hypertension on behavioral variables: Results from the systolic hypertension in the elderly Program. *Archives of Internal Medicine, 154*(19), 2154−2160. https://doi.org/10.1001/archinte.1994.00420190047006.

Ashby, E. L., Miners, J. S., Kehoe, P. G., & Love, S. (2016). Effects of hypertension and anti-hypertensive treatment on amyloid-B plaque load and AB-synthesizing and AB-degrading enzymes in frontal cortex. *Journal of Alzheimer's Disease, 50*(4), 1191−1203. https://doi.org/10.3233/JAD-150831.

Bermejo-Pareja, F., Benito-León, J., Louis, E. D., Trincado, R., Carro, E., Villarejo, A., et al. (2010). Risk of incident dementia in drug-untreated arterial hypertension: A population-based study. *Journal of Alzheimer's Disease, 22*(3), 949−958. https://doi.org/10.3233/JAD-2010-101110.

Carnevale, D., Mascio, G., D'Andrea, I., Fardella, V., Bell, R. D., Branchi, I., et al. (2012). Hypertension induces brain β-amyloid accumulation, cognitive impairment, and memory deterioration through activation of receptor for advanced glycation end products in brain vasculature. *Hypertension, 60*(1), 188−197. https://doi.org/10.1161/HYPERTENSIONAHA.112.195511.

Diener, H. C., Sacco, R. L., Yusuf, S., Cotton, D., Ôunpuu, S., Lawton, W. A., et al. (2008). Effects of aspirin plus extended-release dipyridamole versus clopidogrel and telmisartan on disability and cognitive function after recurrent stroke in patients with ischaemic stroke in the Prevention Regimen for Effectively Avoiding Second Strokes (PRoFE. *The Lancet Neurology, 7*(10), 875−884. https://doi.org/10.1016/S1474-4422(08)70198-4.

Faraco, G., & Iadecola, C. (November, 2013). Hypertension: A harbinger of stroke and dementia. *Hypertension.* https://doi.org/10.1161/HYPERTENSIONAHA.113.01063.

Forette, F., Seux, M. L., Staessen, J. A., Thijs, L., Babarskiene, M. R., Babeanu, S., et al.Systolic hypertension in Europe, I. (2002). the prevention of dementia with antihypertensive treatment: New evidence from the systolic hypertension in Europe (Syst-Eur) study [see comment][erratum appears in Arch Intern Med. 2003 Jan 27;163(2):241.] *Archives of Internal Medicine, 162*(18), 2046−2052.

Forette, F., Seux, M. L., Staessen, J. A., Thijs, L., Birkenhäger, W. H., Babarskiene, M. R., et al. (1998). Prevention of dementia in randomised double-blind placebo-controlled Systolic Hypertension in Europe (Syst-Eur) trial. *Lancet, 352*(9137), 1347−1351. https://doi.org/10.1016/S0140-6736(98)03086-4.

Glodzik, L., Rusinek, H., Pirraglia, E., McHugh, P., Tsui, W., Williams, S., et al. (2014). Blood pressure decrease correlates with tau pathology and memory decline in hypertensive elderly. *Neurobiology of Aging, 35*(1), 64–71. https://doi.org/10.1016/j.neurobiolaging.2013.06.011.

Gottesman, R. F., Albert, M. S., Alonso, A., Coker, L. H., Coresh, J., Davis, S. M., et al. (2017). Associations between midlife vascular risk factors and 25-year incident dementia in the atherosclerosis risk in communities (ARIC) cohort. *JAMA Neurology, 388*(10046), 797–805. https://doi.org/10.1001/jamaneurol.2017.1658.

Gottesman, R. F., Coresh, J., Catellier, D. J., Sharrett, A. R., Rose, K. M., Coker, L. H., et al. (2010). Blood pressure and white-matter disease progression in a biethnic cohort: Atherosclerosis risk in communities (ARIC) study. *Stroke, 41*(1), 3–8. https://doi.org/10.1161/STROKEAHA.109.566992.

Gottesman, R. F., Schneider, A. L. C., Albert, M., Alonso, A., Bandeen-Roche, K., Coker, L., et al. (2014). Midlife hypertension and 20-year cognitive change. *JAMA Neurology, 21287*(10), 1–10. https://doi.org/10.1001/jamaneurol.2014.1646.

Guo, Z., Fratiglioni, L., Winblad, B., & Viitanen, M. (1997). Blood pressure and performance on the Mini-Mental State Examination in the very old. Cross-sectional and longitudinal data from the Kungsholmen Project. *American Journal of Epidemiology, 145*(12), 1106–1113.

Hayden, K. M., Zandi, P. P., Lyketsos, C. G., Khachaturian, A. S., Bastian, L. A., Charoonruk, G., et al. (2006). Vascular risk factors for incident Alzheimer disease and vascular dementia: The Cache County Study. *Alzheimer Disease and Associated Disorders, 20*(2), 93–100. https://doi.org/10.1097/01.wad.0000213814.43047.86.

Holmstedt, C. A., Turan, T. N., & Chimowitz, M. I. (2013). Atherosclerotic intracranial arterial stenosis: Risk factors, diagnosis, and treatment. *The lancet Neurology, 12*(11), 1106–1114. https://doi.org/10.1016/S1474-4422(13)70195-9.

Jennings, J. R., Mendelson, D. N., Muldoon, M. F., Ryan, C. M., Gianaros, P. J., Raz, N., et al. (2012). Regional grey matter shrinks in hypertensive individuals despite successful lowering of blood pressure. *Journal of Human Hypertension, 26*(5), 295–305. https://doi.org/10.1038/jhh.2011.31.

Kivipelto, M., Helkala, E. L., Laakso, M. P., Hänninen, T., Hallikainen, M., Alhainen, K., et al. (2002). Apolipoprotein e4 allele, elevated midlife total cholesterol level, and high midlife systolic blood pressure are independent risk factors for late-life Alzheimer disease. *Annals of Internal Medicine, 137*(3), 149–155. https://doi.org/10.7326/0003-4819-137-3-200208060-00006 ([pii]).

Koike, M. A., Green, K. N., Blurton-Jones, M., & Laferla, F. M. (2010). Oligemic hypoperfusion differentially affects tau and amyloid-{beta}. *American Journal Of Pathology, 177*(1), 300–310. https://doi.org/10.2353/ajpath.2010.090750.

Langbaum, J. B. S., Chen, K., Launer, L. J., Fleisher, A. S., Lee, W., Liu, X., et al. (2012). Blood pressure is associated with higher brain amyloid burden and lower glucose metabolism in healthy late middle-age persons. *Neurobiology of Aging, 33*(4). https://doi.org/10.1016/j.neurobiolaging.2011.06.020, 827.e11-9.

Launer, L. J., Masaki, K., Petrovitch, H., Foley, D., & Havlik, R. J. (1995). The association between midlife blood pressure levels and late-life cognitive function. The Honolulu-Asia Aging Study. *Journal of the American Medical Association, 274*(23), 1846–1851.

Launer, L. J., Ross, G. W., Petrovitch, H., Masaki, K., Foley, D., White, L. R., et al. (2000). Midlife blood pressure and dementia: The Honolulu-Asia aging study. *Neurobiology of Aging, 21*(1), 49–55. https://doi.org/10.1016/S0197-4580(00)00096-8.

Li, Z., Cao, Y., Li, L., Liang, Y., Tian, X., Mo, N., et al. (2014). Prophylactic angiotensin type 1 receptor antagonism confers neuroprotection in an aged rat model of postoperative cognitive dysfunction. *Biochemical and Biophysical Research Communications, 449*(1), 74–80. https://doi.org/10.1016/j.bbrc.2014.04.153.

Lindsay, J., Laurin, D., Verreault, R., Hébert, R., Helliwell, B., Hill, G. B., et al. (2002). Risk factors for Alzheimer's disease: A prospective analysis from the Canadian study of health and aging. *American Journal of Epidemiology, 156*(5), 445–453. https://doi.org/10.1093/aje/kwf074.

Li, G., Rhew, I. C., Shofer, J. B., Kukull, W. A., Breitner, J. C. S., Peskind, E., et al. (2007). Age-varying association between blood pressure and risk of dementia in those aged 65 and older: A community-based prospective cohort study. *Journal of the American Geriatrics Society, 55*(8), 1161–1167. https://doi.org/10.1111/j.1532-5415.2007.01233.x.

Lithell, H., Hansson, L., Skoog, I., Elmfeldt, D., Hofman, A., Olofsson, B., et al. (2003). The study on cognition and prognosis in the elderly (SCOPE): Principal results of a randomized double-blind intervention trial. *Journal of Hypertension, 21*(5), 875—886. https://doi.org/10.1097/01.hjh.0000059028.82022.89.

Maillard, P., Seshadri, S., Beiser, A., Himali, J. J., Au, R., Fletcher, E., et al. (2012). Effects of systolic blood pressure on white-matter integrity in young adults in the framingham heart study: A cross-sectional study. *The Lancet Neurology, 11*(12), 1039—1047. https://doi.org/10.1016/S1474-4422(12)70241-7.

Marpillat, N. L., Macquin-Mavier, I., Tropeano, A.-I., Bachoud-Levi, A.-C., & Maison, P. (2013). Antihypertensive classes, cognitive decline and incidence of dementia: A network meta-analysis. *Journal of Hypertension, 31*(6), 1073—1082. https://doi.org/10.1097/HJH.0b013e3283603f53.

McGuinness, B., Todd, S., Passmore, P., & Bullock, R. (2009). Blood pressure lowering in patients without prior cerebrovascular disease for prevention of cognitive impairment and dementia. *Cochrane Database of Systematic Reviews, 4.* https://doi.org/10.1002/14651858.CD004034.pub3.

Muller, M., Sigurdsson, S., Kjartansson, O., Aspelund, T., Lopez, O. L., Jonnson, P. V., et al. (2014). Joint effect of mid- and late-life blood pressure on the brain: The AGES-Reykjavik Study. *Neurology, 82*(24), 2187—2195. https://doi.org/10.1212/WNL.0000000000000517.

Muller, M., Tang, M.-X., Schupf, N., Manly, J. J., Mayeux, R., & Luchsinger, J. A. (2007). Metabolic syndrome and dementia risk in a multiethnic elderly cohort. *Dementia and Geriatric Cognitive Disorders, 24*(3), 185—192. https://doi.org/10.1159/000105927.

Muller, M., Van Der Graaf, Y., Visseren, F. L., Mali, W. P. T. M., & Geerlings, M. I. (2012). Hypertension and longitudinal changes in cerebral blood flow: The SMART-MR study. *Annals of Neurology, 71*(6), 825—833. https://doi.org/10.1002/ana.23554.

Pandav, R., Dodge, H. H., DeKosky, S. T., & Ganguli, M. (2003). Blood pressure and cognitive impairment in India and the United States: A cross-national epidemiological study. *Archives of Neurology, 60*(8), 1123—1128. https://doi.org/10.1001/archneur.60.8.1123.

Pantoni, L. (2010). Cerebral small vessel disease: From pathogenesis and clinical characteristics to therapeutic challenges. *The lancet Neurology, 9*(7), 689—701. https://doi.org/10.1016/S1474-4422(10)70104-6.

Peters, R., Beckett, N., Forette, F., Tuomilehto, J., Clarke, R., Ritchie, C., et al. (2008). Incident dementia and blood pressure lowering in the hypertension in the very elderly trial cognitive function assessment (HYVET-COG): A double-blind, placebo controlled trial. *The Lancet Neurology, 7*(8), 683—689. https://doi.org/10.1016/S1474-4422(08)70143-1.

Posner, H. B., Tang, M.-X., Luchsinger, J., Lantigua, R., Stern, Y., & Mayeux, R. (2002). The relationship of hypertension in the elderly to AD, vascular dementia, and cognitive function. *Neurology, 58*(8), 1175—1181.

Power, M. C., Tchetgen, E. J. T., Sparrow, D., Schwartz, J., & Weisskopf, M. G. (2013). Blood pressure and cognition: Factors that may account for their inconsistent association. *Epidemiology, 24*(6), 886—893. https://doi.org/10.1097/EDE.0b013e3182a7121c.

Power, M. C., Weuve, J., Gagne, J. J., McQueen, M. B., Viswanathan, A., & Blacker, D. (2011). The association between blood pressure and incident Alzheimer disease: A systematic review and meta-analysis. *Epidemiology, 22*(5), 646—659. https://doi.org/10.1097/EDE.0b013e31822708b5.

Prince, M. J., Bird, A. S., Blizard, R. A., & Mann, A. H. (1996). Is the cognitive function of older patients affected by antihypertensive treatment? Results from 54 months of the medical research council's trial of hypertension in older adults. *BMJ, 312*(7034), 801—805. https://doi.org/10.1136/bmj.312.7034.801.

Qu, Q., Qiao, J., Han, J., Yang, J., Guo, F., Luo, G., et al. (2005). The incidence of dementia among elderly people in Xi' an, China. *Zhonghua Liu Xing Bing Xue Za Zhi = Zhonghua Liuxingbingxue Zazhi, 26*(7), 529—532.

Raffaitin, C., Gin, H., Empana, J.-P., Helmer, C., Berr, C., Tzourio, C., et al. (2009). Metabolic syndrome and risk for incident Alzheimer's disease or vascular dementia: The three-city study. *Diabetes Care, 32*(1), 169—174. https://doi.org/10.2337/dc08-0272.

Reitz, C., Tang, M.-X., Manly, J., Mayeux, R., & Luchsinger, J. A. (2007). Hypertension and the risk of mild cognitive impairment. *Archives of Neurology, 64*(12), 1734—1740. https://doi.org/10.1001/archneur.64.12.1734.

Rönnemaa, E., Zethelius, B., Lannfelt, L., & Kilander, L. (2011). Vascular risk factors and dementia: 40-Year follow-up of a population-based cohort. *Dementia and Geriatric Cognitive Disorders, 31*(6), 460–466. https://doi.org/10.1159/000330020.

Ruitenberg, A., Skoog, I., Ott, A., Aevarsson, O., Witteman, J. C. M., Lernfelt, B., et al. (2001). Blood pressure and risk of dementia: Results from the Rotterdam study and the Gothenburg H-70 study. *Dementia and Geriatric Cognitive Disorders, 12*(1), 33–39. https://doi.org/10.1159/000051233.

Schneider, J. A., Arvanitakis, Z., Bang, W., & Bennett, D. A. (2007). Mixed brain pathologies account for most dementia cases in community-dwelling older persons. *Neurology, 69*(24), 2197–2204. https://doi.org/10.1212/01.wnl.0000271090.28148.24.

Shah, R. C., Wilson, R. S., Bienias, J. L., Arvanitakis, Z., Evans, D. A., & Bennett, D. A. (2005). Relation of blood pressure to risk of incident Alzheimer's disease and change in global cognitive function in older persons. *Neuroepidemiology, 26*(1), 30–36. https://doi.org/10.1159/000089235.

Shams, S., Martola, J., Granberg, T., Li, X., Shams, M., Fereshtehnejad, S. M., et al. (2015). Cerebral microbleeds: Different prevalence, topography, and risk factors depending on dementia diagnosis-the Karolinska imaging dementia study. *American Journal of Neuroradiology, 36*(4), 661–666. https://doi.org/10.3174/ajnr.A4176.

Swan, G. E., Carmelli, D., & Larue, A. (1998). Systolic blood pressure tracking over 25 to 30 years and cognitive performance in older adults. *Stroke, 29*(11), 2334–2340. https://doi.org/10.1161/01.STR.29.11.2334.

Torika, N., Asraf, K., Roasso, E., Danon, A., & Fleisher-Berkovich, S. (2016). Angiotensin converting enzyme inhibitors ameliorate brain inflammation associated with microglial activation: Possible implications for Alzheimer's disease. *Journal of Neuroimmune Pharmacology, 11*(4), 774–785. https://doi.org/10.1007/s11481-016-9703-8.

Tsivgoulis, G., Alexandrov, A. V., Wadley, V. G., Unverzagt, F. W., Go, R. C. P., Moy, C. S., et al. (2009). Association of higher diastolic blood pressure levels with cognitive impairment. *Neurology, 73*(8), 589–595. https://doi.org/10.1212/WNL.0b013e3181b38969.

Tzourio, C., Anderson, C., Chapman, N., Woodward, M., Neal, B., MacMahon, S., et al. (2003). Effects of blood pressure lowering with perindopril and indapamide therapy on dementia and cognitive decline in patients with cerebrovascular disease. *Archives of Internal Medicine, 163*(9), 1069–1075. https://doi.org/10.1001/archinte.163.9.1069.

van Velsen, E. F. S., Vernooij, M. W., Vrooman, H. A., van der Lugt, A., Breteler, M. M. B., Hofman, A., et al. (2013). Brain cortical thickness in the general elderly population: The Rotterdam Scan Study. *Neuroscience Letters, 550*, 189–194. https://doi.org/10.1016/j.neulet.2013.06.063.

Williamson, J. D., Pajewski, N. M., Auchus, A. P., Bryan, R. N., Chelune, G., Cheung, A. K., et al. (2019). Effect of intensive vs standard blood pressure control on probable dementia: A randomized clinical trial. *JAMA, 321*, 553. https://doi.org/10.1001/JAMA.2018.21442.

Whelton, P. K., Carey, R. M., Aronow, W. S., Casey, D. E., Collins, K. J., Dennison Himmelfarb, C., et al. (2018). 2017 ACC/AHA/AAPA/ABC/ACPM/AGS/APhA/ASH/ASPC/NMA/PCNA guideline for the prevention, detection, evaluation, and management of high blood pressure in adults: Executive summary. *Journal of the American College of Cardiology, 71*(19), 2199–2269. https://doi.org/10.1016/j.jacc.2017.11.005.

Wong, L. J., Kupferman, J. C., Prohovnik, I., Kirkham, F. J., Goodman, S., Paterno, K., et al. (2011). Hypertension impairs vascular reactivity in the pediatric brain. *Stroke, 42*(7), 1834–1838. https://doi.org/10.1161/STROKEAHA.110.607606.

Yamada, M., Kasagi, F., Sasaki, H., Masunari, N., Mimori, Y., & Suzuki, G. (2003). Association between dementia and midlife risk factors: The radiation effects research foundation adult health study. *Journal of the American Geriatrics Society, 51*(3), 410–414.

Zlokovic, B. V. (2011). Neurovascular pathways to neurodegeneration in Alzheimer's disease and other disorders. *Nature Reviews Neuroscience, 12*(12), 723–738. https://doi.org/10.1038/nrn3114.

CHAPTER 9

Genetics of dementia: a focus on Alzheimer's disease

Francesca Fernandez[1], Jessica L. Andrews[2]

[1]Faculty of Health Sciences, School of Behavioural and Health Sciences, Australian Catholic University, Nudgee, QLD, Australia; [2]Office of the DVC Research, The University of Sydney, Sydney, NSW, Australia

List of abbreviations

ABCA7 ATP-binding cassette subfamily A member 7

AD Alzheimer's disease

ADAM 10 ADAM metallopeptidase domain 10

AICD intracellular cytoplasmic/C-terminal domain

APOE apolipoprotein E

APP amyloid precursor protein

Aβ amyloid-β peptide

BIN1 bridging integrator-1

CD2AP CD2-associated protein

CD33 Siglec-3

CLU clusterin

CR1 complement receptor type 1

CTFβ carboxyterminal fragment-β

CTFα carboxyterminal fragment-α

DSG2 desmoglein 2

ECHDC3 enoyl-CoA hydratase domain containing 3

EPHA1 ephrin type-A receptor

FAD familial AD

FBXL7 F-box and leucine-rich repeat protein 7

GWAS genome-wide association study

HLA-DRB1 HLA class II histocompatibility antigen, DRB1 beta chain

HLA-DRB5 HLA class II histocompatibility antigen, DRB5 beta chain

INPP5D inositol polyphosphate-5-phosphatase D

MEF2C myocyte enhancer factor 2C

MS4A membrane-spanning 4-domain family, subfamily A

MTHFD1L monofunctional C1-tetrahydrofolate synthase

NME8 NME family member 8 COBL—Cordon-bleu protein

PFDN1 Prefoldin subunit 1

PICALM phosphatidylinositol-binding clathrin assembly protein

PLCG2 phospholipase C gamma 2

PTK2K PTK2 protein tyrosine kinase 2

RANBP2 RAN binding protein 2

sAPPβ secreted amino-terminals APPβ

sAPPα secreted amino-terminals APPα

SNP single-nucleotide polymorphism

Diagnosis and Management in Dementia
ISBN 978-0-12-815854-8, https://doi.org/10.1016/B978-0-12-815854-8.00009-4

SORL1 sortilin-related receptor 1
TREM2 triggering receptor expressed on myeloid cells 2
UNC5C Unc-5 netrin receptor C
USP6NL USP6 N-terminal-like

Introduction

Dementia is a major problem among the world's aging population, affecting more than 6.5% of people aged over 65 years and more than 22% of people over the age of 85 worldwide. In 2015, dementia was reported to affect an estimated 46.8 million people worldwide according to the World Alzheimer Report, with the incidence of dementia reaching epidemic proportions; worldwide projections are that the number of people living with dementia will reach 75 million by 2030 and 131.5 million by 2050 (Prince et al., 2015). Primary manifestations of dementia include cognitive dysfunction, significant disruptions to thoughts, and other intellectual impairments associated with a severe decline in emotional control and social behavior. Dementia can be caused by many brain-related diseases including schizophrenia, attention deficit hyperactivity disorder, Parkinson's disease, and vascular dementia. The most common and well-known cause of dementia is Alzheimer's disease (AD), which is responsible for about 70% of dementia cases. AD is a progressive neurodegenerative disease clinically characterized by gradual cognitive decline including loss of memory, orientation, and reasoning. AD is pathologically characterized by the presence of neurofibrillary tangles (NFTs) and amyloid plaques in the brain. For the last 2 decades, research on AD has been extensive; however, a simple, definitive diagnostic test has yet to be developed (Goedert & Spillantini, 2006). When AD symptoms are mild, clinical diagnosis can be quite challenging and easily overlooked (Hampel et al., 2004); it can only be diagnosed when the pathology has been clinically established, which usually occurs when progressive cognitive deficits affect a person's ability to cope with the functional demands of social or professional life. There is therefore great interest in developing a reliable method for early detection of AD (Hampel et al., 2004; Sunderland, Hampel, Takeda, Putnam, & Cohen, 2006).

To date a large and increasing number of research studies have focused on investigating biomarker candidates for AD (Bailey, 2007; Hampel et al., 2004). Several neurochemical biomarkers and biomarkers visible on magnetic resonance imaging (MRI) are being evaluated for sensitivity and specificity in the detection of AD pathology (Bailey, 2007). In addition to questions regarding the reliability of these markers, it is worth noting that the lumbar punctures necessary for extracting the cerebral spinal fluid (CSF) used for identifying biomarkers in AD are relatively invasive and not well tolerated as a screening method in elderly populations, the primary demographic presenting with AD. Imaging procedures such as MRI are also costly. Despite the large number of promising results, neurochemical and imaging-based markers have so far not been established in routine clinical diagnoses of AD (Sunderland et al., 2006). Currently, the only definitive method of diagnosis is by postmortem biochemical analyses involving the quantification of amyloid plaques and NFTs in the affected brain regions, which remains as the "gold standard" for AD diagnosis (Dubois et al., 2007).

Genetic markers are the most promising biomarkers for AD diagnosis (Bertram, McQueen, Mullin, Blacker, & Tanzi, 2007). With continual advances in technology as well as the discovery of the human genome, genetic research has progressed exceptionally over the last 2 decades. The results of this research have shown both types of AD (early-onset and late-onset) have genetic links (Bertram, 2009) (Fig. 9.1).

Development of early-onset AD starts between 30 and 60 years of age; this is a rare form of AD affecting only 5% of all AD sufferers (Fadil et al., 2009). This form of AD is also known as familial AD (FAD) due to its high prevalence of inheritance. FAD is caused by a number of different genetic mutations in *amyloid precursor protein (APP), presenilin-1 (PSEN1)*, or *presenilin-2 (PSEN2)* genes (Goate et al., 1991; Levy-Lahad et al., 1995; Rogaev et al., 1995; Sherrington et al., 1995). FAD genetic mutations induce increases in amyloid pathway activity, generating amyloid-β peptides (Aβ) resulting from APP cleavage by β- and γ-secretase enzymes (Fig. 9.2). Aβ accumulation and oligomerization

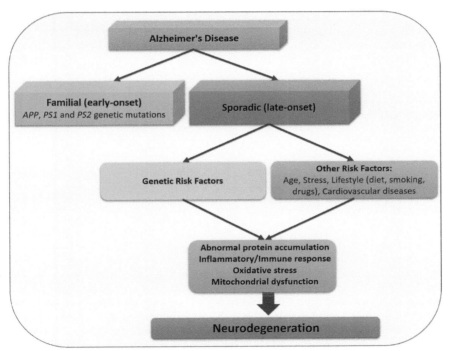

Figure 9.1 *Risk factors for Alzheimer's disease (AD).* In contrast to familial AD, the development of sporadic AD relies on genetic and environmental factors. A combination of genetic and other risk factors influences AD pathogenesis, resulting in neurodegenerative processes.

results in the formation of amyloid plaques in the brain (Fig. 9.2). Even if only one genetic mutation is inherited, the child will have a 50% chance of developing FAD if one parent is affected by FAD (this is classified as an autosomal-dominant genetic disease).

In most people affected by it, AD develops after 60 years of age (late-onset or sporadic AD; Fig. 9.1). Interestingly, most people suffering late-onset AD do not share the genetic mutations identified in FAD (Tanzi & Bertram, 2005). To date, over 600 genes have been proposed as "genetic risk factors" in late-onset AD, but only one of them—a form of the *apolipoprotein E (APOE)* gene—has been validated so far (Allen & Chiu, 2008). The *APOE* gene codes for a protein capable of assisting in the transport of cholesterol in the blood (Fig. 9.3D). Four forms/alleles of the *APOE* gene have been identified,

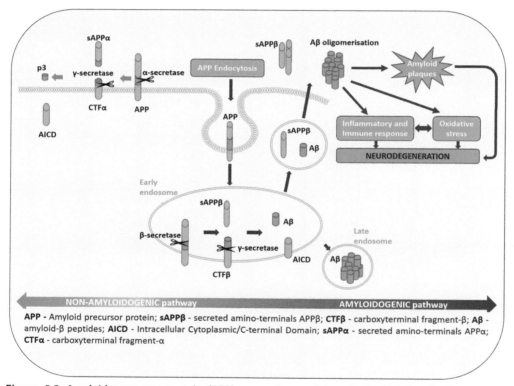

APP - Amyloid precursor protein; **sAPPβ** - secreted amino-terminals APPβ; **CTFβ** - carboxyterminal fragment-β; **Aβ** - amyloid-β peptides; **AICD** - Intracellular Cytoplasmic/C-terminal Domain; **sAPPα** - secreted amino-terminals APPα; **CTFα** - carboxyterminal fragment-α

Figure 9.2 *Amyloid precursor protein (APP) processing and its role in Alzheimer's disease (AD) pathophysiology.* Neuronal APP can be processed through nonamyloidogenic or amyloidogenic pathways, with the latter more active in AD brains. APP is endocytosed before undergoing cleavage by β-secretase into sAPPβ and CTFβ. CTFβ is then cleaved by γ-secretase into Aβ and AICD fragments. Aβ and sAPPβ are then exocytosed back into the extracellular space. Through this process Aβ accumulates, forming oligomers, leading to the generation of amyloid plaques in the brain. The nonamyloidogenic pathway involves APP cleavage by α-secretase; this generates sAPPα and CTFα, with CTFα undergoing further cleavage by γ-secretase, resulting in p3 and AICD fragments. All by-products of this pathway are nonpathogenic.

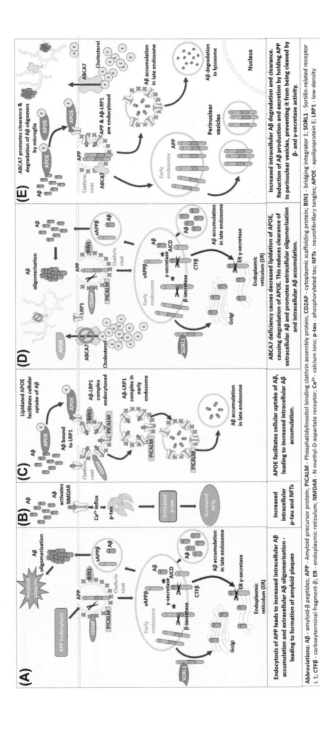

Figure 9.3 *Mechanisms leading to imbalances of amyloid-β peptides (Aβ) and tau in Alzheimer's disease (AD) pathophysiology.* (A) APP endocytosis plays a critical role in the accumulation of Aβ. PICALM, CD2AP, and BIN1 are clathrin-binding proteins that influence APP internalization and subsequent cleavage by early-endosomal β-secretase/γ-secretase, leading to increases in intracellular/extracellular Aβ and the production of amyloid plaques. SORL1 facilitates transport of APP from early endosomes to the Golgi for additional Aβ production. CTFβ can also be transported to the ER where it undergoes cleavage by ER γ-secretase to form Aβ. (B) Extracellular Aβ (previously secreted) can facilitate NMDAR activation leading to an influx of Ca²⁺ leading to accumulation of intracellular p-tau and production of NFTs. (C) APOE binds directly to extracellular Aβ and LRP1. Aβ–LRP1 binding initiates PICALM/clathrin-mediated endocytosis of Aβ–LRP1 complexes. After internalization PICALM remains with Aβ for trafficking. Aβ then accumulates intracellularly in endosomes. (D) ABCA7 modulates the phagocytotic activity of microglia. Dysfunction/deficiency of microglial ABCA7 reduces clearance of Aβ oligomers. Dysfunction/deficiency of neuronal ABCA7 reduces cholesterol transfer to APOE, promoting Aβ production and aggregation. (E) Expression of ABCA7 in microglia induces Aβ oligomer clearance. Neuronal ABCA7 allows transport of cholesterol across the membranes, which when binding with APOE4 facilitates Aβ uptake and degradation. ABCA7 also regulates APP processing and inhibits Aβ secretion by preventing APP cleavage.

but only *APOE* allele E4 (*APOE4*) seems to be present—in 40% of people with late-onset AD (Bu, 2009). Several studies have confirmed that *APOE4* increases the risk of developing AD, but the mechanism of how this happens is not yet understood; however, some people with *APOE4* will never develop the disease (Bu, 2009). Further research is necessary to fully understand the implications of *APOE* in the genetics of AD.

In this chapter, we discuss the genetics of late-onset AD. We first provide a quick overview of current candidate genes for late-onset AD and then propose a paradigm for the role of genes underlying cognitive processes in the genetics of AD.

Genetics of Alzheimer's disease related to its pathophysiology

Over 20,000 studies have been published on the genetics of AD so far, and more than 690 genes have been reported as associated with this pathology (Bertram et al., 2007). Genetic methodologies and analyses have advanced from pedigree-linkage and specific candidate-based association studies to genome-wide association studies (GWASs) and next-generation sequencing (NGS) (Zhu, Tan, Tan, & Yu, 2017). Although the genetic risk factors for late-onset AD are yet to be clearly established, the characterization of novel vulnerability genes for AD have facilitated our understanding of the mechanisms underlying its pathophysiology and have assisted with the development of new avenues for therapies. Here we provide a brief update of the major genes that have been identified and most strongly associated with AD (through significance in GWASs and replicated genetic studies) in relation to their roles in the pathophysiology of sporadic AD (Fig. 9.4 and Table 9.1).

Genetic linkage studies were first performed to identify chromosomal regions associated with AD, leading to the characterization of *APP*, *PSEN1*, and *PSEN2* genes in early-onset AD (Tanzi & Bertram, 2005); this subsequently led to the identification of *APOE4* as a risk factor for late-onset AD (Verghese et al., 2011). To date, *APOE4* remains the strongest genetic risk factor for sporadic AD, with the risk increasing up to 15-fold when two *APOE4* alleles are present; however, *APOE2* has been shown to be protective against AD (Giri et al., 2016). Furthermore, people with AD who carry two *APOE4* alleles have been reported to have high levels of microtubule-associated protein t-tau (total-tau) and p-tau181 (phosphorylated-tau181) in their CSF, both of which are well-established hallmarks of AD pathophysiology (Han et al., 2010) (Fig. 9.3B). *APOE* encodes for apolipoprotein E, which has a role primarily in lipid transport but can also bind to Aβ, having differential effects on the aggregation and/or clearance of Aβ depending on the APOE isoform(s) present (Fig. 9.3E). On one hand, APOE4 can inhibit Aβ clearance, leading to increased Aβ deposition in senile plaques (Kok et al., 2009) (Fig. 9.3C and D), while on the other, more highly lipidated forms of APOE (E2 and E3) are more efficient at inducing intracellular degradation of Aβ (Jiang et al., 2016). APOE4 also promotes Aβ-induced neuroinflammation and increased tau

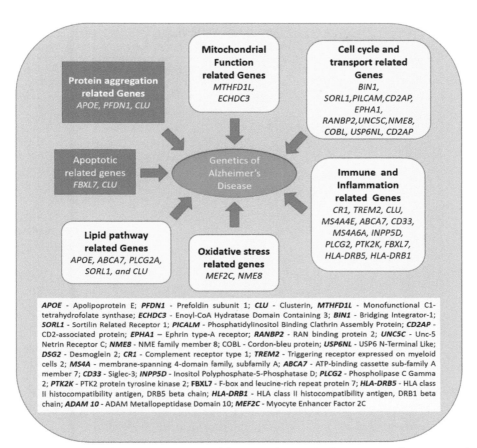

Figure 9.4 *Genetic risk factors for sporadic Alzheimer's disease (AD).* Summary of genes associated with sporadic AD reported by genome wide association studies ($P < 10^{-8}$) and/or replicated genetic studies, organized according to their functions in AD pathophysiology.

Table 9.1 Genes/proteins associated with Alzheimer's disease and their functions.

Gene	Location	Encoded protein	Major functions	References
CR1	1q32.2	Complement receptor 1	Aβ clearance and immune response	Zhu et al. (2017b)
TREM2	6p21.1	Triggering receptor expressed on myeloid cells 2	Immune response	Jonsson et al. (2013)
CLU	8p21.1	Clusterin	Apoptotic process, lipid pathways, inflammation, Aβ aggregation	Haight et al. (2018)

Continued

Table 9.1 Genes/proteins associated with Alzheimer's disease and their functions.—cont'd

Gene	Location	Encoded protein	Major functions	References
MS4A	11q12.2	Membrane spanning 4-domains A4E	Signal transduction, immune function	Ebbert et al. (2016)
BIN1	2q14.3	Bridging integrator 1	Vesicle mediated transport, APP trafficking transport	Chapuis et al. (2013)
FBXL7	5p15.1	F-box and leucine-rich repeat protein 7	Ubiquitination	Freudenberg-Hua et al. (2018)
PFDN1	5q31.3	Prefoldin protein 1	Stabilizes newly synthesized proteins	Freudenberg-Hua et al. (2018)
HLA-DRB	6p21.32	HLA class II beta chain paralogues	Immune response	Hamza et al. (2010)
ABCA7	19p13.3	ABC transporter member 7	Lipid transport and immune response	Steinberg et al. (2015)
MTFHD1L	6q25.1	Methylenetetrahydrofolate dehydrogenase (NADP + dependent) 1	Mitochondrial function	Freudenberg-Hua et al. (2018)
NME8	7p14.1	Thioredoxin domain-containing protein 3	Oxidation process, cell proliferation and differentiation	Liu et al. (2014)
COBL	7p12.1	Cordon-bleu WH2 repeat protein	Actin cytoskeleton and neuronal morphogenesis regulation	Freudenberg-Hua et al. (2018)
ZCWPW1	7q22.1	Zinc finger, CW type with PWWP domain 1	Epigenetic regulation	Allen et al. (2015)
USP6NL	10p14	USP6 N-terminal-like protein	Vesicle trafficking	Jun et al. (2017)
ECHDC3	10p14	Enoyl-CoA hydratase domain-containing protein 3	Mitochondrial function	Jun et al. (2017)
FRMD4	10p13	FERM domain-containing protein	Epithelial cell polarization, tauopathy	Ruiz et al. (2014)

Table 9.1 Genes/proteins associated with Alzheimer's disease and their functions.—cont'd

Gene	Location	Encoded protein	Major functions	References
CELF1	11p11-2	CUG-BP, Elav-like family	Regulation of RNA processing in the nucleus and cytoplasm	Rosenthal, Barmada, Wang, Demirci, & Kamboh, 2014
PICALM	11q14.2	Phosphatidylinositol binding clathrin assembly protein	Vesicle-mediated transport	Schjeide et al. (2011)
SORL 1	11q24.1	Sortilin-related receptor, L1	Vesicle trafficking and APP processing	Cuenco et al. (2008)
SLC24A4	14q32.12	Sodium/potassium/calcium exchanger 4	Ion transport	Larsson et al. (2011)
BZRAP1	17q22	Benzodiazepine receptor (peripheral) associated protein 1	Neurotransmitter interactions	Freudenberg-Hua et al. (2018)
ATP5H/ KCTD2	17q25.1	ATP synthase peripheral stalk subunit H	Mitochondrial function	Boada et al. (2014)
APOE	19q13.32	Apolipoprotein E	Lipid metabolism and protein clearance/aggregation	Roses (1996)
CD33	19q13.41	CD antigen	Vesicle-mediated transport and immune system	Bradshaw et al. (2013)
CASS4	20q13.31	Cas scaffolding protein family member 4	Cell migration and adhesion	Rosenthal et al. (2014)
CD2AP	6p12.3	CD2 associated protein	Actin cytoskeleton	Shulman et al. (2013)
DSG2	18q12.1	Desmoglein 2	Cell adhesion	Lambert et al. (2013)
INPP5D	2q37.1	Inositol polyphosphate-5-phosphatase D	Cell proliferation and survival	Lambert et al. (2013)
ADAM10	15q21.3	A disintegrin and metalloproteinase domain-containing protein 10	Protein aggregation and cell adhesion	Marioni et al. (2018)
EPHA 1	7q34	Ephrin receptor A1	Synaptic development and immune response	Lambert et al. (2013)

kinase activity (resulting in phosphorylation of tau) leading to increased neuroinflammatory response and memory deficits (Jiang et al., 2016) (Fig. 9.5).

When bound to Aβ, APOE docks to low-density lipoprotein receptor-related protein 1 (LRP1), leading to its endocytosis into the neuron in the presence of clathrin (Fig. 9.3C). Clathrin coating and protein shuffling are conducted by proteins including bridging integrator 1 (BIN1), phosphatidylinositol-binding clathrin assembly (PICALM), and CD2-associated protein (CD2), the genes for which have all been associated with late-onset AD (Table 9.1) (Freudenberg-Hua, Li, & Davies, 2018).

Seven isoforms of *BIN1* are known to be expressed in the brain, coding for membrane adaptors that form complexes with clathrin and AP2/α-adaptin, leading to synaptic vesicle endocytosis (Moustafa et al., 2018). *BIN1* expression has been reported to be elevated in postmortem AD brains and has been associated with increased tau aggregation/Aβ oligomer formation but not with Aβ expression levels in earlier stages of disease (Chapuis et al., 2013). Further investigations are necessary to clearly uncover the role of *BIN1* in AD pathophysiology.

Similarly, PICALM, which is mostly expressed in neurons, participates in clathrin-mediated endocytosis and intracellular trafficking. Single-nucleotide polymorphisms

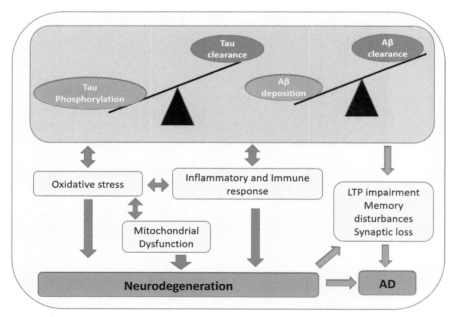

Figure 9.5 *Pathways involved in Alzheimer's disease (AD) pathophysiology.* An imbalance of tau and amyloid-β peptide (Aβ) accumulation induces activation of inflammatory and immune responses along with increased oxidative stress, leading to neurodegeneration. These imbalances also disturb molecular pathways underlying long-term potentiation (LTP), leading to synaptic loss and memory impairment.

(SNPs; rs3851179, rs541458) within *PICALM* have been reported as associated with a neuroprotective effect in AD (Harold et al., 2009), with SNP rs541458 being correlated with decreased levels of Aβ42 in the CSF of people with AD (Schjeide et al., 2011). PIC-ALM is known to play a key regulatory role in the internalization of APP (Fig. 9.3C) as well as in the clearance of Aβ and tau; it is currently being investigated as a potential avenue for novel AD therapies (Zhao et al., 2015).

CD2AP encodes for the cytoplasmic scaffolding protein (CD2) involved in actin cytoskeleton regulation, intracellular trafficking and synaptic endocytosis of receptors such as epidermal growth factor receptor, and apoptosis (Lynch et al., 2003). Association between *CD2AP* SNPs (rs9296559, rs9349407) and sporadic AD has been reported in several studies (Hollingworth et al., 2011; Naj et al., 2011). Prevalence of mutations in *CD2AP* has also been correlated with the presence of amyloid plaques in AD (Hollingworth et al., 2011).

Similar to *CD2AP, MS4A6A/MS4A4E* and *CD33* are genes highly expressed in myeloid cells in the brain that have been associated with AD through GWASs (Table 9.1) and have been correlated with AD symptoms including Braak tangles and amyloid plaques (Griciuc et al., 2013; Karch et al., 2012). Neuronal membrane-spanning 4-domain family, subfamily A (MS4A) protein (also expressed in inflammatory monocytes) has been shown to have a role in the modulation of intracellular calcium, contributing to AD pathophysiology (Yu et al., 2015). The CD33 (sialic acid binding Ig-like lectin 3) protein is expressed in microglia and has significant roles in neuroinflammatory pathways and receptor-mediated endocytosis independent of clathrin (Croker et al., 2008) (Fig. 9.5). Both neuroinflammation and immune system dysfunction have been shown to have significant involvement in the pathogenesis of AD. As illustrated in Fig. 9.4, many genes that encode for inflammatory and/or immune system proteins have been characterized as having an association with AD.

CLU encodes for clusterin (apolipoprotein J), which regulates lipid transport, apoptosis, and cellular interactions. Several SNPs present in *CLU* (rs11136000, rs9331888, rs2279590, rs7982, rs7012010) have been associated with a neuroprotective effect against sporadic AD (Harold et al., 2009; Hollingworth et al., 2011; Naj et al., 2011). Since CLU can bind with Aβ and subsequently pass though the blood—brain barrier, it can modulate the aggregation and toxicity of Aβ (Weinstein et al., 2016). In contrast to CLU, *FBXL7*, which encodes for F-box and leucine-rich repeat protein 7, constitutes one of the subunits of E3 ubiquitin protein ligase and displays proapoptotic activity. A SNP in *FBXL7* (rs75002042) was associated with an increased risk of developing AD (Tosto et al., 2015). CLU also can regulate the complement system, which is highly involved in inflammatory responses. *CR1* encodes for complement receptor type 1 protein and has been associated with AD pathophysiology through genetic association, GWASs, and meta-analysis studies (Moustafa et al., 2018). Similar to CLU, the ephrin receptor A1 protein coded by *EPHA1* is highly involved in immune response. Both rs11767557 and rs11771145 in *EPHA1* have been associated with reduced risk of late-onset AD (Hollingworth et al., 2011; Lambert et al., 2013; Naj et al., 2011).

ABCA7 encodes for ATP-binding cassette transporter A7 protein, which also has roles in regulating immune responses and lipid transport (Moustafa et al., 2018). The ABCA7 protein not only modulates the phagocytotic activity of macrophages/microglia but also reduces Aβ production and aggregation and regulates cholesterol transfer to APOE (Chan et al., 2008) (Fig, 9.3D and E). Several polymorphisms (rs3764650, rs115550680, rs4147929, rs3752246, rs142076058) have been associated with sporadic AD (Aikawa et al., 2018).

TREM2 encodes for the triggering receptor expressed on myeloid cells 2 protein, which is highly expressed on microglia and facilitates phagocytotic activity and downregulation of inflammatory responses (Jonsson et al., 2013). Mutations in *TREM2* including rs75932628 were reported to be associated with increased risk of developing late-onset AD (Jonsson et al., 2013).

Mitochondria provide the main source of energy for all cells including neurons; any dysfunction of these organelles disturbs aerobic respiratory processes and can lead to neuronal death, a characteristic symptom of AD (Hawking, 2016). Variants in the mitochondrial gene *MTHFD1L,* along with mutations in genes involved in neuronal oxidative stress *(MEF2C, NME8)*, have been associated with increased genetic susceptibility to late-onset AD (Ma et al., 2012; Moustafa et al., 2018).

Sortilin-related receptor L1 (SORL1), encoded by *SORL1,* is involved in vesicle trafficking and lipid pathways. SORL1 promotes endocytosis of APP, resulting in the generation of intraendosomal Aβ (Fig. 9.3A). Variants in this gene have been reported as associated with late-onset AD in different ethnicities and with AD symptomatology (including hippocampal atrophy and increased CSF Aβ levels) (Moustafa et al., 2018).

Decades of research and advances in molecular biology technologies and statistical analyses have allowed for the characterization of genes associated with AD (Table 9.1, Fig. 9.4). These genes can be roughly classified into six main pathways underlying AD pathophysiology and development: protein aggregation, mitochondrial dysfunction, apoptotic processes, immune/inflammatory responses, oxidative stress processes, and lipid metabolism and transport (Fig. 9.4). Despite the identification of these genetic variants in AD through linkage studies *(APOE)*, candidate gene studies/GWASs, and NGS *(TREM2, UNC5C,* a netrin receptor implicated in axonal transport, and *ADAM10* involved in increased nonamyloidogenic pathway and α-secretase activity), the heritability of sporadic AD remains to be uncovered. Several limitations have been identified in genetic linkage and candidate gene studies for sporadic AD, mainly due to small cohort sizes, false-positive outcomes, and vulnerability of locus heterogeneity. Although most genes identified in AD have failed to explain heritability for sporadic AD, they play a crucial role in gaining a better understanding of not only the genetic complexity underlying sporadic AD but also the possible molecular mechanisms responsible for the neurodegenerative symptoms of AD. By improving the body of knowledge about AD pathophysiology and exploring AD phenotypes such as CSF biomarkers, imaging data, and neuropsychological assessments in line with identified genetic variants, potential effective therapies will be unveiled.

Memory impairment is one of the main AD endophenotypes, along with disturbances in executive functioning, language, attention, and affect. In the next section, genes associated with cognitive dysfunction in the context of AD will be briefly discussed.

Genes related to cognition involved in Alzheimer's disease

Progressive deterioration of higher order functions such as memory, affective changes, and behavioral changes (e.g., repetitive behavior) are hallmarks indicating the development of AD (Goedert & Spillantini, 2006). Memory is mediated by a set of neural mechanisms allowing individuals to encode, consolidate, retain, and retrieve information. It is a critical function providing people with a sense of self that links past, present, and future. Different memory types have been defined according to their function and nature— short-term memory versus long-term memory or implicit versus declarative memory (Kandel, 2006). Declarative memory includes semantic memory (storage of context-independent information) and episodic memory (storage of specific information related to a context, generally time and location) (Kandel, 2006). People with AD first exhibit disruption in their semantic memory, usually illustrated by the difficulty of naming items and general verbal fluency (Jahn, 2013). In the context of AD, disturbances in semantic memory reflect the presence of neuropathological dysfunction (such as loss of synaptic connections and decreases in dendritic density) in both temporal and frontal lobes (Starr et al., 2005). Located in the temporal lobe, the hippocampal formation is critical for the encoding and consolidation of information from short-term to long-term memory (Kandel, 2006). Deterioration of declarative memory in people with AD reflects damage and atrophy of the hippocampus, which occurs in the early stages of AD development (Drebing et al., 1994), whereas damage in the frontal lobes reflects disturbances in working memory (executive functioning, attention) in people with AD (Kalpouzos et al., 2005).

Damage to memory-related brain regions has been associated with gene mutations and alterations in the expression of candidate genes for sporadic AD. *APOE4* has been correlated with both episodic memory decline and atrophy of the temporal lobes (Wolk et al., 2011), particularly in the hippocampal region (Kerchner et al., 2014). *APOE4* carriers also showed decreased gray matter volume with age and increased amyloid load associated with impaired glucose metabolism; furthermore, they showed cerebral amyloid angiopathy in their temporal lobes, reflecting the struggle for local brain tissue to maintain adequate functions for cell survival (Ramanan et al., 2014). Deposition of Aβ plaques in the AD brain was also correlated with a decline in both episodic memory and hippocampal volume (Mormino et al., 2009). Excessive accumulation of senile plaques and NFTs in the temporal lobe compared with other brain regions reflects the progression of cognitive decline in AD (Wolk et al., 2011). Interactions between *PICALM* (rs3851179) and *APOE4* were also found to be associated with cognitive impairment and temporal lobe atrophy in AD brains (Morgen et al., 2014).

Regarding immune-related genetic variants, a variant in *NME8* (rs2718058) was reported to be associated with cognitive dysfunction along with atrophy of the hippocampal region and increased CSF tau levels (Liu et al., 2014). *CD33* microglial gene expression was positively correlated with the presence of amyloid plaques and cognitive dysfunction (Griciuc et al., 2013). Interestingly, the *EPHA1* variant rs11771145 showed a protective effect against sporadic AD and was correlated with functional modifications of the hippocampal formation and temporal lobe (Wang et al., 2015).

Following whole-genome sequencing, a rare APP variant (rs63750847) was also characterized, demonstrating a neuroprotective role for AD in an elderly Icelandic population, illustrated by a decreased level of cognitive decline as well as the decreased production and aggregation of Aβ compared with that in people with AD (Jonsson et al., 2012).

Learning and memory processes are supported by molecular networks such as the glutamatergic system, including metabotropic (AMPA) and ionotropic (NMDA) receptors, which are responsible for long-term potentiation (LTP). These processes occur exclusively in the hippocampus and underlie synaptic plasticity (Kandel, 2006). The accumulation of both Aβ oligomers and phosphorylated-tau in the hippocampus has been associated with decreased LTP and dysfunctional neuronal communication (Palop & Mucke, 2010) (Fig. 9.5).

To date, many candidate gene association studies have been performed, and genetic variants have been implicated in many different genes related to neurobiological processes underlying learning and memory such as *BDNF, HTR2A, COMT, CLSTN2, KIBRA, CAMTA1, PSD,* and *CPEB3*. Genetic associations with AD have been previously reported for these genes that are mainly involved in neurotransmitter systems. However, for some, these associations did not retain their significance in GWAS analysis or showed controversial results in replicated studies (Rogaeva & Schmitt-Ulms, 2016). Future studies will be required for further investigating the role of cognitive-related genes in sporadic AD.

Conclusion

AD is a complex neurodegenerative disease with a strong genetic component. Here we have reviewed the main genes known to be involved in the genetic vulnerability of sporadic AD and in relation to AD pathophysiology: *APOE, BIN1, PICALM, CD2AP, MSA4A, CLU, CR1, CD 33, TREM2, FBXL7, EPHA1, MTHFD1L ABC7, MEF2C, NME8,* and *SORL1*. We have also discussed the role of some of these genes in cognitive decline as observed in people with AD. Gaining a better understanding of the genetic profiles of people with AD in relation to their cognitive impairment and neuronal dysfunction is critical to the future design and assessment of efficient therapies for late-onset AD.

Although genetic components continue to be important risk factors for sporadic AD, this pathology remains multifactorial with complex interactions between genetic, epigenetic, and environmental factors. Epigenetic mechanisms are reversible and can alter levels of gene expression through modulation of gene transcription, leading to changes in phenotypes such as altered behaviors observed in people with AD.

Better understanding genetic and epigenetic markers for AD will allow us to predict the efficiency of therapies based on individual genetic profiles and allow for better diagnostic screening for early-stage AD. Overall, with advances in genetic technology and a better understanding of AD genetic markers, the development of future therapies for AD seems promising.

Definitions

Candidate gene studies: examining the association of one or more genetic markers in a single gene in relation to a disease and/or trait

Genes: sequence of DNA encoding specific proteins

Genetic mutations: alterations in the sequence and/or type of nucleotide in the DNA

Genetic risk factors: DNA sequence responsible for the phenotypical expression of a disease or trait

Genome-wide association studies: a study examining the entire genome (every gene from an individual) of everyone within the study (usually case–control) to determine whether specific genetic regions are associated with a disease or trait

Linkage studies: investigating the position/location of genes associated with a disease or trait relative to specific chromosomal regions within families

Sequencing: genetic technology used to identify the order of specific nucleotides of tested DNA

Single-nucleotide polymorphism: change of only *one* nucleotide in the DNA sequence

Key facts of cognition

- Cognition is the ability to integrate information from all your senses, creating a mental picture of yourself and the world around you.
- Learning, memory, attention, and the formation of knowledge are all processes involved in cognition.

- Cognitive dysfunction is disruption to thoughts, attention, learning, and memory processes and is a symptom of many disorders affecting the brain including schizophrenia, attention deficit hyperactivity disorder, epilepsy, Parkinson's disease, and Alzheimer's disease.
- The hippocampus is a seahorse-shaped brain region responsible for converting short-term memory into long-term memory.
- M. H. famously had both hippocampi removed during drastic surgery to treat severe epilepsy; he lived for over 50 years with no hippocampi, assisting researchers studying cognitive disorders for the remainder of his life.

Summary points

- This chapter focuses on providing a brief overview of genes associated with sporadic (late-onset) Alzheimer's disease (AD).
- Despite decades of research, no current and reliable test is currently available for the diagnosis of AD.
- Focus on genetic biomarkers as diagnostic tools is promising.
- Genes associated with sporadic AD through linkage, association, and genome-wide association study case studies are briefly described.
- Genes underlying cognitive function are discussed in the context of AD.

References

Aikawa, T., Holm, M.-L., Kanekiyo, T. (2018). ABCA7 and pathogenic pathways of Alzheimer's disease. *Brain Sciences, 8*.

Allen, P. B., Chiu, D. T. (2008). Alzheimer's disease protein Abeta1-42 does not disrupt isolated synaptic vesicles. *Biochimica et Biophysica Acta, 1782*, 326–334.

Allen, M., Kachadoorian, M., Carrasquillo, M. M., Karhade, A., Manly, L., Burgess, J. D., et al. (2015). Late-onset Alzheimer disease risk variants mark brain regulatory loci. *Neurol. Genet., 1*, e15.

Bailey, P. (2007). Biological markers in Alzheimer's disease. *Can. J. Neurol. Sci. J. Can. Sci. Neurol., 34*(Suppl. 1), S72–S76.

Bertram, L. (2009). Alzheimer's disease genetics current status and future perspectives. *International Review of Neurobiology, 84*, 167–184.

Bertram, L., McQueen, M. B., Mullin, K., Blacker, D., Tanzi, R. E. (2007). Systematic meta-analyses of Alzheimer disease genetic association studies: The AlzGene database. *Nature Genetics, 39*, 17–23.

Boada, M., Antúnez, C., Ramírez-Lorca, R., DeStefano, A. L., González-Pérez, A., Gayán, J., et al. (2014). ATP5H/KCTD2 locus is associated with Alzheimer's disease risk. *Molecular Psychiatry, 19*, 682–687.

Bradshaw, E. M., Chibnik, L. B., Keenan, B. T., Ottoboni, L., Raj, T., Tang, A., et al. (2013). CD33 Alzheimer's disease locus: Altered monocyte function and amyloid biology. *Nature Neuroscience, 16*, 848–850.

Bu, G. (2009). Apolipoprotein E and its receptors in Alzheimer's disease: Pathways, pathogenesis and therapy. *Nature Reviews Neuroscience, 10*, 333–344.

Chan, S. L., Kim, W. S., Kwok, J. B., Hill, A. F., Cappai, R., Rye, K.-A., et al. (2008). ATP-binding cassette transporter A7 regulates processing of amyloid precursor protein in vitro. *Journal of Neurochemistry, 106*, 793–804.

Chapuis, J., Hansmannel, F., Gistelinck, M., Mounier, A., Van Cauwenberghe, C., Kolen, K. V., et al. (2013). Increased expression of BIN1 mediates Alzheimer genetic risk by modulating tau pathology. *Molecular Psychiatry, 18*, 1225—1234.

Croker, B. A., Lawson, B. R., Rutschmann, S., Berger, M., Eidenschenk, C., Blasius, A. L., et al. (2008). Inflammation and autoimmunity caused by a SHP1 mutation depend on IL-1, MyD88, and a microbial trigger. *Proceedings of the National Academy of Sciences of the United States of America, 105*, 15028—15033.

Cuenco, K., Lunetta, K. L., Baldwin, C. T., McKee, A. C., Guo, J., Cupples, L. A., et al. (2008). Association of distinct variants in SORL1 with cerebrovascular and neurodegenerative changes related to Alzheimer disease. *Archives of Neurology, 65*, 1640—1648.

Drebing, C. E., Moore, L. H., Cummings, J. L., Gorp, W. G. V., Hinkin, C., Perlman, S. L., et al. (1994). Patterns of neuropsychological performance among forms of subcortical dementia: A case study approach. *Cognitive and Behavioral Neurology, 7*, 57—66.

Dubois, B., Feldman, H. H., Jacova, C., Dekosky, S. T., Barberger-Gateau, P., Cummings, J., et al. (2007). Research criteria for the diagnosis of Alzheimer's disease: Revising the NINCDS-ADRDA criteria. *The Lancet Neurology, 6*, 734—746.

Ebbert, M. T. W., Boehme, K. L., Wadsworth, M. E., Staley, L. A., Alzheimer's Disease Neuroimaging Initiative, Alzheimer's Disease Genetics Consortium, Mukherjee, S., et al. (2016). Interaction between variants in CLU and MS4A4E modulates Alzheimer's disease risk. *Alzheimers Dement. J. Alzheimers Assoc., 12*, 121—129.

Fadil, H., Borazanci, A., Ait Ben Haddou, E., Yahyaoui, M., Korniychuk, E., Jaffe, S. L., et al. (2009). Early onset dementia. *International Review of Neurobiology, 84*, 245—262.

Freudenberg-Hua, Y., Li, W., Davies, P. (2018). The role of genetics in advancing precision medicine for Alzheimer's disease-A narrative review. *Frontiers of Medicine, 5*, 108.

Giri, M., Zhang, M., Lü, Y. (2016). Genes associated with Alzheimer's disease: An overview and current status. *Clinical Interventions in Aging, 11*, 665—681.

Goate, A., Chartier-Harlin, M. C., Mullan, M., Brown, J., Crawford, F., Fidani, L., et al. (1991). Segregation of a missense mutation in the amyloid precursor protein gene with familial Alzheimer's disease. *Nature, 349*, 704—706.

Goedert, M., Spillantini, M. G. (2006). A century of Alzheimer's disease. *Science, 314*, 777—781.

Griciuc, A., Serrano-Pozo, A., Parrado, A. R., Lesinski, A. N., Asselin, C. N., Mullin, K., et al. (2013). Alzheimer's disease risk gene CD33 inhibits microglial uptake of amyloid beta. *Neuron, 78*, 631—643.

Haight, T., Bryan, R. N., Meirelles, O., Tracy, R., Fornage, M., Richard, M., et al. (2018). Associations of plasma clusterin and Alzheimer's disease-related MRI markers in adults at mid-life: The CARDIA Brain MRI sub-study. *PloS One, 13*, e0190478.

Hampel, H., Mitchell, A., Blennow, K., Frank, R. A., Brettschneider, S., Weller, L., et al. (2004). Core biological marker candidates of Alzheimer's disease - perspectives for diagnosis, prediction of outcome and reflection of biological activity. *Journal of Neural Transmission Vienna Austria, 111*, 247—272, 1996.

Hamza, T. H., Zabetian, C. P., Tenesa, A., Laederach, A., Montimurro, J., Yearout, D., et al. (2010). Common genetic variation in the HLA region is associated with late-onset sporadic Parkinson's disease. *Nature Genetics, 42*, 781—785.

Han, M.-R., Schellenberg, G. D., Wang, L.-S., & Alzheimer's Disease Neuroimaging Initiative. (2010). Genome-wide association reveals genetic effects on human $A\beta42$ and τ protein levels in cerebrospinal fluids: A case control study. *BMC Neurology, 10*, 90.

Harold, D., Abraham, R., Hollingworth, P., Sims, R., Gerrish, A., Hamshere, M. L., et al. (2009). Genome-wide association study identifies variants at CLU and PICALM associated with Alzheimer's disease. *Nature Genetics, 41*, 1088—1093.

Hawking, Z. L. (2016). Alzheimer's disease: The role of mitochondrial dysfunction and potential new therapies. *Bioscience Horizons International Journal of Students Research, 9*.

Hollingworth, P., Harold, D., Sims, R., Gerrish, A., Lambert, J.-C., Carrasquillo, M. M., et al. (2011). Common variants at ABCA7, MS4A6A/MS4A4E, EPHA1, CD33 and CD2AP are associated with Alzheimer's disease. *Nature Genetics, 43*, 429—435.

Jahn, H. (2013). Memory loss in Alzheimer's disease. *Dialogues in Clinical Neuroscience, 15*, 445—454.

Jiang, T., Zhang, Y.-D., Chen, Q., Gao, Q., Zhu, X.-C., Zhou, J.-S., et al. (2016). TREM2 modifies microglial phenotype and provides neuroprotection in P301S tau transgenic mice. *Neuropharmacology, 105*, 196—206.

Jonsson, T., Atwal, J. K., Steinberg, S., Snaedal, J., Jonsson, P. V., Bjornsson, S., et al. (2012). A mutation in APP protects against Alzheimer's disease and age-related cognitive decline. *Nature, 488*, 96—99.

Jonsson, T., Stefansson, H., Steinberg, S., Jonsdottir, I., Jonsson, P. V., Snaedal, J., et al. (2013). Variant of TREM2 associated with the risk of Alzheimer's disease. *New England Journal of Medicine, 368*, 107—116.

Jun, G. R., Chung, J., Mez, J., Barber, R., Beecham, G. W., Bennett, D. A., et al. (2017). Transethnic genome-wide scan identifies novel Alzheimer's disease loci. *Alzheimers Dementia Journal of the Alzheimer's Association, 13*, 727—738.

Kalpouzos, G., Eustache, F., de la Sayette, V., Viader, F., Chételat, G., Desgranges, B. (2005). Working memory and FDG-PET dissociate early and late onset Alzheimer disease patients. *Journal of Neurology, 252*, 548—558.

Kandel, E. R. (2006). *Search of memory : the emergence of a new science of mind.* New York : W. W. Norton & Company.

Karch, C. M., Jeng, A. T., Nowotny, P., Cady, J., Cruchaga, C., Goate, A. M. (2012). Expression of novel Alzheimer's disease risk genes in control and Alzheimer's disease brains. *PloS One, 7*, e50976.

Kerchner, G. A., Berdnik, D., Shen, J. C., Bernstein, J. D., Fenesy, M. C., Deutsch, G. K., et al. (2014). APOE ε4 worsens hippocampal CA1 apical neuropil atrophy and episodic memory. *Neurology, 82*, 691—697.

Kok, E., Haikonen, S., Luoto, T., Huhtala, H., Goebeler, S., Haapasalo, H., et al. (2009). Apolipoprotein E-dependent accumulation of Alzheimer disease-related lesions begins in middle age. *Annals of Neurology, 65*, 650—657.

Lambert, J. C., Ibrahim-Verbaas, C. A., Harold, D., Naj, A. C., Sims, R., Bellenguez, C., et al. (2013). Meta-analysis of 74,046 individuals identifies 11 new susceptibility loci for Alzheimer's disease. *Nature Genetics, 45*, 1452—1458.

Larsson, M., Duffy, D. L., Zhu, G., Liu, J. Z., Macgregor, S., McRae, A. F., et al. (2011). GWAS findings for human iris patterns: Associations with variants in genes that influence normal neuronal pattern development. *The American Journal of Human Genetics, 89*, 334—343.

Levy-Lahad, E., Wasco, W., Poorkaj, P., Romano, D. M., Oshima, J., Pettingell, W. H., et al. (1995). Candidate gene for the chromosome 1 familial Alzheimer's disease locus. *Science, 269*, 973—977.

Liu, Y., Yu, J.-T., Wang, H.-F., Hao, X.-K., Yang, Y.-F., Jiang, T., et al. (2014). Association between NME8 locus polymorphism and cognitive decline, cerebrospinal fluid and neuroimaging biomarkers in Alzheimer's disease. *PloS One, 9*, e114777.

Lynch, D. K., Winata, S. C., Lyons, R. J., Hughes, W. E., Lehrbach, G. M., Wasinger, V., et al. (2003). A Cortactin-CD2-associated protein (CD2AP) complex provides a novel link between epidermal growth factor receptor endocytosis and the actin cytoskeleton. *Journal of Biological Chemistry, 278*, 21805—21813.

Marioni, R. E., Harris, S. E., Zhang, Q., McRae, A. F., Hagenaars, S. P., Hill, W. D., et al. (2018). GWAS on family history of Alzheimer's disease. *Translational Psychiatry, 8*, 99.

Ma, X.-Y., Yu, J.-T., Wu, Z.-C., Zhang, Q., Liu, Q.-Y., Wang, H.-F., et al. (2012). Replication of the MTHFD1L gene association with late-onset Alzheimer's disease in a Northern Han Chinese population. *Journal of Alzheimer's Disease, 29*, 521—525.

Morgen, K., Ramirez, A., Frölich, L., Tost, H., Plichta, M. M., Kölsch, H., et al. (2014). Genetic interaction of PICALM and APOE is associated with brain atrophy and cognitive impairment in Alzheimer's disease. *Alzheimers Dementia Journal of Alzheimers Association, 10*, S269—S276.

Mormino, E. C., Kluth, J. T., Madison, C. M., Rabinovici, G. D., Baker, S. L., Miller, B. L., et al. (2009). Episodic memory loss is related to hippocampal-mediated beta-amyloid deposition in elderly subjects. *Brain Journal of Neurology, 132*, 1310—1323.

Moustafa, A. A., Hassan, M., Hewedi, D. H., Hewedi, I., Garami, J. K., Al Ashwal, H., et al. (2018). Genetic underpinnings in Alzheimer's disease - a review. *Reviews in the Neurosciences, 29*, 21—38.

Naj, A. C., Jun, G., Beecham, G. W., Wang, L.-S., Vardarajan, B. N., Buros, J., et al. (2011). Common variants at MS4A4/MS4A6E, CD2AP, CD33 and EPHA1 are associated with late-onset Alzheimer's disease. *Nature Genetics, 43*, 436−441.

Palop, J. J., Mucke, L. (2010). Amyloid-β−induced neuronal dysfunction in Alzheimer's disease: From synapses toward neural networks. *Nature Neuroscience, 13*, 812−818.

Prince, M., Wimo, A., Guerchet, M., Ali, G.-C., Wu, Y.-T., Prina, M., et al. (2015). *World Alzheimer Report 2015, the global impact of dementia: An analysis of prevalence, incidence, cost and trends*.

Ramanan, V. K., Risacher, S. L., Nho, K., Kim, S., Swaminathan, S., Shen, L., et al. (2014). APOE and BCHE as modulators of cerebral amyloid deposition: A florbetapir PET genome-wide association study. *Molecular Psychiatry, 19*, 351−357.

Rogaeva, E., Schmitt-Ulms, G. (2016). Does BDNF Val66Met contribute to preclinical Alzheimer's disease? *Brain Journal of Neurology, 139*, 2586−2589.

Rogaev, E. I., Sherrington, R., Rogaeva, E. A., Levesque, G., Ikeda, M., Liang, Y., et al. (1995). Familial Alzheimer's disease in kindreds with missense mutations in a gene on chromosome 1 related to the Alzheimer's disease type 3 gene. *Nature, 376*, 775−778.

Rosenthal, S. L., Barmada, M. M., Wang, X., Demirci, F. Y., Kamboh, M. I. (2014). Connecting the dots: Potential of data integration to identify regulatory SNPs in late-onset Alzheimer's disease GWAS findings. *PloS One, 9*, e95152.

Roses, A. D. (1996). Apolipoprotein E alleles as risk factors in Alzheimer's disease. *Annual Review of Medicine, 47*, 387−400.

Ruiz, A., Heilmann, S., Becker, T., Hernández, I., Wagner, H., Thelen, M., et al. (2014). Follow-up of loci from the International Genomics of Alzheimer's Disease Project identifies TRIP4 as a novel susceptibility gene. Transl. *Psychiatry, 4*, e358.

Schjeide, B.-M. M., Schnack, C., Lambert, J.-C., Lill, C. M., Kirchheiner, J., Tumani, H., et al. (2011). The role of clusterin, complement receptor 1, and phosphatidylinositol binding clathrin assembly protein in Alzheimer disease risk and cerebrospinal fluid biomarker levels. *Archives of General Psychiatry, 68*, 207−213.

Sherrington, R., Rogaev, E. I., Liang, Y., Rogaeva, E. A., Levesque, G., Ikeda, M., et al. (1995). Cloning of a gene bearing missense mutations in early-onset familial Alzheimer's disease. *Nature, 375*, 754−760.

Shulman, J. M., Chen, K., Keenan, B. T., Chibnik, L. B., Fleisher, A., Thiyyagura, P., et al. (2013). Genetic susceptibility for Alzheimer disease neuritic plaque pathology. *JAMA Neurology, 70*, 1150−1157.

Starr, J. M., Loeffler, B., Abousleiman, Y., Simonotto, E., Marshall, I., Goddard, N., et al. (2005). Episodic and semantic memory tasks activate different brain regions in Alzheimer disease. *Neurology, 65*, 266−269.

Steinberg, S., Stefansson, H., Jonsson, T., Johannsdottir, H., Ingason, A., Helgason, H., et al. (2015). Loss-of-function variants in ABCA7 confer risk of Alzheimer's disease. *Nature Genetics, 47*, 445−447.

Sunderland, T., Hampel, H., Takeda, M., Putnam, K. T., Cohen, R. M. (2006). Biomarkers in the diagnosis of Alzheimer's disease: Are we ready? *Journal of Geriatric Psychiatry and Neurology, 19*, 172−179.

Tanzi, R. E., Bertram, L. (2005). Twenty years of the Alzheimer's disease amyloid hypothesis: A genetic perspective. *Cell, 120*, 545−555.

Tosto, G., Fu, H., Vardarajan, B. N., Lee, J. H., Cheng, R., Reyes-Dumeyer, D., et al. (2015). F-box/LRR-repeat protein 7 is genetically associated with Alzheimer's disease. *Annals of Clinical and Translational Neurology, 2*, 810−820.

Verghese, P. B., Castellano, J. M., Holtzman, D. M. (2011). Apolipoprotein E in Alzheimer's disease and other neurological disorders. *The Lancet Neurology, 10*, 241−252.

Wang, H.-F., Tan, L., Hao, X.-K., Jiang, T., Tan, M.-S., Liu, Y., et al., Alzheimer's Disease Neuroimaging Initiative. (2015). Effect of EPHA1 genetic variation on cerebrospinal fluid and neuroimaging biomarkers in healthy, mild cognitive impairment and Alzheimer's disease cohorts. *Journal of Alzheimer's Disease, 44*, 115−123.

Weinstein, G., Beiser, A. S., Preis, S. R., Courchesne, P., Chouraki, V., Levy, D., et al. (2016). Plasma clusterin levels and risk of dementia, Alzheimer's disease, and stroke. *Alzheimer's & dementia (Amsterdam, Netherlands), 3*, 103−109.

Wolk, D. A., Dunfee, K. L., Dickerson, B. C., Aizenstein, H. J., DeKosky, S. T. (2011). A medial temporal lobe division of labor: Insights from memory in aging and early Alzheimer disease. *Hippocampus, 21,* 461−466.

Yu, L., Chibnik, L. B., Srivastava, G. P., Pochet, N., Yang, J., Xu, J., et al. (2015). Association of Brain DNA methylation in SORL1, ABCA7, HLA-DRB5, SLC24A4, and BIN1 with pathological diagnosis of Alzheimer disease. *JAMA Neurology, 72,* 15−24.

Zhao, Z., Sagare, A. P., Ma, Q., Halliday, M. R., Kong, P., Kisler, K., et al. (2015). Central role for PICALM in amyloid-β blood-brain barrier transcytosis and clearance. *Nature Neuroscience, 18,* 978−987.

Zhu, J.-B., Tan, C.-C., Tan, L., Yu, J.-T. (2017a). State of play in Alzheimer's disease genetics. *Journal of Alzheimer's Disease, 58,* 631−659.

Zhu, X.-C., Wang, H.-F., Jiang, T., Lu, H., Tan, M.-S., Tan, C.-C., et al. (2017b). Effect of CR1 genetic variants on cerebrospinal fluid and neuroimaging biomarkers in healthy, mild cognitive impairment and Alzheimer's disease cohorts. *Molecular Neurobiology, 54,* 551−562.

CHAPTER 10

Clinical and pathological phenotypes in dementia: a focus on autosomal dominant frontotemporal dementia

Innocenzo Rainero, Alessandro Vacca, Elisa Rubino
Aging Brain and Memory Clinic, Department of Neuroscience "Rita Levi Montalcini", University of Torino, Torino, Italy

List of abbreviations

AD Alzheimer's disease
aFTD autosomal dominant frontotemporal dementia
ALS amyotrophic lateral sclerosis
bvFTD behavioral variant frontotemporal dementia
C9orf72 chromosome 9 open reading frame 72 gene
CBD corticobasal syndrome
CHCHD10 coiled-coil—helix—coiled-coil—helix domain-containing 10 gene
CHMP2B chromatin-modifying protein 2B gene
EWS Ewing sarcoma protein
FTD frontotemporal dementia
FTLD frontotemporal lobar degeneration
FUS fused in sarcoma gene
GRN granulin precursor gene
LBD Lewy body dementia
MAPT microtubule-associated protein tau gene
MND motor neuron disease
MT microtubule
NFT neurofibrillary tangle
NII neuronal intranuclear inclusion
OPTN optineurin gene
PNFA progressive nonfluent aphasia
PPA primary progressive aphasia
PSP progressive supranuclear palsy
SD semantic dementia
SQSTM1 sequestosome 1 gene
TAF15 TATA-binding associated factor 15 protein
TARDBP transactive response DNA-binding protein gene
TBK1 TANK-1-binding kinase 1 gene
TDP-43 TAR DNA-binding protein 43
TREM2 triggering receptor expressed on myeloid cells 2 gene
UBQLN2 ubiquilin 2 gene
VCP valosin-containing protein gene

Diagnosis and Management in Dementia
ISBN 978-0-12-815854-8, https://doi.org/10.1016/B978-0-12-815854-8.00010-0

Mini-dictionary of terms

Chromosome 9 open reading frame 72 the protein encoded by this gene plays an important role in the regulation of endosomal trafficking, interacting with Rab proteins, which are involved in autophagy and endocytic transport. Expansion of a GGGGCC repeat in an intronic region of this gene is associated with ALS and FTD.

Frontotemporal dementia FTD is a heterogeneous neurodegenerative disease characterized by progressive impairment in social, behavioral, language, and motor functions.

Frontotemporal lobar degeneration FTLD refers to the pathological neurodegenerative process occurring in the frontal and temporal lobes of the brain.

Granulin precursor the *GRN* gene encodes a protein called granulin (also known as progranulin) that is the precursor of several active peptides, called progranulins. These proteins regulate the growth, division, and survival of neurons. They also play important roles in early embryonic development, regulation of the body's immune system response, and wound healing.

Microtubule-associated protein tau the *MAPT* gene encodes a protein called tau, a protein involved in assembling and stabilizing MTs. MTs are essential for maintaining cell shape, assist in the process of cell division, and are essential for the transport of materials within cells.

Motor neuron disease the term MND relates to a group of diseases of the nervous system that are characterized by steadily progressive deterioration of the motor neurons in the brain, brain stem, and spinal cord.

Primary progressive aphasia primary progressive aphasia (PPA) is a neurodegenerative syndrome caused by progressive impairment of language function.

Progressive nonfluent aphasia PNFA is a form of PPA mainly characterized by agrammatism, laborious speech, alexia, and agraphia, frequently accompanied by apraxia of speech. Language comprehension is relatively preserved.

Semantic dementia this is a subtype of PPA mainly characterized by progressive impairment of confrontation naming and word comprehension.

Introduction

The term "frontotemporal dementia" (FTD) refers to a spectrum of neurodegenerative disorders clinically characterized by insidious and progressive deterioration in behavioral, executive, language, and motor functions (Bang, Spina, & Miller, 2015). The neuropathological correlate of FTD is a selective and progressive atrophy of the frontal and temporal lobes, sometimes asymmetric, defined as frontotemporal lobar degeneration (FTLD) (Lashley et al., 2015).

FTD is an important cause of dementia worldwide, being the third common cause after Alzheimer's disease (AD) and Lewy body dementia. However, FTD is often underdiagnosed, being misdiagnosed with other types of dementia, and, in individuals under 65 years of age, the disorder is at least as prevalent as AD. FTLD accounts for approximately 10% of all pathologically diagnosed dementia cases.

Table 10.1 shows the more common types of dementia and their prevalence.

The disease imposes a great burden on the patient, the family, and society, being associated with substantial direct and indirect costs, diminished quality of life, and increased caregiver burden. In the United States, the economic burden for FTD is approximately

Table 10.1 Common types of dementia and their prevalence.

Type of dementia	Prevalence
Alzheimer's disease	55%—80%
Vascular	20%—30%
Lewy body disease	10%—25%
Frontotemporal	10%—15%

The more common types of dementia and their prevalence are shown.

twice that reported for AD, with an estimated annual direct cost of approximately US$70,000 per patient (Galvin et al., 2015). Progressive changes in the patient's behavior significantly modify the interpersonal relations and induce depression, anxiety, and stress in both primary caregivers and family members. Despite the burden associated with the disease, at present there is no approved therapy for FTD.

Genetics of frontotemporal dementia

Several studies clearly showed that FTD has a large genetic component. More than 40% of FTD patients report a family history of dementia, and approximately one-third of familial cases show an autosomal dominant pattern of inheritance of the phenotype (autosomal dominant FTD, aFTD). Since 2000, several causative and susceptibility genes for FTD have been discovered. Genetic variants of three major genes, microtubule-associated protein tau (*MAPT*), granulin precursor (*GRN*), and chromosome 9 open reading frame 72 (*C9orf72*), account for about half of the aFTD cases. In addition, rare defects in the chromatin-modifying protein 2B (*CHMP2B*), valosin-containing protein (*VCP*), fused in sarcoma (*FUS*), transactive DNA-binding protein (*TARDBP*), sequestosome 1 (*SQSTM1*), optineurin (*OPTN*), ubiquilin 2 (*UBQLN2*), TANK-1 binding kinase 1 (*TBK1*), triggering receptor expressed on myeloid cells 2 (*TREM2*), and coiled-coil—helix—coiled-coil—helix domain-containing 10 (*CHCHD10*) genes have been described in families segregating the FTD phenotype. Fig. 10.1 shows, on the left, the percentage of familial, autosomal dominant, and sporadic FTD cases and, on the right, the approximate percentage of known genes in the genetic forms of FTD.

In 2014, candidate gene association studies as well as genome–wide association studies reported that genetic factors related to immune functions and autophagic pathways may be involved in the pathogenesis of both sporadic and familial FTD (Ferrari et al., 2014). Further genes are expected to be found in the future.

Clinical symptoms of frontotemporal dementia

Clinically, FTD is a highly heterogeneous disease, with signs and symptoms varying greatly in different patients, even within the same family. Generally, patients are

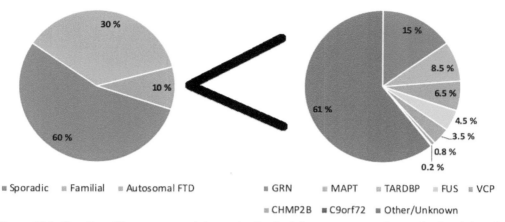

■ Sporadic   ■ Familial   ■ Autosomal FTD   ■ GRN   ■ MAPT   ■ TARDBP   ■ FUS   ■ VCP

■ CHMP2B   ■ C9orf72   ■ Other/Unknown

Figure 10.1 *Genetics of frontotemporal dementia.* On the left, percentages of sporadic, familial, and autosomal dominant frontotemporal dementia; on the right, percentages of unknown and known genetic variants in autosomal dominant frontotemporal dementia.

categorized according to the symptoms at disease onset. However, with the progression of the disease, an overlap between distinct clinical syndromes is frequent. At present, three main clinical syndromes are recognized:

Behavioral variant frontotemporal dementia

This syndrome is characterized by progressive impairment in social and executive skills, with altered emotional responsivity and emergence of a variety of abnormal behaviors, including apathy, disinhibition, obsessions, rituals, and stereotypes. Behavioral variant FTD (bvFTD) develops subtly, and early detection may depend on impairment in social awareness, sexual disturbance, and altered dietary habits. The bvFTD syndrome is currently classified according to the criteria of Rascovsky et al. (2011). These criteria require that, for the diagnosis of *possible* bvFTD, three of the following symptoms must be persistent within the first 3 years of disease: behavioral disinhibition; apathy or inertia; loss of sympathy or empathy; perseverative, stereotyped, or compulsive/ritualistic behavior; hyperorality; and dietary changes. Neuropsychological findings include executive deficits with relative sparing of memory and visuospatial functions. For the diagnosis of *probable* bvFTD, a significant functional decline and neuroimaging findings consistent with bvFTD (i.e., frontal and/or anterior atrophy or hypometabolism) must also be present. Finally, for bvFTD with definite FTLD pathology, neuropathological evidence of FTLD (on biopsy or at postmortem) or the presence of a known pathogenic mutation is needed.

Semantic dementia

This is a highly characteristic syndrome led by progressive impairment of semantic memory. Patients commonly present with fluent speech associated with difficulty in retrieving names and impaired comprehension of word meanings. Early in the course of the disease, the semantic deficit may be well compensated and may emerge only on neuropsychological testing. A more pervasive semantic impairment, also affecting visual information and other nonverbal domains, generally develops later in the disease course, as do behavioral disturbances broadly similar to those in bvFTD.

Progressive nonfluent aphasia

This syndrome is characterized by a progressive impairment in language output with nonfluent speech. In some patients, phonemic or articulatory errors are the dominant feature, whereas in others the syndrome is dominated by expressive agrammatism. These features commonly coexist as the disease evolves. Apraxia of other orofacial movements or swallowing often accompanies speech apraxia. In 2011, Gorno-Tempini et al. published detailed criteria for the classification of patients with the different variants of degenerative language impairment (the so-called primary progressive aphasia).

The main FTD phenotypes significantly overlap with the syndromes of progressive supranuclear palsy (PSP), corticobasal syndrome (CBD), and motor neuron disease (FTD—MND). PSP, also called Steele—Richardson—Olszewski syndrome, is characterized by early postural instability with falls, impairment of vertical gaze, and frontal behavioral changes with marked cognitive impairment. Features of CBD include asymmetric apraxia accompanied by rigidity and myoclonus and classically nonvolitional or "alien" actions of the affected limb. The phenotype may include prominent behavioral or language deficits. In the syndrome of FTD—MND, behavioral or language dysfunction may evolve in parallel with MND. Therefore, all patients with FTD presentations should be evaluated for the appearance of motor neuron signs. At presentation or during the disease course, FTD patients may develop various hyperkinetic or hypokinetic movement disorders, such as symmetrical parkinsonism, chorea, orofacial dyskinesias, myoclonus, and dystonia (Baizabal-Carvallo & Jankovic, 2016). Finally it is of interest to note that a subgroup of FTD patients (phenocopy FTD) do not show progressive functional decline and remain with normal neuroimaging over time.

Neuropathology of frontotemporal dementia

The different FTD syndromes arise from FTLD, a pathological term denoting the degeneration of cortical and subcortical structures within frontal and temporal regions of the brain. Affected structures include the frontoinsular cortices, anterior temporal poles, basal ganglia, substantia nigra, brain stem, and thalamus. In some forms of genetic FTD the cerebellum may also be involved. Macroscopically, brain atrophy predominates in the frontal and temporal lobes, with sparing of the posterior brain regions. At the microscopic examination of the brain, neuronal loss, gliosis, vacuolization of the superficial cortex, and ballooned neurons may be observed.

FTLD shows several abnormal protein inclusions in neurons and glial cells with diverse molecular pathologies. Based on the nature of these inclusions, *three subgroups of FTLD* have been recognized: (1) *FTLD—TAU*, with tau-positive inclusions; (2) *FTLD—TDP*, with tau-negative inclusions containing TAR DNA-binding protein 43 (TDP-43) conjugated with ubiquitin; and (3) *FTLD—FET*, containing FUS protein, Ewing sarcoma (EWS), or TATA-binding associated factor 15 (TAF15). Therefore, the pathological diagnosis requires extensive tissue sampling and immunohistochemistry. FTLD—TDP is considered to be the most common underlying pathology for FTD, FTLD—TAU is considered slightly less common, while the FTLD—FET pathology is rare (around 5% of cases).

Based on the types of inclusions and their distribution throughout the brain, FTLD—TDP can be further subdivided into four types: type A, with many neuronal cytoplasmic inclusions (NCIs) and many dystrophic neurites; type B, with moderate NCIs and few dystrophic neurites; type C, with few NCIs and many long dystrophic neurites; and type D, with few NCIs, many short dystrophic neurites, and many lentiform neuronal intranuclear inclusions. Fig. 10.2 shows the actual subclassification of FTLD subtypes.

In bvFTD, early selective neuronal degeneration of Von Economo neurons and fork cells has been described (Seeley et al., 2006). These neurons are found in the anterior insula, which together with the anterior cingulate cortex, is an early region of degeneration in bvFTD. This selective neuronal loss has been observed in FTLD—TAU, FTLD—TDP, and FTLD—FET, suggesting it occurs irrespective of the abnormal protein.

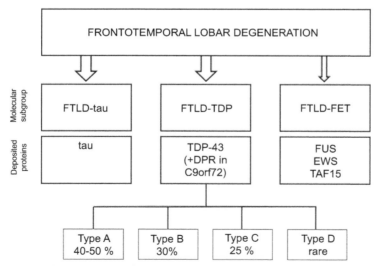

Figure 10.2 *Neuropathological classification of frontotemporal lobar degeneration subtypes.* The actual classification of different frontotemporal lobar degeneration subtypes according to neuropathological characteristics and misfolded proteins is shown.

Genotype—phenotype correlations in autosomal dominant frontotemporal dementia

The correlation between genotype and phenotype is defined as the probability of a distinct mutation being associated with specific clinical characteristics. Genotype and phenotype share a statistical relationship: the more frequently a specific phenotype is observed in association with a certain genotype, the higher is the possibility that an unrelated subject with the same genotype will present the traits or clinical characteristics observed in the population carrying the same genetic variant.

Monogenic diseases are characterized by complex genotype—phenotype correlations. A specific mutation may be associated with a wide spectrum of symptoms whose expression is probably modulated by the effects of additional modifying genes, epigenetic factors, and environmental factors. In addition, according to the type of gene mutation, there may be either loss or gain of function, and both may be partial or total. Finally, several genes have multiple isoforms, and a specific mutation may differentially affect the isoforms.

Analysis of genotype—phenotype correlations in FTD is complicated by a large number of factors, such as the lack of specific biomarkers for the disease, the heterogeneity of the symptoms, and the variability of the clinical characteristics according to the patient's age. Several studies have investigated genotype—phenotype correlation in patients with FTD, with partially conflicting results. The disease has a high rate of misdiagnosis, which makes difficult the genotype—phenotype correlation. In addition, the clinical phenotype of FTD may differ even within the same family, adding potential difficulty in accurate diagnosis.

The purpose of this review is to discuss available data regarding the effects of genetic variants on the clinical and pathological characteristics of the disease in autosomal dominant FTD.

Microtubule-associated protein tau

In 1998, the first causal gene of aFTD was identified. Investigating pedigrees with FTD and parkinsonism (so-called FTDP-17), Hutton et al. found the cosegregation between mutations in the *MAPT* gene and the clinical phenotype. The gene is located on chromosome 17q21.31, and consists of a noncoding exon followed by 14 coding exons. *MAPT* encodes the MAPT protein, a phosphoprotein with several phosphorylation sites. Alternative splicing processes of exons 2, 3, and 10 of the *MAPT* gene generate six main protein isoforms: three isoforms containing three amino acid repeats (3R) and three isoforms with four repeats (4R).

The main function of tau protein is to stabilize and organize the axonal microtubule network. Binding to tubulin, tau promotes the assembly of microtubules (MTs) and also plays a role in modulating vesicle transport mediated by motor proteins along the MTs.

The deposition of abnormally phosphorylated tau proteins in neurons and glial cells is a common feature of several neurodegenerative diseases, collectively called tauopathies, including AD, FTD, PSP, corticobasal syndrome (CBS), and Niemann—Pick disease.

More than 80 different *MAPT* gene mutations and polymorphisms have been reported in patients with FTD, including 44 that are clearly pathogenetic (http://www.molgen.ua.ac.be) in 134 families. Most cases are caused by heterozygous mutation, although rare homozygous mutations have been found. Mutations are clustered from exon 9 to 13, encoding the four MT-binding domains. A substantial number of mutations are located in intron 10, leading to anomalous ratios of 3R to 4R tau. The three most common mutations, the P301L mutation, the mutation involving the splice site in exon 10 (IVS10+16C > T), and the N279K mutation, account for about 50% of the known mutations.

Age at onset of the disease in patients with the *MAPT* gene shows a significant variability, ranging from 40 to 70 years. Clinically, patients with *MAPT* gene mutations show generally the bvFTD phenotype, including disinhibition and obsessive—compulsive behavior. Extrapyramidal signs and symptoms are frequent, resembling the other sporadic tauopathies such as PSP and CBS. In patients carrying *MAPT* mutations, a semantic dementia (SD)-like presentation, associated with features of bvFTD, can also be observed. Although the clinical phenotype is highly variable, early age at onset, parkinsonism, and oculomotor deficits should lead the clinician to suspect *MAPT* mutation in the patient. However, the lack of a clear correlation between *MAPT* gene mutations and clinical features indicates that additional genetic and/or environmental factors can produce significant phenotypic variations.

The neuropathological phenotype varies substantially in morphological characteristics, severity, and distribution, depending on the type of *MAPT* gene mutation. Tau deposits may be abundant in the cerebral cortex, white matter, and some subcortical and brain-stem nuclei. Neurofibrillary tangles (NFTs) or Pick bodies in neurons, as well as astrocytic plaques, tufted astrocytes, and coiled bodies in oligodendroglial cells in the glial pathology may be found. The location of the *MAPT* mutation influences the type of cellular involvement (i.e., neurons, astrocytes, or oligodendroglia) and the type of neuronal inclusion. Mutations in exons 1 and 10 (as well as in the intron following exon 10) are associated with neuronal and glial tau deposition, while mutations in exons 9, 11, 12, and 13 lead to deposits of tau filaments mainly in neurons. Mutations in exons 12 and 13 lead to the formation of NFTs with paired helical filaments.

Granulin precursor gene

In 2006, Baker et al. and Cruts et al. described the presence of *GRN* gene mutations in patients with familial and sporadic FTD (Baker et al., 2006; Cruts et al., 2006). The gene is located on chromosome 17p21.31, consists of 13 exons, and encodes progranulin, a

593-amino-acid-long precursor protein. Cleavage of progranulin, both in the lysosomes and in the extracellular matrix, produces a variety of active granulin peptides.

Progranulin and related peptides are implicated in a wide range of biological processes. Within the central nervous system, progranulin is predominantly expressed at the level of activated microglial cells, pointing to its regulatory role in the inflammatory response in the brain. In addition, a neuroprotective role of GRN has been suggested, although the mechanism of this function is still under investigation.

One hundred seventy-two different *GRN* gene mutations have been described (http://www.molgen.ua.ac.be) worldwide. Among these, 79 are clearly pathogenetic and have been described in 259 unrelated families. In contrast to *MAPT*, these mutations are scattered all over the gene. Generally, *GRN* mutations are null mutations that lead to nonsense-mediated decay of mutant *GRN* mRNA and reduced expression of progranulin. Consequently, mutation carriers can be identified by measuring plasma, serum, and cerebrospinal fluid progranulin concentrations. The most frequently observed *GRN* mutations are the R493X nonsense mutation in exon 11, and the c.-8+5G>C (IVS0+5G>C) substitution in a splicing site in intron 0. Mutation penetrance is very high, with 90% of carriers manifesting symptoms by the age of 75 years.

Clinical phenotypes associated with *GRN* mutations are remarkably variable, both between families with different mutations and among members of the same family. The mean age at onset is around 60 years and the mean duration of disease is about 8 years (range 3—22 years). *GRN* mutations are associated mainly with the bvFTD phenotype, with social withdrawal and apathy being the most common behavioral changes. Intriguingly, a large range of psychiatric symptoms has been included in this phenotype. Hallucinations and delusions are frequently concomitant. At presentation, about one-third of patients show an early isolated language dysfunction, suggestive of a nonfluent type. Episodic memory deficits may also occur, leading to a clinical diagnosis of mild cognitive impairment or, if they occur together with parietal deficits, such as dyscalculia, visuospatial dysfunction, and limb apraxia, to an erroneous diagnosis of AD. Extrapyramidal signs are frequently observed and include parkinsonism, dystonia, and limb apraxia. MND is only a very rare part of the clinical spectrum within GRN families.

The most consistent pathological feature in FTLD patients with *GRN* gene mutations is the presence of ubiquitin lentiform neuronal intranuclear inclusions in the neocortex and striatum. In addition, the neocortex shows moderate-to-severe superficial laminar spongiosis, chronic degenerative changes, and NCIs containing, as a major component, the TDP-43 protein.

Chromosome 9 open reading frame 72 gene

Investigating families segregating a combined FTD and amyotrophic lateral sclerosis phenotype (FTD—ALS), in 2011 two independent research groups discovered that an expansion within the *C9orf72* gene was the cause of the diseases (DeJesus–Hernandez et al., 2011; Renton et al., 2011). The gene is located on 9p21.2, and consists of 12 exons encoding three different transcripts and two protein isoforms. FTD—ALS is related to an expansion (GGGGCC, G4C2) within a noncoding region of the gene. In healthy subjects, the number of repeats is fewer than 10, although 30 is universally accepted as a nonpathological limit. By contrast, in expanded individuals, the number of repeats ranges from a minimum of 400 to several thousands. The function of the C9orf72 protein is currently unclear. First experimental evidence suggests a role for C9orf72 in the regulation of endosomal trafficking and neuronal autophagy.

Several studies have demonstrated that *C9orf72* pathological expansion is a major cause of familial FTD, FTD—ALS, and ALS. At this writing, 336 families segregating the FTD—ALS phenotype with *C9orf72* expansion have been described (http://www.molgen.ua.ac.be). It is interesting to note that, regardless of clinical presentation or ethnic origin, all the patients carrying the *C9orf72* variant inherit the expansion on the same genetic background, suggesting the presence of a founder effect.

The clinical presentation may vary widely even within the same family (FTD or ALS or a combination of both diseases). Age at onset ranges from 27 to 83 years, with an average of 50 years. Considering the FTD spectrum, the most dominant phenotype in these patients is bvFTD associated with ALS, whereas progressive nonfluent aphasia (PNFA) is the least frequent. Psychosis and obsessive—compulsive disorders are common at the onset of the disease (Galimberti et al., 2013). Memory deficits are very frequent (up to 50%—65% of cases), presenting the clinician with a challenging differential diagnosis with AD. On examination, cerebellar and extrapyramidal signs may also be present.

The neuropathology associated with the *C9orf72* mutation is a combination of FTLD—TDP and ALS. Postmortem examination usually shows TDP-43-positive inclusions in the basal ganglia, substantia nigra, hippocampus, extramotor cerebral cortex, and lower motor neurons of the brain stem and spinal cord. Nevertheless, the majority of reports agree that there is a specific pattern of neuropathological characteristics with p62-positive inclusions. These characteristic neuronal inclusions can be found in several neuroanatomical sites, but the hippocampal pyramidal neurons and cerebellar granule cell layer seem to be the most affected.

Chromatin-modifying protein 2B gene

In 2008, a mutation in the *CHMP2B* gene on chromosome 3p11.2 was identified in a Danish FTD family (Van der Zee et al., 2008). This gene encodes a component of the highly conserved endosomal sorting complex required for transport III (ESCRT-III), which plays a role in the recycling or degradation of cell surface receptors in endosomal, lysosomal, and autophagic degradation pathways. *CHMP2B* has six exons and is expressed in neurons of all major brain regions.

The age at the onset of symptoms in FTD patients carrying *CHMP2B* mutations is between 46 and 65 years, with an average of 58 years. Clinically, early behavioral changes are a common onset symptom, suggesting a bvFTD variant, but a minority of patients have been reported to show a particular progressive aphasia characterized by a spontaneous reduction of speech and preserved reading and repetition skills. Later in the disease course, parkinsonism, dystonia, pyramidal signs, and myoclonus can be observed. Notably, at-risk subjects with *CHMP2B* mutations show cognitive changes dominated by executive dysfunctions years before they fulfill the diagnostic criteria of FTD.

Histological observation shows enlarged vacuoles in the frontal cortex and parietal, temporal, and occipital neurons, probably due to altered endosome—lysosome fusion and autophagic mechanism impairment. Cytoplasmic inclusions are negative for tau, TDP-43, and FUS.

Valosin-containing protein

Mutations in *VCP*, a 17-exon protein-coding gene located on chromosome 9p13.3, were identified when studying families presenting with inclusion body myopathy with Paget's disease of bone and FTD (IBMPFD), a complex clinical picture characterized by muscle weakness due to inclusion body myopathy (IBM), osteolytic lesions compatible with Paget's disease of bone (PDB), and autosomal dominant FTD (Watts et al., 2004). The VCP protein is a member of a family of proteins involved in vesicle transport and fusion and in proteasome function. It also plays a role in dendritic spinogenesis and in several cellular events related to mitosis and ubiquitin-dependent protein degradation.

At this writing, 19 mutations have been described in this gene, identified in 49 independent families, and the R155H mutation, found in exon 5, is the most frequent. However, mutations in *VCP* are rare and account for less than 1% of cases of aFTD. The penetrance is incomplete for the three typical clinical conditions and patients may present only one clinical phenotype. bvFTD and SD are the FTD subtypes most frequently reported. Generally, symptoms occur in the sixth decade of life in 25%—30% of IBMPFD patients and penetrance of the three clinical conditions (IBM, PDB, FTLD) is independent of the underlying mutation.

Pathologically, FTLD patients carrying *VCP* mutations show several cytoplasmic and nuclear intraneuronal inclusions and dystrophic neurites, consistent with a TDP-43 proteinopathy.

Sequestosome 1

Mutations in the *SQSTM1* gene were initially identified as a cause of PDB, a common disorder of bone metabolism characterized by an increased tissue turnover. More recently, mutations in *SQSTM1* have been found in both sporadic and familial ALS patients and in FTD patients (Rubino et al., 2012).

SQSTM1 is an 11-exon gene located on 5q35 and it encodes p62, a 440-amino-acid adapter protein containing several protein—protein interaction domains with multiple functions in signal transduction. p62 also acts as a transport factor that directs ubiquitinated proteins to the autophagic degradation or proteasome pathway, playing a key role in the formation of ubiquitin-positive protein inclusions in neurons with defective autophagy. In addition, pathological studies in humans have demonstrated increased immunoreactivity for p62 in several neurodegenerative diseases, such as AD, Lewy body dementia, FTLD, Parkinson's disease, and Huntington's disease (Geetha et al., 2012).

SQSTM1 gene mutations in FTD patients have been confirmed by several studies. However, at this writing, no clear genotype—phenotype correlations have been demonstrated. Gene mutations have been associated with both bvFTD and SD phenotypes. In 2018, homozygous mutations in *SQSTM1* were linked to a complex clinical phenotype, including ataxia, cognitive decline, gaze palsy, dyskinesia, and dystonia (Muto et al., 2018). It seems appropriate to check patients with PDB for signs and symptoms of both dementia and MND.

Detailed histopathology demonstrated that mutations in *SQSTM1* are related to widespread neuronal and glial phospho-TDP-43 pathology. Autoptic investigations have shown that FTD or ALS carriers of *C9ORF72* expansions have a larger number of p62-positive inclusions. According to a recent description of a new role of p62 in maintaining mitochondrial integrity, a portion of p62 localizes directly within the mitochondria and stabilizes electron transport by forming heterogeneous protein complexes.

Fused in sarcoma gene

FUS is a highly conserved, ubiquitously expressed protein-coding gene located on 16p11.2. The gene encodes the FUS protein, a 526-amino-acid protein component of the heterogeneous nuclear ribonucleoprotein complex and a member of the FET protein family, which also includes EWS and TAF15. These are nuclear multifunctional DNA/RNA-binding proteins, implicated in cellular processes that include regulation of gene expression, maintenance of genomic integrity, and mRNA/microRNA processing. FUS is present in both the nucleus and the cytoplasm, but in neurons there is proportionally more in the nucleus, with smaller amounts in the cytoplasm, and the expression in the glia can even be exclusively nuclear. In FTD, the ability of FUS to shuttle to the nucleus is impaired because of an unspecified transporting-mediated nuclear import defect; this results in cytoplasmic accumulation of the full-length protein in stress granules. Purified FUS is extremely prone to aggregation and it aggregates more rapidly than TDP-43.

FUS mutations were discovered to be the cause of about 3% of familial ALS cases (Kwiatkowski et al., 2009), and few cases of familial FTD. Disease-causing mutations, mostly located in the carboxy-terminus of the protein, seem to cause protein cytoplasmic mislocalization. This could happen through two different mechanisms: a defect in transporting, reducing the efficiency of nuclear import of all FET proteins, or certain unknown posttranslational modifications of FET proteins, decreasing their solubility. Similar to TDP-43, both a loss of function and a gain of toxic properties via their seques-tration in aggregates are plausible. FTD—FUS should be suspected in patients with onset of symptoms before 40 years of age, with no family history of FTD, and with atrophy of the frontoinsular and cingulate cortex and of the head of the caudate nucleus on the neuroimaging study.

Ubiquilin 2 gene

UBQLN2 is an intronless gene on Xp11.21, discovered in 2011 as a cause of X-linked ALS and ALS/dementia (Deng et al., 2011). UBQLN2 protein is a member of the ubiq-uilin family of proteins that are characterized by the presence of an N-terminal ubiquitin-like domain and a C-terminal ubiquitin-associated domain.

UBQLN2 mutations are rare in Central European FTD and ALS patients, but they seem to be more frequent in the sporadic disease forms. Clinically, mutated *UBQLN2* can give rise to the full ALS—FTD continuum. Immunohistochemistry studies reported that mutated UBQLN2 accumulates in neuronal inclusions in the brain and spinal cord. Importantly, the *UBQLN2* gene could be involved in ALS-related dementia, even without *UBQLN2* mutations. In addition, UBQLN2 pathology has been found in asso-ciation with inclusion bodies in synucleinopathies and polyglutamine diseases, leading to the hypothesis that it may be a common downstream mechanism in neurodegenerative diseases.

Optineurin gene

In 2010, homozygous deletions of exon 5 of the gene encoding OPTN (*OPTN)* were described in Japanese siblings affected by ALS (Maruyama et al., 2010). *OPTN* is a 16-exon gene located on 10p13, encoding the coiled-coil-containing protein OPTN. This gene is involved in cellular morphogenesis, membrane and vesicle trafficking, and transcription activation through its interactions with the RAB8, huntingtin, and tran-scription factor IIIA proteins. Like other proteins already described, OPTN also seems to be involved in protein degradation via autophagy.

The pathomechanism causing the disease may be different depending on the recessive and dominant nature of the underlying mutation. For example, it is thought that ALS due to a recessive mutation might show a loss of function resulting from nonsense-mediated mRNA decay of transcription; in any case, it is very difficult to confirm any hypothesis because the lack of autopsy material in these recessive families precludes extensive histopathological analyses. Intriguingly, OPTN has also been linked to PDB. OPTN is colocalized with TDP-43 in the characteristic inclusion bodies of sporadic ALS.

Triggering receptor expressed on myeloid cells 2 gene

The *TREM2* gene is located on 6p21.1, and its product is a membrane protein that forms a receptor signaling complex with the TYRO protein tyrosine kinase—binding protein. TREM2 plays a role in the immune response and may be involved in chronic inflammation. Defects in this gene are one cause of a genetic syndrome named Nasu–Hakola disease (NHD), also known as polycystic lipomembranous osteodysplasia with sclerosing leukoencephalopathy, which is characterized by early-onset progressive dementia associated with sclerosing leukoencephalopathy and systemic bone cysts (Madry et al., 2007). The early onset of dementia as well as the marked involvement of the frontal regions in NHD are features resembling FTD. A study identified different homozygous mutations in *TREM2* in three Turkish probands among 44 identified with FTLD spectrum-like disease (Guerreiro et al., 2013). These three patients presented with behavioral changes and subsequent cognitive impairment and motor features, but without any bone cysts or bone-associated phenotypes.

As of this writing, 14 different mutations have been identified in *TREM2* in FTD patients, but an increased frequency of rare heterozygous *TREM2* variations has also been detected in AD patients. *TREM2* mutations might contribute to the neurodegenerative process leading to an altered immune response with extensive inflammation or defective microglial function or survival.

FTD patients carrying *TREM2* genetic variants frequently show atypical clinical signs such as epilepsy, parkinsonism, early parietal and hippocampus involvement, and corpus callosum thickness on brain MRI.

TANK-binding kinase 1 gene

The *TBK1* gene, located on chromosome 12, encodes a serine/threonine kinase that plays a relevant role in regulating inflammatory responses to foreign agents. In addition, TBK1 also has a major role in autophagy and mitophagy. Several mutations in this gene have been described in families showing FTD and ALS, mainly in Belgium (Freischmidt el al., 2015). More recently, the phenotypic spectrum was extended, including also PSP-like and cerebellar syndromes (Wilke et al., 2018). Postmortem investigations generally show FTLD—TDP type A pathology, but with unusual features of numerous TDP-43-positive neuritic structures at the cerebral cortex/subcortical white matter junction.

Coiled-coil—helix—coiled-coil—helix domain-containing 10 gene

In 2014, mutations in the *CHCHD10* gene were identified in a large family with a complex phenotype variably associating FTD with ALS, cerebellar ataxia, myopathy, and

Table 10.2 Frequency of various frontotemporal dementia gene mutations.

Gene	No. Mutations	No. Families
MAPT	44 (8.51%)	134 (8.14%)
GRN	79 (15.28%)	259 (15.74%)
C9orf72	1 (0.19%)	336 (20.41%)
CHMP2B	4 (0.77%)	5 (0.30%)
VCP	19 (3.68%)	49 (2.98%)
FUS	23 (4.45%)	49 (2.98%)
TARDBP	33 (6.38%)	134 (8.14%)
TBK1	28 (5.42%)	45 (2.73%)

The numbers of pathogenetic mutations as well as families as of 2020 described according to the Alzheimer Disease and Frontotemporal Dementia Mutation database (http://www.molgen.ua.ac.be).

hearing impairment (Bannwarth et al., 2014). This gene, located on chromosome 22q11.23, encodes a protein located in the mitochondrial intermembrane space and involved in mitochondrial genome stability and maintenance of cristae junctions. The frequency of mutations is generally low (2%—3%) in European countries but slightly higher in China (7.5%). The age at onset of the disease is around 50 years and the clinical presentation is highly heterogeneous, ranging from bvFTD to SD and ALS. Preliminary studies showed that CHCHD10 dysfunction induces cytoplasmic TDP-43 accumulation, resulting in mitochondrial and synaptic damage.

Double mutants in frontotemporal dementia

Studies have found double mutations in families with FTD, which may further increase our understanding of the underlying processes involved in this disease. van Blitterswijk et al. (2013) detected *C9orf72* repeat expansions in approximately 2% of North American and Italian families harboring *GRN* or *MAPT* gene mutations. Additional patients with double mutations have been described. These findings suggest that the cooccurrence of two pathogenic mutations could contribute to the pleiotropy that is detected in FTD patients and that genetic counselors should take into account this phenomenon when advising patients and their family members.

Table 10.2 shows the known percentages of various gene mutations in aFTD and the number of families described so far.

Conclusions

A large number of studies have investigated the genotype—phenotype correlations in aFTD patients carrying different gene mutations. At this writing, due to the clinical, pathological, and genetic heterogeneity of FTD, no clear genotype—phenotype correlation is available. However, there are initial suggestions that may help the clinician in requesting genetic counseling. Patients with *TARDP* gene mutations in addition to the bvFTD phenotype frequently present with complex motor symptoms. Mutations in *GRN* are frequently characterized by psychiatric symptoms, associated with language dysfunction and parkinsonism. Finally, patients carrying *C9orf72* repeat expansion present with behavioral disturbances, psychotic symptoms, and MND. Additional, larger studies are warranted in the next years to better elucidate the complex genotype—phenotype correlations in autosomal dominant FTD.

Key facts

- FTD is the third most common cause of degenerative dementia worldwide. However, the disease is still underrecognized and underdiagnosed.
- The clinical picture of FTD is highly heterogeneous, with the large majority of patients showing behavioral abnormalities. Approximately 20% of patients present, at disease onset, with language dysfunction.
- In approximately 40% of patients a familial history of dementia is present. However, in only 10% of cases the disease shows a clear autosomal dominant inheritance of the phenotype (aFTD).
- In patients with aFTD, genetic variants in several genes have been described, with more than 20 different genes potentially involved in the disease.
- Genetic variants in *MAPT*, *GRN*, and *C9orf72* genes are the mutations more frequently reported.
- Genotype—phenotype studies are currently investigating the relationship between clinical symptoms of the disease and genetic variants.
- At this writing, no clear genotype—phenotype correlations have been demonstrated in aFTD. However, specific clinical symptoms may help the clinician in selecting molecular genetic investigations.
- Additional studies are needed to better elucidate the genotype—phenotype correlation in FTD.

Summary points

- A correlation between genotype and phenotype is defined as a significant probability of a distinct gene variant being associated with a particular physical feature or abnormality.
- Studies of genotype—phenotype correlations in patients with FTD have provided, so far, no clear relationship.
- However, there are some preliminary studies that may help clinicians in requesting molecular genetic investigation.

- Most FTD patients with a *MAPT* gene mutation present with the behavioral variant of the disease and frequently have extrapyramidal symptoms.
- Patients with FTD who carry a *GRN* mutation are characterized by psychiatric symptoms, like hallucinations and delusions, and may develop memory impairment and language dysfunctions.
- *C9orf72* gene expansion in patients with FTD is frequently associated with symptoms of motor neuron disease, psychiatric symptoms, and behavioral disturbances.

References

Baizabal-Carvallo, J. F., & Jankovic, J. (2016). Parkinsonism, movement disorders and genetics in frontotemporal dementia. *Nature Reviews Neurology, 12*, 175—185.

Baker, M., Mackenzie, I. R., Pickering-Brown, S. M., Gass, J., Rademakers, R., Lindholm, C., et al. (2006). Mutations in progranulin cause tau-negative frontotemporal dementia linked to chromosome 17. *Nature, 442*, 916—919.

Bang, J., Spina, S., & Miller, B. L. (2015). Frontotemporal dementia. *Lancet, 386*, 1672—1683.

Bannwarth, S., Ait-El-Mkadem, S., Chaussenot, A., Genin, E. C., Lacas-Gervais, S., Fragaki, K., et al. (2014). A mitochondrial origin for frontotemporal dementia and amyotrophic lateral sclerosis through CHCHD10 involvement. *Brain, 137*, 2329—2345.

van Blitterswijk, M., Baker, M. C., DeJesus-Hernandez, M., Ghidoni, R., Benussi, L., Finger, E., et al. (2013). C9ORF72 repeat expansions in cases with previously identified pathogenic mutations. *Neurology, 81*, 1332—1341.

Cruts, M., Gijselinck, I., van der Zee, J., Engelborghs, S., Wils, H., Pirici, D., et al. (2006). Null mutations in progranulin cause ubiquitin-positive frontotemporal dementia linked to chromosome 17q21. *Nature, 442*, 920—924.

DeJesus-Hernandez, M., Mackenzie, I. R., Boeve, B. F., Boxer, A. L., Baker, M., Rutherford, N. J., et al. (2011). Expanded GGGGCC hexanucleotide repeat in noncoding region of C9ORF72 causes chromosome 9p-linked FTD and ALS. *Neuron, 72*, 245—256.

Deng, H. X., Chen, W., Hong, S. T., Boycott, K. M., Gorrie, G. H., Siddique, N., et al. (2011). Mutations in UBQLN2 cause dominant X-linked juvenile and adult-onset ALS and ALS/dementia. *Nature, 477*, 211—215.

Ferrari, R., Hernandez, D. G., Nalls, M. A., Rohrer, J. D., Ramasamy, A., Kwok, J. B., et al. (2014). Frontotemporal dementia and its subtypes: A genomewide association study. *The Lancet Neurology, 13*, 686—689.

Freischmidt, A., Wieland, T., Richter, B., Ruf, W., Schaeffer, V., Müller, K., et al. (2015). Haploinsufficiency of TBK1 causes familial ALS and frontotemporal dementia. *Nature Neuroscience, 18*, 631—636.

Galimberti, D., Fenoglio, C., Serpente, M., Villa, C., Bonsi, R., Arighi, A., et al. (2013). Autosomal dominant frontotemporal lobar degeneration due to the C9ORF72 hexanucleotide repeat expansion: Late-onset psychotic clinical presentation. *Biological Psychiatry, 74*, 384—391.

Galvin, J. E., Howard, D. H., Denny, S. S., Dickinson, S., Tatton, N., et al. (2015). The social and economic burden of frontotemporal degeneration. *Neurology, 89*, 2049—2056.

Geetha, T., Vishwaprakash, N., Sycheva, M., Babu, J. R., et al. (2012). Sequestosome 1/p62: Across diseases. *Biomarkers, 17*, 99—103.

Gorno-Tempini, M. L., Hillis, A. E., Weintraub, S., Kertesz, A., Mendez, M., Cappa, S. F., et al. (2011). Classification of primary progressive aphasia and its variants. *Neurology, 76*, 1006—1014.

Guerreiro, R. J., Lohmann, E., Brás, J. M., Gibbs, J. R., Rohrer, J. D., Gurunlian, N., et al. (2013). Using exome sequencing to reveal mutatins in TREM2 presenting as a frontotemporal dementia-like syndrome without bone involvement. *JAMA Neurology, 70*, 78—84.

Hutton, M., Lendon, C. L., Rizzu, P., Baker, M., Froelich, S., Houlden, H., et al. (1998). Association of missense and 5'-splice-site mutations in tau within the inherited dementia FTDP-17. *Nature, 393,* 702–705.

Kwiatkowski, T. J., Jr., Bosco, D. A., Leclerc, A. L., Tamrazian, E., Vanderburg, C. R., Russ, C., et al. (2009). Mutations in the FUS/TLS gene on chromosome 16 cause familial amyotrophic lateral sclerosis. *Science, 323,* 1205–1208.

Lashley, T., Rohrer, J. D., Mead, S., Revesz, T., et al. (2015). Review: An update on clinical, genetic and pathological aspects of frontotemporal lobar degenerations. *Neuropathology and Applied Neurobiology, 41,* 858–881.

Madry, H., Prudlo, J., Grgic, A., Freyschmidt, J., et al. (2007). Nasu-hakola disease (PLOSL): Report of five cases and review of the literature. *Clinical Orthopaedics and Related Research, 454,* 262–269.

Maruyama, H., Morino, H., Ito, H., Izumi, Y., Kato, H., Watanabe, Y., et al. (2010). Mutations of optineurin in amyotrophic lateral sclerosis. *Nature, 465,* 223–236.

Muto, V., Flex, E., Kupchinsky, Z., Primiano, G., Galehdari, H., Dehghani, M., et al. (2018). Biallelic SQSTM1 mutations in early-onset, variably progressive neurodegeneration. *Neurology, 91,* e319–e330.

Rascovsky, K., Hodges, J. R., Knopman, D., Mendez, M. F., Kramer, J. H., Neuhaus, J., et al. (2011). Sensitivity of revised diagnostic criteria for the behavioural variant of frontotemporal dementia. *Brain, 134,* 2456–2477.

Renton, A. E., Majounie, E., Waite, A., Simón-Sánchez, J., Rollinson, S., Gibbs, J. R., et al. (2011). A hexanucleotide repeat expansion in C9ORF72 is the cause of chromosome 9p21-linked ALS-FTD. *Neuron, 72,* 257–268.

Rubino, E., Rainero, I., Chiò, A., Rogaeva, E., Galimberti, D., Fenoglio, P., et al. (2012). SQSTM1 mutations in frontotemporal lobar degeneration and amyotrophic lateral sclerosis. *Neurology, 79,* 1556–1562.

Seeley, W. W., Carlin, D. A., Allman, J. M., Macedo, M. N., Bush, C., Miller, B. L., et al. (2006). Early frontotemporal dementia targets neurons unique to apes and humans. *Annals of Neurology, 60,* 660–667.

Van der Zee, J., Urwin, H., Engelborghs, S., Bruyland, M., Vandenberghe, R., Dermaut, B., et al. (2008). CHMP2B C-truncating mutations in frontotemporal lobar degeneration are associated with an aberrant endosomal phenotype in vitro. *Human Molecular Genetics, 17,* 313–322.

Watts, G. D., Wymer, J., Kovach, M. J., Mehta, S. G., Mumm, S., Darvish, D., et al. (2004). Inclusion body myopathy associated with Paget disease of bone and frontotemporal dementia is caused by mutant valosin-containing protein. *Nature Genetics, 36,* 377–381.

Wilke, C., Baets, J., De Bleecker, J. L., Deconinck, T., Biskup, S., & Hayer, S. N (2018). Beyond ALS and FTD: The phenotypic spectrum of TBK1 mutations includes PSP-like and cerebellar phenotypes. *Neurobiology of Aging, 62,* 244.e9.

CHAPTER 11

Environmental and genetic risk factors for dementia

Vanesa Bellou[1], Lazaros Belbasis[1], Evangelos Evangelou[1,2]
[1]Department of Hygiene and Epidemiology, University of Ioannina Medical School, Ioannina, Greece; [2]Department of Epidemiology and Biostatistics, School of Public Health, Imperial College London, London, United Kingdom

List of abbreviations

BMI body mass index
CI confidence interval
GWAS genome-wide association study
MR Mendelian randomization
OR odds ratio
RCT randomized clinical trial
RR risk ratio
SNP single-nucleotide polymorphism
WHO World Health Organization
WMHs white matter hyperintensities

Mini-dictionary of terms

Bias A bias is a systematic error in the study design or statistical analysis resulting in an incorrect estimate of the true effect of an exposure on the outcome of interest. Biases are broadly divided into selection bias, information bias, and confounding.

Between-study heterogeneity Between-study heterogeneity is a statistical term used in meta-analyses to describe the observed variation of effect sizes among different studies of a meta-analysis.

Confounding Confounding is a phenomenon leading to the observation of a spurious association that is not causal. The observed association is caused by a confounder that is a third factor associated with both the disease of interest and the risk factor of interest.

Effect size An effect size is a quantitative measure of the impact of a factor on an outcome. Commonly used metrics of effect size for dichotomous exposures are odds ratio, risk ratio, and hazard ratio.

Field synopsis A field synopsis is a regularly updated snapshot of the current state of knowledge about genetic associations and common diseases.

Genome-wide association study A genome-wide association study is an epidemiological study design that examines the association between hundreds of thousands of genetic variants and risk for a disease in hundreds of thousands of individuals.

Heritability Heritability is a term used to describe the degree of variation in a phenotype as a result of genetic variation between individuals of the population.

Mendelian randomization study A Mendelian randomization study is a novel epidemiological study design that exploits findings from genetic association studies to examine the potential causal effect of an exposure on risk for a disease.

Diagnosis and Management in Dementia
ISBN 978-0-12-815854-8, https://doi.org/10.1016/B978-0-12-815854-8.00011-2

Meta-analysis   Meta-analysis is the process of quantitative synthesis of findings from multiple studies on the same research question.

Single nucleotide polymorphism   A single-nucleotide polymorphism is a type of variation in the genome that occurs when a single nucleotide at a specific genetic position is changed for another nucleotide.

Venice criteria   Venice criteria are a set of methodological criteria derived for the assessment of epidemiological credibility in genetic association studies.

Umbrella review   An umbrella review is a systematic search and critical appraisal of systematic reviews and meta-analyses on a specific research topic.

Introduction

Dementia is a clinical syndrome describing a wide range of symptoms affecting a patient's memory, behavior, and verbal or nonverbal communication. Dementia is categorized in subtypes that vary in clinical manifestation, pathophysiology, prognosis, and treatment (Burns and Iliffe, 2009; Robinson, Tang and Taylor, 2015). It is a very common and debilitating disease that has a significant impact not only on individual quality of life but also on public health, resulting in enormous healthcare expenditures. To elaborate, it is estimated that 47 million people worldwide were living with dementia in 2015. This number will nearly double every 20 years, reaching 75 million people in 2030 and 132 million people in 2050. Through meta-analysis of the available evidence, it is estimated that the incidence of dementia is over 9.9 million worldwide, implying that a new case occurs every 3.2 s. The economic burden of caring for dementia is more than US$800 billion per year globally and is expected to reach US$2 trillion by 2030 (Alzheimer's Disease International, 2015).

Indeed, the significant effect of dementia on public health is depicted by the World Health Organization (WHO) announcing the launch of the Global Dementia Observatory on December 2017 to provide a constant monitoring service for data related to dementia planning around the world. The WHO Global Action Plan on the Public Health Response to Dementia lays out the framework for action, but the warning signs are growing ever stronger (The Lancet, 2017).

Currently, there is neither an effective disease-modifying treatment nor an effective preventive strategy for dementia. All available medications alleviate symptoms only modestly. So the focus is steered toward primary prevention of the disease to reduce the size of the affected population (Baumgart et al., 2015; Yaffe, 2018). The first step for planning a preventive and screening strategy for dementia is the identification of a set of credible risk factors. Thus, the present chapter aims to summarize the current advances in the field of genetic and environmental epidemiology of dementia based on the findings of high-quality and large-scale epidemiological studies. We particularly focus on Alzheimer's disease and vascular dementia, which are the most prevalent forms of dementia.

Why are the risk factors of dementia important?

Before discussing the evidence on risk factors for dementia, it is important to address the significance and implications of the identification of these risk factors. Dementia is a medical disorder that commonly remains undiagnosed. This is attributed to the common misconception that dementia is a natural consequence of aging. Also, many patients refrain from seeking medical attention for their symptoms. Fewer than half of patients with dementia have a formal diagnosis. Also, when diagnosis is made, it typically occurs late, during an advanced stage of the disease. Earlier diagnosis of dementia might allow disease modification or improve an individual's quality of life through the application of pharmacotherapy during less advanced stages (Alzheimer's Disease International, 2011; Department of Health Working Group, 2009). Moreover, timely diagnosis reserves time for consideration of future plans of care, which may subsequently improve quality of life for patients and their families (Bradford et al., 2009; Dubois et al., 2016).

However, already approved medications for dementia have shown only modest beneficial effects, and none of them alter the progression of disease. Keeping that in mind, prevention of dementia is the most straightforward approach to tackle the increasing incidence and burden of disease. Despite dementia being considered a disorder of the elderly, the underlying brain pathology begins in midlife but remains clinically silent. This fact suggests a window of opportunity to intervene by addressing dementia risk factors in middle age (Frankish and Horton, 2017). The most straightforward approach for prevention is the identification of modifiable environmental risk factors. Even if effective drugs arise in the future, the need for an effective prevention strategy will remain, given the rising rate of new cases of dementia. Also, a risk prediction model could be created based on these risk factors to detect individuals at high risk for developing dementia.

Moreover, the contribution of genetic risk factors in dementia etiology should not be overlooked. One quarter of people over 55 years of age have a family history of dementia. For the majority of them, the family history is due to genetically complex disease, where many genetic variations of small effect interact to increase the risk of dementia (Loy et al., 2014). A minority of dementia cases are an attribute of an autosomal-dominant pattern of transmission. Thus, efforts should be made to identify genetic risk factors for dementia (Fenoglio et al., 2018). Elucidation of genetic variants associated with increased risk for dementia leads to two major applications. First, genetic associations offer a better understanding of biological pathways leading to the pathogenesis of disease. Second, these associations could serve as potential drug targets to modify the clinical course of dementia and thus prevent or delay its development (Van Cauwenberghe, Van Broeckhoven and Sleegers, 2016).

Environment-wide mapping of risk factors

A large variety of environmental factors have been examined as putative risk factors for dementia. A recently published umbrella review aimed to summarize and critically appraise the published meta-analyses of observational studies of this research field (Bellou et al., 2017). In this effort, a total of 76 unique associations were identified, and these associations were categorized based on the definition of outcome (i.e., all types of dementia, Alzheimer's disease, and vascular dementia). The examined risk factors covered a wide range of dietary patterns, psychological and social factors, medications, and comorbid conditions. The majority of meta-analyses examined risk factors for Alzheimer's disease, whereas only a small portion of meta-analyses focused on risk factors for vascular dementia.

Additionally, in this umbrella review, a thorough assessment of epidemiological credibility was performed based on prespecified methodological criteria (Bellou et al., 2017). This assessment was based on the level of significance, sample size, presence of between-study heterogeneity, and presence of systematic biases. These criteria also have been applied for the identification of robust risk factors for other neurological conditions (Belbasis, Bellou, et al., 2015; Belbasis, Bellou and Evangelou, 2016; Bellou et al., 2016), whereas a similar assessment of epidemiological credibility has been proposed for the meta-analyses of genetic association studies about a decade ago (Ioannidis et al., 2008). For dementia, the factors that presented high epidemiological credibility were the following: type 2 diabetes mellitus, late-life depression and depression at any age, benzodiazepine use, physical activity, frequency of social contacts, and history of cancer. Table 11.1 presents the associations supported by high epidemiologic credibility for risk of dementia as presented in this umbrella review. In the following sections, we discuss the credibility and potential biases of these associations from an epidemiological perspective.

Environmental risk factors

Type 2 diabetes mellitus has a strong relationship with all types of dementia (RR, 1.60; 95% CI, 1.43 to 1.79), as observed in a published meta-analysis of cohort studies (Gudala et al., 2013). This association was also observed for both Alzheimer's disease (RR, 1.94; 95% CI, 1.39 to 1.72) and vascular dementia (RR, 2.28; 95% CI, 1.96 to 2.66), and it is supported by a large number of prospective cohort studies (Gudala et al., 2013). The observed effect was larger for vascular dementia. However, it is not clear whether this association is genuine, and we should consider the potential of shared environmental and genetic determinants for dementia and type 2 diabetes mellitus. A recently published umbrella review of meta-analyses focused on environmental risk factors for type 2 diabetes mellitus and provides insights about the credibility of these associations (Bellou et al., 2018). Indeed, some environmental factors, including physical activity, educational status, and obesity, present a statistically significant association with both risk for dementia and risk for type 2 diabetes mellitus.

Another factor that has been associated with all types of dementia is depression. A multitude of observational studies have examined the association between depression and dementia and investigated whether the age of clinical presentation of depression affects subsequent development of dementia (da Silva et al., 2013; Diniz et al., 2013).

Table 11.1 Associations of environmental factors with risk for dementia supported by high epidemiological credibility.

Risk factor	Level of comparison	Effect size metric	RE summary effect size (95% CI)	P value	95% PI	I^2	SSE/ESS
All types of dementia							
Benzodiazepine use	Ever versus Never	RR	1.49 (1.30–1.72)	2.7×10^{-8}	1.03 to 2.17	35.1	No/No
Depression at any age	Yes versus No	RR	1.99 (1.84–2.16)	8.0×10^{-62}	1.65 to 2.40	27.8	No/No
Late-life depression	Yes versus No	RR	1.85 (1.67–2.05)	3.1×10^{-32}	1.66 to 2.06	0	No/No
Frequency of social contact	Low level versus High level	RR	1.57 (1.32–1.85)	1.9×10^{-7}	1.27 to 1.93	0	No/No
Type 2 diabetes mellitus	Yes versus No	RR	1.60 (1.43–1.79)	5.4×10^{-17}	1.05 to 2.44	72.3	No/No
Alzheimer's disease							
Late-life depression	Yes versus No	RR	1.65 (1.42–1.92)	4.8×10^{-11}	1.36 to 1.99	2.2	No/No
Type 2 diabetes mellitus	Yes versus No	RR	1.54 (1.39–1.72)	3.1×10^{-15}	1.37 to 1.73	0	No/No
Cancer	Yes versus No	HR	0.62 (0.53–0.74)	4.6×10^{-8}	0.50 to 0.78	0	Yes/No
Depression at any age	Yes versus No	RR	1.77 (1.48–2.13)	6.0×10^{-10}	0.86 to 3.66	69.6	Yes/Yes
Physical activity	High level versus Low level	HR	0.62 (0.52–0.72)	5.0×10^{-9}	0.51 to 0.75	0	Yes/No
Vascular dementia							
Type 2 diabetes mellitus	Yes versus No	RR	2.28 (1.94–2.66)	1.1×10^{-24}	1.91 to 2.71	0	No/No

CI, confidence interval; ESS, excess statistical significance; HR, hazard ratio; OR, odds ratio; PI, prediction interval; RE, random effects; RR, risk ratio; SSE, small-study effects.

Both early-life and late-life depression were associated with dementia, but the association is stronger when considering only late-life depression rather than depression at any age. A variety of mechanisms have been proposed to explain this finding. Depression may act as a prodrome or a clinical manifestation in the early pathway of dementia preceding clinical diagnosis. In this case, this association could be considered a result of reverse causation. An alternative explanation is the vascular depression hypothesis, which postulates that vascular disease could be the underlying link between depression and dementia by inflicting subclinical long-term structural damage in the brain. Another possible hypothesis is that depressive symptomatology could be a defensive mechanism of the individual that perceives the beginning of cognitive impairment (Bellou et al., 2017; Byers and Yaffe, 2011).

Another association that has emerged based on multiple prospective cohort studies is the one between the use of benzodiazepines and risk for dementia (Zhong et al., 2015). Consumption of benzodiazepines could lower cognitive reserve capacity in the long run, which might reduce the resiliency of the brain against early-phase brain damage by soliciting accessory neuronal networks (Bellou et al., 2017). However, a cohort study that examined the possibility of a dose—response association did not find such a relationship (Gray et al., 2016). This observation goes against a causal relationship. A possible explanation could be reverse causation, meaning that patients with dementia in an early stage experience anxiety, depressive symptoms, and trouble sleeping, which could lead them to use benzodiazepines (Bellou et al., 2017).

Large longitudinal studies have found that frequency of social contacts is another important risk factor for dementia (Kuiper et al., 2015). A low level of social contacts was associated with an RR of 1.57 (95% CI, 1.32 to 1.85) for all types of dementia. Although other factors related to social relationships have been examined, including loneliness, satisfaction with social network, social network size, and social participation, these factors presented either a weak or nonsignificant effect on the risk for dementia (Bellou et al., 2017).

The association of a history of cancer with Alzheimer's disease is also supported by high epidemiologic credibility (Bellou et al., 2017). The meta-analysis of observational studies showed a lower risk for Alzheimer's disease in individuals with a history of cancer. However, patients with cancer generally live shorter lives than the general population does, so this association could be attributed to the competing risks between risk for death and risk for Alzheimer's disease (Zhang et al., 2015). Moreover, the observational studies that examined this association used models adjusting mainly for age and sex but did not perform adjustment for other potential confounders, like history of depression and type 2 diabetes mellitus. These limitations bring into question the credibility of this association (Bellou et al., 2017).

Evidence from prospective cohort studies indicated that physical activity has a protective effect against the development of dementia (Blondell, Hammersley-Mather & Veerman, 2014). Also, physical activity has been examined as a potential protective factor against vascular dementia (Aarsland et al., 2010). Although increased physical activity led to a lower risk for vascular dementia, only a small number of epidemiological studies, based on a relatively small number of cases, examined this association. The beneficial effect of physical activity against cognitive decline has been examined in randomized clinical trials, but current evidence is scarce (Brasure et al., 2018).

Some additional risk factors received much attention, but they presented only suggestive evidence based on the published umbrella review (Bellou et al., 2017). Among them are obesity, level of educational attainment, nonsteroidal antiinflammatory drugs, and statins. A notorious risk factor that has been examined extensively for an association with dementia is body mass index (BMI), which constitutes a measure of obesity. A large number of epidemiological studies have examined the effect of middle-life and late-life BMI on the risk for dementia (Bellou et al., 2017). Middle-life obesity is associated with an increased risk for dementia with an RR of 1.91 (95% CI, 1.40 to 2.62) (Loef and Walach, 2013). However, there is not adequate epidemiological evidence that obesity in late life contributes to the development of dementia (Bellou et al., 2017; Pedditizi, Peters and Beckett, 2016).

A basic hypothesis in the pathogenesis of dementia is the cognitive reserve hypothesis. Brain reserve refers to the ability of the brain to withstand the changes inflicted by age and disease without manifesting cognitive impairment. A multitude of biological and epidemiological studies have depicted that physical and mental stimulation has a protective effect against cognitive deterioration (Fratiglioni and Wang, 2007). Level of educational attainment, socioeconomic status, social network, leisure activities, and work complexity are factors that help preserve cognitive function in old age (Fratiglioni and Wang, 2007). Higher level of education, higher socioeconomic status, high frequency of social contacts, high work complexity, and high physical activity lowered the risk for developing dementia, and all of these associations could be mediated by increased cognitive function (Bellou et al., 2017).

Two medication classes presented only suggestive evidence for an association with dementia. Both statins and NSAIDs depicted a protective effect against developing dementia. The association between statin use and risk for dementia has been examined in 12 observational studies including approximately 38,000 cases (Richardson et al., 2013). However, this association was accompanied by large between-study heterogeneity and presence of small-study effects, indicating that the observed effect could be inflated (Bellou et al., 2017). Furthermore, a meta-analysis examined the association between NSAIDs and risk for Alzheimer's disease (Wang et al., 2015). However, this association was also characterized by large between-study heterogeneity. Of note, these inferences are based only on observational studies. Considering that the gold standard study design to make inferences for drugs is the randomized clinical trial (RCT), these associations should be interpreted with caution because they could be the result of confounding and systematic biases.

A range of other factors has been proposed for an association with dementia. However, after assessing available evidence, these associations did not fulfill enough criteria to reach high credibility (Bellou et al., 2017). Nutritional factors, including dietary intake of vitamins as well as alcohol and coffee consumption, did not affect the risk for dementia. Also, evidence from observational studies indicate that medications, including antihypertensive drugs, statins, aspirin, and NSAIDs, do not alter the risk for dementia. Also, some comorbid conditions (i.e., atrial fibrillation, hypertension, traumatic brain injury, and rheumatoid arthritis) either had weak evidence or did not have a statistically significant association with risk for dementia (Bellou et al., 2017).

Genetic risk factors

Genetic polymorphisms participate in the pathogenesis of complex diseases. The identification of genetic risk factors is mainly based on two study designs, candidate gene association studies and genome-wide association studies (GWASs). Two field synopses summarized the findings of candidate gene association studies for Alzheimer's disease and vascular dementia (Belbasis, Panagiotou, et al., 2015). A field synopsis for Alzheimer's disease summarized the available genetic information from 789 publications reporting on 802 different polymorphisms in 277 genes (Bertram et al., 2007). In this effort, apart from APOE-ε4, 20 polymorphisms located in 13 genes yielded a statistically significant effect and high epidemiological credibility, which was assessed through Venice criteria (Ioannidis et al., 2008). The findings of this systematic approach have been presented online through the AlzGene platform (http://www.alzgene.org). The last update was performed on April 2011, and it led to the inclusion of 1395 publications reporting on 2973 polymorphisms in 695 genes.

In a similar systematic approach, researchers collected the candidate gene association studies on genetic associations for vascular impairment including vascular dementia (Dwyer et al., 2013). However, this effort was not accompanied by an online database. In this field synopsis, 104 eligible publications were captured reporting on 72 different polymorphisms in 47 genes. The APOE-ε4 variant and an MTHFR single-nucleotide polymorphism (SNP) (rs1801133) were associated with a higher risk for vascular dementia, whereas the APOE-ε2 variant had a protective effect.

Large-scale collaborative GWASs have significantly advanced the knowledge regarding the genetic determinants of late-onset Alzheimer's disease by the identification of at least 20 additional genetic loci (Van Cauwenberghe et al., 2016). The susceptibility genes are linked to a number of biological pathways: endosomal vesicle cycling, immune and inflammatory response, cholesterol and lipid metabolism, cytoskeletal function and axonal transport, cell migration, neural development, and synaptic function. Also, the associated genes are linked with β-amyloid cascade and/or tau pathology (Van Cauwenberghe et al., 2016). About 30 GWASs for Alzheimer's disease are available through the GWAS Catalog, a curated collection of all published GWASs, and they have examined the effects of genetic polymorphisms on the risk for Alzheimer's disease (MacArthur et al., 2017).

A major genetic risk factor for Alzheimer's disease is the APOE genotype, which accounts for approximately 25% of explained genetic variance (Ridge et al., 2016). Three major allelic variants at a single gene locus (ε2, ε3, and ε4) have been identified in the APOE genotype, and these variants encode for different isoforms (ApoE2, ApoE3, ApoE4) (Van Cauwenberghe et al., 2016). The APOE-ε4 allele increases risk in familial and sporadic early-onset and late-onset Alzheimer's disease. The APOE-ε2 allele is thought to be protective and to delay onset age (Van Cauwenberghe et al., 2016). No risk loci linked with Alzheimer's disease have an effect on the magnitude of APOE-ε4, because the effect estimates range from an odds ratio (OR) of 1.1 to 2.0 per risk allele.

Findings from GWASs have permitted estimation of the proportion of the phenotypic variance of AD attributable to genetic risk alleles (Cuyvers and Sleegers, 2016). Early studies estimated that about 24% of total phenotypic variance can be explained by directly genotyped SNPs (Lee et al., 2013), and about 33% of total phenotypic variance can be explained by the combination of directly genotyped SNPs and HapMap-imputed SNPs (Ridge et al., 2013). A family-based analysis in the UK Biobank revealed that the estimated heritability for Alzheimer's disease is close to 35% (Muñoz et al., 2016).

The genetic determinants of vascular dementia are poorly understood, and studies on its genetic basis are scarce (Ikram et al., 2017). Candidate gene approaches did not identify robust associations (Dwyer et al., 2013; Ikram et al., 2017). A GWAS for vascular dementia identified a SNP near the androgen receptor gene on the X chromosome (Schrijvers et al., 2012). Furthermore, indirect information for the genetics of vascular dementia could be derived from GWASs for white matter hyperintensities (WMHs). WMHs constitute a magnetic resonance imaging marker of small-vessel cerebral disease associated with increased risk for dementia (Ikram et al., 2017; Prins and Scheltens, 2015). A field synopsis summarized the candidate gene association studies for WMHs and failed to identify any credible associations (Paternoster, Chen and Sudlow, 2009). Meta-analyses of GWASs have identified some SNPs associated with increased risk for WMHs (Fornage et al., 2011; Verhaaren et al., 2015).

Apart from biological pathways leading to the development of dementia, findings from GWASs offer additional important insights. As already mentioned, prospective cohort studies indicated an increased risk for Alzheimer's disease and all types of dementia in individuals diagnosed with depression in late life. A recently published study assessed the presence of shared genetic architecture between Alzheimer's disease and major depressive disorder by applying two complementary techniques, linkage disequilibrium regression and polygenic profile scoring (Gibson et al., 2017). Both approaches revealed no evidence for overlap in the polygenic architecture of major depressive disorder and Alzheimer's disease. Additionally, an analysis of shared heritability in common disorders of the brain found no genetic correlation between Alzheimer's disease and a number of neurological and psychiatric conditions (Anttila et al., 2017). However, educational attainment and intelligence were genetically correlated with Alzheimer's disease (Anttila et al., 2017).

Moreover, cardiovascular diseases are considered to have an obvious role in vascular dementia, and further, vascular pathology is considered important in the pathogenesis of Alzheimer's disease. A recently published study examined whether this association was observed due to genetic overlap between cardiovascular diseases and Alzheimer's disease (Karlsson et al., 2017). This study found that a genetic risk score predicting risk for coronary artery disease was not associated with risk for developing Alzheimer's disease. Also, a gene-based analysis showed that there is no genetic overlap between AD and coronary artery disease. The same gene-based analysis indicated that Alzheimer's disease has some genetic overlap with BMI, total cholesterol, high-density lipoprotein, low-density lipoprotein, and triglycerides. However, pathway analyses found six common lipid-related pathways between Alzheimer's disease and coronary artery disease (Karlsson et al., 2017). These pathways pertained to statin pathway, chylomicron-mediated lipid transport, lipoprotein metabolism, retinoid metabolism and transport, lipid digestion, mobilization and transport, and visual phototransduction. Using gene-based analysis, this study also indicated no genetic overlap between Alzheimer's disease and type 2 diabetes (Karlsson et al., 2017). Furthermore, an analysis based on linkage disequilibrium regression did not find evidence of a genetic correlation between coronary artery disease and Alzheimer's disease (Bulik-Sullivan et al., 2015).

Mendelian randomization studies

Recent advances in understanding the genetic determinants of chronic diseases and traits led to the conceptualization of Mendelian randomization (MR) studies. The MR approach utilizes Mendel's second law within an epidemiological setting (Davey Smith and Ebrahim, 2003). The main idea of an MR study is to examine exposure—disease associations by exploiting the random assignment of genes as a means of reducing confounding and systematic biases (Davey Smith and Ebrahim, 2003).

The MR approach has been used to test the causal association between some environmental factors and Alzheimer's disease. The most recently published MR study examined the potential causal association of 22 potentially modifiable risk factors with Alzheimer's disease (Larsson et al., 2017). The researchers used summary genetic data from four GWASs for Alzheimer's disease. The examined risk factors were divided to educational and intelligence characteristics, lifestyle and dietary factors, cardiometabolic factors, and inflammatory factors. Among them, only level of educational attainment had a P-value lower than the Bonferroni-corrected threshold and presented a potential causal association with Alzheimer's risk. The OR was 0.89 (95% CI, 0.84 to 0.93) per year of education completed. Also, the authors claimed a suggestive association for the risk factors that were nominally significant. These factors pertained to intelligence, smoking quantity, coffee consumption, and serum vitamin D. However, the findings of the MR study are not definitive because the genetic determinants of the examined exposures are not fully uncovered. Table 11.2 presents the findings of the aforementioned MR study for Alzheimer's disease.

Biases in observational studies for dementia

The identification of credible risk factors for dementia is a quite challenging process, because a number of obstacles and biases hinder the conduct of observational studies in this field. These obstacles could be distinguished into two categories, challenges that are common in many research areas and those that are more profound in dementia research (Weuve et al., 2015).

Confounding is a common bias in observational studies. Although some statistical approaches have been proposed to deal with confounders, residual confounding due to unmeasured confounders could still lead to false-positive findings in observational research. The "gold standard" study design in the identification of causal associations is RCT, because the randomization tends to attenuate the differences in the distribution of potential confounding factors between the group of exposed and non-exposed participants. However, the design of preventive RCTs for dementia is a challenging process, because the ideal RCT would start an intervention at midlife and would have a long follow-up. Also, it is unethical to design a randomized experiment for some research questions.

Furthermore, a major issue in dementia research is the definition of disease. Currently, different diagnostic criteria are used to diagnose dementia, and these criteria present varying validity and reliability (Weuve et al., 2015). Thus, uncertainty in diagnostic criteria and measurement error in neuropsychiatric assessments hinder the valid identification of dementia cases in observational studies.

Mortality and dementia have several risk factors in common, which results in higher mortality in patients with dementia. The product of this association is that people truncated during the follow-up period by death may have developed dementia but remained undiagnosed due to periodic follow-up. This association results in a competing risk between dementia and death in cohort studies and creates survival bias, which might produce a biased effect between exposure and risk for disease (Alzheimer's Disease International, 2014; Weuve et al., 2015).

During the last 2 decades, systematic review and meta-analysis of observational studies has emerged as a reliable approach toward the identification of credible associations. However, biases are still present in the level of meta-analysis. This observation is reflected in the findings of the umbrella review for risk factors of dementia. The majority of the meta-analyses presented large between-study heterogeneity, and this fact limits the credibility of the observed associations (Bellou et al., 2017). The origins of between-study

Table 11.2 Findings from the most recently published Mendelian randomization study for Alzheimer's disease.

Risk factor	Level of comparison	N of SNPs	Effect size metric	Effect size (95% CI)	P value
Years of education	Per 1-year increase	152	OR	0.89 (0.84–0.93)	2.4×10^{-6}
College/university	Yes versus No	32	OR	0.74 (0.63–0.86)	8.0×10^{-5}
Intelligence	Per 1-SD increase	16	OR	0.73 (0.57–0.93)	0.01
Smoking quantity	Per 10 cigarettes/ day increase	4	OR	0.69 (0.49–0.99)	0.04
Alcohol consumption	Per 1 drink/ week	3	OR	0.72 (0.50–1.04)	0.08
Coffee consumption	Per 1 cup/day	5	OR	1.26 (1.05–1.51)	0.01
25-Hydroxyvitamin D	Per 20% increase	4	OR	0.92 (0.85–0.98)	0.01
Serum folate	Per 1-SD increase	2	OR	0.98 (0.72–1.33)	0.89
Serum vitamin B_{12}	Per 1-SD increase	7	OR	1.11 (0.95–1.30)	0.18
Total homocysteine	Per 1-SD increase	18	OR	0.99 (0.88–1.11)	0.86
BMI	Per 1-SD increase	76	OR	1.05 (0.91–1.21)	0.51
Waist-to-hip ratio	Per 1-SD increase	38	OR	1.18 (0.97–1.45)	0.10
Type 2 diabetes	Yes versus No	50	OR	1.02 (0.97–1.07)	0.49
Fasting glucose	Per 1-SD increase	36	OR	1.14 (0.99–1.32)	0.07
Fasting insulin	Per 1-SD increase	19	OR	1.13 (0.85–1.51)	0.40
Systolic blood pressure	Per 1-SD increase	93	OR	0.94 (0.77–1.14)	0.51
Diastolic blood pressure	Per 1-SD increase	105	OR	0.96 (0.79–1.16)	0.65
HDL cholesterol	Per 1-SD increase	70	OR	0.98 (0.90–1.07)	0.64
LDL cholesterol	Per 1-SD increase	56	OR	1.07 (0.98–1.17)	0.14
Total cholesterol	Per 1-SD increase	73	OR	1.03 (0.94–1.12)	0.54
Triglycerides	Per 1-SD increase	40	OR	0.96 (0.87–1.06)	0.40
C-reactive protein	Per 1-SD increase	17	OR	1.04 (0.94–1.17)	0.44

BMI, body mass index; CI, confidence interval; LDL, low–density lipoprotein; OR, odds ratio; SD, standard deviation; SNPs, single-nucleotide polymorphisms.

heterogeneity could be identified in the differences in study design among the observational studies in dementia research. This observation could be the result of mixing patients with different subtypes of dementia in studies examining all types of dementia. On the contrary, in observational studies limited to one subtype of dementia, heterogeneity could arise from diagnostic misclassification or inclusion of mixed dementia cases. There is still no consensus in diagnostic criteria used for dementia diagnosis (Bellou et al., 2017). Furthermore, in the published umbrella review, small-study effect and excess significance bias were present in many meta-analyses, causing false-positive associations or inflated effect estimates. The phenomenon of small-study effect and excess significance bias is caused by the selective publication of studies showing statistically significant effects.

Conclusions

As already mentioned, the identification of credible risk factors for dementia is an important step toward disease prevention and early identification of patients. Meta-analyses of prospective cohort studies indicate strong hints that type 2 diabetes mellitus, depression, benzodiazepine use, frequency of social contacts, physical activity, and history of cancer are credible risk factors for dementia. Also, some additional factors, such as obesity and educational attainment, are supported by suggestive evidence. However, it is not yet clear whether these factors constitute causal factors in the pathways of the development of dementia. Indeed, these associations are based on observational studies. Even if an effort is made to modify these risk factors, the observed associations might not be translated into large preventive benefits for dementia (Bellou et al., 2017; Prasad et al., 2013). Although some large-scale efforts have examined the genetic determinants of dementia, only a small number of SNPs have been identified. Evidence from GWASs suggest that Alzheimer's disease does not share common genetic variants with major depressive disorder or cardiovascular diseases. MR analyses have only showed an association between genetically higher educational attainment and risk for Alzheimer's disease.

Summary points

- This chapter focuses on the environmental and genetic risk factors for dementia.
- Dementia is a clinically heterogeneous neurodegenerative disease that has a tremendous impact on patient quality of life and has major public health consequences.
- Prevention of dementia by adressing modifiable environmental risk factors is considered the most efficient method to stop the increasing incidence and burden of disease.
- Credible environmental risk factors for development of dementia are type 2 diabetes mellitus, depression, benzodiazepines use, physical activity, frequency of social contacts, and history of cancer.

- The presence of biases in current literature inflates or attenuates the observed effect size of examined factors. Such biases include recall bias, reverse causation, confounding, misclassification of disease, and competing risk of death.
- Genome-wide association studies identified 20 genetic loci associated with increased risk for Alzheimer's disease.
- A genome-wide association study for vascular dementia identified only one single-nucleotide polymorphism located near the androgen receptor gene on the X chromosome.
- Findings from genome-wide association studies do not support the hypothesis of shared genetic determinants between dementia and depression or cardiovascular diseases.
- A Mendelian randomization study suggested a potentially causal association between genetically determined educational attainment and risk for Alzheimer's disease.

Acknowledgment

Lazaros Belbasis and Vanesa Bellou are supported by Ph.D. scholarships funded by the IKY Greek State Scholarships Foundation.

References

Aarsland, D., et al. (2010). Is physical activity a potential preventive factor for vascular dementia? A systematic review. *Aging and Mental Health, 14*(4), 386–395. https://doi.org/10.1080/13607860903586136.

Alzheimer's Disease International. (2011). *World Alzheimer Report 2011: The benefits of early diagnosis and intervention.*

Alzheimer's Disease International. (2014). *World alzheimer report 2014: Dementia and risk reduction.*

Alzheimer's Disease International. (2015). *World Alzheimer's Report 2015: The global impact of dementia* (London).

Anttila, V., et al. (2017). Analysis of shared heritability in common disorders of the brain. *bioRxiv.* https://doi.org/10.1101/048991.

Baumgart, M., et al. (2015). Summary of the evidence on modifiable risk factors for cognitive decline and dementia: A population-based perspective. *Alzheimer's and Dementia, 11*(6), 718–726. https://doi.org/10.1016/j.jalz.2015.05.016.

Belbasis, L., Bellou, V., et al. (2015). Environmental risk factors and multiple sclerosis: An umbrella review of systematic reviews and meta-analyses. *The Lancet Neurology, 14*(3), 263–273. https://doi.org/10.1016/S1474-4422(14)70267-4.

Belbasis, L., Bellou, V., & Evangelou, E. (2016). Environmental risk factors and amyotrophic lateral sclerosis: An umbrella review and critical assessment of current evidence from systematic reviews and meta-analyses of observational studies. *Neuroepidemiology, 46*(2), 96–105. https://doi.org/10.1159/000443146.

Belbasis, L., Panagiotou, O. A., et al. (2015). A systematic appraisal of field synopses in genetic epidemiology: A HuGE review. *American Journal of Epidemiology, 181*(1), 1–16. https://doi.org/10.1093/aje/kwu249.

Bellou, V., et al. (2016). Environmental risk factors and Parkinson's disease: An umbrella review of meta-analyses. *Parkinsonism and Related Disorders, 23*, 1–9. https://doi.org/10.1016/j.parkreldis.2015.12.008.

Bellou, V., et al. (2017). Systematic evaluation of the associations between environmental risk factors and dementia: An umbrella review of systematic reviews and meta-analyses. *Alzheimer's and Dementia: The Journal of the Alzheimer's Association, 13*(4), 406–418. https://doi.org/10.1016/j.jalz.2016.07.152.

Bellou, V., et al. (2018). Risk factors for type 2 diabetes mellitus: An exposure-wide umbrella review of meta-analyses. *PLoS One, 13*(3), e0194127. https://doi.org/10.1371/journal.pone.0194127.

Bertram, L., et al. (2007). Systematic meta-analyses of Alzheimer disease genetic association studies: The AlzGene database. *Nature Genetics, 39*(1), 17–23. https://doi.org/10.1038/ng1934.

Blondell, S. J., Hammersley-Mather, R., & Veerman, J. (2014). Does physical activity prevent cognitive decline and dementia?: A systematic review and meta-analysis of longitudinal studies. *BMC Public Health, 14*(1), 510. https://doi.org/10.1186/1471-2458-14-510.

Bradford, A., et al. (2009). Missed and delayed diagnosis of dementia in primary care. *Alzheimer Disease and Associated Disorders, 23*(4), 306–314. https://doi.org/10.1097/WAD.0b013e3181a6bebc.

Brasure, M., et al. (2018). Physical activity interventions in preventing cognitive decline and Alzheimer-type dementia. *Annals of Internal Medicine, 168*(1), 30. https://doi.org/10.7326/M17-1528.

Bulik-Sullivan, B., et al. (2015). An atlas of genetic correlations across human diseases and traits. *Nature Genetics, 47*(11), 1236–1241. https://doi.org/10.1038/ng.3406.

Burns, A., & Iliffe, S. (2009). Dementia. *BMJ, 338*(feb05 1), b75. https://doi.org/10.1136/bmj.b75.

Byers, A. L., & Yaffe, K. (2011). Depression and risk of developing dementia. *Nature Reviews Neurology, 7*(6), 323–331. https://doi.org/10.1038/nrneurol.2011.60.

Cuyvers, E., & Sleegers, K. (2016). Genetic variations underlying Alzheimer's disease: Evidence from genome-wide association studies and beyond. *The Lancet. Neurology, 15*(8), 857–868. https://doi.org/10.1016/S1474-4422(16)00127-7.

Davey Smith, G., & Ebrahim, S. (2003). "Mendelian randomization": Can genetic epidemiology contribute to understanding environmental determinants of disease?*. *International Journal of Epidemiology, 32*(1), 1–22. https://doi.org/10.1093/ije/dyg070.

Department of Health Working Group. (2009). *Living well with dementia: A national dementia strategy.* Edited by D. of H. W. Group. London. United Kingdom: Department of Health.

Diniz, B. S., et al. (2013). Late-life depression and risk of vascular dementia and Alzheimer's disease: Systematic review and meta-analysis of community-based cohort studies. *British Journal of Psychiatry: The Journal of Mental Science, 202*(5), 329–335. https://doi.org/10.1192/bjp.bp.112.118307.

Dubois, B., et al. (2016). Timely diagnosis for Alzheimer's disease: A literature review on benefits and challenges. *Journal of Alzheimer's Disease: JAD, 49*(3), 617–631. https://doi.org/10.3233/JAD-150692.

Dwyer, R., et al. (2013). Using Alzgene-like approaches to investigate susceptibility genes for vascular cognitive impairment. *Journal of Alzheimer's Disease, 34*(1), 145–154. https://doi.org/10.3233/JAD-121069.

Fenoglio, C., et al. (2018). Role of genetics and epigenetics in the pathogenesis of Alzheimer's disease and frontotemporal dementia. *Journal of Alzheimer's Disease, 62*(3), 913–932. https://doi.org/10.3233/JAD-170702.

Fornage, M., et al. (2011). Genome-wide association studies of cerebral white matter lesion burden: The CHARGE consortium. *Annals of Neurology, 69*(6), 928–939. https://doi.org/10.1002/ana.22403.

Frankish, H., & Horton, R. (2017). Prevention and management of dementia: A priority for public health. *The Lancet, 390*(10113), 2614–2615. https://doi.org/10.1016/S0140-6736(17)31756-7.

Fratiglioni, L., & Wang, H.-X. (2007). Brain reserve hypothesis in dementia. *Journal of Alzheimer's Disease, 12*(1), 11–22.

Gibson, J., et al. (2017). 'Assessing the presence of shared genetic architecture between Alzheimer's disease and major depressive disorder using genome-wide association data. *Translational Psychiatry, 7*(4), e1094. https://doi.org/10.1038/tp.2017.49.

Gray, S. L., et al. (2016). Benzodiazepine use and risk of incident dementia or cognitive decline: Prospective population based study. *BMJ (Clinical Research Ed.), 352*, i90. Available from: http://www.ncbi.nlm.nih.gov/pubmed/26837813.

Gudala, K., et al. (2013). Diabetes mellitus and risk of dementia: A meta-analysis of prospective observational studies. *Journal of Diabetes Investigation, 4*(6), 640–650. https://doi.org/10.1111/jdi.12087.

Ikram, M. A., et al. (2017). Genetics of vascular dementia - review from the ICVD working group. *BMC Medicine, 15*(1), 48. https://doi.org/10.1186/s12916-017-0813-9.

Ioannidis, J. P. A., et al. (2008). Assessment of cumulative evidence on genetic associations: Interim guidelines. *International Journal of Epidemiology, 37*(1), 120–132. https://doi.org/10.1093/ije/dym159.

Karlsson, I. K., et al. (2017). Genetic susceptibility to cardiovascular disease and risk of dementia. *Translational Psychiatry, 7*(5), e1142. https://doi.org/10.1038/tp.2017.110.

Kuiper, J. S., et al. (2015). Social relationships and risk of dementia: A systematic review and meta-analysis of longitudinal cohort studies. *Ageing Research Reviews, 22*, 39–57. https://doi.org/10.1016/j.arr.2015. 04.006.

Larsson, S. C., et al. (2017). Modifiable pathways in Alzheimer's disease: Mendelian randomisation analysis. *BMJ (Clinical Research Ed.), 359*, j5375. Available from: http://www.ncbi.nlm.nih.gov/pubmed/ 29212772.

Lee, S. H., et al. (2013). 'Estimation and partitioning of polygenic variation captured by common SNPs for Alzheimer's disease, multiple sclerosis and endometriosis. *Human Molecular Genetics, 22*(4), 832–841. https://doi.org/10.1093/hmg/dds491.

Loef, M., & Walach, H. (2013). Midlife obesity and dementia: meta-analysis and adjusted forecast of dementia prevalence in the United States and China. *Obesity (Silver Spring, Md.), 21*(1), E51–E55. https:// doi.org/10.1002/oby.20037.

Loy, C. T., et al. (2014). Genetics of dementia. *The Lancet, 383*(9919), 828–840. https://doi.org/10.1016/ S0140-6736(13)60630-3.

MacArthur, J., et al. (2017). The new NHGRI-EBI Catalog of published genome-wide association studies (GWAS Catalog). *Nucleic Acids Research, 45*(D1), D896–D901. https://doi.org/10.1093/nar/gkw1133.

Muñoz, M., et al. (2016). Evaluating the contribution of genetics and familial shared environment to common disease using the UK Biobank. *Nature Genetics, 48*(9), 980–983. https://doi.org/10.1038/ ng.3618.

Paternoster, L., Chen, W., & Sudlow, C. L. M. (2009). Genetic determinants of white matter hyperintensities on brain scans: A systematic assessment of 19 candidate gene polymorphisms in 46 studies in 19,000 subjects. *Stroke, 40*(6), 2020–2026. https://doi.org/10.1161/STROKEAHA.108.542050.

Pedditizi, E., Peters, R., & Beckett, N. (2016). The risk of overweight/obesity in mid-life and late life for the development of dementia: A systematic review and meta-analysis of longitudinal studies. *Age and Ageing, 45*(1), 14–21. https://doi.org/10.1093/ageing/afv151.

Prasad, V., et al. (2013). Observational studies often make clinical practice recommendations: An empirical evaluation of authors' attitudes'. *Journal of Clinical Epidemiology, 66*(4), 361–366.e4. https://doi.org/ 10.1016/j.jclinepi.2012.11.005.

Prins, N. D., & Scheltens, P. (2015). White matter hyperintensities, cognitive impairment and dementia: An update. *Nature Reviews Neurology, 11*(3), 157–165. https://doi.org/10.1038/nrneurol.2015.10.

Richardson, K., et al. (2013). Statins and cognitive function: A systematic review. *Annals of Internal Medicine, 159*(10), 688–697. https://doi.org/10.7326/0003-4819-159-10-201311190-00007.

Ridge, P. G., et al. (2013). Alzheimer's disease: Analyzing the missing heritability. *PLoS One, 8*(11), e79771. https://doi.org/10.1371/journal.pone.0079771.

Ridge, P. G., et al. (2016). Assessment of the genetic variance of late-onset Alzheimer's disease. *Neurobiology of Aging, 41*, 200.e13–200.e20. https://doi.org/10.1016/j.neurobiolaging.2016.02.024.

Robinson, L., Tang, E., & Taylor, J.-P. (2015). Dementia: Timely diagnosis and early intervention. *BMJ, 350*. https://doi.org/10.1136/bmj.h3029. h3029–h3029.

Schrijvers, E. M. C., et al. (2012). Genome-wide association study of vascular dementia. *Stroke, 43*(2), 315–319. https://doi.org/10.1161/STROKEAHA.111.628768.

da Silva, J., et al. (2013). Affective disorders and risk of developing dementia: Systematic review. *The British Journal of Psychiatry, 202*(3), 177–186. https://doi.org/10.1192/bjp.bp.111.101931.

The Lancet. (2017). Dementia burden coming into focus. *The Lancet, 390*(10113), 2606. https://doi.org/ 10.1016/S0140-6736(17)33304-4.

Van Cauwenberghe, C., Van Broeckhoven, C., & Sleegers, K. (2016). The genetic landscape of Alzheimer disease: Clinical implications and perspectives. *Genetics in Medicine, 18*(5), 421–430. https://doi.org/ 10.1038/gim.2015.117.

Verhaaren, B. F. J., et al. (2015). Multiethnic genome-wide association study of cerebral white matter hyperintensities on MRI. *Circulation Cardiovascular Genetics, 8*(2), 398–409. https://doi.org/10.1161/ CIRCGENETICS.114.000858.

Wang, J., et al. (2015). Anti-inflammatory drugs and risk of Alzheimer's disease: An updated systematic review and meta-analysis. *Journal of Alzheimer's Disease, 44*(2), 385–396. https://doi.org/10.3233/JAD-141506.

Weuve, J., et al. (2015). Guidelines for reporting methodological challenges and evaluating potential bias in dementia research. *Alzheimer's and Dementia, 11*(9), 1098–1109. https://doi.org/10.1016/j.jalz.2015.06.1885.

Yaffe, K. (2018). Modifiable risk factors and prevention of dementia. *JAMA Internal Medicine, 178*(2), 281. https://doi.org/10.1001/jamainternmed.2017.7299.

Zhang, Q., et al. (2015). Inverse relationship between cancer and Alzheimer's disease: A systemic review meta-analysis. *Neurological Sciences: Official Journal of the Italian Neurological Society And of the Italian Society of Clinical Neurophysiology, 36*(11), 1987–1994. https://doi.org/10.1007/s10072-015-2282-2.

Zhong, G., et al. (2015). Association between benzodiazepine use and dementia: A meta-analysis. *PLoS One, 10*(5), e0127836. https://doi.org/10.1371/journal.pone.0127836. Edited by A. Aleman.

CHAPTER 12

Lipids, brain ageing, dementia, and lipidomics

Anne Poljak[1,2,3], Braidy Nady[1], Wong Matthew Wai Kin[1], Yue Liu[1], Mahboobeh Housseini[1], Sachdev Perminder Singh[1,4]

[1]Centre for Healthy Brain Ageing (CHeBA), School of Psychiatry, University of New South Wales, Sydney, New South Wales, Australia; [2]Bioanalytical Mass Spectrometry Facility (BMSF), University of New South Wales, Sydney, New South Wales, Australia; [3]School of Medical Sciences, University of New South Wales, Sydney, New South Wales, Australia; [4]Neuropsychiatric Institute, Euroa Centre, Prince of Wales Hospital, Sydney, New South Wales, Australia

List of abbreviations

AD Alzheimer disease
APO apolipoprotein
Aβ amyloid β
BACE1 beta-amyloid cleavage enzyme
BBB blood—brain barrier
CNS central nervous system
CSF cerebrospinal fluid
DAG diacylglycerol
DHA docosahexaenoic acid
EPA eicosapentaenoic acid
HDL high-density lipoprotein
HNE 4-hydroxynonenal
LDL low-density lipoprotein
PC phosphatidylcholine
PE phosphatidylethanolamine
PKC protein kinase C
PLA2 phospholipase A2
PS phosphatidylserine
PUFAs polyunsaturated fatty acids
SNP single nucleotide polymorphism
VLDL very low-density lipoprotein

Mini-dictionary of terms

Plasma Lipidome The full complement of lipids in plasma.
Oxidative Stress Damage to biological molecules (e.g., protein, lipids, DNA) resulting from attack by free radicals.
Proteotoxicity Cellular damage resulting from malformed/structurally compromised proteins.

Diagnosis and Management in Dementia
ISBN 978-0-12-815854-8, https://doi.org/10.1016/B978-0-12-815854-8.00012-4

Protein-Lipid Interactome Complex associations/assemblies formed between proteins and lipids, including, but not exclusively, high- and low-density lipoprotein particles, lipid rafts, and membrane bilayers.

Cognitive Decline A decline in cognitive ability that often occurs with ageing.

Neuroinflammation Activation of the brain's endogenous immune system (glial cells), which can happen in response to infection, trauma, toxins, etc., but at a subclinical level is also detected in a variety of age-related neurodegenerative diseases.

Synaptic Signaling Transmission of chemical signals between nerve cells via neurotransmitter molecules.

Lipoprotein Particles Particles of 500—600 nm size, found in blood and CSF, and composites of phospholipids, fatty acids, cholesterol, and lipophilic proteins, particularly apolipoproteins. They vary in size, density, and composition, and are named according to their relative density: very low-density lipoprotein (VLDL), low-density lipoprotein (LDL), and high-density lipoprotein (HDL).

Introduction

Importance of lipids in the brain: Lipids comprise over 50% of brain's dry weight, making it one of the most lipid-rich organs, second only to adipose tissue. Brain function is critically dependent on the multiple structural and functional roles that lipids serve, including signaling, maintenance of membrane structure and integrity, myelin sheath formation, neurotransmission, and protein associations in both plasma and brain. Brain physiological processes such as neuroplasticity, synaptic and mitochondrial function, lipid raft formation, and cell signaling are critically dependent on lipid composition. Additionally, lipid by-products, particularly metabolites of arachidonic acid and lipid products of oxidative stress (Fig. 12.1), are drivers of inflammation (Weiser, Butt, & Mohajeri, 2016). Lipids are known to play a vital role in brain development during gestation and childhood, but much less is known about how lipid profiles change with ageing and the role lipids play in the mechanisms of brain ageing.

Relevance of plasma lipids to brain function: Brain lipid and phospholipid content is dependent upon essential dietary polyunsaturated fatty acids (PUFAs), such as docosahexaenoic acid (DHA), for which de novo synthesis is negligible. Essential PUFAs are delivered from blood into the brain, either as free fatty acids or in the form of phospholipids such as lysophosphatidylcholine, through the blood—brain barrier (BBB) by both passive and active transport mechanisms. Some transport proteins involved are fatty acid—binding protein, endothelial lipases, sodium-dependent lysophosphatidylcholine symporters, and apolipoproteins. Consequently, the plasma lipidome, and the plethora of proteins that regulate it, have a considerable impact on the brain lipid environment, vascular function, and inflammatory processes. The plasma lipid profile also represents a lifestyle modifiable factor that can play a decisive role in the state of brain health and maintenance of cognitive function during ageing. A considerable proportion of plasma proteins are lipophilic and associate with lipids and lipoprotein particles (Gordon et al., 2013) and Table 12.2. Strong correlations have been shown between plasma apolipoproteins, cholesterol, and plasma lipoprotein particles (Song et al., 2012).

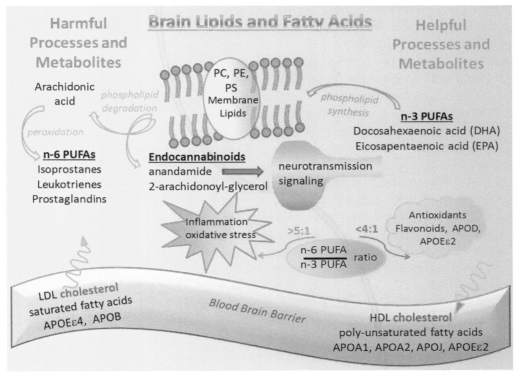

Figure 12.1 *The pleiotropic roles of brain lipids.* Lipids are transported across the blood–brain barrier by a variety of passive and active mechanisms, aided by plasma constituents such as LDL, HDL and interactions with lipophilic proteins such as apolipoproteins. Within the brain, lipids undergo a diversity of metabolic processes and subserve structural roles such as the cell membrane phospholipids, which are synthesized from dietary n-3 polyunsaturated fatty acids (n-3 PUFAs). Phospholipid degradation and peroxidation in turn produces a plethora of functional derivatives, including n-6 polyunsaturated fatty acids (n-6 PUFAs), which are mediators of inflammation, and endocannabinoids, which are lipid-signaling molecules and neuromodulators. The n-6 PUFA/n-3 PUFA ratio is an indicator of whether the brain environment is in an inflammatory or antioxidant state.

Plasma lipids

Lipids comprise a large portion of plasma metabolites, but apart from routinely measured cholesterol and triglycerides, detailed descriptions of the plasma lipidome have been lacking until the last decade. The introduction of mass spectrometry platforms in lipidomic analysis and the pioneering work of the LIPID Metabolites and Pathways Strategy (LIPID MAPS) consortium (www.lipidmaps.org) has greatly enhanced our knowledge of plasma lipids and introduced a six-category plasma lipid classification system: glycerolipids, glycerophospholipids, fatty acids, sphingolipids, sterols, and prenols (Quehenberger & Dennis, 2011). There are up to 100,000 lipids in the human lipidome (Shevchenko & Simons, 2010), suggesting a greater lipid diversity than found among proteins.

This diversity is related to variability in lipid structure, which can vary by size, carbon chain length, double bonds, and polarity of the head group relative to the more hydrophobic fatty acid tails (Brugger, 2014). It is also not uncommon for lipids of the same class to share the same molecular weight (isobaric species), which makes identification a daunting task. Nevertheless, this extreme structural diversity of lipids is of interest to researchers since lipids play many complex roles in health and disease.

Most circulating lipids are either derived from the diet (Fernandez & West, 2005) or directly synthesized by the liver. Plasma lipids generally do not solubilize well in blood and are transported through the bloodstream by associating with apolipoproteins in lipoprotein complexes, allowing redistribution to cells and tissues. While there is evidence that some lipids do cross the BBB, via ApoA-I (Stukas et al., 2014), or specific transporters (Nguyen et al., 2014), the distribution and synthesis of apolipoproteins is different between the central nervous system and the periphery (Mahley, 2016). For instance, ApoB is not present in CSF, and ApoE is the most abundant apolipoprotein in the central nervous system, but is produced locally by astrocytes and microglia. Thus, for a majority of lipids, the peripheral and central compartments regulate lipid storage and metabolism independently.

Plasma lipids have already been implicated in various health conditions, including metabolic syndrome, cardiovascular disease, cancer, preeclampsia, and sepsis (Meikle, Wong, Barlow, & Kingwell, 2014) (see Table 12.1), but are also of interest as potential biomarkers for early detection, diagnosis, or prognosis of neurological disorders, such as dementia (Wong, Braidy, Poljak, Pickford, et al., 2017; Wong, Braidy, Poljak, & Sachdev, 2017). In particular, a variety of lipid indices have been established as indicators of health or cardiovascular risk (Table 12.2).

Plasma lipids and the brain over the lifespan

Plasma lipids tend to increase from young adulthood to middle and old age (Lawton et al., 2008). One study found that individuals in the 58—66 years age bracket had lipid profiles corresponding to increased cardiovascular risk relative to individuals aged 25—32 years (Okecka-Szymanska, 2011). Another study focusing on similar age groups found significant associations of some lipids with age (Ishikawa et al., 2014), though age-related differences were more prominent among females with substantial increases (>1.5-fold) in many species of triglycerides. Changes in plasma lipids among much older subjects (>65 years) are still under investigation; there is a suggestion that centenarians could have a reversal in trend of lipids to reflect profiles of younger subjects (Jove et al., 2017), including reduced peroxidizability of fatty acids. This is salient since longevity in mammals was previously associated with a reduction in long-chain free fatty acids, especially polyunsaturated fatty acids (Jove et al., 2013). In centenarians, phospholipids and sphingolipids could signify a lipidome related to longevity (Montoliu et al., 2014). Plasmalogens, or ethyl phospholipids, were also increased, and have been known

Table 12.1 Plasma lipids as clinical biomarkers of disease.

	Lipid measure	Biomarker of conditions and/or diseases	Levels	Common assay approaches
Cholesterol	Total cholesterol	Hypercholesterolemia, cardiovascular risk	<200 mg/dL optimal >240 mg/dL elevated and risk factor for cardiovascular disease	Enzymatic assay, ultracentrifugation, electrophoresis, chemical precipitation, magnetic beads, homogeneous assays
	High-density lipoprotein cholesterol (HDL–C)	Decreased level associated with cardiovascular disease, tangier disease	Deficiency level; <20 mg/dL	
	Low-density lipoprotein cholesterol (LDL–C)	Familial combined hyperlipidemia, familial hypercholesterolemia, dysbetalipoproteinemia, phytosterolemia, cerebrotendinous xanthomatosis	Elevated level; >160 mg/dL optimal level; <100 mg/dL	
	Very low-density lipoprotein cholesterol (VLDL–C)	As for LDL–C		
	Small dense LDL–C (sdLDL–C)	Atherogenic dyslipidemia, cardiovascular disease, diabetes	Elevated level; >50 mg/dL Normal = 0.8 mmol/L diabetes = 1.45 mmol/L	
Fatty acids	Triglycerides (TG)	Hypertriglyceridemia, cardiovascular disease, insulin resistance	Extreme elevated level; >1000 mg/dL elevated level; >150 mg/dL	Enzymatic assay, mass spectrometry (LCMS or GCMS)

Continued

Table 12.1 Plasma lipids as clinical biomarkers of disease.—cont'd

	Lipid measure	Biomarker of conditions and/or diseases	Levels	Common assay approaches
	Plasma α-linolenic acid (ALA)	Cardiovascular risk	<12.0 μg/L increased cardiovascular risk >30.0 μg/L optimal	
	Plasma eicosapentaenoic acid (EPA)	Cardiovascular risk	<10.0 μg/L increased cardiovascular risk > 50.0 μg/L optimal >150.0 μg/L significantly decreased cardiovascular risk	
	Plasma docosahexaenoic acid (DHA)	Cardiovascular risk	<45.0 μg/L increased cardiovascular risk >100.0 μg/L optimal	
Lipoproteins	lipoprotein(a); Lp(a)	High levels associate with premature cardiovascular disease, stroke, genetic dyslipidemia, defects in blood clotting, diabetes, autoimmune diseases	Cutoffs not definitively established, >30 mg/dL suggested risk factor	Gel electrophoresis, ELISA
	Apolipoprotein A1	Assessing cardiovascular disease risk, HDL function and deficiency, tangier disease, lecithin: cholesteryl acyltransferase (LCAT) deficiency, hepatic lipase and cholesteryl ester transfer protein deficiency, cancer	Normal = 1.5 g/L decreased levels associated with disease	
	Apolipoprotein B	Abetalipoproteinemia, hypobetalipoproteinemia	Normal = 1 g/L increased levels associated with disease	

| **Phospholipid** | Cardiolipin | Barth syndrome, age-related neurodegenerative diseases, fatty liver disease, cardiovascular disease, tangier disease, cancer, diabetes, antiphospholipid syndrome | Normal range = 9.1–24.2 µg/mL in plasma | Thin layer chromatography (TLC), high-pressure liquid chromatography (HPLC), mass spectrometry (MS), enzymatic with fluorimetric detection |
| **Microbial lipids (gram-positive bacteria)** | Lipopolysaccharide (LPS) endotoxins
Lipoteichoic acid (LTA)
Lipoarabinomannan (LAM) | Sepsis, infection
Sepsis, infection
Tuberculosis detection | Detection of any level is indicative of infection | Bioassays such as the traditional limulus amebocyte lysate (LAL) bioassay, ELISA, LCMSMS |

A variety of lipid measures that are in current use for the diagnosis of a variety of conditions and diseases are summarized. This is a list of common measures, but not an exhaustive list.

The table has not previously been published, but the content is based on previously published work Deguchi, H., Fernandez, J. A., Hackeng, T. M., Banka, C. L., & Griffin, J. H. (2000). Cardiolipin is a normal component of human plasma lipoproteins. *Proceedings of the National Academy of Sciences of the United States of America, 97*(4), 1743–1748; Inaku, K. O., Ogunkeye, O. O., Abbiyesuku, F. M., Chuhwak, E. K., Isichei, C. O., Imoh, L. C., et al. (2017). Elevation of small, dense low density lipoprotein cholesterol–a possible antecedent of atherogenic lipoprotein phenotype in type 2 diabetes patients in Jos, North-Central Nigeria. *BMC Clinical Pathology, 17*, 26. doi: 10.1186/s12907-017-0065-9; Schaefer, E. J., Tsunoda, F., Diffenderfer, M., Polisecki, E., Thai, N., & Asztalos, B. (2000). The measurement of lipids, lipoproteins, apolipoproteins, fatty acids, and sterols, and next generation sequencing for the diagnosis and treatment of lipid disorders. In L. J. De Groot, G. Chrousos, K. Dungan, K. R. Feingold, A. Grossman, J. M. Hershman, et al. (Eds.), Endotext. *South Dartmouth (MA)*.

Table 12.2 Lipid indices of disease risk.

Lipid index	Level and associated health risk/benefit
Fatty acid index	>33% increased cardiovascular risk
	<30% optimal value and decreased cardiovascular risk
Transfatty acid index	>0.80% increased cardiovascular risk
Monounsaturated fatty acid index	<19% decreased, associated with increased cardiovascular risk
	>22% optimal value and decreased cardiovascular risk
Omega-6 fatty acid index	<41% decreased
	>46% increased
Unsaturated/saturated fatty acid ratio	<2.0 increased cardiovascular risk
	>2.25 optimal
Omega-3 fatty acid index	<1.85% increased cardiovascular risk
	>4.50% optimal
Total cholesterol (TC)/HDL-C ratio	>5 increased risk of cardiovascular disease, insulin resistance, and metabolism syndrome
	<5 decreased health risk
Triglycerides (TG)/HDL-C ratio	>3 increased risk of insulin resistance, cardiovascular disease
	<1.1 optimal
LDL-C/HDL-C ratio	>2.5 increased risk of cardiovascular disease
n-6 PUFA/n-3 PUFA ratio	>5 risk factor for obesity, inflammation
	<2–4 optimal

A variety of lipid indices of disease risk have been established, based on studies of large populations. The unsaturated/saturated fatty acid ratio is calculated by dividing the sum of beneficial fatty acids by the sum of harmful fatty acids. The omega-3 fatty acid index is calculated as (EPA + DHA)%/total fatty acids.
This table has not previously been published and is a summary of some lipid indices used to determine health risks. The levels are based on previously published work Arsenault, B. J., Rana, J. S., Stroes, E. S., Despres, J. P., Shah, P. K., Kastelein, J. J., et al. (2009). Beyond low-density lipoprotein cholesterol: Respective contributions of non-high-density lipoprotein cholesterol levels, triglycerides, and the total cholesterol/high-density lipoprotein cholesterol ratio to coronary heart disease risk in apparently healthy men and women. *Journal of the American College of Cardiology, 55*(1), 35–41. doi: 10.1016/j.jacc.2009.07.057; Baez-Duarte, B. G., Zamora-Ginez, I., Gonzalez-Duarte, R., Torres-Rasgado, E., Ruiz-Vivanco, G., Perez-Fuentes, R., et al. (2017). Triglyceride/high-density lipoprotein cholesterol (TG/HDL-C) index as a reference criterion of risk for metabolic syndrome (MetS) and low insulin sensitivity in apparently healthy subjects. *Gaceta Médica de México, 153*(2), 152–158; Schaefer, E. J., Tsunoda, F., Diffenderfer, M., Polisecki, E., Thai, N., & Asztalos, B. (2000). The measurement of lipids, lipoproteins, apolipoproteins, fatty acids, and sterols, and next generation sequencing for the diagnosis and treatment of lipid disorders. In L. J. De Groot, G. Chrousos, K. Dungan, K. R. Feingold, A. Grossman, J. M. Hershman, et al. (Eds.), *Endotext. South Dartmouth (MA)*; Simopoulos, A. P. (2016). An increase in the omega-6/omega-3 fatty acid ratio increases the risk for obesity. *Nutrients, 8*(3), 128. doi: 10.3390/nu8030128.

to act as scavengers for reactive oxygen species (Lessig & Fuchs, 2009). The lipidome of extreme-aged individuals may therefore be uniquely protective against oxidative stress and inflammation (Montoliu et al., 2014). Brain ageing is accompanied by a marked decrease in membrane lipids including phospholipids, cerebrosides, and sulfatides, which starts from the age of 20, and is greatly accelerated after the age of 80. Other age-related changes include decreased cholesterol in the human hippocampus and cortex (Soderberg, Edlund, Kristensson, & Dallner, 1990), as well as total brain gangliosides (Svennerholm, Bostrom, Jungbjer, & Olsson, 1994). These lipids are important components of cell membrane microdomains known as lipid rafts (Simons & Gerl, 2010), which are less fluid, detergent-resistant regions, which facilitate anchoring of key proteins involved in signal transduction. The altered physicochemical properties associated with lipid rafts could also directly modulate the function of ion channels or proteins involved in neuro-transmission and vesicular fusion (Dart, 2010). Thus any major changes to the distribution or function of lipid rafts and membrane fluidity could affect neuronal function and synaptic signaling in older age (Colin et al., 2016). Additionally, lipid rafts are regions that control amyloid beta processing via the beta-amyloid cleavage enzyme (BACE1), and altered raft composition may have a role in AD (Fabelo et al., 2014).

The role of lipids in brain ageing and cognition

In humans and nonhuman primates, normal ageing is associated with cognitive decline and is correlated with synaptic loss, decreased postsynaptic density, and increased abundance of nonsynaptic buttons (Morrison & Baxter, 2012). Furthermore, increased inflammation and oxidative stress and mitochondrial dysfunction are also age-associated processes that impact nervous and immune tissues (Braidy et al., 2014). Lipids can impact synaptic function (Marza et al., 2008) and membrane plasticity (Pinot et al., 2014), and in the central nervous system such physiological effects may affect cortical function (Stutz & Horvath, 2016). Furthermore, lipids both mediate and modulate neuroinflammation (Farooqui, Horrocks, & Farooqui, 2007). The n-3 PUFAs are partic-ularly important, as they are the most abundant substrates for membrane phospholipid synthesis. A meta-analysis (Yurko-Mauro, Alexander, & Van Elswyk, 2015) showed episodic memory improvements in older adults taking dietary supplements of DHA and EPA. Dietary enrichment with n-3 PUFAs as well as DHA/EPA supplements has been suggested to confer cognitive benefits (Dyall, 2017), whereas lower plasma levels of DHA/PC-DHA may have adverse consequences for cognitive function in older age groups (Tan et al., 2012). Plasma HDL levels correlate with cognitive function in advanced old age (Atzmon et al., 2002). By contrast, LDL-cholesterol has adverse effects on vascular and cognitive function (Ma et al., 2017), as do some of the metabolites of arachidonic acid, which are oxidation and membrane breakdown by-products, and drivers of inflammation (Keleshian, Modi, Rapoport, & Rao, 2013). Since much of the brain lipid milieu depends on dietary sources and/or BBB transport/clearance mech-anisms, they may also be altered either for good or harm by lifestyle variables such as diet, exercise, and gut microbiota (Tucker, 2016).

Lipid-derived metabolites and neurotoxicity

It is well established that lipid peroxidation can alter the structure and function of polyunsaturated fatty acids in membrane phospholipids. For instance, oxidative stress–induced lipid peroxidation can stimulate the production of bioactive aldehydes such as 4-hydroxyalkenals and malondialdehyde. Common products of lipid peroxidation, 4-hydroxynonenal (HNE) malondialdehyde, have been associated with lower neuronal viability in several in vitro and in vivo studies (Pizzimenti et al., 2013). Oxidative stress–mediated lipoperoxidation has also been shown to decrease intracellular reduced glutathione levels and increase its oxidized form (Liu et al., 2000). Increased levels of HNE have been shown to increase mitochondrial permeability and cytochrome c release, culminating in apoptotic cell death by stimulating the activity of several caspase enzymes (Mahr et al., 2014). Enhanced activity of glutathione transferase A4-4 activity and a subsequent increase in glutathionyl-HNE have been shown to protect against HNE toxicity and may likely represent a mechanism of tolerance (McElhanon et al., 2013). Cardiolipin is an anionic phospholipid that can inhibit the activity of glutathione transferase A4-4. It represents a crucial component of mitochondrial membranes, and increased oxidative stress can lead to a decline in the levels of this lipid (Moon et al., 2014). Oxidation of cardiolipin stimulates the release of fatty acids from phospholipids by PLA2. This subsequently leads to mitochondrial dysfunction and disintegration of the outer mitochondrial membrane, leading to the release of cytochrome c from the mitochondria and eventually cell death via an apoptotic mechanism (Vladimirov, Proskurnina, & Alekseev, 2013). Consequently, PLA2 inhibitors have been shown to ameliorate cell death in vitro. Phosphatidylserine (PS) has also been shown to regulate apoptotic signaling, and reduced levels of PS and increased neuronal cell death have been reported following exposure to oxidative stress (Akbar, Baick, Calderon, Wen, & Kim, 2006). Docosahexaenoic acid (DHA: 22:6n-3) has been shown to prevent neuronal apoptosis by enhancing accumulation of PS. Increased PLA2 activity and oxidation-mediated HNE production can reduce PS levels (Akbar et al., 2006).

Ceramides have been associated with mitochondrial impairment and induction of apoptotic pathways in several neurodegenerative disorders (Novgorodov, Gudz, & Obeid, 2008).Ceramides are formed in the central nervous system (CNS) by either a de novo synthesis or are a product of sphingomyelin hydrolysis. The neuronal expression of serine palmitoyltransferase was elevated in caspase 3-positive neurons following exposure to oxidative stress (Saito et al., 2010). This suggests that de novo ceramide synthesis is associated with apoptotic neurodegeneration in the CNS. While there are some protective benefits of ceramide synthase 6 (CerS6), this enzyme is involved in the production of C16-ceramides and, consequently, sphingomyelin and glucosylceramides. Sphingomyelin is necessary for the production of diacylglycerol, which can activate protein kinase C, and prevent apoptosis (Novgorodov et al., 2011).

While some lipids have been shown to promote neurotoxicity, others have demonstrated protective effects against oxidation. For instance, cholesterol is increased in neuronal membranes exposed to oxidative stress (Barcelo-Coblijn, Wold, Ren, & Murphy, 2013). Cholesterol is essential for the maintenance of membrane rigidity, and an increase in the cholesterol represents a compensatory mechanism to protect against oxidative damage to the cell membrane. Oxidation of cardiolipin leads to loss of membrane stability and poor rigidity.

Phosphatidylethanolamine is another important lipid that serves as a substrate for acyltransferases and N-acylphosphoethanolamine (NAPE). The levels of NAPE in the cell membrane are regulated by oxidative stress and inflammation (Subbanna, Shivakumar, Psychoyos, Xie, & Basavarajappa, 2013). Importantly, NAPE is a key precursor for the synthesis of N-acylethanolamines, which have important roles in learning and memory, neuroinflammation, oxidative stress, neuroprotection, and neurogenesis. Treatment with palmitoylethanolamine has been shown to protect neurons against oxidative stress, and apoptosis in vitro, and improve mouse survival in a murine model for chronic constriction injury (Duncan, Chapman, & Koulen, 2009). Additionally, acylethanolamines, and palmitoylethanolamine in particular, have important neuroprotective roles. Acylethanolamines are mainly distributed in the mitochondria. It is noteworthy to mention that treatment with ϖ type-3 unsaturated fatty acids and DHA exerts neuroprotection by enhancing the formation of acylethanolamine (Akbar et al., 2006). This suggests that increased levels of acylethanolamine can attenuate cellular damage due to oxidative stress.

Protein-lipid interactome and the role of lipids in proteotoxicity

Recent work shows that altered plasma lipid profiles precede diseases and disorders of brain ageing (Mapstone et al., 2014). By contrast, age-related brain pathologies are generally represented in terms of protein dysregulation, such as the amyloid, tau and presenilins in dementia, synuclein in Parkinson's, huntingtin in Huntington's, and mutant superoxide dismutase in amyotrophic lateral sclerosis. Interestingly, most of these proteins are associated with membranes, such as plasma membrane, vesicles, microtubules, or the cytoskeleton, with the exception of superoxide dismutase, which is an antioxidant. We propose that at a molecular level, a variety of lipid-protein interactions are the drivers of either successful processes or adverse events in ageing. Considering protein dysregulation in isolation, as is typically done in the literature, this is inadequate to a full understanding of the mechanisms that drive "amyloidosis" and diseases driven by protein structure dysregulation. This is because most of these proteins exist in lipid-rich environments, such as within or near cell membranes, and therefore lipids and proteins are partners in structural function or dysfunction. The main lipophilic component of plasma is comprised of the lipoprotein particles (LDL, HDL, VLDL, chylomicrons), which are assemblies of proteins and lipids, generally understood to be the vehicles for cholesterol and fatty acid transport. Previous studies have shown associations between apolipoproteins and cholesterol, HDL, and LDL (Song et al., 2012). Lipidation of APOE and APOA1 (i.e., HDL apolipoproteins) by ABCA1 may regulate clearance of misfolded/toxic proteins such as Aβ (Fan, Donkin, & Wellington, 2009), whereas plasma-derived lipids such as docosahexaenoic acid may inhibit Aβ aggregate formation (Grimm et al., 2011). Altered lipid composition may also trigger the unfolded protein response that ensures proteostasis (Hou & Taubert, 2014). Furthermore, mutations in lipid-associated genes can alter ligand-binding properties of the expressed protein, as in the case of fatty acid binding protein (FABP) (Shimamoto et al., 2014).

Fatty acids in ageing and dementia

Fatty acids (FAs) are essential structural constituents of cell membranes as well as many vital signaling biomolecules. The molecular structure of phospholipids and glycolipids that form significant parts of all cell membranes is predominantly made up of FAs. Both saturated and unsaturated fatty acids have important biological roles. With their relatively higher melting point, saturated fatty acids are solid at room temperature and as cellular membrane components, reduce membrane fluidity and permeability. They also increase blood LDL-cholesterol level, which in turn increases risk of vascular disease. By contrast, unsaturated fatty acids have at least one double bond, and mediate a variety of health benefits by: (A) reducing blood LDL-cholesterol, (B) increasing HDL-cholesterol, (C) lowering the risk of vascular disease (Tutino, Caruso, De Leonardis, De Nunzio, & Notarnicola, 2017), and (D) controlling blood sugar and insulin sensitivity and reducing inflammation (Bjermo et al., 2012). Most fatty acids can be synthesized within the body, except for linoleic acid and alpha-linolenic acid (omega-6 and omega-3 fats, respectfully), which are essential dietary fatty acids.

Fatty acids, particularly polyunsaturated FAs, are necessary for the normal development of the CNS (Zarate, El Jaber-Vazdekis, Tejera, Perez, & Rodriguez, 2017). The brain has a unique distribution of fatty acids, including large amounts of DHA (22:6n-3), arachidonic acid (AA, 20:4n-6) and palmitate (16:0), and much lower levels of other omega-3 PUFAs (Dyall, 2015). PUFAs are an integral part of several brain functions, including neurogenesis, neuron inflammation, and production of neurotransmitters (Balanza-Martinez et al., 2011). Furthermore, a variety of psychiatric diseases, including schizophrenia, depression, Parkinson's disease, and Alzheimer disease, are associated with PUFA deficiency (Sinn & Howe, 2008). Late-life dementias are on a significant upward trajectory, doubling every 5 years after age 65 (Qiu, Kivipelto, & von Strauss, 2009) with both environmental risk factors and genetic vulnerability playing significant roles (Cole, Ma, & Frautschy, 2009). Consequently, dietary factors, such as omega-3 fatty acids and antioxidants, may play an important role in moderating risk of cognitive decline, maintaining healthy neuronal function and cognitive reserve and minimizing dementia risk (Fotuhi et al., 2008). In particular, DHA protects nervous system function by multiple mechanisms, including reduction of arachidonic acid metabolite levels and increasing trophic signal transduction (Calon et al., 2004; Cole et al., 2009). DHA may also have a role in AD by reducing amyloid β (Aβ) production and accumulation and suppressing several signal transduction pathways induced by Aβ (Cole & Frautschy, 2010; Love, 2004; Mucke & Pitas, 2004). Furthermore, fatty acids may modulate protein misfolding in other neurodegenerative diseases, including Parkinson's disease (Sharon et al., 2003). Drug and/or dietary interventions targeting omega-3 fatty acids and associated pathways may therefore be a cost-effective approach with fewer side effects that can be used for prevention of dementia and other CNS diseases (Augustin et al., 2018), however, more work is needed, as a number of clinical trials utilizing omega-3 fatty acids have also been unsuccessful.

Lipids in dementia and dementia biomarkers

At present, the role of lipids as suitable biomarkers for AD remains to be elucidated. For instance, it remains unclear whether cholesterol constitutes a causative event in AD. However, several studies have shown that Aβ can perturb cholesterol homeostasis (Zarrouk, Nury, Hammami, & Lizard, 2015), and hypercholesterolemia in middle age represents a risk factor for the development of AD (Solomon, Kivipelto, Wolozin, Zhou, & Whitmer, 2009). APOE-ε4 represents the strongest genetic risk factor for AD compared to the more common ε3 allele, while the ε2 allele reduces the risk of developing AD (C. C. Liu, Liu, Kanekiyo, Xu, & Bu, 2013). APOE-ε4 has also been shown to promote amyloidogenesis. One study that examined the expression of apoE2, apoE3, and apoE4 in humans and mice showed that the amyloid load varied with E4 > E3 > E2 (Holtzman et al., 2000), while clearance of soluble Aβ trended E4 < E3 < E2 (Castellano et al., 2011). Another study found no association between cholesteryl ester transfer protein (CETP) gene polymorphisms and AD (Li et al., 2014). CETP is an important enzyme involved in cholesterol metabolism and facilitates the transfer of cholesteryl esters from HDL-cholesterol to LDL-cholesterol utilizing triacylglycerol (Barter & Kastelein, 2006).

Furthermore, oxysterols may also be involved in the pathogenesis of AD, due to its roles in neuroinflammation, Aβ accumulation, and cell death (Gamba et al., 2015). One oxysterol derivative, 24(S)- hydroxycholesterol (24S-OHC), may represent an important biomarker of AD (Jeitner, Voloshyna, & Reiss, 2011). 24S-OHC is an enzymatically oxidized cholesterol product produced in the brain, and neurons in particular, via the neuronal enzyme CYP46A1 (Lund, Guileyardo, & Russell, 1999). Impaired 24S−OHC synthesis inhibits the cholesterol biosynthetic pathway, leading to reduced levels of several metabolites including geranylgeraniol diphosphate, which is necessary for learning and memory and synaptic plasticity (Russell, Halford, Ramirez, Shah, & Kotti, 2009). Therefore, altered levels of 24S-OHC in CSF and/or in the plasma of AD patients may represent a useful biomarker of impaired cholesterol homeostasis. Either decreased or increased concentrations of 24S-OHC have been previously reported in both the plasma and CSF of AD patients and correlated with the expression levels of CYP46A1. Reduced levels of 24S-OHC in the plasma of AD patients correlated with increased brain atrophy, likely to be due to a decline in metabolically active neurons (Zuliani et al., 2011). The presence of the APOE-ε4 allele was associated with reduced levels of 24SOHC (Leoni & Caccia, 2011). Recently, other members of the oxysterol family, such as alkylaminooxysterols, have been associated with the pathogenesis of AD. Alkylaminooxysterols are formed from the aminolysis of 5α, 6α—epoxycholesterol with several other naturally occurring amines, including histamine, putrescine, spermidine, or spermine (de Medina, Paillasse, Payre, Silvente-Poirot, & Poirot, 2009). Dendrogenin B (5alpha-Hydroxy6beta-[3-(4-aminobutylamino)propylamino]cholest-7-en3beta-ol) has been shown to stimulate neurite outgrowth in several cell lines, and enhance neuronal survival at nanomolar concentrations (de Medina et al., 2009). Dendrogenin B (5alpha-Hydroxy6beta-[3-(4-aminobutylamino)propylamino]cholest-7-en3beta-ol) has been shown to stimulate neurite outgrowth in several cell lines, and enhance neuronal survival at nanomolar concentrations (de Medina et al., 2009).

Moreover, condensed membrane nanodomains or microdomains, otherwise known as lipid rafts, which are enriched in sphingolipids and cholesterol, represent major sites for the binding and oligomerization of amyloidogenic proteins. Ganglioside M1 has been shown to promote the release of Aβ into the extracellular medium (Lemkul & Bevan, 2011). The enzyme sphingomyelin synthase (which is required for GM1 production) promotes Aβ generation, and inhibition of enzyme activity impairs Aβ production in a dose- and time-dependent manner. The insertion of Aβ monomers into lipid monolayers are also regulated by electrostatic interactions between Aβ and phospholipid head groups, which enhance the cytotoxic effects of Aβ in neurons (Hsiao, Fu, Hill, Halliday, & Kim, 2013).

Furthermore, ceramide and ceramide 1-phosphate, sphingosine and sphingosine 1-phosphate are also involved in oxidative stress, neuroinflammation, and/or cell death, reported in AD (Farooqui, Ong, & Farooqui, 2010). For example, cardiolipin profile in the brain of 3xTg-AD mice is associated with early synaptic mitochondrial dysfunction in AD (Monteiro-Cardoso et al., 2015). Similarly, selective deficiency of ethanolamine plasmalogens and choline plasmalogens has been reported in postmortem brain samples from patients with AD (Igarashi et al., 2011). Low amounts of circulating plasmalogens have also been reported in serum from clinically and pathologically diagnosed AD subjects, and the rate of decline correlated with the severity of dementia (Wood et al., 2010). Plasmalogens are highly susceptible to oxidative stress, and therefore, it is not surprising that increased oxidative stress in AD may contribute to the observed decrease in plasmalogen levels.

At present, plasma or serum biomarkers for AD are of considerable interest in the diagnosis of AD, especially because blood-based biomarkers are noninvasive and can be performed routinely in the clinic. Circulating phospholipids have been associated with the development of AD (Mapstone et al., 2014). In particular, C18 ceramide and total sphingomyelins have been positively correlated with the levels of Aβ and tau in the CSF of a group of cognitively normal individuals that have a confirmed parental history of AD (Mielke et al., 2014). Serum sphingomyelin and ceramides also correlated well with memory impairment (Mielke et al., 2010). The levels of several choline phospholipids have been reported to be subnormal in AD plasma, and deficit in cognitively normal people that rapidly develop clinical MCI and AD within 2–3 years (Mapstone et al., 2014). Interestingly, a panel of metabolites including phosphatidylcholines (PCs), (PC diacyl (aa) C36:6, PC aa C38:0, PC aa C38:6, PC aa C40:1, PC aa C40:2, PC aa C40:6, PC acyl-alkyl (ae) C40:6), lysophosphatidylcholine (lysoPC a C18:2), and acylcarnitines (ACs) (Propionyl AC (C3) and C16:1-OH) were reported to be in deficit in the plasma of participants that have developed MCI and later AD compared to the nonconverter group (Mapstone et al., 2014), though replicating results in independent cohorts is required before a set of reliable lipid biomarkers can be instituted clinically for predictive or diagnostic context of use. Taken together, these studies suggest that the plasma lipidome may be a reliable and relatively inexpensive means of identifying subjects with MCI/AD, including potentially several years before clinical onset. Whether restoration of the levels of altered lipids to "healthy" levels can improve clinical onset and severity of AD remains to be evaluated.

Key facts

Key facts of plasma lipidomics

- Plasma comprises >10,000 individual lipid species, with heterogeneity arising from their structural diversity, not only in molecular weight differences but also in chain length, numbers of double bonds, and head groups.
- Lipidomics experiments are generally performed by chromatography on a C18 column, in-line to a mass spectrometer, which identifies lipid mass and structural subcomponents, using a technique called collision-induced dissociation. The entire process is often abbreviated as LCMSMS, standing for liquid chromatography with two tandem mass spectrometry events (molecular ion determination followed by fragment mass determination).
- Lipids in plasma can travel freely (such as free fatty acids), but are most often lipids associated with lipoprotein particles, which are protein-lipid-cholesterol composites of diverse size, range, and density and go by names such as VLDL, HDL, LDL, and chylomicrons (see Fig. 12.2).

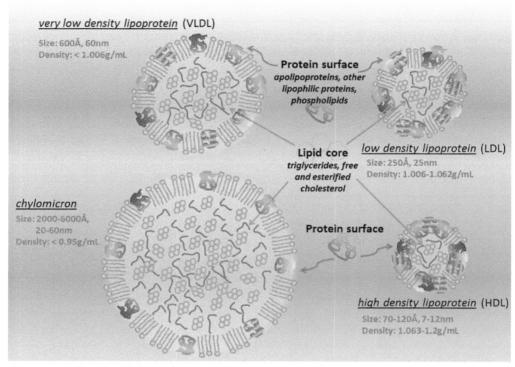

Figure 12.2 *Diversity of plasma lipoprotein particles.* Phospholipid particles are comprised of a lipid core (cholesterol, triacylglycerol) and a lipophilic surface comprising proteins (predominantly apolipoproteins, but also other lipophilic plasma proteins) and phospholipids. They are named by their density, such as high-density lipoprotein (HDL), which is relatively protein rich, higher density but smaller in diameter, and low-density lipoprotein (LDL), which has lower density than HDL but a larger diameter. Particle sizes and densities are on a continuum, but by convention are partitioned into a smaller number of categories, with commonly discussed variants shown in the schematic.

- Most fats can be synthesized in the body except for two essential fatty acids, viz. linoleic and α–linolenic (one of the 3 omega 3 fatty acids, the others being the longer-chain DHA and EPA), which must be obtained from the diet. Synthesis of the longer-chain fatty acids DHA and EPA may be impaired during ageing.
- Free fatty acids (FFAs) impact glucose metabolism by inhibiting insulin–stimulated glucose uptake and glycogen synthesis. Therefore, alterations in plasma levels of FFAs may play a role in the pathogenesis of dementia.

Key facts on the function of lipids in the brain

- Phospholipids form the membrane bilayer of cells, the composition of which can vary across cell types, and in which a variety of lipophilic membrane proteins are imbedded.

- Oxidation processes, often resulting from disease processes and poorly regulated metabolism, can damage lipid membranes, creating lipid peroxides and amplifying further damage to lipids and also other biopolymers such as DNA and proteins.
- Some metabolites or degradation products of lipids, such as arachidonic acid, are within pathways generating metabolites that have regulatory functions in their own right, such as isoprostanes, prostaglandins, and endocannabinoids.

Summary points

- This chapter focuses on the plasma lipidome and its association with brain function during ageing and its role in age-related neurodegenerative diseases such as dementia.
- Mechanisms by which plasma lipids may affect brain function are discussed, including changes to lipid profiles, associations with cognitive function, and interactions with specific proteins.
- What is known about changes to lipid expression profiles with ageing and in dementia is outlined.
- The roles of lipids in neurotoxicity and proteotoxicity are discussed.
- The roles of plasma fatty acids, impact on brain function, and their potential as biomarkers are reviewed.
- Lipids are examined as potential biomarkers of dementia.

Acknowledgments

The authors thank Dr. Hamid Roshan for his help with typographical editing of parts of the manuscript. This work was supported by a National Health and Medical Research Council of Australia Program Grant (APP1054544) to Prof. Perminder Sachdev. Dr. Nady Braidy is the recipient of an ARC DECRA Postdoctoral Fellowship at the University of New South Wales. Mr. Matthew Wai Kin Wong is the recipient of an Australian Postgraduate Award (APA) from the Australian Commonwealth Government, in support of his PhD candidature.

References

Akbar, M., Baick, J., Calderon, F., Wen, Z., & Kim, H. Y. (2006). Ethanol promotes neuronal apoptosis by inhibiting phosphatidylserine accumulation. *Journal of Neuroscience Research, 83*(3), 432—440. https://doi.org/10.1002/jnr.20744.

Arsenault, B. J., Rana, J. S., Stroes, E. S., Despres, J. P., Shah, P. K., Kastelein, J. J., et al. (2009). Beyond low-density lipoprotein cholesterol: Respective contributions of non-high-density lipoprotein cholesterol levels, triglycerides, and the total cholesterol/high-density lipoprotein cholesterol ratio to coronary heart disease risk in apparently healthy men and women. *Journal of the American College of Cardiology, 55*(1), 35—41. https://doi.org/10.1016/j.jacc.2009.07.057.

Atzmon, G., Gabriely, I., Greiner, W., Davidson, D., Schechter, C., & Barzilai, N. (2002). Plasma HDL levels highly correlate with cognitive function in exceptional longevity. *The Journals of Gerontology. Series A, Biological Sciences and Medical Sciences, 57*(11), M712—M715.

Augustin, K., Khabbush, A., Williams, S., Eaton, S., Orford, M., Cross, J. H., et al. (2018). Mechanisms of action for the medium-chain triglyceride ketogenic diet in neurological and metabolic disorders. *The Lancet Neurology, 17*(1), 84—93. https://doi.org/10.1016/S1474-4422(17)30408-8.

Baez-Duarte, B. G., Zamora-Ginez, I., Gonzalez-Duarte, R., Torres-Rasgado, E., Ruiz-Vivanco, G., Perez-Fuentes, R., et al. (2017). Triglyceride/high-density lipoprotein cholesterol (TG/HDL-C) index as a reference criterion of risk for metabolic syndrome (MetS) and low insulin sensitivity in apparently healthy subjects. *Gaceta Médica de México, 153*(2), 152—158.

Balanza-Martinez, V., Fries, G. R., Colpo, G. D., Silveira, P. P., Portella, A. K., Tabares-Seisdedos, R., et al. (2011). Therapeutic use of omega-3 fatty acids in bipolar disorder. *Expert Review of Neurotherapeutics, 11*(7), 1029—1047. https://doi.org/10.1586/ern.11.42.

Barcelo-Coblijn, G., Wold, L. E., Ren, J., & Murphy, E. J. (2013). Prenatal ethanol exposure increases brain cholesterol content in adult rats. *Lipids, 48*(11), 1059—1068. https://doi.org/10.1007/s11745-013-3821-3.

Barter, P. J., & Kastelein, J. J. (2006). Targeting cholesteryl ester transfer protein for the prevention and management of cardiovascular disease. *Journal of the American College of Cardiology, 47*(3), 492—499. https://doi.org/10.1016/j.jacc.2005.09.042.

Bjermo, H., Iggman, D., Kullberg, J., Dahlman, I., Johansson, L., Persson, L., et al. (2012). Effects of n-6 PUFAs compared with SFAs on liver fat, lipoproteins, and inflammation in abdominal obesity: A randomized controlled trial. *American Journal of Clinical Nutrition, 95*(5), 1003—1012. https://doi.org/10.3945/ajcn.111.030114.

Braidy, N., Poljak, A., Grant, R., Jayasena, T., Mansour, H., Chan-Ling, T., et al. (2014). Mapping NAD(+) metabolism in the brain of Wistar rats: Potential targets for influencing brain senescence. *Biogerontology, 15*(2), 177—198. https://doi.org/10.1007/s10522-013-9489-5.

Brugger, B. (2014). Lipidomics: Analysis of the lipid composition of cells and subcellular organelles by electrospray ionization mass spectrometry. *Annual Review of Biochemistry, 83*, 79—98. https://doi.org/10.1146/annurev-biochem-060713-035324.

Calon, F., Lim, G. P., Yang, F., Morihara, T., Teter, B., Ubeda, O., et al. (2004). Docosahexaenoic acid protects from dendritic pathology in an Alzheimer's disease mouse model. *Neuron, 43*(5), 633—645. https://doi.org/10.1016/j.neuron.2004.08.013.

Castellano, J. M., Kim, J., Stewart, F. R., Jiang, H., DeMattos, R. B., Patterson, B. W., et al. (2011). Human apoE isoforms differentially regulate brain amyloid-beta peptide clearance. *Science Translational Medicine, 3*(89). https://doi.org/10.1126/scitranslmed.3002156, 89ra57.

Cole, G. M., & Frautschy, S. A. (2010). DHA may prevent age-related dementia. *Journal of Nutrition, 140*(4), 869—874. https://doi.org/10.3945/jn.109.113910.

Cole, G. M., Ma, Q. L., & Frautschy, S. A. (2009). Omega-3 fatty acids and dementia. *Prostaglandins Leukotrienes and Essential Fatty Acids, 81*(2—3), 213—221. https://doi.org/10.1016/j.plefa.2009.05.015.

Colin, J., Gregory-Pauron, L., Lanhers, M. C., Claudepierre, T., Corbier, C., Yen, F. T., et al. (2016). Membrane raft domains and remodeling in aging brain. *Biochimie, 130*, 178—187. https://doi.org/10.1016/j.biochi.2016.08.014.

Dart, C. (2010). Lipid microdomains and the regulation of ion channel function. *The Journal of Physiology, 588*(Pt 17), 3169—3178. https://doi.org/10.1113/jphysiol.2010.191585.

Deguchi, H., Fernandez, J. A., Hackeng, T. M., Banka, C. L., & Griffin, J. H. (2000). Cardiolipin is a normal component of human plasma lipoproteins. *Proceedings of the National Academy of Sciences of the United States of America, 97*(4), 1743—1748.

Duncan, R. S., Chapman, K. D., & Koulen, P. (2009). The neuroprotective properties of palmitoylethanolamine against oxidative stress in a neuronal cell line. *Molecular Neurodegeneration, 4*, 50. https://doi.org/10.1186/1750-1326-4-50.

Dyall, S. C. (2015). Long-chain omega-3 fatty acids and the brain: A review of the independent and shared effects of EPA, DPA and DHA. *Frontiers in Aging Neuroscience, 7*, 52. https://doi.org/10.3389/fnagi.2015.00052.

Dyall, S. C. (2017). Interplay between n-3 and n-6 long-chain polyunsaturated fatty acids and the endocannabinoid system in brain protection and repair. *Lipids, 52*(11), 885—900. https://doi.org/10.1007/s11745-017-4292-8.

Fabelo, N., Martin, V., Marin, R., Moreno, D., Ferrer, I., & Diaz, M. (2014). Altered lipid composition in cortical lipid rafts occurs at early stages of sporadic Alzheimer's disease and facilitates APP/BACE1 interactions. *Neurobiology of Aging, 35*(8), 1801−1812. https://doi.org/10.1016/j.neurobiolaging.2014.02.005.

Fan, J., Donkin, J., & Wellington, C. (2009). Greasing the wheels of Abeta clearance in Alzheimer's disease: The role of lipids and apolipoprotein E. *BioFactors, 35*(3), 239−248. https://doi.org/10.1002/biof.37.

Farooqui, A. A., Horrocks, L. A., & Farooqui, T. (2007). Modulation of inflammation in brain: A matter of fat. *Journal of Neurochemistry, 101*(3), 577−599. https://doi.org/10.1111/j.1471-4159.2006.04371.x.

Farooqui, A. A., Ong, W. Y., & Farooqui, T. (2010). Lipid mediators in the nucleus: Their potential contribution to Alzheimer's disease. *Biochimica et Biophysica Acta, 1801*(8), 906−916. https://doi.org/10.1016/j.bbalip.2010.02.002.

Fernandez, M. L., & West, K. L. (2005). Mechanisms by which dietary fatty acids modulate plasma lipids. *Journal of Nutrition, 135*(9), 2075−2078. https://doi.org/10.1093/jn/135.9.2075.

Fotuhi, M., Zandi, P. P., Hayden, K. M., Khachaturian, A. S., Szekely, C. A., Wengreen, H., et al. (2008). Better cognitive performance in elderly taking antioxidant vitamins E and C supplements in combination with nonsteroidal anti-inflammatory drugs: The Cache County study. *Alzheimers Dement, 4*(3), 223−227. https://doi.org/10.1016/j.jalz.2008.01.004.

Gamba, P., Testa, G., Gargiulo, S., Staurenghi, E., Poli, G., & Leonarduzzi, G. (2015). Oxidized cholesterol as the driving force behind the development of Alzheimer's disease. *Frontiers in Aging Neuroscience, 7*, 119. https://doi.org/10.3389/fnagi.2015.00119.

Gordon, S. M., Deng, J., Tomann, A. B., Shah, A. S., Lu, L. J., & Davidson, W. S. (2013). Multi-dimensional co-separation analysis reveals protein-protein interactions defining plasma lipoprotein subspecies. *Molecular and Cellular Proteomics, 12*(11), 3123−3134. https://doi.org/10.1074/mcp.M113.028134.

Grimm, M. O., Kuchenbecker, J., Grosgen, S., Burg, V. K., Hundsdorfer, B., Rothhaar, T. L., et al. (2011). Docosahexaenoic acid reduces amyloid beta production via multiple pleiotropic mechanisms. *Journal of Biological Chemistry, 286*(16), 14028−14039. https://doi.org/10.1074/jbc.M110.182329.

Holtzman, D. M., Bales, K. R., Tenkova, T., Fagan, A. M., Parsadanian, M., Sartorius, L. J., et al. (2000). Apolipoprotein E isoform-dependent amyloid deposition and neuritic degeneration in a mouse model of Alzheimer's disease. *Proceedings of the National Academy of Sciences of the United States of America, 97*(6), 2892−2897. https://doi.org/10.1073/pnas.050004797.

Hou, N. S., & Taubert, S. (2014). Membrane lipids and the endoplasmic reticulum unfolded protein response: An interesting relationship. *Worm, 3*(3). https://doi.org/10.4161/21624046.2014.962405. e962405.

Hsiao, J. H., Fu, Y., Hill, A. F., Halliday, G. M., & Kim, W. S. (2013). Elevation in sphingomyelin synthase activity is associated with increases in amyloid-beta peptide generation. *PLoS One, 8*(8). https://doi.org/10.1371/journal.pone.0074016. e74016.

Igarashi, M., Ma, K., Gao, F., Kim, H. W., Rapoport, S. I., & Rao, J. S. (2011). Disturbed choline plasmalogen and phospholipid fatty acid concentrations in Alzheimer's disease prefrontal cortex. *Journal of the Alzheimers Disease, 24*(3), 507−517. https://doi.org/10.3233/JAD-2011-101608.

Inaku, K. O., Ogunkeye, O. O., Abbiyesuku, F. M., Chuhwak, E. K., Isichei, C. O., Imoh, L. C., et al. (2017). Elevation of small, dense low density lipoprotein cholesterol-a possible antecedent of atherogenic lipoprotein phenotype in type 2 diabetes patients in Jos, North-Central Nigeria. *BMC Clinical Pathology, 17*, 26. https://doi.org/10.1186/s12907-017-0065-9.

Ishikawa, M., Maekawa, K., Saito, K., Senoo, Y., Urata, M., Murayama, M., et al. (2014). Plasma and serum lipidomics of healthy white adults shows characteristic profiles by subjects' gender and age. *PLoS One, 9*(3). https://doi.org/10.1371/journal.pone.0091806. e91806.

Jeitner, T. M., Voloshyna, I., & Reiss, A. B. (2011). Oxysterol derivatives of cholesterol in neurodegenerative disorders. *Current Medicinal Chemistry, 18*(10), 1515−1525.

Jove, M., Naudi, A., Aledo, J. C., Cabre, R., Ayala, V., Portero-Otin, M., et al. (2013). Plasma long-chain free fatty acids predict mammalian longevity. *Scientific Reports, 3*, 3346. https://doi.org/10.1038/srep03346.

Jove, M., Naudi, A., Gambini, J., Borras, C., Cabre, R., Portero-Otin, M., et al. (2017). A stress-resistant lipidomic signature confers extreme longevity to humans. *The Journals of Gerontology. Serirs A, Biological Sciences and Medical Sciences, 72*(1), 30–37. https://doi.org/10.1093/gerona/glw048.

Keleshian, V. L., Modi, H. R., Rapoport, S. I., & Rao, J. S. (2013). Aging is associated with altered inflammatory, arachidonic acid cascade, and synaptic markers, influenced by epigenetic modifications, in the human frontal cortex. *Journal of Neurochemistry, 125*(1), 63–73. https://doi.org/10.1111/jnc.12153.

Lawton, K. A., Berger, A., Mitchell, M., Milgram, K. E., Evans, A. M., Guo, L., et al. (2008). Analysis of the adult human plasma metabolome. *Pharmacogenomics, 9*(4), 383–397. https://doi.org/10.2217/14622416.9.4.383.

Lemkul, J. A., & Bevan, D. R. (2011). Lipid composition influences the release of Alzheimer's amyloid beta-peptide from membranes. *Protein Science, 20*(9), 1530–1545. https://doi.org/10.1002/pro.678.

Leoni, V., & Caccia, C. (2011). Oxysterols as biomarkers in neurodegenerative diseases. *Chemistry and Physics of Lipids, 164*(6), 515–524. https://doi.org/10.1016/j.chemphyslip.2011.04.002.

Lessig, J., & Fuchs, B. (2009). Plasmalogens in biological systems: Their role in oxidative processes in biological membranes, their contribution to pathological processes and aging and plasmalogen analysis. *Current Medicinal Chemistry, 16*(16), 2021–2041.

Li, Q., Huang, P., He, Q. C., Lin, Q. Z., Wu, J., & Yin, R. X. (2014). Association between the CETP polymorphisms and the risk of Alzheimer's disease, carotid atherosclerosis, longevity, and the efficacy of statin therapy. *Neurobiology of Aging, 35*(6), 1513 e1513–1523. https://doi.org/10.1016/j.neurobiolaging.2013.12.032.

Liu, W., Kato, M., Akhand, A. A., Hayakawa, A., Suzuki, H., Miyata, T., et al. (2000). 4-hydroxynonenal induces a cellular redox status-related activation of the caspase cascade for apoptotic cell death. *Journal of Cell Science, 113*(Pt 4), 635–641.

Liu, C. C., Liu, C. C., Kanekiyo, T., Xu, H., & Bu, G. (2013). Apolipoprotein E and Alzheimer disease: Risk, mechanisms and therapy. *Nature Reviews Neurology, 9*(2), 106–118. https://doi.org/10.1038/nrneurol.2012.263.

Love, R. (2004). Good fats prevent dendritic damage in mouse model of AD. *The Lancet Neurology, 3*(11), 636.

Lund, E. G., Guileyardo, J. M., & Russell, D. W. (1999). cDNA cloning of cholesterol 24-hydroxylase, a mediator of cholesterol homeostasis in the brain. *Proceedings of the National Academy of Sciences of the United States of America, 96*(13), 7238–7243.

Mahley, R. W. (2016). Central nervous system lipoproteins: ApoE and regulation of cholesterol metabolism. *Arteriosclerosis, Thrombosis, and Vascular Biology, 36*(7), 1305–1315. https://doi.org/10.1161/ATVBAHA.116.307023.

Mahr, A., Batteux, F., Tubiana, S., Goulvestre, C., Wolff, M., Papo, T., et al. (2014). Brief report: Prevalence of antineutrophil cytoplasmic antibodies in infective endocarditis. *Arthritis and Rheumatism, 66*(6), 1672–1677. https://doi.org/10.1002/art.38389.

Mapstone, M., Cheema, A. K., Fiandaca, M. S., Zhong, X., Mhyre, T. R., MacArthur, L. H., et al. (2014). Plasma phospholipids identify antecedent memory impairment in older adults. *Nature Medicine, 20*(4), 415–418. https://doi.org/10.1038/nm.3466.

Marza, E., Long, T., Saiardi, A., Sumakovic, M., Eimer, S., Hall, D. H., et al. (2008). Polyunsaturated fatty acids influence synaptojanin localization to regulate synaptic vesicle recycling. *Molecular Biology of the Cell, 19*(3), 833–842. https://doi.org/10.1091/mbc.e07-07-0719.

Ma, C., Yin, Z., Zhu, P., Luo, J., Shi, X., & Gao, X. (2017). Blood cholesterol in late-life and cognitive decline: A longitudinal study of the Chinese elderly. *Molecular Neurodegeneration, 12*(1), 24. https://doi.org/10.1186/s13024-017-0167-y.

McElhanon, K. E., Bose, C., Sharma, R., Wu, L., Awasthi, Y. C., & Singh, S. P. (2013). Gsta4 null mouse embryonic fibroblasts exhibit enhanced sensitivity to oxidants: Role of 4-hydroxynonenal in oxidant toxicity. *Open Journal of Apoptosis, 2*(1). https://doi.org/10.4236/ojapo.2013.21001.

de Medina, P., Paillasse, M. R., Payre, B., Silvente-Poirot, S., & Poirot, M. (2009). Synthesis of new alkylaminooxysterols with potent cell differentiating activities: Identification of leads for the treatment of cancer and neurodegenerative diseases. *Journal of Medicinal Chemistry, 52*(23), 7765–7777. https://doi.org/10.1021/jm901063e.

Meikle, P. J., Wong, G., Barlow, C. K., & Kingwell, B. A. (2014). Lipidomics: Potential role in risk prediction and therapeutic monitoring for diabetes and cardiovascular disease. *Pharmacology and Therapeutics, 143*(1), 12—23. https://doi.org/10.1016/j.pharmthera.2014.02.001.

Mielke, M. M., Bandaru, V. V., Haughey, N. J., Rabins, P. V., Lyketsos, C. G., & Carlson, M. C. (2010). Serum sphingomyelins and ceramides are early predictors of memory impairment. *Neurobiology of Aging, 31*(1), 17—24. https://doi.org/10.1016/j.neurobiolaging.2008.03.011.

Mielke, M. M., Haughey, N. J., Bandaru, V. V. R., Zetterberg, H., Blennow, K., Andreasson, U., et al. (2014). Cerebrospinal fluid sphingolipids, beta-amyloid, and tau in adults at risk for Alzheimer's disease. *Neurobiology of Aging, 35*(11), 2486—2494. https://doi.org/10.1016/j.neurobiolaging.2014.05.019.

Monteiro-Cardoso, V. F., Oliveira, M. M., Melo, T., Domingues, M. R., Moreira, P. I., Ferreiro, E., et al. (2015). Cardiolipin profile changes are associated to the early synaptic mitochondrial dysfunction in Alzheimer's disease. *Journal of Alzheimers Disease, 43*(4), 1375—1392. https://doi.org/10.3233/JAD-141002.

Montoliu, I., Scherer, M., Beguelin, F., DaSilva, L., Mari, D., Salvioli, S., et al. (2014). Serum profiling of healthy aging identifies phospho- and sphingolipid species as markers of human longevity. *Aging, 6*(1), 9—25. https://doi.org/10.18632/aging.100630.

Moon, K. H., Tajuddin, N., Brown, J., 3rd, Neafsey, E. J., Kim, H. Y., & Collins, M. A. (2014). Phospholipase A2, oxidative stress, and neurodegeneration in binge ethanol-treated organotypic slice cultures of developing rat brain. *Alcoholism: Clinical and Experimental Research, 38*(1), 161—169. https://doi.org/10.1111/acer.12221.

Morrison, J. H., & Baxter, M. G. (2012). The ageing cortical synapse: Hallmarks and implications for cognitive decline. *Nature Reviews Neuroscience, 13*(4), 240—250. https://doi.org/10.1038/nrn3200.

Mucke, L., & Pitas, R. E. (2004). Food for thought: Essential fatty acid protects against neuronal deficits in transgenic mouse model of AD. *Neuron, 43*(5), 596—599. https://doi.org/10.1016/j.neuron.2004.08.025.

Nguyen, L. N., Ma, D., Shui, G., Wong, P., Cazenave-Gassiot, A., Zhang, X., et al. (2014). Mfsd2a is a transporter for the essential omega-3 fatty acid docosahexaenoic acid. *Nature, 509*(7501), 503—506. https://doi.org/10.1038/nature13241.

Novgorodov, S. A., Chudakova, D. A., Wheeler, B. W., Bielawski, J., Kindy, M. S., Obeid, L. M., et al. (2011). Developmentally regulated ceramide synthase 6 increases mitochondrial Ca^{2+} loading capacity and promotes apoptosis. *Journal of Biological Chemistry, 286*(6), 4644—4658. https://doi.org/10.1074/jbc.M110.164392.

Novgorodov, S. A., Gudz, T. I., & Obeid, L. M. (2008). Long-chain ceramide is a potent inhibitor of the mitochondrial permeability transition pore. *Journal of Biological Chemistry, 283*(36), 24707—24717. https://doi.org/10.1074/jbc.M801810200.

Okecka-Szymanska, J. (2011). Effects of age, gender and physical activity on plasma lipid profile. *Biomedical Human Kinetics, 3*(1), 1—9. https://doi.org/10.2478/v10101-011-0001-x.

Pinot, M., Vanni, S., Pagnotta, S., Lacas-Gervais, S., Payet, L. A., Ferreira, T., et al. (2014). Lipid cell biology. Polyunsaturated phospholipids facilitate membrane deformation and fission by endocytic proteins. *Science, 345*(6197), 693—697. https://doi.org/10.1126/science.1255288.

Pizzimenti, S., Ciamporcero, E., Daga, M., Pettazzoni, P., Arcaro, A., Cetrangolo, G., et al. (2013). Interaction of aldehydes derived from lipid peroxidation and membrane proteins. *Frontiers in Physiology, 4*, 242. https://doi.org/10.3389/fphys.2013.00242.

Qiu, C., Kivipelto, M., & von Strauss, E. (2009). Epidemiology of Alzheimer's disease: Occurrence, determinants, and strategies toward intervention. *Dialogues in Clinical Neuroscience, 11*(2), 111—128.

Quehenberger, O., & Dennis, E. A. (2011). The human plasma lipidome. *New England Journal of Medicine, 365*(19), 1812—1823. https://doi.org/10.1056/NEJMra1104901.

Russell, D. W., Halford, R. W., Ramirez, D. M., Shah, R., & Kotti, T. (2009). Cholesterol 24-hydroxylase: An enzyme of cholesterol turnover in the brain. *Annual Review of Biochemistry, 78*, 1017—1040. https://doi.org/10.1146/annurev.biochem.78.072407.103859.

Saito, M., Chakraborty, G., Hegde, M., Ohsie, J., Paik, S. M., Vadasz, C., et al. (2010). Involvement of ceramide in ethanol-induced apoptotic neurodegeneration in the neonatal mouse brain. *Journal of Neurochemistry, 115*(1), 168—177. https://doi.org/10.1111/j.1471-4159.2010.06913.x.

Schaefer, E. J., Tsunoda, F., Diffenderfer, M., Polisecki, E., Thai, N., & Asztalos, B. (2000). The measurement of lipids, lipoproteins, apolipoproteins, fatty acids, and sterols, and next generation sequencing for the diagnosis and treatment of lipid disorders. In L. J. De Groot, G. Chrousos, K. Dungan, K. R. Feingold, A. Grossman, J. M. Hershman, et al. (Eds.), *Endotext*. South Dartmouth (MA).

Sharon, R., Bar-Joseph, I., Frosch, M. P., Walsh, D. M., Hamilton, J. A., & Selkoe, D. J. (2003). The formation of highly soluble oligomers of alpha-synuclein is regulated by fatty acids and enhanced in Parkinson's disease. *Neuron, 37*(4), 583–595.

Shevchenko, A., & Simons, K. (2010). Lipidomics: Coming to grips with lipid diversity. *Nature Reviews Molecular Cell Biology, 11*(8), 593–598. https://doi.org/10.1038/nrm2934.

Shimamoto, C., Ohnishi, T., Maekawa, M., Watanabe, A., Ohba, H., Arai, R., et al. (2014). Functional characterization of FABP3, 5 and 7 gene variants identified in schizophrenia and autism spectrum disorder and mouse behavioral studies. *Human Molecular Genetics, 23*(24), 6495–6511. https://doi.org/10.1093/hmg/ddu369.

Simons, K., & Gerl, M. J. (2010). Revitalizing membrane rafts: New tools and insights. *Nature Reviews Molecular Cell Biology, 11*(10), 688–699. https://doi.org/10.1038/nrm2977.

Simopoulos, A. P. (2016). An increase in the omega-6/omega-3 fatty acid ratio increases the risk for obesity. *Nutrients, 8*(3), 128. https://doi.org/10.3390/nu8030128.

Sinn, N., & Howe, P. R. C. (2008). Mental health benefits of omega-3 fatty acids may be mediated by improvements in cerebral vascular function. *Bioscience Hypotheses, 1*(2), 103–108.

Soderberg, M., Edlund, C., Kristensson, K., & Dallner, G. (1990). Lipid compositions of different regions of the human brain during aging. *Journal of Neurochemistry, 54*(2), 415–423.

Solomon, A., Kivipelto, M., Wolozin, B., Zhou, J., & Whitmer, R. A. (2009). Midlife serum cholesterol and increased risk of Alzheimer's and vascular dementia three decades later. *Dementia and Geriatric Cognitive Disorders, 28*(1), 75–80. https://doi.org/10.1159/000231980.

Song, F., Poljak, A., Crawford, J., Kochan, N. A., Wen, W., Cameron, B., et al. (2012). Plasma apolipoprotein levels are associated with cognitive status and decline in a community cohort of older individuals. *PLoS One, 7*(6). https://doi.org/10.1371/journal.pone.0034078. e34078.

Stukas, S., Robert, J., Lee, M., Kulic, I., Carr, M., Tourigny, K., et al. (2014). Intravenously injected human apolipoprotein A-I rapidly enters the central nervous system via the choroid plexus. *Journal of the American Heart Association, 3*(6). https://doi.org/10.1161/JAHA.114.001156. e001156.

Stutz, B., & Horvath, T. L. (2016). Synaptic lipids in cortical function and psychiatric disorders. *EMBO Molecular Medicine, 8*(1), 3–5. https://doi.org/10.15252/emmm.201505749.

Subbanna, S., Shivakumar, M., Psychoyos, D., Xie, S., & Basavarajappa, B. S. (2013). Anandamide-CB1 receptor signaling contributes to postnatal ethanol-induced neonatal neurodegeneration, adult synaptic, and memory deficits. *Journal of Neuroscience, 33*(15), 6350–6366. https://doi.org/10.1523/JNEUROSCI.3786-12.2013.

Svennerholm, L., Bostrom, K., Jungbjer, B., & Olsson, L. (1994). Membrane lipids of adult human brain: Lipid composition of frontal and temporal lobe in subjects of age 20 to 100 years. *Journal of Neurochemistry, 63*(5), 1802–1811.

Tan, Z. S., Harris, W. S., Beiser, A. S., Au, R., Himali, J. J., Debette, S., et al. (2012). Red blood cell omega-3 fatty acid levels and markers of accelerated brain aging. *Neurology, 78*(9), 658–664. https://doi.org/10.1212/WNL.0b013e318249f6a9.

Tucker, K. L. (2016). Nutrient intake, nutritional status, and cognitive function with aging. *Annals of the New York Academy of Sciences, 1367*(1), 38–49. https://doi.org/10.1111/nyas.13062.

Tutino, V., Caruso, M. G., De Leonardis, G., De Nunzio, V., & Notarnicola, M. (2017). Tissue fatty acid profile is differently modulated from olive oil and omega-3 polyunsaturated fatty acids in ApcMin/+ mice. *Endocrine, Metabolic and Immune Disorders - Drug Targets, 17*(4), 303–308. https://doi.org/10.2174/1871530317666170911161623.

Vladimirov, Y. A., Proskurnina, E. V., & Alekseev, A. V. (2013). Molecular mechanisms of apoptosis. structure of cytochrome c-cardiolipin complex. *Biochemistry, 78*(10), 1086–1097. https://doi.org/10.1134/S0006297913100027.

Weiser, M. J., Butt, C. M., & Mohajeri, M. H. (2016). Docosahexaenoic acid and cognition throughout the lifespan. *Nutrients, 8*(2), 99. https://doi.org/10.3390/nu8020099.

Wong, M. W., Braidy, N., Poljak, A., Pickford, R., Thambisetty, M., & Sachdev, P. S. (2017a). Dysregulation of lipids in Alzheimer's disease and their role as potential biomarkers. *Alzheimers Dement, 13*(7), 810–827. https://doi.org/10.1016/j.jalz.2017.01.008.

Wong, M. W., Braidy, N., Poljak, A., & Sachdev, P. S. (2017b). The application of lipidomics to biomarker research and pathomechanisms in Alzheimer's disease. *Current Opinion in Psychiatry, 30*(2), 136–144. https://doi.org/10.1097/YCO.0000000000000303.

Wood, P. L., Mankidy, R., Ritchie, S., Heath, D., Wood, J. A., Flax, J., et al. (2010). Circulating plasmalogen levels and Alzheimer disease assessment scale-cognitive scores in Alzheimer patients. *Journal of Psychiatry and Neuroscience, 35*(1), 59–62.

Yurko-Mauro, K., Alexander, D. D., & Van Elswyk, M. E. (2015). Docosahexaenoic acid and adult memory: A systematic review and meta-analysis. *PLoS One, 10*(3). https://doi.org/10.1371/journal.pone.0120391. e0120391.

Zarate, R., El Jaber-Vazdekis, N., Tejera, N., Perez, J. A., & Rodriguez, C. (2017). Significance of long chain polyunsaturated fatty acids in human health. *Clinical and Translational Medicine, 6*(1), 25. https://doi.org/10.1186/s40169-017-0153-6.

Zarrouk, A., Nury, T., Hammami, M., & Lizard, G. (2015). Oligomerized amyloid-β1-40 peptide favors cholesterol, oxysterol, and fatty acid accumulation in human neuronal SK-N-be cells. *International Journal of Clinical Medicine, 6*, 813–824. http://creativecommons.org/licenses/by/4.0/.

Zuliani, G., Donnorso, M. P., Bosi, C., Passaro, A., Dalla Nora, E., Zurlo, A., et al. (2011). Plasma 24S-hydroxycholesterol levels in elderly subjects with late onset Alzheimer's disease or vascular dementia: A case-control study. *BMC Neurology, 11*, 121. https://doi.org/10.1186/1471-2377-11-121.

CHAPTER 13

The 5×FAD mouse model of Alzheimer's disease

Caroline Ismeurt[1], Patrizia Giannoni[2], Sylvie Claeysen[1]

[1]IGF, Univ Montpellier, CNRS, INSERM, Montpellier, France; [2]EA7352 CHROME, University of Nîmes, Nîmes, France

List of abbreviations

AD Alzheimer's disease
APP amyloid precursor protein
Aβ amyloid-β peptide
BACE1 β-secretase
BDNF brain-derived neurotrophic factor
CAA cerebral amyloid angiopathy
FAD familial Alzheimer's disease
GFAP glial fibrillary acidic protein
Iba1 ionized calcium-binding adapter molecule 1
IL interleukin
LTP long-term potentiation
NFT neurofibrillary tangle
PHFs paired helical filaments
PSEN1 presenilin 1
TGFβ transforming growth factor β
TLR Toll-like receptor
TNFα tumor necrosis factor α

Mini-dictionary of terms

Amyloid plaques extracellular aggregates mainly composed of neurotoxic amyloid peptides and found in postmortem brains of AD patients. With aging, plaques can also appear in the brains of cognitively healthy people.

Neurofibrillary tangles intracellular fibrillary structures composed of hyperphosphorylated tau protein. These aggregates are a marker of AD but are also found in tauopathies.

Complement system part of the immune system and corresponds to plasma proteins that recognize pathogens and participate in their elimination.

Novel object recognition test memory test including a training phase in which the mouse can explore two similar objects and a restitution phase in which one of the objects is replaced by a novel object. The amount of time between the training and the restitution phase allows analysis of the short-term memory (e.g., 1 h) or the long-term memory (e.g., 24 h).

Pericytes perivascular cells involved in the regulation of capillary blood flow and maintenance of the BBB.

Diagnosis and Management in Dementia
ISBN 978-0-12-815854-8, https://doi.org/10.1016/B978-0-12-815854-8.00013-6

Introduction

Many transgenic mouse models have been developed to reproduce the numerous brain features of Alzheimer's disease (AD), ranging from amyloid-β (Aβ) peptide deposition to neurofibrillary tangles (NFTs), neuroinflammation and synaptic alterations, and neuronal loss (Selkoe and Schenk, 2003). As wild-type mice develop neither plaques nor tangles in their life span, the introduction of human genes bearing mutations involved in familial forms of the disease was a key step in recapitulating AD pathology (Oakley et al., 2006). The heterologous mouse line Tg6799, also named 5×FAD, expresses five familial AD mutations under the control of the neuron-specific *thy1* promoter: three mutations on the human amyloid protein precursor (APP695; London, Swedish, and Florida) and two mutations on human presenilin-1 (PS1; M146L and L286V) (Fig. 13.1). No human tau transgene is expressed in this mouse line. The 5×FAD mouse line is the transgenic mouse model of AD presenting the earliest onset and most intense amyloid pathology. Interestingly, this mouse model recapitulates all the neuropathological features associated with AD: amyloid plaques, gliosis, cognitive impairment, neurotransmission defects, synaptic loss, and neurodegeneration (Fig. 13.2). The only missing features are the tau tangles, even if some reports are contradictory on this point. The AD-like pathology in 5×FAD mice can be divided into three phases as in human AD. The asymptomatic phase, around 2 months of age, is characterized by the presence of amyloid plaques and appearance of gliosis (Fig. 13.2). The prodromal phase, with extensive amyloid deposits and gliosis, is defined by the appearance of cognitive deficits around 4 months of age. The symptomatic phase starts around 6 months of age and over, with extreme amyloid pathology and associated neuroinflammation, marked cognitive deficits, and neurotransmission impairments. Finally, late AD-like stages, or "severe AD-like pathology," are characterized by synaptic and neuronal loss around 9 months of age. This chapter describes all the neuropathological and behavioral features of the 5×FAD mouse model to highlight its interest for fundamental research and preclinical studies.

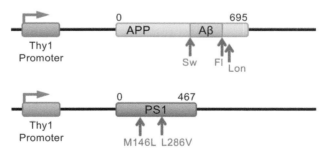

Figure 13.1 Schematic representation of the 5×FAD transgenes. *Arrows* indicate the familial Alzheimer's disease mutations in human amyloid precursor protein (*APP*) and presenilin-1 (*PS1*) transgenes that are under the control of the neuron-specific *thy1* promotor. Coding regions are depicted. For a more complete view of the untranslated regions see Oakley et al. (2006). *Aβ*, amyloid-β peptide.

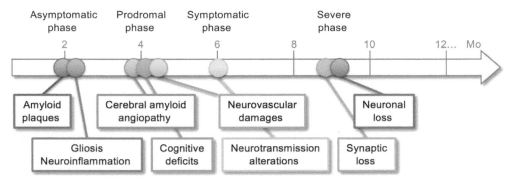

Figure 13.2 Schematic representation of pathology time course in the 5×FAD mouse model. Key Alzheimer's disease features that are reproduced by the mouse *line* are summarized in accordance with the time scale. The putative correspondence with clinical disease is indicated. As it is controversial in the model, tau pathology is not indicated in this picture.

Amyloid pathology

The overproduction and aggregation of Aβ in the brain are major pathological hallmarks of AD. Aβ is produced and released in the extracellular domain through the amyloidogenic pathway of APP processing. This pathway involves two successive cleavages by proteases: BACE1 and γ-secretase. The cleavage by γ-secretase is not precise and leads to several Aβ peptides with different lengths, including $A\beta_{1-40}$ (40 amino acids) and $A\beta_{1-42}$ (42 amino acids), the two main Aβ isoforms. In the 5×FAD mouse model, the three FAD mutations on the human APP transgene are located around the cleavage sites of β- and γ-proteases and are involved in the overproduction of either total Aβ or specifically $A\beta_{1-42}$ peptides. The London (V717I) and Florida (I716V) mutations, located around the γ-secretase cleavage site, have been reported to induce an increase in the $A\beta_{1-42}/A\beta_{1-40}$ ratio (De Jonghe et al., 2001; Herl et al., 2009). The Swedish mutation (K670N/M671L), close to the BACE1 cleavage site, has been reported to induce an increase in total Aβ (Hsiao et al., 1996). The two FAD mutations present on the PSEN1 transgene (catalytic subunit of γ-secretase) have been shown to also increase the $A\beta_{1-42}/A\beta_{1-40}$ ratio (Citron et al., 1997; Scheuner et al., 1996). Due to a constitutive expression of the transgene, an overproduction of Aβ is detectable as early as 1.5 months of age in 5×FAD mouse brains. This amyloid accumulation corresponds mainly to soluble Aβ peptides ($A\beta_{1-40}$ and $A\beta_{1-42}$). At the age of 2 months, Aβ oligomers aggregate in Aβ fibrils and diffuse amyloid plaques, while forming dense amyloid plaques later on. The amyloid pathology begins in the deep cortical layers of the brain and subiculum. Then, with aging of the mice, plaques spread to cover much of the cortex and hippocampus (Schaeffer et al., 2011) (Fig. 13.3). This increase in plaque number and density reaches a plateau at 10 months of age in male 5×FAD mice, whereas in females the maximal density is not even reached at 14 months of age (Bhattacharya, Haertel,

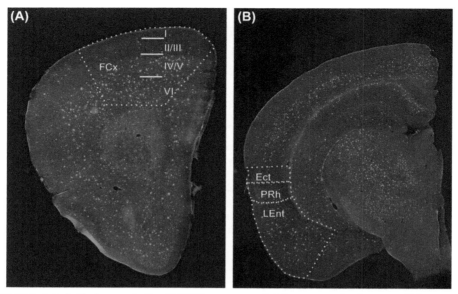

Figure 13.3 Amyloid plaques in the 5×FAD mouse model. Representative images of 30-μm sagittal brain sections in which plaques are stained with thioflavin-S in (A) the frontal cortex and (B) the entorhinal cortex. *I–VI*, cortical layers I to VI; *Ect*, ectorhinal area; *FCx*, frontal cortex; *LEnt*, entorhinal area, lateral part; *PRh*, perirhinal area.

Maelicke, & Montag, 2014). This amyloid load has deleterious effects on the brain, including the development of a persistent neuroinflammation, neuronal loss, cognitive deficits, and inhibition of long-term potentiation (LTP).

Neuroinflammation

Resident microglia and astrocytes, which are innate immune cells of the brain, are clustered around amyloid plaques (Perlmutter, Barron, & Chui, 1990) and are mediators of the neuroinflammation commonly observed in AD (Fig. 13.4). In a healthy brain, the microglia are in a resting state and survey the surrounding area. In the 5×FAD mouse brain, the Aβ pathology induces the activation of microglia into an antiinflammatory state, called M2. This activation is mediated mainly by Toll-like receptor 2, which recognizes $A\beta_{1-42}$ (Liu et al., 2012). M2 microglia release antiinflammatory cytokines such as interleukin-10 (IL-10) and transforming growth factor β (Orihuela, McPherson, & Harry, 2016) and have a phagocytic activity that contributes to the clearance of Aβ oligomers. Nevertheless, as the amyloid pathology is persistent, microglial cells fail to clear all Aβ peptides and transit chronically into a proinflammatory state, called M1. Several cytokines, such as IL-6, IL-β, and tumor necrosis factor α, are released, inducing a persistent neuroinflammation, which aggravates brain damage (Spangenberg et al., 2016). This proinflammatory environment is believed to appear around the age of 2 months, when the amyloid deposits have arisen. However, the microglial activation may occur

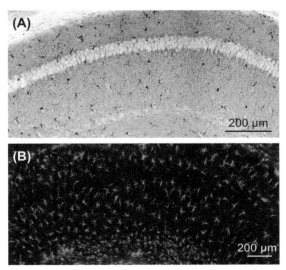

Figure 13.4 Microgliosis and astrogliosis in the 5×FAD mouse model. Representative images of 30-μm sagittal brain sections in the CA1 region of the hippocampus. (A) Ionized calcium-binding adapter molecule 1 staining of microglia. (B) Glial fibrillary acidic protein staining of astrocytes.

earlier (Boza-Serrano, Yang, Paulus, & Deierborg, 2018). The astrocytes, the second type of immune cells in the brain, are activated in a manner proportional to the level of $A\beta_{1-42}$ and amyloid plaques (Oakley et al., 2006). This activation is mediated by C1q, one component of the complement system, and induces the transition of astrocytes into reactive astrocytes (Fonseca, Zhou, Botto, & Tenner, 2004). The latter have a phagocytic activity to clear Aβ peptides and produce neurotrophic factors such as brain-derived neurotrophic factor (Kimura, Takahashi, Tashiro, & Terao, 2006; Wyss-Coray et al., 2003). However, reactive astrocytes also produce proinflammatory cytokines (Johnstone, Gearing, & Miller, 1999), and with aging they present impairments in clearing Aβ peptides and in neuronal support (Iram et al., 2016). This alteration in astrocyte function participates in the maintenance of the chronic neuroinflammation and brain damage.

Behavioral aspects of the model

The 5×FAD mouse model develops different types of memory deficits around the age of 4 months, when amyloid plaques and neuroinflammation are present (Oakley et al., 2006). In many tests involving hippocampus-dependent memory function, such as the Morris water maze, the Y maze, fear conditioning, and object or social recognition, 5×FAD mice have demonstrated impaired performance (Hongpaisan, Sun, & Alkon, 2011; Jawhar et al., 2011). Working memory (short or long term) is altered in 5×FAD mice as early as 4 months of age (in the B6/SJL background) (Giannoni et al., 2013) (Fig. 13.5). An impairment of frontal cortex function has also been demonstrated at 4 months of age using the olfactory H maze and the delayed reaction paradigm

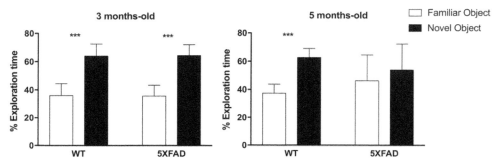

Figure 13.5 Learning performance of 3- and 5-month-old wild-type (*WT*) and 5×FAD mice assessed with the novel object recognition test. The 5×FAD mice are deficient at 5 months of age as indicated by the lower percentage of time devoted to the exploration of the novel object in comparison with the WT mice. Retention interval 24 h, 5–8 mice per group. Data are the means ± SEM. ***$P < .001$.

(Girard et al., 2013). Associative learning memory is also affected in this mouse model, but only around the age of 6 months (Baranger et al., 2017; Roddick, Schellinck, & Brown, 2014). Cognitive performance of 5×FAD mice in conditioned taste aversion and contextual fear conditioning paradigms is impaired around 6 and 9 months of age, respectively (Devi and Ohno, 2010). As noticed in The Jackson Laboratory depository (https://www.jax.org/strain/006554) and in comparison with the original B6/SJL—5 ×FAD strain, the phenotype in the 5×FAD mice is less pronounced when the mice are backcrossed on the C57BL/6J background, the more commonly used laboratory mouse strain to perform cognitive assessment. Indeed, we have experienced an around 1-month delay in the phenotype when using C57BL/6J—5×FAD mice compared with the original B6/SJL—5×FAD mice (Oakley et al., 2006). Amyloid pathology is delayed, and thus the first cognitive deficits appear around 4.5—5 months of age. The status of the animal housing facility is another parameter to consider. It has recently been demonstrated that gut microbiota composition has an impact on the amyloid pathology development and that AD mice bred in conventional microbiota have more amyloid in the brain than AD mice born in a germ-free environment (Brandscheid et al., 2017; Harach et al., 2017). Consequently, if the mice are bred in specific-pathogen-free or specific- and opportunist-pathogen-free facilities, the behavioral phenotypes could be delayed due to a less aggressive amyloid pathology.

Synaptic plasticity and long-term potentiation

According to the amyloid cascade hypothesis, amyloid accumulation in the AD brain induces numerous neuropathological changes that precede neurodegeneration (Hardy & Selkoe, 2002). LTP is a perfect readout of cellular communication processes that underlie learning and memory. In 5×FAD mice, phenotypic differences between plaque deposits in the cortex and hippocampus (more dense plaques in the latter) were linked to differences in LTP in these areas (Crouzin et al., 2013). Within the hippocampus of 5×FAD mice, an intense astrogliosis in the dentate gyrus, compared with the CA1 region, was associated with enhanced tonic GABA (γ-aminobutyric acid) current mediated by α5-GABA$_A$ receptors in dentate granule cells (Wu, Guo, Gearing, & Chen, 2014). Blockade of these GABA$_A$ receptors rescued LTP deficits and restored working memory performances in the Y-maze test. Restoration of LTP in 5×FAD mice is thus an important parameter to consider in the evaluation tests of candidate molecules or actors as promising therapeutic targets against AD (Baranger et al., 2016; MacPherson et al., 2017).

Vascular pathology

In recent years, the impact of the cerebral vasculature on the pathophysiology of AD has gained attention. Indeed, a link has been evidenced between AD development and vascular risk factors such as hypertension, obesity, type 2 diabetes, stroke, and atherosclerosis (Cechetto, Hachinski, & Whitehead, 2008). An efficient brain vasculature is necessary for the correct delivery of nutrients and oxygen, but it is also fundamental for its role of detoxification and barrier. Brain vessels help in clearing possible toxic elements, such as the amyloid peptides responsible for plaque formation, and modulate the substances that access the brain thanks to the blood—brain barrier (BBB). This finely organized structure is composed of tight junctions between endothelial cells that strictly control the movements of molecules between the blood and the brain. Several elements, including aging and genetic and epigenetic factors, but also a number of pathologies, determine an impairment of brain vasculature efficiency, with consequent accumulation of toxic species and alterations in brain homeostasis. The 5×FAD mouse model has been demonstrated to accurately reproduce the alterations in brain vasculature typical of AD presymptomatic and symptomatic stages (Giannoni et al., 2016). Thanks to a longitudinal study conducted with in vivo two-photon microscopy associated with classical postmortem analysis, it was evidenced that 5×FAD mice accumulate amyloid around vessels in what is defined as cerebral amyloid angiopathy (CAA), starting at 3 months of age and accumulating over time (Fig. 13.6A). The study confirms a preceding observation of BBB disruption (Kook et al., 2012) near amyloid deposits (Fig. 13.6A—C), also associated with matrix metalloprotease secretion. Indeed, although the association between CAA and cognitive performance is still discussed, the majority of human AD cases show this feature. Another set of cells participating in the neurovascular unit and controlling vessel efficiency and contraction, the pericytes, is also affected in 5×FAD mice. Their expression level changes over time, becoming more important at 9 and 12 months of age, considered late-stage AD in this model. To note, their increased level of expression seems associated with their redistribution, although further studies are needed to better characterize this phenomenon. Pericytes have been proposed as major players in the reparation

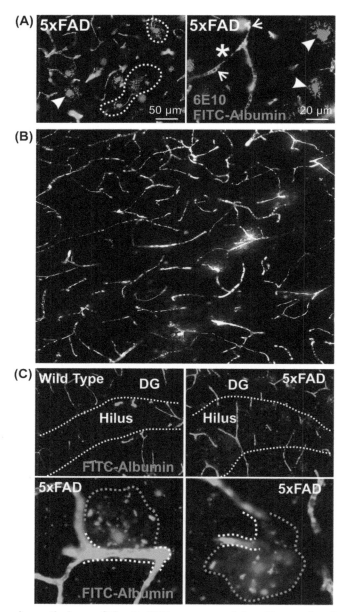

Figure 13.6 Vascular pathology of the 5×FAD mouse model at 12 months of age. (A) Amyloid accumulation is stained using 6E10 anti-Aβ antibody; vasculature is evidenced using fluorescent albumin staining (FITC-albumin). With aging, 5×FAD mice present vascular amyloid deposits (*arrows*) and intense parenchymal Aβ deposits, including dense plaques (*arrowheads*) and diffuse plaques (*dots*). Vascular deposits of Aβ induce vessel damages (*star*) as evidenced by leakages of FITC-albumin staining. (B) Larger view of cortical vasculature and leakages of FITC-albumin. (C) No leakages are detectable in wild-type aged mice (upper left) compared with 5×FAD of the same age (upper right); dots delimit the hilus area of the hippocampus. Bottom: higher magnification of vessel damage and leakage in 5×FAD mice; dots delimit the damaged vessels and associated leakages. *Aβ*, amyloid-β peptide; *DG*, dentate gyrus. *(Courtesy of Dr. Nicola Marchi (IGF, Montpellier).)*

process after a brain insult and are currently under investigation both in human AD cases and in mouse models, as they represent a well-promising target of intervention. Finally, 5×FAD mice reproduce one of the main aspects contributing to AD development: vascular inflammation. A strong microglial activation associated with astrogliosis is evidenced starting at 3 months of age around brain vessels and in strict association with amyloid accumulation and vessel damage. Clusters of reactive inflammatory cells were found correlating with disruption of tight junctions detected by leakages of fluorescein isothiocyanate—albumin. Changes in the expression of inflammation markers (e.g., IL-6) have also been observed in this model (Gurel et al., 2018), probably contributing to vascular inflammation. As for the human pathology, vascular inflammation aggravates with aging in the 5×FAD mouse model.

Neuronal loss/synaptic dysfunctions

The 5×FAD mouse model reproduces the neuronal loss and synaptic dysfunctions that characterize human AD. This feature is particularly interesting, as many other AD animal models do not present any neuronal damages. Interestingly, neuronal loss is particularly evident in the hippocampus and in cortex areas involved in memory formation and processing (Eimer and Vassar, 2013). A profound loss of layer 5 pyramidal neurons is observed in 5×FAD mice starting at 12 months of age (Oakley et al., 2006), while other alterations such as axonal dystrophy and loss of spines and basal dendrites (Buskila, Crowe, & Ellis-Davies, 2013) appear much earlier, around 4—6 months. As behavioral deficits appear around 4 months in this model, such early changes are thought to influence cognitive performance of the mice before late-stage neuronal death occurs. Thanks to electrophysiological studies of such neurons, a direct link between intraneuronal Aβ accumulation, which precedes plaque formation, and neuronal loss has been hypothesized. Presynaptic vesicular machinery, as well as postsynaptic effects, has been highlighted, probably due to the impact of Aβ peptides on glutamatergic conductance. A general consensus is present in the actual literature that considers the presence of amyloid as responsible for synaptic dysfunctions (Marchetti and Marie, 2011) and neuronal loss (Eimer and Vassar, 2013).

Tau pathology

Although the 5×FAD model is considered one of the most exhaustive concerning the reproduction of AD signs, the presence of NFTs is still discussed. While the first description of the 5×FAD model reported the lack of NFTs (Oakley et al., 2006), other more recent papers show a positive staining of paired helical filament tau sites Ser202 and Thr205 with the AT8 antibody, and identify NFTs in the cortex and hippocampus of female 5×FAD mice 6—8 months of age (Tohda, Urano, Umezaki, Nemere, & Kuboyama, 2012). As the antibody used was directed against the human tau, this supposes a cross-reaction with the mouse form. Hyperphosphorylation of tau seems confirmed in 5×FAD mice (Grinan-Ferre et al., 2016), as stated by mRNA analysis. However, the work of Maarouf and colleagues evidenced no NFTs identified with AT8 antibody (Maarouf et al., 2013). Possible causes of heterogeneity in the observed results could be the age, sex, and genetic background of the animals, as well as methodological and analysis differences (immunohistochemistry protocol, microscope, etc.). Discrepancies among results point to the need of further investigations.

Gender issue

In the human population, the prevalence of AD is higher among women than among men, with a 2:1 women/men ratio. A detailed review of current knowledge on sex-related differences in human AD has been conducted (Ferretti et al., 2018). This excellent work pointed out sex differences in AD patients not only from an epidemiological point of view, but also in clinical manifestations, biomarker patterns, risk factors, and treatment performance. Clearly, researchers have now to consider gender issues in the deciphering of molecular and mechanistic comprehension of AD in preclinical research. A higher Aβ production and deposition has been reported in female 5×FAD mice compared with male mice of the same age (Bhattacharya et al., 2014; Oakley et al., 2006). This feature was also reported in other transgenic AD mouse models such as APP/PS1 or 3×Tg-AD (Callahan et al., 2001; Wang, Tanila, Puolivali, Kadish, & van Groen, 2003). Of interest, despite an increased Aβ burden, female 5×FAD mice (SJL background strain) have a similar life span compared with males, whereas in 3×Tg-AD mice an early mortality in males has been reported (Rae and Brown, 2015). External environment can also induce gender-related pathology, as stress increases Aβ levels in female 5×FAD but not in males (Devi, Alldred, Ginsberg, & Ohno, 2010). Female 5×FAD mice present better olfactory performance than males in olfactory working memory tasks (Roddick et al., 2014). However, odor detection performance may vary in 5×FAD depending on the concentration of the odorant molecules (Roddick, Roberts, Schellinck, & Brown, 2016). As in humans, treatment efficiency can differ between male and female 5×FAD mice. Vitamin D supplementation is efficient only when administered during the symptomatic phase for the females (Landel, Millet, Baranger, Loriod, & Feron, 2016) and only in the early asymptomatic phase for the males (Morello et al., 2018). We also experienced differences in treatment efficiency when a serotonin type 4 receptor agonist was chronically administered to 5×FAD mice. Two-month treatments are sufficient to prevent cognitive deficits in females, whereas longer treatments were necessary to reach similar effects in males (Baranger et al., 2017; Giannoni et al., 2013). Collectively, there is a crucial need to systematically investigate sex differences in mouse models of AD with adapted sizes of groups to obtain an accurate statistical power.

Drawbacks of the 5×FAD mouse model

In this transgenic mouse, the extremely aggressive Aβ pathology, with extensive extracellular plaque formation beginning at 2 months of age, is definitely not in accordance with the time course of human sporadic AD. The slow progressive pathophysiological development of clinical AD in patients is not accurately reproduced in this transgenic mouse model. New models, using injections of adeno-associated viruses in rodents, have been developed to mimic more precisely the initial stages and progression of AD in patients, by controlling the expression in time and space of the transgenes (Audrain et al., 2016).

Advantages of the 5×FAD mouse model

The extreme rapidity of the amyloid neuropathology and associated behavioral deficits make this transgenic mouse a proper research model for early-onset AD (Jawhar, Trawicka, Jenneckens, Bayer, & Wirths, 2012; Oakley et al., 2006). Moreover, the appearance of cognitive deficits as early as 4 months of age renders the 5×FAD mice suitable for evaluating the preventive potential of chronic treatments with disease-modifying molecules (Baranger et al., 2017; Giannoni et al., 2013). Indeed, if the chronic administration of a compound is able to slow Aβ production and delay the appearance of cognitive impairments in 5×FAD mice, which produce the highest level of amyloid, the same compound should be efficient in transgenic mouse models presenting moderate Aβ accumulation. In addition, such a rapidity in obtaining relevant and measurable features in the model induces a reduction in time (and costs) involving animal housing and protocol-related expenses. The ethical aspect of such improvement is a particularly important argument promoting the use of this mouse model.

Conclusion

To conclude, 5×FAD mice represent the more rapid and intense transgenic mouse model of AD, in terms of amyloid pathology. This model recapitulates numerous features of clinical AD in human patients, such as amyloid deposits, gliosis, synaptic degeneration, working memory impairment, reduced anxiety, and selective neuronal loss. Despite controversies due to the early and aggressive plaque development phenotype, this model remains of great interest to preclinically evaluate the disease-modifying potential of new compounds against AD.

Key facts of neurovascular pathology

- A healthy vascular system is mandatory for proper functioning of the brain.
- Amyloid accumulation around vessels (CAA) induces vessel damage in AD.

- Vessel breaks in AD are evidenced by leaks of immunoglobulins in the brain.
- An intense vascular inflammation accompanies vessel damage.
- The neurovascular unit is remodeled in AD with redistribution of the pericytes.

Summary points

- The 5×FAD mouse model reproduces the majority of AD hallmarks, including amyloid plaques, neuronal death, synaptic dysfunctions, CAA, and vessel alterations associated with inflammation.
- As in human AD, the gender issue should be considered in the 5×FAD mouse model, since the pathology is more aggressive in females than in males.
- The rapid and aggressive progression of the pathology in 5×FAD mice is not in accordance with the slow development of sporadic AD in humans.
- The 5×FAD mouse line is a proper research model for early-onset AD.
- This animal model enables a time reduction in the experimental protocols and a diminution of the research costs.

Acknowledgments

The authors thank Dr. Nicola Marchi for the gift of vasculature images. This work was supported by CNRS, INSERM, Montpellier University, the French Research National Agency ANR (ANR-12-BSV4-008-01 ADAMGUARD), the Association France Alzheimer, the Fondation Vaincre Alzheimer, Région Languedoc Roussillon, and the Fondation Plan Alzheimer. We are grateful for the confocal MRI platform at Montpellier, France.

References

Audrain, M., Fol, R., Dutar, P., Potier, B., Billard, J. M., Flament, J., et al. (2016). Alzheimer's disease-like APP processing in wild-type mice identifies synaptic defects as initial steps of disease progression. *Molecular Neurodegeneration, 11*, 5.

Baranger, K., Giannoni, P., Girard, S. D., Girot, S., Gaven, F., Stephan, D., et al. (2017). Chronic treatments with a 5-HT4 receptor agonist decrease amyloid pathology in the entorhinal cortex and learning and memory deficits in the 5xFAD mouse model of Alzheimer's disease. *Neuropharmacology, 126*, 128—141.

Baranger, K., Marchalant, Y., Bonnet, A. E., Crouzin, N., Carrete, A., Paumier, J. M., et al. (2016). MT5-MMP is a new pro-amyloidogenic proteinase that promotes amyloid pathology and cognitive decline in a transgenic mouse model of Alzheimer's disease. *Cellular and Molecular Life Sciences, 73*, 217—236.

Bhattacharya, S., Haertel, C., Maelicke, A., & Montag, D. (2014). Galantamine slows down plaque formation and behavioral decline in the 5XFAD mouse model of Alzheimer's disease. *PLoS One, 9*. e89454.

Boza-Serrano, A., Yang, Y., Paulus, A., & Deierborg, T. (2018). Innate immune alterations are elicited in microglial cells before plaque deposition in the Alzheimer's disease mouse model 5xFAD. *Scientific Reports, 8*, 1550.

Brandscheid, C., Schuck, F., Reinhardt, S., Schafer, K. H., Pietrzik, C. U., Grimm, M., et al. (2017). Altered gut microbiome composition and tryptic activity of the 5xFAD Alzheimer's mouse model. *Journal of Alzheimer's Disease, 56*, 775—788.

Buskila, Y., Crowe, S. E., & Ellis-Davies, G. C. (2013). Synaptic deficits in layer 5 neurons precede overt structural decay in 5xFAD mice. *Neuroscience, 254*, 152—159.

Callahan, M. J., Lipinski, W. J., Bian, F., Durham, R. A., Pack, A., & Walker, L. C. (2001). Augmented senile plaque load in aged female beta-amyloid precursor protein-transgenic mice. *American Journal of Pathology, 158,* 1173—1177.

Cechetto, D. F., Hachinski, V., & Whitehead, S. N. (2008). Vascular risk factors and Alzheimer's disease. *Expert Review of Neurotherapeutics, 8,* 743—750.

Citron, M., Westaway, D., Xia, W., Carlson, G., Diehl, T., Levesque, G., et al. (1997). Mutant presenilins of Alzheimer's disease increase production of 42-residue amyloid beta-protein in both transfected cells and transgenic mice. *Nature Medicine, 3,* 67—72.

Crouzin, N., Baranger, K., Cavalier, M., Marchalant, Y., Cohen-Solal, C., Roman, F. S., et al. (2013). Area-specific alterations of synaptic plasticity in the 5XFAD mouse model of Alzheimer's disease: Dissociation between somatosensory cortex and hippocampus. *PLoS One, 8,* e74667.

De Jonghe, C., Esselens, C., Kumar-Singh, S., Craessaerts, K., Serneels, S., Checler, F., et al. (2001). Pathogenic APP mutations near the gamma-secretase cleavage site differentially affect Abeta secretion and APP C-terminal fragment stability. *Human Molecular Genetics, 10,* 1665—1671.

Devi, L., Alldred, M. J., Ginsberg, S. D., & Ohno, M. (2010). Sex- and brain region-specific acceleration of beta-amyloidogenesis following behavioral stress in a mouse model of Alzheimer's disease. *Molecular Brain, 3,* 34.

Devi, L., & Ohno, M. (2010). Genetic reductions of beta-site amyloid precursor protein-cleaving enzyme 1 and amyloid-beta ameliorate impairment of conditioned taste aversion memory in 5XFAD Alzheimer's disease model mice. *European Journal of Neuroscience, 31,* 110—118.

Eimer, W. A., & Vassar, R. (2013). Neuron loss in the 5XFAD mouse model of Alzheimer's disease correlates with intraneuronal Abeta42 accumulation and Caspase-3 activation. *Molecular Neurodegeneration, 8,* 2.

Ferretti, M. T., Iulita, M. F., Cavedo, E., Chiesa, P. A., Schumacher Dimech, A., Santuccione Chadha, A., et al. (2018). Sex differences in Alzheimer disease - the gateway to precision medicine. *Nature Reviews Neurology, 14,* 457—469.

Fonseca, M. I., Zhou, J., Botto, M., & Tenner, A. J. (2004). Absence of C1q leads to less neuropathology in transgenic mouse models of Alzheimer's disease. *Journal of Neuroscience, 24,* 6457—6465.

Giannoni, P., Arango-Lievano, M., Neves, I. D., Rousset, M. C., Baranger, K., Rivera, S., et al. (2016). Cerebrovascular pathology during the progression of experimental Alzheimer's disease. *Neurobiology of Disease, 88,* 107—117.

Giannoni, P., Gaven, F., de Bundel, D., Baranger, K., Marchetti-Gauthier, E., Roman, F. S., et al. (2013). Early administration of RS 67333, a specific 5-HT4 receptor agonist, prevents amyloidogenesis and behavioral deficits in the 5XFAD mouse model of Alzheimer's disease. *Frontiers in Aging Neuroscience, 5,* 96.

Girard, S. D., Baranger, K., Gauthier, C., Jacquet, M., Bernard, A., Escoffier, G., et al. (2013). Evidence for early cognitive impairment related to frontal cortex in the 5XFAD mouse model of Alzheimer's disease. *Journal of Alzheimer's Disease, 33,* 781—796.

Grinan-Ferre, C., Sarroca, S., Ivanova, A., Puigoriol-Illamola, D., Aguado, F., Camins, A., et al. (2016). Epigenetic mechanisms underlying cognitive impairment and Alzheimer disease hallmarks in 5XFAD mice. *Aging (Albany NY), 8,* 664—684.

Gurel, B., Cansev, M., Sevinc, C., Kelestemur, S., Ocalan, B., Cakir, A., et al. (2018). Early stage alterations in CA1 extracellular region proteins indicate dysregulation of IL6 and iron homeostasis in the 5XFAD Alzheimer's disease mouse model. *Journal of Alzheimer's Disease, 61,* 1399—1410.

Harach, T., Marungruang, N., Duthilleul, N., Cheatham, V., Mc Coy, K. D., Frisoni, G., et al. (2017). Reduction of Abeta amyloid pathology in APPPS1 transgenic mice in the absence of gut microbiota. *Scientific Reports, 7,* 41802.

Hardy, J., & Selkoe, D. J. (2002). The amyloid hypothesis of Alzheimer's disease: progress and problems on the road to therapeutics. *Science, 297,* 353—356.

Herl, L., Thomas, A. V., Lill, C. M., Banks, M., Deng, A., Jones, P. B., et al. (2009). Mutations in amyloid precursor protein affect its interactions with presenilin/gamma-secretase. *Molecular and Cellular Neuroscience, 41,* 166—174.

Hongpaisan, J., Sun, M. K., & Alkon, D. L. (2011). PKC epsilon activation prevents synaptic loss, Abeta elevation, and cognitive deficits in Alzheimer's disease transgenic mice. *Journal of Neuroscience, 31,* 630—643.

Hsiao, K., Chapman, P., Nilsen, S., Eckman, C., Harigaya, Y., Younkin, S., et al. (1996). Correlative memory deficits, Abeta elevation, and amyloid plaques in transgenic mice. *Science, 274*, 99–102.

Iram, T., Trudler, D., Kain, D., Kanner, S., Galron, R., Vassar, R., et al. (2016). Astrocytes from old Alzheimer's disease mice are impaired in Abeta uptake and in neuroprotection. *Neurobiology of Disease, 96*, 84–94.

Jawhar, S., Trawicka, A., Jenneckens, C., Bayer, T. A., & Wirths, O. (2012). Motor deficits, neuron loss, and reduced anxiety coinciding with axonal degeneration and intraneuronal Abeta aggregation in the 5XFAD mouse model of Alzheimer's disease. *Neurobiology of Aging, 33*, 196 e29-40.

Jawhar, S., Wirths, O., Schilling, S., Graubner, S., Demuth, H. U., & Bayer, T. A. (2011). Overexpression of glutaminyl cyclase, the enzyme responsible for pyroglutamate A{beta} formation, induces behavioral deficits, and glutaminyl cyclase knock-out rescues the behavioral phenotype in 5XFAD mice. *Journal of Biological Chemistry, 286*, 4454–4460.

Johnstone, M., Gearing, A. J., & Miller, K. M. (1999). A central role for astrocytes in the inflammatory response to beta-amyloid; chemokines, cytokines and reactive oxygen species are produced. *Journal of Neuroimmunology, 93*, 182–193.

Kimura, N., Takahashi, M., Tashiro, T., & Terao, K. (2006). Amyloid beta up-regulates brain-derived neurotrophic factor production from astrocytes: Rescue from amyloid beta-related neuritic degeneration. *Journal of Neuroscience Research, 84*, 782–789.

Kook, S. Y., Hong, H. S., Moon, M., Ha, C. M., Chang, S., & Mook-Jung, I. (2012). Abeta(1)(-)(4)(2)-RAGE interaction disrupts tight junctions of the blood-brain barrier via Ca(2)(+)-calcineurin signaling. *Journal of Neuroscience, 32*, 8845–8854.

Landel, V., Millet, P., Baranger, K., Loriod, B., & Feron, F. (2016). Vitamin D interacts with Esr1 and Igf1 to regulate molecular pathways relevant to Alzheimer's disease. *Molecular Neurodegeneration, 11*, 22.

Liu, S., Liu, Y., Hao, W., Wolf, L., Kiliaan, A. J., Penke, B., et al. (2012). TLR2 is a primary receptor for Alzheimer's amyloid beta peptide to trigger neuroinflammatory activation. *The Journal of Immunology, 188*, 1098–1107.

Maarouf, C. L., Kokjohn, T. A., Whiteside, C. M., Macias, M. P., Kalback, W. M., Sabbagh, M. N., et al. (2013). Molecular differences and similarities between Alzheimer's disease and the 5XFAD transgenic mouse model of amyloidosis. *Biochemistry Insights, 6*, 1–10.

MacPherson, K. P., Sompol, P., Kannarkat, G. T., Chang, J., Sniffen, L., Wildner, M. E., et al. (2017). Peripheral administration of the soluble TNF inhibitor XPro1595 modifies brain immune cell profiles, decreases beta-amyloid plaque load, and rescues impaired long-term potentiation in 5xFAD mice. *Neurobiology of Disease, 102*, 81–95.

Marchetti, C., & Marie, H. (2011). Hippocampal synaptic plasticity in Alzheimer's disease: What have we learned so far from transgenic models? *Reviews in the Neurosciences, 22*, 373–402.

Morello, M., Landel, V., Lacassagne, E., Baranger, K., Annweiler, C., Feron, F., et al. (2018). Vitamin D improves neurogenesis and cognition in a mouse model of Alzheimer's disease. *Molecular Neurobiology.*

Oakley, H., Cole, S. L., Logan, S., Maus, E., Shao, P., Craft, J., et al. (2006). Intraneuronal beta-amyloid aggregates, neurodegeneration, and neuron loss in transgenic mice with five familial Alzheimer's disease mutations: Potential factors in amyloid plaque formation. *Journal of Neuroscience, 26*, 10129–10140.

Orihuela, R., McPherson, C. A., & Harry, G. J. (2016). Microglial M1/M2 polarization and metabolic states. *British Journal of Pharmacology, 173*, 649–665.

Perlmutter, L. S., Barron, E., & Chui, H. C. (1990). Morphologic association between microglia and senile plaque amyloid in Alzheimer's disease. *Neuroscience Letters, 119*, 32–36.

Rae, E. A., & Brown, R. E. (2015). The problem of genotype and sex differences in life expectancy in transgenic AD mice. *Neuroscience and Biobehavioral Reviews, 57*, 238–251.

Roddick, K. M., Roberts, A. D., Schellinck, H. M., & Brown, R. E. (2016). Sex and genotype differences in odor detection in the 3xTg-AD and 5XFAD mouse models of Alzheimer's disease at 6 Months of age. *Chemical Senses, 41*, 433–440.

Roddick, K. M., Schellinck, H. M., & Brown, R. E. (2014). Olfactory delayed matching to sample performance in mice: Sex differences in the 5XFAD mouse model of Alzheimer's disease. *Behavioural Brain Research, 270*, 165–170.

Schaeffer, E. L., Figueiro, M., & Gattaz, W. F. (2011). Insights into Alzheimer disease pathogenesis from studies in transgenic animal models. *Clinics, 66*(Suppl. 1), 45–54.

Scheuner, D., Eckman, C., Jensen, M., Song, X., Citron, M., Suzuki, N., et al. (1996). Secreted amyloid beta-protein similar to that in the senile plaques of Alzheimer's disease is increased in vivo by the presenilin 1 and 2 and APP mutations linked to familial Alzheimer's disease. *Nature Medicine, 2*, 864–870.

Selkoe, D. J., & Schenk, D. (2003). Alzheimer's disease: Molecular understanding predicts amyloid-based therapeutics. *Annual Review of Pharmacology and Toxicology, 43*, 545–584.

Spangenberg, E. E., Lee, R. J., Najafi, A. R., Rice, R. A., Elmore, M. R., Blurton-Jones, M., et al. (2016). Eliminating microglia in Alzheimer's mice prevents neuronal loss without modulating amyloid-beta pathology. *Brain, 139*, 1265–1281.

Tohda, C., Urano, T., Umezaki, M., Nemere, I., & Kuboyama, T. (2012). Diosgenin is an exogenous activator of 1,25D(3)-MARRS/Pdia3/ERp57 and improves Alzheimer's disease pathologies in 5XFAD mice. *Scientific Reports, 2*, 535.

Wang, J., Tanila, H., Puolivali, J., Kadish, I., & van Groen, T. (2003). Gender differences in the amount and deposition of amyloidbeta in APPswe and PS1 double transgenic mice. *Neurobiology of Disease, 14*, 318–327.

Wu, Z., Guo, Z., Gearing, M., & Chen, G. (2014). Tonic inhibition in dentate gyrus impairs long-term potentiation and memory in an Alzheimer's [corrected] disease model. *Nature Communications, 5*, 4159.

Wyss-Coray, T., Loike, J. D., Brionne, T. C., Lu, E., Anankov, R., Yan, F., et al. (2003). Adult mouse astrocytes degrade amyloid-beta in vitro and in situ. *Nature Medicine, 9*, 453–457.

PART II

Biomarkers, psychometric instruments and diagnosis

CHAPTER 14

Use of cerebrospinal fluid in diagnosis of dementia

Angelo Nuti, Cecilia Carlesi, Elena Caldarazzo Ienco, Gabriele Cipriani, Paolo Del Dotto
Neurology Unit, Versilia Hospital, Camaiore (Lucca), Italy

List of abbreviations

Abeta Amyloid-beta peptide
Abeta40 Abeta1-40
Abeta42 Abeta1-42
AD Alzheimer disease
alpha syn alpha-synuclein
APP Amyloid precursor protein
BACE1 Beta-secretase
BBB Blood—brain barrier
C9ORF72 chromosome 9 opening reading frame 72
CCL2 C-C chemokine ligand 2
CNS central nervous system
CJD Creutzfeldt—Jakob disease
CSF Cerebrospinal fluid
DLB Dementia with Lewy bodies
FTD Frontotemporal dementia
MCI Mild cognitive impairment
MMP Matrix metalloproteinase
NF-L Neurofilament light
NINCDS-ADRDA National Institute of Neurological and Communicative Disorders and Stroke - Alzheimer's Disease and Related Disorders Association workgroup
p-Tau Phosphorylated tau
PDD Parkinson's disease dementia
t-Tau Total tau
TDP-43 TAR DNA-binding protein 43 Kda
TIMP-1 Tissue inhibitor of metalloproteinase 1
TREM-2 triggering receptor expressed on myeloid cells
VAD Vascular dementia
VCI Vascular cognitive impairment
WML White matter lesions

Diagnosis and Management in Dementia
ISBN 978-0-12-815854-8, https://doi.org/10.1016/B978-0-12-815854-8.00014-8

Mini-dictionary of terms

Alzheimer disease Alzheimer disease is an age-dependent neurodegenerative disorder and the most common cause of dementia with aging.

Amyloid plaque Amyloid plaque is one of the major lesions found in the brain of patients with Alzheimer disease and consists of an Abeta amyloid core with a corona of argyrophilic axonal and dendritic processes, Abeta amyloid fibrils, glial cell processes, and microglia cells.

Biomarker biomarker defines a biological process or disease characteristic that is objectively measured.

Blood—brain barrier The blood—brain barrier is a selectivity permeability barrier that separates the brain from the central nervous system.

Dementia with Lewy bodies Dementia with Lewy bodies is a primary degenerative dementia sharing features of Parkinson's disease and most often also of Alzheimer disease, characterized histopathologically by intraneuronal Lewy bodies.

Frontotemporal dementia Frontotemporal dementia is a complex multifactorial disorder characterized by progressive degeneration of the frontal and/or temporal lobes.

Neurofibrillary tangles Neurofibrillary tangles are intracellular inclusions composed of ubiquinate and phosphorylated protein tau.

Parkinson's disease dementia Parkinson's disease dementia is a complex clinical picture characterized by the cardinal parkinsonian features (resting tremors, bradykinesia, rigidity, and postural abnormalities) plus neurocognitive disorders.

Tau protein Tau protein is a microtubule-binding protein mainly located in a thin unmyelinated cortical axon, where it stabilizes microtubules and participates in axonal transport.

Vascular cognitive impairment Vascular cognitive impairment is an umbrella term encompassing all forms of cognitive dysfunction associated with cerebrovascular disease.

Introduction

In general a biomarker defines a biological process or disease characteristic that is objectively measured (Biomarkers Definition Working Group, 2001). In clinical neurology, cerebrospinal fluid (CSF) biomarkers are used for diagnostic purposes and it is possible to measure CSF biomarkers related to accumulations of proteins, neuronal injury, inflammation, and other neuropathological processes (Mattsson, Grigoriou, & Zetterberg, 2017).

An ideal clinical assessment of dementia comprises a mental status examination, cognitive battery, informant interview, neurological examination, laboratory tests, and neuroimaging to exclude alternative diagnosis and to give a positive diagnosis. The diagnosis of a specific type of dementia can be made using clinical criteria and can be supplemented by biomarkers (Zetterberg, Rhorer, & Schott, 2017, pp. 85—97).

CSF may be used in diagnosis of dementia to exclude neuroinflammatory and neuro-infectious disorders measuring CSF—to—serum albumin ratio, IgG and IgM indices, and oligoclonal bands (Zetterberg et al., 2017). Moreover, molecular biomarkers of several dementia disorders may be measured in CSF.

Alzheimer disease

Alzheimer disease (AD) is a progressive neurodegenerative disease with cognitive, behavioral, and functional abnormalities. AD is the most common type of dementia; it accounts for approximately 70% of cases of progressive cognitive impairment in aged individuals.

Age is the single most important risk factor, and the prevalence of AD doubles every 5 years after the age of 60 and reaches 40% after 90 (Dubois & Upenskaya-Cadoz, 2016, pp. 353—360).

AD is characterized microscopically by:

(1) amyloid/neuritique plaque, deposits of the amyloid-beta peptide (Abeta) in cerebral parenchyma and in blood vessels as cerebral amyloid angiopathy

(2) intraneuronal neurofibrillary tangles, composed of aggregates and often abnormally hyperphosphorylated tau protein, which is a microtubule-associated protein.

Additionally, neuronal loss in specific brain regions, synaptic dysfunction and loss, and microglial activation have been reported (Masters, 2017, pp. 470—485).

Amyloid beta-related biomarkers

Abeta peptides

Amyloid beta (abeta) peptides are derived from the type 1 transmembrane amyloid precursor protein (APP), ubiquitously expressed by the neurons in the central nervous system (CNS). APP is cleaved by three proteolytic enzymes: alpha secretase, beta secretase (BACE1), and gamma secretase. The cleavage combined (amyloidogenic pathway) of BACE1 and gamma secretase primarily leads to production of Abeta1-42 (Abeta42) and Abeta1-40 (Abeta40). CSF Abeta42 is reduced in CSF of AD patients and correlates inversely with Abeta brain accumulation, both in neuropathology and in vivo PET imaging studies. This reduction may be due to Abeta42 sequestration in senile plaques in the brain, but hypothetically can be due to other mechanisms such as altered release of Abeta42, formation of Abeta42 oligomers not detected by common assays, binding of Abeta42 to other proteins which alter common assays, or intracellular Abeta accumulation (Mattsson et al., 2017).

Low CSF Abeta42 level in absence of plaque formation has been reported in neuroinflammatory diseases: bacterial meningitis, multiple sclerosis, neuroborreliosis, and human immunodeficiency virus—associated dementia (Zetterberg et al., 2017).

Moreover, reduced levels of CSF Abeta42 has been observed also in some cases of vascular dementia (VAD), frontotemporal dementia (FTD), dementia with Lewy bodies (LBDs), multiple system atrophy, Creutzfeldt—Jakob disease (CJD), and amyotrophic lateral sclerosis. In some of these cases the reduction of CSF Abeta42 levels reflects concomitant AD pathology (mixed dementia), but in other cases it is also possible that the reduction of CSF Abeta42 occurs by other mechanisms, for example, as a consequence of white matter pathology (Mattsson et al., 2017).

APP processing produces other Abeta peptides and, in particular, the isoform Abeta40, which is relatively unchanged in AD patients. Instead, the ratio of Abeta42: Abeta40 has been suggested to be a better CSF indicator of Abeta pathology than Abeta42 alone (Mattsson et al., 2017; Zetterberg et al., 2017).

Abeta-generating enzymes

A soluble, measurable form of BACE1 exists in CSF. The CSF activity of BACE1 has been shown to be higher in some studies in AD patients and also in mild cognitive impairment (MCI) cases, but other studies have reported negative results. It is possible that this increase of BACE1 CSF activity occurs only in early phases of the disease (Lista, Zetterberg, O'Briant, Blennow, & Hampel, 2017, pp. 528–538).

Another processing enzyme of APP is gamma secretase. It has more than 100 known substrates, one of which is alcadein, which is present in CSF; it may have the potential to be a useful marker for studying the activity of gamma secretase in CSF (Mattsson et al., 2017).

sAPP peptides

APP cleavage also gives rise to soluble N-terminal fragments called soluble alphaAPP (after alpha secretase cleavage) and soluble betaAPP (after betaAPP secretase cleavage). The studies have shown conflicting results; some studies showed increased levels of these soluble fragments in CSF and other studies showed no variations with respect to nondemented controls (Mattsson et al., 2017).

Abeta oligomers

Soluble oligomers of Abeta may be more toxic than fibrillar Abeta aggregates. Several studies have measured CSF levels of Abeta oligomers, but these are difficult to quantify and characterize and the results are variable; some studies reported elevated levels and other studies reported unaltered CSF levels of Abeta oligomers (Lista et al., 2017).

TAU-related biomarkers

Tau protein is a neuronal protein mainly localized in thin unmyelinated cortical axons where it stabilizes microtubules and takes part in axonal transport. In the adult human brain six isoforms are expressed by alternative splicing from the MAPT gene, which leads to tau isoforms with three (3R-tau) or four (4R-tau) microtubule-binding repeat domains.

Total tau

CSF total tau (t-Tau) is clearly elevated in AD patients, a finding that has been reported in hundreds of research papers (Mattsson et al., 2017; Zetterberg et al., 2017). CSF t-Tau correlates with hippocampal atrophy (Wang et al., 2012) and gray matter degeneration (Glodzik et al., 2012). The CSF concentration of t-Tau reflects ongoing axonal degeneration and neuronal injury and may predict the clinical course of AD (Wallin et al., 2010). CSF t-Tau is also found to be increased in all neurological diseases with significant neuronal loss; the highest concentrations are seen in severe and rapid neurodegeneration such as in CJD (Mattsson et al., 2017).

Posphorylated tau

Posphorylated tau (p-Tau) reflects a process of phosphorylation of protein tau at specific threonine or serine residues. This process leads to an alteration on chemical-physical properties of tau proteins, which forms intraneuronal aggregates of neurofibrillary tangles. CSF p-Tau levels are reduced in AD patients and correlates with neurofibrillary tangle pathology in AD. However, is not clear why CSF p-Tau levels are reduced only in AD patients and not in other tauopathies (Mattsson et al., 2017; Zetterberg et al., 2017).

CSF p-Tau levels are also elevated, but in the absence of tangle formation, in other three conditions: (1) during normal brain development (Mattsson, Sävman, Osterlundh, Blennow, & Zetterberg, 2010); (2) herpes encephalitis (Grahn et al., 2013); and (3) superficial CNS siderosis (Ikeda et al., 2010).

Tau isoforms

The study of specific isoforms of tau may provide additional information in AD. A study using a specific assay for 3-4/R-Tau isoform showed a selective reduction of 4R-Tau in AD and progressive supranuclear palsy patients with respect to controls and lower levels of 4R-tau in AD cases compared with Parkinson's disease (PD) patients with dementia (Luk et al., 2012).

Clinical use of CSF biomarkers in AD

Many studies have found that AD patients have about 50% reduction of CSF Abeta42 levels and several times increased CSF t-Tau and p-Tau levels compared to nondemented controls (Bloudek, Spackman, Blankenburg, & Sullivan, 2011). Moreover, multiple studies suggest that the combined CSF test for t-Tau, p-Tau, and Abeta42 outperforms each biomarker used alone with diagnostic sensitivity and specificity >90% in single-center studies (Zetterberg et al., 2017). Thus CSF biomarkers offer significant clinical diagnostic accuracy in AD and are now incorporated into new criteria for the diagnosis of AD (Jack et al., 2018).

Recent data have shown that it is possible to identify changes in CSF biomarkers for AD decades before patients become demented and these changes may develop in a specific sequence, as presented in the "dynamic biomarker model" (Jack et al., 2013). In this model CSF Abeta42 is thought to change before CSF t-Tau and p-Tau because the initial pathological event in the amyloid cascade hypothesis is the aggregation of Abeta peptides, and only after the occurrence of tau pathology, neuroinflammation, synaptic dysfunction, and neuronal loss (Hardy & Selkoe, 2002). Moreover, most studies show that CSF AD biomarkers are essentially stable once patients have converted to AD and the biomarkers trajectories are sigmoid, suggesting that they have reached a plateau phase (Mattsson et al., 2017; Zetterberg et al., 2017).

Then AD can be identified by CSF biomarkers prior to clinical symptomatology and the earliest CSF modification is the reduction of the CSF Abeta42 levels. Numerous studies have verified that CSF Abeta42 levels are highly predictive of future AD in cases of MCI and in cases of cognitively normal control (Zetterberg et al., 2017). In MCI patients that later progress to AD dementia, CSF Abeta42, t-Tau and P- are altered with sensitivity and specificity of 70%—90% compared to patients who developed other dementias or patients who remain stable (Visser et al., 2009).

The typical clinical presentation of AD is the amnestic form, but more rarely some AD cases may present with variants such as cortical posterior atrophy, logopenic aphasia, or frontal variant of AD. In general, CSF biomarkers do not differ between these clinical presentations of AD (Mattsson et al., 2017).

In the National Institute of Neurological and Communicative Disorders and Stroke and the Alzheimer's Disease and Related Disorders Association (NINCDS-ADRDA) concept of AD (McKhann et al., 1984) diagnosis cannot be made clinically and needs a postmortem confirmation to be ascertained. Therefore a clinical diagnosis of AD can only be probable. For the International Working Group the availability in vivo of specific biomarkers of AD pathology has moved the definition of AD from a clinicopathological entity to a clinicobiological entity. As biomarkers can be considered as surrogate markers of histopathological changes, the clinical diagnosis can now be established in vivo, with high level of specificity and predictive validity (Dubois & Uspenskaya-Cadoz, 2016).

Synaptic biomarkers

Synaptic loss is a characteristic of AD. Several presynaptic and postsynaptic proteins, such as rab3A, synaptotagmin, growth-associated protein 43, synaptosomal-associated protein 25, and neurogranin have been found in CSF. The best established CSF synaptic biomarker is neurogranin. The CSF levels of neurogranin are increased in AD patients, with correlation with tau protein CSF concentration. The neurogranin in CSF AD patients is increased in a disease-specific manner. In fact, whereas AD patients have marked increase in CSF neurogranin levels, FTD patients show low concentration (Mattsson et al., 2017).

CSF biomarkers of microglial activation

In AD there is some evidence that there is activation of microglial cells, in particular the M1 phenotype of the microglia, the native macrophages of the brain, and neuroinflammation is a crucial mechanism in the pathophysiology of AD.

The CSF activity of chitotriosidase, an enzyme expressed in CNS by activated macrophages, is elevated in AD patients compared to nondemented controls (Watabe-Rudolph et al., 2012). YKL-40 is a glycoprotein analoge to chitotriosidase without enzyme activity and is expressed both in microglia and astrocytes. The CSF levels of YKL-40 can predict progression from prodromal MCI to AD dementia. Moreover, a positive association between CSF YKL-40 and other biomarkers of neurodegeneration-in particular t–Tau protein– has been reported during the asymptomatic preclinical stage of AD (Baldacci, Lista, Cavedo, Bonuccelli, & Hampel, 2017).

Another microglial CSF biomarker is the C–C chemokine ligand 2 (CCL2), which is produced by microglial cells. This is a ligand for C-C chemokine receptor 2, expressed on monocytes, and it is important for recruitment of these cells in the inflammatory response. Higher CSF levels of CCL2 have been reported in AD patients compared with nondemented controls (Corrêa, Starling, Teixeira, Caramelli, & Silva, 2011) as well as in the MCI phase (Galimberti et al., 2006).

CD14 is a protein surface mostly expressed by macrophages and performs a significative role in the innate immune response of the CNS. Yin et al. (2009) have reported higher CSF levels of CD14 in AD patients and in PD patients versus healthy controls. The microglial cells in the CNS selectively express the secreted ectodomain of TREM2 and CSF concentration of TREM2 is increased in AD and correlates with markers of neuronal injury such as CSF level of t-Tau and p-Tau (Suàrez-Calvet et al., 2016). The increase of CSF level of TREM2 seems to be disease specific (Piccio et al., 2016).

Vascular cognitive impairment

Vascular cognitive impairment (VCI) is an umbrella term that encompasses all forms of mild to severe cognitive dysfunction due to a variety of cerebrovascular causes. Both small and large vessel disease can provoke VCI and vascular dementia (VAD). VAD due to large artery disease is often the result of multiple infarction (multi–infarct dementia) and from a clinical point of view presents a stepwise course. Small-vessel disease is much more common and may produce either lacunae without white matter lesions (WMLs) or extensive white matter damage with and without lacunae (Rosenberg, Bjerke, & Wallin, 2014). Cerebrovascular disease can induce blood—brain barrier (BBB) impairment, breakdown of extracellular matrix, and white matter damage, which may be reflected in the CSF (Zetterberg et al., 2017).

Biomarkers of blood—brain barrier

The BBB is a selective permeability barrier that separates the circulating blood from the CNS. Recent data indicates that the dysfunction of the BBB may play a role in the pathogenesis of VAD. The CSF—serum albumin ratio is the gold standard fluid biomarker in CSF for measuring BBB integrity. To calculate this ratio it is necessary to measure albumin in CSF and serum collected simultaneously. This ratio increases with advancing age and further increases in patients with VCI as compared to AD and with worsening of WMLs (Wallin et al., 2017). Recent data suggests that there is a lack of association between CSF—serum albumin ratio and AD biomarkers, suggesting that BBB dysfunction is not inherent to AD but might represent, if present in a patient with AD, concomitant cerebrovascular pathology (Skillbäck et al., 2017).

Biomarkers of extracellular matrix breakdown

Matrix metalloproteinases (MMPs) are a family of enzymes active at the level of the extracellular matrix, at the cell surface, and intracellularly. There are 26 families of MMPS, but only the types 2, 3, 7, 9, 10, and 12 are active on the brain (Wallin et al., 2017). Elevated levels of MMP 9 are reported in CSF of patients with VCI and MD, but not in AD. Also the CSF levels of tissue inhibitor of metalloproteinase 1 (TIMP-1) are found elevated in VAD patients but not in AD patients, and the TIMP-1 CSF concentration as been shown to be correlated with the albumine ratio only in VAD patients (Bjerke et al., 2011).

Biomarkers of white matter damage

Markers of white matter injury include: myelin basic protein, a major structural constituent of the myelin sheath; sulfatide, a glycophospholipid also found in myelin sheath; and neurofilament light (NF-L), a protein expressed in large-caliber myelinated axons. These biomarkers are elevated in the CSF of VAD and in subcortical VCI, but this data is not specific for VCI because these biomarkers are elevated in other CNS diseases that involve white matter, such as multiple sclerosis (Zetterberg et al., 2017).

Frontotemporal dementia

FTD is an umbrella term that covers clinically, genetically, and pathologically heterogeneous neurodegenerative disorders that preferentially affect the frontal and anterior temporal lobes. It can be subdivided into three clinical subtypes: behavioral-variant FTD, progressive nonfluent aphasia, and semantic dementia. It can overlap with motor neuron disease and parkinsonism disorders, in particular, corticobasal syndrome and progressive supranuclear palsy. Family history of dementia and related disorders can be found in 40%—50% of patients with FTD, but only 10%—40% will have an autosomal dominant pattern of inheritance. The more frequent mutations (80% of familial FTD cases) are mutations in progranulin (*GRN*), microtubule-associated protein tau, and C9ORF72 genes. Neuropathologically, FTD is characterized by proteinaceous inclusions usually containing an abnormal form of one of two protein tau or TAR DNA binding protein 43 kDA (TDP-43), although less frequently some patients have inclusions containing fused in sarcoma protein.

CSF neurofilament

In FTD, NF-L CSF levels have been shown to be higher compared to AD (Skillbäck et al., 2014), but the elevated CSF levels of NF-L do not help to discriminate between FTD and other non-AD dementia, such as VAD. In FTD, high NF-L CSF levels correlate with neuropsychological performance and atrophy and with a shorter survival (Scherling et al., 2014).

CSF TDP-43

The intraneuronal inclusion positive for TDP-43 accounts for 50% of TDP cases, and several studies have explored fluid biomarkers related to TDP-43 in CSF of FTD patients. CSF levels of TDP-43 may be increased in patients with FTD and with amyotrophic lateral sclerosis, but the results varied for different FTD mutations and for different TDP variants (Mattsson et al. 2017). Moreover, unfortunately most of the protein measurable in the CSF appears to be blood-derived and so blood-based TDP43 may contaminate CSF measurement and its CSF concentration does not accurately reflect the neuropathology (Zetterberg et al., 2017).

CSF progranulin

Up to 10% of FTD cases have mutations of *GRN* gene, which encodes for the secreted protein progranulin. Progranulin concentration has been reported to be reduced in CSF and plasma in *GRN* mutation carriers but not in other FTD cases (Ghidoni, Paterlini, & Benussi, 2012). This data suggests progranulin determination as a screening test with high sensitivity and specificity for mutation carriers versus controls or versus patients with other dementias.

CSF AD biomarkers

A small proportion of patients with behavioral variants of FTD have positive AD CSF biomarkers, but generally CSF biomarkers for AD are normal in FTD patients. Combining the result of the normality of the classic AD biomarkers (Abeta42, t-Tau, p-Tau) with increase of CSF NF-L concentration the diagnostic accuracy for FTD has a specificity of 94%—100% and a sensitivity of 75%—86% with respect to AD and cognitively normal controls (de Jong et al., 2007) (Table 14.1).

Synucleinopathies

Parkinson's disease dementia

PD is the second most common neurodegenerative disorder and the best known movement disorder, especially in elderly people. It is caused by accumulation of protein alpha-synuclein (alpha-syn) in dopaminergic neurons, provoking neuronal cell death.

Although considered primarily a motor disorder, nonmotor symptoms accompany PD from its early stages and can be present even before the manifestation of motor symptomatology. Both prevalence and incidence of dementia are increased in PD. In a systematic review, the point prevalence of dementia was found as 24%—31%, and 3%—4% of all dementias were estimated to be due to PD dementia (PDD) (Aarsland, Zaccai, & Brayne, 2005).

Table 14.1 CSF biomarkers profile in different types of cognitive decline.

Condition	Total-TAU	Phospho-TAU	Aβ42
Normal aging	Normal	Normal	Normal
Alzheimer's disease	Moderate/marked increase	Moderate/marked increase	Moderate/marked decrease
Frontotemporal dementia	Normal/mild increase	Normal/mild increase	Normal/mild decrease
Parkinson's disease	Normal	Normal	Normal
Dementia with Lewy body	Normal/mild increase	Normal	Mild/moderate decrease
Creutzfeldt–Jakob disease	Very marked increase	Normal/mild moderate increase	Moderate/marked decrease
Vascular dementia	Normal/moderate increase	Normal	Normal/mild decrease

Discrepant data have been reported regarding the data of total alpha-syn in the CSF in PD, ranging from higher to unaltered and lower (Llorens, Schmitz, Ferrer, & Zerr, 2016). However, based on two meta-analyses, the current consensus is that alpha-syn CSF levels are significantly lower in PD patients compared to controls and AD cases, and are similar to those in dementia with Lewy bodies (DLB) and multiple system atrophy (Gao et al., 2015; Sako, Murakami, Izumi, & Kaji, 2014). Regarding PDD, two recent studies found a correlation between increased CSF alpha-syn levels and cognitive decline over 8 years' follow-up (Hall et al., 2015; Stewart et al., 2014).

Abeta and tau proteins, previously described as biomarkers for AD, were studied in PD as potential biomarkers for cognitive decline. CSF tau/Abeta42 and Abeta42 levels are able to predict cognitive decline in PD cases or discriminate PDD from nondemented PD patients (Parnetti et al., 2008).

Dementia with lewy bodies

In DLB the aggregates of alpha-syn are present not only in substantia nigra but also in cortical areas, causing cognitive dysfunction and dementia. In addition, clinically parkinsonian features may occur. Moreover, most patients neuropathologically have some degree of AD pathology, in particular, amyloid plaques.

The concomitant AD pathology is reflected in the CSF as decreased levels of Abeta42 (Marques, Van Rumund, Kuiperij, & Verbeek, 2017, pp. 85–97). Decreased CSF alpha-syn levels have been observed in DLB patients when compared to controls and AD cases, and lower levels of alpha-synuclein correlate with DLB duration (Llorens et al., 2016).

Recently, in a group of DLB patients elevated CSF–serum albumin ratios have been reported, but the clinical significance of this data is unknown (Llorens et al., 2015).

Key facts

- Cerebrospinal fluid biomarkers represent a useful tool for diagnosis of dementia syndrome.
- Alzheimer disease is the most prevalent form of dementia.
- Definite diagnosis of Alzheimer disease requires an autopsy confirmation.
- Pathological CSF biomarkers can be considered as surrogate markers of Alzheimer disease pathology.
- The availability of these biomarkers has moved the definition of Alzheimer disease from a clinicopathological entity to a clinicobiological entity.
- Now clinical diagnosis of Alzheimer disease can be established in vivo.
- Other biomarkers can be studied in other dementia syndromes, such as alpha-synuclein in the group of the synucleinopathies.

Summary points

- Overall diagnosis of dementia is clinical but diagnosis of specific types of dementia can be made using clinical criteria supplemented by biomarkers.
- It is possible to measure cerebrospinal fluid biomarkers related to the presence of aggregates of proteins, neuroinflammation, index of neuronal injuries, and other pathological processes.
- The classical cerebrospinal fluid AD signature is: high level of total and phosphorylated protein tau and reduced concentration of Abeta42 compared to nondemented controls.
- Cerebrovascular disease seems to provoke abnormalities in functioning of the blood—brain barrier, breakdown of the extracellular matrix, and white matter damage, which is reflected in abnormalities in the CSF.
- No pathognomonic CSF abnormalities are reported in patients with frontotemporal dementia.
- Cerebrospinal fluid alpha-synuclein levels are typically reduced in dementia with Lewy bodies and in Parkinson's disease dementia.

References

Aarsland, D., Zaccai, J., & Brayne, C. (2005). A systematic review of prevalence studies of dementia in Parkinson's disease. *Movement Disorders, 20*, 1255—1263.

Baldacci, F., Lista, S., Cavedo, E., Bonuccelli, U., & Hampel, H. (2017). Diagnostic function of the neuro-inflammatory biomarker YKL-40 in Alzheimer's disease and other neurodegenerative diseases. *Expert Review of Proteomics, 14*, 285—299.

Biomarkers Definitions Working Group. (2001). Biomarkers and surrogate endpoints: Preferred definitions and conceptual framework. *Clinical Pharmacology and Therapeutics, 69*, 89—95.

Bjerke, M., Zetterberg, H., Edman, Å., Blennow, K., Wallin, A., & Andreasson, U. (2011). Cerebrospinal fluid matrix metalloproteinases and tissue inhibitor of metalloproteinases in combination with subcortical and cortical biomarkers in vascular dementia and Alzheimer's disease. *Journal of Alzheimer's Disease, 27*, 665—676.

Bloudek, L. M., Spackman, D. E., Blankenburg, M., & Sullivan, S. D. (2011). Review and meta-analysis of biomarkers and diagnostic imaging in Alzheimer's disease. *Journal of Alzheimer's Disease, 26*, 627—645.

Corrêa, J. D., Starling, D., Teixeira, A. L., Caramelli, P., & Silva, T. A. (2011). Chemokines in CSF of Alzheimer's disease patients. *Arquivos de Neuro-Psiquiatria, 69*, 455—459.

Dubois, B., & Upenskaya-Cadoz, O. (2016). Changing Concepts and new definitions for Alzheimer's disease. In M. Husai, & J. M. Schott (Eds.), *Oxford texbook of cognitive neurology and dementia* (pp. 353—360). Oxford: Oxford University Press.

Galimberti, D., Schoonenboom, N., Scheltens, P., Fenoglio, C., Bouwman, F., Venturelli, E., et al. (2006). Intrathecal chemokine synthesis in mild cognitive impairment and Alzheimer disease. *Archives of Neurology, 63*, 538—543.

Gao, L., Tang, H., Nie, K., Wang, L., Zhao, J., Gan, R., et al. (2015). Cerebrospinal fluid alpha-synuclein as a biomarker for Parkinson's disease diagnosis: A systematic review and meta-analysis. *International Journal of Neuroscience, 12*, 645—654.

Ghidoni, R., Paterlini, A., & Benussi, L. (2012). Circulating progranulin as a biomarker for neurodegenerative diseases. *American Journal of Neurodegenerative Disease, 1*, 180—190.

Glodzik, L., Mosconi, L., Tsui, W., de Santi, S., Zinkowski, R., Pirraglia, E., et al. (2012). Alzheimer's disease markers, hypertension, and gray matter damage in normal elderly. *Neurobiology of Aging, 33*, 1215—1227.

Grahn, A., Hagberg, L., Nilsson, S., Blennow, K., Zetterberg, H., & Studahl, M. (2013). Cerebrospinal fluid biomarkers in patients with varicella-zoster virus CNS infections. *Journal of Neurology, 260*, 1813—1821.

Hall, S., Surova, Y., Öhrfelt, A., Zetterberg, H., Lindqvist, D., & Hansson, O. (2015). CSF biomarkers and clinical progression of Parkinson disease. *Neurology, 84*, 57—63.

Hardy, J., & Selkoe, D. J. (2002). The amyloid hypothesis of Alzheimer's disease: Progress and problems on the road to therapeutics. *Science, 297*, 353—356.

Ikeda, T., Noto, D., Noguchi-Shinohara, M., Ono, K., Takahashi, K., Ishida, C., et al. (2010). CSF tau protein is a useful marker for effective treatment of superficial siderosis of the central nervous system: Two case reports. *Clinical Neurology and Neurosurgery, 112*, 62—64.

Jack, C. R., Jr., Bennett, D. A., Blennow, K., Carrillo, M. C., Dunn, B., Haeberlein, S. B., et al. (2018). NIA-AA research framework. *Alzheimer's and dementia: the Journal of the Alzheimer's Association, 14*, 535—562.

Jack, C. R., Jr., Knopman, D. S., Jagust, W. J., Petersen, R. C., Weiner, M. W., Aisen, P. S., et al. (2013). Tracking pathophysiological processes in Alzheimer's disease: An updated hypothetical model of dynamic biomarkers. *The Lancet Neurology, 12*, 207—216.

de Jong, D., Jansen, R. W., Pijnenburg, Y. A., van Geel, W. J., Borm, G. F., Kremer, H. P., et al. (2007). CSF neurofilament proteins in the differential diagnosis of dementia. *Journal of Neurology, Neurosurgery, and Psychiatry, 78*, 936—938.

Lista, S., Zetterberg, H., O'Briant, S. E., Blennow, K., & Hampel, H. (2017). Blood and cerebrospinal fluid biomarkers in Alzheimer's disease. In D. Ames, J. T. O'Brien, & A. Burns (Eds.), *Dementia, fifth edition* (pp. 528—538). Boca Raton: CRC Press.

Llorens, F., Schmitz, M., Ferrer, I., & Zerr, I. (2016). CSF biomarkers in neurodegenerative and vascular dementias. *Progress in Neurobiology, 138*, 36—53.

Llorens, F., Schmitz, M., Gloeckner, S. F., Kaerst, L., Hermann, P., Schmidt, C., et al. (2015). Increased albumin CSF/serum ratio in dementia with Lewy bodies. *Journal of the Neurological Sciences, 358*, 398—403.

Luk, C., Compta, Y., Magdalinou, N., Martí, M. J., Hondhamuni, G., Zetterberg, H., et al. (2012). Development and assessment of sensitive immuno-PCR assays for the quantification of cerebrospinal fluid three- and four-repeat tau isoforms in tauopathies. *Journal of Neurochemistry, 123*, 96—405.

Marques, T. N., Van Rumund, A., Kuiperij, H. B., & Verbeek, M. M. (2017). Biomarkers in cerebrospinal fluid for synucleinopathies, tauopathies, and other neurodegenerative disorders. In F. Deisenhammer, C. E. Teunissen, & H. Tumani (Eds.), *Cerebrospinal fluid in neurologic disorders: Vol 146. Handbook of clinical neurology* (pp. 85—97). Amsterdam: Elsevier (3rd series).

Masters, C. L. (2017). The neuropathology of Alzheimer's disease. In D. Ames, J. T. O'Brien, & A. Burns (Eds.), *Dementia, fifth edition* (pp. 470—485). Boca Raton: CRC Press.

Mattsson, N., Grigoriou, S., & Zetterberg, H. (2017). Fluid biomarkers in Alzheimer's disease and fronto-temporal dementia. In G. Galimberti, & E. Scarpini (Eds.), *Neurodegenerative diseases* (pp. 221–252). Cham: Springer.

Mattsson, N., Sävman, K., Osterlundh, G., Blennow, K., & Zetterberg, H. (2010). Converging molecular pathways in human neural development and degeneration. *Neuroscience Research, 66*, 30–32.

McKhann, G., Drachman, D., Folstein, M., Katzman, R., Price, D., & Stadlan, E. M. (1984). Clinical diagnosis of Alzheimer's disease: Report of the NINCDS-ADRDA work group under the auspices of Department of Health and Human Services Task Force on Alzheimer's disease. *Neurology, 34*, 939–944.

Parnetti, L., Tiraboschi, P., Lanari, A., Peducci, M., Padiglioni, C., D'Amore, C., et al. (2008). Cerebrospinal fluid biomarkers in Parkinson's disease with dementia and dementia with Lewy bodies. *Biological Psychiatry, 64*, 850–855.

Piccio, L., Deming, Y., Del-Águila, J. L., Ghezzi, L., Holtzman, D. M., Fagan, A. M., et al. (2016). Cerebrospinal fluid soluble TREM2 is higher in Alzheimer disease and associated with mutation status. *Acta Neuropathologica, 131*, 925–933.

Rosenberg, G. A., Bjerke, M., & Wallin, A. (2014). Multimodal markers of inflammation in the subcortical ischemic vascular disease type of vascular cognitive impairment. *Stroke, 45*, 1531–1538.

Sako, W., Murakami, N., Izumi, Y., & Kaji, R. (2014). Reduced alpha-synuclein in cerebrospinal fluid in synucleinopathies: Evidence from a meta-analysis. *Movement disorders: Official Journal of the Movement Disorder Society, 29*, 1599–1605.

Scherling, C. S., Hall, T., Berisha, F., Klepac, K., Karydas, A., Coppola, G., et al. (2014). Cerebrospinal fluid neurofilament concentration reflects disease severity in frontotemporal degeneration. *Annals of Neurology, 75*, 116–126.

Skillbäck, T., Delsing, L., Synnergren, J., Mattsson, N., Janelidze, S., Nägga, K., et al. (2017). CSF/serum albumin ratio in dementias: A cross-sectional study on 1861 patients. *Neurobiology of Aging, 59*, 1–9.

Skillbäck, T., Farahmand, B., Bartlett, J. W., Rosén, C., Mattsson, N., Nägga, K., et al. (2014). CSF neurofilament light differs in neurodegenerative diseases and predicts severity and survival. *Neurology, 18*, 1945–1953.

Stewart, T., Liu, C., Ginghina, C., Cain, K. C., Auinger, P., Cholerton, B., et al. (2014). Cerebrospinal fluid α-synuclein predicts cognitive decline in Parkinson disease progression in the DATATOP cohort. *American Journal of Pathology, 184*, 966–975.

Suárez-Calvet, M., Kleinberger, G., Araque Caballero, M.Á., Brendel, M., Rominger, A., Alcolea, D., et al. (2016). sTREM2 cerebrospinal fluid levels are a potential biomarker for microglia activity in early-stage Alzheimer's disease and associate with neuronal injury markers. *EMBO Molecular Medicine, 2*, 466–476.

Visser, P. J., Verhey, F., Knol, D. L., Scheltens, P., Wahlund, L. O., Freund-Levi, Y., et al. (2009). Prevalence and prognostic value of CSF markers of Alzheimer's disease pathology in patients with subjective cognitive impairment or mild cognitive impairment in the DESCRIPA study: A prospective cohort study. *The Lancet Neurology, 8*, 619–627.

Wallin, A. K., Blennow, K., Zetterberg, H., Londos, E., Minthon, L., & Hansson, O. (2010). CSF biomarkers predict a more malignant outcome in Alzheimer disease. *Neurology, 11*, 1531–1537.

Wallin, A., Kapaki, E., Boban, M., Engelborghs, S., Hermann, D. M., Huisa, B., et al. (2017). Biochemical markers in vascular cognitive impairment associated with subcortical small vessel disease - a consensus report. *BMC Neurology, 23*, 102–109.

Wang, L., Fagan, A. M., Shah, A. R., Beg, M. F., Csernansky, J. G., Morris, J. C., et al. (2012). Cerebrospinal fluid proteins predict longitudinal hippocampal degeneration in early-stage dementia of the Alzheimer type. *Alzheimer Disease and Associated Disorders, 2*, 314–321.

Watabe-Rudolph, M., Song, Z., Lausser, L., Schnack, C., Begus-Nahrmann, Y., Scheithauer, M. O., et al. (2012). Chitinase enzyme activity in CSF is a powerful biomarker of Alzheimer disease. *Neurology, 78*, 569–577.

Yin, G. N., Jeon, H., Lee, S., Lee, H. W., Cho, J. Y., & Suk, K. (2009). Role of soluble CD14 in cerebrospinal fluid as a regulator of glial functions. *Journal of Neuroscience Research, 15*(87), 2578–2590.

Zetterberg, H., Rohrer, J. D., & Schott, J. M. (2017). Cerebrospinal fluids in dementias. In F. Deisenhammer, C. E. Teunissen, & H. Tumani (Eds.), *Cerebrospinal fluid in neurologic disorders: Vol 146. Handbook of clinical neurology* (pp. 85–97). Amsterdam: Elsevier (3rd series).

CHAPTER 15

Salivary biomarkers in Alzheimer's disease

Jessica L. Andrews[1], Francesca Fernandez[2]
[1]Office of the DVC Research, The University of Sydney, Sydney, NSW, Australia; [2]Faculty of Health Sciences, School of Behavioural and Health Sciences, Australian Catholic University, Nudgee, QLD, Australia

List of abbreviations

AChE acetylcholinesterase
AICD intracellular cytoplasmic/C-terminal domain
APP amyloid precursor protein
Aβ peptide amyloid-β peptide
CTFβ carboxy-terminal fragment-β
CTFα carboxy-terminal fragment-α
ELISA enzyme-linked immunosorbent assay
FUPLC MS faster ultra-performance liquid chromatography mass spectrometry
HRP horseradish peroxidase
MALDI-TOF MS matrix-assisted laser desorption/ionization time-of-flight mass spectrometry
p-tau phosphorylated tau
sAPPβ secreted amino-terminal APPβ
sAPPα secreted amino-terminal APPα
SDS−PAGE sodium dodecyl sulfate−polyacrylamide gel electrophoresis
^{1}H NMR proton nuclear magnetic resonance

Mini-dictionary of terms

Biomarkers biological parameters indicating the presence of a state, condition, and/or disease.
ELISA a laboratory technique using antibodies that is used to quantify proteins of interest in samples.
Gland a structure in the human body able to secrete a biological fluid.
Saliva watery fluid present in the mouth of individuals necessary for the protection of teeth and for carbohydrate digestion, which also has enzymatic and antibacterial functions.
Western blot a method of quantification using specific molecules that bind to the targeted protein in a sample.

Introduction

Alzheimer's disease (AD) is a devastating neurodegenerative disorder of the central nervous system and is the most common and well-known cause of dementia, resulting in about 70% of dementia cases worldwide. The neurodegeneration in AD results in cognitive dysfunction and is physiologically caused by a combination of dysfunctional mechanisms and pathological abnormalities in the brain, primarily in the temporal

Diagnosis and Management in Dementia
ISBN 978-0-12-815854-8, https://doi.org/10.1016/B978-0-12-815854-8.00015-X

lobe, including the hippocampus, the brain region responsible for the consolidation of information from short-term to long-term memory (Kandel, 2006). These dysfunctional mechanisms include neuronal death caused by apoptosis, mitochondrial dysfunction, immune and inflammatory responses, oxidative stress, and disruptions to cell-cycle and cell-transport pathways; pathological abnormalities include the aggregation and accumulation of toxic molecules (neurofibrillary tangles and amyloid plaques) both intra- and extraneuronally.

Currently AD can be diagnosed only when the pathology has been clinically established (Knopman et al., 2001). This task is quite difficult to perform at early stages of the disease when symptoms are mild; therefore, AD is usually diagnosed only once the cognitive deficits resulting from the pathology have begun to significantly affect the person's ability to cope with the demands of his or her social and/or professional life. Early detection of neurodegenerative disorders such as AD is critical, to be able to provide treatments to patients as early as possible, when they are most likely to be effective (DeKosky and Marek, 2003). A large number of studies have investigated biomarker candidates as a method of detecting AD earlier in its development. By definition, a biomarker is a substance, physiological characteristic, or gene that indicates the presence of disease, a physiological abnormality, or a psychological condition.

Current methods for biomarker detection of AD involve lumbar punctures to collect cerebrospinal fluid (CSF), a protective cushioning fluid surrounding the brain and spinal cord. Sampling CSF for biomarker analysis can provide information about the physiological status of the brain; however, the lumbar puncture procedure required for collection of CSF is costly, invasive, and quite painful, especially for elderly populations who are the primary demographic for AD (Evans, 1998). Therefore, an accurate and less invasive method for biomarker analysis for the detection of AD is required.

Saliva is a watery physiological fluid that contains many important substances such as mucus, electrolytes, enzymes, and antibacterial agents, which functions to protect teeth and other oral tissues and aid in the digestion of food (Mandel, 1987; Schenkels, Veerman, & Nieuw Amerongen, 1995). There are three main salivary glands responsible for the secretion of saliva in the mouth: the parotid gland, sublingual gland, and submandibular gland (Fig. 15.1). These glands interface with the central nervous system via their connection to the glossopharyngeal and facial cranial nerves, which innervate the parotid gland and sublingual and submandibular glands, respectively (Moore, Dalley, & Agur, 2013). Due to this interface between the salivary glands and the central nervous system biomarkers found in saliva may prove to be a useful tool in determining normal and pathological physiologies of the central nervous system (Shi et al., 2011). Furthermore, saliva is easy to collect (Fig. 15.2); its collection is noninvasive and does not require expert training, unlike current methods of collecting CSF for biomarker analysis of neurological disorders. Using salivary biomarkers as a possible avenue for the diagnosis of AD and many brain-related diseases, including schizophrenia, attention-deficit

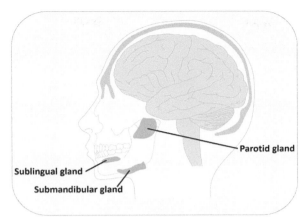

Figure 15.1 *Location of salivary glands.* There are three main salivary glands responsible for the secretion of saliva in the mouth: the parotid gland, sublingual gland, and submandibular gland.

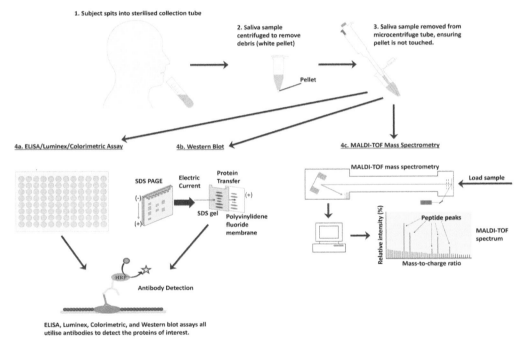

Figure 15.2 *Collection and processing of saliva samples for biomarker analysis.* Collection and processing of saliva samples from subjects is quick, easy, and painless for the subject, making saliva an excellent choice of biofluid for the assessment of diagnostic markers for Alzheimer's disease. Once collected, saliva samples are processed via a number of biochemical assays, including ELISA, Luminex, colorimetric, and western blot assays, plus MALDI-TOF mass spectrometry. *ELISA*, enzyme-linked immunosorbent assay; *HRP*, horseradish peroxidase; *MALDI-TOF*, matrix-assisted laser desorption/ionization time-of-flight; *SDS–PAGE*, sodium dodecyl sulfate–polyacrylamide gel electrophoresis.

hyperactivity disorder, Parkinson's disease, and vascular dementia, would allow for earlier diagnosis of the disease, allowing for earlier treatment interventions for these devastating neurological disorders.

Salivary biomarkers are established for the diagnosis and prediction of the progression of a number of diseases (Streckfus & Bigler, 2002); however, as of this writing, despite a large number of promising results, none have been established in the routine clinical diagnosis of AD. This chapter will provide an overview of the different types of salivary biomarkers that have been examined in AD and their methods of analysis, and then discuss novel promising salivary biomarkers for AD, genetic biomarkers.

Types of salivary biomarkers in Alzheimer's disease and methods of analysis

The etiology of AD is pathologically characterized by the presence of neurofibrillary tangles and amyloid plaques in the brain, which can be attributed to two molecules: the amyloid-β peptide (Aβ), which has a normal physiological role in memory formation, lipid homeostasis, and neuron outgrowth and activity, and the microtubule-associated protein tau, which has a role in maintaining neuronal structure and intracellular transport.

In AD, Aβ is formed through the cleavage of amyloid precursor protein by β- and γ-secretase enzymes (Fig. 15.3). The amyloid-β42 isoform (Aβ42) is known to be more neurotoxic than other isoforms of Aβ, and the amyloid-β40 isoform (Aβ40) is known to be a stabilizer for Aβ42. Extracellular Aβ42 accumulation and oligomerization result in the formation of amyloid plaques in the brain (Fig. 15.3).

A study by Bermejo-Pareja, Antequera, Vargas, Molina, and Carro (2010) used enzyme-linked immunosorbent assay (ELISA) to examine and quantify levels of Aβ proteins extracted from saliva samples of AD patients compared with two sets of controls: elderly healthy controls and people with Parkinson's disease (PD). Patients with mild AD showed a statistically significant increase in salivary Aβ42 levels compared with their controls while no difference in salivary Aβ42 levels was reported between PD patients and their respective controls, indicating that the increase in salivary Aβ42 levels in mild AD patients was specific to the AD pathology and not neurodegeneration in general. It is important to note here that, while Aβ42 levels were increased, there was a very large standard deviation, and therefore these results might be considered with caution. Contrasting results by Kim, Choi, Song, and Song (2014) obtained from both an antibody-based magnetic nanoparticle assay and a highly sensitive ELISA have demonstrated that salivary Aβ42 levels increase with disease progression, ranging from subjects with mild cognitive impairment (MCI) right up to those with severe AD; it must also be noted that the authors make mention of the high standard deviation in their study. Despite this, results showed a significant threefold increase in both Aβ42 and Aβ40 levels

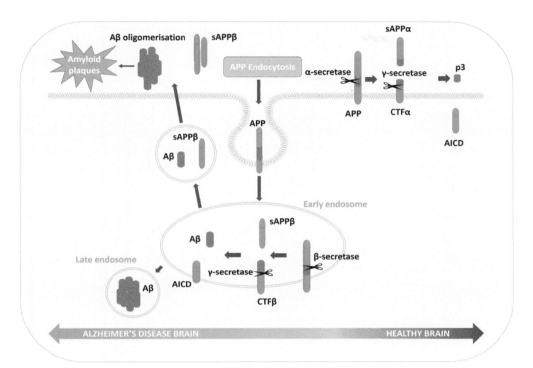

Figure 15.3 *Amyloid precursor protein (APP) processing in Alzheimer's disease.* In Alzheimer's disease, neuronal APP is endocytosed before undergoing cleavage by β-secretase into sAPPβ and CTFβ. CTFβ cleavage by γ-secretase results in Aβ and AICD fragments. Aβ and sAPPβ are exocytosed back into the extracellular space, resulting in extracellular accumulation of Aβ and formation of oligomers, leading to the generation of amyloid plaques in the brain. In the nonpathologic brain APP is cleaved by α-secretase, generating sAPPα and CTFα. CTFα is further cleaved by γ-secretase, forming p3 and AICD fragments. All by-products of this pathway are nonpathogenic. *Aβ peptide*, amyloid-β peptide; *AICD*, intracellular cytoplasmic/C-terminal domain; *CTFα*, carboxy-terminal fragment-α; *CTFβ*, carboxy-terminal fragment-β; *sAPPα*, secreted amino-terminal APPα; *sAPPβ*, secreted amino-terminal APPβ. [1]Bermejo-Pareja et al. (2010); [2]Kim et al. (2014); [3]Lee et al. (2017); [4]Shi et al. (2011); [5]Sayer et al. (2004); [6]Bakhtiari et al. (2017); [7]Boston et al. (2008); [8]Carro et al. (2017); [9]Su et al. (2008); [10]Ashton et al. (2018); [11]Pekeles et al. (2018); [12]Liang et al. (2015); [13]Yilmaz et al. (2017).

in the saliva of people with severe AD compared with subjects with MCI (Kim et al., 2014). Furthermore, there was a 10-fold increase in Aβ42 and Aβ40 levels in the saliva of people with severe AD compared with the healthy controls (Kim et al., 2014). A more recent study by Lee, Guo, Kennedy, McGeer, and McGeer (2017) used ELISA to confirm the results of Bermejo-Pareja et al. (2010) and Kim et al. (2014), showing a significant increase in Aβ42 levels in the saliva of people with AD compared with healthy controls, importantly, with a much smaller standard error than the previous two studies. While Lee et al. (2017) did not examine the different stages of AD severity, they did, however, additionally test three cases of people classified as pre-AD. These three pre-AD cases had a very high risk of developing AD, two due to a strong family history and one who carried a *presenilin 1* genetic mutation known to result in a 100% risk of developing AD (Lee et al., 2017). These three individual cases had very high salivary Aβ42 levels, with the authors determining the risk of these three subjects developing AD being definite, rather than possible (Lee et al., 2017). The results of all of these studies are in contrast to the findings of the wider body of literature regarding the widely reported CSF Aβ42 levels, which are consistently reported to be decreased in people with AD (Hampel et al., 2004; Moonis et al., 2005; Riemenschneider et al., 2002), despite Aβ42 levels being increased in the brains of people with AD in the form of amyloid plaques (Grimmer et al., 2009). Interestingly, a metaanalysis of 14 studies reported that plasma levels of Aβ42, which were largely reported to be lower in people with AD compared with healthy controls, were not a strong predictor of AD symptomatology; this may be due to the lack of consistency in the methodologies used to measure Aβ42 in the plasma (Koyama et al., 2012).

The results of the abovementioned three studies indicate that high salivary levels of Aβ42 may prove to be a predictive biomarker for the development of AD. Increasing the stability of Aβ42 in saliva by reducing its aggregation (by the addition of thioflavin S to samples) and limiting the presence of bacteria (by the addition of sodium azide to collected samples) will greatly improve the establishment of a reliable and reproducible standard salivary test for AD (Lee et al., 2017).

Microtubule-associated protein tau has a role in maintaining neuronal structure and intracellular transport; in AD, tau is abnormally phosphorylated. High levels of total tau (t-tau) and phosphorylated tau (p-tau) have been found intraneuronally and in the CSF of people with AD (Han, Schellenberg, Wang, & Alzheimer's Disease Neuroimaging Initiative, 2010). Furthermore, tau is known to work interdependently with Aβ42 in the pathophysiology of AD, enhancing their cytotoxic functions (Bloom, 2014).

Shi et al. (2011) detected t-tau and p-tau in the saliva of all tested AD and healthy control subjects using matrix-assisted laser desorption ionization—time-of-flight—tandem mass spectrometry and quantified t-tau and p-tau using highly sensitive Luminex assays (Table 15.1, Fig. 15.4). Normalized t-tau levels were unchanged between AD and control subjects, and normalized p-tau was increased in AD compared with control subjects, although not at a statistically significant level. When the ratio of p-tau/t-tau was examined, this ratio was significantly increased in the AD subjects compared with the control group. A more recent study by Pekeles et al. (2018) examined p-tau/t-tau ratios utilizing specific antibodies for four different p-tau phosphorylation sites (threonine 181, serine 396, serine 404, and a combined antibody for serine 400/threonine 403/serine 404 sites) in the saliva of two separate batches of subjects using western blotting (Table 15.1, Fig. 15.4). In batch 1 it was found that AD subjects have increased p-tau/t-tau levels at three of the four tested phosphorylation sites compared with controls, and in batch 2 it was found that

Table 15.1 Summary of results of tested protein markers in saliva of Alzheimer's disease.

AD marker tested	Number of cases/controls	Analysis method	Result	References
Aβ42	70 AD subjects 56 elderly controls 51 PD subjects (controls)	• ELISA	• Increase in Aβ42 in mild AD compared with controls • No difference in Aβ42 levels between control groups	Bermejo-Pareja et al. (2010)
	28 AD subjects 17 controls	• Antibody-based magnetic nanoparticle assay	• Aβ42 increases with AD disease progression	Kim et al. (2014)
	7 AD subjects 3 pre-AD subjects 27 controls	• ELISA • ELISA	• Aβ42 is increased in severe AD compared with controls • Aβ42 is increased in AD compared with controls	Lee et al. (2017)
Tau	21 AD subjects 38 healthy controls	• MALDI-TOF MS • Luminex	• Increased ratio of p-tau/t-tau in AD compared with controls • Nonsignificant increase in p-tau in AD compared with controls	Shi et al. (2011)
	Batch 1: 55 MCI subjects 46 AD subjects 47 healthy elderly controls	• Western blot	• Increased ratio of p-tau/t-tau in AD compared with healthy elderly controls	Pekeles et al. (2018)
	Batch 2: 16 FTD subjects 41 AD subjects 44 healthy elderly controls 12 neurology subjects 76 young controls	• Western blot	• Increased ratio of p-tau/t-tau in AD and FTD compared with healthy elderly controls	
	68 MCI subjects 53 AD subjects 160 healthy elderly controls	• Human t-tau assay on an HD–1 Simoa analyzer	• No difference in t-tau concentrations in AD compared with controls	Ashton et al. (2018)

Continued

Table 15.1 Summary of results of tested protein markers in saliva of Alzheimer's disease.—cont'd

AD marker tested	Number of cases/controls	Analysis method	Result	References
AChE	22 mild dementia subjects 14 AD subjects 11 healthy controls	• Ellman colorimetric assay • Western blot	• Decreased AChE activity in the AD subjects nonresponsive to AChE inhibitors	Sayer et al. (2004)
	15 AD subjects 13 VD subjects 13 control subjects	• Ellman colorimetric assay	• Nonsignificant decrease in AChE activity in AD and VD compared with controls	Boston et al. (2008)
	15 AD subjects 15 healthy elderly controls	• Ellman colorimetric assay	• Nonsignificant decrease in AChE activity in AD compared with controls	Bakhtiari et al. (2017)
Lactoferrin	80 AD subjects 44 aMCI subjects 59 PD subjects (controls) 91 healthy elderly controls	• MALDI-TOF MS • ELISA	• Decreased lactoferrin in aMCI and AD subjects compared with healthy elderly controls • Lactoferrin levels decreased with increasing severity of disease; aMCI>AD. • Lactoferrin levels in PD were higher than levels of the control group • Positive correlation between salivary lactoferrin and CSF Aβ42 levels • Negative correlation between salivary lactoferrin and CSF t-tau levels	Carro et al. (2017)
Protein carbonyls	21 MCI subjects 15 AD subjects 30 healthy controls	• ELISA	• No difference in protein carbonyls between AD or MCI and control groups • Peak of protein carbonyl levels for all groups at 2:00 p.m.	Su et al. (2008)

Metabolites	256 AD subjects 218 healthy controls	• FUPLC MS	• Sphinganine-1-phosphate, ornithine, and phenyllactic acid were increased in AD compared with controls • Inosine, 3-dehydrocarnitine, and hypoxanthine were decreased in AD compared with controls • Specificity and accuracy of Sphinganine-1-phosphate, ornithine, and phenyllactic acid were determined to be sufficient to distinguish AD from control subjects	Liang et al. (2015)
	8 MCI subjects 9 AD subjects 12 healthy controls	• ¹H NMR spectroscopy	• Accurately quantified and identified 57 metabolites in saliva • Alterations in concentration of 22 metabolites in MCI and AD subjects compared with controls • Specificity and sensitivity of galactose, imidazole, and acetone using regression models made these the strongest predictive markers to distinguish between MCI and control subjects	Yilmaz et al. (2017)

Aβ, amyloid-β peptide; *AChE*, acetylcholinesterase; *AD*, Alzheimer's disease; *aMCI*, amnestic mild cognitive impairment; *CSF*, cerebrospinal fluid; *ELISA*, enzyme-linked immunosorbent assay; *FTD*, frontotemporal dementia; *FUPLC MS*, faster ultra-performance liquid chromatography mass spectrometry; *MALDI-TOF MS*, matrix-assisted laser desorption/ionization time-of-flight mass spectrometry; *MCI*, mild cognitive impairment; *¹H NMR*, proton nuclear magnetic resonance; *PD*, Parkinson's disease; *p-tau*, phosphorylated tau; *t-tau*, total tau; *VD*, vascular dementia.

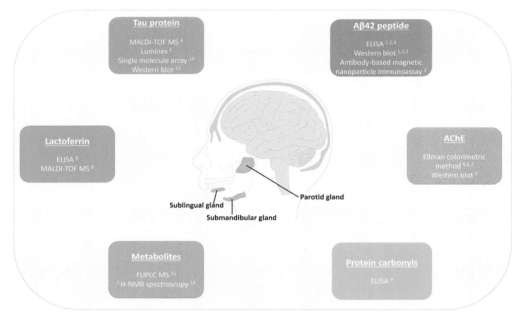

Figure 15.4 *Salivary biomarkers examined in Alzheimer's disease and their methods of analysis.* There are six main biomarkers that have been examined in Alzheimer's disease: Aβ42 peptide, tau protein, AChE, lactoferrin, protein carbonyls, and metabolites. These have all been examined using a variety of biochemical assays. *Aβ42 peptide*, amyloid-β42 peptide; *AChE*, acetylcholinesterase; *ELISA*, enzyme-linked immunosorbent assay; *FUPLC MS*, faster ultra-performance liquid chromatography mass spectrometry; *MALDI-TOF MS*, matrix-assisted laser desorption/ionization time-of-flight mass spectrometry; *¹H NMR*, proton nuclear magnetic resonance.

p-tau/t-tau levels were increased at the S396 phosphorylation site in AD and fronto-temporal dementia subjects compared with healthy elderly controls. Another recent study by Ashton et al. (2018) examined salivary t-tau using a human total tau assay on an HD-1 Simoa analyzer (a digital form of ELISA) in people with MCI and AD compared with healthy controls. Similar to Shi et al. (2011), they also found no significant difference in levels of t-tau in the two diagnostic groups examined compared with the healthy controls (Ashton et al., 2018). While further study is required to assess the viability of tau as a salivary biomarker for AD, the aforementioned studies have determined that tau is detectable and quantifiable in saliva samples by several biochemical assays. Interestingly, tau was also found to be elevated in buccal cells of AD subjects, supporting the potentiality of this biomarker in the diagnosis of AD pathophysiology (François, Leifert, Martins, Thomas, & Fenech, 2014).

A number of other biomarkers have been tested in the saliva of people with AD, including acetylcholinesterase (AChE), lactoferrin, and protein carbonyls (Table 15.1, Fig. 15.4).

AChE inhibitors are one of the first lines of treatment in the early stages of AD, since cholinergic neurons are some of the first to degenerate. Sayer, Law, Connelly, and Breen (2004) investigated salivary AChE activity and AChE protein expression in subjects with mild dementia who responded to AChE therapy, people with AD who did not respond to AChE therapy, and matched elderly controls. AChE activity was decreased in the AD subjects who were nonresponsive to AChE inhibitors, and the expression of the AChE enzyme in saliva was confirmed by western blot analysis. More recent studies by Boston, Gopalkaje, Manning, Middleton, and Loxley (2008) and Bakhtiari et al. (2017) have found similar results, whereby AChE enzymatic activity tended to decrease in saliva samples from AD compared with control subjects; however, these results were not statistically significant (Table 15.1, Fig. 15.4). Further work is required to determine if salivary AChE activity levels can be used as a diagnostic biomarker for AD.

The antimicrobial peptide lactoferrin is known to be able to bind to Aβ and contributes to the regulation of immune and inflammatory responses in the brain (Gifford, Hunter, & Vogel, 2005; Holmes et al., 2009; Wang, Sato, Zhao, & Tooyama, 2010). Due to these characteristics, lactoferrin has been examined as a potential biomarker for AD. Carro et al. (2017) identified lactoferrin in salivary samples and quantified it using ELISA in a cohort totaling 274 subjects (Table 15.1, Fig. 15.4). They found salivary lactoferrin levels to be decreased in subjects with amnestic MCI and AD compared with healthy elderly controls, as well as PD controls, and that levels decreased with increasing disease severity. Positive and negative correlations were also found between salivary lactoferrin and CSF levels of Aβ42 and t-tau, respectively (Carro et al., 2017). This study presents lactoferrin as a promising novel biomarker for the diagnosis of AD, but requires further work in a larger population.

Oxidative stress processes are known to underlie AD pathophysiology. Since protein carbonyls are a well-accepted marker of protein oxidation, Su et al. (2008) examined levels of total carbonylated proteins in the saliva of people with AD and MCI compared with a control group control using ELISA. While there were no significant differences in protein carbonyl levels in AD or MCI compared with controls, it was found that protein carbonyl levels were highest at 2:00 p.m. for all tested groups. This is the first study to report significant diurnal fluctuations in salivary redox markers. The lack of change in total carbonylated proteins in the saliva of people with AD and MCI indicates that protein carbonyls do not appear to be in a state of oxidative stress in the examined disease states; therefore, protein carbonyls would not be a viable biomarker option for AD. Due to the limited number of tested individuals in this study, further studies using larger populations are necessary to better investigate the potential for protein carbonyls to be recognized as biomarkers for AD.

Finally, metabolomics, the study of small molecules resulting from metabolic activity, appears to be a novel and promising area of research for exploring potential biomarkers for AD. Liang et al. (2015) used faster ultra-performance liquid chromatography mass spectrometry to examine the presence and levels of metabolites in the saliva of AD and control subjects (Table 15.1, Fig. 15.4). They found that levels of sphinganine-1-phosphate, ornithine, and phenyllactic acid were increased in AD compared with controls, and that inosine, 3-dehydrocarnitine, and hypoxanthine were decreased in AD compared with controls. The researchers also showed that sphinganine-1-phosphate, ornithine, and phenyllactic acid were detected with sufficient accuracy and specificity to be able to distinguish between AD and control subjects (Liang et al., 2015). More recently, Yilmaz et al. (2017) used proton nuclear magnetic resonance spectroscopy to accurately identify and quantify 57 metabolites in saliva of MCI, AD, and control subjects. Alterations in the concentrations of 22 metabolites were reported in MCI and AD subjects compared with controls, and galactose, imidazole, and acetone were identified as strong predictive markers to distinguish between MCI and control subjects. Further studies are necessary to establish the potential of metabolite analysis as biomarkers for AD.

Another type of biomarker that has not yet been discussed in this review are genetic biomarkers of AD. The next section will cover the potential for this type of biomarker in saliva for the potential diagnosis of AD.

Genetic and epigenetic salivary biomarkers in Alzheimer's disease: a promising avenue for diagnosis

Genetic markers are the most promising biomarkers for the diagnosis of AD (Reitz and Mayeux, 2009). With advances in technology and in the sequencing of the human genome, genetic research has progressed explosively for the last two decades. However, the majority of genetic and epigenetic studies have been performed using brain tissues from AD patients compared with matched controls, and limited work has been reported using saliva. Using whole-genome transcriptome arrays (Fig. 15.5), we have characterized the expression of the whole genome in saliva of people in varying stages of AD compared with age-matched controls, and have established different profiles of gene expression according to the disease progression. To our knowledge, we are the first to examine the whole genome in the saliva of AD patients, in the quest to find novel biomarkers for the diagnosis of this devastating neurodegenerative disease. At the time of this writing, we are in the process of validating our results using real-time PCR (Fig. 15.5).

Prominent epigenetic markers that are measurable in saliva include DNA methylation. Epigenetics is the study of molecular modifications in cells, modulating gene expression in response to environmental changes that have occurred. DNA methylation (one of the best understood epigenetic markers) can act as an interface between genetic and environmental factors, effectively "switching" genes on and off and playing an essential role in the development and differentiation of cells in response to environmental events. Like gene expression, as mentioned earlier with regard to transcriptome arrays, DNA methylation can be measured in a genome-wide manner in saliva using high-throughput technologies. Salivary DNA has already been used to study

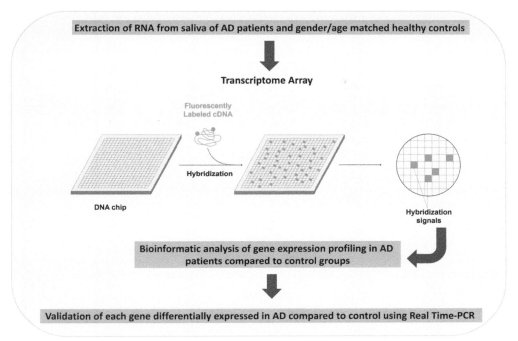

Figure 15.5 *Detection of genetic biomarkers in Alzheimer's disease (AD).* The use of transcriptome arrays to uncover potential genetic biomarkers in a case—control elderly population and validation of significantly differentially expressed genes through individual real-time PCR is depicted.

methylation changes in relation to systemic conditions, such as cancer pathologies (Wren, Shirtcliff, & Drury, 2015). Interestingly, salivary DNA methylation patterns have also been found to correlate with the methylation expression profile reported across different regions of the brain in cohorts of healthy adults and people with schizophrenia (Wren et al., 2015). Methylation profiles in the brain were also found to be more reflective of DNA methylation profiles in saliva compared with those found in blood samples (Smith et al., 2015).

Altogether, little is known about the characterization of genetic and epigenetic biomarkers for AD in saliva. With the success of saliva-based diagnostic tools for cancer (based on transcriptomic studies) and the reduced cost of high-throughput genetic analysis technologies, there is no doubt that reliable genetic and/or epigenetic markers for AD will be discovered in the near future.

Conclusion

Unlike blood and CSF, saliva is the most noninvasive means to collect biological fluid for the detection of potential biomarkers for the diagnosis of neurological pathologies.

However, only a limited number of studies have explored this avenue of research in the context of AD diagnosis. One limitation encountered in the examination of saliva of people with AD is the effects of current anticholinergic medications for AD; these drugs are known to reduce the production of saliva, making its collection a little more challenging for some people with AD. Since salivary parameters can change rapidly in response to physiological and/or pathological stimuli, additional studies are urgently needed to further identify reliable and reproducible biomarkers for AD pathophysiology. The establishment of a specific diagnostic tool will also benefit future discovery of effective therapy for this devastating neurodegenerative disease.

Key facts of salivary biomarkers

- Humans produce around 1.5–2 L of saliva each day.
- Saliva can contain over 70 types of bacteria.
- Heavy cell phone users have a significantly increased salivary flow rate compared with more restrained cell phone users.
- Saliva contains its own painkiller, opiorphin, which is six times more powerful than morphine.
- Like the kidney, salivary glands and their related ducts can contain "stones," small solid balls formed from crystallized chemicals in the saliva.

Summary points

- Saliva is a powerful biofluid that is able to respond dynamically to physiological and pathological states.
- Protein biomarkers are the most studied types of biomarkers in the investigation of potential biomarkers for Alzheimer's disease.
- Both amyloid-β42 and tau proteins currently represent the most promising biomarkers for Alzheimer's disease.
- Further studies are required to identify genetic and epigenetic biomarkers for Alzheimer's disease.
- The characterization of specific biomarkers for Alzheimer's disease will improve not only the diagnosis of this pathology but also its therapy.

References

Ashton, N. J., Ide, M., Schöll, M., Blennow, K., Lovestone, S., Hye, A., et al. (2018). No association of salivary total tau concentration with Alzheimer's disease. *Neurobiology of Aging, 70*, 125–127.
Bakhtiari, S., Moghadam, N. B., Ehsani, M., Mortazavi, H., Sabour, S., & Bakhshi, M. (2017). Can salivary acetylcholinesterase be a diagnostic biomarker for Alzheimer? *Journal of Clinical and Diagnostic Research, 11*, ZC58–ZC60.

Bermejo-Pareja, F., Antequera, D., Vargas, T., Molina, J. A., & Carro, E. (2010). Saliva levels of Abeta1-42 as potential biomarker of Alzheimer's disease: A pilot study. *BMC Neurology, 10*, 108.

Bloom, G. S. (2014). Amyloid-β and tau: The trigger and bullet in Alzheimer disease pathogenesis. *Journal of the American Medical Association Neurology, 71*, 505—508.

Boston, P. F., Gopalkaje, K., Manning, L., Middleton, L., & Loxley, M. (2008). Developing a simple laboratory test for Alzheimer's disease: Measuring acetylcholinesterase in saliva - a pilot study. *International Journal of Geriatric Psychiatry, 23*, 439—440.

Carro, E., Bartolomé, F., Bermejo-Pareja, F., Villarejo-Galende, A., Molina, J. A., Ortiz, P., et al. (2017). Early diagnosis of mild cognitive impairment and Alzheimer's disease based on salivary lactoferrin. *Alzheimer's and Dementia: Diagnosis, Assessment and Disease Monitoring, 8*, 131—138.

DeKosky, S. T., & Marek, K. (2003). Looking backward to move forward: Early detection of neurodegenerative disorders. *Science, 302*, 830—834.

Evans, R. W. (1998). Complications of lumbar puncture. *Neurologic Clinics, 16*, 83—105.

François, M., Leifert, W., Martins, R., Thomas, P., & Fenech, M. (2014). Biomarkers of Alzheimer's disease risk in peripheral tissues; focus on buccal cells. *Current Alzheimer Research, 11*, 519—531.

Gifford, J. L., Hunter, H. N., & Vogel, H. J. (2005). Lactoferricin. *Cellular and Molecular Life Sciences, 62*, 2588.

Grimmer, T., Riemenschneider, M., Förstl, H., Henriksen, G., Klunk, W. E., Mathis, C. A., et al. (2009). Beta amyloid in Alzheimer's disease: Increased deposition in brain is reflected in reduced concentration in cerebrospinal fluid. *Biological Psychiatry, 65*, 927—934.

Hampel, H., Teipel, S. J., Fuchsberger, T., Andreasen, N., Wiltfang, J., Otto, M., et al. (2004). Value of CSF β-amyloid1-42 and tau as predictors of Alzheimer's disease in patients with mild cognitive impairment. *Molecular Psychiatry, 9*, 705—710.

Han, M.-R., Schellenberg, G. D., Wang, L.-S., & Alzheimer's Disease Neuroimaging Initiative. (2010). Genome-wide association reveals genetic effects on human Aβ42 and τ protein levels in cerebrospinal fluids: A case control study. *BMC Neurology, 10*, 90.

Holmes, C., Cunningham, C., Zotova, E., Woolford, J., Dean, C., Kerr, S., et al. (2009). Systemic inflammation and disease progression in Alzheimer disease. *Neurology, 73*, 768—774.

Kandel, E. R. (2006). *Search of memory : the emergence of a new science of mind*. New York : W. W. Norton & Company.

Kim, C.-B., Choi, Y. Y., Song, W. K., & Song, K.-B. (2014). Antibody-based magnetic nanoparticle immunoassay for quantification of Alzheimer's disease pathogenic factor. *Journal of Biomedical Optics, 19*, 051205.

Knopman, D. S., DeKosky, S. T., Cummings, J. L., Chui, H., Corey-Bloom, J., Relkin, N., et al. (2001). Practice parameter: Diagnosis of dementia (an evidence-based review). Report of the Quality Standards Subcommittee of the American Academy of Neurology. *Neurology, 56*, 1143—1153.

Koyama, A., Okereke, O. I., Yang, T., Blacker, D., Selkoe, D. J., & Grodstein, F. (2012). Plasma amyloid-β as a predictor of dementia and cognitive decline: A systematic review and meta-analysis. *Archives of Neurology, 69*, 824—831.

Lee, M., Guo, J.-P., Kennedy, K., McGeer, E. G., & McGeer, P. L. (2017). A method for diagnosing Alzheimer's disease based on salivary amyloid-β protein 42 levels. *Journal of Alzheimer's Disease, 55*, 1175—1182.

Liang, Q., Liu, H., Zhang, T., Jiang, Y., Xing, H., & Zhang, A. (2015). Metabolomics-based screening of salivary biomarkers for early diagnosis of Alzheimer's disease. *RSC Advances, 5*, 96074—96079.

Mandel, I. D. (1987). The functions of saliva. *Journal of Dental Research, 66*, 623—627.

Moonis, M., Swearer, J. M., Dayaw, M. P. E., St, G.-H., Rogaeva, E., Kawarai, T., et al. (2005). Familial Alzheimer disease: Decreases in CSF Aβ42 levels precede cognitive decline. *Neurology, 65*, 323—325.

Moore, K. L., Dalley, A. F., & Agur, A. M. R. (2013). *Clinically oriented anatomy*. Lippincott Williams & Wilkins.

Pekeles, H., Qureshi, H. Y., Paudel, H. K., Schipper, H. M., Gornistky, M., & Chertkow, H. (2018). Development and validation of a salivary tau biomarker in Alzheimer's disease. *Alzheimer's and Dementia: Diagnosis, Assessment and Disease Monitoring, 11*, 53—60.

Reitz, C., & Mayeux, R. (2009). Use of genetic variation as biomarkers for Alzheimer's disease. *Annals of the New York Academy of Sciences, 1180*, 75—96.

Riemenschneider, M., Lautenschlager, N., Wagenpfeil, S., Diehl, J., Drzezga, A., & Kurz, A. (2002). Cerebrospinal fluid tau and β-amyloid 42 proteins identify Alzheimer disease in subjects with mild cognitive impairment. *Archives of Neurology, 59*, 1729—1734.

Sayer, R., Law, E., Connelly, P. J., & Breen, K. C. (2004). Association of a salivary acetylcholinesterase with Alzheimer's disease and response to cholinesterase inhibitors. *Clinical Biochemistry, 37*, 98—104.

Schenkels, L. C., Veerman, E. C., & Nieuw Amerongen, A. V. (1995). Biochemical composition of human saliva in relation to other mucosal fluids. *Critical Reviews in Oral Biology and Medicine: An Official Publication of the American Association of Oral Biologists, 6*, 161—175.

Shi, M., Sui, Y.-T., Peskind, E. R., Li, G., Hwang, H., Devic, I., et al. (2011). Salivary tau species are potential biomarkers of Alzheimer's disease. *Journal of Alzheimer's Disease, 27*, 299—305.

Smith, A. K., Kilaru, V., Klengel, T., Mercer, K. B., Bradley, B., Conneely, K. N., et al. (2015). DNA extracted from saliva for methylation studies of psychiatric traits: Evidence tissue specificity and relatedness to brain. *American Journal of Medical Genetics. Part B, Neuropsychiatric Genetics, 168*, 36—44.

Streckfus, C. F., & Bigler, L. R. (2002). Saliva as a diagnostic fluid. *Oral Diseases, 8*, 69—76.

Su, H., Gornitsky, M., Geng, G., Velly, A. M., Chertkow, H., & Schipper, H. M. (2008). Diurnal variations in salivary protein carbonyl levels in normal and cognitively impaired human subjects. *Age, 30*, 1—9.

Wang, L., Sato, H., Zhao, S., & Tooyama, I. (2010). Deposition of lactoferrin in fibrillar-type senile plaques in the brains of transgenic mouse models of Alzheimer's disease. *Neuroscience Letters, 481*, 164—167.

Wren, M. E., Shirtcliff, E. A., & Drury, S. S. (2015). Not all biofluids are created equal: Chewing over salivary diagnostics and the epigenome. *Clinical Therapeutics, 37*, 529—539.

Yilmaz, A., Geddes, T., Han, B., Bahado-Singh, R. O., Wilson, G. D., Imam, K., et al. (2017). Diagnostic biomarkers of Alzheimer's disease as identified in saliva using 1H NMR-based metabolomics. *Journal of Alzheimer's Disease, 58*, 355—359.

CHAPTER 16

Diacylglycerols: biomarkers of a sustained immune response in proteinopathies

Paul L. Wood[1], Randall J. Woltjer[2]

[1]Metabolomics Unit, Associate Dean of Research, College of Veterinary Medicine, Lincoln Memorial University, Harrogate, TN, United States; [2]Department of Neurology, Oregon Health Science University and Portland VA Medical Center, Portland, OR, United States

List of abbreviations

DAGs Diacylglycerols
PLC Phospholipase C
PLD Phospholipase D

Mini-dictionary of terms

Lipid nomenclature (e.g., phosphatidylcholine 36:5) Glycerol backbone with a phosphocholine substitution from a hydroxyl group and two fatty acids from the other hydroxyl groups equivalent to 36 carbons and 5 double bonds; the major form of this would be the two fatty acids 16:0 and 20:5.
Proteinopathies These include AD where there is deposition of amyloid and neurofibrillary tangles as well PD and LBD where there is deposition of synuclein. However, all of these disorders are often characterized by multiple neuropathologies (Bennett, 2017).

Introduction

While a number of proteinopathies are associated with the development of dementia, the roles of these protein deposits on clinical outcomes are far from being understood. Clinical proteinopathies include Alzheimer's disease (AD), Parkinson's disease (PD), and Lewy body disease (LBD). Studies of these disorders, however, have revealed that proteinopathies occur with significant incidence in elderly persons who demonstrate no cognitive deficits (Crawford, Bjorklund, Taglialatela, & Gomer, 2012; Jellinger, 1995; Kramer et al., 2011; Maarouf et al., 2011; Wood, Barnette, Kaye, Quinn, & Woltjer, 2015; Wood, Locke et al., 2015). These individuals, who have been coined nondemented with significant Alzheimer's neuropathology (NDAN) are being more intensively studied in an attempt to define their resistance factors and/or cognitive reserve (Bennett, 2017).

In addition to abnormal protein deposition in proteinopathies (Walker et al., 2015), these diseases are characterized by both systemic inflammation (Bettcher et al., 2018;

Diagnosis and Management in Dementia
ISBN 978-0-12-815854-8, https://doi.org/10.1016/B978-0-12-815854-8.00016-1

Busse et al., 2017; Corlier et al., 2018; King & Thomas, 2017; Lai et al., 2017; Swardfager et al., 2017; Zuliani et al., 2007) and neuroinflammation (Bachstetter et al., 2015; Wood, 2003; Wood et al., 1993). Of particular interest, neuroinflammation has been demonstrated in AD and mild cognitive impairment (MCI) patients utilizing in vivo imaging (Knezevic & Mizrahi, 2018; Parbo et al., 2017). Since MCI is the precursor to the onset of clinical AD, these data indicate that neuroinflammation occurs early in the disease process and that this immune response is sustained. In marked contrast, there appears to be minimal neuroinflammation in the brains of NDAN subjects (Crawford et al., 2012).

Lipidomics of proteinopathies

The question that we have asked is what changes in the lipidome occur during this sustained immune response. In this regard, we and other laboratories have demonstrated elevated levels of diacylglycerols (DAGs) in the plasma (González-Domínguez, García-Barrera, & Gómez-Ariza, 2015; Wood, Phillipps, Woltjer, Kaye, & Quinn, 2014; Wood, Barnette et al., 2015; Wood, Locke et al., 2015; Wood, Medicherla et al., 2015) and frontal cortex (Chan et al., 2012; Lam et al., 2014; Wood et al., 2014; Wood, Barnette et al., 2015; Wood, Locke et al., 2015) of AD patients. Of particular interest, DAG levels were also increased in the frontal cortex (Barbash et al., 2017; Wood, Barnette et al., 2015; Wood, Locke et al., 2015) and plasma (Wood, Locke et al., 2015; Wood, Medicherla et al., 2015) of MCI patients, while they were not elevated in the cortex of NDAN subjects (Barbash et al., 2017; Wood et al., 2014; Wood, Barnette et al., 2015). These DAG data support previous findings suggesting minimal neuroinflammation in the brains of NDAN subjects (Crawford et al., 2012).

In our laboratory, high-resolution mass spectrometric analyses of AD brain (Fig. 16.1) and plasma (Fig. 16.2) DAG levels demonstrated a general increase in most DAG levels,

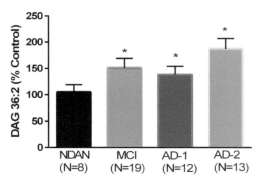

Figure 16.1 Frontal cortex levels of DAG 36:2 expressed as % of controls (67 ± 12 years, N = 15). *AD-1*, Alzheimer's; 76 ± 7 years; *AD-2*, Alzheimer's, 91 ± 5 years; *MCI*, mild cognitive impairment; 91 ± 8 years; *NDAN*, nondemented with AD neuropathology; 92 ± 4 years). * = *P* < .05.

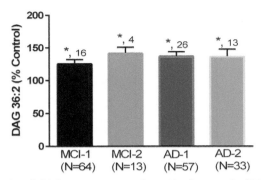

Figure 16.2 Plasma levels of DAG 36:2 expressed as % of controls (67 ± 12 yrs, N = 15). *AD*-1, moderate Alzheimer's; *AD-2*, severe Alzheimer's; *MCI-1*, moderate mild cognitive impairment; *MCI-2*, severe MCI. *Number = $P < .05$, % of subjects with levels greater than 1 SD above control values.

with DAG 36:2 presented as a prototypic example. Of importance, an analysis of the fatty acid composition of individual DAGs revealed no alterations in the fatty acid composition of the augmented DAGs in the AD frontal cortex (Wood, Medicherla et al., 2015).

Elevated DAGs have been found in the frontal cortex of PD (Cheng et al., 2011; Wood, Tippireddy, Feriante, Woltjer et al., 2018) and LBD (Wood, Tippireddy et al., 2018) subjects. In parallel with the augmented DAG levels we also measured decreases in the levels of phosphatidylcholines and lysophosphatidic acids, both precursors of DAGs (Figs. 16.3 and 16.4). These data contrast with our previous AD data for the frontal cortex, where we monitored decrements in phosphatidylethanolamines but not phosphatidylcholines (Wood, Barnette et al., 2015). These data suggest that multiple alterations in glycerophospholipid metabolism can result in elevated DAG levels in proteinopathies.

DAG levels are under tight regulation because they serve complex structural and signal transduction roles (Brose, Betz, & Wegmeyer, 2004; Carrasco & Mérida, 2007). Physiological levels of free DAGs are maintained mainly via the actions of DAG kinase, an enzyme that is decreased in AD (Schlam et al., 2013). In contrast, phospholipase C (PLC) (Shimohama et al., 1998) and phospholipase D (PLD) (Wang, Yu, & Tan, 2014), enzymes that can augment DAG levels, are both increased in AD. However, both enzymes constitute large families of discrete isozymes that remain to be more fully characterized in proteinopathies. The pathological relevance of elevated DAG levels still remains to be more extensively defined. However, alterations in the levels of DAGs in membranes may lead to altered membrane function, whereas free DAGs can dramatically alter synaptic function (Brose et al., 2004; Carrasco & Mérida., 2007).

Diacylglycerols and immune responses

The potential roles of DAGs in immune regulation can be complex with both PLC (Xiao, Kawakami, & Kawakami, 2013; Zhang, Dhillon, Prasad, & Markesbery, 1998)

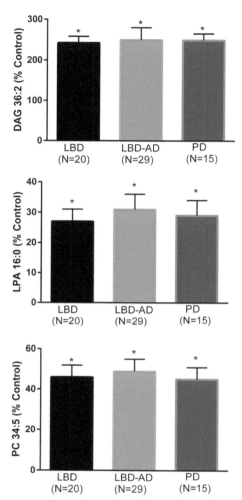

Figure 16.3 Frontal cortex levels of DAG 36:2, lysophosphatidic acid (LPA 16:0), and phosphatidylcholine (PC 36:5) expressed as % of controls (N = 43). *LBD*, Lewy body disease with intermediate AD pathology; N = 20); *LBD-AD*, Lewy body disease with AD pathology; N = 29; *PD*, Parkinson's disease; N = 15. *$P < .01$.

and PLD (Chien, Chen, Chien, Yeh, & Lu, 2004; Serrander, Fällman, & Stendahl, 1996) being upregulated during immune response. In this regard, we have monitored elevated serum levels of DAG associated with a sustained immune response to both an active bacterial infection (Leptospirosis) and an immunization against the bacteria in horses (Wood, Steinman et al., 2018). In this equine study, we also monitored elevated levels of cyclic phosphatidic acid 16:0, a PLD product. These data suggest that PLC/PLD lipid products may be useful biomarkers of a sustained immune

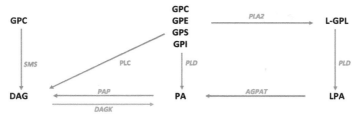

Figure 16.4 Schematic presentation of lipid pathways that modulate DAG levels. *AGPAT*, acylglycerophosphate acyltransferase; *DAG*, diacylglycerols; *DAGK*, DAG kinase; *GPC*, diacylglycero-phosphocholine; *GPE*, diacylglycerophosphoethanolamine; *GPI*, diacylglycerophosphoinositol; *GPS*, diacylglycerophosphoserine; *L-GPL*, lyso(monoacyl)-glycerophospholipid; *LPA*, lysophosphatidic acid; *PA*, phosphatidic acid; *PAP*, PA phosphatase; *PLA2*, phospholipase A2; *PLC*, phospholipase C; *PLD*, phospholipase D.

response. In the case of our Lewy body disease study (Wood, Tippireddy et al., 2018), we could not detect cyclic phosphatidic acid 16:0 in the control frontal cortex, while detectable levels were monitored in the disease groups: controls = 0 of 432; LBD = 16 of 20; LBD with AD = 17 of 29; and PD = 11 of 15.

The potential negative roles of systemic inflammation and neuroinflammation in proteinopathies have been championed by the McGeers based on historical data demonstrating that long-term antiinflammatory therapy (e.g., NSAIDs) decreases the risk of developing AD (in t' Veld et al., 2001; McGeer, Schulzer, & McGeer, 1996; McGeer, Rogers, McGeer, 2016). However, prospective studies of interventions with NSAIDs in established AD patients have failed to demonstrate any clinical benefit (McGeer et al., 2016). This has led to the conclusion that earlier antiinflammatory intervention is needed, presumably in early MCI patients (Imbimbo, Solfrizzi, & Panza, 2010; Lichtenstein, Carriba, Masgrau, Pujol, & Galea, 2010; McGeer et al., 2016). To this end we propose that such a study of early NSAID intervention in MCI be funded by the NIH to determine the utility of this approach. Furthermore, only MCI patients with an established immune response should be enrolled in the study. Parameters to include could be DAGs, cyclic phosphatidic acids, and cytokines.

Key facts

- Proteinopathies are associated with a sustained immune response as reflected by unresolved neuroinflammation and systemic inflammation.
- Elevated DAG levels in proteinopathies are lipid biomarkers of this sustained immune response in proteinopathies.
- A prospective study of the ability of NSAIDs to delay or block the progression of MCI to AD should be undertaken utilizing biomarkers of immune activation in MCI patients.

Summary points

- Proteinopathies are characterized by a sustained immune response.
- Lipid biomarkers of this sustained immune response are DAGs.
- Elevated DAG levels are most likely the combinatorial result of increased PLC and PLD activities and decrements in DAG kinase activity.
- It is a subset of patients with proteinopathies that have sustained systemic inflammation and neuroinflammation.
- Retrospective data have indicated that chronic antiinflammatory therapy delays the onset of AD, while acute antiinflammatory therapy has no clinical benefit in AD patients.
- A prospective analysis of the ability of NSAIDs to delay the onset of AD in MCI patients is warranted in MCI subjects demonstrating sustained systemic inflammation.

References

Bachstetter, A. D., Van Eldik, L. J., Schmitt, F. A., Neltner, J. H., Ighodaro, E. T., Webster, S. J., et al. (2015). Disease-related microglia heterogeneity in the hippocampus of Alzheimer's disease, dementia with Lewy bodies, and hippocampal sclerosis of aging. *Acta Neuropathologica Communications, 3*, 32.

Barbash, S., Garfinkel, B. P., Maoz, R., Simchovitz, A., Nadorp, B., Guffanti, A., et al. (2017). Alzheimer's brains show inter-related changes in RNA and lipid metabolism. *Neurobiology of Disease, 106*, 1—13.

Bennett, D. A. (2017). Mixed pathologies and neural reserve: Implications of complexity for Alzheimer disease drug discovery. *PLOS Medicine, 14*. e1002256.

Bettcher, B. M., Johnson, S. C., Fitch, R., Casaletto, K. B., Heffernan, K. S., Asthana, S., et al. (2018). Cerebrospinal fluid and plasma levels of inflammation differentially relate to CNS markers of Alzheimer's disease pathology and neuronal damage. *Journal of Alzheimer's Disease, 62*, 385—397.

Brose, N., Betz, A., & Wegmeyer, H. (2004). Divergent and convergent signaling by the diacylglycerol second messenger pathway in mammals. *Current Opinion in Neurobiology, 14*, 328—340.

Busse, M., Michler, E., von Hoff, F., Dobrowolny, H., Hartig, R., Frodl, T., et al. (2017). Alterations in the peripheral immune system in dementia. *Journal of Alzheimer's Disease, 58*, 1303—1313.

Carrasco, S., & Mérida, I. (2007). Diacylglycerol, when simplicity becomes complex. *Trends in Biochemical Sciences, 32*, 27—36.

Chan, R. B., Oliveira, T. G., Cortes, E. P., Honig, L. S., Duff, K. E., Small, S. A., et al. (2012). Comparative lipidomic analysis of mouse and human brain with Alzheimer disease. *Journal of Biological Chemistry, 287*, 2678—2688.

Cheng, D., Jenner, A. M., Shui, G., Cheong, W. F., Mitchell, T. W., Nealon, J. R., et al. (2011). Lipid pathway alterations in Parkinson's disease primary visual cortex. *PLOS One, 6*. e17299.

Chien, E. J., Chen, C. C., Chien, C. H., Yeh, T. P., & Lu, L. M. (2004). Activation and up-regulation of phospholipase D expression by lipopolysaccharide in human peripheral T cells. *The Chinese Journal of Physiology, 47*, 203—209.

Corlier, F., Hafzalla, G., Faskowitz, J., Kuller, L. H., Becker, J. T., Lopez, O. L., et al. (2018). Systemic inflammation as a predictor of brain aging: Contributions of physical activity, metabolic risk, and genetic risk. *NeuroImage, 172*, 118—129.

Crawford, J. R., Bjorklund, N. L., Taglialatela, G., & Gomer, R. H. (2012). Brain serum amyloid P levels are reduced in individuals that lack dementia while having Alzheimer's disease neuropathology. *Neurochemical Research, 37*, 795—801.

González-Domínguez, R., García-Barrera, T., & Gómez-Ariza, J. L. (2015). Application of a novel metabolomic approach based on atmospheric pressure photoionization mass spectrometry using flow injection analysis for the study of Alzheimer's disease. *Talanta, 131*, 480—489.

Imbimbo, B. P., Solfrizzi, V., & Panza, F. (2010). Are NSAIDs useful to treat Alzheimer's disease or mild cognitive impairment? *Frontiers in Aging Neuroscience, 2*. pii: 19.

Jellinger, K. A. (1995). Alzheimer's changes in non-demented and demented patients. *Acta Neuropathologica, 89*, 112−113.

King, E., & Thomas, A. (2017). Systemic inflammation in Lewy body diseases: A systematic review. *Alzheimer Disease and Associated Disorders, 31*, 346−356.

Knezevic, D., & Mizrahi, R. (2018). Molecular imaging of neuroinflammation in Alzheimer's disease and mild cognitive impairment. *Progress In Neuro-Psychopharmacology and Biological Psychiatry, 80*, 123−131.

Kramer, P. L., Xu, H., Woltjer, R. L., Westaway, S. K., Clark, D., Erten-Lyons, D., et al. (2011). Alzheimer disease pathology in cognitively healthy elderly: A genome-wide study. *Neurobiology of Aging, 32*, 2113−2122.

Lai, K. S. P., Liu, C. S., Rau, A., Lanctôt, K. L., Köhler, C. A., Pakosh, M., et al. (2017). Peripheral inflammatory markers in Alzheimer's disease: A systematic review and meta-analysis of 175 studies. *Journal of Neurology Neurosurgery and Psychiatry, 88*, 876−882.

Lam, S. M., Wang, Y., Duan, X., Wenk, M. R., Kalaria, R. N., Chen, C. P., et al. (2014). Brain lipidomes of subcortical ischemic vascular dementia and mixed dementia. *Neurobiology of Aging, 35*, 2369−2381.

Lichtenstein, M. P., Carriba, P., Masgrau, R., Pujol, A., & Galea, E. (2010). Staging anti-inflammatory therapy in Alzheimer's disease. *Frontiers in Aging Neuroscience, 2*, 142.

Maarouf, C. L., Daugs, I. D., Kokjohn, T. A., Walker, D. G., Hunter, J. M., Kruchowsky, J. C., et al. (2011). Alzheimer's disease and non-demented high pathology control nonagenarians: Comparing and contrasting the biochemistry of cognitively successful aging. *PLOS One, 6*. e27291.

McGeer, P. L., Rogers, J., & McGeer, E. G. (2016). Inflammation, antiinflammatory agents, and Alzheimer's disease: The last 22 years. *Journal of Alzheimer's Disease, 54*, 853−857.

McGeer, P. L., Schulzer, M., & McGeer, E. G. (1996). Arthritis and anti-inflammatory agents as possible protective factors for Alzheimer's disease: A review of 17 epidemiologic studies. *Neurology, 47*, 425−432.

Parbo, P., Ismail, R., Hansen, K. V., Amidi, A., Mårup, F. H., Gottrup, H., et al. (2017). Brain inflammation accompanies amyloid in the majority of mild cognitive impairment cases due to Alzheimer's disease. *Brain, 140*, 2002−2011.

Schlam, D., Bohdanowicz, M., Chatgilialoglu, A., Steinberg, B. E., Ueyama, T., Du, G., et al. (2013). Diacylglycerol kinases terminate diacylglycerol signaling during the respiratory burst leading to heterogeneous phagosomal NADPH oxidase activation. *Journal of Biological Chemistry, 288*, 23090−23104.

Serrander, L., Fällman, M., & Stendahl, O. (1996). Activation of phospholipase D is an early event in integrin-mediated signalling leading to phagocytosis in human neutrophils. *Inflammation, 20*, 439−450.

Shimohama, S., Sasaki, Y., Fujimoto, S., Kamiya, S., Taniguchi, T., Takenawa, T., et al. (1998). Phospholipase C isozymes in the human brain and their changes in Alzheimer's disease. *Neuroscience, 82*, 999−1007.

Swardfager, W., Yu, D., Ramirez, J., Cogo-Moreira, H., Szilagyi, G., Holmes, M. F., et al. (2017). Peripheral inflammatory markers indicate microstructural damage within periventricular white matter hyperintensities in Alzheimer's disease: A preliminary report. *Alzheimer's and dementia: diagnosis, assessment and disease monitoring, 7*, 56−60.

in t'Veld, B. A., Ruitenberg, A., Hofman, A., Launer, L. J., van Duijn, C. M., Stijnen, T., et al. (2001). Nonsteroidal antiinflammatory drugs and the risk of Alzheimer's disease. *New England Journal of Medicine, 345*, 1515−1521.

Walker, L., McAleese, K. E., Thomas, A. J., Johnson, M., Martin-Ruiz, C., Parker, C., et al. (2015). Neuropathologically mixed Alzheimer's and Lewy body disease: Burden of pathological protein aggregates differs between clinical phenotypes. *Acta Neuropathologica, 129*, 729−748.

Wang, J., Yu, J. T., & Tan, L. (2014). PLD3 in Alzheimer's disease. *Molecular Neurobiology, 51*, 480−486.

Wood, P. L. (2003). Microglia: Roles of microglia in chronic neurodegenerative diseases. In P. L. Wood (Ed.), *Neuroinflammation: Mechanisms and management* (2nd ed., pp. 3−27). Totowa, NJ: Humana Press. ISBN: 978-1-4684-9720-5, ISBN: 978-1-59259-297-5 (eBook).

Wood, P. L., Barnette, B. L., Kaye, J. A., Quinn, J. F., & Woltjer, R. L. (2015). Non-targeted lipidomics of CSF and frontal cortex gray and white matter in control. *Mild Cognitive, 27*, 270−278.

Wood, P. L., Locke, V. A., Herling, P., Passaro, A., Vigna, G. B., Volpato, S., et al. (2015). Targeted lipidomics distinguishes patient subgroups in mild cognitive impairment (MCI) and late onset Alzheimer's disease (LOAD). *BBA Clinical, 5*, 25—28.

Wood, P. L., Medicherla, S., Sheikh, N., Terry, B., Phillipps, A., Kaye, J. A., et al. (2015). Targeted lipidomics of fontal cortex and plasma diacylglycerols (DAG) in mild cognitive impairment and Alzheimer's disease: Validation of DAG accumulation early in the pathophysiology of Alzheimer's disease. *Journal of Alzheimer's Disease, 48*, 537—546.

Wood, P. L., Phillipps, A., Woltjer, R. L., Kaye, J. A., & Quinn, J. F. (2014). Increased lysophosphatidylethanolamine and diacylglycerol levels in Alzheimer's disease plasma. *JSM Alzheimer's Disease and Related Dementia, 1*, 1001.

Wood, P. L., Steinman, M., Erol, E., Carter, C., Christmann, U., & Verma, A. (2018). Lipidomic analysis of immune activation in equine leptospirosis and leptospira-vaccinated horses. *PLOS One, 13*. e0193424.

Wood, P. L., Tippireddy, S., Feriante, J., & Woltjer, R. L. (2018). Augmented frontal cortex diacylglycerol levels in Parkinson's disease and Lewy body disease. *PLOS One, 13*. e0191815.

Wood, J. A., Wood, P. L., Ryan, R., Graff-Radford, N. R., Pilapil, C., Robitaille, Y., et al. (1993). Cytokine indices in Alzheimer's temporal cortex: No changes in mature IL-1 beta or IL-1RA but increases in the associated acute phase proteins IL-6, alpha 2-macroglobulin and C-reactive protein. *Brain Research, 629*, 245—252.

Xiao, W., Kawakami, Y., & Kawakami, T. (2013). Immune regulation by phospholipase C-β isoforms. *Immunologic Research, 56*, 9—19.

Zhang, D., Dhillon, H., Prasad, M. R., & Markesbery, W. R. (1998). Regional levels of brain phospholipase C-gamma in Alzheimer's disease. *Brain Research, 811*, 161—165.

Zuliani, G., Ranzini, M., Guerra, G., Rossi, L., Munari, M. R., Zurlo, A., et al. (2007). Plasma cytokines profile in older subjects with late onset Alzheimer's disease or vascular dementia. *Journal of Psychiatric Research, 41*, 686—693.

CHAPTER 17

Butyrylcholinesterase as a biomarker in Alzheimer's disease

Drew R. DeBay[1,2], Sultan Darvesh[1,2]
[1]Department of Medical Neuroscience, Dalhousie University, Halifax, NS, Canada; [2]Department of Medicine (Neurology), Dalhousie University, Halifax, NS, Canada

List of abbreviations

ACh acetylcholine
AChE acetylcholinesterase
AD Alzheimer's disease
Aβ amyloid-beta
BChE butyrylcholinesterase
BCHE butyrylcholinesterase gene
CH1 high-affinity choline transporter
ChAT choline acetyltransferase
CSF cerebrospinal fluid
mAChR muscarinic ACh receptor
MRI Magnetic Resonance Imaging
nAChR nicotinic ACh receptor
PET Positron Emission Tomography
SPECT Single Photon Emission Computed Tomography
VAChT vesicular acetylcholine transporter

Minidictionary of terms

Acetylcholinesterase (AChE) A serine hydrolase enzyme that coregulates the duration of action of the neurotransmitter acetylcholine.

Biomarker An objectively measured characteristic evaluated as an indicator of biological processes (normal or pathogenic) or pharmacologic responses to a therapeutic intervention.

Butyrylcholinesterase (BChE) A serine hydrolase enzyme that coregulates the duration of action of the neurotransmitter acetylcholine.

Magnetic Resonance Imaging An imaging technique using large magnetic fields and radio waves to measure and localize signal from hydrogen atoms in the body to generate detailed anatomical images.

Positron Emission Tomography (PET) A nuclear medicine imaging technique that detects and localizes coincident high-energy photons generated by a positron-emitting radiotracer, providing images of the radiotracer distribution in the body.

Sensitivity The probability of a diagnostic test to correctly classify an individual with a disorder as positive.

Diagnosis and Management in Dementia
ISBN 978-0-12-815854-8, https://doi.org/10.1016/B978-0-12-815854-8.00017-3
263

Single Photon Emission Computed Tomography (SPECT) A nuclear medicine imaging technique that detects and localizes high-energy gamma rays emitted from a radiotracer, providing images of radiotracer distribution in the body.

Specificity The probability of a diagnostic test to correctly classify an individual without a disorder as negative.

Alzheimer's disease

Alzheimer's disease (AD) is a progressive neurodegenerative disorder of insidious onset and is the most common cause of dementia (Scheltens et al., 2016). The prevalence of AD and concomitant socioeconomic economic burdens are predicted to dramatically rise over the next decades (Scheltens et al., 2016). AD remains a clinical diagnosis based on acquired impairment in various cognitive domains (memory, language, visuospatial, or executive functions) that interfere with activities of daily living (McKhann et al., 2011). Considerable overlap of symptomatology exists between AD and other non-AD dementias, making the reliability of a clinical diagnosis of AD insufficient (sensitivity = 81%, specificity = 70%) (Knopman et al., 2001). A definitive diagnosis of AD requires the clinically observed presence of dementia during life and disease-defining hallmarks at autopsy, namely, extracellular neuritic plaques containing β-amyloid (Aβ) and intracellular neurofibrillary tangles (NFTs) containing hyperphosphorylated tau (Montine et al., 2012). Aβ and NFTs are necessary criteria but not alone sufficient for an AD diagnosis, as they lack specificity since up to $\sim$30% of cognitively normal older individuals also have this pathology in the brain (Jansen et al., 2015; Snowdon & Nun, 2003). Nevertheless, characterizing Aβ and NFT biomarkers in the living brain has helped improve the accuracy of AD diagnosis during life (Johnson et al., 2013).

A shift in research focus toward a biological (biomarker-based) definition of AD has been suggested recently. This includes a focus on the A, T, (N) framework; that is, the presence of amyloid, tau and neurodegeneration (Jack et al., 2018), which is currently evaluated with cerebrospinal fluid (CSF)-based and quantitative molecular imaging approaches using Positron Emission Tomography (PET) and Single Photon Emission Computed Tomography (SPECT) imaging.

β-Amyloid as an AD biomarker

Amyloid plaques are extracellular aggregates of β-Amyloid (Aβ) peptide, with 39–43 amino acid residues resulting from γ- and β-secretase proteolytic cleavage of the amyloid precursor protein (APP) (Villemagne, 2016). An imbalance of production, and clearance of Aβ_{42} oligomer in particular, has been implicated in AD, whereby Aβ_{42} and soluble fragments aggregate into large fibrils and insoluble extracellular plaques. This amyloid cascade is thought to be an early event that initiates the pathogenesis of AD (Selkoe & Hardy, 2016).

Cerebrospinal fluid Aβ biomarkers

Cerebrospinal fluid (CSF) Aβ biomarkers have been evaluated in AD, reflecting the homeostatic balance between Aβ production and clearance in the brain. In AD, a marked decrease in CSF $A\beta_{42}$ and $A\beta_{42}/A\beta_{40}$ ratio has been shown consistently with a mean decrease to approximately 56% of cognitively normal individuals (Olsson et al., 2016). It has been suggested that this is due to $A\beta_{42}$ aggregates becoming sequestered in Aβ plaques in the brain resulting in lower amounts remaining in the CSF (Blennow & Zetterberg, 2018).

Aβ imaging biomarkers

The pursuit of Aβ molecular imaging has taken cues from the chemistry of established histopathological dyes (Villemagne, 2016). A derivative of thioflavin T, N-methyl-$[^{11}C]$2-(4′-methylaminophenyl)-6-hydroxybenzothiazole (^{11}C-PiB) (Klunk et al., 2004) was the first to show high affinity and selectivity for Aβ plaques (Villemagne, 2016).

Subsequently, three additional ^{18}F-labeled Aβ tracers have been approved for brain imaging by the US Food and Drug Administration, to capitalize on its longer half-life ($t_{1/2}$, 110 min) and more widespread utility globally. This includes ^{18}F-florbetapir, ^{18}F-flutemetamol, and ^{18}F-florbetaban (Villemagne, 2016). These agents represent a major leap forward for molecular imaging of Aβ during life as an adjunct test in AD. However, the lack of specificity of Aβ in AD limits their diagnostic value as $\sim$30% of cognitively normal elderly individuals have Aβ pathology at autopsy (Snowdon & Nun, 2003). Aβ imaging as an AD biomarker provides an average sensitivity of 90% and specificity of 85% (Morris et al., 2016) (Table 17.1).

Tau as an AD biomarker

In addition to Aβ, tau neurofibrillary tangles (NFTs) are required for the diagnosis of AD, though they are not exclusive to this disease. Tau (tubulin-associated unit) is a microtubule-stabilizing protein, important for the neuronal cytoskeleton and for axonal transport (Iqbal, Liu, & Gong, 2016). Six isoforms of tau exist in humans, classified by the microtubule-binding domain repeats, namely, 3 or 4 repeats (3 and 4R, respectively) (Iqbal et al., 2016). Abnormal tauopathies occur when hyperphosphorylation of tau generates conformational changes (misfolding and aggregation), which, in turn, prevents normal microtubule binding. This promotes axonal destabilization, transport impairment and degeneration, ultimately leading to neuronal dysfunction (Iqbal et al., 2016). Insoluble paired helical filaments, a major component of NFTs, also arise from tau hyperphosphorylation and are a common feature in the AD brain (Iqbal et al., 2016).

Table 17.1 Performance summary of current AD biomarkers.

Diagnostic modality	Sensitivity (%)	Specificity (%)	References
[a]Clinical diagnosis	81	70	Knopmann et al. (2001)
[b]MRI	83	89	Bloudek et al. (2011)
[c]^{18}FDG–PET	90	89	Bloudek et al. (2011)
[d]SPECT perfusion	80	85	Bloudek et al. (2011)
[e]Aβ-PET	90	85	Morris et al. (2016)
[f]tau-PET	100	86	Wang et al. (2016)
[g]CSF Aβ$_{42}$	80	82	Bloudek et al. (2011)
[h]CSF P-tau	80	83	Bloudek et al. (2011)
[i]CSF T-tau	82	90	Bloudek et al. (2011)
[j]BChE neuropathology	100	100	Macdonald et al. (2017)

Diagnostic performance summary (sensitivity and specificity) of current AD biomarkers (AD vs. cognitively normal meta-analyses).
[a]Clinical diagnosis using neuropathology "gold standard."
[b]Magnetic Resonance Imaging (MRI)—based hippocampal atrophy.
[c]^{18}Fluorodeoxyglucose (^{18}FDG)-Positron Emission Tomography (PET) temporoparietal hypometabolism.
[d]Single Photon Emission Computed Tomography (SPECT) perfusion using pooled^{99m}Tc-HMPAO,^{99m}Tc-ECD or^{123}IMP imaging.
[e]Amyloid (Aβ)-PET using pooled^{18}florbetapir,^{18}flutemetamol or^{18}florbetaben imaging data.
[f]Tau-PET using radiotracer^{18}F-AV1451.
[g]Cerebrospinal fluid (CSF) Aβ$_{42}$.
[h]CSF Phosphorylated (P)-tau.
[i]CSF Total (T)-tau.
[j]Butyrylcholinesterase (BChE) neuropathological quantification.
Adapted from Morris, E., Chalkidou, A., Hammers, A., Peacock, J., Summers, J., & Keevil, S. (2016). Diagnostic accuracy of (18)F amyloid PET tracers for the diagnosis of Alzheimer's disease: A systematic review and meta-analysis. *European Journal of Nuclear Medicine and Molecular Imaging, 43*(2), 374—385. doi:10.1007/s00259-015-3228-x and Bloudek, L. M., Spackman, D. E., Blankenburg, M., & Sullivan, S. D. (2011). Review and meta-analysis of biomarkers and diagnostic imaging in Alzheimer's disease. *Journal of Alzheimer's Disorder, 26*(4), 627—645. doi:10.3233/JAD-2011-110458.

Tauopathies with both 3 and 4R isoforms include AD, Down syndrome, and chronic traumatic encephalopathy. Tauopathies that possess the 4R isoform include progressive supranuclear palsy, corticobasal degeneration (CBD), and argyrophilic grain disease, while Pick's disease (Frontotemporal Dementia (FTD)-Tau) expresses a 3R isoform (Villemagne & Okamura, 2016). Despite this widespread and varied presence of tau in numerous tauopathies, tau biomarkers continue to be considered a contributor for AD diagnosis.

Cerebrospinal fluid tau biomarkers

Over the last two decades, assays for cerebrospinal fluid (CSF) tau have been developed, including the total tau (T-tau) assay, recognizing all six isoforms of tau. T-tau is a reflection of an overall state of neurodegeneration rather than a direct marker of pathophysiological process in AD (Blennow & Zetterberg, 2018). In AD, T-tau is elevated 2.5-fold that of cognitively normal individuals, reflecting heightened neurodegeneration (Olsson et al., 2016). However, increased T-tau is also observed in several other neurodegenerative disorders (Zetterberg, 2017).

Of particular relevance in AD, the measurement of CSF phosphorylated-tau (P-tau) targets tau protein residues possessing the same post-translational phosphorylation observed in the AD brain, namely, hyperphosphorylation at threonine-181, threonine-231, and serine-199 residues (Blennow & Zetterberg, 2018). Individuals with AD show increased CSF P-tau concentrations approximately 1.9-fold compared to cognitively normal individuals (Olsson et al., 2016); however, to date, only modest associations with direct measures of neurofibrillary pathology have been documented (Zetterberg, 2017).

Taken together, current fluid biomarkers in AD have been consolidated into an Alzheimer CSF profile, namely, decreased CSF $A\beta_{42}$ concentrations with increased CSF levels of T-tau and P-tau (Table 17.1).

Tau imaging biomarkers

Tau as an imaging biomarker is in a very early stage of development. A benzimidazole derivative, 7-(6-[^{18}F]fluoropyridin-3yl)-5H-pyrido[4,3-b]indole (^{18}F-AV1451), has shown high affinity for tau and minimal binding to white matter in the brain (Villemagne & Okamura, 2016). However, "off-target" binding has been identified in non-tau regions of the brain including retention seen in the basal ganglia, anterior midbrain, venous sinuses, and the choroid plexus (Villemagne & Okamura, 2016). Brain imaging of tau with ^{18}F —AV1451 has been shown to distinguish AD from cognitively normal controls (sensitivity = 100%, specificity = 86%) (Wang et al., 2016) (Table 17.1). The presence of NFTs in a number of non-AD tauopathies represents a major limitation for the use of such agents for specific AD diagnosis.

Neurodegeneration

Neurodegenerative biomarkers in AD includes detection of brain atrophy in the medial temporal lobe with Magnetic Resonance Imaging (MRI) (Frisoni, Fox, Jack, Scheltens, & Thompson, 2010) as well as measurement of impaired cerebral metabolism and perfusion in temporoparietal regions with ^{18}Flurodeoxyglucose PET (FDG-PET) (Mosconi, 2005) and ^{99m}Tc-Exametazime SPECT (Yeo, Lim, Khan, & Pal, 2013), respectively. Moderate diagnostic performance has been achieved with these modalities (sensitivity = 83%, specificity = 85%) for MRI (sensitivity = 83%, specificity = 86%) for FDG-PET and (sensitivity = 80%, specificity = 85%) for SPECT (Bloudek, Spackman, Blankenburg, & Sullivan, 2011) (Table 17.1). Neurodegenerative biomarkers do not inform the cause for neurodegeneration, and the lack of AD specificity provides limited predictive value for a definitive AD diagnosis. Consequently, though considerable progress has been made characterizing these AD biomarkers, they currently must be used with caution and in conjunction with clinical history and cognitive testing, to support the clinical diagnosis of AD (Johnson et al., 2013).

The limitations of Aβ, tau, and neurodegeneration biomarkers notwithstanding, the AT(N) classification system has provided an opportunity to formalize our current understanding of AD progression at a biological level. Biomarker development beyond AT(N) will be important to augment existing approaches for AD diagnosis.

The cholinergic system and the AD cholinergic hypothesis

The cholinergic system is responsible for essential functions in the central nervous system (CNS). One of the major cholinergic pathways in the brain originates in the cholinergic complex of the basal forebrain and forms the major projections to cortical and subcortical regions important in memory and other cognitive functions (Mesulam, 2013). The observation that the central cholinergic system is one of the first neuronal networks affected in AD (Davies & Maloney, 1976) led to development of the cholinergic hypothesis of memory dysfunction. The cholinergic hypothesis posits that loss of cholinergic function significantly contributes to the cognitive decline associated with AD (Coyle, Price, & DeLong, 1983).

Neurons that synthesize and release the neurotransmitter acetylcholine (ACh) are said to be cholinergic. Components of the cholinergic system include the ACh-synthesizing enzyme choline acetyltransferase (ChAT), high-affinity choline transporters (CHT1) that permit uptake of choline at the presynaptic terminal of neurons, vesicular acetylcholine transporter (VAChT) to transport ACh into vesicles, and cholinoceptive muscarinic (mAChR) and nicotinic (nAChR) ACh receptors for synaptic signaling and modulatory enzymes acetylcholinesterase (AChE) and butyrylcholinesterase (BChE) (Giacobini & Pepeu, 2006). This chapter focuses on cholinesterases with a particular emphasis on BChE.

Cholinesterases: BChE and AChE

BChE and AChE are enzymes that coregulate cholinergic neurotransmission in the brain by controlling the concentration of the neurotransmitter ACh (Darvesh, Hopkins, & Geula, 2003; Silver, 1974). BChE is an enzyme of the hydrolase class, utilizing a serine residue as its catalytic center (Silver, 1974). While AChE is highly specific for ACh, BChE shows broader substrate repertoire and can hydrolyze a wide variety of esters and related compounds (Lockridge, 2015). In addition to enzymatic activity, BChE is involved in the development of the nervous system (Layer, 1983), detoxification, and drug metabolism (Lockridge, 2015). Importantly, BChE has been shown to interact with other proteins, such as β-amyloid (Mesulam & Geula, 1994) and NFTs (Moran, Mufson, & Gomez-Ramos, 1994).

BChE is a 574 amino acid residue glycoprotein. As a hydrolase enzyme, BChE has the serine (S198:human BChE numbering of mature enzyme), glutamate (E325), and histidine (H438) residues at the catalytic site (Lockridge et al., 1987; Nicolet et al., 2003). This catalytic triad is located near the bottom of a 20 Å deep active site gorge, where substrates such as choline esters react with serine to generate a tetrahedral intermediate that collapses to expel alcohol to form an acylated enzyme that undergoes hydrolysis to release the enzyme and product (Nicolet, Lockridge, Masson, Fontecilla-Camps, & Nachon, 2003; Sussman et al., 1991). The active site gorge of BChE is large ($\sim$ 500 Å) relative to AChE ($\sim$ 300 Å) and can accommodate larger substrate and inhibitor ligands than AChE (Saxena, Redman, Jiang, Lockridge, & Doctor, 1997).

BChE exists in different molecular forms that have significance in health and disease. BChE monomers and oligomers are comprised of identical catalytic subunits (Massoulie, Pezzementi, Bon, Krejci, & Vallette, 1993) and include the G_1 monomeric (globular) form, the G_2 dimeric form comprised of two monomers joined by a disulfide bridge at identical cysteine 571 residues (Lockridge et al., 1987), and the G_4 globular tetramer, containing two G_2 forms held together by hydrophobic interactions (Lockridge et al., 1987). Asymmetric G_4 membrane-bound tetramers also exist that are tethered to membranes either by a proline-rich membrane anchor (PRiMA) (Massoulie et al., 1993) or by triple helical collagen-tailed anchor in single- (A_4), double- (A_8), or triple-tetramer configurations (Massoulie et al., 1993). The G_1 form of the enzyme is prevalent during embryonic development (Arendt, Bruckner, Lange, & Bigl, 1992), while the G_4 form predominates in the mature healthy brain (Arendt et al., 1992). Interestingly, there is a reversal of the molecular form to embryonic G_1 form of the enzyme in AD (Arendt et al., 1992).

Cholinesterase distribution in normal and AD brain

In the healthy brain, AChE is found in neurons and neuropil whereas BChE is found in distinct neurons in regions such as the amygdala, hippocampus, and thalamus as well as in glia and white matter (Darvesh et al., 2003). However, BChE is found at relatively low concentrations in the cerebral cortex in the normal brain (Macdonald et al., 2017). In the AD brain, AChE levels are observed to be reduced (Perry et al., 1978) while those of BChE are markedly increased (Perry et al., 1978) or remain relatively constant (Darvesh, Reid, & Martin, 2010). Importantly, whereas there is very little BChE activity in the cerebral cortex in cognitively normal individuals, this distribution changes in AD and BChE preferentially associates with Aβ plaques and NFTs in the cerebral cortex in a highly AD-specific manner (Figs. 17.2–17.3) (Darvesh et al., 2010; Macdonald et al., 2017; Mesulam & Geula, 1994), the significance of which is not entirely clear. In AD, the brain ratio of BChE to AChE increases from BChE/AChE = 0.2 in normal brain to BChE/AChE = 11.0 in the AD brain (Perry, 1980). Additionally, BChE in the cerebral cortex is also virtually absent in other forms of dementia (Macdonald et al., 2017).

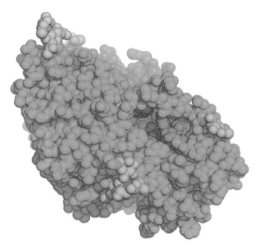

Figure 17.1 Crystal structure of human Butyrylcholinesterase (BChE). 3-Dimensional rendering of the monomeric crystal structure of human butyrylcholinesterase (BChE) enzyme. BChE is a glycoprotein with an active site (red) containing catalytic triad serine (S198), glutamate (E325), and histidine (H438). Other BChE amino acids shown are in green and glycosylated regions are in blue. *(Image generated in PyMOL (https://pymol.org) using crystal structure from Protein Data Bank (https://www.rcsb. org).)*

Butyrylcholinesterase as an AD biomarker

The preferential association of BChE with AD neuropathology positions BChE as a viable diagnostic target in AD. A recent study compared human brain tissue at autopsy from AD brains, cognitively normal older individuals and cognitively normal older individuals with Aβ plaques (Macdonald et al., 2017) (Fig. 17.2). The diagnostic performance of Aβ versus BChE quantification in predicting AD was evaluated, using clinicopathological criteria as the gold standard. A virtual absence of BChE in non-AD orbitofrontal cortex was seen, and significantly elevated BChE was observed in AD brains only. BChE quantification in these brains provided better overall diagnostic performance (sensitivity/specificity = 100%/100%, receiver operating characteristic (ROC) area under curve (AUC) = 1.0, diagnostic accuracy = 100%) outperforming Aβ quantification (sensitivity/specificity = 100%/85.7%, ROC AUC = 0.98, diagnostic accuracy = 90%) (Macdonald et al., 2017). Importantly, BChE was virtually absent in other common dementias, including CBD, FTD-Tau, dementia with Lewy bodies, and vascular dementia (Macdonald et al., 2017). The superior specificity of BChE distinguishes plaques associated with AD from those found in the cognitively normal individuals with Aβ plaques. BChE is a sensitive and specific biomarker with high predictive value in AD and a valuable diagnostic target that could enhance the current AD biomarker armamentarium.

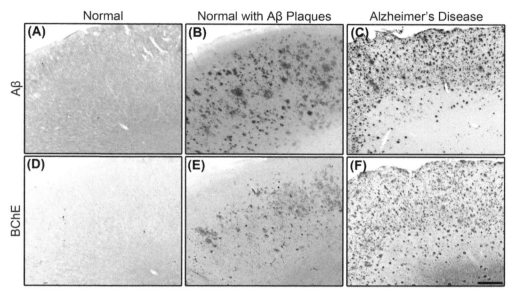

Figure 17.2 Butyrylcholinesterase (BChE) and Aβ plaques in Alzheimer's disease. Orbitofrontal cortex photomicrographs from cognitively normal (A,D), cognitively normal with Aβ plaques (B,E), and AD (C,F) brains stained for Aβ (A,B,C) immunohistochemistry, and butyrylcholinesterase (BChE, (D,E,F) histochemistry. Note, lack of BChE staining in normal orbitofrontal cortex (D), paucity of BChE activity in cognitively normal with Aβ plaques (E), and significant BChE activity in AD (F). Scale bar = 500 um *(Representative unpublished photomicrographs adapted from Macdonald, I. R., Maxwell, S. P., Reid, G. A., Cash, M. K., DeBay, D. R., & Darvesh, S. (2017). Quantification of butyrylcholinesterase activity as a sensitive and specific biomarker of Alzheimer's disease. Journal of Alzheimer's Disorder, 58(2), 491–505. doi:10.3233/JAD-170164)*

BChE CSF biomarkers

BChE activity in the CSF has been evaluated in AD and compared in relation to AD neuropathology and cognitive assessments. BChE activity in CSF has been observed to decrease in AD (Darreh-Shori, Brimijoin, Kadir, Almkvist, & Nordberg, 2006) in contrast to the well-established increase of BChE-associated plaques in the cerebral cortex in AD. BChE CSF levels have also been shown to be 40%–60% lower in carriers of one or two *APOE4* (AD genetic risk factor) alleles and directly correlate with brain metabolism ([18]FDG-PET) and cognitive function (Darreh-Shori et al., 2006). This reciprocal relationship between decreased BChE in the CSF and increased BChE in the cortex in AD has been suggested to mark the incorporation of BChE into Aβ plaques in the brain and as such may be a predictive biomarker for AD.

BChE as an imaging biomarker

Molecular imaging of the cholinergic system has generated a number of radioligands that target VAChT, cholinergic receptors mAChR, nAChR, as well as AChE and BChE cholinesterases (Volkow et al., 2001). Different strategies have been adopted for imaging of cholinesterases in the brain. The metabolic trapping principle (Kikuchi et al., 2007) is a concept that has been successfully used for imaging of AChE, whereby a radioligand that crosses the blood—brain barrier (BBB) acts as a substrate for AChE and will be hydrolyzed by AChE into hydrophilic products that will be trapped inside the brain, labeling areas where the AChE enzyme is present (Kikuchi et al., 2007). The enzymatic trapping principle shows promise for BChE radioligands. This approach relies on incorporating the radionuclide in the portion of a BChE ligand that remains attached to the longer-lived acyl enzyme intermediate, providing a more stable radiolabeled complex to accurately localize BChE in the brain (Darvesh, 2013).

Several PET agents targeting AChE (Namba et al., 2002) and BChE (Kikuchi et al., 2004; Kuhl et al., 2006; Snyder et al., 2001) have been tested for human brain imaging of cholinesterases in neurodegenerative disorders. In particular, AChE imaging with N-[^{11}C]methylpiperidin-4-yl acetate (^{11}C-MP4A) and ^{11}C-MP4P are established radiotracers in the assessment of AChE. AChE PET imaging was able to successfully characterize AChE activity in vivo as a therapeutic monitoring tool after treatment with cholinesterase inhibitor donepezil in the AD brain (Ota et al., 2010).

The quest for visualization of BChE associated with AD plaques is still an ongoing pursuit. Previous studies have shown that chemical entities containing N-methylpyrrolidinol or N-methylpiperidinol moiety readily enter the brain (Kikuchi et al., 2004; Kuhl et al., 2006; Roivainen et al., 2004); however, an increase in radioligand uptake in regions typically associated with cholinesterase AD plaques was not clearly established. For example, in vivo testing of 1-^{11}C-methyl-4-piperidinyl n-butyrate (Kuhl et al., 2006), a BChE-specific ester substrate, demonstrated rapid clearance (Roivainen et al., 2004) and overall decreased uptake in AD brain, contrary to the known histochemical and enzymatic distribution studies of BChE (Mesulam & Geula, 1994).

A recent *in vitro* study comparing radioligand uptake in AD brain tissue from a cognitively normal brain with Aβ plaques using autoradiography revealed that an Aβ imaging agent, ^{123}I-IMPY, did not distinguish between the two tissue types, while a cholinesterase imaging agent, 4-^{123}I-Iodophenylcarbamate (^{123}I-PIP), selectively labeled plaques in the cortex of the AD brain (Macdonald, Reid, Pottie, Martin, & Darvesh, 2016). This demonstrates that cholinesterase imaging is a viable approach for the specific detection of AD pathology. Although ^{123}I-PIP was effective as an AD imaging agent in autoradiographic analysis, it was unable to cross the BBB to detect AD pathology in a living AD mouse model.

Other prospective ^{123}I SPECT radioligands have been generated also containing the N-methylpiperidinol moiety, which, again, rely on enzyme (BChE) trapping of the radiolabel (Darvesh, 2013) in an attempt to prolong ligand-enzyme latencies (Macdonald et al., 2011, 2016). To prevent early loss of radiolabel from the BChE—substrate complex, incorporation of a radioactive marker on the acyl moiety, instead of the alcohol portion (first leaving group) of the ester, gives the radioactive marker longer time to remain associated with the enzyme. Specificity of BChE over AChE was achieved by increasing the size of the acyl group from the natural AChE substrate, ACh. The aromatic ring bearing an iodine atom is amenable to exchange with [123]I (Macdonald et al., 2011). Earlier autoradiographic evidence was generated showing that one such iodobenzoate derivative, 1-methylpiperidin-4yl 4-iodobenzoate, when injected into a rat, enters rodent brain and labels areas known to exhibit BChE activity in histochemical studies (Macdonald et al., 2011).

In recent studies, the BChE-specific substrate ligand, N-methylpiperidinyl-4-[123]Iodo benzoate, readily entered the brain and provided sequential SPECT images of BChE that could differentiate a 5XFAD mouse model of AD from a WT counterpart (Figs. 17.3—17.4) (DeBay et al., 2017). Therefore, imaging of BChE in the cortex represents a promising diagnostic marker for AD that has the potential for early disease detection in the living brain, not yet realized by current Aβ or NFT imaging efforts.

The functional role of BChE in AD

The exact role of BChE in AD remains unclear. A putative role is in the maturation of Aβ plaques (Mesulam & Geula, 1994). Furthermore, in a genome-wide association study, BChE, in conjunction with AD risk factor *APOE4*, was found to be a chief determinant of Aβ deposition in the AD brain (Ramanan et al., 2014).

Additionally, among the over 60 known genetic variants of BChE, certain genetic variants cause reduction of BChE activity, providing an opportunity to evaluate the association of altered BChE expression to AD progression. However, the influence of BChE genetic polymorphism in AD remains controversial. The most common *BCHE*-K variant sees a BChE reduction of 30% (Lockridge, 2015). While some studies have reported a protective effect of *BCHE*-K or no effect, others have suggested increased AD risk, particularly when associated with genetic risk factor *APOE4* (Wang et al., 2015).

As in human AD, mouse models of AD (5XFAD) also accumulate BChE-associated Aβ plaques (Darvesh & Reid, 2016). BChE knockout mice have been generated (5XFAD/BChE-KO), and interestingly, these mice see a significant reduction in fibrillar Aβ plaque deposition compared to the 5XFAD mice, especially in males (Fig. 17.5)

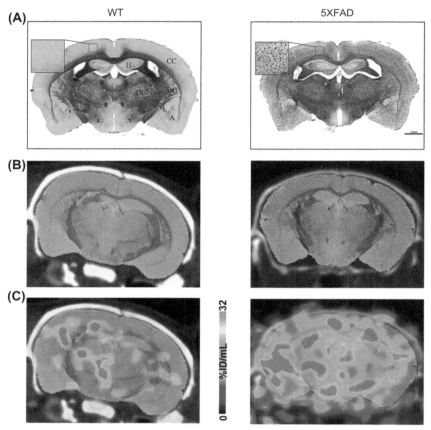

Figure 17.3 Molecular imaging of butyrylcholinesterase using SPECT radiotracer *N*-methylpiperi-dinyl-4-^{123}Iodo-benzoate. Comparison of BChE activity (midcoronal level) for WT (left) and 5XFAD (right) mice detected with SPECT radiotracer *N*-methylpiperidinyl-4-^{123}Iodo-benzoate. (A) Representative BChE histochemistry in WT and 5XFAD. Scant BChE observed in cerebral cortex of WT and marked elevation of BChE in cerebral cortex of 5XFAD mice. Significant accumulation of BChE in subcortical regions also apparent in AD. Scale bar = 1 mm, 100 mm (inset). (B) Representative CT/MRI for WT and 5XFAD brains. (C) 10 min post-injection SPECT and coregistered CT/MRI images. Marked retention in the 5XFAD cortex evident in the 5XFAD brain compared to WT and to a lesser extent subcortically. Image intensities expressed as percent injected dose/mL (%ID/mL) and set to a common scale of 0%ID/mL–32%ID/mL. Abbreviations: A, amygdala; BChE, butyrylcholinesterase; BG, basal ganglia; CC, cerebral cortex; CT, computed tomography; H, hippocampal formation; MR, magnetic resonance; SPECT, single photon emission computed tomography; Th, thalamus; WT, wild type; 5XFAD, B6SJL-Tg(APPSwFlLon, PSEN1*M146 L*L286 V) 6799Vas/Mmjax mouse strain. *(Representative unpublished photomicrographs adapted from DeBay, D. R., Reid, G. A., Pottie, I. R., Martin, E., Bowen, C. V., & Darvesh, S. (2017). Targeting butyrylcholinesterase for preclinical single photon emission computed tomography (SPECT) imaging of Alzheimer's disease. Alzheimers Dement (N Y), 3(2), 166–176. doi:10.1016/j.trci.2017.01.005.)*

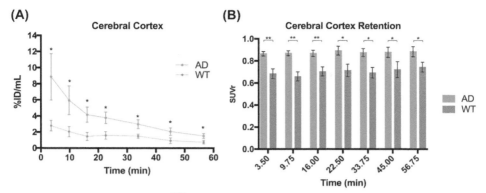

Figure 17.4 *N*-methylpiperidinyl-4-^{123}Iodo-benzoate BChE radiotracer retention. (A) BChE-specific radiotracer, *N*-methylpiperidinyl-4-^{123}Iodo-benzoate retention in Alzheimer's (5XFAD) mouse brain. Time-activity curves (% injected dose/mL) for 5XFAD (red) and WT (blue) mice (mean ± SEM). Significantly greater retention in cerebral cortex observed in 5XFAD compared to WT brain over scan duration (*, $P < .05$). (B) Cortical retention index (relative standardized uptake value (SUVr), normalized to whole brain) was up to 31% greater in 5XFAD brains versus WT (*, $P < .05$; **, $P < .01$). *(Adapted from data published in DeBay, D. R., Reid, G. A., Pottie, I. R., Martin, E., Bowen, C. V., & Darvesh, S. (2017). Targeting butyrylcholinesterase for preclinical single photon emission computed tomography (SPECT) imaging of Alzheimer's disease. Alzheimers Dement (N Y), 3(2), 166–176. doi:10.1016/j.trci.2017.01.005.)*

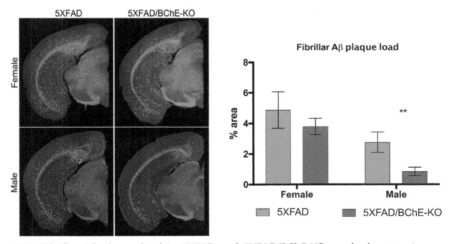

Figure 17.5 Fibrillar Aβ plaque load in 5XFAD and 5XFAD/BChE-KO cerebral cortex. Images (left) showing thioflavin-S staining of fibrillar Aβ in female and male 5XFAD and 5XFAD/BChE-KO mice at 6 months. Bar graphs (right) indicate percentage of cerebral cortex area covered with fibrillar Aβ pathology in female and male mice, aged 6 months, presented as mean ± SD (**, $P < .01$). *(Adapted from data published in Reid G.A. and Darvesh S., Butyrylcholinesterase-knockout reduces brain deposition of fibrillar beta-amyloid in an Alzheimer mouse model. Neuroscience, 298, 2015, 424-435. doi:10.1016/j.neuroscience.2015.04.039.)*

(Darvesh & Reid, 2016). Taken together, these data suggest that BChE may have mechanistic significance in the pathogenesis of AD. Regardless of whether or not BChE plays a causal role in AD development, this does not preclude its utility as a highly sensitive and specific biomarker of the disease.

Conclusions

The cholinergic system is a critical aspect of the proper functioning of the CNS, and cholinergic deficits have been shown to be inextricably linked with cognitive impairment and Aβ and tau deposition in AD. BChE is an important diagnostic target in AD and could be adopted into the AD biomarker armamentarium.

Conflict of interest

The corresponding author is a scientific cofounder and stockholder in Treventis Corporation, a biotech company developing diagnostic and therapeutic agents for Alzheimer's disease and is listed as inventor on patents related to this work that is assigned to Treventis Corporation.

Key facts of Alzheimer disease

- Alzheimer's disease (AD) is the most common cause of dementia
- A definitive AD diagnosis can only occur after death at autopsy
- Features of AD include the deposition of amyloid plaques (Aβ) and tau neurofibrillary tangles (NFTs) as well as neurodegeneration
- Current Aβ and NFT biomarkers include cerebrospinal fluid (CSF)-based and molecular-imaging approaches to visualize Aβ and NFT in the brain
- A major limitation of these biomarkers is that ~30% of cognitively normal individuals also have Aβ and NFTs
- Biomarkers of the cholinergic system may enhance current diagnostic approaches in AD

Summary points

- This chapter focuses on biomarkers of the cholinergic system, one of the first neuronal networks that degenerates in AD
- The enzyme butyrylcholinesterase (BChE) is elevated in the AD brain, preferentially associating with AD pathology in the cerebral cortex
- BChE is a highly sensitive and specific biomarker and an important diagnostic target in AD
- Brain imaging of BChE could greatly enhance the accuracy and timely detection of AD

References

Arendt, T., Bruckner, M. K., Lange, M., & Bigl, V. (1992). Changes in acetylcholinesterase and butyrylcho-linesterase in Alzheimer's disease resemble embryonic development–a study of molecular forms. *Neurochemistry International, 21*(3), 381–396.

Blennow, K., & Zetterberg, H. (2018). Biomarkers for Alzheimer's disease: Current status and prospects for the future. *Journal of Internal Medicine.* https://doi.org/10.1111/joim.12816.

Bloudek, L. M., Spackman, D. E., Blankenburg, M., & Sullivan, S. D. (2011). Review and meta-analysis of biomarkers and diagnostic imaging in Alzheimer's disease. *Journal of Alzheimer's Disorder, 26*(4), 627–645. https://doi.org/10.3233/JAD-2011-110458.

Coyle, J. T., Price, D. L., & DeLong, M. R. (1983). Alzheimer's disease: A disorder of cortical cholinergic innervation. *Science, 219*(4589), 1184–1190.

Darreh-Shori, T., Brimijoin, S., Kadir, A., Almkvist, O., & Nordberg, A. (2006). Differential CSF butyryl-cholinesterase levels in Alzheimer's disease patients with the ApoE epsilon4 allele, in relation to cognitive function and cerebral glucose metabolism. *Neurobiology of Disease, 24*(2), 326–333. https://doi.org/10.1016/j.nbd.2006.07.013.

Darvesh, S. (2013). Butyrylcholinesterase radioligands to image Alzheimer's disease brain. *Chemico-Biological Interactions, 203*(1), 354–357. https://doi.org/10.1016/j.cbi.2012.08.009.

Darvesh, S., Hopkins, D. A., & Geula, C. (2003). Neurobiology of butyrylcholinesterase. *Nature Reviews Neuroscience, 4*(2), 131–138. https://doi.org/10.1038/nrn1035.

Darvesh, S., & Reid, G. A. (2016). Reduced fibrillar beta-amyloid in subcortical structures in a butyrylcholinesterase-knockout Alzheimer disease mouse model. *Chemico-Biological Interactions.* https://doi.org/10.1016/j.cbi.2016.04.022.

Darvesh, S., Reid, G. A., & Martin, E. (2010). Biochemical and histochemical comparison of cholinesterases in normal and Alzheimer brain tissues. *Current Alzheimer Research, 7*(5), 386–400.

Davies, P., & Maloney, A. J. (1976). Selective loss of central cholinergic neurons in Alzheimer's disease. *Lancet, 2*(8000), 1403.

DeBay, D. R., Reid, G. A., Pottie, I. R., Martin, E., Bowen, C. V., & Darvesh, S. (2017). Targeting butyrylcholinesterase for preclinical single photon emission computed tomography (SPECT) imaging of Alzheimer's disease. *Alzheimers Dement (N Y), 3*(2), 166–176. https://doi.org/10.1016/j.trci.2017.01.005.

Frisoni, G. B., Fox, N. C., Jack, C. R., Jr., Scheltens, P., & Thompson, P. M. (2010). The clinical use of structural MRI in Alzheimer disease. *Nature Reviews Neurology, 6*(2), 67–77. https://doi.org/10.1038/nrneurol.2009.215.

Giacobini, E., & Pepeu, G. (2006). *The brain cholinergic system in health and disease.* Abingdon, Oxon England: Informa Healthcare.

Iqbal, K., Liu, F., & Gong, C. X. (2016). Tau and neurodegenerative disease: The story so far. *Nature Reviews Neurology, 12*(1), 15–27. https://doi.org/10.1038/nrneurol.2015.225.

Jack, C. R., Jr., Bennett, D. A., Blennow, K., Carrillo, M. C., Dunn, B., Haeberlein, S. B., et al. (2018). NIA-AA Research Framework: Toward a biological definition of Alzheimer's disease. *Alzheimers Dement, 14*(4), 535–562. https://doi.org/10.1016/j.jalz.2018.02.018.

Jansen, W. J., Ossenkoppele, R., Knol, D. L., Tijms, B. M., Scheltens, P., Verhey, F. R., et al. (2015). Prevalence of cerebral amyloid pathology in persons without dementia: A meta-analysis. *Journal of the American Medical Association, 313*(19), 1924–1938. https://doi.org/10.1001/jama.2015.4668.

Johnson, K. A., Minoshima, S., Bohnen, N. I., Donohoe, K. J., Foster, N. L., Herscovitch, P., et al. (2013). Appropriate use criteria for amyloid PET: A report of the amyloid imaging task force, the society of nuclear medicine and molecular imaging, and the Alzheimer's association. *Alzheimers Dement, 9*(1). https://doi.org/10.1016/j.jalz.2013.01.002. e-1-16.

Kikuchi, T., Okamura, T., Fukushi, K., Takahashi, K., Toyohara, J., Okada, M., et al. (2007). Cerebral acetylcholinesterase imaging: Development of the radioprobes. *Current Topics in Medicinal Chemistry, 7*(18), 1790–1799.

Kikuchi, T., Zhang, M. R., Ikota, N., Fukushi, K., Okamura, T., Suzuki, K., et al. (2004). N-[18F]fluoroe-thylpiperidin-4-ylmethyl butyrate: A novel radiotracer for quantifying brain butyrylcholinesterase activity by positron emission tomography. *Bioorganic and Medicinal Chemistry Letters, 14*(8), 1927–1930. https://doi.org/10.1016/j.bmcl.2004.01.080.

Klunk, W. E., Engler, H., Nordberg, A., Wang, Y., Blomqvist, G., Holt, D. P., et al. (2004). Imaging brain amyloid in Alzheimer's disease with Pittsburgh Compound-B. *Annals of Neurology, 55*(3), 306–319. https://doi.org/10.1002/ana.20009.

Knopman, D. S., DeKosky, S. T., Cummings, J. L., Chui, H., Corey-Bloom, J., Relkin, N., et al. (2001). Practice parameter: Diagnosis of dementia (an evidence-based review). Report of the quality standards subcommittee of the American Academy of Neurology. *Neurology, 56*(9), 1143–1153.

Kuhl, D. E., Koeppe, R. A., Snyder, S. E., Minoshima, S., Frey, K. A., & Kilbourn, M. R. (2006). In vivo butyrylcholinesterase activity is not increased in Alzheimer's disease synapses. *Annals of Neurology, 59*(1), 13–20. https://doi.org/10.1002/ana.20672.

Layer, P. G. (1983). Comparative localization of acetylcholinesterase and pseudocholinesterase during morphogenesis of the chicken brain. *Proceedings of the National Academy of Sciences of the United States of America, 80*(20), 6413–6417.

Lockridge, O. (2015). Review of human butyrylcholinesterase structure, function, genetic variants, history of use in the clinic, and potential therapeutic uses. *Pharmacology and Therapeutics, 148*, 34–46. https://doi.org/10.1016/j.pharmthera.2014.11.011.

Lockridge, O., Bartels, C. F., Vaughan, T. A., Wong, C. K., Norton, S. E., & Johnson, L. L. (1987). Complete amino acid sequence of human serum cholinesterase. *Journal of Biological Chemistry, 262*(2), 549–557.

Macdonald, I. R., Maxwell, S. P., Reid, G. A., Cash, M. K., DeBay, D. R., & Darvesh, S. (2017). Quantification of butyrylcholinesterase activity as a sensitive and specific biomarker of Alzheimer's disease. *Journal of Alzheimer's Disorder, 58*(2), 491–505. https://doi.org/10.3233/JAD-170164.

Macdonald, I. R., Reid, G. A., Joy, E. E., Pottie, I. R., Matte, G., Burrell, S., et al. (2011). Synthesis and preliminary evaluation of piperidinyl and pyrrolidinyl iodobenzoates as imaging agents for butyrylcholinesterase. *Molecular Imaging and Biology, 13*(6), 1250–1261. https://doi.org/10.1007/s11307-010-0448-0.

Macdonald, I. R., Reid, G. A., Pottie, I. R., Martin, E., & Darvesh, S. (2016). Synthesis and preliminary evaluation of Phenyl 4-123I-iodophenylcarbamate for visualization of cholinesterases associated with Alzheimer disease pathology. *Journal of Nuclear Medicine, 57*(2), 297–302. https://doi.org/10.2967/jnumed.115.162032.

Massoulie, J., Pezzementi, L., Bon, S., Krejci, E., & Vallette, F. M. (1993). Molecular and cellular biology of cholinesterases. *Progress in Neurobiology, 41*(1), 31–91.

McKhann, G. M., Knopman, D. S., Chertkow, H., Hyman, B. T., Jack, C. R., Jr., Kawas, C. H., et al. (2011). The diagnosis of dementia due to Alzheimer's disease: Recommendations from the National Institute on Aging-Alzheimer's Association workgroups on diagnostic guidelines for Alzheimer's disease. *Alzheimers Dement, 7*(3), 263–269. https://doi.org/10.1016/j.jalz.2011.03.005.

Mesulam, M. M. (2013). Cholinergic circuitry of the human nucleus basalis and its fate in Alzheimer's disease. *Journal of Comparative Neurology, 521*(18), 4124–4144. https://doi.org/10.1002/cne.23415.

Mesulam, M. M., & Geula, C. (1994). Butyrylcholinesterase reactivity differentiates the amyloid plaques of aging from those of dementia. *Annals of Neurology, 36*(5), 722–727. https://doi.org/10.1002/ana.410360506.

Montine, T. J., Phelps, C. H., Beach, T. G., Bigio, E. H., Cairns, N. J., Dickson, D. W., et al. (2012). National Institute on aging-Alzheimer's association guidelines for the neuropathologic assessment of Alzheimer's disease: A practical approach. *Acta Neuropathologica, 123*(1), 1–11. https://doi.org/10.1007/s00401-011-0910-3.

Moran, M. A., Mufson, E. J., & Gomez-Ramos, P. (1994). Cholinesterases colocalize with sites of neurofibrillary degeneration in aged and Alzheimer's brains. *Acta Neuropathologica, 87*(3), 284–292.

Morris, E., Chalkidou, A., Hammers, A., Peacock, J., Summers, J., & Keevil, S. (2016). Diagnostic accuracy of (18)F amyloid PET tracers for the diagnosis of Alzheimer's disease: A systematic review and meta-analysis. *European Journal of Nuclear Medicine and Molecular Imaging, 43*(2), 374–385. https://doi.org/10.1007/s00259-015-3228-x.

Mosconi, L. (2005). Brain glucose metabolism in the early and specific diagnosis of Alzheimer's disease. FDG-PET studies in MCI and AD. *European Journal of Nuclear Medicine and Molecular Imaging, 32*(4), 486–510. https://doi.org/10.1007/s00259-005-1762-7.

Namba, H., Fukushi, K., Nagatsuka, S., Iyo, M., Shinotoh, H., Tanada, S., et al. (2002). Positron emission tomography: Quantitative measurement of brain acetylcholinesterase activity using radiolabeled substrates. *Methods, 27*(3), 242–250.

Nicolet, Y., Lockridge, O., Masson, P., Fontecilla-Camps, J. C., & Nachon, F. (2003). Crystal structure of human butyrylcholinesterase and of its complexes with substrate and products. *Journal of Biological Chemistry, 278*(42), 41141–41147. https://doi.org/10.1074/jbc.M210241200.

Olsson, B., Lautner, R., Andreasson, U., Ohrfelt, A., Portelius, E., Bjerke, M., et al. (2016). CSF and blood biomarkers for the diagnosis of Alzheimer's disease: A systematic review and meta-analysis. *The Lancet Neurology, 15*(7), 673–684. https://doi.org/10.1016/S1474-4422(16)00070-3.

Ota, T., Shinotoh, H., Fukushi, K., Kikuchi, T., Sato, K., Tanaka, N., et al. (2010). Estimation of plasma IC50 of donepezil for cerebral acetylcholinesterase inhibition in patients with Alzheimer disease using positron emission tomography. *Clinical Neuropharmacology, 33*(2), 74–78. https://doi.org/10.1097/WNF.0b013e3181c71be9.

Perry, E. K. (1980). The cholinergic system in old age and Alzheimer's disease. *Age and Ageing, 9*(1), 1–8.

Perry, E. K., Tomlinson, B. E., Blessed, G., Bergmann, K., Gibson, P. H., & Perry, R. H. (1978). Correlation of cholinergic abnormalities with senile plaques and mental test scores in senile dementia. *British Medical Journal, 2*(6150), 1457–1459.

Ramanan, V. K., Risacher, S. L., Nho, K., Kim, S., Swaminathan, S., Shen, L., et al. (2014). APOE and BCHE as modulators of cerebral amyloid deposition: A florbetapir PET genome-wide association study. *Molecular Psychiatry, 19*(3), 351–357. https://doi.org/10.1038/mp.2013.19.

Roivainen, A., Rinne, J., Virta, J., Jarvenpaa, T., Salomaki, S., Yu, M., et al. (2004). Biodistribution and blood metabolism of 1-11C-methyl-4-piperidinyl n-butyrate in humans: An imaging agent for in vivo assessment of butyrylcholinesterase activity with PET. *Journal of Nuclear Medicine, 45*(12), 2032–2039.

Saxena, A., Redman, A. M., Jiang, X., Lockridge, O., & Doctor, B. P. (1997). Differences in active site gorge dimensions of cholinesterases revealed by binding of inhibitors to human butyrylcholinesterase. *Biochemistry, 36*(48), 14642–14651. https://doi.org/10.1021/bi971425+.

Scheltens, P., Blennow, K., Breteler, M. M., de Strooper, B., Frisoni, G. B., Salloway, S., et al. (2016). Alzheimer's disease. *Lancet, 388*(10043), 505–517. https://doi.org/10.1016/S0140-6736(15)01124-1.

Selkoe, D. J., & Hardy, J. (2016). The amyloid hypothesis of Alzheimer's disease at 25 years. *EMBO Molecular Medicine, 8*(6), 595–608. https://doi.org/10.15252/emmm.201606210.

Silver, A. (1974). *The Biology of Cholinesterases*. Amsterdam: North Holland Publishing.

Snowdon, D. A., & Nun, S. (2003). Healthy aging and dementia: Findings from the Nun study. *Annals of Internal Medicine, 139*(5 Pt 2), 450–454.

Snyder, S. E., Gunupudi, N., Sherman, P. S., Butch, E. R., Skaddan, M. B., Kilbourn, M. R., et al. (2001). Radiolabeled cholinesterase substrates: In vitro methods for determining structure-activity relationships and identification of a positron emission tomography radiopharmaceutical for in vivo measurement of butyrylcholinesterase activity. *Journal of Cerebral Blood Flow and Metabolism, 21*(2), 132–143. https://doi.org/10.1097/00004647-200102000-00004.

Sussman, J. L., Harel, M., Frolow, F., Oefner, C., Goldman, A., Toker, L., et al. (1991). Atomic structure of acetylcholinesterase from *Torpedo californica*: A prototypic acetylcholine-binding protein. *Science, 253*(5022), 872–879.

Villemagne, V. L. (2016). Amyloid imaging: Past, present and future perspectives. *Ageing Research Reviews, 30*, 95–106. https://doi.org/10.1016/j.arr.2016.01.005.

Villemagne, V. L., & Okamura, N. (2016). Tau imaging in the study of ageing, Alzheimer's disease, and other neurodegenerative conditions. *Current Opinion in Neurobiology, 36*, 43–51. https://doi.org/10.1016/j.conb.2015.09.002.

Volkow, N. D., Ding, Y. S., Fowler, J. S., & Gatley, S. J. (2001). Imaging brain cholinergic activity with positron emission tomography: its role in the evaluation of cholinergic treatments in Alzheimer's dementia. *Biological Psychiatry, 49*(3), 211−220.

Wang, L., Benzinger, T. L., Su, Y., Christensen, J., Friedrichsen, K., Aldea, P., et al. (2016). Evaluation of tau imaging in staging Alzheimer disease and revealing interactions between beta-amyloid and tauopathy. *Journal of the American Medical Association Neurology, 73*(9), 1070−1077. https://doi.org/10.1001/jamaneurol.2016.2078.

Wang, Z., Jiang, Y., Wang, X., Du, Y., Xiao, D., Deng, Y., et al. (2015). Butyrylcholinesterase K variant and Alzheimer's disease risk: A meta-analysis. *Medical Science Monitor, 21*, 1408−1413. https://doi.org/10.12659/MSM.892982.

Yeo, J. M., Lim, X., Khan, Z., & Pal, S. (2013). Systematic review of the diagnostic utility of SPECT imaging in dementia. *European Archives of Psychiatry and Clinical Neuroscience, 263*(7), 539−552. https://doi.org/10.1007/s00406-013-0426-z.

Zetterberg, H. (2017). Review: Tau in biofluids - relation to pathology, imaging and clinical features. *Neuropathology and Applied Neurobiology, 43*(3), 194−199. https://doi.org/10.1111/nan.12378.

CHAPTER 18

Blood brain-derived neurotrophic factor as a biomarker of Alzheimer's disease

Marta Balietti
Center of Neurobiology of Aging, IRCCS INRCA, Ancona, Italy

List of abbreviations

AD Alzheimer's disease
ADAS-Cog Alzheimer's Disease Assessment Scale—Cognitive Subscale
BDNF brain-derived neurotrophic factor
CDR Clinical Dementia Rating scale
LTD long-term depression
LTP long-term potentiation
MCI mild cognitive impairment
MMSE Mini-Mental State Examination
SNP single-nucleotide polymorphism

Mini-dictionary of terms

Biomarker molecule easy to collect, not time consuming nor expensive to assess, able to mirror the changes induced by the pathology in the affected organs. It can be used for diagnosis, prognosis, and/or therapy monitoring.
Brain-derived neurotrophic factor a neurotrophin primarily synthesized by neurons and involved in key processes of the developing and mature nervous systems, such as cell survival/differentiation, synaptic function, and energetic needs.
Confounding variables factors able to modulate blood BDNF level irrespective of AD.
Modulating approaches pharmacological and nonpharmacological protocols, used as potential preventive/therapeutic interventions in AD, that can change blood BDNF concentration.
Polymorphism genetic variation that can be found in more than 1% of the population.

Introduction

One of the main challenges in Alzheimer's disease (AD) research is the identification of biomarkers useful in the early diagnosis, ideally in the presymptomatic stage. Despite a wide effort, no molecule has proven to be conclusive. A detailed analysis of the possible AD biomarkers is beyond the aim of this chapter; however, Table 18.1 summarizes some of the "candidates."

Diagnosis and Management in Dementia
ISBN 978-0-12-815854-8, https://doi.org/10.1016/B978-0-12-815854-8.00018-5

Table 18.1 Some of possible biomarkers of Alzheimer's disease.

Matrix	Biomarker	AD-related change
Brain	**Pittsburgh compound B positron emission tomography**	↑
CSF	**Aβ$_{1-42}$**	↓
CSF	**Total TAU**	↑
CSF	**Phosphorylated TAU**	↑
CSF	Neurogranin	↑
Blood	MicroRNA miR-107	↓
Blood	Plasma neurofilament light	↑
Blood	Platelet amyloid precursor protein isoform ratio	↓
Blood	Products of lipid peroxidation	↑
Blood	Vascular cell adhesion molecule-1	↑
Blood	Brain-derived neurotrophic factor	Details in the text

The clinically recommended biomarkers are shown in boldface. *CSF*, cerebrospinal fluid.

Brain-derived neurotrophic factor (BDNF) belongs—together with nerve growth factor, neurotrophin-3, and neurotrophin-4—to the neurotrophins family, a group of highly homologous proteins with key roles in the developing and adult nervous systems. BDNF is synthesized as a large propeptide that, after a posttranslational dimerization, can be either secreted or processed to the mature molecule; both precursor and mature forms are active extracellular signals (Costa, Perestrelo, & Almeida, 2018; Lessmann & Brigadski, 2009; Mizui, Ishikawa, Kumanogoh, & Kojima, 2016) (Fig. 18.1). Evidence exists that cerebral alterations in BDNF production and signaling may be involved in the etiopathogenesis of AD. Most studies report decreased mRNA and protein levels in the brain of AD patients, especially in areas involved in learning and memory and targeted by cholinergic fibers; interestingly, some authors have evidenced a localized increased amount, suggesting a compensatory mechanism against amyloid-β-induced neurotoxicity in the early stage of the pathology (Song, Yu, & Tan, 2015; Tanila, 2017). On this basis, a mounting attention has been paid to the potential use of blood BDNF as an AD biomarker. Indeed, in addition to peripheral sources (Laske et al., 2006), circulating BDNF derives from the brain, and human and animal studies have evidenced a correlation between blood and brain levels as well as between blood level and cerebral phenomena (Driscoll et al., 2012; Erickson et al., 2010; Hwang et al., 2015; Klein et al., 2011; Lang, Hellweg, Seifert, Schubert, & Galliant, 2007) (Fig. 18.2).

Blood BDNF in Alzheimer's disease: possible reasons for literature inconsistency

According to the literature, AD patients have higher, lower, or even unchanged blood BDNF levels in comparison with healthy elderly controls (Table 18.2). This inconsistency highly limits the biomarker reliability and can be due, at least in part, to

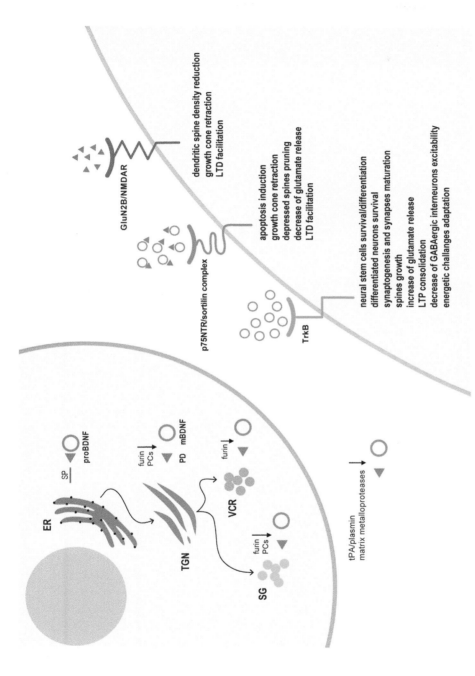

Figure 18.1 *Sketch of neuronal brain-derived neurotrophic factor (BDNF) synthesis and functions.* BDNF is synthesized in the lumen of the endoplasmic reticulum (*ER*) as a precursor composed of a signal peptide (*SP*), immediately removed, and a propeptide (*proBDNF*). ProBDNF is cleaved by proteolytic enzymes to produce the mature neurotrophin (*mBDNF*) in the *trans*-Golgi network (*TGN*), the vesicles for constitutive release (*VCR*), the secretory granules (*SG*), and the extracellular matrix. Part of proBDNF remains unprocessed and is secreted via the constitutive release pathway (VCR) or, preferentially, via activity-regulated exocytosis (SG). mBDNF binds TrkB, while proBDNF binds p75NTR complexed with sortilin; the prodomain (*PD*) is also active by binding GluN2B containing NMDA receptors (*NMDAR*). *LTD*, long-term depression; *LTP*, long-term potentiation; *PCs*, protein convertases; *tPA*, tissue plasminogen activator.

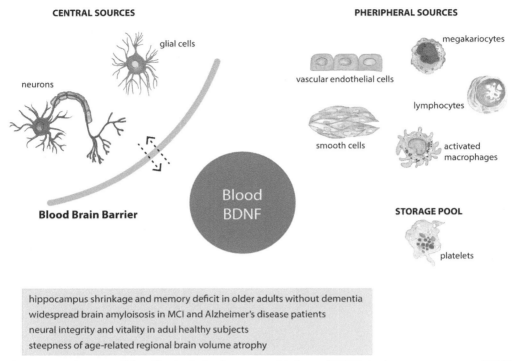

Figure 18.2 *Sources of blood brain-derived neurotrophic factor (*BDNF*).* In the nervous system, BDNF is mainly produced by neurons, but glial cells are also significant sources. BDNF can cross the blood–brain barrier in both directions through a saturable transport system (Pan, Banks, Fasold, Bluth, & Kastin, 1998). Peripherally, BDNF is synthesized by lymphocytes, activated macrophages, smooth cells, vascular endothelial cells, and megakaryocytes, while platelet BDNF is the major storage pool. In the box are listed the cerebral phenomena correlated with blood BDNF level. *MCI*, mild cognitive impairment.

nonhomogeneous patient enrollment criteria and methodological biases (Balietti, Giuli, & Conti, 2018; Kim et al., 2017). A first, key, issue is the possible dependency of blood BDNF amount on AD staging, with an early, compensatory increase followed by a late decrease (Balietti et al., 2017; Konukoglu, Andican, Firtina, Erkol, & Kurt, 2012; Laske et al., 2006; Platenick et al., 2014). Even if not all findings support this hypothesis (Angelucci et al., 2010; Faria et al., 2014; O'Bryant et al., 2009), the strict definition of patient staging may certainly ameliorate data comparability. AD and healthy subjects should also be age and gender balanced, in that BDNF seems to change during aging (Lommatzsch et al., 2005; Passaro et al., 2015; Platenik et al., 2014) and differences exist between males and females (Alvarez, Aleixandre, Linares, Masliah, & Moessler, 2014; Komulainen et al., 2008; Lommatzsch et al., 2005; Ziegenhorn et al., 2007). Unfortunately, numerous studies did not correctly manage these influencing factors (Angelucci et al., 2010; Leyhe, Stransky, Eschweiler, Buchkremer, & Laske, 2008; O'Bryant et al., 2011; Passaro et al.,

Table 18.2 Summary of studies that compared blood brain-derived neurotrophic factor levels of Alzheimer's disease patients and healthy elderly subjects.

		BDNF level[a]	AD (n)	HC (n)	P
Angelucci et al. (2010)	Higher	Mild AD, 6061.6 ± 1335 vs. 5166.9 ± 1387 pg/mL[b,d]	54	27	<.01
		Moderate to severe AD, 6082.7 ± 1458 vs. 5166.9 ± 1387 pg/mL[b,d]	35		<.01
Faria et al. (2014)	Lower	2545.3, 1497.4–4153.4 vs. 1503.8, 802.3–2378.4 pg/mL[c,e]	50	56	<.001
Yasutake et al. (2006)		14.73 ± 5.88 vs. 19.72 ± 7.53 ng/mL[b,f]	60	33	.002
Laske et al. (2007)		18.6 vs. 21.3 ng/mL[b,g]	27	28	.041
Leyhe et al. (2008)		19.2 ± 3.7 vs. 23.2 ± 3.7 ng/mL[b,f]	19	20	.015
Lee et al. (2009)		22.9 ± 5.0 vs. 27.9 ± 6.9 ng/mL[b,f]	47	39	<.01
Forlenza et al. (2010)		581.9 ± 379.4 vs. 777.5 ± 467.8 pg/L[b,f]	30	59	.001
Gezen-Ak et al. (2013)		na	52	32	<.05
Ventriglia et al. (2013)		33.16 ± 12.4 vs. 39.89 ± 9.48 ng/mL[b,f]	266	169	<.001
Coelho et al. (2014)		255 ± 116.2 vs. 353 ± 169.9 pg/mL[c,f]	21	18	.04
Liu et al. (2015)		97.03 ± 66.98 vs. 139.15 ± 102.18 ng/mL[b,f]	110	120	<.001
Passaro et al. (2015)		264.7 ± 14.7 vs. 383.9 ± 204.6 pg/mL[c,f]	44	47	.002
Janel et al. (2017)		2.00 ± 0.80 vs. 3.23 ± 1.25 ng/mL[c,f]	84	36	<.001
O'Bryant et al. (2009)	No difference	23.5, 0.8–42.9 vs. 23.8, 0.1–36.7 ng/mL[b,e]	99	99	.82
O'Bryant et al. (2011)		31.46, 6.40–43.0 vs. 30.96, 2.10–43.0 ng/mL[b,e]	198	201	.53
Woolley et al. (2012)		na	34	38	ns
Sonali et al. (2013)		12268.3 ± 7099.9 vs. 9362.833 ± 5883.3 pg/mL[b,f]	63	63	.07
Alvarez et al. (2014)		15.16 ± 9.48 vs. 16.73 ± 11.83 ng/mL[b,f]	252	38	ns

AD, Alzheimer's disease; *BDNF*, brain-derived neurotrophic factor; *HC*, healthy elderly subjects; *na*, numeric data not available; *ns*, not significant.
[a] AD versus HC.
[b] Serum.
[c] Plasma.
[d] Mean ± SEM.
[e] Median and interquartile range values.
[f] Mean ± SD.
[g] Median values.

2015). A widely underrated variable is lifestyle, despite it being well known that alcohol consumption, smoking, physical exercise, and eating behavior (Balietti et al., 2018) can modulate blood BDNF regardless of dementia. Only a few authors analyzed at least one of these habits (Angelucci et al., 2010; Coelho et al., 2014; Gezen-AK et al., 2013; Laske et al., 2006; Platenik et al., 2014) and no authors considered more than two habits (Balietti et al., 2017). Most of the elderly, demented but also cognitively healthy, consume several medications each day. Some of these medications may change circulating BDNF level, either directly (i.e., acetylcholinesterase inhibitors, antidepressants, benzodiazepines, lipid-lowering drugs, antidiabetics) or by influencing platelet functioning (i.e., nonsteroidal antiinflammatories, anticoagulants, antihypertensives) (Allard et al., 2016; Leyhe et al., 2008; Polyakova et al., 2015; Ventriglia et al., 2013; Zhang et al., 2017). Several studies did not assess the medications panel (Coelho et al., 2014; Forlenza et al., 2010; Gezen-AK et al., 2013; Janel et al., 2017; Laske et al., 2007, 2006; Lee et al., 2009; Liu et al., 2015; O'Bryant et al., 2011; Passaro et al., 2015, 2009; Platenik et al., 2014; Sonali, Tripathi, Sagar, & Vivekanandhan, 2013; Yasutake, Kuroda, Yanagawa, Okamura, & Yoneda, 2006), only a few took into account medication use in both AD patients and controls (Alvarez et al., 2014; Angelucci et al., 2010; Balietti et al., 2017; Faria et al., 2014; Laske et al., 2006), and even fewer considered both psychoactive and nonpsychoactive drugs (Alvarez et al., 2014; Angelucci et al., 2010; Balietti et al., 2017; Leyhe et al., 2008). The high susceptibility of AD patients to depression (Modrego, 2010) may represent a further potential bias. It has been widely reported that depression induces a significant reduction in blood BDNF (Molendijk et al., 2014); nonetheless, several studies completely missed the evaluation of depressive symptoms (Faria et al., 2014; Jenel et al., 2017; Konukoglu et al., 2012; Liu et al., 2015; O'Bryant et al., 2011, 2009; Sonali et al., 2013; Ventriglia et al., 2013), thus creating a possible overlap between depression- and AD-linked changes. An interesting open question is, serum or plasma? Serum BDNF is released from platelets during experimental clotting (Fujimura et al., 2002), while plasma BDNF represents the fraction that freely crosses the blood—brain barrier. During AD, platelets undergo numerous changes (Balietti et al., 2018) and these changes can differentially influence the BDNF amount contained in the two matrices (Serra-Millas, 2016). At this writing, we do not have enough data to choose between serum and plasma; in contrast, it is plausible to hypothesize that they can provide complementary information, serum BDNF mirroring a long-term storage pool and plasma BDNF representing the bioactive form. Finally, there are methodological features to consider: (1) although blood BDNF follows a circadian rhythm (Pluchino et al., 2009), several studies did not clearly state if the time of blood sampling was always the same and what it was, exactly (Angelucci et al., 2010; Coelho et al., 2014; Faria et al., 2014; Gezen-Ak et al., 2013; Jenel et al., 2017; O'Bryant et al., 2011, 2009; Sonali et al., 2013; Woolley et al., 2012); (2) BDNF concentration is negatively affected by storage duration and influenced by storage temperature (Trajkovska et al., 2007); nonetheless, information is almost never provided regarding this key factor; and (3) blood BDNF is nonnormally distributed (Komulainen et al., 2008; Trajkovska et al., 2007; Ziegenhorn et al., 2007), but this variable characteristic is rarely controlled (Alvarez et al., 204; Balietti et al., 2017; Faria et al., 2014; Gezen-Ak et al., 2013; Jenel et al., 2017; Liu et al., 2015; Platenik et al., 2014) and parametric statistics is often, wrongly, applied.

To really determine if serum/plasma BDNF could be a reliable biomarker for AD, a wide work of standardization should be the starting point. Fig. 18.3 schematizes the possible influencing factors and concise, practical advice to manage them is provided in the legend.

Polymorphisms of the BDNF *gene*

The human BDNF *gene* has a complex structure composed of several untranslated and one coding exon. It has been widely studied to investigate possible associations between its variations and psychiatric/neurological diseases and, despite thousands of polymorphisms being found, only some of them seem involved in AD. A substitution at codon 66 (196G/A) is a frequent variant that alters the amino acid sequence of the protein pro-region, i.e., change of valine to methionine (Val66Met). It is the most widely analyzed polymorphism in AD, but a consensus about its role has not been achieved. Investigations found that Met carriers do not have higher risk for AD; others showed a significant association between Val66Met and AD, with ethnic and gender differences, and some others even suggested that Val/Val homozygotes have higher susceptibility to the pathology (Song et al., 2015). The influence of Val66Met polymorphism on cognition is also controversial, but clues suggest that Met BDNF worsens performance in specific domains (Savitz, Solms, & Ramesar, 2006). Literature inconsistency also exists for the other single-nucleotide polymorphisms (SNPs), which may be due to ethnic differences. Indeed, a recent metaanalysis proved that the C270T SNP significantly increases the risk of AD in Asians but not in Africans and Europeans (Ji et al., 2015), whereas 11757 G/C polymorphism associates with Chinese but not with Italian AD patients (Boiocchi et al., 2013).

But what about the influence of these polymorphisms on blood BDNF? The issue has been barely considered and exclusively for Val66Met. The few evidences reported that no significant differences in blood BDNF level exist between Val/Val and Met carriers in AD (Forlenza et al., 2010; Kim et al., 2015) or in mild cognitive impairment (MCI) (Yu et al., 2008) patients, thus suggesting that this SNP does not influence the circulating BDNF amount. However, the lack of enough experimental data does not allow us to draw sound conclusions and may contribute to the general inconsistency.

Several experimental/scientific issues need to be clarified to define the possible influence, and eventual proper control, of BDNF gene polymorphisms on blood

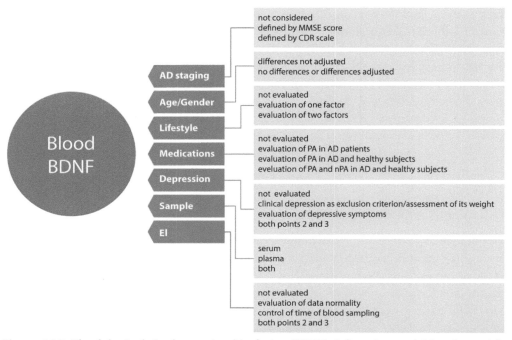

Figure 18.3 *Blood brain-derived neurotrophic factor (*BDNF*) influencing variables that might contribute to literature inconsistency.* The variables that can modulate blood BDNF are reported inside the arrows; next to each arrow, the approaches applied in the published studies are shown. (1) Alzheimer's disease (*AD*) staging: Classify AD patients as mild, moderate, and severe (Clinical Dementia Rating scale [*CDR*]: 1, 2, 3) or as early, middle, and late (Mini-Mental State Examination [*MMSE*] score: ≥ 20, 20–10, <10). (2) Age/gender: If control and AD cohorts have age and gender ratio differences, apply dedicated statistical methods (e.g., model adjustment). (3) Lifestyle: (a) Compare physical activity rates of healthy and AD subjects (e.g., using the Physical Activity Scale for the Elderly) and if the rates are different, consider it as a covariate; (b) classify healthy and AD subjects as current, former, and nonsmokers; for current smokers, define number of cigarettes and age at which they began to smoke; for former smokers, consider the abstinence period as a covariate; (c) include alcohol addiction in the exclusion criteria; (d) subdivide healthy and AD individuals according to their body mass index (BMI) (i.e., moderate/severe thinness, BMI ≤ 17 kg/m^2; underweight, $17 <$ BMI < 18.5 kg/m^2; normal weight, $18.5 \leq$ BMI < 25.0 kg/m^2; overweight, $25.0 \leq$ BMI < 30.0 kg/m^2; obesity, BMI ≥ 30.0 kg/m^2; obesity class 3, BMI > 40.0 kg/m^2). (4) Medications: Analyze both psychoactive (*PA*) and nonpsychoactive (*nPA*) drugs in both AD and cognitively healthy subjects and then treat them as potentially interfering variables. (5) Depression: Exclude subjects with a diagnosis of depression or weight the possible influence; furthermore, evaluate the depressive symptoms of enrolled subjects (preferentially, using the Geriatric Depression Scale) and treat them as potentially interfering variables. (6) Sample: Only one study analyzed both serum and plasma (Laske et al., 2006); thus, experiments are needed to establish if one matrix better mirrors the brain or if the two matrices provide complementary information. (7) Experimental issues (*EI*): (a) Verify normality of blood BDNF concentration and apply nonparametric statistics or logarithmic transformation if the variable is not normally distributed; (b) strictly control the time of blood sampling to avoid changes due to the circadian rhythm; (c) apply a limited, uniform storage time to avoid the artificial reduction of blood BDNF concentration.

BDNF level: (1) the use of reliable sample sizes for the investigation of a complete ethnic panel; (2) the evaluation of possible interactions among polymorphisms (e.g., combining 196G/A, 270C/T, and 11757G/C SNPs, Boiocchi and colleagues (2013) showed that specific haplotypes could be novel AD markers); and (3) the combination of BDNF polymorphisms with other genetic variants, such as the ε4 allele of apolipoprotein E (ApoE).

Blood BDNF and "preventive/therapeutic" approaches for Alzheimer's disease

Lack of data uniformity limits the use of blood BDNF as an AD biomarker, especially in monitoring potentially preventive/therapeutic approaches. At this writing, only a few studies have been published in this field. The effect of physical exercise is the most widely explored intervention, even if it has been primarily evaluated in MCI. The literature nearly consistently reports an upregulating effect on blood BDNF, but specific variables might influence the result. Nascimento et al. (2015) showed that only MCI subjects with the Val/Val genotype increased plasma BDNF level after a 16-week multimodal program, whereas MCI subjects with at least one Met allele did not show a change in concentration. Another genotypic characteristic may play a role: MCI African Americans who were ApoE non-ε4 carriers or ApoE ε4 carriers evidenced different percentage changes in serum BDNF after a 6-month aerobic exercise program, ε4 carriers being less responsive (Allard et al., 2017). A gender-related difference was also found, with a serum BDNF reduction in MCI women and a serum BDNF increase in MCI men induced by a 6-month aerobic protocol (Baker et al., 2010). The capacity of physical exercise to modulate blood BDNF does not necessarily need long-lasting approaches. In MCI subjects, a single bout of aerobic exercise was sufficient to trigger a significant, short-lived upregulation of serum BDNF (Tsai, Ukropec, Ukropcova, & Pai, 2018), and after a submaximal graded test on a treadmill, AD patients significantly increased plasma BDNF (Coelho et al., 2014). Interestingly, Suzuki et al. (2013) found that higher serum BDNF levels at baseline were significantly correlated with the amelioration rate on the AD Assessment Scale—Cognitive Subscale (ADAS-Cog) induced in MCI patients by a 6-month multicomponent program, thus providing one of the few pieces of evidence of an existing link among physical exercise, cognitive performance, and blood BDNF. Except for one investigation (Coelho et al., 2014), the major limitation of the studies that explored the effect of physical exercise on blood BDNF is the lack of a control group. Indeed, the assessment of the baseline level of blood BDNF in MCI/AD patients in comparison to cognitively healthy elderly controls is unavoidable to understand the real meaning of the up- or downregulation induced by the intervention.

The possible use of blood BDNF as a biomarker has been evaluated for another nonpharmacological approach, i.e., cognitive stimulation (CS). In MCI subjects, 12-week comprehensive multimodal interventions induced an increase in serum BDNF levels significantly correlated with improvement in different cognitive domains, among which is the modified ADAS-Cog, suggesting the involvement of BDNF-linked brain plasticity in the mechanisms underlying cognitive changes (Jeong et al., 2016). A correlation among CS, cognitive performance, and blood BDNF was also found in AD patients. Immediately after a 10-week comprehensive intervention, both ADAS-Cog score and plasma BDNF level, which was higher at baseline than in healthy controls, were reduced, with higher initial ADAS-Cog scores associated with greater plasma BDNF reductions (Balietti et al., 2017).

Pharmacological strategies have provided interesting, but inconsistent, information. A 15-month administration of donepezil, an acetylcholinesterase inhibitor, increased serum BDNF levels of AD patients to the values of healthy controls, but the effect on cognition was less clear. Indeed, the Mini–Mental State Examination score significantly decreased after the treatment, even if the absence, for ethical reasons, of a placebo group did not allow the authors to establish if donepezil was ineffective on cognition or was able to slow down the impairment rate (Leyhe et al., 2008). A nearly opposite result was evidenced after 2-month treatment with escitalopram, an antidepressant of the selective serotonin reuptake inhibitor class: patients with a diagnosis of possible mild neurocognitive disorder due to AD evidenced a general improvement of clinical parameters and no changes in plasma BDNF level (Levada, Cherednichenko, Trailin, & Troyan, 2016, p. 4095723). Lack of correspondence between cognitive modifications and blood BDNF changes weakens the possibility of using the peripheral neurotrophin to monitor the effect of these drugs. Interestingly, lithium, a medication applied in the therapy of affective disorders, not only upregulated serum BDNF amount in AD patients to the level of the healthy controls, but had a positive effect on ADAS-Cog score, even if no correlation was evidenced between BDNF and ADAS-Cog score changes (Leyhe et al., 2009).

In summary, encouraging data have been collected on the potentialities of blood BDNF in "therapy" monitoring, but much remains to be done. In our opinion, the first step should be, again, the adoption of uniform enrolling criteria and experimental settings. Indeed, not all the possible confounding factors were uniformly managed and this may create differences among cohorts, reducing reliability and generalizability (Table 18.3).

Table 18.3 Summary of studies that evaluated blood BDNF as a biomarker of "preventive/therapeutic" approaches: strengths and weaknesses.

		HE	CP	LS	DS	D	EI
Baker et al. (2010)	Physical exercise	—	—	pe	—	—	ts
Suzuki et al. (2013)		—	Yes	—	Yes	—	ts
Coelho et al. (2014)		Yes	—	pe eb	Yes	—	—

Table 18.3 Summary of studies that evaluated blood BDNF as a biomarker of "preventive/therapeutic" approaches: strengths and weaknesses.—cont'd

		HE	CP	LS	DS	D	EI
Nascimento et al. (2015)		—	—	pe eb	Yes	p	vn
Allard et al. (2017)		—	—	ac eb	Yes	p^a np^a	ts vn
Tsai et al. (2018)		—	—	eb	Yes	—	ts vn
Jeong et al. (2016)	CS	—	Yes	ac	Yes	p	—
Balietti et al. (2017)		Yes	Yes	s ac	Yes	p np	ts vn
Leyhe et al. (2008)	Drugs	Yes	—	—	Yes	p^a np^a	ts
Leyhe et al. (2009)		Yes	—	ac	—	p np	ts
Levada et al. (2016)		Yes	—	—	—	—	—

CP, correlation with cognitive performance; CS, cognitive stimulation; D, drug influence (p, psychoactive medications; np, nonpsychoactive medications); DS, depressive symptoms assessment; eb, eating behavior by body mass index; EI, experimental issues (ts, controlled time of sampling; vn, variable normality); HE, presence of a healthy control group; LS, lifestyle evaluation (ac, alcohol consumption; pe, physical exercise rate; s, smoking).
[a]Incomplete panel.

Conclusion

A biomarker of a neurological disorder should be easy to collect, not time consuming nor expensive to test, able to mirror the cerebral changes due to the pathology, useful to define risk factors, and capable of reflecting the positive effects induced by preventive/therapeutic approaches. Blood BDNF certainly accomplishes some of these requirements but also has numerous limits that, at this writing, reduce its usefulness in clinical practice. Rigorous definition of enrollment criteria (especially regarding AD staging and depression influence), standardized management of confounding factors, complete analyses of polymorphism influence, and better design of studies evaluating the effects of pharmacological and nonpharmacological interventions should certainly create the basis to reliably establish the possible introduction of blood BDNF dosage in the diagnostic phase as well as in the therapeutic follow-up.

Key facts of brain-derived neurotrophic factor and Alzheimer's disease

- BDNF is a neurotrophin discovered in 1982. Physiologically, it promotes neuron survival and differentiation, regulates long-term potentiation/long-term depression, and supports adaptation to energetic challenge.

- Alterations in brain BDNF metabolism have been reported in several pathologies, such as depression and Huntington's disease. In AD, results are less widely accepted, but an early, compensatory increased production seems to be followed by a significant decrease.
- As blood BDNF level may correlate with the brain amount and reflect neural phenomena, its use as a diagnostic/therapeutic follow-up biomarker in AD has been proposed.
- Numerous variables can influence blood BDNF level irrespective of dementia (e.g., depressive symptoms, lifestyle, medications), and lack of proper control has contributed to conflicting data; higher, lower, and unchanged values have been reported in comparison with healthy controls.
- A large work of standardization is urgently needed to define the soundness of blood BDNF in clinical practice.

Summary points

- Evidence exists that production and signaling of BDNF are impaired in the brain of AD patients.
- A mounting attention has been paid to the possible use of blood BDNF as an AD biomarker, but literature inconsistency limits the clinical potentialities.
- Lack of results uniformity could be due, at least in part, to nonhomogeneous patient enrollment criteria as well as to methodological biases.
- The variables that can influence blood BDNF levels regardless of AD (e.g., medications, lifestyle, depression, sample type) were analyzed and practical advice to manage them was provided.
- The possible influence of BDNF gene polymorphisms was also considered.
- Finally, the available data regarding blood BDNF reliability in monitoring potentially preventive/therapeutic approaches were reviewed, stressing strengths and weaknesses.

Acknowledgment

The author gratefully thanks Letizia Tisba and Marzio Marcellini for graphic assistance.

References

Allard, J. S., Ntekim, O., Johnson, S. P., Ngwa, J. S., Bond, V., Pinder, D., et al. (2017). APOEε4 impacts up-regulation of brain-derived neurotrophic factor after a six-month stretch and aerobic exercise intervention in mild cognitively impaired elderly African Americans: A pilot study. *Experimental Gerontology, 87*, 129–136.

Allard, J. S., Perez, E. J., Fukui, K., Carpenter, P., Ingram, D. K., & de Cabo, R. (2016). Prolonged metformin treatment leads to reduced transcription of Nrf2 and neurotrophic factors without cognitive impairment in older C57BL/6J mice. *Behavioural Brain Research, 301*, 1–9.

Alvarez, A., Aleixandre, M., Linares, C., Masliah, E., & Moessler, H. (2014). Apathy and APOE4 are associated with reduced BDNF levels in Alzheimer's disease. *Journal of Alzheimer's Disease, 42*, 1347—1355.

Angelucci, F., Spalletta, G., di Iulio, F., Ciaramella, A., Salani, F., Colantoni, L., et al. (2010). Alzheimer's disease (AD) and Mild Cognitive Impairment (MCI) patients are characterized by increased BDNF serum levels. *Current Alzheimer Research, 7*, 15—20.

Baker, L. D., Frank, L. L., Foster-Schubert, K., Green, P. S., Wilkinson, C. W., McTiernan, A., et al. (2010). Effects of aerobic exercise on mild cognitive impairment: A controlled trial. *Archives of Neurology, 67*, 71—79.

Balietti, M., Giuli, C., & Conti, F. (2018). Peripheral blood brain-derived neurotrophic factor as a biomarker of Alzheimer's disease: Are there methodological biases? *Molecular Neurobiology.* https://doi.org/10.1007/s12035-017-0866-y.

Balietti, M., Giuli, C., Fattoretti, P., Fabbietti, P., Papa, R., Postacchini, D., et al. (2017). Effect of a comprehensive intervention on plasma BDNF in patients with Alzheimer's disease. *Journal of Alzheimer's Disease, 57*, 37—43.

Boiocchi, C., Maggioli, E., Zorzetto, M., Sinforiani, E., Cereda, C., Ricevuti, G., et al. (2013). Brain-derived neurotrophic factor gene variants and Alzheimer disease: An association study in an Alzheimer disease Italian population. *Rejuvenation Research, 16*, 57—66.

Coelho, F. G., Vital, T. M., Stein, A. M., Arantes, F. J., Rueda, A. V., Camarini, R., et al. (2014). Acute aerobic exercise increases brain-derived neurotrophic factor levels in elderly with Alzheimer's disease. *Journal of Alzheimer's Disease, 39*, 401—408.

Costa, R. O., Perestrelo, T., & Almeida, R. D. (2018). PROneurotrophins and CONSequences. *Molecular Neurobiology, 55*, 2934—2951.

Driscoll, I., Martin, B., An, Y., Maudsley, S., Ferrucci, L., Mattson, M. P., et al. (2012). Plasma BDNF is associated with age-related white matter atrophy but not with cognitive function in older, non-demented adults. *PloS One, 7*, e35217.

Erickson, K. I., Prakash, R. S., Voss, M. W., Chaddock, L., Heo, S., McLaren, M., et al. (2010). Brain-derived neurotrophic factor is associated with age-related decline in hippocampal volume. *Journal of Neuroscience, 30*, 5368—5375.

Faria, M. C., Goncalves, G. S., Rocha, N. P., Moraes, E. N., Bicalho, M. A., Gualberto Cintra, M.,T., et al. (2014). Increased plasma levels of BDNF and inflammatory markers in Alzheimer's disease. *Journal of Psychiatric Research, 53*, 166—172.

Forlenza, O. V., Diniz, B. S., Teixeira, A. L., Ojopi, E. B., Talib, L. L., Mondonca, V. A., et al. (2010). Effect of brain-derived neurotrophic factor Val66Met polymorphism and serum levels on the progression of mild cognitive impairment. *World Journal of Biological Psychiatry, 11*, 774—780.

Fujimura, H., Altar, C. A., Chen, R., Nakamura, T., Nakahashi, T., Kambayashi, J., et al. (2002). Brain-derived neurotrophic factor is stored in human platelets and released by agonist stimulation. *Thrombosis and Haemostasis, 87*, 728—734.

Gezen-Ak, D., Dursun, E., Hanağasi, H., Bilgiç, B., Lohman, E., Araz, Ö. S., et al. (2013). BDNF, TNFα, HSP90, CFH, and IL-10 serum levels in patients with early or late onset Alzheimer's disease or mild cognitive impairment. *Journal of Alzheimer's Disease, 37*, 185—195.

Hwang, K. S., Lazaris, A. S., Eastman, J. A., Teng, E., Thompson, P. M., Gylys, K. H., et al. (2015). Plasma BDNF levels associate with Pittsburgh compound B binding in the brain. *Alzheimer's and Dementia, 1*, 187—193.

Janel, N., Alexopoulos, P., Badel, A., Lamari, F., Camproux, A. C., Lagarde, J., et al. (2017). Combined assessment of DYRK1A, BDNF and homocysteine levels as diagnostic marker for Alzheimer's disease. *Translational Psychiatry, 7*, e1154.

Jeong, J. H., Na, H. R., Choi, S. H., Kim, J., Na, D. L., Seo, S. W., et al. (2016). Group- and home-based cognitive intervention for patients with mild cognitive impairment: A randomized controlled trial. *Psychotherapy and Psychosomatics, 85*, 198—207.

Ji, H., Dai, D., Wang, Y., Jiang, D., Zhou, X., Lin, P., et al. (2015). Association of BDNF and BCHE with Alzheimer's disease: Meta-analysis based on 56 genetic case-control studies of 12,563 cases and 12,622 controls. *Experimental and Therapeutic Medicine, 9*, 1831—1840.

Kim, A., Fagan, A. M., Goate, A. M., Benzinger, T. L., Morris, J. C., Head, D., et al. (2015). Lack of an association of BDNF Val66Met polymorphism and plasma BDNF with hippocampal volume and memory. *Cognitive, Affective, and Behavioral Neuroscience, 15*, 625−643.

Kim, B. Y., Lee, S. H., Graham, P. L., Angelucci, F., Lucia, A., Pareje-Galeano, H., et al. (2017). Peripheral brain-derived neurotrophic factor levels in Alzheimer's disease and mild cognitive impairment: A comprehensive systematic review and meta-analysis. *Molecular Neurobiology, 54*, 7297−7311.

Klein, A. B., Williamson, R., Santini, M. A., Clemmensen, C., Ettrup, A., Rios, M., et al. (2011). Blood BDNF concentrations reflect brain-tissue BDNF levels across species. *International Journal of Neuropsychopharmacology, 14*, 347−353.

Komulainen, P., Pedersen, M., Hänninen, T., Bruunsgaard, H., Lakka, T. A., Kivipelto, M., et al. (2008). BDNF is a novel marker of cognitive function in ageing women: The DR's EXTRA study. *Neurobiology of Learning and Memory, 90*, 596−603.

Konukoglu, D., Andican, G., Fırtına, S., Erkol, G., & Kurt, A. (2012). Serum brain-derived neurotrophic factor, nerve growth factor and neurotrophin-3 levels in dementia. *Acta Neurologica Belgica, 112*, 255−260.

Lang, U. E., Hellweg, R., Seifert, F., Schubert, F., & Gallinat, J. (2007). Correlation between serum brain-derived neurotrophic factor level and an in vivo marker of cortical integrity. *Biological Psychiatry, 62*, 530−535.

Laske, C., Stransky, E., Leyhe, T., Eschweiler, G. W., Maetzler, W., Wittorf, A., et al. (2007). BDNF serum and CSF concentrations in Alzheimer's disease, normal pressure hydrocephalus and healthy controls. *Journal of Psychiatric Research, 14*, 387−394.

Laske, C., Stransky, E., Leyhe, T., Eschweiler, G. W., Schott, K., Langer, H., et al. (2006). Decreased brain-derived neurotrophic factor (BDNF)- and beta-thromboglobulin (beta-TG)- blood levels in Alzheimer's disease. *Journal of Thrombosis and Haemostasis, 96*, 102−103.

Laske, C., Stransky, E., Leyhe, T., Eschweiler, G. W., Wittorf, A., Richartz, E., et al. (2006). Stage-dependent BDNF serum concentrations in Alzheimer's disease. *Journal of Neural Transmission, 113*, 1217−1224.

Lee, J. G., Shin, B. S., You, Y. S., Kim, J. E., Yoon, S. W., Jeon, D. W., et al. (2009). Decreased serum brain-derived neurotrophic factor levels in elderly Korean with dementia. *Psychiatry Investigation, 6*, 299−305.

Lessmann, V., & Brigadski, T. (2009). Mechanisms, locations, and kinetics of synaptic BDNF secretion: An update. *Neuroscience Research, 65*, 11−22.

Levada, O. A., Cherednichenko, N. V., Trailin, A. V., & Troyan, A. S. (2016). *Plasma brain-derived neurotrophic factor as a biomarker for the main types of mild neurocognitive disorders and treatment efficacy: A preliminary study.* Disease Markers, 2016.

Leyhe, T., Eschweiler, G. W., Stransky, E., Gasser, T., Annas, P., Basun, H., et al. (2009). Increase of BDNF serum concentration in lithium treated patients with early Alzheimer's disease. *Journal of Alzheimer's Disease, 16*, 649−656.

Leyhe, T., Stransky, E., Eschweiler, G. W., Buchkremer, G., & Laske, C. (2008). Increase of BDNF serum concentration during donepezil treatment of patients with early Alzheimer's disease. *European Archives of Psychiatry and Clinical Neuroscience, 258*, 124−128.

Liu, Y. H., Jiao, S. S., Wang, Y. R., Bu, X. L., Yao, X. Q., Xiang, Y., et al. (2015). Associations between ApoEε4 carrier status and serum BDNF levels—new insights into the molecular mechanism of ApoEε4 actions in Alzheimer's disease. *Molecular Neurobiology, 51*, 1271−1277.

Lommatzsch, M., Zingler, D., Schuhbaeck, K., Schloetcke, K., Zingler, C., Schuff-Werner, P., et al. (2005). The impact of age, weight and gender on BDNF levels in human platelets and plasma. *Neurobiology of Aging, 26*, 115−123.

Mizui, T., Ishikawa, Y., Kumanogoh, H., & Kojima, M. (2016). Neurobiological actions by three distinct subtypes of brain-derived neurotrophic factor: Multi-ligand model of growth factor signaling. *Pharmacological Research, 105*, 93−98.

Modrego, P. J. (2010). Depression in Alzheimer's disease. Pathophysiology, diagnosis, and treatment. *Journal of Alzheimer's Disease, 21*, 1077−1087.

Molendijk, M. L., Spinhoven, P., Polak, M., Bus, B. A., Penninx, B. W., & Elzinga, B. M. (2014). Serum BDNF concentrations as peripheral manifestations of depression: Evidence from a systematic review and meta-analyses on 179 associations (N=9484). *Molecular Psychiatry, 19*, 791–800.

Nascimento, C. M., Pereira, J. R., Pires de Andrade, L., Garuffi, M., Ayan, C., Kerr, D. S., et al. (2015). Physical exercise improves peripheral BDNF levels and cognitive functions in mild cognitive impairment elderly with different bdnf Val66Met genotypes. *Journal of Alzheimer's Disease, 43*, 81–91.

O'Bryant, S. E., Hobson, V., Hall, J. R., Waring, S. C., Chan, W., Massman, P., et al. (2009). Brain-derived neurotrophic factor levels in Alzheimer's disease. *Journal of Alzheimer's Disease, 17*, 337–341.

O'Bryant, S. E., Hobson, V. L., Hall, J. R., Barber, R. C., Zhang, S., Johnson, L., et al. (2011). Serum brain-derived neurotrophic factor levels are specifically associated with memory performance among Alzheimer's disease cases. *Dementia and Geriatric Cognitive Disorders, 31*, 31–36.

Pan, W., Banks, W. A., Fasold, M. B., Bluth, J., & Kastin, A. J. (1998). Transport of brain-derived neurotrophic factor across the blood-brain barrier. *Neuropharmacology, 37*, 1553–1561.

Passaro, A., Dalla Nora, E., Morieri, M. L., Soavi, C., Sanz, J. M., Zurlo, A., et al. (2015). Brain-derived neurotrophic factor plasma levels: Relationship with dementia and diabetes in the elderly population. *The Journals of Gerontology. Series A, Biological Sciences and Medical Sciences, 70*, 294–302.

Pláteník, J., Fišar, Z., Buchal, R., Jirák, R., Kitzlerová, E., Zvěřová, M., et al. (2014). GSK3β, CREB, and BDNF in peripheral blood of patients with Alzheimer's disease and depression. *Progress in Neuro-Psychopharmacology and Biological Psychiatry, 50*, 83–93.

Pluchino, N., Cubeddu, A., Begliuomini, S., Merlini, S., Giannini, A., Bucci, F., et al. (2009). Daily variation of brain-derived neurotrophic factor and cortisol in women with normal menstrual cycles, undergoing oral contraception and in post menopause. *Human Reproduction, 24*, 2303–2309.

Polyakova, M., Stuke, K., Schuemberg, K., Mueller, K., Schoenknecht, P., & Schroeter, M. L. (2015). BDNF as a biomarker for successful treatment of mood disorders: A systematic & quantitative meta-analysis. *Journal of Affective Disorders, 174*, 432–440.

Savitz, J., Solms, M., & Ramesar, R. (2006). The molecular genetics of cognition: Dopamine, COMT and BDNF. *Genes, Brain and Behavior, 4*, 311–328.

Serra-Millàs, M. (2016). Are the changes in the peripheral brain-derived neurotrophic factor levels due to platelet activation? *World Journal of Psychiatry, 6*, 84–101.

Sonali, N., Tripathi, M., Sagar, R., & Vivekanandhan, S. (2013). Val66Met polymorphism and BDNF levels in Alzheimer's disease patients in North Indian population. *International Journal of Neuroscience, 123*, 409–416.

Song, J. H., Yu, J. T., & Tan, L. (2015). Brain-Derived neurotrophic factor in Alzheimer's disease: Risk, mechanisms, and therapy. *Molecular Neurobiology, 52*, 1477–1493.

Suzuki, T., Shimada, H., Makizako, H., Doi, T., Yoshida, D., Ito, K., et al. (2013). Randomized controlled trial of multicomponent exercise in older adults with mild cognitive impairment. *PloS One, 8*, e61483.

Tanila, H. (2017). The role of BDNF in Alzheimer's disease. *Neurobiology of Disease, 97*, 114–118.

Trajkovska, V., Marcussen, A. B., Vinberg, M., Hartvig, P., Aznar, S., & Knudsen, G. M. (2007). Measurements of brain-derived neurotrophic factor: Methodological aspects and demographical data. *Brain Research Bulletin, 73*, 143–149.

Tsai, C. L., Ukropec, J., Ukropcová, B., & Pai, M. C. (2018). An acute bout of aerobic or strength exercise specifically modifies circulating exerkine levels and neurocognitive functions in elderly individuals with mild cognitive impairment. *NeuroImage: Clinical, 17*, 272–284.

Ventriglia, M., Zanardini, R., Bonomini, C., Zanetti, O., Volpe, D., Pasqualetti, P., et al. (2013). Serum brain-derived neurotrophic factor levels in different neurological diseases. *BioMed Research International, 2013*, 9D1D82.

Woolley, J. D., Strobl, E. V., Shelly, W. B., Karydas, A. M., Robin Ketelle, R. N., Wolkowitz, O. M., et al. (2012). BDNF serum concentrations show no relationship with diagnostic group or medication status in neurodegenerative disease. *Current Alzheimer Research, 9*, 815–821.

Yasutake, C., Kuroda, K., Yanagawa, T., Okamura, T., & Yoneda, H. (2006). Serum BDNF, TNF-alpha and IL-1beta levels in dementia patients: Comparison between Alzheimer's disease and vascular dementia. *European Archives of Psychiatry and Clinical Neuroscience, 256*, 402–406.

Yu, H., Zhang, Z., Shi, Y., Bai, F., Xie, C., Qian, Y., et al. (2008). Association study of the decreased serum BDNF concentrations in amnestic mild cognitive impairment and the Val66Met polymorphism in Chinese Han. *Journal of Clinical Psychiatry, 69*, 1104–1111.

Zhang, J., Mu, X., Breker, D. A., Li, Y., Gao, Z., & Huang, Y. (2017). Atorvastatin treatment is associated with increased BDNF level and improved functional recovery after atherothrombotic stroke. *International Journal of Neuroscience, 127*, 92–97.

Ziegenhorn, A. A., Schulte-Herbrüggen, O., Danker-Hopfe, H., Malbranc, M., Hartung, H. D., Anders, D., et al. (2007). Serum neurotrophins—a study on the time course and influencing factors in a large old age sample. *Neurobiology of Aging, 28*, 1436–1445.

CHAPTER 19

The neuroscience of dementia: diagnosis and management methods of amyloid positron emission tomography imaging and its application to the Alzheimer's disease spectrum

Shizuo Hatashita

Department of Neurology, Shonan Atsugi Hospital, Atsugi, Japan

List of abbreviations

[11C] Carbon-11
[18F] Fluorine-18
AA Alzheimer's association
AD Alzheimer's disease
AIT Amyloid imaging task force
AUC Appropriate use criteria
Aβ Amyloid-beta
CERAD Consortium to Establish a Registry for Alzheimer's Disease
CL Centiloid scale
CN Cognitively normal
DVR Distribution volume ratio
HC Healthy control
MRI Magnetic resonance imaging
NIA National Institute on Aging
PET Positron emission tomography
PIB Pittsburgh compound B
ROIs Regions of interest
SoT Standard of truth
SUV Standardized uptake value
SUVR Standardized uptake value ratio

Mini-dictionary of terms

Amyloid-beta Peptide of 36—43 amino acids that derive from the amyloid precursor protein
Positron emission tomography A nuclear medicine imaging technique detecting pairs of gamma rays emitted indirectly by a positron-emitting radionuclide

Diagnosis and Management in Dementia
ISBN 978-0-12-815854-8, https://doi.org/10.1016/B978-0-12-815854-8.00019-7

Carbon-11 A radioisotope of carbon for the radioactive labeling of molecules

Fluorine-18 A fluorine radioisotope that is an important source of positrons

Standardized uptake value A nuclear medicine term and ratio of image-derived radioactivity concentration and whole-body concentration of injected radioactivity

Introduction

Alzheimer's disease (AD) is characterized by progressive cognitive impairment initially confined to the episodic memory system. The neuropathological features of AD constitute the extracellular deposition of amyloid-beta (Aβ) aggregates (senile plaques) and intracellular inclusions of hyperphosphorylated tau aggregates (neurofibrillary tangles). The criteria for the clinical diagnosis of AD were defined by a National Institute of Neurological and Communicative Disorders and Stroke (NINCDS) and Alzheimer's Disease and Related Disorders Association (ADRDA) working group (Mckhann et al., 1984). These criteria have been widely adopted and used for almost 30 years. AD has generally been diagnosed only by clinical assessments and histopathology confirmed by examination of the postmortem brain. Therefore, this clinical—pathological correspondence is not always consistent.

The advent of amyloid positron emission tomography (PET) imaging during the last decade has enabled the detection of human brain Aβ deposition in living people. The development of amyloid PET imaging for AD has led to improved diagnostics and accelerated the development of new therapies. The National Institute on Aging (NIA)—Alzheimer's Association (AA) working group proposed new diagnostic criteria for the dementia phase of AD, the symptomatic predementia phase, and the asymptomatic preclinical phase, which are incorporated with biomarkers of the underlying pathological process in addition to clinical symptoms (Jack et al., 2011). This disease framework of AD with biomarkers is an important advance in clinical and pathophysiological progression. The major biomarkers for AD include Aβ biomarkers of amyloid PET imaging and cerebrospinal fluid (CSF) Aβ 42, and neurodegeneration biomarkers of CSF tau, fluorodeoxyglucose PET imaging and structure magnetic resonance imaging (MRI). At present, tau PET imaging has been developed as a new biomarker. Among these biomarkers, amyloid PET imaging plays a major role in defining the pathophysiological process of AD because it is a noninvasive neuroimaging method capable of quantifying biological processes based on the dynamic distribution of Aβ deposition.

The primary PET molecular agent for detecting fibrillar Aβ in vivo is the carbon-11 ([11C]) labeled thioflavin T derivative, which is known as [11C]-Pittsburgh compound B ([11C]-PIB). Amyloid PET imaging with [11C]-PIB has been used in many different subjects throughout the world and has demonstrated the usefulness of assessing the Aβ plaque status of subjects. Nearly all patients who have clinical diagnostic criteria for AD dementia have been distinguished from healthy control (HC) by amyloid PET imaging with [11C]-PIB (Klunk et al., 2004). Furthermore, the in vivo retention of

PIB measured by PET imaging has been shown to be directly related to Aβ-containing amyloid plaques in the post—mortem AD brain (Ikonomovic et al., 2008). We have already reported that a diagnostic framework with Aβ deposition by [11C]-PIB PET in different clinical stages of AD allows for an earlier and more specific AD diagnosis process (Hatashita & Yamasaki, 2010). Thus, amyloid PET imaging with [11C]-PIB, which has a high affinity for fibrillar Aβ, can detect Aβ deposition in the brain and is a distinctive and reliable biomarker of AD. Following its early success, several fluorine-18 ([18F])-radiolabeled Aβ-selective radiopharmaceuticals have been developed for more routine clinical usefulness than [11C]-PIB. These include [18F]-florbetapir, [18F]-flutemetamol, [18F]-florbetaben, and [18F]-NAV4694. All exhibit selective binding with a high affinity to fibrillar Aβ in the brain similar to that of [11C]-PIB. Amyloid PET imaging with these [18F]-labeled Aβ tracers can discriminate AD subjects from HCs with high accuracy and is widely used in clinical practice (Barthel et al., 2011; Camus et al., 2012; Hatashita et al., 2014; Rowe et al., 2013).

This chapter focuses on amyloid PET imaging in the AD spectrum. We describe the historical background, pathological validation, imaging procedure, data analysis, and diagnostic accuracy for amyloid PET imaging. We also demonstrate the clinical application of amyloid PET imaging.

Historical background of amyloid positron emission tomography imaging

Beginning in the early 1990s, Mathis and his group began work on a suitable PET radio-tracer to image Aβ plaques in the brains of AD in their laboratories at the University of Pittsburgh. The laboratory studies were conducted starting with Congo red derivatives (histological dyes) and finally with thioflavin-T in late 1999. The neutral thioflavin-T derivatives (benzothiazole-aniline or BTA derivatives) were synthesized and bound selectively with high affinity to aggregated fibrillar Aβ in human AD brain tissues. The [11C]-labeled BTA compound, [11C] 6-OH-BTA-1, was assessed as having good in vivo brain uptake and clearance in the nonhuman primate brain (Mathis et al., 2003). The Pittsburgh group approached the Uppsala group regarding collaborative imaging studies. The Uppsala group named 6-OH-BAT-1 "Pittsburgh compound B" or PIB. The first human study with [11C]-PIB PET was performed on a subject with memory problems at the Uppsala PET center in 2002. The [11C]-PIB PET images clearly showed that the pattern of PIB retention matched the regional distribution of Aβ deposits reported from postmortem studies (Klunk et al., 2004). The [11C]-PIB PET imaging demonstrated an ability to distinguish between patients with AD and HC and then to relate [11C]-PIB retention to postmortem pathological findings (Ikonomovic et al., 2008). [11C]-PIB PET imaging has been extensively used in clinical research, trial, and practice for AD. However, [11C]-PIB PET tracer can only be used in large PET centers with their own on-site cyclotron and radiopharmacy facilities due to the 20 min half-life of [11C].

The [18F]-labeled amyloid PET radiopharmaceuticals with a 110 min half-life have been developed with a growing potential for clinical and research purposes. They are more suitable radioisotopes for widespread clinical use and allow distribution from a production site to multiple PET centers. These [18F]-labeled Aβ imaging ligands are [18F]-florbetapir, [18F]-flutemetamol, [18F]-florbetaben, and [18F]-NAV4694. Of these, [18F]-flutemetamol is a benzothiazole derivative and [18F]-NAV4694 is a benzofuran derivative; these two ligands are close structural analogues of [11C]-PiB. Both [18F] florbetapir and [18F]florbetaben are stilbene derivatives. All the ligands have shown selective binding to aggregated fibrillar Aβ in the brain tissue of AD patients in vitro. [18F]-labeled PET imaging has been demonstrated to accurately detect Aβ deposition in the brain and differentiate patients with AD from HC, similar to results from [11C]-PIB PET imaging (Barthel et al., 2011; Camus et al., 2012; Hatashita et al., 2014; Rowe et al., 2013). The Food and Drug Administration and European Medicines Agency approved [18F]-florbetapir (Eli Lilly, Amyvid) in 2012, [18F]-flutemetamol (GE Healthcare, Vizamyl) in 2013, and [18F]-florbetaben (Piramal Imaging Limited, Neuraceq) in 2014 for use in clinical practice, while [18F]-NAV4694 (AZD4694) (AstraZeneca) is currently in phase 3 trials. It is highly likely that the [18F]-labeled PET imaging will replace [11C]-PIB PET imaging in general clinical practice.

Pathological validation of amyloid positron emission tomography imaging

For the detection of Aβ pathology in amyloid PET imaging, a postmortem study with [11C]-PIB PET imaging has demonstrated that PIB binding is highly selective for insoluble fibrillar Aβ (Ikonomovic et al., 2008). Three [18F]-labeled tracers of [18F]-florbetapir, [18F]-flutemetamol, and [18F]-florbetaben have also been demonstrated to bind to Aβ plaques in postmortem human brain sections with AD using autoradiography, Bielschowsky silver stains, and immunohistochemistry correlation studies (Clark et al., 2012). The target of Aβ deposition in amyloid PET imaging includes all forms of Aβ plaques. Fibrillar Aβ may appear in varying amounts in the different plaque types, which include fleecy, amorphous, diffuse, compact, cored, or neuritic. Cored and neuritic plaques typically have large amounts of fibrillar Aβ, whereas fleecy and amorphous plaques have very little. In contrast, diffuse plaques have widely varying amounts of fibrillar Aβ. The currently available Aβ tracer in amyloid PET image is not directly representative of oligomers because of its low concentration relative to insoluble Aβ fibrils. Amyloid PET imaging can detect cored plaque with large amounts of fibrillar Aβ.

Postmortem histopathology is the most appropriate standard of truth (SoT) for obtaining regulatory approval for the detection of neuritic Aβ deposition in the brain.

Table 19.1 Diagnostic accuracy in amyloid positron emission tomography (PET) imaging radiotracers.

Radiotracer	[18F]-Florbetapir Amyvid	[18F]-Flutemetamol Vizamyl	[18F]-Florbetaben NeuraCeq	[11C]-PIB
Visual reading of PET imaging with autopsy-based SoT (CERAD score)				
Sensitivity (%)	92	91	100	83[a]
Specificity (%)	100	87	91	100[a]
Visual interpretation of PET imaging with discriminating AD from HC				
Sensitivity (%)	95	97	80	100
Specificity (%)	95	85	91	73
Quantitative assessment of PET imaging with discriminating AD from HC				
Sensitivity (%)	92	97	85	100
Specificity (%)	90	85	91	92

AD, Alzheimer's disease; *CERAD*, Consortium to Establish a Registry for Alzheimer's Disease; *HC*, healthy control; *PIB*, Pittsburgh compound B; *SoT*, standard of truth.
[a]Quantitative assessment with CERAD score.

The visual or quantitative assessments for clinical use of amyloid PET imaging are compared with the neuropathological assessment of the absence or presence of Aβ plaques according to the criteria of the Consortium to Establish a Registry for Alzheimer's Disease (CERAD) as the SoT and evaluated as the pathological validation of clinical use with amyloid PET imaging (Table 19.1) (Mirra et al., 1991). When the [11C]-PIB PET imaging for 50 individuals was compared with postmortem amyloid burden, the quantitative assessment of amyloid positivity had a sensitivity of 83% and specificity of 100% for the CERAD criteria (Villeneuve et al., 2015). Furthermore, blinded visual interpretation of the [18F]-flutemetamol images exhibited a high sensitivity of 91% and specificity of 87% using the original CERAD in an autopsy cohort of 106 subjects (Salloway et al., 2017). Visual assessment of [18F]-florbetaben PET images according to CERAD SoT showed a sensitivity of 100% and specificity of 91% in a multicenter phase 3 trial (Sabri et al., 2015). Also, in 59 participants exhibiting a range from normal to advanced dementia who had an autopsy within 2 years of PET imaging, the sensitivity and specificity of [18F]-florbetapir PET imaging to CERAD SoT were 92% and 100%, respectively (Clark et al., 2012). Thus, [18F]-labeled PET imaging, in addition to [11C]-PIB PET imaging, could also be used to distinguish individuals with no or sparse amyloid plaques from those with moderate to frequent plaques.

Positron emission tomography imaging procedure

In [11C]-PIB PET imaging, [11C]-PIB is injected intravenously as a bolus with a mean dose of 370−555 megabecquerels (MBq). Dynamic PET scanning is performed for

Table 19.2 Amyloid positron emission tomography (PET) imaging radiotracers.

Radiotracer	[18F]-florbetapir Amyvid	[18F]-flutemetamol Vizamyl	[18F]-florbetaben NeuraCeq	[11C]-PIB
Parent molecule recommended PET scan	Stilbene	Benzothiazole	Stilbene	Thioflavin-T
Dose (MBq)	370	185	300	370—555
Start time (min)	30—50	60—120	45—130	0—40
Scan duration (min)	10	10—20	15—20	20—90
Image display scale	Black—white	Color	Black—white	Color
Approved date for clinical use				
FDA	Apr. 2012	Oct. 2013	Mar. 2014	

FDA, Food and Drug Administration; *MBq*, megabecquerels; *min*, minutes; *PIB*, Pittsburgh compound B.

60 min using a predetermined acquisition protocol (31 frames: 4×15 s, 8×30 s, 9×60 s, 2×120 s, 8×300 s) or for 90 min (34 frames: 4×15 s, 8×30 s, 9×60 s, 2×120 s, 8×300 s, 3×600 s). In contrast, static PET scanning is performed from 40—60 min or 50—70 min after [11C]-PIB injection.

The procedures for [18F]-labeled PET imaging are performed according to different protocols, and the time of initiation and the duration of the PET scan vary depending on the administered Aβ tracer (Table 19.2). For [18F]-florbetapir PET scan, a 10 min scan is acquired starting 30—50 min after injection of 370 MBq [18F]-florbetapir, while [18F]-florbetaben PET scan is for 15—20 min starting 45—130 min after 300 MBq [18F]-florbetaben, and [18F]-flutemetamol PET scan is for 10—20 min starting 60—120 min after injection of 185 MBq [18F]-flutemetamol.

All scans are performed on a PET or PET/CT scanner. The general process of image acquisition begins with the patient lying still in a supine position on the scanner bed. The patient's head is secured firmly in the head folder with tape or a self-adhering wrap across the forehead to minimize motion artifacts during the scan. Based on the PET imaging protocol in three-dimensional mode, transmission acquisition is performed for 10 min without administration. An iterative or filtered back-projection reconstruction algorithm is performed with a slice thickness of 2—4 mm, matrix size of 128×128 with pixel sizes of approximately 2 mm. A post—smoothing Gaussian filter with a fill width a half-maximum of 3—5 mm is generally applied. Attenuation correction, appropriate scatter, and energy corrections are applied as with any PET scan. At scanning completion, the images are reconstructed in the transaxial planes using an interactive reconstruction process. Data are reconstructed using the vendor's reconstruction software.

Data analysis

[11C]-Pittsburgh compound B data analysis

MRI is obtained for PET image coregistration and to facilitate anatomical localization of regions of interest (ROIs) used in the analysis of the PET data. Coregistration of the PET images with the MRI is performed with statistical parameter mapping, PMOD, or another software package. The ROIs are manually drawn on the coregistered MR image in each subject or automatically with a single-subject MRI template.

The PIB retention is evaluated with standardized uptake value (SUV), regional standardized uptake value ratio (SUVR), binding potential (BP) or distribution volume ratio (DVR) measures, and these different parametric [11C]-PIB images are generated. The cerebellar gray matter is primarily used as reference tissue. The DVR, which is determined through Logan graphical analysis as simplified dynamic method, is the most widely used kinetic analysis method for [11C]-PIB PET (Lopresti et al., 2005). The linear Logan graphical analysis with image-derived cerebellar cortex, is used with simplified PIB PET data over 60—90 min intervals, and the regional DVR value for each cortical region is calculated. In contrast, the SUVR is evaluated between 40 and 70 min after [11C]-PIB injection with a short time window (20—30 min) and is normalized to the reference region. A 20 min window over 40—60 min is subsequently verified by fully quantitative studies and has been demonstrated to be a good substitute for the dynamic methods. The quantitative threshold of amyloid positivity in brain tissue could be done independently at each facility, because the timing of scan acquisition, duration of the acquisition, and choice of reference region differ.

For clinical purposes, visual interpretation of [11C]-PIB images is sufficient rather than deriving quantitative measures. Reading experience has an impact on visual interpretation of DVR image, and extensive training may be necessary to accommodate the difficulty of reading SUV and SUVR images. In general, [11C]-PIB PET images use a rainbow color scale normalized by setting cerebellar white matter to yellow, and [11C]-PIB PET images are rated as either positive or negative. A positive scan shows PIB retention in more than one cortical brain region, while a negative scan shows PIB retention predominantly in white matter but not in any of the cortical regions. DVR or BP images have shown the highest interreader and intermethod agreement in visual interpretation of [11C] PIB images in clinical practice. [11C]-PIB PET DVR image is the best in evaluating accumulation of PIB retention among the parameter images (Fig. 19.1).

[18F]-Labeled positron emission tomography data analysis

Visual interpretation of [18F]-labeled PET imaging applies almost exclusively to use in clinical practice. The visual reading depends on the observer's experience, because there is not a clear cutoff value between normal and pathological findings. Therefore, the

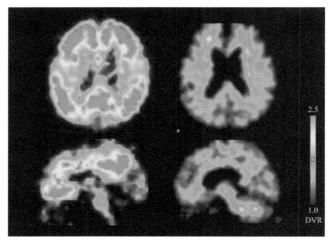

Figure 19.1 Representative axial (upper) and sagittal (lower) [11C]-PIB PET DVR images of AD patient (left) and HC subjects (right). Typical positive image from AD patient shows extensive [11C]-PIB uptake in cortical regions, while negative image from HC subject shows mild nonspecific [11C]-PIB uptake in white matter.

[18F]-labeled PET image should be interpreted by readers who have completed an electronic or in-person training program. Both [18F]-florbetapir and [18F]-florbetaben PET images are displayed for evaluation in gray scale, whereas [18F]-flutemetamol PET images are read using a color scale. Visual reading for a screening into amyloid-positive or amyloid-negative subtypes can be relatively standardized. Given an [18F]-labeled PET SUVR image, it is easy to interpret an Aβ deposition in an [18F]-labeled SUVR image (Fig. 19.2). Positive scans show at least one cortical region with increased radioactivity in cortical gray matter and/or with loss of the normally distinct gray—white matter contrast. The negative scans show more radioactivity in white matter than in gray matter, creating clear gray—white matter contrast. When an MRI is available, the reader can examine the MRI to clarify the relationship between [18F] tracer uptake and gray matter anatomy. Atrophy may affect the interpretability of scans, particularly in the frontal, temporal and parietal lobes. Some scans may be difficult to interpret due to image noise, suboptimal patient positioning, or smoothing of the reconstructed image.

Quantitative analysis of [18F]-labeled PET images is widely used for research studies but currently may not generally be incorporated into clinical use. However, quantitative analysis can be more useful than visual interpretation when the detection of small amounts of Aβ deposition is necessary. Quantification is mostly performed by SUVR normalized to a reference region such as the whole cerebellum, cerebellar gray matter, or pons. The SUVR in the regional cortical region is evaluated with an MRI

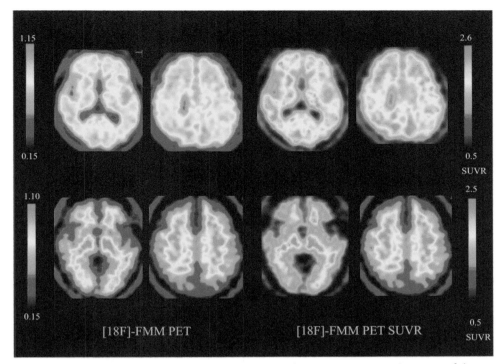

1.15

2.6

0.15

0.5
SUVR

1.10

2.5

0.15

0.5
SUVR

[18F]-FMM PET   [18F]-FMM PET SUVR

Figure 19.2 Representative [18F]-flutemetamol (FMM) and [18F]-FMM PET SUVR images of the same patients with typical positive (upper) and negative scan (lower). The typical positive image on [18F]-FMM PET shows the cortical uptake with a loss of the gray—white matter demarcation, while that of the [18F]-FMM SUVR image shows distinctly high uptake in gray matter. The negative image shows more uptake in white matter than in gray matter.

template-based approach or the PET template-based approach. Both approaches accurately assess the cortical distribution of [18F]-labeled amyloid radiotracers and are used for SUVR calculation. The MRI-based quantification uses PET and corresponding MRI data for better brain segmentation. There is the manual and automated ROI delineation for PET imaging. In the manual ROI delineation method, each subject's native PET and MRI data are coregistered, and ROIs are hand drawn on each coregistered MRI. Manual ROI delineation is generally the standard approach for the extraction of radioactivity from PET imaging. The automated ROI sampling approach is often needed for efficient and standardized sampling of large data sets and/or longitudinal follow-up. A single subject MRI template is used for automated ROI delineation. Each subject's native PET and MRI data are coregistered, and then spatially normalized to the MRI template. In contrast, the PET template-based quantification uses a PET

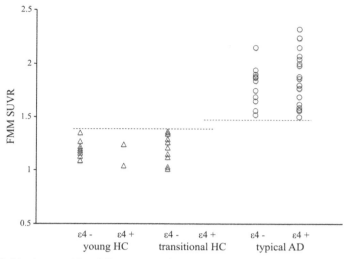

Figure 19.3 Individual cortical [18F]-flutemetamol (FMM) SUVR values in young HC (≤45 years of age) and transitional HC subjects (46–55 years of age) and AD patients with apolipoprotein E (APOE) ε4 (ε4+) and without APOE ε4 (ε4−). The upper dotted line indicates the threshold of FMM-positive SUVR, while the lower line indicates for FMM-negative SUVR. *(Data are from Hatashita, S., Yamasaki, H., Suzuki, Y., Tanaka, K., Wakebe, D., & Hayakawa, H. (2014). [18F]-Flutemetamol amyloid-beta PET imaging compared with [11C]-PIB across the spectrum of Alzheimer's disease.* European Journal of Nuclear Medicine and Molecular Imaging, *41, 290–300, with permission from the publishers.)*

template and does not require an MRI image, and this approach is used more clinically and is also easier to use. Analyzing Aβ images without an MRI may be a complex problem due to anatomical information, partial volume effects, and variability of intensity distribution across the brain. However, PET-only methods are quite important in the clinical setting because the Aβ assessment is available immediately after the scan.

Positive or negative diagnostic labels require that cutoff values are applied to continuous biological phenomena, but accepted standards for quantitative analysis of amyloid PET imaging are lacking. Quantification of PET imaging must rely on local laboratory specific standards. Imaging laboratories must derive a normal range for their method and each Aβ radiotracer. In our laboratory, the quantitative threshold of SUVR between AD and HC has been primarily based on young HC subjects (≤45 years of age) with negative scans and AD patients with typical positive scans (Fig. 19.3) (Hatashita et al., 2014).

Centiloid standardization method

Quantitative assessments of regional Aβ deposition with these [18F]-labeled PET tracers have been variable. Results of quantitative analysis are influenced by the timing of scan acquisition after administration of the Aβ tracer, duration of the acquisition, image reconstruction algorithms, partial volume correction, choice and extent of cortical regions, and the quantitative analysis method used. An international working party of Aβ imaging researchers has developed a method to standardize quantitative Aβ imaging measures by scaling the outcome to the Centiloid scale (CL) (Klunk et al., 2015). The Centiloid project has standardized quantitative Aβ imaging outcomes to a common scaled unit ranging from 0 to 100, independent of the individual Aβ tracer used. It is then possible to derive an additional linear equation to convert the result obtained from any preferred in-house analysis method to CL units by analysis of PIB images that have been analyzed by the standard CL method. The widespread use of the Centiloid standardization method will facilitate direct comparison of results across laboratories, clear definition of cutoffs for amyloid-positivity, further representation of longitudinal change, and direct comparison of different tracers.

Diagnostic accuracy of amyloid positron emission tomography imaging

The diagnostic accuracy of amyloid PET imaging is evaluated with visual and quantitative analysis (Table 19.1). The sensitivity and specificity for distinguishing patients with AD from HC subjects has been evaluated for each tracer in research studies. The visual analysis of the [11C]-PIB PET imaging has been demonstrated to discriminate AD patients from older HC subjects (≥74 years of age) with a sensitivity of 100% and specificity of 73%. Quantitative [11C]-PIB analysis with regional DVR showed a sensitivity of 100% and a specificity of 92% (Ng et al., 2007). In [18F]-labeled PET imaging, the visual assessment of [18F]-florbetapir PET imaging showed a sensitivity of 95% and specificity of 95% for distinguishing patients with probable AD from elderly HC, and the quantitative assessment of the global cortex SUVR showed a sensitivity of 92% and specificity of 90% (Camus et al., 2012). The visual assessment of [18F]-florbetaben PET images showed a sensitivity of 80% and specificity of 91% (Barthel et al., 2011). Lineal discriminant quantitative analysis of regional SUVR yielded a sensitivity of 85% and specificity of 91%. Furthermore, the visual assessment of [18F]-flutemetamol images discriminated AD patients from older HC subjects (≥56 years of age) with a sensitivity of 97% and specificity of 85% and from young and transitional HC subjects (≤55 years of age) with a specificity of 100%, being similar to the quantitative assessment of cortical regional SUVR (Hatashita et al., 2014). There are no noticeable differences in sensitivity and specificity among the different [18F]-labeled ligands. These findings suggest that the three [18F]-labeled PET imagings discriminated AD patients from older HC subjects with sufficient sensitivity and specificity. The visual interpretation of amyloid PET imaging is sufficiently accurate compared with quantitative assessment, and amyloid PET imaging can differentiate AD patients from HC subjects with high accuracy in routine clinical practice.

Clinical application

Use of amyloid positron emission tomography imaging in Alzheimer's disease dementia

Amyloid PET imaging detects AD pathophysiology but is not intended to make a clinical diagnosis of AD dementia. Clinical evaluation is still important for making more accurate diagnoses of AD dementia with amyloid PET. An amyloid PET imaging study for AD dementia has demonstrated that 328 (96%) of 341 patients with AD dementia diagnosed by NINCDS—ADRDA criteria were amyloid PET positive in 14 specialized centers plus the Alzheimer's Disease Neuroimaging Initiative (Klunk, 2011). Furthermore, diagnoses of probable AD dementia made using standard criteria in a specialized center in the United States have been confirmed by autopsy in over 95% of cases (Mayeux et al., 1998). If the dementia expert can sufficiently assess the core clinical criteria for AD dementia, the diagnosis can be accurate without the use of amyloid PET imaging. In contrast, the accurate diagnosis rate for probable AD dementia confirmed by autopsy dropped to near 70% in less specialized settings (Knopman et al., 2001). These findings suggest that amyloid PET imaging can be required in routine clinical practice to make diagnosis more accurate if a general medical physician evaluates for AD dementia with core clinical criteria. The use of amyloid PET imaging is useful for general clinical practice as an optional clinical tool to enhance certainty of diagnosis for AD dementia.

Prognostic value of amyloid positron emission tomography imaging

The Aβ deposition in the brain is likely to occur 10—15 years before cognitive impairment and is a very early event in the full spectrum of the AD pathophysiological process. The asymptomatic and symptomatic predementia phases of the AD pathophysiological process are a part of a continuum of clinical and biological AD. The NIA—AA working group has proposed the diagnostic criteria for the symptomatic predementia phase of AD, referring to an MCI due to AD, which could be used to identify individuals with AD pathophysiological processes as the primary cause of their progressive cognitive dysfunction. It has been evaluated whether the MCI patients with positive amyloid PET scan can progress to AD dementia. A community-based study has shown that 161 (58%) of 277 MCI patients were amyloid positive at baseline amyloid PET imaging (Klunk, 2011). In a longitudinal study, 53 (52%) of 101 MCI patients with amyloid positive scans converted to AD dementia within 1—3 years. In addition, we have recently demonstrated, using [11C]-PIB PET imaging, that 28 (72%) of 39 MCI patients with Aβ deposition converted to AD dementia within 3—8 years (Hatashita & Wakebe, 2017). Patients with MCI can progress to AD dementia if Aβ deposition is identified by amyloid PET imaging. Cognitively normal (CN) individuals with Aβ deposition, who are defined as preclinical AD by NIA—AA criteria, have also been evaluated to progress to MCI or AD dementia. Knopman et al. (2012) described that the rate of progression to MCI was 18% over a short follow-up of 15 months in 90 CN participants with amyloid positivity (preclinical AD stages 1—3). Villemagne et al. (2011) reported that 8 (25%) of 32 healthy individuals with positive Aβ PET scans progressed to MCI within 3 years. Our recent study has found that 12 (55%) of 22 subjects with preclinical AD with positive [11C]-PIB scan progressed to MCI within 7 years, of whom two subjects (9%) progressed to AD dementia (Hatashita & Wakebe, 2019). A higher proportion of the CN subjects with Aβ deposition might progress to MCI or AD dementia although some older individuals with preclinical AD may not become symptomatic during their lifetime. Therefore, it would be of great value to use amyloid PET imaging to predict the progression from predementia phase to AD dementia.

Appropriate use criteria

The use and application of clinical amyloid PET imaging requires careful consideration based on scientific and economic effects. The Amyloid Imaging Task Force (AIT) of the Society of Nuclear Medicine and Molecular Imaging and the AA recommended appropriate use criteria (AUC) for amyloid PET imaging (Johanson et al. 2013). The AIT outlined three clinical indications for the clinical use of amyloid PET imaging based on the following characteristics: (one) patients with persistent or progressive unexplained MCI; (two) patients with core clinical criteria for possible AD with an atypical presentation or suspected mixed etiology; and (three) patients with atypically early-age-onset progressive cognitive decline. A review study evaluating the utility of AUC for amyloid imaging has demonstrated that amyloid PET scanning improves diagnostic accuracy and results in therapeutic changes (Apostolova et al., 2016). In contrast, a recent study with AUC for amyloid PET imaging reported no difference in any outcomes of clinical utility between AIT-appropriate and AIT-inappropriate groups (Altomare et al., 2018). A consensus agreement on the appropriate use of the amyloid PET imaging has not yet been reached. Amyloid PET imaging could have an impact on changes in diagnosis and treatment regardless of whether the patients meet the AUC criteria or not.

Conclusion

Even general medical physicians who are not experts in dementia can diagnose AD dementia with more certainty if they evaluate the clinical core criteria of AD and identify Aβ deposition using amyloid PET imaging. Furthermore, in both the asymptomatic and symptomatic predementia phases of AD, amyloid positivity using amyloid PET imaging could predict progression to AD dementia. Amyloid PET imaging can identify the status of Aβ deposition in the underlying AD pathophysiology, increase diagnostic certainty, and alter management. This approach could improve the diagnose and management for patients with memory loss or cognitive dysfunction.

Key facts of predementia phase of Alzheimer's disease

- The NIA—AA working group has referred to the diagnostic criteria for the symptomatic predementia phase of AD as MCI due to AD.
- MCI due to AD is diagnosed with certainty in MCI patients based on core clinical criteria for MCI with Aβ deposition using amyloid PET imaging.
- Patients defined as MCI due to AD could progress to AD dementia.
- An asymptomatic predementia phase of AD has been demonstrated as preclinical AD in CN individuals who have Aβ deposition.
- Subjects with preclinical AD defined by amyloid PET imaging are at a higher risk for progression to MCI or AD dementia.

Summary points

- [11C]-PIB amyloid PET imaging detects Aβ deposition in the brain and is a reliable biomarker of AD.
- [18F]-labeled amyloid PET imaging, including [18F]-florbetapir, [18F]-flutemetamol, and [18F]-florbetaben, has been approved and replaces [11C]-PIB PET imaging in clinical practice.
- Amyloid PET imaging can distinguish individuals with no or sparse amyloid plaques from those with moderate to frequent plaques.
- AD patients can differentiate from HC subjects with a high sensitivity and specificity by amyloid PET imaging.
- Amyloid PET imaging identifies the status of the Aβ deposition of the underlying AD pathophysiology, increases diagnostic certainty, and alters management.

References

Altomare, G., Ferrari, C., Festari, C., Guerra, U. P., Muscio, C., Padovani, A., et al. (2018). Quantitative appraisal of the amyloid imaging taskforce appropriate use criteria for amyloid-PET. *Alzheimers Dement*. https://doi.org/10.1016/j.jalz.2018.02.022.

Apostolova, L. G., Haider, J. M., Groukasian, N., Rabinovici, G. D., Chetelat, G., Ringman, J. M., et al. (2016). Critical review of the Appropriate use criteria for amyloid imaging: Effect on diagnosis and patient care. *Alzheimers Dement, 5*, 15—22.

Barthel, H., Gertz, H. J., Dresel, S., Peters, O., Bartenstein, P., Burger, K., et al. (2011). Cerebral amyloid-beta PET with florbetaben (18F) in patients with Alzheimer's disease and healthy controls: A multicenter phase 2 diagnostic study. *The Lancet Neurology, 10*, 424—435.

Camus, V., Payoux, P., Barre, L., Desgranges, B., Voisin, T., Tauber, C., et al. (2012). Using PET with 18F-AV-45 (florbetapir) to quantify brain amyloid load in a clinical environment. *European Journal of Nuclear Medicine, 39*, 621—631.

Clark, C. M., Pontecorvo, M. J., Beach, T. G., Bedell, B. J., Coleman, R. E., Doraiswamy, P. M., et al. (2012). Cerebral PET with florbetapir compared with neuropathology at autopsy for detection of neurotic amyloid-beta plaques: A prospective cohort study. *The Lancet Neurol, 11*, 669—678.

Hatashita, S., & Wakebe, D. (2017). Amyloid-β deposition and long-term progression in mild cognitive impairment due to Alzheimer's disease defined with amyloid PET imaging. *Journal of Alzheimer's Disease, 57,* 765—773.

Hatashita, S., & Yamasaki, H. (2010). Clinically different stages of Alzheimer's disease associated by amyloid deposition with [11C]-PIB PET imaging. *Journal of Alzheimer's Disease, 21,* 995—1003.

Hatashita, S., Yamasaki, H., Suzuki, Y., Tanaka, K., Wakebe, D., & Hayakawa, H. (2014). [18F]-Flutemetamol amyloid-beta PET imaging compared with [11C]-PIB across the spectrum of Alzheimer's disease. *European Journal of Nuclear Medicine and Molecular Imaging, 41,* 290—300.

Hatashita, S., & Wakebe, D. (2019). Amyloid β deposition and glucose metabolism on the long-term progression of preclinical Alzheimer's disease. *Future Sci OA, 5,* FSO356. https://doi.org/10.4155/fosa-0069.

Ikonomovic, M. D., Klunk, W. E., Abrahamson, E. E., Mathis, C. A., Price, J. C., Tsopelas, N. D., et al. (2008). Post-mortem correlates of in vivo PiB-PET amyloid imaging in a typical case of Alzheimer's disease. *Brain, 131,* 1630—1645.

Jack, C. R., Jr., Albert, M., Knopman, D. S., Mckhann, G. M., Sperling, R. A., Carillo, M., et al. (2011). Introduction to revised criteria for the diagnosis of Alzheimer's disease: National Institute on Aging and Alzheimer's Association Workgroups. *Alzheimers and Dementia, 7,* 257—262.

Johanson, K. A., Minoshima, S., Bohnen, N. I., Donoche, K. J., Forter, N. L., Hercovitch, P., et al. (2013). Appropriate use criteria for amyloid PET: A report of the amyloid imaging task force, the society of nuclear medicine and molecular imaging, and the Alzheimer's association. *Journal of Nuclear Medicine, 54,* 476—490.

Klunk, W. E. (2011). Amyloid imaging as a biomarker for cerebral B-amyloidosis and risk-prediction for Alzheimer dementia. *Neurobiology of Aging, 32,* S20—S36.

Klunk, W. E., Engler, H., Nordberg, A., Wang, Y., Blomqvist, G., Holt, D. P., et al. (2004). Imaging brain amyloid in Alzheimer' disease with pittsburgh compound-B. *Annals of Neurology, 55,* 306—319.

Klunk, W. E., Koeppe, R. A., Price, J. C., Benzinger, T., Devous Sr, M. D., Jagust, W., et al. (2015). The centiloid project: Standardizing quantitative amyloid plaque estimation by PET. *Alzheimers and Dementia, 11,* 1—15.

Knopman, D. S., DeKosky, S. T., Cummings, J. L., Chui, H., Corey-Bloom, J., Relkin, N., et al. (2001). Practice paeameter: Diagnosis of dementia (an evidence-based review). Report of the quality standards subcommittee of the American academy of Neurology. *Neurology, 56,* 1143—1153.

Knopman, D. S., Jack, C. R., Jr., Wiste, H. J., Weigand, S. D., Vemuri, P., Lowe, V., et al. (2012). Short-term clinical outcomes for stages of NIA-AA preclinical Alzheimer disease. *Neurology, 78,* 1576—1582.

Lopresti, B. J., Klunk, W. E., Mathis, C. A., Hoge, J. A., Ziolko, S. K., Lu, X., et al. (2005). Simplified quantification of Pittsburgh compound B amyloid imaging PET studies: A comparative analysis. *Journal of Nuclear Medicine, 46,* 1959—1972.

Mathis, C. A., Wang, Y., Holt, D. P., Huang, G. F., Debnath, M. L., & Klunk, W. E. (2003). Synthesis and evaluation of [11C]-labeled 6-substituted 2-arylbenzothiazoles as amyloid imaging agents. *Journal of Medicinal Chemistry, 46,* 2740—2754.

Mayeux, R., Saunders, A. M., Shea, S., Mirra, S., Evans, D., Roses, A. D., et al. (1998). Utility of the apolipoprotein E genotype in the diagnosis of Alzheimer's disease. Alzheimer's disease centers consorium on Apolipoprotein E and Alzheimer's disease. *New England Journal of Medicine, 338,* 506—511.

Mckhann, G. M., Drachman, D. A., Folstein, M., Katzman, R., Price, D., & Stadlan, E. M. (1984). Clinical diagnosis of Alzheimer's disease-report of the NINCDS-ADRDA work group under the auspices of Department of Health and Human services task force on Alzheimer's disease. *Neurology, 34,* 939—944.

Mirra, S. S., Heymann, A., Mckeel, D., Sumi, S. M., Crain, B. J., Brownlee, L. M., et al. (1991). The criteria of the consortium to establish a registry for AD (CERAD). Part II standardization of the neuropathologic assessment of Alzheimer's disease. *Neurology, 41,* 479—486.

Ng, S., Villemagne, V. L., Berlangieri, S., Lee, S. T., Cherk, M., Gong, S. J., et al. (2007). Visual assessment verus quantitative assessment of [11C]-PIB PET and [18F]-FDG PET for detection of Alzheimer's disease. *Journal of Nuclear Medicine, 48,* 547—552.

Rowe, C. C., Pejoska, S., Mulligan, R. S., Jones, G., Chan, J. G., Svensson, S., et al. (2013). Head-to head comparison of [11C]-PIB and [18F]-AZD4694 (NAV4694) for beta-amyloid imaging in aging and dementia. *Journal of Nuclear Medicine, 54*, 880–886.

Sabri, O., Sabbagh, M. N., Seibyl, J., Barthel, H., Akatsu, H., Ouchi, Y., et al. (2015). Florbetaben PET imaging to detect amyloid-beta plaques in Alzheimer's disease: Phase 3 study. *Alzheimers and Dementia, 11*, 964–974.

Salloway, S., Gamez, J. E., Singh, U., Sadowsky, C. H., Villena, T., Sabbagh, M. N., et al. (2017). Performance of [18F]-Flutemetamol amyloid imaging against the neuritic plaque component of CERAD and the current (2012) NIA-AA recommendation for the neuropathologic diagnosis of Alzheimer's disease. *Alzheimers and Dementia (Amst), 9*, 25–34.

Villemagne, V. L., Pike, K. E., Chetelat, G., Ellis, K. A., Mulligan, R. S., Bourgeat, P., et al. (2011). Longitudinal assessment of Aβ and cognition in aging and Alzheimer's disease. *Annals of Neurology, 69*, 181–192.

Villeneuve, S., Rabinovici, G. D., Cohn-Sheehy, B. I., Madison, C., Ayakta, N., Ghosh, P. M., et al. (2015). Existing Pittsburgh compound-B positron emission tomography thresholds are too high: statistical and pathologic evaluation. *Brain, 138*, 2020–2033.

CHAPTER 20

Diffusion-weighted imaging (DWI) tractography and Alzheimer's disease

Nicola Amoroso

Dipartimento Interateneo di Fisica "M. Merlin", Università degli studi di Bari "A. Moro", Istituto Nazionale di Fisica Nucleare - Sez. di Bari, Bari, Italy

List of abbreviations

AD Alzheimer's disease
DWI diffusion-weighted imaging
MRI Magnetic Resonance Imaging
MCI mild cognitive impairment
NIH National Institutes of Health
ROC curve receiver operating characteristic curve

Mini-dictionary of terms

Complex networks "complex networks" usually refers to mathematical objects called graphs. A complex network is defined by the collection of elementary constituents, the nodes; connections between the nodes, the edges; and measures of intensity associated with each node, the weights.

Deep learning this is a novel learning approach based on multilevel representations of the patterns to learn. It is particularly suitable for image analysis; in fact, images are complex objects allowing different representations according, for example, to the dimensional scale they are observed with.

Double dipping this is also referred to as circular analysis. It consists in using the same data for both selection and classification purposes. The main drawback of this practice is the lack of generalization affecting the results, in particular for classification studies, the overoptimistic estimation of the performance.

Human connectome this is a map of the neural connections in the brain, estimated with diffusion imaging techniques. The map is built by measuring the diffusion properties of water in white matter fibers and thus reconstructing how different regions of the brain are connected.

Receiver operating characteristic curve this is a graphical plot representing the classification accuracy in terms of two variables, true positive and false positive rates. True positive rate is also known as sensitivity and specificity. False positive rate can be calculated in terms of specificity, which is by definition its complement to 1 (specificity = 1 − false positive rate). The area under this curve is a measure of accuracy. The main advantage of this measure is that it does not require any thresholding as it is directly given by classification scores. Of course, classification scores have to be binarized to provide an accuracy measure.

Supervised learning the machine-learning practice to learn models using labeled data is supervised learning. The basic idea of supervised learning is that it is possible to learn an unknown function of predictor variables, which (usually) results in a continuous output, with the performance for the given observation to belong to one class. In contrast, unsupervised approaches can learn models without labeled data.

Diagnosis and Management in Dementia
ISBN 978-0-12-815854-8, https://doi.org/10.1016/B978-0-12-815854-8.00020-3
313

Introduction

Magnetic resonance imaging (MRI) provides useful biomarkers for the early diagnosis of Alzheimer's disease (AD). Nowadays, a cascade model of AD (Jack et al., 2010), according to which the disease progression can last for decades, is widely accepted. It is important to design biomarkers able to detect its early signs in order to develop drugs or disease-modifying therapies. In particular, MRI can provide measures of brain atrophy and a quantitative evaluation of the neuronal tissue loss. Both cross-sectional (Cuingnet et al., 2011; de Vos et al., 2016; Frisoni, Fox, Jack, Scheltens, & Thompson, 2010; Tangaro, Fanizzi, Amoroso, & Bellotti, 2017) and longitudinal (Chincarini et al., 2016; Leung et al., 2010; Spulber et al., 2013) studies have outlined the importance and effectiveness of MRI-based biomarkers for early detection of AD.

MRI measures have been used extensively to provide accurate discrimination of AD, especially using machine-learning approaches. Several studies have explored different image-processing approaches, feature-selection algorithms, and learning strategies, not to mention that these studies used different data sets; therefore, it is really difficult to compare the published results or to evaluate how they would perform in clinical practice. This is why a number of international challenges have been promoted in recent years; interestingly, the presented results could not achieve the performance of previously published works (Allen et al., 2016; Amoroso, Diacono et al., 2018; Bron et al., 2015).

There is strong evidence that reliable and accurate early diagnosis of AD should rely on decision support systems, combining different measures and imaging modalities (de Vos et al., 2016; Dyrba et al., 2013; Zhang et al., 2014). In particular, scientific studies based on diffusion-weighted imaging (DWI) brain scans are receiving an increasing importance, probably because they present a natural framework to investigate diseases related to brain connectivity, such as AD.

DWI provides contrast information about the molecular motion of water. The basic idea behind DWI is the application of two opposite gradient pulses for all points of a specific direction, let us say the x direction. If water molecules have no net movement, the two gradients' effect is null; on the contrary, if they have a net movement in the x direction, the effect of the gradients is no longer null and a difference proportional to the water displacement in that direction can be obtained. In this way a three-dimensional map of water movement can be constructed; this map accounts essentially for water diffusion, the principal mechanism for water displacement within the brain. Thus, DWI provides potentially unique information on the viability of brain tissue (Schaefer, Grant, & Gonzalez, 2000). This information becomes crucial when dealing with AD effects.

White matter axons play a fundamental role in cognitive functions and all cerebral functions are at their full potential only when myelination is complete. Studies by Roher et al. (2002) demonstrated that AD is characterized by white matter degeneration. As white matter integrity can be investigated only by DWI inspection, this imaging becomes strategic to define novel biomarkers and to monitor AD onset.

DWI measures supporting Alzheimer's disease diagnosis

The first attempts to perform AD classification using DWI-based measures adopted voxel-wise statistics, such as fractional anisotropy and mean diffusivity. Fractional anisotropy quantifies the asymmetry of water diffusion processes; when water molecules flow along a white matter fiber their anisotropy is high because they tend to move along a privileged direction (Fig. 20.1). The assumption behind the use of fractional anisotropy is that white matter integrity is proportional to anisotropy (Landman et al., 2007; Le Bihan et al., 2001). Mean diffusivity measures how water molecules are free to move in a tissue. Typically, mean diffusivity is low in white and gray matter tissues and it is high in cerebrospinal fluid; an increment of mean diffusivity in white matter could be a sign of myelin disruption(Alexander et al., 2011).

Multivariate approaches based on fractional anisotropy and mean diffusivity maps were investigated. Independent component analysis was explored by Ouyang et al. (2015) to reveal fractional anisotropy changes in the bilateral corona radiate and in the corpus callosum; region of interest—based approaches were used to classify patients with AD using support vector machines (Dyrba, Grothe, Kirste, & Teipel, 2015; Metro et al., 2012) and elastic net regression (Schouten et al., 2016); a random forest analysis, including an evaluation of double dipping bias, was presented by Maggi pinto et al. (2017). It is worthwhile noting that supervised approaches would need extremely large training sets to be fully representative and for their results to be sufficiently robust and generalizable. In addition, these techniques rely on the underlying hypothesis that disease patterns are voxel-wise: basically, the disease affects the brain by modifying the voxel intensity. Thanks to these changes, it is possible to detect the disease onset. On the other hand, it is possible to pursue a holistic approach that looks for disease biomarkers from a global perspective.

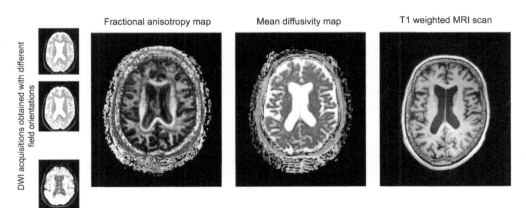

Figure 20.1 *Diffusion-weighted imaging (DWI) data, fractional anisotropy, and mean diffusivity maps compared with a T1 MRI scan.* DWI data are acquired varying the magnetic field orientation to improve diffusion map reconstruction. Fractional anisotropy (left) and mean diffusivity (middle) are examples of information retrieved from DWI. See for comparison the T1-weighted acquisition (right).

This second approach was suggested in seminal works showing how the whole brain and its organization could be accurately described in terms of complex networks (Bassett & Bellmore 2006; Bellmore & Sporns 2009): nodes, accounting for the elements constituting the system of interest, and edges, the interactions occurring between the nodes. Nowadays, network neuroscience represents a consolidated framework to describe, characterize, analyze, and investigate the human brain. The most intriguing challenge remains, probably, the use of network descriptions for prediction/diagnostic purposes (Bassett & Sporns 2017). According to this perspective, studies have investigated the possibility of using DWI tractography, not only to model the brain, but also to design effective diagnosis support systems.

Algorithms for tractography

The potential of tractography has been widely recognized (Perkel, 2013). The mapping of anatomical connections, the construction of the so called "human connectome," might help researchers to understand the normal processes occurring inside the brain and how they change according to learning, feelings, or aging. Moreover, understanding the human connectome could be fundamental to deepen our understanding of several diseases, including AD, schizophrenia, or autism-related pathologies. These were the main goals motivating the huge investments in research of recent years, such as the $40 million funding provided by the US NIH for the Human Connectome Project (http://www.humanconnectomeproject.org).

The first and pivotal challenge to face deals with the design and implementation of efficient and accurate algorithms to reconstruct white matter fibers from DWI scans. In fact, there are different algorithms and models allowing tractography, but it is not clear which ones should be preferred and why. Tractography algorithms should be able to track the paths of white matter fibers and how they kiss or cross one another. For this goal, the most common model is based on the diffusion tensor; this tensor indicates at each voxel the number and the orientation of each distinct fiber passing through it (Huisman, 2010). The construction of this tensor requires at least the acquisition of six distinct DWI images. The diffusion tensor is a symmetric 3×3 matrix whose diagonal elements D_{ii} are the diffusion coefficients measured along the three x, y, and z axes, and the remaining D_{ij} elements measure the correlation of random motions between each pair of principal directions (as a consequence the tensor must be symmetric and knowledge of the six elements uniquely defines it). The problem is that this tensor does not provide any information about the number of fibers crossing the voxel, but it can only summarize the net effect. Accordingly, it can result in misleading fiber reconstructions.

Several solutions have been proposed. High-angular-resolution diffusion imaging is one possibility. This strategy is based on using several tens of DWI scans to improve the diffusion tensor estimation (Assaf & Basser, 2005; Behrens et al., 2003; Descoteaux, Deriche, Knosche, & Anwander, 2009; Tuch, 2004; Özarslan, Shepherd, Vemuri, -Blackband, & Mareci, 2006). The proposed solutions are twofold. On one hand are deterministic algorithms: deterministic algorithms follow the fibers according to indications yielded by the diffusion tensor. On the other hand, probabilistic algorithms repeat deterministic computations many times with slightly different initial conditions and then they average the results; thus, they obtain a connectivity map whose elements are the probabilities that a given voxel is connected to another one. The wide availability of different algorithms is both an advantage and a curse: they allow the detection of distinct aspects with different pros, but, at the same time, their heterogeneity prevents a fair and objective comparison of different studies and findings. A noticeable study performed a comparison of 10 different tractography algorithms (Fillard et al., 2011); more recently, the performance of nine different algorithms was evaluated for AD classification (Zhan et al., 2015). A short sample of the latest results (at the time of writing, limited to NC—AD classification) is presented in Table 20.1.

Although consistent works have brought significant improvements to the algorithms for tractography reconstruction, there is evidence that anatomical accuracy of brain connections derived from them is still limited (Thomas et al., 2014). Tractography results are strongly dependent on the adopted parameters; more importantly, their lack of reliability is not related to image quality. In the end, even if diffusion MRI remains a fundamental tool for mapping the human connectome and offers important measures to evaluate the patients, there is room for improvement. Nonetheless, tractography-based studies confirm the enormous informative content of this kind of imaging and the possibility of developing more and more accurate diagnosis support systems for AD.

Table 20.1 Recent tractography results about AD classification.

Authors	Data	Performance
Bouts et al. (2018)	35 NC—30 AD	AUC 94%
Amoroso, Monaco, Tangaro, and Neuroimaging Initiative (2017)	52 NC—47 AD	AUC 95% (cross-validation)
Shouten et al. (2017)	173 NC—77 AD	AUC 92% (cross-validation)
Shouten et al. (2016)	173 NC—77 AD	AUC 95% (cross-validation)
Ebadi et al. (2017)	15 NC—15 AD	ACC 80% (cross-validation)

The recent development of novel algorithms for tractography leads to promising results for Alzheimer's disease (*AD*) classification. *AUC*, area under the curve; *NC*, normal controls.

From brain tractography to complex networks

A common processing pipeline for tractography usually consists of three steps:

- Preprocessing: eddy correction is usually performed; in addition, depending on the registration algorithm used, other preprocessing steps could be required, as for example, skull stripping.
- Registration: each subject in the examined cohort is registered to a reference template. Some studies coregister T1-weighted images and the corresponding non-diffusion-weighted images ($b = 0$ s/mm^2) and then perform the registration to the reference template.
- Tractography: region-of-interest analysis is performed in the reference space. Commonly adopted templates are provided of atlases, labeled maps with anatomical regions denoted. Accordingly, the T1 scans are segmented and white matter stream-lines are computed using the reference atlas. The regions segmented in the atlas are used as "seed regions." In other words, for each voxel of this region the deterministic or probabilistic reconstruction is performed. Thus, for each pair of regions it is possible to determine the number of traits connecting them.

The result of these processes is a subject-specific connectivity matrix. In fact, tractography results can be stored in a squared matrix, whose columns and rows are equals to the regions of interest presented in the atlas, and each element of the matrix (i,j) represents the number of streamlines connecting the region i to the region j. In principle, this matrix could be asymmetric, but making it symmetric is a common practice.

This representation is typical of an undirected weighted graph. In fact, by definition, each column/row of the matrix, which is called, for instance, the adjacency matrix, represents a node of the graph; each nonnull element (i,j) of the matrix represents an edge, while the number of traits stored there is the weight related to that edge. In the end, this description provides three distinct sets: N, the set of nodes; E, the set of edges; and W, the set of weights; by definition, a mathematical object consisting of these sets is a graph $G = G(N,E,W)$. A schematic overview is presented in (Fig. 20.2).

This mathematical description is rich in attributes and quantitative measures, which can be promptly used to characterize brain connectivity and, eventually, to learn patterns discriminating AD patients from controls (Rubinov, & Sporns 2010). Let us provide a list of possible attributes; the interested reader can refer to the excellent work by Boccaletti et al. (2006) for a comprehensive overview. Strength may represent the size, the intensity of connections; higher strengths usually denote nodes whose centrality, i.e., importance, in the network is relevant. Another centrality measure is betweenness; by definition, betweenness of a node measures the number of paths within the networks passing through it (note the difference between strength and betweenness). These measures do not take into account characteristics of other nodes; this is why other centrality measures have been introduced, such as inverse participation. Inverse participation

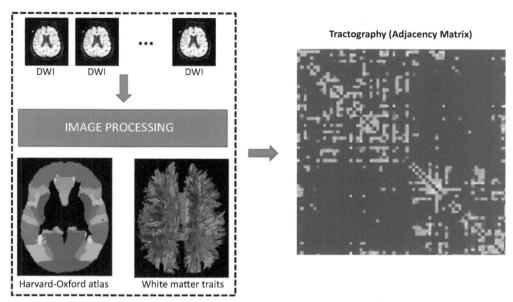

Figure 20.2 *Schematic pipeline for brain tractography.* Starting from diffusion-weighted imaging (*DWI*) scans, it is possible to reconstruct white matter fibers and to track which regions of a suitable atlas (the Harvard—Oxford, for example) are connected. The final result is a weighted matrix, called the adjacency matrix, whose elements represents the intensity of pairwise connections.

measures the strength of a node and compares it with the strength distribution of neighbor nodes (hint: if all the nodes of a network have the same strength, then this cannot be considered an important attribute anymore). Centrality measures provide an importance measure for each node of the network; however, it is somehow important to characterize and describe the whole network, too.

It is important to evaluate the structural efficiency of a network. A path in a network is the walk connecting two (distinct) nodes. Shortest paths are very important in communication networks, as they provide the most efficient way to steer information. Thus, the average shortest path length provides a measure of how efficient the communication flow within the network is. Moreover, another important structural measure for networks is clustering. Clustering measures the probability that when two nodes are both connected to a third node, then they are connected, too; this is the reason clustering is also known as transitivity. The combination of average shortest path length and clustering can yield important information about the network organization; for example, they can be used to evaluate the so-called small worldness of the network or, in contrast, its randomness.

All the previously mentioned measures, and others, which were excluded for brevity, contribute to providing a quantitative evaluation of the characteristics of the network describing the subject's brain and, therefore, of his or her connectome. These measures can be used to outline differences between groups (Lo et al., 2010; Tijms, Series, Willshaw, & Lawrie, 2012) or individual subject classification (Cui et al., 2012; Prasad, Joshi, Nir, Toga, & Thompson, 2015; Soman et al., 2016). Of course, a prominent role for these investigations is played by machine learning.

Machine learning for diagnosis support systems

Machine-learning approaches have gained increasing attention in recent years, especially for neuroimaging applications to brain pathologies. MRI data provide a huge base of knowledge; however, there are no analytical approaches or models that allow the direct exploitation of this informative content. In contrast, a more and more commonly adopted strategy consists in letting models and knowledge arise from the data themselves. In particular, machine-learning approaches allow the observation, data driven, of important features related to the observed disease; they provide a compact framework for storing information and an accurate tool to support clinical practice. Machine-learning approaches can manage and combine heterogeneous data, for example, acquired by different imaging sources. A not-secondary aspect to consider is the possibility of obtaining data through machine-learning diagnostic scores, which can naturally be considered probabilistic measures of disease severity. Finally, machine-learning models are able to incorporate novel information when available; thus, they are particularly suitable for managing the huge amount of neuroimaging data collected by hospitals and publicly available data sets.

Many recent studies have outlined group differences between cohorts of controls and patients. This approach is mainly based on measuring an indicator for the two cohorts and then evaluating whether significant statistical differences are present by the means of P values. Even if important knowledge has been acquired in this way, these studies are not able to provide a direct answer to the question: "Is this subject affected by the disease?" In contrast, classification provides a direct answer to this question; its reliability is usually measured in terms of accuracy or area under the receiver operating characteristic curve.

Machine-learning studies are often composed of two parts: feature selection and classification. The feature selection phase consists in determining the predictor variables, which can help in classification and disregarding confounding factors. This crucial study should be performed without any knowledge of the clinical labels to prevent the so-called double-dipping bias. A common procedure to avoid this bias is cross-validation. By separating training and validation sets, one can use training to learn the most informative features and train the classification model, then classification is performed on the validation set to obtain unbiased accuracy estimations. However, it is not rare to see neuroimaging studies combining feature selection and classification without distinction of training and validation set (Arbabshirani, Plis, Sui, & Calhoun, 2017).

This bias arises because feature selection is usually performed with statistical test comparing the distribution of a predictor variable in the patient and control cohorts. Thus, if no separation of training and validation test is performed, feature selection is performed, including in the learning phase the knowledge of clinical labels. Of course, this helps classification and leads to overestimation of the classification performance (Kriegeskorte, Simmons, Bellgowan, & Baker, 2009). The same considerations hold for more sophisticated, multivariate feature selection strategies. The golden rule remains to separate data from training and validation sets; best practices require cross-validation to be performed many times with random initializations to gain statistical robustness.

Classification is the second crucial phase of machine-learning analyses. Different classification strategies have been proposed. However, support vector machines, random forests, and deep learning deserve a particular mention. Support vector machines (Drucker, Wu, & Vapnik, 1999) are a state-of-the-art machine-learning method. It is based on the idea that classification and regression tasks can optimally be solved in high- or infinite-dimensional spaces defining the so-called support vectors. The algorithm explores the feature space to determine the optimal hyperplane separating the two populations to classify. Support vector machines are widely adopted because of their great versatility and classification accuracy. Random forests (Breiman, 2001) is another classification method widely adopted; it is based on the random construction of classification trees, which form then a forest. The classification is obtained by averaging the vote of each tree. The great advantage of random forests is their simplicity. In fact, they are really easy to tune, having a small number of parameters to be explored; basically, optimal results are achieved just by tuning the number of trees and the number of variables to be sampled when growing a tree. Finally, deep-learning (LeCun, Bengio, & Hinton, 2015) techniques deserve to be mentioned because they represent a radical change. Deep-learning strategies are designed to learn patterns, especially from imaging data, without a supervised knowledge. Thanks to the increasing availability of neuroimaging data, they pave the way to novel investigations and further comprehension of brain pathologies. They could probably represent the standard of future approaches.

Conclusions

This chapter was aimed at providing a short yet exhaustive description of recent approaches to DWI tractography and commonly adopted techniques to exploit the informative content of this imaging technique, with particular attention paid to the complex network framework and to developing diagnosis support systems. Of course, the main goal for these studies is the design and implementation of software libraries and diagnostic tools helpful for clinical practice. For what concerns AD patients, the main challenge remains to reach accurate and reliable classification performance; a not-secondary goal is the design of tools supporting further comprehension of the disease and its mechanisms.

Another crucial point deals with mild cognitive impairment (MCI) subjects. MCI is considered a prodromal AD phase. Of course, the release of reliable software for characterization and classification of MCI patients would have very important consequences; in this regard, it is worth noting that many studies are investigating possible solutions, but a consolidated and accepted framework is far from being obtained. Finally, neurodegenerative diseases, such as AD, share common features, especially concerning their negative effects on brain connectivity. Thus, the development of effective tools for AD could bring significant improvements and further comprehension of other diseases.

Key facts of DWI tractography and Alzheimer's disease classification

- DWI scans allow connectome reconstruction and, therefore, the investigation of brain connectivity.
- Neurodegenerative diseases affect brain connectivity. This is true in particular for AD, and this is why DWI can play a fundamental role in AD early diagnosis.
- State-of-the-art approaches are based on the evaluation of DWI-related features, such as fractional anisotropy and mean diffusivity.
- Novel approaches are focused on brain tractography.
- Complex networks allow a comprehensive modeling of the brain and a quantitative description of brain connectivity.
- The fundamental elements of a complex network description are the nodes of the network, its basic constituents; the edges, its connections; and the weights, the intensity related to each connection.
- Machine-learning approaches can suitably include the base of knowledge extracted from DWI data; this is also true for complex network features. Supervised learning is the most commonly adopted framework, but novel deep-learning techniques are now being investigated.

Summary points

- This chapter summarizes the uses of DWI data for AD assessment.
- Particular attention is given to tractography and related brain connectivity measures.
- The exploration of brain connectivity features with machine-learning methods is a major trend in current research.
- However, tractography algorithms must be used with attention, as they require fundamental improvements to become reliable tools.
- Supervised classifiers, such as support vector machines and random forests, allow an accurate classification of controls and AD patients.
- Future research will be more focused on unsupervised learning strategies, especially deep learning.

References

Alexander, A. L., Hurley, S. A., Samsonov, A. A., Adluru, N., Hosseinbor, A. P., Mossahebi, P., et al. (2011). Characterization of cerebral white matter properties using quantitative magnetic resonance imaging stains. *Brain Connectivity, 1*(6), 423–446.

Allen, G. I., Amoroso, N., Anghel, C., Balagurusamy, V., Bare, C. J., Beaton, D., et al. (2016). Crowd-sourced estimation of cognitive decline and resilience in Alzheimer's disease. *Alzheimer's and Dementia: The Journal of the Alzheimer's Association, 12*(6), 645–653.

Amoroso, N., Diacono, D., Fanizzi, A., La Rocca, M., Monaco, A., Lombardi, A., et al. (2018). Deep learning reveals Alzheimer's disease onset in MCI subjects: Results from an international challenge. *Journal of Neuroscience Methods, 302*, 3–9.

Amoroso, N., Monaco, A., Tangaro, S., & Neuroimaging Initiative. (2017). Topological measurements of DWI tractography for Alzheimer's disease detection. *Computational and Mathematical Methods in Medicine.*

Arbabshirani, M. R., Plis, S., Sui, J., & Calhoun, V. D. (2017). Single subject prediction of brain disorders in neuroimaging: Promises and pitfalls. *Neuroimage, 145*, 137–165.

Assaf, Y., & Basser, P. J. (2005). Composite hindered and restricted model of diffusion (CHARMED) MR imaging of the human brain. *Neuroimage, 27*(1), 48–58.

Bassett, D. S., & Bullmore, E. D. (2006). Small-world brain networks. *The Neuroscientist, 12*(6), 512–523.

Bassett, D. S., & Sporns, O. (2017). Network neuroscience. *Nature Neuroscience, 20*(3), 353.

Behrens, T. E., Woolrich, M. W., Jenkinson, M., Johansen-Berg, H., Nunes, R. G., Clare, S., et al. (2003). Characterization and propagation of uncertainty in diffusion-weighted MR imaging. *Magnetic Resonance in Medicine, 50*(5), 1077–1088.

Boccaletti, S., Latora, V., Moreno, Y., Chavez, M., & Hwang, D. U. (2006). Complex networks: Structure and dynamics. *Physics Reports, 424*(4–5), 175–308.

Bouts, M. J., Möller, C., Hafkemeijer, A., van Swieten, J. C., Dopper, E., van der Flier, W. M., et al. (2018). Single subject classification of Alzheimer's disease and behavioral variant frontotemporal dementia using anatomical, diffusion tensor, and resting-state functional magnetic resonance imaging. *Journal of Alzheimer's Disease*, 1–13.

Breiman, L. (2001). Random forests. *Machine Learning, 45*(1), 5–32.

Bron, E. E., Smits, M., Van Der Flier, W. M., Vrenken, H., Barkhof, F., Scheltens, P., et al. (2015). Standardized evaluation of algorithms for computer-aided diagnosis of dementia based on structural MRI: The CAD Dementia challenge. *Neuroimage, 111*, 562–579.

Bullmore, E., & Sporns, O. (2009). Complex brain networks: Graph theoretical analysis of structural and functional systems. *Nature Reviews Neuroscience, 10*(3), 186.

Chincarini, A., Sensi, F., Rei, L., Gemme, G., Squarcia, S., Longo, R., et al. (2016). Integrating longitudinal information in hippocampal volume measurements for the early detection of Alzheimer's disease. *Neuroimage, 125*, 834–847.

Cuingnet, R., Gerardin, E., Tessieras, J., Auzias, G., Lehéricy, S., Habert, M. O., et al. (2011). Automatic classification of patients with Alzheimer's disease from structural MRI: A comparison of ten methods using the ADNI database. *Neuroimage, 56*(2), 766–781.

Cui, Y., Wen, W., Lipnicki, D. M., Beg, M. F., Jin, J. S., Luo, S., et al. (2012). Automated detection of amnestic mild cognitive impairment in community-dwelling elderly adults: A combined spatial atrophy and white matter alteration approach. *Neuroimage, 59*(2), 1209–1217.

Descoteaux, M., Deriche, R., Knosche, T. R., & Anwander, A. (2009). Deterministic and probabilistic tractography based on complex fibre orientation distributions. *IEEE Transactions on Medical Imaging, 28*(2), 269–286.

Drucker, H., Wu, D., & Vapnik, V. N. (1999). Support vector machines for spam categorization. *IEEE Transactions on Neural Networks, 10*(5), 1048–1054.

Dyrba, M., Ewers, M., Wegrzyn, M., Kilimann, I., Plant, C., Oswald, A., et al. (2013). Robust automated detection of microstructural white matter degeneration in Alzheimer's disease using machine learning classification of multicenter DTI data. *PloS One, 8*(5), e64925.

Dyrba, M., Grothe, M., Kirste, T., & Teipel, S. J. (2015). Multimodal analysis of functional and structural disconnection in Alzheimer's disease using multiple kernel SVM. *Human Brain Mapping, 36*(6), 2118–2131.

Ebadi, A., Dalboni da Rocha, J. L., Nagaraju, D. B., Tovar-Moll, F., Bramati, I., Coutinho, G., et al. (2017). Ensemble classification of Alzheimer's disease and mild cognitive impairment based on complex graph measures from diffusion tensor images. *Frontiers in Neuroscience, 11*, 56.

Fillard, P., Descoteaux, M., Goh, A., Gouttard, S., Jeurissen, B., Malcolm, J., et al. (2011). Quantitative evaluation of 10 tractography algorithms on a realistic diffusion MR phantom. *Neuroimage, 56*(1), 220–234.

Frisoni, G. B., Fox, N. C., Jack, C. R., Jr., Scheltens, P., & Thompson, P. M. (2010). The clinical use of structural MRI in Alzheimer disease. *Nature Reviews Neurology, 6*(2), 67.

Huisman, T. A. G. M. (2010). Diffusion-weighted and diffusion tensor imaging of the brain, made easy. *Cancer Imaging, 10*(1A), S163.

Jack, C. R., Jr., Knopman, D. S., Jagust, W. J., Shaw, L. M., Aisen, P. S., Weiner, M. W., et al. (2010). Hypothetical model of dynamic biomarkers of the Alzheimer's pathological cascade. *The Lancet Neurology, 9*(1), 119–128.

Kriegeskorte, N., Simmons, W. K., Bellgowan, P. S., & Baker, C. I. (2009). Circular analysis in systems neuroscience: The dangers of double dipping. *Nature Neuroscience, 12*(5), 535.

Landman, B. A., Farrell, J. A., Jones, C. K., Smith, S. A., Prince, J. L., & Mori, S. (2007). Effects of diffusion weighting schemes on the reproducibility of DTI-derived fractional anisotropy, mean diffusivity, and principal eigenvector measurements at 1.5 T. *Neuroimage, 36*(4), 1123–1138.

Le Bihan, D., Mangin, J. F., Poupon, C., Clark, C. A., Pappata, S., Molko, N., et al. (2001). Diffusion tensor imaging: concepts and applications. *Journal of Magnetic Resonance Imaging, 13*(4), 534–546.

LeCun, Y., Bengio, Y., & Hinton, G. (2015). Deep learning. *Nature, 521*(7553), 436.

Leung, K. K., Barnes, J., Ridgway, G. R., Bartlett, J. W., Clarkson, M. J., Macdonald, K., et al. (2010). Automated cross-sectional and longitudinal hippocampal volume measurement in mild cognitive impairment and Alzheimer's disease. *Neuroimage, 51*(4), 1345–1359.

Lo, C. Y., Wang, P. N., Chou, K. H., Wang, J., He, Y., & Lin, C. P. (2010). Diffusion tensor tractography reveals abnormal topological organization in structural cortical networks in Alzheimer's disease. *Journal of Neuroscience, 30*(50), 16876–16885.

Maggipinto, T., Bellotti, R., Amoroso, N., Diacono, D., Donvito, G., Lella, E., et al. (2017). DTI measurements for Alzheimer's classification. *Physics in Medicine and Biology, 62*(6), 2361.

Mesrob, L., Sarazin, M., Hahn-Barma, V., de Souza, L. C., Dubois, B., Gallinari, P., et al. (2012). DTI and structural MRI classification in Alzheimer's disease. *Advances in Molecular Imaging, 2*(02), 12.

Ouyang, X., Chen, K., Yao, L., Wu, X., Zhang, J., Li, K., et al. (2015). Independent component analysis-based identification of covariance patterns of microstructural white matter damage in Alzheimer's disease. *PloS One, 10*(3), e0119714.

Özarslan, E., Shepherd, T. M., Vemuri, B. C., Blackband, S. J., & Mareci, T. H. (2006). Resolution of complex tissue microarchitecture using the diffusion orientation transform (DOT). *Neuroimage, 31*(3), 1086–1103.

Perkel, J. M. (2013). Life science technologies: This is your brain: Mapping the connectome. *Science, 339*(6117), 350–352.

Prasad, G., Joshi, S. H., Nir, T. M., Toga, A. W., & Thompson, P. M. (2015). Brain connectivity and novel network measures for Alzheimer's disease classification. *Neurobiology of Aging, 36*, S121–S131.

Roher, A. E., Weiss, N., Kokjohn, T. A., Kuo, Y. M., Kalback, W., Anthony, J., et al. (2002). Increased Aβ peptides and reduced cholesterol and myelin proteins characterize white matter degeneration in Alzheimer's disease. *Biochemistry, 41*(37), 11080–11090.

Rubinov, M., & Sporns, O. (2010). Complex network measures of brain connectivity: Uses and interpretations. *Neuroimage, 52*(3), 1059–1069.

Schaefer, P. W., Grant, P. E., & Gonzalez, R. G. (2000). Diffusion-weighted MR imaging of the brain. *Radiology, 217*(2), 331–345.

Schouten, T. M., Koini, M., de Vos, F., Seiler, S., de Rooij, M., Lechner, A., et al. (2017). Individual classification of Alzheimer's disease with diffusion magnetic resonance imaging. *Neuroimage, 152*, 476—481.

Schouten, T. M., Koini, M., de Vos, F., Seiler, S., van der Grond, J., Lechner, A., et al. (2016). Combining anatomical, diffusion, and resting state functional magnetic resonance imaging for individual classification of mild and moderate Alzheimer's disease. *Neuroimage: Clinical, 11*, 46—51.

Soman, S., Prasad, G., Hitchner, E., Massaband, P., Moseley, M. E., Zhou, W., et al. (2016). Brain structural connectivity distinguishes patients at risk for cognitive decline after carotid interventions. *Human Brain Mapping, 37*(6), 2185—2194.

Spulber, G., Simmons, A., Muehlboeck, J. S., Mecocci, P., Vellas, B., Tsolaki, M., et al. (2013). An MRI-based index to measure the severity of Alzheimer's disease-like structural pattern in subjects with mild cognitive impairment. *Journal of internal medicine, 273*(4), 396—409.

Tangaro, S., Fanizzi, A., Amoroso, N., & Bellotti, R. (2017). A fuzzy-based system reveals Alzheimer's Disease onset in subjects with Mild Cognitive Impairment. *Physica Medica: European Journal of Medical Physics, 38*, 36—44.

Thomas, C., Frank, Q. Y., Irfanoglu, M. O., Modi, P., Saleem, K. S., Leopold, D. A., et al. (2014). Anatomical accuracy of brain connections derived from diffusion MRI tractography is inherently limited. *Proceedings of the National Academy of Sciences, 111*(46), 16574—16579.

Tijms, B. M., Seriès, P., Willshaw, D. J., & Lawrie, S. M. (2012). Similarity-based extraction of individual networks from gray matter MRI scans. *Cerebral Cortex, 22*(7), 1530—1541.

Tuch, D. S. (2004). Q-ball imaging. *Magnetic Resonance in Medicine, 52*(6), 1358—1372.

de Vos, F., Schouten, T. M., Hafkemeijer, A., Dopper, E. G., van Swieten, J. C., de Rooij, M., et al. (2016). Combining multiple anatomical MRI measures improves Alzheimer's disease classification. *Human Brain Mapping, 37*(5), 1920—1929.

Zhang, Y., Wang, S., & Dong, Z. (2014). Classification of Alzheimer disease based on structural magnetic resonance imaging by kernel support vector machine decision tree. *Progress In Electromagnetics Research, 144*, 171—184.

Zhan, L., Zhou, J., Wang, Y., Jin, Y., Jahanshad, N., Prasad, G., et al. (2015). Comparison of nine tractography algorithms for detecting abnormal structural brain networks in Alzheimer's disease. *Frontiers in Aging Neuroscience, 7*, 48.

CHAPTER 21

Transcranial magnetic stimulation in the cortical exploration of dementia

Thanuja Dharmadasa[1], William Huynh[1,2], Matthew C. Kiernan[3]
[1]Brain and Mind Centre, The University of Sydney, Sydney, NSW, Australia; [2]Prince of Wales Hospital, Randwick, NSW, Australia; [3]Bushell Chair of Neurology Department of Neurology Royal Prince Alfred Hospital, Sydney, NSW, Australia

List of abbreviations

AD Alzheimer's disease
ALS amyotrophic lateral sclerosis
bvFTD behavioral variant frontotemporal dementia
CBD corticobasal degeneration
CS conditioning stimulus
CSP cortical silent period
CST corticospinal tract
D direct wave
DLPFC dorsolateral prefrontal cortex
EMG electromyography
FTLD frontotemporal lobar degeneration
I indirect wave
ICF intracortical facilitation
IFG inferior frontal gyrus
IHI interhemispheric inhibition
ISI interstimulus interval
iSP ipsilateral silent period
LICI long-interval intracortical inhibition
M1 primary motor cortex
MCI mild cognitive impairment
MEP motor-evoked potential
MT motor threshold
PD Parkinson's disease
PNFA progressive nonfluent aphasia
PSP progressive supranuclear palsy
RMT resting motor threshold
SD semantic dementia
SICI short-interval intracortical inhibition
TDP43 TAR-DNA binding protein 43
TMS transcranial magnetic stimulation
TS test stimulus
TT-TMS threshold tracking transcranial magnetic stimulation

Diagnosis and Management in Dementia
ISBN 978-0-12-815854-8, https://doi.org/10.1016/B978-0-12-815854-8.00021-5

Mini-dictionary of terms

Transcranial magnetic stimulation (TMS) a noninvasive technique that uses a brief magnetic pulse (generated via a coil placed on the scalp) to induce electric current flow in a targeted region of the human brain

Single-pulse TMS uses a single-pulse of TMS to generate electrical activity in the cortex

Paired-pulse TMS describes the TMS method that delivers two successive magnetic pulses to the same area separated by a short time interval; this method is mainly used to explore intracortical networks (inhibitory or excitatory)

Cortical relates to the outer layer of the cerebrum (mainly gray matter), the cerebral cortex

Subcortical relates to the region immediately below the cortex, mostly associated with structures within the white matter (such as the thalamus, basal ganglia, and brain stem)

Primary motor cortex the key region of the motor cortex responsible for voluntary movement and the main brain area targeted in diagnostic TMS protocols

Cortical excitability describes the level of neuronal excitability (either inhibited or excited) within the brain and is mainly determined by interactions between cellular receptors and neurotransmitters

Introduction

Early recognition, diagnosis, and differentiation among dementia subtypes continues to rely primarily on clinical evaluation and remains challenging. This has accelerated the search to better understand the biological basis of disease, reinforcing the need for objective biomarkers to facilitate screening, diagnosis, and treatment (Sosa–Ortiz, Acosta–Castillo, & Prince, 2012). In this context, neurophysiological techniques such as transcranial magnetic stimulation (TMS) have offered an opportunity for functional assessment of the cerebral cortex as a valuable tool in the understanding of primary and secondary dementias.

Since its original description more than three decades ago (Barker, Jalinous, & Freeston, 1985), TMS has undergone significant evolution as a noninvasive neurostimulation and neuromodulation technique, providing precious insight into the functional integrity of brain pathways. Its main application has been investigation of the complex neuronal networks of the primary motor cortex (M1), which is influenced by both inhibitory and excitatory mechanisms (Vucic & Kiernan, 2017). TMS studies have supported the clinical, neuroimaging, and neuropathological involvement of motor networks in dementia (Filippi et al., 2012), unmasking aspects of the neurobiochemical milieu to differentiate the glutamatergic, GABAergic, and cholinergic contribution to specific disease subtypes. This chapter will present an overview of the various TMS techniques and cortical parameters that have been instrumental to understanding dementia, focusing on insights into the pathophysiological and biological networks of disease across cortical, cortico-subcortical, and subcortical dementias classified according to pathology.

Transcranial magnetic stimulation: principles and techniques

The technique of TMS uses a transient magnetic field to induce an electric current in the cortex (Rossini et al., 2015) according to Faraday's law of electromagnetic induction. This magnetic field is generated through a stimulating coil held over a subject's head that painlessly and noninvasively penetrates the skull without attenuation (Fig. 21.1). Depending on stimulation intensity and coil type used, the electromagnetic force can stimulate neurons at 1.5—3.0 cm beneath the scalp (Roth & Basser, 1990). Several theoretical models have been postulated to explain the exact effect of this electromagnetic field on biological tissue, with studies in both animals and humans conferring that TMS generates a corticomotoneuronal volley of direct (D) and indirect (I) waves occurring at intervals of 1.5—2.5 ms (Di Lazzaro et al., 1998). D waves are thought to represent the activation of corticospinal axons and are only recruited at high intensities, while I waves seem to be activated at lower intensities and are mediated by a more complex interaction between cortical excitatory and inhibitory neurons (Di Lazzaro et al., 2012). TMS delivered over the primary motor cortex is thought to activate pyramidal

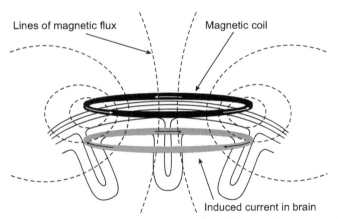

Figure 21.1 *Magnetic coil current flow direction and induced electrical current in the brain.* A magnetic impulse is produced in the magnetic coil *(dark gray circle)*, which is placed tangentially to the scalp surface (with the direction of current flow in the coil shown as the *white arrow* within the *dark gray circle*). This produces a magnetic field with lines of flux *(dashed lines* of magnetic flux) perpendicular to the plane of the coil. A perpendicular electrical field is induced, generating electrical current in the brain *(light gray circle; arrow shows direction)*. *(Reproduced with permission Rossini, P., Burke, D., Chen, R., Cohen, L., Daskalakis, Z., Di Iorio, R., et al. (2015). Non-invasive electrical and magnetic stimulation of the brain, spinal cord, roots and peripheral nerves: Basic principles and procedures for routine clinical and research application. an updated report from an I.F.C.N. Committee.* Clinical Neurophysiology, 126, *1071—1107.)*

neurons (Betz cells) transsynaptically via I waves (Chen et al., 2008), but the exact neural circuitries evoked remain to be determined. Significantly, modeling studies have suggested a simultaneous activation of subcortical white matter, which may be of particular influence due to the strong connections between white matter cortico-cortical axons to total corticospinal output (Laakso, Hirata, & Ugawa, 2014). This is of clear relevance in the neurophysiological evaluation of the complex cortical networks in dementia, in which connectivity across motor and nonmotor areas remains critical to disease manifestation. These complex neural circuits are critically dependent on both excitatory and inhibitory interneuronal systems facilitated by cellular receptor and neurotransmitter interactions (Robinson, 1992). Excitation is primarily mediated by glutamate/NMDA receptor interaction, while inhibition is facilitated by γ-aminobutyric acid (GABA)/GABA$_{A/B}$ receptor action (Ziemann, 2003).

Single-pulse transcranial magnetic stimulation
Motor threshold
Motor threshold (MT) is influenced by the glutamatergic system and is a global measure of corticospinal neuronal membrane excitability, representing the overall ease by which corticomotoneurons are excited. Resting motor threshold (RMT) is determined by the lowest intensity of motor cortex stimulation necessary to produce a motor-evoked potential (MEP) amplitude in a target muscle (Rossini et al., 2015). Typically, lower MTs are generated by a higher density of corticomotoneuronal projections onto motor neurons (such as in the cortical hand region) and/or in the presence of excessive glutamate activity (Ziemann, 2003).

Motor-evoked potential
The MEP is generated following a single pulse of TMS delivered to the primary motor cortex, with the response recorded using surface electrodes attached to the contralateral target muscle at rest (e.g., the abductor pollicis muscle of the hand). MEPs display distinct physiological characteristics to those of the MT. First, MEPs assess cortical neurons that are comparatively distant from the TMS magnetic field of stimulation and are therefore typically less excitable, requiring a higher threshold (Chen et al., 2008). Furthermore, the modulation of MEP amplitude in response to changes in neurotransmission (i.e., suppression in response to GABAergic transmission and increases with glutamatergic and noradrenergic neurotransmission) occurs independently of MT changes, supporting the theory that separate biological mechanisms underpin the generation of these variables (Boroojerdi, Battaglia, Muellbacher, & Cohen, 2001; Ziemann, 2003).

Cortical silent period

The CSP represents a transient period of electrical silence that interrupts surface electromyographic (EMG) activity after a suprathreshold TMS stimulus is applied to the contralateral M1 during voluntary contraction of a target muscle. CSP duration lasts a few hundred milliseconds and increases with stimulus intensity. The underlying physiological mechanisms are a complex interaction of both spinal and intracortical M1 inhibitory processes, but the cortical contribution is the most predominant influence. CSP therefore essentially represents a measure of intracortical inhibition and changes in intracortical excitability, mediated mainly via GABA receptor transmission (Cantello, Gianelli, Civardi, & Mutani, 1992). Pharmacological studies have further suggested $GABA_A$ receptor activation when shorter CSPs are elicited (typically at lower stimulus intensities) and $GABA_B$ receptor activation with CSP durations of >100 ms (at higher stimulus intensities) (Rossini et al., 2015; Ziemann, 2003).

An ipsilateral silent period (iSP) can also be recorded when a TMS pulse is delivered ipsilaterally to M1, which typically has a shorter duration of approximately 30 ms compared with that of the CSP. This phenomenon is most likely conveyed through corpus callosal networks and is thought to reflect transcallosal inhibition (Ferbert et al., 1992).

Paired-pulse transcranial magnetic stimulation

Novel paradigms have been developed to facilitate exploration of inhibitory and excitatory interneuronal circuits within the cortex. The paired-pulse technique involves applying two TMS impulses—a subthreshold "conditioning" stimulus (CS) followed by a suprathreshold "test" stimulus (TS) that is modulated by the effect of the former (Kujirai et al., 1993). By varying the time interval between the pair of pulses (the interstimulus interval, ISI), a number of parameters can be determined (Table 21.1). This commonly includes short-interval intracortical inhibition (SICI), intracortical facilitation, and long-interval intracortical inhibition (LICI). These variables have been increasingly recognized as important outcome measures in several neurological conditions, including dementia, and can also be applied to study nonmotor areas.

Table 21.1 Paired-pulse transcranial magnetic stimulation parameters.

	Conditioning (first) stimulus	Test (second) stimulus	Interstimulus interval	Neurotransmitter
SICI	Subthreshold	Suprathreshold	1—7 ms	$GABA_A$
ICF	Subthreshold	Suprathreshold	10—30 ms	Glutamate, norepinephrine
SAI	Peripheral (median) nerve	Suprathreshold	19—50 ms	Acetylcholine

Summary of paired-pulse TMS methods applied to M1. *GABA*, γ-aminobutyric acid; *ICF*, intracortical facilitation; *ms*, milliseconds; *SAI*, short-latency afferent inhibition; *SICI*, short-interval intracortical inhibition.

Short-interval intracortical inhibition and intracortical facilitation

Two distinct phases have been identified at ISI≤1 ms and 3 ms using the threshold tracking TMS (TT-TMS) technique, with the 3 ms response mediated by postsynaptic inhibition via GABA$_A$ receptors (Di Lazzaro, Pilato, Dileone, Ranieri et al., 2006). The first phase remains less well defined but may partly relate to refractoriness of local cortical axons and synaptic activation of different cortical inhibitory circuits (Roshan, Paradiso, & Chen, 2003). Other neurotransmitter mechanisms can also regulate SICI, including dopamine agonists and noradrenergic antagonists (Ziemann, 2003). The exact physiological process underscoring ICF remains poorly understood, but neural circuits within the cerebral cortex distinct to those mediating SICI seem to guide this response (Di Lazzaro, Pilato, Dileone, Saturno et al., 2006).

Long-interval intracortical inhibition and interhemispheric inhibition

LICI is a separate cortical phenomenon probably mediated by GABA$_B$ receptors on inhibitory interneurons and occurring between an ISI of 50–200 ms (McDonnell, Orekhov, & Ziemann, 2006; Valls-Sole, Pascual-Leone, Wassermann, & Hallett, 1992). Specifically, it refers to the inhibition of the MEP response when a suprathreshold CS is delivered before the TS (Di Lazzaro, Oliviero, Mazzone, et al., 2002).

Transcallosal pathways and interhemisp7heric interactions can also be explored using paired-pulse paradigms by applying single stimuli sequentially to different regions of the brain. Commonly, a suprathreshold CS is delivered to M1 followed by a suprathreshold TS to the contralateral M1, which results in interhemispheric inhibition (IHI). IHI has been suggested to be mediated by postsynaptic GABA$_B$ receptors, and is largely produced by interhemispheric excitatory pathways through the corpus callosum and local inhibitory pathways in the target M1 (Ni et al., 2009). However, subcortical mechanisms and nonmotor areas are also contributory, and therefore IHI represents a more widespread network that extends to involves areas that include the dorsal premotor cortex, somatosensory cortex and dorsolateral prefrontal cortex (Ni et al., 2009).

Short-latency afferent inhibition

A TMS-induced MEP response can also be inhibited when preceded by a contralateral peripheral cutaneous stimulation at short latencies, a phenomenon referred to as short-latency afferent inhibition (SAI). The specific technique involves prestimulating a peripheral sensory nerve (e.g., median nerve) in the contralateral limb prior to delivering a TMS stimulus to M1, and recording from the corresponding muscle (e.g., abductor pollicis brevis) (Fig. 21.2). For median nerve stimulation, this inhibitory modulation of M1 commonly occurs at an ISI of 20 ms (Tokimura et al., 2000). Maximum inhibition is seen when the sensory impulse is delivered at the N20 latency of the somatosensory cortical potential evoked from that nerve (Bikmullina, Kicic, Carlson, & Nikulin, 2009). Pharmacological studies have supported that this inhibition is primarily due to primary involvement of central cholinergic activity (Di Lazzaro, Pilato, Dileone, Tonali, & Ziemann, 2005).

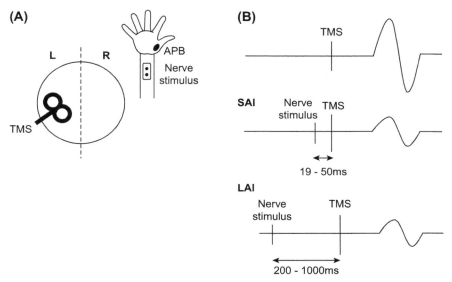

Figure 21.2 *Principles of short-afferent inhibition and long-afferent inhibition.* (A) The right median nerve is stimulated at the wrist (nerve stimulus) prior to generating a TMS pulse over the contralateral (left) M1. SAI is recorded via surface EMG electrodes placed over the target muscle (APB shown in diagram). (B) Top panel—an unconditioned MEP response produced by a single TMS pulse; middle panel: SAI induced by stimulating a peripheral nerve between ∼19 and 50 ms prior to the TMS pulse, inhibiting MEP response; bottom pane: LAI induced by stimulating a peripheral nerve between ∼200 and 1000 ms prior to the TMS pulse, inhibiting MEP response. *APB*, abductor pollicis brevis; *L*, left; *LAI*, long-afferent inhibition; *ms*, milliseconds; *R*, right; *SAI*, short-afferent inhibition; *TMS*, transcranial magnetic stimulation. *(Reproduced with permission Turco CV., El-Sayes J., Savoi MJ., Fassett HJ., Locke MB., Nelson AJ., Short-and long-latency afferent inhibition; uses, mechanisms and influencing factors, Brain Stimul. 11, 2018, 59-74.)*

Understanding dementia using transcranial magnetic stimulation: key diagnostic and pathophysiologic findings

TMS has allowed dementia subtypes to be segregated based on their neurobiological substrates, evidencing distinct pathogenic networks for each disease. The following section will describe the key findings for the common dementia subtypes. A simple diagnostic algorithm is also shown in Fig. 21.3, highlighting some of the main features discussed in the following sections. In addition, key abnormalities are summarized as listed in Table 21.2.

Cortical dementias

Alzheimer's disease

As the most common cause of primary dementia and with an estimated quadruple-fold rise in prevalence by 2050, the timely diagnosis and management of Alzheimer's disease (AD) is of critical importance (Brookmeyer, Johnson, Ziegler-Graham, & Arrighi, 2007). The clinical profile of AD remains central to understanding the regions of cortical

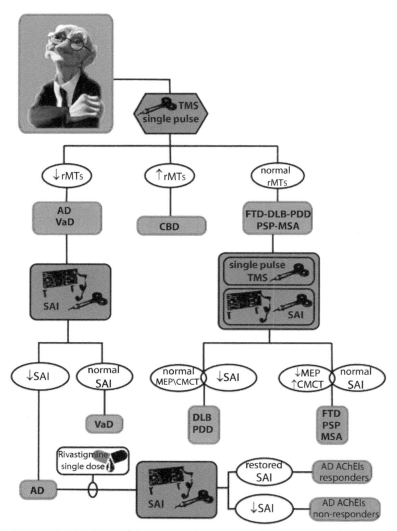

Figure 21.3 *Diagnostic algorithm of dementia subtypes guided by transcranial magnetic stimulation (TMS) abnormalities.* A simple diagnostic algorithm to differentiate key dementia subtypes based on TMS parameters. *AChEI*, acetylcholinesterase inhibitors; *AD*, Alzheimer's disease; *CBD*, corticobasal degeneration; *DLB*, dementia with Lewy bodies; *FTD*, frontotemporal dementia; *MSA*, multiple system atrophy; *PDD*, Parkinson's disease dementia; *PSP*, progressive supranuclear palsy; *RMT*, resting motor threshold; *VaD*, vascular dementia. *(Reproduced with permission Cantone, M., Di Pino, G., Capone, F., Piombo, M., Chiarello, D., Cheeran, B, et al. (2014). The contribution of transcranial magnetic stimulation in the diagnosis and in the management of dementia.* Clinical Neurophysiology, 125, *1509–1532.)*

Table 21.2 Summary of transcranial magnetic stimulation parameter abnormalities across common dementia subtypes.

	RMT (%)	MEP amplitude (%)	CSP duration (ms)	SICI (%)	ICF (%)	SAI (%)
Cortical dementias						
Alzheimer's disease	↓	↑	Normal	↓ (or normal)	Normal	↓
Frontotemporal dementia	Normal	Absent (or ↓)	Normal	↓	Normal	Normal
Cortico-subcortical dementias						
Dementia with Lewy bodies	Normal	Normal	N/A	Normal (or ↓)	N/A	↓
Corticobasal degeneration	↑	Normal	↓	↓	↓	–
Subcortical dementias						
Parkinson's disease dementia	Normal	Normal	↓	↓	Normal	↓
Progressive supranuclear palsy	Normal	↑	↓	↓	Normal	Normal

The main TMS abnormalities identified across the dementia subtypes. *CSP*, cortical silent period; *MEP*, motor-evoked potential; *N/A*, not assessed; *RMT*, resting motor threshold; *SAI*, short-latency afferent inhibition; *SICI*, short-interval intracortical inhibition.

involvement during disease progression. Earlier stages of the disease are typically heralded by nonmotor symptoms such as short-term memory loss, language deficits, and mood and behavioral changes, consistent with the histopathological changes predominantly reported in nonmotor areas (such as the hippocampus, the entorhinal and cingulate cortices, and the medial temporal lobe). More advanced stages, however, show clear involvement of motor pathways, indicated clinically by symptoms such as gait impairment, hypokinesia, and rigidity (Fischer et al., 2007) and consistent with late evidence of M1 involvement histopathologically (Suva et al., 1999). Imaging studies in some patients with a mild cognitive impairment (MCI) subtype have also shown M1 volumetric change, which has been associated with a deterioration of gait (Annweiler et al., 2013). Furthermore, observation of altered corticomotor connectivity systems in early disease cements the motor pathway as an integral network underpinning the progressive neurodegenerative changes that unfold.

   TMS studies of M1 have suggested that three main neurophysiological changes occur in AD: (one) a global increase in motor cortex excitability with reduction of RMT, (two) an impairment in the cholinergic system with reduction of SAI, and (three) early involvement of CC pathways, with abnormalities in iSP (Cantone et al., 2014). In the first instance, RMT changes seem to occur in a bimodal pattern with an initial reduction in early disease, suggesting a process of cortical excitability, followed by a gradual increase in more advanced stages, consistent with neuronal degeneration in the motor cortex (Pennisi et al., 2002). This pattern of dysfunction has also shown persistence despite institution of pharmacological treatment (Pennisi et al., 2002), potentially suggesting an intrinsic cortical network failure. Further support for cortical hyperexcitability has been implied through studies showing a reduction of SICI in AD, but this finding has been inconsistent across the literature, with many studies conversely reporting normal SICI (Hoeppner et al., 2012). Parameters such as ICF and CSP are also reportedly normal in this disease (Di Lazzaro, Oliviero, Tonali, et al., 2002).

Dysfunction of central cholinergic systems in AD has been implicated in several TMS studies over the last decade that demonstrate a reduction in SAI (Cantone et al., 2014). For patients with mild cognitive impairment, these SAI changes are only significant in the amnestic subgroup, highlighting its potential as a biomarker to identify the subgroup at risk of developing dementia. Patients with AD also show normalization of SAI after administration of anticholinesterase agents (such as rivastigmine). Such findings emphasize the prognostic potential of SAI and its possible utility as a therapeutic marker. Lastly, abnormalities in iSP (i.e., delayed onset and increased duration) (Hoeppner et al., 2012) have provided important evidence for the early functional involvement of the corpus callosum and transcallosal pathways in AD, supported by structural DTI studies (Fjell & Walhovd, 2010).

Frontotemporal lobar degeneration

Frontotemporal lobar degeneration (FTLD) is a group of heterogenous disorders that includes three main clinical syndromes: (one) behavioral variant frontotemporal degeneration (bvFTD), the most common subtype accounting for more than 50% of cases, (two) progressive nonfluent aphasia (PNFA), and (three) semantic dementia (SD) (Mann & Snowden, 2017). In keeping with this variable clinical spectrum, findings from TMS and neuropathological studies have revealed a heterogeneity of cortical dysfunction, underlying complex pathophysiological mechanisms. The involvement of the motor system has been highlighted by the clinical, genetic, and histopathologic association of FTLD with primary motor disorders such as Parkinson's disease (PD) and amyotrophic lateral sclerosis (ALS). Moreover, FTLD and ALS are now recognized to exist as part of a disease continuum (Burrell et al., 2016; Dharmadasa et al., 2017) sharing a common genetic mutation (C9ORF72) (Snowden et al., 2013), characteristic TAR-DNA binding protein 43 (TDP43) neuropathological inclusions (Neumann et al., 2006) and overlapping imaging abnormalities (Dharmadasa et al., 2018; Lillo et al., 2012). Although ALS primarily represents a motor systems failure involving both upper and lower motor neurons, cognitive dysfunction is now recognized to occur in up to 50% of patients, with 10%—15% meeting the criteria for FTLD (Dharmadasa et al., 2017). Similarly, a large proportion of FTLD patients develop motor dysfunction during their disease, with approximately 12% meeting criteria for ALS (Burrell, Kiernan, Vucic, & Hodges, 2011). Paired-pulse TMS techniques suggest cortical hyperexcitability as a key feature in ALS patients, specifically marked by a significant reduction or absence in SICI that is postulated to occur via glutamate-mediated excitotoxic mechanisms (Vucic, Ziemann, Eisen, Hallett, & Kiernan, 2013).These TMS abnormalities can be identified early in the disease, prior to clinical signs of upper motor neuron dysfunction, and are able to reliably differentiate ALS from disease mimics (Menon et al., 2015; Vucic, & Kiernan, 2006).

The clear involvement of motor pathways in FTLD may signify pathological spread of neurodegeneration across central networks, starting from the frontal cortex and extending to M1 (Burrell et al., 2011). TMS studies support abnormalities of central motor circuitries even in patients without clinical features of motor dysfunction, typified by changes in MEP (absent or reduced amplitude with an increased latency) and prolongation of CMCT (Di Lazzaro, Pilato, Dileone, Saturno, et al., 2006). These findings are predominantly apparent in bvFTD and SD subgroups (Burrell et al., 2011). A reduction of SICI has also been reported in patients with primary progressive aphasia, accompanied by some lower motor neuron change in keeping with an ALS overlap, but to a lesser degree than that seen in classic ALS. No changes in SICI have been found in the other subgroups of FTLD, and no changes in CSP or MT have been reported in FTLD overall (Pierantozzi et al., 2004), suggesting that although there is a degree of overlap with ALS, the underlying FTLD neural network is cortically distinct. Furthermore, the lack of central cholinergic activity changes (as evidenced by a normal SAI) and demonstration of normal MT and ICF additionally support a network distinct from that of AD (Pierantozzi et al., 2004).

Cortico-subcortical dementias
Dementia with lewy bodies

Dementia with Lewy bodies (LBD) is an alpha-synucleinopathy with the characteristic clinical features of impaired attention, fluctuating mental state, complex hallucinations, and loss of visuospatial ability (McKeith et al., 2005), commonly associated with the motor manifestations of parkinsonism. The major finding from the exploration of the LBD motor cortex via TMS is that of central cholinergic deficits, as suggested by a reduction of SAI (Di Lazzaro et al., 2007). Additionally, patients with more significant reductions in cholinergic activity are more likely to have visual hallucinations (Marra et al., 2012), with the severity of such symptoms also correlating with the degree of cortical excitability in the visual cortex (Taylor et al., 2011).

Corticobasal degeneration

Marked TMS abnormalities have been described in the rarer neurodegenerative syndrome of corticobasal degeneration (CBD), in which patients develop a frontal-type dementia during disease evolution. At clinical onset, CBD typically manifests as a unilateral "alien limb", characterized by upper limb apraxia and akinetic hypertonia and supported by a lack of dopamine responsiveness (Grijalvo-Perez, & Litvan, 2014). TMS studies of the motor cortex have revealed an underlying global inhibitory dysfunction demonstrated by an increase in MT and reduction of SICI and CSP duration (Burrell, Hornberger, Vucic, Kiernan, & Hodges, 2014). There has also been an absence of iSP reported, suggesting a functional impairment of transcallosal pathways (Kuhn et al., 2004) and supporting the pathologic basis for global network dysfunction.

Subcortical dementias

Parkinson's disease dementia

Although PD is predominated by motor manifestations, features such as cognitive impairment and behavioral and personality changes are now well recognized (Jellinger, 2012). Several TMS studies in PD patients with a degree of cognitive impairment have shown dramatic decreases in SAI corresponding to the clinically affected side (Celebi, Temucin, Elibol, & Saka, 2012), suggesting that development of cognitive dysfunction is secondary to cholinergic system degeneration. Conversely, such findings on TMS may potentially be useful in identifying the subgroup of PD patients at risk of developing dementia, but this needs further investigation. Other abnormalities described using TMS have implied an impairment of intracortical inhibition, with reduction in SICI, CSP, and iSP corresponding to the clinically affected side (Mackinnon, Gilley, Weis-McNulty, & Simuni, 2005). Such changes in the motor cortical networks of PD may represent the neurobiological processes of disease, which is supported by imaging findings of widespread atrophy as well as the changes to motor cortex metabolism secondary to dopamine depletion (Lindenbach, & Bishop, 2013).

Progressive supranuclear palsy

Progressive supranuclear palsy (PSP) is commonly misdiagnosed for PD due to clinical similarities at early onset that include akinetic rigidity, gait disturbances, and dystonia (Ling, 2016). Impairment of voluntary eye movements is a key differentiating feature among these patients. Distinct to PD, TMS studies have pointed to impairment of callosal function as the major pathological process in PSP, as evidenced by prolongation of iSP (Kuhn et al., 2004; Whittstock et al., 2013). Reduced intracortical M1 inhibition also features as a key mechanism in both pathologies, evidenced by large MEP amplitudes, reduced SICI, and reduced CSP correlating with disease progression (Kuhn et al., 2004).

Conclusion

The utility of TMS in exploration of the dementia spectrum has provided critical insights into the neuropathological basis of these common diseases. Specific abnormalities of TMS parameters have enabled a dynamic understanding of the complex cortical networks involved in the pathophysiological process of dementia subtypes and of the neurochemical and neurobiological landscape that differentiates them. Recognition of the earliest cortical changes using this tool may also enable more accurate and timely diagnoses of specific stages or subtypes of dementia. Furthermore, the potential for presymptomatic identification of "at risk" patients greatly increases the potential efficacy of earlier therapeutic interventions. With an accelerating clinical trials pipeline in dementia, TMS thus remains a pivotal tool that can aid in the accurate diagnosis, differentiation, and biological monitoring of therapeutic agents. The development of potential cortical biomarkers would be a powerful adjunct to enable targeted treatment strategies in subtypes.

Key facts of transcranial magnetic stimulation

- Transcranial magnetic stimulation (TMS) is a noninvasive technique that can assess pathways of the central nervous system in vivo including the brain, spinal cord, and nerve roots.
- TMS uses the principle of Faraday's law of electromagnetic induction to generate electrical current in the brain using painless magnetic fields. It is predominantly used to stimulate the motor cortex and motor pathways and to assess complex intracortical networks.
- Prior to TMS, investigation of the cortex occurred through transcranial electrical stimulation techniques (TES), which involved painful high-voltage electrical stimulation through the scalp.
- TMS was first introduced in 1985 due to the intolerance of TES and other electrical methods. Guidelines for the use of TMS were first published more than 20 years ago (1994).
- TMS technique requires a storing capacitor to generate a magnetic field in addition to a stimulating coil. Different coils have been developed to look at specific brain regions of interest, and some can stimulate deep brain targets at a depth of up to approximately 3 cm.
- Both TES and TMS stimulate axons (not cell bodies) of neurons. TMS stimulation of the motor cortex activates neurons both directly and indirectly through cortical networks. The exact circuits that are activated remain to be fully elucidated.
- TMS outcome parameters of the motor cortex are now well established and can identify central motor pathway involvement early in disease. As such, TMS has been shown to be of prognostic and diagnostic significance and has particularly expanded the understanding of neurodegenerative conditions such as amyotrophic lateral sclerosis.

Summary points

- This chapter focuses on the use of transcranial magnetic stimulation (TMS) applied to the primary motor cortex of the brain to further unmask the pathophysiology of dementia subtypes.

- Using TMS, Alzheimer's disease has been characterized by global motor cortex excitability, impairment of central cholinergic systems, and early involvement of the corpus callosal pathways.

- TMS studies of Frontotemporal lobar degeneration have evidenced heterogenous abnormalities of central motor networks but lacking central cholinergic involvement, in distinct contrast to AD.

- The major TMS finding in dementia with Lewy bodies is visual cortex excitability and central cholinergic deficits, with the degree of deficit correlating with the likelihood of developing visual hallucinations.

- Corticobasal degeneration is a rare syndrome, and TMS studies show underlying global inhibitory dysfunction through the involvement of transcallosal networks.

- The development of cognitive changes in Parkinson's disease has been linked to cholinergic system degeneration using TMS. This remains to be further validated.

- TMS in the progressive supranuclear palsy subtype demonstrates impairment of corpus callosal function as its major pathological process, distinguishing it from the clinically similar Parkinson's disease.

References

Annweiler, C., Beauchet, O., Bartha, R., Wells, J., Borrie, J., & Hachinski, V. (2013). Motor cortex and gait in mild cognitive impairment: A magnetic resonance spectroscopy and volumetric imaging study. *Brain, 136*, 859–871.

Barker, A., Jalinous, R., & Freeston, I. (1985). Non-invasive magnetic stimulation of human motor cortex. *The Lancet, 1*, 1106–1107.

Bikmullina, R., Kicic, D., Carlson, S., & Nikulin, V. (2009). Sensory afferent inhibition within and between limbs in humans. *Clinical Neurophysiology, 120*, 610–618.

Boroojerdi, B., Battaglia, F., Muellbacher, W., & Cohen, L. (2001). Mechanisms influencing stimulus-response properties of the human corticospinal system. *Clinical Neurophysiology, 112*, 931–937.

Brookmeyer, R., Johnson, E., Ziegler-Graham, K., & Arrighi, H. (2007). Forecasting the global burden of Alzheimer's disease. *Alzheimer's Dementia, 3*, 186–191.

Burrell, J. R., Halliday, G. M., Kril, J. J., Ittner, L. M., Gotz, J., Kiernan, M. C., et al. (2016). The frontotemporal dementia-motor neuron disease continuum. *The Lancet, 388*(10047), 919–931. https://doi.org/10.1016/S0140-6736(16)00737-6.

Burrell, J., Hornberger, M., Vucic, S., Kiernan, M., & Hodges, J. (2014). Apraxia and motor dysfunction in corticobasal syndrome. *PLoS, 9*.

Burrell, J., Kiernan, M., Vucic, S., & Hodges, J. (2011). Motor neuron dysfunction in frontotemporal dementia. *Brain, 134*, 13.

Cantello, R., Gianelli, M., Civardi, C., & Mutani, R. (1992). Magnetic brain stimulation: The silent period after the motor evoked potential. *Neurology, 42*, 1951–1959.

Cantone, M., Di Pino, G., Capone, F., Piombo, M., Chiarello, D., Cheeran, B., et al. (2014). The contribution of transcranial magnetic stimulation in the diagnosis and in the management of dementia. *Clinical Neurophysiology, 125*, 1509−1532.

Celebi, O., Temucin, C., Elibol, B., & Saka, E. (2012). Short latency afferent inhibition in Parkinson's disease patients with dementia. *Movement Disorders, 27*, 1052−1055.

Chen, R., Cros, D., Curra, A., Di Lazzaro, V., Lefaucheur, J., Magistris, M., et al. (2008). The clinical diagnostic utility of transcranial magnetic stimulation: Report of an IFCN committee. *Clinical Neurophysiology, 119*, 504−532.

Dharmadasa, T., Henderson, R., Talman, P., Macdonell, R., Mathers, S., Schultz, D., et al. (2017). Motor neurone disease: Progress and challenges. *Medical Journal of Australia, 206*(8), 357−362.

Dharmadasa, T., Huynh, W., Tsugawa, J., Shimatani, Y., Ma, Y., & Kiernan, M. (2018). Implications of structural and functional brain changes in amyotrophic lateral sclerosis. *Expert Rev Neurother, 18*, 407−419.

Di Lazzaro, V., Oliviero, A., Profice, P., Saturno, E., Pilato, F., Insola, A., et al. (1998). Comparison of descending volleys evoked by transcranial magnetic and electric stimulation in conscious humans. *Electroencephalography and Clinical Neurophysiology, 109*, 5.

Di Lazzaro, V., Oliviero, A., Mazzone, P., Pilato, F., Saturno, E., Insola, A., et al. (2002). Direct demonstration of long latency cortico-cortical inhibition in normal subjects and in a patient with vascular parkinsonism. *Clinical Neurophysiology, 113*, 1673−1679.

DiLazzaro, V., Oliviero, A., Tonali, P., Marra, C., Daniele, A., Profice, P., et al. (2002). Noninvasive in vivo assessment of cholinergic cortical circuits in AD using transcranial magnetic stimulation. *Neurology, 59*, 392−397.

Di Lazzaro, V., Pilato, F., Dileone, M., Ranieri, F., Ricci, V., Profice, P., et al. (2006). GABAA receptor subtype specific enhancement of inhibition in human motor cortex. *The Journal of Physiology, 575*, 721−726.

Di Lazzaro, V., Pilato, F., Dileone, M., Saturno, E., Oliviero, A., & Marra, C. (2006). In vivo cholinergic circuit evaluation in frontotemporal and Alzheimer's dementias. *Neurology, 66*, 1111−1113.

Di Lazzaro, V., Pilato, F., Dileone, M., Saturno, E., Profice, P., & Marra, C. (2007). Functional evaluation of cerebral cortex in dementia with lewy bodies. *Neuroimage, 37*, 422−429.

Di Lazzaro, V., Pilato, F., Dileone, M., Tonali, P., & Ziemann, U. (2005). Dissociated effects of diazepam and lorazepam on short latency afferent inhibition. *The Journal of Physiology, 569*, 315−323.

Di Lazzaro, V., Profice, P., Ranieri, F., Capone, F., Dileone, M., Oliviero, A., et al. (2012). I-wave origin and modulation. *Brain Stimulation, 5*, 512−525.

Ferbert, A., Priori, A., Rothwell, J., Day, B., Colebatch, J., & Marsden, C. (1992). Interhemispheric inhibition of the human motor cortex. *The Journal of Physiology, 453*, 525−546.

Filippi, M., Agosta, F., Barkhof, F., Dubois, B., Fox, N., Frisoni, G., et al. (2012). The use of neuroimaging in the diagnosis of dementia. *European Journal of Neurology, 19*, 1487−1501.

Fischer, P., Jungwirth, S., Zehetmayer, S., Weissgram, S., Hoenigschnable, S., Gelpi, E., et al. (2007). Conversion from subtypes of mild cognitive impairment to Alzheimer dementia. *Neurology, 68*, 288−291.

Fjell, A., & Walhovd, K. (2010). Structural brain changes in aging: courses, causes and cogntiive consequences. *Reviews in the Neurosciences, 21*, 187−222.

Grijalvo-Perez, A., & Litvan, I. (2014). Corticobasal degeneration. *Seminars in Neurology, 34*, 160−173.

Hoeppner, J., Wegrzyn, M., Thome, J., Bauer, A., Oltmann, I., Buchmann, J., et al. (2012). Intra- and intercortical motor excitabiltiy in Alzheimer's disease. *Journal of Neural Transmission, 119*, 605−612.

Jellinger, K. (2012). Neurobiology of cognitive impairment in Parkinson's disease. *Expert Review of Neurotherapeutics, 12*, 1451−1466.

Kuhn, A., Grosse, P., Holtz, K., Brown, P., Meyer, B., & Kupsh, A. (2004). Patterns of abnormal motor cortex excitability in atypical parkinsonian syndromes. *Clinical Neurophysiology, 115*, 1786−1795.

Kujirai, T., Caramia, M., Rothwell, J., Day, B., Thompson, P., Ferbert, A., et al. (1993). Corticocortical inhibition in human motor cortex. *The Journal of Physiology, 471*, 501−519.

Laakso, I., Hirata, A., & Ugawa, Y. (2014). Effects of coil orientation on the electric field induced by TMS over hte hand motor area. *Physics in Medicine and Biology, 59*, 203−218.

Lillo, P., Mioshi, E., Burrell, J., Kiernan, M., Hodges, J., & Hornberger, M. (2012). Grey and white matter changes across the amyotrophic lateral sclerosis-frontotemporal dementia continuum. *PLoS One, 7,* e43993.

Lindenbach, D., & Bishop, C. (2013). Critical involvement of the motor cortex in the pathophysiology and treatment of Parkinson's disease. *Neuroscience and Biobehavioral Reviews, 37,* 2737—2750.

Ling, H. (2016). Clinical approach to progressive supranuclear palsy. *Journal of Movement Disorders, 9,* 3—13.

Mackinnon, C., Gilley, E., Weis-McNulty, A., & Simuni, T. (2005). Pathways mediating abnormal intracortical inhibition in Parkinson's disease. *Annals of Neurology, 58,* 516—524.

Mann, D., & Snowden, J. (2017). Frontotemporal lobar degeneration: Pathogenesis, pathology and pathways to phenotype. *Brain Pathology, 27*(6), 723—736.

Marra, C., Quaranta, D., Profice, P., Pilato, F., Capone, F., Iodice, F., et al. (2012). Central cholinergic dysfunction measured "in vivo" correlates with different behavioral disorders in Alzheimer's disease and dementia with Lewy body. *Brain Stimulation, 5,* 533—538.

McDonnell, M., Orekhov, Y., & Ziemann, U. (2006). The role of GABAB receptors in intracortical inhibition in the human motor cortex. *Experimental Brain Research, 173,* 86—93.

McKeith, I., Dickson, D., Lowe, J., Emre, M., O'Brien, J., & Feldman, H. (2005). Diagnosis and management of dementia with Lewy bodies: Third report of the DLB consortium. *Neurology, 65,* 1863—1872.

Menon, P., Geevasinga, N., Yiannikas, C., Howells, J., Kiernan, M., & Vucic, S. (2015). Sensitivity and specificity of threshold tracking transcranial magnetic stimulation for diagnosis of amyotrophic lateral sclerosis: a prospective study. *The Lancet Neurology, 14*(5), 7.

Neumann, M., Sampathu, D., Kwong, L., Truax, A., Micsenyi, M., Chou, T., et al. (2006). Ubiquitinated TDP-43 in frontotemporal lobar degeneration and amyotrophic lateral sclerosis. *Science, 314,* 130—133.

Ni, Z., Gunraj, C., Nelson, A., Yeh, I., Castillo, G., Hoque, T., et al. (2009). Two phases of interhemispheric inhibition between motor related cortical areas and the primary motor cortex in human. *Cerebral Cortex, 19,* 1654—1655.

Pennisi, G., Alagona, G., Ferri, R., Greco, S., Santonocito, D., Pappalardo, A., et al. (2002). Motor cortex excitability in Alzheimer disease: One year follow-up study. *Neuroscience Letters, 329,* 293—296.

Pierantozzi, M., Panella, M., Palmieri, M., Koch, G., Giordano, A., Marciani, M., et al. (2004). Different TMS patterns of intracortical inhibition in early onset Alzheimer dementia and frontotemporal dementia. *Clinical Neurophysiology, 115*(10), 2410—2418.

Robinson, D. (1992). Implications of neural networks for how we think about brain function. *Behavioral and Brain Sciences, 13,* 644—655.

Roshan, L., Paradiso, G., & Chen, R. (2003). Two phases of short-interval intracortical inhibition. *Experimental Brain Research, 151,* 330—337.

Rossini, P., Burke, D., Chen, R., Cohen, L., Daskalakis, Z., Di Iorio, R., et al. (2015). Non-invasive electrical and magnetic stimulation of the brain, spinal cord, roots and peripheral nerves: Basic principles and procedures for routine clinical and research application. an updated report from an I.F.C.N. Committee. *Clinical Neurophysiology, 126,* 1071—1107.

Roth, B., & Basser, P. (1990). A model of the stimulation of a nerve fiber by electromagnetic induction. *IEEE Transactions on Biomedical Engineering, 37,* 588—597.

Snowden, J., Harris, J., Richardson, A., Rollinson, S., Thompson, J., Neary, D., et al. (2013). Frontotemporal dementia with amyotrophic lateral sclerosis: A clinical comparison of patients with and without repeat expansions in C9orf72. *Amyotrophic Lateral Sclerosis Frontotemporal Degeneration, 14,* 172—176.

Sosa-Ortiz, A., Acosta-Castillo, I., & Prince, M. (2012). Epidemiology of dementias and Alzheimer's disease. *Archives of Medical Research, 43,* 600—608.

Suva, D., Favre, I., Kraftsik, R., Esteban, M., Lobrinus, A., & Miklossy, J. (1999). Primary motor cortex involvement in ALzheimer disease. *Journal of Neuropathology and Experimental Neurology, 58,* 1125—1134.

Taylor, J., Firbank, M., Barnett, N., Pearce, S., Livingstone, A., Mosimann, U., et al. (2011). Visual hallucinations in dementia with Lewy bodes: Trancranial magentic stimulation study. *British Journal of Psychiatry, 199,* 492—500.

Tokimura, H., Lazzaro, V., Tokumura, Y., Oliviero, A., Profice, P., Insola, A., et al. (2000). Short latency inhibition of human hand motor cortex by somatosensory input from the hand. *The Journal of Physiology, 523,* 503—513.

Turco, C. V., El-Sayes, J., Savoi, M. J., Fassett, H. J., Locke, M. B., & Nelson, A. J. (2018). Short-and long-latency afferent inhibition; uses, mechanisms and influencing factors. *Brain Stimul, 11*, 59−74.

Valls-Sole, J., Pascual-Leone, A., Wassermann, E., & Hallett, M. (1992). Human motor evoked responses to paired transcranial magentic stimuli. *Electroencephalography and Clinical Neurophysiology, 85*, 355−364.

Vucic, S., & Kiernan, M. (2006). Novel threshold tracking techniques suggest that cortical hyperexcitability is an early feature of motor neuron disease. *Brain, 129*, 11.

Vucic, S., & Kiernan, M. (2017). Transcranial magnetic stimulation for the assessment of neurodegenerative disease. *Neurotherapeutics, 14*, 91−106.

Vucic, S., Ziemann, U., Eisen, A., Hallett, M., & Kiernan, M. (2013). Transcranial magnetic stimulation and amyotrophic lateral sclerosis: Pathophysiological insights. *Jounal of Neurology Neurosurgery and Psychiatry, 84*, 10.

Whittstock, M., Pohley, I., Walter, U., Grossmann, A., Benecke, R., & Wolters, A. (2013). Interhemispheric inhibition in different phenotypes of progressive supranuclear palsy. *Journal of Neural Transmission, 120*, 453−461.

Ziemann, U. (2003). Pharmacology of TMS. *Supplements to Clinical Neurophysiology, 56*, 1264−1275.

CHAPTER 22

The role of retinal imaging in Alzheimer's disease

Victor T.T. Chan, Carol Y. Cheung*
Department of Ophthalmology and Visual Sciences, The Chinese University of Hong Kong, Shatin, Hong Kong

List of abbreviations

AD Alzheimer's disease
CNS central nervous system
GC-IPL macular ganglion cell—inner plexiform layer
GCC macular ganglion cell complex
MCI mild cognitive impairment
RGC retinal ganglion cell
RNFL retinal nerve fiber layer
SD-OCT spectral-domain optical coherence tomography

Mini-dictionary of terms

Spectral-domain optical coherence tomography (SD-OCT) The SD-OCT captures the layered retinal structure in detail using a spectrometer in the detection arm of an interferometer. The signals detected are then converted to depth information with Fourier transformation. The SD-OCT can image the layered retinal structure with a biopsy-like resolution (~ 5 μm) and retinal layers can be segmented by built-in software.

Ganglion cell—inner plexiform layer (GC-IPL) The GC-IPL is a combined layer of ganglion cell layer and inner plexiform layer, which represent the cell bodies and dendrites of RGCs, respectively. Its thickness is usually measured at the macula, where the majority of RGC bodies are located. As the boundary between the ganglion cell layer and the inner plexiform layer is not distinguishable in conventional SD-OCT, they are usually reported together as GC-IPL.

Macular ganglion cell complex (GCC) The macular GCC is composed of the RNFL and the GC-IPL and represents the RGCs in the macular region.

Retinal nerve fiber layer (RNFL) The RNFL represents the axons of RGCs. It is usually measured at the peripapillary area, where the axons merge to form the optic nerve.

Fundus photography Fundus photography is routinely used in clinical settings of ophthalmology to document the ophthalmoscopic appearance of 30 to 50 degrees of the retina. It consists of a specialized low-power microscope and a camera, and operates based on the optical principle of an indirect ophthalmoscope, which has independent light paths for the observation and illumination systems.

*Senior author.

Diagnosis and Management in Dementia
ISBN 978-0-12-815854-8, https://doi.org/10.1016/B978-0-12-815854-8.00022-7
345

Central retinal artery equivalent (CRAE) A parameter used to summarize the average diameter of retinal arterioles, which is estimated from the widths of blood columns of the arterioles using the revised Parr—Hubbard formula (Knudtson et al., 2003).

Central retinal venule equivalent (CRVE) A parameter used to summarize the average diameter of retinal venules, which is estimated from the widths of blood columns of the venules using the revised Parr—Hubbard formula (Knudtson et al., 2003).

Fractal dimension A type of retinal vascular geometric parameter that measures the branching complexity of the retinal vascular network. It is computed using the "box-counting method," which segments an image using squares with different side lengths in each segmentation and plots the number of squares containing retinal vessels against the side length of the squares to obtain the slope (i.e., fractal dimension). Smaller values of fractal dimension indicate a sparser retinal vasculature, while larger values of fractal dimension indicate a more complex branching pattern of the retinal vasculature.

Tortuosity A type of retinal vascular geometric parameter that summarizes the curviness or straightness of retinal vessels. It is computed using the integral of the curvature square along the retinal vessels and the values are normalized by the total length of the vessels (Cheung, Zheng et al., 2011; Sasongko et al., 2010). Larger values of tortuosity indicate more tortuous retinal vessels.

Bifurcation parameters A collection of parameters that summarize the relationship between parent and daughter vessels at a bifurcation. Examples include retinal vascular branching angle, angular asymmetry, asymmetry ratio, and asymmetry factor (Cheung, Tay et al., 2011).

Introduction

With an increase in the aging population, the incidence of age-related dementia is expected to rise substantially. Current evidence suggests the pathological changes of Alzheimer's disease (AD), which is the most common cause of age-related dementia, begin years before the onset of cognitive decline, providing opportunities for disease detection and timely intervention. Hence, there has been great interest in discovering biomarkers to facilitate early detection of AD.

During embryonic development, the retina and optic nerve develop as direct extensions of the diencephalon and share numerous similarities with the brain, including anatomical structure, immunological nature, and physiological properties (Table 22.1). Being a part of the central nervous system (CNS), the retina has thus long been regarded as a "window" to detect changes in the brain *in vivo* because of the transparency of the eye and the similarities between the retina and the brain. Hence, it has been proposed that changes in the retina mirror degenerative changes of AD occurring in the brain.

The retinal neuronal structure and vasculature can now be imaged noninvasively and efficiently using spectral-domain optical coherence tomography (SD-OCT) and fundus photography, respectively (Chan et al., 2017, 2019; Cheung, Ikram, Chen, & Wong, 2017). Since 2010, studies have identified a spectrum of retinal neuronal and vascular changes in AD, which show potential to be applied clinically in AD. This chapter aims to provide an overview of the retinal neuronal and vascular changes in AD, and discuss their potential clinical applications.

Table 22.1 The similarities between the retina and the brain.

Similarities	Description
Embryological origin	The retina and optic nerve are derived from the diencephalon during development. When the optic vesicles extend toward the surface ectoderm, the connections to the forebrain form the optic stalks, which are eventually filled with the ganglion cell fibers to become the optic nerves. Hence, the retina preserves its connection with the brain (such as lateral geniculate nucleus in the thalamus and the superior colliculus in the midbrain) via the optic nerve fibers.
Cytological structure	The retina is composed of different layers of neurons that are interconnected through synapses. Similar to other neurons in the CNS, the RGCs are composed of a cell body, dendrites, and an axon. The axons of RGCs are covered with myelin sheaths produced by oligodendrocytes and are ensheathed by all three layers of meninges.
Response to insult	Insult to the axons of RGCs results in a similar response compared with other CNS axons, including retro- and anterograde axonal degeneration, myelin destruction, scar formation, and creation of a neurotoxic environment that leads to secondary degeneration (Levkovitch-Verbin et al., 2001, 2003).
Limited regenerative ability	Like other CNS neurons, the axons of RGCs have limited regenerative ability due to gliosis, absence of growth-promoting factors, and presence of molecules that restrict the outgrowth of neurites (such as myelin debris, reticulon-4, myelin-associated glycoprotein, and oligodendrocyte myelin glycoprotein) (Filbin, 2003; Fischer, He, & Benowitz, 2004; Lingor et al., 2007; Rolls et al., 2008; Schwab, 2004; Silver & Miller, 2004).
Immunological properties	Both the retina and other parts of the CNS are considered immune-privileged sites (Streilein, 2003). Due to the similar collection of cell surface molecules, immunoregulatory molecules, and cytokines, the retina produces an immune response that resembles that of other parts of the CNS (Kaur, Foulds, & Ling, 2008; Streilein, 2003). In addition, the aqueous humor of the eye contains antiinflammatory and immunoregulatory molecules that can also be found in the cerebrospinal fluid (Taylor & Streilein, 1996; Wilbanks & Wayne Streilein, 1992).
Presence of blood barrier	The inner blood—retinal barrier is regarded as the retinal counterpart of the blood—brain barrier. The endothelial cells in the blood—retinal barrier are nonfenestrated, firmly connected by tight junctions, and surrounded by end-feet from glial and Müller cells (Kaur et al., 2008).
Vascular properties	The retinal vasculature shares similarities with the cerebral circulation, including autoregulatory mechanisms, barrier function, end arterioles without anastomoses, and relatively low-flow systems with high oxygen extraction (Patton et al., 2005).

The similarities between the retina and the brain are listed. Due to these similarities, the retina has been considered as a window to assess neuronal and vascular changes in the central nervous system, including those occurring in Alzheimer's disease. *RGCs*, retinal ganglion cells.

Retinal neuronal changes in Alzheimer's disease

The neuronal architecture of the retina is organized into distinct layers, with each layer representing a specific component of retinal neurons (Fig. 22.1). For instance, the axons of retinal ganglion cells (RGCs) can be quantified by the retinal nerve fiber layer (RNFL), while their cell bodies and dendrites comprise the macular ganglion cell—inner plexiform layer (GC-IPL). The layered retinal neuronal structure can be imaged noninvasively, accurately (with resolution of ~3 μm), and efficiently (<1.7 s per 3D scan) by SD-OCT, which captures the retina with a spectrometer and converts the signals to depth information using Fourier transformation.

Differences in the retinal neuronal layers have been observed in AD. Previous studies showed that thicknesses of the RNFL and the GC-IPL were significantly reduced in AD compared with cognitively normal controls (Chan et al., 2019; Cheung et al., 2015; Choi, Park, & Kim, 2016; Cunha et al., 2016; denHaan, Verbraak, Visser, & Bouwman, 2017) (Fig. 22.1). Consistently, patients with AD also have reduced thickness of the macular ganglion cell complex, which is composed of the RNFL and GC-IPL (Chan et al., 2019; Bayhan, Aslan Bayhan, Celikbilek, Tanık, & Gürdal, 2015; Cunha et al., 2016; Marziani et al., 2013). Two large-scale population-based studies, the UK Biobank Study and the Rotterdam Study, also demonstrated that thinning of RNFL was associated with a significantly increased risk of developing both cognitive decline (Ko et al., 2018) and AD dementia (Mutlu et al., 2018). Thinning of the RNFL and the GC-IPL in AD indicates decreased numbers of axons and cell body—dendrite complexes of the RGCs, respectively. Furthermore, macular thickness and macular volume were also significantly reduced in AD compared with cognitively normal controls (Chan et al., 2019; Choi et al., 2016; denHaan et al., 2017; Gao, Liu, Li, Bai, & Liu, 2015).

The findings of retinal neuronal reduction in AD favor the hypothesis that retinal neuronal loss reflects neurodegenerative changes in the brain of AD patients. Consistently, histopathologic studies have reported significant loss of RGCs and their axons in ocular tissues of both animal models and human subjects with AD (Blanks, Schmidt et al., 1996; Blanks, Torigoe, Hinton, & Blanks, 1996). Clinical studies also observed close associations between visual symptoms and AD. Two possible mechanisms have been hypothesized in an attempt to explain the retinal neuronal loss in AD. The first hypothesis suggested that retrograde degeneration of the optic nerve occurs secondary to the pathological events affecting the visual pathway in the cerebrum (Ascaso et al., 2014), particularly in the subcortical visual centers (Leuba & Saini, 1995). Meanwhile, others hold the view that AD-induced neurodegeneration occurs simultaneously in the retina and the cerebrum due to the presence of amyloid-β (Aβ) plaques, fibrillar tau, and AD-specific signs of neuroinflammation in ocular tissues of both patients and animal models of AD (Goldstein et al., 2003; Kayabasi, Sergott, & Rispoli, 2014; Koronyo et al., 2017; Koronyo-Hamaoui et al., 2011). However, most studies in the current literature have a cross-sectional design and do not assess the longitudinal changes in retinal neuronal layers in AD. Further studies are warranted to explore whether the retinal neuronal loss is directly caused by the pathology of AD or whether the magnitude of the retinal neuronal loss correlates with the extent of neurodegenerative changes in the brain.

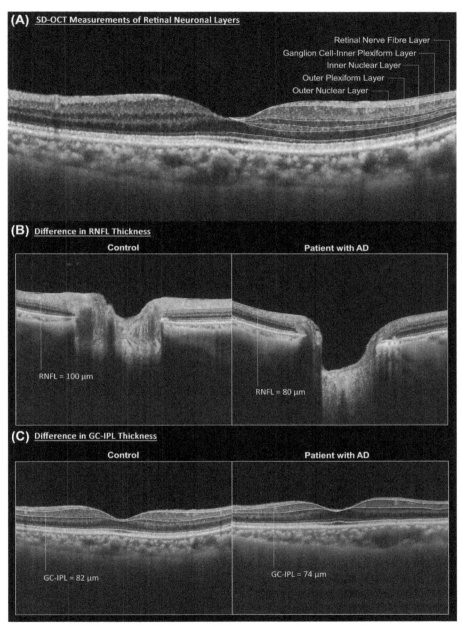

Figure 22.1 *An example of retinal neuronal changes in Alzheimer's disease (AD) dementia.* (A) The neuronal architecture of the retina is organized into distinct layers, with each layer representing a specific component of retinal neurons. Hence, changes in thicknesses of the retinal neuronal layers provide information on retinal neuronal injury. (B) The retinal nerve fiber layer (*RNFL*), consisting of axons of the retinal ganglion cells (RGCs), is thinner in AD compared with controls. (C) The ganglion cell—inner plexiform layer (*GC-IPL*), which comprises RGC cell bodies and dendrites, is also thinner in AD compared with controls. *SD-OCT*, spectral-domain optical coherence tomography.

Retinal vascular changes in Alzheimer's disease

The homologies between retinal and cerebral circulation (Table 22.1) allow assessment of vascular changes in the CNS via visualization of the retinal vasculature, which can be achieved easily with fundus photography. Previous studies have identified a spectrum of retinal vascular changes in AD, namely (1) qualitative retinal lesions, (2) retinal vascular calibers, and (3) retinal vascular geometric parameters.

While fundus photography allows direct identification of qualitative retinal lesions (e.g., microaneurysms, hemorrhages, cotton wool spots, and hard exudates) (Fig. 22.2), quantitative properties of the retinal vasculature (e.g., retinal vascular calibers and retinal vascular geometric parameters) can also be measured reliably using computer-assisted analysis programs. One of the most widely used methods of measuring quantitative retinal vascular changes was initially developed in the Atherosclerosis Risk in Communities (ARIC) study by Hubbard et al., who estimated the retinal vascular calibers by measuring the blood column diameters of retinal vessels within 0.5–1 disc diameter from the margin of the optic disc (referred to as the standard zone; Fig. 22.3) (Hubbard et al., 1999). Subsequently, Cheung et al. (2010) demonstrated that inclusion of peripheral vessels within 0.5–2 disc diameters from the disc margin (referred to as the extended zone; Fig. 22.3) provides a better representation of the retinal vasculature and allows more accurate assessment of the vascular geometric properties. Currently, the retinal vascular calibers are assessed in both standard and extended zones, whereas the retinal vascular geometric parameters are assessed only in the extended zone. The average calibers of the retinal arterioles and venules are summarized as central retinal artery equivalent (CRAE) and central retinal venule equivalent (CRVE) using the revised Parr–Hubbard formula (Knudtson et al., 2003).

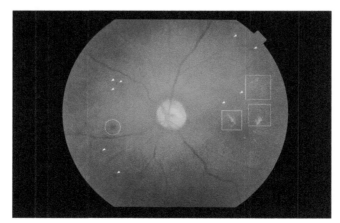

Figure 22.2 *Examples of qualitative retinal lesions.* Qualitative retinal lesions, such as hemorrhages (indicated by *white circle*), microaneurysms (indicated by *white arrowheads*), and hard exudate (indicated by *white squares*), can be directly identified from retinal photographs.

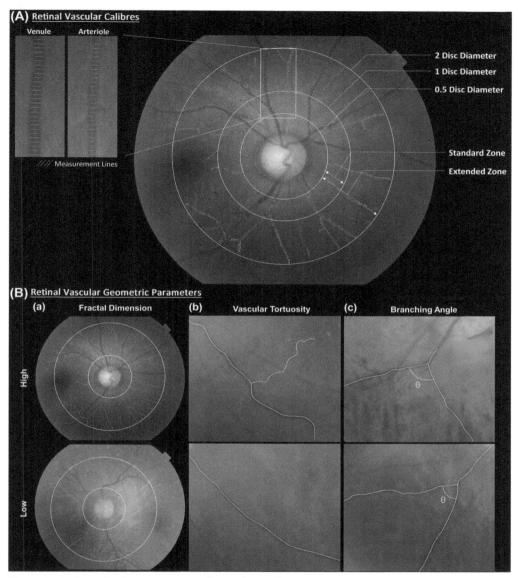

Figure 22.3 *Quantitative assessment of retinal vasculature.* (A) The quantitative retinal vascular parameters are measured in the standard zone (0.5–1 disc diameter away from the disc margin) and the extended zone (0.5–2 disc diameters away from the disc margin). The retinal vascular calibers can be estimated from the measurement lines that span the width of blood columns of retinal vessels and summarized as central retinal artery equivalent (CRAE) and central retinal venule equivalent (CRVE), using the revised Parr–Hubbard formula (Knudtson et al., 2003). CRAE and CRVE are assessed in both the standard and the extended zones. (B) The retinal vascular geometric parameters, which are assessed in the extended zone only, capture the geometrical differences of the retinal vasculature and reflect the optimality and efficiency of blood circulation in the retinal vascular network. (a) The retinal fractal dimension reflects the branching complexity of the retinal vascular network. Smaller values of fractal dimension indicate a sparser branching pattern of the retinal vasculature. (b) The retinal vascular tortuosity reflects the average straightness of the retinal vessels. Larger values of tortuosity indicate more tortuous retinal vessels. (c) The retinal vascular branching angle summarizes the average angle sustained between the two daughter vessels at each bifurcation. It has been proposed that sub-optimal branching angle reflects decreased energy deficiency of blood distribution.

Several studies have explored associations between AD and qualitative retinal lesions, such as microaneurysms, hemorrhages, hard exudates, and cotton wool spots (Cheung et al., 2017). Cross-sectional data from the Rotterdam Study showed that qualitative retinal lesions were associated with AD (Schrijvers et al., 2012). The Cardiovascular Health Study showed that qualitative retinal lesions were associated with AD only in subjects with hypertension and in subjects without diabetes (Baker et al., 2007). However, the AGES-Reykjavik Study reported no associations between qualitative retinal lesions and AD (Qiu et al., 2010). In addition, longitudinal data from the Rotterdam Study showed no associations between qualitative retinal lesions and risk of developing AD (Schrijvers et al., 2012).

Alterations in retinal vascular calibers in AD have also been shown in previous studies, although the direction of changes remains controversial. Frost et al. (2013) and Cheung et al. (2014) found that both CRAE and CRVE were reduced in AD. Conversely, William et al. (2015) reported an increase of CRAE in AD and did not find a significant difference in CRVE compared with control subjects. Furthermore, the associations between changes in CRAE and AD became statistically nonsignificant after adjusting for cardiovascular risk factors (Cheung et al., 2014; Williams et al., 2015).

Studies have demonstrated associations between changes in retinal vascular geometric parameters and dementia. These parameters captured geometrical differences in the retinal vasculature in AD and may reflect decreased optimality and efficiency of blood circulation in the retinal vascular network, as suggested by Murray's principle (Murray, 1926a,b). Current literature has consistently shown that decreased retinal fractal dimension (i.e., a sparser retinal vascular network) is associated with AD (Cheung et al., 2014; Frost et al., 2013; Williams et al., 2015). The pathological basis of a less complex retinal vascular network could be the result of retinal vascular rarefaction and collapse (Hammes et al., 2011). In addition, previous studies also reported changes in retinal vascular tortuosity in AD (Cheung et al., 2014; Frost et al., 2013; Williams et al., 2015). Although both increased (Cheung et al., 2014) and decreased (Frost et al., 2013; Williams et al., 2015) retinal vascular tortuosity were reported, the finding of increased retinal vascular tortuosity in AD showed agreement with previous observations in cerebrovascular disease, stroke, and hypertension (Benitez-Aguirre et al., 2011; Ong et al., 2014; Owen et al., 2011). Furthermore, suboptimal bifurcation parameters were also implicated in AD, such as larger venular branching asymmetry factor (i.e., the ratio of the squares of the two branching venules widths) (Frost et al., 2013).

The retinal vascular changes in AD provided unique insights into the role of microvascular or small-vessel disease in AD, in particular information distinct from neuroimaging measures. Due to the homology between retinal and cerebral vasculature, the vascular changes observed in AD (such as decreased vessel calibers, decreased fractal dimension, and increased vascular tortuosity) may mirror vascular changes in the cerebral circulation of AD subjects. For instance, decreased retinal fractal dimension in AD coincides with reduced vessel density in AD brain, which occurs due to sequestration of vascular endothelial growth factor with Aβ (Brown & Thore, 2011). Similarly, pathological studies also reported more tortuous cerebral arterioles in AD, resembling vascular changes in the retina (Hunter et al., 2012). See Tables 22.2 and 22.3 for summaries of the neuronal and vascular changes in the retina in AD.

Table 22.2 Retinal vascular changes in Alzheimer's disease.

Measure	Description	Current understanding
Central retinal artery equivalent (CRAE)	Represents the average diameter of retinal arterioles	While Cheung et al. (2014) and Frost et al. (2013) reported a reduction of CRAE in AD, William et al. (2015) observed an increase of CRAE in AD
Central retinal venule equivalent (CRVE)	Represents the average diameter of retinal venules	CRVE is significantly reduced in AD, suggesting that retinal venular narrowing is associated with AD
Fractal dimension	Smaller value indicates a sparser branching pattern of the retinal vasculature	Retinal fractal dimension is significantly reduced in AD, suggesting that a sparser retinal vascular network is associated with AD
Tortuosity	Larger value indicates more tortuous retinal vessels	Both increased and decreased retinal vascular tortuosity was observed in patients with AD; specific change in tortuosity in AD remains to be determined
Bifurcation parameters	Estimate the optimality of branching of the retinal vasculature	Asymmetry factor, which is the ratio of the squares of the two branching venules' widths, is significantly larger in patients with AD

The retinal vasculature can be imaged easily using fundus photography, which allows measurement of the quantitative vascular properties. Among these parameters, retinal vascular calibers and geometric parameters (e.g., fractal dimension, tortuosity, and bifurcation parameters) have been shown to be associated with AD. This table summarizes the associations between retinal vascular changes and AD reported in previous studies. *AD*, Alzheimer's disease.

Table 22.3 Retinal neuronal changes in Alzheimer's disease.

Measure	Description	Current understanding
Peripapillary RNFL thickness	Represents number of axons of the RGCs	Peripapillary RNFL is significantly thinner in AD, suggesting a reduction in number of RGC axons
Macular GC-IPL thickness	Represents number of cell bodies and dendrites of the RGCs	Macular GC-IPL is significantly thinner in AD, suggesting a reduction in number of RGC cell bodies and dendrites
Macular GCC thickness	Represents number of RGCs	Macular GCC is significantly thinner in AD, suggesting a reduction in number of RGC
Macular thickness	Represents the total number of neurons in the entire layer of macular	Macular thickness is significantly thinner in AD, suggesting a reduction in retinal neurons
Macular volume	Represents the total number of neurons in the entire layer of macular	Macular volume is significantly reduced in AD, suggesting a reduction in retinal neurons

The retinal structure including neuronal layers can be assessed noninvasively and with high resolution using spectral-domain optical coherence tomography. Previous studies have identified several retinal neuronal changes in AD, which are summarized in this table. *AD*, Alzheimer's disease; *GCC*, ganglion cell complex; *GC-IPL*, ganglion cell—inner plexiform layer; *RGC*, retinal ganglion cell; *RNFL*, retinal nerve fiber layer.

Associations between retinal changes with amyloid-β burden and markers of cerebral small-vessel diseases

A few studies have investigated associations between retinal imaging measures and amyloid burden, the upstream pathology of AD (Frost et al., 2013; Golzan et al., 2017). While Frost et al. (2013) observed higher venular branching asymmetry factor and arteriolar length-to-diameter ratio in healthy subjects with high amyloid plaque burden, Golzan et al. (2017) found no correlations between the thickness of the RGC layer and neocortical amyloid burden.

In addition, there are also associations between retinal vascular changes and markers of cerebral small-vessel disease, an important risk factor for cognitive decline and dementia (Pantoni, 2010). For instance, lacunar infarcts have been associated with qualitative retinal lesions, decreased CRAE, increased CRVE, and increased retinal fractal dimension in patients without stroke (Cheung et al., 2010; Cooper et al., 2006; Ikram et al., 2006). Similarly, white matter lesion and its progression were also associated with qualitative retinal lesions, decreased CRAE, increased CRVE, and suboptical retinal bifurcation parameters (Baker et al., 2010; Cheung et al., 2010; Doubal, DeHaan et al., 2010; Doubal, MacGillivray et al., 2010; Haan et al., 2012; Hughes et al., 2016; Ikram et al., 2006; Longstreth et al., 2007; Qiu et al., 2009). In addition, it has also been reported that increased retinal arteriolar tortuosity and decreased retinal arteriolar fractal dimension were associated with cerebral microbleed, another marker of cerebral small-vessel disease (Hilal et al., 2014). These associations were independent of vascular risk factors and other markers of cerebral small-vessel disease (e.g., white matter lesion and lacunar infarcts), suggesting that retinal arteriolar tortuosity and arteriolar fractal dimension may be early markers of cerebral small-vessel disease.

Potential role of retinal imaging as a stratification tool of Alzheimer's disease

Identifying asymptomatic individuals at higher risk of AD is the first step toward the success of preventing or slowing down dementia. Current approaches to detecting the preclinical phase of AD mainly rely on detection of Aβ and tau in positron emission tomography (PET) imaging and cerebrospinal fluid (CSF) and assessment of neurodegeneration or neuronal injury using neuroimaging (such as [18F]fluorodeoxyglucose-PET and structural MRI) (Jack et al., 2016). However, data from longitudinal studies suggest that an abnormal level of amyloid burden is not sufficient to cause cognitive decline (Burnham et al., 2016; Mormino, Betensky, Hedden, Schultz, Amariglio et al., 2014; Mormino, Betensky, Hedden, Schultz, Ward et al., 2014) and, consistently, approximately 30% of clinically normal elderly subjects have signs of elevated Aβ burden on autopsy examination (Bennett et al., 2006; Price & Morris, 1999). In addition, the cost of neuroimaging and the invasiveness of obtaining CSF have hindered the application of these biomarkers in population screening for AD.

Retinal imaging can potentially be used as a risk stratification tool to identify asymptomatic individuals with higher risk of AD. Compared with neuroimaging and testing of CSF, retinal imaging allows noninvasive and efficient (<1 min per scan) visualization of neuronal structure and the vasculature of the CNS with biopsy-like resolution. Retinal imaging is also relatively accessible compared with PET scan and requires much less training to interpret the results. Furthermore, retinal imaging provides unique information on the extent of neuronal and vascular damage in AD, which cannot be assessed in detail using neuroimaging techniques available in most hospitals.

Current limitations of retinal imaging

Several major challenges should be sufficiently addressed before translating retinal imaging into a stratification tool clinically. The retinal neuronal and vascular differences in AD are only modest and have not been consistently observed. This could be partly explained by the heterogeneities between studies, such as severity of cognitive impairment, number of statistically adjusted confounders, and differences in imaging and analysis methods. Since the current clinical diagnostic criteria often fail to differentiate between AD and non-AD dementia (Beach, Monsell, Phillips, & Kukull, 2012), the inclusion of subjects with non-AD dementia may contribute to the variability of the outcomes. Further studies with a well-characterized study population may provide better characterization of the effects of AD on the retinal neuronal and vascular structure. In addition, retinal neuronal and vascular changes may have different underlying substrates compared with AD and other types of dementia, limiting the specificity of retinal measurements. For instance, thinning of the RNFL has also been reported in other neurodegenerative diseases, such as glaucoma, multiple sclerosis, and Parkinson's disease (London, Benhar, & Schwartz, 2013). Retinal vascular changes have also been described in hypertension and diabetes, which are common diseases in the elderly population.

Recent advancements

Advances in artificial intelligence provide a platform for improving the interpretation of retinal images. The deep learning algorithm, which is a subset of artificial intelligence, can now be used to develop a convolutional neural network (CNN) to recognize an image with a specific pattern and its variants. Since AD-specific features in retinal images can be automatically identified by the CNN in the feature extraction stage and do not involve any objective judgment, AD-specific features that have been neglected by previous studies may be discovered. The method of deep learning has been applied in other fields of ophthalmology to automatically detect diabetic retinopathy from fundus photographs with high sensitivity and specificity (Ting et al., 2017). In the near future, deep learning techniques may provide solutions to resolve the sophisticated neuronal and vascular changes in AD and improve the specificity of retinal imaging measurements by identifying AD-specific features.

On the other hand, the retinal vasculature can now be imaged at the capillary level using the state-of-the-art OCT-angiography (OCT-A). Measurements of OCT-A may improve the sensitivity and specificity of retinal imaging because microvascular changes are expected to be earlier events than macrovascular changes, and the intricate capillary network may manifest AD-specific changes. New studies have demonstrated associations between AD and OCT-A measurements, such as increased foveal avascular zone and decreased retinal vessel density, in line with previous findings on fundus photography (Bulut et al., 2017).

The advent of ultra-wide-field retinal imaging also enables assessment of more peripheral retinal vasculature. The principle of ultra-wide-field retinal imaging is based on confocal laser scanning microscopy, combined with a concave elliptical mirror. Compared with traditional fundus photography, it can capture up to 200 degrees of the retina without pupil dilatation (Kernt et al., 2012) and provide a more comprehensive representation of the retinal vasculature (Fig. 22.4). It is possible that decreased arteriolar fractal dimension and increased peripheral hard drusen deposition are potential biomarkers of AD (Csincsik et al., 2018).

Conclusion

Being an extension of the brain, the retina is an excellent location to study the effects of dementia. Previous studies have demonstrated a spectrum of retinal neuronal and vascular changes in AD. These retinal changes provided unique insights into the effects of AD on the CNS neuronal structure and vasculature. Clinically, retinal imaging could provide a complementary approach to neuroimaging and cognitive assessment and may be employed as a risk stratification tool to identify individuals with higher risk of AD. Further studies are required to validate the role of retinal imaging measurements in AD and address the current limitations with the aid of recent advancements.

Key facts of retinal imaging

- A landmark discovery in ophthalmology, SD-OCT can image the three-dimensional retinal structure, including neuronal layers, noninvasively and reliably within seconds. SD-OCT has been widely used to detect eye diseases (e.g., glaucoma and age–related macular degeneration) and to study other neurodegenerative diseases, including AD and multiple sclerosis.

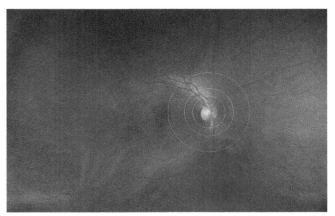

Figure 22.4 *Retinal vasculature captured by ultra-wide-field retinal imaging.* Compared with traditional fundus photography, ultra-wide-field retinal imaging can capture up to 200 degrees of the retina without pupil dilatation and provide a more comprehensive representation of the retinal vasculature. It is expected that more Alzheimer's disease—specific retinal changes can be extracted from the peripheral retinal vasculature.

- Fundus photography is routinely used in clinical settings of ophthalmology to document the ophthalmoscopic appearance of 30 to 50 degrees of the retina. It consists of a specialized low-power microscope and a camera, and operates based on the optical principle of an indirect ophthalmoscope, which has independent light paths for the observation and illumination systems.
- Quantitative properties of the retinal vasculature can be measured using computer-assisted programs, such as Singapore "I" Vessel Assessment (SIVA; National University of Singapore, Singapore), Vessel Assessment and Measurement Platform for Images of the Retina (VAMPIRE), and Retinal Analysis (RA; University of Wisconsin).
- The computer-assisted programs can detect the location of the optic disc, generate skeletonized vessel tracings, and measure quantitative retinal vascular parameters automatically within seconds. However, trained graders are required to examine the accuracy of these vessel tracings and, if necessary, correct inaccurate tracings manually.
- AD is associated with a number of visual symptoms, including inferior hemifield loss, poorer color discrimination, deteriorated ocular motor function, and decreased contrast sensitivity (Javaid, Brenton, Guo, & Cordeiro, 2016).

Summary points

- Our retina shares prominent similarities with the brain, including embryological origin, anatomical structure, and physiological properties. Hence, neuronal and vascular changes in the retina may reflect similar changes in the brain and provide unique information on the extent of neurodegenerative and cerebrovascular damage in AD.
- The retinal structure, including neuronal layers, can be assessed noninvasively and with high resolution using SD-OCT. Previous studies have shown that patients with AD had significant reductions in macular thickness and volume, as well as thicknesses of peripapillary RNFL and macular GC-IPL.
- The retinal vasculature can also be imaged using fundus photography, which allows direct identification of qualitative retinal lesions (e.g., microaneurysms, hemorrhages, cotton wool spots, and hard exudates) and measurement of quantitative properties of the retinal vasculature (e.g., vascular calibers and a spectrum of retinal vascular geometric parameters) using computer-assisted analysis programs. Previous studies have reported the presence of qualitative retinal lesions (e.g., microaneurysms, hemorrhages, cotton wool spots, and hard exudates) and changes in retinal vascular

calibers and retinal vascular geometric parameters (e.g., fractal dimension, tortuosity, and bifurcation parameters) in subjects with AD or cognitive impairment.

- Alterations in the retinal neuronal structure and vasculature in AD are potential biomarkers that can be used to facilitate stratification of individuals with higher risk for AD.
- Recent advancements in artificial intelligence, OCT-A, and ultra-wide-field retinal imaging are expected to facilitate more breakthroughs in this promising field.

References

Ascaso, F. J., Cruz, N., Modrego, P. J., Lopez-Anton, R., Santabárbara, J., Pascual, L. F., et al. (2014). Retinal alterations in mild cognitive impairment and Alzheimer's disease: An optical coherence tomography study. *Journal of Neurology, 261*(8), 1522—1530. https://doi.org/10.1007/s00415-014-7374-z.

Baker, M. L., Marino Larsen, E. K., Kuller, L. H., Klein, R., Klein, B. E. K., Siscovick, D. S., et al. (2007). Retinal microvascular signs, cognitive function, and dementia in older persons: The cardiovascular health study. *Stroke, 38*(7), 2041—2047. https://doi.org/10.1161/STROKEAHA.107.483586.

Baker, M. L., Wang, J. J., Liew, G., Hand, P. J., DeSilva, D. A., Lindley, R. I., et al. (2010). Differential associations of cortical and subcortical cerebral atrophy with retinal vascular signs in patients with acute stroke. *Stroke, 41*(10), 2143—2150. https://doi.org/10.1161/STROKEAHA.110.594317.

Bayhan, H. A., Aslan Bayhan, S., Celikbilek, A., Tanık, N., & Gürdal, C. (2015). Evaluation of the chorioretinal thickness changes in Alzheimer's disease using spectral-domain optical coherence tomography. *Clinical and Experimental Ophthalmology, 43*(2), 145—151. https://doi.org/10.1111/ceo.12386.

Beach, T. G., Monsell, S. E., Phillips, L. E., & Kukull, W. (2012). Accuracy of the clinical diagnosis of alzheimer disease at National Institute on Aging Alzheimer Disease Centers, 2005-2010. *Journal of Neuropathology and Experimental Neurology, 71*(4), 266—273. https://doi.org/10.1097/NEN.0b013e31824b211b.

Benitez-Aguirre, P., Craig, M. E., Sasongko, M. B., Jenkins, A. J., Wong, T. Y., Wang, J. J., et al. (2011). Retinal vascular geometry predicts incident retinopathy in young people with type 1 diabetes: A prospective cohort study from adolescence. *Diabetes Care, 34*(7), 1622—1627. https://doi.org/10.2337/dc10-2419.

Bennett, D. A., Schneider, J. A., Arvanitakis, Z., Kelly, J. F., Aggarwal, N. T., Shah, R. C., et al. (2006). Neuropathology of older persons without cognitive impairment from two community-based studies. *Neurology, 66*(12), 1837—1844. https://doi.org/10.1212/01.wnl.0000219668.47116.e6.

Blanks, J. C., Schmidt, S. Y., Torigoe, Y., Porrello, K. V., Hinton, D. R., & Blanks, R. H. I. (1996). Retinal pathology in Alzheimer's disease. II. Regional neuron loss and glial changes in GCL. *Neurobiology of Aging, 17*(3), 385—395. https://doi.org/10.1016/0197-4580(96)00009-7.

Blanks, J. C., Torigoe, Y., Hinton, D. R., & Blanks, R. H. I. (1996b). Retinal pathology in Alzheimer's disease. I. Ganglion cell loss in foveal/parafoveal retina. *Neurobiology of Aging, 17*(3), 377—384. https://doi.org/10.1016/0197-4580(96)00010-3.

Brown, W. R., & Thore, C. R. (2011). Review: Cerebral microvascular pathology in ageing and neurodegeneration. *Neuropathology and Applied Neurobiology, 37*(1), 56—74. https://doi.org/10.1111/j.1365-2990.2010.01139.x.

Bulut, M., Kurtulus, F., Gözkaya, O., Erol, M. K., Cengiz, A., Akıdan, M., et al. (2017). Evaluation of optical coherence tomography angiographic findings in Alzheimer's type dementia. *British Journal of Ophthalmology*. https://doi.org/10.1136/bjophthalmol-2017-310476. bjophthalmol-2017-310476.

Burnham, S. C., Bourgeat, P., Doré, V., Savage, G., Brown, B., Laws, S., et al. (2016). Clinical and cognitive trajectories in cognitively healthy elderly individuals with suspected non-Alzheimer's disease pathophysiology (SNAP) or Alzheimer's disease pathology: A longitudinal study. *The Lancet Neurology, 15*(10), 1044—1053. https://doi.org/10.1016/S1474-4422(16)30125-9.

Chan, V. T. T., Tso, T. H. K., Tang, F., Tham, C., Mok, V., Chen, C., et al. (2017). Using retinal imaging to study dementia. *Journal of Visualized Experiments, 129*(129). https://doi.org/10.3791/56137. e56137–e56137.

Chan, V. T. T., Sun, Z., Tang, S., Chen, L. J., Wong, A., Tham, C. C., et al. (2019). Spectral-domain OCT measurements in Alzheimer's disease: A systematic review and meta-analysis. *Ophthalmology, 126*(4), 497–510.

Cheung, C. Y., Ikram, M. K., Chen, C., & Wong, T. Y. (2017). Imaging retina to study dementia and stroke. *Progress in Retinal and Eye Research*. https://doi.org/10.1016/j.preteyeres.2017.01.001.

Cheung, N., Mosley, T., Islam, A., Kawasaki, R., Sharrett, A. R., Klein, R., et al. (2010). Retinal microvascular abnormalities and subclinical magnetic resonance imaging brain infarct: A prospective study. *Brain, 133*(7), 1987–1993. https://doi.org/10.1093/brain/awq127.

Cheung, C. Y., Ong, Y. T., Hilal, S., Ikram, M. K., Low, S., Ong, Y. L., et al. (2015). Retinal ganglion cell analysis using high-definition optical coherence tomography in patients with mild cognitive impairment and Alzheimer's disease. *Journal of Alzheimer's Disease, 45*(1), 45–56. https://doi.org/10.3233/JAD-141659.

Cheung, C. Y., Ong, Y. T., Ikram, M. K., Ong, S. Y., Li, X., Hilal, S., et al. (2014). Microvascular network alterations in the retina of patients with Alzheimer's disease. *Alzheimer's and Dementia, 10*(2), 135–142. https://doi.org/10.1016/j.jalz.2013.06.009.

Cheung, C. Y., Tay, W. T., Mitchell, P., Wang, J. J., Hsu, W., Lee, M. L., et al. (2011). Quantitative and qualitative retinal microvascular characteristics and blood pressure. *Journal of Hypertension, 29*(7), 1380–1391. https://doi.org/10.1097/HJH.0b013e328347266c.

Cheung, C. Y., Zheng, Y., Hsu, W., Lee, M. L., Lau, Q. P., Mitchell, P., et al. (2011). Retinal vascular tortuosity, blood pressure, and cardiovascular risk factors. *Ophthalmology, 118*(5), 812–818. https://doi.org/10.1016/j.ophtha.2010.08.045.

Choi, S. H., Park, S. J., & Kim, N. R. (2016). Macular ganglion cell -inner plexiform layer thickness is associated with clinical progression in mild cognitive impairment and Alzheimers disease. *PLoS One, 11*(9), e0162202. https://doi.org/10.1371/journal.pone.0162202.

Cooper, L. S., Wong, T. Y., Klein, R., Sharrett, A. R., Bryan, R. N., Hubbard, L. D., et al. (2006). Retinal microvascular abnormalities and MRI-defined subclinical cerebral infarction: The atherosclerosis risk in communities study. *Stroke, 37*(1), 82–86. https://doi.org/10.1161/01.STR.0000195134.04355.e5.

Csincsik, L., MacGillivray, T. J., Flynn, E., Pellegrini, E., Papanastasiou, G., Barzegar-Befroei, N., et al. (2018). Peripheral retinal imaging biomarkers for Alzheimer's disease: A pilot study. *Ophthalmic Research*. https://doi.org/10.1159/000487053.

Cunha, L. P., Lopes, L. C., Costa-Cunha, L. V. F., Costa, C. F., Pires, L. A., Almeida, A. L. M., et al. (2016). Macular thickness measurements with frequency domain-OCT for quantification of retinal neural loss and its correlation with cognitive impairment in Alzheimer's disease. *PLoS One, 11*(4), e0153830. https://doi.org/10.1371/journal.pone.0153830.

denHaan, J., Verbraak, F. D., Visser, P. J., & Bouwman, F. H. (2017). Retinal thickness in Alzheimer's disease: A systematic review and meta-analysis. *Alzheimer's and Dementia (Amsterdam, Netherlands, 6,* 162–170. https://doi.org/10.1016/j.dadm.2016.12.014.

Doubal, F. N., DeHaan, R., MacGillivray, T. J., Cohn-Hokke, P. E., Dhillon, B., Dennis, M. S., et al. (2010). Retinal arteriolar geometry is associated with cerebral white matter hyperintensities on magnetic resonance imaging. *International Journal of Stroke, 5*(6), 434–439. https://doi.org/10.1111/j.1747-4949.2010.00483.x.

Doubal, F. N., MacGillivray, T. J., Patton, N., Dhillon, B., Dennis, M. S., & Wardlaw, J. M. (2010b). Fractal analysis of retinal vessels suggests that a distinct vasculopathy causes lacunar stroke. *Neurology, 74*(14), 1102–1107. https://doi.org/10.1212/WNL.0b013e3181d7d8b4.

Filbin, M. T. (2003). Myelin-associated inhibitors of axonal regeneration in the adult mammalian CNS. *Nature Reviews Neuroscience, 4*(9), 703–713. https://doi.org/10.1038/nrn1195.

Fischer, D., He, Z., & Benowitz, L. I. (2004). Counteracting the Nogo receptor enhances optic nerve regeneration if retinal ganglion cells are in an active growth state. *Journal of Neuroscience: The Official Journal of the Society for Neuroscience, 24*(7), 1646–1651. https://doi.org/10.1523/JNEUROSCI.5119-03.2004.

Frost, S., Kanagasingam, Y., Sohrabi, H., Vignarajan, J., Bourgeat, P., Salvado, O., et al. (2013). Retinal vascular biomarkers for early detection and monitoring of Alzheimer's disease. *Translational Psychiatry, 3*(2), e233. https://doi.org/10.1038/tp.2012.150.

Gao, L., Liu, Y., Li, X., Bai, Q., & Liu, P. (2015). Abnormal retinal nerve fiber layer thickness and macula lutea in patients with mild cognitive impairment and Alzheimer's disease. *Archives of Gerontology and Geriatrics, 60*(1), 162−167. https://doi.org/10.1016/j.archger.2014.10.011.

Goldstein, L. E., Muffat, J. A., Cherny, R. A., Moir, R. D., Ericsson, M. H., Huang, X., et al. (2003). Cytosolic β-amyloid deposition and supranuclear cataracts in lenses from people with Alzheimer's disease. *Lancet, 361*(9365), 1258−1265. https://doi.org/10.1016/S0140-6736(03)12981-9.

Golzan, S. M., Goozee, K., Georgevsky, D., Avolio, A., Chatterjee, P., Shen, K., et al. (2017). Retinal vascular and structural changes are associated with amyloid burden in the elderly: Ophthalmic biomarkers of preclinical Alzheimer's disease. *Alzheimer's Research and Therapy, 9*(1). https://doi.org/10.1186/s13195-017-0239-9.

Haan, M., Espeland, M. A., Klein, B. E., Casanova, R., Gaussoin, S. A., Jackson, R. D., et al. (2012). Cognitive function and retinal and ischemic brain changes: The Women's health initiative. *Neurology, 78*(13), 942−949. https://doi.org/10.1212/WNL.0b013e31824d9655.

Hammes, H.-P., Feng, Y., Pfister, F., Brownlee, M., Calcutt, N., Cooper, M., et al. (2011). Diabetic retinopathy: Targeting vasoregression. *Diabetes, 60*(1), 9−16. https://doi.org/10.2337/db10-0454.

Hilal, S., Ong, Y. T., Cheung, C. Y., Tan, C. S., Venketasubramanian, N., Niessen, W. J., et al. (2014). Microvascular network alterations in retina of subjects with cerebral small vessel disease. *Neuroscience Letters, 577*, 95−100. https://doi.org/10.1016/j.neulet.2014.06.024.

Hubbard, L. D., Brothers, R. J., King, W. N., Clegg, L. X., Klein, R., Cooper, L. S., et al. (1999). Methods for evaluation of retinal microvascular abnormalities associated with hypertension/sclerosis in the Atherosclerosis Risk in Communities Study. *Ophthalmology, 106*(12), 2269−2280. https://doi.org/10.1016/S0161-6420(99)90525-0.

Hughes, A. D., Falaschetti, E., Witt, N., Wijetunge, S., Thom, S. A. M. G., Tillin, T., et al. (2016). Association of retinopathy and retinal microvascular abnormalities with stroke and cerebrovascular disease. *Stroke, 47*(11), 2862−2864. https://doi.org/10.1161/STROKEAHA.116.014998.

Hunter, J. M., Kwan, J., Malek-Ahmadi, M., Maarouf, C. L., Kokjohn, T. A., Belden, C., et al. (2012). Morphological and pathological evolution of the brain microcirculation in aging and Alzheimer's disease. *PLoS One, 7*(5). https://doi.org/10.1371/journal.pone.0036893.

Ikram, M. K., DeJong, F. J., VanDijk, E. J., Prins, N. D., Hofman, A., Breteler, M. M. B., et al. (2006). Retinal vessel diameters and cerebral small vessel disease: The Rotterdam Scan Study. *Brain, 129*(1), 182−188. https://doi.org/10.1093/brain/awh688.

Jack, C. R., Bennett, D. A., Blennow, K., Carrillo, M. C., Feldman, H. H., Frisoni, G. B., et al. (2016). A/T/N: An unbiased descriptive classification scheme for Alzheimer disease biomarkers. *Neurology*. https://doi.org/10.1212/WNL.0000000000002923.

Javaid, F. Z., Brenton, J., Guo, L., & Cordeiro, M. F. (2016). Visual and ocular manifestations of Alzheimer's disease and their use as biomarkers for diagnosis and progression. *Frontiers in Neurology*. https://doi.org/10.3389/fneur.2016.00055.

Kaur, C., Foulds, W. S., & Ling, E. A. (2008). Blood-retinal barrier in hypoxic ischaemic conditions: Basic concepts, clinical features and management. *Progress in Retinal and Eye Research*. https://doi.org/10.1016/j.preteyeres.2008.09.003.

Kayabasi, U., Sergott, R. C., & Rispoli, M. (2014). Retinal examination for the diagnosis of Alzheimer's disease. *International Journal of Ophthalmic Practice, 03*(04). https://doi.org/10.4172/2324-8599.1000145.

Kernt, M., Hadi, I., Pinter, F., Seidensticker, F., Hirneiss, C., Haritoglou, C., et al. (2012). Assessment of diabetic retinopathy using nonmydriatic ultra-widefield scanning laser ophthalmoscopy (Optomap) compared with ETDRS 7-field stereo photography. *Diabetes Care, 35*(12), 2459−2463. https://doi.org/10.2337/dc12-0346.

Knudtson, M. D., Lee, K. E., Hubbard, L. D., Wong, T. Y., Klein, R., & Klein, B. E. K. (2003). Revised formulas for summarizing retinal vessel diameters. *Current Eye Research, 27*(3), 143−149. https://doi.org/10.1076/ceyr.27.3.143.16049.

Ko, F., Muthy, Z. A., Gallacher, J., Sudlow, C., Rees, G., Yang, Q., et al. (2018). Association of retinal nerve fiber layer thinning with current and future cognitive decline: A study using optical coherence tomography. *JAMA neurology, 75*(10), 1198–1205.

Koronyo-Hamaoui, M., Koronyo, Y., Ljubimov, A. V., Miller, C. A., Ko, M. K., Black, K. L., et al. (2011). Identification of amyloid plaques in retinas from Alzheimer's patients and noninvasive in vivo optical imaging of retinal plaques in a mouse model. *NeuroImage, 54*(Suppl. 1), S204–S217. https://doi.org/10.1016/j.neuroimage.2010.06.020.

Koronyo, Y., Biggs, D., Barron, E., Boyer, D. S., Pearlman, J. A., Au, W. J., et al. (2017). Retinal amyloid pathology and proof-of-concept imaging trial in Alzheimer's disease. *JCI Insight, 2*(16), 196–208. https://doi.org/10.1172/jci.insight.93621.

Leuba, G., & Saini, K. (1995). Pathology of subcortical visual centres in relation to cortical degeneration in Alzheimer's disease. *Neuropathology and Applied Neurobiology, 21*(5), 410–422. https://doi.org/10.1111/j.1365-2990.1995.tb01078.x.

Levkovitch-Verbin, H., Quigley, H. A., Kerrigan-Baumrind, L. A., D'Anna, S. A., Kerrigan, D., & Pease, M. E. (2001). Optic nerve transection in monkeys may result in secondary degeneration of retinal ganglion cells. *Investigative Ophthalmology and Visual Science, 42*(5), 975–982.

Levkovitch-Verbin, H., Quigley, H. A., Martin, K. R., Zack, D. J., Pease, M. E., & Valenta, D. F. (2003). A model to study differences between primary and secondary degeneration of retinal ganglion cells in rats by partial optic nerve transection. *Investigative Ophthalmology and Visual Science, 44*(8), 3388–3393. https://doi.org/10.1167/iovs.02-0646.

Lingor, P., Teusch, N., Schwarz, K., Mueller, R., Mack, H., Bähr, M., et al. (2007). Inhibition of Rho kinase (ROCK) increases neurite outgrowth on chondroitin sulphate proteoglycan in vitro and axonal regeneration in the adult optic nerve in vivo. *Journal of Neurochemistry, 103*(1), 181–189. https://doi.org/10.1111/j.1471-4159.2007.04756.x.

London, A., Benhar, I., & &Schwartz, M. (2013). The retina as a window to the brain - from eye research to CNS disorders. *Nature Reviews Neurology.* https://doi.org/10.1038/nrneurol.2012.227.

Longstreth, W. T., Larsen, E. K. M., Klein, R., Wong, T. Y., Sharrett, A. R., Lefkowitz, D., et al. (2007). Associations between findings on cranial magnetic resonance imaging and retinal photography in the elderly: The cardiovascular health study. *American Journal of Epidemiology, 165*(1), 78–84. https://doi.org/10.1093/aje/kwj350.

Marziani, E., Pomati, S., Ramolfo, P., Cigada, M., Giani, A., Mariani, C., et al. (2013). Evaluation of retinal nerve fiber layer and ganglion cell layer thickness in Alzheimer's disease using spectral-domain optical coherence tomography. *Investigative Opthalmology and Visual Science, 54*(9), 5953. https://doi.org/10.1167/iovs.13-12046.

Mormino, E. C., Betensky, R. A., Hedden, T., Schultz, A. P., Amariglio, R. E., Rentz, D. M., et al. (2014). Synergistic effect of β-amyloid and neurodegeneration on cognitive decline in clinically normal individuals. *Journal of the American Medical Association Neurology, 71*(11), 1379–1385. https://doi.org/10.1001/jamaneurol.2014.2031.

Mormino, E. C., Betensky, R. A., Hedden, T., Schultz, A. P., Ward, A., Huijbers, W., et al. (2014). Amyloid and APOE ε4 interact to influence short-term decline in preclinical Alzheimer disease. *Neurology, 82*(20), 1760–1767. https://doi.org/10.1212/WNL.0000000000000431.

Mutlu, U., Colijn, J. M., Ikram, M. A., Bonnemaijer, P. W. M., Licher, S., Wolters, F. J., et al. (2018). Association of retinal neurodegeneration on optical coherence tomography with dementia: a population-based study. *JAMA Neurology, 75*(100), 1256–1263.

Murray, C. D. (1926a). The physiological principle of minimal work. I. The vascular system and the cost of blood volume. *Proceedings of the National Academy of Sciences United States of America, 12*, 207–214. https://doi.org/10.1085/jgp.9.6.835.

Murray, C. D. (1926b). The physiological principle of minimum work applied to the angle of branching of arteries. *The Journal of General Physiology,* (4), 835–841. https://doi.org/10.1103/PhysRevC.71.064610.

Ong, Y. T., Hilal, S., Cheung, C. Y., Xu, X., Chen, C., Venketasubramanian, N., et al. (2014). Retinal vascular fractals and cognitive impairment. *Dementia and Geriatric Cognitive Disorders Extra, 4*(2), 305–313. https://doi.org/10.1159/000363286.

Owen, C. G., Rudnicka, A. R., Nightingale, C. M., Mullen, R., Barman, S. A., Sattar, N., et al. (2011). Retinal arteriolar tortuosity and cardiovascular risk factors in a multi-ethnic population study of 10-year-old children; the child heart and health study in England (CHASE). *Arteriosclerosis, Thrombosis, and Vascular Biology, 31*(8), 1933—1938. https://doi.org/10.1161/ATVBAHA.111.225219.

Pantoni, L. (2010). Cerebral small vessel disease: From pathogenesis and clinical characteristics to therapeutic challenges. *The Lancet Neurology, 9*(7), 689—701. https://doi.org/10.1016/S1474-4422(10)70104-6.

Patton, N., Aslam, T., Macgillivray, T., Pattie, A., Deary, I. J., & Dhillon, B. (2005). Retinal vascular image analysis as a potential screening tool for cerebrovascular disease: A rationale based on homology between cerebral and retinal microvasculatures. *Journal of Anatomy, 206*(4), 319—348. https://doi.org/10.1111/j.1469-7580.2005.00395.x.

Price, J. L., & Morris, J. C. (1999). Tangles and plaques in nondemented aging and "preclinical" Alzheimer's disease. *Annals of Neurology, 45*(3), 358—368. https://doi.org/10.1002/1531-8249(199903)45:3<358::AID-ANA12>3.0.CO;2-X.

Qiu, C., Cotch, M. F., Sigurdsson, S., Jonsson, P. V., Jonsdottir, M. K., Sveinbjörnsdottir, S., et al. (2010). Cerebral microbleeds, retinopathy, and dementia: The AGES-Reykjavik Study. *Neurology, 75*(24), 2221—2228. https://doi.org/10.1212/WNL.0b013e3182020349.

Qiu, C., Cotch, M. F., Sigurdsson, S., Klein, R., Jonasson, F., Klein, B. E. K., et al. (2009). Microvascular lesions in the brain and retina: The age, gene/environment susceptibility-Reykjavik study. *Annals of Neurology, 65*(5), 569—576. https://doi.org/10.1002/ana.21614.

Rolls, A., Shechter, R., London, A., Segev, Y., Jacob-Hirsch, J., Amariglio, N., et al. (2008). Two faces of chondroitin sulfate proteoglycan in spinal cord repair: A role in microglia/macrophage activation. *PLoS Medicine, 5*(8), 1262—1277. https://doi.org/10.1371/journal.pmed.0050171.

Sasongko, M. B., Wang, J. J., Donaghue, K. C., Cheung, N., Benitez-Aguirre, P., Jenkins, A., et al. (2010). Alterations in retinal microvascular geometry in young type 1 diabetes. *Diabetes Care, 33*(6), 1331—1336. https://doi.org/10.2337/dc10-0055.

Schrijvers, E. M. C., Buitendijk, G. H. S., Ikram, M. K., Koudstaal, P. J., Hofman, A., Vingerling, J. R., et al. (2012). Retinopathy and risk of dementia: The rotterdam study. *Neurology, 79*(4), 365—370. https://doi.org/10.1212/WNL.0b013e318260cd7e.

Schwab, M. E. (2004). Nogo and axon regeneration. *Current Opinion in Neurobiology.* https://doi.org/10.1016/j.conb.2004.01.004.

Silver, J., & Miller, J. H. (2004). Regeneration beyond the glial scar. *Nature Reviews Neuroscience.* https://doi.org/10.1038/nrn1326.

Streilein, J. W. (2003). Ocular immune privilege: Therapeutic opportunities from an experiment of nature. *Nature Reviews Immunology.* https://doi.org/10.1038/nri1224.

Taylor, A. W., & Streilein, J. W. (1996). Inhibition of antigen-stimulated effector T cells by human cerebrospinal fluid. *Neuroimmunomodulation, 3*(2—3), 112—118.

Ting, D. S. W., Cheung, C. Y., Lim, G., Tan, G. S. W., Quang, N. D., Gan, A., et al. (2017). Development and validation of a deep learning system for diabetic retinopathy and related eye diseases using retinal images from multiethnic populations with diabetes. *Journal of the American Medical Association, 318*(22), 2211—2223. https://doi.org/10.1001/jama.2017.18152.

Wilbanks, G. A., & Wayne Streilein, J. (1992). Fluids from immune privileged sites endow macrophages with the capacity to induce antigen-specific immune deviation via a mechanism involving transforming growth factor-beta. *European Journal of Immunology, 22*(4), 1031—1036. https://doi.org/10.1002/eji.1830220423.

Williams, M. A., McGowan, A. J., Cardwell, C. R., Cheung, C. Y., Craig, D., Passmore, P., et al. (2015). Retinal microvascular network attenuation in Alzheimer's disease. *Alzheimer's and Dementia: Diagnosis, Assessment and Disease Monitoring, 1*(2), 229—235. https://doi.org/10.1016/j.dadm.2015.04.001.

CHAPTER 23

Prediction of Alzheimer's disease

Athanasios Alexiou[1,2], Stylianos Chatzichronis[2,3], Ghulam Md Ashraf[4]

[1]Novel Global Community Educational Foundation, Hebersham, NSW, Australia; [2]AFNP Med Austria, Wien, Austria; [3]National and Kapodistrian University of Athens, Department of Informatics and Telecommunications, Bioinformatics Program, Zografou, Greece; [4]King Fahd Medical Research Center, King Abdulaziz University, Jeddah, Saudi Arabia

List of abbreviations

^{18}F-AV-45 Florbetapir-fluorine-18
^{31}P-PMRS Phosphorus MR spectroscopy
AD Alzheimer's disease
ADAS-Cog13 Alzheimer's Disease Assessment Scale-Cognitive Subtest
ADLs Activities of daily living
ADNI Alzheimer's Disease Neuroimaging Initiative
APP$_{669-711}$ Amyloid-β Precursor Protein
AV45-PET Florbetapir-fluorine-18 positron emission tomography
Aβ Amyloid beta
CDRSB Clinical dementia rating sum of boxes score
CSF Cerebrospinal fluid
DAT Alzheimer's-type dementia
DLB Dementia with Lewy bodies
FDG-PET Fludeoxyglucose positron emission tomography
GP Gaussian process
IADLs Instrumental activities of daily living
mAChR Muscarinic acetylcholine receptor antagonist
MCI Mild cognitive impairment
MCMC Markov chain Monte Carlo
MKL Multiple kernel learning
MRI Magnetic resonance imaging
PET Positron emission tomography
pGP Personalized Gaussian process
PHS Polygenic hazard score
ROC Receiver operating characteristic
rsEEG Resting state eyes-closed electroencephalographic
SD Standard deviation
SPECT Single-photon emission computed tomography
SVMs Support vector machines
VPH Virtual physiological human
T1-MRI T1 Weighted Magnetic Resonance Imaging

Diagnosis and Management in Dementia
ISBN 978-0-12-815854-8, https://doi.org/10.1016/B978-0-12-815854-8.00023-9

Mini-dictionary of terms

- Geriatric assessment programs are multidimensional diagnostic processes for the identification of functional and pathophysiological problems in the life of a frail person through long-term use of a clinical decision support system.
- An advanced geriatrics assessment includes evaluation of the following systems: audition, balance, blood pressure, bones, cardiovascular, central nervous, endocrine, gastrointestinal, genitourinary, hematologic, immune, muscle, peripheral nervous, pulmonary, regulation, renal, smell, temperature, thirst, and vision.
- A Bayesian network uses a directed acyclic graph (G) with (N) nodes and a joint distribution of the form $P(N) = \prod_{n \in N} p(n \vee pa(n))$ in order to represent a set of variables $n \in N$ and their conditional dependencies $pa(n)$.
- Machine learning combines sophisticated algorithms and statistical methods for optimized data analysis through supervised and unsupervised learning.
- A null model is a graph that matches one specific graph in parts of its structural features.
- T1-MRI outputs T1 images that show emphasis in displaying FAT-type tissues.

Introduction

Recent studies have shown that the first pathophysiological symptoms of Alzheimer's disease (AD) could start at a very young age, making it difficult to determine the right time for appropriate inhibitory therapy. Medications targeting dementia or depression, or the use of advanced nanotechnological products, could reduce cognitive decline or neuropsychiatric symptoms in AD, but unfortunately they do not offer a holistic treatment. Given the complexity of the disease, many researchers worldwide have focused on alternative solutions such as reliable early diagnosis or prediction in an effort to investigate a broad range of clinical trial results and formulate a relationship between lesions and symptoms as well as diagnosis, monitoring, and disease management tools. Scientists using the collective manner of the virtual physiological human (VPH) framework are establishing medical expert systems based on machine learning techniques that offer clinical support for investigating a large number of neurodegenerative diseases like AD, age-related dementia, and human frailty. Clinical prediction of AD is currently based on clinical expert systems such as cognitive-behavioral testing (Fig. 23.1) within mild cognitive impairment (MCI), the combination and evaluation of cerebrospinal fluid (CSF) and plasma biomarkers, neuroimaging analyzing software, advanced statistical methods, and virtual reality programs that simulate and evaluate virtual activities of daily life. A standardized procedure for the development of an integrative multistage VPH model for the accurate evaluation of AD progression must also include advanced monitoring of the efficacy of treatment and improved cost-effectiveness of diagnostic protocols in the general population. Several components like neuropsychological test scores, molecular data, imaging and electrophysiological data, CSF and plasma biomarkers, medical history

Figure 23.1 *Cognitive and functional assessment in groups with subjective cognitive decline, mild cognitive impairment, and dementia.* The most common cognitive assessment procedures applied in clinical or individualized testing for Alzheimer's disease. These methods mainly target groups with subjective cognitive decline, mild cognitive impairment, or dementia.

and demographic details, and many other risk factors must be modeled within an ontology database (Ascoli, Donohue, & Halavi, 2007; Bota, Dong, & Swanson 2005; Fensel, 2000; Lein et al., 2007; Martone et al., 2002, 2003, 2007; Muller, Kenny, & Sternberg, 2004; Wang, Williams, & Manly, 2003) that can provide data storage and manipulation to extract an accurate clinical decision for AD prediction.

Geriatric assessment programs

By using interdisciplinary assessment scales, geriatricians manage to recognize chronic diseases like temporary functional disabilities or other risk factors and comorbidities that frail patients usually present at 65 years and over. Several systems must be measured by using the necessary activity-of-daily-living (ADL) and instrumental-activity-of-daily-living (IADL) methods. In the literature we can access many published screening tests involving frailty and geriatric clinical medicine (Appels & Scherder, 2010; Brodaty, Low, Gibson, & Burns, 2006; Buys, 2000; Cassel, 2003; Hancock & Larner, 2011; Mathuranth, Nestor, Berrios, Rakowicz, & Hodges, 2000; Moniz-Cook et al., 2008; Wright & Shay, 2000; Zekry et al., 2008) from eating, dressing, bathing, transferring, and toileting up to heavier housework, managing finances, and telephoning activities. To recognize a decline in cognitive function or depression, the Folstein Mini-Mental State Examination, the Yesavage Geriatric Depression Scale, and the PHQ-9 are well-known procedures. While several clinical trials for elderly medication therapies led to novel therapeutic products like cholinesterase inhibitors for AD and many other afflictions, medication errors are a common problem that may be present in many different phases of the process. Therapeutic failure in frail persons can also occur due to prescription errors, difficulties in doctor—patient communication, and treatment dispensing and administration. According to these principles, we can mention a few of the main principles and questions to be addressed for clinical decisions generated from an expert system for accurate geriatric pharmacotherapy:

- Is medication necessary?
- Is medication adjusted for age?
- Do new medications have any adverse effects related to other medications or diseases?
- Have the initial doses been adjusted for the frail patient?
- Has the geriatrician defined the therapeutic endpoints?

According to the *Geriatric Medicine Survival Handbook* (Misiaszek, 2008), an advanced and accurate geriatric evaluation consists of several tests as well as the physical examination of the frail person, which can be summarized as follows and stored in a biological database for further modeling and formulation:

- Personal and family medical history
- Brief biography
- Current living situation
- Medication review
- Adverse drug reactions like allergies, intolerabilities, or other side effects
- Functional inquiry regarding ADLs, IADLs, or other support in daily living
- Physical examination
- Review of individual human body systems
- Cognitive testing and behavior
- Laboratory and imaging tests
- Neurodegeneration

Predicting Alzheimer's disease development

Alzheimer's disease prediction is a nontrivial task in the scientific community but is still considered ambiguous because of difficulty using conventional methods to distinguish the phases over which the disease develops. Recent studies have mainly focused on image analysis methods for identifying biomarkers of AD prediction due to the noninvasiveness of those methods. Machine learning has been a tool for prediction for various scientific fields for automated prediction. Thus, although early studies did not include such a process, it is widely used in AD prediction, especially in imaging biomarkers. Other studies follow alternative but felicitous approaches to predict specific stages of AD-like modeling (Table 23.1).

Table 23.1 Methods targeting Alzheimer's disease prediction.

Prediction test	Benefits
Beyond classical regression model (Teipel et al., 2015)	High accuracy, validation in large data, 6 −31 months of prediction
Comparison of cognitive criteria and associated neuroimaging biomarkers (Callahan et al., 2015)	High accuracy, 24 months of prediction, low cost
Free and cued selective reminding test (Di Stefano et al., 2015)	Validation in large clinical trials, minimal false-negative values
Delay of gamma responses (Basar et al., 2016)	Discrimination of healthy and AD event-related and sensory-evoked oscillations, discovery of 100 ms latency
Bayesian model (Alexiou et al., 2017; Matzavinos et al., 2017)	High accuracy, efficient use of all up-to-now validated biomarkers
Cortical neural synchronization mechanisms in MCI patients (Babiloni et al., 2017)	Discovery of low individual alpha frequency peak, low alpha in posterior and high delta in parental
EEG machine learning (Simpraga et al., 2017)	High accuracy, better understanding of brain cholinergic events using scopolamine
Machine learning in amyloid imaging (Mathotaarachchi et al., 2017)	High accuracy, 24 months prediction
Null longitudinal model (Gavidia-Bovadilla, Kanaan-Izquierdo, Mataró-Serrat, & Perera-Lluna 2017)	High accuracy, prediction 1.9 years earlier for females and 1.4 years earlier for males than prediction times from usual methods
Quantification of MRI deformation (Long et al., 2017)	High accuracy
pGP for AD prediction (Peterson et al., 2017)	Experimentally has been shown that pGP models are more efficient from population-level models in AD prediction
Predicting AD with PHS (Desikan et al., 2017)	High accuracy, large dataset
Changes in phosphate metabolism (Rijpma et al., 2018)	Discovering that energy metabolism changes in mild AD
Petri net modeling (Ashraf et al., 2018)	Discovery of role of calpain and its inhibitor calpastatin in the progression of a neural pathway associated with AD.
Subnetwork selection and graph kernel principal component analysis based on minimum spanning tree brain functional network (Cui et al., 2018)	High accuracy in classifying subjects with AD
Checking alternations in FDG-PET images (Popuri et al., 2018)	Validated use of FDG-PET DAT score as biomarker
Predicting AD using cerebral blood flow in xenon-enhanced CT (Sase et al., 2018)	Efficient classification of AD and MCI
QEEG with Lewy bodies (Stylianou et al., 2018)	High accuracy

Continued

Table 23.1 Methods targeting Alzheimer's disease prediction.—cont'd

Prediction test	Benefits
Regional association with tau deposition (Swinford, Risacher, Charil, Schwarz, & Saykin, 2018)	Discovery of self-memory related with tau in frontal region and informative-memory with tau in parietal area
Multimodal MRI and high-throughput brain phenotyping (Wang et al., 2018)	High accuracy
Selecting ROIs using multiple kernel learning (Rondina et al., 2017)	High accuracy
High-performance amyloid-β biomarkers (Nakamura et al., 2018)	High accuracy, discovery of the biomarker $APP_{669-711}/AB_{1-42}$

Methods for prediction of Alzheimer's disease development or progression and their advantages related to accuracy; identification of new potential biomarkers and the period of prediction.

Imaging prediction methods

A contemporary imaging method was created to predict whether patients with MCI will develop AD (Teipel, Kurth, Krause, & Grothe, 2015) using data from the Alzheimer's Disease Neuroimaging Initiative (ADNI) database. In particular, it used regularized regression in the form of both penalized and unpenalized regression models to extract necessary information from florbetapir-fluorine-18 positron emission tomography (AV45-PET), fludeoxyglucose positron emission tomography (FDG-PET), and magnetic resonance imaging (MRI) images. In general, experiments showed that models using penalties for overfitting such as the logistic regression and the penalized Cox model produce better results than those of similar unpenalized approaches. Moreover, an emphasis is shown for partial volume effect correction in positron emission tomography (PET) images because it was noted that this correction plays an optimal predictive role in the results produced in FDG-PET when used in contrast to AV45-PET. Another method utilizing data from the ADNI database combines similar methods to achieve better predictions for identifying patients with AD and assesses how the disease progresses over the following 24 months (Callahan, Ramirez, Berezuk, Duchesne, & Black, 2015). It mainly focuses on the different standard deviations (SDs) that appear in data sets using either one or two memory tests each time. This has shown that only when stricter thresholds are applied to single memory tests do they show results similar to those of two measured tests regarding quality, while emphasis is given to the use of at least two memory tests due to a 10% increase in optimal prediction accuracy. Further, improvements were abetted by the use of apolipoprotein ε4 status and whole brain atrophy to attain precision. Also, having specific clinical restrictions, an additional memory test could achieve meliorated results. In short, high accuracy has been proved, regardless of biomarkers, when memory is assessed using delayed recall of a short story and a word list using a cutoff of <-1 SD below normative references.

Machine learning and advanced statistical prediction methods

In later research for AD prediction, the use of the free and cued selective reminding test was applied to examine patients diagnosed with subjective cognitive decline or MCI who are likely to develop AD (Di Stefano et al., 2015). The elder groups compared in this study were selected from the GuidAge Trial, and evaluation of the findings was done using receiver operating characteristic (ROC) plots. A cutoff was used to identify subjects who seemed to develop AD with minimal risk. The results revealed that the test was quite sufficient for large clinical trials. However, the study was transacted with a lack of physiopathological markers for valuation of AD that could be utilized to strengthen verification of results. Additional research estimated AD stages, primarily checking gamma waves associated with AD (Basar, Emek-Savas, Güntekina, & Yener, 2016). After analyzing three gamma frequency bands within four time windows in AD subjects, a delay was noticed in event-related and sensory-evoked oscillations of at least 100 ms.

Other approaches have used simulation to achieve AD prediction. A recent study presents a hybrid Bayesian model (Alexiou, Mantzavinos, Greig, & Kamal, 2017; Mantzavinos & Alexiou, 2017) offering modeling and a combination of several biomarkers and risk factors to predict AD at prodromal stages (Fig. 23.2). Markov chain Monte Carlo (MCMC) and Bayesian theory are the fundamental elements of research for predicting the development of the disease. This approach successfully incorporates all known biomarkers into a single probabilistic model to achieve an approximate 100% accuracy excluding the MCMC error; this is an invasive method that accounts for current knowledge. Similarly, an EEG study checked differences in resting state eyes-closed electroencephalographic (rsEEG) rhythms between AD MCI and control subjects using data from the Informal European Consortium (Babiloni et al., 2017). It concluded that different features might exist among rsEEG rhythms including cortical neural synchronization at delta and alpha frequencies underpinning brain arousal and vigilance in quiet wakefulness.

Machine learning techniques and EEG have been combined to limit the development of AD (Simpraga et al., 2017). It has been hypothesized that deficits of cholinergic signaling contribute to EEG slowing in AD, which is also supported by the reversal of EEG slowing by cholinergic drugs. Using data from healthy subjects receiving scopolamine, scientists developed an index of the muscarinic acetylcholine receptor antagonist (mAChR) including 14 EEG biomarkers. This mAChR index yielded higher classification performance than that of any single EEG biomarker, with cross-validated accuracy, sensitivity, specificity, and precision ranging from 88% to 92%. The mAChR index also discriminated healthy elderly patients from those with AD; however, an index optimized for AD pathophysiology provided a better classification. Another recent study combines machine learning algorithms with PET (Mathotaarachchi et al., 2017), targeting patients with AD in primary stages to recognize possible developments over the upcoming 24 months. The novelty of this algorithm is based on the successful automated separation

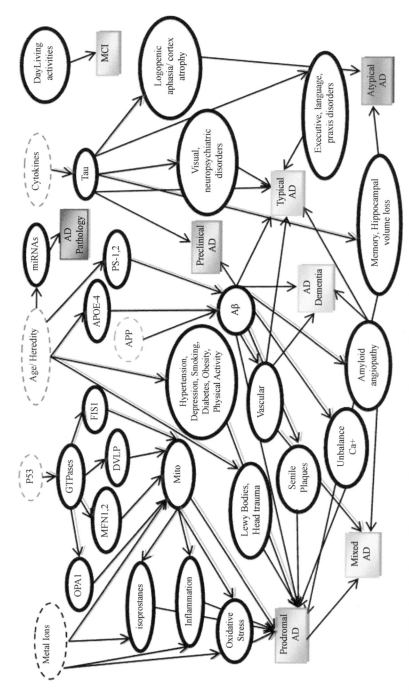

Figure 23.2 Alzheimer's disease biomarkers imported to a prediction model taken from the Creative Commons Attribution License (CC BY) study of Mantzavinos et al. (2017). The most commonly used biomarkers related to Alzheimer's disease early diagnosis and prediction expressed through a Bayesian network (Mantzavinos et al., 2017).

of stable MCI and progressive MCI subjects using random undersampling random forest classifiers, which until recently was a challenging task. Images from PET scans and corresponding CSF biomarkers from the ADNI database were used as well. This analytic method experimentally proved to be the most efficient compared with previously known algorithms, reaching 84% precision and a high 0.906 ROC criterion.

Gavidillia-Bovadilla et al. (2017) has created null aging-based models for early and efficient prediction of subjects with AD. Firstly, the authors supported the idea of finding changes in healthy subjects that had occurred in the brain due to aging. Then they constructed a linear mixed-effects model based on changes in 166 AD biomarkers originated from ADNI database MRI images. Afterward, they identified significant variant and quasivariant brain regions over time to use to create the aged-based null model. Support vector machine (SVM) models have also played a significant role, as they have been used to classify subjects and for early detection of the transformation of MCI to AD. What has been found are reductions in most cortical volumes and thicknesses in subcortical regions, including greater atrophy in the hippocampus. Likewise, as compared with standard clinical diagnosis methods, SVM classifiers predicted the conversion of AD to occur 1.9 years earlier for females (ACC: 72.5%) and 1.4 years earlier for males (ACC: 69%).

Another method, the personalized Gaussian process (GP) model (Peterson, Rudovic, Guerrero, & Picard, 2017), uses a personalized Gaussian process (pGP) model to predict the critical metrics of AD progression (mini-mental state examination, ADAS-Cog13, clinical dementia rating sum-of-boxes score, and CSF) based on each patient's previous visits. The authors exploit the advantage, noted in previous studies, of the probabilistic consistency of such models in clinical analyses when a nonparametric regression approach is followed. They create a model taking multimodal data from other patients' visits following the basic Gaussian process regression. Later, this model is adopted sequentially over time to a new patient using domain–adaptive GPs to form a patient pGP. It is shown that the development of such personalized models is more efficient in prediction than are conventional models using Gaussian process regression. However, more significant improvements are needed, as in this research the prediction is limited to one visit forward.

Cognitive and genomics methods

Another innovative classifying technique for AD prediction is oriented to the quantification of MRI deformation (Long, Chen, Jiang, & Zhang, 2017). It is structured within two main phases, where one distinguishes AD and MCI from healthy elders and the second predicts AD conversion in MCI patients by computing and analyzing the regional morphological differences of brain between groups. Distance between each pair of subjects was quantified from a symmetric diffeomorphic registration followed by an embedding algorithm and a learning approach for classification. Accuracy percentages for classification were 96.5% mild AD versus control group, 88.99% for separating progressive MCI and stable MCI, and a corresponding 91.74% for progressive MCI and control group. Finally, a large deformation in the hippocampus and amygdala in patients with progressive MCI has been noted, while diffusive morphological changes in the whole-brain gray matter were prominent in identifying mild or moderate AD patients. A drawback of the algorithm is that reaching high levels of accuracy demands enormous computational time, and as the authors suggest, a desirable equilibrium should be found.

A different approach is focused on merging stats from single-nucleotide polymorphisms and apolipoproteins related to AD into a polygenic hazard score (PHS) (Desikal et al., 2017). By integrating population-based incidence proportion and genome-wide data into a genetic epidemiology framework, scientists have developed a PHS for quantifying age-associated risk for developing AD. A unique study is investigating brain -phospholipids and energy metabolism in AD by 31P MRS (Rijpma, van der Graaf, Meulenbroek, Olde Rikkert, & Heerschap, 2018). Increased phosphocreatine levels were found in regions that show early degeneration in AD, but not in the ACC, a region known to be involved in this disease at a later stage. Together with increased pH, this indicates that energy metabolism is altered in mild AD. With the development of Petri Nets models, it is feasible to understand how the neuronal pathways work and in that way modify the process of AD (Ashraf, Ahmad, Ali, & Ul-Haq 2018). The primary role of the Calpain's inhibitor calpastatin in the development of AD has been analyzed. The AD prediction problem recently has been enhanced by a complicated but effective method classifying AD, MCI, and healthy subjects using data from the ADNI database (Cui et al., 2018). It involves minimum spanning tree, gSpan, graph kernel principal component analysis, and SVM models. It achieves 98.3%, 91.3%, and 77.3% for MCI versus NC, AD versus NC, and AD versus MCI, respectively, but needs to be applied to more substantial data.

Checking alternations in FDG-PET images in the form of a score constructs a new way of predicting Alzheimer's-type dementia (DAT) (Popuri et al., 2018). In that way, the metabolism of the patient is being watched. A score defined as FDG-PET DAT score is created that ranges from 0 to 1, showing the probability of a subject to suffer from DAT. Another imaging method includes xenon-enhanced tomography to successfully classify mild AD subjects from control groups (Sase, Yamamoto, Kawashima, Tan, & Sawa, 2018). Using QEEG measurements in a study oriented to find differences between dementia with Lewy bodies (DLB), AD, Parkinson's disease dementia, and healthy subjects (Stylianou et al., 2018), it was shown that this method can efficiently distinguish AD from DLB in patients. This kind of diagnosis is supported by high percentages of accuracy, specificity, and sensitivity.

Additional studies for prediction of AD include associated examination proteins. Notably, Swiford checked for a relationship between self- and informant-memory concerns and tau aggregation using ^{18}F-flortaucipir PET data from the ADNI database (Swinford et al., 2018). It was noted that self-memory is related to tau located in the frontal brain regions.

Multimodal MRI and high-throughput brain phenotyping has also been applied in AD prediction (Wang et al., 2018). Scientists tested the utility of structural MRI and diffusion MRI as imaging markers of AD using high-throughput brain phenotyping including morphometry and whole-brain tractography, and machine learning analytics for classification. The resulting machine learning model was trained on large-scale brain phenotypes (n = 34,646) classified as AD, MCI, and SMC with unprecedented accuracy (AD/SMC: 97% accuracy; MCI/SMC: 83% accuracy; AD/MCI: 98% accuracy) with strict iterative nested tenfold cross-validation. Another interesting approach applies multiple kernel learning (MKL) on brain regions of interest (Rondina et al., 2017) to predict AD patients with high accuracy. In fact, imaging data have been taken from the fMRI Atlas project in the form of T1-MRI, ^{18}F-FDG-PET, and regional cerebral blood flow single-photon emission computed tomography (SPECT). Two approaches have been used, an SVM model dealing with the whole-brain and MKL-region of interest (MKL-ROI) methods, with their performances compared. MKL-ROI seems more efficient in ^{18}F-FDG-PET images (92%, 5% accuracy). In later research, a novel formula was discovered for identifying biomarkers related to early-stage AD using blood samples from PET images to examine amyloid-β in plasma (Nakamura et al., 2018). Immunoprecipitation mass spectrometry was used, due to its effectiveness, to measure Aβ in plasma and Pittsburgh compound B-PET for evaluation of results. The biomarkers discovered were the ratios $APP_{669-711}/A\beta_{1-42}$ and $A\beta_{1-40}/A\beta_{1-42}$. Due to the use of large, different, and multiple data sets, those biomarkers show great potential but need further validation to be applied in clinical trials.

Conclusion

With more than 35,000,000 cases worldwide, AD is among the six most deadly diseases. Unfortunately, AD fear has remained because of failed clinical trials for new effective drugs based mainly on the amyloid hypothesis. Therefore, the scientific community has adopted and integrated artificial intelligence methods within the framework of AD management, targeting early diagnosis through accurate prediction. It seems at present that the best treatment is prevention.

Key facts of predicting Alzheimer's disease

- Many disorders are not caused by aging itself and do not occur in every elderly patient, such as anemia, cardiovascular diseases, cancer, stroke, diabetes mellitus, hypothyroidism, osteoporosis, prostate disease, sexual dysfunction, vision loss, and free radicals.
- Various risk factors, functional status, physical activation, nutritional habits, baseline demographic data, medication history, family history, and other comorbidity factors must be recorded for an accurate AD prediction test.

- The assessment of cognitive function usually includes mini-mental state examination, clinical dementia rating, medical history, physical examination, Hachinski ischemic scale, the geriatric depression scale, functional assessment questionnaire, the neuropsychiatric inventory questionnaire and the Alzheimer's disease assessment scale-cognitive (ADAS-COG).
- Many patients with mild cognitive impairment display the same morphological changes as those of AD patients and show no symptom progression, whereas others eventually develop other types of dementia. Thus, it is not certain whether patients with mild cognitive impairment will convert this impairment to AD.
- The majority of the modern predictive methods are focused on machine learning using either Electroencephalography (EEG) data or imaging data from MRI, positron emission tomography (PET), and SPECT due to the high accuracy and non-invasiveness of those methods.
- In vitro scanning tunneling microscopy detects amyloid beta, two-photon Rayleigh scattering assay techniques can be used for tau detection, and in vivo MRI and optical imaging can be used for detecting amyloid beta plaques.

Summary points

- This chapter focuses on tools and methods to predict AD development.
- AD is classified as prodromal AD, AD dementia, typical AD, atypical AD, mixed AD, preclinical AD, Alzheimer's pathology, and MCI.
- Hippocampus volume or medial temporal lobe atrophy in MRI and temporoparietal/precuneus hypometabolism or hypoperfusion on PET or SPECT biomarkers of neurological injury are considered biomarkers with high efficacy.
- When the two biomarkers amyloid beta and tau/phosphorylated tau protein are positively measured, the probability of AD development increases.
- Geriatric assessment programs offer a holistic management of human frailty regarding diagnosis and proper pharmacotherapy for clinicians.

Acknowledgments

Athanasios Alexiou gratefully acknowledges the facilities provided by AFNP Med in Austria. Ghulam Md Ashraf thanks the Almighty Allah and gratefully acknowledges the facilities provided by King Fahd Medical Research Center (KFMRC), King Abdulaziz University, Jeddah, Saudi Arabia.

References

Alexiou, A., Mantzavinos, V., Greig, N. H., & Kamal, M. A. (2017). A Bayesian model for prediction and early diagnosis of Alzheimer's disease. *Frontiers in Aging Neuroscience, 31*(9), 77.

Appels, B., & Scherder, E. (2010). The diagnostic accuracy of dementia-screening instruments with an administration time of 10 to 45 minutes for use in secondary care: A systematic review. *American Journal of Alzheimer's Disease and Other Dementias, 25*, 301–316.

Ascoli, G. A., Donohue, D. E., & Halavi, M. (2007). NeuroMorpho. Org: A central resource for neuronal morphologies. *The Journal of Neuroscience, 27*, 9247–9251.

Ashraf, J., Ahmad, J., Ali, A., & Ul-Haq, Z. (2018). Analyzing the behavior of neuronal pathways in Alzheimer's disease using Petri net modeling approach. *Frontiers in Neuroinformatics, 12*, 26. https://doi.org/10.3389/fninf.2018.00026.

Babiloni, C., Del Percio, C., Lizio, R., Noce, G., Cordone, S., Lopez, S., et al. (2017). Abnormalities of cortical neural synchronization mechanisms in subjects with mild cognitive impairment due to Alzheimer's and Parkinson's diseases: An EEG study. *Journal of Alzheimer's Disease, 59*(1), 339–358.

Basar, E., Emek-Savas, D. D., Güntekina, B., & Yener, G. G. (2016). Delay of cognitive gamma responses in Alzheimer's disease. *NeuroImage: Clinical, 11*, 106–115.

Bota, M., Dong, H. W., & Swanson, L. W. (2005). Brain architecture management system. *Neuroinformatics, 3*, 15–48.

Brodaty, H., Low, L., Gibson, L., & Burns, K. (2006). What is the best dementia screening instrument for general practitioners to use? *American Journal of Geriatric Psychiatry, 14*, 391–400.

Buys, C. H. (2000). Telomeres, telomerase, and cancer. *New England Journal of Medicine, 342*, 1282–1283.

Callahan, B. L., Ramirez, J., Berezuk, C., Duchesne, S., & Black, S. E. (2015). Predicting Alzheimer's disease development: A comparison of cognitive criteria and associated neuroimaging biomarkers. *Alzheimer's Research and Therapy, 7*, 68.

Cassel, C. K. (2003). *Geriatric medicine. An evidence-based approach* (4th ed.). New York: Springer.

Cui, X., Xiang, J., Guo, H., Yin, G., Zhang, H., Lan, F., et al. (2018). Classification of Alzheimer's disease, mild cognitive impairment, and normal controls with subnetwork selection and graph Kernel Principal Component Analysis based on minimum spanning tree brain functional network. *Frontiers in Computational Neuroscience, 12*, 31.

Desikan, R. S., Fan, C. C., Wang, Y., Schork, A. J., Cabral, H. J., Cupples, L. A., et al. (2017). Genetic assessment of age-associated Alzheimer disease risk: Development and validation of a polygenic hazard score. *PLoS Medicine, 14*(3), e1002258. https://doi.org/10.1371/journal.pmed.1002258.

Di Stefano, F., Epelbaum, S., Coley, N., Cantet, C., Ousset, P. J., Hampel, H., et al. (2015). Prediction of Alzheimer's disease dementia: Data from the GuidAge prevention trial. *Journal of Alzheimer's Disease, 48*, 793–804.

Fensel, D. (2000). *Ontologies: Silver bullet for knowledge management and electronic commerce.* Springer-Verlag Berlin Heidelberg.

Gavidia-Bovadilla, G., Kanaan-Izquierdo, S., Mataró-Serrat, M., & Perera-Lluna, A. (2017). Early prediction of Alzheimer's disease using null longitudinal model-based classifiers. *PLoS One, 12*(1), e0168011.

Hancock, P., & Larner, A. (2011). Test your memory test: Diagnostic utility in a memory clinic population. *International Journal of Geriatric Psychiatry, 25*, 976–980.

Lein, E. S., et al. (2007). Genome-wide atlas of gene expression in the adult mouse brain. *Nature, 445*(7124), 168–176.

Long, X., Chen, L., Jiang, C., & Zhang, L. (2017). Prediction and classification of Alzheimer disease based on quantification of MRI deformation. *PLoS One, 12*(3), e0173372.

Mantzavinos, V., & Alexiou, A. (2017). Biomarkers for Alzheimer's disease diagnosis. *Current Alzheimer Research, 14*(11), 1149–1154.

Martone, M. E., Gupta, A., Wong, M., Qian, X., Sosinsky, G., Ludaescher, B., et al. (2002). A cell centered database for electron tomographic data. *Journal of Structural Biology, 138*, 145–155.

Martone, M. E., Zaslavsky, I., Gupta, A., Memon, A., Tran, J., Wong, W., et al. (2007). The smart atlas: Spatial and semantic strategies for multiscale integration of brain data. In A. Burger, et al. (Eds.), *Anatomy ontologies for Bioinformatics: Principles and practice.* London: Springer-Verlag.

Martone, M. E., Zhang, S., Gupta, A., Qian, X., He, H., Price, D. L., et al. (2003). The cell-centered database: A database for multiscale structural and protein localization data from light and electron microscopy. *Neuroinformatics, 1*, 379–395.

Mathotaarachchi, S., Pascoal, T. A., Shin, M., Benedet, A. L., Kang, M. S., Beaudry, T., et al. (2017). Identifying incipient dementia individuals using machine learning and amyloid imaging. *Neurobiology of Aging, 59*, 80–90.

Mathuranth, P., Nestor, P., Berrios, G., Rakowicz, W., & Hodges, J. R. (2000). A brief cognitive test battery to differentiate Alzheimer's disease and frontotemporal dementia. *Neurology, 55*, 1613–1620.

Misiaszek, B. C. (2008). *Geriatric medicine survival Handbook.* Michael G. DeGroote School of Medicine at McMaster University.

Moniz-Cook, E., Verooij-Dassen, M., Woods, R., Verhey, F., Chattat, R., DE Vugt, M., et al. (2008). A European consensus on outcome measures for psychosocial intervention research in dementia care. *Aging and Mental Health, 12*(1), 14—29.

Muller, H. M., Kenny, E. E., & Sternberg, P. W. (2004). Textpresso: An ontology-based information retrieval and extraction system for biological literatures. *PLoS Biology, 2,* e309.

Nakamura, A., Kaneko, N., Villemagne, V. L., Kato, T., Doecke, J., Doré, V., et al. (2018). High performance plasma amyloid-β biomarkers for Alzheimer's disease. *Nature, 554*(7691), 249—254.

Peterson, K., Rudovic, O., Guerrero, R., & Picard, R. W. (2017). *Personalized Gaussian processes for future prediction of Alzheimer's disease progression* (arXiv preprint arXiv:1712.00181).

Popuri, K., Balachandar, R., Alpert, K., Lu, D., Bhalla, M., Mackenzie, I. R., et al. (2018). Development and validation of a novel dementia of Alzheimer's type (DAT) score based on metabolism FDG-PET imaging. *NeuroImage Clinical, 18,* 802—813.

Rijpma, A., van der Graaf, M., Meulenbroek, O., Olde Rikkert, M. G. M., & Heerschap, A. (2018). Altered brain high-energy phosphate metabolism in mild Alzheimer's disease: A 3-dimensional 31P MR spectroscopic imaging study. *NeuroImage Clinical, 18,* 254—261.

Rondina, J. M., Ferreira, L. K., de Souza Duran, F. L., Kubo, R., Ono, C. R., Leite, C. C., et al. (2017). Selecting the most relevant brain regions to discriminate Alzheimer's disease patients from healthy controls using multiple kernel learning: A comparison across functional and structural imaging modalities and atlases. *NeuroImage Clinical, 17,* 628—641.

Sase, S., Yamamoto, H., Kawashima, E., Tan, X., & Sawa, Y. (2018). Discrimination between patients with mild Alzheimer's disease and healthy subjects based on cerebral blood flow images of the lateral views in xenon-enhanced computed tomography. *Psychogeriatrics, 18*(1), 3—12.

Simpraga, S., Alvarez-Jimenez, R., Mansvelder, H. D., van Gerven, J. M. A., Groeneveld, G. J., Poil, S. S., et al. (2017). EEG machine learning for accurate detection of cholinergic intervention and Alzheimer's disease. *Scientific Reports, 7*(1), 5775.

Stylianou, M., Murphy, N., Peraza, L. R., Graziadio, S., Cromarty, R., Killen, A., et al. (2018). Quantitative electroencephalography as a marker of cognitive fluctuations in dementia with Lewy bodies and an aid to differential diagnosis. *Clinical Neurophysiology, 129*(6), 1209—1220.

Swinford, C. G., Risacher, S. L., Charil, A., Schwarz, A. J., & Saykin, A. J. (2018). Memory concerns in the early Alzheimer's disease prodrome: Regional association with tau deposition. *Alzheimers Dement, 10,* 322—331.

Teipel, J. S., Kurth, J., Krause, B., & Grothe, M. J. (2015). The relative importance of imaging markers for the prediction of Alzheimer's disease dementia in mild cognitive impairment - beyond classical regression. *NeuroImage Clinical, 8,* 583—593.

Wang, J., Williams, R. W., & Manly, K. F. (2003). WebQTL: Web-based complex trait analysis. *Neuroinformatics, 1*(10), 299—308.

Wang, Y., Xu, C., Lee, S., Stern, Y., Kim, J. H., Yoo, S., et al. (2018). Accurate prediction of Alzheimer's disease using multi-modal MRI and high-throughput brain phenotyping. *bioRxiv,* 255141.

Wright, W. E., & Shay, J. W. (2000). Telomere dynamics in cancer progressionand prevention: Fundamental differences in humanand mouse telomere biology. *Nature Medicine, 6,* 849—851.

Zekry, D., Herrmann, F., Gradjean, R., Meynet, M. P., Michel, J. P., Gold, G., et al. (2008). Demented versus non-demented very old inpatients: The same comorbidities but poorer functional and nutritional status. *Age and Ageing, 37,* 83—89.

CHAPTER 24

Addenbrooke's Cognitive Examination

Jordi A. Matias-Guiu

Department of Neurology, Hospital Clínico San Carlos, San Carlos Institute for Health Research (IdISSC), Universidad Complutense, Madrid, Spain

List of abbreviations

ACE Addenbrooke's Cognitive Examination (any version)
ACE-I Addenbrooke's Cognitive Examination (first version)
ACE-III Addenbrooke's Cognitive Examination (third version)
ACE-R Addenbrooke's Cognitive Examination Revised

Mini-dictionary of terms

Mild cognitive impairment intermediate stage between normal aging and dementia, which may be a prodromal stage in the development of dementia. Cognitive performance is below normal limits, but patients are independent in daily living activities and do not fulfill the criteria for dementia.

Alzheimer's disease the most frequent cause of dementia. It is a neurodegenerative disease characterized by neurodegeneration and deposition of amyloid and tau. Several clinical presentations are possible, but memory loss is the most frequent.

Frontotemporal dementia includes several clinicopathological entities characterized by neurodegeneration of the frontal and/or temporal lobes. There are three main clinical subtypes: behavioral variant, nonfluent aphasia, and semantic dementia.

Primary progressive aphasia clinical syndrome characterized by progressive language deterioration due to neurodegeneration of language brain systems. Three subtypes are currently recognized: nonfluent, semantic, and logopenic.

Parkinson's disease neurodegenerative disorder affecting mainly the motor system. It is produced by degeneration of dopaminergic neurons in the "substantia nigra."

Amyotrophic lateral sclerosis neurodegenerative disease affecting mainly the motor system (first and second motor neurons). Predicted survival is short. It shares some clinical and pathological findings with frontotemporal dementia.

Posterior cortical atrophy clinical syndrome characterized by progressive impairment of visuospatial and visuoperceptive abilities. It is regarded as a clinical variant of Alzheimer's disease.

Progressive supranuclear palsy neurodegenerative disorder affecting multiple brain regions. Initially, it was considered an aggressive variant of Parkinson's disease with early falls and oculomotor abnormalities, but today, several phenotypes of presentation are recognized, including nonfluent progressive aphasia.

Corticobasal degeneration neurodegenerative disorder affecting the cerebral cortex and basal ganglia. It usually presents with asymmetric parkinsonism, aphasia, apraxia, alien hand syndrome, and frontal-lobe cognitive impairment.

Vascular cognitive impairment refers to any kind of cognitive impairment caused by brain vascular damage. It is considered the second cause of dementia. It may be produced by several mechanisms.

Diagnosis and Management in Dementia
ISBN 978-0-12-815854-8, https://doi.org/10.1016/B978-0-12-815854-8.00024-0

Introduction

The incidence and prevalence of cognitive disorders have increased considerably due to population aging. In this setting, memory complaints are becoming a frequent cause of consultation in primary and specialized care. Formerly, the diagnosis of neurodegenerative disorders was usually achieved in advanced stages. In fact, diagnosis was mainly directed to exclude potentially reversible causes, such as tumors, hydrocephalus, etc. However, over the past years, we have aimed to diagnose patients in earlier stages, and the differential diagnosis between distinct neurodegenerative diseases has become more important.

Neuropsychological assessment has emerged as an essential and irreplaceable tool for the diagnosis of cognitive disorders. In this regard, neuropsychological evaluations are usually arranged in steps. The first step comprises *screening*. This aims to obtain a dichotomous outcome: normal *versus* impaired. However, it should not provide a specific diagnosis. Although the Mini-Mental State Examination (MMSE) is considered the gold standard, it has several pitfalls and limitations, making new cognitive screening tests necessary (Carnero-Pardo, 2014). Cognitive screening is very important because cognitive symptoms may be very frequent in the general population. In addition, the diagnosis of neurodegenerative disorders is usually delayed. Several tests have been proposed for screening, including the MMSE, Montreal Cognitive Assessment (MoCA), Memory Impairment Screen (MIS), clock test, Rowland Universal Dementia Assessment Scale (RUDAS), Test Your Memory, etc. (Tsoi, Chan, Hirai, Wong, & Kwok, 2015) (Table 24.1). Some of them evaluate several cognitive domains (MMSE, MoCA, RUDAS, ...), while others are mainly focused on memory (MIS, Test Your Memory, ...). Although they present important differences, these tests have several tasks in common, which are considered the most useful for screening, for instance, temporal orientation and verbal fluency or the learning of a list of words, which are included in MMSE, RUDAS, or MoCA, among others.

Table 24.1 Main cognitive screening tests in dementia.

Clock drawing test
Cross-cultural dementia screening
DemTect
General practitioner assessment of cognition
Memory impairment screen
Mini-mental state examination
Montreal cognitive assessment
Test your memory
Rowland universal dementia assessment
Six-item cognitive impairment test

The choice of one test or another is determined by several factors, including the clinical setting and the specific patient (Velayudhan et al., 2014). The first factor regards the clinical question we need to answer. For instance, do we want to screen for dementia? Or do we want to screen for mild cognitive impairment? Or are we seeking the differential diagnosis between two entities (Alzheimer's disease vs. frontotemporal dementia, for instance)? Or are we maybe interested in obtaining a cognitive profile of different cognitive functions? Second, what is our clinical setting (primary care, specialized care, memory clinic)? How much time do we have to examine the patient? And third, what are the patient's characteristics that we want to examine (age, educational level, sensory or motor disabilities, etc.)?

The second step in neuropsychological assessment comprises a *general cognitive assessment*, which evaluates the main cognitive domains and allows us to obtain a cognitive profile. This profile may be very suggestive of each particular neurodegenerative disorder in an adequate setting, because each neurodegenerative disorder tends to impair more or less specific brain regions and systems in the early stages. This step is conducted using some neuropsychological batteries or joining several tests evaluating multiple cognitive domains. However, this step of neuropsychological assessment is often time consuming (Lezak, Howieson, Bigler, & Tranel, 2012).

In this chapter, we will review Addenbrooke's Cognitive Examination (ACE), which is a useful test that has been validated in several languages, countries, and cultures around the world over the past years (Mirza, Panagioti, Waheed, & Waheed, 2017; Wang et al., 2017).

Development, versions, and main characteristics

ACE was initially developed as a modified MMSE, with the addition of some further tasks to evaluate memory, language, constructive praxis, and verbal fluency. Since the publication of the first version in the year 2000 (Mathuranah, Nestor, Barrios, Rakowicz, & Hodges, 2000), two additional versions have been developed (a revised [ACE-R] and a third version [ACE-III]), as well as an abbreviated variant, the Mini-ACE. ACE-III is the first version not including all the items of the MMSE, and it is currently the recommended version (Hsieh, Schubert, Hoon, Mioshi, & Hodges, 2013). Omission of MMSE items was necessary because of limiting copyright restrictions. The duration of administration of the different versions of ACE is approximately 15 min, and only pen and paper are required. An online version has been developed, with free autoscoring and autoreporting (http://www.acemobile.org) (Hodges & Larner, 2017; Newman et al., 2017).

The test scores out of a total of 100 points. Interestingly, since ACE-R, specific scores for cognitive domains have been provided: attention/orientation, memory, verbal fluency, language, and visuospatial abilities (Table 24.2).

The last modification developed for ACE has been the Mini-ACE. This test was developed using Mokken scaling analysis in a cohort of patients with different types of

Table 24.2 Items and scoring of Addenbrooke's Cognitive Examination, third revision.

	Total score: 100	Domains
Temporal orientation	5	Attention and orientation domain (18)
Spatial orientation	5	
Registration of three words	3	
Serial subtraction	5	
Free recall of three items previously registered	3	Memory domain (26)
Encoding and learning of a name and address	7	
Retrograde memory	4	
Free recall of name and address (at the end of the test)	7	
Recognition of name and address (at the end of the test)	5	
Words beginning with "p"	7	Verbal fluency (14)
Animals	7	
Comprehension of a complex command	3	Language domain (26)
Writing two sentences	2	
Repetition of complex words	2	
Repetition of two sentences	2	
Confrontation naming	12	
Semantic task	4	
Reading irregular words	1	
Copy of intersecting infinity loops	1	Visuospatial domain (16)
Copy of a cube	1	
Drawing a clock	5	
Counting dots	4	
Identifying incomplete letter	4	

Left column shows the description of each task, middle column the scoring of tasks, and right column the domain to which they belong and the total score of each cognitive domain.

neurodegenerative dementia assessed with ACE-III and further validated in an independent sample (Hsieh et al., 2014). This allowed the selection of ACE-III items with the highest ability to discriminate between patients and controls. Accordingly, time orientation, animal verbal fluency, clock drawing, and the learning and recall of a name and address were the items included in the Mini-ACE (Table 24.3). This test has been demonstrated to be a very useful tool for screening, even more so than the whole ACE-III. This is because the Mini-ACE includes only items that truly discriminate between cognitively impaired and unimpaired subjects, eliminating those that, although clinically interesting, are not diagnostic (Matías-Guiu, Cortés-Martínez et al., 2017). Furthermore, the Mini-ACE is a very brief test, as it can be administered in 5 min. In addition, scoring of the Mini-ACE may be calculated in cases in which ACE-III is administered completely (Matías-Guiu & Fernández-Bobadilla, 2016).

Table 24.3 Items and scoring of Mini-Addenbrooke's Cognitive Examination.

	Mini-ACE (30 points)
Temporal orientation	4 (season orientation not included)
Encoding and learning of a name and address	7
Animals	7
Drawing a clock	5
Free recall of name and address	7

Left column shows the description of the tasks, and right column the total score of each task. *Mini-ACE*, Mini-Addenbrooke's Cognitive Examination.

Diagnostic properties and interpretation

Several studies have been performed supporting the use and analysis of the diagnostic and psychometric properties of ACE and its subsequent versions. Reliability was high according to Cronbach's α coefficient, and convergent validity with other neuropsychological tests, such as the MMSE, was adequate. Importantly, area under the curve, sensitivity, specificity, and predictive values for diagnosis of dementia, mild cognitive impairment, and different cognitive disorders were adequate in different studies.

In the first study of ACE-I (Mathuranath et al., 2000), two cutoff points were suggested for identifying dementia, with sensitivity of .93 for 88/100 and specificity of .96 for 83/100. Patients with frontotemporal dementia and Alzheimer's disease were also compared, and statistically significant differences were obtained for several tasks. On that basis, an index known as the VLOM ratio was suggested to distinguish between frontotemporal dementia and Alzheimer's disease:

$$VLOM = (verbal\ fluency + language)/(orientation + delayed\ recall).$$

In this index, higher VLOM (>3.2) would be suggestive of Alzheimer's disease, while in frontotemporal dementia the VLOM ratio would be lower (<2.2). These results emphasized the applicability of ACE as a general test for screening but, especially, focusing on other causes of cognitive impairment beyond Alzheimer's disease.

Later, a second version named Addenbrooke's Cognitive Examination—Revised was developed. In this version, several items were grouped to provide additional scores informing about attention/orientation, memory, verbal fluency, language, and visuospatial abilities (Mioshi, Dawson, Mitchell, Arnold, & Hodges, 2006). In 2013, the validation of the third version (ACE-III) was published. This version is very similar to ACE-R, but items belonging to the MMSE were replaced.

ACE is a very rich test, since it truly provides helpful information for knowing neuropsychological status. Some items may provide different signs, for instance, constructive praxis tasks or language items, among others. Qualitative assessment of cognitive performance may provide key data in the diagnostic process in some cases and may be useful in deducing the reason for failure during test performance (Díaz-Orueta, Blanco-Campal, & Burke, 2018).

In addition, the test gives several scores. First, it provides a total score, for which cutoff points could be applied, and allows drawing a cognitive profile using the scores for each cognitive domain. The comparison between cognitive domains in the same patient enables one to search for dissociations between cognitive functions, which are relevant for the diagnosis of different causes of dementias. For instance, a case with impairment of memory domain, with preservation of other cognitive domains, may be suggestive of a prodromal stage of Alzheimer's disease in the appropriate clinical context (Fig. 24.1). In contrast, a patient with impairment in verbal fluency may suggest executive dysfunction in the context of a behavioral variant frontotemporal dementia (Fig. 24.2). Impairment of verbal fluency may also suggest the diagnosis of primary progressive aphasia, especially if the language score is also altered. Also, analysis of specific language tasks may help in the differential diagnosis and monitoring of primary progressive aphasia subtypes (Leyton, Hornberger, Mioshi, & Hodges, 2010).

The five score domains of ACE-III have been highly correlated with standardized neuropsychological tests specific for each cognitive domain. For instance, memory domain of ACE-III was correlated with the Free and Cued Selective Reminding Test (r = 0.806 for total recall and total delayed recall) and Rey Auditory Verbal Learning Test (r = 0.59) in two independent studies (Matías-Guiu, Valles-Salgado et al., 2017; Hsieh et al., 2013). In this regard, the memory domain showed an area under the curve of .906 to detect impairment in the Free and Cued Selective Reminding Test. This is especially meaningful if we consider that the Free and Cued Selective Reminding Test is the recommended neuropsychological test for the diagnosis of prodromal Alzheimer's disease according to the International Working Group (Dubois et al., 2014). Similarly, the language domain was correlated with the Boston Naming Test (r = 0.744). In contrast, visuospatial domain and attention domain were only moderately correlated with the specific tests.

Overall, ACE cognitive domains allow an approximation of each patient's cognitive profile, in a similar way compared with conducting a full neuropsychological assessment (Table 24.4). In addition to score domains, the assessment of specific items and patterns of responses also improves the etiological diagnosis (McGrory, Starr, Shenkin, Austin, & Hodges, 2015). However, interpretation of the test is partially limited by the effects of demographic factors. Specifically, ACE scores are mainly influenced by age and education. Moreover, different validation studies have reported heterogeneous cutoff points, suggesting evident differences between languages and cultures. Interestingly, the diagnostic capacity of ACE-III improved with the use of normative data instead of raw scores (Matías-Guiu, Fernández-Bobadilla et al., 2016). For these reasons, normative data are recommended to adequately interpret the test results. Unfortunately, only a few studies have provided normative data until now.

69 year-old, 18 years of formal education	Raw score	Age- and education-adjusted scaled score																
		<1	1	2	3-5	6-10	11-18	19-28	29-40	41-59	60-71	72-81	82-89	90-94	95-97	98	99	>99
ACE-III (total score)	83	2	3	4	5	6	7	8	9	10	11	12	13	14	15	16	17	18
Attention / Orientation	14	2	3	4	5	6	7	8	9	10	11	12	13	14	15	16	17	18
Memory	15	2	3	4	5	6	7	8	9	10	11	12	13	14	15	16	17	18
Verbal fluency	12	2	3	4	5	6	7	8	9	10	11	12	13	14	15	16	17	18
Language	26	2	3	4	5	6	7	8	9	10	11	12	13	14	15	16	17	18
Visuospatial	16	2	3	4	5	6	7	8	9	10	11	12	13	14	15	16	17	18

Figure 24.1 Cognitive profile of a patient with prodromal Alzheimer's disease. Memory and attention/orientation domains are impaired, while the other domains are within normal limits. Memory domain score is the earliest score to be affected in typical presentation of Alzheimer's disease.

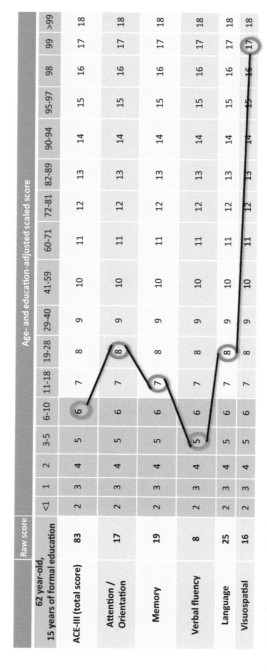

Figure 24.2 Cognitive profile of a patient with behavioral variant frontotemporal dementia. Verbal fluency is impaired, while the other domains are within normal limits. This may be suggestive of executive dysfunction, especially if words beginning with "p" are more affected than animal verbal fluency.

Table 24.4 Summary of interpretation of ACE-III scores and Mini-ACE.

ACE-III (total score)	– Detection of Alzheimer's disease dementias – Discrimination between Alzheimer's disease CDR 0.5 and CDR 1 – Follow-up – Useful for the detection of non–Alzheimer's disease dementias
ACE-III (memory domain)	– Detection of Alzheimer's disease at early stage (CDR 0.5) – Detection of Alzheimer's disease at stage CDR 1
ACE-III (verbal fluency)	– May be the first score impaired in early stages of primary progressive aphasia – May be indicative of executive dysfunction
ACE-III (visuospatial)	– Very useful in detecting posterior cortical atrophy and other dementias with visuospatial impairment
Mini-ACE	– The most sensitive test for screening, especially in Alzheimer's disease

Main uses for each score of ACE-III are shown. *ACE*, Addenbrooke's Cognitive Examination; *CDR*, clinical dementia rating.

Metaanalysis

A metaanalysis covering studies published between 2002 and 2010 found 45 studies about ACE and ACE-R, and nine papers were included in the metaanalysis (Crawford, Whitnall, Robertson, & Evans, 2012). The authors concluded that ACE and ACE-R were reliable for detecting cognitive impairment, but more studies examining the differential diagnosis between dementia subtypes and detection of mild cognitive impairment would be needed.

A subsequent metaanalysis published by Larner and Mitchell in 2014 concluded a superior diagnostic accuracy of ACE-R, in comparison to the MMSE and ACE-I. Sensitivity and specificity of ACE-R was higher than those of the MMSE in both low- and high-prevalence settings, such as primary care and memory clinics, respectively (Larner & Mitchell, 2014).

Clinical and research use

ACE has mainly been validated in three clinical scenarios: (1) diagnosis of dementia, (2) diagnosis of mild cognitive impairment, and (3) differential diagnosis between Alzheimer's disease and frontotemporal dementia.

In addition, ACE has been validated or applied in other conditions. In this regard, it has been specifically validated in primary progressive aphasia (Leyton et al., 2010). ACE includes several language tasks, which may be useful in the classification of primary progressive aphasia into nonfluent, semantic, and logopenic variants. Accordingly, a semantic index was suggested to distinguish between Alzheimer's disease and semantic dementia (Davis, Dawson, Mioshi, Erzingçlioglu, & Hodges, 2008).

ACE has also been applied for the detection of vascular cognitive impairment (Lees et al., 2017; Pendlebury, Mariz, Bull, Mehta, & Rothwell, 2012). The relatively exhaustive assessment of language in ACE is an important advantage over other screening tools after stroke. However, the strong influence of language on the score may also limit the applicability of the test in some cases (Lees et al., 2017).

ACE has also been validated for the detection of cognitive impairment associated with Parkinson's disease and the differential diagnosis of Parkinson's disease and atypical parkinsonisms (progressive supranuclear palsy and corticobasal degeneration) (McColgan et al., 2012; Reyes et al., 2009; Rittman et al., 2013). However, some studies have found that other tools may be more accurate for screening for cognitive impairment associated with Parkinson's disease than ACE (Kaszas et al., 2012; Komadina et al., 2011).

Furthermore, ACE-III has also been used for other conditions, such as amyotrophic lateral sclerosis (Matías-Guiu, Pytel et al., 2016; Xu, Alruwaili, Henderson, & McCombe, 2017), in the differential diagnosis between depression and dementia (Dudas, Berrios, & Hodges, 2005), and for posterior cortical atrophy (Ahmed et al., 2016), among others.

Comparison with other screening tests

Most studies validating ACE have compared it with the MMSE, usually demonstrating the superiority of ACE. However, a few studies have compared ACE with other screening cognitive tests beyond the MMSE. In 2017, in a study performed by our group, ACE-III achieved the highest diagnostic accuracy in comparison with the MMSE, MoCA, RUDAS, and MIS for the diagnosis of Alzheimer's disease at the clinical dementia rating stage 1 (Matías-Guiu, Cortés-Martínez et al., 2017). Another study from 2017 has shown the superiority of ACE-III compared with the MMSE and MoCA in the identification of everyday functioning impairment in mild to moderate dementia (Giebel & Challis, 2017). Previously, an improvement of 16% in the diagnosis of Alzheimer's disease using ACE-R instead of the MMSE was estimated in an interesting study using data from the national dementia register of Scotland (Law et al., 2013).

Conversely, in a study performed in United Kingdom, the Mini-ACE was compared with the MoCA (Larner, 2017). In this study, area under the curve was excellent for both for the diagnosis of dementia, but the MoCA achieved a better diagnostic capacity than the Mini-ACE for mild cognitive impairment. Thus, further studies are necessary to clarify the specific role and advantages of ACE in comparison with other screening tools.

Concluding remarks

ACE and its subsequent versions, ACE-R and ACE-III, have become one of the most important cognitive screening tests in clinical practice and research (Fig. 24.3). Several studies have validated their usefulness in multiple causes of cognitive impairment and different settings.

ACE plays a dual role in cognitive assessment (Matías-Guiu, Fernández-Bobadilla, & Cortés-Martínez, 2018). On one hand, it is very useful to screen, especially with the reduced version of the Mini-ACE. This reduced version is shorter than the MMSE and other brief tests and has proved to have very good diagnostic properties. On the other hand, it serves as a brief neuropsychological battery assessing the main cognitive domains, which is useful in the differential diagnosis of cognitive disorders and to approximate the cognitive profile of patients or to guide a more comprehensive neuropsychological examination.

ACE has demonstrated several advantages over other cognitive tests. But there are no fully optimal tests, and ACE also has some disadvantages or limitations. For instance, executive functioning is relatively unexamined, and social cognition, which is becoming increasingly important in the assessment of frontal-lobe dementias, is not examined. Furthermore, demographic factors strongly influence test performance, and thus, normative data are highly recommended to achieve an adequate interpretation (Matías-Guiu et al., 2015; Nieto, Galtier, Hernández, Velasco, & Barroso, 2016). In turn, in ACE-III, the MMSE score is not

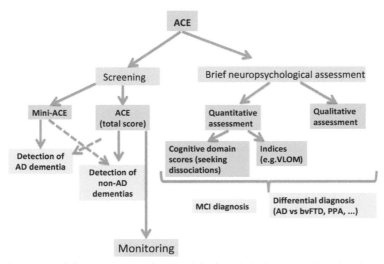

Figure 24.3 Summary of the main uses of *ACE*, Addenbrooke's Cognitive Examination . ACE may be used for screening (especially using total score and the Mini-ACE) and monitoring (mainly with total score) and as an approximation of a full neuropsychological examination, which may allow a differential diagnosis (using qualitative interpretation and quantitative assessment with cognitive domain scores). *AD*, Alzheimer's disease; *bvFTD*, behavioral variant frontotemporal dementia; *MCI*, mild cognitive impairment; *PPA*, primary progressive aphasia.

incorporated, which may be important in some settings. These limitations may be overcome by using different methods, such as combining ACE-III with other cognitive tests or behavioral scales (Hancock & Larner, 2009), or using conversion methods between ACE-III and the MMSE (Matías-Guiu, Pytel et al., 2017).

Key facts

Key facts of neuropsychological findings in Alzheimer's disease in ACE

- Episodic memory dysfunction is usually the first neuropsychological deficit in Alzheimer's disease.
- It may be demonstrated by trying to remember a name and address.
- Other sensitive markers are animal verbal fluency and temporal orientation.

Key facts of neuropsychological findings in primary progressive aphasia in ACE

- ACE includes several items evaluating language: visual confrontation naming, a semantic task, writing, reading irregular words, comprehension of a complex command, repetition of complex words, and repetition of two sentences. Verbal fluencies are also examined.
- Some findings may suggest a specific variant of primary progressive aphasia.
- Other items depend on language. For this reason, a very low score in ACE in a patient with little or no functional impairment may indicate an aphasic disorder.

Key facts of neuropsychological findings in behavioral variant frontotemporal dementia in ACE

- Patients with frontotemporal dementia in the early stages are usually oriented, but they show executive and/or language impairment. Other cognitive functions, such as visuospatial and memory, are usually more preserved.
- Neuropsychological findings in frontotemporal dementia are heterogeneous, however, and some patients may also show memory dysfunction.
- Cognitive impairment may appear later than behavioral disorder in frontotemporal dementia. For this reason, scoring above normal limits does not necessarily exclude a diagnosis of frontotemporal dementia.

Summary points

- ACE is a cognitive screening test widely used for the diagnosis of dementia and neurodegenerative disorders.
- Three versions have been developed (ACE, ACE-R, and ACE-III), and a short tool (Mini-ACE).

- ACE provides a total score and five additional scores on the following cognitive domains: attention and orientation, memory, verbal fluency, language, and visuospatial abilities.
- ACE serves both as a screening test and as a brief neuropsychological battery.
- ACE is influenced by demographic and cultural factors, and accordingly, normative data are recommended for an optimal interpretation.
- ACE has mainly been validated in three clinical scenarios: (1) diagnosis of dementia, (2) diagnosis of mild cognitive impairment, and (3) differential diagnosis between Alzheimer's disease and frontotemporal dementia.
- Other conditions in which ACE has been used are vascular cognitive impairment, primary progressive aphasia, parkinsonian syndromes, and amyotrophic lateral sclerosis, among others.

References

Ahmed, S., Baker, I., Husain, M., Thompson, S., Kipps, C., Hornberger, M., et al. (2016). Memory impairment at initial clinical presentation in posterior cortical atrophy. *Journal of Alzheimer's Disease, 52*, 1245—1250.

Carnero-Pardo, C. (2014). Should the mini-mental state examination be retired? *Neurología, 29*, 473—481.

Crawford, S., Whitnall, L., Robertson, L., & Evans, J. J. (2012). A systematic review of the accuracy and clinical utility of the Addenbrooke's Cognitive Examination and the Addenbrooke's Cognitive Examination-Revised in the diagnosis of dementia. *International Journal of Geriatric Psychiatry, 27*, 659—669.

Davis, R. R., Dawson, K., Mioshi, E., Erzinçlioglu, S., & Hodges, J. R. (2008). Differentiation of semantic dementia and Alzheimer's disease using the Addenbrooke's Cognitive Examination (ACE). *International Journal of Geriatric Psychiatry, 23*, 370—375.

Díaz-Orueta, U., Blanco-Campal, A., & Burke, T. (2018). Rapid review of cognitive screening instruments in MCI: Proposal for a process-based approach modification of overlapping tasks in select widely used instruments. *International Psychogeriatrics, 30*, 663—672.

Dubois, B., Feldman, H. H., Jacova, C., Hampel, H., Molinuevo, J. L., Blennow, K., et al. (2014). Advancing research diagnostic criteria for Alzheimer's disease: The IWG-2 criteria. *The Lancet Neurology, 13*, 614—629.

Dudas, R. B., Berrios, G. E., & Hodges, J. R. (2005). The Addenbrooke's Cognitive Examination (ACE) in the differential diagnosis of early dementias versus affective disorder. *American Journal of Geriatric Psychiatry, 13*, 218—226.

Giebel, G. M., & Challis, D. (2017). Sensitivity of the Mini-Mental State Examination, Montreal Cognitive Assessment and the Addenbrooke's Cognitive Examination III to everyday activity impairments in dementia: An exploratory study. *International Journal of Geriatric Psychiatry, 32*, 1085—1093.

Hancock, P., & Larner, A. J. (2009). Diagnostic utility of the Informant Questionnaire on Cognitive Decline in the Elderly (IQCODE) and its combination with the Addenbrooke's Cognitive Examination-Revised (ACE-R) in a memory clinic-based population. *International Psychogeriatrics, 21*, 526—530.

Hodges, J. R., & Larner, A. J. (2017). Addenbrooke's Cognitive Examination: ACE, ACE-R, ACE-III, ACEapp, and M-ACE. In A. J. Larner (Ed.), *Cognitive screening instruments*. Springer International Publishing Switzerland.

Hsieh, S., McGrory, S., Leslie, F., Dawson, K., Ahmed, S., Butler, C. R., et al. (2014). The mini-Addenbrooke's Cognitive Examination: A new assessment tool for dementia. *Dementia and Geriatric Cognitive Disorders, 39*, 1—11.

Hsieh, S., Schubert, S., Hoon, C., Mioshi, E., & Hodges, J. R. (2013). Validation of the Addenbrooke's Cognitive Examination III in frontotemporal dementia and Alzheimer's disease. *Dementia and Geriatric Cognitive Disorders, 36*, 242—250.

Kaszas, B., Kovacs, N., Balas, I., Kallai, J., Aschermann, Z., Kerekes, Z., et al. (2012). Sensitivity and specificity of Addenbrooke's Cognitive Examination, mattias dementia rating scale, frontal assessment battery and mini mental state examination for diagnosing dementia in Parkinson's disease. *Parkinsonism & Related Disorders, 18,* 554–556.

Komadina, N. C., Terpening, Z., Huang, Y., Halliday, G. M., Naismith, S. L., & Lewis, S. J. (2011). Utility and limitations of Addenbrooke's Cognitice Examination-Revised for detecting mild cognitive impairment in Parkinson's disease. *Dementia and Geriatric Cognitive Disorders, 31,* 349–357.

Larner, A. J. (2017). MACE versus MoCA: Equivalence or superiority? Pragmatic diagnostic test accuracy study. *International Psychogeriatrics, 29,* 931–937.

Larner, A. J., & Mitchell, A. J. (2014). A meta-analysis of the accuracy of the Addenbrooke's Cognitive Examination (ACE) and the Addenbrooke's Cognitive Examination-Revised (ACE-R) in the detection of dementia. *International Psychogeriatrics, 26,* 555–563.

Law, E., Connelly, P. J., Randall, E., McNeill, C., Fox, H. C., Parra, M. A., et al. (2013). Does the Addenbrooke's Cognitive Examination-Revised add to the Mini-Mental State Examination in established Alzheimer disease? Results from a national dementia research register. *International Journal of Geriatric Psychiatry, 28,* 351–355.

Lees, R. A., Hendry Ba, K., Broomfield, N., Stott, D., Larner, A. J., & Quinn, T. J. (2017). Cognitive assessment in stroke: Feasibility and test properties using differing approaches to scoring of incomplete items. *International Journal of Geriatric Psychiatry, 32,* 1072–1078.

Leyton, C. R., Hornberger, M., Mioshi, E., & Hodges, J. R. (2010). Application of Addenbrooke's Cognitive Examination to diagnosis and monitoring of primary progressive aphasia. *Dementia and Geriatric Cognitive Disorders, 29,* 504–509.

Lezak, M. D., Howieson, D. B., Bigler, E. D., & Tranel, D. (2012). *Neuropsychological assessment* (5th ed.). New York: Oxford University Press.

Mathuranath, P. S., Nestor, P. J., Barrios, G. E., Rakowicz, W., & Hodges, J. R. (2000). A brief cognitive test battery to differentiate Alzheimer's disease and frontotemporal dementia. *Neurology, 55,* 1613–1620.

Matías-Guiu, J. A., Cortés-Martínez, A., Valles-Salgado, M., Rognoni, T., Fernández-Matarrubia, M., Moreno-Ramos, T., et al. (2017). Addenbrooke's Cognitive Examination III: Diagnostic utility for mild cognitive impairment and dementia and correlation with standardized neuropsychological tests. *International Psychogeriatrics, 29,* 105–113.

Matías-Guiu, J. A., & Fernandez-Bobadilla, R. (2016). Validation of the Spanish-language version of mini-Addenbrooke's Cognitive Examination as a dementia screening tool. *Neurología, 31,* 646–648.

Matías-Guiu, J. A., Fernández-Bobadilla, R., & Cortés-Martínez, A. (2018). Addenbrooke's Cognitive Examination III: A neuropsychological test useful to screen and obtain a cognitive profile. *Neurología, 33,* 140.

Matías-Guiu, J. A., Fernández-Bobadilla, R., Fernández-Oliveira, A., Valles-Salgado, M., Rognoni, T., Cortés-Martínez, A., et al. (2016). Normative data for the Spanish version of the Addenbrooke's Cognitive Examination III. *Dementia and Geriatric Cognitive Disorders, 41,* 243–250.

Matías-Guiu, J. A., Fernández de Bobadilla, R., Escudero, G., Pérez-Pérez, J., Cortés, A., Morenas-Rodríguez, E., et al. (2015). Validation of the Spanish version of Addenbrooke's Cognitive Examination III for diagnosing dementia. *Neurología, 30,* 545–551.

Matías-Guiu, J. A., Pytel, V., Cabrera-Martín, M. N., Galán, L., Valles-Salgado, M., Guerrero, A., et al. (2016). Amyloid- and FDG-PET imaging in amyotrophic lateral sclerosis. *European Journal of Nuclear Medicine and Molecular Imaging, 43,* 2050–2060.

Matías-Guiu, J. A., Pytel, V., Cortés-Martínez, A., Valles-Salgado, M., Rognoni, T., Moreno-Ramos, T., et al. (2017). Conversion between Addenbrooke's Cognitive Examination III and Mini-Mental State Examination. *International Psychogeriatrics.* https://doi.org/10.1017/S104161021700268X.

Matías-Guiu, J. A., Valles-Salgado, M., Rognoni, T., Hamre-Gil, F., Moreno-Ramos, T., & Matías-Guiu, J. (2017). Comparative diagnostic accuracy of the ACE-III, MIS, MMSE, MoCA, and RUDAS for screening of Alzheimer disease. *Dementia and Geriatric Cognitive Disorders, 43,* 237–246.

McColgan, P., Evans, J. R., Breen, D. P., Mason, S. L., Barker, R. A., & Williams-Gray, C. H. (2012). Addenbrooke's Cognitive Examination-Revised for mild cognitive impairment in Parkinson's disease. *Movement Disorders, 27,* 1173–1177.

McGrory, S., Starr, J. M., Shenkin, S. D., Austin, E. J., & Hodges, J. R. (2015). Does the order of item difficulty of the Addenbrooke's Cognitive Examination add anything to subdomain scores in the clinical assessment of dementia? *Dementia and Geriatric Cognitive Disorders, 5,* 155–169.

Mioshi, E., Dawson, K., Mitchell, J., Arnold, R., & Hodges, J. R. (2006). The Addenbrooke's Cognitive Examination revised (ACE-R): A brief cognitive test battery for dementia screening. *International Journal of Geriatric Psychiatry, 21,* 1078–1085.

Mirza, N., Panagioti, M., Waheed, M. W., & Waheed, W. (2017). Reporting of the translation and cultural adaptation procedures of the Addenbrooke's Cognitive Examination version III ACE-III) and its predecessors: A systematic review. *BMC Medical Research Methodology,* 141.

Newman, C. G. J., Bevins, A. D., Zajicek, J. P., Hodges, J. R., Vuillermoz, E., Dickenson, J. M., et al. (2017). Improving the quality of cognitive screening assessments: ACEmobile, an iPad-based version of the Addenbrooke's Cognitive Examination-III. *Alzheimers Dementia (Amsterdam), 10,* 182–187.

Nieto, A., Galtier, I., Hernández, E., Velasco, P., & Barroso, J. (2016). Addenbrooke's Cognitive Examination-Revised: Effects of education and age. Normative data for the Spanish speaking population. *Archives of Clinical Neuropsychology, 7,* 811–818.

Pendlebury, S. T., Mariz, J., Bull, L., Mehta, Z., & Rothwell, P. M. (2012). MoCA, ACE-R, and MMSE versus the national Institute of neurological disorders and stroke-Canadian stroke network vascular cognitive impairment harmonization standards neuropsychological battery after TIA and stroke. *Stroke, 43,* 464–469.

Reyes, M. A., Pérez-Lloret, S., Roldán Gerschcovich, E., Martin, M. E., Leiguarda, R., & Merello, M. (2009). Addenbrooke's cognitive examination validation in Parkinson's disease. *European Journal of Neurology, 16,* 142–147.

Rittman, T., Ghosh, B. C., McColgan, P., Breen, D. P., Evans, J., Williams-Gray, C. H., et al. (2013). The Addenbrooke's Cognitive Examination for the differential diagnosis and longitudinal assessment of patients with parkinsonian disorders. *Journal of Neurology, Neurosurgery, and Psychiatry, 84,* 544–551.

Tsoi, K. K., Chan, J. Y., Hirai, H. W., Wong, S. Y., & Kwok, T. C. (2015). Cognitive tests to detect dementia: A systematic review and meta-analysis. *JAMA Internal Medicine, 175,* 1450–1458.

Velyudhan, L., Ryu, S. H., Raczek, M., Philpot, M., Lindesay, J., Critchfield, M., et al. (2014). Review of brief cognitive tests for patients with suspected dementia. *International Psychogeriatrics, 26,* 1247–1264.

Wang, B. R., Ou, Z., Gu, X. H., Wei, C. S., Xu, J., & Shi, J. Q. (2017). Validation of the Chinese version of Addenbrooke's Cognitive Examination III for diagnosing dementia. *International Journal of Geriatric Psychiatry.* https://doi.org/10.1002/gps.4680.

Xu, Z., Alruwaili, A. R. S., Henderson, R. D., & McCombe, P. A. (2017). Screening for cognitive and behavioural impairment in amyotrophic lateral sclerosis: Frequency of abnormality and effect on survival. *Journal of the Neurological Sciences, 376,* 16–23.

CHAPTER 25

Beyond the cutoffs: a Bayesian approach to the use of the Montreal Cognitive Assessment as a screening tool for mild cognitive impairment and dementia

Andrea Bosco[1], Alessandro O. Caffò[1], Giuseppina Spano[1,2], Antonella Lopez[1]

[1]Department of Education Science, Psychology, Communication, Università degli Studi di Bari "Aldo Moro", Bari, Italy;
[2]Department of Agro-Environmental and Territorial Sciences, Università degli Studi di Bari "Aldo Moro", Bari, Italy

List of abbreviations

ACE-III Addenbrooke's Cognitive Examination III

AD Alzheimer's disease

AUC area under the curve

CI confidence interval

DA discriminant analysis

DSM *Diagnostic and Statistical Manual of Mental Disorders*

ICD International Classification of Diseases

J Youden index

LR− negative likelihood ratio

LR likelihood ratio

LR+ positive likelihood ratio

MCI mild cognitive impairment

MIS Memory impairment screen

MMSE Mini-Mental State Examination

MoCA Montreal Cognitive Assessment

NINCDS−ADRDA National Institute of Neurological and Communicative Disorders and Stroke and the Alzheimer's Disease and Related Disorders Association

NPP negative predictive powers

PPP positive predictive powers

QUADAS-2 Quality Assessment of Diagnostic Accuracy Studies

ROC receiver operating characteristics

RUDAS Rowland Universal Dementia Assessment Scale

SD standard deviation

Se sensitivity

Sp specificity

Diagnosis and Management in Dementia
ISBN 978-0-12-815854-8, https://doi.org/10.1016/B978-0-12-815854-8.00025-2

Mini-dictionary of terms

AUC: AUC is the abbreviation for area under the curve. An area of 1 represents the highest diagnostic accuracy, while a value of 0.5 represents a test in which diagnostic accuracy is null.

Cutoff score: the cutoff represents the threshold value of a test. In a psychometric test, it represents the score beyond which the presence of a characteristic is determined or fixed.

Likelihood ratios: the likelihood ratios (LR) express the times that it is more (or less) likely that the result of the test, positive (+) or negative (−), rises from the population in which the illness is present rather than absent. The LR+ may vary between 1 and $+\infty$, while the LR− may vary between 0 and 1, with 1 representing no discrimination.

Negative predictive power: this represents the proportion of people who do not show the characteristic among those with a negative result from the test.

Positive predictive power: this represents the proportion of people who possess the characteristic among those with a positive result from the test.

ROC curve: ROC is the abbreviation for receiver operating characteristic. It is a graphical method that shows the accuracy of a diagnostic test.

Sensitivity: sensitivity is the proportion of the patients with a known positive condition for which the predicted condition is positive.

Specificity: specificity is the proportion of the patients with a known negative condition for which the predicted condition is negative.

Youden index: this is a combinatory index of sensitivity and specificity at a cutoff point t: $J(t) = Se(t) + Sp(t) - 1$.

The Montreal Cognitive Assessment as a screening tool for mild cognitive impairment and dementia

The Montreal Cognitive Assessment (MoCA) has had great success since its first validation by Nasreddine and colleagues on a Canadian sample of elderly people in 2005. The MoCA was proposed as a cognitive screening tool, most suitable for detecting people with mild cognitive impairment (MCI). In the original study, the authors showed that by applying a cutoff of 26, the Mini-Mental State Examination (MMSE; Folstein, Folstein, & McHugh, 1975) showed 18% sensitivity in identifying patients with MCI, while the sensitivity of the MoCA was 90%. Using the same cutoff of 26, the MoCA showed 100% sensitivity in detecting participants with Alzheimer's disease (AD), compared with a sensitivity of 78% for the MMSE. Therefore, the reason for the widespread use of this screening tool was its favorable performance compared with the test it was compared with, i.e., the MMSE in identifying the prodromal stage of AD, that is, MCI. Increasing attention to the evolution of normal and pathological aging and the need for brief and practical tools with powerful diagnostic properties explains the great success of the MoCA.

The original MoCA consists of a series of items that measure different cognitive domains, namely memory, visuospatial skills, attention and working memory, language, and orientation in time and place. The administration lasts only 10 min and can be conducted by professionals at multiple levels of expertise, due to simple and clear instructions. There are different versions of the test. The most commonly used, in both clinical and research settings, is the full paper-and-pencil version, but some other versions are available at www.mocatest.org/ (Wong et al., 2015).

The test has been validated in normal and in several clinical populations, other than MCI and AD (for a complete review, see Julayanont & Nasreddine, 2017). In addition, the full-version MoCA has been translated and validated in several languages and dialects. In fact, a 2017 review by O'Driscoll and Shaikh (2017) analyzed and described the cross-cultural differences that influence the validity of the MoCA.

The MoCA has the merit of overcoming the limitations of other cognitive screening tests. In a 2017 study by Matías-Guiu et al. (2017), the authors compared the diagnostic accuracy of the MMSE, the MoCA, Addenbrooke's Cognitive Examination III (ACE-III; Hsieh, Schubert, Hoon, Mioshi, & Hodges, 2013), the Memory Impairment Screen (MIS; Buschke et al., 1999), and the Rowland Universal Dementia Assessment Scale (Storey, Rowland, Conforti, & Dickson, 2004) for screening AD. These tests have similar characteristics, with the exception of the MIS, which is specific to the memory domain. Overall, all tests achieved satisfactory diagnostic accuracy, as measured by receiver operating characteristics (ROC) analysis. In fact, the area under the curve (AUC) for all tests exceeded 0.850. The highest AUC in this study was 0.897 for ACE-III, while the lowest AUC was 0.856. The strong diagnostic power of ACE-III confirmed the results of a previous metaanalysis by Tsoi, Chan, Hirai, Wong, and Kwok (2015). In this study, the ability to identify MCI was also investigated and the MoCA appeared to be completely comparable to the MMSE, confirming the ability of the MoCA to identify the first signs of cognitive impairment (Table 25.1).

The exploration of the diagnostic power of the MoCA for discriminating between normal aging, MCI, and AD and other dementias continues to be under assessment. Alongside the numerous validations and the search for specific cutoffs for each population, an in-depth study on the MoCA subtests was carried out (Cecato, Martinelli, Izbicki, Yassuda, & Aprahamian, 2016). The study showed that some items were more effective in discriminating early signs of cognitive impairment and, thus, leading to the diagnosis of probable MCI, i.e., inverse digits, serial seven, phrases, verbal fluency, abstraction, and episodic memory (word recall). Moreover, the tests of clock numbers and hands, rhino naming, serial seven, word recall, and orientation were useful in identifying more severe impairments, which characterized the progression from MCI to AD.

In conclusion, the MoCA appears to be an easy, brief, and reliable cognitive screening tool, for both clinical and research purposes (Yokomizo, Simon, & de Campos Bottino, 2014), since it seems to meet the 16 criteria for a good diagnostic tool screening developed by Milne, Culverwell, Guss, Tuppen, and Whelton (2008) and summarized in four key domains: (1) practicality ([a] time it would take for clinicians to familiarize themselves with the measure, [b] cost of the instrument, [c] availability); (2) feasibility ([d] acceptance by patients, [e] acceptance by clinicians, [f] ease of administering, [g] ease of scoring, [h] application duration, [i] ease of interpreting the scores); (3) range of applicability ([j] applicability to a wide age range, [k] sensitivity to different educational levels, [l] sensitivity to language and culture, [m] applicability to different types of dementia); and (4) psychometric properties ([n] validity and reliability, [o] specificity, [p] sensitivity).

The clinical use of Montreal Cognitive Assessment cutoff values

It would be very difficult to list all the validation studies of the MoCA as well as describing all the results obtained, and moreover this is not the aim of the present study. We instead focus on results related to the diagnostic accuracy of MoCA for patients with dementia, in most cases with AD, and with MCI, taking into account a certain number of reviews and systematic reviews that have been published from 2010 to 2018.

Mitchell and Malladi (2010) proposed a metaanalysis of the diagnostic accuracy of all brief multidomain tests alternative to the MMSE (Folstein et al., 1975) for the detection of dementia. They included the original MoCA paper by Nasreddine et al. (2005), who performed a series of ROC curve analyses for all tests considered. Results showed that MoCA was in the top five tests for detecting dementia and achieved a satisfactory screening performance in specialist settings, where the prevalence of dementia is generally higher than in community settings. Regarding early dementia and MCI, MoCA diagnostic performance was not statistically different compared with MMSE.

In a review of reviews, Yokomizo et al. (2014) included MoCA among a number of brief instruments alternative to MMSE to be used in the primary care context for the screening of dementia, considering different factors such as application time, sensitivity, specificity, and number of studies, also suggesting that the professional should be able to get some expertise in choosing the appropriate instrument to use, alone or in combination, depending on the profile of the patient and on the administration setting.

A systematic review by Davis et al. (2015) examined the diagnostic accuracy of the MoCA for the diagnosis of AD and other dementias. Seven studies were included in the final sample, on the basis of a very detailed quality assessment according to the Quality Assessment of Diagnostic Accuracy Studies-2 criteria (Whiting et al., 2011). Two studies collected data from community settings and five studies from secondary care settings (three in memory clinics, two in hospital clinics); no studies with data collection from primary care settings were included. There were 9422 participants in total, although all studies but one recruited only small samples. The prevalence of dementia ranged from 22% to 54% in care settings studies, and from 5% to 10% in population samples. Sensitivity varied from 0.93 (95% confidence interval [CI] 0.68−1.00) to 1.00 (95% CI 0.92−1.00), specificity from 0.51 (95% CI 0.34−0.69) to 0.87 (95% CI 0.72−0.96), and cutoff scores from 18 to 26 in secondary care populations. In community-based populations sensitivity varied from 0.77 (95% CI 0.69−0.84) to 0.98 (95% CI 0.98−0.99), specificity from 0.57 (95% CI 0.47−0.67) to 0.52 (95% CI 0.51−0.53), and cutoff scores from 20 to 25. Authors reported that in four of seven studies the recommended cutoff score of 26 or over indicating normal cognition (Nasreddine et al., 2005) yielded high sensitivity but very poor specificity, and concluded that cutoff scores lower than 26 might be more accurate for detecting dementia and that further studies conducted with high-quality standards are required to confirm this prediction.

Tsoi et al. (2017) performed a metaanalytic study on the accuracy of recall tests for the detection of patients with MCI. MoCA was included among widely used tests based on recall items. Thirty-five studies employing MoCA were included in the sample, for a total of 2107 (35.6%) patients with MCI identified from 5922 participants. Sensitivity varied from 0.64 (95% CI 0.49−0.78) to 1.00 (95% CI 0.29−1.00), with a pooled value of 0.83 (95% CI 0.80−0.86), specificity from 0.27 (95% CI 0.12−0.46) to 0.95 (95% CI 0.87−0.99), with a pooled value of 0.75 (95% CI 0.69−0.80). The heterogeneity among studies was large, with I-square statistic for sensitivity and specificity of 0.69 and 0.92, respectively. The hierarchical summary ROC curve resulted in a diagnostic odds ratio of 14.8, with an AUC value of 87% (95% CI 84%−89%). Cutoff scores used for the detection of MCI ranged from 20 to 29. Authors recommended the MoCA be used for MCI detection, since it was better than the MMSE and other multi-domain tests.

O'Driscoll and Shaikh (2017) proposed a systematic review taking into account the cross-cultural applicability of MoCA for differentiating dementias comparing AD, MCI patients' dementias, and normal controls. Eighteen studies, for a total of 12,670 participants, investigated the diagnostic accuracy of MoCA in differentiating patients with dementia, i.e., AD, dementia with Lewy bodies, mixed-type dementia, behavioral variant frontotemporal dementia, vascular dementia, and major neurocognitive disorders, from healthy aged individuals. Sensitivity varied from 0.75 (95% CI 0.53−0.90) to 1.00 (95% CI 0.95−1.00), specificity from 0.57 (95% CI 0.47−0.67) to 1.00 (95% CI 0.91−1.00). Cutoff scores used by the studies ranged from 13 to 26. Twenty-six studies, for a total of 19,060 participants, investigated the diagnostic accuracy of MoCA in differentiating patients with MCI from healthy aged individuals. Sensitivity ranged from 0.54 (95% CI 0.39−0.69) to 0.96 (95% CI 0.91−0.99), specificity from 0.19 (95% CI 0.19−0.33) to 0.97 (95% CI 0.91−0.99). Cutoff scores used in these studies ranged from 17 to 26. Seven studies, for a total of 1087 participants, investigated the diagnostic accuracy of MoCA in differentiating patients with dementia from patients with MCI. Sensitivity varied from 0.81 (95% CI 0.65−0.92) to 1.00 (95% CI 0.91−1.00), specificity from 0.69 (95% CI 0.57−0.79) to 0.98 (95% CI 0.88−1.00). The cutoff scores used in these studies ranged from 15 to 25. The authors concluded that there was a wide range of suggested cutoff scores for detecting patients with both dementia and MCI, and that poor methodological rigor had affected operating characteristics and diagnostic accuracy, suggesting the use of MoCA as a screening and not as a diagnostic tool.

Carson, Leach and Murphy (2018) proposed an examination of MoCA cutoff scores for discriminating patients with MCI from healthy aged individuals. Nine studies were included in the metaanalysis, for a total of 478 (53.1%) patients with MCI identified from 1048 participants. Cutoff scores varied from 13 to 30. The authors reported that a cutoff score of 23 provided the best diagnostic accuracy across a range of parameters included in the metaanalysis and recommended the use of this score to screen for patients with MCI.

Thus, several sources of heterogeneity characterized the studies presented and affected the variability of cutoff scores and operating characteristics. A nonexhaustive list of possible moderator variables that can account for heterogeneity in systematic reviews should comprise: (1) type of study conducted (e.g., prevalence and/or incidence studies, prospective or retrospective cohort studies, cross-sectional or longitudinal studies, case—control or population studies); (2) setting for selection and recruitment of patients and healthy aged participants (e.g., community or primary/secondary care facilities), which can affect the prevalence of dementia and MCI; (3) criteria used as the gold standard for diagnosis of dementia and MCI (e.g., *Diagnostic and Statistical Manual of Mental Disorders* [DSM] or International Classification of Diseases criteria for dementia; National Institute of Neurological and Communicative Disorders and Stroke—Alzheimer's Disease and Related Disorders Association or other criteria for AD; Petersen or DSM-V criteria for MCI); (4) country, which may have a differential impact on the prevalence and incidence of AD and MCI, on the socioeconomic status of participants, on access to national health services, and on quality of public and private facilities for neurodegenerative diseases, as well as differences within the same country (e.g., urban/rural contexts, more or less polluted areas, etc.); and (5) different versions of the MoCA used (e.g., paper and pencil, telephonic, electronic, etc.).

Behind the cutoff: statistical methods for calculating diagnostic thresholds

Clinical diagnostic tests are used to evaluate the presence or the absence of a disease. In the diagnostic process it is necessary to determine or redetermine the clinical condition of the patient, predicted from a specific test, not just using clinical experience, but also with reliable and consistent statistical methods. This is possible with the determination of an optimal cutoff (see Fig. 25.1) (Bosco et al., 2017; Sharma, 2014).

Several methods have been developed to determine a cutoff point for a clinical test, to classify patients as positive or negative for the disease. A list of the most accredited methods (Kim, Choi, Chung, Rha, & Kim, 2004) is provided below.

Mean $\pm$ *2SD method*: Once an adequate sample of a population of interest is recruited, and the diagnostic test is run, the mean and standard deviation (SD) of the scores can be computed. An interval, obtained by subtracting/adding two times the SD from the mean predicts that the chance of a test score falling outside this interval will be less than 5% (see Fig. 25.2).

The 90th percentile: An individual's performance may be considered at risk if it falls within the 90th percentile of the sample data. The k-th percentile splits the data distribution into two parts: the lower k% part of the data and the upper part, which includes the remaining data points (see Fig. 25.3).

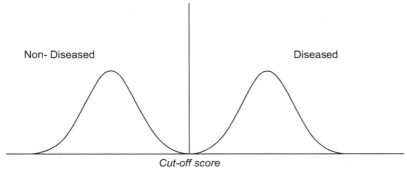

Figure 25.1 *Nonoverlapping test distributions.* Two nonoverlapping normal distributions of a test are shown. It is the ideal condition in discriminating "diseased" from "nondiseased" individuals.

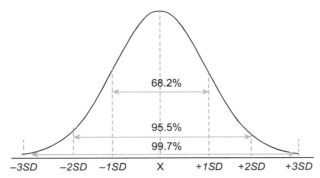

Figure 25.2 *Mean ± 2SD method.* Mean ± 1, 2, and 3 SD intervals is shown. The 2SD interval is usually considered appropriate as the cutoff for a test.

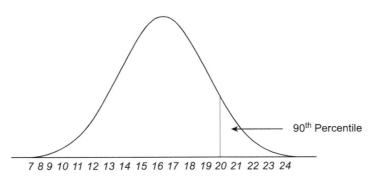

Figure 25.3 *The 90th percentile method.* The 90th percentile method divides the scores into two parts: those that are less than or equal to the score corresponding to the 90th percentile and those that exceed it.

Logistic regression: Logistic regression is a statistical method that can be used to predict the presence or absence of a disease, taking into account a series of independent variables. This method predicts the posttest probability of disease.

Discriminant function analysis: Discriminant analysis is a statistical multivariate procedure and can be considered as a posterior procedure of a multivariate analysis of variance. It is used to classify positive and negative cases with observed diagnostic test values (Fig. 25.4). The discriminant score is the output of the analysis and predicts group membership for every case. It is considered to be a cutoff point that classifies patients from normal to diseased (in ascending order). This is a suitable method adopted to determine cutoff scores when more than one diagnostic test is administered at a time. The predictors have a normal distribution and the outcomes are measured as a continuous variable. It is also used in place of logistic regression, given that the two methods yield the same conclusion.

Receiver operating characteristic curve: The most popular method to predict cutoff values is the ROC (DeLong, DeLong, & Clarke-Pearson, 1988, pp. 837–845; Metz, 1978) curve analysis based on sensitivity and specificity (see Fig. 25.5), as can be seen in the earlier paragraph on the MoCA cutoff scores. Under ideal conditions it should be possible to distinguish all diseased patients from those who are disease free. Under real conditions, there are four possible outcomes: true positive and negative, and false positive and negative. The accuracy of a diagnostic test can be measured by the probability of identifying a true positive case, namely, sensitivity, and the probability of identifying a true negative, namely, specificity. The ROC curve is a graph of the sensitivity (y axis) versus 1−specificity (x axis).

To identify the optimal cutoff point for the purpose of diagnosis, the Youden index (J) is used (Fig. 25.5).

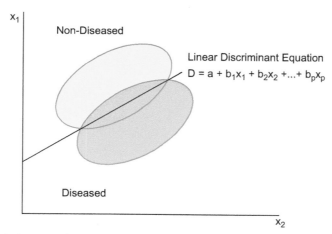

Figure 25.4 *Discriminant analysis method.* The graph usually used for representing the discriminant equation is shown.

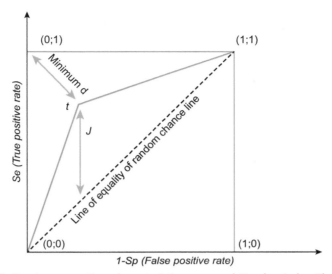

Figure 25.5 ROC, Receiver operating characteristic curve and Youden index. The ROC curve is a graphical method representing the accuracy of a diagnostic test. The Youden is a combinatory index based on sensitivity (*Se*) and specificity (*Sp*) at a given cutoff point, according to the following formula: $J(t) = Se(t) + Sp(t) - 1$.

Choosing a suitable cutoff calculation method is of paramount importance to use a test successfully. To sidestep this, the researcher could also consider the prevalence of the disease to make a diagnostic decision, using Bayesian decision analysis. In the next paragraphs, a Bayesian approach will be discussed; adding information about how a test result combined with the prevalence of the disease would change the probability of having or not having a disease.

A Bayesian approach to the use of clinical cutoff in the Montreal Cognitive Assessment

The use of a test of a diagnostic nature should always be supplemented by its operational characteristics, i.e., the degree of its diagnostic/screening reliability. The previous paragraphs highlighted the role of sensitivity and specificity as indicators of the diagnostic accuracy (e.g., Altman & Bland, 1994) of the test and, in particular, of its aptitude for minimizing false negatives and false positives.

In addition to the operational characteristics of the test, we can also calculate other indices that are, in principle, useful for the diagnostic process: the positive and negative predictive powers. Unfortunately, these measures are not sufficiently effective indicators of predictive validity due to their well-known dependence on the prevalence of the illness (e.g., Baldessarini, Finklestein, & Arana, 1983). Paradoxically, at the level of diagnosis in the clinical setting, the use of predictive powers can be misleading. The so-called likelihood ratios are, however, extremely useful for these purposes. Likelihood ratios are based on conditional statistics known as Bayesian statistics, whose cornerstone is the Bayes theorem (Eq. 25.1):

$$P(\text{Illness} \mid \text{Test}) = \frac{P(\text{Test} \mid \text{Illness}) \cdot P(\text{Illness})}{P(\text{Test})}, \tag{25.1}$$

where Test is a certain result obtained by a diagnostic test (e.g., MoCA score under the intended cutoff score) and Illness is the intended state of disease (e.g., MCI, AD), and these are considered to be events occurring with a certain probability, P(Test) and P(Illness), respectively. Moreover, $P(\text{Test}) \neq 0$. P(Illness|Test) denotes the conditional probability of the event Illness (the condition being studied) given a certain result of the Test, as the product of P(Illness), the former corresponding to the prevalence of the Illness in the appropriate population and the conditional probabilities of a certain result on the Test, given the event of Illness: P(Illness|Test). The product is standardized by the probability of obtaining an intended result on the test: P(Test). The ratio in the equation can easily be viewed as based on likelihood ratios (positive likelihood ratio [LR$^+$] and negative likelihood ratio [LR$^-$]). The higher the LR$^+$ and the more it tends to infinity, the more the test, in the case of a positive result, will be discriminative and therefore important from a diagnostic point of view. In contrast, the closer the LR$^-$ is to zero, the more the test, in the case of a negative result, is important for excluding the diagnosis. It is easy to demonstrate (but it is not the aim of the present chapter) that this calculation can be shortened by the following Eqs. (25.2 and 25.3):

$$LR^+ = \frac{\text{Sensitivity}}{(1 - \text{Specificity})}, \tag{25.2}$$

$$LR^- = \frac{(1 - \text{Sensitivity})}{(\text{Specificity})}, \tag{25.3}$$

where sensitivity and specificity are the operational characteristics obtained by the test in previous studies.

The product of a certain LR and the prevalence (previously transformed in ODDS) of the intended illness returns the so-called posterior probability, allowing the clinician to decide whether the result is diagnostically satisfactory (e.g., setting a 5% statistical criterion). Actually, it is not possible to directly multiply the LRs for the prevalence expressed in terms of proportions. It is necessary to transform the proportion into the corresponding ODDS. The following equation transforms prevalence into ODDS (Eq. 25.4):

$$ODDS = \frac{\text{Prevalence}}{(1 - \text{Prevalence})}. \tag{25.4}$$

It is possible to return to a probability, according to the following inverse Eq. (25.5):

$$\text{Posterior probability} = \frac{\text{ODDS}}{(1 + \text{ODDS})}. \tag{25.5}$$

According to our aim, we try to assess the role that MoCA can play in making a probabilistic decision regarding the likelihood that a person may or may not have MCI or dementia (Fig. 25.6). In this regard, the review by O'Driscoll and Shaikh (2017) is very comprehensive. That review does not provide LR values; therefore, we

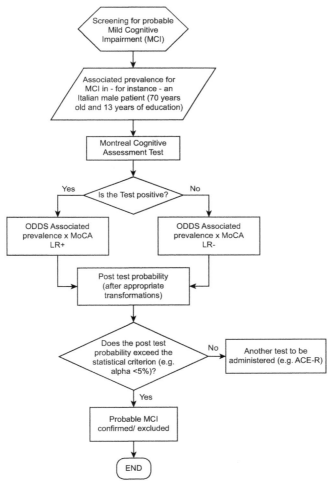

Figure 25.6 *Bayesian diagnostic decision algorithm.* The flow chart of a screening process according to the Bayesian approach is shown. *ACE-R*, Addenbrooke's Cognitive Examination—Revised; *LR+*, positive likelihood ratio; *LR−*, negative likelihood ratio; *MoCA*, Montreal Cognitive Assessment.

can provide this supplement as an example. As a starting point, Prince et al. (2013) suggested that the prevalence of dementia in the general population (≥ 60) can be estimated in the range of 5%—7%. The review of Roberts and Knopman (2013) shows a wide range of prevalence for MCI, with values ranging from 2.4% (i.e., Fisk, Merry, & Rockwood, 2003) to values 20 times greater, that is, as high as 42% (i.e., Artero et al., 2008), for people ≥ 65 years of age.

Montreal Cognitive Assessment likelihood ratios for detecting dementia

First of all, a series of methodological decisions was made: (1) we decided to focus on AD studies. (2) The highest Youden index score was used as a criterion in choosing the best study among those showing the same cutoff. We found eight studies showing MoCA cutoff scores ranging from 14 to 26. The LR+ ranged from 4.9 to 44.0, and the LR− ranged from 0.001 to 0.143. Accordingly, we calculated the posttest probabilities by multiplying the ODDS of the prevalence (Eq. 25.4) with LRs, both positive and negative. The posttest probability for positive test results varied between 0.205 and 0.698 (prevalence 0.05) and between 0.269 and 0.768 (prevalence 0.07), while, for negative test results, it varied between 0.0001 and 0.01 (prevalence 0.05) and between 0.0001 and 0.001 (prevalence 0.07).

Therefore, the best solution is associated with a MoCA cutoff of 22 for the diagnosis of dementia, as emerged in the study of Costa et al. (2012). That is, a person who received an assessment based on MoCA with a prior probability of having dementia between 5% and 7%, and with a value of less than or equal to 22, would have a 10—12 times higher probability of having a dementia. On the other hand, if he or she gets a score above 22, he or she would have 30 times less posttest probability with respect to the prior probability, that is, the initial prevalence.

Montreal Cognitive Assessment likelihood ratios for detecting mild cognitive impairment

We made the same methodological decisions described in the previous paragraphs. Ten studies showing MoCA cutoff scores ranging from 18 to 28 were finally selected to handle the efficacy of the MoCA in detecting people with MCI. LR+ values ranged between 1.102 and 30.333, and the LR− ranged from 0.053 to 0.902. Posttest probability for a test with positive results varied between 0.026 and 0.427 (prevalence 0.024) and between 0.444 and 0.956 (prevalence 0.42). Posttest probabilities as an effect of a negative test result varied between 0.0013 and 0.0217 (prevalence 0.024) and between 0.037 and 0.395 (prevalence 0.42).

In the detection of MCI, the best MoCA cutoff seemed to be 23, as emerged in the study of Tan et al. (2015). A person who received an assessment based on MoCA with a prior probability of having an MCI ranging between 2.4% and 42%, and who got a value of less than or equal to 23, would have up to 17 times greater probability (associated with a prevalence of 2.4%) of having an MCI. On the other hand, if he or she gets a score above 23, he or she would have a reduction in posttest probability of up to 12 times (associated with a prevalence of 2.4%) with respect to the prior probability, that is, the initial prevalence.

Concluding remarks

This chapter aimed to provide an overview of the MoCA as a screening/initial diagnosis tool for MCI and dementia and to discuss the importance of using LRs in diagnostic

Table 25.1 The most widespread screening tools for cognitive efficiency, first Author, year of publication, test length and cognitive domains evaluated are shown.

Name	Abbreviation	Authors	Year	No. items	Domains
Addenbrooke's Cognitive Examination III	ACE-III	Hsieh et al.	2013	21	Attention, memory, fluency, language, visuospatial abilities
Alzheimer's Disease Assessment Scale—cognitive subscale	ADAS-cog	Graham et al.	2004	11	Orientation, 10-word list learning task, 12-word recognition task, recall of instructions, comprehension, naming, word finding difficulty, language, praxis
Memory Impairment Screen	MIS	Buschke et al.	1999	4	Memory
Mini-Mental State Examination	MMSE	Folstein et al.	1975	30	Orientation, memory, attention and calculation, language, visual constructive abilities
Montreal Cognitive Assessment	MoCA	Nasreddine et al.	2005	30	Visuospatial abilities, language, memory, attention, executive functions, orientation
Rowland Universal Dementia Assessment Scale	RUDAS	Storey	2004	6	Memory, body orientation, visuospatial abilities, judgment, language

practice in neurology, geriatrics, and neuropsychology, primarily. Some concluding remarks arise from this study.

(1) The wide variety of operational characteristic values in several national validation studies raises questions regarding methodological issues and highlights the need for a clearer distinction between case—control and population studies. To solve this kind of methodological difficulty it would be necessary to undertake multicenter and multinational cross-sectional studies.

(2) (From our study, it would seem that the best performance in terms of LR is obtained by two independent studies that establish a single point of difference between the optimal cutoff for dementia (i.e., 22) and that for MCI (i.e., 23). This result casts some doubts on the possibility that only one point in the test will differentiate AD and MCI compared with samples of healthy participants. A direct comparison between MCI and AD remains open but clinically and methodologically more difficult to perform. This concern could be resolved with longitudinal studies on the same cohorts of elderly people tested for a long time along the different trajectories that their cognitive efficiency will take over the years.

Key facts

- The MoCA is a brief test of global cognitive functioning.
- The MoCA takes about 10 min to administer, and its total score ranges from 0 to 30.
- The MoCA assesses short-term memory, visuospatial function, executive function, attention, concentration and working memory, language, and orientation.
- The MoCA was originally proposed in 2005 by Nasreddine and colleagues as a brief screening tool for the diagnosis of MCI, and then extended also for the diagnosis of dementia.
- The MoCA seems to be a promising alternative to the MMSE, because of its sensitivity to early detection of dementia and MCI.
- Nowadays the MoCA is also employed as a screening tool for cognitive impairment in a great range of conditions other than age-related cognitive decline, such as Parkinson's disease, traumatic brain injuries, stroke, epilepsy, Korsakoff syndrome, etc.
- There are multiple versions of MoCA, such as paper and pencil, electronic, haptic, mini-version, etc., all available on www.mocatest.org.

Summary points

- A positive result from a MoCA test does not necessarily mean illness.
- Sensitivity and specificity are important indicators but are not sufficient for diagnosis.
- LRs are useful for diagnosis.

- The combination of LRs and specific prevalence returns the posterior probability of a diagnosis.
- Multicenter, multinational, and longitudinal studies are needed to further improve the MoCA as a screening tool.

References

Altman, D. G., & Bland, J. M. (1994). Diagnostic tests. 1: Sensitivity and specificity. *British Medical Journal, 308*(6943), 1552.

Artero, S., Ancelin, M. L., Portet, F., Dupuy, A., Berr, C., Dartigues, J. F., et al. (2008). Risk profiles for mild cognitive impairment and progression to dementia are gender specific. *Journal of Neurology, Neurosurgery and Psychiatry, 79*(9), 979—984.

Baldessarini, R. J., Finklestein, S., & Arana, G. W. (1983). The predictive power of diagnostic tests and the effect of prevalence of illness. *ArchGen Psychiatry, 40*, 569—573.

Bosco, A., Spano, G., Caffò, A. O., Lopez, A., Grattagliano, I., Saracino, G., et al. (2017). Italians do it worse. Montreal Cognitive Assessment (MoCA) optimal cut-off scores for people with probable Alzheimer's disease and with probable cognitive impairment. *Aging Clinical and Experimental Research, 29*(6), 1113—1120.

Buschke, H., Kuslansky, G., Katz, M., Stewart, W. F., Sliwinski, M. J., Eckholdt, H. M., et al. (1999). Screening for dementia with the memory impairment screen. *Neurology, 52*(2), 231—238.

Carson, N., Leach, L., & Murphy, K. J. (2018). A re-examination of Montreal Cognitive Assessment (MoCA) cutoff scores. *International Journal of Geriatric Psychiatry, 33*(2), 379—388.

Cecato, J. F., Martinelli, J. E., Izbicki, R., Yassuda, M. S., & Aprahamian, I. (2016). A subtest analysis of the Montreal Cognitive Assessment (MoCA): Which subtests can best discriminate between healthy controls, mild cognitive impairment and Alzheimer's disease? *International Psychogeriatrics, 28*(5), 825—832.

Costa, A. S., Fimm, B., Friesen, P., Soundjock, H., Rottschy, C., Gross, T., et al. (2012). Alternate-form reliability of the Montreal Cognitive Assessment screening test in a clinical setting. *Dementia and Geriatric Cognitive Disorders, 33*(6), 379—384.

Davis, D. H., Creavin, S. T., Yip, J. L., Noel-Storr, A. H., Brayne, C., & Cullum, S. (2015). *Montreal Cognitive Assessment for the diagnosis of Alzheimer's disease and other dementias*. The Cochrane Library.

DeLong, E. R., DeLong, D. M., & Clarke-Pearson, D. L. (1988). *Comparing the areas under two or more correlated receiver operating characteristic curves: A nonparametric approach*. Biometrics.

Fisk, J. D., Merry, H. R., & Rockwood, K. (2003). Variations in case definition affect prevalence but not outcomes of mild cognitive impairment. *Neurology, 61*(9), 1179—1184.

Folstein, M. F., Folstein, S. E., & McHugh, P. R. (1975). "Mini-mental state": A practical method for grading the cognitive state of patients for the clinician. *Journal of Psychiatric Research, 12*(3), 189—198.

Graham, D. P., Cully, J. A., Snow, A. L., Massman, P., & Doody, R. (2004). The Alzheimer's disease assessment scale-cognitive subscale: Normative data for older adult controls. *Alzheimer Disease & Associated Disorders, 18*(4), 236—240.

Hsieh, S., Schubert, S., Hoon, C., Mioshi, E., & Hodges, J. R. (2013). Validation of the Addenbrooke's Cognitive Examination III in frontotemporal dementia and Alzheimer's disease. *Dementia and Geriatric Cognitive Disorders, 36*(3—4), 242—250.

Julayanont, P., & Nasreddine, Z. S. (2017). Montreal Cognitive Assessment (MoCA): Concept and clinical review. In *Cognitive screening instruments* (pp. 139—195). Cham: Springer.

Kim, I., Choi, Y. H., Chung, H. C., Rha, S. Y., & Kim, B. S. (2004). Statistical method of determining a cut off value between normal and disease groups. *Bulletin of Informatics and Cybernetics, 36*(1), 63—72.

Matías-Guiu, J. A., Valles-Salgado, M., Rognoni, T., Hamre-Gil, F., Moreno-Ramos, T., & Matías-Guiu, J. (2017). Comparative diagnostic accuracy of the ACE-III, MIS, MMSE, MoCA, and RUDAS for screening of Alzheimer disease. *Dementia and Geriatric Cognitive Disorders, 43*(5—6), 237—246.

Metz, C. E. (1978). Basic principles of ROC analysis. *Seminars in Nuclear Medicine, 8*(4), 283—298. Elsevier.

Milne, A., Culverwell, A., Guss, R., Tuppen, J., & Whelton, R. (2008). Screening for dementia in primary care: A review of the use, efficacy and quality of measures. *International Psychogeriatrics, 20*(5), 911−926.

Mitchell, A. J., & Malladi, S. (2010). Screening and case finding tools for the detection of dementia. Part I: Evidence-based meta-analysis of multidomain tests. *The American Journal of Geriatric Psychiatry, 18*(9), 759−782.

Nasreddine, Z. S., Phillips, N. A., Bédirian, V., Charbonneau, S., Whitehead, V., Collin, I., et al. (2005). The montreal cognitive assessment, MoCA: A brief screening tool for mild cognitive impairment. *Journal of the American Geriatrics Society, 53*(4), 695−699.

O'Driscoll, C., & Shaikh, M. (2017). Cross-cultural applicability of the Montreal Cognitive Assessment (MoCA): A systematic review. *Journal of Alzheimer's Disease, 58*(3), 789−801.

Prince, M., Bryce, R., Albanese, E., Wimo, A., Ribeiro, W., & Ferri, C. P. (2013). The global prevalence of dementia: A systematic review and metaanalysis. *Alzheimer's and Dementia: The Journal of the Alzheimer's Association, 9*(1), 63−75.

Roberts, R., & Knopman, D. S. (2013). Classification and epidemiology of MCI. *Clinics in Geriatric Medicine, 29*(4), 753−772.

Sharma, B. (2014). *Right choice of a method for determination of cut-off values: A statistical tool for a diagnostic test.*

Storey, J. E., Rowland, J. T., Conforti, D. A., & Dickson, H. G. (2004). The Rowland universal dementia assessment scale (RUDAS): A multicultural cognitive assessment scale. *International Psychogeriatrics, 16*(1), 13−31.

Tan, J. P., Li, N., Gao, J., Wang, L. N., Zhao, Y. M., Yu, B. C., et al. (2015). Optimal cutoff scores for dementia and mild cognitive impairment of the Montreal Cognitive Assessment among elderly and oldest-old Chinese population. *Journal of Alzheimer's Disease, 43*(4), 1403−1412.

Tsoi, K. K., Chan, J. Y., Hirai, H. W., Wong, S. Y., & Kwok, T. C. (2015). Cognitive tests to detect dementia: A systematic review and meta-analysis. *Journal of the American Medical Association Internal Medicine, 175*(9), 1450−1458.

Tsoi, K. K., Chan, J. Y., Hirai, H. W., Wong, A., Mok, V. C., Lam, L. C., et al. (2017). Recall tests are effective to detect mild cognitive impairment: A systematic review and meta-analysis of 108 diagnostic studies. *Journal of the American Medical Directors Association, 18*(9) (807-e17).

Whiting, P. F., Rutjes, A. W., Westwood, M. E., Mallett, S., Deeks, J. J., Reitsma, J. B., et al. (2011). QUADAS-2: A revised tool for the quality assessment of diagnostic accuracy studies. *Annals of Internal Medicine, 155*(8), 529−536.

Wong, A., Nyenhuis, D., Black, S. E., Law, L. S., Lo, E. S., Kwan, P. W., et al. (2015). Montreal Cognitive Assessment 5-minute protocol is a brief, valid, reliable, and feasible cognitive screen for telephone administration. *Stroke, 46*(4), 1059−1064.

Yokomizo, J. E., Simon, S. S., & de Campos Bottino, C. M. (2014). Cognitive screening for dementia in primary care: A systematic review. *International Psychogeriatrics, 26*(11), 1783−1804.

CHAPTER 26

Utility of ALBA screening instrument: the prodromal phase of dementia with Lewy bodies

Waleska Berríos, Angel Golimstok
Department of Neurology, Italian Hospital of Buenos Aires, Buenos Aires, Argentina

List of abbreviations

AD Alzheimer's disease
ASI ALBA Screening Instrument
DAT dopamine transporter
DLB Dementia with Lewy bodies
DSM Diagnostic and Statistical Manual for Mental Disorders
IRBD idiopathic RBD
LBD Lewy body disease
LBs Lewy bodies
MCI mild cognitive impairment
MIBG myocardial meta-iodobenzylguanidine
MNCD major neurocognitive disorder
mNCD mild neurocognitive disorder
MPS mild parkinsonian signs
PD Parkinson's disease
PDD Parkinson's disease dementia
PET positron emission tomography
RBD REM sleep behavior disorder
REM rapid eye movement
SPECT single-photon emission computed tomography
VH visual hallucinations
αSyn α-synuclein

Mini-dictionary of terms

Autonomic nervous system Is the branch of the nervous system that works without conscious control.
Biomarker Is a broad subcategory of medical signs that are objective indications of medical state observed from outside the patient, which can be measured accurately and reproducibly.
Dementia Is the long-term gradual decrease in the cognitive abilities affecting a person's daily functioning.
Lewy bodies Are abnormal aggregates of protein that develop inside neurons, that are found in Parkinson's disease, the Lewy body dementias (Parkinson's disease dementia and dementia with Lewy bodies), and some other disorders.
Parkinsonism Is a clinical syndrome characterized by tremor, bradykinesia, rigidity, and postural instability.

Diagnosis and Management in Dementia
ISBN 978-0-12-815854-8, https://doi.org/10.1016/B978-0-12-815854-8.00026-4

Prodrome Is an early sign or symptom (or set of signs and symptoms) that often indicates the onset of a disease before more diagnostically specific signs and symptoms develop. It is derived from the Greek word *prodromos*, meaning "running before."

Rapid eye movement sleep Is one of the phases of sleep in mammals and birds, which is characterized by random/rapid movement of the eyes, associated with low muscle tone throughout the body, and usually vivid dreaming.

Introduction

Dementia with Lewy bodies (DLB) is the second most frequent neurodegenerative disorder following Alzheimer disease (AD) with a prevalence rate of 5% in the elderly and up to 30% of all dementias (Zaccai, McCracken & Brayne, 2005) and is characterized by symptoms such as cognitive fluctuation, spontaneous and progressive motor symptoms of Parkinsonism, recurrent visual hallucinations, rapid eye movement (REM) sleep disorders, autonomic symptoms, and neuroleptic sensitivity (McKeith et al., 2005). Recently, the *Diagnostic and Statistical Manual for Mental Disorders* (DSM-5) included as entities dementia with Lewy bodies as "Major Neurocognitive Disorder (MNCD) due to Lewy Body Disease (LBD)" and nondemented patients as "Mild Neurocognitive Disorder (mNCD) with Lewy Bodies" (American Psychiatric Association, 2013).

The clinical features of DLB and Parkinson's disease dementia (PDD) are similar (Emre et al., 2007; Metzler-Baddeley, 2007). Based on the third international consensus, DLB is diagnosed when cognitive impairment precedes parkinsonism or begins within a year of parkinsonism. PDD is diagnosed when parkinsonism precedes cognitive impairment by more than 1 year (McKeith et al., 2005). DSM-5 recognized also PDD as "Major and Mild Neurocognitive Disorder due to Parkinson's disease (PD)" (American Psychiatric Association, 2013).

Both types of disorder, the DLB and PDD, shared the pathology of α-synuclein (αSyn) and pathological lesions as Lewy bodies (LBs) and Lewy neurites (Jellinger, 2009). At the end stage of these disorders, the distribution of pathological lesions is very similar (Beach et al., 2009), suggesting they could be clinical variants of the same disease (Aarsland, Ballard, & Halliday, 2004). The earlier cortical involvement in DLB and earlier brainstem involvement in PDD is a pathological difference between DLB and PDD that could explain different sequences of clinical features, i.e., motor symptoms followed by cognitive-behavioral impairment in PDD, and inversely in DLB (Halliday, Hely, Reid, & Morris, 2008; Jellinger, 2004).

The December 2005 recommendations of the DLB Consortium on diagnosis and management have been useful for diagnosis and management of disease in clinical and research practice, but sensitivity is still insufficient with a percentage of underdiagnosis, with cases often misdiagnosed as AD (McKeith et al., 2005). The revised DLB criteria from the fourth DLB Consortium presented in 2017 maintained the structure of previous version, incorporating new developments and recognizing the interest in detecting early stage disease, anticipating that a similar report to that for the dementia phase was being developed for the prodromal stage (McKeith et al., 2017).

In this revised version, the criteria for clinical diagnosis were improved, clearly differentiating between clinical characteristics and diagnostic biomarkers.

While clinical signs and symptoms were classified as central or supportive, the same was done with the biomarkers, as indicative or supportive, according to their diagnostic specificity and the available evidence.

The support items help in clinical decision-making, providing evidence for a diagnosis of DLB, even when they are not conclusive by themselves.

The main change of the clinical criteria has been the reassignment of REM sleep behavior disorder (RBD) from the suggestive to the core features. Also, neuroleptic sensitivity, and low–dopamine transporter (DAT) imaging, have been reassigned in this new version.

These revised criteria shown in Table 26.1 yielded categories of probable and possible DLB, describing the most typical clinical presentations of DLB.

The prodromal phase has acquired great importance in dementia due to the search for an early therapeutic intervention. In LBD, this stage is characterized by cognitive, motor, and behavioral clinical features, and also by biomarkers detected in laboratory, neurophysiological, or image tests (Chiba et al., 2012; Fujishiro, Nakamura, Sato, & Iseki, 2015). A biomarker is an emerging concept, which means "a characteristic that is objectively measured and evaluated as an indicator of normal biological processes, pathogenic processes, or pharmacologic responses to a therapeutic intervention" (Biomarkers Definitions Working Group, 2001). The detection of biomarkers in DLB would be a step required to define candidates to implement preventive strategies in order to avoid or delay the progression of the disease.

In this chapter, we describe the main characteristics of the prodromal phase of DLB and a screening instrument designed and validated by our team, that is useful, even at this stage of the disease.

Characteristics of prodromal DLB

Features of prodromal DLB (Fig. 26.1) can be divided into the following items: (1) Cognitive symptoms, (2) Behavioral symptoms, (3) Sleep phenomena, (4) Autonomic dysfunction, (5) Physical symptoms, and (6) Olfactory dysfunction.

Cognitive symptoms

Although it has been reported in two longitudinal studies that cognitive fluctuations are not frequently found in the prodromal phase (Jicha et al., 2010; Molano et al., 2009), it is

Table 26.1 New criteria for the diagnosis of dementia with Lewy bodies.

An **essential** criterion is the diagnosis of dementia, defined as a cognitive decline on one or more cognitive domains and/or behavioral changes resulting in impairment of normal daily function

Memory impairment would not be remarkable in the early stages but is usually evident with progression

Neuropsychological test is characterized by deficit of attention, executive function, and visuoperceptual ability, even at early stage

Core clinical features (note that the first three typically occur early and may persist along the course of disease)

Fluctuating cognition (pronounced variations in attention and alertness)
Recurrent visual hallucinations (these are typically well formed and detailed)
REM sleep behavior disorder (may precede cognitive decline)
One or more spontaneous cardinal features of parkinsonism such as bradykinesia, rest tremor, or rigidity

Supportive clinical features

Dangerous sensitivity to antipsychotic agents; postural instability; frequent falls; syncope or other transient episodes of unresponsiveness; severe autonomic dysfunction, urinary incontinence; hypersomnia; hyposmia; hallucinations in other modalities; systematized delusions; apathy, anxiety, and depression.

Indicative biomarkers

SPECT/PET: Reduced dopamine transporter uptake in basal ganglia
123iodine-MIBG myocardial scintigraphy: low uptake
REM sleep without atonia confirmed by polysomnography

Supportive biomarkers

Computed Tomography/Magnetic Resonance Imaging scan show relative preservation of medial temporal lobe structures

Evidence of generalized low uptake on SPECT/PET perfusion/metabolism scan with reduced occipital activity, the cingulate island sign on FDG-PET imaging

Electroencephalogram shows prominent posterior slow-wave activity with periodic fluctuations in the pre-alpha/theta range

Diagnosis of PROBABLE DLB:

  a. Evidence of two or more core clinical features of DLB, with or without indicative biomarkers, or
  b. Evidence of one core clinical feature plus one or more indicative biomarkers
Important: Probable DLB should not be diagnosed based on biomarkers alone

Diagnosis of POSSIBLE DLB:

  a. Evidence of one core clinical feature of DLB alone, or
  b. Evidence of one or more indicative biomarkers without core clinical features

Table 26.1 New criteria for the diagnosis of dementia with Lewy bodies.—cont'd

DLB is less likely:

a. When other physical illness or brain disorder, e.g., cerebrovascular disease, can explain the clinical picture partially or totally, although these do not exclude a DLB diagnosis and may serve to indicate mixed or multiple pathologies contributing to the clinical presentation, or

b. If parkinsonian features are the only core clinical feature and appear for the first time at a stage of severe dementia.

Pay attention: DLB should be diagnosed when dementia occurs before or concurrently with parkinsonism as well as Parkinson's disease dementia (PDD) should be diagnosed when dementia occurs in the context of well-established PD. In research studies in which distinction needs to be made between DLB and PDD, the arbitrary 1-year rule between the onset of dementia and parkinsonism continues to be helpful and recommended.

It is also recommended to use the term "Lewy body disease" in the common clinical practice.

These are the revised criteria for the clinical diagnosis of dementia with Lewy bodies according to the fourth consensus (2017).
Adapted from McKeith, I.G., Boeve, B.F., Dickson, D.W., Halliday, G., Taylor, J.-P., Weintraub, D., et al. 2017. Diagnosis and management of dementia with Lewy bodies: fourth consensus report of the DLB Consortium. *Neurology,* *89*(1), 88—100. http://doi.org/10.1212/WNL.0000000000004058.

possible that the scarce detection at this stage is due to the fact that caregivers would not be aware of these types of symptoms, and therefore they do not report them on a scale.

In the study of Jicha et al. (2010), which used "operational definitions for symptoms," and defined "fluctuations" as "descriptions of episodic confusion, or staring spells not associated with seizure-like activity or known history of seizure disorder," one-third of the mild cognitive impairment (MCI) patients with DLB and none of the MCI-AD subjects exhibited this symptom.

The data of that study showed that "fluctuations" are associated with further DLB and not with AD, suggesting that careful attention to this feature in the prodromal phase may help to predict a conversion to this type of dementia.

In a recent paper, Ferman et al. (2013) reported that patients with nonamnestic MCI, most of them with concomitant attention and/or visuospatial impairment, were 10 times more likely to develop clinically probable DLB; whereas those with amnestic MCI were 10 times more likely to progress to probable AD. In the DLB group, the progression was particularly evident in patients with nonamnestic multiple domain impairment.

Comparing the neuropsychological profile in MCI-DLB patients with respect to those with MCI-AD, frontal-executive and visuospatial functions were worse in the first group, while verbal recognition memory impairment was greater in the later (Yoon, Kim, Moon, Yong, & Hong, 2015). Similar results were described in a retrospective study at MCI stage comparing the same both groups. Only the number of pentagon angles and the digit backward test were significantly correlated with the diagnosis of DLB, suggesting a more specific dorsal visual stream impairment in MCI-DLB with respect to the ventral visual stream, which seems to be involved in both DLB and AD.

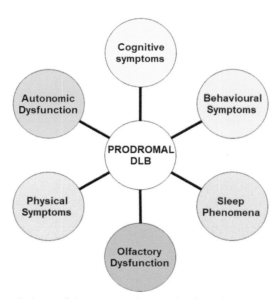

Figure 26.1 he prodromal phase of dementia with Lewy bodies. These are the main components of the prodromal phase of dementia with Lewy bodies.

This group found that the number of pentagon angles, examined with the Qualitative Scoring MMSE Pentagon Test, differentiated DLB from AD with a specificity of 96% and sensitivity of 42%, as well as with a positive predictive value of 91% and a negative predictive value of 63% (Cagnin et al., 2015). A study of RBD patients who developed a neurodegenerative disorder (PD/DLB) in 35% of cases during the follow-up and were pair-matched with healthy controls reported a lower performance in attention and executive functions 6 years before the diagnosis. In this study the Trail Making Test (part B) was the best to detect early prodromal DLB, whereas verbal fluency (semantic) and verbal episodic learning tests seemed to be useful for monitoring changes over time in this disorder (Marchand et al., 2018).

Behavioral symptoms

The presence of visual hallucinations (VHs) is one of the core criteria for clinical diagnosis in DLB, and differently from AD, this symptom may be present initially in the evolution of disease. In about half the cases with DLB, VHs may precede the memory impairment (Auning et al., 2011; Chiba et al., 2012; Fujishiro, Iseki, Nakamura et al., 2013). Despite not being described in the literature yet, in our experience, patients with prodromal DLB usually report having had sporadic episodes of VH for several years before the onset of diagnosis. However, there are some single reports about others behavioral symptoms, like polymodal hallucinations, preceding in many years the full clinical DLB syndrome (Abbate et al., 2014).

Future studies that quantify and characterize the presence of VH in the early prodromal phases are needed to decide the inclusion in clinical criteria for this stage.

Anxiety and depression were described to be present in around one-quarter of patients with diagnosis of the disease (Auning et al., 2011; Chiba et al., 2012). It is very difficult to know if depression is a prodromal feature or only a risk factor. Recent research that used low uptake of myocardial meta-iodobenzylguanidine (MIBG) to diagnose nonclinical LBD showed that this test can be a predictor of progression to DLB when it is combined with major depressive disorder (Fujishiro, Iseki, Nakamura et al., 2013).

However, depressive symptoms are not specific, and may be difficult to use as a biomarker for early diagnosis.

Sleep phenomena

RBD, the core feature recently included in DLB clinical criteria, is a sleep disorder (parasomnia) in which patients appear to physically act out dreams during REM sleep. The behaviors include talking, shouting, grabbing, kicking, punching, or jumping from the bed. Sometimes violent behaviors can occur and be associated with injury. RBD is a well-known prodromal symptom that can precede alpha-synucleinopathies by years or even decades. Some longitudinal studies of RBD have shown that more than 80%—90% of cases would develop a synucleinopathy, i.e., PD, PDD, DLB or multiple system atrophy (Chiba et al., 2012; Fujishiro et al., 2015; Iranzo et al., 2013; Marchand et al., 2018; Schenck, Boeve, & Mahowald, 2013). A very recent multicenter study showed that 41% of idiopathic RBD (IRBD) patients developed defined neurodegenerative synucleinopathy after 5 years of follow-up and were more likely to report family history of dementia, to have autonomic and motor symptoms, and higher use of clonazepam as well. They did not find differences in risk factors between those converted to dementia and parkinsonism (Postuma et al., 2015). IRBD is diagnosed when it occurs in the absence of another neurologic disease. However, several IRBD studies have found subtle cognitive abnormalities of visuospatial abilities, verbal memory, attention, and executive functioning, which are the characteristic cognitive domains affected in DLB (Ferini—Strambi et al., 2004; Massicotte-Marquez et al., 2008; Postuma, Gagnon, & Montplaisir, 2008).

Furthermore, other studies in IRBD found multiple subtle motor, autonomic, visual, and olfactory signs that allow us to assume that this entity is a prodromal phase of synucleinopathy (Lanfranchi, Fradette, Gagnon, Colombo, & Montplaisir, 2007; Postuma, Lang, Massicotte-Marquez, & Montplaisir, 2006; Stiasny-Kolster et al., 2004).

Thus, RBD is a very strong feature of prodromal synucleinopathy. This provides several opportunities for early intervention in future research (Postuma, Gagnon, & Montplaisir, 2013a, p. 3).

Autonomic dysfunction

According to previous reports, dysautonomia would be almost as common as RBD in α-synucleinopathy patients. Thus, dysautonomia may be a prodromal feature, like RBD. The most frequent and early complaint described to happen before motor symptoms in PD (Abbott et al., 2001), and also before cognitive and behavioral symptoms in DLB, is constipation (Kaufmann, Nahm, Purohit, & Wolfe, 2004). Furthermore, IRBD has been associated with subtle sympathetic dysautonomia expressed as abnormal cardiac MIBG (Miyamoto, Miyamoto, Inoue, Usui, Suzuki, & Hirata, 2006).

In a retrospective study, 74% of patients classified as PD, PD plus MCI, or PDD and DLB or MCI plus concurrent parkinsonism experienced dysautonomia predominantly expressed as postural fall in blood pressure, urinary incontinence or retention, and constipation (Claassen et al., 2010). Also, orthostatic hypotension predicted conversion to LBD in around 60% up to 3 years before diagnosis in a group of IRBD patients (Postuma, Gagnon, Pelletier, & Montplaisir, 2013b).

Several reports described the presence of LBs in enteric nervous system, parasympathetic and sympathetic ganglia in PD (Takeda, Yamazaki, Miyakawa, & Arai, 1993; Wakabayashi, Takahashi, Ohama, & Ikuta, 1990), and other locations including hypothalamus (Hague, Lento, Morgello, Caro, & Kaufmann, 1997).

Furthermore, neuropathological studies in large cohorts of asymptomatic aged cases have shown LB pathology in autonomic system (Minguez-Castellanos et al., 2007; Oinas et al., 2010).

Thus, it is hypothesized that there is a strong and early association between synucleinopathy and dysautonomia. However, we still do not know in detail the characteristics of this association, and autonomic symptoms are nonspecific markers for the prodromal phase. The MIBG scintigraphy as biomarker should be studied in future series to define its potential utility as a predictor in this phase.

Other autonomic symptoms less frequently reported, but more specific, may be considered if they are present (Ferini-Strambi et al., 2014).

Physical symptoms

Mild parkinsonian signs (MPSs) are common in the elderly population and have been associated with PD among other pathologies. Parkinsonism is one of the core features of DLB and may be a presenting symptom in around one out of every four patients (Auning et al., 2011). MPSs were reported to be associated with depression (Uemura, Wada-Isoe, Nakashita, & Nakashima, 2013), hyposmia (Louis, Marder, Tabert, & Devanand, 2008), slight cognitive impairment (Louis et al., 2005), dementia (Louis, Tang, & Schupf, 2010), an increased risk of mortality (Zhou, Duan, Sun, Yan, & Ren, 2010), and more recently with PD, sharing many risk factors and prodromal markers of PD (Lerche et al., 2014). All these associations allow us to hypothesize that

MPS could be the clinical expression of a prodromal DLB/PDD. Patients with mild PDD have more significant motor deficits than DLB patients, mostly resting tremor and bradykinesia. While the resting tremor predominates in PDD, the severity of postural and intention tremor was similar in both patients groups (Petrova, Mehrabian-Spasova, Aarsland, Raycheva, & Traykov, 2015).

Tremor, previously reported to be common in DLB, was characterized as mixed tremors, i.e., rest and postural/action tremor (Onofrj et al., 2013). Longitudinal studies to define the utility of MPS as a prodromal clinical marker of DLB are needed for the future.

Olfactory dysfunction

Olfactory symptoms in PD are well established and may represent one of the earliest signs of the disease. Hyposmia may be detected years before the development of motor symptomatology (Ponsen et al., 2004; Ross et al., 2008).

Olfactory dysfunction (anosmia/hyposmia) may be associated to IRBD being a feature of prodromal synucleinopathy (Miyamoto et al., 2010). However, it is difficult to consider this disorder as a specific prodromal feature of DLB because many reports showed that olfactory impairment is associated to other dementias (Duff, McCaffrey, & Solomon, 2002) and also with ageing (Murphy et al., 2002).

On the basis of these evidences, it is possible to consider smell dysfunction as a prodromal sign in DLB but in combination with other symptoms to improve its specificity.

Biomarkers

To define the potential biomarkers in DLB, it is possible to follow the natural process found in the stages of Braak et al. and the model described by Jack et al. (2010) in AD— first, markers of abnormal protein deposition, followed by markers of neuronal injury, structural brain changes, cognitive impairment, and finally, daily living function scales.

Biomarkers of αSyn in vivo are only in a phase of development (Yu et al., 2012) and there is not clear evidence of reduced cerebrospinal fluid αSyn in DLB (Kasuga, Nishizawa, & Ikeuchi, 2012). Blood biomarkers in DLB are in the same situation of AD, and are not available. To our knowledge, studies examining such biomarkers in DLB are lacking.

Biomarkers of neuronal injury and systems dysfunction are available with a growing level of evidence. MIBG scintigraphy is a promising technique based on the use of an analogous of noradrenaline to identify presynaptic sympathetic nerve terminals in the heart, but it is still not validated as a prodromal biomarker, as we commented above.

Assessment of the integrity of dopaminergic circuits in the basal ganglia can be made using a range of radiolabeled ligands, detected with single-photon emission computed tomography (SPECT) or positron emission tomography (PET) (Brooks, 2016). The SPECT of presynaptic DAT, which is responsible for reuptake of dopamine by the presynaptic nerve terminal, with 123I-FP-CIT (ioflupane), is the most frequently used for the differentiation of parkinsonian syndromes from other causes of tremor.

PET also has been used with various radiolabeled ligands to evaluate dopaminergic transmission in the basal ganglia (Brooks, 2016); a major disadvantage of both PET and SPECT, however, is the exposure of subjects to ionizing radiation.

In prodromal phase there are very few studies. Iranzo et al. (2013) followed patients with IRBD and found that more than 90% of them converted to LBD by 14 years and those who were disease-free had decreased striatal DAT uptake.

Another marker of substantia nigra pathology is enlarged hyperechogenicity on ultrasound. One study showed that a baseline hyperechogenicity in substantia nigra increased 20.6 times the risk to develop PD in 5 years of follow-up (Berg et al., 2013). In a study of IRBD, substantia nigra hyperechogenicity was found in 36% of participants, compared with 11% of controls (Iranzo et al., 2010). Another biomarker of neuronal death or dysfunction is 18 F-fluorodeoxyglucose PET. The unique pattern that seems to be predictive of conversion to DLB in MCI patients was baseline occipital hypometabolism (Fujishiro, Iseki, Kasanuki, et al., 2013).

Then, inexpensive and noninvasive biomarkers with higher sensitivity and specificity are needed in the future.

Recently in Argentina, a brief Spanish clinical instrument for screening of LBD was validated as useful even in the prodromal stage. The ALBA Screening Instrument (ASI) has very good sensitivity and specificity (Garcia Basalo et al., 2017). The ASI consists of 17 items divided into three sections (Table 26.2):

(1) The Clinical Questionnaire, consisting of 11 questions, is directed to the patient about possible symptoms associated with the disease. The first four questions are related to the presence of dream enactment behaviors, present in almost all patients with RBD. The fifth and sixth questions are designed to screen VH and the seventh and eighth questions are similar to the core question for RBD in the Mayo Clinic Questionnaire. The last three questions are based on cognitive criteria for DLB.

(2) A brief physical exam to confirm the presence or absence of postural and resting tremor.

(3) A cognitive assessment consisting of four tests that evaluated: (a) Category Verbal Fluency, (b) Working Memory, (c) Executive functions, and (d) Visuospatial functions.

The maximum score is 24 points, with a higher probability of LBD diagnosis as the scale score increases.

ASI is a reliable and valid tool for the assessment of LBD that can be administered by nonmedical neuropsychological staff in about 10 min.

Table 26.2 ALBA screening instrument (ASI): assessment protocol.

Clinical questionnaire

Ask the questions of the questionnaire as they are written. Ideally, the questionnaire must be conducted in the presence of a family caregiver. Questions 4, 6, 7, and 8 are scored two points if the answer is yes, as long as the information comes from the caregiver. However, if affirmative answers to these questions are from the patients, the score must be "1."

1.	Do you have, or have you had, any nightmares in recent times that you recall?	Yes — No	/1
2.	Do dream images seem vivid to you? (Do they seem very real?)	Yes — No	/1
3.	Are the vivid dreams frequent?	Yes — No	/1
4.	Have you ever woken up and tried searching for a dream character or some object that you were dreaming of?	Yes — No	/2
5.	Have you felt that a person close to you, or a family member that has passed away, or somebody from your childhood is currently in your home?	Yes — No	/1
6.	Have you seen an image of any of these people, even fleetingly, although they are not really there?	Yes — No	/2
7.	Have you ever heard the patient talking or screaming during sleep? (Caregiver) Do you know if you have ever spoken or screamed while sleeping? (Patient alone)	Yes — No	/2
8.	Have you ever seen the patient acting out his/her dreams while sleeping? (Punching, flailing of arms, kicking)? (Caregiver). Do you know if you have ever acted out dreams (Punching, flailing of arms, kicking)? (Patient alone)	Yes — No	/2
9.	Have you experienced forgetfulness in recent times?	Yes — No	/1
10.	Have you had any episodes of confusion, even during a hospital stay?	Yes — No	/2
11.	Could you say that your mental status changes from time to time or on different days, some better and some worse?	Yes — No	/1

Physical examination

For the assessment of both items 12 and 13, the presence of tremor must be clear and evident at first sight. If the patient argues that the tremor occurs due to medication or another factor, nevertheless please assign a score of "1."

12.	Check for postural tremor, asking patient to stretch hands before him/her and spread the fingers.	Yes — No	/1

Continued

Table 26.2 ALBA screening instrument (ASI): assessment protocol.—cont'd

13. Check for rest tremor, asking patient to put hands at rest on lap or table.	Yes — No	/1

Cognitive examination (Yes = incorrect; No = correct)

Every cognitive item must be scored "1" or "0" depending on an incorrect or correct performance, respectively, for each item.

- Item 14 (category fluency): Score "1" if the patient names less than 12 animals (e.g.,11).
- Item 15 (Inverse Digit Span). Score "0" points for the sequence correctly repeated. (The correct response for this trial is 7-2-8-5.)
- Item 16 (Executive and visuoconstructional skill: Clock-test). Score: One point is allocated for each of the following three criteria: Circle (0 points): The clock face must be a circle with only minor distortion acceptable (e.g., slight imperfection on closing the circle); numbers (0 points): All clock numbers must be present with no additional numbers; numbers must be in the correct order and placed in the approximate quadrants on the clock face; roman numerals are acceptable but numbers cannot be placed outside the circle contour; hands (0 points): The patient must draw two hands jointly, indicating the correct time; the hour hand must be clearly shorter than the minute hand; hands must be centered within the clock face with their junction close to the clock center. A point is assigned for a given element if any of the above criteria are not met.
- Item 17 (Visuoconstructional Skill: Pentagon copy) Score: To be acceptable, there must be two figures that appear to be intersecting. At least one of the figures must have five angles. Tremor, rotation, relative size, and symmetry are ignored. Consider only lines that the patient has drawn (ignore any lines from the original model that the patient may have included as part of the copy). If the patient has drawn more than one copy, grade the best copy.

14. Category fluency. Say: "I want you to name for me as many animals as you can. You will have 60 s before I tell you to stop. Do you have any questions?"	Yes — No	/1
15. Inverse Digit Span. Ask the patient to repeat the following series in reverse order: 5-8-2-7. Read at a speed of one digit per second.	Yes — No	/1
16. Executive and Visuoconstructional Skill (Clock test). "Draw a circular clock. Put in all the numbers and set the time to 10 past 11."	Circle/Numbers/Hands	/3
17. Visuoconstructional Skill (Pentagon copy). Ask the patient to copy the interlocking pentagons:	Yes — No	/1

Total score: /24.
Screening tool designed for Lewy body dementia, including 17 items about clinical symptoms, physical exam to detect tremor (postural or rest), and cognitive assessment.

The first five questions of the scale, evaluating (1) the presence of nightmares, (2) vivid dreams, (3) the frequency of vivid dreams, (4) searching for someone or something from the dream upon wakening, and (5) the feeling of someone else present, were the more relevant items for diagnosis.

Future research should focus on validating a combination of biomarkers and clinical brief instruments for early diagnosis of prodromal phase of DLB in order to achieve therapeutic interventions that modify the course of the disease.

Key facts of dementia with Lewy bodies

- Dementia with Lewy bodies is a very frequent type of brain disease.
- Fluctuations in cognitive function, hand tremors and/or mild rigidity in face, arms or legs (parkinsonism), recurrent visual hallucinations, dreams enactment that is called REM sleep behavior disorder, e.g., punching, shouting, kicking during sleep, and autonomic symptoms as hypotension, constipation, etc. are clinical characteristics of DLB.
- The presymptomatic phase of DLB (previous to the cognitive impairment) could be characterized by the presence of RBD, deficit in attention – executive function, visuospatial abilities, attentional, some psychiatric symptoms, orthostatic hypotension, mild parkinsonism, smell dysfunction, and alterations in DAT scan.
- The ALBA Screening Instrument is a new valid scale to screen symptoms of DLB even in early stage.

Summary points

- Dementia with Lewy bodies is the second most frequent neurodegenerative disorder.
- Clinical characteristics of DLB consist of symptoms such as cognitive fluctuation, spontaneous and progressive motor symptoms of Parkinsonism, recurrent visual hallucinations, REM sleep behavior disorders, autonomic symptoms, and others.
- Prodromal phase of DLB could be featured by REM sleep behavior disorder, frontal-executive and visuospatial dysfunctions, some psychiatric symptoms, orthostatic hypotension, mild parkinsonian signs, hyposmia, and decreased striatal DAT uptake.
- ALBA Screening Instrument is a reliable and valid tool that consists of 17 items for the assessment of LBD even in prodromal phase, which can be administered by nonmedical neuropsychological staff in about 10 min.
- Future research should focus on validating biomarkers for early diagnosis of DLB.

References

Aarsland, D., Ballard, C. G., & Halliday, G. (2004). Are Parkinson's disease with dementia and dementia with Lewy bodies the same entity? *Journal of Geriatric Psychiatry and Neurology, 17*(3), 137–145.

Abbate, C., Trimarchi, P. D., Inglese, S., Viti, N., Cantatore, A., De, L. A., et al. (2014). Preclinical polymodal hallucinations for 13 years before dementia with Lewy bodies. *Behavioural Neurology, 2014*, 694296–694296.

Abbott, R. D., Petrovitch, H., White, L. R., Masaki, K. H., Tanner, C. M., Curb, J. D., et al. (2001). Frequency of bowel movements and the future risk of Parkinson's disease. *Neurology, 57*(3), 456–462.

American Psychiatric Association. (2013). *Diagnostic and statistical manual of mental disorders* (5th ed.). Washington, D.C: American Psychiatric Association.

Auning, E., Rongve, A., Fladby, T., Booij, J., Hortobágyi, T., Siepel, F. J., et al. (2011). Early and presenting symptoms of dementia with Lewy bodies. *Dementia and Geriatric Cognitive Disorders, 32*(3), 202–208.

Beach, T. G., Adler, C. H., Lue, L., Sue, L. I., Bachalakuri, J., Henry-Watson, J., et al. (2009). Unified staging system for Lewy body disorders: Correlation with nigrostriatal degeneration, cognitive impairment and motor dysfunction. *Acta Neuropathologica, 117*(6), 613–634.

Berg, D., Behnke, S., Seppi, K., Godau, J., Lerche, S., Mahlknecht, P., et al. (2013). Enlarged hyperechogenic substantia nigra as a risk marker for Parkinson's disease. *Movement Disorders, 28*(2), 216–219.

Biomarkers Definitions Working Group. (2001). Biomarkers and surrogate endpoints: Preferred definitions and conceptual framework. *Clinical Pharmacology and Therapeutics, 69*(3), 89–95.

Brooks, D. J. (2016). Molecular imaging of dopamine transporters. *Ageing Research Reviews, 30*, 114–121.

Cagnin, A., Bussè, C., Gardini, S., Jelcic, N., Guzzo, C., Gnoato, F., et al. (2015). Clinical and cognitive phenotype of mild cognitive impairment evolving to dementia with Lewy bodies. *Dementia and Geriatric Cognitive Disorders Extra, 5*(3), 442–449.

Chiba, Y., Fujishiro, H., Iseki, E., Ota, K., Kasanuki, K., Hirayasu, Y., et al. (2012). Retrospective survey of prodromal symptoms in dementia with Lewy bodies: Comparison with Alzheimer's disease. *Dementia and Geriatric Cognitive Disorders, 33*(4), 273–281.

Claassen, D. O., Josephs, K. A., Ahlskog, J. E., Silber, M. H., Tippmann-Peikert, M., & Boeve, B. F. (2010). REM sleep behavior disorder preceding other aspects of synucleinopathies by up to half a century. *Neurology, 75*(6), 494–499.

Duff, K., McCaffrey, R. J., & Solomon, G. S. (2002). The pocket smell test: Successfully discriminating probable Alzheimer's dementia from vascular dementia and major depression. *Journal of Neuropsychiatry and Clinical Neurosciences, 14*(2), 197–201.

Emre, M., Aarsland, D., Brown, R., Burn, D. J., Duyckaerts, C., Mizuno, Y., et al. (2007). Clinical diagnostic criteria for dementia associated with Parkinson's disease. *Movement Disorders, 22*(12), 1689–1707.

Ferini-Strambi, L., Oertel, W., Dauvilliers, Y., Postuma, R. B., Marelli, S., Iranzo, A., et al. (2014). Autonomic symptoms in idiopathic REM behavior disorder: A multicentre case–control study. *Journal of Neurology, 261*(6), 1112–1118.

Ferini–Strambi, L., Di Gioia, M. R., Castronovo, V., Oldani, A., Zucconi, M., & Cappa, S. F. (2004). Neuropsychological assessment in idiopathic REM sleep behavior disorder (RBD) Does the idiopathic form of RBD really exist? *Neurology, 62*(1), 41–45.

Ferman, T. J., Smith, G. E., Kantarci, K., Boeve, B. F., Pankratz, V. S., Dickson, D. W., et al. (2013). Nonamnestic mild cognitive impairment progresses to dementia with Lewy bodies. *Neurology, 81*(23), 2032–2038.

Fujishiro, H., Iseki, E., Kasanuki, K., Chiba, Y., Ota, K., Murayama, N., et al. (2013a). A follow up study of non-demented patients with primary visual cortical hypometabolism: prodromal dementia with Lewy bodies. *Journal of Neurological Sciences, 334*(1), 48–54.

Fujishiro, H., Iseki, E., Nakamura, S., Kasanuki, K., Chiba, Y., Ota, K., et al. (2013b). Dementia with Lewy bodies: Early diagnostic challenges. *Psychogeriatrics, 13*(2), 128–138.

Fujishiro, H., Nakamura, S., Sato, K., & Iseki, E. (2015). Prodromal dementia with Lewy bodies. *Geriatrics and Gerontology International, 15*(7), 817–826.

Garcia Basalo, M. M., Fernandez, M. C., Ojea Quintana, M., Rojas, J. I., Garcia Basalo, M. J., Bogliotti, E., et al. (2017). ALBA screening instrument (ASI): A brief screening tool for Lewy body dementia. *Archives of Gerontology and Geriatrics, 70*, 67–75.

Hague, K., Lento, P., Morgello, S., Caro, S., & Kaufmann, H. (1997). The distribution of Lewy bodies in pure autonomic failure: Autopsy findings and review of the literature. *Acta Neuropathologica, 94*(2), 192–196.

Halliday, G., Hely, M., Reid, W., & Morris, J. (2008). The progression of pathology in longitudinally followed patients with Parkinson's disease. *Acta Neuropathologica, 115*(4), 409–415.

Iranzo, A., Lomeña, F., Stockner, H., Valldeoriola, F., Vilaseca, I., Salamero, M., et al. (2010). Decreased striatal dopamine transporter uptake and substantia nigra hyperechogenicity as risk markers of synucleinopathy in patients with idiopathic rapid-eye-movement sleep behaviour disorder: a prospective study. *The Lancet Neurology, 9*(11), 1070–1077.

Iranzo, A., Tolosa, E., Gelpi, E., Molinuevo, J. L., Valldeoriola, F., Serradell, M., et al. (2013). Neurodegenerative disease status and post-mortem pathology in idiopathic rapid-eye-movement sleep behaviour disorder: An observational cohort study. *The Lancet Neurology, 12*(5), 443–453.

Jack, C. R., Jr., Knopman, D. S., Jagust, W. J., Shaw, L. M., Aisen, P. S., Weiner, M. W., et al. (2010). Hypothetical model of dynamic biomarkers of the Alzheimer's pathological cascade. *The Lancet Neurology, 9*(1), 119–128.

Jellinger, K. A. (2004). Lewy body-related α-synucleinopathy in the aged human brain. *Journal of Neural Transmission, 111*(10–11), 1219–1235.

Jellinger, K. A. (2009). Formation and development of Lewy pathology: A critical update. *Journal of Neurology, 256*(3), 270–279.

Jicha, G. A., Schmitt, F. A., Abner, E., Nelson, P. T., Cooper, G. E., Smith, C. D., et al. (2010). Prodromal clinical manifestations of neuropathologically confirmed Lewy body disease. *Neurobiology of Aging, 31*(10), 1805–1813.

Kasuga, K., Nishizawa, M., & Ikeuchi, T. (2012). α-Synuclein as CSF and blood biomarker of dementia with Lewy bodies. *International Journal of Alzheimer's Disease, 2012*, 437025–437025.

Kaufmann, H., Nahm, K., Purohit, D., & Wolfe, D. (2004). Autonomic failure as the initial presentation of Parkinson disease and dementia with Lewy bodies. *Neurology, 63*(6), 1093–1095.

Lanfranchi, P. A., Fradette, L., Gagnon, J. F., Colombo, R., & Montplaisir, J. (2007). Cardiac autonomic regulation during sleep in idiopathic REM sleep behavior disorder. *Sleep, 30*(8), 1019–1025.

Lerche, S., Hobert, M., Brockmann, K., Wurster, I., Gaenslen, A., Hasmann, S., et al. (2014). Mild parkinsonian signs in the elderly—is there an association with PD? Crossectional findings in 992 individuals. *PLoS One, 9*(3), e92878.

Louis, E. D., Marder, K., Tabert, M. H., & Devanand, D. P. (2008). Mild Parkinsonian signs are associated with lower olfactory test scores in the community-dwelling elderly. *Movement Disorders, 23*(4), 524–530.

Louis, E. D., Schupf, N., Manly, J., Marder, K., Tang, M. X., & Mayeux, R. (2005). Association between mild parkinsonian signs and mild cognitive impairment in a community. *Neurology, 64*(7), 1157–1161.

Louis, E. D., Tang, M. X., & Schupf, N. (2010). Mild parkinsonian signs are associated with increased risk of dementia in a prospective, population-based study of elders. *Movement Disorders, 25*(2), 172–178.

Marchand, D. G., Postuma, R. B., Escudier, F., De Roy, J., Pelletier, A., Montplaisir, J., et al. (April 17, 2018). How does dementia with Lewy bodies start? Prodromal cognitive changes in REM sleep behavior disorder. *Annals of Neurology.* https://doi.org/10.1002/ana.25239 (Epub ahead of print).

Massicotte-Marquez, J., Décary, A., Gagnon, J. F., Vendette, M., Mathieu, A., Postuma, R. B., et al. (2008). Executive dysfunction and memory impairment in idiopathic REM sleep behavior disorder. *Neurology, 70*(15), 1250–1257.

McKeith, I. G., Boeve, B. F., Dickson, D. W., Halliday, G., Taylor, J. P., Weintraub, D., et al. (2017). Diagnosis and management of dementia with Lewy bodies Fourth consensus report of the DLB Consortium. *Neurology, 89*(1), 88–100.

McKeith, I. G., Dickson, D. W., Lowe, J., Emre, M., O'brien, J. T., Feldman, H., et al. (2005). Diagnosis and management of dementia with Lewy bodies third report of the DLB consortium. *Neurology, 65*(12), 1863–1872.

Metzler-Baddeley, C. (2007). A review of cognitive impairments in dementia with Lewy bodies relative to Alzheimer's disease and Parkinson's disease with dementia. *Cortex, 43*(5), 583–600.

Minguez-Castellanos, A., Chamorro, C. E., Escamilla-Sevilla, F., Ortega-Moreno, A., Rebollo, A. C., Gomez-Rio, M., et al. (2007). Do α-synuclein aggregates in autonomic plexuses predate Lewy body disorders? A cohort study. *Neurology, 68*(23), 2012–2018.

Miyamoto, T., Miyamoto, M., Inoue, Y., Usui, Y., Suzuki, K., & Hirata, K. (2006). Reduced cardiac 123I-MIBG scintigraphy in idiopathic REM sleep behavior disorder. *Neurology, 67*(12), 2236–2238.

Miyamoto, T., Miyamoto, M., Iwanami, M., Hirata, K., Kobayashi, M., Nakamura, M., et al. (2010). Olfactory dysfunction in idiopathic REM sleep behavior disorder. *Sleep Medicine, 11*(5), 458–461.

Molano, J., Boeve, B., Ferman, T., Smith, G., Parisi, J., Dickson, D., et al. (2009). Mild cognitive impairment associated with limbic and neocortical Lewy body disease: A clinicopathological study. *Brain, 133*(2), 540–556.

Murphy, C., Schubert, C. R., Cruickshanks, K. J., Klein, B. E., Klein, R., & Nondahl, D. M. (2002). Prevalence of olfactory impairment in older adults. *JAMA, 288*(18), 2307–2312.

Oinas, M., Paetau, A., Myllykangas, L., Notkola, I. L., Kalimo, H., & Polvikoski, T. (2010). α-Synuclein pathology in the spinal cord autonomic nuclei associates with α-synuclein pathology in the brain: A population-based vantaa 85+ study. *Acta Neuropathologica, 119*(6), 715–722.

Onofrj, M., Varanese, S., Bonanni, L., Taylor, J. P., Antonini, A., Valente, E. M., et al. (2013). Cohort study of prevalence and phenomenology of tremor in dementia with Lewy bodies. *Journal of Neurology, 260*(7), 1731–1742.

Petrova, M., Mehrabian-Spasova, S., Aarsland, D., Raycheva, M., & Traykov, L. (2015). Clinical and neuropsychological differences between mild Parkinson's disease dementia and dementia with Lewy bodies. *Dementia and Geriatric Cognitive Disorders Extra, 5*(2), 212–220.

Ponsen, M. M., Stoffers, D., Booij, J., van Eck-Smit, B. L., Wolters, E. C., & Berendse, H. W. (2004). Idiopathic hyposmia as a preclinical sign of Parkinson's disease. *Annals of Neurology, 56*(2), 173–181.

Postuma, R. B., Gagnon, J. F., & Montplaisir, J. (2008). Cognition in REM sleep behavior disorder—A window into preclinical dementia? *Sleep Medicine, 9*(4), 341–342.

Postuma, R. B., Gagnon, J. F., & Montplaisir, J. Y. (2013a). *REM sleep behavior disorder and prodromal neurodegeneration-where are we headed?* (p. 3) New York, NY): Tremor and other hyperkinetic movements.

Postuma, R. B., Gagnon, J. F., Pelletier, A., & Montplaisir, J. (2013b). Prodromal autonomic symptoms and signs in Parkinson's disease and dementia with Lewy bodies. *Movement Disorders, 28*(5), 597–604.

Postuma, R. B., Iranzo, A., Hogl, B., Arnulf, I., Ferini-Strambi, L., Manni, R., et al. (2015). Risk factors for neurodegeneration in idiopathic rapid eye movement sleep behavior disorder: A multicenter study. *Annals of Neurology, 77*(5), 830–839.

Postuma, R. B., Lang, A. E., Massicotte-Marquez, J., & Montplaisir, J. (2006). Potential early markers of Parkinson disease in idiopathic REM sleep behavior disorder. *Neurology, 66*(6), 845–851.

Ross, G. W., Petrovitch, H., Abbott, R. D., Tanner, C. M., Popper, J., Masaki, K., et al. (2008). Association of olfactory dysfunction with risk for future Parkinson's disease. *Annals of Neurology, 63*(2), 167–173.

Schenck, C. H., Boeve, B. F., & Mahowald, M. W. (2013). Delayed emergence of a parkinsonian disorder or dementia in 81% of older men initially diagnosed with idiopathic rapid eye movement sleep behavior disorder: A 16-year update on a previously reported series. *Sleep Medicine, 14*(8), 744–748.

Stiasny-Kolster, K., Doerr, Y., Möller, J. C., Höffken, H., Behr, T. M., Oertel, W. H., et al. (2004). Combination of 'idiopathic' REM sleep behaviour disorder and olfactory dysfunction as possible indicator for α-synucleinopathy demonstrated by dopamine transporter FP-CIT-SPECT. *Brain, 128*(1), 126–137.

Takeda, S., Yamazaki, K., Miyakawa, T., & Arai, H. (1993). Parkinson's disease with involvement of the parasympathetic ganglia. *Acta Neuropathologica, 86*(4), 397–398.

Uemura, Y., Wada-Isoe, K., Nakashita, S., & Nakashima, K. (2013). Depression and cognitive impairment in patients with mild parkinsonian signs. *Acta Neurologica Scandinavica, 128*(3), 153–159.

Wakabayashi, K., Takahashi, H., Ohama, E., & Ikuta, F. (1990). Parkinson's disease: An immunohistochemical study of Lewy body-containing neurons in the enteric nervous system. *Acta Neuropathologica, 79*(6), 581–583.

Yoon, J. H., Kim, M., Moon, S. Y., Yong, S. W., & Hong, J. M. (2015). Olfactory function and neuropsychological profile to differentiate dementia with Lewy bodies from Alzheimer's disease in patients with mild cognitive impairment: A 5-year follow-up study. *Journal of Neurological Sciences, 355*(1), 174—179.

Yu, L., Cui, J., Padakanti, P. K., Engel, L., Bagchi, D. P., Kotzbauer, P. T., et al. (2012). Synthesis and in vitro evaluation of α-synuclein ligands. *Bioorganic and Medicinal Chemistry, 20*(15), 4625—4634.

Zaccai, J., McCracken, C., & Brayne, C. (2005). A systematic review of prevalence and incidence studies of dementia with Lewy bodies. *Age and Ageing, 34*(6), 561—566.

Zhou, G., Duan, L., Sun, F., Yan, B., & Ren, S. (2010). Association between mild parkinsonian signs and mortality in an elderly male cohort in China. *Journal of Clinical Neuroscience, 17*(2), 173—176.

CHAPTER 27

The quick mild cognitive impairment screen and applications to dementia

Rónán O'Caoimh[1,4], D. William Molloy[2], Roger Clarnette[3]

[1]Clinical Sciences Institute, National University of Ireland Galway, Galway City, Ireland; [2]Centre for Gerontology and Rehabilitation, St. Finbarr's Hospital, University College Cork, Cork City, Ireland; [3]Medical School, University of Western Australia, Crawley, WA, Australia; [4]Department of Geriatric Medicine, Mercy University Hospital, Cork City, Ireland

List of abbreviations

ABCS 135 AB Cognitive Screen 135
MMSE Mini-Mental State Examination
MoCA Montreal Cognitive Assessment
Qmci Quick Mild Cognitive Impairment screening tool

Introduction

Cognitive impairment is a common clinical presentation among older people. It is associated with significant disability and affects quality of life for those affected, as well as for their family members, and has significant economic implications (Wimo, Jönsson, Bond, Prince, & Winblad, 2013). The prevalence of dementia is continuing to increase worldwide (Prince et al., 2013), although trends suggest that the incidence is stabilizing or even falling in some regions (Wu et al., 2017). Cognitive impairment is usually due to degenerative brain disease, and Alzheimer's disease (AD) is the commonest cause; the prevalence of other subtypes is lower and influenced by referral patterns and the diagnostic accuracy of testing (Brunnström, Gustafson, Passant, & Englund, 2009). Despite a lack of effective therapies, identification of cognitive deficits when impairment is at a mild stage is considered important (Borson et al., 2013). Identifying when mild cognitive impairment (MCI) progresses to dementia is particularly important to time the introduction of cognitive stabilizers, including cholinesterase inhibitors and N-methyl-D-aspartate receptor antagonists (Tricco et al., 2013; Cooper, Li, Lyketsos, & Livingston, 2013). Other benefits include the ability to plan for future care needs, to make adjustments to work/life balance, and the potential to prescribe effective disease-modifying therapies if and when these become available (Solomon & Murphy, 2005). There is merit, therefore, in cognitive screening for symptomatic older adults, albeit evidence is limited and opinion divided (Le Couteur, Doust, Creasey, & Brayne, 2013; Riley McCarten, 2013).

Diagnosis and Management in Dementia
ISBN 978-0-12-815854-8, https://doi.org/10.1016/B978-0-12-815854-8.00027-6

429

We have previously described three limitations of screening for cognitive impairment including dementia (O'Caoimh et al., 2017). These include the low diagnostic accuracy, including suboptimal sensitivity and specificity, of short cognitive screening instruments; the variable effects of education and aging on cutoff scores; and the potential harm of erroneous diagnostic classifications (O'Caoimh et al., 2017). In addition, screening tests should not be considered as a substitute for more detailed neuropsychological assessment (Coen, Robertson, Kenny, & King-Kallimanis, 2016). Caution is therefore needed in interpreting cognitive screening scores when they are applied in a population of asymptomatic older people, where the evidence for the usefulness of screening in the absence of symptoms is low (Pottie et al., 2016). However, these instruments also have utility in providing rapid initial assessment of those with cognitive complaints (Wojtowicz and Larner, 2017) and those with MCI (Larner, 2016). In a review of screening instruments, Cullen and colleagues found that of 39 instruments examined, none were ideally suited to all clinical circumstances (Cullen et al., 2007).

Main text

The Quick Mild Cognitive Impairment (Q*mci*) screen (see Fig. 27.1) was developed in an attempt to overcome the problem of established instruments by minimizing floor and ceiling effects to discriminate between normal cognition, MCI, and dementia, so that one test could, in theory, be adaptable across the spectrum of cognitive aging and different causes of dementia (O'Caoimh et al., 2012; O'Caoimh and Molloy, 2017). Being able to distinguish between MCI and dementia is important, as experimental treatments increasingly target the prodromal stage of AD (Cummings, Lee, Mortsdorf, Ritter, & Zhong, 2017). The Q*mci* screen is based on the AB Cognitive Screen 135 (ABCS 135), an instrument that was developed to better differentiate between MCI and dementia (Molloy, Standish, & Lewis, 2005; Standish, Molloy, Cunje, & Lewis, 2007). The original validation data were derived from comparisons of a referred clinic population over 55 years of age, comprising 124 individuals with MCI and 111 with normal cognition (Molloy, Standish, & Lewis, 2005). The ABCS 135 assessed five domains: orientation, five-word registration, delayed word recall, clock drawing, and semantic fluency. A subsequent study examining the subtests of the ABCS 135 showed that the test could be administered in less than 5 min and showed good discrimination between those with MCI and dementia, although it was noted that several domains had suboptimal accuracy in distinguishing MCI from controls with normal cognition (Standish, Molloy, Cunje, & Lewis, 2007). The ABCS 135 was subsequently revised and refined to improve accuracy in identifying mild changes in cognition, specifically to improve the ability to separate normal cognition and subjective cognitive impairment from MCI and dementia, to create the Q*mci* screen. This included a reweighting of some domains and the addition of a new subtest (O'Caoimh, Foley et al., 2012).

The Quick *Mild Cognitive Impairment* (Q*mci*) screen

Name: DOB: Gender: Years in Education: Date: Time:

2. Orientation 🕐 Ten seconds for each answer. 🖉 Give **2 points for correct answer**, 1 if attempted but incorrect, 0 if no attempt.	What country is this? ___ / 2 What year is this? ___ / 2 What month is this? ___ / 2 What is todays date? ___ / 2 What day of the week is it? ___ / 2				**Score** ___/10

2. Word Registration

"I am going to say 5 words. After I have said these 5 words, repeat them back to me."

🕐 30 seconds.
🖉 Give 1 point per word repeated, in any order, no hints. When finished, say......
"Remember these words because I'll ask you to recall them later."

	Dog	Rain	Butter	Love	Door	
Alternates						
	Cat	Dark	Pepper	Fear	Bed	___/5
	Rat	Heat	Bread	Round	Chair	

3. Clock Drawing

"Draw a clock face and set the time to 'ten past eleven'."
(circle provided over page)
🕐 One minute.
🖉 Give 1 mark for each number, 1 for each hand & 1 for the pivot correctly placed or close to their ideal location. Lose 1 mark for each number duplicated or, if greater than 12.

Score:	Numbers	Correct	+ _____ / 12	
		Errors	- _____	
	Hands		+ _____ / 2	
	Pivot		+ _____ / 1	
	Total		+ _____ / 15	___ / 15

4. Delayed Recall

"A few minutes ago, I said five words. Please name as many words as you can remember."
🕐 30 seconds.
🖉 Recall in any order, within 30 seconds, giving 4 points per word, no hints.

	Dog	Rain	Butter	Love	Door	
Alternates						
	Cat	Dark	Pepper	Fear	Bed	___ / 20
	Rat	Heat	Bread	Round	Chair	

5. Verbal Fluency

"Name as many animals as you can in one minute."
🕐 One minute.
🖉 Give half a point per animal named; to a maximum of 40. Accept all 'creatures' including birds, fish, insects etc.

Alternative forms include: **fruit & veg** or **towns & cities**.
Score 0.5 x number of animals =
List here, in 'shorthand' if required:

___ / 20

6. Logical Memory

"I am going to read you a short story. When I am finished tell me as much of the story as you can."
🕐 One minute.
Give **2 points per highlighted word**, recalled exactly, immediately within 30 seconds, in any order, no hints. Two alternatives are provided.

Story 1		Alternative version 1		Alternative version 2	
The red fox	It was a hot	The brown dog	It was a cold	The white hen	It was a warm
ran across the	May	Ran across	October day.	Walked across	September afternoon.
Ploughed field.	morning.	the metal bridge.	Ripe	the concrete road.	Dry
	Fragrant Blossoms		Apples		Leaves
It was chased by	were forming on	It was hunting	were hanging on	It was followed by	were blowing in
a brown dog.	the bushes.	a white rabbit.	the trees.	a black cat.	the wind.

___ / 30

Administered by: _____   Total score ___ / 100*

*Cognitive Impairment is suggested if the score <62/100 but requires adjustment for age and education.

Figure 27.1 Scoring sheet for the Quick Mild Cognitive Impairment (Q*mci*) screen. *(Originally published at https://academic.oup.com/ageing/article/41/5/624/47050.)*

The Q*mci* screen was validated in a sample of 965 patients and caregivers (providing normal controls) referred to four memory clinics in Canada (O'Caoimh et al., 2012). It is scored out of 100 points, with a score of 0 indicating likely severe impairment and a score of 100 suggesting normal cognition. It was designed so that a score of 50 or below would identify those with dementia (O'Caoimh et al., 2013). Cutoff scores do, however, vary, depending on age and educational status (O'Caoimh, Timmons, & Molloy, 2016). A new subtest, logical memory, a test of immediate verbal recall of a brief story and assessing episodic memory, was added to the ABCS 135 and the scoring reweighted, giving the Q*mci* screen six domains in total: orientation (10 points), registration of five words (5 points), clock drawing (15 points), word recall (20 points), verbal fluency (20 points), and logical memory (30 points). The domains were weighted according to their utility in measuring change at different stages of cognitive decline. Logical memory is the most sensitive domain for detecting MCI (O'Caoimh et al., 2013). This added 30 s to the administration, although the median time remains approximately 5 min (O'Caoimh et al., 2013).

The Q*mci* screen was developed and subsequently validated by utilizing data from a number of sources, including memory clinics. While the original validation sample was collected in Canada (O'Caoimh et al., 2012, 2013), the Q*mci* screen has since been externally validated in Ireland (O'Caoimh et al., 2016), Australia (Clarnette et al., 2016, 2018), Turkey (Yavuz et al., 2016, 2017), Italy (Iavarone et al., 2018), China (Xu et al., 2017), Portugal (Dos Santos et al., 2019) and the Netherlands (Bunt et al., 2015). It has also been validated in different subtypes of MCI and dementia in those attending memory clinics (O'Caoimh, Timmons, & Molloy, 2013; O'Caoimh & Molloy, 2013; O'Caoimh & Molloy, 2015; O'Caoimh et al., 2016; O'Caoimh & Molloy, 2019). Further, it has been validated in clinical trials (O'Caoimh et al., 2014), the community (Clarnette et al., 2018), general practice (O'Caoimh, Cadoo et al., 2015), and a rehabilitation hospital (O'Caoimh, Timmons et al., 2013). Most recently it was used to detect cognitive impairment at high altitude using a modified environmental version, the eQ*mci* (Phillips, Basnyat, Chang, Swenson, & Harris, 2017).

The Q*mci* screen has been shown to be useful in distinguishing between diseases that cause cognitive impairment (O'Caoimh & Molloy, 2019; O'Caoimh, Cadoo et al., 2015). An assessment of the accuracy of the Q*mci* screen total score and subtests showed that the Q*mci* was more accurate in identifying AD than vascular dementia, with area under the curve of receiver operating characteristic curves of 0.97 (95% confidence interval 0.96–0.97) and 0.87 (95% confidence interval 0.84–0.91), respectively (O'Caoimh, Cadoo et al., 2015). The Q*mci* screen was also significantly more accurate than the Mini-Mental State Examination (MMSE) and nonsignificantly more accurate than the Montreal Cognitive Assessment (MoCA) in distinguishing dementia in those with Parkinson's disease attending a movement disorder clinic in Ireland (O'Caoimh, Foley, Trawley et al., 2012). The Q*mci* screen has similar (non-statistically significantly different) accuracy compared with the MoCA in separating MCI from dementia across a range of dementia subtypes (including AD, vascular, frontotemporal, Lewy body, and Parkinson's disease) attending a geriatric memory clinic, but with a shorter administration time (median of 4.5 vs. 9.5 min, respectively, $W = 121$, $P < .001$) (O'Caoimh et al., 2016). The *Qmci* has also been validated against noncognitive measures, correlating with assessments of functional status (Lawton—Brody activities of daily living scale) and a global rating of cognition (the Clinical Dementia Rating Scale) (O'Caoimh et al., 2014a) important in correctly identifying dementia in clinical practice.

While useful across the cognitive spectrum, the *Qmci* screen, as the name suggests, is most useful in separating prodromal dementia from normal cognition and subjective cognitive disorders. The sensitivity and specificity, as well as the positive predictive value and negative predictive value of the *Qmci* screen in differentiating MCI from those with normal cognition (healthy volunteers) compared with the Standardized MMSE (SMMSE), are presented in Table 27.1 (unpublished data). At the recommended cutoff for separating MCI from normal (<67/100), it had a sensitivity of 70% and specificity of 82%. This compares with a sensitivity of only 5% for the SMMSE at its commonly used cutoff of <24/30. Similarly, the psychometric properties of the *Qmci* screen in differentiating MCI from those with subjective memory complaints compared with the MoCA are presented in Table 27.2 (unpublished data), showing that at its recommended cutoff for MCI (<65/100), it had a sensitivity of 80% and specificity of 68%.

Regarding other psychometric properties, its interrater reliability (IRR) is well established in different clinical settings and populations, ranging from 0.77 in an Irish rehabilitation unit (O'Caoimh, McKeogh et al., 2013) to 0.90 in a Turkish geriatric clinic (Yavuz et al., 2017). Test—retest reliability is high, ranging from 0.86 (O'Caoimh et al., 2013) to 0.91 (Yavuz et al, 2017), depending on the setting. Examining dementia specifically, IRR was also strong (0.84), and there was moderate test—retest reliability (0.67) (Yavuz et al., 2017). The use of cutoff points to identify the likelihood of specific clinical syndromes, i.e., normal cognition but subjective cognitive symptoms, MCI, or dementia, should only guide the clinician, as these distinctions are determined by thorough clinical assessment (Larner, 2015a). Nevertheless, cut points are useful in helping to determine the likelihood of a clinical syndrome (Larner, 2015b), although it is known that cognitive screening instruments are affected by age and education (Crum, Anthony, Bassett, & Folstein, 1991). While cut points indicate optimal points of sensitivity and specificity, these also differ by age, education, and the method by which the cutoff score is derived (O'Caoimh et al., 2017). For the differentiation of MCI from dementia (non-differentiated), the optimal cutoff using Youden's index was <45/100 overall, rising to <50/100 for younger patients (≤75 years) with more education (≥12 years) (O'Caoimh et al., 2014b; O'Caoimh et al., 2017).

The *Qmci* screen is, however, a new instrument and the number of validation studies available is still low. While it has been externally validated in several languages and countries, these are usually restricted to single sites. Further, while its use in separating

Table 27.1 Sensitivity, specificity, positive predictive value, and negative predictive value, with 95% CIs, for different Quick Mild Cognitive Impairment screen and Standardized Mini-Mental State Examination cutoff scores for mild cognitive impairment (without adjustment for age or education) versus normal controls (healthy volunteers with no cognitive complaints).

Cognitive screen	Cutoff score	Sensitivity (95% CI)	Specificity (95% CI)	PPV (95% CI)	NPV (95% CI)	False positive (95% CI)	False negative (95% CI)
				Prevalence of MCI 19.7%			
Qmci screen							
Youden's index (J)							
0.51	<74	91% (85%–95%)	60% (56%–64%)	36% (31%–41%)	96% (94%–98%)	64% (59%–69%)	4% (2%–6%)
0.55	<72	88% (82%–93%)	67% (63%–70%)	39% (34%–45%)	96% (93%–97%)	61% (55%–66%)	4% (3%–7%)
0.55	<70	81% (73%–86%)	74% (71%–78%)	43% (38%–49%)	94% (91%–96%)	57% (51%–62%)	6% (4%–9%)
0.56	<69[b]	79% (72%–85%)	77% (73%–80%)	46% (40%–52%)	94% (91%–96%)	54% (48%–60%)	6% (4%–9%)
0.53	<68	73% (65%–80%)	80% (76%–83%)	47% (41%–54%)	92% (90%–94%)	53% (46%–59%)	8% (6%–10%)
0.52	<67[a]	70% (62%–77%)	82% (79%–85%)	49% (42%–55%)	92% (89%–94%)	51% (45%–58%)	8% (6%–11%)
0.52	<66	68% (60%–75%)	84% (81%–87%)	51% (44%–58%)	91% (89%–93%)	49% (42%–56%)	9% (7%–11%)
0.48	<65	61% (53%–69%)	87% (84%–89%)	54% (46%–61%)	90% (87%–92%)	46% (39%–54%)	10% (8%–13%)
0.47	<64	57% (49%–65%)	90% (87%–92%)	57% (49%–65%)	90% (87%–92%)	43% (35%–51%)	10% (8%–13%)
0.41	<62	48% (40%–56%)	93% (91%–95%)	63% (54%–72%)	88% (85%–90%)	37% (28%–46%)	12% (10%–15%)
0.37	<60	43% (35%–51%)	94% (92%–96%)	64% (54%–73%)	87% (84%–89%)	36% (27%–46%)	13% (11%–16%)
SMMSE							
Youden's index (J)							
0.20	<30	79% (72%–85%)	41% (37%–45%)	25% (21%–29%)	89% (85%–92%)	75% (71%–79%)	11% (8%–15%)
0.27	<29	55% (47%–63%)	72% (68%–75%)	32% (27%–38%)	87% (83%–89%)	68% (62%–73%)	13% (11%–17%)
0.24	<28	34% (27%–42%)	90% (87%–92%)	45% (36%–54%)	85% (82%–87%)	55% (46%–65%)	15% (13%–18%)
0.09	<26	10% (6%–16%)	99% (98%–100%)	70% (47%–86%)	82% (79%–84%)	30% (14%–53%)	18% (16%–21%)
0.05	<24	5% (2%–9%)	100% (99%–100%)	78% (40%–96%)	81% (78%–84%)	22% (4%–60%)	19% (16%–22%)

CI, 95% confidence interval; MCI, mild cognitive impairment; NPV, negative predictive value; PPV, positive predictive value; Qmci, Quick Mild Cognitive Impairment screen; SMMSE, Standardized Mini-Mental State Examination.

[a]Cutoff for MCI (vs. normal) selected from O'Caoimh et al. (2017).

[b]Most appropriate based on Youden's index.

Table 27.2 Sensitivity, specificity, positive predictive value, and negative predictive value, with 95% CIs, for different Quick Mild Cognitive Impairment screen and Montreal Cognitive Assessment cutoff scores for mild cognitive impairment (without adjustment for age or education) versus subjective memory complaints.

Cognitive screen (reference)	Cutoff score	Sensitivity (95% CI)	Specificity (95% CI)	PPV (95% CI)	NPV (95% CI)	False positive (95% CI)	False negative (95% CI)
				Prevalence of MCI of 57%			
Qmci screen							
Youden's index (J)							
0.47	<66	83% (74%–90%)	64% (52%–76%)	76% (66%–83%)	64% (61%–84%)	24% (17%–34%)	26% (16%–39%)
0.48	<65[a]	80% (71%–87%)	68% (56%–78%)	77% (68%–85%)	72% (60%–82%)	23% (15%–32%)	28% (18%–40%)
0.47	<64	74% (64%–82%)	73% (62%–83%)	79% (69%–86%)	68% (56%–78%)	21% (14%–31%)	32% (22%–44%)
0.48	<63	69% (59%–78%)	79% (67%–87%)	81% (71%–89%)	66% (55%–75%)	19% (11%–29%)	34% (25%–45%)
0.45	<62	64% (54%–73%)	81% (70%–89%)	82% (72%–90%)	63% (52%–72%)	18% (10%–28%)	37% (28%–48%)
0.43	<61	59% (49%–69%)	84% (73%–91%)	83% (72%–91%)	61% (50%–70%)	17% (9%–28%)	39% (30%–50%)
0.44	<60	56% (46%–66%)	88% (78%–94%)	86% (75%–93%)	60% (50%–69%)	14% (7%–25%)	40% (31%–50%)
0.38	<58	47% (37%–57%)	91% (81%–96%)	87% (75%–94%)	56% (46%–65%)	13% (6%–26%)	44% (35%–54%)
0.35	<56	43% (33%–53%)	92% (83%–97%)	88% (75%–95%)	54% (45%–63%)	12% (4%–25%)	46% (37%–55%)
MoCA							
Youden's index (J)							
0.27	<26	86% (77%–92%)	41% (30%–53%)	66% (58%–74%)	69% (53%–81%)	34% (26%–42%)	31% (19%–47%)
0.30	<25	73% (63%–81%)	57% (45%–69%)	70% (60%–78%)	61% (49%–73%)	30% (22%–40%)	39% (27%–61%)
0.39	<24	68% (58%–77%)	71% (59%–80%)	76% (66%–84%)	62% (51%–72%)	24% (16%–34%)	38% (28%–49%)
0.41	<23	57% (47%–67%)	84% (73%–91%)	83% (72%–90%)	59% (49%–69%)	17% (10%–28%)	41% (31%–51%)
0.33	<22[b]	44% (34%–54%)	89% (80%–95%)	85% (71%–93%)	54% (45%–63%)	15% (7%–29%)	46% (37%–55%)

CI, 95% confidence interval; MCI, mild cognitive impairment; MoCA, Montreal Cognitive Assessment; NPV, negative predictive value; PPV, positive predictive value; Qmci, Quick Mild Cognitive Impairment screen.

[a]Cutoff for MCI (vs. normal) selected from O'Caoimh et al. (2017).

[b]Cutoff for MCI selected from Freitas et al. (2013).

MCI from normal cognition or those with subjective complaints is relatively large, few studies have studied the use of the Qmci in different dementia subtypes. Hence, it is not known if it is useful in all of these, particularly as dementia progresses from mild to more advanced stages. Further study is now needed to demonstrate these and compare them with commonly used instruments such as the MMSE and MoCA. In addition, cutoff scores need to be examined and developed in other samples and settings. Concerning these, while their utility is limited by context and uncertainty regarding the optimal method of establishing test cut points, in everyday clinical practice they will remain useful in guiding further investigation of cognitive symptoms. The Qmci has already shown its usefulness across the cognitive spectrum with few ceiling or floor effects compared with more commonly used instruments. For example, the SMMSE (Molloy, Alemayehu, & Roberts, 1991) and MoCA (Nasreddine et al., 2005) have a much narrower range of scores, such that a few points difference can compel a clinician to consider a patient to be normal or impaired; for both of these tests, patients who are clinically normal, are mildly impaired, or have established dementia can all be in the 26—30 points range (Molloy & O'Caoimh, 2017).

Our investigations indicate that cutoff scores generated by using Youden's index (Youden, 1950) are marginally more accurate than using the maximal accuracy method (Larner, 2015b). Overall, data currently available show that the best cutoff for cognitive impairment for the Qmci screen (MCI or dementia) is <62/100, which gives a sensitivity of 83% and specificity of 87% (O'Caoimh et al., 2017). The ability to accurately identify early cognitive change while being able to measure decline over time is one of the main strengths of the instrument. The recognition that MCI is a useful clinical entity and that its identification can be standardized has reduced the importance of identifying dementia as the most critical clinical cognitive syndrome (Petersen, 2004). The most commonly used instrument to screen for MCI remains the MoCA. Traditionally, a score of <26/30 is deemed indicative of cognitive impairment (Nasreddine et al., 2005). However, this score has been shown to have poor specificity (McLennan, Mathias, Brennan, & Stewart, 2011; Luis et al., 2009), thus resulting in a high number of false positives. For some older individuals with relatively low educational status, a score below 26/30 does not represent cognitive impairment. Data for the Qmci screen derived from a clinic population indicates that the optimum cutoff for MCI using the MoCA is ≤23 (Clarnette et al., 2016; O'Caoimh et al., 2016). Similarly, lower cutoff scores for MCI for the MoCA have been found in other studies (Larner, 2012; Freitas, Simoes, Alves, & Santana, 2013). Our data show that the Qmci screen has accuracy similar to that of the MoCA but better specificity. Its shorter administration time and arguably simpler instructions suggest that the Qmci screen may be a better instrument to use in clinical practice, although, as said, further study is required to build on the evidence and to compare with the already extensive data collected for the MMSE and MoCA among many other more established short cognitive screening instruments.

In summary, the Q*mci* screen is a new, versatile, and brief cognitive screening instrument that is useful in identifying and separating normal cognition and subjective cognitive decline from MCI and dementia. Given the recent focus on identifying neurodegenerative disease when symptoms are mild, the use of valid and reliable instruments that facilitate this is a priority for clinicians and researchers. The fact that identification of biomarkers for AD will remain for now relatively expensive and largely inaccessible for clinicians, further emphasizes this need. The Q*mci* screen is validated in different cognitive impairment syndromes, including in different unselected patients with dementia attending outpatient clinics, and has good to excellent diagnostic accuracy and moderate to strong reliability in a range of clinical settings. More research is now required to further examine its psychometric properties in different dementia subtypes, its utility in different languages and cultures, and its predictive accuracy over time.

Key facts of brief cognitive assessment and the Q*mci* screen

- Accurate brief tests of cognitive function guide further investigation and aid in diagnosis.
- There is a lack of evidence that brief cognitive screens are useful for screening in asymptomatic populations.
- Alzheimer's disease is the commonest cause of cognitive impairment among the elderly.
- Brief cognitive tests should cover all important cognitive domains.
- The most important function of brief cognitive testing is to facilitate the distinction between dementia and normal and mild cognitive impairment.

Summary points

- The Q*mci* screen has been validated in numerous clinical settings, languages and countries.
- The Q*mci* screen is more accurate than the MMSE and has similar accuracy to the MoCA.
- The Q*mci* screen can be completed in 5 min.
- The Q*mci* screen is an accurate test to use in a busy clinical practice.

References

Borson, S., Frank, L., Bayley, P. J., et al. (2013). Improving dementia care: The role of screening and detection of cognitive impairment. *Alzheimer's and Dementia, 9*(2), 151–159.

Brunnström, H., Gustafson, L., Passant, U., & Englund, E. (2009). Prevalence of dementia subtypes: A 30-year retrospective survey of neuropathological reports. *Archives of Gerontology and Geriatrics, 49*(1), 146–149.

Bunt, S., O'Caoimh, R., Krijnen, W. P., et al. (2015). Validation of the Dutch version of the quick mild cognitive impairment screen (Qmci-D). *BMC Geriatrics, 15*(115). https://doi.org/10.1186/s12877-015-0113-1.

Clarnette, R., Goh, M., Bharadwaj, S., et al. (2018). Screening for cognitive impairment in an Australian aged care assessment team as part of comprehensive geriatric assessment. *Aging, Neuropsychology, and Cognition.* https://doi.org/10.1080/13825585.2018.1439447.

Clarnette, R., O'Caoimh, R., Antony, D., et al. (2016). Comparison of the quick mild cognitive impairment (Qmci) screen to the montreal cognitive assessment (MoCA) in an Australian geriatrics clinic. *International Journal of Geriatric Psychiatry, 32*(6), 643–649.

Coen, R. F., Robertson, D. A., Kenny, R. A., & King-Kallimanis, B. L. (2016). Strengths and limitations of the MoCA for assessing cognitive functioning: Findings from a large representative sample of Irish older adults. *Journal of Geriatric Psychiatry and Neurology, 29*(1), 18–24.

Cooper, C., Li, R., Lyketsos, C., & Livingston, G. (2013). Treatment for mild cognitive impairment: Systematic review. *The British Journal of Psychiatry, 203*(4), 255–264.

Crum, R. M., Anthony, J. C., Bassett, S. S., & Folstein, M. F. (1991). Population- based norms for the Mini-Mental State Examination by age and educational level. *Journal of the American Medical Association, 269,* 2386–2391.

Cullen, B., O'Neill, B., Evans, J. J., et al. (2007). A review of screening tests for cognitive impairment. *Journal of Neurology Neurosurgery and Psychiatry, 78,* 790–799.

Cummings, J., Lee, G., Mortsdorf, T., Ritter, A., & Zhong, K. (2017). Alzheimer's disease drug development pipeline. *Alzheimer's and Dementia: Translational Research and Clinical Interventions, 3*(3), 367–384.

Dos Santos, P. M., O'Caoimh, R., Svendrovski, A., Casanovas, C., Pernas, F. O., Illario, M., Molloy, W., & Paul, C. (2019). The RAPid COmmunity COGnitive screening programme (RAPCOG): Developing the Portuguese version of the quick mild cognitive impairment (Qmci-P) screen as part of the EIP on AHA twinning scheme. *Translational Medicine@ UniSa, 19,* 82.

Freitas, S., Simoes, M. R., Alves, L., & Santana, I. (2013). Montreal cognitive assessment: Validation study for mild cognitive impairment and Alzheimer disease. *Alzheimer Disease and Associated Disorders, 27*(1), 37–43.

Iavarone, A., Carpinelli Mazzi, M., Russo, G., et al. (2018). The Italian version of the quick mild cognitive impairment (Qmci-I) screen: Normative study on 307 healthy subjects. *Aging Clinical and Experimental Research.* https://doi.org/10.1007/s40520-018-0981-2. Jun 8.

Larner, A. J. (2012). Screening utility of the montreal cognitive assessment (MoCA): In place of – or as well as – the MMSE? *International Psychogeriatrics, 24*(3), 391–396.

Larner, A. J. (2015a). Introduction. In A. J. Larner (Ed.), *Diagnostic test accuracy studies in dementia* (pp. 1–17). Springer International Publishing Switzerland.

Larner, A. J. (2015b). Optimising the cutoffs of cognitive screening instruments in pragmatic diagnostic accuracy studies: Maximising accuracy or the Youden index? *Dementia and Geriatric Cognitive Disorders, 39,* 167–175.

Larner, A. J. (2016). Cognitive screening instruments for the diagnosis of mild cognitive impairment. *Progress in Neurology and Psychiatry, 20*(2), 21–26.

Le Couteur, D., Doust, J. A., Creasey, H., & Brayne, C. (2013). Political drive to screen for pre-dementia: Not evidence based and ignores the harms of diagnosis. *BMJ, 347,* f5125.

Luis, C. A., Keegan, A. P., & Mullan, M. (2009). Cross validation of the Montreal Cognitive Assessment in community dwelling older adults in the Southeastern US. *International Journal of Geriatric Psychiatry, 24*(2), 197–201.

McLennan, S. N., Mathias, J. L., Brennan, L. C., & Stewart, S. (2011). Validity of the Montreal Cognitive Assessment (MoCA) as a screening test for mild cognitive impairment (MCI) in a cardiovascular population. *Journal of Geriatric Psychiatry and Neurology, 24*(1), 33–38.

Molloy, D. W., Alemayehu, E., & Roberts, R. (1991). Reliability of a standardised mini-mental state examination compared with the traditional mini-mental state examination. *The American Journal of Psychiatry, 148*(1), 102–105.

Molloy, D. W., & O'Caoimh, R. (2017). *The quick guide.* Waterford, Ireland: Newgrange Press.

Molloy, D. W., Standish, T., & Lewis, D. L. (2005). Screening for mild cognitive impairment: Comparing the MMSE and the ABCS. *The Canadian Journal of Psychiatry, 50*(1), 52−58.

Nasreddine, Z. S., Phillips, N. A., Bedirian, V., et al. (2005). The montreal cognitive assessment, MoCA: A brief screening tool for mild cognitive impairment. *Journal of the American Geriatrics Society, 53*(4), 695−699.

O'Caoimh, R., Cadoo, S., Russell, A., et al. (2015). Comparison of three short cognitive screening instruments for Mild Cognitive Impairment and Dementia in General practice. *Irish Ageing Studies Review, 6*(1), 295.

O' Caoimh, R., Foley, M. J., Trawley, S., et al. (2012). Screening cognitive impairment in a movement disorder clinic: Comparison of the montreal cognitive assessment to the SMMSE. *Irish Journal of Medical Science, 181*(S(7)), 228.

O'Caoimh, R., Gao, Y., Eustace, J., & Molloy, D. W. (2014b). Cognitive screening tests need to be adjusted for age and education in patients presenting with symptomatic memory loss. *Irish Journal of Medical Science, 183S*(7), 314.

O'Caoimh, R., Gao, Y., Gallagher, P., et al. (2013). Which part of the quick mild cognitive impairment screen (Qmci) discriminates between normal cognition, mild cognitive impairment and dementia? *Age and Ageing, 42*, 324−330.

O'Caoimh, R., Gao, Y., McGlade, C., et al. (2012). Comparison of the quick mild cognitive impairment (Qmci) screen and the SMMSE in screening for mild cognitive impairment. *Age and Ageing, 41*(5), 624−629.

O'Caoimh, R., Gao, Y., Svendvski, A., et al. (2017). Comparing approaches to optimize cut-off scores for short cognitive screening instruments in mild cognitive impairment and dementia. *Journal of Alzheimer's Disease, 57*, 123−133.

O'Caoimh, R., McKeogh, J., Daly, B., et al. (2013). Screening for cognitive impairment in a hospital rehabilitation unit. *Irish Journal of Medical Science, 182S*(6), 286.

O'Caoimh, R., & Molloy, D. W. (2013). Diagnosing vascular mild cognitive impairment with atrial fibrillation remains a challenge. *Heart, 99*(11), 819.

O'Caoimh, R., & Molloy, D. W. (2015). Accuracy of cognitive screening instruments in Alzheimer's disease and other dementia subtypes. *Irish Ageing Studies Review, 6*(1), 292.

O'Caoimh, R., & Molloy, D. W. (2017). The quick mild cognitive impairment screen (Qmci). In *Cognitive screening instruments* (pp. 255−272). Cham: Springer.

O'Caoimh, R., Sato, S., Wall, J., et al. (2015). Potential for a "memory gym" intervention to delay conversion of mild cognitive impairment to dementia. *Journal of the American Medical Directors Association.* https://doi.org/10.1016/j.jamda.2015.01.081. pii:S1525-8610(15)00082-1.

O'Caoimh, R., Svendrovski, A., Johnston, B., et al. (2014a). The quick mild cognitive impairment screen correlated with the standardised Alzheimer's assessment scale- cognitive section in clinical trials. *Journal of Clinical Epidemiology, 67*, 87−92.

O'Caoimh, R., Timmons, S., & Molloy, D. W. (2013). Comparison of the quick mild cognitive impairment screen (Qmci) to the montreal cognitive Asssessment. *Irish Journal of Medical Science, 182S*(6), 286.

O'Caoimh, R., Timmons, S., & Molloy, D. W. (2016). Screening for mild cognitive impairment: Comparison of "MCI specific" screening instruments. *Journal of Alzheimer's Disease, 51*, 619−629.

O'Caoimh, R., & Molloy, D. W. (2019). Comparing the diagnostic accuracy of two cognitive screening instruments in different dementia subtypes and clinical depression. *Diagnostics, 9*(3), 93.

Petersen, R. C. (2004). Mild cognitive impairment as a diagnostic entity. *Journal of Internal Medicine, 256*, 183−194.

Phillips, L., Basnyat, B., Chang, Y., Swenson, E. R., & Harris, N. S. (2017). Findings of cognitive impairment at high altitude: Relationships to acetazolamide use and acute mountain sickness. *High Altitude Medicine and Biology, 18*(2), 121−127.

Pottie, K., Rahal, R., Jaramillo, A., et al. (2016). Recommendations on screening for cognitive impairment in older adults. *Canadian Medical Association Journal, 188*(1), 37−46.

Prince, M., Bryce, R., Albanese, E., et al. (2013). The global prevalence of dementia: A systematic review and metaanalysis. *Alzheimer's and Dementia, 9*(1), 63−75.

Riley McCarten, J. (2013). The case for screening for cognitive impairment in older adults. *Journal of the American Geriatrics Society, 61*(7), 1203—1205.

Solomon, P. R., & Murphy, C. A. (2005). Should we screen for Alzheimer's disease? A review of the evidence for and against screening Alzheimer's disease in primary care practice. *Geriatrics, 60*, 26—31.

Standish, T., Molloy, D. W., Cunje, A., & Lewis, D. (2007). Do the ABCS 135 short cognitive screen and its subtests discriminate between normal cognition, mild cognitive impairment and dementia? *International Journal of Geriatric Psychiatry, 22*, 189—194.

Tricco, A. C., Soobiah, C., Berliner, S., et al. (2013). Efficacy and safety of cognitive enhancers for patients with mild cognitive impairment: A systematic review and meta-analysis. *Canadian Medical Association Journal, 185*(16), 1393—1401.

Wimo, A., Jönsson, L., Bond, J., Prince, M., & Winblad, B.((2013). The worldwide economic impact of dementia 2010. *Alzheimer's and Dementia, 9*(1), 1.

Wojtowicz, A., & Larner, A. J. (2017). Diagnostic test accuracy of cognitive screeners in older people. *Progress in Neurology and Psychiatry, 21*(1), 17—21.

Wu, Y. T., Beiser, A. S., Breteler, M. M., et al. (2017). The changing prevalence and incidence of dementia over time—current evidence. *Nature Reviews Neurology, 13*(6), 327.

Xu, Y., Li, X., Chen, Z., et al. (2017). Development of the Chinese version of the quick mild cognitive impairment (Qmci-CN). *Age and Ageing, 46*(S3), 57.

Yavuz, B. B., Hacer, D. V., O'Caoimh, R., et al. (2016). Validation of the Turkish version of the quick mild cognitive impairment screen (Qmci-TR). *European Geriatric Medicine, 7*(S1), S69—S70.

Yavuz, B. B., Varan, H. D., O'Caoimh, R., et al. (2017). Validation of the Turkish version of the quick mild cognitive impairment screen. *American Journal of Alzheimer's Disease and Other Dementias, 32*(3), 145—156.

Youden, W. J. (1950). Index for rating diagnostic tests. *Cancer, 3*(1), 32—35.

CHAPTER 28

The usefulness of evaluating performance of activities of daily living in the diagnosis of mild cognitive disorders

Patricia De Vriendt, Elise Cornelis, Ellen Gorus

Department Gerontology, Vrije Universiteit Brussel, Brussels, Belgium

List of abbreviations

a advanced
ADLs activities of daily living
b basic
CHP cognitively healthy person
i instrumental
MCI mild cognitive impairment
MD mild dementia
NCD neurocognitive disorder

Mini-dictionary of terms

Activities of daily living ADLs are all the activities a person performs on a daily basis that allow him or her to live autonomously as well as being integrated into his or her environment, fulfilling a determined social role.

Advanced ADLs these refer to all the other and more complex activities not included in the basic and housekeeping types.

Basic ADLs these refer to self-care (washing, dressing, etc.).

Instrumental ADLs these refer to housekeeping activities.

Mild cognitive impairment this is the early phase of cognitive deterioration, generally used to refer to a transitional zone between normal cognitive functioning and clinically probable dementia.

Dementia and mild cognitive impairment

Today, 47 million people live with dementia worldwide, but since increasing age is the most important risk factor for dementia, its prevalence will increase dramatically, to more than 82 million by 2030 and 152 million by 2050 (World Health Organization, 2017). Dementia is a progressive clinical syndrome characterized by cognitive and functional decline and the occurrence of behavioral and psychological symptoms, affecting the autonomy and participation of older persons in the community. In old age many

Diagnosis and Management in Dementia
ISBN 978-0-12-815854-8, https://doi.org/10.1016/B978-0-12-815854-8.00028-8

conditions in addition to dementia can cause cognitive deterioration. Therefore, an extensive multidisciplinary (at least medical, cognitive, psychological, and functional evaluation) diagnostic workup is needed to increase the validity of the diagnosis. Since, as of this writing, dementia cannot be resolved or cured, interventions have to focus on improving and optimizing functioning. These are most effective when they start as early as possible, which means that early diagnosis is pivotal. The concept of mild cognitive impairment (MCI) has been developed to capture the early phase of cognitive deterioration and is generally used to refer to a transitional zone between normal cognitive functioning and clinically probable dementia (Winblad et al., 2004). However, MCI is a heterogeneous concept in its clinical presentation and its progression to dementia; mainly, amnestic MCI has a high risk of dementia, but some persons remain stable or even revert to normal cognition. Since boundaries between normal aging, MCI, and mild dementia are vague, discussion about the MCI criteria and their operationalization is ongoing. The differentiation between mild and major neurocognitive disorders (NCDs), referring to the new version of the *Diagnostic Statistical Manual of Mental Disorders*, may be a step in the right direction (American Psychiatric Association, 2013). The extent to which cognitive decline interferes with everyday functioning determines the distinction between mild and major NCDs. In major NCD or dementia, cognitive impairment causes dependence in everyday functioning, while in mild NCD, which comes close to MCI, individuals remain autonomous, although subtle problems may already occur in complex activities of daily living (ADLs).

There has been increased interest in understanding the cognitive correlates of everyday functioning, especially in mild cognitive disorders. It has consistently been found that greater cognitive impairment is associated with increased levels of functional dependency. Several studies have shown a significant relationship between everyday functioning and memory (Farias et al., 2008; Yeh, Tsang, et al., 2011), executive functioning (McAlister, Schmitter-Edgecombe, & Lamb, 2016; Monaci & Morris, 2012), and complex reasoning (Beaver & Schmitter-Edgecombe, 2017). However, the strength of the relationship varies across studies. A review (Royall et al., 2007) exploring the cognitive correlates of functional status found that the variance in daily functioning that can be specifically attributed to cognition is modest. On average, cognition explained 21% of variance in functional outcomes, with a median of 15.9%. This finding has been confirmed in a sample of MCI patients (Reppermund et al., 2011). In this study, even though all cognitive domains were correlated with instrumental (i-)ADLs, these correlations were very small (ranging from −.053 to −.155). In a longitudinal study (Monaci & Morris, 2012), it was shown that, while at baseline cognitive function was robustly associated with i-ADLs but not with basic (b-)ADLs, this relationship diminished over time and was not present at follow-up. Consequently, early detection of subtle functional limitations and understanding their cognitive and noncognitive correlates is essential. Therefore, assessment of ADLs is paramount to determine the degree of impairment in everyday functioning and the underlying reasons of impairment and subsequently to underpin accurate diagnostic classification in NCDs. On top of that, evidence shows that ADL disability may increase the risk for incident dementia. Therefore, an evaluation of ADLs might be useful not just as a diagnostic tool but also as an indicator of the risk for future dementia (Farias et al., 2017; Fauth et al., 2013; McAlister et al., 2016).

The concept of activities of daily living

Despite the widespread use of the concept of ADLs, there is no generally accepted definition nor typology. Therefore, it is useful to clarify how it is conceptualized in this chapter. ADLs refer to "the everyday activities that people do as individuals, in families, and within communities to occupy time and bring meaning and purpose to life," including activities one needs to, wants to, and is expected to do. In the geriatric literature, ADLs commonly refer to the ability to perform everyday tasks and are seen as a measurement of disability and independence (Reppermund et al., 2011). That is to say, ADLs are all the activities a person performs on a daily basis that allow him or her to live autonomously as well as to be integrated into his or her environment, fulfilling a determined social role (Devi, 2018). Daily functioning is complex, resulting from an interplay of different factors requiring an optimal combination of cognitive, motor, and psychological skills, as well as the appropriate environmental conditions. Therefore, a good ADL evaluation should rely on an apt disentangling of underlying skills. It is a major challenge—in particular in old age—to detect which deficient skills cause the limited functioning.

Several typologies are available to describe ADLs, but the concept of Reuben et al. (Reuben, Laliberte, Hiris, & Mor, 1990) seems the most appropriate to use in cognitive problems in a geriatric population (De Vriendt et al., 2012). ADLs are divided into three types, stratified according to difficulty, complexity, and vulnerability to cognitive decline. The b-ADLs are the activities that meet basic physiological and self-maintenance needs, such as personal hygiene, dressing, and eating, traditionally described in the Katz index (Katz, Ford, Moskowitz, Jackson, & Jaffe, 1963), mastered early in life and relatively uniform across people and cultures. The i-ADLs include, as described in the Lawton scale (Lawton & Brody, 1969), more complex activities, essential to maintain independent living, such as cooking, shopping, and managing medication. So far, this ADL typology is commonly used in literature. However, one level is often missing in the literature and measurement tools, namely the advanced (a-)ADLs, also called complex or extended ADLs. The a-ADLs cover those activities that are volitional, influenced by cultural and motivational factors, expressing a personal engagement in satisfying activities that are beyond what is needed to be independent. Examples of a-ADLs are leisure activities, self-development activities, or (semi-)professional work.

The measurement of activities of daily living

Report-based scales

Report-based scales or questionnaires are cheap, easy, quick, and the most direct and simple method for gathering information, including the perception of the person him- or herself, which is pivotal for choosing treatment options later on (Sikkes & Rotrou, 2014). However, self-report questionnaires have poor relations with objective measures of cognition and weak reliability in individuals without self-awareness. Therefore, for individuals with cognitive disorders, an informant report seems to be more superior, as patients may over- or underestimate their functional limitations (Puente, Terry, Faraco, Brown, & Miller, 2014). Informant report (i.e., relative, close friend, proxy) is considered more reliable, providing a good representation of everyday performance (Schmitter-Edgecombe & Parsey, 2014), and has better correlations with objective cognitive measures (Tsang, Diamond, Mowszowski, Lewis, & Naismith, 2012). However, informant report might also be prone to error, since reporter bias has been found due to caregiver burden and distress (Puente et al., 2014; Schmitter-Edgecombe & Parsey, 2014). In addition, most report-based tools have poor diagnostic validity for subtle functional impairments in early cognitive decline (Fieo, Austin, Starr, & Deary, 2011; Jekel et al., 2015).

Currently, frequently used report-based evaluations for b- and i-ADLs are the Katz index (Katz et al., 1963) and the Lawton scale (Lawton & Brody, 1969), respectively. Although widespread, both scales have poorly described psychometric properties, their scoring systems are not sensitive enough to detect subtle deficits, and they do not identify causes of limitations in ADLs (Koskas et al., 2014; Paula et al., 2014). Other report-based evaluations have been developed, such as the Functional Activities Questionnaire (Pfeffer, Kurosaki, Harrah, Chance, & Filos, 1982), the Alzheimer's Disease Cooperative Study ADL scale (Galasko et al., 1997), the Everyday Cognition scale (Farias et al., 2008), and the Everyday Technology Use Questionnaire (Malinowsky, Nygard, & Kottorp, 2011). These instruments are promising scales for use in mild cognitive disorders. However, they evaluate a mixed spectrum of self-care, household, and other activities, or assess only one specific activity, or assess everyday cognitive abilities such as memory, language, or attention.

Performance-based scales

Since report-based evaluations have been criticized, performance-based scales have been developed. In this case, individuals are asked to perform activities while being observed by an assessor in an attempt to estimate actual abilities equivalent to those performed in the home environment. In this way, performance-based scales provide a more objective measure of ADLs. However, these evaluations are often performed in artificial environments, such as a hospital, and outside the daily routine in which individuals normally perform their ADLs, which hampers accurate assessment. Also, these evaluations can be influenced by the motivation and behavior of the patient and the assessor, and they are (more) time consuming and expensive to administer. Moreover, only an observation of a small excerpt of everyday functioning is conducted, and this is likely not representative of global independence.

The Assessment of Motor and Process Skills (Fisher & Jones, 2012) is regarded as one of the best measures of functional performance. Unfortunately, it is time consuming and the compulsory training is expensive, which limits its usability, as shown by a study in Belgium. Another performance-based tool is the Erlangen Test of Activities of Daily Living in Mild Dementia or Mild Cognitive Impairment (Luttenberger, Reppermund, Schmiedeberg-Sohn, Book, & Graessel, 2016), with good psychometric properties and short administration time, but its international use is limited as it has been validated only in a German sample. Other well-known performance-based instruments are the Direct Assessment of Functional Status (Loewenstein et al., 1989) and the Texas Functional Living scale (Cullum et al., 2001), useful at the dementia stage, but not at the stage of MCI. Recently, some performance-based ADL scales have tried to capture functional limitations in MCI, such as the Sydney Test of Activities of Daily Living in Memory Disorders (Reppermund et al., 2017) and the Naturalistic Action Test (Giovannetti, Libon, Buxbaum, & Schwartz, 2002), both valid, reliable, and time- and resource-efficient measures and successful in discriminating MCI from normal cognition and mild dementia.

Shortcomings in the evaluations of activities of daily living in mild cognitive disorders

Despite the importance of the criterion concerning ADLs in mild cognitive problems, there is not yet a consensus on how to evaluate ADLs. The tools used in the earliest studies mainly involved a dichotomous scoring. The majority of the reported measures evaluated only the outcome of a task, by rating the success or failure in completing the task. The process of completing a task (e.g., instance errors, duration, etc.) is not reflected in the scores, while it is evident that this information could be of vital importance and meaningful for diagnostic purposes. In an attempt to improve the diagnostic value, other items, expected to be relevant, were added, e.g., executive functions, thus broadening their scope, but also resulting in an amalgam of measured concepts. Since there is sufficient knowledge that the correlation between neuropsychological functioning and ADLs does not equal "one," they should be evaluated separately. In addition, in particular when detecting cognitive problems, it is important to make a distinction between underlying causes of impairment. It should be ascertained that limitations are due to cognitive deficits and not to comorbidities or physical impairments often seen in old age. Evaluations of ADLs often also entail gender-dependent tasks, and then the question is how to evaluate activities one does not perform. Clear-cut guidelines on how to deal with irrelevant activities are lacking. The International Classification of Functioning, Disability and Health (ICF) (World Health Organization, 2001)—considered the current reference in assessment—offers in this regard two possibilities. One can evaluate "capacity," meaning an individual's ability to execute a task or an action. This indicates the highest level of functioning of a person in a given domain at a given moment in a standardized environment. Or, one can evaluate "performance," what an individual does in his or her current environment. The majority of the tools are not aligned with these definitions. Only a few instruments evaluate the more complex ADLs, or a-ADLs, and a comprehensive one is absent. Finally, the question remains, which activities may be impaired and to what extent, to make the diagnosis of MCI. This discussion is important since controversy exists whether functional limitations are part of MCI or already announce a conversion toward dementia.

State of the art: what do we know already?

There is enough evidence to warrant the hypothesis that people with MCI experience subtle problems in everyday life, which is different from dementia. Moreover, in populations with a low educational level or a multicultural background—in which the diagnosis of cognitive disorders might be influenced by, e.g., language and education— an ADL evaluation is a more ecological approach than neuropsychological assessment (Iavarone et al., 2007).

The process of functional decline in mild cognitive disorders: degradation of functional abilities occurs in a stepwise and hierarchical manner

A functional continuum emerges, ranging from optimal functioning in successful aging to age-related functional decline, subtle functional problems (first in a- and later in i-ADLs) in MCI, moderate problems (seen in a- and i-ADLs), and finally functional problems on all levels of ADLs, including b-ADLs, in severe dementia. So, the degradation of functional abilities occurs in a stepwise hierarchical manner, with b-ADLs affected after i- and a-ADLs. Functional decline occurs over the course of MCI and cumulative changes accompany the conversion to dementia (Lechowski et al., 2010; Morris, Berg, Fries, Steel, & Howard, 2013; Schmitter-Edgecombe & Parsey, 2014). Although functional changes in MCI are interposed between healthy aging and dementia, an overlap can be observed on both ends of the ADL spectrum, implying that the distinction between normal aging and MCI, on one hand, and MCI and dementia, on the other hand, is quite difficult. Statistically it is possible to make a distinction between groups, but at the individual level it is more difficult to make a prediction, since functional status varies between persons. That is to say, individual slopes of decline should be taken into account.

The number of activities varies across stages of cognitive decline

Significant differences are seen between the total numbers of activities—also named as activity capital—performed by cognitively healthy older adults, persons with MCI, and persons with (mild) dementia, with the last performing significantly fewer activities (De Vriendt et al., 2012, 2013; De Vriendt, Mets, Petrovic, & Gorus, 2015). A more frequent engagement in activities and a slower rate of decline in the variety of activities are associated with a lower risk of progressing from mild to severe cognitive impairment over time (Hughes, Flatt, Fu, Chang, & Ganguli, 2013). Performing activities such as computer use, playing games, reading books, craft activities, and social activities is associated with a significantly reduced risk of incident MCI (Geda et al., 2011; Krell-Roesch et al., 2018).

Of course, no causal relationship has been evidenced in this studies and it remains unclear which mediating factors play a role in the association between activities and the decreased risk of incident MCI. Nevertheless, other studies found that doing voluntary work and engaging in leisure activities lead to a slower cognitive decline and are protective for developing dementia (Griep et al., 2017). Mainly cognitive, intellectual, or mentally stimulating activities are associated with a decreased risk of cognitive decline (Akbaraly et al., 2009; Marioni et al., 2015; Then et al., 2013).

Certain activities are more sensitive to functional decline than others

The type of limited activities might also give clinicians guidance for diagnosing mild cognitive disorders, since some activities are more sensitive to cognitive decline than others. Since the oldest studies mainly focused on b- and i-ADLs, the best evidence is available for these activities. MCI patients demonstrate problems in i-ADLs (Farias et al., 2006; Gold, 2012; Pedrosa et al., 2010; Peres et al., 2008; Yeh, Lin, et al., 2011): use of the telephone, handling finances, taking care of medications, using public transportation, meal preparation, housework. Particularly, activities relying on everyday memory (e.g., remembering a few shopping items out of a list), planning and organization, and initiating and carrying them out appeared to be impaired (e.g., finding personal belongings, keeping appointments). Last, at the level of a-ADLs, it was observed that people with MCI have more difficulties with computer use, playing games, reading books, crafts (Geda et al., 2011), handwriting (Werner, Rosenblum, Bar-On, Heinik, & Korczyn, 2006), and driving (Pedrosa et al., 2010; Wadley et al., 2009). It has also been shown that the overall ability to manage everyday technology is more decreased in mild dementia or MCI compared with older adults without known cognitive impairment (Nygard & Starkhammar, 2007; Nygard & Winblad, 2006).

Subtle decrease in the quality of performance

In addition, the quality of the performance can be of interest when trying to make a distinction between cognitively healthy, MCI, or mild dementia. While in dementia, quality of performance obviously decreases, with a loss of autonomy as a consequence, mild functional deficits in MCI are characterized by inefficient task execution, poorly sequenced task steps, and inaccurate object selection; however, this does not preclude accomplishment of the primary task goals (Giovannetti et al., 2002). Patients with MCI make mostly commission errors (e.g., a step or subtask is performed more than once, or the action is completed without the necessary tools), demonstrating a pattern of errors suggestive of preserved task knowledge with difficulties applying this knowledge to the efficient execution of ADLs. They perform complex activities worse and show less persistence and more fatigue, disorientation, confusion, decreased efficiency, slower performance, and repetition of minor mistakes (Ahn et al., 2009; Wadley et al., 2009). Difficulties with initiating or taking interest in a particular task and with planning and organization are also reported (Yeh, Lin, et al., 2011). In addition, the restriction in i-ADLs is associated with a much lower chance to return to normal cognition (Ahn et al., 2009).

Conclusion

Based on the existing evidence, it might be recommended to start the ADL evaluation by exploring the patient's life history and daily living activities to gain insight into the patient's activity capital. However, starting from "fixed" questions and using typologies and lists of activities have the insidious risk of "one size fits all" questionnaires. Therefore, these standardized measures should comprise a personal component, exploring the perception of the individual and that of his or her relatives. To uncover, these "narratives," questionnaires should start by focusing on what the patient considers relevant "activities." Activities are considered relevant if they are currently performed or have been performed previously, and an ADL evaluation should target only these activities. This is in line with the ICF's performance construct. Subsequently, in-depth information about how these relevant activities are performed and—in case of a limitation—about the underlying cause(s) of limitation is needed. Fig. 28.1 illustrates this assessment process.

However, a typical problem is that these "personalized" measures are not psychometrically sound because of the high degree of variability they allow. Therefore, sophisticated indices should be developed and validated while applying sound statistical strategies. When taking the b-, i-, and a-ADL tool of De Vriendt et al. (2012, 2013, 2015) and Cornelis, Gorus, Beyer, Bautmans and De Vriendt (2017), Cornelis, Gorus, Van Weverbergh, Beyer and De Vriendt (2018) as an example, it is possible to combine both objective and subjective rates in one comprehensive index and validate them. The instruments showed good to excellent content, face, construct, convergent, and discriminative validity. In conclusion, even within personalized ADL evaluations, both standardization and personalization are possible, and excellent validity should be the aim.

Finally, this method of ADL evaluation might be considered time consuming at the start of the clinical process. Therefore, standardized manuals allowing enough personalization should be developed in a way that guarantees easy handling in clinical practice.

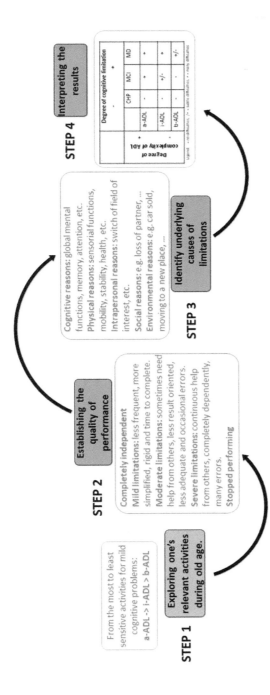

Figure 28.1 Steps in the assessment of performance of activities of daily living (ADL). The sequence of steps one should undertake in an evaluation of daily functioning to underpin a diagnosis of (mild) cognitive disorder is shown. a-, advanced; b-, basic; i-, instrumental; CHP, cognitively healthy person; MCI, mild cognitive impairment; MD, mild dementia.

Moreover, clinicians should be convinced that personalized evaluation is probably time saving when considering the entire span of the care process and offers more valid diagnosis when embedded in a multidisciplinary workup.

Key facts of everyday functioning in mild cognitive disorders

- MCI is seen as a transitional zone between normal cognitive aging and dementia. MCI causes cognitive changes.
- In dementia, cognitive impairment causes dependence in everyday functioning. In MCI, individuals remain autonomous, although subtle problems may occur in performing complex ADLs.
- Therefore, next to evaluating cognitive decline, it is also important to evaluate everyday functioning when diagnosing persons with cognitive disorders such as MCI and dementia.
- An accurate evaluation of ADLs may contribute to a more accurate diagnosis, but is also valuable in setting up nonpharmacological treatments of cognitive disorders.
- Existing ADL evaluations often show poor psychometric properties.

Summary points

- This chapter focuses on the diagnosis of dementia as a multidisciplinary diagnostic workup and the extent to which cognitive decline interferes with everyday functioning.
- The ADL evaluation should start by exploring the patient's activity capital, since the number of activities differs between cognitively healthy older adults, persons with MCI, and persons with (mild) dementia, with the first having more activities than the last.
- Subsequently, in-depth information about how these activities are performed should be collected, since certain activities are more sensitive to cognitive decline than others.
- Finally, the extent of the limitation and underlying causes should be clarified.

References

Ahn, I. S., Kim, J. H., Kim, S., Chung, J. W., Kim, H., Kang, H. S., et al. (2009). Impairment of instrumental activities of daily living in patients with mild cognitive impairment. *Psychiatry Investigation, 6*(3), 180—184. https://doi.org/10.4306/pi.2009.6.3.180.

Akbaraly, T. N., Portet, F., Fustinoni, S., Dartigues, J. F., Artero, S., Rouaud, O., et al. (2009). Leisure activities and the risk of dementia in the elderly: Results from the three-city study. *Neurology, 73*(11), 854—861. https://doi.org/10.1212/WNL.0b013e3181b7849b.

American Psychiatric Association. (2013). *DSM-5: Diagnostic and statistical manual of mental disorders* (5th ed.). Washington & DC: APA Association.

Beaver, J., & Schmitter-Edgecombe, M. (2017). Multiple types of memory and everyday functional assessment in older adults. *Archives of Clinical Neuropsychology, 32*(4), 413–426. https://doi.org/10.1093/arclin/acx016.

Cornelis, E., Gorus, E., Beyer, I., Bautmans, I., & De Vriendt, P. (2017). Early diagnosis of mild cognitive impairment and mild dementia through basic and instrumental activities of daily living: Development of a new evaluation tool. *PLoS Medicine, 14*(3), e1002250. https://doi.org/10.1371/journal.pmed.1002250.

Cornelis, E., Gorus, E., Van Weverbergh, K., Beyer, I., & De Vriendt, P. (2018). Convergent and concurrent validity of a report- versus performance-based evaluation of everyday functioning in the diagnosis of cognitive disorders in a geriatric population. *International Psychogeriatrics, 1*–12. https://doi.org/10.1017/S1041610218000327.

Cullum, C. M., Saine, K., Chan, L. D., Martin-Cook, K., Gray, K. F., & Weiner, M. F. (2001). Performance-Based instrument to assess functional capacity in dementia: The Texas Functional Living Scale. *Cognitive and Behavioral Neurology, 14*(2), 103–108.

De Vriendt, P., Gorus, E., Cornelis, E., Bautmans, I., Petrovic, M., & Mets, T. (2013). The advanced activities of daily living: A tool allowing the evaluation of subtle functional decline in mild cognitive impairment. *The Journal of Nutrition, Health and Aging, 17*(1), 64–71. https://doi.org/10.1007/s12603-012-0381-9.

De Vriendt, P., Gorus, E., Cornelis, E., Velghe, A., Petrovic, M., & Mets, T. (2012). The process of decline in advanced activities of daily living: A qualitative explorative study in mild cognitive impairment. *International Psychogeriatrics, 24*(6), 974–986. https://doi.org/10.1017/s1041610211002766.

De Vriendt, P., Mets, T., Petrovic, M., & Gorus, E. (2015). Discriminative power of the advanced activities of daily living (a-ADL) tool in the diagnosis of mild cognitive impairment in an older population. *International Psychogeriatrics, 27*(9), 1419–1427. https://doi.org/10.1017/s1041610215000563.

Devi, J. (2018). The scales of functional assessment of activities of daily living in geriatrics. *Age and Ageing.* https://doi.org/10.1093/ageing/afy050.

Farias, S. T., Lau, K., Harvey, D., Denny, K. G., Barba, C., & Mefford, A. N. (2017). Early functional limitations in cognitively normal older adults predict diagnostic conversion to mild cognitive impairment. *Journal of the American Geriatrics Society, 65*(6), 1152–1158. https://doi.org/10.1111/jgs.14835.

Farias, S. T., Mungas, D., Reed, B. R., Cahn-Weiner, D., Jagust, W., Baynes, K., et al. (2008). The measurement of everyday cognition (ECog): Scale development and psychometric properties. *Neuropsychology, 22*(4), 531–544. https://doi.org/10.1037/0894-4105.22.4.531.

Farias, S. T., Mungas, D., Reed, B. R., Harvey, D., Cahn-Weiner, D., & Decarli, C. (2006). MCI is associated with deficits in everyday functioning. *Alzheimer Disease and Associated Disorders, 20*(4), 217–223. https://doi.org/10.1097/01.wad.0000213849.51495.d9.

Fauth, E. B., Schwartz, S., Tschanz, J. T., Ostbye, T., Corcoran, C., & Norton, M. C. (2013). Baseline disability in activities of daily living predicts dementia risk even after controlling for baseline global cognitive ability and depressive symptoms. *International Journal of Geriatric Psychiatry, 28*(6), 597–606. https://doi.org/10.1002/gps.3865.

Fieo, R. A., Austin, E. J., Starr, J. M., & Deary, I. J. (2011). Calibrating ADL-IADL scales to improve measurement accuracy and to extend the disability construct into the preclinical range: A systematic review. *BMC Geriatrics, 11*, 42. https://doi.org/10.1186/1471-2318-11-42.

Fisher, A. G., & Jones, L. B. (2012). Assessment of motor and process skills. *Development, standardization, and administration manual vol. 1.* Fort Collins, CO: Three Star Press, Inc.

Galasko, D., Bennett, D., Sano, M., Ernesto, C., Thomas, R., Grundman, M., et al. (1997). An inventory to assess activities of daily living for clinical trials in Alzheimer's disease. The Alzheimer's Disease Cooperative Study. *Alzheimer Disease and Associated Disorders, 11*(Suppl. 2), S33–S39.

Geda, Y. E., Topazian, H. M., Roberts, L. A., Roberts, R. O., Knopman, D. S., Pankratz, V. S., et al. (2011). Engaging in cognitive activities, aging, and mild cognitive impairment: A population-based study. *Journal of Neuropsychiatry and Clinical Neurosciences, 23*(2), 149–154. https://doi.org/10.1176/appi.neuropsych.23.2.149.

Giovannetti, T., Libon, D. J., Buxbaum, L. J., & Schwartz, M. F. (2002). Naturalistic action impairments in dementia. *Neuropsychologia, 40*(8), 1220–1232.

Gold, D. A. (2012). An examination of instrumental activities of daily living assessment in older adults and mild cognitive impairment. *Journal of Clinical and Experimental Neuropsychology, 34*(1), 11–34. https://doi.org/10.1080/13803395.2011.614598.

Griep, Y., Hanson, L. M., Vantilborgh, T., Janssens, L., Jones, S. K., & Hyde, M. (2017). Can volunteering in later life reduce the risk of dementia? A 5-year longitudinal study among volunteering and non-volunteering retired seniors. *PloS One, 12*(3), e0173885. https://doi.org/10.1371/journal.pone.0173885.

Hughes, T. F., Flatt, J. D., Fu, B., Chang, C. C., & Ganguli, M. (2013). Engagement in social activities and progression from mild to severe cognitive impairment: The MYHAT study. *International Psychogeriatrics, 25*(4), 587–595. https://doi.org/10.1017/S1041610212002086.

Iavarone, A., Milan, G., Vargas, G., Lamenza, F., De Falco, C., Gallotta, G., et al. (2007). Role of functional performance in diagnosis of dementia in elderly people with low educational level living in Southern Italy. *Aging Clinical and Experimental Research, 19*(2), 104–109.

Jekel, K., Damian, M., Wattmo, C., Hausner, L., Bullock, R., Connelly, P. J., et al. (2015). Mild cognitive impairment and deficits in instrumental activities of daily living: A systematic review. *Alzheimer's Research and Therapy, 7*(1), 17. https://doi.org/10.1186/s13195-015-0099-0.

Katz, S., Ford, A. B., Moskowitz, R. W., Jackson, B. A., & Jaffe, M. W. (1963). Studies of illness in the aged. The index of ADL: A standardized measure of biological and psychosocial function. *Journal of the American Medical Association, 185*, 914–919.

Koskas, P., Henry-Feugeas, M. C., Feugeas, J. P., Poissonnet, A., Pons-Peyneau, C., Wolmark, Y., et al. (2014). The Lawton instrumental activities daily living/activities daily living scales: A sensitive test to Alzheimer disease in community-dwelling elderly people? *Journal of Geriatric Psychiatry and Neurology, 27*(2), 85–93. https://doi.org/10.1177/0891988714522694.

Krell-Roesch, J., Feder, N. T., Roberts, R. O., Mielke, M. M., Christianson, T. J., Knopman, D. S., et al. (2018). Leisure-time physical activity and the risk of incident dementia: The Mayo clinic study of aging. *Journal of Alzheimer's Disease, 63*(1), 149–155. https://doi.org/10.3233/JAD-171141.

Lawton, M. P., & Brody, E. M. (1969). Assessment of older people: Self-maintaining and instrumental activities of daily living. *The Gerontologist, 9*(3), 179–186.

Lechowski, L., Van Pradelles, S., Le Crane, M., d'Arailh, L., Tortrat, D., Teillet, L., et al. (2010). Patterns of loss of basic activities of daily living in Alzheimer patients: A cross-sectional study of the French REAL cohort. *Dementia and Geriatric Cognitive Disorders, 29*(1), 46–54. https://doi.org/10.1159/000264632.

Loewenstein, D. A., Amigo, E., Duara, R., Guterman, A., Hurwitz, D., Berkowitz, N., et al. (1989). A new scale for the assessment of functional status in Alzheimer's disease and related disorders. *Journal of Gerontology, 44*(4), P114–P121.

Luttenberger, K., Reppermund, S., Schmiedeberg-Sohn, A., Book, S., & Graessel, E. (2016). Validation of the Erlangen test of activities of daily living in persons with mild dementia or mild cognitive impairment (ETAM). *BMC Geriatrics, 16*, 111. https://doi.org/10.1186/s12877-016-0271-9.

Malinowsky, C., Nygard, L., & Kottorp, A. (2011). Psychometric evaluation of a new assessment of the ability to manage technology in everyday life. *Scandinavian Journal of Occupational Therapy, 18*(1), 26–35. https://doi.org/10.3109/11038120903420606.

Marioni, R. E., Proust-Lima, C., Amieva, H., Brayne, C., Matthews, F. E., Dartigues, J. F., et al. (2015). Social activity, cognitive decline and dementia risk: A 20-year prospective cohort study. *BMC Public Health, 15*, 1089. https://doi.org/10.1186/s12889-015-2426-6.

McAlister, C., Schmitter-Edgecombe, M., & Lamb, R. (2016). Examination of variables that may affect the relationship between cognition and functional status in individuals with mild cognitive impairment: A meta-analysis. *Archives of Clinical Neuropsychology, 31*(2), 123–147. https://doi.org/10.1093/arclin/acv089.

Monaci, L., & Morris, R. G. (2012). Neuropsychological screening performance and the association with activities of daily living and instrumental activities of daily living in dementia: Baseline and 18- to 24-month follow-up. *International Journal of Geriatric Psychiatry, 27*(2), 197–204. https://doi.org/10.1002/gps.2709.

Morris, J. N., Berg, K., Fries, B. E., Steel, K., & Howard, E. P. (2013). Scaling functional status within the interRAI suite of assessment instruments. *BMC Geriatrics, 13*, 128. https://doi.org/10.1186/1471-2318-13-128.

Nygard, L., & Starkhammar, S. (2007). The use of everyday technology by people with dementia living alone: Mapping out the difficulties. *Aging and Mental Health, 11*(2), 144–155. https://doi.org/10.1080/13607860600844168.

Nygard, L., & Winblad, B. (2006). Measuring long term effects and changes in the daily activities of people with dementia. *The Journal of Nutrition, Health and Aging, 10*(2), 137–138.

Paula, J. J., Bertola, L., Avila, R. T., Assis Lde, O., Albuquerque, M., Bicalho, M. A., et al. (2014). Development, validity, and reliability of the general activities of daily living scale: A multidimensional measure of activities of daily living for older people. *Revista Brasileira de Psiquiatria, 36*(2), 143–152. https://doi.org/10.1590/1516-4446-2012-1003.

Pedrosa, H., De Sa, A., Guerreiro, M., Maroco, J., Simoes, M. R., Galasko, D., et al. (2010). Functional evaluation distinguishes MCI patients from healthy elderly people–the ADCS/MCI/ADL scale. *The Journal of Nutrition, Health and Aging, 14*(8), 703–709.

Peres, K., Helmer, C., Amieva, H., Orgogozo, J. M., Rouch, I., Dartigues, J. F., et al. (2008). Natural history of decline in instrumental activities of daily living performance over the 10 years preceding the clinical diagnosis of dementia: A prospective population-based study. *Journal of the American Geriatrics Society, 56*(1), 37–44. https://doi.org/10.1111/j.1532-5415.2007.01499.x.

Pfeffer, R. I., Kurosaki, T. T., Harrah, C. H., Jr., Chance, J. M., & Filos, S. (1982). Measurement of functional activities in older adults in the community. *Journal of Gerontology, 37*(3), 323–329.

Puente, A. N., Terry, D. P., Faraco, C. C., Brown, C. L., & Miller, L. S. (2014). Functional impairment in mild cognitive impairment evidenced using performance-based measurement. *Journal of Geriatric Psychiatry and Neurology, 27*(4), 253–258. https://doi.org/10.1177/0891988714532016.

Reppermund, S., Birch, R. C., Crawford, J. D., Wesson, J., Draper, B., Kochan, N. A., et al. (2017). Performance-based assessment of instrumental activities of daily living: Validation of the Sydney test of activities of daily living in memory disorders (STAM). *Journal of the American Medical Directors Association, 18*(2), 117–122. https://doi.org/10.1016/j.jamda.2016.08.007.

Reppermund, S., Sachdev, P. S., Crawford, J., Kochan, N. A., Slavin, M. J., Kang, K., et al. (2011). The relationship of neuropsychological function to instrumental activities of daily living in mild cognitive impairment. *International Journal of Geriatric Psychiatry, 26*(8), 843–852. https://doi.org/10.1002/gps.2612.

Reuben, D. B., Laliberte, L., Hiris, J., & Mor, V. (1990). A hierarchical exercise scale to measure function at the Advanced Activities of Daily Living (AADL) level. *Journal of the American Geriatrics Society, 38*(8), 855–861.

Royall, D. R., Lauterbach, E. C., Kaufer, D., Malloy, P., Coburn, K. L., Black, K. J., et al. (2007). The cognitive correlates of functional status: A review from the committee on research of the American Neuropsychiatric Association. *Journal of Neuropsychiatry and Clinical Neurosciences, 19*(3), 249–265. https://doi.org/10.1176/jnp.2007.19.3.249.

Schmitter-Edgecombe, M., & Parsey, C. M. (2014). Assessment of functional change and cognitive correlates in the progression from healthy cognitive aging to dementia. *Neuropsychology, 28*(6), 881–893. https://doi.org/10.1037/neu0000109.

Sikkes, S. A., & Rotrou, J. (2014). A qualitative review of instrumental activities of daily living in dementia: what's cooking? *Neurodegenerative Disease Management, 4*(5), 393–400. https://doi.org/10.2217/nmt.14.24.

Then, F. S., Luppa, M., Schroeter, M. L., Konig, H. H., Angermeyer, M. C., & Riedel-Heller, S. G. (2013). Enriched environment at work and the incidence of dementia: Results of the Leipzig longitudinal study of the aged (LEILA 75+). *PloS One, 8*(7), e70906. https://doi.org/10.1371/journal.pone.0070906.

Tsang, R. S., Diamond, K., Mowszowski, L., Lewis, S. J., & Naismith, S. L. (2012). Using informant reports to detect cognitive decline in mild cognitive impairment. *International Psychogeriatrics, 24*(6), 967–973. https://doi.org/10.1017/S1041610211002900.

Wadley, V. G., Okonkwo, O., Crowe, M., Vance, D. E., Elgin, J. M., Ball, K. K., et al. (2009). Mild cognitive impairment and everyday function: An investigation of driving performance. *Journal of Geriatric Psychiatry and Neurology, 22*(2), 87–94. https://doi.org/10.1177/0891988708328215.

Werner, P., Rosenblum, S., Bar-On, G., Heinik, J., & Korczyn, A. (2006). Handwriting process variables discriminating mild Alzheimer's disease and mild cognitive impairment. *Journals of Gerontology Series B: Psychological Sciences and Social Sciences, 61*(4), P228–P236.

Winblad, B., Palmer, K., Kivipelto, M., Jelic, V., Fratiglioni, L., Wahlund, L. O., et al. (2004). Mild cognitive impairment–beyond controversies, towards a consensus: Report of the international working group on mild cognitive impairment. *Journal of Internal Medicine, 256*(3), 240–246. https://doi.org/10.1111/j.1365-2796.2004.01380.x.

WorldHealthOrganization. (2001). *International classification of functioning, disability and health.*

World Health Organization. (2017). *Dementia.* Retrieved from http://www.who.int/mediacentre/factsheets/fs362/en/.

Yeh, Y. C., Lin, K. N., Chen, W. T., Lin, C. Y., Chen, T. B., & Wang, P. N. (2011a). Functional disability profiles in amnestic mild cognitive impairment. *Dementia and Geriatric Cognitive Disorders, 31*(3), 225–232. https://doi.org/10.1159/000326910.

Yeh, Y. C., Tsang, H. Y., Lin, P. Y., Kuo, Y. T., Yen, C. F., Chen, C. C., et al. (2011b). Subtypes of mild cognitive impairment among the elderly with major depressive disorder in remission. *American Journal of Geriatric Psychiatry, 19*(11), 923–931. https://doi.org/10.1097/JGP.0b013e318202clc6.

PART III

Pharmacological treatments for dementia

CHAPTER 29

Cholinesterase inhibitors in dementias: an overview

Patrizia Mecocci, Lucia Paolacci, Virginia Boccardi
Department of Medicine, Institute of Gerontology and Geriatrics, University of Perugia, Perugia, Italy

List of abbreviations

ACh Acetylcholine
AD Alzheimer's dementia
ADAS-Cog Alzheimer's disease assessment scale, cognitive subscale
ADL Activity of daily living
APP Amyloid precursor protein
Aβ β-amyloid
BACE1 β-secretase 1
BID Bis in die (twice a day)
BPSD Behavioral and psychological symptoms of dementia
ChAT Choline acetyltransferase
ChEIs Cholinesterase inhibitors
DLB Dementia with Lewy bodies
FDA Food and drug administration
FTD Frontotemporal dementia
MCI Mild cognitive impairment
MMSE Mini-mental state examination
MTDL Multitarget-directed ligand
NH Nursing home
NMDA N-methyl-D-aspartate
NSAID Nonsteroidal antiinflammatory drug
OD Omne in die (once a day)
PDD Parkinson's disease dementia
PET Positron emission tomography
PKC Protein kinase C
VaD Vascular dementia

Mini-dictionary of terms

Alzheimer's dementia This is the most common form of dementia, and it is marked histologically by neuronal degeneration in specific areas of the cerebral cortex and by the presence of neurofibrillary tangles and plaques containing β-amyloid.
Choline acetyltransferase This is the enzyme responsible for the synthesis of the neurotransmitter ACh.
Cholinesterase inhibitors Drugs that inhibit the choline acetyltransferase enzyme, increasing both ACh levels and duration, leading to benefits on cognitive and behavioral symptoms and functional status.

Diagnosis and Management in Dementia
ISBN 978-0-12-815854-8, https://doi.org/10.1016/B978-0-12-815854-8.00029-X

M receptors These are ACh receptors that form G-protein-coupled receptor complexes in the cell membranes, involved in learning and memory functions.

N receptors These are ionotropic receptors that respond to ACh and nicotine in a brain cell and other tissues, involved in neurodegenerative disease and neurogenesis in ischemic stroke.

Oldest old They represent the part of the population age 85 years or older, a distinct clinical and pathological entity. Several studies found that their response to ChEI is very encouraging.

Introduction

Dementia is a progressive brain disease marked by multiple cognitive deficits (such as impairment of memory, language, attention, executive functions, reasoning and judgment, and visual perception) that lead to a progressive reduction in functional status. Alzheimer's disease is the most common cause of dementia, with a prevalence of 48 million persons worldwide (Prince, Comas-Herrera, Knapp, Guerchet, & Karagiannidou, 2016). The new lexicon for defining dementia (Dubois et al., 2010) established that the term "disease" involves all the pathophysiological processes that begin many years before the onset of clinical symptoms, whereas the term "dementia" is more appropriate when clinical symptoms emerge; here we will often refer to Alzheimer's dementia (AD). Overall, dementias are caused by neuronal death in different brain areas, and clinical manifestation appears evident when a neuronal loss in the brain reaches a threshold; in AD this is localized in the hippocampus and cerebral neocortex and mainly involves cholinergic neurons. In recent years, many efforts have been spent in understanding the causes of dementia with the aim of developing new treatments. The most corroborated theories as of this writing are presented in Fig. 29.1, but there are many other possible mechanisms and processes yet to be identified.

The amyloid cascade hypothesis is the most accepted, and many drugs have been developed upon such a theory, but it is now clear that β-amyloid (Aβ) fragment production alone appears not sufficient to produce the clinical expression of AD, considering that a great amyloid burden has also been found in brains of old and very old subjects without dementia. A hypothetical model of AD sequence leading to cognitive impairment postulates that Aβ accumulation is an "upstream" event in the cascade that is associated with "downstream" synaptic dysfunction, neurodegeneration, and eventual neuronal loss. Due to the poor clarity of the detailed pathogenesis of AD and of other dementias, an effective therapy has not yet been developed. So far, there are no curative treatments for dementia, but research has directed its efforts toward disease-modifying (trying to change the underlying disease process, producing an enduring effect on the clinical course of the AD) and symptomatic drugs. Only these last have been included in phase IV of the AD treatment development pipeline (Tables 29.1 and 29.2).

In this chapter, we aim at presenting the role of cholinesterase inhibitors (ChEIs), the most corroborated drugs for treating AD since 1990. We will focus on their usefulness, limits, and further capabilities in light of new evidence such as the development of multitarget directed therapy.

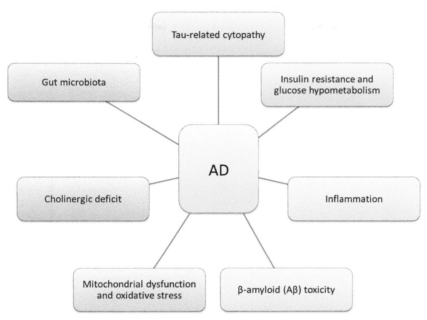

Figure 29.1 *Potential pathways involved in Alzheimer's disease pathogenesis and treatment targets.* Many pathways are involved in AD pathogenesis, beyond β-amyloid toxicity and tau cytopathy. Corroborated hypotheses are cholinergic dysfunction, oxidative stress, and inflammation, while new pathways have been proposed involving gut microbiota activity and insulin resistance. All these pathways could be treatment targets. *AD*, Alzheimer's disease.

Cholinergic dysfunctions and cholinesterase inhibitor rationale

In the second half of the 20th century, several studies on neuroanatomy and neurotransmitters based on biochemistry, immunohistochemistry, and pharmacology discovered the most important neurotransmitters and their receptors, linking their modification to specific neurologic diseases. In the same years the cholinergic deficit hypothesis of AD was conceptualized (Bartus, Dean, Beer, & Lippa, 1982), and it is still a pillar of dementia research in light of the responsiveness of the disease to therapy with ChEIs. The cholinergic system plays a crucial role in cognitive functions, in particular within the hippocampus and other brain regions associated with memory and learning functions. Cholinergic neurons are widely distributed in the brain, and they are organized both in specific areas and in projections as reported in Fig. 29.2A, B.

In subjects with AD, the level of acetylcholine (ACh) released by cholinergic synapses becomes progressively lower due to the progressive alteration of the terminal part of the nerve, which becomes unable to transmit the nerve impulse at the synaptic level. This causes a progressive deterioration of cognitive functions. Moreover, AD postmortem brains showed reduced choline uptake, reduced ACh release, and presynaptic cholinergic

Table 29.1 Available pharmacological treatment of dementia.

Symptom	Drug class	Main active	Recommendations
Cognitive	Cholinesterase inhibitors	Donepezil Rivastigmine Galantamine	Use cholinesterase inhibitors in the treatment of mild to moderate Alzheimer's disease; donepezil is approved also for severe dementia. Cholinesterase inhibitors should not be stopped just because the point of severe dementia has been reached. Switch one cholinesterase inhibitor to another if the first is not tolerated or effective. Use rivastigmine and donepezil in dementia with Lewy bodies and Parkinson's disease dementia. Cholinesterase inhibitors could be useful also in mixed vascular dementia, but not in frontotemporal dementia.
	NMDA antagonist	Memantine	Use memantine in moderate to severe Alzheimer's disease. Memantine could produce global improvements in Lewy body dementia and mixed vascular dementia, but not in frontotemporal dementia.
Noncognitive	Antipsychotics	Atypical (risperidone, olanzapine, quetiapine)	Use lowest efficacious dosage; pay attention to sedation, orthostasis, and extrapyramidal symptoms.

Table 29.1 Available pharmacological treatment of dementia.—cont'd

Symptom	Drug class	Main active	Recommendations
		Typical (haloperidol, fluphenazine, thiothixene)	They have more side effects than atypical drugs and they can cause an increased risk of cerebrovascular events in elderly patients with dementia-related psychosis.
	Antidepressants	SSRIs (fluoxetine, paroxetine, sertraline, citalopram, fluvoxamine)	SSRIs may help some behavioral aspects of frontotemporal dementia, but do not improve cognition. Typical side effects include sweating, tremors, nervousness, insomnia or somnolence, dizziness, and various gastrointestinal and sexual disturbances.
		Heterocyclic and noncyclic antidepressant agents (nefazodone and mirtazapine)	Effective, especially in patients with associated anxiety. Mirtazapine also improves appetite and weight gain.
	Anxiolytics	Benzodiazepines (llorazepam, oxazepam, temazepam, triazolam)	Useful in insomnia, anxiety, and agitation, but regular use can lead to addiction, tolerance, depression, and confusion or paradoxic agitations.
	Mood-stabilizing (antiagitation) drugs	Trazodone Carbamazepine Divalproex sodium	Useful in place of or added to antipsychotic agents for control of severe behavioral symptoms.

Currently available treatments of dementia include drugs for cognitive and noncognitive symptoms (behavioral and mood disorders). *NMDA*, N-methyl-D-aspartate; *SSRI*, selective serotonin reuptake inhibitor.
Modified from O'Brien, J.T., Holmes, C., Jones, M., Jones, R., Livingston, G., McKeith, I., et al., (2017). Clinical practice with anti-dementia drugs: A revised (third) consensus statement from the British association for psychopharmacology. *Journal of Psychopharmacology*, 31, 147–168 and Sadowsky C.H., & Galvin, J.E. (2012). Guidelines for the management of cognitive and behavioral problems in dementia. *The Journal of the American Board of Family Medicine*, 25, 350–366.

Table 29.2 Characteristics of available cholinesterase inhibitors.

Drug	Formulation	Dosing	Half-life; metabolism/ elimination	FDA approval	Demonstrated efficacy
Donepezil	Immediate release tablets	5 mg OD 10 mg OD	$T^{1}/_{2}$ 90 h Liver, kidney	5 mg: mild to moderate 10 mg: mild, moderate, severe 23 mg: moderate, severe	Entire clinical spectrum of the disease: cognitive symptoms (mild to severe), behavior, and mood
	Sustained release tablets	23 mg OD			
	Orally disintegrating tablets	5 or 10 mg OD			
Rivastigmine	Capsule	1.5 mg BID 3 mg BID 4.5 mg BID 6 mg BID	$T^{1}/_{2}$ 2 h Kidney	Mild to moderate AD	Cognition, mood behavior
	Patch	4.6 mg/ 24 h 9.5 mg/ 24 h 13.3 mg/ 24 h		Mild, moderate, severe AD; mild, moderate PDD	
Galantamine	Immediate release	4 mg BID 8 mg BID 12 mg BID	$T^{1}/_{2}$ 5—7 h Liver, kidney	Mild, moderate AD	Cognition, mood behavior
	Extended release	8 mg OD 16 mg OD 24 mg OD	$T^{1}/_{2}$ 5—7 h released immediately and 12 h later Liver, kidney		

Cholinesterase inhibitor formulation, dosing, metabolism, FDA approval, and efficacy are shown. *AD*, Alzheimer's dementia; *BID*, twice a day; *OD*, once a day; *PDD*, Dementia in Parkinson's disease.

deficit that correlate with diminished cognitive function and accumulation of neuritic plaques and neurofibrillary tangles (Tata, Velluto, D'Angelo, & Reale, 2014).

Data suggest that cholinergic dysfunction can occur during release or processing; it concerns an altered coupling between muscarinic cholinergic receptors and second-messenger systems, with diminished levels of both ACh and choline acetyltransferase activity. Moreover, an unbalanced reduction of muscarinic receptors has been reported, with preferred preservation of postsynaptic receptors (M1, M3) versus a substantial reduction in nicotinic and muscarinic M2 receptors (mainly presynaptic). New evidence also suggested that cholinergic dysfunction in AD is closely associated with Aβ and tau toxicity through muscarinic (mostly M1) and nicotinic 7N receptors. M1 receptors are the primary muscarinic subtype, involved in learning and memory functions; they have been shown to promote amyloid precursor protein (APP) cleavage and a decrease in Aβ accumulation by modulating protein kinase C and β-secretase 1 activity. Despite M1 receptor expression appearing unaltered in AD patients, it has been proposed that Aβ fragment may induce an uncoupling of M1 receptors from Gq protein, altering their function. Another significant change observed in AD regards 7 nicotinic receptors, which bind with high affinity to Aβ fragments, leading to neurotoxic effects, an intracellular accumulation of Aβ, an impairment of 7 receptor activity, a decrease in cholinergic neurotransmission, and ultimately increased levels of tau phosphorylation. Conversely, nicotinic ligands decrease Aβ internalization and reduce cholinergic dysfunction (Confaloni, Tosto & Tata, 2016).

Based on this notion, it seems reasonable and useful to use drugs that can enhance or "potentiate" the cholinergic system. ChEIs represent the first drugs approved by the US Food and Drug Administration: they inhibit the cholinesterase enzyme that catalyzes the hydrolysis of the neurotransmitter ACh into choline and acetic acid, ultimately increasing both its levels and its duration. *Tacrine* was the first drug approved in 1993 as a potent and clinically useful therapy. However, due to its hepatotoxicity, it was soon withdrawn from the pharmaceutical market in 1998, but it provided evidence that the symptoms of the AD could be significantly reduced and it confirmed the link between cholinergic activity and cognitive symptoms of AD (Bartus, 2000). Other ChEIs, namely *donepezil, rivastigmine*, and *galantamine*, were later introduced, with superior pharmacokinetic and pharmacodynamic properties and low incidence of adverse effects compared with tacrine. They can be divided into two groups: irreversible and reversible inhibitors. Rivastigmine is a slowly reversible acetyl- and butyryl-ChEI that acts by binding the esterasic part of the active site. Butyrylcholinesterase is similar to acetylcholinesterase, and it is linked to Aβ and neurofibrillary tangle deposition and toxicity. Rivastigmine inhibits both enzymes, leading to a better cholinergic neurotransmission; furthermore, it is also able to selectively inhibit the monomeric form of acetylcholinesterase, which represents only a small amount of this enzyme in normal brain (due to a more significant amount of tetrameric structure), but the predominant form in AD. Furthermore, it can inhibit

(A)

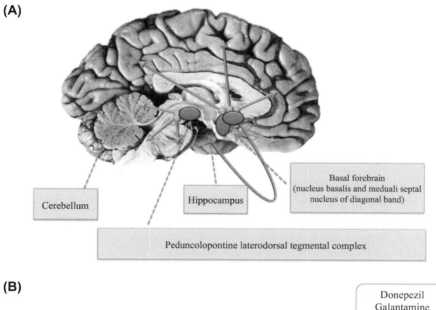

Cerebellum

Hippocampus

Basal forebrain
(nucleus basalis and meduali septal
nucleus of diagonal band)

Peduncolopontine laterodorsal tegmental complex

(B)

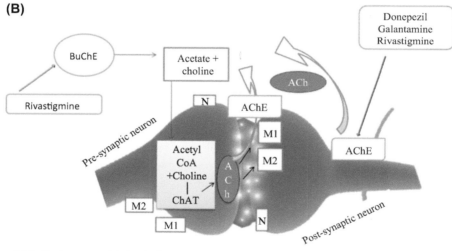

Figure 29.2 *Brain areas rich in cholinergic neurons and principal cholinergic projections and mechanism of action of cholinesterase inhibitors.* (A) Important cholinergic systems originate in the basal forebrain (a complex of subcortical nuclei projecting to brain areas critical to cognition) and brain stem, projecting to the limbic system and cerebral cortex. The sites of greater acetylcholine synthesis in the brain are the interpeduncular nucleus and pedunculopontine complex (located in the midbrain). (B) Choline acetyltransferase enzyme (*ChAT*) catalyzes the transfer of an acetyl group from acetyl-CoA to choline for synthesis of acetylcholine (*ACh*), while acetylcholinesterase (*AChE*) and butyrylcholinesterase (*BuChE*) terminate its function, catalyzing ACh hydrolysis. Inhibitors of the latter enzymes improve ACh availability in the brain. M1, muscarinic receptor type 1; M2, muscarinic receptor type 2; N, nicotinic receptor.

acetylcholinesterase specifically in the cortex and hippocampus (Confaloni et al., 2016). Donepezil and galantamine are reversible ChEIs, and they selectively target only acetyl-cholinesterase. Galantamine is also known to show a positive allosteric modulation of 7 receptors, thus potentiating the ACh effects and possibly reducing the Aβ/7 nicotinic receptor binding, as previously described.

Acetylcholinesterase inhibitor therapy: the five W's

Who could benefit from therapy?

Guidelines suggest starting treatment when the diagnosis of AD is postulated (Sadowsky & Galvin, 2012) by these criteria:
- progressive cognitive impairment, including typical amnesia, severe enough to inter-fere with daily functioning;
- pathophysiological injury or topographical biomarkers;
- possible AD autosomal dominant mutation (i.e., presenilin or APP).

Subjects with milder disease at the initiation of ChEI therapy at optimal doses will better maintain cognition and activities of daily living (ADLs) for a longer time, support-ing the importance of early diagnosis and thus initiation of antidementia drugs at a milder stage of the disease (Wattmo & Wallin, 2017). A systematic analysis of double-blind, placebo-controlled trials of ChEIs in patients with mild to moderate dementia demon-strated a significant cognitive improvement at 6 months and 1 year (Birks, 2006). In addi-tion to their effects on cognition, these agents have also shown beneficial effects on measures of behavior, ADLs, and global subject function. A 2017 systematic review and metaanalysis that analyzed clinical trials involving over 16,000 patients found that ChEIs improved functional outcomes (Blanco-Silvente et al., 2017), an essential goal, since guidelines acknowledge that preventing or delaying further loss of ADL function is crucial for AD treatment (Farlow & Cummings, 2007). Donepezil and galantamine demonstrated a significant effect in behavioral symptoms reduction (Dyer, Harrison, Laver, Whitehead, & Crotty, 2018). This effect could be related to the so-called "cholin-ergic deficiency syndrome". It is based on the idea that treatment with anticholinergic drugs can cause restlessness, confusion, perceptual distortions, anxiety, and illusions or visual hallucinations, so that enhancing the cholinergic system could be an excellent strat-egy to reduce these symptoms (Lemstra, Richard, & van Goo, 2007). Probably linked to their functional, cognitive, and behavioral effects, ChEIs have also been shown to reduce AD caregiver stress burden because of their positive effects on behavioral symptoms (Hashimoto, Yatabe, Kaneda, Honda, & Ikeda, 2009) and ADLs (Feldman et al., 2003).

A peculiar mention pertains to the treatment of very old patients (85 years old or more) with dementia, who often do not yield to therapy. Younger age is associated with better anatomical and functional conditions of the brain that may positively present a better short-term effect of treatment (Bullock et al., 2006). The exclusion from drug prescription on the basis of age is not appropriate, since it has been demonstrated that the oldest age is associated with a slower progression of the disease, independent of the initial response, during the subsequent years of treatment, compared with subjects with an initial positive response (Boccardi et al., 2017; Droogsma et al., 2015; Wattmo, Paulsson, Minthon, & Londos, 2013). Thus, age per se cannot be considered as a limit in starting and maintaining therapy.

Which drug should be prescribed?

There are many studies of each drug individually reporting on safety and efficacy, but only relatively few "head-to-head" open-label trials comparing different ChEIs (e.g., Abolfazli, Ghazanshahi, & Nazeman, 2008; Shimizu, Kanetaka, Hirose, Sakurai, & Hanyu, 2015; Wouters, Dautzenberg, Thissen, & Dautzenberg, 2010) (Fig. 29.3). Tricco et al. (2018) have examined the comparative effectiveness and safety of cognitive enhancers (AChEIs and memantine) in a systematic review involving nearly 150 studies, but all of them were drug versus placebo. They found that donepezil and memantine, donepezil, and

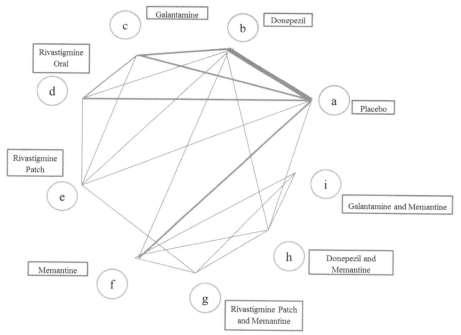

Figure 29.3 *Network metaanalysis of current Alzheimer's treatments.* In this network metaanalysis, the authors considered 41 studies comparing nine treatments: the thickness of the lines is proportional to the number of direct comparisons between each pair of treatments. *(Modified from Jackson, D., Veroniki, A.A., Law, M., Tricco, A.C., & Baker, R. (2017). Paule-Mandel estimators for network meta-analysis with random inconsistency effects.* Research Synthesis Methods, 8, 416–434.)

galantamine were the most effective agents for AD regarding main outcomes (cognitive status, especially for donepezil; global function, especially for galantamine; behavior problems, mortality, and adverse events). However, many studies are difficult to interpret because they were conducted using open-label design, so in a *real-world* setting there is no reason to believe that one drug is better than another, and the choice should be made on the basis of the individual tolerability, better availability, and daily consumption (see Table 29.2) (Solomon & Budson, 2016).

When should therapy start and stop?

It is clear that treatment should be initiated as soon as possible after diagnosis of dementia (Fig. 29.4). By focusing on the several stages of dementia, doubts can be raised regarding prescribing medicine in the very initial (*start*) or very late stages (*stop*). An important issue lies in the treatment of subjects with mild cognitive impairment (MCI), who have a probability of dementia conversion of 15% in 2 years (Petersen et al., 2018). Unfortunately, several studies concerned the idea that ChEIs are not efficacious at this stage of the disease. In particular, donepezil does not show any reduction in progression to possible or probable AD after 3 years of follow-up (Petersent et al., 2005), nor a slower progression on various cognitive scales (Doody et al., 2009; Salloway et al., 2004). Data on galantamine are more restricted, but it was concluded that it is ineffective for reducing progression to dementia over 24 months (Winblad et al., 2008). Again, a study of 48 months in patients with MCI treated with rivastigmine did not show a reduction in the rate of progression to possible or probable AD (Feldman et al., 2007). Therefore, according to this evidence, in MCI, clinicians may choose to offer ChEIs, specifying that it is an off-label prescription not currently backed by empirical evidence (Petersen et al., 2018). In later stages, it is still possible to start therapy as long as there is a function that one wishes to preserve. In this case, clinicians and patients should expect a slower decline rather than improvement or stabilization, as seen from placebo-comparative studies (Solomon & Budson, 2016). Another debated issue is when to stop therapy. Some guidelines recommend discontinuation when AD becomes too severe (MMSE score <10), in line with the marketing authorization (NICE, 2016). However, the effects of medications may be unpredictable, especially over long durations of treatment, since cognitive and behavioral impairments change during the disease course. Overall, it is recommended to discontinue therapy when the patient is no longer able to enjoy any aspect of life: in this case, the goal of treatment will be to help the patient spend his or her last months or years of living with comfort and dignity. The literature concluded that ChEI discontinuation might deleteriously affect cognition and neuropsychiatric symptoms and argued against interruption in patients with severe AD (O'Regan, Lanctot, Mazereeuw, & Herrmann, 2015) without a proper motivation. Moreover, a modest reduction in caregivers' psychological symptoms has been seen for patients treated even in severe stages. This suggests the possibility that caregivers living with treated patients perceived treatment benefits (Howard et al., 2012) or that the act of giving medication—regardless of its material effects—may signify that they care for the patient. In conclusion, it seems reasonable to apply an individualized approach to discontinuation while engaging patients and families in treatment decisions (Renn et al., 2018).

Where is it appropriate to use these drugs (home, nursing homes, hospital)?

The usefulness of ChEIs at home has already been extensively discussed regarding cognitive and behavioral symptoms. Another gain associated with more extended periods of treatment with ChEIs is the decreased risk for nursing home (NH) placement (Beusterien et al., 2004; Lopez et al., 2005) with associated personal, social, and economic benefits. Costs related to the care of AD patients are very high, estimated from \$82,125 to \$92,378 per year for NH care, as shown by the Alzheimer's Association (2017). For this reason, delaying NH admission may have an impact on the overall costs associated with the disease (Lopez et al., 2009). The principal factors driving NH admission are the severity of cognitive impairment, dependency in ADLs, and prominent behavioral and

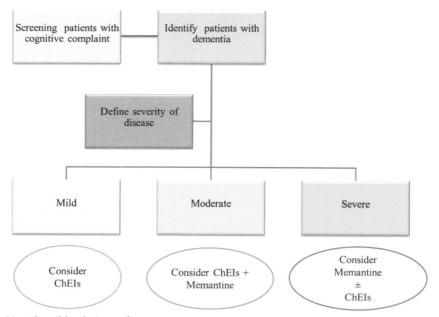

Figure 29.4 *Possible choices of treatment and management of Alzheimer's dementia.* A flowchart representing possible choices of treatment based on the severity of the disease is shown. Cholinesterase inhibitors (*ChEIs*) should be the first choice in patients with mild–moderate disease and they could be added to memantine for better disease management in severe stages.

psychological symptoms of dementia (BPSD). Although ChEIs did not affect the survival time in NHs (Wattmo, Londos, & Minthon, 2016), they are very useful to delay the placement in this setting: a retrospective analysis of a large US medical claims database showed that over a 27-month follow-up period, more patients who were not treated with ChEIs were placed in NHs (11.0%) than were those who received therapy (Beusterien et al., 2004). Unfortunately, in NHs these drugs are often discontinued (Parsons, Briesacher, Givens, Chen, & Tjia, 2011), leading to adverse behavioral changes, increasing the severity of aggressive behaviors, and decreasing time spent engaging in leisure-related activities. The reason for this discontinuation is not fully clear; a survey conducted between a group of hospice medical directors in the United States highlighted that they did not consider ChEIs or memantine to be effective in this type of person, recommending discontinuing these therapies to families at the time of hospice admission (Shega, Ellner, Lau, & Maxwell, 2009).

Indeed, ChEIs seem to be useful also in acute care, even if in hospitals they are often stopped in the deprescribing process. Except for real side effects, discontinuing these drugs is not recommended also because it could worsen BPSD as well as precipitating delirium. Although there is no clear evidence that ChEIs are effective in preventing delirium compared with placebo (Siddiqi et al., 2016), it is known that the main neurochemical correlate of delirium could lie in a decreased cholinergic activity.

Why is it important to prescribe these drugs?

Collectively, prescribing ChEIs is safe, and they are well tolerated and efficacious. ChEIs are tolerated by more than 90% of patients (Solomon & Budson, 2016), and the most common side effects are due to muscarinic cholinergic receptors in the gut (Table 29.3). However, gastrointestinal side effects often attenuate in a few days and only in a small

Table 29.3 Common side effects of cholinesterase inhibitors.

Drug	Side effects	Notes
Donepezil	Nausea, vomiting, and diarrhea	They can be reduced taking medication with food
	Agitation	Generally it subsides after few weeks of therapy
Rivastigmine	Nausea, vomiting and diarrhea, weight loss, abdominal pain	They can be reduced by using the patch
	Headaches, fatigue, anxiety, agitation	
Galantamine	Nausea, vomiting, and diarrhea	Titrate dosage gradually and administer drug with meals

Cholinesterase inhibitors' common side effects affect the gastrointestinal tract and behavior. They are limited by gradually titrating, reducing dosage, or using the patch formulation available for rivastigmine (the last is useful only to treat gastrointestinal symptoms).

percentage of cases do they seriously need a reduction of the dose or even the interruption of therapy. Other common effects are dizziness, insomnia, and headache, and studies have reported a rare slowing of heart rate (that can lead to syncope or require a pacemaker). Most common side effects can be controlled with some techniques. For example, the rivastigmine patch is useful when patients do not tolerate the oral drug; equally, it is possible to try evening administration to reduce nausea or loss of appetite. In contrast, if there is night agitation or vivid dreams, it is possible to take the drug in the morning. Dizziness, headache, and fasciculation can be avoided by dose reduction. Serious adverse events (cardiac, rash, enteral bleeding) require discontinuing the drug and a thorough evaluation of the patient. At the same time, cautions must be employed in certain conditions such as asthma and chronic obstructive pulmonary disease (they can cause bronchospasm or increase secretions); gastrointestinal bleeding history or peptic ulcer disease, as well as concomitant use of nonsteroidal antiinflammatory drugs; urinary obstruction; and renal or hepatic impairment, which can raise ChEI levels (Khoury, Rajamanickam, & Grossberg, 2018; Solomon & Budson, 2016).

Cholinesterase inhibitors in other dementias

Vascular dementia (VaD) represents the second most common type of dementia. In older patients, in particular, the combination of VaD and AD is very common, known as mixed dementia. Evidence showed that VaD is also associated with cholinergic deficit, suggesting that ChEIs may be beneficial also in its treatment (Erkinjuntti, Román, Gauthier, Feldman, & Rockwood, 2004). Two Cochrane Reviews support the role of donepezil (Malouf & Birks, 2004) and rivastigmine (Birks, McGuinness, & Craig, 2013) in VaD, showing improvements in cognitive function and ADLs, as well as more global measures of behavioral symptoms and caregiver stress (the last two only for rivastigmine) (Kandiah et al., 2017; Kavirajan & Schneider, 2007).

Dementia with Lewy bodies (DLB) and Parkinson's disease dementia (PDD) are two entities of major neurocognitive disorders with still unknown etiology characterized by cognitive and motor symptoms. Findings from an in vivo positron emission tomography imaging study showed high cortical AChE deficit in these patients (Bohnen et al., 2003). Here again, robust evidence supports the use of ChEIs to treat the cognitive and psychiatric symptoms of these disorders, since they may produce a reduction in apathy, visual hallucinations, and delusions in both disorders (Galasko, 2017). In particular some studies have shown that rivastigmine provides significant benefits to cognition, particularly in executive function and attention; ADLs; and neuropsychiatric symptoms, especially hallucinations and delusions (Kandiah et al., 2017), but it was associated with greater risk of adverse events (gastrointestinal, tremor, somnolence, dizziness) in both DLB and PDD (Stinton et al. 2015).

On the other hand, some disorders typically do not benefit from ChEIs, for example, frontotemporal dementia (FTD) and bipolar disorders, due to the risk of behavioral symptoms exacerbation (Solomon & Budson, 2016). Evidence has demonstrated not clinically significant efficacy in treating FTD patients, and the drugs may hasten cognitive decline or worsen behavioral symptoms (Mocellin, Scholes, Walterfang, Looi, & Velakoulis, 2015). This is likely due to a lack of consistent evidence of a cholinergic deficit in FTD without joint pathology of other major neurocognitive disorders (Chow, 2005). However, at least one study reported some efficacy in improving behavioral symptoms with galantamine treatment in an aphasic subgroup of FTD without a preponderance of serious adverse events, although no effect was found in the overall treatment of the general FTD subject population (Kertesz et al., 2008).

Novel targeting strategies

New strategies for the creation of "multitarget-directed ligands" have emerged, also mixing existing compounds with novel molecules that might enhance the efficacy of currently available treatments. These drugs aim to simultaneously address several critical pathophysiological processes as described earlier (Terry, Callahan, & Hernandez, 2015). Studies have investigated other mechanisms of the action of ChEIs regarding disease-modifying effects (Caccamo et al., 2006), focusing on new molecules potentially able to activate complementary pathways of muscarinic and nicotine receptors. Referring to the role of M1 muscarinic subtype receptors in learning and memory, beyond their effects on ACh release, they can promote APP processing, attenuating plaque and tangle formation. Donepezil, rivastigmine, and galantamine showed, in preclinical studies of in vivo and in vitro models of AD, several neuroprotective mechanisms, such as contrasting Aβ toxicity, tau pathology, neuroinflammation, glutamate-induced toxicity, and vascular pathology, via many pathways, including the 7 nicotinic receptors, the 1-phosphatidylinositol-3-phosphate 5-kinase—Akt pathway (involved in metabolism, growth, proliferation, survival, transcription, and protein synthesis), and the inhibition of glycogen synthesis kinase-3 (which has a role in tau hyperphosphorylation) (Kim et al., 2017). Unfortunately, none of these studies concretized a compound capable of opposing the course of the disease. Some novel M1 ligands (i.e., alvameline, xanomeline, cevimeline) have been tested in clinical trials, but although they showed improvement in memory and cognition in AD patients, clinical results were unsatisfactory mainly due to their side effects (Confaloni et al., 2016). Other ligands that selectively alter the function and activity of nicotinic receptors have also been discovered. Several nicotinic receptor agonists and modulators have thus been characterized (i.e., encenicline/EVP 6124, ABT 126, TC-1734/AZD3480, GLN-1062), but none of them have proceeded past phase III (Solomon & Budson, 2016).

Key facts on acetylcholine and dementias

- ACh is a neurotransmitter released in neuromuscular junctions, in the autonomic ganglia, and in the brain, where it influences neuronal excitability and presynaptic release of neurotransmitters and coordinates the activity of groups of neurons.
- Production of ACh is catalyzed by choline acetyltransferase enzyme, while acetylcholinesterase and butyrylcholinesterase terminate its function, catalyzing ACh hydrolysis.
- ACh binds muscarinic and nicotinic receptors (either pre- or postsynaptic) and modifies in various ways brain activity, increasing neuronal firing and contributing to long-term potentiation.
- Due to the role of these receptors, ACh plays a role in cognition, mainly in learning and memory.
- Dysfunctions of central cholinergic systems are a core feature in Alzheimer's disease and other dementias. Due to degeneration of cholinergic neurons, a reduction of cholinergic tone in the cortex and hippocampus occurs, leading to cognitive dysfunction.

Summary points

- Dementia is a progressive brain disease marked by multiple cognitive deficits that lead to a gradual reduction in functional status.
- The most common form of dementia is Alzheimer's disease, and the US FDA has approved ChEIs for treatment.
- ChEIs are symptomatic drugs, but there is evidence that they could also have some disease-modifying effects.
- ChEIs are efficacious and well tolerated.
- It is recommended to try ChEIs in other dementias, i.e., VaD, DLB, and PDD, whereas they seem not useful in FTD and dementia associated with bipolar disorder.

References

Abolfazli, S., Ghazanshahi, M., & Nazeman, M. (2008). Effects of 6-months treatment with donepezil and rivastigmine on results of neuropsychological tests of MMSE, NPI, Clock and Bender in patients with Alzheimer's disease. *Acta Medica Iranica, 46,* 99—104.

Bartus, R. T. (2000). In neurodegenerative diseases, models, and treatment strategies: Lessons learned and lessons forgotten a generation following the cholinergic hypothesis. *Experimental Neurology, 163,* 495—529.

Bartus, R. T., Dean, R. L., Beer, B., & Lippa, A. S. (1982). The cholinergic hypothesis of geriatric memory dysfunction. *Science, 217,* 408—414.

Beusterien, K. M., Thomas, S. K., Gause, D., Kimel, M., Arcona, S., & Mirski, D. (2004). Impact of rivastigmine use on the risk of nursing home placement in a US sample. *CNS Drugs, 18,* 1143—1148.

Birks, J. (2006). Cholinesterase inhibitors for Alzheimer's disease. *Cochrane Database of Systematic Reviews, 1,* CD005593.

Birks, J., McGuinness, B., & Craig, D. (2013). Rivastigmine for vascular cognitive impairment. *Cochrane Database of Systematic Reviews, 5,* CD004744.

Blanco-Silvente, L., Castells, X., Saez, M., Barceló, M. A., Garre-Olmo, J., Vilalta-Franch, J., et al. (2017). Discontinuation, efficacy, and safety of cholinesterase inhibitors for Alzheimer's disease: A meta-analysis and meta-regression of 43 randomized clinical trials enrolling 16106 patients. *International Journal of Neuropsychopharmacology, 20,* 519—528.

Boccardi, V., Baroni, M., Smirne, N., Clodomiro, A., Ercolani, S., Longo, A., et al. (2017). Short-term response is not predictive of long-term response to acetylcholinesterase inhibitors in old age subjects with Alzheimer's disease: A "Real World" study. *Journal of Alzheimer's Disease, 56,* 239—248.

Bohnen, N. I., Kaufer, D. I., Ivanco, L. S., Lopresti, B., Koeppe, R. A., Davis, J. G., et al. (2003). Cortical cholinergic function is more severely affected in Parkinsonian dementia than in Alzheimer disease: An in vivo positron emission tomographic study. *Archives of Neurology, 60,* 1745—1748.

Bullock, R., Bergman, H., Touchon, J., Gambina, G., He, Y., Nagel, J., et al. (2006). Effect of age on response to rivastigmine or donepezil in patients with Alzheimer's disease. *Current Medical Research and Opinion, 22,* 483—494.

Caccamo, A., Oddo, S., Billings, L. M., Green, K. N., Martinez-Coria, H., Fisher, A., et al. (2006). M1 receptors play a central role in modulating AD-like pathology in transgenic mice. *Neuron, 49,* 671—682.

Chow, T. W. (2005). Treatment approaches to symptoms associated with frontotemporal degeneration. *Current Psychiatry Reports, 7,* 376—380.

Confaloni, A., Tosto, G., & Tata, A. M. (2016). Promising therapies for Alzheimer's disease. *Current Pharmaceutical Design, 22,* 2050—2056.

Doody, R. S., Ferris, S. H., Salloway, S., Sun, Y., Goldman, R., Watkins, W. E., et al. (2009). Donepezil treatment of patients with MCI: A 48-week randomized, placebo-controlled trial. *Neurology, 72,* 1555—1561.

Droogsma, E., Van Asselt, D., Diekhuis, M., Veeger, N., Van der Hooft, C., & De Deyen, P. P. (2015). Initial cognitive response to cholinesterase inhibitors and subsequent long-term course in patients with mild Alzheimer's disease. *International Psychogeriatrics, 27,* 1323—1333.

Dubois, B., Feldman, H. H., Jacova, C., Cummings, J. L., Dekosky, S. T., Barberger- Gateau, P., et al. (2010). Revising the definition of Alzheimer's disease: A new lexicon. *The Lancet Neurology, 9,* 1118—1127.

Dyer, S. M., Harrison, S. L., Laver, K., Whitehead, C., & Crotty, M. (2018). An overview of systematic reviews of pharmacological and non-pharmacological interventions for the treatment of behavioral and psychological symptoms of dementia. *International Psychogeriatrics, 30,* 295—309.

Erkinjuntti, T., Román, G., Gauthier, S., Feldman, H., & Rockwood, K. (2004). Emerging therapies for vascular dementia and vascular cognitive impairment. *Stroke, 35,* 1010—1017.

Farlow, M. R., & Cummings, J. L. (2007). Effective pharmacologic management of Alzheimer's disease. *The American Journal of Medicine, 120,* 388—397.

Feldman, H. H., Ferris, S., Winblad, B., Sfikas, N., Mancione, L., He, Y., et al. (2007). Effect of rivastigmine on delay to diagnosis of Alzheimer's disease from mild cognitive impairment: The InDDEx study. *The Lancet Neurology, 6,* 501—512.

Feldman, H., Gauthier, S., Hecker, J., Vellas, B., Emir, B., Mastey, V., et al. (2003). Efficacy of donepezil on maintenance of activities of daily living in patients with moderate to severe Alzheimer's disease and the effect on caregiver burden. *Journal of the American Geriatrics Society, 51,* 737—744.

Galasko, D. (2017). Lewy body disorders. *Neurologic Clinics, 35,* 325—338.

Hashimoto, M., Yatabe, Y., Kaneda, K., Honda, K., & Ikeda, M. (2009). Impact of donepezil hydrochloride on the care burden of family caregivers of patients with Alzheimer's disease. *Psychogeriatrics, 9,* 196—203.

Howard, R., McShane, R., Lindesay, J., Ritchie, C., Baldwin, A., Barber, R., et al. (2012). Donepezil and memantine for moderate-to-severe Alzheimer's disease. *New England Journal of Medicine, 366,* 893—903.

Jackson, D., Veroniki, A. A., Law, M., Tricco, A. C., & Baker, R. (2017). Paule-Mandel estimators for network meta-analysis with random inconsistency effects. *Research Synthesis Methods, 8,* 416—434.

Kandiah, N., Pai, M. C., Senanarong, V., Looi, I., Ampil, E., Park, K. W., et al. (2017). Rivastigmine: The advantages of dual inhibition of acetylcholinesterase and butyrylcholinesterase and its role in subcortical vascular dementia and Parkinson's disease dementia. *Clinical Interventions in Aging, 12,* 697—707.

Kavirajan, H., & Schneider, L. S. (2007). Efficacy and adverse effects of cholinesterase inhibitors and memantine in vascular dementia: A meta-analysis of randomised controlled trials. *The Lancet Neurology, 6*, 782—792.

Kertesz, A., Morlog, D., Light, M., Blair, M., Davidson, W., Jesso, S., et al. (2008). Galantamine in frontotemporal dementia and primary progressive aphasia. *Dementia and Geriatric Cognitive Disorders, 25*, 178—185.

Khoury, R., Rajamanickam, J., & Grossberg, G. T. (2018). An update on the safety of current therapies for Alzheimer's disease: Focus on rivastigmine. *Therapeutic Advances in Drug Safety, 9*, 171—178.

Kim, S. H., Kandiah, N., Hsu, J. L., Suthisisang, C., Udommongkol, C., & Dash, A. (2017). Beyond symptomatic effects: Potential of donepezil as a neuroprotective agent and disease modifier in Alzheimer's disease. *British Journal of Pharmacology, 174*, 4224—4232.

Lemstra, A. W., Richard, E., & van Goo, W. A. (2007). Cholinesterase inhibitors in dementia: Yes, no, or maybe? *Age and Ageing, 36*, 625—627.

Lopez, O. L., Becker, J. T., Saxton, J., Sweet, R. A., Klunk, W., & DeKosky, S. T. (2005). Alteration of a clinically meaningful outcome in the natural history of Alzheimer's disease by cholinesterase inhibition. *Journal of the American Geriatrics Society, 53*, 83—87.

Lopez, O. L., Becker, J. T., Wahed, A. S., Saxton, J., Sweet, R. A., Wolk, D. A., et al. (2009). Long-term effects of the concomitant use of memantine with cholinesterase inhibition in Alzheimer disease. *Journal of Neurology, Neurosurgery and Psychiatry, 80*, 600—607.

Malouf, R., & Birks, J. (2004). Donepezil for vascular cognitive impairment. *Cochrane Database of Systematic Reviews, 1*, CD004395.

Mocellin, R., Scholes, A., Walterfang, M., Looi, J. C., & Velakoulis, D. (2015). Clinical update on frontotemporal dementia: Diagnosis and treatment. *Australasian Psychiatry, 23*, 481—487.

National Institute for Health and Care Excellence (NICE). (2016). *Donepezil, galantamine, rivastigmine, and memantine for the treatment of Alzheimer's disease (CG42).*

O'Brien, J. T., Holmes, C., Jones, M., Jones, R., Livingston, G., McKeith, I., et al. (2017). Clinical practice with anti-dementia drugs: A revised (third) consensus statement from The British Association for Psychopharmacology. *Journal of Psychopharmacology, 31*, 147—168.

O'Regan, J., Lanctot, K. L., Mazereeuw, G., & Herrmann, N. (2015). Cholinesterase inhibitor discontinuation in patients with Alzheimer's disease: A meta-analysis of randomized controlled trials. *Journal of Clinical Psychiatry, 76*, 1424—1431.

Parsons, C., Briesacher, B. A., Givens, J. L., Chen, Y., & Tjia, J. (2011). Cholinesterase inhibitor and memantine use in newly admitted nursing home residents with dementia. *Journal of the American Geriatrics Society, 59*, 1253—1259.

Petersen, R. C., Lopez, O., Armstrong, M. J., Getchius, T. S. D., Ganguli, M., Gloss, D., et al. (2018). Practice guideline update summary: Mild cognitive impairment: Report of the guideline development, dissemination, and implementation subcommittee of the American academy of neurology. *Neurology, 90*, 126—135.

Petersen, R. C., Thomas, R. G., Grundman, M., Bennett, D., Doody, R., Ferris, S., et al. (2005). Vitamin E and donepezil for the treatment of mild cognitive impairment. *New England Journal of Medicine, 352*, 2379—2388.

Prince, M., Comas-Herrera, A., Knapp, M., Guerchet, M., & Karagiannidou, M. (2016). *World Alzheimer report 2016: Improving healthcare for people living with dementia: Coverage, quality and costs now and in the future.* London, UK: Alzheimer's Disease International (ADI).

Renn, B. N., Asghar-Ali, A. A., Thielke, S., Catic, A., Martini, S. R., Mitchell, B. G., et al. (2018). A systematic review of practice guidelines and recommendations for discontinuation of cholinesterase inhibitors in dementia. *American Journal of Geriatric Psychiatry, 26*, 134—147.

Sadowsky, C. H., & Galvin, J. E. (2012). Guidelines for the management of cognitive and behavioral problems in dementia. *The Journal of the American Board of Family Medicine, 25*, 350—366.

Salloway, S., Ferris, S., Kluger, A., Goldman, R., Griesing, T., Kumar, D., et al. (2004). Efficacy of donepezil in mild cognitive impairment: A randomized placebo-controlled trial. *Neurology, 63*, 651—657.

Shega, J. W., Ellner, L., Lau, D. T., & Maxwell, T. L. (2009). Cholinesterase inhibitor and N-methyl-D-aspartic acid receptor antagonist use in older adults with end-stage dementia: A survey of hospice medical directors. *Journal of Palliative Medicine, 12*, 779–783.

Shimizu, S., Kanetaka, H., Hirose, D., Sakurai, H., & Hanyu, H. (2015). Differential effects of acetylcholinesterase inhibitors on clinical responses and cerebral blood flow changes in patients with Alzheimer's disease: A 12-month, randomized, and open-label trial. *Dementia and Geriatric Cognitive Disorders Extra, 5*, 135–146.

Siddiqi, N., Harrison, J. K., Clegg, A., Teale, E. A., Young, J., Taylor, J., et al. (2016). Interventions for preventing delirium in hospitalised non-ICU patients. *Cochrane Database of Systematic Reviews, 3*, CD005563.

Solomon, P. R., & Budson, A. E. (2016). Cholinesterase inhibitors. In A. E. Budson, & P. R. Solomon (Eds.), *Memory loss, Alzheimer's disease, and dementia: A practical guide for clinicians* (2nd ed., pp. 160–173). Elselvier Inc.

Stinton, C., McKeith, I., Taylor, J. P., Lafortune, L., Mioshi, E., Mak, E., et al. (2015). Pharmacological management of Lewy body dementia: A systematic review and meta-analysis. *American Journal of Psychiatry, 172*, 731–742.

Tata, M. A., Velluto, L., D'Angelo, C., & Reale, M. (2014). Cholinergic system dysfunction and neurodegenerative diseases: Cause or effect? *CNS and Neurological Disorders – Drug Targets, 13*, 1294–1303.

Terry, A. V., Jr., Callahan, P. M., & Hernandez, C. M. (2015). Nicotinic ligands as multifunctional agents for the treatment of neuropsychiatric disorders. *Biochemical Pharmacology, 97*, 388–398.

Tricco, A. C., Ashoor, H. M., Soobiah, C., Rios, P., Veroniki, A. A., Hamid, J. S., et al. (2018). Comparative effectiveness and safety of cognitive enhancers for treating Alzheimer's disease: Systematic review and network meta-analysis. *Journal of the American Geriatrics Society, 66*, 170–178.

Wattmo, C., Londos, E., & Minthon, L. (2016). Cholinesterase inhibitors do not alter the length of stay in nursing homes among patients with Alzheimer's disease: A prospective, observational study of factors affecting survival time from admission to death. *BMC Neurology, 16*, 156.

Wattmo, C., Paulsson, E., Minthon, L., & Londos, E. (2013). A longitudinal study of risk factors for community-based home help services in Alzheimer's disease: The influence of cholinesterase inhibitor therapy. *Clinical Interventions in Aging, 8*, 329–339.

Wattmo, C., & Wallin, Å. K. (2017). Early-versus late-onset Alzheimer disease: Long-term functional outcomes, nursing home placement, and risk factors for rate of progression. *Dementia and Geriatric Cognitive Disorders Extra, 7*, 172–187.

Winblad, B., Gauthier, S., Scinto, L., Feldman, H., Wilcock, G. K., Truyen, L., et al. (2008). Safety and efficacy of galantamine in subjects with mild cognitive impairment. *Neurology, 70*, 2024–2035.

Wouters, C. J., Dautzenberg, L., Thissen, A., & Dautzenberg, P. L. (2010). Oral galantamine versus rivastigmine transdermal patch: A descriptive study at a memory clinic in The Netherlands. *Tijdschrift voor Gerontologie en Geriatrie, 41*, 146–150.

CHAPTER 30

Choline-containing phospholipids and treatment of adult-onset dementia disorders

Francesco Amenta[1], Gopi Battineni[2], Enea Traini[2], Graziano Pallotta[1]

[1]Clinical Research, Telemedicine and Telepharmacy Centre, School of Medicinal and Health Products Sciences, University of Camerino, Camerino, Italy; [2]Telemedicine and Telepharmacy Center, School of Pharmacological Sciences and Health Products, University of Camerino, Camerino, Italy

List of abbreviations

ACh Acetylcholine
AChR Acetylcholine receptor
AD Alzheimer's disease
CCPL Choline-containing phospholipids
ChAT Choline acetyltransferase
ChE Cholinesterase
GPC Choline alphoscerate
PL Phospholipids

Mini-dictionary of terms

Acetylcholine ACh is an organic chemical that acts as a neurotransmitter in the brain and in the body of many different animals, including humans.

Acetyltransferase acetyltransferase (also knows as *transacetylase*) is a kind of transferase enzyme that transfers an acetyl group.

Cholinergic neurons a cholinergic neuron is a nerve cell that primarily uses ACh as a neurotransmitter to send its messages. Many neurological systems in our body are cholinergic. Cholinergic neurons represent the primary source of ACh for the cerebral cortex.

Neurotransmitter neurotransmitters are endogenous chemicals that allow neurotransmission. They are a type of chemical messenger that transmits signals across a chemical synapse from one neuron to another neuron, muscle cell, or gland cell.

Phospholipids PLs are lipids that are a major component of all cell membranes. They can form lipid bilayers thanks to their amphiphilic characteristic. The structure of the PL molecule commonly consists of two hydrophobic fatty acid "tails" and a hydrophilic "head" consisting of a phosphate group.

Introduction

Choline, the main precursor of the brain and autonomic neurotransmitter acetylcholine (ACh), is a quaternary ammonium salt (Fig. 30.1). Choline is an essential component of various membrane phospholipids (PLs) and contributes to the structural integrity of cell membranes. Choline–containing PLs (CCPLs) include phosphatidylcholine (PC),

Diagnosis and Management in Dementia
ISBN 978-0-12-815854-8, https://doi.org/10.1016/B978-0-12-815854-8.00030-6

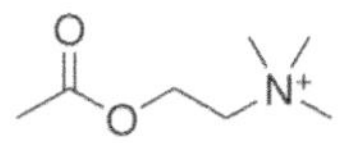

Figure 30.1 Acetylcholine. The structural formula of acetylcholine is shown.

phosphatidylserine, sphingomyelin (SM), cytidine 5′-diphosphocholine (CDP—choline or citicoline), and choline alphoscerate (GPC). PC is the major PL in most eukaryotic cells and is involved in SM synthesis, choline/choline metabolite regeneration, and fatty acid/GPC formation. Moreover, the role of CCPLs in the biosynthesis of various neurotransmitters, ACh included, has been extensively demonstrated.

This chapter summarizes the main preclinical and clinical data on the use of CCPLs in the treatment of adult-onset dementia and suggests that, if properly used, these compounds could still have a place in the pharmacotherapy of dementia disorders.

Cholinergic neurotransmission in Alzheimer's disease and the cholinergic approach in the treatment of Alzheimer's disease

Dr. Aloysius Alzheimer observed feeble dementia in his patient Auguste Deter in 1901, and there are still no specific treatments for the disease, although several approaches have been proposed to alleviate its cognitive and noncognitive symptoms (Amenta, Parnetti, Gallai, & Wallin, 2001; Parnetti, Mignini, Tomassoni, Traini, & Amenta, 2007).

ACh, the first neurotransmitter ever identified, is largely diffused in the central, peripheral, and autonomic nervous systems. In the central nervous system, the cholinergic neurons are widely distributed. Nerve cell bodies are located in the spinal cord, hindbrain, medial habenula, mesopontine region, basal forebrain, striatum, olfactory tubercle, and islands of Calleja complex. From these nerve cell bodies originate projections innervating almost all brain areas.

The cholinergic system plays a role in learning and memory processes. Endogenous ACh released by neurons located in the basal forebrain (the *nucleus basalis magnocellularis* of Meynert and in the septal nuclei) is involved in the modulation of acquisition, encoding, consolidation, extinction, and retrieval of memory. Cholinergic projections originating from the nucleus basalis magnocellularis supply the cerebral cortex, whereas those originating from septal nuclei supply the hippocampus. The hippocampus is a cerebral region involved in a variety of functions, including learning and memory processes. It plays an important role in the elaboration of mapping and working memory, in the improvement of selective attention, and in the inhibition of inappropriate behavioral responses. The hippocampus shows a high sensitivity to aging and is involved in the pathogenesis of learning and memory dysfunction occurring with age and in adult-onset dementia disorders (Amenta et al., 2001; Parnetti et al., 2007).

The important role of the cholinergic neurons originating from the nucleus basalis of Meynert and from septal nuclei in memory is highlighted by the fact that a specific degeneration of these neurons takes place in Alzheimer's disease (AD) and contributes to the memory loss exhibited by AD patients (Amenta et al., 2001). The observation of a loss of the ACh biosynthetic enzyme choline acetyltransferase (ChAT) in the cerebral cortex of AD patients stimulated the development of cholinergic strategies to counter cognitive dysfunction typical of adult-onset dementia, including AD itself (Amenta et al., 2001).

The observation that administration of the muscarinic antagonist scopolamine to healthy young subjects induced a cognitive impairment resembling that found in adult-onset dementia, and the subsequent demonstration of a remarkable decrease in ChAT activity in the cerebral cortex and hippocampus in AD patients, contributed to the development of the cholinergic hypothesis of geriatric memory dysfunctions (Bartus, Dean, Pontecorvo, & Flicker, 1985). The cholinergic system is not the only neurotransmitter system affected in adult-onset cognitive impairment, AD included. However, changes in cholinergic function are implicated in the pathogenesis of learning and memory alterations occurring in adult-onset dementia (Avery, Baker, & Asthana, 1997; Muir, 1997). An analysis of the involvement of cholinergic receptors in cognitive functions has shown that central muscarinic and nicotinic cholinergic receptors might be involved in learning and memory through complex mechanisms. Studies of the brain of aged subjects or of patients suffering from AD have shown a marked loss of ChAT and of nicotinic cholinergic receptors (Davies & Maloney, 1976; Hellström-Lindahl, Mousavi, Zhang, Ravid, & Nordberg, 1999). A correlation between the loss of cortical cholinergic synapses and cognitive decline was demonstrated, as well as the existence of a close relationship between this loss and the decrease in high-affinity cholinergic receptors. The different pharmacological treatments proposed or tested included intervention with ACh precursors, stimulation of ACh release, use of muscarinic or nicotinic receptor agonists, and acetylcholinesterase (AChE) inhibition (Amenta et al., 2001; Parnetti et al., 2007).

Choline-containing phospholipids and the precursor-loading approach for treating age-related dementia disorders

PLs (i.e., PC, phosphatidylserine, phosphatidylethanolamine, and SM) play an important role in cellular constitution and metabolism and are relevant components of various cell membranes (i.e., cells, mitochondria, endoplasmic reticulum, Golgi apparatus, peroxisomes, and lysosomes). Choline is an essential nutrient with a complex role in the body. Choline is necessary for the biosynthesis of the neurotransmitter ACh, cell-membrane signaling PLs, lipid transport (lipoproteins), and methyl-group metabolism (plasmatic homocysteine reduction). Choline is also the major dietary source of methyl groups via the synthesis of S-adenosylmethionine (Tayebati & Amenta, 2013).

Cholinergic precursor-loading therapy was the first approach tried to restore deficient cholinergic neurotransmission and to relieve cognitive impairment in dementia disorders.

Given the controversial nature of the existence of a direct role of choline in ACh release, and given that an increase in brain free choline does not always imply an increase in brain ACh (Tayebati & Amenta, 2013), it was CCPLs that were initially proposed for treating adult-onset dementia disorders.

ACh is a neurotransmitter derived from choline and acetyl-coenzyme A (Fig. 30.2). The biosynthesis is catalyzed by the enzyme ChAT, and the biosynthesis of the cosubstrate acetyl-coenzyme A is not specific for cholinergic neurons. Nervous tissue is unable to synthesize choline, which derives from the diet and is delivered to neurons through the bloodstream. ACh released from cholinergic synapses is hydrolyzed by AChE into choline and acetyl-coenzyme A (Fig. 30.2). Approximately 50% of choline derived from ACh hydrolysis is recovered through a high-affinity transporter. Neurons require, therefore, a further supply of choline to synthesize ACh (Amenta et al., 2001).

Cholinergic neurons probably have a particular avidity for choline, and it has been hypothesized that when the provision of choline is insufficient, neurons obtain it by hydrolyzing membrane PLs. This mechanism, known as autocannibalism, may render cholinergic neurons more susceptible to injury and may contribute to cholinergic neuron degeneration (Wurtman, Blusztajn, & Ulus, 1990).

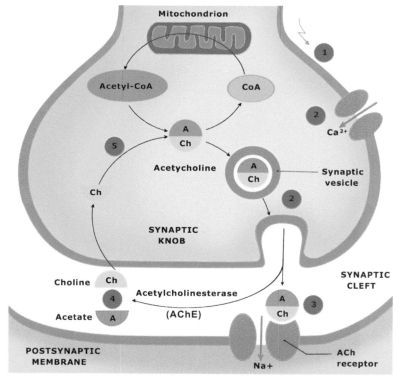

Figure 30.2 Cholinergic synapsis. Acetylcholine processing in a cholinergic synapse is shown. After the release, acetylcholine is broken down by the enzyme acetylcholinesterase.

The CCPLs proposed for the treatment of age-related dementia disorders include lecithin, phosphatidylserine, CDP—choline, and GPC. The main evidence of their activity is reviewed next.

Lecithin

Lecithin is a CCPL representing the main source of choline. Lecithin has been shown to increase serum choline levels more effectively than orally administered choline (Wurtman et al., 1990). Apparently, lecithin may accelerate ACh synthesis in the brain through enhanced availability of choline. Lecithin has been tried out in the treatment of dementia, alone or in combination with an acetylcholinesterase inhibitors (AChEI). Lecithin is largely available as a nutraceutical.

The role of this compound in dementia and in cognitive impairment was the focus of a Cochrane Review in 2009 (Higgins & Flicker, 2003). In the field of dementia and cognitive impairment, lecithin was investigated in 21 studies. Eight studies used the compound alone, others in association with tacrine, physostigmine, and piracetam.

The Cochrane analysis selected 12 randomized trials involving patients with AD (265 patients), parkinsonian dementia (21 patients), and individual memory problems (90 patients). No trials reported any clear clinical benefit of lecithin for AD or parkinsonian dementia. Only a few trials have led to data for meta-analyses. The only statistically significant result was in favor of placebo for adverse events, based on one single trial. Relevant results in favor of lecithin were obtained in a trial of subjects with individual memory problems (Higgins & Flicker, 2003).

Phosphatidylserine

Phosphatidylserine (Fig. 30.3) is a CCPL largely diffused in animal tissues (brain, liver, kidney, heart, and spleen), bluefish, soy, and egg yolk. Phosphatidylserine is a nutraceutical that was originally obtained from the distillation of bovine brains, which show a high

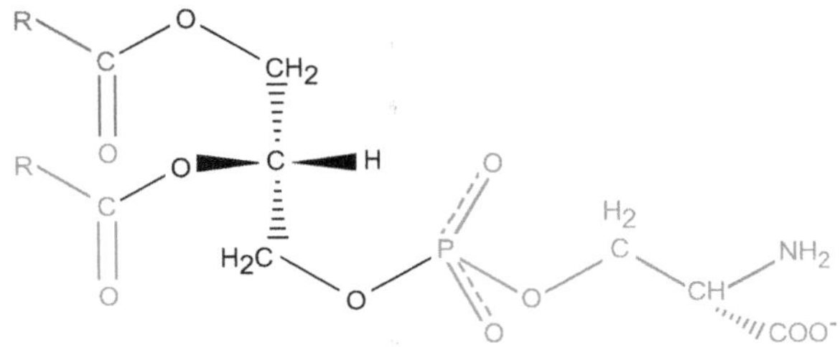

Figure 30.3 Phosphatidylserine. The structural formula of phosphatidylserine is shown. The different components of the phosphatidylserines are colored as described below: Blue and green, variable fatty acid groups; black, glycerol; red, phosphate; purple, serine.

concentration of this PL. After the epidemics of bovine spongiform encephalopathy, soy lecithin became the main source of phosphatidylserine.

Supplementation of this PL is based on the observation that reduced levels of DHA—phosphatidylserine in the cerebral cortex are associated with conversion of mild cognitive impairment to overt dementia (Olivera-Pueyo & Pelegrín-Valero, 2017). It has been suggested that phosphatidylserine content decreases in aging in parallel to the increase in cholesterol in neuronal membranes. These changes modify cell membrane viscosity by reducing enzymatic catalysis with consequent higher susceptibility of nerve cells to injury (Olivera-Pueyo & Pelegrín-Valero, 2017).

Seven clinical trials have investigated the effects of phosphatidylserine administration in AD. These studies, performed between 1986 and 1994, suggested some beneficial effects of the compound, though without providing conclusive evidence. More recent studies using phosphatidylserine as a supplement in patients with mild cognitive impairment or very early stage dementia have reported significant improvement in learning, memory, and verbal fluency, as well as visual learning, attention, communication, initiative, and socialization. Another study has reported reduced apathy, increased motivation and interest, and improved memory (Olivera-Pueyo & Pelegrín-Valero, 2017).

The limited number of patients investigated in the aforementioned studies and the lack of established diagnostic criteria do not allow one to draw a certain conclusion about a possible role of phosphatidylserine as a therapeutic agent in dementia.

CDP-choline

CDP—choline, or citicoline, is a CCPL composed of cytidine and choline linked together by a diphosphate bridge (Fig. 30.4). Citicoline is the international nonproprietary name of CDP—choline. Citicoline is marketed as a prescription-only drug in several European countries and in Japan, and as a nutraceutical in the United States and in Europe. CDP—choline is an intermediate in the synthesis of PC in Kennedy's pathway. PC is a cell membrane component that is degraded during cerebral ischemia to free fatty acids and free radicals (Parnetti et al., 2007). The functions of CDP—choline include the repair of the neural membrane through the synthesis of PC, the reduction of accumulated

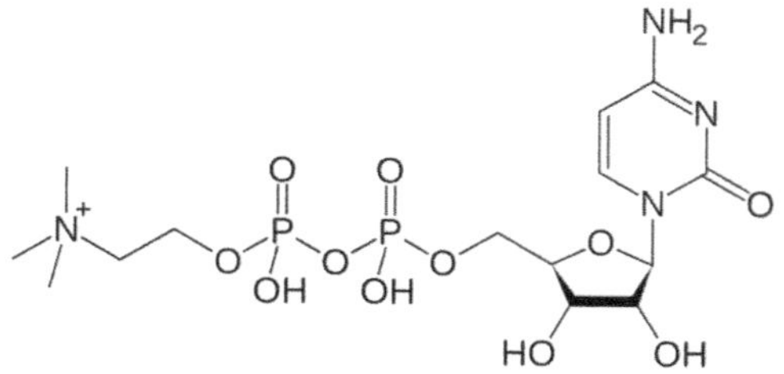

Figure 30.4 Citicoline. The structural formula of citicoline is shown.

fat responsible for increased cognitive deficit, and an increase in ACh levels. In poststroke patients, a neuroprotective effect of CDP—choline, contributing to significant improvement in temporal orientation, attention tasks, and executive function, has been reported, along with an ability, albeit in an experimental setting, to assist in neural repair. Citicoline reduced the duration of hospitalization and improved the recovery of neurological function and level of consciousness (Colucci et al., 2012). The activity of citicoline on the cognitive domain was also analyzed in a meta-analysis evaluating 13 studies carried out from the 1970s to the early 2000s, which demonstrated the effectiveness of the compound in the treatment of cognitive, emotional, and behavioral disorders (Colucci et al., 2012). These effects were attributed to the effect of citicoline on metabolic activation and were noticeable in patients with dementia of degenerative and/or vascular origin and in patients with cerebrovascular disease (Colucci et al., 2012).

More recent studies have reviewed the pharmacological profile, kinetics, and possible new uses of citicoline (Gareri et al., 2015). The CITIRIVAD Study investigated the effects of a combined treatment with the AChEI rivastigmine and citicoline in AD and in mixed dementia. This showed the greater effectiveness of a combined administration of citicoline plus rivastigmine (an AChEI) versus AChEI alone (Castagna, Cotroneo, Ruotolo, & Gareri, 2016). A 2017 trial investigated the effects of an oral 9-month treatment with citicoline in addition to AChEI treatment, compared with the AChEI alone, in AD patients (Citicholinage Study). An increase in the Mini-Mental State Examination (MMSE) scores was noticeable in CDP—choline-treated patients (Gareri et al., 2017). The Citicholinage Study concluded that coadministration of citicoline and AChEI supports the combined administration in managing the disease, by slowing disease progression (Gareri et al., 2017).

The effectiveness of citicoline was proposed in patients with traumatic brain injury, probably due to its effect on edema (Colucci et al., 2012). However, more recent studies did not confirm this effect. Citicoline is a compound that has been available in the pharmaceutical market since the beginning of the 1980s; more recently it has also been marketed as a dietary supplement. Several studies have demonstrated positive effects of the compound on cognition, whereas other investigations failed to confirm the positive results in the cognitive domain. In view of these discrepancies, additional clinical studies are necessary to confirm the potential benefits of citicoline in the treatment of adult-onset dementia disorders.

Choline alphoscerate

GPC, or α-glycerylphosphorylcholine (Fig. 30.5), is a semisynthetic derivative of lecithin. Following oral administration, it is converted to phosphorylcholine, a metabolically active form of choline able to reach the cholinergic nerve terminals, where it increases ACh synthesis, levels, and release (Amenta & Tayebati, 2008; Amenta et al., 2014).

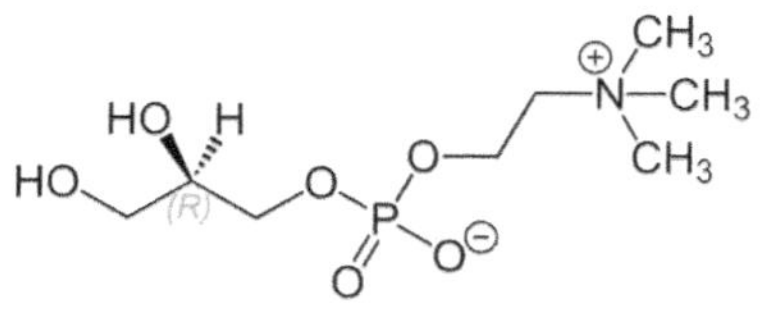

Figure 30.5 Choline alphoscerate. The structural formula of α-GPC (choline alphoscerate) is shown.

Although GPC has been on the pharmaceutical market since 1987, its interest seems to have suffered since the introduction of the therapy with cholinesterase inhibitors. Since 2010 there has been renewed attention paid to this compound, with preclinical studies, clinical investigations, and review articles published in the literature (Traini, Bramanti, & Amenta, 2013).

GPC interferes with brain PL metabolism and increases brain choline and ACh levels and release in rat hippocampus, facilitates learning and memory in experimental animals, improves brain transduction mechanisms, and decreases the age-dependent structural changes occurring in the rat frontal cortex and hippocampus. Moreover, this compound contributes to anabolic processes responsible for membrane PL and glycerolipid synthesis, positively influencing membrane fluidity. GPC was shown to ameliorate cognitive deficit in experimental models of aging brain and to reverse mnemonic deficits induced by scopolamine administration (Traini et al., 2013).

A restorative role of GPC on the central cholinergic system was documented by studies performed on old rodents. Neuroprotective effects of GPC were also reported in a rodent model of altered cholinergic neurotransmission caused by lesions of the nucleus basalis magnocellularis, which constitutes the main source of cholinergic innervation of the cerebral neocortex. A positive effect of treatment with GPC on hippocampus microanatomy and glial reaction was documented in spontaneously hypertensive rats. In a test of the cholinergic precursors lecithin, CDP—choline, and GPC, the last elicited the most relevant stimulation of vesicular ACh and choline transporters in the same model of brain vascular injury. This suggests that GPC is an enhancer of central cholinergic neurotransmission (Traini et al., 2013).

The activity of CCPLs on brain PL biosynthesis may influence brain metabolism and different neurotransmitter systems (Fig. 30.6). Based on the observation that the CCPL CDP—choline has a monoaminergic profile, an investigation was carried out on the activity of GPC on brain dopamine and serotonin levels and on dopamine plasma membrane transporter, vesicular monoamine transporters 1 and 2, serotonin transporter, and norepinephrine transporter (Traini et al., 2013). Administration of the compound increased dopamine levels in frontal cortex and cerebellum and serotonin levels in frontal cortex and striatum. It also stimulated the dopamine plasma membrane transporter in the frontal cortex and cerebellum. This investigation concluded that GPC also possesses a monoaminergic profile and interferes to some extent with brain monoamine transporters (Traini et al., 2013).

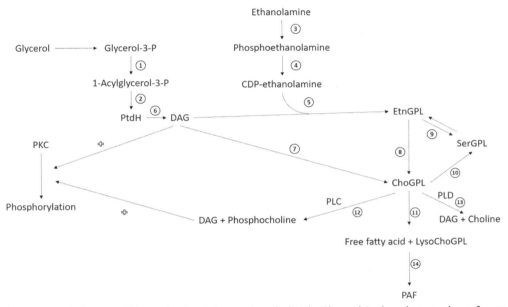

Figure 30.6 Pathways of biosynthesis of glycerophospholipids. Glycerol-3-phosphate acyltransferase (1); lysophosphatidic acid acyltransferase (2); ethanolamine kinase (3); ethanolamine cytidylyltransferase (4); CDP—phosphoethanolamine cytidylyltransferase (5); phosphatidic acid phosphatase (6); CDP—phosphocholine cytidylyltransferase (7); phosphatidylethanolamine methyltransferase (8); phosphatidylserine decarboxylase (9); phosphatidylserine synthase (10); phospholipase A2 (11); phospholipase C (*PLC*) (12); phospholipase D (*PLD*) (13); PAF synthesizing enzymes (14). *CDP—ethanolamine*, cytidine 5'-diphosphoethanolamine; *ChoGPL*, choline glycerophospholipid; *DAG*, diacylglycerol; *EtnGPL*, ethanolamine glycerophospholipid; *LysoChoGPL*, choline lysoglycerophospholipid; *PKC*, protein kinase C; *PtdH*, phosphatidic acid; *SerGPL*, serine glycerophospholipid.

The majority of clinical studies available on the effect of GPC on cognitive function in neurodegenerative and cerebrovascular disorders were detailed in two review articles from our group (Parnetti et al., 2007; Traini et al., 2013). Studies published before 2001 investigated 1570 patients, of which 854 were in controlled trials. Patients examined were affected by dementia of degenerative, vascular, or combined origin, such as dementia of the Alzheimer's type, vascular dementia, and acute cerebrovascular diseases (transitory ischemic attack and stroke) (Amenta et al., 2001). Test batteries for assessing the effect of GPC on cognitive domains were primarily the MMSE for disorders of neurodegenerative origin and the Sandoz Clinical Assessment Geriatric scale for disorders of vascular origin (e.g., vascular dementia) (Traini et al., 2013). The activity of GPC was also investigated in 789 patients with cognitive impairment of vascular origin. Three homogeneous–case trials evaluated 408 patients, while three combined–case trials included 381 patients with vascular dementia. Treatment with GPC improved memory and attention, as well as affective and somatic symptoms (fatigue, vertigo). The effects of GPC were greater than those of placebo and of the same extent as or superior to those of reference compounds (Amenta et al., 2001).

Preclinical studies showed that the association of GPC with (acetyl)cholinesterase inhibitors enhances the effects of both drugs on cholinergic neurotransmission. This prompted the development in Italy of an independent trial known as "Effect of association between a cholinesterase inhibitor and GPC on cognitive deficits in AD associated with cerebrovascular impairment" (ASCOMALVA). This controlled, randomized, and double-blind multicenter (two neurological units in Naples and Mantua) study was designed to assess whether the association between the AChEI donepezil (at the daily dose of 10 mg) and GPC (at the daily dose of 1200 mg) was accompanied by changes in MMSE; Alzheimer's Disease Assessment Scale, Cognitive subscale (ADAS-Cog); Basic Activities of Daily Living; Instrumental Activities of Daily Living; and Neuropsychiatric Inventory (NPI) scores. The last included the evaluation of measures of severity and of caregiver stress.

The patients were between 56 and 91 years of age (mean 75 ± 10 years) and were included in the protocol with an MMSE score between 15 and 23. They showed ischemic brain damage documented by neuroimaging (MRI and CT scan), with a score ≥ 2 in at least one subfield (white matter or basal ganglia), according to the new rating scale for Age-Related White Matter Changes. Recruited patients were then randomly allotted to an active treatment group (donepezil + GPC) or to a reference treatment group (donepezil + placebo) and were treated for 48 months; examination took place at recruitment and after 3, 6, 9, 12, 18, 24, 36, and 48 months of treatment. Consistent with literature data, in patients allotted to the reference treatment group (donepezil + placebo), a slight time-dependent worsening of MMSE and ADAS-Cog scores was found. Treatment with donepezil + GPC (active treatment) countered the decline in MMSE and ADAS-Cog scores. The effect of the association on psychometric tests was statistically significant after 12 months of treatment (Figs. 30.7 and 30.8) (Amenta et al., 2014).

Another aspect explored by the ASCOMALVA trial was the influence of GPC on apathy. Apathy is a common symptom in AD and has a significant impact on several outcomes. No treatment has proven to be effective against apathy, although the administration of cholinesterase inhibitors has been associated with modest improvements in the short term. The ASCOMALVA trial measured apathy at baseline and at 3, 6, 9, 12, 18, and 24 months, through the apathy subtest of the NPI, in 113 mild–moderate AD patients. Two matched groups were compared: group A (n = 56 subjects), treated with donepezil plus GPC, and group 2B (n = 57 subjects), treated with donepezil alone. The combination of donepezil plus GPC was more effective than donepezil alone in countering symptoms of apathy in AD. This suggests that the availability in the brain of a higher amount of ACh may affect apathy in AD subjects with spared executive functions (Amenta & Tayebati, 2008).

Behavioral and psychological symptoms of dementia (BPSD) are a group of psychological reactions, psychiatric symptoms, and behaviors commonly found in AD.

(A)

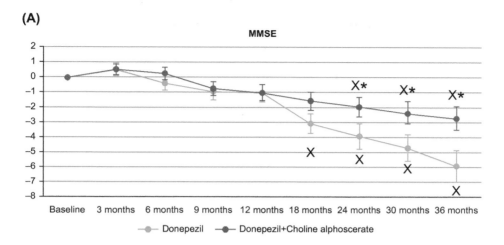

(B)

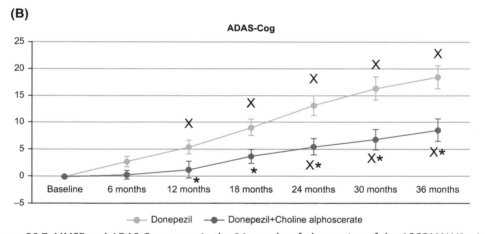

Figure 30.7 MMSE and ADAS-Cog scores in the 36 months of observation of the ASCOMALVA trial. The first results of the ASCOMALVA trial after 12 months of treatment are shown, including evaluation of cognitive tests. Data are the means of the difference in the scores from baseline ± SEM × $P > .05$ versus baseline; *$P < .05$ of donepezil versus associated therapy. *ADAS-Cog*, Alzheimer's Disease Assessment Scale, Cognitive subscale; *MMSE*, Mini-Mental State Examination.

Four clusters of BPSD have been described: mood disorders (depression, anxiety, and apathy), psychotic symptoms (delusions and hallucinations), aberrant motor behaviors (pacing, wandering, and other purposeless behaviors), and inappropriate behaviors (agitation, disinhibition, and euphoria). Most of them are attributed to ACh deficiency. The ASCOMALVA trial investigated the influence of the addition of GPC to donepezil on BPSD at the baseline and after 24 months in 113 mild–moderate AD patients. BPSD were analyzed through the NPI. NPI data revealed a significant decrease in BPSD severity and caregiver distress in patients of group A compared with group B. Mood

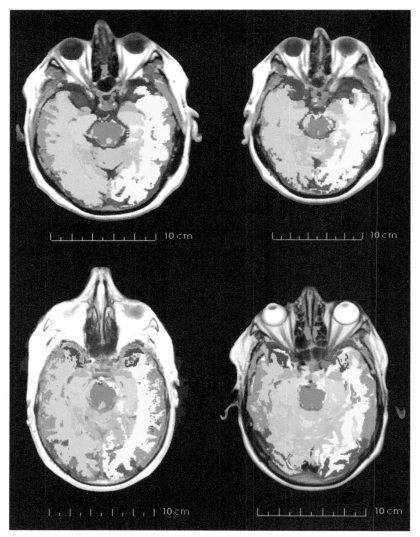

Figure 30.8 Nuclear magnetic resonance images of two patients from the ASCOMALVA study. Images from the ASCOMALVA study, relating to two patients belonging respectively to the donepezil + placebo group (first row) and to the donepezil + choline alphoscerate group (second row) are shown. The hippocampal area is indicated in the parceling from the yellow area. *(The images are original and are modified MRI used in morphometric evaluation of brain volume.)*

disorders (depression, anxiety, and apathy) were significantly decreased in subjects treated with donepezil and GPC, while their severity and frequency were increased in the other group. In conclusion, patients treated with donepezil plus GPC showed a lower level of behavioral disturbances than subjects treated with donepezil only, suggesting that the combination may have beneficial effects (Carotenuto et al., 2017).

A further contribution of the ASCOMALVA trial was the analysis of the influence of a treatment with GPC associated with a treatment with donepezil on brain atrophy in AD. Cerebral atrophy is a common feature of neurodegenerative disorders, such as AD. This pathology includes a loss of gyri and sulci in the temporal lobe and parietal lobe and in parts of the frontal cortex and of the cingulate gyrus. Participants of the ASCOMALVA trial underwent yearly MRI for diagnostic purposes. In 56 patients who achieved 3 years of therapy, brain MRI was analyzed by voxel morphometry techniques. After 3 years of treatment, in patients treated with donepezil plus GPC, a reduction in the volume loss of the gray matter (with a concomitant increase in the volume of the ventriculi and cerebrospinal fluid space) was observed, compared with the reference group, treated with donepezil only. The areas in which brain atrophy was sensitive to combination treatment with GPC were the frontal and temporal lobes, hippocampus, amygdala, and basal ganglia. In other areas, no significant differences were noticeable between the two groups. Morphological data were confirmed by neuropsychological assessment performed alongside the trial.

In conclusion, a cholinergic precursor-loading strategy with GPC in combination with donepezil counters to some extent the atrophy occurring in some brain areas of AD patients. The parallel observation of an improvement in cognitive and functional tests in those patients suggests that morphological changes observed may also have functional relevance.

Conclusions and future perspectives

One option for improving impaired cerebrocortical cholinergic neurotransmission consists in the inhibition of endogenous ACh degradation by using ChE inhibitors (Amenta et al., 2001; Parnetti et al., 2007).

Unfortunately, based on available data from clinical trials, the usefulness of cholinergic precursors in the treatment of AD appears doubtful. The dietary supplementation of free choline, for instance, does not increase the synthesis or the release of ACh in the brain (Amenta & Tayebati, 2008; Sigala et al., 1992; Traini et al., 2013). However, some CCPLs involved in ACh synthesis play an important role in cholinergic neurotransmission, and increase the availability and release of ACh. These PLs include phosphatidylserine, CDP—choline, and GPC. Among these CCPLs, GPC is the compound inducing the most marked release of ACh in animal models. Clinical trials performed with CDP—choline and GPC have shown that these compounds have a more pronounced activity than choline GPC. The results of clinical trials with these compounds were, however, obtained on relatively small numbers of patients. Nonetheless, the results are encouraging to further studies on these molecules.

The ASCOMALVA trial is also of interest as it demonstrated that a cholinergic precursor-loading strategy with GPC and donepezil counters to some extent the atrophy occurring in some brain areas of AD patients. The parallel observation of an improvement in cognitive and functional tests in those patients suggests that morphological changes observed may also have functional relevance.

It is worth mentioning that the main effects of CCPLs were observed after 1—3 years of treatment. This suggests a reconsideration of the time of observation needed for identifying the therapeutic effects of drugs in a brain disorder as complex and as long lasting as AD.

Key facts about choline and phospholipids

- Choline and PLs are important, not only for structural purposes within the lipid bilayer, but also for preventing certain diseases and general disorders at the cellular level.
- PLs are present in various foods, such as milk, chicken eggs, soy, fish eggs, and sunflower seeds.
- PC, phosphatidylserine, and phosphatidylethanolamine are three of the numerous forms of PLs with specific and specialized functions.
- PC is the most abundant PL in the human body; it performs functions that aid in neural processes, including memory, reasoning, and learning.
- Phosphatidylserine is produced by almost every cell in the body, but also needs to be included in the diet to ensure that all cells receive an adequate supply.
- Phosphatidylethanolamine is known for playing a significant role in cellular membrane formation, cognition, and memory.
- A proper daily dietary intake of choline is necessary to decrease the risk of many problems such as fatty liver, liver damage, cognitive decline, nerve damage, fatigue, and memory loss.
- Choline is responsible for good cognitive abilities. Therefore, diets consisting of fish, leafy green vegetables, and soy are essential for maintaining good health.

Summary points

- Cholinergic neurons, for example, ACh-delivering neurons, are essentially associated with the pathogenesis of AD.
- Phosphatidylserine and, in addition, CDP—choline increase ACh substance and discharge.
- ACh and choline are very important for memory and intellectual capacity.
- Choline takes part in various neurochemical processes. It is an antecedent and a metabolite of ACh, it has a role in single-carbon digestion, and it is a fundamental segment of various film PLs.
- Hypoxia/ischemia leads to hydrolysis of PC and a further breakdown of glycerophosphocholine.
- An expansion of free choline does not generally imply expansion of ACh.

Acknowledgment

The support of MDM SpA (Monza, Italy) in the realization of the ASCOMALVA trial is gratefully acknowledged.

References

Amenta, F., Carotenuto, A., Fasanaro, A. M., Rea, R., & Traini, E. (2014). The ASCOMALVA (association between the cholinesterase inhibitor donepezil and the cholinergic precursor choline alphoscerate in Alzheimer's disease) trial: Interim results after two years of treatment. *Journal of Alzheimer's Disease, 42*(Suppl. 3), S281–S288.

Amenta, F., Parnetti, L., Gallai, V., & Wallin, A. (2001). Treatment of cognitive dysfunction associated with Alzheimer's disease with cholinergic precursors. Ineffective treatments or inappropriate approaches? *Mechanism of Ageing and Development, 122*(16), 2025–2040. https://doi.org/10.1016/S0047-6374(01)00310-4.

Amenta, F., & Tayebati, S. (2008). Pathways of acetylcholine synthesis, transport and release as targets for treatment of adult-onset cognitive dysfunction. *Current Medicinal Chemistry, 15*(5), 488–498. Review. PubMed PMID: 18289004.

Avery, E. E., Baker, L. D., & Asthana, S. (December 1997). Potential role of muscarinic agonists in Alzheimer's disease. *Drugs and Aging, 11*(6), 450–459. Review. PubMed PMID: 9413702.

Bartus, R. T., Dean, R. L., Pontecorvo, M. J., & Flicker, C. (1985). The cholinergic hypothesis: A historical overview, current perspective, and future directions. *Annals of the New York Academy of Sciences.* https://doi.org/10.1111/j.1749-6632.1985.tb37600.x.

Carotenuto, A., Rea, R., Traini, E., Fasanaro, A. M., Ricci, G., Manzo, V., et al. (2017). The effect of the association between donepezil and choline alphoscerate on behavioral disturbances in Alzheimer's disease: Interim results of the ASCOMALVA trial. *Journal of Alzheimer's Disease, 56*(2), 805–815.

Castagna, A., Cotroneo, A. M., Ruotolo, G., & Gareri, P. (2016). The CITIRIVAD study: CITIcoline plus RIVAstigmine in elderly patients affected with dementia study. *Clinical Drug Investigation, 36*, 1059–1065.

Colucci, L., Bosco, M., Rosario Ziello, A., Rea, R., Amenta, F., & Fasanaro, A. M. (2012). Effectiveness of nootropic drugs with cholinergic activity in treatment of cognitive deficit: A review. *Journal of Experimental Pharmacology, 4*, 163–172. https://doi.org/10.2147/JEP.S35326. Published 2012 Dec 11.

Davies, P., & Maloney, A. J. F. (1976). Selective loss OF central cholinergic neurons in Alzheimer's disease. *The Lancet.* https://doi.org/10.1016/S0140-6736(76)91936-X.

Gareri, P., Castagna, A., Cotroneo, A. M., Putignano, D., Conforti, R., Santamaria, F., et al. (2017). The citicholinage study: Citicoline plus cholinesterase inhibitors in aged patients affected with Alzheimer's disease study. *Journal of Alzheimer's Disease, 56*, 557–565.

Gareri, P., Castagna, A., Cotroneo, A. M., Putignano, S., De Sarro, G., & Bruni, A. C. (2015). The role of citicoline in cognitive impairment: Pharmacological characteristics, possible advantages, and doubts for an old drug with new perspectives. *Clinical Interventions in Aging, 10*, 1421–1429.

Hellström-Lindahl, E., Mousavi, M., Zhang, X., Ravid, R., & Nordberg, A. (March 20, 1999). Regional distribution of nicotinic receptor subunit mRNAs in human brain: Comparison between Alzheimer and normal brain. *Molecular Brain Research, 66*(1–2), 94–103. PubMed PMID: 10095081.

Higgins, J. P., & Flicker, L. (2003). Lecithin for dementia and cognitive impairment. *Cochrane Database of Systematic Reviews, 3*, CD001015. Review. PubMed PMID: 12917896.

Muir, J. L. (April 1997). Acetylcholine, aging, and Alzheimer's disease. *Pharmacology Biochemistry and Behavior, 56*(4), 687–696. Review. PubMed PMID: 9130295.

Olivera-Pueyo, J., & Pelegrín-Valero, C. (September 2017). Dietary supplements for cognitive impairment. *Actas Españolas de Psiquiatría, 45*(Suppl.), 37–47. Epub 2017 Sep 1. PubMed PMID: 29171642.

Parnetti, L., Mignini, F., Tomassoni, D., Traini, E., & Amenta, F. (2007). Cholinergic precursors in the treatment of cognitive impairment of vascular origin: Ineffective approaches or need for re-evaluation? *Journal of the Neurological Sciences.* https://doi.org/10.1016/j.jns.2007.01.043.

Sigala, S., Imperato, A., Rizzonelli, P., Casolini, P., Missale, C., & Spano, P. (1992). L-α-glycerylphorylcholine antagonizes scopolamine-induced amnesia and enhances hippocampal cholinergic transmission in the rat. *European Journal of Pharmacology.* https://doi.org/10.1016/0014-2999(92)90392-H.

Tayebati, S. K., & Amenta, F. (2013). Choline-containing phospholipids: Relevance to brain functional pathways. *Clinical Chemistry and Laboratory Medicine.* https://doi.org/10.1515/cclm-2012-0559.

Traini, E., Bramanti, V., & Amenta, F. (2013). Choline alphoscerate (alpha-glyceryl-phosphoryl-choline) an old choline- containing phospholipid with a still interesting profile as cognition enhancing agent. *Current Alzheimer Research.* https://doi.org/10.2174/15672050113106660173.

Wurtman, R. J., Blussztajn, J. K., & Ulus, I. H. (1990). Advances in neurology. In R. J. Wurtman (Ed.), *Alzheimer disease* (p. 117). New York: Raven Press.

Further readings

Uney, J. B., Jones, G. M., Rebeiro, A., & Levy, R. (March 15, 1992). The effect of long-term high dose lecithin on erythrocyte choline transport in Alzheimer patients. *Biological Psychiatry, 31*(6), 630—633. PubMed PMID: 1581445.

Barbeau, A. (February 1978). Emerging treatments: Replacement therapy with choline or lecithin in neurological diseases. *The Canadian Journal of Neurological Sciences, 5*(1), 157—160. Review. PubMed PMID: 148319.

Berry, I. R., & Borkan, L. (1985). Biochemistry of Alzheimer's disease: A report. *Psychopharmacology Bulletin, 21*(2), 347—355. Review. PubMed PMID: 2860697.

Boller, F., & Forette, F. (1989). Alzheimer's disease and THA: A review of the cholinergic theory and of preliminary results. *Biomedicine & Pharmacotherapy, 43*(7), 487—491. Review. PubMed PMID: 2684292.

Crook, T., Petrie, W., Wells, C., & Massari, D. C. (1992). Effects of phosphatidylserine in Alzheimer's disease. *Psychopharmacology Bulletin, 28*(1), 61—66. PubMed PMID: 1609044.

Di Perri, R., Coppola, G., Ambrosio, L. A., Grasso, A., Puca, F. M., & Rizzo, M. (1991). A multicentre trial to evaluate the efficacy and tolerability of alpha-glycerylphosphorylcholine versus cytosine diphosphocholine in patients with vascular dementia. Jul—Aug *Journal of International Medical Research, 19*(4), 330—341. PubMed PMID: 1916007.

Engel, R. R., Satzger, W., Günther, W., Kathmann, N., Bove, D., Gerke, S., et al. (June 1992). Double-blind cross-over study of phosphatidylserine vs. placebo in patients with early dementia of the Alzheimer type. *European Neuropsychopharmacology, 2*(2), 149—155. PubMed PMID: 1633433.

Forssell, L. G., Hellström, A., Ericsson, K., & Winblad, B. (1989). Early stages of late onset Alzheimer's disease. Diagnostic criteria, protein metabolism, precursor loading effects, neurochemical and neuropsychological applications. *Acta Neurologica Scandinavica - Supplement, 121*, 1—95. PubMed PMID: 2565644.

Fünfgeld, E. W., Baggen, M., Nedwidek, P., Richstein, B., & Mistlberger, G. (1989). Double-blind study with phosphatidylserine (PS) in parkinsonian patients with senile dementia of Alzheimer's type (SDAT). *Progress in Clinical and Biological Research, 317*, 1235—1246. PubMed PMID: 2690093.

Gauthier, S., Bouchard, R., Bacher, Y., Bailey, P., Bergman, H., Carrier, L., et al. (November 1989). Progress report on the Canadian Multicentre Trial of tetrahydroaminoacridine with lecithin in Alzheimer's disease. *The Canadian Journal of Neurological Sciences, 16*(4 Suppl. l), 543—546. PubMed PMID: 2680009.

Gottfries, C. G. (October 1989). Pharmacological treatment strategies in dementia disorders. *Pharmacopsychiatry, 22*(Suppl. 2), 129—134. Review. PubMed PMID: 2690151.

Growdon, J. H. (October 1978). Effects of choline on tardive dyskinesia and other movement disorders. *Psychopharmacology Bulletin, 14*(4), 55—56. PubMed PMID: 151871.

Kanof, P. D., Mohs, R. C., Gross, J., Davidson, M., Bierer, L. M., & Davis, K. L. (1991). Platelet phospholipid synthesis in Alzheimer's disease. Jan—Feb *Neurobiology of Aging, 12*(1), 65—69. PubMed PMID: 2002885.

Miller, B. L., Read, S., Tang, C., & Jenden, D. (1991). Differences in red blood cell choline and lipid-bound choline between patients with Alzheimer disease and control subjects. Jan—Feb *Neurobiology of Aging, 12*(1), 61—64. PubMed PMID: 2002884.

Miller, B. L. (April 1991). A review of chemical issues in 1H NMR spectroscopy: N-acetyl-L-aspartate, creatine and choline. *NMR in Biomedicine, 4*(2), 47—52. Review. PubMed PMID: 1650241.

Nitsch, R. M., Blusztajn, J. K., Pittas, A. G., Slack, B. E., Growdon, J. H., & Wurtman, R. J. (March 1, 1992). Evidence for a membrane defect in Alzheimer disease brain. *Proceedings of the National Academy of Sciences of the United States of America, 89*(5), 1671—1675. PubMed PMID: 1311847; PubMed Central PMCID: PMC48514.

Perry, E. K., Perry, R. H., & Tomlinson, B. E. (July 30, 1977). Dietary lecithin supplements in dementia of Alzheimer type. *Lancet, 2*(8031), 242—243. PubMed PMID: 69845.

Peters, B. H., & Levin, H. S. (September 1979). Effects of physostigmine and lecithin on memory in Alzheimer disease. *Annals of Neurology, 6*(3), 219—221. PubMed PMID: 534419.

Schwartz, J. H. (1991). Chemical messengers: Small molecules and peptides. In E. R. Kandel, J. H. Schwartz, & T. M. Jessel (Eds.), *Principles of neural science* (3rd ed., pp. 213—224). New York: Elsevier.

Siow, B. L. (April 1985). Cerebral ageing, neurotransmitters and therapeutic implications. *Singapore Medical Journal, 26*(2), 151—153. PubMed PMID: 2863876.

Söderberg, M., Edlund, C., Kristensson, K., & Dallner, G. (June 1991). Fatty acid composition of brain phospholipids in aging and in Alzheimer's disease. *Lipids, 26*(6), 421—425. PubMed PMID: 1881238.

Tudorache, B., Lupulescu, R., Dutan, I., & Sârbulescu, A. (1990). Assessment of various psychopharmacological combinations in the treatment of presenile and senile primary degenerative dementia. Oct—Dec *Romanian Journal of Neurology and Psychiatry, 28*(4), 277—294. PubMed PMID: 2100154.

CHAPTER 31

Donepezil in the treatment of Alzheimer's disease

Gabriella Marucci[1], Michele Moruzzi[1], Francesco Amenta[2]

[1]School of Medicinal and Health Sciences Products, University of Camerino, Camerino, Italy; [2]Clinical Research, Telemedicine and Telepharmacy Centre, School of Medicinal and Health Products Sciences, University of Camerino, Camerino, Italy

List of abbreviations

ACh Acetylcholine
AChE Acetylcholinesterase
AD Alzheimer's disease
Aβ Aβ-amyloid peptide
BuChE Butyrylcholinesterase
ChE cholinesterase
ChE-Is (acetyl)-cholinesterase inhibitors
CI confidence interval
FDA US Food and Drug Administration
MD mean difference

Mini-dictionary of terms

AChE An enzyme involved in the hydrolysis of the neurotransmitter acetylcholine. It has anionic and esteratic sites.
Catalytic triad of AChE This is the active site of the enzyme (the so-called esteratic site) constituted of three amino acids: histidine 440, glutamate 327, and serine 200.
Hybrid compound The combination of two different and independently acting compounds or parts of compounds.
Hyperactive phosphorylation The introduction of phosphoryl groups in a molecule.
Pharmacophore An essential part of the molecule interacting with a specific biological target.
Structure—activity relationships Relationship between the molecule structure and its biological activity.

Introduction

The hypothesis that altered brain cholinergic transmission has a key role in the impairment of cognitive function in adult-onset dementia has stimulated research of possible therapeutic approaches for countering cholinergic neurotransmission deficits. This hypothesis was corroborated by the observation that administration of the muscarinic receptor antagonist scopolamine to healthy subjects induced a cognitive impairment resembling that found in adult-onset dementia (Amenta, Parnetti, Gallai, & Wallin,

Diagnosis and Management in Dementia
ISBN 978-0-12-815854-8, https://doi.org/10.1016/B978-0-12-815854-8.00031-8

2001). The demonstration of a decrease of the ACh biosynthetic enzyme choline acetyltransferase in the cerebral cortex and hippocampus in AD individuals (Amenta et al., 2001) contributed to the success of the cholinergic hypothesis of geriatric memory dysfunction. Although the cholinergic system is not the only neurotransmitter system affected in adult-onset cognitive impairment, changes in cholinergic function are implicated in the pathogenesis of learning and memory changes occurring in adult-onset dementia disorders (Amenta et al., 2001).

One approach to the symptomatic treatment of cognitive and behavioral symptoms of AD was based on the inhibition of ACh-degrading enzyme cholinesterases AChE and butyrylcholinesterase (BuChE) by a class of drugs known as (acetyl)–cholinesterase inhibitors (ChE-Is). This inhibition increases neurotransmitter levels, thus enhancing deficient brain cholinergic neurotransmission. ChE-Is are the first registered drugs with a specific indication for symptomatic treatment of AD (Table 31.1).

The development of these drugs started in the mid-1970s with the observation that the ChE-I physostigmine had positive effects on memory function in young and aged normal subjects. ChE-Is were shown to provide temporarily modest improvement in symptoms of AD and to stabilize or temporarily slow the decline of cognitive function and functional ability. ChE-Is are called *nonspecific*, or simply ChE-Is, when they inhibit AChE, BuChE, and other cholinesterases; they are called *specific* when they inhibit AChE only. These drugs also can be classified as reversible, pseudoirreversible, or irreversible based on the degree of enzyme inhibition (Giacobini, 1998).

Tacrine (tetrahydroaminoacridine) was the first agent approved by the FDA, in 1993, for the treatment of symptoms of AD. It is a reversible, nonspecific ChE-I featuring variable absorption, extensive distribution and central nervous system penetration. One key drawback of tacrine is its hepatotoxicity. In spite of the potential interest of tacrine, its efficacy for symptoms of dementia remains controversial (Amenta et al., 2001).

Donepezil is a piperidine-based, second-generation reversible inhibitor of AChE, displaying negligible effect on nonspecific ChEs. It is the second FDA-approved AChE inhibitor for the symptomatic treatment of AD (Rogers, Mohs, & Dody, 1998). The pharmacokinetics of donepezil are linear in the range of therapeutic dosage. Side effects of donepezil are remarkably lower than with those of tacrine. Donepezil has a relatively slow clearance and a long elimination half-life, which allows once-daily dosing. The other two ChE-Is approved by the FDA and European Medicines Agency and still in use are rivastigmine, a pseudoirreversible carbamate-selective AChE-I and BuChE inhibitor, and galantamine, which is both an AChE-I and an allosteric modulator of nicotinic receptors to increase ACh release. The latter was introduced in Sweden in 2000 (Amenta et al., 2001). Table 31.2 lists the names and indications of compounds proposed to support treatment of adult-onset dementia.

This chapter summarizes the main aspects of the medicinal chemistry, preclinical studies, and therapeutic use of donepezil.

Table 31.1 Licensed drugs for treating Alzheimer's disease.

Name	Indication	Countries of license	Pharmaceutical form	Recommended dosage
Galantamine hydrobromide	Alzheimer's disease—dementia (mild to moderate)	34	Capsule, extended-release/solution/tablet	*Oral solution and Tablets:* 4 mg twice day After 4 weeks increase at 8 mg twice a day After 4 weeks at 8 mg increase at 12 mg twice a day *Extended Release:* 16–24 mg/day
Donepezil hydrochloride	Alzheimer's disease—dementia (mild to moderate)	38	Tablet/tablet, disintegrating	*Mild to moderate:* 10 mg/day *Moderate to severe:* 10 or 23 mg/day
Rivastigmine tartrate	Alzheimer's disease—dementia (mild to severe)	33	Oral capsule/oral solution and transdermal patch	*Transdermal patch:* 4.6 mg/24 h After 4 weeks increase to 9.5 mg/24 h To maintain increase to 13.3/24 h Oral route: 1.5 mg orally twice daily for 4 weeks Increase of 1.5 mg twice daily every 4 weeks up to a maximum of 6 mgtwice daily
Memantine hydrochloride	Dementia	39	Capsule, extended release, solution/tablet	*Immediate Release:* Week 1: 5 mg orally once a day Week 2: 10 mg orally/day Week 3: 15 mg orally/day Week 4: 20 mg orally/day *Extended Release:* Dose: 7 mg orally once a day Maintenance and maximum dose: 28 mg orally once a day
Memantine hydrochloride/donepezil hydrochloride	Alzheimer's disease—dementia (moderate to severe)	2	Capsule, delayed-release	Memantine hydrochloride 28 mg/donepezil 10 mg orally once daily in the evening

Table 31.2 Drugs with evidence favoring efficacy in treating some symptoms of Alzheimer's disease.

Name	Indication	Countries of license	Pharmaceutical form	Recommended dosage
Carbamazepine	Agitation–dementia	30	Chewable tablet, immediate-release tablet, extended-release capsule, extended-release tablet, suspension.	Begin with 200 mg orally twice daily. May adjust in increments of 200 mg/day to achieve optimum clinical response.
Ergoloid mesylates	Dementia, symptomatic treatment of age-related dementia	4	Capsule, tablet, and liquid	Capsule and tablet usual dose 1 mg three times daily. The liquid is dosed as 1 mL (1 mg) three times daily.
Isoxsuprine hydrochloride	Dementia	18	Oral route	The recommended dose is 10–20 mg three or four times daily.
Olanzapine	Agitation, acute dementia	39	Oral route	The initial dose is 10 or 15 mg once daily, and doses may be increased at intervals of not less than 24 h, by 5 mg daily.

Medicinal chemistry of donepezil

To overcome the disadvantages of physostigmine (short-acting action) and tacrine (hepatotoxicity), a new AChE-I was proposed that is structurally different from the above-mentioned compounds. A derivative of indanone, also known as E2020, was developed by Sugimoto and coworkers at Eisai Research Laboratory in Japan starting in 1983. This molecule represents an effective AChE-I and is extensively used for the treatment of AD. The patent expired in November 2010, and the compound is now produced as a generic formulation and marketed by many companies worldwide.

The first synthesis of the compound was published in 1995, and only later the code E2020 was changed to donepezil (Fig. 31.1) (Sugimoto, Iimura, Yamanishi, & Yamatsu, 1995).

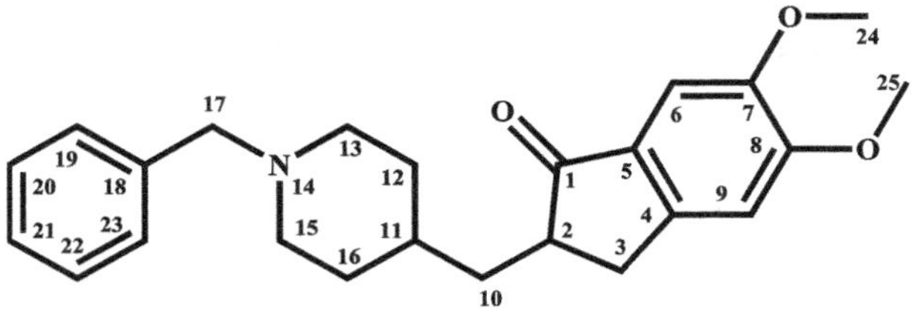

Figure 31.1 Donepezil structure.

The Sugimoto paper reported the synthesis of a series of indanone derivatives substituted with various groups. Among them, the 2-((1-benzylpiperazine-4-yr) methyl)-5,6-dimethoxy-2,3-dihydro-1H-in-1-1 (donepezil) exhibited a high AChE inhibitory activity in vitro, with an $IC_{50} = 5.7$ nM, the highest of the whole series. The compound displayed a selective affinity 1250 times greater for AChE than for BuChE. The synthesis was performed by the aldol condensation/dehydration between the indanone (I) and the piperidine derivative (II), leading to an unsaturated precursor (III) that was hydrogenated to yield the desired compound, donepezil.

The overall process yield was 27.4%, and large-scale production would have faced obstacles such as subzero temperature ($-78°C$) and the need for hazardous chemicals during synthesis. The same authors continued their synthetic work to identify other AChE-Is, but none of the newly synthesized compounds was able to outstrip the activity of donepezil (Sugimoto, Yamanishi, Iimura, & Kawakami, 2000).

Generally, pharmaceutical research in industry aims to develop products of the highest quality at reasonable cost. Efforts were therefore made to obtain a synthetic pathway for donepezil that was economically affordable, had high yields, and could easily be transferred to a large-scale industrial context (Devries, 1997; Gutman, Shkolnik, Tishin, Nisnevich, & Zaltzman, 2000; Yoichi, 2001). New synthetic processes were developed but still involved problematic operations in a larger-scale context or required the use of expensive chemical precursors (Stephen, 1997). A cheaper synthetic process for donepezil was not published until 2006 (Chandrashekar et al., 2006). This approach, using commercially available and affordable materials, showed an overall yield of 56.0%. This procedure avoided using a platinum catalyzer and relied instead on a less expensive one based on palladium carbon (Pd/C). Niphade et al. (2008) developed a new "improved, cost-effective and scalable process for donepezil hydrochloride" via condensation of the 5,6-dimethoxy-1-indanone with 1-benzylpiperidine-4-carboxaldehyde using sodium hydroxide, in the presence of tetrabutyl ammonium bromide in a mixed-solvent water/toluene, to give 1-benzyl-4-(5,6-dimethoxy-1-oxidant-2-ylidenemethyl) piperidine. The advantages of this synthesis were the removal of hazardous chemicals like n-butyl lithium, cryogenic temperatures, and purification via chromatography columns.

Another group achieved significant cost reductions in the synthesis of donepezil using very small quantities of solvent and avoiding hazardous reagents. This required the synthesis of a new soluble intermediate, the 6-dimethoxy-3-pyridine-4-ylspiro(indene-2,2-oxirane)-1(3H)-1, using only three isolation and drying steps with a high overall yield of 60% (Dubey, Kharbanda, Dubey, & Mathela, 2010). To obtain a more effective and safer drug with improved pharmacokinetics, pharmacodynamics, and toxicity profiles, other groups synthesized AChE-Is structurally related to donepezil. These attempts were not successful, as all synthesized compounds had AChE-I activity that was lower or at least similar to donepezil with comparable pharmacokinetics, pharmacodynamics, and toxicity profiles (Gabr & Abdel-Raziq, 2018b; Ismail, Kamel, Mohamed, & Faggal, 2012; Li et al., 2016; Mishra et al., 2017).

These studies allow considerations of structure—activity relationships. In particular, the two methoxyl groups on the indanone structure can be replaced by two-acetamide groups or other electron-attractor groups without loss of activity. Replacement of the methylene spacer between the indanone structure and the piperidine moiety by an ethyl or propyl chain leads to molecules with good anti-AChE activity. Compounds with a benzisoxazole moiety instead of the indanone molecule were found to also act as AChE-Is. Similar results were obtained by replacing the benzyl group with a naphthol. The presence of the piperidine ring seems to be essential, since its replacement with various heterocycles caused a loss of activity (Ismail et al., 2012; Li et al., 2016; Mishra et al., 2017; Gabr & Abdel-Raziq, 2018a).

AD is a multifactorial disease, and impaired cholinergic neurotransmission is a consequence rather than a cause of other phenomena; it is therefore difficult to tackle the progression of the disease by administering AChE-Is only. Medicinal chemistry research has slowly shifted from identifying selective AChE-Is toward BuChE-Is or nonselective ChE-Is. This was suggested by the observation of high BuChE levels in cortical regions of patients with severe AD (Perry, Perry, Blessed, & Tomlinson, 1978). In recent years, new theories have been proposed concerning the onset and progression of AD based on Aβ-amyloid peptide (Aβ) and tau. These include the hypotheses that neurotoxic agents, oxidative stress, iron overload, and cholesterol levels in neuronal rafts trigger anomalous signaling cascades that promote tau hyperphosphorylation.

If several actors are involved, it would be impossible for any ChE-I developed in the future to be a sufficient therapeutic agent against AD. Research should instead develop hybrid compounds to treat AD by targeting cholinergic neurotransmission, such as AChE-Is, but also to inhibit Aβ formation and deposition and decrease oxidative stress. Shen et al. (2008) obtained a new series of 2-phenoxy-indan-1-1 derivatives with an alkylamine side chain. These compounds had AChE and BuChE as targets but also displayed antioxidant activity. Among them, AP2238 was demonstrated to interact with AChE and inhibit the proaggregation of AChE on Aβ (Piazzi et al., 2003, Fig. 31.3). A series of hybrid compounds bearing fragments of donepezil and AP2238 were subsequently designed and synthesized (Rizzo et al., 2010) (Fig. 31.2). Some displayed good anti-AChE activity and inhibited Aβ aggregation similarly to the effects of donepezil.

Other enzymes playing an important role in the progression of AD are monoamine oxidases (MAOs) that catalyze the oxidative deamination of monoamines. This reaction produces hydrogen peroxide implicated in the generation of toxic species based on oxygenated radicals. A series of hybrid molecules was developed, derived from donepezil, propargylamine, and 8-hydroxyquinoline, and named DPH (Wang et al., 2014) (Fig. 31.3).

The most promising compound of this series was racemic α-aminonitrile-4-(1-benzylpiperazine-4-yl)-2-(((8-hydroxyquinoline-5-yl)methyl) (prop-2-yn-1-yl)amino) butane-nitrile (DPH6), which irreversibly blocked MAOs A/B and resulted in an

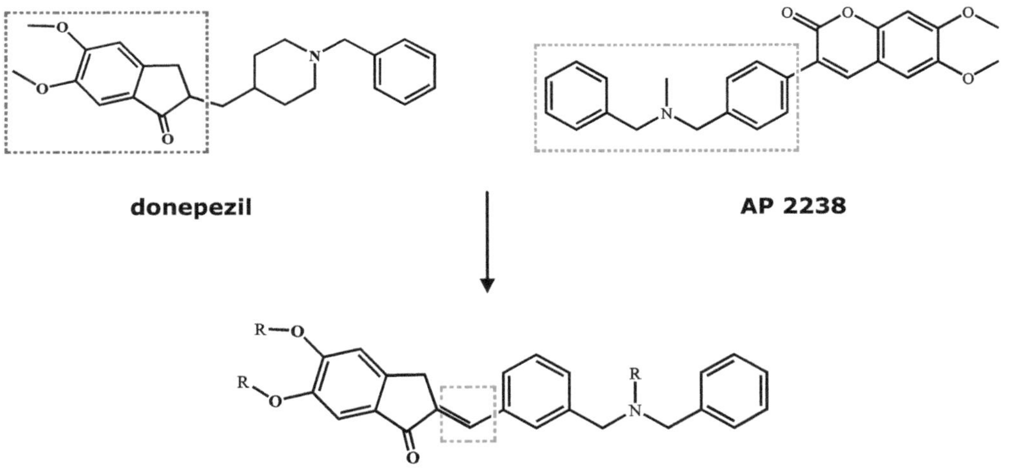

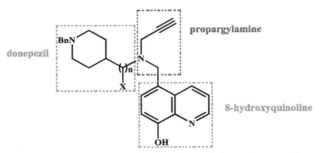

Figure 31.2 General structure of hybrid compounds bearing pharmacophoric fragments of donepezil and AP2238.

Figure 31.3 General structure of hybrid molecules derived from donepezil, propargylamine, and 8-hydroxyquinoline.

AChE–I interacting with complex biometals Cu(II), Zn(II), and Fe(III). This compound is less toxic than donepezil and improves cognitive function. Following the same approach, Wu et al. (2017) produced a number of hybrid molecules, of which some inhibited AChE and BuChE in nanomolar concentration and MAO-A in the micromolar range.

The donepezil–huprine hybrid AVCRI104P4 synthesized by Sola et al. (2015) is a potential drug candidate for AD (Fig. 31.4). It advantageously displays inhibitory activities against AChE (low nanomolar range), BuChE, Aβ aggregation, and β–secretase BACE-1 (submicromolar or low micromolar range).

The combination of donepezil with clioquinol, which chelates redox–active metals, resulted in a series of molecules with multitargeting activities, selectively inhibiting BuChE at micromolar concentrations and preventing Aβ self-aggregation. Some of these molecules chelate copper (II) and zinc (II) (Prati et al., 2016). This approach resulted in a series of new hydroxypyridinone-benzofuran hybrids (Hiremathad et al., 2017) designed by conjugating the donepezil benzylpiperidine part with the indanone moiety. The last one is a metal chelator, radical scavenger, and inhibitor of Aβ aggregation.

Unfortunately, the results obtained were disappointing, as all newly synthesized molecules were less active than the parent compound in spite of having higher Aβ aggregation inhibitory activity, metal chelating capacity, and radical scavenging activity. Dias Viegas et al. (2018) developed a new series of N-benzyl-piperidine-aryl-acylhydrazones hybrid derivatives (Fig. 31.5). The hybrid was synthesized by combining the N-benzyl-piperidine subunit of donepezil with a hydroxy-piperidine fragment of the AChE-I LASSBio-767 using an acylhydrazone linker. Only one compound of this series exhibited AChE inhibitory activity like donepezil as well as antiinflammatory activity countering Aβ formation.

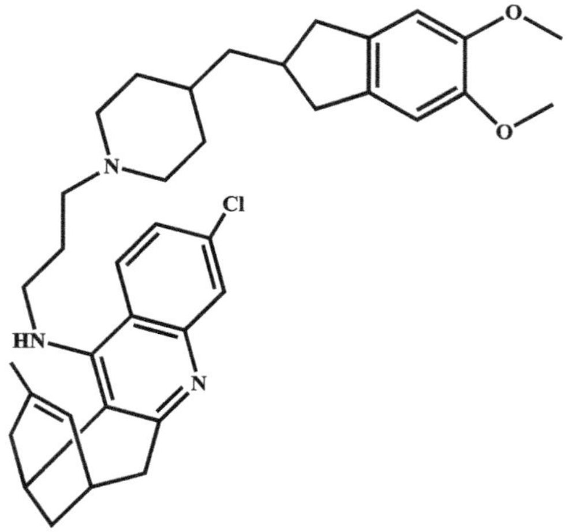

Figure 31.4 AVCRI104P4 structure.

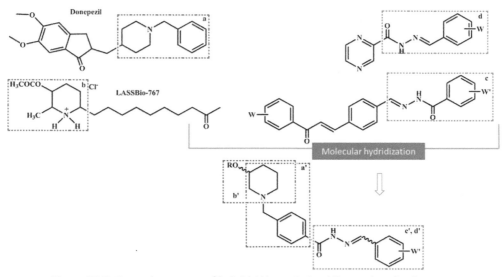

Figure 31.5 General structure of hybrid *N*-benzyl-piperidine-aryl-acylhydrazones.

Two other hybrids were recently synthesized containing the indanone-piperidine moiety of donepezil as well as alpha-lipoic acid (Terra et al., 2018). Alpha-lipoic acid is an antioxidant agent with neuroprotective activity (Amenta et al., 2018; Jacobson & Sabbagh, 2008). One of these hybrids displayed moderate inhibitory AChE activity and a more pronounced inhibitory activity on BuChE. Interestingly, this hybrid also had good antioxidant properties, more pronounced than those of alpha-lipoic acid itself (Terra et al., 2018). The *N*-benzyl-piperidine-aryl-acylhydrazone hybrid was considered a potential anti-AD agent.

In conclusion, much effort has been spent to improve the activity, the pharmacokinetics, pharmacodynamics, and toxicity profiles of donepezil. So far, donepezil remains the best among cognate compounds developed and tested, and the only one of this class that is in clinical use.

Preclinical studies of donepezil

For over 2 decades, preclinical studies have analyzed the pharmacological profile of ChE-Is including donepezil (Jacobson & Sabbagh, 2008). These investigations have observed that the properties of donepezil include a role in processing amyloid precursor protein and Aβ neurotoxicity, mitigating the effects of cholinergic deafferentiation and oxidative stress, influencing AChE isoform expression, upregulating cerebrocortical nicotinic receptors, and inhibiting excitotoxic injury. Moreover, donepezil modulates different neurotransmitter systems by interacting with alpha-1 adrenergic receptors, improves cerebral blood flow and activity-flow coupling, enhances neuronal plasticity, and reduces levels of proinflammatory cytokines (Jacobson & Sabbagh, 2008).

In terms of age-related memory impairment, experiments with aged male Fisher 344 rats (Yuede, Dong, & Csernanskya, 2007) showed that water-maze acquisition and retention was improved with a daily dose of donepezil (0.75, but not 0.375 mg/kg) beginning 4 days before testing and continuing for 15 days, compared with saline-treated age-matched controls (Hernandez et al., 2006). A significant improvement in water-maze performance with doses of 0.25 and 0.5 mg/kg was also reported (Abe et al., 2003). Subcutaneous administration of a comparable dose of donepezil (0.695 mg/kg/day) for 3 weeks before testing for the subsequent 2 weeks did not improve radial-arm-maze performance (Barnes et al., 2000).

Lesions of different brain areas using several approaches were carried out to assess the effects of donepezil on deficits in hippocampal-dependent memory tasks. The low dose of 0.1 mg/kg was ineffective (Xu, Chen, Yanai, Huang, & Wei, 2002). Male Wistar rats exposed to cerebral ischemia and intracerebroventricular Aβ infusion showed improvement in a radial-arm-maze task with 3.0 mg/kg donepezil given acutely before testing (Iwasaki et al., 2006). In the model of ibotenic acid lesions of the entorhinal cortex, donepezil was effective in improving acquisition in a water-maze task at doses of 0.3 and 3.0 mg/kg, with better performance compared with that of untreated lesioned animals, although it did not reach the levels observed in sham controls. No effects were seen on retention in this model (Spowart-Manning & van der Staay, 2005).

Collectively the above studies on the effects of donepezil on hippocampal-dependent memory tests showed positive results, suggesting that the dose and paradigm used to assess memory impairment are important factors in showing the "efficacy" of the drug (Yuede et al., 2007).

Donepezil and neuroprotection

Preclinical studies on models of AD suggest a neuroprotective activity of donepezil independent of ChE inhibition (Kim et al., 2017). It was suggested that this neuroprotection might depend on the attenuation of Aβ-induced toxicity via α7nAChRs and the PI3K-Akt pathway.

A study performed on rat cortical neurons exposed to oxygen-glucose deprivation, using lactate dehydrogenase release as a marker of neuronal damage, showed that donepezil has a protective effect against injury induced by oxygen-glucose deprivation. It was hypothesized that this compound may protect from progressive degeneration of cortical neuronal cells in AD (Zhou, Fu, & Tang, 2001). Moreover, donepezil protects from glutamate-induced neurotoxicity through a direct or indirect stimulation of nicotinic cholinergic receptors before the activation of nitric oxide and MAP kinase (Takada-Takatori et al., 2008). Other studies have confirmed the antiinflammatory effect of donepezil in suppressing IL-1β and cyclooxygenase-2 expression in the brain and spleen, suggesting that donepezil directly prevents systemic inflammation (Kim et al., 2017). Antiamnesic and neuroprotective effects against Aβ-induced toxicity were also demonstrated (Meunier, Ieni, & Maurice, 2006).

In conclusion, a neuroprotective and antiinflammatory activity of donepezil was shown in models relevant to AD (Aβ-induced cell death, oxygen-glucose deprivation, glutamate-induced cell death). These properties may be behind the suggested role of the compound against progressive degeneration of brain neurons.

Clinical studies on donepezil

Donepezil was introduced in the US market in 1996 and was subsequently licensed in most Western countries. The US FDA approved it for use in mild, moderate, and severe AD. In most countries it is licensed for the symptomatic treatment of mild to moderate forms of AD. In terms of commercially available pharmaceutical formulations, donepezil was first sold in tablet form. Liquid and transdermal formulation were subsequently developed. The recommended oral dose is 5 mg once a day, increasing to 10 mg once a day after at least 1 month of treatment. In 2010, the FDA approved a 23 mg once-a-day tablet of donepezil for symptomatic treatment of severe forms of AD.

Numerous clinical trials were conducted on the activity and effectiveness of donepezil in AD. This topic was reviewed in a recent Cochrane meta-analysis that examined 30 studies involving 8257 participants (Birks & Harvey, 2018). Twenty-eight studies reported results allowing a meta-analysis. The majority of trials lasted about 6 months or less, while one small trial continued for 52 weeks using donepezil tablets at a dosage of 5 or 10 mg. Two studies out of 28 tested a slow-release oral formulation at a dose of 23 mg/day. After 26 weeks of treatment, compared with placebo, donepezil gave rise to better outcomes of cognitive function when measured by the Alzheimer's disease assessment scale-cognitive (ADAS-Cog) scale (score range 0−70, mean difference, or MD: −2.67, 95%, confidence interval, or CI: −3.31 to −2.02, 1130 participants, five studies), by mini-mental state examination (score range 0−30, MD: 1.05, 95% CI: 0.73 to 1.37, 1757 participants, seven studies) and by severe impairment battery (score range 0−100, MD: 5.92, 95% CI: 4.53 to 7.31, 1348 participants, five studies). Results were summarized for MDs for continuous outcomes, with 95% CIs, or in descriptive format when meta-analysis was not possible. Donepezil was also associated with a better function, measured with the Alzheimer's disease cooperative study activities of daily living score for severe Alzheimer's disease (MD: 1.03, 95%, CI: 0.21 to 1.85, 733 participants, three studies). A higher proportion of participants treated with donepezil experienced improvement on the clinician-rated global impression of change scale (odds ratio: 1.92, 95%, CI: 1.54 to 2.39, 1674 participants, six studies). No differences between donepezil and placebo were reported for behavioral symptoms measured by the neuropsychiatric inventory (MD: −1.62, 95%, CI: −3.43 to 0.19, 1035 participants, four studies) or by the behavioral pathology in Alzheimer's disease scale (MD: 0.4, 95% CI −1.28 to 2.08, 194 participants, one study), or by quality of life (MD: −2.79, 95% CI −8.15 to 2.56, 815 participants, two studies).

After 6 months of treatment, donepezil shows modest but significant benefits. The side effects reported are mainly mild, but they can induce people to stop treatment. Stabilization of cognitive performance or ability to maintain activities of daily living are important clinical end points. Yet clinical studies did not demonstrate any effect of donepezil on quality of life. The usefulness of the above clinical trials is limited, however, by a short observation time compared with the longer-term evolution of AD.

In summary, donepezil has shown some activity in the treatment of cognitive symptoms in mild to moderate stages of AD, which corresponds to its regulatory indication. More data are still required from longer-term clinical studies examining measures of disease progression or the amount of time before needing full-time care (Birks & Harvey, 2018).

In 2010 the FDA approved a higher-dose (23 mg/daily) donepezil formulation for the treatment of patients suffering from moderate-to-severe AD. This authorization was granted on the strength of positive results from a phase three clinical trial that compared switching to donepezil 23 mg/day against continuing treatment with a daily dose of 10 mg/day of the same drug (Adlimoghaddam, Neuendorff, Roy, & Albensi, 2018). Clinical results with increasing doses of donepezil (5, 10, and 23 mg/day) suggest that a higher dosage of donepezil improves cognition and preserves function in individuals with severe AD. Adverse effects occurred more frequently as the dose of donepezil increased, leading some patients to discontinue treatment (Adlimoghaddam et al., 2018). These results do not allow the drawing of definitive conclusions about the practical usefulness of quite high doses of the compound during the advanced stages of AD.

Other off-label uses of donepezil in the treatment of adult-onset dementia (Kumar & Sharma, 2018) include

- *Lewy body dementia*: some studies have reported benefits of donepezil for the treatment of cognitive and behavioral symptoms in Lewy body dementia.
- *Traumatic brain injury*: a few studies have suggested improvement in memory dysfunction in patients with traumatic brain injury treated with donepezil.
- *Dementia associated with Parkinson's disease*: some evidence suggests that donepezil can improve cognition, executive function, and global status in Parkinson's disease dementia.

Suggestions of off-label treatment with donepezil have focused, however, on vascular dementia (VaD). VaD is a common type of dementia caused by reduced blood flow to the brain. Donepezil has been tested in two 24-week, double-blind, randomized, placebo-controlled trials (n = 1219) (Wilkinson et al., 2010; Román et al., 2010). In the former study (n = 603; mean age 73.9 years), patients receiving donepezil 5 mg or 10 mg/daily displayed significant improvements in cognition compared with placebo on the ADAS-Cog cognitive subscale ($P = .001$ for donepezil 5 mg and $P < .001$ for donepezil 10 mg) (Wilkinson et al., 2010). In a second study (n = 616; mean age 75 years), both donepezil groups demonstrated cognitive improvements on the ADAS-Cog cognitive subscale compared with placebo (donepezil 5 mg, $P = .003$; donepezil

10 mg, $P = .0002$) (Román et al., 2010). In a multicenter, open-label, 30-week extension study (n = 885) of the two above trials, cognitive improvements were observed in the double-blind phase (Román et al., 2010). These improvements were sustained and remained above baseline for the whole follow-up period. In the open-label extension phase, patients who had completed the initial double-blind studies received donepezil 5 mg/day for the initial 6 weeks, then 10 mg/day for a total of 30 weeks regardless of the medication received during the double-blind studies. A mean reduction of between 0.6 and 1.15 points in the ADAS-Cog subscale (primary study outcome) from baseline score (double-blind study entry) to week 54 was recorded for patients receiving donepezil at either dose during the entire 54 week period. For patients receiving placebo in the double-blind studies, no significant improvement in ADAS-Cog scores was observed, and scores remained steady and near baseline values throughout the 30-week extension study. Overall, the ADAS-Cog results indicated a difference in the level of cognitive function between early and delayed initiation of donepezil, although the reasons for this finding remain unclear (Wilkinson et al., 2003). Donepezil may improve cognition in patients with VaD but not overall global functioning (Kumar & Sharma, 2018).

Conclusions

Donepezil is the second ChE-I introduced into the market. In spite of notable efforts by medicinal chemists, no new molecule has been developed showing better activity than that of the starting compound. Donepezil was developed to increase brain ACh levels by reversibly inhibiting the neurotransmitter catabolic enzyme AChE. Growing evidence indicates that donepezil does not act solely at the neurotransmitter level but displays a range of other effects. This drug affects several cellular and molecular processes involved in the pathophysiology of AD, and this eclectic profile is probably the basis of its activity in ameliorating cognitive symptoms, and to a lesser extent behavioral problems occurring in AD. Early and long-lasting treatment with donepezil is considered to preserve cognitive function more effectively than delayed treatment. Yet, the clinical benefits of donepezil administration are regarded as small and may not be clinically meaningful.

Key facts of Alzheimer's treatment

- The acetylcholinesterase (AChE)-I tacrine (Cognex) was the first drug approved by the US Food and Drug Administration (FDA) in 1993 for symptomatic treatment of mild and moderate stages of Alzheimer's disease (AD).
- Donepezil (Aricept) was approved by the FDA in 1996 and is currently used in mild and moderate stages of AD.
- Memantine is an NMDA receptor antagonist authorized in 2000 for the treatment of moderate-severe forms of AD.

Summary points

- The cholinergic hypothesis of geriatric memory dysfunction was proposed in 1980. The hypothesis assumes that cognitive deficits occurring in AD are due to degeneration of cholinergic neurons of the basal forebrain. This loss impairs brain cholinergic neurotransmission.
- Based on the reduced cholinergic tone in cognitive dysfunction, research was directed to identify therapeutic strategies to counter impaired brain cholinergic neurotransmission. One of these strategies was the introduction of compounds slowing the degradation of the neurotransmitter acetylcholine by inhibiting the acetylcholine (ACh)-degrading enzymes AChE/cholinesterase (ChE).
- Donepezil effectively inhibits AChE and was extensively used for the treatment of mild and moderate stages of AD. Several parent molecules of donepezil were developed while attempting to improve the properties of the native drug, but with no success.
- AD is a multifactorial disease in which the impaired cholinergic neurotransmission represents the consequence but not the cause of the observed cognitive dysfunction. Besides increasing acetylcholine levels by inhibiting AChE, donepezil protects against neurotoxic products involved in AD pathogenesis, upregulates nicotinic receptors, and may counter the progression of hippocampal atrophy.

Acknowledgments

The support of MDM SpA (Monza, Italy) to the completion of this chapter is gratefully acknowledged.

References

Abe, Y., Aoyagi, A., Hara, T., Abe, K., Yamazaki, R., Kumagae, Y., et al. (2003). Pharmacological characterization of RS-1259, an orally active dual inhibitor of acetylcholinesterase and serotonin transporter, in rodents: Possible treatment of Alzheimer's disease. *Journal of Pharmacological Sciences, 93*, 95—105.

Adlimoghaddam, A., Neuendorff, M., Roy, B., & Albensi, B. C. (2018). A review of clinical treatment considerations of donepezil in severe Alzheimer's disease. *CNS Neuroscience and Therapeutics*, 1—13.

Amenta, F., Buccioni, M., Dal Ben, D., Lambertucci, C., Martì Navia, A., Ngouadjeu Ngnintedem, M. A., et al. (2018). Ex-vivo absorption study of lysine R-lipoate salt, a new pharmaceutical form of R-ALA. *European Journal of Pharmaceutical Sciences, 15*(118), 200—207.

Amenta, F., Parnetti, L., Gallai, V., & Wallin, A. (2001). Treatment of cognitive dysfunction associated with Alzheimer's disease with cholinergic precursors. Ineffective treatments or inappropriate approaches? *Mechanism of Ageing and Development, 122*, 2025—2040.

Barnes, C. A., Meltzer, J., Houston, F., Orr, G., McGann, K., & Wenk, G. L. (2000). Chronic treatment of old rats with donepezil or galantamine: Effects on memory, hippocampal plasticity and nicotinic receptors. *Neuroscience, 99*, 17—23.

Birks, J. S., & Harvey, R. J. (2018). Donepezil for dementia due to Alzheimer's disease. *Cochrane Database of Systematic Reviews, 6.* https://doi.org/10.1002/14651858.CD001190.pub3.

Chandrashekar, R. E., Naveenkumar, K., Subrahmanyeswara, R. C., Pravinchandra, J. V., Venkataraman, S., Himabindu, V., et al. (2006). New synthesis of donepezil through palladium-catalyzed hydrogenation approach. *Synthetic Communications, 36*(2), 169—174.

Devries, K. M. (1997). *Processes and intermediates for preparing 1-benzyl-4-((5,6-dimethoxy-1-indanon)-2-yl) methylpiperidine*. WO patent 22584.

Dias Viegas, F. P., de Freitas Silva, M., Divino da Rocha, M., Castelli, M. R., Riquiel, M. M., Machado, R. P., et al. (2018). Design, synthesis and pharmacological evaluation of N-benzyl-piperidinyl-aryl-acylhydrazone derivatives as donepezil hybrids:Discovery of novel multi-target anti-alzheimer prototype drug candidates European. *Journal of Medicinal Chemistry, 10*(147), 48—65.

Dubey, S. K., Kharbanda, M., Dubey, S. K., & Mathela, C. S. (2010). New commercially viable synthetic route for DonepezilHydrochloride:anti-Alzheimer's drug. *Chemical and Pharmaceutical Bulletin, 58*, 1157—1160.

Gabr, M. T., & Abdel-Raziq, M. S. (2018a). Design and synthesis of donepezil analogues as dual AChE and BACE-1 inhibitors. *Bioorganic Chemistry, 80*, 245—252.

Gabr, M. T., & Abdel-Raziq, M. S. (2018b). Structure-based design, synthesis, and evaluation of structurally rigid donepezil analogues as dual AChE and BACE-1 inhibitors. *Bioorganic and Medicinal Chemistry Letters, 28*(17), 2910—2913. https://doi.org/10.1016/j.bmcl.2018.07.019.

Giacobini, E. (1998). Cholinesterase inhibitors for Alzheimer's disease: From tacrine to future applications. *Neurochemistry International, 32*, 413—419.

Gutman, L. A., Shkolnik, E., Tishin, B., Nisnevich, G., & Zaltzman, I. (2000). *Process and intermediates for production of donepezil and related compounds*. WO patent 09483.

Hernandez, C. M., Gearhart, D. A., Parikh, V., Hohnadel, E. J., Davis, L. W., Middlemore, M. L., et al. (2006). Comparison of Galantamine and Donepezil for effects on nerve growth factor, cholinergic markers, and memory performance in aged rats. *Journal of Pharmacology and Experimental Therapeutics, 316*, 679—694.

Hiremathad, A., Chand, K., Tolayan, L., Rajeshwari-Keri, R. S., Esteves, A. R., Cardoso, S. M., et al. (2017). Hydroxypyridinone-benzofuran hybrids with potential protective roles for Alzheimer's disease therapy. *Journal of Inorganic Biochemistry, 179*, 82—96.

Ismail, M. M., Kamel, M. M., Mohamed, L. W., & Faggal, S. I. (2012). Synthesis of new indole derivatives structurally related to donepezil and their biological evaluation as acetylcholinesterase inhibitors. *Molecules, 25*(17), 4811—4823.

Iwasaki, K., Egashira, N., Hatip-Al-Khatib, I., Akiyoshi, Y., Arai, T., Takagaki, Y., et al. (2006). Cerebral ischemia combined with β-amyloid impairs spatial memory in the eight-arm radial maze task in rats. *Brain Research, 1097*, 216—233.

Jacobson, S. A., & Sabbagh, M. N. (2008). Donepezil: Potential neuroprotective and disease-modifying effects. *Expert Opinion on Drug Metabolism and Toxicology, 4*, 1363—1369.

Kim, S. H., Kandiah, N., Hsu, J. L., Suthisisang, C., Udommongkol, C., & Dash, A. (2017). Beyond symptomatic effects: Potential of donepezil as a neuroprotective agent and disease modifier in Alzheimer's disease. *British Journal of Pharmacology, 174*, 4224—4232.

Kumar, A., & Sharma, S. (2018). *Donepezil*. StatPearls Publishing.

Li, F., Wang, Z. M., Wu, J. J., Wang, J., Xie, S. S., Lan, J. S., et al. (2016). Synthesis and pharmacological evaluation of donepezil-based-agents as new cholinesterase/monoamine oxidase inhibitors for the potential application against Alzheimer's disease. *Journal of Enzyme Inhibition and Medicinal Chemistry, 31*(S3), 41—53.

Meunier, J., Ieni, J., & Maurice, T. (2006). The anti-amnesic and neuroprotective effects of donepezil against amyloid β25-35 peptide-induced toxicity in mice involve an interaction with the σ1receptor. *British Journal of Pharmacology, 149*, 998—1012.

Mishra, C. B., Kumari, S., Manral, A., Prakash, A., Saini, V., Lynn, A. M., et al. (2017). Design, synthesis, in-silico and biological evaluation of novel donepezil derivatives as multi-target-directed ligands for the treatment of Alzheimer's disease. *European Journal of Medicinal Chemistry, 5*(125), 736—750.

Niphade, N., Mali, A., Jagtap, K., Ojha, R. C., Vankawala, P. J., & Mathad, V. T. (2008). An improved and efficient process for the production of donepezil hydrochloride: Substitution of sodium hydroxide for n-butyl lithium via phase transfer catalysis. *Organic Process Research and Development, 12*, 731—735.

Perry, E. K., Perry, R. H., Blessed, G., & Tomlinson, B. E. (1978). Changes in brain cholinesterases in senile dementia of Alzheimer type. *Neuropathology and Applied Neurobiology, 4*, 273—277.

Piazzi, L., Rampa, A., Bisi, A., Gobbi, S., Belluti, F., Cavalli, A., et al. (2003). 3-(4-{[Benzyl(Methyl) Amino]-Methyl}Phenyl)-6,7-Dimethoxy-2H-2-Chromenone(AP2238) inhibits both acetylcholinesterase and acetylcholinesterase-induced beta-amyloid aggregation a dual function lead for Alzheimer's disease therapy. *Journal of Medicinal Chemistry, 46*, 2279—2282.

Prati, F., Bergamini, C., Fato, R., Soukup, O., Korabecny, J., Andrisano, V., et al. (2016). Novel 8-hydroxyquinoline derivatives as multitarget compounds for the treatment of Alzheimer's disease. *ChemMedChem, 20*(11), 1284−1295.

Rizzo, S., Bartolini, M., Ceccarini, L., Piazzi, L., Gobb, i S., Cavalli, A., et al. (2010). Targeting Alzheimer's disease:Novel indanone hybrids bearing a pharmacophoric fragment of AP2238. *Bioorganic and Medicinal Chemistry, 1*(18), 1749−1760.

Rogers, S. L., Mohs, R., & Dody, R. S. (1998). Aricept:a well-tolerated and clinically effective treatment for the symptoms of AD. Results from world-wide clinical trials. *Alzheimer's Research, 1*(Suppl. 1), S13−S14.

Román, G. C., Salloway, S., Black, S. E., Royall, D. R., Decarli, C., Weiner, M. W., et al. (2010). Randomized, placebo-controlled, clinical trial of donepezil in vascular dementia: Differential effects by hippocampal size. *Stroke, 41*, 1213−1221.

Shen, Y., Sheng, R., Zhang, J., He, Q., Yang, B., & Hu, Y. (2008). 2-Phenoxy-indan-1-one derivatives as acetylcholinesterase inhibitors:a study on the importance of modifications at the side chain on the activity. *Bioorganic and Medicinal Chemistry, 15*(16), 7646−7653.

Sola, I., Viayna, E., Gómez, T., Galdeano, C., Cassina, M., Camps, P., et al. (2015). Multigram synthesis and in vivo efficacy studies of a novel multitarget anti-Alzheimer's compound. *Molecules, 10*(20), 4492−4515.

Spowart-Manning, L., & van der Staay, F. J. (2005). Spatial discrimination deficits by excitotoxic lesions in the Morris water escape task. *Behavioural Brain Research, 156*, 269−276.

Stephen, L. (1997). *Process for the preparation of benzyl-PiperidylmethylIndanones.* U.S. Patent 5,606,064.

Sugimoto, H., Iimura, Y., Yamanishi, Y., & Yamatsu, K. (1995). Síntesis and structure-activity relationships of acetylcholinesterase inhibitors:1-benzyl-4-[(5,6-dimethoxy-1-oxoindan-2-yl]methylpiperidinehydrochloride and related compounds. *Journal of Medicinal Chemistry, 38*, 4821−4829.

Sugimoto, H., Yamanishi, Y., Iimura, Y., & Kawakami, Y. (2000). Donepezil hydrochloride (E2020) and other acetylcholinesterase inhibitors. *Current Medicinal Chemistry, 7*, 303−339.

Takada-Takatori, Y., Kume, T., Ohgi, Y., Izumi, Y., Niidome, T., Fujii, T., et al. (2008). Mechanism of neuroprotection by donepezil pretreatment in rat cortical neurons chronically treated with donepezil. *Journal of Neuroscience Research, 86*, 3575−3583.

Terra, B. S., da Silva, P. H. C., Tramarin, A., Franco, L. L., da Cunha, E. F. F., Macedo Junior, F., et al. (2018). Two novel donepezil-lipoic acid hybrids: Synthesis, anticholinesterase and antioxidant activities and theoretical studies. *Journal of the Brazilian Chemical Society, 29*, 738−747.

Wang, L., Esteban, G., Ojima, M., Bautista-Aguilera, O. M., Inokuchi, T., Moraleda, I., et al. (2014). Donepezil+propargylamine+8-hydroxyquinoline hybrids as new multifunctional metal-chelators, ChE and MAO inhibitors for the potential treatment of Alzheimer's disease. *European Journal of Medicinal Chemistry, 10*(80), 543−561.

Wilkinson, D., Róman, G., Salloway, S., Hecker, J., Boundy, K., Kumar, D., et al. (2003). Donepezil in vascular dementia. A randomized, placebo- controlled study. *Neurology, 61*, 479−486.

Wu, M. Y., Esteban, G., Brogi, S., Shionoya, M., Wang, L., Campiani, G., et al. (2017). Donepezil-like multifunctional agents:Design, synthesis, molecular modeling and biological evaluation. *European Journal of Medicinal Chemistry, 4*(121), 864−879.

Xu, A. J., Chen, Z., Yanai, K., Huang, Y. W., & Wei, E. Q. (2002). Effect of 3-[1-(phenylmethyl)-4-piperidinyl]-1-(2,3,4,5-tetrahydro-1H-1-benzazepin-8-yl)-1-propanone fumarate, a novel acetylcholinesterase inhibitor, on spatial cognitive impairment induced by chronic cerebral hypoperfusion in rats. *Neuroscience Letters, 331*, 33−36.

Yoichi, I. (2001). *Process for production of donepezil derivative.* US patent 6,252,081.

Yuede, C. M., Dong, H., & Csernanskya, J. G. (2007). Anti-dementia drugs and hippocampal-dependent memory in rodents. *Behavioural Pharmacology, 18*, 347−363.

Zhou, J., Fu, Y., & Tang, X. C. (2001). Huperzine A and donepezil protect rat pheochromocytoma cells against oxygen-glucose deprivation. *Neuroscience Letters, 306*, 53−56.

CHAPTER 32

Memantine and Alzheimer's disease: a comprehensive review

Sergio del Río-Sancho

Instituto de Ciencias Biomédicas, Departamento de Farmacia, Facultad de Ciencias de la Salud, Universidad CEU Cardenal Herrera, CEU Universities, Valencia, Spain

List of abbreviations

Ach acetylcholine
AChE acetylcholinesterase
AD Alzheimer's disease
Aβ β-amyloid peptide
ChEIs cholinesterase inhibitors
NFTs tau neurofibrillary tangles
NMDAR N-methyl-D-aspartate receptor
US-FDA US Food and Drug Administration

Mini-dictionary of terms

Alzheimer's treatment Unfortunately, there is no treatment that acts on the underlying cause of the disease at present. Therefore, current AD treatments aim to counteract manifestations of the disease as well as delay its progression and the appearance of symptoms.
Binding site Area on the neuron where a molecule interacts to induce a response.
Memantine Memantine is the only medication with an effect on the glutamatergic system that has been approved for the treatment of AD.
Moderate-to-severe stage Middle to late stage of the progressive degeneration of neurons that occurs during the development of AD.
Uncompetitive antagonist Molecule that blocks the response of an agonist by binding to a different site than its opposite. The activation of the receptor by the agonist is mandatory for the antagonist to access its binding site.

Introduction

Alzheimer's disease (AD) is a degenerative neuronal disease and the most common form of dementia. Dementia is generally characterized by a decline in cognitive function characterized by a group of symptoms such as memory loss and the reduction of linguistic and logical thinking capacities as well as other cognitive skills. These symptoms severely affect a patient's ability to fulfill routine tasks, from remembering recent conversations or events to, in more severe cases, swallowing and walking (Alzheimer's Association, 2017).

Diagnosis and Management in Dementia
ISBN 978-0-12-815854-8, https://doi.org/10.1016/B978-0-12-815854-8.00032-X

AD was first described in 1906 by the German psychiatrist and neurologist Alois Alzheimer, and in 1976 it was presented as the principal cause of precocious dementia in the world as well as a major health issue (Katzman, 1976). As of 2017 it is considered the principal cause of death that cannot be prevented, delayed, or cured. According to data compiled by the "Alzheimer's Association" about the US population (2017), 1 of 10 people over 65 years of age is affected by AD (5.3 million people in total), which in 2017 alone created a cost of $259 billion for health care. It is important to note that the population costs related to the 15 million caregivers that assist patients, either volunteer or family members, are not included.

With a pathology of hereditary or sporadic origin (in 98% of cases), it is the interaction of multiple risk factors (i.e., age, the consumption of hypercaloric diets, and sedentary lifestyle) and the interaction of genetical and environmental factors that are supposed to lead to the development of AD (Folch et al., 2018).

The progressive degeneration of neurons during the development of the disease provokes the exacerbation of the above-mentioned symptoms. According to severity of symptoms, three stages of AD are recognized by clinicians: mild (early stage), moderate (middle stage), and severe (late stage). Progression of the disease differs with each patient; however, the loss of brain mass leads to patient death, on average, within 5—10 years of onset (Anand, Patience, Sharma, & Khurana, 2017).

Mechanisms of Alzheimer's disease

Several mechanisms have been identified involving the decline of neuronal activity and the appearance of AD. Namely, the presence of senile plaques and Tau neurofibrillary tangles (NFTs) (considered the neuropathological hallmarks of AD), and the cholinergic and glutamatergic systems (Fig. 32.1).

Senile plaques are composed of extracellular deposits of neurotoxic β-amyloid peptide (Aβ), which are formed due to an alteration of the usual activity of the amyloid cascade. Aβ peptides are believed to be responsible for the activation of cationic channels controlled by N-methyl-D-aspartate receptors (NMDARs) and in the formation of NFTs. In summary, the elevated level of Aβ peptide in the brain leads to synaptic dysfunction and neuronal death, either due to its accumulation or its activity (Kim et al., 2017; Olivares et al., 2012).

On the contrary, NFTs comprise mainly intracellular deposits of hyperphosphorylated Tau protein. The aberrant hyperphosphorylation of protein leads to abnormal associations with axonal tubulin, which either hampers the physiological functionality of the axons or precipitates and causes neuronal death by apoptosis (Anand et al., 2017; Tiraboschi, Hansen, hal, & Corey-Bloom, 2004).

In terms of physiological conditions, acetylcholine (Ach) is an essential neurotransmitter responsible for cognitive function and learning. An increased activity of

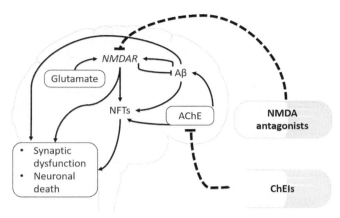

Figure 32.1 *Mechanisms and treatments of Alzheimer's disease (AD).* Schematic representation of the main factors that lead to the degeneration of neurons during the development of AD and the therapeutic target of the two main pharmacological treatments.

acetylcholinesterase (AChE) has been reported in AD patients, which results in a reduction of the concentration and function of Ach. Additionally, this enzyme is related to the formation of senile plaques and NFTs (Singh et al., 2013).

With respect to the glutamatergic system, a massive release of glutamate provokes increased activation of NMDARs, causes extensive influx of Ca^{2+} into the cell, and causes the generation of free radicals. Ultimately, the cytotoxicity generated leads to NFT formation, synaptic dysfunction, and neuronal death (Thomas & Grossberg, 2009).

Treatments for Alzheimer's disease

Unfortunately, current treatments for AD do not address the underlying cause of the disease. Therefore, AD treatments focus on counteracting the manifestation of the disease, delaying both its progression and the appearance of symptoms. In general, the different treatments approved by the US Food and Drug Administration (US-FDA) are classified into two groups that regulate the levels of the two neurotransmitters in the brain, Ach and glutamate (Fig. 32.1) (Greig, 2015).

Cholinesterase inhibitors (ChEIs) are drugs designed to raise Ach levels in the brain and are especially useful in patients with mild to moderate AD. This group includes donepezil, galantamine, and rivastigmine.

The second group consists of NMDAR antagonists, which aim to alleviate hyperexcitation of the receptor. They are also useful in other aspects of the disease as discussed below. Within this group we find memantine (1-amino-3,5-dimethyladamantane; Fig. 32.2), an adamantane derivate that has been approved by the US-FDA for commercialization in treating moderate to severe AD.

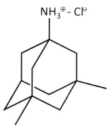

$NH_3^{\oplus} - Cl^{\ominus}$

Figure 32.2 *Structure of memantine hydrochloride.* Molecular structure of memantine hydrochloride.

Memantine: mechanism of action

Under normal synaptic transmission conditions, NMDARs in the postsynaptic neuron are activated by binding with glutamate, thereby producing depolarization. The depolarization results in the exit of magnesium ion from the NMDAR channel and entry of cations into the neuron (Fig. 32.3). The NMDAR channels are only open for a few milliseconds, because although the entry of cations is essential for learning and memory, its excess can lead to excitotoxicity (Lipton, 2004).

Memantine is considered an uncompetitive, low-affinity and open–channel blocker because under resting conditions, memantine is not able to displace the Mg^{2+} to bind to or near the magnesium-binding site. In fact, it needs prior activation of the NMDAR by glutamate (resulting in the opening of the channel) before it can gain access to its binding site. In contrast to other potent NMDAR blockers (i.e., MK 801), its action only occurs when the channel is open for a longer time than usual—that is, under pathological conditions—but being able to leave the NMDAR channel upon strong synaptic depolarization (Tampi & van Dyck, 2007).

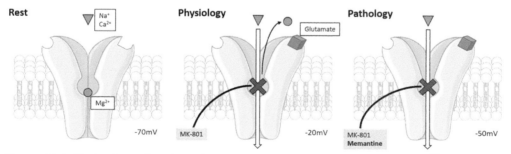

Figure 32.3 *Mechanism of action of memantine.* Under resting conditions, Mg^{2+} blocks the N-methyl-D-aspartate receptor (NMDAR) channel, thus preventing the entry of cations in the neuron. This situation is reversed while on normal synaptic transmission, as the binding of the NMDAR with glutamate activates the channel and expels the Mg^{2+}. When the channel fails to deactivate, memantine acts as an open-channel blocker. In contrast with other antagonists, memantine is able to leave the channel upon strong synaptic depolarization. *(This figure was created with images adapted from Servier Medical Art by Servier. Original images are licensed under a Creative Commons Attribution 3.0 Unported License.)*

In addition, the blocking/unblocking kinetics of memantine make it a very well-tolerated medication. Most NMDAR antagonists present side effects such as hallucinations and schizophrenia-type symptoms. However, for memantine, these side effects were only observed sporadically (Lipton, 2006).

History of memantine

The timeline of memantine's development and use in AD is illustrated in Fig. 32.4. Memantine was first synthesized and patented in 1968 by Eli Lilly and Company for the treatment of diabetes due to its efficacy in reducing blood sugar levels (Gerzon, Krumkalns, Brindle, Marshall, & Root, 1963). Despite a negative outcome for its intended purpose, pharmaceutical company Merz, in collaboration with Neurobiological Technologies, Inc., recognized its activity at the central nervous system level and requested (under the name D145) German patent protection as a potential treatment for several neurological diseases (Parkinson's disease, spasticity, and other cerebral disturbances) in 1972. The patent protection was granted in 1975 followed by a US patent in 1978 (Balazs, Bridges, & Cotman, 2005).

In 1983, Wesemann et al. postulated an incorrect mechanism of action of INN-memantine, suggesting its influence on the metabolism of different neurotransmitters (dopamine, noradrenaline, and serotonin). However, they did not consider a difference in concentration of 100-fold between the in vitro and the in vivo data (Parsons, Danysz, & Quack, 1999; Wesemann, Sontag, & Maj, 1983). The discovery about its activity as an antagonist of NMDAR did not occur until 1989, when the German company Merz first described its action as a potent blocker of NMDAR channels (Bormann, 1989). At present, it is known whether memantine is an uncompetitive antagonist rather than a potent blocker. Nevertheless, these results allowed it to achieve patent protection for the treatment of cerebral ischemia and AD under the trade name Axura (Witt, Macdonald, & Kirkpatrick, 2004).

The commercialization of the first treatment for AD was approved in 1996. It was Aricept (Donepezil) by Eisai Co., an AChE inhibitor shown to be active in mild to moderate stages of the disease (Cacabelos, 2007). Not much later, in 1999, the results of the first major study demonstrating the clinical efficacy and safety of memantine as monotherapy in moderately severe to severe primary dementia were published (Winblad & Poritis, 1999). Based on favorable data regarding the risk/benefit of memantine given by the Committee for Proprietary Medicinal Products (CPMP), the European Agency for the Evaluation of Medical Products (EMEA) approved its use in 2002, marketed as Ebixa, to H. Lundbeck A/S for the treatment of moderately severe to severe AD (Kilpatrick & Tilbrook, 2002). In October 2003, Forest Laboratories, Inc. announced approval by the US Food and Drug Administration (US-FDA) (Witt et al., 2004) under the name Namenda IR, resulting in the only noncholinergic medication for the treatment of AD (Keiski, 2017).

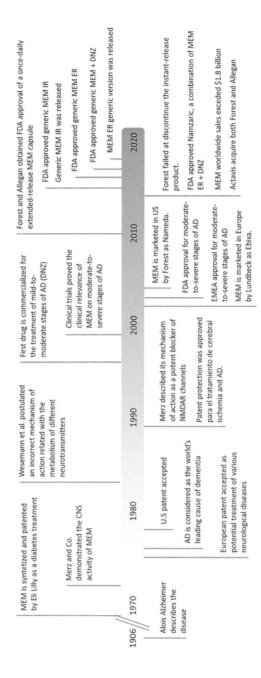

Figure 32.4 Overview of the history of memantine. Timeline of memantine and its use in AD. *AD,* Alzheimer's disease; *DNZ,* donepezil; *EMEA,* European Agency for the Evaluation of Medical Products; *ER,* extended-release; *FDA,* Food and Drug Administration; *IR,* immediate-release; *MEM,* memantine.

In 2010, Forest Laboratories obtained US-FDA approval for Namenda XR, a once-daily extended-release (ER) memantine capsule. The new product represented a substantial improvement in treatment adherence compared with that of the "immediate-release" formulation (IR). This could be attributed to the more convenient dosing regimen (e.g., a lower pill burden) and the facilitation of administration to patients with difficulties in swallowing (del Río-Sancho et al., 2012, 2017; Plosker, 2015).

Four years later, in February 2014, Forest Laboratories announced their interest in discontinuing Namenda IR and only formulating the ER version, probably because of the upcoming 2015 expiration of the patent exclusivity period for memantine IR. The courts decided against the proposal, claiming that given the absence of generics in 2014, patients would be forced to switch to ER (Deardorff & Grossberg, 2016). At the end of the year, Forest Laboratories and Allergan US obtained US-FDA approbation for Namzaric, a once-daily fixed-dose combination of memantine with donepezil for the treatment of moderate to severe AD. The product launch and memantine's resulting $1.8 billion in worldwide sales encouraged Actavis to acquire both Forest and Allergan, leaving Actavis as the major supplier of Namenda and Namzaric (Alam, Lingenfelter, Bender, & Lindsley, 2017).

As mentioned, the patent exclusivity period expired in 2015, leading to the release of generic versions of the drug. In August 2015, the US-FDA approved the generic version of memantine IR tablets. This was followed by US-FDA approval for a generic version of Memantine ER (2016), the approval of a generic version of memantine and Donepezil ER capsules (2017), and more recently the release of a generic version of memantine ER (2018).

Therapeutic efficacy: clinical trials

Since AD cannot be prevented, delayed, or cured, the therapeutic efficacy of memantine is evaluated according to different outcomes. Some of the main rated outcomes and evaluation scales used are (one) evaluation of the degree of deterioration by studying patient ability to perform activities of daily living [scale: Alzheimer's Disease Cooperative Study-Activities of the Daily Living Inventory (ADCS -ADL) (Galasko, Schmitt, Thomas, Jin, & Bennett, 2005)]; (two) degree of patient cognition, either independently [scales: Alzheimer's Disease Assessment Scale-Cognitive Subscale (Rosen, Terry, Fuld, Katzman, & Peck, 1980), Mini-Mental State Examination (MMSE) (Folstein, Folstein, & McHugh, 1975), and Severe Impairment Battery (SIB) (Schmitt et al., 1997)] or in combination with function and behavior [scale: Behavioral Rating Scale for Geriatric Patients (BGP) (van der Kam & Hoeksma, 1989)]; (three) evaluation of the general outcome of the patient according to the global impression of change [scales: Clinician's Global Impression of Change and Clinician's Interview-Based Impression of Change with Caregiver Input (Schneider et al., 1997)]; (four) global functioning [scales: Functional Assessment Staging Scale (FAST) (Sclan & Reisberg, 1992) and Global Deterioration Scale

(Reisberg, Ferris, de Leon, & Crook, 1982)]; and (five) neuropsychiatric symptoms [scale: Neuropsychiatric Inventory (Cummings et al., 1994)].

Monotherapy clinical studies

The first major study in which clinical efficacy and safety of memantine was evaluated as a monotherapy for moderately severe to severe primary dementia (AD and vascular dementia) was published in 1999 by Winblad and Poritis (1999) (Table 32.1). The results of the study highlighted significant and consistent clinical relevance in terms of cognition and care dependence. In 2003, Reisberg et al. (2003) published the results of the first study evaluating the efficacy of memantine in outpatients with moderate to severe AD. The study showed improved cognition and better global assessment in the memantine-treated group, and thus it concluded that patients receiving memantine had a better outcome than those receiving placebo. The positive results observed in both clinical trials were crucial to the approval of the use of memantine by the EMEA (in 2002) and the US-FDA (in 2003), since the study by Reisberg et al. covered data collected from August 1998 to April 1999.

In 2007, Van Dyck et al. published the first study following the approval and commercialization of memantine in Europe and the United States (van Dyck, Tariot, Meyers, & Malca Resnick, 2007). However, compared with previous studies, no significant treatment benefits in terms of cognition, outcome, or deterioration were observed. These surprising and opposing results were explained by a violation of parametric statistical assumptions or other unexplained factors.

Similar negative results were published in 2012 and 2013, where a lack of significant improvement in agitation (Fox et al., 2012) and inconclusive results in cognition (Wang et al., 2013) were reported in patients with moderate to severe AD. Furthermore, the results of the clinical trials in which memantine was evaluated as a monotherapy for the treatment of mild to moderate AD were also found very inconsistent. In areas like cognition and global change efficacy can vary between trials, and improvements in deterioration were not observed in any of them (Bakchine & Loft, 2008; Peskind et al., 2006; Schmidt et al., 2008).

In 2011, the results of a Phase 2 clinical trial in Japanese outpatients by Kitamura, Homma, and Nakamura (2011) were published. Among the two doses evaluated (10–20 mg/day), the former did not lead to improvements in terms of cognition or global functioning. Moreover, none reduced the deterioration or improved the general outcome of patients. These results were in clear contrast to those published by Winblad and Poritis in 1999, where an improved global impression of change of patients was reported. These discrepancies could be due to the type of scale used—e.g., Winblad and Poritis used the BGP scale to evaluate cognition and function whereas Kitamura et al. used FAST, MMSE, and SIB scales. Later the same year, results of the continuation of the above-mentioned study (Phase 3) were published by Nakamura et al., concluding similar negative results (2011). To determine whether these differences with respect to previously published

Table 32.1 Clinical trials in which memantine was administered as monotherapy.

Study	Subject sample	Dosage	Duration (weeks)	Deterioration	Cognition	Global change	Global functioning	Neuropsychiatric Symptoms	Other
Moderate to severe									
Winblad and Poritis (1999)	166	10 mg/day	12		✓✓	✓✓			
Reisberg et al. (2003)	252	10 mg, bid	28	✓	✓✓	✓✓			
van Dyck et al. (2007)	350	2 × 5 mg, bid	24	×	×	×			
Kitamura et al. (2011)	315	10 mg/day 10 mg, bid	24	× ×	× ✓✓	× ×	× ✓✓		
Nakamura et al. (2011)	432	10 mg, bid	24	×	✓✓	×		✓✓	
Fox et al. (2012)	153	10 mg, bid	12		✓✓	×		×	
Wang et al. (2013)	26	10 mg, bid	24		✓/×				

Continued

Table 32.1 Clinical trials in which memantine was administered as monotherapy.—cont'd

Study	Subject sample	Dosage	Duration (weeks)	Outcome measures						
				Deterioration	Cognition	Global change	Global functioning	Neuropsychiatric Symptoms	Other
Mild to moderate									
Peskind et al. (2006)	403	10 mg, bid	24	×	✓	✓		✓	
Bakchine and Loft (2008)	470	10 mg, bid	24	×	×	×			
Schmidt et al. (2008)	36	10 mg, bid	52	×	✓				× Brain volume

Comprehensive review of the main clinical studies where memantine was administered as monotherapy. Outcome measures are presented highlighting the presence (✓) or absence (×) of significant treatment benefits.

clinical trials were due to the different lifestyle between Japan and Western countries or only to a lack of statistical power, a pooled analysis encompassing both publications was published in 2014. The results showed better outcomes for memantine compared with those of placebo in terms of cognition as well as behavioral and psychological symptoms (including activity disturbances and aggressiveness) (Nakamura et al., 2014).

Finally, a meta-analysis of the different clinical trials was published in 2017. The meta-analysis demonstrated that using memantine as monotherapy results in significant improvement in cognitive function in patients at all levels of AD severity as well as a reduction in behavioral disturbances in moderate to severe AD (Kishi et al., 2017).

Combination therapy clinical studies

Due to the positive effects observed by memantine and by ChEIs in moderate to severe AD, different studies considered the possibility of evaluating the clinical benefit combining both strategies (Table 32.2) (Tampi & van Dyck, 2007).

The first clinical trial to evaluate the combined effect of memantine with ChEIs (in this case, donepezil) was published by Tariot et al. in (2004). The results of this first study were very promising because they determined that combined therapy improved the cognition, activities of daily living, global outcome, and behavior of the patients. Additionally, the combination of both medications could result in the positive results obtained by Cretu, Szalontay, Chirita, and Chirita (2008), in which treatment with memantine was found to be well tolerated. Moreover, it reduced agitation and aggression, irritability, and appetite eating disturbances in patients who were agitated at baseline.

Four years later, Howard et al. published the results of a more extensive study including a higher patient number, but unfortunately reported less promising results. According to their study, no significant benefits were found for the combination of donepezil and memantine over donepezil alone, neither to cognition nor to neuropsychiatric symptoms (Howard et al., 2012). In the following years, studies evaluated the combination of memantine with donepezil or other ChEIs, giving disparate and inconclusive results for both communication skills and neuropsychiatric symptoms (Grossberg et al., 2013; Herrmann, Gauthier, Boneva, & Lemming, 2013; Saxton et al., 2012).

This trend continued until 2014, at which time Araki et al. published the results of a twice-longer study than Howard et al. and demonstrated that the combination of memantine and donepezil improved clinical symptoms, overall cognitive function, and behavioral and psychological symptoms of dementia (Araki et al., 2014). At the end of the year, the US-FDA approved a once-daily fixed-dose combination of memantine and donepezil for treatment of moderate to severe AD.

However, with regard to treatment of mild to moderate AD, combination therapy (with either ChEIs or antioxidants) did not result in significant improvements in any outcome measure evaluated (Ashford et al., 2011; Choi et al., 2011; Dysken et al., 2014; Peters et al., 2015; Porsteinsson, Grossberg, Mintzer, & Olin, 2008).

Table 32.2 Clinical trials in which memantine was administered as combination therapy.

Study	Subject sample	Dosage	Duration (weeks)	Deterioration	Cognition	Global change	Global functioning	Neuropsychiatric symptoms	Other
Moderate to severe									
Tariot et al. (2004)	24	MEM 2 × 5 mg, bid + DNZ NR	24	✓	✓	✓		✓	
Cretu et al. (2008)	43	MEM 10 mg, bid + DNZ NR	24					✓	
Howard et al. (2012)	295	MEM 10 mg, bid + DNZ 10 mg	52		×			×	
Saxton et al. (2012)	265	MEM 10 mg, bid + ChEI	12	×					✓/× Communication
Grossberg et al. (2013)	677	MEM 28 mg od + ChEIs NR	24		✓	✓		✓	
Herrmann et al. (2013)	369	MEM 20 mg/day + ChEIs NR	24					×	
Araki et al. (2014)	37	MEM 20 mg/day + DNZ NR	24		✓	✓		✓	

Mild to moderate

Study	N	Treatment	Weeks						
Porsteinsson et al. (2008)	433	MEM 20 mg, od + ChEI	24		×	×	×	×	× Neuronal tissue volume
Ashford et al. (2011)	17	MEM 10 mg, bid + DNZ NR	52		×				
Choi et al. (2011)	172	MEM 20 mg/day + RIV patch 17.6 mg	16	×	×			×	
Dysken et al. (2014)	613	MEM 20 mg/day + vitamin E 2000 IU/d	~27	×	×		×	×	
Peters et al. (2015)	232	MEM 10 mg, bid + GAL 24 mg	52		×		×	×	

Comprehensive review of the main clinical studies where memantine was administered as combination therapy. Outcome measures are presented highlighting the presence (✔) or absence (×) of significant treatment benefits. *ChEIs*, cholinesterase inhibitors; *DNZ*, donepezil; *GAL*, galantamine; *MEM*, memantine; *NR*, not reported; *RIV*, rivastigmine.

Concluding remarks

Based on the results of clinical trials, memantine has been shown to be an effective and well-tolerated treatment for moderate AD. Due to its unique mechanism of action, its administration should be considered the treatment of choice for moderate to severe AD patients until the discovery of a cure for the disease. With regards to combination therapy with ChEIs, only the association with donepezil was found superior to monotherapy with ChEIs. Overall, either as monotherapy or in combination with donepezil, memantine results in significant improvement for patients and relief for caregivers.

Key facts of memantine

- Multiple factors may lead to neuronal activity decline and the appearance of AD.
- Among those, the cholinergic and glutamatergic systems play an important role in development of the disease.
- Memantine was originally synthesized and patented in 1968 by Eli Lilly and Company but for the treatment of diabetes.
- Its effect on the glutamatergic system was first described by Merz in 1989.
- Since 2003, it is the only noncholinergic medication approved by the US-FDA for the treatment of AD.

Summary points

- This chapter describes the role of memantine on the treatment of AD.
- Among approved medications for the treatment of AD, memantine is the only drug that acts on the glutamatergic system.
- Moreover, its blocking/unblocking kinetics make it a very well-tolerated medication.
- In addition, memantine's history illustrates its evolution in consonance with the needs of moderate to severe AD patients.
- Either as monotherapy or in combination with donepezil, memantine results in significant improvement for patients and relief for caregivers.

References

Alam, S., Lingenfelter, K. S., Bender, A. M., & Lindsley, C. W. (2017). Classics in chemical neuroscience: Memantine. *ACS Chemical Neuroscience, 8*(9), 1823—1829.

Alzheimer's Association. (2017). 2017 Alzheimer's disease facts and figures. *Alzheimer's and Dementia, 13*(4), 325—373.

Anand, A., Patience, A. A., Sharma, N., & Khurana, N. (2017). The present and future of pharmacotherapy of Alzheimer's disease: A comprehensive review. *European Journal of Pharmacology, 815*, 364—375.

Araki, T., Wake, R., Miyaoka, T., Kawakami, K., Nagahama, M., Furuya, M., et al. (2014). The effects of combine treatment of memantine and donepezil on Alzheimer's disease patients and its relationship with cerebral blood flow in the prefrontal area. *International Journal of Geriatric Psychiatry, 29*(9), 881—889.

Ashford, J. W., Adamson, M., Beale, T., La, D., Hernandez, B., Noda, A., et al. (2011). MR spectroscopy for assessment of memantine treatment in mild to moderate Alzheimer dementia. *Journal of Alzheimer's Disease, 26*(Suppl. 3), 331−336.

Bakchine, S., & Loft, H. (2008). Memantine treatment in patients with mild to moderate Alzheimer's disease: Results of a randomised, double-blind, placebo-controlled 6-month study. *Journal of Alzheimer's Disease, 13*(1), 97−107.

Balazs, R., Bridges, R. J., & Cotman, C. W. (2005). *Excitatory amino acid transmission in health and disease*. New York: Oxford University Press.

Bormann, J. (1989). Memantine is a potent blocker of N-methyl-D-aspartate (NMDA) receptor channels. *European Journal of Pharmacology, 166*(3), 591−592.

Cacabelos, R. (2007). Donepezil in Alzheimer's disease: From conventional trials to pharmacogenetics. *Neuropsychiatric Disease and Treatment, 3*(3), 303−333.

Choi, S. H., Park, K. W., Na, D. L., Han, H. J., Kim, E. J., Shim, Y. S., et al. (2011). Tolerability and efficacy of memantine add-on therapy to rivastigmine transdermal patches in mild to moderate Alzheimer's disease: A multicenter, randomized, open-label, parallel-group study. *Current Medical Research and Opinion, 27*(7), 1375−1383.

Cretu, O., Szalontay, A. S., Chirita, R., & Chirita, V. (2008). Effect of memantine treatment on patients with moderate-to-severe Alzheimer's disease treated with donepezil. *Revista Medico-Chirurgicala a Societati de Medicii si Naturalisti din Iasi, 112*(3), 641−645.

Cummings, J. L., Mega, M., Gray, K., Rosenberg-Thompson, S., Carusi, D. A., & Gornbein, J. (1994). The neuropsychiatric inventory: Comprehensive assessment of psychopathology in dementia. *Neurology, 44*(12), 2308−2314.

Deardorff, W. J., & Grossberg, G. T. (2016). A fixed-dose combination of memantine extended-release and donepezil in the treatment of moderate-to-severe Alzheimer's disease. *Drug Design, Development and Therapy, 10*, 3267−3279.

van Dyck, C. H., Tariot, P. N., Meyers, B., & Malca Resnick, E. (2007). A 24-week randomized, controlled trial of memantine in patients with moderate-to-severe Alzheimer disease. *Alzheimer Disease and Associated Disorders, 21*(2), 136−143.

Dysken, M. W., Sano, M., Asthana, S., Vertrees, J. E., Pallaki, M., Llorente, M., et al. (2014). Effect of vitamin E and memantine on functional decline in Alzheimer disease: The TEAM-AD VA cooperative randomized trial. *JAMA, 311*(1), 33−44.

Folch, J., Busquets, O., Ettcheto, M., Sanchez-Lopez, E., Castro-Torres, R. D., Verdaguer, E., et al. (2018). Memantine for the treatment of dementia: A review on its current and future Applications. *Journal of Alzheimer's Disease, 62*(3), 1223−1240.

Folstein, M. F., Folstein, S. E., & McHugh, P. R. (1975). "Mini-mental state". A practical method for grading the cognitive state of patients for the clinician. *Journal of Psychaitric Research, 12*(3), 189−198.

Fox, C., Crugel, M., Maidment, I., Auestad, B. H., Coulton, S., Treloar, A., et al. (2012). Efficacy of memantine for agitation in Alzheimer's dementia: A randomised double-blind placebo controlled trial. *PloS One, 7*(5), e35185.

Galasko, D., Schmitt, F., Thomas, R., Jin, S., & Bennett, D. (2005). Detailed assessment of activities of daily living in moderate to severe Alzheimer's disease. *Journal of the International Neuropsychological Society, 11*(4), 446−453.

Gerzon, K., Krumkalns, E. V., Brindle, R. L., Marshall, F. J., & Root, M. A. (1963). The Adamantyl group in medicinal Agents. I. Hypoglycemic N-Arylsulfonyl-N'-adamantylureas. *Journal of Medicinal Chemistry, 6*(6), 760−763.

Greig, S. L. (2015). Memantine ER/donepezil: A review in Alzheimer's disease. *CNS Drugs, 29*(11), 963−970.

Grossberg, G. T., Manes, F., Allegri, R. F., Gutierrez-Robledo, L. M., Gloger, S., Xie, L., et al. (2013). The safety, tolerability, and efficacy of once-daily memantine (28 mg): A multinational, randomized, double-blind, placebo-controlled trial in patients with moderate-to-severe Alzheimer's disease taking cholinesterase inhibitors. *CNS Drugs, 27*(6), 469−478.

Herrmann, N., Gauthier, S., Boneva, N., & Lemming, O. M. (2013). A randomized, double-blind, placebo-controlled trial of memantine in a behaviorally enriched sample of patients with moderate-to-severe Alzheimer's disease. *International Psychogeriatrics, 25*(6), 919−927.

Howard, R., McShane, R., Lindesay, J., Ritchie, C., Baldwin, A., Barber, R., et al. (2012). Donepezil and memantine for moderate-to-severe Alzheimer's disease. *New England Journal of Medicine, 366*(10), 893–903.

van der Kam, P., & Hoeksma, B. H. (1989). The usefulness of BOP and SIVIS (ADL and behavior rating scales) for the estimation of workload in a psychogeriatric nursing home. Results of a time-standard study. *Tijdschrift voor Gerontologie en Geriatrie, 20*(4), 159–166.

Katzman, R. (1976). Editorial: The prevalence and malignancy of Alzheimer disease. A major killer. *Archives of Neurology, 33*(4), 217–218.

Keiski, M. A. (2017). Chapter 16 - Memantine: A safe and tolerable NMDA antagonist with potential benefits in traumatic brain injury A2. In K. Heidenreich (Ed.), *New therapeutics for traumatic brain Injury* (pp. 253–271). San Diego: Academic Press.

Kilpatrick, G. J., & Tilbrook, G. S. (2002). Memantine. Merz. *Current Opinion in Investigational Drugs, 3*(5), 798–806.

Kim, S. H., Kandiah, N., Hsu, J. L., Suthisisang, C., Udommongkol, C., & Dash, A. (2017). Beyond symptomatic effects: Potential of donepezil as a neuroprotective agent and disease modifier in Alzheimer's disease. *British Journal of Pharmacology, 174*(23), 4224–4232.

Kishi, T., Matsunaga, S., Oya, K., Nomura, I., Ikuta, T., & Iwata, N. (2017). Memantine for Alzheimer's disease: An updated systematic review and meta-analysis. *Journal of Alzheimer's Disease, 60*(2), 401–425.

Kitamura, S., Homma, A., & Nakamura, Y. (2011). Late phase II study of mementine hydrochloride, a new NMDA receptor antagonist, in patients with moderate to severe Alzheimer's disease. *Japanese Journal of Geriatric Psychiatry, 22*(4), 453–463.

Lipton, S. A. (2004). Paradigm shift in NMDA receptor antagonist drug development: Molecular mechanism of uncompetitive inhibition by memantine in the treatment of Alzheimer's disease and other neurologic disorders. *Journal of Alzheimer's Disease, 6*(6 Suppl.), S61–S74.

Lipton, S. A. (2006). Paradigm shift in neuroprotection by NMDA receptor blockade: Memantine and beyond. *Nature Reviews Drug Discovery, 5*(2), 160–170.

Nakamura, Y., Homma, A., Kitamura, S., & Yoshimura, I., III (2011). Phase III study of memantine hydrochloride, a new NMDA receptor antagonist, in patients with moderate to severe Alzheimer's disease—efficacy and safety. *Japanese Journal of Geriatric Psychiatry, 22*(4), 464–473.

Nakamura, Y., Kitamura, S., Homma, A., Shiosakai, K., & Matsui, D. (2014). Efficacy and safety of memantine in patients with moderate-to-severe Alzheimer's disease: Results of a pooled analysis of two randomized, double-blind, placebo-controlled trials in Japan. *Expert Opinion on Pharmacotherapy, 15*(7), 913–925.

Olivares, D., Deshpande, V. K., Shi, Y., Lahiri, D. K., Greig, N. H., Rogers, J. T., et al. (2012). N-methyl D-aspartate (NMDA) receptor antagonists and memantine treatment for Alzheimer's disease, vascular dementia and Parkinson's disease. *Current Alzheimer Research, 9*(6), 746–758.

Parsons, C. G., Danysz, W., & Quack, G. (1999). Memantine is a clinically well tolerated N-methyl-d-aspartate (NMDA) receptor antagonist—a review of preclinical data. *Neuropharmacology, 38*(6), 735–767.

Peskind, E. R., Potkin, S. G., Pomara, N., Ott, B. R., Graham, S. M., Olin, J. T., et al. (2006). Memantine treatment in mild to moderate Alzheimer disease: A 24-week randomized, controlled trial. *American Journal of Geriatric Psychiatry, 14*(8), 704–715.

Peters, O., Fuentes, M., Joachim, L. K., Jessen, F., Luckhaus, C., Kornhuber, J., et al. (2015). Combined treatment with memantine and galantamine-CR compared with galantamine-CR only in antidementia drug naïve patients with mild-to-moderate Alzheimer's disease. *Alzheimer's and Dementia: Translational Research and Clinical Interventions, 1*(3), 198–204.

Plosker, G. L. (2015). Memantine extended release (28 mg once daily): A review of its use in Alzheimer's disease. *Drugs, 75*(8), 887–897.

Porsteinsson, A. P., Grossberg, G. T., Mintzer, J., & Olin, J. T. (2008). Memantine treatment in patients with mild to moderate Alzheimer's disease already receiving a cholinesterase inhibitor: A randomized, double-blind, placebo-controlled trial. *Current Alzheimer Research, 5*(1), 83–89.

Reisberg, B., Doody, R., Stoffler, A., Schmitt, F., Ferris, S., & Mobius, H. J. (2003). Memantine in moderate-to-severe Alzheimer's disease. *New England Journal of Medicine, 348*(14), 1333–1341.

Reisberg, B., Ferris, S. H., de Leon, M. J., & Crook, T. (1982). The global deterioration scale for assessment of primary degenerative dementia. *American Journal of Psychiatry, 139*(9), 1136–1139.

del Río-Sancho, S., Serna-Jimenez, C. E., Calatayud-Pascual, A., Balaguer-Fernandez, C., Femenia-Font, A., Merino, V., et al. (2012). Transdermal absorption of memantine. Effect of chemical enhancers, iontophoresis and role of enhancer lipophilicity. *European Journal of Pharmaceutics and Biopharmaceutics, 82*(1), 164—170.

del Río-Sancho, S., Serna-Jimenez, C. E., Sebastian-Morello, M., Calatayud-Pascual, M. A., Balaguer-Fernandez, C., Femenia-Font, A., et al. (2017). Transdermal therapeutic systems for memantine delivery. Comparison of passive and iontophoretic transport. *International Journal of Pharmaceutics, 517*(1—2), 104—111.

Rosen, W. G., Terry, R. D., Fuld, P. A., Katzman, R., & Peck, A. (1980). Pathological verification of ischemic score in differentiation of dementias. *Annals of Neurology, 7*(5), 486—488.

Saxton, J., Hofbauer, R. K., Woodward, M., Gilchrist, N. L., Potocnik, F., Hsu, H. A., et al. (2012). Memantine and functional communication in Alzheimer's disease: Results of a 12-week, international, randomized clinical trial. *Journal of Alzheimer's Disease, 28*(1), 109—118.

Schmidt, R., Ropele, S., Pendl, B., Ofner, P., Enzinger, C., Schmidt, H., et al. (2008). Longitudinal multi-modal imaging in mild to moderate Alzheimer disease: A pilot study with memantine. *Journal of Neurology, Neurosurgery, and Psychiatry, 79*(12), 1312—1317.

Schmitt, F. A., Ashford, W., Ernesto, C., Saxton, J., Schneider, L. S., Clark, C. M., et al. (1997). The severe impairment battery: Concurrent validity and the assessment of longitudinal change in Alzheimer's disease. The Alzheimer's disease cooperative study. *Alzheimer Disease and Associated Disorders, 11*(Suppl. 2), S51—S56.

Schneider, L. S., Olin, J. T., Doody, R. S., Clark, C. M., Morris, J. C., Reisberg, B., et al. (1997). Validity and reliability of the Alzheimer's disease cooperative study-clinical global impression of change. The Alzheimer's disease cooperative study. *Alzheimer Disease and Associated Disorders, 11*(Suppl. 2), S22—S32.

Sclan, S. G., & Reisberg, B. (1992). Functional assessment staging (FAST) in Alzheimer's disease: Reliability, validity, and ordinality. *International Psychogeriatrics, 4*(Suppl. 1), 55—69.

Singh, M., Kaur, M., Kukreja, H., Chugh, R., Silakari, O., & Singh, D. (2013). Acetylcholinesterase inhibitors as Alzheimer therapy: From nerve toxins to neuroprotection. *European Journal of Medicinal Chemistry, 70*, 165—188.

Tampi, R. R., & van Dyck, C. H. (2007). Memantine: Efficacy and safety in mild-to-severe Alzheimer's disease. *Neuropsychiatric Disease and Treatment, 3*(2), 245—258.

Tariot, P. N., Farlow, M. R., Grossberg, G. T., Graham, S. M., McDonald, S., & Gergel, I. (2004). Memantine treatment in patients with moderate to severe Alzheimer disease already receiving donepezil: A randomized controlled trial. *JAMA, 291*(3), 317—324.

Thomas, S. J., & Grossberg, G. T. (2009). Memantine: A review of studies into its safety and efficacy in treating Alzheimer's disease and other dementias. *Clinical Interventions in Aging, 4*, 367—377.

Tiraboschi, P., Hansen, L. A., Thal, L. J., & Corey-Bloom, J. (2004). The importance of neuritic plaques and tangles to the development and evolution of AD. *Neurology, 62*(11), 1984—1989.

Wang, T., Huang, Q., Reiman, E. M., Chen, K., Li, X., Li, G., et al. (2013). Effects of memantine on clinical ratings, fluorodeoxyglucose positron emission tomography measurements, and cerebrospinal fluid assays in patients with moderate to severe Alzheimer dementia: A 24-week, randomized, clinical trial. *Journal of Clinical Psychopharmacology, 33*(5), 636—642.

Wesemann, W., Sontag, K., & Maj, J. (1983). Pharmacodynamics and pharmacokinetics of memantine. *Arzneimittel Forschung, 33*(8), 1122—1134.

Winblad, B., & Poritis, N. (1999). Memantine in severe dementia: Results of the 9M-best study (benefit and efficacy in severely demented patients during treatment with memantine). *International Journal of Geriatric Psychiatry, 14*(2), 135—146.

Witt, A., Macdonald, N., & Kirkpatrick, P. (2004). Memantine hydrochloride. Nature reviews. *Drug Discovery, 3*(2), 109—110.

CHAPTER 33

A new neuroprotective strategy for the drug therapy of Parkinson's disease: the Ca^{2+}/cAMP signaling as therapeutic target

Afonso Caricati-Neto[1], Fúlvio Alexandre Scorza[2], Leandro Bueno Bergantin[3]

[1]Head of Laboratory of Autonomic and Cardiovascular Pharmacology, Department of Pharmacology, Universidade Federal de São Paulo (UNIFESP), São Paulo, SP, Brazil; [2]Department of Neurology/Neurosurgery, Universidade Federal de São Paulo (UNIFESP), São Paulo, SP, Brazil; [3]Department of Pharmacology, Universidade Federal de São Paulo (UNIFESP), São Paulo, SP, Brazil

List of abbreviations

[Ca^{2+}]c Cytosolic Ca^{2+} concentration
[cAMP]c Cytosolic cAMP concentration
AC Adenylyl cyclase
cAMP Cyclic adenosine monophosphate
CCB Ca^{2+}channel blockers
cGMP Cyclic guanosine monophosphate
CREB cAMP response element binding protein
DHP Dihydropyridine
Epac Exchange protein activated by cAMP
ER Endoplasmic reticulum
GC Guanylyl cyclase
HCN Hyperpolarization-activated cyclic nucleotide-gated ion channels
IP3 Inositol trisphosphate
IP3R Inositol trisphosphate receptors
MCU Mitochondrial Ca^{2+} uniporter
mNCX Mitochondrial Na$^+$/Ca^{2+}-ATPase
MPTP Mitochondrial permeability transition pore
NMDA N-methyl-D-aspartate
PDE Phosphodiesterases
PKA cAMP-dependent protein kinase
PKG cGMP-dependent protein kinase
PMCA Plasma membrane Ca^{2+}-ATPase
RyR Ryanodine receptors
VACC Voltage-activated Ca^{2+} channels

Diagnosis and Management in Dementia
ISBN 978-0-12-815854-8, https://doi.org/10.1016/B978-0-12-815854-8.00033-1

Mini-dictionary of terms

Calpains Proteins belonging to the family of Ca^{2+}-dependent, nonlysosomal cysteine proteases (proteolytic enzymes) expressed ubiquitously in mammals and many other organisms.

Dysregulations of Ca^{2+} homeostasis The dysregulations of Ca^{2+} concentrations in cells and living organisms.

Forskolin An activator of adenylyl cyclases that increases the concentrations of cAMP.

James Parkinson Parkinson was the first person to describe individuals with symptoms of the disease that bears his name.

Targets Biological molecules that can be regulated by medicines to produce a therapeutic effect.

Introduction

Initially described by the English surgeon James Parkinson in 1817, Parkinson's disease (PD) is the second most common neurodegenerative disease in developed countries, and is a major medical, social, financial, and scientific problem worldwide. In 2015, PD affected 6.2 million people and resulted in about 117,400 deaths worldwide (Salat et al., 2016). This neurodegenerative disease is strongly associated with aging, exponentially increasing in incidence above the age of 65. Estimates indicate that the incidence of PD will dramatically increase with the increment of the life expectancy worldwide. Actually, the worldwide population over 65 years of age is estimated around 8.5%, but studies performed by the World Health Organization and U.S. National Institutes of Health indicate that this number will double by 2050 (Salat et al., 2016). Despite this worrying scenario and enormous research efforts, PD is still incurable and only symptomatic relief drugs are available.

PD is clinically diagnosed based on the presence of typical motor dysfunctions that include bradykinesia, rigidity, abnormal posture, and resting tremor. These dysfunctions result from the degeneration and death of dopamine (DA) neurons in the substantia nigra (SN) pars compacta (Salat et al., 2016). This neuronal loss is responsible for a striatal DA deficiency associated to intracellular inclusions, containing aggregates of α-synuclein protein (α-syn) called Lewy body, commonly found in the brains of patients with clinical PD (Poewe et al., 2017). Although there are signs of a distributed neuropathology, as judged by Lewy body formation, the motor symptoms of PD are the first to clinically appear and are clearly linked to the degeneration and death of SN DA neurons (Braak et al., 2004; Poewe et al., 2017). Attenuation of these dysfunctions by drugs that produce an increment of dopaminergic neurotransmission has been used to confirm the clinical diagnosis of PD (Hayes et al., 2010; Salat et al., 2016). PD is also associated with many nonmotor symptoms that add to an overall disability. Thus, early falls or autonomic symptoms, accompanied by a positive response to PD medicines, should raise evidence about the diagnosis (Salat et al., 2016). Functional imaging of the dopaminergic cerebral areas obtained by a single photon emission computed tomography, or positron emission tomography (PET), can be very useful in the diagnosis of early PD (Salat et al., 2016).

PET studies examining the rate of decline in the DA-producing cells suggest that humans have already lost 50%—70% of their SN DA neurons before they develop motor symptoms, and it has been estimated that the duration of this "presymptomatic" phase is about 5 years (Salat et al., 2016).

Unfortunately, there is no cure for PD. Current pharmacotherapy is based in a compensation of the progressive deficit of DA in SN by administering its precursor 3,4-dihydroxyphenylalanine (L-DOPA) and/or agonists of DA receptors. Combined therapy using L-DOPA and DOPA-decarboxylase inhibitors, to stimulate the synaptic neurotransmission by DA in SN, is currently the most efficient pharmacotherapy for PD (Hayes et al., 2010). Anticholinergic drugs are also used to attenuate the motor dysfunctions in PD. The use of 200 mg of L-DOPA three times daily, compared with an initial dose of 100 mg three times daily, provides slightly greater benefit for reducing the motor symptoms in PD, but at the cost of earlier wearing-off symptoms and dyskinesias (Hayes et al., 2010). Although this therapeutic strategy reveals the centrality of SN DA neurons in the motor dysfunctions related to PD, it is ineffective to interrupt or prevent the progression of this disease (see Table 33.1).

The underlying molecular pathogenesis of PD involves multiple pathways and mechanisms: α-syn proteostasis, mitochondrial function, oxidative stress, neuronal Ca^{2+} homeostasis, axonal transport, and neuroinflammation (Poewe et al., 2017). In the past 2 decades, an increasing amount of evidence has indicated that an age-related deregulation of neuronal Ca^{2+} homeostasis is intimately involved in the pathogenesis of PD and other neurodegenerative diseases (Caricati-Neto, García, & Bergantin, 2015; Pchitskaya, Popugaeva, & Bezprozvanny, 2018). In addition, epidemiological studies have demonstrated a strong linkage between a deregulation of neuronal Ca^{2+} homeostasis and the

Table 33.1 List of main medicines for current therapy to treat Parkinson's disease, including comparison of drugs (efficacy) for treating symptoms.

Main medicines to treat Parkinson's disease (PD) and their mechanism of action

Levodopa (LD) — DOPA decarboxylase enzyme substrate
Bromocriptine (BR) — dopamine receptors agonist
MAO-B inhibitors — monoamine oxidase type B inhibitors
COMT inhibitors — catechol-O-methyltransferase inhibitors
Amantadine — antiinfluenza medication that has demonstrated benefits in alleviating PD symptoms
Anticholinergics — blockers of brain receptors for acetylcholine
Efficacy
"The data revealed no evidence to support the use of early BR/LD combination therapy as a strategy to prevent or delay the onset of motor complications in the treatment of PD." Extracted from "Bromocriptine/levodopa combined versus levodopa alone for early Parkinson's disease." (Van Hilten et al., 2007).

risk of developing PD (Gudala, Kanukula, & Bansal, 2015; Pasternak et al., 2012). Thus, the precise understanding of the pathophysiological role of the deregulation of neuronal Ca^{2+} homeostasis in PD could promote important advances in the development of new therapeutic strategies for PD.

This chapter will outline the perspectives of the pharmacological modulation of neuronal signaling mediated by Ca^{2+}, and other intracellular messengers, as a new neuroprotective strategy for the drug therapy of PD.

Pathophysiological role of neuronal Ca^{2+} homeostasis in PD

As the neurons use Ca^{2+} as a messenger in multiple cellular responses, these cells require an extremely precise spatial-temporal control of Ca^{2+}-dependent processes to efficiently perform the neuronal responses. This control involves several cellular mechanisms, which include proteins (Ca^{2+} channels, transporters, and buffers), intracellular messengers (IP_3, cAMP, and others), and subcellular organelles. When neurons are stimulated, depolarization of the plasma membrane opens voltage-activated Ca^{2+} channels (VACC) allowing a rapid Ca^{2+} influx that transiently elevates the cytosolic Ca^{2+} concentration ($[Ca^{2+}]_c$) (Berridge, 2013; Caricati-Neto et al., 2013). The neurons express multiple VACC types, including the dihydropyridine (DHP)-sensitive or L-type (Cav1.1, Cav1.2, Cav1.3, Cav1.4) and DHP-insensitive (Cav2.1, Cav2.2, Cav2.3) high-voltage-activated Ca^{2+} channels, and also the low-voltage-activated or T-type (Cav3.1, Cav3.2, Cav3.3) Ca^{2+} channels. Ca^{2+} influx through L-type VACC stimulates a Ca^{2+} release from the endoplasmic reticulum (ER) through Ca^{2+} channels regulated by ryanodine receptors (RyR), amplifying $[Ca^{2+}]_c$. ER Ca^{2+} channels regulated by IP3 receptors (IP3R) also control a Ca^{2+} flux from the ER into the cytosol. These mechanisms increase $[Ca^{2+}]_c$, triggering multiple neuronal responses, including neurotransmitter release, neuronal growth, gene expression, apoptosis, synaptic plasticity, cell survival, and others (Berridge, 2013; Caricati-Neto et al., 2013).

After stimulation, the $[Ca^{2+}]_c$ returns to the basal state due to the action of several Ca^{2+}-sensitive proteins involved in the sequestration, extrusion, and buffering of Ca^{2+} (Berridge, 2013; Caricati-Neto et al., 2013). Calmodulin buffers Ca^{2+} into the cytosol, sarco-endoplasmic Ca^{2+}-ATPase (SERCA) sequestrates Ca^{2+} from the cytosol into ER, and plasma membrane Ca^{2+}-ATPase (PMCA) and Na^+/Ca^{2+}-ATPase exchanger (NCX) extrude Ca^{2+} from the cytosol into an extracellular space. Mitochondria is a vital organelle to neuronal functions and survival by supplying energy, and affects cell physiology via Ca^{2+}, ROS, and signaling proteins (Berridge, 2013; Caricati-Neto et al., 2013; Pchitskaya et al., 2018). Ca^{2+} influx into mitochondrial matrix is mainly regulated by a mitochondrial Ca^{2+} uniporter (MCU), and its efflux is regulated by a mitochondrial

Na^+/Ca^{2+}-ATPase (mNCX). Ca^{2+} influx into mitochondrial matrix mediated by MCU stimulates the mitochondrial metabolism in neurons, increasing energy production in the form of ATP. However, mitochondrial Ca^{2+} overload due to a deregulation of neuronal Ca^{2+} homeostasis activates a mitochondrial permeability transition pore (MPTP), leading to a neuronal death (Michel, Hirsch, & Hunot, 2016; Pchitskaya et al., 2018; Wang et al., 2017).

Deregulation of neuronal Ca^{2+} homeostasis and its implications for PD

Several studies have supported that an enhanced Ca^{2+} influx through L-type VACC, an impaired ability of mitochondria to buffer or cycle Ca^{2+}, and a perturbed Ca^{2+} regulation in ER Ca^{2+} stores, are importantly implicated in the death of SN DA neurons and motor dysfunctions related to PD (Caricati-Neto et al., 2015; Gudala et al., 2015; Ilijic, Guzman, & Surmeier, 2011; Mattson & Bezprozvanny, 2008; Michel et al., 2016; Pchitskaya et al., 2018; Swart & Hurley, 2016). Evidence has come from the observation that the involvement of L-type VACC during autonomous pacemaking elevates the sensitivity of SN DA neurons to mitochondrial toxins used to produce animal models of PD, suggesting that a Ca^{2+} influx via L-type VACC is an important determinant factor of selective vulnerability of these DA neurons (Surmeier et al., 2011). Studies using brain obtained from PD patients, as well as animal PD models, have shown the involvement of an upregulation of L-type VACC, especially Cav1.2 and Cav1.3 subtypes, in the degeneration and death of SN DA neurons (Ilijic et al., 2011; Swart & Hurley, 2016). In addition, a pharmacological blockade of L-type VACC by Ca^{2+} channel blockers (CCBs) prevented an L-type VACC upregulation and DA neuron depletion in the SN DA neurons, and partly restored the DA content in the striatum (Ilijic et al., 2011; Swart & Hurley, 2016). This linkage between L-type VACC dysfunctions and the risk of developing PD also has been supported by epidemiological studies (Gudala et al., 2015; Pasternak et al., 2012).

The increment of Ca^{2+} influx in DA neurons due to an L-type VACC upregulation increases a mitochondrial oxidant stress, and this stress is exacerbated by a deletion of DJ-1, a gene associated with an early onset, recessive form of PD (Surmeier et al., 2011). This increased Ca^{2+} influx in SN DA neurons drives a sustained feed-forward stimulation of mitochondrial oxidative phosphorylation (Michel et al., 2016; Surmeier et al., 2011). Although this design helps to prevent a bioenergetic failure when an activity needs to be sustained, it leads to a basal mitochondrial oxidant stress (Michel et al., 2016; Surmeier et al., 2011). Over decades, this basal oxidant stress could compromise mitochondrial function and increase the mitophagy, resulting in an increased vulnerability to other proteostatic stressors, such as an elevated α-syn expression (Michel et al., 2016; Surmeier et al., 2011). In addition, a mitochondrial Ca^{2+} overload in response to a physiological Ca^{2+} stimulation due to a reduced mNCX activity, which leads to a delayed Ca^{2+} efflux, has been implicated in the pathogenesis of familial form of PD associated to a PINK-1 deficiency (Ludtmann & Abramov, 2018). These findings support that a mitochondrial collapse and a metabolic stress in SN DA neurons due to a deregulation of neuronal Ca^{2+} homeostasis represent central converging trigger factors for an idiopathic and familial PD, and a potential therapeutic target of PD.

Unlike the vast majority of neurons in the brain, adult SN DA neurons are autonomously activated, generating broad, regular slow action potentials in the absence of a synaptic input. This pacemaking activity can be very important in maintaining ambient DA levels in regions that are innervated by these DA neurons, particularly the striatum. While most neurons rely exclusively on monovalent cation channels to drive a pacemaking, SN DA neurons also engage ion channels that allow a Ca^{2+} influx, leading to an increased $[Ca^{2+}]_c$ that contributes to an increment of intracellular Ca^{2+} signaling and neurodegeneration (Michel et al., 2016; Pchitskaya et al., 2018). This increased $[Ca^{2+}]_c$ may be due to changes in the discharge activity in DA neurons and multiple PD-related events, such as α-syn aggregation, and mitochondrial and ER dysfunctions (Michel et al., 2016; Pchitskaya et al., 2018).

Deregulation of neuronal Ca^{2+} homeostasis: molecular mechanisms

It was suggested that DA neuron survival may be compromised in situations where levels of Ca^{2+} exceed the upper or fall below the lower limit of the physiological range in the cytosol and subcellular organelles (Blandini et al., 2004). In addition, a cytosolic Ca^{2+} overload in DA neurons may result from a sustained engagement of N-methyl-D-aspartate (NMDA) glutamate receptors, as a consequence of the overactivity of subthalamic nucleus glutamatergic inputs (Blandini et al., 2004). In addition, glutamate is extracellularly increased in the vicinity of SN DA neurons (Blandini et al., 2004). In fact, an excitotoxic process may occur through a neuronal Ca^{2+} overload, then leading to neurodegenerative events, including nitrosative and oxidative stress, processes which can be enhanced by glutamate. Indeed, a mitochondrial bioenergetic collapse associated to PD may reduce the Mg^{2+} blockade of the NMDA channel pore, and as consequence may increase the sensitivity of DA neurons to a glutamate-mediated excitotoxic stress (Blandini et al., 2004).

SN DA neurons are also characterized by a Ca^{2+}-dependent pacemaking, an autonomous mode of discharge that elevates $[Ca^{2+}]_c$ due to a Ca^{2+} influx through L-type VACC (Guzman et al., 2010; Michel et al., 2016). This activity, which serves to maintain a basal DA tone in the striatum, may confer a specific vulnerability to SN DA neurons, which have a low intrinsic Ca^{2+} buffering capacity (Guzman et al., 2010; Michel et al., 2016). Indeed, a Ca^{2+} influx through an autonomous pacemaking results in an increase of basal mitochondrial oxidative stress in SN DA neurons caused by a Ca^{2+} overload (Guzman et al., 2010; Michel et al., 2016). One of the consequences of this stress is to reduce the bioenergetic reserve capacity of DA neurons, which in turn makes them particularly vulnerable in the conditions of an elevated metabolic demand (Guzman et al., 2010). This mitochondrial oxidative stress generated by a Ca^{2+} influx during a pacemaking is enhanced in SN DA neurons that lack the PD gene DJ-1, in agreement with the putative role of this gene in regulating oxidant defenses (Guzman et al., 2010). This stress was also increased in perinuclear and dendritic compartments of DA neurons, exhibiting intracellular α-syn Lewy bodies, like inclusions, formed in vitro by recruitment of preformed fibrils of α-syn (Subramaniam et al., 2014). Similarly, a progressive and sustained increase in the spike rate in mouse SN DA neurons overexpressing A53T mutant α-syn was also shown (Subramaniam et al., 2014). These findings support the direct involvement of a deregulated proteostasis and mitochondrial collapse in PD pathogenesis. This scenario suggests that a diminution of Ca^{2+} influx through the L-type VACC not only can reduce a basal mitochondrial oxidant stress in SN DA neurons but also can protect these neurons from damage caused by mitochondrial toxins, such as MPP^+ and rotenone, that collapse mitochondrial respiration and energy production (Guzman et al., 2010; Michel et al., 2016; Wang et al., 2017).

Regardless of the nature of the mechanism actually causing a neuronal Ca^{2+} overload, the capacity of the mitochondria to handle Ca^{2+} probably remains crucial to prevent neurodegenerative diseases (Ludtmann & Abramov, 2018). Evidence suggests that a mitochondrial Ca^{2+} overload in response to a physiological Ca^{2+} stimulation, detected in the familial form of PD associated to PINK-1 deficiency, is intimately related to a delayed mitochondrial Ca^{2+} efflux due to reduced mNCX activity (Ludtmann & Abramov, 2018). Thus, mitochondrial Ca^{2+} overload and ensuing PD-neurodegenerative events may directly result from increased Ca^{2+} transport from the cytosol into the mitochondria via MCU, combined with delayed Ca^{2+} efflux from mitochondria due to reduced mNCX activity (Ludtmann & Abramov, 2018). Often in close apposition to the ER, mitochondria can also accumulate Ca^{2+} into the matrix through the coordinated activation of ER Ca^{2+} channels regulated by IP_3R receptors and MCU (Ludtmann & Abramov, 2018). Therefore, mitochondrial Ca^{2+} overload and mitochondrial oxidative stress in DA neurons may also be due to the fact that too much Ca^{2+} is transported from cytosol into the ER via SERCA, an energy costly process that may as such further amplify a mitochondrial collapse in a vicious cycle (Guzman et al., 2010; Michel et al., 2016; Wang et al., 2017).

Studies performed in aged mice, that express disease-relevant levels of a wild-type α-syn from the complete human SNCA locus, showed decreased discharge frequencies in SN DA neurons (Janezic et al., 2013). Considering healthy DA neurons present an absence of overt aggregation pathology, one may assume that α-syn preaggregates were the cause of a reduced spiking, and that this deficit may represent an early biomarker of neurodegeneration (Janezic et al., 2013). One way to reconcile this set of results with data implicating an activity-dependent Ca^{2+} influx in DA neuronal death would be to assume that DA neurons go successively into degeneration through hypo- and hyperactive phases, during which they endure a Ca^{2+} deficiency and a Ca^{2+} overload, respectively. This proposal has been supported by studies using the MitoPark mouse, a genetic mitochondrial model of PD in which silent DA neurons physiologically appear less compromised than the hyperactive ones (Good et al., 2011). The contribution of activity, and Ca^{2+} deficits, to the death of DA neurons may imply that these neurons have a fundamental need for proteins regulated by Ca^{2+}. Thus, PI3K may represent one of the cytosolic proteins required for cellular survival (Ries et al., 2009).

Ca^{2+} stored in the ER may also enable DA neurons to regulate their own survival. In particular, elevating basal $[Ca^{2+}]_c$ due to an activation of ER Ca^{2+} channels regulated by RyR provided protection to midbrain DA neurons in several culture paradigms where neurodegeneration is either spontaneous or induced by a trophic support deprivation or MPP^+ intoxication (Guerreiro et al., 2008). The beneficial action of ER Ca^{2+} mobilization was attributed to the action of cytosolic Ca^{2+} on a putative protein target required for DA cell survival, but it was also proposed that the resultant decrease in ER Ca^{2+} load may be beneficial by limiting mitochondrial Ca^{2+} overload, occurring through ER Ca^{2+} mobilization. However, reducing Ca^{2+} shuttling from ER into the mitochondria through a blockade of ER IP_3R or MCU was detrimental for DA neurons, indicating that a Ca^{2+} transport between these two subcellular organelles maintains the mitochondrial bioenergetic machinery functional in DA neurons, and ultimately preserves their survival (Calì et al., 2012). Thus, an early and concomitant depletion in mitochondrial Ca^{2+}, and ATP, preceded DA neuronal loss in midbrain cultures exposed to the mitochondrial toxin MPP^+ (Calì et al., 2012). These findings need to be in parallel with data showing that the PD protein α-syn facilitates mitochondrial Ca^{2+} transients elicited by IP_3R activation, whereas α-syn loss of function, in addition to reducing Ca^{2+} fluxes into the mitochondria, also results in an increased autophagy (Calì et al., 2012). In accordance with these findings, α-syn was reported to associate with mitochondria-associated ER membranes, a structural and functional distinct subdomain of the ER (Guardia-Laguarta et al., 2014). It was shown that other PD-associated proteins, DJ-1 and parkin, facilitate ER-mitochondria tethering upon a stimulation of IP_3R, whereas mutated forms of DJ-1 and parkin siRNA (small interfering RNA) impaired this process (Ottolini et al., 2013).

Increase of $[Ca^{2+}]_c$ also stimulates DA biosynthesis in DA neurons through the Ca^{2+}-dependent activation of tyrosine hydroxylase (TH). TH catalyzes the hydroxylation of tyrosine to L-DOPA and is the rate-limiting enzyme of DA biosynthesis in DA neurons (Rittenhouse & Zigmond, 1999). Thus, neuronal Ca^{2+} overload produces neuronal damage due to a toxic impact of L-DOPA via an autoxidation of the DA or posttranslational modifications of α-syn (Mosharov et al., 2009). In addition, an increase of $[Ca^{2+}]_c$ may also exert adverse effects for DA neurons through an activation of calpains, a family of Ca^{2+}-dependent cysteine proteases known to induce a cleavage of cdk5, a protein that has a facilitating role in DA cell death (Dufty et al., 2007). In fact, the presence of C-terminal calpain-cleaved α-syn has been detected in brain of PD patients (Dufty et al., 2007). Overexpression of the endogenous and specific inhibitor of calpain, such as calpastatin, importantly reduced α-syn-positive aggregates in human A30P α-syn transgenic mice (Diepenbroek et al., 2014).

In summary, the synergistic interactions between α-syn, Ca^{2+}, and DA may lead to an imbalanced protein turnover and selective susceptibility of SN DA neurons to death. Pharmacological modulation of any of these toxicity mediators can be beneficial for the survival of SN DA neurons, providing multiple opportunities for targeted drug interventions aimed at modifying the PD progression. Thus, the age-related deregulation of neuronal Ca^{2+} homeostasis, which leads to degeneration and death of SN DA neurons, increasing the risk of developing PD, could be adequately attenuated by drugs that modulate the neuronal Ca^{2+} signaling (Caricati-Neto et al., 2015). The following diagram summarizes the previous discussion (Diagram 33.1).

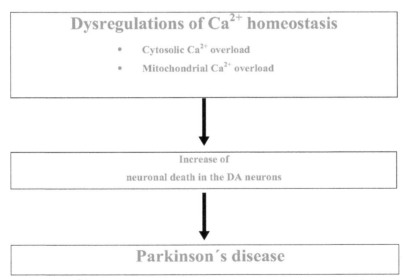

Diagram 33.1 The Ca^{2+} signaling dysregulations and their consequences: Parkinson's disease.

Neuronal Ca^{2+} signaling as a new neuroprotective target for PD therapy

Synaptic terminals are particularly vulnerable to a Ca^{2+}-mediated degeneration because they experience repeated bouts of Ca^{2+} influx and have unusual high energy requirements to support their ion-homeostatic and signaling systems. A potential clue to the vulnerability of SN DA neurons is their increasing reliance on Ca^{2+} channels to maintain an autonomous activity with age. This reliance could pose a sustained metabolic stress on mitochondria, accelerating cellular ageing and death. Ca^{2+} influx underlying autonomous activity in DA neurons is closely related to the L-type VACC, and thus pharmacological modulation of these Ca^{2+} channels by CCB could force these neurons to revert to a juvenile Ca^{2+} signaling to generate an autonomous activity and maintain the functional integrity of mitochondria (Caricati-Neto et al., 2015). This reversion produced by CCB could confer protection against a neurotoxic stimulus that produces neuronal degeneration, pointing to a potential neuroprotective strategy with a drug class that has been safely used in human beings for decades (Caricati-Neto et al., 2015). Thus, due to their physiological importance in synaptic plasticity and their involvement in the degeneration and death of SN DA neurons, the L-type VACC have become a strategic target in PD pharmacotherapy (Caricati-Neto et al., 2015; Pchitskaya et al., 2018).

Ca^{2+} channel blockers as new neuroprotective strategies for PD therapy: epidemiological aspects

Almost 50 years after the discovery that drugs like nifedipine, verapamil, and diltiazem exert their vasorelaxation and cardiodepressant effects by selectively blocking Ca^{2+} influx through L-type VACC in vascular and cardiac myocytes, it was proposed that these CCBs could be used to treat neurodegenerative diseases due to their neuroprotective effects. Numerous studies have confirmed this proposal (Caricati-Neto et al., 2015; Gudala et al., 2015; Pchitskaya et al., 2018; Swart & Hurley, 2016; Wang et al., 2017). Experimental studies have shown that isradipine, a nifedipine analogous, prevented against an MPTP-induced L-type VACC upregulation (especially Cav1.2 and Cav1.3 subunits) and DA neuron depletion in the SN DA neurons, and partly restored the DA content in the striatum (Pchitskaya et al., 2018; Wang et al., 2017). Isradipine also acts as a potent activator of antiageing neuroprotective protein sirtuin 1 (Mai et al., 2009), which appears to regulate autophagy function. Isradipine produced a dose-dependent sparing of DA fibers and cell bodies at concentrations achievable in humans (Ilijic et al., 2011). A phase II clinical trial showed that isradipine was safely tolerated by patients with PD, and a phase III trial is currently underway to determine whether treatment with isradipine is neuroprotective, and therefore able to slow the progression of PD (Swart & Hurley, 2016). Analyzing data from PubMed, EBSCO, and the Cochrane library, a recent meta-analysis involving 11,941 PD cases clearly showed a strong association between the L-type CCB use and lower risk of PD (Gudala et al., 2015). This study showed that distinct classes of L-type CCB, including DHP (nifedipine derivates) and non-DHP (verapamil and diltiazem derivates), significantly reduced the PD incidence (Gudala et al., 2015). Another historical cohort study, involving 461,984 person-years of about a decade follow-up, demonstrated the association between CCB use and lower risk of PD (Pasternak et al., 2012). In addition, L-type CCB users aged >65 years showed lower PD incidence (Pasternak et al., 2012). Among patients with PD, L-type CCB use was associated with reduced risk of death (Pasternak et al., 2012). These studies support that the use of L-type CCB is associated with a reduced risk of PD, particularly in older patients, and with reduced mortality among patients with PD.

Ca^{2+}/cAMP signaling as new neuroprotective targets for PD therapy

Recent studies suggest that the neuroprotective effects of L-type CCB can be significantly potentiated by drugs that increase the intracellular levels of cyclic adenosine monophosphate (cAMP), classified as cAMP-enhancer compounds, such as phosphodiesterase (PDE) inhibitors and adenylyl cyclase (AC) activators (Bergantin, 2017a,b, 2018; Bergantin & Caricati-Neto, 2016a–d, 2017; Caricati-Neto et al., 2015, 2017; Caricati-Neto & Bergantin, 2016a–b, 2017a–g, Table 33.2).

Ca^{2+} and cAMP are important messengers in signal transduction cascades and essential in cellular signaling in a variety of neuronal functions, and the functional interaction between the intracellular signaling pathways mediated by Ca^{2+} and cAMP (Ca^{2+}/cAMP signaling) are involved in several neuronal responses, including neurotransmitter release, neuronal plasticity, and survival (Caricati-Neto et al., 2015; Caricati-Neto & Bergantin, 2017a–e). Thus, this potentiation of neuroprotective effects of L-type CCB results from the pharmacological modulation of Ca^{2+}/cAMP signaling (Caricati-Neto et al., 2015;

Table 33.2 CCB and cAMP-enhancer compounds.

Ca^{2+} channel blockers (CCB)	cAMP-enhancer compounds
1. Verapamil	1. Rolipram
2. Nifedipine	2. 3-isobutyl-1-methylxanthine (IBMX)
3. Diltiazem	3. Forskolin
4. Isradipine	4. Aminophylline
5. Amlodipine	5. Theophylline
6. Nicardipine	6. Paraxanthine

Caricati-Neto & Bergantin, 2017a—e). As the AC activity is finely modulated by a cytosolic Ca^{2+}, the reduction of $[Ca^{2+}]_c$ produced by a partial blockade of L–type VACC by CCB (in low concentration) increments the AC activity and cAMP biosynthesis, increasing cAMP-mediated neuronal responses due to an elevation of $[cAMP]_c$ (Caricati-Neto et al., 2015; Caricati-Neto & Bergantin, 2017a—e). This increment of $[cAMP]_c$ increases the activity of cAMP-dependent protein kinase (PKA), which in turn increases the phosphorylation of cAMP response element binding protein (CREB) (Caricati-Neto et al., 2015; Caricati-Neto & Bergantin, 2017a—e). CREB is a transcription factor associated with several neuronal functions, including memory and synaptic plasticity (Caricati-Neto et al., 2015; Caricati-Neto & Bergantin, 2017a—e). In response to signals at the cell surface, CREB is phosphorylated in the nucleus by various protein kinases via secondary messengers, such as cAMP and/or Ca^{2+}, for regulating specific genes (Caricati-Neto et al., 2015; Caricati-Neto & Bergantin, 2017a—e). As the PDE regulates $[cAMP]_c$, PDE inhibitors potentiate the cAMP/PKA/CREB signaling, preserving the neuronal function and survival (Caricati-Neto et al., 2015; Caricati-Neto & Bergantin, 2017a—e).

Over the past decades, the involvement of an age-related deregulation of synthesis, execution, and/or degradation of cyclic nucleotide signaling in neurodegenerative diseases aroused the interest of scientists in the role of cyclic nucleotide signaling in neuronal survival (Kelly, 2018; Li et al., 2011). This deregulation includes: abnormal activity of AC, cAMP, PKA, CREB, exchange protein activated by cAMP (Epac), hyperpolarization-activated cyclic nucleotide-gated ion channels (HCNs), soluble and particulate guanylyl cyclase (GC), cGMP, cGMP-dependent protein kinase (PKG), and PDE (Kelly, 2018; Li et al., 2011). PDEs are fundamental enzymes that belong to signal transduction cascade, hydrolyzing cAMP and/or cGMP. PDEs are classified into 11 subtypes, as cAMP (PDE 4,7,8), cGMP (PDE 5,6,9), and double-specific PDE (PDE 1,2,3,10,11) (Kelly, 2018). The PDE4s are essential regulators of cAMP abundance in the central nervous system through their ability to regulate PKA activity, the CREB phosphorylation, and other important elements of signal transduction (Kelly, 2018; Li et al., 2011). CREB signaling has been recently involved in several brain pathological conditions, including neurodegenerative disorders (Kelly, 2018; Li et al., 2011). The increase of $[cAMP]_c$ induced by PDE4 inhibition enhances the CREB phosphorylation, and transcription of proteins related to synaptic plasticity and neuronal survival, producing neuroprotection (Caricati-Neto et al., 2015; Caricati-Neto & Bergantin, 2017a—e; Kelly, 2018; Li et al., 2011). Thus, the pharmacological modulation of neuronal Ca^{2+}/cAMP signaling by the combined use of L-type CCB and PDE4 inhibitors (see Fig. 33.1) could prevent neuronal dysfunction and death by an attenuation of cytosolic Ca^{2+} overload, and stimulation of cellular survival pathways regulated by the cAMP/PKA/CREB signaling pathway (Caricati-Neto et al., 2015; Caricati-Neto & Bergantin, 2017a—d). Additionally, this pharmacological strategy could attenuate or prevent a motor deficit due to an increase in central DA neurotransmission caused by an increment in DA release from DA neurons (Caricati-Neto et al., 2015; Caricati-Neto & Bergantin, 2017a—e) (Fig. 33.1).

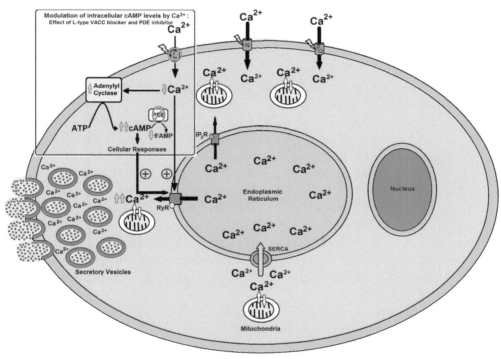

Neuroendocrine Cell   Designed by Caricati-Neto and Bergantin

Figure 33.1 Theoretical model of the pharmacological modulation of neuronal Ca^{2+}/cAMP signaling mediated by L-type CCB and PDE inhibitors. *(Adapted from Caricati-Neto, A., & Bergantin, L. B. (2017a). A new hope for the therapy of Alzheimer's and Parkinson's diseases: Pharmacological modulation of neural Ca^{2+}/cAMP signaling interaction. Brain and Nervous System Current Research, 1(1), 1—2; Caricati-Neto, A., & Bergantin, L. B. (2017b). A new hope for the therapy of the neurodegenerative diseases: Pharmacological modulation of neural Ca^{2+}/cAMP signaling interaction. Journal of Pharmacology Research Drug Design, 1(1), 1—2; Caricati-Neto, A., & Bergantin, L. B. (2017c). Novel therapeutics from old pharmaceuticals: The Ca^{2+}/cAMP signalling interaction as a new pharmacological target for treatment of diseases related to aging. Journal of Neurological Neurodisability, 1(4), 1—12; Caricati-Neto, A., & Bergantin, L. B. (2017d). Pharmaceutical intervention on Ca^{2+}/cAMP signaling interaction: Benefits for combating neuro-degeneration and diseases related to aging. International Journal of Human Anatomy, 1(1), 1—6; Caricati-Neto, A., & Bergantin, L. B. (2017e). Perspectives of the pharmacological modulation of Ca^{2+}/cAMP signaling interaction as a new neuroprotector therapeutic strategy for amyotrophic lateral sclerosis (ALS). Biomedical Journal of Scientific and Technical Research, 1, 1.)*

Key facts about James Parkinson

- He first described the disease that would bear his name as paralysis agitans or shaking palsy.
- He was born on April 11, 1755, in London, England.
- The first volume of his *Organic Remains of a Former World* was published in 1804.
- His alma mater is The London Hospital.
- He died on December 21, 1824, after a stroke that interfered with his speech.

Summary points

- Degeneration and death of SN DA neurons caused by synergistic interactions between α-syn, Ca^{2+}, and DA are the neuropathological hallmarks of PD.
- Pharmacological modulation of any of these toxicity mediators can be beneficial for the survival of these neurons, providing multiple opportunities for targeted drug interventions aimed at modifying PD progression.
- Pharmacological modulation of neuronal Ca^{2+}/cAMP signaling by the combined use of L-type CCB and PDE4 inhibitors could prevent neuronal dysfunction and death by an attenuation of cytosolic Ca^{2+} overload and stimulation of cellular survival pathways regulated by the cAMP/PKA/CREB signaling pathway.
- Pharmacological modulation of neuronal Ca^{2+}/cAMP signaling could also attenuate or prevent motor deficit due to an increment in DA release from DA neurons.
- An increase of $[Ca^{2+}]_c$ may also exert adverse effects for DA neurons through an activation of calpains.
- Several studies have supported that an enhanced Ca^{2+} influx through L-type VACC, an impaired ability of mitochondria to buffer or cycle Ca^{2+}, and a perturbed Ca^{2+} regulation in ER Ca^{2+} stores are importantly implicated in the death of SN DA neurons and motor dysfunctions related to PD.

References

Bergantin, L. B. (2017a). Neurodegenerative diseases: Where to go from now? Thought provoking through Ca^{2+}/cAMP signaling interaction. *Brain Disorders and Therapy, 6,* e125. https://doi.org/10.4172/2168-975X.1000e125.

Bergantin, L. B. (2017b). Neurological disorders: Is there a horizon? Emerging ideas from the interaction between Ca^{2+} and camp signaling pathways. *Journal of Neurological Disorder, 5,* e124. https://doi.org/10.4172/2329-6895.1000e124.

Bergantin, L. B. (2018). Dysregulation of intracellular Ca^{2+} and camp signalling:plausible targets for neurological disorders. *Journal of Neurological Disorder, 6,* e125. https://doi.org/10.4172/2329-6895.1000e125.

Bergantin, L. B., & Caricati-Neto, A. (2016a). *From discovering "calcium paradox" to Ca^{2+}/cAMP interaction: Impact in human health and disease* (1st ed.). Riga: Scholars Press.

Bergantin, L. B., & Caricati-Neto, A. (2016b). Challenges for the pharmacological treatment of neurological and psychiatric disorders: Implications of the Ca^{2+}/cAMP intracellular signalling interaction. *European Journal of Pharmacology, 788,* 255—260.

Bergantin, L. B., & Caricati-Neto, A. (2016c). Recent advances in pharmacotherapy of neurological and psychiatric disorders promoted by discovery of the role of Ca^{2+}/cAMP signaling interaction in the neurotransmission and neuroprotection. *Advance Pharmaceutical Journal, 1*(3), 66.

Bergantin, L. B., & Caricati-Neto, A. (2016d). Novel insights for therapy of Parkinson's disease: Pharmacological modulation of the Ca^{2+}/cAMP signalling interaction. *Austin Neurology and Neurosciences, 1*(2), 1009.

Bergantin, L. B., & Caricati-Neto, A. (2017). Emerging concepts for neuroscience field from Ca^{2+}/cAMP signalling interaction. *The Journal of Neurology and Experimental Neuroscience, 3*(1), 29—32. https://doi.org/10.17756/jnen.2017-024.

Berridge, M. J. (2013). Dysregulation of neural calcium signaling in Alzheimer disease, bipolar disorder and schizophrenia. *Prion, 7*(1), 2—13.

Blandini, F., Braunewell, K. H., Manahan-Vaughan, D., et al. (2004). Neurodegeneration and energy metabolism: From chemistry to clinics. *Cell Death and Differentiation, 11*, 479—484.

Braak, H., Ghebremedhin, E., Rub, U., et al. (2004). Stages in the development of Parkinson's disease-related pathology. *Cell and Tissue Research, 318*, 121—134.

Calí, T., Ottolini, D., Negro, A., et al. (2012). α-Synuclein controls mitochondrial calcium homeostasis by enhancing endoplasmic reticulum-mitochondria interactions. *Journal of Biological Chemistry, 287*, 17914—17929.

Caricati-Neto, A., & Bergantin, L. B. (2016a). New therapeutic strategy of Alzheimer's and Parkinson's diseases: Pharmacological modulation of neural Ca^{2+}/cAMP intracellular signaling interaction. *Asian Journal of Pharmacy and Pharmacology, 2*(6), 136—143.

Caricati-Neto, A., & Bergantin, L. B. (2016b). Recent discovery of the role of Ca^{2+}/cAMP signaling interaction in the neurotransmission and neuroprotection and its impact in pharmacotherapy of neurological and psychiatric disorders. *Asian Journal of Pharmacy and Pharmacology, 3*, 45.

Caricati-Neto, A., & Bergantin, L. B. (2017a). A new hope for the therapy of Alzheimer's and Parkinson's diseases: Pharmacological modulation of neural Ca^{2+}/cAMP signaling interaction. *Brain and Nervous System Current Research, 1*(1), 1—2.

Caricati-Neto, A., & Bergantin, L. B. (2017b). A new hope for the therapy of the neurodegenerative diseases: Pharmacological modulation of neural Ca^{2+}/cAMP signaling interaction. *Journal of Pharmacology Research Drug Design, 1*(1), 1—2.

Caricati-Neto, A., & Bergantin, L. B. (2017c). Novel therapeutics from old pharmaceuticals: The Ca^{2+}/cAMP signalling interaction as a new pharmacological target for treatment of diseases related to aging. *Journal of Neurological Neurodisability, 1*(4), 1—12.

Caricati-Neto, A., & Bergantin, L. B. (2017d). Pharmaceutical intervention on Ca^{2+}/cAMP signaling interaction: Benefits for combating neurodegeneration and diseases related to aging. *International Journal of Human Anatomy, 1*(1), 1—6.

Caricati-Neto, A., & Bergantin, L. B. (2017e). Perspectives of the pharmacological modulation of Ca^{2+}/cAMP signaling interaction as a new neuroprotector therapeutic strategy for amyotrophic lateral sclerosis (ALS). *Biomedical Journal of Scientific and Technical Research, 1*, 1.

Caricati-Neto, A., & Bergantin, L. B. (2017f). From a "eureka insight" to a novel potential therapeutic target to treat Parkinson's disease: The Ca^{2+}/camp signalling interaction. *Journal of Systems and Integrative Neuroscience, 4*. https://doi.org/10.15761/JSIN.1000187.

Caricati-Neto, A., & Bergantin, L. B. (2017g). The passion of a scientific discovery: The "calcium paradox" due to Ca^{2+}/camp interaction. *Journal of Systems and Integrative Neuroscience, 3*. https://doi.org/10.15761/JSIN.1000186.

Caricati-Neto, A., García, A. G., & Bergantin, L. B. (2015). Pharmacological implications of the Ca^{2+}/cAMP signalling interaction: From risk for antihypertensive therapy to potential beneficial for neurological and psychiatric disorders. *Pharmacology Research and Perspectives, 3*(5), e00181.

Caricati-Neto, A., Padin, J. F., Silva-Junior, E. D., et al. (2013). Novel features on the regulation by mitochondria of calcium and secretion transients in chromaffin cells challenged with acetylcholine at 37 degrees C. *Physics Reports, 1*(7), e00182.

Caricati-Neto, A., Scorza, F. A., Scorza, C. A., Cysneiros, R. M., Menezes- Rodrigues, F. S., et al. (2017). Sudden unexpected death in Parkinson's disease and the pharmacological modulation of the Ca^{2+}/

cAMP signaling interaction: A shot of good news. *Brain Disorders and Therapy, 6,* 231. https://doi.org/10.4172/2168-975X.1000231.

Diepenbroek, M., Casadei, N., Esmer, H., et al. (2014). Overexpression of the calpain-specific inhibitor calpastatin reduces human alpha-Synuclein processing, aggregation and synaptic impairment in [A30P]αSyn transgenic mice. *Human Molecular Genetics, 23,* 3975—3989.

Dufty, B. M., Warner, L. R., Hou, S. T., et al. (2007). Calpain cleavage of alpha-synuclein: Connecting proteolytic processing to disease linked aggregation. *American Journal of Pathology, 170,* 1725—1738.

Good, C. H., Hoffman, A. F., Hoffer, B. J., et al. (2011). Impaired nigrostriatal function precedes behavioral deficits in a genetic mitochondrial model of Parkinson's disease. *The FASEB Journal, 25,* 1333—1344.

Guardia-Laguarta, C., Area-Gomez, E., Rüb, C., et al. (2014). α-Synuclein is localized to mitochondria-associated ER membranes. *Journal of Neuroscience, 34,* 249—259.

Gudala, K., Kanukula, R., & Bansal, D. (2015). Reduced risk of Parkinson's disease in users of calcium channel blockers: A meta-analysis. *International Journal of Chronic Diseases,* 697404.

Guerreiro, S., Toulorge, D., Hirsch, E., et al. (2008). Paraxanthine, the primary metabolite of caffeine, provides protection against dopaminergic cell death via stimulation of ryanodine receptor channels. *Molecular Pharmacology, 74,* 980—989.

Guzman, J., Sanchez-Padilla, J., Wokosin, D., et al. (2010). Oxidant stress evoked by pacemaking in dopaminergic neurons is attenuated by DJ-1. *Nature, 468,* 696—700.

Hayes, M. W., Fung, V. S., Kimber, T. E., et al. (2010). Current concepts in the management of Parkinson disease. *Medical Journal of Australia, 192*(3), 144—149.

Ilijic, E., Guzman, J. N., & Surmeier, D. J. (2011). The L-type channel antagonist isradipine is neuroprotective in a mouse model of Parkinson's disease. *Neurobiology of Disease, 43*(2), 364—371.

Janezic, S., Threlfell, S., Dodson, P., et al. (2013). Deficits in dopaminergic transmission precede neuron loss and dysfunction in a new Parkinson model. *Proceedings of the National Academy of Sciences of the United States of America, 110,* E4016—E4025.

Kelly, M. P. (2018). Cyclic nucleotide signaling changes associated with normal aging and age-related diseases of the brain. *Cellular Signalling, 42,* 281—291.

Li, Y., Cheng, Y., Huang, Y., et al. (2011). Phosphodiesterase-4D knock-out and RNA interference-mediated knock-down enhance memory and increase hippocampal neurogenesis via increased cAMP signaling. *Journal of Neuroscience, 31,* 172—183.

Ludtmann, M. H., & Abramov, A. Y. (2018). Mitochondrial calcium imbalance in Parkinson's disease. *Neuroscience Letters, 663,* 86—90.

Mai, A., Valente, S., Meade, S., et al. (2009). Study of 1,4-dihydropyridine structural scaffold: Discovery of novel sirtuin activators and inhibitors. *Journal of Medicinal Chemistry, 52,* 5496—5504.

Mattson, M. P., & Bezprozvanny, I. (2008). Neuronal calcium mishandling and the pathogenesis of Alzheimer's disease. *Trends in Neurosciences, 31*(9), 454—463.

Michel, P., Hirsch, E., & Hunot, S. (2016). Understanding dopaminergic cell death pathways in Parkinson disease. *Neuron, 90*(4), 675—691.

Mosharov, E., Larsen, K., Kanter, E., et al. (2009). Interplay between cytosolic dopamine, calcium, and alpha-synuclein causes selective death of substantia nigra neurons. *Neuron, 62*(2), 218—229.

Ottolini, D., Calì, T., Negro, A., et al. (2013). The Parkinson disease related protein DJ-1 counteracts mitochondrial impairment induced by the tumor suppressor protein p53 by enhancing endoplasmic reticulum-mitochondria tethering. *Human Molecular Genetics, 22,* 2152—2168.

Pasternak, B., Svanström, H., Nielsen, N., et al. (2012). Use of calcium channel blockers and Parkinson's disease. *American Journal of Epidemiology, 175*(7), 627—635.

Pchitskaya, E., Popugaeva, E., & Bezprozvanny, I. (2018). Calcium signaling and molecular mechanisms underlying neurodegenerative diseases. *Cell Calcium, 70,* 87—94.

Poewe, W., Seppi, K., Tanner, C., et al. (2017). Parkinson disease. *Nature Reviews Disease Primers, 3,* 17013.

Ries, V., Cheng, H., Baohan, A., et al. (2009). Regulation of the postnatal development of dopamine neurons of the substantia nigra in vivo by Akt/protein kinase B. *Journal of Neurochemistry, 110,* 23—33.

Rittenhouse, A., & Zigmond, R. (1999). Role of N- and L-type calcium channels in depolarization-induced activation of tyrosine hydroxylase and release of norepinephrine by sympathetic cell bodies and nerve terminals. *Journal of Neurobiology, 40*, 137–148.

Salat, D., Noyce, A., Schrag, A., et al. (2016). Challenges of modifying disease progression in prediagnostic Parkinson's disease. *The Lancet Neurology, 15*(6), 637–648.

Subramaniam, M., Althof, D., Gispert, S., et al. (2014). Mutant α-synuclein enhances firing frequencies in dopamine substantia nigra neurons by oxidative impairment of A-type potassium channels. *Journal of Neuroscience, 34*, 13586–13599.

Surmeier, J., Guzman, J., Sanchez-Padilla, J., et al. (2011). The role of calcium and mitochondrial oxidant stress in the loss of substantia nigra pars compacta dopaminergic neurons in Parkinson's disease. *Neuroscience, 198*, 221–231.

Swart, T., & Hurley, M. (2016). Calcium channel antagonists as disease-modifying therapy for Parkinson's disease: Therapeutic rationale and current status. *CNS Drugs, 30*(12), 1127–1135.

Van Hilten, Ramaker, C. C., Stowe, R., & Ives, N. (2007). *Bromocriptine/levodopa combined versus levodopa alone for early Parkinson's disease.* Cochrane Library. https://doi.org/10.1002/14651858.CD003634.pub2.

Wang, Q. M., Xu, Y., Liu, S., et al. (2017). Isradipine attenuates MPTP-induced dopamine neuron degeneration by inhibiting up-regulation of L-type calcium channels and iron accumulation in the substantia nigra of mice. *Oncotarget, 8*(29), 47284–47295.

CHAPTER 34

Pitfalls and possible solutions for research and development of dementia therapies

Thomas Müller
Department of Neurology, St. Joseph Hospital Berlin-Weissensee, Berlin, Germany

Introduction

The most frequent chronic neurodegenerative brain disorder is dementia, particularly Alzheimer's disease (AD). In the past years, many compounds failed in clinical programs for the development of new treatment paradigms. Disillusionment and frustration is high in clinical researchers, their patients, and the pharmaceutical industry (Müller & Foley, 2016). This chapter aims to discuss probable fundamental causes for this negative development and to provide suggestions for a way out of this dilemma of clinical research and drug approval following an initial comprehensive review of introductory current perceptions of dementia.

Dementia is a syndrome

In the clinic, dementia is characterized by cognitive deterioration with progression of short-term memory loss, orientation problems, and reasoning as main important clinical symptoms. They are supplemented by further individually varying psychopathological symptoms. Out of the spectrum of dementia forms, AD is looked upon as the most common form of dementia.

AD accounts for 60%—70% of all cases of dementia according to epidemiological trials (Emmerzaal, Kiliaan, & Gustafson, 2015; Takizawa, Thompson, Van, Faure, & Maier, 2015). Current translational research focus on "pure" AD forms. However, the common dementia syndrome is mostly a mixture between vascular dementia and AD on top of additional manifestations of further infectious or metabolic morbidities, i.e., diabetes, hypertension, etc., in clinical practice (Fig. 34.1). Intensity of the dementia syndrome may fluctuate depending on the quality and adherence to therapies for these additional disorders (Fig. 34.1).

Diagnosis and Management in Dementia
ISBN 978-0-12-815854-8, https://doi.org/10.1016/B978-0-12-815854-8.00034-3

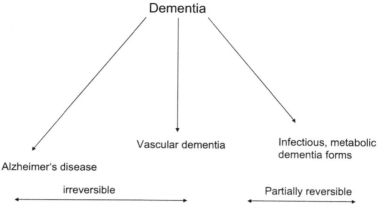

Figure 34.1 Most common dementia subtypes.

Dementia and clinical maintenance of patients

Increase in life expectancy will cause a dramatic elevation of the occurrence of dementia forms in the next decades. There are estimates that number of dementia patients will double every 20 years worldwide with 30% as mild, 40% as moderate, and 30% as severe cases in the patient population (Emmerzaal et al., 2015; Takizawa et al., 2015). The expected burden for the health care and social systems will be enormous. Development of effective drug therapies will become more and more necessary. Currently, there is ongoing research on specific therapies using the academic well-defined diagnosis criteria based on neuropathology findings (Müller, 2018). These diagnostic features often only partially and artificially reflect the whole spectrum of signs observed in the clinic. Current drug treatment in dementia patients focuses on symptomatic amelioration of quality of life for patients and their caregivers. Quite often one only tries to delay the need to stay in a nursing home. Standardized guidelines based on so-called evidence-based analyses of trials are of limited value (Müller, 2018). Off-label use of drugs is common.

Diversity between experimental findings and the clinic of dementia

To date, it is far from clear whether neuropathology findings, such as Lewy body accumulation, β-amyloid or tau protein enrichment play an active role in the ongoing chronic disease process itself in affected neurons (Müller, 2018). These processes may also just reflect well-wrapped protein garbage as a consequence of disease-affected neurons (Sian-Hulsmann, Monoranu, Strobel, & Riederer, 2015). Generally, chronic neurodegenerative processes result from different metabolic cascades. It is well known that they finally end up in cell death inducing events and well-described mechanisms. These processes induce individual different signs in the clinic. Hypotheses based on neuropathology findings support the concepts of protein misfolding. However, more generally, occurrence of protein misfolding is a first-line defense process. It involves protein refolding, which is mediated by chaperone proteins (Müller, 2018). Process failures cause protein degradation and/or accumulation. If this refolding/degradation machinery cannot prevent misfolded proteins, a stress response is activated involving upregulation of refolding and degradation processes (Kumar, Jha, Jha, Ramani, & Ambasta, 2015). Too severe stress by protein misfolding activates cell death programs (Fig. 34.2). Accordingly, currently tested experimental, potential therapeutic strategies include reduction of protein misfolding, repair of misfolded proteins, and facilitation of degradation of proteins particularly when they are so damaged without any chance for repair. However, there is a certain capacity of the human brain to compensate these initial events for considerable intervals before the clinical onset of initial mild and unspecific symptoms of the neurodegenerative disease (Müller, 2018). Thus, initiation and rate of disease progression varies individually. Preclinical and experimental researchers still primarily focus on these processes and the associated acetylcholine deficit in their models, which do not reflect the variability observed in the clinic. These drawbacks of experimental research may represent main causes, why therapeutic interventions on tau and β-amyloid metabolism failed in clinical trials. Now morphological and functional imaging techniques, such as magnetic resonance imaging tomography or positron emission tomography with various radiotracers, are increasingly used as instrumental diagnostic tools and potential biomarkers for disease progression. One radiotracer currently used is $[^{18}F]$-AV-45 (Florbetapir) for the measurement of Aβ-plaque density in AD patients. It is believed that these functional imaging techniques with specific radiotracers will help to interrogate the biological mechanisms of disease initiation, progression, and quantification of successful, potential future disease-modifying therapies. However, these kinds of imaging techniques only focus on visualization of pure forms of AD, which are rare and heterogeneous in clinical practice as aforementioned (Müller & Foley, 2016).

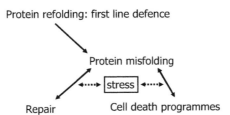

Figure 34.2 Simplified scheme for repair of misfolded protein and induction of cell death programs.

Past and current concepts for disease modification in AD patients

Aggregated $A\beta_{1-42}$ was used in $A\beta$ animal models as an approach for active immunization. Plaques diminished and cognitive deficits improved. However, the subsequently performed clinical vaccination trials were stopped due to onset of meningoencephalitis in up to 6% of the participants. There was decline of cognitive deterioration in vaccinated patients with a high titer and accordingly lowered amyloid load found in postmortem findings. No association to the corresponding amyloid exposure appeared. As the next step, the shift from active to passive immunization was performed according to findings in AD models. This approach was regarded as safer, with the ability to be better controlled, and thus more efficacious. The phase III study programs with two anti-$A\beta$ monoclonal antibodies, Bapineuzumab and Solanezumab, failed again. Thus, currently the only way to counteract AD onset or to modify the course of AD is prevention of the major risk factors for AD (Doody et al., 2014; Lannfelt et al., 2014; Li, Liu, Wang, & Jiang, 2013; Matsubara et al., 2013; Salloway et al., 2014). Arteriosclerosis disease, type II diabetes, midlife hypertension, midlife obesity, smoking, and physical inactivity are looked upon as the most common ones. The preventive, more educational therapeutic lifestyle approaches particularly make sense in subjects with a certain genetic risk factor to develop AD. The most known one is the ApoE polymorphism. There are three major isoforms, ApoE2, ApoE3, and ApoE4 of this gene, which impact the risks for developing AD. Carriers of the homozygous ApoE4 allele have the highest AD risk, ApoE3 is considered as normal, and ApoE2 is looked upon as protective. Accordingly, preclinical researchers have developed genetically predisposed models of AD (Müller Foley, 2016). The occurrence of pure genetic forms of AD is rare. However, the current tendency is to screen and test novel drugs in animals, such as knockout mice designed according to the observed genetic predisposition or defect. However, these mice sometimes do not develop all or at least even the essential dementia features. Therefore, their suitability may be limited to serve as an animal model for screening of new drugs.

Testing of symptomatic and disease-modifying treatments for dementia: Are new concepts necessary?

Fig. 34.3 describes some of the current most influential players on study designs. All of them have particular interests and concerns. In the past, the role of authorities and ethical committees was less influential, whereas the role of physicians, who are involved in the daily maintenance of dementia patients, was more pronounced. These physicians often discovered effects of drugs by clinical observation. Approval of a drug was easier. Accordingly, more therapeutic approaches survived for the real world. One of the currently marketed and used compounds for AD and other dementia forms is the

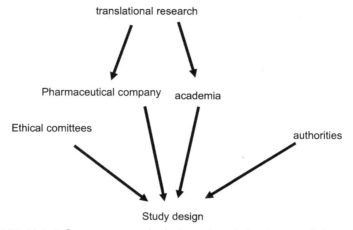

Figure 34.3 Main influencers on study designs: the missing impact of the real world.

NMDA antagonist memantine, which is now also available as a delayed-release formulation. This drug has a predominant vigilance enhancing effect and was suggested by neurologists who were involved in the daily care of AD patients (Davey, 2014). Also on the market are acetyl cholinesterase inhibitors, which are agents that aim to compensate for the cholinergic deficit in AD patients. All these compounds are currently approved. They are used either alone or in combination, but they only provide a limited symptomatic benefit on cognition and associated features, i.e., activities of daily living (Anderson & Egge, 2014; Davey, 2014; De la Torre, 2014; Tai et al., 2014; Villeneuve, Brisson, Marchant, & Gaudet, 2014). Since the onset and the signs of AD are heterogeneous and pure forms of AD are rare, one may focus on the most common dementia form, which is a mixture between AD and vascular dementia. These patient populations are more suitable to reflect the real world. Frequently, demented patients also have additional disorders. They may aggravate or even induce cognitive dysfunction or falsify or imitate cognitive deterioration, i.e., in the case of additional onset of depression or apathy. Clinical trials on cognition enhancing or disease modifying drugs mostly employ cognitive abilities measuring semistandardized neuropsychological tests (Larner, 2016). They aim to demonstrate the effect of the tested drug. They produce false outcomes, when the general health condition of the study participants worsens or improves during the course of the trial. Moreover, performance of these tests suffers from additional factors, such as motivation, education, and personality features.

In terms of disease modification, particularly in the case of testing of biological drugs, one may assume that monoclonal antibodies do not probably interfere with the function of affected neurons or interact with their downstream second-messenger systems. These drugs just aim to remove extracellularly located pathologic and/or misfolded proteins.

To date, it is far from clear whether these proteins play a role in the ongoing chronic disease process itself in the affected neurons at all or whether they just represent well-wrapped protein garbage as consequence of affected neurons. The complexity of dementia and its therapy also results from the additional appearance of various kinds of nonmotor symptoms, e.g., depression. Therefore the translational approach to test compounds, which were successful in models of AD with their focus on cognition only, had to fail in the clinical study world despite huge efforts on standardization of neuropsychological assessment tools (Larner, 2016). The limitations of many of these animal models and thus the performed experimental investigations are the focus on cholinergic neurotransmission. The models often only investigate one mechanism of neuronal cell death. Experimental researchers neglect that chronic neurodegenerative disease is a result of each other complementing metabolic events ending up in a cascade in conjunction with an accelerated ageing process. The final steps are cell death inducing events and individual different clinical signs of the neurodegenerative disease (Müller & Foley, 2016). There is a certain capacity of the human brain to compensate these initial events for a considerable interval before the clinical onset of initial mild and unspecific symptoms of the neurodegenerative disease. This so-called "neuroplasticity" phenomenon also impacts as aforementioned the velocity of disease progression also differs in an individually different manner. In summary, this heterogeneous and individ-ually different disease progression in combination with relative short trial periods may also contribute to a failure of trials on disease modification in AD. Mortality or increase of life expectancy, caregiver burden, or delay of transfer to nursing homes may be more robust clinical endpoints in terms of disease modification. However, these endpoints would demand longer study durations. There are some potential solutions or failure risk-reducing approaches out of this scenario (Fig. 34.4).

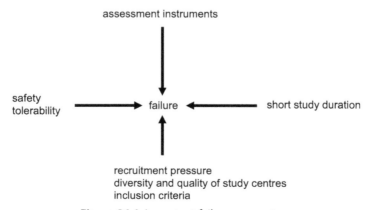

Figure 34.4 Important failure parameters.

Potential approaches out of this dilemma

Longer intervals of placebo control may limit the influence of natural fluctuations of disease progression. Use of biological markers may counteract the investigator bias to a certain extent. They will lead the way to prodromal diagnosis of dementia and then disease-modifying therapies. There are some well-known genetic forms of dementia syndromes. One approach would be to enrich the study population with only well-defined genetic forms of dementia. It may reduce the heterogeneity of disease signs and disease course, and probably of cell death mechanisms. However, it may also limit the approval of the drug to this specific population, but this should not be decided by the authorities. Instead the therapy should be approved, i.e., with an expedited procedure similar to the granted breakthrough status in the United States. The responsibility for a decision on the value of a therapy should be transferred to the real world again. Clinical practice shows that so-called "dementia at-risk individuals" are often not interested in predictive genetic examinations. It is obvious that a positive outcome may cause heavy burden for their future life. Both lack of cure and of effective disease-modifying therapeutic strategies in dementia also reduce individuals' motivation for testing and participation. Thus recruitment for trials aiming to expand the healthy interval before the conversion to mild cognitive impairment or AD dementia in individuals with a genetically determined risk will be problematic. One way out of this dilemma would be to offer genetic testing to these "dementia at-risk individuals" without telling them the outcome and keep them blind with their consent (Fig. 34.5). In this scenario, negative subjects may serve as healthy controls. Demented ApoE gene carriers may either receive the therapeutic intervention or placebo. Optionally one can even include a delayed start design. There will be a lot of concerns on this pragmatic strategy for such a novel study design, for instance, by ethical naysayers, health politicians, lawyers, and data protection officers. However, these opinion makers should respect individual decisions and pragmatic procedures. The "dementia at-risk population" will probably have less doubt, since they have nothing to lose. Moreover it should still remain their own individual responsibility and benefit-risk calculation whether they will participate in a placebo arm and stay in the whole study or not. Such designs will ease recruitment of larger cohorts and enable ApoE stratification, sex, and age (Fig. 34.5). Study outcomes will allow consistent statements and conclusions on a possible disease-modifying effect of the intervention. A further precondition will be longer study durations over several years, which is a drawback for the financing pharmaceutical company or institution. However, the current practice of the evaluation of therapeutic strategies for disease modification often suffers from too short study intervals, too small cohorts, and the focus on the translation of results from experimental models in the clinic. This concept created failures in the past and this will continue in the future. These attempts neglect the abovementioned capacity of the human brain to compensate initial dementia events

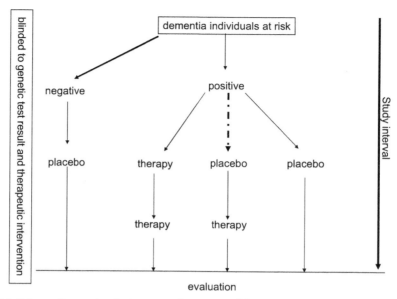

Figure 34.5 Schematic study design on disease modification. *Arrow,* condition necessary for performance of delayed start design. *(Modified from Müller, T. (2017). Investigational agents for the management of Huntington's disease.* Expert Opinion on Investigational Drugs, 26, 175–185.)

for a considerable, individual different interval before the clinical onset of initial mild and unspecific cognitive and depressive symptoms, i.e., cognitive decline, apathy, and depression.

In terms of determination of novel symptomatic treatments, better scoring instruments are needed. The employed rating scales for evaluation of drug efficacy predominantly focus on cognitive abilities in AD. Use of these scales is enforced by the authorities to achieve approval for the tested compound. As an example, the MiniMental State Examination is used as a selection criterion, but its scores have a considerable bias by the educational level of the patient (Larner, 2016). Generally, rating instruments like the neuropsychiatric inventory lack assessment of psychopathological phenomena in relation to exogenous influencing factors, i.e., well-being of caregivers etc. (Larner, 2016). Clinicians point out that noncognition-related signs are often essential limiting for quality of life. As a result, they investigate the efficacy of already available compounds often in observational or naturalistic small trials. These outcomes are not considered as essential by the authorities-driven evidence-based-medicine classification of trials. Additionally, most of the used assessment instruments are not objective. They are biased by the attitude and habits of the investigator. Cheap, easy-to-apply biological markers do not exist and need to be established (Fig. 34.6). One underestimates that the rating situation and the stress for the patient often cause an insufficient appraisal of the tested compound.

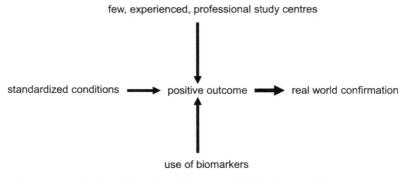

Figure 34.6 Real world confirmation scenario following positive outcomes.

The safety hysteria

Clinical researchers increasingly point out the importance of the so-called nocebo effect. This means that patient experiences a side effect once being informed of its potential occurrence (Kleine-Borgmann & Bingel, 2018; Ren & Xu, 2018). In the clinical research scenario, the side effect profile and the tolerability of a tested compound appear to have more or at least the same importance than its efficacy. In clinical practice, however, the application of a compound is often the result of a careful benefit-risk evaluation performed by the prescribing physician and the increasingly well-informed, mature patient. Nowadays compounds often only reach the market when they are well tolerated and safe. Thus a provocative hypothesis may suggest that only better placebos are currently developed for treatment of dementia. In summary, there is need to improve the whole scenario for intervention testing and approval. Currently, the existing system incapacitates patients and their prescribing physicians. Both have to discuss and to calculate the benefit—risk ratio of a therapeutic drug approach on an individual basis together and not the authorities. Their role should be restricted to inform the community on possible side effects of the therapy.

Possible solutions

A way out of this dilemma may be to inaugurate trials with standardized use of more objectively measuring instruments only in a few study units with a well-trained staff under specific, more constant condition. These organizations should recruit and represent networks of outpatient units, which overlook hundreds of patients for trials (Fig. 34.6). They should work independently of pharmaceutical companies and authorities to avoid corruption and abuse. Such a commercially financed but independent working and more professional organization of clinical research would enable better control of data quality with more profound training of the staff. Functional imaging biomarkers should be mandatory (Fig. 34.6). Their use will reduce patient numbers to demonstrate an effect. One should allow a financial compensation for patients when they participate in trials. This scenario works in healthy volunteers, but there are ethical concerns in patients. Generally, with the still growing impact of ethical committees, respectively drug authorities must be reduced again. There is some indirect evidence that such a scenario would be successful. Many tested compounds had successful phase II trial out comes before entering phase III of development. Phase II as proof-of-concept study was performed with a few investigators in a certain region. These trials demonstrate the efficacy, safety, and tolerability of an intervention. Then authorities and pharmaceutical companies themselves often perform global phase III trials with hundreds of patients. Consequently, the trial quality and experience increasingly varies between participating investigators and induces a negative bias. Since there is an enormous recruitment pressure in this scenario, not optimally suitable patients are often also included. This resulting vicious circle ends up in the failure of the tested compound. Here again, an alternative approach would be to go for better quality instead of quantity (Müller & Foley, 2016). This contradicts marketing strategies of pharmaceutical companies and demands of the authorities. However, this system with smaller studies performed with a more experimental approach to demonstrate positive benefits of a compound would also reduce the risk of failures (Fig. 34.4). Costs for drug developments will reduce. Consecutive high pricing of drugs will also decrease. However, such a cheaper research and development scenario is not enabled by drug-approving authorities currently. On the contrary, European countries already demand that pharmaceutical companies prove the advantage of the new compound over existing, respectively generic drugs. In the short run, this reduces drug costs in the health care system. However, it also probably slows development of therapies in the long run. As a result the current conditions will not solve the future dementia problem in the ageing population with a rise of life expectancy in the industrialized counties in particular in the short run. An additional approach is to protect the new compound longer than 10 years. This process would reduce the pressure for the pharmaceutical company to earn the investment in the new drug within 10 years before it becomes generic (Müller Foley, 2016).

Key facts

- Dementia as syndrome is the most frequent chronic neurodegenerative brain disorder.
- It will cause an enormous future burden for health care and social systems worldwide.
- As disease entity, features and progression of dementia differ individually.
- Diagnosis is too late due to missing, easy-to-apply biomarkers.
- Symptomatic drug treatment needs an individually adapted and not a standardized regime.

- Available animal models do not reflect the different dementia syndromes.
- Suitability of these models for translational research is limited but overestimated in experimental research.
- Common instruments for efficacy evaluation are not sensitive and specific enough.
- Regulatory authorities lack experience with patient maintenance.

Summary points

- Burden of dementia is the future problem for health care.
- Diagnosis is too late, easy-to-apply biomarkers are needed.
- The dementia syndrome is heterogeneous disease entity.
- Conditions for research and approval of new therapies of dementia are too complex.
- Future high drug costs are enforced for return of investment.
- Safety and tolerability concerns block innovative groundbreaking treatment possibilities.

References

Anderson, L. A., & Egge, R. (2014). Expanding efforts to address Alzheimer's disease: The healthy brain initiative. *Alzheimer's and Dementia: The Journal of the Alzheimer's Association, 10*, 453−456.

Davey, D. A. (2014). Alzheimer's disease and vascular dementia: One potentially preventable and modifiable disease? Part II: Management, prevention and future perspective. *Neurodegenerative Disease Management, 4*, 261−270.

De la Torre, J. C. (2014). Detection, prevention, and pre-clinical treatment of Alzheimer's disease. *Journal of Alzheimer's Disease, 42*, 431−442.

Doody, R. S., Thomas, R. G., Farlow, M., Iwatsubo, T., Vellas, B., Joffe, S., et al. (2014). Phase 3 trials of solanezumab for mild-to-moderate Alzheimer's disease. *New England Journal of Medicine, 370*, 311−321.

Emmerzaal, T. L., Kiliaan, A. J., & Gustafson, D. R. (2015). 2003−2013: A decade of body mass index, Alzheimer's disease, and dementia. *Journal of Alzheimer's Disease, 43*, 739−755.

Kleine-Borgmann, J., & Bingel, U. (2018). Nocebo effects: Neurobiological mechanisms and strategies for prevention and optimizing treatment. *International Review of Neurobiology, 138*, 271−283.

Kumar, P., Jha, N. K., Jha, S. K., Ramani, K., & Ambasta, R. K. (2015). Tau phosphorylation, molecular chaperones, and ubiquitin E3 ligase: Clinical relevance in Alzheimer's disease. *Journal of Alzheimer's Disease, 43*, 341−361.

Lannfelt, L., Moller, C., Basun, H., Osswald, G., Sehlin, D., Satlin, A., et al. (2014). Perspectives on future Alzheimer therapies: Amyloid-beta protofibrils - a new target for immunotherapy with BAN2401 in Alzheimer's disease. *Alzheimer's Research and Therapy, 6*, 16.

Larner, A. J. (2016). Correlation or limits of Agreement? Applying the bland-Altman approach to the comparison of cognitive screening instruments. *Dementia and Geriatric Cognitive Disorders, 42*, 247−254.

Li, Y., Liu, Y., Wang, Z., & Jiang, Y. (2013). Clinical trials of amyloid-based immunotherapy for Alzheimer's disease: End of beginning or beginning of end? *Expert Opinion on Biological Therapy, 13*, 1515−1522.

Matsubara, E., Takamura, A., Okamoto, Y., Oono, H., Nakata, T., Wakasaya, Y., et al. (2013). Disease modifying therapies for Alzheimer's disease targeting Abeta oligomers: Implications for therapeutic mechanisms. *BioMed Research International, 2013*, 984041.

Müller, T. (2017). Investigational agents for the management of Huntington's disease. *Expert Opinion on Investigational Drugs, 26*, 175−185.

Müller, T. (2018). Editorial regarding: Practical treatment of Lewy body disease in the clinic: Patient and physician perspectives. *Neurology and Therapy, 7*, 1−3.

Müller, T., & Foley, P. (2016). Clinical drug research in chronic central neurodegenerative disorders. *Expert Review of Neurotherapeutics, 16*, 497–504.

Ren, Y., & Xu, F. (2018). How patients should be counseled on adverse drug reactions: Avoiding the nocebo effect. *Research in Social and Administrative Pharmacy, 14*(7), 705.

Salloway, S., Sperling, R., Fox, N. C., Blennow, K., Klunk, W., Raskind, M., et al. (2014). Two phase 3 trials of bapineuzumab in mild-to-moderate Alzheimer's disease. *New England Journal of Medicine, 370*, 322–333.

Sian-Hulsmann, J., Monoranu, C., Strobel, S., & Riederer, P. (2015). Lewy bodies: A spectator or salient killer? *CNS and Neurological Disorders - Drug Targets, 14*, 947–955.

Tai, L. M., Koster, K. P., Luo, J., Lee, S. H., Wang, Y. T., Collins, N. C., et al. (2014). Amyloid-beta pathology and APOE genotype modulate retinoid X receptor agonist activity in vivo. *Journal of Biological Chemistry, 289*, 30535–30555.

Takizawa, C., Thompson, P. L., Van, W. A., Faure, C., & Maier, W. C. (2015). Epidemiological and economic burden of Alzheimer's disease: A systematic literature review of data across Europe and the United States of America. *Journal of Alzheimer's Disease, 43*, 1271–1284.

Villeneuve, S., Brisson, D., Marchant, N. L., & Gaudet, D. (2014). The potential applications of Apolipoprotein E in personalized medicine. *Frontiers in Aging Neuroscience, 6*, 154.

CHAPTER 35

All-trans retinoic acid in Alzheimer's disease

Siamak Beheshti

Department of Plant and Animal Biology, Faculty of Biological Science and Technology, University of Isfahan, Isfahan, Iran

List of abbreviations

AD Alzheimer's disease
APP Amyloid precursor protein
APP-C99 Amyloid precursor protein C-terminal fragment
ATRA All-trans retinoic acid
Aβ Amyloid beta
BACE-1 β-site amyloid precursor protein cleaving enzyme 1
DNA Deoxyribonucleic acid
ERK Extracellular signal-regulated kinase
fAβ Fibrillar amyloid beta
GSK3β Glycogen synthase kinase-3 beta
IL-6 Interleukin-6
LPS Lipopolysaccharide
MAP Microtubule-associated protein
PS1 Presenilin 1
RARs Retinoic acid receptors
RXRs Retinoid X receptors

Mini-dictionary of terms

Amyloid Beta Amyloid beta represents peptides of 36—43 amino acids that are the main constituents of the amyloid plaques found in the brains of AD patients.

APP APP is an abbreviation for amyloid precursor protein. It is an integral membrane protein that is expressed in many tissues and concentrated in the synapses of neurons.

Neuroinflammation Neuroinflammation is inflammation of the nervous tissue that may be originated in response to infection, traumatic brain injury, toxic metabolites, or autoimmunity.

Neurotransmission Neurotransmission is the process by which signaling molecules named neurotransmitters are released by the axon terminal of a neuron (the presynaptic neuron) and bind to their receptors on another neuron (the postsynaptic neuron).

Tau Protein Tau is a microtubule-associated protein that interacts with tubulin and promotes its assembly into microtubules and stabilizes the microtubule network.

Diagnosis and Management in Dementia
ISBN 978-0-12-815854-8, https://doi.org/10.1016/B978-0-12-815854-8.00035-5
559

Introduction

Alzheimer's disease (AD) is an irreversible neurodegenerative disease. It is the second most prevalent neurological disorder after stroke. AD has two main neuropathological marks, the accumulation of amyloid beta (Aβ) in extracellular plaques and the prevalence of intracellular neurofibrillary tangles (Serrano-Pozo, Frosch, Masliah, & Hyman, 2011). There are a limited number of drugs for AD therapy. Moreover, most increase neural transmission and do not modify the disease process (Ono & Yamada, 2012).

All-trans retinoic acid (ATRA) is an active metabolite of vitamin A in the brain (Fig. 35.1). It is synthesized by retinol in cortex, amygdala, hypothalamus, hippocampus, striatum, and related brain regions (Obulesu, Dowlathabad, & Bramhachari, 2011). ATRA controls neurogenesis and neuronal survival and helps preserve neuronal plasticity and cognitive function in later life. It acts via specific retinoic acid receptors (RARs) and retinoid X receptors (RXRs), each with three subtypes: α, β, and γ (Olson & Mello, 2010). A substantial number of studies have argued that retinoid signaling might have beneficial effects in AD. AD patients have been reported to have low serum and plasma concentrations of vitamin A and β-carotene (Bourdel-Marchasson et al., 2001; Jimenez-Jimenez et al., 1999; Zaman et al., 1992). Likewise, among people aged 65 and older, higher beta-carotene plasma levels were associated with better memory performance (Perrig, Perrig, & Stahelin, 1997). Moreover, the transport and function of retinoic acid was shown to be defective in the AD brain (Goodman & Pardee, 2003). Vitamin A deficiency caused Aβ accumulation in rats (Corcoran, So, & Maden, 2004) and produced a severe deficit in spatial learning and memory (Hernandez-Pinto, Puebla-Jimenez, & Arilla-Ferreiro, 2006).

Due to the increasing number of studies showing the relevance of retinoid signaling in the pathogenesis of AD, recent findings on the potential prophylactic and therapeutic effects of retinoid signaling in AD will be discussed in this chapter.

Retinoids in animals

Animals cannot synthetize retinoids. Instead, they convert plant or microorganism carotenoids to retinoids. These comprise all-trans retinoic acid, 13-cis retinoic acid, 13-cis-4-oxo retinoic acid, all-trans-4-oxo retinoic acid, and all-trans-retinoyl-β-glucuronide (Barua & Sidell, 2004; Wyss & Bucheli, 1997). ATRA is the most abundant form of

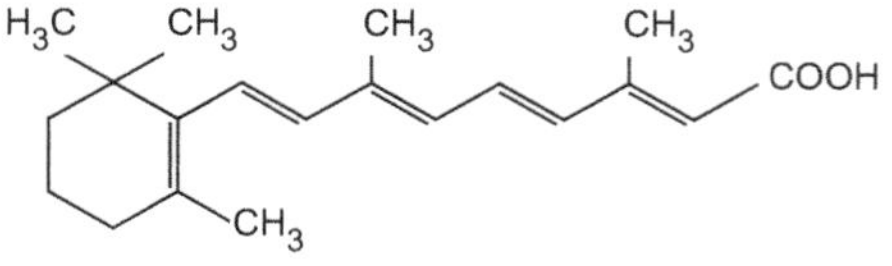

Figure 35.1 Chemical structure of all-trans retinoic acid. The structure of all-trans retinoic acid indicates a cyclic end group, a polyene side chain, and a polar end group.

retinoid. The majority of these retinoids are transported in plasma bound to albumin. Retinoids are present in plasma at nanomolar concentrations (about 5–10 nM). The levels of most of these retinoids are reliant on the intake of vitamin A and rise two to four times after ingestion of a large amount of vitamin A.

Retinoid receptors and their distribution in the brain

Different retinoid receptors have been identified so far. A retinoic acid receptor (RARα) that belongs to the family of steroid/thyroid hormone receptors was cloned in 1987. This family of nuclear receptors are ligand-dependent transcription factors that regulate gene expression by binding to short DNA sequences (hormone-responsive elements or enhancers) in the vicinities of target genes. Thereafter, two additional RARs (RARβ and RARγ) were identified (Blomhoff & Blomhoff, 2006). Later, a new subfamily of RARs (called RXRα, RXRβ, and RXRγ) was cloned. The primary sequences of these receptors differ substantially from those of the RARs (Szanto et al., 2004). Hence, six different genes coding for nuclear RARs have been cloned to date. The expression of these six receptors varies substantially between cells. Nuclear RARs function as heterodimers (one RAR complexed with one RXR) and possibly also as a homodimer of two RXRs, binding to DNA sequences called RAR elements or retinoid X response elements located within the promoter of target genes. In vitro binding studies have demonstrated that all-trans retinoic acid and 9-cis retinoic acid, but not 13-cis retinoic acid, are high-affinity ligands for RARs, whereas only 9-cis retinoic acid binds with high affinity to RXRs (Soprano, Qin, & Soprano, 2004). However, the physiological role of 9-cis retinoic acid has been questioned. Thus, the most important ligand for the RAR-RXR heterodimer seems to be all-trans retinoic acid binding to the RAR heterodimer partner (Blomhoff & Blomhoff, 2006).

Retinoid receptors were detected throughout the central nervous system including spinal cord, cortex, amygdala, hypothalamus, hippocampus, striatum, and associated brain regions (Lane & Bailey, 2005). In the hippocampus, RARα, β, and γ were detected in neurons of the dentate gyrus, while a smaller number of neurons within the CA_1 region expressed these receptors. This contrasts with mouse, in which RARγ is the only receptor present in the dentate gyrus while in the CA_1/CA_3 subfields, RARα and γ are expressed. A limited number of papers have described retinoic acid receptors in the human brain. RARα and RARγ were reported in approximately 12% of the granule neurons of human dentate gyrus and RARα in some neurons of the hippocampal subfields (Rioux & Arnold, 2005). The high expression of both receptors and retinoic acid synthetic enzymes compared with mouse models implies a crucial function in the human hippocampus.

The effect of retinoid signaling on amyloid-β

Amyloid-β peptide is a natural metabolic product (Shoji et al., 1992). It comprised 36—43 amino acids, though Aβ peptides of 40 and 42 amino acids (Aβ40 and Aβ42) predominate (De Strooper & Annaert, 2000). Aβ peptides originate from proteolysis of amyloid-β precursor protein (APP) by the consecutive enzymatic actions of β-site amyloid precursor protein cleaving enzyme 1 (BACE-1) and γ-secretase (Iizuka et al., 1996; Vassar, 2014). Aβ40 is the most abundant species normally formed. However, an imbalance in synthesis and clearance, which leads to accumulation of aggregated Aβ42 peptides, is thought to be the starting cause in AD (Saido, 2013).

Aggregation of Aβ and its subsequent deposition as extracellular amyloid plaques in the brain is a major hallmark of AD (Mroczko, Groblewska, Litman-Zawadzka, Kornhuber, & Lewczuk, 2018). Aggregation of Aβ is a self-assembly process that starts with monomeric protein and involves transient intermediates and finalizes with the creation of Aβ fibrils. The created fibrils then deposit as plaques in the brain (Hardy & Allsop, 1991). Oligomers and protofibrils are the main intermediate species and have higher toxicity than mature fibrils. Accordingly, there is a major role for these soluble aggregates in the pathogenesis of AD (Verma, Vats, & Taneja, 2015).

Studies have clarified that retinoid signaling affects Aβ content in the brain. RARα signaling in vitro prevented both intracellular and extracellular Aβ accumulation and prevented Aβ-induced neuronal cell death in cortical cultures (Jarvis et al., 2010). All-trans retinoic acid regulated all secretases (α, β, and γ) in an antiamyloidogenic sense at the levels of transcription, translation, and activation (Koryakina, Aeberhard, Kiefer, Hamburger, & Kuenzi, 2009). The RAR agonist tamibarotene (Am80) reduced insoluble Aβ levels in brains of APP 23 mice (Kawahara et al., 2009). Ding et al. (2008) tested the effect of systemic administration of ATRA on Aβ deposition in APP/presenilin 1 (PS1) double-transgenic mice that exhibit Aβ plaques as early as 2.5 months of age. ATRA attenuated Aβ levels in both the frontal cortex and the hippocampus. There was also a significant decrease in Aβ deposition. Vitamin A and beta-carotene dose-dependently inhibited formation of fibrillar Aβ (fAβ) from fresh Aβ, as well as their extension. Moreover, they dose-dependently destabilized preformed fAβs (Ono et al., 2004). Retinoic acid treatment of cells resulted in significant inhibition of γ-secretase-mediated processing of the amyloid precursor protein C-terminal fragment (APP-C99). Retinoic acid-elicited signaling was found to significantly increase accumulation of APP-C99 and decrease production of secreted Aβ40. In addition, retinoic acid-induced inhibition of γ-secretase activity was found to be mediated through significant activation of extracellular signal-regulated kinases (ERK1/2). Treatment of cells with the specific ERK inhibitor PD98059 completely abolished retinoic acid-mediated inhibition of γ-secretase. Consistent with these findings, retinoic acid was observed to inhibit secretase-mediated proteolysis of full-length APP. Retinoic acid inhibited γ-secretase through nuclear RARα and RXRα (Kapoor et al., 2013).

Besides the effect of retinoid signaling on the content of Aβ, some studies have also indicated that Aβ could affect retinoic acid synthesis. Aβ downregulated RARα signaling that inhibited the synthesis of the endogenous ligand, retinoic acid. This effect could be reversed by a RARα agonist. RARα signaling promoted Aβ clearance and prevented tau phosphorylation (Goncalves et al., 2013). All these evidence support the idea that retinoids have a crucial role in Aβ clearance in the brain (Fig. 35.2 and Table 35.1).

Role of retinoids on tau phosphorylation

Tau is a microtubule-associated protein (MAP). It is the main MAP of a mature neuron. The other two neuronal MAPs are MAP1 and MAP2 (Goedert, Spillantini, Jakes, Rutherford, & Crowther, 1989). Tau interacts with tubulin and promotes its assembly into microtubules and stabilizes the microtubule network (Weingarten, Lockwood, Hwo, & Kirschner, 1975). This activity of tau is regulated by its degree of phosphorylation (Lindwall & Cole, 1984). The normal adult human brain has 2–3 mol phosphate/mole of tau protein (Kopke et al., 1993). Hyperphosphorylation of tau reduces its biological activity. In the AD brain, tau is ~threefold to fourfoldmore hyperphosphorylated than in the normal adult brain tau, and in this hyperphosphorylated state it is polymerized, forming neurofibrillary tangles, which leads to neurodegeneration (Iqbal & Grundke-Iqbal, 2008). Due to the crucial role of tau in the

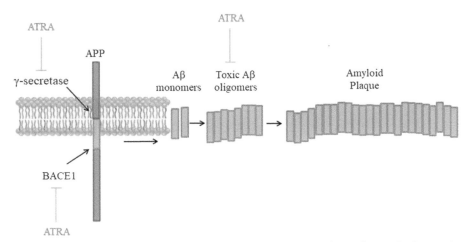

Figure 35.2 Effects of all-trans retinoic acid on the amyloidogenic pathway for amyloid-β production. All-trans retinoic acid has the capacity to modulate brain amyloid-β content by acting on different players in the amyloidogenic pathway of amyloid-β production. It is shown to inhibit γ-secretase, β-site amyloid precursor protein cleaving enzyme 1, or amyloid-β fibrils at different levels. ATRA, all-trans retinoic acid; APP, amyloid precursor protein; Aβ, amyloid-β; BACE-1, β-site amyloid precursor protein cleaving enzyme 1.

Table 35.1 The effects of retinoid signaling on amyloid-β content and tau phosphorylation.

Retinoid	Effect	Authors	Year
All-trans retinoic acid	Regulation of α, β, and γ secretases at the levels of transcription, translation, and activation	Koryakina et al.	2009
All-trans retinoic acid	Attenuation of amyloid-β levels in the frontal cortex and hippocampus of APP/PS1 double-transgenic mice	Ding et al.	2008
All-trans retinoic acid	Prevention of intracellular and extracellular amyloid-β accumulation in cortical cultures	Jarvis et al.	2010
Am80	Reduction of insoluble amyloid-β levels in the brain	Kawahara et al.	2009
Vitamins A and β-carotene	Inhibition of fibrillar amyloid-β formation from fresh amyloid-β	Ono et al.	2004
AM 580	Prevention of tau phosphorylation	Goncalves et al.	2013

The potential impact of retinoid signaling on amyloid-β content and tau phosphorylation is shown. AM80 and AM580 are retinoic acid receptor-α agonists.

pathogenesis of AD, inhibition of abnormal hyperphosphorylation of tau offers a hopeful therapeutic target for AD (Iqbal, Liu, Gong, & Grundke-Iqbal, 2010). In this regard, RARα signaling has been shown to prevent tau phosphorylation (Table 35.1; Goncalves et al., 2013).

Glycogen synthase kinase-3β (GSK3β) is a key enzyme in the regulation of cell cycle. In neurons it plays a central role in the regulation of tau phosphorylation (Hernandez & Avila, 2008). Studies have demonstrated that the deregulation of GSK3β activity is involved in several other pathological events associated with AD, like increased production of the Aβ peptide, induction of apoptosis, and impaired neurogenesis and synaptic plasticity (Lippa & Morris, 2006). It was shown that miR-138 is increased in AD models and its overexpression activated GSK-3β and increased tau phosphorylation in human embryonic kidney 293/tau cells. Furthermore, RARα was shown to be a direct target of miR-138, and supplement of RARα substantially suppressed GSK-3β activity, and reduced tau phosphorylation induced by miR-138. It seems that miR-138 promotes tau phosphorylation by targeting the RARα/GSK-3β pathway (Wang, Tan et al., 2015).

Antineuroinflammatory activity of retinoids

An inflammatory response is present in both AD patients and animal models of AD (Johnston, Boutin, & Allan, 2011). Studies have shown antiinflammatory roles of retinoids in neurodegenerative disorders (Kuenzli, Tran, & Saurat, 2004). ATRA regressed the mediators of ethanol induced neuroinflammation by decreasing oxidative stress and regulating the expression of nuclear factor kappa-B and sirtuin 1 (Priyanka, Syam Das, Thushara, Rauf, & Indira, 2018).

Microglia clear fAβ through phagocytosis of amyloid fibrils and bulky Aβ aggregates. However, this process is blocked in the presence of inflammatory cytokines that leads to inactivation of the microglia phagocytosis (Weitz & Town, 2012). Retinoids have been shown to inhibit production of proinflammatory cytokine interleukin-6 (IL-6) (Zitnik et al., 1994). Accordingly, downregulation of IL-6 by retinoids may be a useful therapeutic strategy against AD. Retinoids also prevent lipopolysaccharide (LPS)-induced or Aγ-induced tumor necrosis factor-α production and inhibit expression of inducible nitric oxide synthase in activated microglia (Carratu, Marasco, Signorile, Scuderi, & Steardo, 2012; Dheen, Jun, Yan, Tay, & Ling, 2005). It was indicated that RAR agonist AM80 has antiinflammatory responses in an LPS-induced inflammation model in vivo. AM80 could increase the production of brain-derived neurotrophic factor that could provide neuroprotection (Katsuki et al., 2009). ATRA downregulated the expression of BACE-1 in the brains of Tg2576 mice (mice that overexpress a mutant form of APP (isoform 695) with the Swedish mutation (KM670/671NL), resulting in elevated levels of Aβ and ultimately amyloid plaques) and in mice fed a high-fat diet, conditions that are associated with a neuroinflammatory response (Wang, Chen et al., 2015). Consequently, retinoids appear to have noteworthy potential in inhibiting inflammatory responses and promoting amyloid clearance in AD (Fig. 35.3).

Role of retinoids on neurotransmission in Alzheimer's disease

It is well known that many neurotransmitter systems, especially the cholinergic and catecholaminergic systems, are disturbed in AD (St George-Hyslop, 2000; Trillo et al., 2013;

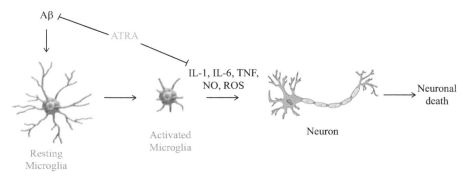

Figure 35.3 Proposed antineuroinflammatiory activity of all-trans retinoic acid. Activated microglia produce a variety of inflammatory cytokines, which in turn inhibit the phagocytic activity of microglia. All-trans retinoic acid can inhibit some of these cytokines; thereby show anti-neuroinflammatory activity. *Aβ*, amyloid-β; *ATRA*, all-trans retinoic acid; *IL-1*, interleukin-1; *IL-6*, interleukin-6; *TNF*, tumor necrosis factor; *NO*, nitric oxide; *ROS*, reactive oxygen species.

Wenk, 2003). Degeneration of cholinergic neurons in the basal forebrain that project to the neocortex, hippocampus, and amygdala is a hallmark of AD (Trillo et al., 2013). Administrations of cholinesterase inhibitors (which prevent the breakdown of acetylcholine and prolong its action in the brain) stimulate learning and memory processes in animals. Meanwhile, administration of these drugs is presently used to treat symptoms of AD in humans (Santucci, Kanof, & Haroutunian, 1989). Retinoids, which are significantly decreased in AD brain (Goodman, 2006), can provide an alternative treatment for AD symptoms, as they have trophic effects on cholinergic neurons. Activation of RARα upregulates expression of choline acetyltransferase and vesicular acetylcholine transporter protein, which helps to transport acetylcholine into synaptic vesicles (Berse & Blusztajn, 1995). Retinoids also upregulate acetylcholine and choline acetyltransferase mRNAs (Goodman, 2006; Pedersen, Berse, Schuler, Wainer, & Blusztajn, 1995).

AD also disrupts monoaminergic systems, including the noradrenergic and dopaminergic systems (Trillo et al., 2013). The locus coeruleus is one of the central noradrenergic nuclei of the brain. It has widespread projections to the cortex and limbic system and shows significant degeneration in AD (Mann, Lincoln, Yates, Stamp, & Toper, 1980). The levels of tyrosine hydroxylase (the rate-limiting enzyme for both norepinephrine and dopamine synthesis) and dopamine-β-hydroxylase (the enzyme required for the synthesis of norepinephrine) are reduced in AD (Iversen et al., 1983). In addition, AD is associated with reduced levels of norepinephrine in the cortex (Reinikainen et al., 1988; Storga, Vrecko, Birkmayer, & Reibnegger, 1996) and dopamine levels in the cortex, amygdala, and striatum (Pinessi et al., 1987), and dopamine receptors in the striatum (Cross, Crow, Ferrier, Johnson, & Markakis, 1984; Pizzolato et al., 1996). As retinoids control the expression of tyrosine hydroxylase and dopamine β-hydroxylase (Kim, Hong, LeDoux, & Kim, 2001) and can also modulate the expression of dopamine D2 receptors (Samad, Krezel, Chambon, & Borrelli, 1997), their trophic properties on both noradrenergic and dopaminergic systems may aid in relieving AD symptoms (Fig. 35.4).

Retinoid signaling and animal models of Alzheimer's disease

Much evidence indicates the beneficial effects of retinoid signaling in animal models of AD. Some of these studies show a protective action and others indicate a therapeutic action (Table 35.2). It was shown that administration of all-trans retinoic acid improved the memory deficits induced by streptozotocin in mice. Meanwhile, it restored acetylcholinesterase activity and attenuated oxidative alterations and brain myeloperoxidase level. It also reduced amyloid deposition in brain (Sodhi & Singh, 2013). In hemizygous amyloid-β precursor protein 23 mice, which overexpress human-type AβPP, it was shown that coadministration of tamibarotene (Am80), RAR α, β agonist, and specific RXR pan agonist HX630 significantly improved

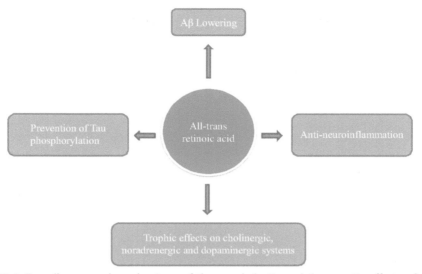

Figure 35.4 Overall proposed mechanisms of the prophylactic and therapeutic effects of all-trans retinoic acid on Alzheimer's disease. Schematic drawing on some of the potential mechanisms proposed for the beneficial effects of all-trans retinoic acid in Alzheimer's disease. $A\beta$, Amyloid-β.

Table 35.2 Studies that evaluated the role of retinoid signaling in animal models of Alzheimer's disease.

Retinoid	Effect	Animal model of Alzheimer's disease	Authors	Year
All-trans retinoic acid	Prophylactic and therapeutic action	Amyloid precursor protein and presenilin 1 double-transgenic mice	Ding et al.	2008
All-trans retinoic acid	Prophylactic action	Intracerebroventricular injection of amyloid-β	Beheshti et al.	2017
All-trans retinoic acid	Therapeutic action	Intracerebroventricular injection of streptozotocin	Sodhi and Singh	2013
Am80 and HX630	Therapeutic action	8.5 month–old amyloid-β protein precursor 23 mice	Kawahara et al.	2014
All-trans retinoic acid	Preventive action	Lipopolysaccharide-induced memory impairment	Behairi et al.	2016
All-trans retinoic acid	Recovery of adult neurogenesis in the hippocampus	APPSwe/PS1M146V/ tauP301L (3 × Tg) mice	Takamura et al.	2017

Prophylactic and/or therapeutic effects of all-trans retinoic acid or other retinoids are shown in some animal models of Alzheimer's disease. AM80 and HX630 are retinoic acid receptor -α and retinoid X receptor agonists, respectively. APPswe/PS1M146V/tauP301L (3 × Tg) mice, harbor a Psen1 PS1M146V mutation and the coinjected APPSwe and tauP301L transgenes.

memory deficits in the Morris water maze and reduced the level of insoluble Aβ peptide in the brain. However, administration of either agent alone produced no effect (Kawahara et al., 2014). ATRA showed decreased activation of microglia and astrocytes, attenuated neuronal degeneration, and improved spatial learning and memory in APP and PS1 double-transgenic mice, a well-established AD mouse model (Ding et al., 2008). Chronic intracerebroventricular injection of ATRA had a prophylactic effect in the amyloid-β model of AD (Beheshti & Soleimanipour, 2017). ATRA pretreatment prevented LPS-induced deleterious effects on memory in aged rats (Behairi et al., 2016). ATRA inhibited activation of microglia, which recovered adult neurogenesis in the hippocampus in a mouse model of AD (Takamura, Watamura, Nikkuni, & Ohshima, 2017).

Conclusion

Taken together, it appears that retinoid signaling has crucial prophylactic as well as therapeutic effects on AD. As AD is a multifactorial disease, targeted therapeutic agents lead to relatively modest benefits. Retinoids are capable of acting upon multiple AD-associated targets including Aβ and tau reduction, acetylcholine activation, and anti-inflammation. Therefore, retinoids appear to make up a practical solution, a single drug with numerous targets to treat multifactorial AD.

Key facts of retinoids

- Retinoids are a class of compounds chemically related to vitamin A.
- The basic structure of the hydrophobic retinoid molecule contains a cyclic end group, a polyene side chain, and a polar end group.
- The main sources of retinoids from the diet are plant pigments such as carotenes and retinyl esters derived from animal sources.
- The most abundant form of retinoids is all-trans retinoic acid.
- Retinoids are used in the treatment of a variety of diseases and might also be useful in the treatment of AD.

Summary points

- All-trans retinoic acid is an active metabolite of vitamin A in the brain.
- Retinoid signaling is involved in the pathogenesis of AD.
- The transport and function of retinoic acid was shown to be defective in the AD brain.
- Retinoid signaling affects amyloid-β content, tau phosphorylation, neurotransmission, and neuroinflammation in the brain.
- Therefore, retinoids appear to be make up a practical solution as a single drug with numerous targets to treat multifactorial AD.

References

Barua, A. B., & Sidell, N. (2004). Retinoyl β-glucuronide: A biologically active interesting retinoid. *Journal of Nutrition, 134*(1), 286–289.

Behairi, N., Belkhelfa, M., Rafa, H., Labsi, M., Deghbar, N., Bouzid, N., et al. (2016). All-trans retinoic acid (ATRA) prevents lipopolysaccharide-induced neuroinflammation, amyloidogenesis and memory impairment in aged rats. *Journal of Neuroimmunology, 300*, 21–29.

Beheshti, S., & Soleimanipour, A. (2017). Prophylactic effect of all-trans retinoic acid in an amyloid-β rat model of Alzheimer's disease. *Physiology and Pharmacology, 21*(1), 34–43.

Berse, B., & Blusztajn, J. K. (1995). Coordinated up-regulation of choline acetyltransferase and vesicular acetylcholine transporter gene expression by the retinoic acid receptor alpha, cAMP, and leukemia inhibitory factor/ciliary neurotrophic factor signaling pathways in a murine septal cell line. *Journal of Biological Chemistry, 270*(38), 22101–22104.

Blomhoff, R., & Blomhoff, H. K. (2006). Overview of retinoid metabolism and function. *Journal of Neurobiology, 66*(7), 606–630.

Bourdel-Marchasson, I., Delmas-Beauvieux, M. C., Peuchant, E., Richard-Harston, S., Decamps, A., Reignier, B., et al. (2001). Antioxidant defences and oxidative stress markers in erythrocytes and plasma from normally nourished elderly Alzheimer patients. *Age and Ageing, 30*(3), 235–241.

Carratu, M. R., Marasco, C., Signorile, A., Scuderi, C., & Steardo, L. (2012). Are retinoids a promise for Alzheimer's disease management? *Current Medicinal Chemistry, 19*(36), 6119–6125.

Corcoran, J. P., So, P. L., & Maden, M. (2004). Disruption of the retinoid signalling pathway causes a deposition of amyloid-β in the adult rat brain. *European Journal of Neuroscience, 20*(4), 896–902.

Cross, A. J., Crow, T. J., Ferrier, I. N., Johnson, J. A., & Markakis, D. (1984). Striatal dopamine receptors in Alzheimer-type dementia. *Neuroscience Letters, 52*(1–2), 1–6.

De Strooper, B., & Annaert, W. (2000). Proteolytic processing and cell biological functions of the amyloid precursor protein. *Journal of Cell Science, 113*(11), 1857–1870.

Dheen, S. T., Jun, Y., Yan, Z., Tay, S. S., & Ling, E. A. (2005). Retinoic acid inhibits expression of TNF-α and iNOS in activated rat microglia. *Glia, 50*(1), 21–31.

Ding, Y., Qiao, A., Wang, Z., Goodwin, J. S., Lee, E. S., Block, M. L., et al. (2008). Retinoic acid attenuates β-amyloid deposition and rescues memory deficits in an Alzheimer's disease transgenic mouse model. *Journal of Neuroscience, 28*(45), 11622–11634.

Goedert, M., Spillantini, M. G., Jakes, R., Rutherford, D., & Crowther, R. A. (1989). Multiple isoforms of human microtubule-associated protein tau: Sequences and localization in neurofibrillary tangles of Alzheimer's disease. *Neuron, 3*(4), 519–526.

Goncalves, M. B., Clarke, E., Hobbs, C., Malmqvist, T., Deacon, R., Jack, J., et al. (2013). Amyloid-β inhibits retinoic acid synthesis exacerbating Alzheimer disease pathology which can be attenuated by an retinoic acid receptor alpha agonist. *European Journal of Neuroscience, 37*(7), 1182–1192.

Goodman, A. B. (2006). Retinoid receptors, transporters, and metabolizers as therapeutic targets in late onset Alzheimer disease. *Journal of Cellular Physiology, 209*(3), 598–603.

Goodman, A. B., & Pardee, A. B. (2003). Evidence for defective retinoid transport and function in late onset Alzheimer's disease. *Proceedings of the National Academy of Sciences of the United States of America, 100*(5), 2901–2905.

Hardy, J., & Allsop, D. (1991). Amyloid deposition as the central event in the aetiology of Alzheimer's disease. *Trends in Pharmacological Sciences, 12*(10), 383–388.

Hernandez-Pinto, A. M., Puebla-Jimenez, L., & Arilla-Ferreiro, E. (2006). A vitamin A-free diet results in impairment of the rat hippocampal somatostatinergic system. *Neuroscience, 141*(2), 851–861.

Hernandez, F., & Avila, J. (2008). The role of glycogen synthase kinase 3 in the early stages of Alzheimers' disease. *FEBS Letters, 582*(28), 3848–3854.

Iizuka, T., Shoji, M., Kawarabayashi, T., Sato, M., Kobayashi, T., Tada, N., et al. (1996). Intracellular generation of amyloid-β protein from amyloid-β protein precursor fragment by direct cleavage with β- and γ-secretase. *Biochemical and Biophysical Research Communications, 218*(1), 238–242.

Iqbal, K., & Grundke-Iqbal, I. (2008). Alzheimer neurofibrillary degeneration: Significance, etiopathogenesis, therapeutics and prevention. *Journal of Cellular and Molecular Medicine, 12*(1), 38–55.

Iqbal, K., Liu, F., Gong, C. X., & Grundke-Iqbal, I. (2010). Tau in Alzheimer disease and related tauopathies. *Current Alzheimer Research, 7*(8), 656−664.

Iversen, L. L., Rossor, M. N., Reynolds, G. P., Hills, R., Roth, M., Mountjoy, C. Q., et al. (1983). Loss of pigmented dopamine-beta-hydroxylase positive cells from locus coeruleus in senile dementia of Alzheimer's type. *Neuroscience Letters, 39*(1), 95−100.

Jarvis, C. I., Goncalves, M. B., Clarke, E., Dogruel, M., Kalindjian, S. B., Thomas, S. A., et al. (2010). Retinoic acid receptor-alpha signalling antagonizes both intracellular and extracellular amyloid-β production and prevents neuronal cell death caused by amyloid-β. *European Journal of Neuroscience, 32*(8), 1246−1255.

Jimenez-Jimenez, F. J., Molina, J. A., de Bustos, F., Orti-Pareja, M., Benito-Leon, J., Tallon-Barranco, A., et al. (1999). Serum levels of β-carotene, α-carotene and vitamin A in patients with Alzheimer's disease. *European Journal of Neurology, 6*(4), 495−497.

Johnston, H., Boutin, H., & Allan, S. M. (2011). Assessing the contribution of inflammation in models of Alzheimer's disease. *Biochemical Society Transactions, 39*(4), 886−890.

Kapoor, A., Wang, B. J., Hsu, W. M., Chang, M. Y., Liang, S. M., & Liao, Y. F. (2013). Retinoic acid-elicited RARα/RXRα signaling attenuates Aβ production by directly inhibiting γ-secretase-mediated cleavage of amyloid precursor protein. *ACS Chemical Neuroscience, 4*(7), 1093−1100.

Katsuki, H., Kurimoto, E., Takemori, S., Kurauchi, Y., Hisatsune, A., Isohama, Y., et al. (2009). Retinoic acid receptor stimulation protects midbrain dopaminergic neurons from inflammatory degeneration via BDNF-mediated signaling. *Journal of Neurochemistry, 110*(2), 707−718.

Kawahara, K., Nishi, K., Suenobu, M., Ohtsuka, H., Maeda, A., Nagatomo, K., et al. (2009). Oral administration of synthetic retinoid Am80 (Tamibarotene) decreases brain β-amyloid peptides in APP23 mice. *Biological and Pharmaceutical Bulletin, 32*(7), 1307−1309.

Kawahara, K., Suenobu, M., Ohtsuka, H., Kuniyasu, A., Sugimoto, Y., Nakagomi, M., et al. (2014). Cooperative therapeutic action of retinoic acid receptor and retinoid x receptor agonists in a mouse model of Alzheimer's disease. *Journal of Alzheimer's Disease, 42*(2), 587−605.

Kim, H. S., Hong, S. J., LeDoux, M. S., & Kim, K. S. (2001). Regulation of the tyrosine hydroxylase and dopamine β-hydroxylase genes by the transcription factor AP-2. *Journal of Neurochemistry, 76*(1), 280−294.

Kopke, E., Tung, Y. C., Shaikh, S., Alonso, A. C., Iqbal, K., & Grundke-Iqbal, I. (1993). Microtubule-associated protein tau. Abnormal phosphorylation of a non-paired helical filament pool in Alzheimer disease. *Journal of Biological Chemistry, 268*(32), 24374−24384.

Koryakina, A., Aeberhard, J., Kiefer, S., Hamburger, M., & Kuenzi, P. (2009). Regulation of secretases by all-trans-retinoic acid. *The FEBS Journal, 276*(9), 2645−2655.

Kuenzli, S., Tran, C., & Saurat, J. H. (2004). Retinoid receptors in inflammatory responses: A potential target for pharmacology. *Current Drug Targets. Inflammation and Allergy, 3*(4), 355−360.

Lane, M. A., & Bailey, S. J. (2005). Role of retinoid signalling in the adult brain. *Progress in Neurobiology, 75*(4), 275−293.

Lindwall, G., & Cole, R. D. (1984). Phosphorylation affects the ability of tau protein to promote microtubule assembly. *Journal of Biological Chemistry, 259*(8), 5301−5305.

Lippa, C. F., & Morris, J. C. (2006). Alzheimer neuropathology in nondemented aging: Keeping mind over matter. *Neurology, 66*(12), 1801−1802.

Mann, D. M., Lincoln, J., Yates, P. O., Stamp, J. E., & Toper, S. (1980). Changes in the monoamine containing neurones of the human CNS in senile dementia. *British Journal of Psychiatry, 136*, 533−541.

Mroczko, B., Groblewska, M., Litman-Zawadzka, A., Kornhuber, J., & Lewczuk, P. (2018). Cellular receptors of amyloid-β oligomers (AβOs) in Alzheimer's Disease. *International Journal of Molecular Sciences, 19*(7).

Obulesu, M., Dowlathabad, M. R., & Bramhachari, P. V. (2011). Carotenoids and Alzheimer's disease: An insight into therapeutic role of retinoids in animal models. *Neurochemistry International, 59*(5), 535−541.

Olson, C. R., & Mello, C. V. (2010). Significance of vitamin A to brain function, behavior and learning. *Molecular Nutrition and Food Research, 54*(4), 489−495.

Ono, K., & Yamada, M. (2012). Vitamin A and Alzheimer's disease. *Geriatrics and Gerontology International, 12*(2), 180−188.

Ono, K., Yoshiike, Y., Takashima, A., Hasegawa, K., Naiki, H., & Yamada, M. (2004). Vitamin A exhibits potent antiamyloidogenic and fibril-destabilizing effects in vitro. *Experimental Neurology, 189*(2), 380—392.

Pedersen, W. A., Berse, B., Schuler, U., Wainer, B. H., & Blusztajn, J. K. (1995). All-trans- and 9-cis-retinoic acid enhance the cholinergic properties of a murine septal cell line: Evidence that the effects are mediated by activation of retinoic acid receptor-alpha. *Journal of Neurochemistry, 65*(1), 50—58.

Perrig, W. J., Perrig, P., & Stahelin, H. B. (1997). The relation between antioxidants and memory performance in the old and very old. *Journal of the American Geriatrics Society, 45*(6), 718—724.

Pinessi, L., Rainero, I., De Gennaro, T., Gentile, S., Portaleone, P., & Bergamasco, B. (1987). Biogenic amines in cerebrospinal fluid and plasma of patients with dementia of Alzheimer type. *Functional Neurology, 2*(1), 51—58.

Pizzolato, G., Chierichetti, F., Fabbri, M., Cagnin, A., Dam, M., Ferlin, G., et al. (1996). Reduced striatal dopamine receptors in Alzheimer's disease: Single photon emission tomography study with the D2 tracer [123I]-IBZM. *Neurology, 47*(4), 1065—1068.

Priyanka, S. H., Syam Das, S., Thushara, A. J., Rauf, A. A., & Indira, M. (2018). All-trans retinoic acid attenuates markers of neuroinflammation in rat brain by modulation of SIRT1 and NFkB. *Neurochemistry Research, 43*(9), 1791—1801.

Reinikainen, K. J., Paljarvi, L., Huuskonen, M., Soininen, H., Laakso, M., & Riekkinen, P. J. (1988). A post-mortem study of noradrenergic, serotonergic and GABAergic neurons in Alzheimer's disease. *Journal of the Neurological Sciences, 84*(1), 101—116.

Rioux, L., & Arnold, S. E. (2005). The expression of retinoic acid receptor alpha is increased in the granule cells of the dentate gyrus in schizophrenia. *Psychiatry Research, 133*(1), 13—21.

Saido, T. C. (2013). Metabolism of amyloid-β peptide and pathogenesis of Alzheimer's disease. *Proceedings of the Japan Academy. Series B, Physical and Biological Sciences, 89*(7), 321—339.

Samad, T. A., Krezel, W., Chambon, P., & Borrelli, E. (1997). Regulation of dopaminergic pathways by retinoids: Activation of the D2 receptor promoter by members of the retinoic acid receptor-retinoid X receptor family. *Proceedings of the National Academy of Sciences of the United States of America, 94*(26), 14349—14354.

Santucci, A. C., Kanof, P. D., & Haroutunian, V. (1989). Effect of physostigmine on memory consolidation and retrieval processes in intact and nucleus basalis-lesioned rats. *Psychopharmacology, 99*(1), 70—74.

Serrano-Pozo, A., Frosch, M. P., Masliah, E., & Hyman, B. T. (2011). Neuropathological alterations in Alzheimer disease. *Cold Spring Harbor Perspectives in Medicine, 1*(1). a006189.

Shoji, M., Golde, T. E., Ghiso, J., Cheung, T. T., Estus, S., Shaffer, L. M., et al. (1992). Production of the Alzheimer amyloid beta protein by normal proteolytic processing. *Science, 258*(5079), 126—129.

Sodhi, R. K., & Singh, N. (2013). All-trans retinoic acid rescues memory deficits and neuropathological changes in mouse model of streptozotocin-induced dementia of Alzheimer's type. *Progress in Neuropsychopharmacology and Biological Psychiatry, 40*, 38—46.

Soprano, D. R., Qin, P., & Soprano, K. J. (2004). Retinoic acid receptors and cancers. *Annual Review of Nutrition, 24*, 201—221.

St George-Hyslop, P. H. (2000). Piecing together Alzheimer's. *Scientific American, 283*(6), 76—83.

Storga, D., Vrecko, K., Birkmayer, J. G., & Reibnegger, G. (1996). Monoaminergic neurotransmitters, their precursors and metabolites in brains of Alzheimer patients. *Neuroscience Letters, 203*(1), 29—32.

Szanto, A., Narkar, V., Shen, Q., Uray, I. P., Davies, P. J., & Nagy, L. (2004). Retinoid X receptors: X-Ploring their (patho)physiological functions. *Cell Death and Differentiation, 11*(Suppl. 2), S126—S143.

Takamura, R., Watamura, N., Nikkuni, M., & Ohshima, T. (2017). All-trans retinoic acid improved impaired proliferation of neural stem cells and suppressed microglial activation in the hippocampus in an Alzheimer's mouse model. *Journal of Neuroscience Research, 95*(3), 897—906.

Trillo, L., Das, D., Hsieh, W., Medina, B., Moghadam, S., Lin, B., et al. (2013). Ascending monoaminergic systems alterations in Alzheimer's disease. translating basic science into clinical care. *Neuroscience and Biobehavioral Reviews, 37*(8), 1363—1379.

Vassar, R. (2014). BACE1 inhibitor drugs in clinical trials for Alzheimer's disease. *Alzheimer's Research and Therapy, 6*(9), 89.

Verma, M., Vats, A., & Taneja, V. (2015). Toxic species in amyloid disorders: Oligomers or mature fibrils. *Annals of Indian Acadmy of Neurology, 18*(2), 138—145.

Wang, R., Chen, S., Liu, Y., Diao, S., Xue, Y., You, X., et al. (2015). All-trans-retinoic acid reduces BACE1 expression under inflammatory conditions via modulation of nuclear factor kB (NFkB) signaling. *Journal of Biological Chemistry, 290*(37), 22532—22542.

Wang, X., Tan, L., Lu, Y., Peng, J., Zhu, Y., Zhang, Y., et al. (2015). MicroRNA-138 promotes tau phosphorylation by targeting retinoic acid receptor alpha. *FEBS Letters, 589*(6), 726—729.

Weingarten, M. D., Lockwood, A. H., Hwo, S. Y., & Kirschner, M. W. (1975). A protein factor essential for microtubule assembly. *Proceedings of the National Academy of Sciences of the United States of America, 72*(5), 1858—1862.

Weitz, T. M., & Town, T. (2012). Microglia in Alzheimer's disease: It's all about context. *International Journal of Alzheimer's Disease, 2012*, 314185.

Wenk, G. L. (2003). Neuropathologic changes in Alzheimer's disease. *The Journal of Clinical Psychiatry, 64*(Suppl. 9), 7—10.

Wyss, R., & Bucheli, F. (1997). Determination of endogenous levels of 13-cis-retinoic acid (isotretinoin), all-trans-retinoic acid (tretinoin) and their 4-oxo metabolites in human and animal plasma by high-performance liquid chromatography with automated column switching and ultraviolet detection. *Journal of Chromatography. B, Biomedical Sciences and Applications, 700*(1—2), 31—47.

Zaman, Z., Roche, S., Fielden, P., Frost, P. G., Niriella, D. C., & Cayley, A. C. (1992). Plasma concentrations of vitamins A and E and carotenoids in Alzheimer's disease. *Age and Ageing, 21*(2), 91—94.

Zitnik, R. J., Kotloff, R. M., Latifpour, J., Zheng, T., Whiting, N. L., Schwalb, J., et al. (1994). Retinoic acid inhibition of IL-1-induced IL-6 production by human lung fibroblasts. *The Journal of Immunology, 152*(3), 1419—1427.

CHAPTER 36

Dementia and usage of N-methyl-D-aspartate receptor antagonists

B.E. Glynn-Servedio

Clinical Pharmacy Specialist—Ambulatory Care, Durham VA Health Care System, Raleigh 1 Community-Based Outpatient Clinic, Raleigh, NC, United States

List of abbreviations

AChE acetylcholinesterase
AD Alzheimer's disease
ADAS-cog Alzheimer's disease assessment scale—cognitive subscale
ADL Bristol Activities of Daily Living
AGS American Geriatrics Society
CGI clinical global impression
CIBIC-plus clinician's interview-based impression of change with caregiver input
DLB dementia with Lewy bodies
FAST functional assessment scale
FTD frontotemporal dementia
GDS global deterioration scale
MMSE mini-mental state examination
NMDA N-methyl-D-aspartate
NPI neuropsychiatric inventory
PDD Parkinson's disease dementia

Mini-dictionary of terms

Excitotoxicity A pathological process by which neurons are damaged or destroyed by excessive stimulation by glutamate or other excitatory substances.

Glutamate The most abundant neurotransmitter in the nervous system and the principal excitatory neurotransmitter. It is also a precursor to the neurotransmitter gamma-aminobutyric acid (GABA).

Ionotropic receptors Ionotropic receptors are transmembrane receptors that cause the opening or closing of an ion channel in response to a specific ligand binding event. These receptors are so named because they allow ions (typically sodium, potassium, chloride, and calcium) to travel in and out of a cell.

Polypharmacy The concomitant use of multiple medications. It is more common in elderly patients who often have more medical conditions requiring medication use. Generally accepted to mean the use of more than five medications simultaneously or the use of potentially inappropriate medications.

Synaptic plasticity The ability of nervous system synapses to strengthen or weaken over time, typically in response to changes in activity level. It is thought to be one of the neurochemical foundations of learning and memory. Can occur in both excitatory and inhibitory synapses.

Diagnosis and Management in Dementia
ISBN 978-0-12-815854-8, https://doi.org/10.1016/B978-0-12-815854-8.00036-7

Introduction

Along with acetylcholinesterase (AChE) inhibitors, *N*-methyl-D-aspartate (NMDA) receptor antagonists are one of the mainstays of pharmacologic treatment for many types of dementia worldwide.

The NMDA receptor is an ionotropic glutamate receptor found in nerve cells; it is so named because the agonist molecule NMDA binds selectively to the NMDA receptor rather than to other glutamate receptors. The receptor is activated when bound to glutamate (the primary excitatory amino acid neurotransmitter in neurons) and glycine (see Fig. 36.1). Activation allows positively charged ions to flow through the cell membrane. Extracellular Mg^{2+} and Zn^{2+} ions can bind to specific sites on the receptor and block passage of other cations through the open ion channel. However, depolarization of the cell allows a voltage-dependent flow of Na^+ and Ca^{2+} into the cell and K^+ out of the cell. The NMDA receptor plays an important function in controlling synaptic plasticity, a cellular mechanism for learning and memory. Ca^{2+} flow through the receptor is thought to be a critical step in this process. Overactivation of NMDA receptors, which causes Mg^{2+} to unblock the ion channel, causes excessive influx of Ca^{2+} and can lead to excitotoxicity. Excitotoxicity may be involved in neurodegenerative disorders including Alzheimer's, Parkinson's, and Huntington's diseases. An uncompetitive antagonist blocks the NMDA receptor's ion channel only when it is excessively open, which blocks the potentially excitotoxic activity and influx of calcium ions while preserving physiological NMDA receptor activity.

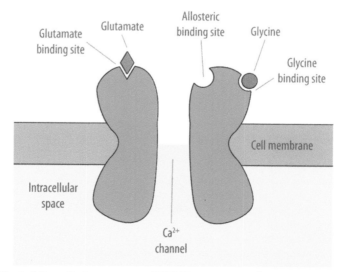

Figure 36.1 *Activated N-methyl-D-aspartate (NMDA) receptor.* A depiction of the NMDA receptor, showing the various binding sites. The receptor is activated when bound to both glutamate and glycine.

Memantine is an uncompetitive antagonist of the NMDA receptor. Memantine mimics the effects of Mg^{2+} at physiological pH and therefore acts as an ion channel blocker. During normal receptor activity, the ion channels only remain open for several milliseconds. This does not allow memantine to bind because it has only a low-to-moderate affinity for the NMDA receptor. Memantine can only bind during prolonged activation of the receptor, as in excitotoxic conditions (Parsons et al., 2007).

Memantine is approved and marketed for the treatment of moderate to severe Alzheimer's disease (AD) in the United States, Canada, Europe, China, and several other countries. It is approved and marketed for the treatment of dementia in Germany.

Memantine in Alzheimer's disease

The majority of trials of memantine use have been conducted in patients with AD. There is little evidence to suggest that patients with mild AD benefit from memantine. A 2011 meta-analysis (Schneider, Dagerman, Higgins, & McShane, 2011) looked at three trials that included a total of 431 patients with mild AD—defined as mini-mental state examination (MMSE) scores of 20—23—and 697 patients with moderate AD (defined as MMSE scores of 10—19). Overall, there were no significant differences between the memantine and placebo groups on any of the measured endpoints, both cognitive and behavioral, among the cohort of patients with mild AD.

Memantine does seem to have some benefit in patients with moderate to severe AD, although any benefits seen are typically modest. Trials using memantine often look at a combination of cognitive and behavioral endpoints. See Tables 36.1—36.3 for a summary

Table 36.1 Common cognitive scales used for dementia.

Scale	Estimated time to administer (minutes)	Summary
MMSE[a]	5—10	• Most widely used cognitive scale in clinical practice • Developed as a screening test for dementia but commonly misused as a diagnostic test • Assesses orientation, memory, attention, calculation, language, and visual construction • Scored between 0 and 30; 23—24 is often used as the cut-off for significant cognitive impairment
ADAS-cog[b]	40	• The most commonly used instrument in dementia clinical trials • Assesses all cognitive areas of dementia • 4-point difference between treatment groups is considered clinically significant

[a]Folstein, Folstein, and McHugh (1975).
[b]Rosen, Mohs, and Davis (1984).

Table 36.2 Common functional and behavioral scales used for dementia.

Scale	Estimated time to administer (minutes)	Summary
Bristol ADL scale[a]	15	• Answered by caregiver
		• Addresses 20 daily living activities
NPI[b]	10	• Assesses frequency and severity of a wide range of behaviors, including delusions, depression, agitation, irritability, and apathy

[a]Bucks, Ashworth, Wilcock, and Siegfried (1996).
[b]Cummings et al. (1994).

Table 36.3 Common scales used to measure overall dementia severity.

Scale	Estimated time to administer (minutes)	Summary
CDR[a]	5	• More reliable staging than MMSE
		• Based on caregiver input regarding cognitive and functional tasks
		• Classifies into four stages: Questionable, mild, moderate, severe
GDS[b]	2	• Used for staging dementia
		• Well validated
		• Classifies into seven stages from "no complaints" to "very severe"
CIBIC-plus[c]	10–40	• Global measure of detectable change in cognitive, functional, and behavioral areas
		• Separate interviews with the patient and the caregiver
CGI[d]	5	• Separate scales to rate severity and improvement
		• 7-point scale
		• Frequently used in studies of mental disorders to evaluate effects of medication

[a]Morris (1993).
[b]Reisberg et al. (1982).
[c]Schneider et al. (1997).
[d]Guy (1976).

of common cognitive, behavioral/functional, and overall disease severity scales referenced in this chapter. A 28-week randomized, placebo-controlled trial of 252 noninstitutionalized patients with a mean MMSE score of eight at study entry (MMSE score range of 3–14) found that memantine significantly reduced deterioration on multiple clinical scales (Reisberg et al., 2003). An open-label extension study including 175 of these patients showed benefits for patients previously taking placebo in all efficacy measures compared with their previous rate of decline; it also confirmed the favorable adverse event profile demonstrated in the original study (Reisberg et al., 2006).

Cholinesterase inhibitors and memantine are also often used in combination, particularly in moderate to severe AD. A combination tablet containing memantine and the cholinesterase inhibitor donepezil is marketed in several countries. The DOMINO-AD trial (Howard et al., 2012) looked at 295 community-dwelling patients with moderate to severe AD (baseline MMSE scores ranging from 5 to 13 with a mean score of 8) already treated with donepezil. This study compared four treatment strategies: no pharmacologic therapy (donepezil discontinued), donepezil continued as monotherapy, donepezil continued and memantine added, and memantine monotherapy. After 1 year, patients who continued donepezil (either with or without memantine) showed modest cognitive and functional benefits. The improvement in MMSE scores exceeded the prespecified minimum clinically important difference. However, the difference in Bristol Activities of Daily Living (ADL) Scale scores did not. Those assigned to receive memantine had a higher MMSE score and a lower score on the Bristol ADL Scale, which are both favorable results, compared with those not receiving memantine, but neither difference met the minimum clinically important difference. This trial was terminated early due to slow recruitment. A long-term follow-up of the DOMINO-AD trial looked at nursing home placement during the first year of the original trial and then every 6 months for an additional 3 years (Howard et al., 2015). Memantine use, either alone or in combination with donepezil, had no effect on the rate of nursing home placement over the 4-year period after randomization; the authors did note that this was a secondary outcome of DOMINO-AD and this analysis was not prespecified in the original study protocol.

A much smaller number of trials have been conducted using memantine in non-Alzheimer's dementia.

Memantine in vascular dementia

Excessive NMDA stimulation can be induced by ischemia, which theoretically suggests that NMDA receptor antagonists may be of use in vascular dementia. Memantine has been studied in vascular dementia, but the data are limited. The MMM 300 trial was a 28-week randomized, double-blind trial conducted in 321 European patients with mild to moderate vascular dementia (Orgogozo, Rigaud, Stoffler, Mobius, & Forette, 2002). Patients with probable vascular dementia and an MMSE score of 12—20 were randomized to receive either memantine 20 mg daily or placebo. After 28 weeks, mean Alzheimer's disease assessment scale—cognitive subscale (ADAS-cog) scores were significantly improved in the memantine group compared with those of the placebo group, and clinician's interview-based impression of change with caregiver input (CIBIC-plus) scores were stable between the two groups. The frequency of adverse effects was similar in both groups. The MMM 500 trial was also a 28-week randomized double-blind trial, conducted in 579 patients in the United Kingdom (Wilcock, Mobius, & Stoffler, 2002). Patients with a diagnosis of probable mild to moderate vascular dementia (mean baseline MMSE score of 17.5) were randomized to receive either memantine 20 mg daily or placebo. After 28 weeks, memantine was shown to improve cognition compared with placebo on ADAS-cog scores; no significant difference was seen in clinical global impression (CGI) scores. The overall impression from these two studies is that memantine may have some benefit on cognitive scales but not on clinical global impressions scales nor activities of daily living when used in patients with vascular dementia. Clinicians will often use memantine despite the limited data due to the lack of other treatment options, either with a cholinesterase inhibitor or as monotherapy.

Memantine in Parkinson's disease dementia and dementia with Lewy bodies

There may also be some benefit to using memantine in Parkinson's disease dementia (PDD) and dementia with Lewy bodies (DLB). A randomized trial of memantine 20 mg daily compared with placebo in 72 patients in Europe with PDD or DLB demonstrated an improvement in CGI scores in the memantine group (Aarsland et al., 2009). A similar study conducted in eight countries among 199 patients with mild to moderate PDD or DLB found an improvement in Alzheimer's disease cooperative study scores and neuropsychiatric inventory (NPI) scores in patients with DLB treated with memantine; no such difference was seen in patients with PDD (Emre et al., 2010).

Memantine in frontotemporal dementia

In contrast, memantine is not recommended for use in patients with frontotemporal dementia (FTD). Small open-label studies initially suggested that memantine may improve some behavioral symptoms of FTD. These studies included 3–21 patients with FTD and looked at a variety of clinical and functional scales (Boxer et al., 2009; Diehl-Schmid, Forstyl, Perneczky, Pohl, & Kurtz, 2008; Sharre, Warner, Davis, & Theado-Miller, 2005; Swanberg, 2007). However, two randomized trials failed to confirm these findings. A trial of 49 patients with FTD randomized patients to receive either memantine or placebo (Vercelletto et al., 2011). The primary endpoint was CIBIC-plus score, and several other clinical scales were used to assess secondary endpoints. The mean baseline MMSE score was 25. After 52 weeks, there were no differences in clinical endpoints between the two groups. There were also more patients in the memantine group, compared with the placebo group, who worsened over the study period. A second randomized trial enrolling 81 patients also showed no change in NPI scores or clinical status (CGI scores) when using memantine 20 mg daily compared with placebo (Boxer et al., 2013). Patients treated with memantine in this trial also showed more frequent cognitive adverse events than those taking placebo.

Considerations for duration of treatment with *N*-methyl-ᴅ-aspartate receptor antagonists

With the exception of the DOMINO-AD study discussed earlier, the duration of all available trials is 1 year or less. Therefore, the long-term effects of treatment are unknown. Available clinical guidelines do not offer specific recommendations regarding optimal duration of therapy. In the United States, clinical practice guidelines for the pharmacologic treatment of dementia from the American College of Physicians and the American Academy of Family Physicians (2008) state that any beneficial effect of treatment would generally be observed within 3 months of initiation. They also state that if slowing decline of cognition is no longer a goal, treatment with memantine or an AChE inhibitor is no longer appropriate. The American Academy of Family Physicians treatment algorithm for AD (Winslow et al., 2011) suggests continuing treatment with AChE inhibitors while the patient's condition is stable and adding memantine when the patient's condition deteriorates to moderate to severe AD. It also recommends discontinuing medications if the patient does not adhere to therapy, continues to deteriorate, or develops serious comorbid disease or is terminally ill; this is also true if the patient or caregiver chooses to discontinue treatment. The Fourth Canadian Consensus Conference on the Diagnosis and Treatment of Dementia (2014) suggests that pharmacologic treatment should be discontinued when the patient and/or caregiver makes an informed decision to stop therapy after being advised of the risks and benefits; the patient is nonadherent; the rate of cognitive, functional, or behavioral decline is greater on treatment than prior to initiation of treatment; the patient experiences intolerable side effects; the patient's comorbidities make continued use unacceptably risky or futile (e.g., terminal illness); or the patient's dementia progresses to a stage where there would be no meaningful benefit from continued therapy. It also recommends considering reinstating therapy if an observable decline occurs after discontinuation. United Kingdom guidelines (2018) recommend using memantine as either monotherapy or in combination with an AChE inhibitor in patients with moderate or advanced dementia in specific situations. Guidelines published by the European Federation of the Neurological Societies give a weak recommendation for using memantine in combination with an AChE inhibitor in moderate to severe AD but do not specifically address when to discontinue therapy (Schmidt et al., 2015).

The list of 10 Things Physicians and Patients Should Question (2015), which was developed by the American Geriatrics Society (AGS), recommends that medication should not be prescribed without periodic assessment of perceived cognitive benefits and adverse effects. The 2015 revision includes expanded rationale for this recommendation to state that it is unclear if cognitive changes with available medications are clinically meaningful. The AGS recommends that medications should be discontinued if the desired effects, including stabilization of cognition, are not perceived by the patient, caregiver, and/or clinician within about 12 weeks.

As with all medications, the incidence of adverse events is one of the main considerations for continuing or stopping treatment. This may be especially pertinent in the dementia patient, as factors such as age, frailty, and polypharmacy must be taken into account when reviewing risks versus benefits. The most common adverse effects of memantine are diarrhea, constipation, vomiting, dizziness, confusion, headache, hypertension, and cough. Rare side effects include hallucination, aggressive behavior, fatigue, abdominal pain, urinary incontinence, and dyspnea (Allergan, 2016).

A meta-analysis of 54 randomized placebo-controlled trials looked at patients with cognitive impairment on AChE inhibitors or memantine to determine risk of falls, syncope, and related adverse events (Kim, Brown, Ding, Kiel, & Berry, 2011). The analysis of pooled data from the 14 memantine trials found no significant difference in falls, syncope, or accidental injury. Memantine was associated with a significant decrease in fractures, though the authors noted that these data were extracted from three small unpublished studies.

There are very limited data to support continuation or discontinuation of treatment in patients enrolled in palliative or hospice care for advanced dementia. Advanced dementia is often defined as a global deterioration scale (GDS) score of 7 or a functional assessment scale (FAST) score of 7, indicating loss of all or most verbal and psychomotor skills (Reisberg, 1988; Reisberg, Ferris, de Leon, & Crook, 1982). The majority of clinical trials of patients with advanced dementia exclude patients with a GDS or FAST score of 7. One retrospective study of 10,065 patients over age 65 enrolled in hospice care in the United States found that 21% of patients with end-stage dementia were prescribed cholinesterase inhibitors or memantine at the time of hospice enrollment; this study did not address whether treatment was discontinued at the time of enrollment (Weschules, Maxwell, & Shega, 2008). Clinical practice guidelines for palliative care and hospice do not address if these medications should be continued in patients who are receiving hospice care. A consensus panel of geriatricians identified cholinesterase inhibitors and memantine to be among the medications that may be inappropriate in advanced dementia (Holmes et al., 2008). There are no studies addressing the duration of therapy for memantine in vascular dementia, PDD, or DLB. In the absence of side effects, treatment duration is often lifelong.

Conclusion

Overall, the decision–making process for the use of memantine should be guided by the preference of the patient and/or caregiver. It is reasonable to stop a medication trial if there is no noticeable benefit after the first 3 months of treatment, if significant side effects are noted, or once a patient's dementia progresses to a point where there would be no meaningful benefit from continued therapy. Often, the risks of side effects may outweigh the potential benefits of treatment. It is also reasonable to restart therapy if an observable decline occurs after discontinuation.

Key facts of the development of memantine

- Structurally related to amantadine, which was developed as an antiinfluenza drug but later incidentally found to improve some symptoms of Parkinson's disease (see Fig. 36.2)
- Compound first synthesized by Eli Lilly in 1968 as a potential drug to treat diabetes
- Later discovered to have activity in the central nervous system and was developed by Merz in Germany to treat dementia
- NMDA activity discovered in 1980s after clinical trials had already begun
- First marketed for dementia in Germany starting in 1989 under the brand name Axura
- In 2000, Merz partnered with Forest to develop memantine for the US market under the brand name Namenda (Witt et al., 2004)
- Merz also partnered with Suntory in Japan and Lundbeck for other markets, including Europe, under the brand name Ebixa
- Sales of memantine reached $1.8 billion in 2014 (Quarterly Namenda sales data and retail statistics information, 2014)
- Extended-release formulation launched in 2014
- Also marketed in some countries as a combination drug with the cholinesterase inhibitor donepezil

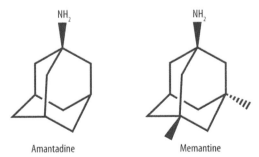

Amantadine   Memantine

Figure 36.2 *Chemical structure of amantadine and memantine.* The chemical structures of the related compounds amantadine (left) and memantine (right), showing structural similarities.

Summary points

- This chapter focuses on the use of NMDA receptor antagonists for the treatment of dementia.
- The NMDA receptor plays an important role in synaptic plasticity, including the functions of memory and learning. Overactivation of NMDA receptors can lead to excitotoxicity and potential neuronal damage.
- Memantine is an uncompetitive NMDA antagonist used for the treatment of certain types of dementia.
- Studies of memantine use have demonstrated some modest benefits in Alzheimer's disease, vascular dementia, Parkinson's disease dementia, and dementia with Lewy bodies.
- Memantine should not be used in frontotemporal dementia. Studies have shown little to no benefit, and it may worsen symptoms.
- Little guidance is available to help determine length of treatment with NMDA receptor antagonists, especially in advanced dementia.
- Use of NMDA receptor antagonists should be guided by patient/caregiver preference, side effects, and progression of disease.

References

Aarsland, D., Ballard, C., Walker, Z., Bostrom, F., Alves, G., Kossakowski, K., et al. (2009). Memantine in patients with Parkinson's disease dementia or dementia with Lewy bodies: A double-blind, placebo-controlled, multicentre trial. *The Lancet Neurology, 8*, 613–618. https://doi.org/10.1016/S1474-4422(09)70146-2.

Allergan, Inc. (2016). *Namenda: Drug label information.* Retrieved from https://dailymed.nlm.nih.gov/dailymed/drugInfo.cfm?setid=b9f27baf-aa2a-443a-9ef5-e002d23407ba.

Boxer, A. L., Knopman, D. S., Kaufer, D. I., Grossman, M., Onyike, C., Graf-Radford, N., et al. (2013). Memantine in patients with frontotemporal lobar degeneration: A multicentre, randomised, double-blind, placebo-controlled trial. *The Lancet Neurology, 12*, 149–156. https://doi.org/10.1016/S1474-4422(12)70320-4.

Boxer, A. L., Lipton, A. M., Womack, K., Meririlees, J., Neuhaus, J., et al. (2009). An open label of memantine treatment in 3 subtypes of frontotemporal lobar degeneration. *Alzheimer Disease and Associated Disorders, 23*, 211–217. https://doi.org/10.1097/WAD.0b013e318197852f.

Bucks, R. S., Ashworth, D. L., Wilcock, G. K., & Siegfried, K. (1996). Assessment of activities of daily living in dementia: Development of the Bristol activities of daily living scale. *Age and Ageing, 25*, 113–120.

Choosing wisely — ten things physicians and patients should question.(April 2015). Retrieved from http://www.choosingwisely.org/societies/american-geriatrics-society.

Cummings, J., Mega, M., Gray, K, Rosenberg-Thompson, S., Carusi, D., & Gornbein, J. (1994). The neuropsychiatric inventory: Comprehensive assessment of psychopathology in dementia. *Neurology, 31*, 2308–2314.

Diehl-Schmid, J., Forstyl, H., Perneczky, R., Pohl, C., & Kurtz, A. (2008). A 6 month, open label study of memantine in patients with frontotemporal dementia. *International Journal of Geriatric Psychiatry, 23*, 754–759. https://doi.org/10.1002/gps.1973.

Emre, M., Tsolaki, M., Bonuccelli, U., Destee, A., Tolosa, E., Kutzelnigg, A., et al. (2010). Memantine for patients with Parkinson's disease dementia or dementia with Lewy bodies: A randomised, double-blind, placebo-controlled trial. *The Lancet Neurology, 9*, 969—977. https://doi.org/10.1016/S1474-4422(10)70194-0.

Folstein, M., Folstein, S., & McHugh, P. (1975). "Mini-Mental state": A practical method for grading the cognitive state of patients for the clinician. *Journal of Psychiatric Research, 12*, 189—198.

Guy, W. (1976). Clinical global impression scale. ECDEU assessment manual for psychopharmacology — revised (DHEW publ No ADM 76-338). In *Rockville, MD: Department of health, education, and welfare, public Health service, alcohol, drug abuse, and mental health administration, NIMH, psychopharmacology research branch, division of extramural research programs* (pp. 218—222).

Holmes, H. M., Sachs, G. A., Shega, J. W., Hougham, G. W., Cox Hayley, D., & Dale, W. (2008). Integrating palliative medicine into the care of persons with advanced dementia: Identifying appropriate medication use. *Journal of the American Geriatrics Society, 56*, 1306—1311. https://doi.org/10.1111/j.1532-5415.2008.01741.x.

Howard, R., McShane, R., Lindesay, J., Ritchie, C., Baldwin, A., et al. (2012). Donepezil and memantine for moderate-to-severe Alzheimer's disease. *New England Journal of Medicine, 366*, 893—903. https://doi.org/10.1056/NEJMoa1106668.

Howard, R., McShane, R., Lindesay, J., Ritchie, C., Baldwin, A., Barber, R., et al. (2015). Nursing home placement in the donepezil and memantine in moderate to severe Alzheimer's disease (DOMINO-AD) trial: Secondary and post-hoc analyses. *The Lancet Neurology, 14*, 1171—1181. https://doi.org/10.1016/S1474-4422(15)00258-6.

Kim, D. H., Brown, R. T., Ding, E. L., Kiel, D. P., & Berry, S. D. (2011). Dementia medications and risk of falls, syncope, and related adverse events: Meta-analysis of randomized controlled trials. *Journal of the American Geriatrics Society, 59*, 1019—1031. https://doi.org/10.1111/j.1532-5415.2011.03450.x.

Kornhuber, J., Weller, M., Schoppmeyer, K., & Riederer, P. (1994). Amantadine and memantine are NMDA receptor antagonists with neuroprotective properties. *Journal of Neural Transmission Supplementum, 43*, 91—104.

Lipton, S. A. (2007). Pathologically activated therapeutics for neuroprotection. *Nature Reviews Neuroscience, 8*, 803—808. https://doi.org/10.1038/nrn2229.

Moore, A., Patterson, C., Lee, L., Vedel, I., & Bergman, H. (2014). Fourth Canadian Consensus Conference on the Diagnosis and Treatment of Dementia: Recommendations for family physicians. *Canadian Family Physician, 60*, 433—438.

Morris, J. C. (1993). The clinical dementia rating (CDR): Current version and scoring rules. *Neurology, 43*, 2412—2414.

National Institute for Health and Care Excellence. (2018). *Dementia: Assessment, management and support for people living with dementia and their carers*. NICE guideline (NG97).

Orgogozo, J. M., Rigaud, A. S., Stoffler, A., Mobius, H. J., & Forette, F. (2002). Efficacy and safety of memantine in patients with mild to moderate vascular dementia: A randomized, placebo-controlled trial (MMM 300). *Stroke, 33*, 1834—1839.

Parsons, C. G., Stöffler, A., & Danysz, W. (2007). Memantine: A NMDA receptor antagonist that improves memory by restoration of homeostasis in the glutamatergic system - too little activation is bad, too much is even worse. *Neuropharmacology, 53*, 699—723. https://doi.org/10.1016/j.neuropharm.2007.07.013.

Qaseem, A., Snow, V., Cross, J. T., Jr., Forciea, M. A., Hopkins, R., Jr., Shekelle, P., et al. (2008). Current pharmacologic treatment of dementia: A clinical practice guideline from the American College of Physicians and the American Academy of Family Physicians. *Annals of Internal Medicine, 148*, 370—378.

Quarterly Namenda sales data and retail statistics information.(February 2014). Retrieved from https://www.drugs.com/stats/namenda.

Reisberg, B. (1988). Functional assessment staging (FAST). *Psychopharmacology Bulletin, 24*, 653—659.

Reisberg, B., Doody, R., Stoffler, A., Schmitt, F., et al. (2003). Memantine in moderate-to-severe Alzheimer's disease. *New England Journal of Medicine, 348*, 1333—1341. https://doi.org/10.1056/NEJMoa013128.

Reisberg, B., Doody, R., Stoffler, A., Schmitt, F., Ferris, S., & Mobius, H. J. (2006). A 24-week open-label extension study of memantine in moderate to severe Alzheimer disease. *Archives of Neurology, 63*, 49—54. https://doi.org/10.1001/archneur.63.1.49.

Reisberg, B., Ferris, S. H., de Leon, M. J., & Crook, T. (1982). The Global Deterioration Scale for assessment of primary degenerative dementia. *American Journal of Psychiatry, 139*, 1136—1139. https://doi.org/10.1176/ajp.139.9.1136.

Rogawski, M. A., & Wenk, G. L. (2003). The neuropharmacological basis for the use of memantine in the treatment of Alzheimer's disease. *CNS Drug Reviews, 9*, 275—308. https://doi.org/10.1111/j.1527-3458.2003.tb00254.x.

Rosen, W., Mohs, R., & Davis, K. (1984). A new rating scale for Alzheimer's disease. *American Journal of Psychiatry, 141*, 1356—1364.

Schmidt, R., Hofer, E., Bouwman, F. H., Buerger, K., Cordonnier, C., Fladby, T., et al. (2015). EFNS-ENS/EAN Guideline on concomitant use of cholinesterase inhibitors and memantine in moderate to severe Alzheimer's disease. *European Journal of Neurology, 22*, 889—898. https://doi.org/10.1111/ene.12707.

Schneider, L. S., Dagerman, K. S., Higgins, J. P., & McShane, R. (2011). Lack of evidence for the efficacy of memantine in mild Alzheimer disease. *Archives of Neurology, 68*, 991—998. https://doi.org/10.1001/archneurol.2011.69.

Schneider, L., Olin, J., Doody, R., Clark, C. M., Morris, J. C., Reisberg, B., et al. (1997). Validity and reliability of the Alzheimer's Disease cooperative study — clinical global impression of change. *Alzheimer Disease and Associated Disorders, 11*, S1—S12.

Sharre, W. D., Warner, J. L., Davis, R. A., & Theado-Miller, N. (2005). Memantine in frontotemporal dementia. *Neurology, 64*, P02077.

Swanberg, M. M. (2007). Memantine for behavioral disturbances in frontotemporal dementia: A case series. *Alzheimer Disease and Associated Disorders, 21*, 164—166. https://doi.org/10.1097/WAD.0b013e318047df5d.

Vercelletto, M., Boutoleau-Bretonnière, C., Volteau, C., Puel, M., Auriacombe, S., Sarazin, M., et al. (2011). Memantine in behavioral variant frontotemporal dementia: Negative results. *Journal of Alzheimer's Disease, 23*, 749. https://doi.org/10.3233/JAD-2010-101632.

Weschules, D. J., Maxwell, T. L., & Shega, J. W. (2008). Acetylcholinesterase inhibitor and N-methyl-D-aspartic acid receptor antagonist use among hospice enrollees with a primary diagnosis of dementia. *Journal of Palliative Medicine, 11*, 738—745.

Wilcock, G., Mobius, H. J., & Stoffler, A. (2002). A double-blind, placebo-controlled multicenter study of memantine in mild to moderate vascular dementia (MMM 500). *International Clinical Psychopharmacology, 17*, 297—305.

Winslow, B. T., Onysko, M. K., Stob, C. M., & Hazlewood, K. A. (2011). Treatment of Alzheimer disease. *American Family Physician, 83*, 1403—1412.

Witt, A., Macdonald, N., & Kirkpatrick, P. (2004). Memantine hydrochloride. *Nature Reviews Drug Discovery, 3*, 109—110.

Xia, P., Chen, H. S., Zhang, D., & Lipton, S. A. (2010). Memantine preferentially blocks extrasynaptic over synaptic NMDA receptor currents in hippocampal autapses. *Journal of Neuroscience, 30*, 11246—11250. https://doi.org/10.1523/JNEUROSCI.2488-10.2010.

CHAPTER 37

Swallowing impairment in Parkinson's disease

Maira Rozenfeld Olchik[1], Marina Padovani[2], Annelise Ayres[3]

[1]Department of Surgery and Orthopedics, Speech Language Pathology Course, Universidade Federal do Rio Grande do Sul, Porto Alegre, Rio Grande do Sul, Brazil; [2]School of Speech-Language Pathology and Audiology, Santa Casa de São Paulo, School of Medical Sciences, São Paulo, Brazil; [3]Postgraduate Program in Health Sciences, Universidade Federal de Ciências da Saúde de Porto Alegre, Porto Alegre, Rio Grande do Sul, Brazil

List of abbreviations

DBS deep brain stimulation
DP Parkinson's disease
FEES fiberoptic endoscopic evaluation of swallowing
MEG magnetoencephalography
SLP speech language pathologist

Minidictionary of terms

Direct therapy Direct therapy involves training the swallowing process with different food textures, volumes, temperatures, and flavors. Compensatory maneuvers may also be practiced.

Dysphagia Dysphagia is characterized by changes in the functioning of any phase of swallowing as a consequence of neurological, mechanical, or psychogenic impairment that present a risk of penetration/aspiration.

Functional evaluation Functional evaluation is the evaluation done by a speech language pathologist (SLP). This assessment should include taking a patient history and conducting a clinical examination in which structural and functional tests are carried out with different food consistencies (liquid, pureed, and solid).

Indirect therapy Indirect therapy includes exercises aimed at modifying the strength, length, and range of motion of structures of the oral cavity, pharynx, and larynx, thus improving swallowing safety.

Objective evaluation Objective evaluation comprises instrumental exams such as fiberoptic endoscopic evaluation of swallowing (FEES) and videofluoroscopy of swallowing, often complementing diagnostic procedures.

Signs and symptoms of dysphagia Signs and symptoms of dysphagia are risk factors that describe unsecured swallowing, including history of aspiration pneumonia; alert state; interaction attention/ability; awareness of the swallowing problem; awareness of secretion; ability to manipulate flows; postural control; fatigability; anatomy and oral, pharyngeal, and laryngeal physiology; orofacial tonus; oral apraxia; orofacial sensitivity; gag pharyngeal contraction; saliva swallowing; cough and hawk; swallowing apraxia; oral residue; delayed swallowing reflex; reduction in laryngeal elevation; wet voice; and multiple swallowing.

Diagnosis and Management in Dementia
ISBN 978-0-12-815854-8, https://doi.org/10.1016/B978-0-12-815854-8.00037-9

Introduction

Swallowing is a coordinated activity that allows the safe, uninterrupted passage of oral contents into the stomach, such as the food bolus, saliva, or secretions. This function is coordinated by the cerebral cortex, brain stem, and brain nerves. It is composed of phases intrinsically related and categorized according to the region of the oropharyngeal passage in which the symptom appears. This can occur during oral preparation, which is voluntary, or during the pharyngeal and esophageal stages, which are involuntary (Inakoa & Albuquerque, 2014; Lind, 2003; Padovani, Moraes, Mangili, & Andrade, 2007; Queiroz, Haguette, & Haguette, 2009; Solazzo et al., 2012). Swallowing occurs approximately 600 times per day in a healthy adult man (Jotz, Angelis, & Barros, 2010) (Fig. 37.1).

Any disorder in the swallowing process that presents a risk of penetration/aspiration is called dysphagia (Inakoa & Albuquerque, 2014; Lind, 2003; Solazzo et al., 2012). Dysphagia is characterized by changes in the functioning of any phase of swallowing as a consequence of neurological, mechanical, or psychogenic impairment (Queiroz et al., 2009). Such a disorder can increase the risk of dehydration, malnutrition, and aspiration pneumonia, leading to death (Inakoa & Albuquerque, 2014; Solazzo et al., 2012; Padovani et al., 2007). In addition, dysphagia negatively impacts patient quality of life; moreover, health complications resulting from dysphagia can inflate medical costs and lengthen hospital stays (Starmer et al., 2012).

Dysphagia can be classified into two distinct types, oropharyngeal and esophageal, according to anatomical region. Oropharyngeal dysphagia refers to difficulties in transport of the alimentary bolus from the mouth to the upper esophageal sphincter. Esophageal dysphagia, on the other hand, refers to disturbances in the passage of the alimentary bolus from the upper esophageal sphincter to the stomach (Jotz et al., 2010; Lind, 2003; Padovani et al., 2007).

Another classification of dysphagia refers to its cause, which may be mechanical due to changes in the structures involved in the swallowing process—or neurological due to damage to the parts of the central nervous system responsible for the swallowing process (Olszewski, 2006).

The SLP is qualified for the evaluation, management, and treatment of oropharyngeal dysphagia (Lind, 2003; Solazzo et al., 2012). This assessment should include taking a patient history and conducting a clinical examination in which structural and functional tests are carried out with different food consistencies (liquid, pureed, and solid). In order to discover noninstrumental clinical evaluations available for the screening and evaluation of dysphagia in patients with Parkinson's disease (PD), we performed a systematic review of scientific evidence. However, we were unable to find approaches regarding clinical screening and assessment instruments for dysphagia in patients with PD that tested food textures and liquid consistencies as well as patient self-perception (Ayres, Scudeiro, & Olchik, 2017).

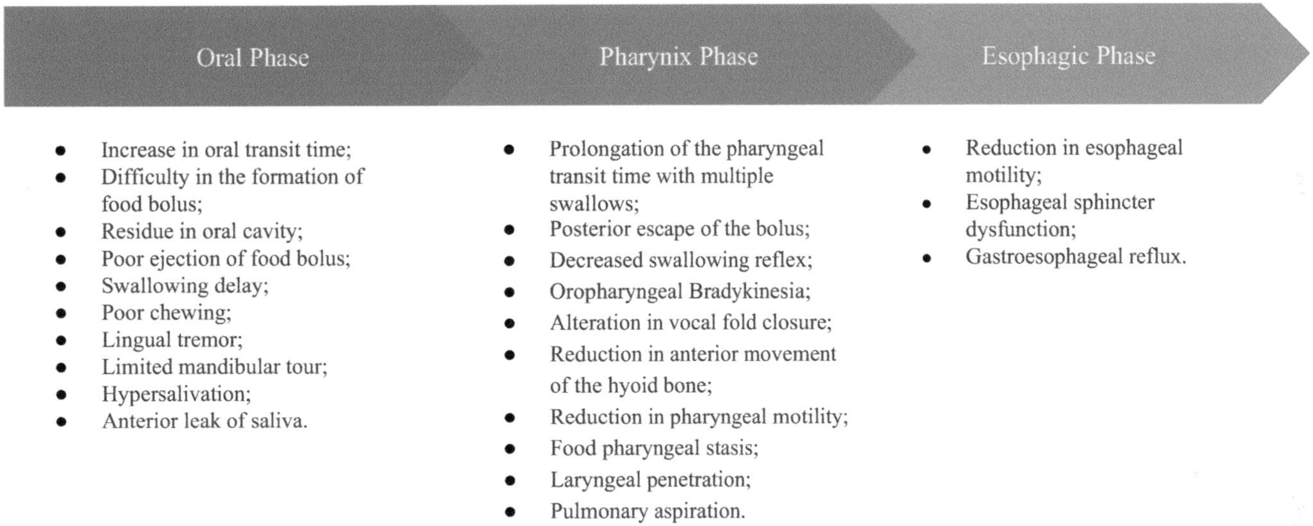

Oral Phase

- Increase in oral transit time;
- Difficulty in the formation of food bolus;
- Residue in oral cavity;
- Poor ejection of food bolus;
- Swallowing delay;
- Poor chewing;
- Lingual tremor;
- Limited mandibular tour;
- Hypersalivation;
- Anterior leak of saliva.

Pharynix Phase

- Prolongation of the pharyngeal transit time with multiple swallows;
- Posterior escape of the bolus;
- Decreased swallowing reflex;
- Oropharyngeal Bradykinesia;
- Alteration in vocal fold closure;
- Reduction in anterior movement of the hyoid bone;
- Reduction in pharyngeal motility;
- Food pharyngeal stasis;
- Laryngeal penetration;
- Pulmonary aspiration.

Esophagic Phase

- Reduction in esophageal motility;
- Esophageal sphincter dysfunction;
- Gastroesophageal reflux.

Figure 37.1 Major signs and symptoms of dysphagia in Parkinson disease.

Instrumental exams such as FEES and videofluoroscopy of swallowing often complement diagnostic procedures. Radiological esophagogram, ultrasound, manometry, electromyography, scintigraphy, and 24-h pH-metry can also be done when necessary (Jotz et al., 2010; Queiroz et al., 2009).

The FEES examination is a low-cost method that is easy to perform bedside without exposing the patient to radiation. A flexible endoscope is introduced through the nose to obtain a view of the pharynx and larynx. During the examination, various food textures dyed with food coloring may be offered in order to better visualize swallowing dynamics. This examination helps determine which food textures are safer for the patient to swallow as well as the maneuvers that work best for speech language therapy (Jotz et al., 2010; Queiroz et al., 2009).

Videofluoroscopy gives a view of the propulsion of the food bolus from the mouth to the esophageal opening. Like the FEES examination, different food consistencies and therapeutic maneuvers can be tested during the assessment (Jotz et al., 2010). It has an advantage over the FEES exam in allowing a better view of the pharyngeal phase and the upper aerodigestive tract in addition to permitting the measurement of any aspirated content. However, its disadvantages are patient exposure to radiation and the intake of barium sulfate. These conditions limit the number of times this type of test can be performed frequently on the same patient (Speyer, 2013) (Fig. 37.2).

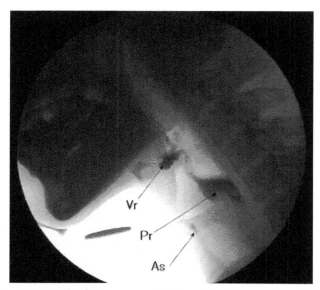

Figure 37.2 Videofluoroscopy with signs of residues and aspiration.

The treatment of oropharyngeal dysphagia involves speech-language rehabilitation via food consistency, food supply, compensatory maneuvers, and protection of the upper airways and mobility, sensitivity, and tonicity exercises for the structures involved in the swallowing process. In addition, in some cases surgical and pharmacological methods may be indicated (Jotz et al., 2010; Lind, 2003).

Oropharyngeal dysphagia in Parkinson's disease

Currently, PD is considered a multisystemic disease that gradually affects multiple components of several functional networks throughout the nervous system (Braak, 2002). This causes a variety of nonmotor symptoms (Suntrup, 2013). These include oropharyngeal dysphagia, which is high in prevalence and an aggravating factor in PD (Kalf, Swart, Bloem, & Munneke, 2012; Luchesi, Kitamura, & Mourão, 2015; Nagaya, Kachi, Yamada, & Igata, 1998; Wintzen et al., 1994).

Dysphagia in PD is characterized by changes in all phases of swallowing (oral, pharyngeal, and esophageal) (Argolo, Sampaio, Pinho, Melo, & Nóbrega, 2015; Kim et al., 2015; Smith, Roddam, & Sheldrick, 2012). The prevalence rates of this symptom found in reports range from 70% to 100% of patients with PD1 (Padovani et al., 2007; Wintzen et al., 1994; Nagaya et al., 1998; Kalf et al., 2012; D'Amelio et al., 2006; Pennington, Snell, Lee, & Walker, 2010; Macleod, Taylor, & Counsell, 2014). In addition, relative risk is 3.2 compared with healthy controls. Dysphagia can occur in varying degrees of severity from the early stages of PD (Sung et al., 2010; Volonte, Porta, & Comi, 2002), but there is still no clear correlation between dysphagia and duration of disease or motor impairment of patients (Lam et al., 2007; Monte, da Silva-Junior, Braga-Neto, Nobre e Souza, & Bruin, 2005).

Dysphagia has a significant impact on quality of life, predisposing patients to recurrent episodes of aspiration pneumonia (Ayres et al., 2017; Mamolar et al., 2017; Michou, Baijens, Rofes, Sanz, & Clave, 2013; Morgante et al., 2000). In addition, it is known that respiratory infection is one of the direct causes of death in patients with PD. Data point to a prevalence of between 30% and 45% of pneumonia cases resulting in the death of patients with PD (D'Amelio et al., 2006; Pennington et al., 2010; Macleod et al., 2014; Pinter et al., 2015). Because such comorbidity is closely linked with immobility and dysphagia, the importance of speech therapy for dysphagia in these individuals has become clear in preventing or delaying the onset of aspiration pneumonia.

The underlying pathophysiology of PD-related dysphagia is still poorly understood. In a study by Suntrup et al. (2013), patient swallowing was assessed with the aid of a FEES examination, a magnetoencephalography (MEG), or an electromyography. Results of the MEG showed that subjects with nondysphagic PD presented focal activation in the caudolateral parts of the primary sensorimotor and premotor cortex and the inferolateral parietal lobe. Activation of the supplementary motor area was significantly reduced in comparison with the group of healthy individuals. In subjects with dysphagic PD, there was a strong overall reduction in activities related to the swallowing task. Moreover, when initiating a swallow, participants with nondysphagic PD showed a prominent change in the lateral direction of the activation peak of the parietal and motor cortex, while activity in the supplemental motor area was significantly reduced. This distinctive pattern was not found in dysphagic patients (Suntrup et al., 2013).

Dysphagia in PD can occur at any stage of swallowing. Swallowing changes will be described more frequently according to each swallowing phase (Altman, Richards, Goldberg, Frucht, & McCabe, 2013; Argolo et al., 2015; Kim et al., 2015; Luchesi, Kitamura, & Mourão, 2013; Mamolar et al., 2017; Nicaretta, Rosso, Mattos, Maliska, & Costa, 2013; Smith et al., 2012; Suntrup et al., 2013) (Fig. 37.3).

The effects of swallowing after deep brain stimulation (DBS) are still inconclusive. Some studies have reported no improvement or decline in swallowing function with DBS. Regarding electrode insertion sites, subthalamic DBS caused more impairment in swallowing than insertion in the *globus pallidus*. However, experimental studies that directly compare swallowing function between these electrode insertion sites, as well as a comparison between unilateral versus bilateral surgery, have not yet been performed (Fuh et al., 1997; Monte et al., 2005; Volkmann et al., 2009).

According to a study by Olchik et al. (2017), there was no significant improvement in swallowing reassessments after DBS. However, according to self-reported questionnaire scores indicative of significant improvement, even small changes in the signs and symptoms of dysphagia had a positive impact on the swallowing quality of life. Furthermore, there was no relation between the patients' motor subtype and swallowing patterns pre- and post-DBS.

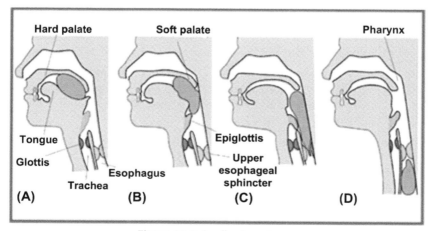

Figure 37.3 Swallowing stages.

Regarding the drug treatment of PD and dysphagia, some studies point to the positive influence of levodopa on the swallowing process. Subjects with higher doses of L-dopa tend to have shorter oral transit times and show better swallowing efficiency (Monte et al., 2005), but these findings are still controversial. Fuh et al. (1997) that pointed an improvement in only 50% of patients after taking l-dopa; the other half did not present changes after medication.

One more important aspect to consider in the follow-up of these patients is the poor self-perception of difficulty in swallowing, which in most cases develops too late. In fact, many patients are diagnosed as dysphagic only after the first episode of aspiration pneumonia (Olanow, Stern, & Sethi, 2009). Manor, Giladi, Cohen, Fliss, and Cohen (2007) published that up to 50% of patients may have objective swallowing disorders that they do not report when questioned, probably because of the frequent lack of perception of the neurological changes brought on by the disease.

Declines in laryngeal sensitivity and cough reflex, increased pharyngeal transport time, and silent aspiration are symptoms that may progress silently until clinical complaints appear. More explicit factors that contribute to the diagnosis of dysphagia occur only in late and advanced stages when rehabilitation options are more restricted (Ertekin et al., 2002; Monteiro et al., 2014).

In corroboration of these data, one study observed a weak correlation between subjective reports of dysphagia and performance in the swallowing test with water (Miller et al., 2009). Thus, it is important to increase the awareness of health professionals to the signs of dysphagia in these patients, since dysphagia in PD may be subclinical or asymptomatic. This demonstrates the plausible idea that patients gradually adapt to dysphagia as a natural consequence of the progression of the disease.

In addition, due to the presence of cognitive impairment or sensory problems, patients with PD may have increased risk of complications due to the underestimation of dysphagia. These data highlight the need for a proactive clinical approach to dysphagia, mainly due to the serious clinical consequences of this symptom (Kalf et al., 2012).

The main examples of such adaptations are a decrease in the size of the food bolus during meals, changes in food textures, and the exclusion of foods that cause greater difficulty while eating. Nonetheless. these modifications may lead to patient malnutrition and dehydration (Kalf et al., 2012).

Treatment of dysphagia in Parkinson's disease

Speech therapy for patients with oropharyngeal dysphagia can be direct or indirect. Direct therapy involves training the swallowing process with different food textures, volumes, temperatures, and flavors. Compensatory maneuvers may also be practiced. Indirect therapy includes exercises aimed at modifying the strength, length, and range of motion of structures of the oral cavity, pharynx, and larynx, thus improving swallowing safety. Among these are tongue strengthening exercises to increase the ejection force of the food bolus, tongue mobility exercises to enhance manipulation of the bolus in the oral cavity, and vocal exercises to increase airway protection through improvement in glottal adduction (Jotz et al., 2010; Lind, 2003).

Considering the serious impact of this symptom on the life of individuals with PD, the number of studies seeking to verify the efficacy of speech language therapy in treating dysphagia in this population has increased. However, the overall number of published reports is still relatively low (Olchik et al., 2017; Pitts et al., 2009; Regan, Walshe, & Tobin, 2010; Smith et al., 2012).

Existing studies have demonstrated the benefits of some specific therapeutic strategies, such as the use of biofeedback during therapeutic intervention (Athukorala, Jone, Sella, & Huckabee, 2014; Felix, Corrêa, & Soares, 2008; Manor, Mootanah, Freud, Giladi, & Cohen, 2013), the use of expiratory muscle strength training for swallowing (Smith et al., 2012), the effects of surface electrical stimulation (Baijens et al., 2012), the effects of myofunctional exercises for swallowing (Luchesi et al., 2015; Argolo, Sampaio, Pinho, Melo, & Nóbrega, 2013), and changes in food consistency (Altman et al., 2013; Luchesi et al., 2013).

In an evidence-based review on swallowing therapy for PD, two workers (Logemann et al., 2008; Robbins et al., 2008) reported on the effect of the chin tuck maneuver. Thus far, Logemann et al. (2008) have conducted the largest study on compensatory approaches for oropharyngeal dysphagia in PD. The authors compared the chin-tuck maneuver with changes in food consistency in the treatment of oropharyngeal dysphagia. The results showed that the postural maneuver was the less-effective strategy in preventing liquid aspiration when compared with changes in food consistency. The same result was found in research carried out by Robbins et al. (2008). This study demonstrated significant differences in interventions to prevent pneumonia. However, these authors suggest that swallowing becomes safer as food consistency changes (with the use of thickened liquids). Other articles show improvements in some swallowing parameters after the investigated treatment strategies (e.g., changes in food texture, tactile thermal oral stimulation, myofunctional exercises, Lee Silverman voice therapy, exercises with biofeedback, and expiratory muscle training), but no significant improvements in global swallowing function were found.

Another systematic review (Russel, Ciucci, Connor, & Schallert, 2010) on speech language therapy for speech and swallowing impairments in patients with PD concluded that there was a shortage of studies on the topic of swallowing problems. Only one article aimed to check the efficacy of a specific therapeutic program for swallowing difficulties. The compiled study found that only three specific methods of speech therapy improved swallowing, namely expired muscle strength training, which improved the cough reflex as well as airway protection; intensive voice therapy, which led to an improvement in tongue mobility and ejection of the food bolus; and intensive speech therapy, which showed evidence of improved swallowing times for solid and liquid consistencies. Even though these treatments did not have the express purpose of treating swallowing impairments, functional progress was nevertheless observed, perhaps due to the recruitment of muscles that are all involved in phonation, speech, and swallowing.

Other new studies have emerged, but small patient samples make it difficult to establish a treatment guideline for patients with PD. Ayres et al., (2017) evaluated the effectiveness of maneuvers in swallowing therapy with PD. A therapeutic program, which consisted of the chin tuck maneuver and guidance on eating practices, was carried out. The chin tuck maneuver improved swallowing performance and self-perception, but not laryngeal function (Ayres et al., 2017) (Table 37.1 and Fig. 37.4).

Wang et al., (2018) have done research on the effect of an easy-to-perform, home-based orolingual exercise program to treat swallowing and breathing coordination

Table 37.1 Swallowing guidelines for Parkinson's disease.

Swallowing orientations

Environmental modifications during mealtimes
  (1) Eat in a quiet place.
  (2) Turn off the television, radio, or any other device that may distract you.
  (3) Stay focused on your meal.
  (4) Do not talk while you chew or swallow.
  (5) Avoid having meals when you are sleepy or tired.
  (6) Try to have meals during ON periods.
Posture during meals
  (1) Sit at a table.
  (2) Keep your torso upright and your head up.
  (3) Avoid lying down immediately after meals.
  (4) Cut or divide food on the plate into small portions.
  (5) Chew food thoroughly before swallowing.
  (6) Swallow more than once until you are sure there is no food left in your mouth.
  (7) If you feel there is food stuck in your throat, swallow more saliva.
Mealtime
  (1) The maximum time you should spend on a main meal (i.e., breakfast, lunch, or dinner) is 30 min.
  (2) Do not eat in a hurry.
Oral Hygiene
  (1) Do not forget to brush your teeth after meals.
  (2) Remember to brush your tongue.
  (3) If you use a dental prosthesis, it should be removed and cleaned with a toothbrush after every meal.
  (4) For patients who do not have teeth, the mouth and tongue should still be cleaned with a toothbrush.

Ayres, A., Jotz, G. P., Rieder, C. R. M., & Olchik, M. R. (2017). Benefit from the chin-down maneuver in the swallowing performance and self-perception of Parkinson's disease patients. *Parkinson's Disease, 2017*, 1–8.

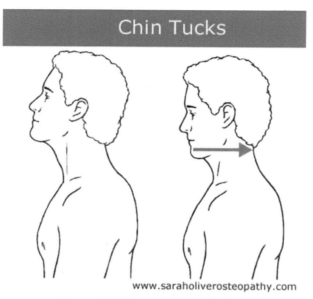

Figure 37.4 Maneuver of swallowing.

in patients with early-stage PD. Park, Oh, Hwang, and Lee (2018) have studied the effect of neuromuscular electrical stimulation therapy applied to the infrahyoid region in combination with forceful swallowing. This is a novel therapeutic approach to dysphagic patients with PD.

Thus, we have observed that although dysphagia is a frequent and high-impact symptom in PD, the body of evidence concerning the efficacy of SLP rehabilitation and management of this symptom in patients with PD is still relatively small. What is more, the vast majority present inconclusive results with little scientific impact. Thus, it is important to continuously search for randomized studies, since several groups continue to work on effective swallowing treatments.

Another important aspect of rehabilitation is the establishment of swallowing guidelines. These support proper management during meals, thereby reducing the risk of penetration/aspiration (Ayres, Jotz, Rieder, & Olchik, 2017).

Conclusion

Dysphagia is a quite prevalent symptom in PD and one of the risk factors for the leading cause of death in PD—aspiration pneumonia. Therefore, it is important to assess for dysphagia in patients with PD regardless of the phase of the disease in order to prevent swallowing disorders in these patients.

Key facts

- Normal swallowing consists of series of coordinated activities that allow for the safe, uninterrupted passage of oral contents into the stomach, such as the food bolus, saliva, or secretions. It is divided into three phases: oral, pharyngeal, and esophageal.
- Concept of dysphagia: To name this difficulty of swallowing, the term dysphagia (from the Greek *dys*—difficulty, and *phagien*—eating) is used. Different concepts are found for the term dysphagia, but the main concept is that dysphagia means disruption of the swallowing process during transportation of the bolus from the oral cavity to the stomach and difficulty in swallowing or any difficulty in swallowing that interferes with transportation of the bolus. It presents different intensities and may lead to malnutrition and dehydration in addition to respiratory complications. Changes in swallowing may occur in the oral, pharyngeal, and esophageal preparatory phases.
- Prevalence and incidence: An estimated 300,000—600,000 new cases of dysphagia occur every year. Dysphagia affects about 6% of the general adult population and survivors of cerebrovascular diseases, resulting in morbidity and mortality in hospitalized patients. The incidence of oropharyngeal dysphagia was reported to be high in patients with PD. Subclinical dysphagia has been reported to appear in early-stage PD.
- Rehabilitation of swallowing: It includes indirect (without food) and direct (with food) swallowing training. There are several techniques, such as thermal stimulation, oral motor exercises, and maneuver, depending on the patient.

Summary points

- This chapter focuses on oropharyngeal dysphagia in PD.
- The prevalence rate of dysphagia in PD ranges from 70% to 100% in patients with the disease, with no direct relation to disease time.
- Dysphagia is one of the causes of pneumonia, which in turn is the main cause of death in PD.
- Dysphagia results in a significant impact on patient quality of life.
- The SLP is qualified to diagnose and treat oropharyngeal dysphagia through modified food texture and liquid consistency, compensatory maneuvers, and airway protection exercises, all aimed at adapting the strength, length, and range of motion of structures involved in the swallowing process. SLPs also provide guidelines for safer ways of offering food to patients.

References

Altman, K. W., Richards, A., Goldberg, L., Frucht, S., & McCabe, D. J. (2013). Dysphagia in stroke, neurodegenerative disease, and advanced dementia. *Otolaryngologic Clinics of North America, 46*(6), 1137—1149.

Argolo, N., Sampaio, M., Pinho, P., Melo, A., & Nóbrega, A. C. (2013). Do swallowing exercises improve swallowing dynamic and quality of life in Parkinson's disease? *NeuroRehabilitation, 32*(4), 949—955. https://doi:10.3233/NRE-130918.

Argolo, N., Sampaio, M., Pinho, P., Melo, A., & Nóbrega, A. C. (2015). Swallowing disorders in Parkinson's disease: Impact of lingual pumping. *International Journal of Language and Communication Disorders, 50*(5), 659—664.

Athukorala, R. P., Jones, R. D., Sella, O., & Huckabee, M. L. (2014). Skill training for swallowing rehabilitation in patients with Parkinson's disease. *Archives of Physical Medicine and Rehabilitation, 95*(7), 1374—1382.

Ayres, A., Jotz, G. P., Rieder, C. R. M., & Olchik, M. R. (2017). Benefit from the chin-down maneuver in the swallowing performance and self-perception of Parkinson's disease patients. *Parkinson's Disease,* (2017), 1—8.

Ayres, A., Scudeiro, L. A. J., & Olchik, M. R. (2017). Instrumentos de avaliação clínica para disfagia orofaríngea na doença de Parkinson: Revisão sistemática. *Audiology Communication Research, 22,* 1—6.

Baijens, L. W. J., Speyer, R., Passos, V. L., Pilz, W., Roodenburg, N., & Clave, P. (2012). The effect of surface electrical stimulation on swallowing in dysphagic Parkinson patients. *Dysphagia, 27,* 528—537.

Braak, H., Del Tredici, K., Bratzke, H., Hamm-Clement, J., Sandmann-Keil, D., & Rüb, U. (2002). Staging of the intracerebral inclusion body pathology associated with idiopathic Parkinson's disease (preclinical and clinical stages). *Journal of Neurology, 249*(3), 1—5.

D'Amelio, M., Ragonese, P., Morgante, L., Reggio, A., Callari, G., Salemi, G., et al. (2006). Long-term survival of Parkinson's disease: A population-based study. *Journal of Neurology, 253*(1), 33—37.

Ertekin, C., Tarlaci, S., Aydogdu, I., Kiylioglu, N., Yuceyar, N., Turman, B., et al. (2002). Electrophysiological evaluation of pharyngeal phase of swallowing in patients with Parkinson's disease. *Movement Disorders, 17,* 942—949.

Felix, V. N., Corrêa, S. M. A., & Soares, R. J. (2008). A therapeutic maneuver for oropharyngeal dysphagia in patients with Parkinson's disease. *Clinics, 63*(5), 661—666.

Fuh, J. L., Lee, R. C., Wang, S. J., Lin, C. H., Wang, P. N., Chiang, J. H., et al. (1997). Swallowing difficulty in Parkinson's disease. *Clinical Neurology and Neurosurgery, 99*(2), 106—112.

Inaoka, C., & Albuquerque, C. (2014). Efetividade da intervenção fonoaudiológica na progressão da alimentação via oral em pacientes com disfagia orofaríngea pós AVE. *Revista CEFAC, 16*(1), 187—196.

Jotz, G. P., Angelis, E. C., & Barros, A. P. B. (2010). *Tratado da deglutição e difagia, no adulto e na criança.* São Paulo: Revinter.

Kalf, J. G., Swart, B. J. M., Bloem, B. R., & Munneke, M. (2012). Prevalence of oropharyngeal dysphagia in Parkinson's disease: A meta-analysis. *Parkinsonism and Related Disorders, 18,* 311—315.

Kim, Y. H., Oh, B. M., Jung, I. Y., Lee, J. C., Lee, G. J., & Han, T. R. (2015). Spatiotemporal characteristics of swallowing in Parkinson's disease. *The Laryngoscope, 125*(2), 389—395.

Lam, K., Lam, F. K., Lau, K. K., Chan, Y. K., Kan, E. Y., Woo, J., et al. (2007). Simple clinical tests may predict severe oropharyngeal dysphagia in Parkinson's disease. *Movement Disorders, 22,* 640—644.

Lind, C. D. (2003). Dysphagia: Evaluation and treatment. *Gastroenterology Clinics of North America, 32*(2), 553—575.

Logemann, J. A., Gensler, G., Robbins, J., Lindblad, A. S., Brandt, D., Hind, J. A., et al. (2008). A randomized study of three interventions for aspiration of thin liquids in patients with dementia or Parkinson's disease. *Journal of Speech, Language, and Hearing Research, 51,* 173—183.

Luchesi, K. F., Kitamura, S., & Mourão, L. F. (2013). Management of dysphagia in Parkinson's disease and amyotrophic lateral sclerosis. *CoDAS, 25*(4), 358—364.

Luchesi, K. F., Kitamura, S., & Mourão, L. F. (2015). Dysphagia progression and swallowing management in Parkinson's disease: An observational study. *Brazilian Journal of Otorhinolaryngology, 81,* 24—30.

Macleod, A. D., Taylor, K. S., & Counsell, C. E. (2014). Mortality in Parkinson's disease: A systematic review and meta-analysis. *Movement Disorders, 29*(13), 1615—1622.

Mamolar, A. S., Santamarina, R. M. L., Granda, M. C. M., Fernández, G. M. J., Sirgo, R. P., & Álvarez, M. C. (2017). Swallowing disorders in Parkinson's disease. *Acta Otorrinolaringologica Española, 68*(1), 15—22.

Manor, Y., Giladi, N., Cohen, A., Fliss, D. M., & Cohen, J. T. (2007). Validation of a swallowing disturbance questionnaire for detecting dysphagia in patients with Parkinson's disease. *Movement Disorders, 22,* 1917—1921.

Manor, Y., Mootanah, R., Freud, D., Giladi, N., & Cohen, J. T. (2013). Video-assisted swallowing therapy for patients with Parkinson's disease. *Parkinsonism and Related Disorders, 19*(2), 207—211.

Michou, E., Baijens, L., Rofes, L., Sanz, P., & Clave, P. (2013). Oropharyngeal swallowing disorders in Parkinson's disease: Revisited. *International Journal of Language and Communication Disorders, 1,* 76—88.

Miller, N., Allcock, L., Hildreth, A. J., Jones, D., Noble, E., & Burn, D. J. (2009). Swallowing problems in Parkinson disease: Frequency and clinical correlates. *Journal of Neurology Neurosurgery and Psychiatry, 80*(9), 1047—1049.

Monte, F. S., da Silva-Junior, F. P., Braga-Neto, P., Nobre e Souza, M. A., & Bruin, V. M. (2005). Swallowing abnormalities and dyskinesia in Parkinson's disease. *Movement Disorders, 20,* 457—462.

Monteiro, L., Souza-Machado, A., Pinho, P., Sampaio, M., Nóbrega, A. C., & Melo, A. (2014). Swallowing impairment and pulmonary dysfunction in Parkinson's disease: The silent threats. *Journal of Neurological Sciences, 339*(1—2), 149—152.

Morgante, L., Salemi, G., Maneghini, F., Di Rosa, A. E., Epifanio, A., Grigoletto, F., et al. (2000). Parkinson disease survival: A population-based study. *Archives of Neurology, 57,* 507—512.

Nagaya, M., Kachi, T., Yamada, T., & Igata, A. (1998). Videofluorographic study of swallowing in Parkinson's disease. *Dysphagia, 13,* 95—100.

Nicaretta, D. H., Rosso, A. L., Mattos, J. P., Maliska, C., & Costa, M. M. B. (2013). Dysphagia and sialorrhea: The relationship to Parkinson's disease. *Arquivos de Gastroenterologia, 50*(1), 42—49.

Olanow, C. W., Stern, M. B., & Sethi, K. (2009). The scientific and clinical basis for the treatment of Parkinson disease. *Neurology, 72,* S1—S136.

Olchik, M. R., Ghisi, M., Ayres, A., Schuh, A. F. S., Oppitz, P., & Rieder, C. R. M. (2017). The impact of deep brain stimulation on the quality of life and swallowing in individuals with Parkinson's disease. *International Archives of Otorhinolaryngology, 22,* 125—129.

Olszewski, J. (2006). Causes, diagnosis and treatment of neurogenic dysphagia as an interdisciplinary clinical problem. *The Polish Otolaryngology, 60*(4), 491—500.

Padovani, A. R., Moraes, D. P., Mangili, L. D., & Andrade, C. R. F. (2007). Protocolo Fonoaudiológico de Avaliação do Risco para Disfagia (PARD). *Revista da Sociedade Brasileira de Fonoaudiologia, 12*(3), 199—205.

Park, J. S., Oh, D. H., Hwang, N. K., & Lee, J. H. (2018). Effects of neuromuscular electrical stimulation in patients with Parkinson's disease and dysphagia: A randomized, single-blind, placebo-controlled trial. *NeuroRehabilitation, 42*(4), 457—463.

Pennington, S., Snell, K., Lee, M., & Walker, R. (2010). The cause of death in idiopathic Parkinson's disease. *Parkinsonism and Related Disorders, 16*(7), 434—437.

Pinter, B., Diem-Zangerl, A., Wenning, G. K., Scherfler, C., Oberaigner, W., Seppi, K., et al. (2015). Mortality in Parkinson's disease: A 38-year follow-up study. *Movement Disorders, 30*(2), 266—269.

Pitts, T., Bolser, D., Rosenbek, J., Troche, M., Okun, M., & Sapienza, C. (2009). Impact of expiratory muscle strength training on voluntary cough and swallow function in Parkinson disease. *Chest Journal, 135,* 1301—1308.

Queiroz, M. A. S., Haguette, R. C. B., & Haguette, E. F. (2009). Achados da videoendoscopia da deglutição em adultos com disfagia orofaríngea neurogênica. *Revista da Sociedade Brasileira de Fonoaudiologia, 14*(4), 454—462.

Regan, J., Walshe, M., & Tobin, W. O. (2010). Immediate effects of thermal—tactile stimulation on timing of swallow in idiopathic Parkinson's disease. *Dysphagia, 25,* 207—215.

Robbins, J., Gensler, G., Hind, J., Logemann, J. A., Lindblad, A. S., Brandt, D., et al. (2008). Comparison of 2 interventions for liquid aspiration on pneumonia incidence: A randomized trial. *Annals of Internal Medicine, 148,* 509—518.

Russell, J. A., Ciucci, M. R., Connor, N. P., & Schallert, T. (2010). Targeted exercise therapy for voice and swallow in persons with Parkinson's disease. *Brain Research Journal, 23*(1341), 3—11.

Smith, S. K., Roddam, H., & Sheldrick, H. (2012). Rehabilitation or compensation: Time for a fresh perspective on speech and language therapy for dysphagia and Parkinson's disease? *International Journal of Language and Communication Disorders, 47*(4), 351–364.

Solazzo, A., Monaco, L., Vecchio, L. D., Tamburrini, S., Iacobellis, F., Berritto, D., et al. (2012). Investigation of compensatory postures with videofluoromanometry in dysphagia patients. *World Journal of Gastroenterology, 18*(23), 2973–2978.

Speyer, R. (2013). Oropharyngeal dysphagia: Screening and assessment. *Otolaryngologic Clinics of North America, 46*(6), 989–1008.

Starmer, H. M., Best, S. R., Agrawal, Y., Chien, W. W., Hillel, A. T., Francis, H. W., et al. (2012). Prevalence, characteristics, and management of swallowing disorders following cerebellopontine angle surgery. *Otolaryngology Head and Neck Surgery, 146*(3), 419–425.

Sung, H. Y., Kim, J. S., Lee, K. S., Kim, Y. I., Song, I. U., Chung, S. W., et al. (2010). The prevalence and patterns of pharyngoesophageal dysmotility in patients with early stage Parkinson's disease. *Movement Disorders, 25*, 2361–2368.

Suntrup, S., Teismann, I., Bejer, J., Suntrup, I., Winkels, M., Mehler, D., et al. (2013). Evidence for adaptive cortical changes in swallowing in Parkinson's disease. *Brain, 136*, 726–738.

Volkmann, J., Albanese, A., Kulisevsky, J., Tornqvist, A. L., Houeto, J. L., Pidoux, B., et al. (2009). Long-term effects of pallidal or subthalamic deep brain stimulation on quality of life in Parkinson's disease. *Movement Disorders, 24*(08), 1154–1161.

Volonte, M. A., Porta, M., & Comi, G. (2002). Clinical assessment of dysphagia in early phases of Parkinson's disease. *Neurological Sciences, 23*, S121–S122.

Wang, C. M., Shieh, W. Y., Ho, C. S., Hu, Y. W., & Wu, Y. R. (2018). Home-based orolingual exercise improves the coordination of swallowing and respiration in early Parkinson disease: A quasi-experimental before-and-after exercise program study. *Frontiers in Neurology, 30*(9), 624.

Wintzen, A. R., Bradrising, U. A., Roos, R. A. C., Vielvoye, J., Liauw, L., & Pauwels, E. K. J. (1994). Dysphagia in ambulant patients with Parkinson's disease: Common, not dangerous. *The Canadian Journal of Neurological Sciences, 21*, 53–55.

CHAPTER 38

Linking astrocytes' exosomes to Alzheimer pathogenesis and therapy

Anna M. Chiarini[1], Ubaldo Armato[1], Claudio Eccher[2], Ilaria Dal Prà[1]

[1]Histology & Embryology Unit, School of Medicine, University of Verona, Verona, Italy; [2]Villa Bianca Hospital, Surgery Unit, Trento, Italy

List of abbreviations

AD Alzheimer disease
ALS Amyotrophic lateral sclerosis
APP Amyloid precursor protein
Aβ Amyloid-β
Aβ-os Amyloid-β oligomers;
Aβs Amyloid-β peptides
BBB Blood—brain barrier
CaSR Calcium-sensing receptor
CNS Central nervous system
CSF Cerebrospinal fluid
ESCRT Endosomal sorting complex required for transport
GFAP Glial fibrillary acidic protein
LOAD Late-onset Alzheimer disease
miRNA MicroRNA
MVBs Multivesicular bodies;
NFTs Neurofibrillary tangles
NO Nitric oxide
nSMase2 Neutral sphingomyelinase2
p-Tau Phosphorylated tau protein
p-Tau-os Phospho-tau oligomers
PHFs Paired helical filaments
SOD1 Superoxide dismutase1
TEMDs Tetraspanin-enriched membrane domains
VEGF-A Vascular endothelial growth factor-A

Mini-dictionary of terms

Amyloid-β (Aβ) Aβ is a peptide of 39—43 amino acids produced by β-secretase-1 cleavage of Amyloid Precursor Protein. Aβ monomers can spontaneously aggregate into neurotoxic oligomers/fibrils.

Calcilytics Pharmaceutical compounds acting as negative allosteric antagonists of the calcium-sensing receptor (CaSR), binding extracellular seven transmembrane domains. Examples: NPS-2143, Calhex-231, Ronacaleret.

Calcium-Sensing Receptor A family C G-protein-coupled receptor whose signaling is activated by Aβ oligomers in human neurons and astrocytes.

Diagnosis and Management in Dementia
ISBN 978-0-12-815854-8, https://doi.org/10.1016/B978-0-12-815854-8.00038-0

Exosomes They are nanovesicles from the endocytic pathway released by the cells into the extracellular environment.

Tau-oligomers Multimeric toxic species of hyperphosphorylated Tau proteins preceding neurofibrillary tangles formation.

Introduction

Alzheimer disease (AD)

The main toxic "AD drivers"

Sporadic (nonfamilial), late-onset Alzheimer's disease (LOAD) is by far the most common kind of dementia in humans. It proceeds very slowly, at first undetected, unfolding as an endocerebral "contagion-like process" following a distinct course from the entorhinal cortex layer II to upper cognitive and other neocortical areas. In this context, "contagion-like process" means the binding of exogenous Aβ oligomers (Aβ-os) to the Ca^{2+}-sensing receptors (CaSRs) of both human cortical neurons and astrocytes (Fig. 38.1). The Aβ-os·CaSR complexes trigger signals whose transduction increases the surplus production, accumulation, and release of neurotoxic Aβ-os and p-Tau (p-Tau-os)—the LOAD "seeds"—plus NO and VEGF-A from neurons and astrocytes

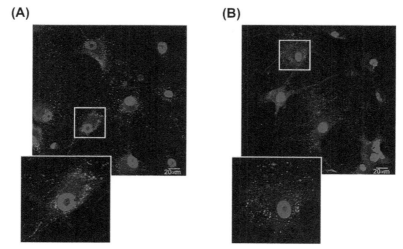

(A)   **(B)**

Figure 38.1 *Aβ·CaSR complexes.* Cortical adult human astrocytes cultured in vitro express the CaSRs whose amino-terminal domains abutting from the plasmalemma bind $Aβ_{25-35}$ and $Aβ_{1-42}$. Such Aβ·CaSR complexes are specifically demonstrated via the in situ proximity ligation assay (Dal Pra et al., 2014). On the nonpermeabilized astrocytes' surfaces, Aβ·CaSR complexes appear as minute red dots later aggregating into increasingly sized patches. DAPI stains the nuclei (blue color). Aβ·CaSR complexes after: (A) 10 min, and (B) 15 min of Aβ exposure. Bars, 20 μm.

Table 38.1 Neurotoxic processes caused by Aβ-os·CaSR signaling in human adult astrocytes.

Process	Effects
CaSR overexpression	Increased sensitivity to toxic Aβ$_{42}$-os.
Buildup of endogenous Aβ$_{42}$-os	Intracellular Aβ$_{42}$-os accumulation enhanced by reduced activities of proteasome and various Aβ-cleaving enzymes.
Increased secretion of endogenous Aβ$_{42}$-os	Oversecreted Aβ$_{42}$-os specifically binds and activates CaSRs of nearby or faraway cells, driving sets of feed-forward recursive vicious loops fostering a progressive extracellular accumulation and diffusion of cytotoxic Aβ$_{42}$-os.
Decreased release of soluble amyloid protein precursor (s-APP)-α	Impairment of s-APP-α neurotrophic/neuroprotective functions.
Increased activity of GSK-3β	Increased intracellular p-Tau-os levels and exosomal p-Tau-os release.
Activation of NO synthase-2	Surplus production/release of NO.
HIF-1α stabilization and nuclear translocation	Excess production/secretion of VEGF-A$_{165}$.

Using as models in vitro cultured human cortical astrocytes and neurons exposed to exogenous Aβ peptides (e.g., Aβ$_{1-42}$, Aβ$_{25-35}$), proved CaSR signaling mediates astrocytes' and neurons' noxious responses to Aβ-exposure (Armato et al., 2013; Dal Prà et al., 2015; Dal Prà et al., 2018; Chiarini, Armato, Gardenal et al., 2017; Chiarini, Armato, Whitfield et al., 2017).

(Armato et al., 2013; Chiarini, Armato, Gardenal, Gui, & Dal Prà, 2017). Thus, the continuous formation of Aβ-os·CaSR complexes recruits through their neurotoxic products increasing numbers of neural cells converting them into novel producers and spreaders of Aβ-os and p-Tau-os—a self-sustaining and self-amplifying propagation LOAD neuropathology (Dal Prà et al., 2015). The neuropathology's slow progression leaves extracellular senile plaques of fibrillar Aβ polymers and intracellular neurofibrillary tangles (NFTs) of p-Tau in its wake (Dal Prà et al., 2015; Dal Prà, Armato, & Chiarini, 2018) (see Table 38.1). Notably, all kinds of cells resident in the human central nervous system (CNS) express the CaSR, and once exposed to exogenous Aβ-os via Aβ-os·CaSR signaling, they can be recruited to promote LOAD. For such reasons, CaSR has recently emerged as an attractive target for treating LOAD (Chiarini, Armato, Whitfield, Dal Pra, 2017).

Astrocytes: brief history

Astrocytes are star-shaped CNS-resident glial cells (Sofroniew & Vinters 2010) (Table 38.2). The term *astrocyte* means star-like cell (Greek *astron*, star, and *kytos*, cell) was coined in 1891 by Michael von Lenhossek, who morphologically analyzed glial cells

Table 38.2 Cell types in central and peripheral nervous system (CNS & PNS).

Location	Cell types	
CNS	Neurons or nerve cells	Neuroglial Cells: Astrocytes Oligodendrocytes Microglia Ependyma
PNS	Ganglion neurons	Schwann cells Satellite cells

The nervous system is a very complex network of (a) neurons, the parenchymal cells that transmit and conduct impulses, and (b) various types of neuroglial cell supporting/protecting neurons.

by the histological techniques developed by Camillo Golgi and Santiago Ramon y Cajal. The most popular astrocyte biomarker, which denotes them under the microscope, is glial fibrillary acidic protein (GFAP), an intermediate filament protein (Eng, 1985). Immunohistochemically labeled GFAP reveals astrocytes as star-like cells with a high number of extended processes. Recently, reactive astrocytes were revealed by injecting them with a dye that would stain the entire cell (Wilhelmsson et al., 2006). Astrocytes form a complex and interconnected network across the brain (Sofroniew & Vinters 2010). A single human astrocyte may have a huge number of connections with neurons; a human protoplasmic astrocyte controls from 270,000 to 2.0 million synapses placed inside its spatial domain. Most importantly, a single astrocyte's branches both envelop a huge number of synapses and touch nearby capillary vessels, thus increasing the blood flow when the same synapses are active (Allen & Eroglu, 2017).

Astrocytes were long assumed to play only supportive roles to neurons involved in synaptic transmission and information processing. However, recent evidence indicates that they (1) are engaged in modulating local cerebral blood flow, neurons metabolism, and water transport; (2) act as neural stem cells in neural development; (3) respond to brain injury; and (4) partake in neural signaling and information processing though not being able to transmit electrical impulses active as neurons do (Sofroniew & Vinters, 2010) (Table 38.3). Astrocytes regulate synaptic transmission by releasing neuroactive substances (e.g., L-glutamate, ATP, adenosine, D-serine, and eicosanoids) and by dynamically controlling the glutamine/glutamate cycle. And, human astrocytes propagate Ca^{2+} waves at speeds of up to 36 μm/s, i.e., 4—10-fold faster than rodents' astrocytes (Emsley & Macklis, 2006; Oberheim et al., 2009).

Human astrocytes

Five classes of structurally and anatomically distinct astrocytes of several subtypes, some of which being typical of the human cerebral cortex, exist in the human CNS (Vasile, Dossi, & Rouach, 2017). *Radial astrocytes*, the first to develop during embryogenesis,

Table 38.3 Astrocytes' main physiological roles.

Physiological processes	Function
Neural development	Guiding developing axons and neuroblasts' migration
	Development and function of synapses
Neurovascular coupling/regulation of local blood flow	Molecular mediators: prostaglandins, NO, and arachidonate production/release
Synaptic environment	Synaptic and interstitial fluid maintenance; ions, pH, and neurotransmitter homeostasis
Synaptic transmission	Release of active molecules: glutamate, purines (ATP, adenosine), GABA, D-serine
CNS metabolism	Glycogen storage; glucose uptake from blood; lactate production/release, etc.

Astrocytes are severalfold more numerous than neurons, develop networks via gap junctions, partake in tripartite synapses, support and shield neurons, and physiologically exchange essential compounds with neurons.

act as scaffolds guiding neurons' migration (Ge & Jia, 2016); they differentiate becoming stellate astrocytes, but after birth persist as such in the cerebellum (Bergmann glia) and the retina (Müller glia). *Interlaminar astrocytes* in cortical layers I and II emit both tangential radial processes and "cable-like" long, vertical processes ending with varicose end-feet contacting vascular walls or in the neuropils; they can propagate Ca^{2+} waves; however, their specific roles are not totally understood (Emsley & Macklis, 2006; Oberheim et al., 2009). *Protoplasmic astrocytes* in cortical layers III and IV are the most numerous type with many ($\sim$200) long branches that in part touch the blood vessels' walls and in part enwrap several thousands of synapses. The end-feet touching the outer walls of arteries and veins form the *glia limitans*, a space for the glymphatic drainage of brain interstitial fluid (lymph). Through such perivascular spaces nutrients reach neurons and glia, while toxic metabolites and soluble Aβs are removed from the CNS (Lliff et al., 2012). On the pia mater inner surface, the end-feet make the CNS a limiting peripheral membrane. *Varicose projection astrocytes*, in cortical layers V and VI, are specifically present only in humans and higher-order primates brains. They emit 1 to 5 processes up to 1 mm long with varicosities every $\sim$10 μm apart, ending on the cerebral vessel walls or in the neuropils (Oberheim et al., 2009). *Fibrous astrocytes* in the white-matter emanate long, thin processes without any branches whose end-feet wrap the axonal nodes of Ranvier or touch the cerebral vessel walls. They help repair brain tissue, especially at the spinal cord, injuries (Hamby & Sofroniew, 2010).

Astrocyte-derived exosomes

All cells in the CNS, including astrocytes and neurons, shed exosomes extracellularly. Exosomes are membrane vesicles $\sim$30–100 nm in diameter. They stem intracellularly from the inward-budding membrane of early endosomes, transforming the latter into multivesicular bodies (MVBs; Fig. 38.2). To form the intraluminal vesicles the endosome's membrane must be organized into distinct tetraspanin (e.g., CD9, CD63, and CD81)-enriched membrane domains (TEMDs). TEMDs comprise proteins easing vesicular fusion and/or fission. TEMDs also include potential ligands for receptor-mediated internalization of exosomes by the recipient cell. The endosomal sorting complex required for transport apparatus proteins also partake in the biogenesis of the exosomes. Once fused with the plasma membrane, MBVs release exosomes with their contents into the extracellular space, which happens under both normal and pathological conditions.

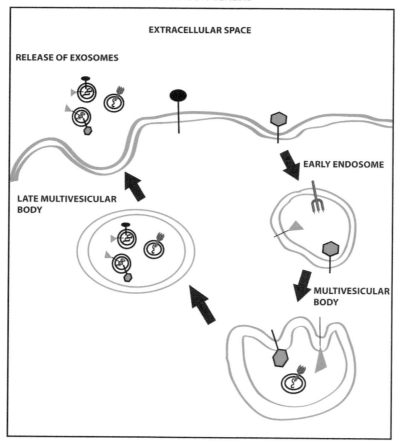

Figure 38.2 *Biogenesis of exosomes.* Exosomes stem from inward budding membranes of early endosomal compartments, which transforms endosomes into MVBs. Upon fusion of MBVs with the plasma membrane, exosomes are released into the extracellular space under both normal and pathological conditions.

Exosomes have a round shape and carry proteins, mRNAs, miRNAs, and lipids, which betray the host cell's functionality. They embody a specific type of intercellular communication by serving as means to deliver or uptake proteins, lipids, and RNA among cells (Mathivanan, Ji, & Simpson, 2010). Neural cells release exosomes under both normal and pathological conditions and have been isolated from the human cerebrospinal fluid (CSF), and their proteome profiled (Street et al., 2012), and from frozen postmortem prefrontal cortices of adult human brain (Banigam et al., 2013).

Although the primary role suggested for exosomes in any tissue, CNS included, was the removal of unnecessary proteins from cells, they function mainly as messengers in cell—cell communication (Janas, Sapoń, Janas, Stowell, & Janas, 2016; Mathivanan et al., 2010; Paolicelli, Bergamini, & Rajendran, 2018; Smalheiser, 2007). Once secreted outside the cells, exosomes can pursue either of these fates: (a) to be endocytosed by neighboring (including the cells of origin) or faraway cells; or (b) to enter the systemic circulation and thence body fluids to be taken up by different tissues. For instance, ((1)) exosomes from different neural cells partake in synaptic excitation by transferring neurotransmission-essential proteins to neurons (Janas et al., 2016; Lachenal et al., 2011); (2) exosomes from glutamate-activated oligodendrocytes modulate metabolism and act neuroprotectively by transferring their cargoes to neurons (Fruhbeis et al., 2013); (3) exosomes mediate the communication between pre- and postsynaptic neurons, helping preserve neural homeostasis while modulating synaptic plasticity (Korkut et al., 2013).

Keeping our attention on astrocytes, Taylor, Robinson, Gifondorwa, Tytell, and Milligan, (2007) and Guescini, Genedani, Stocchi, and Agnati (2010) gave the first evidence of exosomes occurring in astrocyte cultures. Their first work reported that following oxidative stress and hyperthermia, astrocytes from E12 chick lumbar spinal cords shed elevated levels of heat shock protein 70 (Hsp70) within exosomes. Second, they showed that exosomes were vesicular carriers of mitochondrial DNA in cortical astrocyte cultures from neonatal rats. Three years later, Wang et al. (2011) showed cerebral cortex astrocytes-enriched cultures from newborn C57BL/6 mice released exosomes containing synapsin, a synaptic vesicle-associated protein implicated in neural development under conditions of high neuronal activity or cell stress.

In recent years, a growing body of evidence indicated that astrocyte-derived exosomes contribute to CNS physiology and pathophysiology. For instance, astrocytes use exosomes to transfer the excitatory amino acid transporters EAAT-1 and EAAT-2, essential for upkeeping glutamate homeostasis, suggesting that besides their role in cell—cell communication, exosomes partake in glutamate clearance (Gosselin, Meylan, & Decosterd, 2013). Recently, Guitart et al. (2016) showed that murine astrocyte-derived exosomes containing nonpathogenic prion proteins, S3 and P0 proteins, apolipoprotein-E, and laminin receptor dimers, once taken up protected against hypoxic and ischemic stress and improve neuronal survival. Hence, astrocyte-derived exosomes act as intermediaries of astrocyte-neuron communications. The release of exosomal cargo (i.e., proteins, lipids, RNA species) can influence and modify target cells functions. In this regard, Lafourcade, Ramírez, Luarte, Fernández, & Wyneken (2016) suggested a specific role of miRNAs present in astrocyte-derived exosomes as mediators of neuronal maintenance and plasticity, although the evidence of exosome-mediated transfer of astrocytic miRNAs to neurons in vivo is still lacking.

In recent years, several studies have focused on the exosomes released by neuronal and glial cells in several chronic neurodegenerative diseases, suggesting that the disease pathogenesis might result from dysfunctional communication between interconnected neurons and glial cells. Specifically, various studies implicated astrocyte-derived exosomes in promoting the pathogenesis and progression of AD (Chiarini, Armato, Gardenal et al., 2017; Goetzl et al., 2016; Wang et al., 2012), Parkinson's disease (Chistiakov & Chistiakov, 2017), and amyotrophic lateral sclerosis (ALS) (Basso et al., 2013), also showing that changes in astrocyte function are also mirrored in their exosomes. For instance, in murine spinal neuron-astrocyte cocultures used as in vitro models of ALS, superoxide dismutase 1 (SOD1)-mutant astrocytes acted as key drivers of motor neuron death by releasing exosomes containing the mutated SOD1 protein, which is next transferred to spinal motor neurons to induce a selective motor neuron death (Basso et al., 2013).

Understanding the function of astrocyte-derived exosomes in AD

Thus, the molecular contents of exosomes are a "fingerprint" of the releasing cell type and its metabolic and functional status. Neuronal exosomes containing Aβs are implicated in AD pathogenesis (Takahashi et al., 2002). Exosome-associated proteins, like flotillins and Alix, are found in amyloid plaques of AD brains (Rajendran et al., 2006), Hence, astrocytes-derived exosomes are an attractive field of research for diagnostic biomarker discovery in AD, especially since the neuronal uptake of astrocyte-released exosomes was documented (Taylor et al., 2007; Wang et al., 2011).

Astrocyte exosomes could play opposite roles. Neuroprotective roles might include transferring beneficial miRNAs into recipient neural or nonneural cells and removing toxic proteins. Conversely, transferring misfolded proteins, noxious miRNAs or lipids inside exosomes into neighboring recipient cells might help spread the neuropathology. Thus, in AD current studies aim at characterizing astrocyte-derived exosomes either as carriers of pathogenic and/or misfolded Aβs, p-Taus, dysregulated miRNAs, and/or proapoptotic molecules, or as useful carriers of potential biomarkers.

Knowledge about the role(s) of astrocytes-derived exosomes in AD is still scant and is somehow weakened because it derives from astrocytes of differing sources: neonatal mouse primary astrocytes, transgenic familial AD (5xFAD) mouse model, embryo rat astrocytes, neonatal rat primary cortical astrocytes, human cortical adult astrocytes, and human plasma samples. Notably, rodent astrocytes are smaller, have fewer processes, and different Ca^{2+} dynamics when compared to human astrocytes (Vasile et al., 2017). Remarkably, analyzing and comparing transcriptomes from adult mouse and adult human astrocytes revealed significant transcriptional interspecies differences, indicating human astrocytes as the preferred model (Zhang et al., 2016).

Exosomes' neuroprotective roles

Abdullah et al. (2016) demonstrated that astrocyte-enriched cultures from embryo rats constitutively release exosomes with neuroprotective contents; however, $A\beta_{1-42}$ treatment significantly reduced the number of astrocyte-derived exosomes via stimulating c-Jun N-terminal kinase (JNK) signaling pathway. This JNK signaling involvement in exosomes release correlates with findings indicating that alterations of JNK pathways associate with neuronal death in AD (Yarza et al., 2016). In turn, $A\beta$-induced cutbacks of exosomes release would cause $A\beta$s accumulation, worsening their neurotoxicity, suggesting that astrocyte-derived exosomes and microglial cells concur in $A\beta$'s clearance, thereby playing a protective role in AD.

Recent findings implicated rat neonatal cortical astrocytes-derived exosomes in protective responses against AD-associated neuroinflammation obtained by modifying the cargo miRNAs that regulate the synaptic stability and neuronal excitability and not by reducing exosomes release (Chauduri et al., 2018).

Exosomes' neurotoxic role

Wang et al. (2012) used neonatal mouse primary astrocytes and 5xFAD mouse model to investigate neurotoxic roles of astrocyte-derived exosomes in AD. They reported $A\beta$-exposed cortical astrocytes secreted "PAR-4/ceramide-containing lipid vesicles (exosomes)" that triggered apoptosis even in untreated astrocytes. In their experiments, apoptosis was hindered by shRNA-mediated downregulation of PAR-4, a protein sensitizing cells to the sphingolipid ceramide. Moreover, neutral sphingomyelinase 2 (nSMase2)-deficient astrocytes released no exosomes, indicating nSMase2-generated ceramide is critical for exosome release and $A\beta$-induced apoptosis. And, GW4869, an nSMase2 inhibitor, hindered exosome release from murine primary $A\beta_{25-35}$-treated cultures of neurons and astrocytes (Dinkins et al., 2014). They surmised astrocyte-derived exosomes interact with $A\beta$s promoting plaque formation. Further studies pursued a potential therapy for individuals risking developing AD involving the inhibition of exosome secretion (Dinkins et al., 2017). Recent evidence indicated that human adult astrocytes (Chiarini, Armato, Gardenal et al., 2017) and microglia (Asai et al., 2015) secrete exosome-associated p-tau, which might contribute to its spread.

Tau is a microtubule-associated protein that directly binds microtubules and whose primary role was long thought to regulate neuronal axonal microtubule assembly (Guo, Noble, & Hanger, 2017). A broad array of posttranslational modifications target tau, including phosphorylation, acetylation, glycation, nitration, oxidation, sumoylation, ubiquitylation, and truncation (Morris et al., 2015). The hyperphosphorylation of tau is the main trigger of the budding out of NFTs, intraneuronal inclusions composed of tau paired helical filaments (PHFs). NFTs may spread as "seeds" contributing to the diffusion of neurodegeneration in AD (Mudher et al., 2017). Tau has long been considered a neuron-specific intracellular protein. Nevertheless, recent studies revealed that tau can also be found either in neurons-conditioned medium or within exosomes isolated from CSF and plasma of healthy donors and AD patients (for a review, see Guix et al., 2018). The presence of tau in human CSF is likely due to axonal damage and/or neuronal death.

As first evidenced by Wakabayashi et al. (2006), human adult astrocytes also contain tau proteins, which colocalized with Aβs in the *subiculum* and entorhinal cortex of a patient with corticobasal degeneration. They explained their finding as follows: "*the phagocytosis of Aβ coincides with production of phospho-tau in the same reactive astrocytes.*" After that, the relationship between Aβ exposure and p-tau production and release by adult human astrocytes remained elusive. Therefore, we designed a pilot study using as model cultured human adult astrocytes (Armato et al., 2013; Chiarini, Armato, Gardenal et al., 2017).

First, we analyzed tau protein expression in astrocytes via immunofluorescence staining and observed a mostly cytoplasmic diffuse granular tau-immunoreactivity pattern. Next, to identify the several tau isoforms involved, we analyzed *via* Western immunoblotting whole astrocyte lysates from both untreated and $A\beta_{25-35}$-treated cells (Fig. 38.3). Using a *pan*-tau antibody detecting all tau isoforms we recognized, in both untreated and $A\beta_{25-35}$-treated astrocytes, three resolvable tau bands in the 45−60 kDa size range corresponding to 2N4R, 1N3R, and 0N4R tau isoforms, i.e., just the ones found in pre-tangles and NFTs (Espinoza, de Silva, Dickson, & Davies, 2008). Moreover, untreated and Aβ-exposed cortical adult human astrocytes also released tau proteins into the medium and, notably, all the tau proteins human astrocytes released were enclosed within exosomes. Using a p-tau-specific ELISA we demonstrated that (1) physiological p-tau secretion occurred inside exosomes too; (2) exosome-associated p-tau levels increased markedly in $A\beta_{25-35}$-treated versus untreated astrocytes being elicited via Aβ•CaSR signaling; and (3) adding calcilytic NPS 2143 to $A\beta_{25-35}$-exposed astrocytes completely suppressed any exosomal p-tau surge (Fig. 38.3). Our findings suggest that in vivo, such tau/p-tau-containing exosomes would disseminate into the neuropil to be up taken by adjacent neurons, astrocytes, oligodendrocytes, and microglia. And, given astrocytes' higher numbers, Aβ-elicited exosomal p-tau overrelease would worsen neurons' toxic p-tau buildup, favoring their aggregation into pre-tangles and NFTs (Fig. 38.4). Importantly, calcilytics like NPS 2143 have the therapeutic potential to stop AD progression (for review, see Chiarini, Armato, Whitfield et al., 2017) as they can prevent exosomal p-tau-os surges and spread along with $A\beta_{42}$-os oversecretion by human astrocytes and neurons (Armato et al., 2013) and other concomitant neurotoxic effects elicited via Aβ•CaSR signaling (Table 38.1, Fig. 38.4).

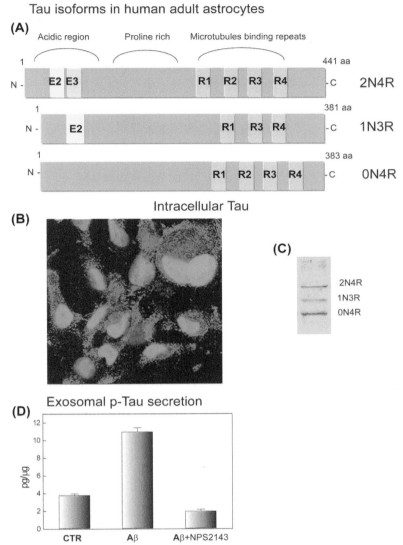

Figure 38.3 *Tau/p-Tau characterization and release of from human adult astrocytes.* (A) Schematic representation of tau isoforms expressed by human adult astrocytes. (B) Immunofluorescence labeling in untreated astrocytes reveals a cytoplasmic diffuse and granular green staining of total tau (antibody HT7). Nuclei are stained blue with DAPI. Magnification, 640×. (C) Typical immunoblot of tau isoforms expressed by human adult astrocytes. (D) Physiologically, p-Tau is released within exosomes and its amount remarkably increases in $A\beta_{25-35}$-treated astrocytes. However, calcilytic NPS 2143 prevents any increase in exosomal p-Tau. Bars are mean values ± SEMs of three duplicate experiments, $P < .01$ vs. CTRs. (For interpretation of the references to color in this figure legend, the reader is referred to the web version of this article).

Alzheimer's disease: increased production of toxic proteins, beta amyloid (Aβ) and phospho Tau (pTau) that spread throughout the brain.

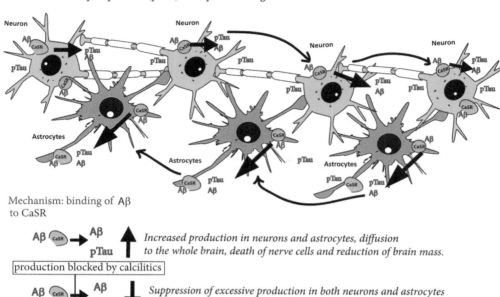

Figure 38.4 *CaSR antagonists (calcilytics) can fully suppress Aβ·CaSR signaling-induced spreading of neurotoxic effects.* Extracellular Aβ$_{42}$-os spread, interact with, and activate the plasma membrane CaSRs of increasing numbers of adjacent astrocytes and neurons, thus recruiting growing numbers of them to overproduce and overrelease Aβ$_{42}$-os and exosomal p-Tau, which spread and attach to or enter nearby or faraway cells. The combined effects of Aβ-os and p-Tau-os are particularly deadly to neurons and detrimental to cognition in vivo. By antagonizing Aβs·CaSR signaling, calcilytics hold the potential to act as novel therapeutics for LOAD, which to be effective, should be given at the earliest LOAD stages to suppress the intrabrain Aβ$_{42}$-os and exosomal p-Tau-os spreading.

Exosomes' roles in the contagion-like propagation of AD are also supported by studies undertaken to identify and characterize human neuron- and astrocyte-derived plasma exosomes from AD patients and heathy subjects (Fiandaca et al., 2015; Goetzl et al., 2016, Goetzl, Schwartz, Abner, Jicha, & Kapogiannis, 2018). As exosomes contribute to the long-distance intercellular communication, they may affect several physiological and pathological processes. Examining the Aβ$_{42}$-generating system constituents in human astrocyte-derived exosomes Goetzl et al. (2016) discovered that levels of γ-secretase, β-secretase/BACE1, soluble (s) amyloid precursor protein (APP)-β, sAPP-α, sAβ$_{42}$, glial-derived neurotrophic factor, p-S^{396}-Tau, and p-T^{181}-Tau were all significantly higher in AD patients vs. controls. Notably, BACE1 and sAPP-β levels were significantly higher in human astrocyte-derived than in neuron-derived exosomes, supporting the pathogenic role astrocytes play in AD. Isolating and characterizing contents of human astrocyte-derived plasma exosomes may also provide diagnostic biomarkers.

Recently, Goetzl et al. (2018) showed that the levels of several terminal complement proteins (C1q, C4b, C3d, factor b, factor d, Bb, C3b, C5b-C9) are significantly higher in AD patients' plasma astrocyte-derived exosomes than in controls. Conversely, in AD patients the exosomal levels of some complement regulatory proteins (CD59, CD46, decay-accelerating factor, complement receptor type1) are lower than in controls. These findings suggest that astrocyte-derived exosomes can also partake in AD's complement-mediated neuroinflammation (Lian et al., 2016). Therefore, it can be posited that astrocyte-derived exosomes convey AD-related systemic signals and their cargoes represent important biomarkers of CNS (dys)function. Further investigations are urgently necessary to characterize plasma astrocyte-derived exosomes with the aims: (1) to identify early AD biomarkers before any neuronal loss occurs, since in preclinical stages AD is asymptomatic; and (2) to monitor their specific cargoes to check the beneficial effects (if any) of therapeutics.

Conclusion

In recent years growing evidence has shown that a contagion-like mechanism underlies the progression of AD neuropathology coupled with senile plaques and NFTs (Eisele et al., 2009; Eisele et al., 2010; Dal Prà et al., 2015). The intracerebral or peripheral inoculation with AD brain homogenate has shown the seeding of Aβ and tau does occur in primates and transgenic mice. Such findings suggested that exosomes may be involved in prion-like spreading of p-Tau-os and in processing APP (Coleman & Hill, 2015). However, using human cortical astrocytes and neuron cultures, we proved that CaSR antagonists (calcilytics), like NPS 2143, do effectively suppress the overproduction and spread of the main AD drivers, Aβ-os and p-Tau-os, otherwise elicited via Aβ42•CaSR signaling, which would prevent neuropathological spread and preserve human cortical neuronal viability and functions in vivo. Thus, given at the preclinical and amnestic minor cognitive impairment (aMCI) stages, calcilytics would stop AD progression, protecting patients' cognitive abilities.

Moreover, it is worth recalling here that exosomes are now considered to be ideal vehicles to deliver therapeutic molecules to previously inaccessible brain regions due to their ability to cross the blood—brain barrier (BBB) bidirectionally. In CNS, perivascular astrocytes partake in the neurovascular units with brain endothelial cells, pericytes, neurons, and microglia (Yamazaki & Kanekiyo, 2017), and by releasing and exchanging exosomes they might pick up beneficial factors that hinder AD onset/progression. In the drug discovery field, new strategies are endeavoring to engineer the exosomes as drug carriers delivering beneficial molecules/therapeutics across the BBB (Armstrong, Holme, & Stevens, 2017).

Key facts of astrocytes in Alzheimer disease

- The World Alzheimer Report announces that worldwide over 50 million people have dementia, estimating 2019s care costs at greater than 1×10^{18}.
- Alzheimer disease (AD) accounts for 50%—70% of all dementia cases.
- Postmortem studies of brains from individuals with AD revealed as hallmarks extracellular amyloid-β plaques and intraneuronal hyperphosphorylated Tau neurofibrillary tangles along with severe neuronal losses.
- Astrocytes are the major central nervous system glial cell type and substantially contribute to human AD development.
- Astrocytes' pathophysiological mechanisms promoting AD have been at least in part clarified.

Key facts of exosomes

- Exosomes are small (30—100 nm diameter) extracellular vesicles mediating cell—cell communications.
- The term "*exosomes*" firstly referred to extracellular vesicles found in biological fluids and released from reticulocytes.
- Many, if not all, types of mammalian cells, including neurons and astrocytes, release exosomes.
- ExoCarta is a database of exosomal protein, lipid, and RNA cargos (www.exocarta.org).
- Recent studies indicate that astrocyte- and neuron-derived exosomes carry amyloid-β and p-Tau, the main drivers of Alzheimer disease.

Summary points

- Astrocytes are neuroglia's main cell type, which actively participate in brain homeostasis maintenance.
- Astrocytes employ various mechanisms to communicate with neuronal networks, and the small extracellular vesicles called exosomes that they release are a novel form of information exchange.
- Recently several studies have proven that astrocyte-released exosomes carry various substances, including tau/p-Tau, amyloid-βs, inflammatory cytokines, and microRNAs.
- Exosomes likely play important roles in several biological and pathological processes within the central nervous system.
- However, the role of astrocyte-derived exosomes and their contents in human neurodegenerative diseases is poorly understood.
- Here, we summarize evidence linking astrocyte-derived exosomes to Alzheimer disease pathogenesis and potential therapy.

References

Abdullah, M., Takase, H., Nunome, M., Enomoto, H., Ito, J., Gong, J. S., et al. (2016). Amyloid-β reduces exosome release from astrocytes by enhancing JNK phosphorylation. *Journal of Alzheimer's Disease, 53*(4), 1433–1441.

Allen, N. J., & Eroglu, C. (2017). Cell biology of astrocyte-synapse interactions. *Neuron, 96*(3), 697–708.

Armato, U., Chiarini, A., Chakravarthy, B., Chioffi, F., Pacchiana, R., Colarusso, E., et al. (2013). Calcium-sensing receptor antagonist (calcilytic) NPS 2143 specifically blocks the increased secretion of endogenous Aβ42 prompted by exogenous fibrillary or soluble Aβ25-35 in human cortical astrocytes and neurons-therapeutic relevance to Alzheimer's disease. *Biochimica et Biophysica Acta, 1832*, 1634–1652.

Armstrong, J. P., Holme, M. N., & Stevens, M. M. (2017). Re-engineering extracellular vesicles as smart nanoscale therapeutics. *ACS Nano, 11*(1), 69–83.

Asai, H., Ikezu, S., Tsunoda, S., Medalla, M., Luebke, J., Haydar, T., et al. (2015). Depletion of microglia and inhibition of exosome synthesis halt tau propagation. *Nature Neuroscience, 18*(11), 1584–1593.

Banigan, M. G., Kao, P. F., Kozubek, J. A., Winslow, A. R., Medina, J., Costa, J., et al. (2013). Differential expression of exosomal microRNAs in prefrontal cortices of schizophrenia and bipolar disorder patients. *PLoS One, 8*(1), e48814.

Basso, M., Pozzi, S., Tortarolo, M., Fiordaliso, F., Bisighini, C., Pasetto, L., et al. (2013). Mutant copper-zinc superoxide dismutase (SOD1) induces protein secretion pathway alterations and exosome release in astrocytes: Implications for disease spreading and motor neuron pathology in amyotrophic lateral sclerosis. *Journal of Biological Chemistry, 288*(22), 15699–15711.

Chaudhuri, A. D., Dastgheyb, R. M., Yoo, S. W., Trout, A., Talbot, C. C., Jr., Hao, H., et al. (2018). TNFα and IL-1β modify the miRNA cargo of astrocyte shed extracellular vesicles to regulate neurotrophic signaling in neurons. *Cell Death and Disease, 9*(3), 363.

Chiarini, A., Armato, U., Gardenal, E., Gui, L., & Dal Prà, I. (2017). Amyloid β-exposed human astrocytes overproduce phospho-tau and overrelease it within exosomes. Effects suppressed by calcilytic NPS 2143-further implications for Alzheimer's therapy. *Frontiers in Neuroscience, 11*, 217.

Chiarini, A., Armato, U., Whitfield, J. F., & Dal Pra, I. (2017). Targeting human astrocytes' calcium-sensing receptors for treatment of Alzheimer's disease. *Current Pharmaceutical Design, 23*(33), 4990–5000.

Chistiakov, D. A., & Chistiakov, A. A. (2017). α-Synuclein-carrying extracellular vesicles in Parkinson's disease: Deadly transmitters. *Acta Neurologica Belgica, 117*, 43.

Coleman, B. M., & Hill, A. F. (2015). Extracellular vesicles — their role in the packaging and spread of misfolded proteins associated with neurodegenerative diseases. *Seminars in Cell & Developmental Biology, 40*, 89–96.

Dal Prà, I., Armato, U., & Chiarini, A. (2018). Astrocytes' role in Alzheimer's disease neurodegeneration. In M. T. Gentile (Ed.), *Astrocyte Physiology and Pathology* (pp. 119–137). London: IntechOpen Limited. https://doi.org/10.5772/intechopen.72974.

Dal Prà, I., Chiarini, A., Gui, L., Chakravarthy, B., Pacchiana, R., Gardenal, E., et al. (2015). Do astrocytes collaborate with neurons in spreading the "infectious" Aβ and Tau drivers of Alzheimer's Disease? *The Neuroscientist, 21*(1), 9–29.

Dal Prà, I., Chiarini, A., Pacchiana, R., Gardenal, E., Chakravarthy, B., Whitfield, J. F., et al. (2014). Calcium-sensing receptors of human astrocyte-neuron teams: Amyloid-β-Driven mediators and therapeutic targets of Alzheimer's disease. *Current Neuropharmacology, 12*(4), 353–364.

Dinkins, M. B., Wang, G., & Bieberich, E. (2017). Sphingolipid-enriched extracellular vesicles and Alzheimer's disease: A decade of research. *Journal of Alzheimer's Disease, 60*(3), 757–768.

Dinkins, M. B., Dasgupta, S., Wang, G., Zhu, G., & Bieberich, E. (2014). Exosome reduction in vivo is associated with lower amyloid plaque load in the 5XFAD mouse model of Alzheimer's disease. *Neurobiology of Aging, 35*(8), 1792–1800.

Eisele, Y. S., Bolmont, T., Heikenwalder, M., Langer, F., Jacobson, L. H., Yan, Z. X., et al. (2009). Induction of cerebral β-amyloidosis: Intracerebral versus systemic Aβ inoculation. *Proceedings of the National Academy of Sciences of the United States of America, 106*(31), 12926–12931.

Eisele, Y. S., Obermüller, U., Heilbronner, G., Baumann, F., Kaeser, S. A., Wolburg, H., et al. (2010). Peripherally applied Aβ-containing inoculates induce cerebral β-amyloidosis. *Science, 330*(6006), 980—982.

Emsley, J. G., & Macklis, J. D. (2006). Astroglial heterogeneity closely reflects the neuronal-defined anatomy of the adult murine CNS. *Neuron Glia Biology, 2*(3), 175—186.

Eng, L. F. (1985). Glial fibrillary acidic protein (GFAP): The major protein of glial intermediate filaments in differentiated astrocytes. *Journal of Neuroimmunology, 8*, 203—214.

Espinoza, M., de Silva, R., Dickson, D. W., & Davies, P. (2008). Differential incorporation of tau isoforms in Alzheimer's disease. *Journal of Alzheimer's Disease, 14*(1), 1—16.

Fiandaca, M. S., Kapogiannis, D., Mapstone, M., Boxer, A., Eitan, E., Schwartz, J. B., et al. (2015). Identification of preclinical Alzheimer's disease by a profile of pathogenic proteins in neurally derived blood exosomes: A case-control study. *Alzheimer's and Dementia, 11*(6), 600—607.e1.

Frühbeis, C., Fröhlich, D., Kuo, W. P., Amphornrat, J., Thilemann, S., Saab, A. S., et al. (2013). Neurotransmitter-triggered transfer of exosomes mediates oligodendrocyte-neuron communication. *PLoS Biology, 11*(7), e1001604.

Ge, W. P., & Jia, G. M. (2016). Local production of astrocytes in the cerebral cortex. *Neuroscience, 323*, 3—9.

Goetzl, E. J., Mustapic, M., Kapogiannis, D., Eitan, E., Lobach, I. V., Goetzl, L., et al. (2016). Cargo proteins of plasma astrocyte-derived exosomes in Alzheimer's disease. *Federation of American Societies for Experimental Biology Journal, 30*(11), 3853—3859.

Goetzl, E. J., Schwartz, J. B., Abner, E. L., Jicha, G. A., & Kapogiannis, D. (2018). High complement levels in astrocyte-derived exosomes of Alzheimer disease. *Annals of Neurology, 83*(3), 544—552.

Gosselin, R.-D., Meylan, P., & Decosterd, I. (2013). Extracellular microvesicles from astrocytes contain functional glutamate transporters: Regulation by protein kinase C and cell activation. *Frontiers in Cellular Neuroscience, 7*, 251.

Guescini, M., Genedani, S., Stocchi, V., & Agnati, L. F. (2010). Astrocytes and Glioblastoma cells release exosomes carrying mtDNA. *Journal of Neural Transmission, 117*, 1.

Guitart, K., Loers, G., Buck, F., Bork, U., Schachner, M., & Kleene, R. (2016). Improvement of neuronal cell survival by astrocyte-derived exosomes under hypoxic and ischemic conditions depends on prion protein. *Glia, 64*(6), 896—910.

Guix, F. X., Corbett, G. T., Cha, D. J., Mustapic, M., Liu, W., Mengel, D., et al. (2018). Detection of aggregation-competent tau in neuron-derived extracellular vesicles. *International Journal of Molecular Sciences, 19*(3). pii: E663.

Guo, T., Noble, W., & Hanger, D. P. (2017). Roles of tau protein in health and disease. *Acta Neuropathologica, 133*(5), 665—704.

Hamby, M. E., & Sofroniew, M. V. (2010). Reactive astrocytes as therapeutic targets for CNS disorders. *Neurotherapeutics, 7*, 494—506.

Janas, A. M., Sapoń, K., Janas, T., Stowell, M. H., & Janas, T. (2016). Exosomes and other extracellular vesicles in neural cells and neurodegenerative diseases. *Biochimica et Biophysica Acta, 1858*(6), 1139—1151.

Korkut, C., Li, Y., Koles, K., Brewer, C., Ashley, J., Yoshihara, M., et al. (2013). Regulation of postsynaptic retrograde signaling by presynaptic exosome release. *Neuron, 77*(6), 1039—1046.

Lachenal, G., Pernet-Gallay, K., Chivet, M., Hemming, F. J., Belly, A., Bodon, G., et al. (2011). Release of exosomes from differentiated neurons and its regulation by synaptic glutamatergic activity. *Molecular and Cellular Neuroscience, 46*(2), 409—418.

Lafourcade, C., Ramírez, J. P., Luarte, A., Fernández, A., & Wyneken, U. (2016). MiRNAs in astrocyte-derived exosomes as possible mediators of neuronal plasticity. *Journal of Experimental Neuroscience, 10*(Suppl. 1), 1—9.

Lian, H., Litvinchuk, A., Chiang, A. C., Aithmitti, N., Jankowsky, J. L., & Zheng, H. (2016). Astrocyte-microglia cross talk through complement activation modulates amyloid pathology in mouse models of Alzheimer's Disease. *Journal of Neuroscience, 36*(2), 577—589.

Lliff, J. J., Wang, M., Liao, Y., Plogg, B. A., Peng, W., Gundersen, G. A., et al. (2012). A paravascular pathway facilitates CSF flow through the brain parenchyma and the clearance of interstitial solutes, including amyloid β. *Science Translational Medicine, 4*, 147ra111.

Mathivanan, S., Ji, H., & Simpson, R. J. (2010). Exosomes: Extracellular organelles important in intercellular communication. *Journal of Proteomics, 73*(10), 1907—1920.

Morris, M., Knudsen, G. M., Maeda, S., Trinidad, J. C., Ioanoviciu, A., Burlingame, A. L., et al. (2015). Tau post-translational modifications in wild-type and human amyloid precursor protein transgenic mice. *Nature Neuroscience, 18*(8), 1183–1189.

Mudher, A., Colin, M., Dujardin, S., Medina, M., Dewachter, I., Alavi Naini, S. M., et al. (2017). What is the evidence that tau pathology spreads through prion-like propagation? *Acta Neuropathologcgica Communications, 5*(1), 99.

Oberheim, N. A., Takano, T., Han, X., He, W., Lin, J. H., Wang, F., et al. (2009). Uniquely hominid features of adult human astrocytes. *Journal of Neuroscience, 29*(10), 3276–3287.

Paolicelli, R. C., Bergamini, G., & Rajendran, L. (2018). Cell-to-cell communication by extracellular vesicles: Focus on microglia. *Neuroscience,* (18), 30254–30259. pii: S0306-4522.

Rajendran, L., Honsho, M., Zahn, T. R., Keller, P., Geiger, K. D., Verkade, P., et al. (2006). Alzheimer's disease beta-amyloid peptides are released in association with exosomes. *Proceedings of the National Academy of Sciences of the United States of America, 103*(30), 11172–11177.

Smalheiser, N. R. (2007). Exosomal transfer of proteins and RNAs at synapses in the nervous system. *Biology Direct, 2*, 35.

Sofroniew, M. V., & Vinters, H. V. (2010). Astrocytes: Biology and pathology. *Acta Neuropathologica, 119*(1), 7–35.

Street, J. M., Barran, P. E., Mackay, C. L., Weidt, S., Balmforth, C., Walsh, T. S., et al. (2012). Identification and proteomic profiling of exosomes in human cerebrospinal fluid. *Journal of Translational Medicine, 10*, 5.

Takahashi, R. H., Milner, T. A., Li, F., Nam, E. E., Edgar, M. A., Yamaguchi, H., et al. (2002). Intraneuronal Alzheimer Aβ42 accumulates in multivesicular bodies and is associated with synaptic pathology. *American Journal of Pathology, 161*, 1869–1879.

Taylor, A. R., Robinson, M. B., Gifondorwa, D. J., Tytell, M., & Milligan, C. E. (2007). Regulation of heat shock protein 70 release in astrocytes: Role of signaling kinases. *Developmental Neurobiology, 67*(13), 1815–1829.

Vasile, F., Dossi, E., & Rouach, N. (2017). Human astrocytes: Structure and functions in the healthy brain. *Brain Structure and Function, 222*(5), 2017–2029.

Wakabayashi, K., Mori, F., Hasegawa, M., Kusumi, T., Yoshimura, I., Takahashi, H., et al. (2006). Co-localization of beta-peptide and phosphorylated tau in astrocytes in a patient with corticobasal degeneration. *Neuropathology, 26*(1), 66–71.

Wang, S., Cesca, F., Loers, G., Schweizer, M., Buck, F., Benfenati, F., et al. (2011). Synapsin I is an oligomannose-carrying glycoprotein, acts as an oligomannose-binding lectin, and promotes neurite outgrowth and neuronal survival when released via glia-derived exosomes. *Journal of Neuroscience, 31*(20), 7275–7290.

Wang, G., Dinkins, M., He, Q., Zhu, G., Poirier, C., Campbell, A., et al. (2012). Astrocytes secrete exosomes enriched with proapoptotic ceramide and prostate apoptosis response 4 (PAR-4): Potential mechanism of apoptosis induction in Alzheimer disease (AD). *Journal of Biological Chemistry, 287*(25), 21384–21395.

Wilhelmsson, U., Bushong, E. A., Price, D. L., Smarr, B. L., Phung, V., Terada, M., et al. (2006). Redefining the concept of reactive astrocytes as cells that remain within their unique domains upon reaction to injury. *Proceedings of the National Academy of Sciences of the United States of America, 103*(46), 17513–17518.

Yamazaki, Y., & Kanekiyo, T. (2017). Blood-brain barrier dysfunction and the pathogenesis of Alzheimer's disease. *International Journal of Molecular Sciences, 18*(9). pii: E1965.

Yarza, R., Vela, S., Solas, M., & Ramirez, M. J. (2016). c-Jun N-terminal kinase (JNK) signaling as a therapeutic target for Alzheimer's disease. *Frontiers in Pharmacology, 6*, 321.

Zhang, Y., Sloan, S. A., Clarke, L. E., Caneda, C., Plaza, C. A., Blumenthal, P. D., et al. (2016). Purification and characterization of progenitor and mature human astrocytes reveals transcriptional and functional differences with mouse. *Neuron, 89*(1), 37–53.

CHAPTER 39

Changing fate: therapeutic mechanisms focused on the switch of amyloid precursor protein processing

Sven Reinhardt, Kristina Endres
Department of Psychiatry and Psychotherapy, Medical Center, Johannes Gutenberg-University of Mainz, Mainz, Germany

List of abbreviations

A-beta Amyloid beta peptide
ADAM10 A disintegrin and metalloproteinase 10
APP Amyloid precursor protein
APPs-alpha Soluble APP cleaved by alpha-secretase
BACE-1 Beta-site of APP cleaving enzyme 1
CSF Cerebrospinal fluid
miR MicroRNA
RAR Retinoic acid receptor
UTR Untranslated region

Mini-dictionary of terms

ADAM10 A disintegrin and metalloproteinase 10, a zinc-dependent metalloproteinase that has been shown to be the major alpha-secretase in neuronal cells.
APPs-alpha Soluble APP cleaved by alpha-secretase
BACE-1 Beta-site of APP cleaving enzyme 1, an aspartic protease that cleaves APP to yield soluble sAPP-beta and A-beta peptides
RAR Retinoic acid receptor, bound to retinoic acid response elements on the DNA of target genes
UTR Untranslated region, the sequence of a gene transcribed but not translated that has an important role in transcriptional regulation

Introduction

The alpha-secretase A disintegrin and metalloproteinase 10 (ADAM10) (Lammich et al., 1999) and the beta-site of APP cleaving enzyme 1 (BACE-1) (Vassar et al., 1999) compete for cleavage of the amyloid precursor protein (APP; see chapter "Implication of alpha- and beta-secretase expression and function in Alzheimer's disease"). Activity of either enzyme leads to production of neurotoxic A-beta peptides (BACE-1) or to secretion of a neurotrophic factor designated APPs-alpha (ADAM10). Approaches aimed at influencing BACE-1 or ADAM10 expression and activity can be deciphered in all stages of the cellular machinery. Beginning with epigenetic modulation of the respective

Diagnosis and Management in Dementia
ISBN 978-0-12-815854-8, https://doi.org/10.1016/B978-0-12-815854-8.00039-2

gene, every intervention point from transcription to translation as well as protein modification, catalytic function, interaction with other molecules, and degradation might be targeted. With this chapter, we do not claim to draw a complete picture of every aspect but try to provide contemporary examples of all stages.

Regulation of protease transcription by transcription factors

The human ADAM10 gene sequence is localized on chromosome 15 (Yamazaki, Mizui, & Tanaka, 1997). Its core promoter has been localized to nucleotides −508 to −300 and the 5′untranslated region (UTR) to 444 bp upstream of the start codon (Prinzen, Muller, Endres, Fahrenholz, & Postina, 2005) (Lammich et al., 2010); the 3′UTR has been identified to stretch to 1254 bp downstream of the stop codon (Augustin et al., 2012). Human BACE-1 is located on chromosome 11. Its promoter sequence is positioned about 2 kb upstream of the start site of the protein coding sequence (Christensen et al., 2004) where bp −1311 to −400 display minimal promoter activity. More recently, the core promoter region has been described to stretch nucleotides −550 to −480 bp relative to the start codon (Xiang et al., 2014). The transcription start site has been found at −691 bp (Zhou & Song, 2006) and the 3′UTR to span 3868 bp with 29 variants, of which some were solely restricted to AD patients (Bettens et al., 2009).

Various transcription factors have been shown to act on secretase promoters; for example, the ER−stress−driven factor XBP1 was found to induce ADAM10 transcription (Reinhardt et al., 2014). With BACE-1, as another example, regulation by NFAT3 has been reported (Mei, Yan, Tan, Zheng, & Situ, 2015). However, knowledge of these regulatory interactions is only of therapeutic value if selective pharmaceutical compounds exist to allow modulation of the respective factor. As an example of such a pharmaceutical substance, we will discuss retinoic acid, which has been implicated in the regulation of both promoters. It was noticed decades ago that alterations in retinoic acid metabolism might play a role in AD development (Goodman & Pardee, 2003). This was confirmed in a meta-analysis concluding that vitamin A—the nutritional precursor for retinoic acid—is reduced in plasma samples of AD patients (Lopes da Silva et al., 2014). Within the ADAM10 promoter, two binding sites for retinoic acid receptors have been identified (RAR-alpha and RAR-beta; Tippmann, Hundt, Schneider, Endres, & Fahrenholz, 2009), and the expression of the secretase has been shown to be inducible by retinoic acid or related molecules in neuroblastoma cells, in primary neurons (Reinhardt et al., 2016), and in mice (e.g., Tippmann et al., 2009). Besides directly stimulating expression of ADAM10 via exogenously applied retinoic acid or liberation of the ligand from intracellular pools via acitretin, induction of RAR-beta via sirtuin 1 evoked by cilostazol has been reported (Lee et al., 2014). Efficacy of RAR-driven ADAM10 stimulation has been demonstrated in a small cohort of moderately to mildly affected AD patients, where acitretin led to an increase in cerebrospinal fluid (CSF) APPs-alpha compared with levels in

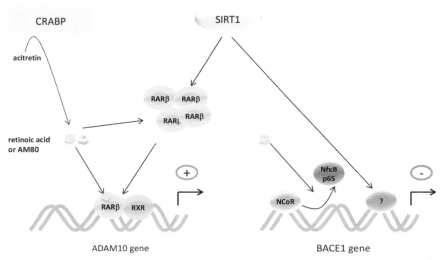

Figure 39.1 Pharmaceutical intervention with secretase balance via retinoic acid and sirtuin 1. All-trans retinoic acid represents a potential Alzheimer's disease therapy by influencing the expressions of BACE-1 and ADAM10. Substances that deliberate retinoic acid from intracellular binding proteins—e.g., cellular retinoic acid binding protein (CRABP)—such as acitretin or direct ligands to RARs like AM80, increase the expression of RARs. RAR-alpha or RAR-beta, together with their respective ligand and the dimerization partner RXR (retinoid X receptor), increase ADAM10 expression. This can also be achieved by elevation of RAR-beta due to sirtuin 1 activation.

placebo-treated patients (Endres et al., 2014). Interestingly, retinoic acid has also been reported to reduce BACE-1 expression within an inflammatory environment by NFκB signaling (Wang et al., 2015), and sirtuin 1 upregulation by the herbal compound fuzhisan also resulted in transcriptional repression of BACE-1 in murine neuroblastoma cells (Gao et al., 2015) (Fig. 39.1).

Translational control of BACE-1 and ADAM10 expression

Another level of regulation of either transcription or translation of both proteases is given by interference with the respective 3′ untranslated region (UTR) RNA sequences. Especially, microRNA (miR) 29 has been found to have an important impact on BACE-1 expression and survival of neuronal cells. miR-29c, a miRNA highly enriched in the brain, has been identified as a suppressor of BACE-1 protein level in SH-SY5Y and HEK293 cells but also in transgenic mice (Zong et al., 2011). Two putative target sites for the miR-29c family were identified in the respective mRNA (Zong et al., 2011) and luciferase-based reporter assay revealed binding of miR-29c to the 3′UTR of human BACE-1 (Lei, Lei, Zhang, Zhang, & Cheng, 2015). Additionally, miR-29c was found downregulated in brains from sporadic AD patients and associated with increased BACE-1 mRNA and protein levels. Interestingly, *in vivo*, brain-specific knockdown of miR-29 in mice resulted in neuronal cell death—e.g., in the hippocampus (Roshan et al., 2014). However, the BACE-1 mRNA level was not elevated in this mouse model.

Measurement of AD-related miRNA expression changes identified miR-107 levels to be decreased significantly, even within the earliest stages of pathology (Wang et al., 2008). A tendency of BACE-1 mRNA levels to increase with decreasing miR-107 levels concurrent with the progression of AD was described, and cell culture reporter assays identified at least one physiological recognition sequence for miR-107 in the 3′UTR of BACE-1 mRNA. Bioinformatical analysis, combining different software tools, predicted miR-107 also would bind to the ADAM10 3′UTR, and this was confirmed by a luciferase-based assay (Augustin et al., 2012). Using miRNA in therapeutic applications might be difficult due to (1) discrepancies of cell culture experiments compared with *in vivo* effects, (2) lack of substrate specificity, and (3) delivery of therapeutic miR-NAs to the brain.

Another important mechanism of translational control is mediated via the respective 5′UTR sequences. In general, translation is mainly regulated by complex secondary structures and the occurrence of upstream open reading frames (ORFs) within the 5′UTR sequence. Several reports indicate that expression of BACE-1 is controlled by its 5′UTR, which comprises 446 nucleotides and is characterized by three alternative ORFs and a mean GC content of about 70% (e.g., Lammich, Schobel, Zimmer, Lichtenthaler, & Haass, 2004). For example, HEK293 cells overexpressing BACE-1 via a construct including the 5′UTR showed a 90% reduction in protein amount as well as in enzyme activity compared with cells transfected with a BACE-1 construct lacking the respective 5′UTR sequence (Lammich et al., 2004). Under these experimental conditions, the BACE-1 mRNA level was not affected, hinting at a mere translational mechanism. Further analysis using UTR-deletion constructs or mutagenesis revealed that the second alternative ORF and the secondary structure of the BACE-1 5′UTR are responsible for the observed translational repression.

In 2006, Marshall, Rattray, and Vaughan were able to demonstrate a posttranscriptional regulation for ADAM10; SH-SY5Y cells cultivated under hypoxic conditions showed a 60% decrease in ADAM10 protein level while quantitation of the respective mRNA showed no significant change. The ADAM10 5′UTR is a GC-rich sequence (69%) with two potential alternative ORFs (Lammich et al., 2010). ADAM10 overexpressing HEK293 cells showed a 3-fold increase in protein amount compared with that of HEK293 cells expressing an ADAM10 construct bearing the 5′UTR sequence, which consequently points out the involvement of the 5′UTR in posttranscriptional regulation (Lammich et al., 2010). This regulation is based on a guanine-quadruplex secondary structure formed at the 5′ end of the ADAM10 5′UTR sequence (Lammich et al., 2011). Such guanine-quadruplexes are known to inhibit the formation and maintenance of the initiator complex during translation.

Protein—protein interactions influencing ADAM10 or BACE-1 activity

Interaction with other proteins is a fundamental feature that finally contributes to regulation of enzyme activity, degradation, or cellular localization. Depletion of the adaptor protein Golgi-localized, gamma adaptin ear–containing, ARF-binding protein 3 (GGA3) has been reported to posttranslationally stabilize BACE-1 (Fig. 39.2A), which consequently leads to increased enzyme activity in a mouse model of cerebral ischemia (Tesco et al., 2007). In AD brain samples, the GGA3 protein level was reported to be significantly decreased and moreover inversely correlated with increased BACE-1 levels (Tesco et al., 2007). This was substantiated by another study, where analysis of frontal cortex

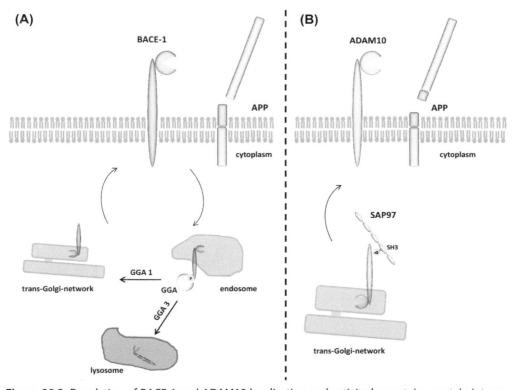

Figure 39.2 Regulation of BACE-1 and ADAM10 localization and activity by protein—protein interactions. (A) GGA proteins are involved in the intracellular trafficking of BACE-1. In the trans-Golgi-network, pro-BACE-1 undergoes maturation, becomes posttranslationally modified, and finally is transported to the plasma membrane. Binding of GGA1 in the endosomes has been reported to recycle BACE-1 back to the cell surface. In contrast, interaction with GGA3 is known to mediate transport into lysosomal structures with subsequent degradation. (B) Interaction with synapse-associated protein-97 (SAP97) is required for synaptic localization of ADAM10. Within a postsynaptic cell, SAP97 binds to the proline-rich domain of the ADAM10 C-terminus. For this interaction, a PKC phosphorylation site within the SAP97 Src homology domain is required.

samples from AD patients revealed about a twofold increase in the level of BACE-1 and a decrease of about 60% in GGA3 (Santosa et al., 2011). In contrast to the aforementioned study, no significant correlation was observed between BACE-1 and GGA3 protein amounts. However, using subcellular fractionation in samples with a low GGA3 status, a strong colocalization of BACE-1 with APP was found in cortical tissue, which would promote the amyloidogenic APP-processing pathway.

It has been reported that BACE-1 is modified with bisecting N-acetylglucosamines, which are enriched in the brain of AD patients (Kizuka et al., 2015). Knockout of the enzyme responsible for this sugar modification in mice led to reduced BACE-1-mediated cleavage of APP, decreased plaque depositions, and improvement of cognitive performance. Such a modification could be implicated in mediating protein–protein interactions, which might influence, for example, the cellular localization of the protease and thereby offer an alternative approach for therapy.

For ADAM10, several proteins have also been reported to influence subcellular localization as well as the activity of the enzyme by protein–protein interaction. It has been described that ADAM10 localized at excitatory synapses is removed from the plasma membrane by clathrin-mediated endocytosis in human hippocampal tissue (Marcello et al., 2013). In contrast, the binding of synapse-associated protein-97 (SAP97) is required for insertion of ADAM10 in the synaptic membrane and also for LTD-induced spine morphogenesis (see Fig. 39.2B; Marcello et al., 2013).

Additionally, several tetraspanins have been identified as interacting with ADAM10 by coimmunoprecipitation experiments (for example, in work by Haining et al., 2012). The C-terminal part of tetraspanin 12 binds to ADAM10 via a palmitoylation-dependent mechanism and further increases nonamyloidogenic processing of APP by promoting enzymatic maturation of the protease (Xu, Sharma, & Hemler, 2009). Prospectively, these findings might lead to novel therapeutic approaches targeting interaction partners of the respective proteases.

Tetraspanins have been described as being involved in maintaining the scaffolding of G-protein-coupled receptors (GPCRs). Interestingly, GPCR ligands were also found to interfere with ADAM10 enzymatic activity. Pituitary adenylate cyclase-activating polypeptide (PACAP) is reduced in cortex tissue of AD patients compared with healthy controls and in 3xTGAD model mice (Han & Liang, et al., 2014; Han & Tang, et al., 2014). In cultured neurons of these animals, treatment with PACAP ameliorated A-beta-induced neurotoxicity. Moreover, long-term application of PACAP stimulated nonamyloidogenic APP processing by increasing ADAM10 activity in a mouse model of AD and improved cognitive performance (Rat et al., 2011).

Even though data obtained in mice with regard to GPCR ligands are promising, clinical studies are missing at this time. In addition, the use of, for example, peptide-based therapeutics such as PACAP might be cost-intensive for broad application.

Lipid-mediated modulation of BACE-1 and ADAM10 activity

Besides protein-based interaction partners, lipids are known to interact with either protease, BACE-1, or ADAM10. Interestingly, both proteins are contrarily regulated by their membrane environment.

In 2003 it had already been hypothesized that APP, which is localized inside lipid rafts, is mainly cleaved by beta-secretase activity, while APP outside rafts undergoes cleavage by the alpha-secretase (Ehehalt, Keller, Haass, Thiele, & Simons, 2003). For BACE-1, an interaction with caveolins and flotillins—general proteins related to lipid rafts—has been described that leads to recruitment of the enzyme into lipid rafts in cultured cells (Hattori et al., 2006). Additionally, palmitoylation of C-terminal cysteine residues mediates anchoring of the enzyme in raft domains as demonstrated by a mutant BACE-1 lacking palmitoylation attachment sites compared with wild-type protein in neuroblastoma cells and rat primary cortical neurons (Motoki et al., 2012). The relationship between localization and activity of BACE-1 has been demonstrated using the BACE-1 inhibitor KMI-574 (Ebina et al., 2009). Incubation with this inhibitory agent resulted in rearrangement of the enzyme localization from rafts to nonraft domains within the membrane of HEK293 cells.

ADAM10 is known to be mainly located outside lipid rafts and alpha-secretase cleavage of APP is mediated in nonraft domains (Kojro, Gimpl, Lammich, Marz, & Fahrenholz, 2001). Targeting of ADAM10 into lipid rafts by an artificial glycosylphosphatidylinositol anchor leads to impaired enzymatic activity in human neuroblastoma cells (e.g., Kojro et al., 2010). It has been reported that the architecture of lipid rafts is altered in AD—microviscosity within these microdomains was increased in the frontal cortex of AD patients compared with healthy controls (Diaz et al., 2015). Moreover, lipid raft microviscosity was positively correlated with BACE-1 levels analyzed in the same brain samples.

Genetic analysis revealed that cholesterol metabolism is causally linked with the onset and/or progression of AD (e.g., Jones et al., 2010). Cholesterol, which plays a crucial role in the formation and maintenance of lipid rafts, is also implicated in regulating BACE-1 and ADAM10 activity; short exposure of primary neurons to cholesterol led to increased proximity of BACE-1 and APP as demonstrated by a fluorescence-based lifetime imaging microscopy technique (Marquer et al., 2011). Longer exposure finally led to endocytosis and consequently to intracellular colocalization. This finding indicates that cholesterol modulates the access of BACE-1 to its substrate APP. This is further consolidated by a study reporting on a cellular model of Niemann-Pick type C, a disease characterized by an increased level of cholesterol. APP and BACE-1 are colocalized intracellularly by a cholesterol-dependent alteration in endocytic trafficking in these cells, which finally leads to an increase in A-beta generation (Malnar et al., 2012). In contrast, it has been previously described that cholesterol depletion leads to enhanced ADAM10 activity in different cellular models. For example, application of the cholesterol-eliminating drug methyl-beta-cyclodextrin resulted in elevated secretion of the ADAM10-dependent cleavage fragment APPs-alpha in several neuronal and nonneuronal cell lines (Kojro et al., 2001).

Besides cholesterol, gangliosides and sphingomyelin are the major components of lipid rafts. These lipids have been described to act as both contributing and preventive endogenous molecules with regard to AD pathogenesis (e.g., Grosgen, Grimm, Friess, & Hartmann, 2010). It has been demonstrated that the amount of the ganglioside GM1 is increased in the brain of AD patients (Pernber, Blennow, Bogdanovic, Mansson, & Blomqvist, 2012). Incubation of rat primary cortical neurons with GM1 enhanced the extracellular A-beta level by decreasing alpha-secretase activity in a protein kinase C (PKC)-dependent manner (Cedazo-Minguez, Wiehager, Winblad, Huttinger, & Cowburn, 2001). Sphingomyelin incorporated into synthetic vesicles increased BACE-1 activity in a cell-free *in vitro* assay (Kalvodova et al., 2005). This was contradicted by a report that inhibition of sphingolipid biosynthesis, which also reduces the sphingomyelin level, by a specific inhibitor of the serine palmitoyltransferase resulted in increased alpha-secretase activity, consequently followed by enhanced secretion of sAPP-alpha in CHO cells (Sawamura et al., 2004).

Epidemiological studies have revealed that dietary intervention, providing the subject with omega-3 polyunsaturated fatty acids such as docosahexaenoic acid, is correlated with reduced risk for age-related cognitive decline. Based on these findings, a clinical trial analyzing the therapeutic and preventive impact of nutritional lipids on the cognitive performance of patients with prodromal AD has been conducted (http://www.lipididiet.eu/). First results indicate that once-daily intake reduced brain shrinkage, particularly in the hippocampus, and preserved the ability of patients to carry out everyday tasks as compared with the placebo group (Soininen et al., 2017).

Small-molecule modulators of BACE-1 and ADAM10 enzymatic activity

A further opportunity for interfering with protease balance is the direct pharmacological modulation of enzymatic activity via small molecules. For ADAM10, such a small-molecule activator targeting the active center of the protease with high selectivity has to our knowledge not yet been described, and its design might be complicated by conservation of the zinc-binding motif throughout the metzincin superfamily.

Conversely, extensive research has been conducted regarding inhibitors of BACE-1 activity; for example, Eli Lilly reported that their inhibitor LY2886721 resulted in robust reduction in cerebral A-beta levels during preclinical and early-phase clinical development (May et al., 2015). This drug also entered a clinical phase II study to prove efficacy in mild cognitively impaired subjects or patients suffering from mild AD. However, the study was terminated by the company due to abnormal biochemical liver parameters as a non-target-related side effect in some subjects (https://clinicaltrials.gov/ct2/show/NCT01561430). The development of novel BACE-1 inhibitors has turned out to be challenging with regard to bioavailability, clearance, and especially blood—brain barrier permeability (Vassar, 2014). For example, several inhibitors, even if efficiently reducing A-beta peptides in rodent AD models, have been shown to be eliminated by the brain efflux transporter p-glycoprotein and consequently lack sufficient cerebral potency (Meredith et al., 2008). Novel small molecule BACE-1 inhibitors have been developed and analyzed in preclinical studies (Vassar, 2014). These drugs were designed to enter the brain with higher efficiency and consequently reduce cerebral A-beta levels in respective AD model mice. Most of these novel potential therapeutic agents have just entered early clinical phase I studies. Only Merck's oral, active-site BACE-1 inhibitor MK-8931 has reached a more advanced stage in clinical research. This drug has been described as well tolerated and decreased CSF A-beta levels in healthy volunteers and patients with mild or moderate AD in phase I clinical trials (Forman et al., 2012). The safety and efficacy of long-term treatment with MK-8931 in cognitively impaired subjects or mildly to moderately affected AD patients are currently being investigated in two parallel phase III clinical trials (NCT01739348, NCT01953601). However, the effect of BACE-1-inhibition on the clinical outcome, in particular on cognitive performance, still remains to be demonstrated.

Hindrances in therapeutic strategies aiming at a switch in amyloid precursor protein processing

As described previously, an essential prerequisite for AD therapy is the ability of the respective drug to overcome the blood—brain barrier. It has been reported that peripheral inhibition of BACE-1, besides a reduction of plasma A-beta level, is not sufficient to affect A-beta in the brain of wild-type or even APP transgenic mice (Georgievska et al., 2015). A second important issue is the starting point of pharmacological intervention. It has been hypothesized that A-beta might initiate cascades of pathological processes already at or even before the onset of disease, which then proceed even if A-beta is therapeutically reduced. Therefore, there is emerging evidence encouraging early treatment in AD therapy. The lack of well-established peripheral biomarkers reflecting the progress of pathogenesis at very early time points might be an additional hindrance in this regard. Third, modulation of the enzymatic activity of both, ADAM10 and BACE-1 might—as with any pharmaceutical intervention—bear potential side effects. Early reports have demonstrated that BACE-1 knockout mice show a rather mild phenotype without severe abnormalities. However, more detailed analyses elucidated a variety of neuronal phenotypes. Several studies demonstrated that BACE-1 is required for axon guidance and that axon mistargeting occurs in BACE-1-null mice (Cao, Rickenbacher, Rodriguez, Moulia, & Albers, 2012). Furthermore, it has been demonstrated that

BACE-1 is required for sensory experience-dependent homeostatic synaptic plasticity in the neocortex (Petrus & Lee, 2014). The identification of novel, physiological relevant substrates of BACE-1 (Dislich et al., 2015) might even add further important cellular functions that have not yet been considered. For ADAM10, greater than 40 substrates have been described based on cell culture experiments (summarized in Endres & Deller, 2017). Consequently, a therapeutic approach targeting ADAM10 might lead to potential side effects in human patients. However, analyzing RNA samples from the brains of ADAM10 transgenic mice using a microarray approach revealed only a mild change in the gene expression profile compared with that of wild-type mice (Prinzen et al., 2009). Moreover, among a few regulated genes, those correlated to inflammation or apoptosis were not overrepresented in ADAM10-overexpressing mice. In contrast, strong overexpression or induction of the protease might interfere with important cellular functions. For instance, mice with a strong enzyme expression rate revealed reduced Akt phosphorylation (Freese, Garratt, Fahrenholz, & Endres, 2009). Therefore, protease-directed therapy must be strictly controlled.

Key facts on sheddases

- The cleavage of cell-surface-located proteins is designated as "shedding."
- Mostly, the large ectodomains of proteins are released by this process.
- Those ectodomains can have various functions acting as growth factors or signaling molecules.
- Sometimes more than one enzyme cleaves a shedded protein, giving rise to a wide variety of different proteolytic products.
- Subsequently to cleavage in close vicinity to the membrane, an additional cleavage within the membrane can occur, leading also to intracellularly released fragments of the protein.

Summary points

- In this chapter, we present approaches to modify BACE-1 and ADAM10 expression and/or activity via endogenous regulators and pharmacotherapy.
- For BACE-1, there is ongoing research regarding novel small-molecule active-site inhibitors.
- For ADAM10, small-molecule activators targeting enzymatic activity directly have so far not been identified.
- However, enhancement of ADAM10 gene expression by, for example, synthetic retinoids such as acitretin represent a promising approach for AD therapy.
- Urgent requirements concerning bioavailability, BBB permeability, and safety still have to be addressed in preclinical and clinical studies.

References

Augustin, R., Endres, K., Reinhardt, S., Kuhn, P. H., Lichtenthaler, S. F., Hansen, J., et al. (2012). Computational identification and experimental validation of microRNAs binding to the Alzheimer-related gene ADAM10. *BMC Medical Genetics, 13*, 35. https://doi.org/10.1186/1471-2350-13-35, 1471-2350-13-35 [pii].

Bettens, K., Brouwers, N., Engelborghs, S., Van, M. H., De Deyn, P. P., Theuns, J., et al. (2009). APP and BACE1 miRNA genetic variability has no major role in risk for Alzheimer disease. *Human Mutation, 30*(8), 1207—1213. https://doi.org/10.1002/humu.21027.

Cao, L., Rickenbacher, G. T., Rodriguez, S., Moulia, T. W., & Albers, M. W. (2012). The precision of axon targeting of mouse olfactory sensory neurons requires the BACE1 protease. *Scientific Reports, 2*, 231. https://doi.org/10.1038/srep00231.

Cedazo-Minguez, A., Wiehager, B., Winblad, B., Huttinger, M., & Cowburn, R. F. (2001). Effects of apolipoprotein E (apoE) isoforms, beta-amyloid (Abeta) and apoE/Abeta complexes on protein kinase C-alpha (PKC-alpha) translocation and amyloid precursor protein (APP) processing in human SH-SY5Y neuroblastoma cells and fibroblasts. *Neurochemistry International, 38*(7), 615—625.

Christensen, M. A., Zhou, W., Qing, H., Lehman, A., Philipsen, S., & Song, W. (2004). Transcriptional regulation of BACE1, the beta-amyloid precursor protein beta-secretase, by Sp1. *Molecular and Cellular Biology, 24*(2), 865—874.

Diaz, M., Fabelo, N., Martin, V., Ferrer, I., Gomez, T., & Marin, R. (2015). Biophysical alterations in lipid rafts from human cerebral cortex associate with increased BACE1/AbetaPP interaction in early stages of Alzheimer's disease. *J Alzheimers Dis, 43*(4), 1185—1198. https://doi.org/10.3233/JAD-141146.

Dislich, B., Wohlrab, F., Bachhuber, T., Muller, S. A., Kuhn, P. H., Hogl, S., et al. (2015). Label-free quantitative proteomics of mouse cerebrospinal fluid detects beta-site APP cleaving enzyme (BACE1) protease substrates in vivo. *Molecular & Cellular Proteomics, 14*(10), 2550—2563. https://doi.org/10.1074/mcp.M114.041533.

Ebina, M., Futai, E., Tanabe, C., Sasagawa, N., Kiso, Y., & Ishiura, S. (2009). Inhibition by KMI-574 leads to dislocalization of BACE1 from lipid rafts. *Journal of Neuroscience Research, 87*(2), 360—368. https://doi.org/10.1002/jnr.21858.

Ehehalt, R., Keller, P., Haass, C., Thiele, C., & Simons, K. (2003). Amyloidogenic processing of the Alzheimer beta-amyloid precursor protein depends on lipid rafts. *The Journal of Cell Biology, 160*(1), 113—123. https://doi.org/10.1083/jcb.200207113.

Endres, K., & Deller, T. (2017). Regulation of alpha-secretase ADAM10 in vitro and in vivo: Genetic, epigenetic, and protein-based mechanisms. *Frontiers in Molecular Neuroscience, 10*, 56. https://doi.org/10.3389/fnmol.2017.00056.

Endres, K., Fahrenholz, F., Lotz, J., Hiemke, C., Teipel, S., Lieb, K., et al. (2014). Increased CSF APPs-alpha levels in patients with Alzheimer disease treated with acitretin. *Neurology, 83*(21), 1930—1935. https://doi.org/10.1212/WNL.0000000000001017. WNL.0000000000001017 [pii].

Forman, M., Tseng, J., Palcza, J., Leempoels, J., Ramael, S., Krishna, G., et al. (2012). The novel BACE inhibitor MK-8931 dramatically lowers CSF A beta peptides in healthy subjects: Results from a rising single dose study. *Neurology, 78*.

Freese, C., Garratt, A. N., Fahrenholz, F., & Endres, K. (2009). The effects of alpha-secretase ADAM10 on the proteolysis of neuregulin-1. *FEBS Journal, 276*(6), 1568—1580. https://doi.org/10.1111/j.1742-4658.2009.06889.x.

Gao, R., Wang, Y., Pan, Q., Huang, G., Li, N., Mou, J., et al. (2015). Fuzhisan, a Chinese herbal medicine, suppresses beta-secretase gene transcription via upregulation of SIRT1 expression in N2a-APP695 cells. *International Journal of Clinical and Experimental Medicine, 8*(5), 7231—7240.

Georgievska, B., Gustavsson, S., Lundkvist, J., Neelissen, J., Eketjall, S., Ramberg, V., et al. (2015). Revisiting the peripheral sink hypothesis: Inhibiting BACE1 activity in the periphery does not alter beta-amyloid levels in the CNS. *Journal of Neurochemistry, 132*(4), 477—486. https://doi.org/10.1111/jnc.12937.

Goodman, A. B., & Pardee, A. B. (2003). Evidence for defective retinoid transport and function in late onset Alzheimer's disease. *Proceedings of the National Academy of Sciences of the United States of America, 100*(5), 2901–2905. https://doi.org/10.1073/pnas.0437937100, 0437937100 [pii].

Grosgen, S., Grimm, M. O., Friess, P., & Hartmann, T. (2010). Role of amyloid beta in lipid homeostasis. *Biochimica et Biophysica Acta, 1801*(8), 966–974. https://doi.org/10.1016/j.bbalip.2010.05.002.

Haining, E. J., Yang, J., Bailey, R. L., Khan, K., Collier, R., Tsai, S., et al. (2012). The TspanC8 subgroup of tetraspanins interacts with A disintegrin and metalloprotease 10 (ADAM10) and regulates its maturation and cell surface expression. *Journal of Biological Chemistry, 287*(47), 39753–39765. https://doi.org/10.1074/jbc.M112.416503.

Han, P., Liang, W., Baxter, L. C., Yin, J., Tang, Z., Beach, T. G., et al. (2014). Pituitary adenylate cyclase-activating polypeptide is reduced in Alzheimer disease. *Neurology, 82*(19), 1724–1728. https://doi.org/10.1212/WNL.0000000000000417.

Han, P., Tang, Z., Yin, J., Maalouf, M., Beach, TG, Reiman, EM, & Shi, J. (2014). Pituitary adenylate cyclase-activating polypeptide protects against β-amyloid toxicity. *Neurobiology of Aging, 35*(9), 2064–2071. https://doi.org/10.1016/j.neurobiolaging.2014.03.022.

Hattori, C., Asai, M., Onishi, H., Sasagawa, N., Hashimoto, Y., Saido, T. C., et al. (2006). BACE1 interacts with lipid raft proteins. *Journal of Neuroscience Research, 84*(4), 912–917. https://doi.org/10.1002/jnr.20981.

Jones, L., Holmans, P. A., Hamshere, M. L., Harold, D., Moskvina, V., Ivanov, D., et al. (2010). Genetic evidence implicates the immune system and cholesterol metabolism in the aetiology of Alzheimer's disease. *PloS One, 5*(11), e13950. https://doi.org/10.1371/journal.pone.0013950.

Kalvodova, L., Kahya, N., Schwille, P., Ehehalt, R., Verkade, P., Drechsel, D., et al. (2005). Lipids as modulators of proteolytic activity of BACE: Involvement of cholesterol, glycosphingolipids, and anionic phospholipids in vitro. *Journal of Biological Chemistry, 280*(44), 36815–36823. https://doi.org/10.1074/jbc.M504484200.

Kizuka, Y., Kitazume, S., Fujinawa, R., Saito, T., Iwata, N., Saido, T. C., et al. (2015). An aberrant sugar modification of BACE1 blocks its lysosomal targeting in Alzheimer's disease. *EMBO Molecular Medicine, 7*(2), 175–189. https://doi.org/10.15252/emmm.201404438.

Kojro, E., Fuger, P., Prinzen, C., Kanarek, A. M., Rat, D., Endres, K., et al. (2010). Statins and the squalene synthase inhibitor zaragozic acid stimulate the non-amyloidogenic pathway of amyloid-beta protein precursor processing by suppression of cholesterol synthesis. *J Alzheimers Dis, 20*(4), 1215–1231. https://doi.org/10.3233/JAD-2010-091621.

Kojro, E., Gimpl, G., Lammich, S., Marz, W., & Fahrenholz, F. (2001). Low cholesterol stimulates the nonamyloidogenic pathway by its effect on the alpha -secretase ADAM 10. *Proceedings of the National Academy of Sciences of the United States of America, 98*(10), 5815–5820. https://doi.org/10.1073/pnas.081612998.

Lammich, S., Buell, D., Zilow, S., Ludwig, A. K., Nuscher, B., Lichtenthaler, S. F., et al. (2010). Expression of the anti-amyloidogenic secretase ADAM10 is suppressed by its 5'-untranslated region. *Journal of Biological Chemistry, 285*(21), 15753–15760. https://doi.org/10.1074/jbc.M110.110742. M110.110742 [pii].

Lammich, S., Kamp, F., Wagner, J., Nuscher, B., Zilow, S., Ludwig, A. K., et al. (2011). Translational repression of the disintegrin and metalloprotease ADAM10 by a stable G-quadruplex secondary structure in its 5'-untranslated region. *Journal of Biological Chemistry, 286*(52), 45063–45072. https://doi.org/10.1074/jbc.M111.296921.

Lammich, S., Kojro, E., Postina, R., Gilbert, S., Pfeiffer, R., Jasionowski, M., et al. (1999). Constitutive and regulated alpha-secretase cleavage of Alzheimer's amyloid precursor protein by a disintegrin metalloprotease. *Proceedings of the National Academy of Sciences of the United States of America, 96*(7), 3922–3927.

Lammich, S., Schobel, S., Zimmer, A. K., Lichtenthaler, S. F., & Haass, C. (2004). Expression of the Alzheimer protease BACE1 is suppressed via its 5'-untranslated region. *EMBO Reports, 5*(6), 620–625. https://doi.org/10.1038/sj.embor.7400166.

Lee, H. R., Shin, H. K., Park, S. Y., Kim, H. Y., Lee, W. S., Rhim, B. Y., et al. (2014). Cilostazol suppresses beta-amyloid production by activating a disintegrin and metalloproteinase 10 via the upregulation of SIRT1-coupled retinoic acid receptor-beta. *Journal of Neuroscience Research, 92*(11), 1581–1590. https://doi.org/10.1002/jnr.23421.

Lei, X., Lei, L., Zhang, Z., Zhang, Z., & Cheng, Y. (2015). Downregulated miR-29c correlates with increased BACE1 expression in sporadic Alzheimer's disease. *International Journal of Clinical and Experimental Pathology, 8*(2), 1565−1574.

Lopes da Silva, S., Vellas, B., Elemans, S., Luchsinger, J., Kamphuis, P., Yaffe, K., et al. (2014). Plasma nutrient status of patients with Alzheimer's disease: Systematic review and meta-analysis. *Alzheimers. Dement, 10*(4), 485−502. https://doi.org/10.1016/j.jalz.2013.05.1771. S1552-5260(13)02464-3 [pii].

Malnar, M., Kosicek, M., Lisica, A., Posavec, M., Krolo, A., Njavro, J., et al. (2012). Cholesterol-depletion corrects APP and BACE1 misstrafficking in NPC1-deficient cells. *Biochimica et Biophysica Acta, 1822*(8), 1270−1283. https://doi.org/10.1016/j.bbadis.2012.04.002.

Marcello, E., Saraceno, C., Musardo, S., Vara, H., de la Fuente, A. G., Pelucchi, S., et al. (2013). Endocytosis of synaptic ADAM10 in neuronal plasticity and Alzheimer's disease. *Journal of Clinical Investigation, 123*(6), 2523−2538. https://doi.org/10.1172/JCI65401.

Marquer, C., Devauges, V., Cossec, J. C., Liot, G., Lecart, S., Saudou, F., et al. (2011). Local cholesterol increase triggers amyloid precursor protein-Bace1 clustering in lipid rafts and rapid endocytosis. *The FASEB Journal, 25*(4), 1295−1305. https://doi.org/10.1096/fj.10-168633.

Marshall, A. J., Rattray, M., & Vaughan, P. F. (2006). Chronic hypoxia in the human neuroblastoma SH-SY5Y causes reduced expression of the putative alpha-secretases, ADAM10 and TACE, without altering their mRNA levels. *Brain Research, 1099*(1), 18−24. https://doi.org/10.1016/j.brainres.2006.05.008.

May, P. C., Willis, B. A., Lowe, S. L., Dean, R. A., Monk, S. A., Cocke, P. J., et al. (2015). The potent BACE1 inhibitor LY2886721 elicits robust central Abeta pharmacodynamic responses in mice, dogs, and humans. *Journal of Neuroscience, 35*(3), 1199−1210. https://doi.org/10.1523/JNEUROSCI.4129-14.2015.

Mei, Z., Yan, P., Tan, X., Zheng, S., & Situ, B. (2015). Transcriptional regulation of BACE1 by NFAT3 leads to enhanced amyloidogenic processing. *Neurochemical Research, 40*(4), 829−836. https://doi.org/10.1007/s11064-015-1533-1.

Meredith, J. E., Jr., Thompson, L. A., Toyn, J. H., Marcin, L., Barten, D. M., Marcinkeviciene, J., et al. (2008). P-glycoprotein efflux and other factors limit brain amyloid beta reduction by beta-site amyloid precursor protein-cleaving enzyme 1 inhibitors in mice. *Journal of Pharmacology and Experimental Therapeutics, 326*(2), 502−513. https://doi.org/10.1124/jpet.108.138974.

Motoki, K., Kume, H., Oda, A., Tamaoka, A., Hosaka, A., Kametani, F., et al. (2012). Neuronal beta-amyloid generation is independent of lipid raft association of beta-secretase BACE1: Analysis with a palmitoylation-deficient mutant. *Brain Behav, 2*(3), 270−282. https://doi.org/10.1002/brb3.52.

Pernber, Z., Blennow, K., Bogdanovic, N., Mansson, J. E., & Blomqvist, M. (2012). Altered distribution of the gangliosides GM1 and GM2 in Alzheimer's disease. *Dementia and Geriatric Cognitive Disorders, 33*(2−3), 174−188. https://doi.org/10.1159/000338181.

Petrus, E., & Lee, H.-K. (2014). BACE1 is necessary for experience-dependent homeostatic synaptic plasticity in visual cortex. *Neural Plasticity, 128631*.

Prinzen, C., Muller, U., Endres, K., Fahrenholz, F., & Postina, R. (2005). Genomic structure and functional characterization of the human ADAM10 promoter. *The FASEB Journal, 19*(11), 1522−1524. https://doi.org/10.1096/fj.04-3619fje, 04-3619fje [pii].

Prinzen, C., Trumbach, D., Wurst, W., Endres, K., Postina, R., & Fahrenholz, F. (2009). Differential gene expression in ADAM10 and mutant ADAM10 transgenic mice. *BMC Genomics, 10*, 66. https://doi.org/10.1186/1471-2164-10-66.

Rat, D., Schmitt, U., Tippmann, F., Dewachter, I., Theunis, C., Wieczerzak, E., et al. (2011). Neuropeptide pituitary adenylate cyclase-activating polypeptide (PACAP) slows down Alzheimer's disease-like pathology in amyloid precursor protein-transgenic mice. *The FASEB Journal, 25*(9), 3208−3218. https://doi.org/10.1096/fj.10-180133.

Reinhardt, S., Grimm, M. O., Stahlmann, C., Hartmann, T., Shudo, K., Tomita, T., et al. (2016). Rescue of hypovitaminosis A induces non-amyloidogenic amyloid precursor protein (APP) processing. *Current Alzheimer Research, 13*(11), 1277−1289.

Reinhardt, S., Schuck, F., Grosgen, S., Riemenschneider, M., Hartmann, T., Postina, R., et al. (2014). Unfolded protein response signaling by transcription factor XBP-1 regulates ADAM10 and is affected in Alzheimer's disease. *The FASEB Journal, 28*(2), 978−997. https://doi.org/10.1096/fj.13-234864. fj.13-234864 [pii].

Roshan, R., Shridhar, S., Sarangdhar, M. A., Banik, A., Chawla, M., Garg, M., et al. (2014). Brain-specific knockdown of miR-29 results in neuronal cell death and ataxia in mice. *RNA, 20*(8), 1287–1297. https://doi.org/10.1261/rna.044008.113. rna.044008.113 [pii].

Santosa, C., Rasche, S., Barakat, A., Bellingham, S. A., Ho, M., Tan, J., et al. (2011). Decreased expression of GGA3 protein in Alzheimer's disease frontal cortex and increased co-distribution of BACE with the amyloid precursor protein. *Neurobiology of Disease, 43*(1), 176–183. https://doi.org/10.1016/j.nbd.2011.03.009.

Sawamura, N., Ko, M., Yu, W., Zou, K., Hanada, K., Suzuki, T., et al. (2004). Modulation of amyloid precursor protein cleavage by cellular sphingolipids. *Journal of Biological Chemistry, 279*(12), 11984–11991. https://doi.org/10.1074/jbc.M309832200.

Soininen, H., Solomon, A., Visser, P. J., Hendrix, S. B., Blennow, K., Kivipelto, M., et al. (2017). 24-month intervention with a specific multinutrient in people with prodromal Alzheimer's disease (LipiDiDiet): A randomised, double-blind, controlled trial. *The Lancet Neurology, 16*(12), 965–975. https://doi.org/10.1016/S1474-4422(17)30332-0.

Tesco, G., Koh, Y. H., Kang, E. L., Cameron, A. N., Das, S., Sena-Esteves, M., et al. (2007). Depletion of GGA3 stabilizes BACE and enhances beta-secretase activity. *Neuron, 54*(5), 721–737. https://doi.org/10.1016/j.neuron.2007.05.012.

Tippmann, F., Hundt, J., Schneider, A., Endres, K., & Fahrenholz, F. (2009). Up-regulation of the alpha-secretase ADAM10 by retinoic acid receptors and acitretin. *The FASEB Journal, 23*(6), 1643–1654. https://doi.org/10.1096/fj.08-121392. fj.08-121392 [pii].

Vassar, R. (2014). BACE1 inhibitor drugs in clinical trials for Alzheimer's disease. *Alzheimer's Research & Therapy, 6*(9), 89. https://doi.org/10.1186/s13195-014-0089-7.

Vassar, R., Bennett, B. D., Babu-Khan, S., Kahn, S., Mendiaz, E. A., Denis, P., et al. (1999). Beta-secretase cleavage of Alzheimer's amyloid precursor protein by the transmembrane aspartic protease BACE. *Science, 286*(5440), 735–741, 7936 [pii].

Wang, R., Chen, S., Liu, Y., Diao, S., Xue, Y., You, X., et al. (2015). All-trans-retinoic acid reduces BACE1 expression under inflammatory conditions via modulation of nuclear factor kappaB (NFkappaB) signaling. *Journal of Biological Chemistry, 290*(37), 22532–22542. https://doi.org/10.1074/jbc.M115.662908. M115.662908 [pii].

Wang, W. X., Rajeev, B. W., Stromberg, A. J., Ren, N., Tang, G., Huang, Q., et al. (2008). The expression of microRNA miR-107 decreases early in Alzheimer's disease and may accelerate disease progression through regulation of beta-site amyloid precursor protein-cleaving enzyme 1. *Journal of Neuroscience, 28*(5), 1213–1223. https://doi.org/10.1523/JNEUROSCI.5065-07.2008, 28/5/1213 [pii].

Xiang, Y., Meng, S., Wang, J., Li, S., Liu, J., Li, H., et al. (2014). Two novel DNA motifs are essential for BACE1 gene transcription. *Scientific Reports, 4*, 6864. https://doi.org/10.1038/srep06864. srep06864 [pii].

Xu, D., Sharma, C., & Hemler, M. E. (2009). Tetraspanin12 regulates ADAM10-dependent cleavage of amyloid precursor protein. *The FASEB Journal, 23*(11), 3674–3681. https://doi.org/10.1096/fj.09-133462.

Yamazaki, K., Mizui, Y., & Tanaka, I. (1997). Radiation hybrid mapping of human ADAM10 gene to chromosome 15. *Genomics, 45*(2), 457–459. https://doi.org/10.1006/geno.1997.4910. S0888-7543(97)94910-7 [pii].

Zhou, W., & Song, W. (2006). Leaky scanning and reinitiation regulate BACE1 gene expression. *Molecular and Cellular Biology, 26*(9), 3353–3364. https://doi.org/10.1128/MCB.26.9.3353-3364.2006, 26/9/3353 [pii].

Zong, Y., Wang, H., Dong, W., Quan, X., Zhu, H., Xu, Y., et al. (2011). miR-29c regulates BACE1 protein expression. *Brain Research, 1395*, 108–115. https://doi.org/10.1016/j.brainres.2011.04.035. S0006-8993(11)00797-9 [pii].

CHAPTER 40

Acetylcholinesterase inhibitory agents in plants and their application to dementia: Alzheimer's disease

Willian Orlando Castillo-Ordoñez, Nohelia Cajas-Salazar

Department of Biology, Research Group Genetic Toxicology and Cytogenetics, Faculty of Natural Sciences and Education, University of Cauca, Popayán, Cauca, Colombia

List of abbreviations

ACh Acetylcholine
AChE Acetylcholinesterase
AD Alzheimer's disease
APP Amyloid precursor protein
BBB Blood—brain barrier
CNS Central nervous system
FDA Food and drug administration
mAChR Muscarinic acetylcholine receptors
nAChR Nicotinic acetylcholine receptors
NMDA N-methyl-D-aspartate
ROS Reactive oxygen species

Mini-dictionary of terms

Cholinergic impairment Alteration of the process involved in the hydrolysis of the acetylcholinesterase enzyme for its degradation, resulting in nervous impulse arrest
Neuritic plaques Extracellular structures formed by amyloid beta peptide deposits
Neurofibrillary tangles Intracellular deposits formed by altered hyperphosphorylated tau protein
Phytochemicals Substances of diverse chemical nature (flavonoids, alkaloids, saponins, estrogens, carbohydrates, etc.) derived from different parts of plants such as seeds, roots, leaves, fruits, and flowers
Synergistic pharmacological effect Property of a combination of substances to exert a biological effect greater than that exerted by each one separately

Introduction

Since ancient times, human health protection has been tightly connected to the discovery and consumption of medicinal plants. These organisms have been shown to be important sources of active compounds that are, in the right (therapeutic) doses, able to modify defense mechanisms of the human body including endocrine signaling, immune response, apoptosis, cell repair, and metabolism. Natural products are thus being intensively

Diagnosis and Management in Dementia
ISBN 978-0-12-815854-8, https://doi.org/10.1016/B978-0-12-815854-8.00040-9

investigated for their clinical use as primary and adjuvant chemotherapies (Horváthová et al., 2012; Melusova, Slamenova, Kozics, Jantova, & Horvathova, 2014; Rodrigues et al., 2015; van Ginkel, Yan, Bhattacharya, Polans, & Kenealey, 2015).

As detailed in previous chapters, AD is a heterogeneous neurodegenerative multifactorial disease, characterized by massive and irreversible damage of brain cells, mainly due to synaptic loss, subsequent neuronal death, and total brain failure (Brookmeyer et al., 2011). Even though considerable research has been in progress to understand the etiological mechanisms of AD and new theories on its causes continue to be formulated, to this point there has not been a general consensus on its pathogenesis. Mounting evidence shows that a decline in DNA repair capacity, accumulation of toxic chemicals from dietary and environmental sources, alteration in insulin metabolism, increasing levels of reactive oxygen species (ROS), chronic inflammation, vascular problems, and altered hormone levels, among other factors, contribute to the progression of the disease and its severity (Brown, Lockwood, & Sonawane, 2005) (see Fig. 40.1).

It is known that pathological manifestations in the AD brain may initiate many years before the development of clinical symptoms; however, understanding where and when molecular alterations occur and biomarkers begin to change has been a challenge to modern medicine, as most studies are conducted in patients with late AD (Carrillo et al., 2013). From a pharmacological point of view, research on the biological activity of natural molecules has claimed a prominent position in AD treatment, and its importance has gathered increased interest from the scientific and medical community. Although lacking, animal and human studies have shown evidence of the ability of some natural products to reduce disease symptoms, while others are able to prevent or even reverse the neurodegenerative progression of the disease (Castillo et al., 2018). Despite this, the few randomized controlled trials conducted so far have not shown efficacy in stopping the progression of the disease. This could be due to the complexity of the disease given by the diverse cumulative effects of genetic and lifestyle risk factors on the mental condition of each individual as well as to the suitability of the molecules to meet physical properties, such as those set out by Lipinski, that predict the therapeutic potential of drugs (Lipinski, Lombardo, Dominy, & Feeney, 1997) (see Fig. 40.2).

In the present chapter, we will focus on describing the existing treatments for AD based on its AChE inhibitory capacity and intended improvement in memory loss. In addition, we provide a list of natural compounds with promising effects for delaying disease progression as well as considerations for suitable treatment points during disease progression based on the cellular mechanisms they are intended to target.

Cholinergic impairment in Alzheimer's disease

Definitive diagnosis of AD can be made only at autopsy, since the presence of neuritic plaques, neurofibrillary tangles, and neuronal loss together with cholinergic impairment

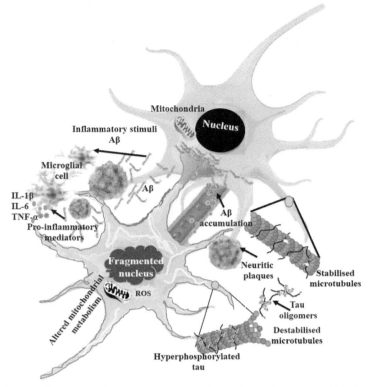

Figure 40.1 *Histopathological alterations observed in Alzheimer's disease.* Alzheimer's disease is characterized by the extracellular accumulation of amyloid β (Aβ) forming neuritic plaques, and hyperphosphorylated tau protein forming intracellular neurofibrillary tangles, ultimately leading to neuronal death. These alterations are accompanied by oxidative stress, mitochondrial depletion, accumulation of Aβ in vascular tissue, and an inflammatory process mediated by microglial activation.

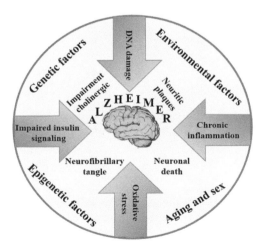

Figure 40.2 *Interaction between risk factors associated with Alzheimer's disease.* Alzheimer's disease is a multifactorial disease caused by the cumulative effect of complex interactions between nonmodifiable (age, sex, and genetic) and modifiable (epigenetic environment) determinants that cause alterations in mental condition throughout the course of life.

need histopathological analysis (see Fig. 40.1). Acetylcholine (ACh) is a low-molecular-weight neurotransmitter that exerts a pivotal role in chemical neurotransmission in the central nervous system (CNS) and peripheral nervous system through its interaction with two classes of receptor proteins: nicotinic acetylcholine receptors (nAChRs) and muscarinic acetylcholine receptors (mAChRs). In the CNS, ACh has a key role in cognitive functions associated with spatial and episodic memory acquisition storage. ACh is stored in cellular vesicles, which release the enzyme when the nerve terminal is depolarized to allow the binding of ACh to its receptor at the synaptic gap. Normally, released ACh has a short half-life due to its hydrolysis at the ester bond by the enzyme acetylcholinesterase (AChE), interrupting impulse transmission at cholinergic synapses (Houghton, Ren, & Howes, 2006) (see Fig. 40.3).

Over the past three decades, ACh inhibition has become a clinical challenge in treating the symptoms of AD, due in part to the complex structure of the active site of the enzyme. AChE is a protein complex with an α/β-hydrolase fold and an overall ellipsoid shape containing a deep groove, called the gorge, that is about 20 Å deep. Hydrolysis of ACh appears to take place at the bottom of the gorge or catalytic site, which consists of the catalytic triad: Ser203, Glu334, and His447 (Atanasova et al., 2015). AChE uses this catalytic triad to improve the ability of Ser to induce a nucleophilic attack on the substrate. Although the hydrolysis process takes place in the base of the gorge, initial binding of ACh is thought to occur at its outer rim in a region called the "peripheral site."

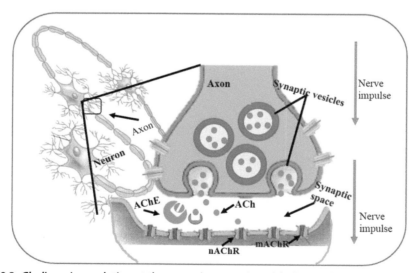

Figure 40.3 *Cholinergic regulation at the synaptic space.* Acetylcholine (ACh) is released from synaptic vesicles of the presynaptic neuron to the synaptic space followed by binding to ACh receptors of the postsynaptic neurons to activate the nervous impulse. The hydrolysis of ACh by AChE interrupts impulse transmission.

Abnormally low levels of ACh cause poor nerve impulse transmission (see Fig. 40.3). Blockade of catabolic degradation of ACh results in increased levels of the neurotransmitter, which may partially correct the cholinergic deficiency seen in AD. At present, knowledge on the structure of AChE is mostly based on work carried out on the enzyme obtained from the electric eel *Torpedo californica*. Although blast alignments of AChE from rat, rabbit, and human have shown that the main residues forming the binding site are highly conserved in AChE from different species, X-ray studies show that binding to AChE may be specific for every AChE inhibitor (Atanasova et al., 2015).

It is still unclear whether cholinergic impairment in AD is a consequence of the loss of cholinergic neurons and AChRs or a direct impact of molecular interactions between amyloid beta (Aβ) peptide and AChRs that leads to dysregulation of the receptors. Studies show that in the AD brain, a severe loss of nAChRs and mAChRs correlates with the severity of the disease at the time of death. Loss of AChRs results in reduced nicotinic and muscarinic cholinergic excitation and additional postsynaptic depolarization, decreasing presynaptic neurotransmitter release and intracellular signaling (Wang et al., 2007).

Therapeutic intervention of Alzheimer's disease based on acetylcholinesterase inhibitors

Tacrine, donepezil, rivastigmine, and galantamine are the only four AChE inhibitors approved thus far by the FDA for the pharmacological treatment of AD. Though they provide palliative treatment, they do not change the course of the disease. In addition, randomized, controlled trials conducted to evaluate rates of improvement observed in AD patients treated with AChE inhibitors have shown variable values ranging from 18% to 48% (Lanctôt et al., 2003).

Tacrine

Tacrine is a synthetic compound known since 1949. It is a nonselective, reversible AChE inhibitor that was approved by the FDA in 1993 and was the first drug marketed for the treatment of AD, under the commercial brand name Cognex. Tacrine is a classical pharmacophore that inhibits both AChE and butyrylcholinesterase (BuChE) at the micromolar scale. Currently, tacrine is rarely used due to its short half-life and the resultant high daily doses needed, which causes hepatotoxicity that requires constant monitoring of liver enzymes levels (Farlow, 2001; Watkins, Zimmerman, Knapp, Gracon, & Lewis, 1994).

Donepezilo

Donepezilo is a piperidine-based agent, a noncompetitive and reversible AChE inhibitor that is chemically unrelated to other AChE inhibitors. It was the first drug to be marketed for the symptomatic treatment of AD in the UK and has been marketed under the commercial brand name Aricept since 1997. Donepezil is highly selective for AChE but shows lower affinity for BuChE; in addition, it exhibits significant antioxidant activity (Atukeren et al., 2017). The most common adverse events reported in clinical trials have been gastrointestinal, including nausea, vomiting, diarrhea, and constipation.

Rivastigmine

Rivastigmine is a noncompetitive AChE and BuChE inhibitor. It binds to both the esteratic and anionic sites of the AChE enzyme, causing carbamylation of its active site, thus preventing the enzyme from metabolizing ACh. Rivastigmine has been marketed since 2000 under the commercial brand name Exelon. While tacrine, donepezil, and galantamine are fast reversible agents, rivastigmine is classified as a pseudoirreversible (very slowly reversible) agent due to its long inhibition of AChE (up to 10 h) despite a very short (2 h) elimination half-life (Desai & Grossberg, 2005). Rivastigmine belongs to the group of compounds called carbamates, which are characterized by lacking binding affinity for dopaminergic, opioid, muscarinic, and nicotinic receptors (Desai & Grossberg, 2001). Clinically, rivastigmine is used for symptomatic treatment of moderate to severe AD stages; however, some authors recommend that a more effective therapeutic strategy would require combining rivastigmine with an N-methyl-D-aspartate (NMDA) receptor antagonist such as memantine, since the glutamatergic damage follows cholinergic impairment. Gastrointestinal side effects associated with rivastigmine consumption include vomiting, diarrhea, nausea, and anorexia.

Galantamine

Unlike the other AChE inhibitors, galantamine has a weak AChE inhibitory effect. Nevertheless, it has a dual-action mechanism because it allosterically modulates nAChRs and inhibits AChE. Additionally, galantamine modulates nonamyloidogenic processing of amyloid precursor protein (APP) by inhibiting beta-site APP cleaving enzyme expression and thus the aggregation and toxicity of Aβ (Li, Wu, Zhang, & Zhang, 2010). This alkaloid, which facilities cholinergic neurotransmission, also exerts antioxidant and antigenotoxic effects that are thought to be due to the removal of reactive oxygen species (Castillo, Aristizabal-Pachon, de Lima Montaldi, Sakamoto-Hojo, & Takahashi, 2016; Ezoulin, Ombetta, Dutertre-Catella, Warnet, & Massicot, 2008). The efficacy of galantamine has been shown for treatment of AD in mild, moderate, and advanced-moderate stages; however, recently its safety and efficacy have been reported in patients with severe AD (Burns et al., 2009). Galantamine was launched in the United States and Europe for the symptom treatment of Alzheimer's disease in 2001, originally as Reminyl, and later as Razadyne in the United States. Side effects associated with galantamine are similar to those of donepezil.

Plants as sources of acetylcholinesterase inhibitors to treat Alzheimer's disease

For centuries, plants and their constituents have been used with advantageous results for medicinal purposes to alleviate common clinical disorders including dementia. Recently, a growing interest has reemerged regarding the use of natural products for the treatment of cognitive impairment in AD and its associated pathologies. Plants and their metabolites are perhaps not only greater in number but also the most relevant group of botanical compounds considered important for developing new substances with the capacity to modulate different pathways deregulated by disease, since complex diseases such as AD are network-related diseases and rarely are caused by a single gene dysfunction or single signaling pathway. The multiple mechanisms involved in the pathogenesis of AD create significant difficulty in producing an effective treatment (Zheng, Fridkin, & Youdim, 2015).

An increasing number of plant metabolites with antioxidant, antigenotoxic, antiinflammatory, and AChE inhibitory activity have been identified. Interestingly, consumption of plants or their metabolites exerts a synergic effect, interfering simultaneously with several deregulated biological circuits in multifactorial diseases as AD (see Fig. 40.4). Moreover, it has been shown that whole-plant extracts can be more effective than their single active compounds due to the synergistic interaction between the active compounds. Experimental studies show that combination of two drugs at subeffective concentration induced a synergistic neuroprotective effect (Romero, Egea, García, & López, 2010). Overall, these findings could inspire an attractive therapeutic strategy based on plants to be explored for patients with neurodegenerative disease, such as AD.

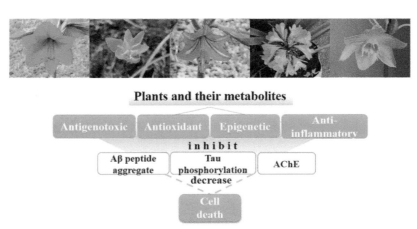

Figure 40.4 *Plants as a source of molecules with therapeutic potential for Alzheimer's disease (AD)*. Plants are an important source of diverse metabolites with the capacity to modulate the genotoxicity, oxidative stress, epigenetic status, inflammation and other cellular alterations observed in AD. Their therapeutic effects are shown to inhibit neuritic plaques, neurofibrillary tangles, cholinergic impairment, and subsequent neuronal death. *(Photographs of Amaryllidaceae flowers are courtesy of Professor Oscar Dario Bermudez, University of Cauca.)*

Amaryllidaceae alkaloids as acetylcholinesterase inhibitors in Alzheimer's disease

Amaryllidaceae is a large family of bulbous plants with roughly 85 genera and 1100 species. It is widely distributed through the tropical and warm regions of the planet with some species native to Andean South America, the Mediterranean, and Southern Africa. Several members of the Amaryllidaceae are natural sources of the previously described galantamine that, due to its ability to penetrate the blood—brain barrier (BBB) and its potential as a cholinergic activator, was used in the 1980s in clinical trials in Western Europe for the treatment of AD. In 1996, Sanochemia Pharmazeutika obtained the first patent on the synthetic processing of galantamine, and in February 2001 the FDA approved galantamine as a therapeutic drug for the treatment of AD.

More than 500 exclusive and structurally diverse alkaloids, namely Amaryllidaceae alkaloids, have been isolated from different members of this family (Jin, 2013). Based on the chemical structure, these alkaloids are classified into nine basic groups including betadine, lycorine, crinine, lycorenine, tazettine, galantamine, narciclasine, cheryline and montanine-type alkaloids (Bastida et al., 2011; Unver, 2007). Several of these ompounds have shown pharmacological activity in vitro with different targets, primarily inhibition of AChE, and simultaneous antioxidant, antigenotoxic, and antiinflammatory action (Adewusi, Fouche, & Steenkamp, 2012; Castillo-Ordóñez et al., 2017; Castillo et al., 2016).

In addition to the Amaryllidaceae family, other plants have been found to be important sources of phytochemical compounds with proven evidence of inhibiting various mechanisms associated with AD. Several experimental studies using biological models, both in vivo and in vitro, have shown evidence that plants or their isolated metabolites can modulate different pathways associated with diseases. In addition to the Amaryllidaceae family, Other plant groups and their compounds have been researched for improving cognition in AD (Bui & Nguyen, 2017). Below, we summarize information on some medicinal plants and their isolated metabolites that have shown potential effects for reducing and preventing symptoms of AD.

Huperzia serrata

This plant contains Huperzine A, a sesquiterpene alkaloid that acts as a reversible AChE inhibitory and NMDA receptor antagonist. In China, it has already been approved as a drug for the symptomatic treatment of AD. Huperzine A has been used in traditional medicine for the treatment of fever and inflammation (Ma, Tan, Zhu, Gang, & Xiao, 2007). Clinical trials have shown that Huperzine A has low or minimal side effects. This is an advantage of Huperzine A compared with other AChE inhibitor drugs for AD treatment.

Ginkgo biloba

This is one the most studied plants for its effect on AD. Among its metabolites are terpenes (ginkgolides and bilobalides), organic acids, and polyphenols (kaempferol, quercetin). Studies have shown that whole ginkgo biloba extract was as effective as the donepezil used as control (Mazza, Capuano, Bria, & Mazza, 2006). Ginkgo biloba extract has been shown in several studies to be a powerful antioxidant. It prevents neuronal death, reverts neuritic dystrophy associated with Aβ, and stimulates neurogenesis. These results suggest a synergistic effect exerted by the various metabolites found in the extract.

Curcuma longa

This is a rhizomatous perennial plant belonging to the Zingiberaceae family. Curcuma has a long history of use as a traditional medicine and food among Asiatic countries. Curcumin is the principal curcuminoid with antioxidant, antitumor, and antiinflammatory properties. Studies have shown that curcumin inhibits Aβ aggregation and Aβ–induced inflammation as well as the activities of β–secretase and AChE (Hamaguchi, Ono, & Yamada, 2010). In addition, it has been reported that curcuma consumption is associated with better cognitive performance in old age, which could be a reason why senior citizens in India, where curcumin is a dietary basic, have a lower prevalence of AD and better cognitive performance (Ng et al., 2006). The exact mechanism by which curcumin exerts its effects are unknown; however, it may be due to its antiinflammatory activity, since inflammation is an important feature in AD.

Resveratrol

Among the molecular pathways altered in AD, epigenetic mechanisms that control gene expression seem to play an important role in the regulation of neuronal differentiation, memory consolidation, and learning during the healthy life span. The reversible nature of epigenetic markers and their modulation by dietary components make them exciting candidates as therapeutic targets (Reuter, Gupta, Park, Goel, & Aggarwal, 2011). Resveratrol, a polyphenol found in grape skin, red wine, nuts, and other plant foods is receiving increasing attention owing to its medical potential in modulating epigenetic pathways associated with ageing. In addition, resveratrol reduces the level of Aβ by inducing the nonamyloidogenic cleavage of APP and enhancing Aβ clearance. Resveratrol also exerts AChE inhibitory activity and decreases cerebrospinal fluid $A\beta_{40}$ and plasma $A\beta_{40}$ levels. The mechanism by which resveratrol exerts a wide range of beneficial effects is not yet clear.

Centella asiatica (L)

This is a plant belonging to the Apiaceae family used in traditional medicine for rejuvenating the neuronal cells and increasing memory. In addition to the neuroprotective effect of *Centella asiatica*, other antiinflammatory, hepatoprotective, immunostimulant, cardioprotective, antidiabetic, and antioxidant biological activities have been reported (Orhan, 2012). Phytochemicals identified from *C. asiatica* include isoprenoids (sesquiterpenes, plant sterols, pentacyclic triterpenoids, and saponins) and phenylpropanoid derivatives (eugenol derivatives, caffeoylquinic acids, and flavonoids). Studies in in vitro models show that *C. asiatica* protects neuronal cells against $A\beta_{40}$-induced neurotoxicity, attenuates mitochondrial dysfunction, and improves antioxidant status (Gray et al., 2018). In addition, *C. asiatica* decreases ROS production and modulates the antioxidative defense system. These activities suggest an important role for this plant in the prevention and treatment of AD. In modern use, *C. asiatica* is typically sold as capsules of the dried herb as a dried extract or as an herbal tincture. It has been reported that supplementation with *C. asiatica* is safe and does not interfere with other concomitant treatments (Cotellese et al., 2018).

Phytooestrogens

These belong to a large group of phenolic compounds found in plants or derived from plant precursors. Phytoestrogens are found in significant amounts in seeds such as soya and flaxseed, vegetables, cereals, tea, chocolate, and fruits. They comprise several classes of chemical compounds (stilbenes, coumestans, isoflavones, ellagitannins, and lignans) that are structurally similar to endogenous estrogens but can have both estrogenic and antiestrogenic effects, with the capacity to bind to the same receptors that our own estrogen does (Landete et al., 2016). Genistein is a major phytoestrogen found in soybeans and has a similar structure to that of human estrogen. The observation that estrogen replacement therapy with phytoestrogens in postmenopausal women is associated with delayed onset and reduced risk of AD aroused interest in genistein in AD. In vitro studies with phytoestrogens show their protective effect against Aβ-induced toxicity and memory and cognition impairments. Genistein can also directly interact with targeted signaling proteins and stabilize their activity to prevent AD (Devi, Shanmuganathan, Manayi, Nabavi, & Nabavi, 2017). At present, the mechanisms by which phytoestrogens exert neuroprotective effects are being elucidated. However, the side effects of genistein on breast cancer cells need to be evaluated, since estrogens are associated with breast and uterine endometrial cell proliferation.

Many studies have been conducted to investigate the effects of total medicinal plant extracts and their isolated active compounds in the treatment of AD. However, the difficulty in demonstrating their strong clinical effect is not surprising given the reduced life expectancy of patients after the clinical manifestation of the disease and because most people use these natural products once the first manifestations of cognitive impairment appears; in other words, treatments would not exert a therapeutic effect, as alterations in multiple pathways implicated in the pathogenesis are already present. On the other hand, lifetime intake of a wide variety of fruits and vegetables may provide a broader degree of protection against the harmful effects of aging and neurodegeneration.

Treatment of Alzheimer's disease: looking to the future

The knowledge gathered about its modifiable and nonmodifiable causes is leading to the formulation of new ways to diagnose and treat the disease and thus search for effective treatments. In addition to the need to provide a unified definition for cognitive health and cognitive impairment, it is also necessary to conceptualize, analyze, and unify emergent theories of the disease, as recent drugs designed based on the classical amyloid hypothesis have failed in clinical trials.

Progress in silico analysis and neuroinformatics tools are accelerating the identification of potential active compounds from plants (See Fig. 40.5). However, the growing number of studies needs to be combined with in vitro and in vivo analysis to evaluate their physiological properties. For instance, only 98% of active molecules from in silico and in vivo analyses do not cross the BBB. At the level of human population studies, the search of early disease biomarkers will allow the stratification of individuals in every stage of pathology, since patients at an early stage of disease could receive appropriate treatments that slow or stop neurodegeneration.

Regarding the use of medicinal plants, successful clinical application in AD requires a clear understanding of the biological pathways involved in the pathophysiology. Several studies in cell and animal models, and to a lesser extent in clinical trials, have shown that plants or their phytochemical compounds are relatively less toxic than synthetic drugs;

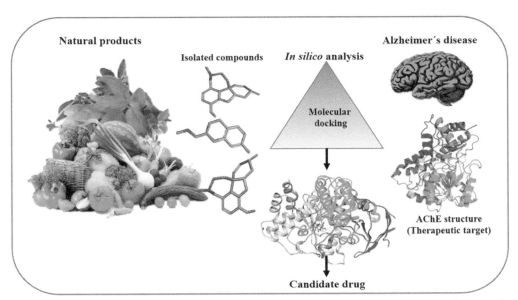

Figure 40.5 *In silico discovery of bioactive molecules as drug candidates for Alzheimer's disease.* The combination of computational and experimental strategies has become of great value in the identification and development of novel promising therapeutic compounds for patients with Alzheimer's disease.

however, in-depth studies are necessary to find new compounds and evaluate adverse and beneficial effects. The multifactorial nature of AD should inspire multifactorial strategies for its prevention and treatment. New approaches such as polypharmacology based on the use of plants could open novel avenues to complex diseases such as Alzheimer's. Plants contain compounds with antiamyloidogenic, anticholinesterase, antioxidant, anti-inflammatory and antiapoptotic properties that could regulate the different altered pathways in AD.

It is clear that an optimal therapeutic strategy for AD should include a combined scheme of phytopharmaceuticals whose multitarget effect can simultaneously regulate the observed cellular and altered pathways during the different stages of the disease. An early intervention should be administered to those individuals with a predisposition to develop the disease due to exposure to known risk factors. For instance, in predisposed individuals, antioxidants (resveratrol, curcumin, flavonoids, and phytoestrogens) should be consumed during the normal cognition stage (see Fig. 40.6). In patients with mild cognition deficiencies who present cholinergic impairments, phytochemical compounds with anticholinergic properties such as galantamine, resveratrol, ginkgo biloba whole extract, and Huperzia serrata whole extract should be considered. Finally, in advanced dementia where both cholinergic impairments and severe neuronal death occur, a combination of natural products with anticholinergic and glutamate modulator activity could preserve the remaining active neurons to slow cognitive and motor impairment.

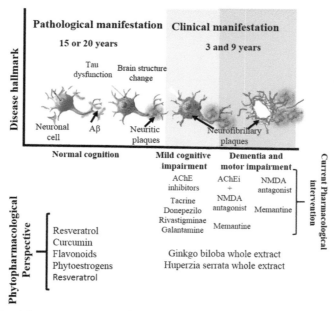

Figure 40.6 *Optimal therapeutic strategy for Alzheimer's disease based on the progression of the disease.* Suitable time for treatment with natural products during the sequence of events of the disease based on the cellular mechanisms they are intended to target.

Key facts of natural products to treat neurodegenerative disorders

- AD is an age-related pathology for which no cure or treatment has been found.
- Neurodegenerative disorders such as AD have a multifactorial etiology, and its pathophysiology is very complex.
- The limited capacity of most researched drugs to cross the BBB is one of the difficulties that has limited identification of an effective treatment for AD.
- Of the drugs approved during the last 30 years, up to 50% are either directly or indirectly derived from natural products.
- The lack of early AD biomarkers and its late diagnosis have limited the knowledge on the therapeutic effectiveness of natural products in each stage of the disease.

Summary points

- This chapter focuses on describing the therapeutic activity of natural AChE inhibitors found in plants.
- Natural products contain metabolites with antioxidant, amyloidogenic, cholinergic, antigenotoxic, antiinflammatory, and estrogenic properties.
- Four AChE inhibitors are currently approved for the palliative treatment of AD.
- Galantamine, a drug commonly used today in the treatment of AD, is an AChE inhibitor obtained from several members of Amaryllidaceae family.

References

Adewusi, E. A., Fouche, G., & Steenkamp, V. (2012). Antioxidant, acetylcholinesterase inhibitory activity and cytotoxicity assessment of the crude extracts of Boophane disticha. *African Journal of Pharmacology and Therapeutics, 1*(3), 78—83.

Atanasova, M., Stavrakov, G., Philipova, I., Zheleva, D., Yordanov, N., & Doytchinova, I. (2015). Galantamine derivatives with indole moiety: Docking, design, synthesis and acetylcholinesterase inhibitory activity. *Bioorganic and Medicinal Chemistry, 23*(17), 5382—5389.

Atukeren, P., Cengiz, M., Yavuzer, H., Gelisgen, R., Altunoglu, E., Oner, S., et al. (2017). The efficacy of donepezil administration on acetylcholinesterase activity and altered redox homeostasis in Alzheimer's disease. *Biomedicine and Pharmacotherapy, 90*, 786—795.

Bastida Armengol, J., Berkov, S., Torras Claveria, L., Pigni, N. B., Andradre, J. P.d., Martínez, V., et al. (2011). Chemical and biological aspects of Amaryllidaceae alkaloids. In D. Muñoz-Torrero (Ed.), *Recent Advances in Pharmaceutical Sciences, 2011, (Chapter 3)* (pp. 65—100).

Brookmeyer, R., Evans, D. A., Hebert, L., Langa, K. M., Heeringa, S. G., Plassman, B. L., et al. (2011). National estimates of the prevalence of Alzheimer's disease in the United States. *Alzheimer's and Dementia, 7*(1), 61—73.

Brown, R. C., Lockwood, A. H., & Sonawane, B. R. (2005). Neurodegenerative diseases: An overview of environmental risk factors. *Environmental Health Perspectives, 113*(9), 1250—1256. https://doi.org/10.1289/ehp.7567.

Bui, T. T., & Nguyen, T. H. (2017). Natural product for the treatment of Alzheimer's disease. *Journal of Basic and Clinical Physiology and Pharmacology, 28*(5), 413—423.

Burns, A., Bernabei, R., Bullock, R., Cruz Jentoft, A. J., Frolich, L., Hock, C., et al. (2009). Safety and efficacy of galantamine (Reminyl) in severe Alzheimer's disease (the SERAD study): A randomised, placebo-controlled, double-blind trial. *The Lancet Neurology, 8*(1), 39—47.

Carrillo, M. C., Brashear, H. R., Logovinsky, V., Ryan, J. M., Feldman, H. H., Siemers, E. R., & Sperling, R. A. (2013). Can we prevent Alzheimer's disease? Secondary "prevention" trials in Alzheimer's disease. *Alzheimer's and Dementia: The Journal of the Alzheimer's Association, 9*(2), 123-131.e121.

Castillo-Ordóñez, W. O., Tamarozzi, E. R., da Silva, G. M., Aristizabal-Pachón, A. F., Sakamoto-Hojo, E. T., Takahashi, C. S., et al. (2017). Exploration of the acetylcholinesterase inhibitory activity of some alkaloids from amaryllidaceae family by molecular docking in silico. *Neurochemical Research, 42*(10), 2826—2830.

Castillo, W. O., Aristizabal-Pachon, A. F., de Lima Montaldi, A. P., Sakamoto-Hojo, E. T., & Takahashi, C. S. (2016). Galanthamine decreases genotoxicity and cell death induced by β-amyloid peptide in SH-SY5Y cell line. *Neurotoxicology, 57*, 291—297.

Castillo, W. O., Aristizabal-Pachon, A. F., Sakamoto-Hojo, E., Gasca, C. A., Cabezas-Fajardo, F. A., & Takahashi, C. (2018). Caliphruria subedentata (Amaryllidaceae) decreases genotoxicity and cell death induced by β-amyloid peptide in sh-sy5y cell line. *Mutation Research: Genetic Toxicology and Environmental Mutagenesis, 836*, 54—61. https://doi.org/10.1016/j.mrgentox.2018.06.010.

Cotellese, R., Hu, S., Belcaro, G., Ledda, A., Feragalli, B., Dugall, M., et al. (2018). *Centella asiatica* (Centellicum®) facilitates the regular healing of surgical scars in subjects at high risk of keloids. *Minerva Chirurgica, 73*(2), 151—156.

Desai, A., & Grossberg, G. (2001). Review of rivastigmine and its clinical applications in Alzheimer's disease and related disorders. *Expert Opinion on Pharmacotherapy, 2*(4), 653—666.

Desai, A., & Grossberg, G. (2005). Rivastigmine for Alzheimer's disease. *Expert Review of Neurotherapeutics, 5*(5), 563—580.

Devi, K. P., Shanmuganathan, B., Manayi, A., Nabavi, S. F., & Nabavi, S. M. (2017). Molecular and therapeutic targets of genistein in Alzheimer's disease. *Molecular Neurobiology, 54*(9), 7028—7041.

Ezoulin, M. J., Ombetta, J.-E., Dutertre-Catella, H., Warnet, J.-M., & Massicot, F. (2008). Antioxidative properties of galantamine on neuronal damage induced by hydrogen peroxide in SK—N—SH cells. *Neurotoxicology, 29*(2), 270—277.

Farlow, M. R. (2001). Pharmacokinetic profiles of current therapiesfor Alzheimer's disease: Implications for switching to galantamine. *Clinical Therapeutics, 23*, A13—A24.

van Ginkel, P. R., Yan, M. B., Bhattacharya, S., Polans, A. S., & Kenealey, J. D. (2015). Natural products induce a G protein-mediated calcium pathway activating p53 in cancer cells. *Toxicology and Applied Pharmacology, 288*(3), 453—462.

Gray, N. E., Magana, A. A., Lak, P., Wright, K. M., Quinn, J., Stevens, J. F., et al. (2018). *Centella asiatica*: Phytochemistry and mechanisms of neuroprotection and cognitive enhancement. *Phytochemistry Reviews, 17*(1), 161—194.

Hamaguchi, T., Ono, K., & Yamada, M. (2010). Curcumin and Alzheimer's disease. *CNS Neuroscience and Therapeutics, 16*(5), 285—297.

Horváthová, E., Kozics, K., Srančíková, A., Hunáková, Ľ., Gálová, E., Ševčovičová, A., et al. (2012). Borneol administration protects primary rat hepatocytes against exogenous oxidative DNA damage. *Mutagenesis, 27*(5), 581—588.

Houghton, P. J., Ren, Y., & Howes, M.-J. (2006). Acetylcholinesterase inhibitors from plants and fungi. *Natural Product Reports, 23*(2), 181—199.

Jin, Z. (2013). Amaryllidaceae and sceletium alkaloids. *Natural Product Reports, 30*(6), 849—868.

Lanctôt, K. L., Herrmann, N., Yau, K. K., Khan, L. R., Liu, B. A., LouLou, M. M., et al. (2003). Efficacy and safety of cholinesterase inhibitors in Alzheimer's disease: A meta-analysis. *Canadian Medical Association Journal, 169*(6), 557—564.

Landete, J., Arqués, J., Medina, M., Gaya, P., de Las Rivas, B., & Muñoz, R. (2016). Bioactivation of phytoestrogens: Intestinal bacteria and health. *Critical Reviews in Food Science and Nutrition, 56*(11), 1826—1843.

Lipinski, C. A., Lombardo, F., Dominy, B. W., & Feeney, P. J. (1997). Experimental and computational approaches to estimate solubility and permeability in drug discovery and development settings. *Advanced Drug Delivery Reviews, 23*(1), 3—25.

Li, Q., Wu, D., Zhang, L., & Zhang, Y. (2010). Effects of galantamine on β-amyloid release and beta-site cleaving enzyme 1 expression in differentiated human neuroblastoma SH-SY5Y cells. *Experimental Gerontology, 45*(11), 842−847.

Ma, X., Tan, C., Zhu, D., Gang, D. R., & Xiao, P. (2007). Huperzine A from Huperzia species—an ethno-pharmacolgical review. *Journal of Ethnopharmacology, 113*(1), 15−34.

Mazza, M., Capuano, A., Bria, P., & Mazza, S. (2006). Ginkgo biloba and donepezil: A comparison in the treatment of Alzheimer's dementia in a randomized placebo-controlled double-blind study. *European Journal of Neurology, 13*(9), 981−985.

Melusova, M., Slamenova, D., Kozics, K., Jantova, S., & Horvathova, E. (2014). Carvacrol and rosemary essential oil manifest cytotoxic, DNA-protective and pro-apoptotic effect having no effect on DNA repair. *Neoplasma, 61*(6), 690−699.

Ng, T.-P., Chiam, P.-C., Lee, T., Chua, H.-C., Lim, L., & Kua, E.-H. (2006). Curry consumption and cognitive function in the elderly. *American Journal of Epidemiology, 164*(9), 898−906.

Orhan, I. E. (2012). *Centella asiatica* (L.) urban: From traditional medicine to modern medicine with neuroprotective potential. *Evidence-based Complementary and Alternative Medicine,* 2012.

Reuter, S., Gupta, S. C., Park, B., Goel, A., & Aggarwal, B. B. (2011). Epigenetic changes induced by curcumin and other natural compounds. *Genes and nutrition, 6*(2), 93−108.

Rodrigues, I. A., Mazotto, A. M., Cardoso, V., Alves, R. L., Amaral, A. C. F., Silva, J. R.d. A., et al. (2015). Natural products: Insights into leishmaniasis inflammatory response. *Mediators of Inflammation,* 12, 2015.

Romero, A., Egea, J., García, A. G., & López, M. G. (2010). Synergistic neuroprotective effect of combined low concentrations of galantamine and melatonin against oxidative stress in SH-SY5Y neuroblastoma cells. *Journal of Pineal Research, 49*(2), 141−148.

Unver, N. (2007). New skeletons and new concepts in Amaryllidaceae alkaloids. *Phytochemistry Reviews, 6*(1), 125−135.

Wang, D., Noda, Y., Zhou, Y., Mouri, A., Mizoguchi, H., Nitta, A., et al. (2007). The allosteric potenti-ation of nicotinic acetylcholine receptors by galantamine ameliorates the cognitive dysfunction in beta amyloid25−35 icv-injected mice: Involvement of dopaminergic systems. *Neuropsychopharmacology, 32*(6), 1261−1271.

Watkins, P. B., Zimmerman, H. J., Knapp, M. J., Gracon, S. I., & Lewis, K. W. (1994). Hepatotoxic effects of tacrine administration in patients with Alzheimer's disease. *Jama, 271*(13), 992−998.

Zheng, H., Fridkin, M., & Youdim, M. (2015). New approaches to treating Alzheimer's disease. *Perspectives in Medicinal Chemistry, 7,* 1.

CHAPTER 41

Removal of blood amyloid-β as an effective and safe therapeutic strategy for Alzheimer's disease

Nobuya Kitaguchi, Kazunori Kawaguchi, Kazuyoshi Sakai
School of Health Sciences, Fujita Health University, Toyoake, Aichi, Japan

List of abbreviations

AD Alzheimer's disease
Aβ β-amyloid, amyloid-β
Aβ$_{1-40}$ 40-amino-acid Aβ
Aβ$_{1-42}$ 42-amino-acid Aβ, aggregates more easily and is more neurotoxic than Aβ$_{1-40}$
CHA charcoal
E-BARS extracorporeal blood Aβ removal systems
HD hemodialysis
HDC hexadecyl-alkylated cellulose particles
PSf polysulfone

Mini-dictionary of terms

Extracorporeal blood circulation Blood is continuously taken from the patient, treated with a certain device(s), and returned to the patient.
Hemodialysis Extracorporeal blood purification for end-stage renal failure patients with dialyzers to remove water, urea, and other small molecules and small proteins.

Introduction: impaired clearance of amyloid-β protein in Alzheimer's disease

One of the major hallmarks of Alzheimer's disease (AD) is the accumulation of amyloid-β protein (Aβ) as senile plaques and an increase in Aβ peptides in the brain (Kuo et al., 1996; Polanco et al., 2018; Selkoe, 2001). There are several Aβ species in the brain and plasma that are approximately 4 kDa in weight, including the 40-amino-acid peptide Aβ$_{1-40}$ and the 42-amino-acid peptide Aβ$_{1-42}$. Aβ$_{1-42}$ aggregates more easily and is more toxic than Aβ$_{1-40}$ (Hung et al., 2008). Aggregated Aβ forms more neurotoxic soluble Aβ oligomers that can cause synapse loss and affect long-term potentiation in hippocampal neurons (Walsh et al., 2002).

Diagnosis and Management in Dementia
ISBN 978-0-12-815854-8, https://doi.org/10.1016/B978-0-12-815854-8.00041-0

Aβ production in the brains of AD patients was reported to be similar to that of normal subjects, yet Aβ clearance from AD brains was approximately 30% lower than in normal subjects (Mawuenyega et al., 2010). It has been hypothesized that it may be possible to treat AD, particularly sporadic cases, by increasing Aβ clearance from the brain. Several methods have been employed to try to enhance brain Aβ clearance. Peripheral administration of anti-Aβ monoclonal antibodies was reported to reduce brain Aβ and maintain cognitive function (e.g., Sevigny et al., 2016). Other Aβ-binding substances, such as peripherally administered gelsolin or the GM1 ganglioside, also reduced brain Aβ (Matsuoka et al., 2003) in an AD mouse model. Further, plasma exchange therapy was shown to be effective in improving cognitive function in AD patients in a phase 2 study (Boada et al., 2009). This method consists of two procedures: removal of plasma that includes Aβ, followed by the intravenous administration of albumin, an Aβ-binding substance, as a supplemental solution. This plasma exchange therapy is currently undergoing a phase 3 trial in AD patients (Boada et al., 2016).

Several factors are capable of transporting Aβ from the brain to the blood, including LRP-1 (lipoprotein receptor–related protein-1), ApoJ (apolipoprotein J), ApoE (apolipoprotein E), and RAGE (receptor for advanced glycation end products) (Bell et al., 2007; DeMattos, Bales, Cummins, Paul, & Holtzman, 2002; Donoghue et al., 2006; Silverberg et al., 2010). RAGE also mediates the transfer of Aβ from the blood into the brain. It has been reported that Aβ in the brain is excreted through the perivascular elimination pathway along the cerebrovascular basement membrane of cerebral capillaries (Carare et al., 2008; Kalaria, Akinyemi, & Ihara, 2012; Maki et al., 2014; Morris, Carare, Schreiber, & Hawkes, 2014; Weller, Djuanda, Yow, & Carare, 2009; Zlokovic, 2011).

Given the observations that suggest that removal of blood Aβ may be linked to reductions in Aβ concentrations in the brain and improved cognitive function, peripheral removal of Aβ may be an effective therapeutic option for AD.

Our hypothesis: extracorporeal blood amyloid-β removal systems may reduce brain amyloid-β

Extracorporeal blood Aβ removal systems (E-BARS) remove Aβ from the body and represent a potential treatment modality for AD (Kawaguchi et al., 2010). E-BARS may also present a safety advantage over drug treatments, as serious adverse reactions associated with drug treatments would be avoided. As shown in Fig. 41.1, E-BARS involve extracorporeal removal of Aβ from the blood using a device. Decreases in the concentration of blood Aβ may then trigger an acceleration of Aβ transport from the brain into the blood.

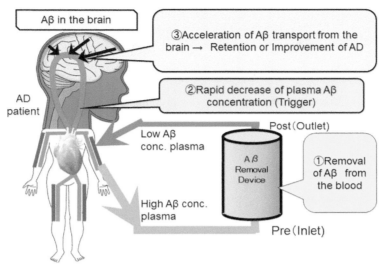

Figure 41.1 Schematic of the therapeutic hypothesis for extracorporeal blood Aβ removal systems in the treatment of Alzheimer's disease (AD). *Aβ*, amyloid-β protein.

Adsorptive materials of devices used in extracorporeal blood amyloid-β removal systems

Hexadecyl-alkylated cellulose particles and charcoal show high amyloid-β-removal activities

To create suitable devices for the removal of blood Aβ, effective materials that adsorb Aβ must be identified. We investigated six adsorbents that are currently in use for blood purification therapy of other intractable diseases, including hexadecyl-alkylated cellulose particles (HDC), which are used to remove β2-microglobulin in carpal tunnel syndrome (Suzuki, Shimazaki, & Kutsuki, 2003); charcoal (CHA), which is used in the treatment of multiple conditions such as hepatic failure; tryptophan-ligated polyvinyl alcohol gel, which is used in the treatment of Guillain—Barré syndrome; cellulose particles ligated with dextran sulfate, used in the treatment of familial hypercholesteremia and systemic lupus erythematosus; and cellulose acetate particles and nonwoven polyethylene terephthalate filters, both of which are used in treating ulcerative colitis.

In studies of the adsorptive properties of these six materials using very high concentrations of synthetic Aβ (100—1000 times higher than those in the blood), HDC and CHA exhibited the highest Aβ-removal activity (Fig. 41.2A) (Kawaguchi et al., 2010). HDC and CHA removed almost 99% of the $A\beta_{1-40}$ and $A\beta_{1-42}$ in batch analyses. Then, minicolumns filled with HDC or CHA were used to evaluate the Aβ-removal activities of these materials in continuous flow using $A\beta_{1-40}$ and $A\beta_{1-42}$ solutions. In these studies, CHA removed >80% of the Aβ after 2 h, whereas HDC removed >90% of the Aβ after 5 h. On the basis of these results, HDC was further evaluated as an Aβ adsorbent.

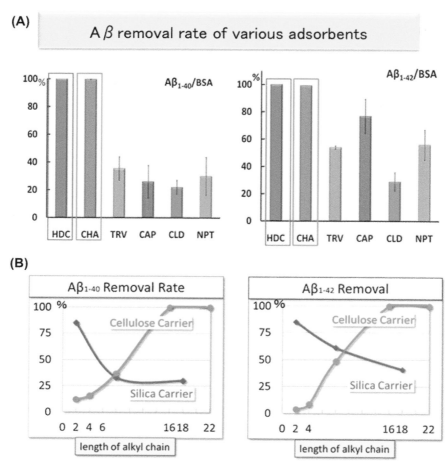

Figure 41.2 Amyloid-β (Aβ) removal properties of various adsorbents. (A) Various medical adsorbents in a 16-h incubation. (B) Adsorbents with various lengths of alkyl chains conjugated to two carrier beads, cellulose and silica. The total incubation time was 1 h. *BSA*, bovine serum albumin; *CAP*, cellulose acetate particles; *CHA*, charcoal; *CLD*, cellulose particles ligated with dextran sulfate; *HDC*, hexadecyl-alkylated cellulose particles; *NPT*, nonwoven polyethylene terephthalate filter; *TRV*, tryptophan-ligated polyvinyl alcohol gel.

Hexadecyl-alkylated cellulose particles effectively removed amyloid-β from human blood

We revealed that HDC are able to remove endogenous Aβ from human blood, as well as synthetic Aβ. Blood samples were collected from hemodialysis patients who underwent blood purification with HDC to remove β2-microglobulin due to carpal tunnel syndrome, a complication associated with hemodialysis. The blood samples were obtained before (pre) and after (post) exposure to the HDC column. The Aβ-removal efficiency, defined as $100 \times (1 - A\beta_{post}/A\beta_{pre})$, of the HDC column was observed to be $51.1\% \pm 6.6\%$ after 1 h of a blood purification session and $46.1\% \pm 6.6\%$ after 4 h for $A\beta_{1-40}$, and $44.9\% \pm 5.0\%$ after 1 h and $38.2\% \pm 5.8\%$ after 4 h for $A\beta_{1-42}$ (Kawaguchi et al., 2013).

Optimization of adsorbent surface properties to improve amyloid-β removal

HDC consist of hydrophobic 16-methylene (C16) ligands attached to cellulose beads. $A\beta$ is also hydrophobic, especially $A\beta_{1-42}$. We sought to optimize the hydrophobic properties of the HDC adsorbent to improve $A\beta$ removal. We tested cellulose and silica beads with different-length alkyl chain ligands as possible adsorbents. With cellulose carriers, longer alkyl chain ligands resulted in higher $A\beta$ removal rates, whereas shorter alkyl chain ligands resulted in higher removal rates with silica beads, as shown in Fig. 41.2B (Kawaguchi et al., 2013). The opposite chain length dependencies of the $A\beta$-removal rates between the cellulose and the silica beads were unexpected. Surface analysis by near-infrared absorption revealed that surface water content is an important factor in $A\beta$ removal (by collaboration with Dr. Takeuchi in Osaka Prefecture University, Kawaguchi et al., 2013). Although the hydrophobicity of the surface is critical for efficient $A\beta$ removal, an appropriate water content is necessary to enable the $A\beta$-containing plasma, a hydrophilic fluid, to contact the surface. Our investigations determined that the best $A\beta$ adsorbent is cellulose beads with C16—C22 alkyl chain ligands.

Hemodialyzers as amyloid-β removal devices in extracorporeal blood amyloid-β removal systems

The most widely used method for blood purification is hemodialysis. Hemodialyzer devices consist of approximately 10,000 hollow fibers with various hydrophilicities (for example, Ronco et al., 2002). Some of the more recently manufactured hemodialyzers have larger pore sizes than older models, to enable 30%—40% of the β2-microglobulin (11,800 Da) to pass through the membranes. The lower molecular weight $A\beta$ (4330 and 4514 Da for $A\beta_{1-40}$ and $A\beta_{1-42}$, respectively) should theoretically pass through the membrane walls from the blood to the dialysate. However, as described later, this is not what occurs.

Hemodialyzers remove blood amyloid-β in hemodialysis sessions

We measured the $A\beta$-removal efficiency of hemodialyzers using hemodialysis samples taken from 57 nondiabetic renal failure patients (ages 59—76 years). The average removal efficiencies were 66.0% and 52.0% at the 1-h time point and 61.1% and 49.2% at the 4-h time point of the hemodialysis sessions for $A\beta_{1-40}$ and $A\beta_{1-42}$, respectively, as shown in Fig. 41.3A (Kato et al., 2012; Kitaguchi et al., 2011). $A\beta_{1-40}$

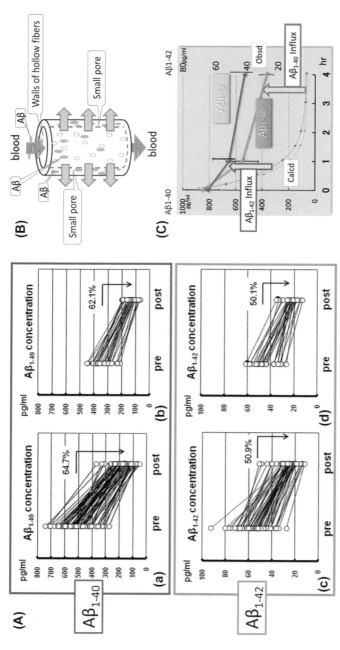

Figure 41.3 Amyloid-β (Aβ) removal with hollow fibers. (A) Aβ removal efficiencies of hemodialyzers during hemodialysis sessions of renal failure patients. Aβ concentrations were measured at pre- (inlet) and post- (outlet) dialyzers at the 1- and 4-h time points of the hemodialysis session. Aβ-removal efficiencies for both Aβ$_{1-40}$ and Aβ$_{1-42}$ were high, approximately 50% or greater. (B) A schematic of the adsorptive filtration system, showing enhanced Aβ adsorption at the inner pores of the hollow fiber walls by enforced pass-through flow. (C) Change in plasma Aβ concentrations in the whole-body circulation during the hemodialysis session. Aβ$_{1-40}$ concentrations are plotted in blue, and Aβ$_{1-42}$ concentrations are plotted in purple. *Calcd*, calculated plasma Aβ concentrations based on a one-compartment model assuming no Aβ influx into the blood; *Obsd*, observed plasma Aβ concentrations. The *arrows* indicate Aβ influx into the blood during the hemodialysis.

exhibited a significantly higher removal efficiency than $A\beta_{1-42}$ at both the 1- and the 4-h time points of each dialysis session ($P < .0001$ for both time points). The sustained removal efficiency during the entire dialysis session suggests the dialyzers had sufficient capacity for $A\beta$ removal during the 4-h treatment. In support of these observations that hemodialyzers remove $A\beta$ from blood, it has been previously reported that $A\beta_{1-42}$ plasma concentrations decrease during hemodialysis (Rubio, Caramelo, Gil, López, & de Yébenes, 2006).

The main mechanism of amyloid-β removal by hemodialyzers is adsorption

The dialyzers used in the hemodialysis sessions described previously contained membranes with large pores that should allow $A\beta$ to pass through the membrane wall to the dialysate. However, very low concentrations of $A\beta$ were detected in the dialysate, which suggests that the main mechanism of $A\beta$ removal during hemodialysis is adsorption, not filtration. This hypothesis was confirmed by in vitro analysis using small fragments of hollow fibers 2−5 mm in length that were incubated in high-concentration solutions of $A\beta$ (40 ng/mL) (Kawaguchi et al., 2016). Filtration does not occur when samples are incubated. The fragments removed >90% of both $A\beta_{1-40}$ and $A\beta_{1-42}$ within 10 min, which strongly suggests adsorption is the major mechanism of $A\beta$ removal during hemodialysis.

Suitable hollow fiber materials for amyloid-β removal

Because $A\beta$ is removed during hemodialysis mainly by adsorption, optimal hemodialyzer membranes for use in $A\beta$ removal were assumed to be hydrophobic rather than hydrophilic. In fact, incubation of $A\beta$ with fragments of hydrophobic hollow fiber materials such as polysulfone (PSf) (for example, Klingel, Ahrenholz, Schwarting, & Röckel, 2002), poly(methyl methacrylate), and polyether polymer alloy removed high concentrations of $A\beta$ over short time periods, whereas hydrophilic materials such as cellulose triacetate and ethylene vinyl alcohol copolymer exhibited lower $A\beta$ removal activities (Kawaguchi et al., 2016). Polyether sulfone had similar $A\beta$ removal activities compared with PSf. Dialyzers must be a little hydrophilic so that blood can easily contact the inner surface of the hollow fibers. Therefore, dialyzers containing materials that are rather hydrophobic but have some hydrophilic character are preferable for $A\beta$ removal.

Other potential methods for use in extracorporeal blood amyloid-β removal systems

We investigated two alternative adsorption systems that could be used in E-BARS to determine if $A\beta$ adsorption could be increased by using hemodialyzers. One alternative adsorption system is an adsorption–accelerating system with low blood flow rates that need not dialysate. Lower blood flow rates of 20 or 50 mL/min resulted in higher Aβ removal efficiencies compared with the higher rate of 200 mL/min, which is the blood flow rate typical of hemodialysis procedures in Japan (Kawaguchi et al., 2016).

Another alternative adsorption system is an adsorptive filtration system (Kitaguchi et al., 2018). In this system, Aβ is adsorbed to the inner pores of the hollow fiber walls, as well as the inside surface of the hollow fibers, by an enforced flow that ensures some of the blood passes through the membrane walls (see Fig. 41.3B). Aβ adsorption was significantly enhanced in this system when approximately 10% of the blood flowed as pass-through flow.

Amyloid-β removal from the blood affects amyloid-β levels in the blood and the brain

The removal of blood amyloid-β during hemodialysis evokes a large influx of amyloid-β into the blood

Dialyzers are effective in removing Aβ from the blood as shown in Fig. 41.3A. Therefore, if no influx of Aβ into the blood is assumed during the hemodialysis procedure, calculations predicted that Aβ concentrations in the blood after a 4-h hemodialysis should have been approximately 10% of the concentrations at the starting point ("Calcd" in Fig. 41.3C, calculated on the basis of a one-compartment model) (Kitaguchi et al., 2011). However, observed Aβ concentrations in the blood ("Obsd" in Fig. 41.3C) were greater than those predicted by the calculations. The differences between the observed and the predicted concentrations were attributed to an influx of Aβ into the blood during the hemodialysis procedure.

The influx of Aβ during a hemodialysis session was estimated to be 32,400 pg/min for $A\beta_{1-40}$ and 3000 pg/min for $A\beta_{1-42}$ based on data collected from 37 nondiabetic hemodialysis patients (Kitaguchi et al., 2011). In a more detailed estimation of Aβ influx during a 4-h hemodialysis in 30 patients, the absolute amounts of 9243 and 719 ng for $A\beta_{1-40}$ and $A\beta_{1-42}$, respectively, were calculated. These values were around five times higher than the level of Aβ in the blood just before hemodialysis, that is, 1952 and 165 ng (Kitaguchi et al., 2015). The amounts of these Aβ influxes were comparable with the total Aβ existing in the Brain (Kitaguchi et al., 2019). A similar influx of Aβ into the blood was also observed during blood Aβ removal in rats using HDC and fragments of hollow fibers (Kitaguchi et al., 2017).

Influxes of amyloid-β into the blood may come from the brain

A likely source of the Aβ that flows into the blood during Aβ removal procedures is the brain (Fig. 41.1). We recently revealed that Aβ accumulation in the brains of a group of patients undergoing hemodialysis was significantly lower than that in age-matched nonhemodialysis controls, as assessed by histopathological studies (Sakai et al., 2016).

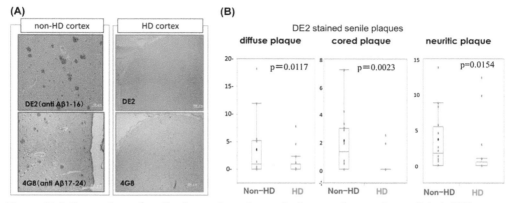

Figure 41.4 Comparison of senile plaques in patients who have undergone hemodialysis (HD) versus those who have not undergone HD. (A) Samples stained with anti-Aβ_{1-16} antibody DE2 (top) and anti-Aβ_{17-24} antibody 4G8 (bottom). (B) The number of diffuse, cored, and neuritic Aβ plaques were significantly lower in HD patients compared with non-HD patients. The senile plaques were investigated in five different fields of 3×10^6 μm^2/field. Non-HD, $n = 16$; HD, $n = 17$.

Three types of senile plaques (cored, diffuse, and neuritic) stained with anti–Aβ antibodies were observed more frequently in nonhemodialysis subjects and were either sparse or not seen at all in hemodialysis patients (Fig. 41.4A). The number of senile plaques stained with the anti-Aβ_{1-16} antibody DE2 in each subject is summarized in Fig. 41.4B for all cortices. The brains of hemodialysis patients exhibited significantly fewer senile plaques than those of nonhemodialysis subjects. This finding was also confirmed for the three plaque types when stained with the anti-Aβ_{17-24} antibody 4G8, that is, diffuse ($P = .0478$), neuritic ($P = .0074$), and cored plaques ($P = .0188$). Further, regarding duration of hemodialysis, patients who underwent hemodialysis for more than 2 years showed significantly fewer neuritic senile plaques with 4G8 compared with nonhemodialysis subjects ($P = .0112$). These histopathological findings suggest that the brain may be one of the sources of the influx of Aβ during the hemodialysis sessions and that repetitive rapid removal of blood Aβ (three times per week) may reduce levels of Aβ in the brain. Removal of Aβ from the brain may be via vessels in the brain, as cerebral amyloid angiopathy is reduced in hemodialysis patients (Sakai et al., unpublished results).

Rat studies demonstrate the influx of amyloid-β from the brain into the blood during blood amyloid-β removal

In studies using rats, we demonstrated that removal of blood Aβ evoked Aβ influx from the brain (cerebrospinal fluid) into the blood (Kawaguchi et al., unpublished results). Because it is difficult to perform extracorporeal blood purifications in transgenic mice because of the small blood volume, we chose to conduct studies of blood Aβ removal procedures in rats. In the rat E-BARS studies, both minicolumns of HDC and fragments of hollow fibers maintained high Aβ-removal efficiencies, around 90%, during a 60-min circulation. Although blood Aβ was effectively removed during the extracorporeal blood purification procedure, the concentration of Aβ in the whole-body blood volume increased 150%−180% compared with blood Aβ concentrations before the purification procedure, for both $Aβ_{1-40}$ and $Aβ_{1-42}$. This increase stopped just after the extracorporeal blood purification procedure was stopped, and the concentrations of blood Aβ levels returned to levels observed before circulation. In contrast to blood Aβ, Aβ levels in cerebrospinal fluid decreased to 60%−70% of preprocedure values for both $Aβ_{1-40}$ and $Aβ_{1-42}$. These findings strongly suggest that blood Aβ removal via extracorporeal blood purification procedures evokes Aβ influxes from the brain into the blood.

Peritoneal dialysis and other methods of removing blood amyloid-β as potential therapeutic options for treating Alzheimer's disease

There are two renal replacement therapies for end-stage renal failure patients: hemodialysis and peritoneal dialysis. In 2017, peritoneal dialysis was shown to be potentially useful in reducing brain Aβ by removal of plasma Aβ in an AD mouse model (Jin et al., 2017). This group also showed that blood purification by parabiosis (the blood circulations were connected between a healthy mouse and an AD model mouse) also reduced brain Aβ in the AD model mouse (Xiang et al., 2015). Further, as described before, plasma exchange with Aβ-free albumin as a supplemental solution was shown to be effective in AD patients in a phase 2 study and is being investigated in a phase 3 trial as of this writing (Boada et al., 2009, 2016).

The aforementioned results strongly suggest that peripheral removal of Aβ, including via E-BARS, may be a useful tool in reducing Aβ in the brain and could be a potentially effective therapy and/or preventive measure for AD (Tholen et al., 2016; Wood, 2017).

Cognitive function is maintained or improved by hemodialysis

As described previously, blood Aβ is effectively removed during hemodialysis, which is hypothesized to evoke the migration of brain Aβ proteins into the blood during hemodialysis. It has been postulated that reductions in Aβ in the brain after hemodialysis may result in improved cognitive function.

Renal failure is known to cause cognitive decline (Yaffe et al., 2010). In our cross-sectional study of approximately 100 subjects, cognitive function as measured by the Mini-Mental State Examination (MMSE) in renal failure patients who did not receive hemodialysis declined as renal function declined. However, once these patients began hemodialysis, MMSE scores began to stabilize, as shown in Fig. 41.5 (Kato et al., 2012). This MMSE change correlated with changes in plasma Aβ concentrations. Prior to the initiation of hemodialysis, plasma concentrations of both $Aβ_{1-40}$ and $Aβ_{1-42}$ increased with a concomitant decline in renal function. However, after initiation of hemodialysis, plasma Aβ concentrations no longer increased.

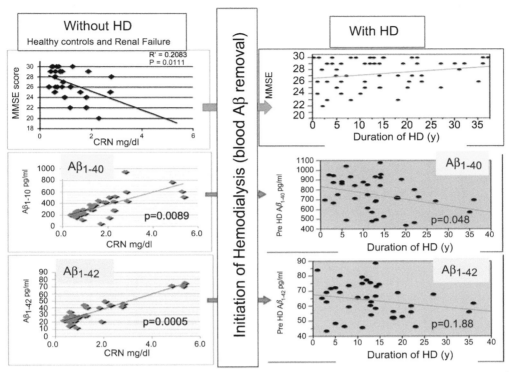

Figure 41.5 Summary of a cross-sectional study of renal failure patients before and after initiation of hemodialysis (HD).The central box indicates initiation of HD. To the left of the central box, data from renal failure patients who had not initiated HD and renal-healthy controls (without HD) are shown. To the right of the central box, data from HD patients (with HD) are shown. Vertical axis: top, the Mini-Mental State Examination (MMSE) score (30 indicates no mistakes); middle, plasma $A\beta_{1-40}$ concentrations; bottom, plasma $A\beta_{1-42}$ concentrations. Plasma for measuring $A\beta$ concentrations after the initiation of HD was sampled at the beginning of each HD session. Horizontal axis: before initiation of HD, plasma creatinine concentrations (CRN), which indicate decline of renal function; after initiation of HD, the duration of HD.

This trend was also confirmed in our prospective studies. In prospective studies with 18- and 36-month intervals, the average MMSE scores did not significantly change, as shown in the top of Fig. 41.6A. However, a patient-by-patient analysis of MMSE changes revealed that most hemodialysis patients maintained or improved their cognitive function, with the exception of patients who had white matter ischemia at baseline (bottom of Fig. 41.6A). This suggests that hemodialysis three times a week, with concomitant $A\beta$ removal from the blood, may have a positive effect on cognitive function but has almost no influence on the cognitive effects of brain ischemia.

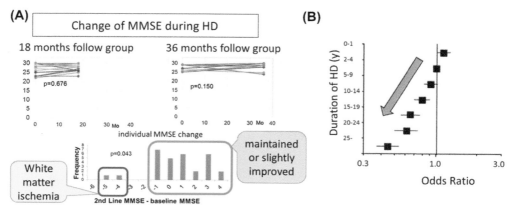

Figure 41.6 Cognitive function and duration of hemodialysis (*HD*). (A) Change in cognitive function of HD patients in prospective studies. Upper left, Mini-Mental State Examination (MMSE) changes over 18 months; upper right, MMSE changes over 36 months; bottom, change in MMSE from baseline for each patient. A change of −1 to 4 is regarded as maintained or improved function. Patients whose MMSE declined by −4 and −5 showed white matter ischemia at baseline by brain computed tomography analysis. (B) Dementia risk (incidence rates of dementia) of nondiabetic HD patients along with duration of HD based on a 1-year prospective study of over 200,000 HD patients in Japan. The incidence of the patients who received HD for 2–4 years is set as the reference (odds ratio is 1.0). Vertical axis, duration of HD. Horizontal axis, odds ratio of dementia risk.

In another 1-year prospective study that utilized a database of over 200,000 hemodialysis patients in Japan, the risk of dementia was analyzed on the basis of onset rates of dementia (Nakai et al., 2018). Patients who had a history of cerebrovascular events were not included. This analysis was done using a multifactor model, including baseline factors from the age-adjusted model plus additional clinical covariates (i.e., gender, comorbidity of diabetes, standardization dialysis dose of urea [Kt/V], albumin level, hemoglobin level, history of myocardial infarction, limb amputation or hip fracture, place of residence, etc.). Lower risk of dementia was observed to be correlated with longer durations of hemodialysis among all subjects. In nondiabetic patients especially, a consistent decrease in dementia risk was correlated with increased duration of hemodialysis (Fig. 41.6B). Diabetic patients showed a weaker association between hemodialysis duration and dementia risk.

Conclusion: pros and cons of extracorporeal blood amyloid-β removal systems

E-BARS have the disadvantage of venipuncture once a week or once a month. However, Aβ proteins are quickly removed from the body by E-BARS (Kitaguchi et al., 2015), as contrasted with anti-Aβ antibody therapy, which sometimes maintains Aβ at high concentrations in the blood for some time (DeMattos et al., 2001; Doody et al., 2014).

On the basis of the findings described in this review, blood Aβ removal via E-BARS or an alternative method may reduce levels of Aβ in the brain and could potentially be a therapeutic option and/or preventive method for AD. Further investigations should be conducted to accelerate future clinical applications.

Key facts about amyloid-β

- Aβ peptides are peptides of approximately 40 amino acids that have strong neurotoxic properties, especially $Aβ_{1-42}$.
- AD does not develop in the absence of Aβ.
- In sporadic AD cases, clearance of Aβ is impaired.
- Administration of Aβ-binding substances, such as anti-Aβ antibodies and gelsolin, reduces brain Aβ, but sometimes causes adverse events.
- Complexes of Aβ and Aβ-binding substances may remain in the blood for a long time.

Summary points

- We hypothesized that E-BARS that rapidly remove Aβ from the blood may accelerate Aβ clearance from the brain.
- Several adsorptive materials, including HDC and hemodialyzers with appropriate membranes, can effectively remove blood Aβ.
- The Aβ concentrations in blood decrease during hemodialysis.
- Removal of blood Aβ evokes a large influx of Aβ into the blood, most likely from the brain.
- Aβ deposition in the brains of patients undergoing hemodialysis has been shown to be significantly lower compared with patients who do not undergo hemodialysis.
- Hemodialysis can lead to stabilized or improved cognitive function, and longer periods of hemodialysis are associated with reduced risk of dementia.
- Blood Aβ removal, including by E-BARS, may reduce brain Aβ and be useful as a therapeutic and/or preventive method for AD.

Acknowledgments

The authors thank the patients who contributed to our study and many collaborators, including Drs. Shigeru Nakai, Masao Kato, Yoshiyuki Hiki, Satoshi Sugiyama, Yukio Yuzawa, Midori Hasegawa, Kazutaka Murakami, Masao Mizuno, Kengo Ito, Takashi Kato, Shinji Ito, Shinji Matsunaga, Hajime Takechi, Toshitaka Nabeshima, Takayoshi Mamiya, Haruyasu Yamaguchi, Hiroshi Kawachi, Yuta Saito, Yukari Murata, Miwa Sakata, Yoshihiro Ota, Naoto Kawamura, and Hirofumi Oka. This work was partly supported by KAKENHI (20509008, 23500531, and 26282126) and the Smoking Research Foundation.

References

Bell, R. D., Sagare, A. P., Friedman, A. E., Bedi, G. S., Holtzman, D. M., Deane, R., et al. (2007). Transport pathways for clearance of human Alzheimer's amyloid beta-peptide and apolipoproteins E and J in the mouse central nervous system. *Journal of Cerebral Blood Flow and Metabolism, 27,* 909–918.

Boada, M., Ortiz, P., Anaya, F., Hernández, I., Muñoz, J., Núñez, L., et al. (2009). Amyloid-targeted therapeutics in Alzheimer's disease: Use of human albumin in plasma exchange as a novel approach for abeta mobilization. *Drug News and Perspectives, 22,* 325–239.

Boada, M., Ramos-Fernández, E., Guivernau, B., Muñoz, F. J., Costa, M., Ortiz, A. M., et al. (2016). Treatment of Alzheimer disease using combination therapy with plasma exchange and haemapheresis with albumin and intravenous immunoglobulin: Rationale and treatment approach of the AMBAR (Alzheimer Management by Albumin Replacement) study. *Neurologia, 31,* 473–481.

Carare, R. O., Bernardes-Silva, M., Newman, T. A., Page, A. M., Nicoll, J. A., Perry, V. H., et al. (2008). Solutes, but not cells, drain from the brain parenchyma along basement membranes of capillaries and arteries: Significance for cerebral amyloid angiopathy and neuroimmunology. *Neuropathology and Applied Neurobiology, 34,* 131–144.

DeMattos, R. B., Bales, K. R., Cummins, D. J., Dodart, J. C., Paul, S. M., & Holtzman, D. M. (2001). Peripheral anti-A beta antibody alters CNS and plasma A beta clearance and decreases brain A beta burden in a mouse model of Alzheimer's disease. *Proceedings of the National Academy of Sciences of the United States of America, 98,* 8850–8855.

DeMattos, R. B., Bales, K. R., Cummins, D. J., Paul, S. M., & Holtzman, D. M. (2002). Brain to plasma amyloid-β efflux: A measure of brain amyloid burden in a mouse model of Alzheimer's disease. *Science, 295,* 2264–2267.

Donahue, J. E., Flaherty, S. L., Johanson, C. E., Duncan, J. A., 3rd, Silverberg, G. D., Miller, M. C., et al. (2006). RAGE, LRP-1, and amyloid-beta protein in Alzheimer's disease. *Acta Neuropathologica, 4,* 405–415.

Doody, R. S., Thomas, R. G., Farlow, M., Iwatsubo, T., Vellas, B., Joffe, S., et al. (2014). Phase 3 trials of solanezumab for mild-to-moderate Alzheimer's disease. *New England Journal of Medicine, 370,* 311–321.

Hung, L. W., Ciccotosto, G. D., Giannakis, E., Tew, D. J., Perez, K., Masters, C. L., et al. (2008). Amyloid-b peptide (Aβ) neurotoxicity is modulated by the rate of peptide aggregation: Aβ dimers and trimers correlate with neurotoxicity. *Journal of Neuroscience, 28,* 11950–11958.

Jin, W.-S., Shen, L.-L., Bu, X.-L., Zhang, W.-W., Chen, S.-H., Huang, Z.-L., et al. (2017). Peritoneal dialysis reduces amyloid-beta plasma levels in humans and attenuates Alzheimer-associated phenotypes in an APP/PS1 mouse model. *Acta Neuropathologica, 134,* 207–220.

Kalaria, R. N., Akinyemi, R., & Ihara, M. (2012). Does vascular pathology contribute to Alzheimer changes? *Journal of Neurological Sciences, 322,* 141–147.

Kato, M., Kawaguchi, K., Nakai, S., Murakami, K., Hori, H., Ohashi, A., et al. (2012). Potential therapeutic system for Alzheimer's disease: Removal of blood abs by hemodialyzers and its effect on the cognitive functions of renal-failure patients. *Journal of Neural Transmission, 119,* 1533–1544.

Kawaguchi, K., Kitaguchi, N., Nakai, S., Murakami, K., Asakura, K., Mutoh, T., et al. (2010). Novel therapeutic approach for Alzheimer's disease by removing amyloid-β protein from the brain with an extracorporeal removal system. *Journal of Artificial Organs, 13,* 31–37.

Kawaguchi, K., Saigusa, A., Yamada, S., Gotoh, T., Nakai, S., Hiki, Y., et al. (2016). Toward the treatment for Alzheimer's disease: Adsorption is primary mechanism of removing amyloid β protein with hollow-fiber dialyzers of the suitable materials, Polysulfone and polymethyl methacrylate. *Journal of Artificial Organs, 19,* 149–158.

Kawaguchi, K., Takeuchi, M., Yamagawa, H., Murakami, K., Nakai, S., Hori, H., et al. (2013). A Potential therapeutic system for Alzheimer's disease using adsorbents with alkyl ligands for removal of blood Amyloid β. *Journal of Artificial Organs, 16,* 211–217.

Kitaguchi, N., Hasegawa, M., Ito, S., Kawaguchi, K., Hiki, Y., Nakai, S., et al. (2015). A prospective study on blood Aβ levels and the cognitive function of patients with hemodialysis: A potential therapeutic strategy for Alzheimer's disease. *Journal of Neural Transmission, 122,* 1593–1607.

Kitaguchi, N., Kawaguchi, K., Nakai, S., Murakami, K., Ito, S., Hoshino, H., et al. (2011). Reduction of Alzheimer's Disease Amyloid-β in plasma by hemodialysis and its relation to cognitive functions. *Blood Purification, 32*, 57—62.

Kitaguchi, N., Kawaguchi, K., Kinomura, J., Hirabayashi, A., Sakata, M., Sakai, K., et al. (2017). Extracorporeal blood Aβ removal system (EBARS) reduced soluble A in the brain by triggering influx into the blood. *Rat studieAlzheimers Dement, 13*(7), P620—P621.

Kitaguchi, N., Kawaguchi, K., Yamazaki, K., Kawachi, H., Sakata, M., Kaneko, M., et al. (2018). Adsorptive filtration systems for effective removal of blood amyloid β: A potential therapy for Alzheimer's disease. *Journal of Artificial Organs, 21*, 220—229.

Kitaguchi, N., Tatebe, H., Sakai, K., Kawaguchi, K., Matsunaga, S., Kitajima, T., et al. (2019). Influx of tau and amyloid-β proteins into the blood during hemodialysis as a therapeutic extracorporeal blood Aβ removal system for Alzheimer's disease. *Journal of Alzheimer's Disease, 69*, 687—707.

Klingel, R., Ahrenholz, P., Schwarting, A., & Röckel, A. (2002). Enhanced functional performance characteristics of a new polysulfone membrane for high-flux hemodialysis. *Blood Purification, 20*, 325—333.

Kuo, Y. M., Emmerling, M. R., Vigo-Pelfrey, C., Kasunic, T. C., Kirkpatrick, J. B., Murdoch, G. H., et al. (1996). Water-soluble Abeta (N-40, N-42) oligomers in normal and Alzheimer disease brains. *Journal of Biological Chemistry, 271*, 4077—4081.

Maki, T., Okamoto, Y., Carare, R. O., Hase, Y., Hattori, Y., Hawkes, C. A., et al. (2014). Phosphodiesterase III inhibitor promotes drainage of cerebrovascular β-amyloid. *Annals of Clinical and Translation Neurology, 1*, 519—533.

Matsuoka, Y., Saito, M., LaFrancois, J., Saito, M., Gaynor, K., Olm, V., et al. (2003). Novel therapeutic approach for the treatment of Alzheimer's disease by peripheral administration of agents with an affinity to β-amyloid. *Journal of Neuroscience, 23*, 29—33.

Mawuenyega, K. G., Sigurdson, W., Ovod, V., Munsell, L., Kasten, T., Morris, J. C., et al. (2010). Decreased clearance of CNS beta-amyloid in Alzheimer's disease. *Science, 330*, 1774.

Morris, A. W., Carare, R. O., Schreiber, S., & Hawkes, C. A. (2014). The cerebrovascular basement membrane: Role in the clearance of β-amyloid and cerebral amyloid angiopathy. *Frontiers in Aging Neuroscience, 6*(1—9), 251.

Nakai, S., Wakai, K., Kanda, E., Kawaguchi, K., Sakai, K., & Kitaguchi, N. (2018). Is hemodialysis itself a risk factor for dementia? An analysis of nationwide registry data of patients on maintenance hemodialysis in Japan. *Renal Replacement Therapy*. https://doi.org/10.1186/s41100-018-0154-y.

Polanco, J. C., Li, C., Bodea, L. G., Martinez-Marmol, R., Meunier, F. A., & Götz, J. (2018). Amyloid-β and tau complexity - towards improved biomarkers and targeted therapies. *Nature Reviews Neurology, 14*, 22—39.

Ronco, C., Bowry, S. K., Brendolan, A., Crepaldi, C., Soffiati, G., Fortunato, A., et al. (2002). Hemodialyzer: From macro-design to membrane nanostructure; the case of the FX-class of hemodialyzers. *Kidney International Supplements, 80*, 126—142.

Rubio, I., Caramelo, C., Gil, A., López, M. D., & de Yébenes, J. G. (2006). Plasma amyloid-beta, Abeta1-42, load is reduced by haemodialysis. *Journal of Alzheimer's Disease, 10*, 439—443.

Sakai, K., Senda, T., Hata, R., Kuroda, M., Hasegawa, M., Kato, M., et al. (2016). Patients that have undergone hemodialysis exhibit lower amyloid deposition in the brain: Evidence supporting a therapeutic strategy for Alzheimer's disease by removal of blood amyloid. *Journal of Alzheimer's Disease, 51*, 997—1002.

Selkoe, D. J. (2001). Alzheimer's disease: Genes, proteins, and therapy. *Physiological Reviews, 81*, 741—766.

Sevigny, J., Chiao, P., Bussière, T., Weinreb, P. H., Williams, L., Maier, M., et al. (2016). The antibody aducanumab reduces Aβ plaques in Alzheimer's disease. *Nature, 537*, 50—56.

Silverberg, G. D., Miller, M. C., Messier, A. A., Majmudar, S., Machan, J. T., Donahue, J. E., et al. (2010). Amyloid deposition and influx transporter expression at the blood-brain barrier increase in normal aging. *Journal of Neuropathology and Experimental Neurology, 69*, 98—108.

Suzuki, K., Shimazaki, M., & Kutsuki, H. (2003). Beta2-microglobulin-selective adsorbent column (Lixelle) for the treatment of dialysis-related amyloidosis. *Therapeutic Apheresis and Dialysis, 7*, 104—107.

Tholen, S., Schmaderer, C., Chmielewski, S., Förstl, H., Heemann, U., Baumann, M., et al. (2016). Reduction of amyloid-β plasma levels by hemodialysis: An anti-amyloid treatment strategy? *Journal of Alzheimer's Disease, 50*, 791–796.

Walsh, D. M., Klyubin, I., Fadeeva, J. V., Cullen, W. K., Anwyl, R., Wolfe, M. S., et al. (2002). Naturally secreted oligomers of amyloid beta protein potently inhibit hippocampal long-term potentiation in vivo. *Nature, 416*, 535–539.

Weller, R. O., Djuanda, E., Yow, H. Y., & Carare, R. O. (2009). Lymphatic drainage of the brain and the pathophysiology of neurological disease. *Acta Neuropathologica, 117*, 1–14.

Wood, H. (2017). Peripheral Aβ clearance — a therapeutic strategy for AD? *Nature Reviews Neurology, 13*, 386.

Xiang, Y., Bu, X.-L., Liu, Y.-H., Zhu, C., Shen, L.-L., Jiao, S.-S., et al. (2015). Physiological amyloid-beta clearance in the periphery and its therapeutic potential for Alzheimer's disease. *Acta Neuropathologica, 130*, 487–499.

Yaffe, K., Ackerson, L., Kurella Tamura, M., Le Blanc, P., Kusek, J. W., Sehgal, A. R., et al. (2010). Chronic kidney disease and congnitive function in older Adults: Findings from the Chronic renal insufficiency cohort cognitive study. *Journal of the American Geriatrics Society, 58*, 338–345.

Zlokovic, B. V. (2011). Neurovascular pathways to neurodegeneration in Alzheimer's disease and other disorders. *Nature Reviews Neuroscience, 12*, 723–738.

PART IV

Non-pharmacological treatments and procedures

CHAPTER 42

Caring for people with dementia in the acute hospital

Robert Briggs, Paul Claffey, Sean P. Kennelly

Centre for Ageing, Neurosciences and the Humanities, Tallaght Hospital, Dublin, Ireland

List of abbreviations

CAM-ICU Confusion assessment method for the intensive care unit
CGA comprehensive geriatric assessment
ED emergency department
IQCODE Informant questionnaire on cognitive decline in the Elderly
MMSE Mini-mental state examination
POPS proactive care of older patients undergoing surgery
PwD people with dementia

Mini-dictionary of terms

Acute hospital The part of the hospital where the patient receives short-term urgent care for a medical illness or injury. It generally refers to the hospital wards and can include medical, surgical, orthopedic wards, etc. The point of entry to the acute hospital is usually via the ED.

Admission avoidance Medical care aimed at preventing admission to the hospital, while still managing the patient's medical illness or injury in a safe, comprehensive manner. This often involves some form of ambulatory care.

Ambulatory care Medical care delivered in an outpatient setting, i.e., not in the acute hospital. It may include outpatient clinics, day hospitals, or care provided in the patient's own home.

Comprehensive geriatric assessment A "multidimensional interdisciplinary diagnostic process focused on determining a frail elderly person's medical, psychological and functional capability in order to develop a coordinated and integrated plan for treatment and long term follow up."

Delirium An acute, fluctuating global cognitive disorder, characterized by inattention and impaired level of consciousness. It is more likely to occur in PwD, especially in the context of acute illness. It is independently associated with poorer outcomes and is also a potent risk factor for future cognitive decline.

Introduction

In 2010 it was estimated that over 35 million people worldwide had dementia, with this figure expected to almost double to 65 million by 2030 (Prince et al., 2013).

People with dementia (PwD) have significantly higher rates of complex medical comorbidity and are more likely to require treatment for acute medical illness. Currently, acute hospital settings such as the emergency department (ED) remain the most frequent entry point to secondary care for acutely unwell PwD, and given these expected

Diagnosis and Management in Dementia
ISBN 978-0-12-815854-8, https://doi.org/10.1016/B978-0-12-815854-8.00042-2

demographic changes it is likely that the coming decades will see increasing numbers of PwD requiring acute medical care.

The acute hospital is currently a challenging environment in which to care for PwD. The vast majority of acute hospitals do not have a specific dementia care pathway or specialist dementia wards (Timmons et al., 2016). Most PwD are regularly treated by physicians who lack confidence in dealing with their complexity (Helm, Balzer, Behncke, Eisemann, & Köpke, 2018) and are cared for by nurses who rate their dementia-specific training as inadequate (Tropea, LoGiudice, Liew, Roberts, & Brand, 2017). Postadmission outcomes are generally poor. Within 6 months of an acute hospital admission, two-thirds of PwD die, are readmitted, or are transferred to a nursing home (Bradshaw et al., 2013).

While this projected increase in the number of PwD presenting to acute care is therefore challenging, it also provides an opportunity to reconfigure acute care so that it caters properly to PwD. The focus of this chapter therefore is to outline important aspects of acute dementia care. We will discuss current approaches to and evidence for identification of PwD in the acute setting, management of delirium, comprehensive geriatric assessment (CGA), dementia-friendly hospital design, and alternative care pathways for PwD, which may help prevent avoidable admission. We begin by discussing the prevalence of dementia within the acute hospital.

Dementia in the acute hospital

Overall, acute hospital service activity currently attributable to dementia care is considerable. PwD present significantly more frequently to the ED than those without dementia, and are more likely to be admitted to the acute hospital after presenting to the ED (LaMantia, Stump, Messina, Miller, & Callahan, 2016).

It is estimated that 2% of all acute hospital inpatient episodes and 10% of hospital bed days involve provision of care to a patient with an established diagnosis of dementia, i.e., not including those with undetected cognitive impairment (Briggs et al., 2016). Studies based in acute hospitals have demonstrated a prevalence of dementia ranging from 21% to 42% (see Table 42.1).

As well as presenting more frequently to the acute setting, PwD also present differently and require expert, nuanced care. For example, the prevalence of dysphagia in PwD is markedly higher (Smith, Kindell, Baldwin, Waterman, & Makin, 2009), increasing the likelihood of aspiration pneumonia and requirement for alternative antibiotic regimes and treatment in the context of respiratory tract infection. Furthermore, swallow can often deteriorate further in the setting of acute illness and should therefore be monitored and assessed by physicians, nurses, and speech and language therapists with sufficient expertise.

Further complexities of caring for PwD in the acute hospital include the higher rates of delirium, medical multimorbidities, and premorbid frailty (Boltz, Lee,

Table 42.1 Studies on prevalence of dementia in the acute hospital.

Author, year	Population/setting	Assessment	Prevalence(%)
Sampson, Blanchard, Jones, Tookman, and King (2009)	617 people ages >70 years, undergoing medical admission to an acute hospital in London, UK	Structured clinical assessment, based on DSM-IV criteria	42
Travers, Byrne, Pachana, Klein, and Gray (2013)	493 people ages ≥70 years, admitted to general medical, general surgical, and orthopedic wards, in four hospitals in Queensland, Australia	MMSE and IQCODE	21
Timmons et al. (2015)	606 people ages ≥70 years, admitted to four acute hospitals in Cork, Ireland	MMSE and IQCODE	25
Briggs et al. (2017)	190 people ages ≥70 years, presenting to acute medical unit in a single acute hospital in Dublin, Ireland	MMSE and AD8	38

Sample of studies estimating prevalence of dementia in acute hospitals. *AD8*, AD8 Dementia Screening Interview; *DSM-IV*, Diagnostic and Statistical Manual of Mental Disorders, fourth edition; *IQCODE*, Informant Questionnaire on Cognitive Decline in the Elderly; *MMSE*, Mini-Mental State Examination.

Chippendale, & Trotta, 2018). PwD who undergo an unscheduled acute hospital admission are also significantly more likely to experience further functional decline during the admission, increasing the risk of nursing home admission, and highlighting the need for coordinated multidisciplinary care (Hartley et al., 2017).

During hospital admission, PwD also have higher rates of preventable complications, such as pressure ulcers, hospital-acquired infections, and metabolic derangements (Bail et al., 2015). It is not surprising, therefore, that admission of PwD is associated with higher cost of care compared with those without dementia (Bail et al., 2015). Furthermore, older people with cognitive impairment have a greater than two-fold increase in length of stay, as well as a 50% increased risk of mortality at 30 days compared with those with intact cognition (Reynish et al., 2017).

Providing high-quality care to PwD is therefore crucial. To achieve this, health care systems should be designed to reliably identify people with cognitive impairment and stream these patients toward appropriate expertize and care pathways within the acute setting.

Identification of cognitive impairment

An initial barrier in caring for patients with dementia in the acute hospital is underdetection of cognitive decline and dementia in acutely unwell older people.

Given the high prevalence of undetected cognitive impairment in the community, it is not surprising that many PwD present to the acute hospital without a preexisting diagnosis (Boustani et al., 2010). In a study based in an acute medical assessment unit, over one-third of patients age 70 years or more met criteria for dementia, yet only one-third of this group had a documented diagnosis prior to admission (Briggs et al., 2017). Furthermore, a UK study demonstrated that almost 40% of patients presenting acutely with delirium had hitherto undiagnosed cognitive impairment (Jackson, MacLullich, Gladman, Lord, & Sheehan, 2016).

Several studies have shown that undiagnosed dementia at the point of admission tends to remain undiagnosed throughout the acute admission (Hustey, Meldon, Smith, & Lex, 2003). There are several potential reasons for this. The emergency setting is not an ideal environment for cognitive assessment due to noise and space constraints, as well as factors such as the burden of the acute illness that has prompted presentation in the first instance. Furthermore, while physicians in the acute setting recognize the importance of screening for cognitive impairment, there is a lack of consensus as to who should actually carry this out (Kennelly et al., 2013). In some cases, there may be a reluctance on the part of the physician, even those with expertize in geriatric medicine, to diagnose dementia for fear of issues such as stigma, lack of confidence in diagnosing dementia, or paternalism (Russ et al., 2012).

Compliance with cognitive assessment in older people admitted to hospitals is therefore generally low (Timmons et al., 2016). Most hospitals do not have a specific screening strategy to detect cognitive impairment in acutely unwell older adults (Timmons et al., 2016), although the Department of Health in the United Kingdom has recently introduced a directive that all older people with unplanned hospital admission must be screened for dementia during the admission (Burn, Fleming, Brayne, Fox, & Bunn, 2018).

Key to any screening strategy for cognitive impairment is selecting an appropriate, accessible assessment tool. The 4AT is a short, reliable tool that can be used to detect cognitive impairment and delirium. It takes less than 2 min to administer, so is suitable for use at the point of entry to the acute hospital and has been validated within an older ED and acute hospital population (O'Sullivan, O'Regan, & Timmons, 2016; Bellelli et al., 2014). Other well-validated tools suitable for the acute setting include the Short Blessed Test and the Six-Item Cognitive Impairment Test (Carpenter et al., 2011; Timmons et al., 2015).

Considering the impact undiagnosed dementia has on acute hospital outcomes in both the short and the longer term, identification of undetected cognitive impairment or dementia should be a priority in the acute setting (Fogg, Meredith, Bridges, Gould, & Griffiths, 2017). This is even more important when we reflect on the fact that almost all patients with dementia wish to be fully informed of their diagnosis (Briggs & O'Neill, 2016) and that disclosing the diagnosis is a prerequisite for accessing specialist care, as well as time to plan for the future.

Impact of hospitalization on cognition

While incipient cognitive decline can often herald a hospital admission, and may even represent the reason for presentation to the acute hospital environment in the first place, there is growing evidence that admission to the acute hospital can have a significant independent adverse effect on cognitive status (Mathews, Arnold, & Epperson, 2014).

A longitudinal study of over 1800 older people demonstrated that after a hospital admission, there was a 1.7- and 3.3-fold increase in the trajectory of decline in executive function and episodic memory, respectively, even after controlling for acute illness severity and prehospital cognitive decline (Wilson et al., 2012). Hospitalization with a critical illness has also been shown to double the risk of dementia during a mean follow-up period of 6 years (Ehlenbach et al., 2010).

When counseling patients regarding future cognitive trajectory, it is therefore important to reference periods of acute illness, as they are often drivers of a step down in cognitive status (see Fig. 42.1).

Possible underlying mechanisms for this association include delirium, medications, depression, and stress. Another important factor in posthospitalization cognitive decline is surgery. Compared with those with no hospitalization, community-dwelling older people who underwent hospitalization for surgery within the prior 12 months performed significantly worse in immediate and delayed recall tasks (O' Brien, O' Leary, Scarlett, O' Hare, & Kenny, 2018). From experience, the risk of perioperative cognitive decline is often not communicated effectively to the patient preoperatively, despite affecting a significant proportion of older surgical patients (Tsai, Sands, & Leung, 2010). There is growing evidence, however, that specific geriatric surgical intervention teams, such as the proactive care of older patients undergoing surgery (POPS) service in the UK, which employ CGA methodology, and intervention can improve outcomes such as survival, length of stay, and dependency levels (Partridge et al., 2017). As yet, however, there is little evidence for the impact of these interventions on cognitive status.

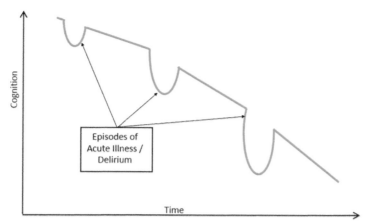

Figure 42.1 *Schema of cognitive trajectory and acute illness in dementia.* A schema showing the impact of episodes of acute illness or delirium on the cognitive trajectory over time is shown.

Delirium

Delirium, an acute, fluctuating global cognitive disorder, characterized by inattention and impaired level of consciousness, is inextricably linked with dementia in the acute setting. PwD are four times more likely to develop delirium during acute illness (Kennedy et al., 2014), and delirium is independently associated with longer length of stay, decline in functional abilities, and higher mortality risk (Inouye, Westendorp, & Saczynski, 2014).

Delirium is also a potent risk factor for future cognitive decline, conferring an almost 9-fold increased risk of incident dementia (Davis et al., 2012). In patients with established Alzheimer's disease, an episode of delirium is associated with a significant acceleration in trajectory of cognitive decline (Fong et al., 2009), with one study demonstrating that individuals with delirium and neuropathological features of dementia declined cognitively at a rate of 0.72 Mini-Mental State Examination points per year faster than age-, sex-, and education-matched controls (Davis et al., 2017). Delirium is also a sign of underlying cognitive vulnerability, and should therefore represent a prompt to evaluate baseline cognitive status once the acute symptoms and illness have resolved, particularly in those with no preexisting diagnosis of dementia.

About 30%—40% of cases of delirium are readily preventable (Inouye et al., 2014), and formal preventative strategies should therefore be a key component in the acute care of all older people, particularly PwD. Further risk factors for incident delirium include higher illness severity, high-risk medications, visual impairment, urinary catheterization, longer length of stay, and electrolyte imbalance, and when PwD are admitted to the acute hospital these factors should be addressed as a priority (Ahmed, Leurent, & Sampson, 2014). Because of these multiple risk factors, and the fact that isolating one cause for delirium in an older patient is often not possible, multicomponent, nonpharmacological interventions represent the most appropriate strategy for delirium prevention, and there is robust evidence that they are effective (Siddiqi et al., 2016). One such intervention, the Hospital Elder Life Program, focusing on six key factors in delirium (cognitive impairment, sleep deprivation, immobility, visual impairment, hearing impairment, and dehydration), significantly reduced delirium incidence, as well as the duration of delirium in those affected (Inouye et al., 1999).

As is the case for dementia, cases of delirium are often undetected in the acute setting, with studies showing that almost half of delirious older medical inpatients are undiagnosed, with even higher rates of missed diagnoses when assessed in the ED (Ryan et al., 2013). Patients with the hypoactive subtype are more likely to be undiagnosed, as symptoms can often be subtle, again highlighting the importance of both the collateral history and standardized cognitive screening in acutely unwell older people. One of the benefits of the 4AT cognitive assessment is that it can be used to detect both delirium and dementia, while the Confusion Assessment Method for the Intensive Care Unit is a well-validated, rapid tool to identify delirium specifically (Wei, Fearing, Sternberg, & Inouye, 2008). Tools such as this take under 5 min to administer on average and should be integrated as part of routine care.

Once delirium is diagnosed, it is important to act quickly, as longer duration of delirium is associated with poorer outcomes (Lee et al., 2018). Delirium should never be a diagnosis in isolation, however, and while treatment is based around similar general multimodal principles of delirium prevention, identifying and treating potential specific triggers, such as infection or culprit medications, is vital. At present, there is no convincing evidence that pharmacological treatment is effective, in either the treatment or the prevention of delirium, and may in fact prolong delirium duration.

Comprehensive geriatric assessment

CGA is the cornerstone of acute medical care of older adults, and any unscheduled presentation of an older person to the acute hospital should prompt a CGA, or at least an assessment of that individual's requirement for a CGA. This is particularly true in the case of PwD, as given the poor outcomes associated with unscheduled attendance in this cohort, they are most likely to derive significant benefit from CGA (see Fig. 42.2).

Key aspects of CGA in PwD include assessments of swallow, gait, and balance; functional status; nutrition; pain; and social circumstances. This requires coordinated multidisciplinary care and team work, and there is some evidence that this care should be delivered in a specialist ward environment (Briggs et al., 2017).

The importance of the collateral or informant history in acutely unwell PwD is also often forgotten. As well as confirming important details around the presenting complaint, it also allows for information to be gleaned regarding premorbid cognitive and functional

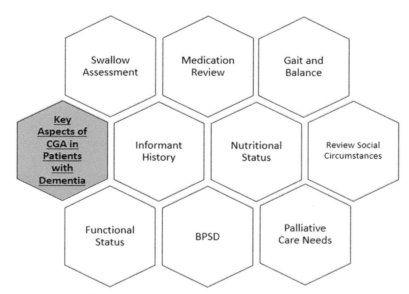

Figure 42.2 *Key aspects of CGA in people with dementia in the acute hospital. BPSD,* behavioral and psychological symptoms in dementia; *CGA,* comprehensive geriatric assessment.

status and social circumstances, as well as offering the carer or family member an opportunity to voice concerns (Briggs & O'Neill, 2016). In the vast majority of cases, consent should be sought from the patient prior to obtaining an informant account, as well as to whether they are agreeable with the sharing of their medical information with the informant (Briggs & O'Neill, 2016).

Polypharmacy is strongly associated with dementia, and PwD are significantly more likely to be prescribed inappropriate medications (Kristensen et al., 2018). When PwD present to acute care, medications should therefore be reviewed and rationalized where appropriate, with particular focus on medicines that adversely affect cognition, such as anticholinergics and antipsychotics.

PwD are more likely to undergo functional decline in the setting of an acute illness, and another key aspect of CGA is identifying any loss of function and planning appropriate individualized rehabilitation strategies to ameliorate this. While cognitive impairment can complicate physical rehabilitation, for example, by reducing carryover, a diagnosis of dementia should not preclude an individual from multidisciplinary rehabilitation. Studies have demonstrated that appropriately selected PwD have similar rates of functional recovery after a period of multidisciplinary rehabilitation following acute illness compared with peers without dementia (Muir-Hunter, Fat, Mackenzie, Wells, & Montero-Odasso, 2016).

An unscheduled admission affords the opportunity to reinforce education around proven strategies to reduce the risk of further cognitive decline, including regular exercise, management of vascular risk factors, moderating alcohol intake, and healthy diet (Briggs, McHale, Fitzhenry, O'Neill, & Kennelly, 2018).

CGA should also inform the need for additional input from specialties such as palliative care or late life psychiatry.

The dementia-friendly hospital

From point of admission to discharge, the physical environment of the acute hospital is also often not suited to caring for PwD.

Given the noise, persistent bright lighting, and limited space, it is not surprising that longer time spent in the ED confers a higher risk of delirium. Royal College of Emergency Medicine guidelines advise that all EDs should provide appropriate dementia-friendly cubicles for assessment and adequate signage and directions to toilets, as well as specific dementia training to all staff members (Royal College of Emergency Medicine 2017). Furthermore, dedicated geriatric EDs, generally located alongside traditional EDs, with physical modifications to lighting, flooring, and visual aids, as well as specific care pathways and staff competencies, are also becoming more prevalent (Hwang & Morrison, 2007).

Once admitted from the ED, the wards of the acute hospital are generally designed to manage acute conditions, with a focus on monitoring, security, and infection control. National audits have demonstrated that most acute hospital wards do not have environmental cues to help PwD to orient themselves, have plain or subtly patterned flooring, have no signs on bathroom doors, and do not label items such as bins or hand dryers (Timmons et al., 2016). Further adaptations that may enable activities of daily living, promote social interaction, and reduce agitation and distress include domestic scale seating and dining areas, availability of personal and self-care items, large-face clocks, and artwork. Most adaptations involved in making a ward dementia friendly are relatively simple and low cost, but there is some initial evidence that they may reduce the use of antipsychotic medications and falls among PwD (Handley, Bunn, & Goodman, 2017).

In addition to a dementia-friendly physical hospital environment, the acute hospital should also aim to cater to the psychological and social needs of PwD. Opportunities for social interaction and engagement should be provided for PwD who are well enough to take part in such activities. For example, music therapy is a proven nonpharmacological strategy to reduce the risk of agitation in PwD in the acute setting (Cheong et al., 2016), while it is also well recognized that social activity has a positive effect on cognitive trajectory (Krueger et al., 2009).

Discharge planning

The importance of coordinated discharge planning is reflected by the fact that 40% of PwD discharged from the acute hospital will be readmitted within the next 3 months

(Draper, Karmel, Gibson, Peut, & Anderson, 2011), and even more so as the quality of discharge planning for unwell older people correlates closely with readmission rates (Bauer, Fitzgerald, Haesler, & Manfrin, 2009).

Carers should be involved in the process where appropriate, although studies suggest that they often feel that communication around discharge planning is poor (Mockford, 2015). Input from social work is often invaluable during this process, especially as a liaison between patient and carers and the medical team, as often a balance must be struck between a methodically planned discharge and ensuring that PwD are not exposed to the risk of hospitalization, such as hospital-acquired infection or deconditioning, any longer than necessary. It is vitally important, however, that any services, such as home care, that are deemed necessary are in place before considering discharge to the community.

After a prolonged admission, it may also be helpful to plan a phased discharge whereby a patient will stay at home overnight initially for one night while retaining his or her bed within the acute hospital, allowing time to address any initial issues around discharge home. It may also be helpful to liaize directly with the relevant public health nurse prior to discharge.

Decisions around discharge can sometimes be fraught, especially in the case of a patient with dementia who wishes to return home despite concerns around the safety of doing so. Care planning meetings can help in this regard but should always involve the patient, as well as giving due priority to his or her wishes. While we obviously have a duty of care to PwD we should always be wary of taking a paternalistic approach in these situations, and aim to balance autonomy with welfare, as generally older people themselves prioritize staying in their own home over minimizing risk of harm (Fahey, Ní Chaoimh, Mulkerrin, Mulkerrin, & O'Keeffe, 2017).

Alternative models of care

In many cases, acutely unwell PwD will require admission to the acute hospital, and admission avoidance strategies would not be in their best interests. However, in some instances admission may be preventable, and given the adverse impact hospital admission can have, exploring different models of care may therefore be beneficial.

Being diagnosed with dementia increases the risk of a potentially preventable hospital admission by almost 80% (Phelan, Borson, Grothaus, Balch, & Larson, 2012). These admissions usually involve ambulatory care—sensitive conditions, such as falls or infections, which could be managed outside of the acute setting if a specific care pathway or model was available.

While many care models focusing on admission avoidance apply to older people in general, given high ED and acute hospital usage, cohorts such as those with dementia and frailty are of particular interest. Models include placing geriatric expertize at the "front door" of the acute hospital, often termed interface geriatrics, with appropriate links to community services or day hospitals to facilitate community-based care (Conroy, Ferguson, Woodard, & Banerjee, 2010) or provision of "hospital at home" for appropriately identified older patients (Shepperd et al., 2008). An alternative model involves CGA of community-based older people deemed to be at risk of future admission or functional decline to reduce future health care use (Edmans, Bradshaw, Franklin, Gladman, & Conroy, 2013).

As yet, however, there is little evidence that interventions such as this are effective in a specific dementia cohort in terms of long-term outcomes. It is likely that alternative models of care such as these will continue to be of interest to policy makers given their potential to relieve pressure on the acute hospital system; however, patient-centered decisions regarding admission should always be at the center of these models, taking into account the risks and benefits of acute hospital admission.

Conclusions

The acute hospital setting presents an extremely challenging environment for a person with dementia and for those providing care to them. It is loud, it frequently involves moving to multiple areas and engaging with multiple individuals, and care commonly involves complicated and uncomfortable procedures. These effects are compounded by the fact that many people with cognitive impairment present to acute services without a formal diagnosis, so no actions are taken to mitigate the potential distress. These limitations have a significant impact on the care and outcomes of hospitalized PwD, but also a broader health care systemic impact with regard to service utilization, cost, and efficiency, given the increased numbers of people living to older age and the amount who live with dementia.

Underlying cognitive impairment has a profound effect on in-hospital complication rates, cost of care, and outcomes such as discharge destination and mortality. Despite this, a substantial proportion of cases of dementia are not identified throughout a hospital admission. This is concerning, especially when we reflect on the fact that the average medical inpatient has contact with over 17 health professionals during his or her time in hospital (Whitt, Harvey, McLeod, & Child, 2007), and that PwD are likely to have a more prolonged length of stay. It is imperative that cognitive impairment is detected in acutely unwell older persons, and steps such as mandatory cognitive screening for all unscheduled admissions of older people will certainly help in this regard.

While PwD are already proportionately one of the key groups served by the acute hospital, demographic changes dictate that coming decades will see an increase in the number of PwD presenting to acute medical services and an enhanced need for multidisciplinary, holistic care focusing on key strategies such as access to CGA, prevention and management of delirium, and timely discharge planning.

PwD regularly present with ambulatory care—sensitive conditions such as infections or noninjurious falls, and the acute hospital must be equipped with appropriate care pathways and community links to manage these conditions while avoiding hospital admission. However, while often attractive to policy makers, admission avoidance will not represent a cure-all for the current issues surrounding acute medical care of older people, particularly those with dementia.

PwD will continue to require admission to the acute hospital and, in fact, in some instances may actually benefit more from in-hospital treatment for conditions such as sepsis, because they are more vulnerable to the consequences of undertreated illness. Making the acute hospital a safer place for PwD must therefore remain an important aim, with a focus on a physical dementia-friendly environment, as well as appropriate dementia-specific training for all staff.

As it stands, acute hospital admission can often signal an inexorable decline in functional and cognitive status in a patient with dementia. This was also previously the case for acute stroke. However, since 2000, reorganization of stroke care, based primarily around systematic change and cohorting of expertise, has led to a reduction in mortality rates by up to 40%. While there are obvious important differences between these two age-related conditions, similar principles should apply, and we should aim to change knowledge and attitudes so that an acute hospital admission for a patient with dementia is seen as a timely intervention aimed at managing an acute illness and at the same time focusing on maintaining functional status, independence, and quality of life.

Key facts of acute dementia care

- The point prevalence of dementia among older people (generally ≥ 70 years) in the acute hospital ranges from 20% to 40% depending on assessment type and wards surveyed.
- Most acute hospitals do not have a specific dementia care pathway.
- In a study based in an acute medical assessment unit, over one-third of patients age 70 years or more met criteria for dementia, yet only one-third of this group had a documented diagnosis prior to admission.
- Within 6 months of an acute hospital admission, two-thirds of PwD die, are readmitted, or are transferred to a nursing home.
- Hospitalization with a critical illness has also been shown to double the risk of dementia during a mean follow-up period of 6 years.
- Delirium is a potent risk factor for future cognitive decline, conferring an almost ninefold increased risk of incident dementia, but up to 40% of cases are preventable.
- Forty percent of patients with dementia discharged from the acute hospital will be readmitted within the next 3 months.

Summary points

- Overall, acute hospital service activity currently attributable to caring for PwD is already considerable, and this is likely to increase significantly in coming decades.
- Unscheduled acute hospital admissions are associated with high in-hospital and 3- to 6- month mortality rates, longer length of stay, and significantly increased risk of discharge to a nursing home.
- Despite this, a substantial proportion of cases of dementia are not identified throughout a hospital admission, and formalized cognitive screening strategies will help in this regard.
- Delirium is inextricably linked with dementia in acute hospitals and is a further independent risk factor for functional decline, mortality, and subsequent accelerated cognitive decline.
- Multicomponent, nonpharmacological interventions represent the most appropriate strategy for delirium prevention, and there is robust evidence that they are effective and cost efficient.
- CGA should be provided to all PwD in the acute hospital.
- Key aspects of CGA in the PwD include assessments of swallow, gait and balance, medications, functional status, nutrition, pain, and social circumstances.
- The importance of coordinated discharge planning is reflected by high rates of representation and readmission among PwD.
- PwD regularly present with ambulatory care—sensitive conditions and the acute hospital must be equipped with appropriate care pathways and community links to manage these conditions while avoiding hospital admission.
- Adopting a "dementia-friendly" hospital design should also be a priority.

References

Ahmed, S., Leurent, B., & Sampson, E. L. (2014). Risk factors for incident delirium among older people in acute hospital medical units: A systematic review and meta-analysis. *Age and Ageing, 43*(3), 326—333.

Bail, K., Goss, J., Draper, B., Berry, H., Karmel, R., & Gibson, D. (2015). The cost of hospital-acquired complications for older people with and without dementia; a retrospective cohort study. *BMC Health Services Research, 15*, 91.

Bauer, M., Fitzgerald, L., Haesler, E., & Manfrin, M. (2009). Hospital discharge planning for frail older people and their family. Are we delivering best practice? A review of the evidence. *Journal of Clinical Nursing, 18*(18), 2539—2546.

Bellelli, G., Morandi, A., Davis, D. H., Mazzola, P., Turco, R., Gentile, S., et al. (2014). Validation of the 4AT, a new instrument for rapid delirium screening: A study in 234 hospitalised older people. *Age and Ageing, 43*(4), 496—502.

Boltz, M., Lee, K. H. H., Chippendale, T., & Trotta, R. L. (2018). Pre-admission functional decline in hospitalized persons with dementia: The influence of family caregiver factors. *Archives of Gerontology and Geriatrics, 74*, 49—54.

Boustani, M., Baker, M. S. S., Campbell, N., Munger, S., Hui, S. L., Castelluccio, P., et al. (2010). Impact and recognition of cognitive impairment among hospitalized elders. *Journal of Hospital Medicine, 5*(2), 69—75.

Bradshaw, L. E., Goldberg, S. E., Lewis, S. A., Whittamore, K., Gladman, J. R., Jones, R. G., et al. (2013). Six-month outcomes following an emergency hospital admission for older adults with co-morbid mental health problems indicate complexity of care needs. *Age and Ageing, 42*(5), 582–588.

Briggs, R., Coary, R., Collins, R., Coughlan, T., O'Neill, D., & Kennelly, S. P. (2016). Acute hospital care: How much activity is attributable to caring for patients with dementia? *Monthly Journal of the Association of Physicians, 109*(1), 41–44.

Briggs, R., Dyer, A., Nabeel, S., Collins, R., Doherty, J., Coughlan, T., et al. (2017). Dementia in the acute hospital: The prevalence and clinical outcomes of acutely unwell patients with dementia. *Monthly Journal of the Association of Physicians, 110*(1), 33–37.

Briggs, R., McHale, C., Fitzhenry, D., O'Neill, D., & Kennelly, S. P. (2018). Dementia, disclosing the diagnosis. *Monthly Journal of the Association of Physicians, 111*(4), 215–216.

Briggs, R., & O'Neill, D. (2016). The informant history: A neglected aspect of clinical education and practice. *International Journal of Medicine, 109*(5), 301–302.

Briggs, R., O'Shea, E., de Siún, A., O'Neill, D., Gallagher, P., Timmons, S., et al. (2017). Does admission to a specialist geriatric medicine ward lead to improvements in aspects of acute medical care for older patients with dementia? *International Journal of Geriatric Psychiatry, 32*(6), 624–632.

Burn, A.-M., Fleming, J., Brayne, C., Fox, C., & Bunn, F. (2018). Dementia case-finding in hospitals: A qualitative study exploring the views of healthcare professionals in English primary care and secondary care. *BMJ Open, 8*(3). e020521.

Carpenter, C. R., Bassett, E. R., Fischer, G. M., Shirshekan, J., Galvin, J. E., & Morris, J. C. (2011). Four sensitive screening tools to detect cognitive dysfunction in geriatric emergency department patients: Brief Alzheimer's screen, short blessed test, ottawa 3DY, and the caregiver-completed AD8. *Academic Emergency Medicine: Official Journal of the Society for Academic Emergency Medicine, 18*(4), 374–384.

Cheong, C. Y., Tan, J. A. Q., Foong, Y.-L., Koh, H. M., Chen, D. Z. Y., Tan, J. J. C., et al. (2016). Creative music therapy in an acute care setting for older patients with delirium and dementia. *Dementia and Geriatric Cognitive Disorders EXTRA, 6*(2), 268–275.

Conroy, S., Ferguson, C., Woodard, J., & Banerjee, J. (2010). Interface geriatrics: Evidence-based care for frail older people with medical crises. *British Journal of Hospital Medicine, 71*(2), 98–101.

Davis, D. H., Muniz Terrera, G., Keage, H., Rahkonen, T., Oinas, M., Matthews, F. E., et al. (2012). Delirium is a strong risk factor for dementia in the oldest-old: A population-based cohort study. *Brain: A Journal of Neurology, 135*(Pt 9), 2809–2816.

Davis, D. H. J., Muniz-Terrera, G., Keage, H. A. D., Stephan, B. C. M., Fleming, J., Ince, P. G., et al. (2017). Association of delirium with cognitive decline in late life: A neuropathologic study of 3 population-based cohort studies. *Journal of the American Medical Association Psychiatry, 74*(3), 244–251.

Draper, B., Karmel, R., Gibson, D., Peut, A., & Anderson, P. (2011). The hospital dementia services project: Age differences in hospital stays for older people with and without dementia. *International Psychogeriatrics, 23*(10), 1649–1658.

Edmans, J., Bradshaw, L., Franklin, M., Gladman, J., & Conroy, S. (2013). Specialist geriatric medical assessment for patients discharged from hospital acute assessment units: Randomised controlled trial. *BMJ (Clinical Research Ed.), 347*.

Ehlenbach, W. J., Hough, C. L., Crane, P. K., Haneuse, S. J., Carson, S. S., Curtis, J. R., et al. (2010). Association between acute care and critical illness hospitalization and cognitive function in older adults. *Journal of the American Medical Association, 303*(8), 763–770.

Fahey, A., Ní Chaoimh, D., Mulkerrin, G. R., Mulkerrin, E. C., & O'Keeffe, S. T. (2017). Deciding about nursing home care in dementia: A conjoint analysis of how older people balance competing goals. *Geriatrics and Gerontology International, 17*(12), 2435–2440.

Fogg, C., Meredith, P., Bridges, J., Gould, G. P., & Griffiths, P. (2017). The relationship between cognitive impairment, mortality and discharge characteristics in a large cohort of older adults with unscheduled admissions to an acute hospital: A retrospective observational study. *Age and Ageing, 46*(5), 794–801.

Fong, T. G., Jones, R. N., Shi, P., Marcantonio, E. R., Yap, L., Rudolph, J. L., et al. (2009). Delirium accelerates cognitive decline in Alzheimer disease. *Neurology, 72*(18), 1570–1575.

Handley, M., Bunn, F., & Goodman, C. (2017). Dementia-friendly interventions to improve the care of people living with dementia admitted to hospitals: A realist review. *BMJ Open, 7*(7).

Hartley, P., Gibbins, N., Saunders, A., Alexander, K., Conroy, E., Dixon, R., et al. (2017). The association between cognitive impairment and functional outcome in hospitalised older patients: A systematic review and meta-analysis. *Age and Ageing, 46*(4), 559−567.

Helm, L., Balzer, K., Behncke, A., Eisemann, N., & Köpke, S. (2018). Patients with dementia in acute care hospitals : A cross-sectional study of physicians' experiences and attitudes. *Zeitschrift fur Gerontologie und Geriatrie, 51*(5), 501−508.

Hustey, F. M., Meldon, S. W., Smith, M. D., & Lex, C. K. (2003). The effect of mental status screening on the care of elderly emergency department patients. *Annals of Emergency Medicine, 41*(5), 678−684.

Hwang, U., & Morrison, R. S. (2007). The geriatric emergency department. *Journal of the American Geriatrics Society, 55*(11), 1873−1876.

Inouye, S. K., Bogardus, S. T., Charpentier, P. A., Leo-Summers, L., Acampora, D., Holford, T. R., et al. (1999). A multicomponent intervention to prevent delirium in hospitalized older patients. *New England Journal of Medicine, 340*(9), 669−676.

Inouye, S. K., Westendorp, R. G. J., & Saczynski, J. S. (2014). Delirium in elderly people. *Lancet, 383*(9920), 911−922.

Jackson, T. A., MacLullich, A. M., Gladman, J. R., Lord, J. M., & Sheehan, B. (2016). Undiagnosed long-term cognitive impairment in acutely hospitalised older medical patients with delirium: A prospective cohort study. *Age and Ageing, 45*(4), 493−499.

Kennedy, M., Enander, R. A., Tadiri, S. P., Wolfe, R. E., Shapiro, N. I., & Marcantonio, E. R. (2014). Delirium risk prediction, health care utilization and mortality of elderly emergency department patients. *Journal of the American Geriatrics Society, 62*(3), 462−469.

Kennelly, S. P., Morley, D., Coughlan, T., Collins, R., Rochford, M., & O'Neill, D. (2013). Knowledge, skills and attitudes of doctors towards assessing cognition in older patients in the emergency department. *Postgraduate Medical Journal, 89*(1049), 137−141.

Kristensen, R. U. U., Nørgaard, A., Jensen-Dahm, C., Gasse, C., Wimberley, T., & Waldemar, G. (2018). Polypharmacy and potentially inappropriate medication in people with dementia: A nationwide study. *Journal of Alzheimer's Disease, 63*(1), 383−394.

Krueger, K. R., Wilson, R. S., Kamenetsky, J. M., Barnes, L. L., Bienias, J. L., & Bennett, D. A. (2009). Social engagement and cognitive function in old age. *Experimental Aging Research, 35*(1), 45−60.

LaMantia, M. A., Stump, T. E., Messina, F. C., Miller, D. K., & Callahan, C. M. (2016). Emergency department use among older adults with dementia. *Alzheimer Disease and Associated Disorders, 30*(1), 35−40.

Lee, H., Ju, J.-W. W., Oh, S.-Y. Y., Kim, J., Jung, C. W. W., & Ryu, H. G. G. (2018). *Impact of timing and duration of postoperative delirium: A retrospective observational study. Surgery*.

Mathews, S. B., Arnold, S. E., & Epperson, C. N. (2014). Hospitalization and cognitive decline: Can the nature of the relationship Be deciphered? *American Journal of Geriatric Psychiatry: Official Journal of the American Association for Geriatric Psychiatry, 22*(5), 465−480.

Mockford, C. (2015). A review of family carers' experiences of hospital discharge for people with dementia, and the rationale for involving service users in health research. *Journal of Healthcare Leadership, 7*, 21−28.

Muir-Hunter, S. W., Fat, G. L., Mackenzie, R., Wells, J., & Montero-Odasso, M. (2016). Defining rehabilitation success in older adults with Dementia—Results from an inpatient geriatric rehabilitation unit. *The Journal of Nutrition, Health and Aging, 20*(4), 439−445.

O' Brien, H., O' Leary, N., Scarlett, S., O' Hare, C., & Kenny, R. A. A. (2018). Hospitalisation and surgery: Are there hidden cognitive consequences? Evidence from the Irish longitudinal study on ageing (TILDA). *Age and Ageing, 47*(3), 408−415.

O'Sullivan, D., O'Regan, N. A., & Timmons, S. (2016). Validity and reliability of the 6-item cognitive impairment test for screening cognitive impairment: A review. *Dementia and Geriatric Cognitive Disorders, 42*(1−2), 42−49.

Partridge, J. S., Harari, D., Martin, F. C., Peacock, J. L., Bell, R., Mohammed, A., et al. (2017). Randomized clinical trial of comprehensive geriatric assessment and optimization in vascular surgery. *British Journal of Surgery, 104*(6), 679−687.

Phelan, E. A., Borson, S., Grothaus, L., Balch, S., & Larson, E. B. (2012). Association of incident dementia with hospitalizations. *Journal of the American Medical Association, 307*(2), 165−172.

Prince, M., Bryce, R., Albanese, E., Wimo, A., Ribeiro, W., & Ferri, C. P. (2013). The global prevalence of dementia: A systematic review and metaanalysis. *Alzheimer's and Dementia : The Journal of the Alzheimer's Association, 9*(1).

Reynish, E. L., Hapca, S. M., De Souza, N., Cvoro, V., Donnan, P. T., & Guthrie, B. (2017). Epidemiology and outcomes of people with dementia, delirium, and unspecified cognitive impairment in the general hospital: Prospective cohort study of 10,014 admissions. *BMC Medicine, 15*, 140.

Royal College of Emergency Medicine. (2017). *Best Practice Guideline; Emergency Department Care*. Available at https://www.rcem.ac.uk/docs/RCEM%20Guidance/RCEM%2050%20Guidance.pdf.

Russ, T. C., Shenkin, S. D., Reynish, E., Ryan, T., Anderson, D., & Maclullich, A. M. (2012). Dementia in acute hospital inpatients: The role of the geriatrician. *Age and Ageing, 41*(3), 282–284.

Ryan, D. J. J., O'Regan, N. A. A., Caoimh, R. O., Clare, J., O'Connor, M., Leonard, M., et al. (2013). Delirium in an adult acute hospital population: Predictors, prevalence and detection. *BMJ Open, 3*(1).

Sampson, E. L., Blanchard, M. R., Jones, L., Tookman, A., & King, M. (2009). Dementia in the acute hospital: Prospective cohort study of prevalence and mortality. *British Journal of Psychiatry : The Journal of Mental Science, 195*(1), 61–66.

Shepperd, S., Doll, H., Angus, R. M., Clarke, M. J., Iliffe, S., Kalra, L., et al. (2008). Admission avoidance hospital at home. *Cochrane Database of Systematic Reviews*, (4).

Siddiqi, N., Harrison, J. K., Clegg, A., Teale, E. A., Young, J., Taylor, J., et al. (2016). Interventions for preventing delirium in hospitalised non-ICU patients. *Cochrane Database of Systematic Reviews, 3*.

Smith, H. A., Kindell, J., Baldwin, R. C., Waterman, D., & Makin, A. J. (2009). Swallowing problems and dementia in acute hospital settings: Practical guidance for the management of dysphagia. *Clinical Medicine, 9*(6), 544–548.

Timmons, S., Manning, E., Barrett, A., Brady, N. M., Browne, V., O'Shea, E., et al. (2015). Dementia in older people admitted to hospital: A regional multi-hospital observational study of prevalence, associations and case recognition. *Age and Ageing, 44*(6), 993–999.

Timmons, S., O'Shea, E., O'Neill, D., Gallagher, P., de Siún, A., McArdle, D., et al. (2016). Acute hospital dementia care: Results from a national audit. *BMC Geriatrics, 16*, 113.

Travers, C., Byrne, G., Pachana, N., Klein, K., & Gray, L. (2013). Prospective observational study of dementia and delirium in the acute hospital setting. *Internal Medicine Journal, 43*(3), 262–269.

Tropea, J., LoGiudice, D., Liew, D., Roberts, C., & Brand, C. (2017). Caring for people with dementia in hospital: Findings from a survey to identify barriers and facilitators to implementing best practice dementia care. *International Psychogeriatrics, 29*(3), 467–474.

Tsai, T. L., Sands, L. P., & Leung, J. M. (2010). An update on postoperative cognitive dysfunction. *Advances in Anesthesia, 28*(1), 269–284.

Wei, L. A., Fearing, M. A., Sternberg, E. J., & Inouye, S. K. (2008). The confusion assessment method (cam): A systematic review of current usage. *Journal of the American Geriatrics Society, 56*(5), 823–830.

Whitt, N., Harvey, R., McLeod, G., & Child, S. (2007). How many health professionals does a patient see during an average hospital stay? *New Zealand Medical Journal, 120*(1253).

Wilson, R. S., Hebert, L. E., Scherr, P. A., Dong, X., Leurgens, S. E., & Evans, D. A. (2012). Cognitive decline after hospitalization in a community population of older persons. *Neurology, 78*(13), 950–956.

CHAPTER 43

Environmental enrichment as a preventative and therapeutic approach to Alzheimer's disease

Kimberley E. Stuart, Anna E. King, James C. Vickers
Wicking Dementia Research and Education Centre, University of Tasmania, Hobart, TAS, Australia

List of abbreviations

AD Alzheimer's disease
APP amyloid precursor protein
BDNF brain-derived neurotrophic factor
EE environmental enrichment
FAD familial Alzheimer's disease
HPA hypothalamic—pituitary—adrenal axis
PS1 presenilin 1
SH standard housing
Wt wild-type

Mini-dictionary of terms

BDNF A widely expressed nerve growth factor of the central nervous system vital for the maintenance, survival, and growth of neurons
Corticosterone The main glucocorticoid of rodents
Environmental enrichment An experimental paradigm in which an animal's housing is designed to promote cognitive, sensory, and physical activity
HPA axis The neuroendocrine adaption component of the stress response, comprising the hypothalamus, pituitary and adrenal glands.
Microglia The brain's immune cells that engulf and remove cells that have undergone apoptosis
Synaptic plasticity A biological process by which specific patterns of synaptic activity result in changes in synaptic strength that contributes to learning and memory

General introduction

Neural plasticity describes the capacity of neurons and neural circuits to undergo structural and functional alterations in response to experience. First posited as early as the 19th century, there is evidence relating to the effect of experience from the environment on the neural system with the idea that a lack of stimulation would cause neural processes to withdraw from each other (Bain, 1872; Cajal, 1894; as cited in Rosenzweig, 1996). In a

Diagnosis and Management in Dementia
ISBN 978-0-12-815854-8, https://doi.org/10.1016/B978-0-12-815854-8.00043-4

time of population aging with concomitant higher vulnerability to neurodegenerative disease and cognitive deterioration, it is critical to understand how to harness and promote neural communication and enhance the plastic capacity of the aging brain.

Animal models of amyloid pathology

A considerable degree of our molecular understanding of Alzheimer's disease (AD) has come from studying familial AD (FAD)-linked mutations, namely the amyloid precursor protein (APP), presenilin 1 (PS1) genes that encode the amyloid precursor protein, and presenilin 1. Many transgenic mouse models used to study the pathogenesis of AD utilize these FAD mutations, which lead to amyloid pathology and memory impairment (Puzzo, Gulisano, Palmeri, & Arancio, 2015). Mouse models with APP and PS1 mutations are considered an early-stage AD model and do not develop the second hallmark pathological feature of AD, neurofibrillary tangles (Garcia-Alloza et al., 2006). Mouse models of amyloid pathology are most commonly used to model AD and will be the focus of this chapter.

Environmental enrichment

Environmental enrichment (EE) is a paradigm applied to animal models to experimentally model the promotion of neural plasticity from experience. The paradigm involves the manipulation of an animal's environment, aimed at encouraging environmental stimulation with objects designed to generate physical, sensory, and cognitive activity. The earliest studies starting in the 1960s reported changes in the neuroanatomy of the healthy rodent brain following exposure to EE, including increased cortical thickness (Van Praag, Kempermann, & Gage, 2000). Heightened brain connectivity and markers of plasticity were subsequently reported, in which increased dendritic branching, spine density, and synaptic contacts have been observed (Hannan, 2014). Living in a stimulating environment has been demonstrated to promote morphological and molecular changes within the healthy central nervous system (CNS) that are thought to modify behavior and heighten cognitive function compared with animals housed in standard housing (SH) conferring minimal stimulation.

Environmental enrichment as an intervention for Alzheimer's disease

Given the range of beneficial effects on the structure and function of the nervous system in healthy animal models, the EE paradigm has been applied to models of neuropathology (for review see Nithianantharajah & Hannan, 2006). A potentially pivotal study by Kamenetz et al. (2003) led to the hypothesis that cognitive stimulation may affect the biochemistry of AD, as neuronal activity was found to bidirectionally control Aβ levels and modulate synaptic plasticity.

From the work of Kamenetz et al. (2003) on neuronal activity and Aβ, Jankowsky, Xu, Fromholt, Gonzales, and Borchelt (2003) proposed to model a functional change in synaptic activity in vivo. In this study, an EE paradigm was employed to promote neuronal activity, and the subsequent effects on Aβ pathology in an APP/PS1 mouse model were assessed. Female mice were placed into EE at an early, presymptomatic stage of the disease. Paradoxical to the reported beneficial effects of EE on the neural system found in healthy rodents and to that of in vitro studies, the authors reported a higher plaque burden in EE-exposed APP/PS1 mice yet increased cognitive function. The authors suggested that higher levels of cognitive stimulation accelerate amyloid pathology, but compensatory mechanisms brought about by enhanced brain function may mitigate these effects. Despite the counterintuitive findings, the authors demonstrated amyloid pathology to be altered in response to experience from the environment.

Since the initial study by Jankowsky et al. (2003), a number of investigations on the effect of EE on a range of transgenic AD mouse models have been performed with variable effects on Aβ neuropathology (Table 43.1). Arendash et al. (2004) approached the enrichment paradigm for an APP mouse model therapeutically by starting EE at an advanced stage of pathology. These mice lived in an EE and were also exposed to a novel environment weekly. At 22 months of age, EE mice performed superiorly on a battery of cognitive tests compared with mice from SH. However, these cognitive improvements were independent of that of Aβ load. The authors concluded, in partial agreement with Jankowsky et al. (2003), that the behavioral benefits of EE to AD may involve a separate mechanism to that of Aβ processing. The findings of the two studies highlight the ambiguous association of Aβ deposition and clinical expression of the disease.

In contrast to the findings of a limited association between EE and Aβ, Lazarov et al. (2005) hypothesized a decrease in Aβ pathology could occur if EE was introduced in early life. At weaning, male APP/PS1 mice were exposed to EE for 5 months. A dramatic reduction in Aβ deposition in neocortex and hippocampus of EE mice compared with that of SH mice was observed. One particularly noteworthy finding of the study was that Aβ burden was inversely correlated with physical activity (time spent on the running wheel). Lazarov et al. further demonstrated that the activity of the Aβ degrading enzyme, neprilysin, was elevated in the brains of EE AD mice compared with Wt and AD SH mice. The authors suggested their contrary findings from that of Jankowsky et al. (2003) could be due to sex differences, as female mice were employed in the Jankowsky et al. study, and female mice have a higher amyloid burden than age-matched males (Wang, Tanila, Puoliväli, Kadish, & Van Groen, 2003). Although conflicting with the previous in vivo studies, Lazarov et al. (2005) provided the first experimental evidence that exposure to EE reduces Aβ burden.

Jankowsky et al. (2005) conducted a follow-up investigation and again found that female EE mice had a 25% higher Aβ load in comparison to SH mice, and agreed that a sexually dimorphic response to EE might be the reason behind the different

Table 43.1 Environmental enrichment interventions for mouse models of amyloid pathology.

First author (publication year)	FAD mouse model	n	Sex	Age entered into EE (months)	Duration of EE (months)	Type of EE	Cognitive function	Aβ pathology
Jankowsky et al. (2003)	APPswe, PSEN1$_{dE9}$	32	F	2	6	Constant with objects rearranged weekly	↑	↑
Arendash et al. (2004)	APP$_{SWE}$	13	Not stated	16	3	Intermittent (3 × /week)	↑	—
Lazarov et al. (2005)	APPswe, PSEN1$_{dE9}$	13	M	1	5	First month constant (3 h/day), then 3 × / week	Not assessed	→
Jankowsky et al. (2005)	APPswe, PSEN1$_{dE9}$	76	F	2	6	Constant	Not assessed	↑
Costa et al. (2007)	APP ± PS1 ±	101	M + F	1	6	Constant + 3 × /week novel exposure	↑	→
Cracchiolo et al. (2007)	APP$_{SWE}$, APP$_{SWE}$PS1$_{M67IL}$	38	M + F	1.5	7	Social/physical/complete + (novelty exposure 3 × /week)	↑	→
Mirochnic et al. (2009)	APP23	38	F	6 & 18	4	Constant with objects rearranged	Not assessed	—
Herring et al. (2011)	APP695	38	F	1	5	Constant	Not assessed	→
Valero et al. (2011)	APP$_{SWE}$	58	F	4	1.5	Constant with objects rearranged weekly	↑	—
Cotel et al. (2012)	APP/PS1 KI	60	F	2	4	Constant with objects rearranged weekly	—	—
Verret et al. (2013)	APP695	112	F	3, 5, 10	2.5	Constant with objects rearranged	↑	Minor ↓
Stuart et al. (2017)	APPswe, PSEN1$_{dE9}$	48	M	6	6	Constant	↑	—
Stuart et al. (2017)	APPswe, PSEN1$_{dE9}$	72	M	6	6	Constant + 3 × /week novel exposure	↑	↑
Huttenrauch (2016)	Tg6799	35	F	1	11	Constant with objects rearranged	—	—

Aβ, beta-amyloid; *APP*, amyloid precursor protein; *EE*, environmental enrichment; *FAD*, familial Alzheimer's disease; *PSEN1*, presenilin 1; ↑, EE increased relative to SH; ↓, EE decreased relative to SH; —, no difference between SH and EE.

pathological outcomes to that of Lazarov et al. (2005). The authors put forward that multiple studies have documented human patients to have intact cognitive function, despite the presence of Aβ plaques at a level sufficient for an AD diagnosis, and further contributed their results to EE working through a mechanism independent to Aβ deposition, such as through building cognitive reserve.

Costa et al. (2007) extended the work from Arendash et al. (2004) and had half of the mice undergo comprehensive behavioral testing, while the other half did not, to control for the potential enriching effect of behavioral testing. Enriched mice outperformed SH mice, and AD mice from EE were performing up to the level of the healthy control mice on cognitive assessments. While nonbehaviorally tested mice showed no difference in Aβ levels, behaviorally tested EE mice demonstrated lower Aβ deposition than nonbehaviorally tested mice. The authors suggested consistent with the findings of Lazarov et al. (2005), that a more demanding EE paradigm may lower Aβ deposition and must induce brain changes that compensate for the damage brought about by Aβ pathology.

What type of environmental enrichment paradigm is efficacious in familial Alzheimer's disease mouse models?

As briefly described, EE appears to lead to inconsistent effects on Aβ pathology (Table 43.1) that may be due to potential confounders, such as the use of different FAD models, age and stage of pathology, gender, type of EE paradigm, length of exposure to EE, and whether EE is intermittent or stable or includes a novelty component.

While positive findings have been produced from introducing EE at an early, presymptomatic age and lasting over disease progression, Herring et al. (2011) set out to examine the effects of living in EE only before or after AD onset and concluded that stimulation delivered either before or after disease onset can reduce Aβ pathology. More recently, studies have provided evidence to the contrary. Verret et al. (2013) reported a reduction in Aβ burden following EE that started in early life but not later in life following the deposition of Aβ. Supporting this finding, an APP/PS1 model exposed to continuous EE following amyloid deposition for 6 months resulted in no alterations to Aβ plaque load (Stuart, King, Fernandez-Martos, Summers, & Vickers, 2017). Similarly, Hüttenrauch et al. (2017) exposed a 5XFAD model to EE at 1 month until 11 months of age and reported no effect on Aβ plaque load, $Aβ_{1-42}$ levels, or Aβ-degrading enzymes. The mouse model employed must too be considered in these differential findings, as Hüttenrauch et al. and Stuart et al. used mouse models with faster and more aggressive pathological development relative to that of Herring and coworkers. Such findings might suggest that intervention initiated before the onset of amyloid deposition may slow amyloidosis; however, once pathology has progressed, EE has little effect on Aβ disease process progression. This research together suggests that EE may be too mild an intervention to support pathological protection into later stages of amyloid pathology but may offer some protection at earlier stages.

Environmental enrichment and amyloid-independent cognitive protection

Studies focused on the association between EE and amyloid pathology have also led to findings regarding amyloid-independent effects. Arendash et al. (2004) and Stuart et al. (2017) reported global cognitive protection in ageing FAD mice exposed to EE despite no reductions in Aβ deposition. Similarly, Jankowsky et al. (2005) and Stuart et al. (2017) reported a counterintuitive increase in Aβ pathology yet a marked attenuation of cognitive dysfunction (Jankowsky et al., 2005) in FAD mice exposed to EE. These findings together might suggest that EE may not directly exert its effects by mitigating Aβ pathology. EE may build cognitive reserve capacity and exert its effects through alternative neural mechanisms that are currently elusive. An understanding of these underlying neural mechanisms will likely be invaluable in developing therapeutics to delay or halt the onset of dementia.

Environmental enrichment as a means of promoting synaptic plasticity in Alzheimer's disease

Synapse loss is currently the strongest neurobiological correlate of cognitive dysfunction in AD (e.g., Arendt, 2009; DeKosky & Scheff, 1990). Evidence of synaptic dysfunction and loss being closely linked with the clinical expression of AD also gives rise to the idea that the synapse should be the focus of therapeutics. AD is an insidious yet slowly progressing disease that emerges clinically many years after disease process inception (Morris et al., 2001). It is conceivable that the process of synaptic degeneration is also a slow-acting process, and there may be a point at which the degeneration of synapses can be rescued. Concurrently, the nervous system has remarkable capacity for plasticity, and while perturbed in AD (Sorrentino, Iuliano, Polverino, Jacini, & Sorrentino, 2014), stimulation from the environment may create heightened capacity for plasticity and subsequent compensation for pathological damage allowing for cognitive maintenance.

EE for healthy wild-type (Wt) rodents has resulted in a spectrum of synaptic alterations. Pre- and postsynaptic changes have been observed through analysis of dendritic spines in the cerebral cortex and hippocampus in which spine densities were increased by EE (for review see Hannan, 2014), and in vivo microscopy has revealed increases in density and turnover of dendritic spines in enriched mice (Jung & Herms, 2014). However, investigations on synaptic alterations following EE in AD mouse models have been more limited. Levi, Jongen-Relo, Feldon, Roses, and Michaelson (2003) investigated EE-induced effects on mice containing the APOE ε4 allele, which is a risk polymorphism for AD (Verghese, Castellano, & Holtzman, 2011), and animals with this genotype did not show cognitive improvement following EE. Moreover, following EE, ε4-carrier mice had reduced hippocampal nerve growth factor and synaptophysin levels compared with non-ε4 carriers exposed to EE. Moreover, EE initiated following Aβ deposition in an APP/PS1 mouse model has been associated with an increase in synaptic puncta in CA1 of the hippocampus with no alterations to Aβ burden (Stuart et al., 2017).

Further evidence of EE increasing plasticity was provided by Costa et al. (2007). Dendritic branching defects observed in FAD mice were not overcome by EE; however, EE was found to induce beneficial changes in gene expression related to neuronal plasticity, which may offer an explanation for the finding of heightened cognitive function in the absence of a reduction in Aβ pathological burden. Herring et al. (2011) also provided evidence for EE upregulating expression of plasticity-associated proteins in an APP mouse model in which AD mice living in EE had increased levels of the plasticity-associated protein Arc and heightened synaptic density relative to that of SH and up to the level of healthy Wt mice. Mirochnic, Wolf, Staufenbiel, and Kempermann (2009) compared with APP mice that lived in SH, EE, or physical activity housing for adult neurogenesis in the hippocampal dentate gyrus at 6 and 18 months. At the 18 month time point, mice living in SH had a higher $Aβ_{1-42}/Aβ_{1-40}$ ratio as well as a lower number of newborn granule cells in the hippocampus dentate gyrus compared with those of EE and physical activity housing conditions. In addition, Lazarov et al. (2005) demonstrated, through microarray analysis, a number of genes that are differentially regulated by EE in an FAD mouse model with elevated expression of genes involved in neurogenesis and neuronal plasticity.

Environmental enrichment, Alzheimer's disease, and brain-derived neurotrophic factor

Alterations in neurotrophins, such as brain-derived neurotrophic factor (BDNF), may provide one underlying mechanism of the association between EE and the cognitive benefit observed in some studies. BDNF is a nerve growth factor expressed throughout the CNS and is vital for the maintenance, survival, and growth of neurons (Mattson, Maudsley, & Martin, 2004). Considerable evidence from in vitro and in vivo experimental studies suggests that BDNF has prosurvival functions on neurons under different pathological conditions (Lu, Nagappan, Guan, Nathan, & Wren, 2013).

BDNF has been implicated in AD, with reduced levels of BDNF found in the hippocampus and frontal and parietal cortices of the AD brain (Hock, Heese, Hulette, Rosenberg, & Otten, 2000). BDNF regulates processing of APP through the nonamyloidogenic pathway, which may yield beneficial effects, such as a reduction in the production of Aβ peptides and the release of the secreted form of APP, which is associated with neuroprotective effects (e.g., Thornton, Vink, Blumbergs, & Van Den Heuvel, 2006). Moreover, studies employing in vitro methods have reported a protective effect of BDNF on $A\beta_{42}$ (Aliaga et al., 2010; Arancibia et al., 2008).

Increases in BDNF expression in the hippocampus of healthy animals following EE have been observed (e.g., Ramírez-Rodríguez et al., 2014) as well as global increases (Ickes et al., 2000). There has been limited work on the relationship between EE and BDNF in FAD models. However, Wolf et al. (2006) reported improved cognitive function following EE in an APP model, putatively accounted for by upregulation of BDNF, and longer-term EE from midlife was associated with increased hippocampal BDNF in an APP/PS1 model (Stuart et al., 2017). There is a wealth of evidence linking increased BDNF expression to physical activity (Erickson, Miller, & Roecklein, 2012), which forms a caveat to the EE research in which it is uncertain whether it is the physical activity alone, rather than cognitive stimulation, that drives the increase in BDNF.

Environmental enrichment, Alzheimer's disease, and microglia

Microglia are widely distributed throughout the brain and constantly survey their environment for pathogens, apoptotic cells, and foreign material (Streit, Mrak, & Griffin, 2004). Microglia are critically involved in generating chronic and self-sustaining neuroinflammation that may act as a key contributor to the pathogenesis of AD (Heneka et al., 2015). However, only more recently have resident macrophages of the brain received attention for their role in CNS dysfunction (Derecki, Katzmarski, Kipnis, & Meyer-Luehmann, 2014).

The dynamics of microglial processes have been found to be regulated by sensory experience and neuronal activity (Tremblay, Lowery, & Majewska, 2010) and have also been implicated in learning processes through their interaction with synaptic contacts. A study in which microglia were ablated in the adult rodent brain led to an impairment in learning and an associated reduction in learning-induced synapse formation in the motor cortex (Parkhurst et al., 2013). These findings together suggest microglia to have a role in experience-dependent synaptic plasticity.

Recent evidence suggests that lifestyle factors such as physical and cognitive activity are involved in promoting microglial health in aging, with reports demonstrating that the neuronal benefits conferred by EE may occur by modulation of innate immune cells including microglia (Ziv et al., 2006). In line with such evidence, a recent study investigated how EE might regulate microglial function in healthy Wt mice injected with oligomeric $A\beta$. Xu et al. (2016) reported that EE suppresses the elevated expression of inflammatory genes that occur in response to oligomeric $A\beta$ and a decrease in uptake of oligomeric $A\beta$ by microglia. The authors reported that EE modulated microglial morphology and density, leading to an increase in microglial resistance to the effects of oligomeric $A\beta$. The study by Xu et al. offered significant insight into one of the underlying mechanisms of the protective effect of EE and that this may be modulated by the brain's innate immune system.

Stress as a potential confounder

Experiencing a stressful stimulus induces activation of the hypothalamic—pituitary—adrenal (HPA) axis, leading to a release of glucocorticoids from the adrenal cortex. Evidence of HPA axis dysfunction in AD comes from the finding that people living with AD typically have increased levels of the stress hormone cortisol in blood (e.g., Csernansky et al., 2006). Elevated levels of the glucocorticoid (corticosterone in rodents) have also been reported in FAD mouse models (Guo, Zheng, & Justice, 2012; Stuart et al., 2017). Whether FAD rodent models have elevated stress hormone levels may act as a confounder to EE studies with Aβ pathological burden as an outcome variable.

The finding of exacerbated Aβ pathology by Jankowsky et al. (2003, 2005) has been suggested to be a result of the EE paradigm inducing stress. Jeong et al. (2011) exposed FAD mice to a stress and EE paradigm and found that EE counteracted the negative effects of stress on AD disease process progression. In line with that of both the Jankowsky studies, Stuart et al. (2017) demonstrated that exposing APP/PS1 mice to EE with novelty following Aβ deposition resulted in an exacerbation of Aβ pathology and elevation of corticosterone levels. Given the highly variable findings of EE on Aβ pathology and evidence that stress plays a major role in disease onset and progression, it is important to consider how it may moderate the relationship between EE, cognition, and pathological features of AD.

Summary

Searching for experimental approaches to promote brain plasticity is exceedingly relevant to our society, in which aging-related diseases are becoming increasingly prominent. Strategies aimed at boosting plasticity, or cognitive reserve, offer a noninvasive alternative approach and avoid the dangerous side effects typical of many pharmacological therapies. Moreover, epidemiological and experimental evidence demonstrates the great potential that cognitive interventions may have in preventing aging-associated cognitive dysfunction. The encouraging data produced from animal enrichment studies offers hope for transferring these results to both human health and disease applications.

While the animal-based literature on EE for healthy animals has a long history and offers promise in staving off neurodegenerative disease, there is much to be elucidated. EE applied to FAD models has focused on amyloid pathology, and findings have been highly variable. The research literature points to more subtle synaptic and cellular changes enhancing the functional capacity of the CNS rather than overt reductions to amyloid pathology. The findings on the relationship between EE and Aβ pathology are contentious, and potential moderating factors such as how EE affects the brain's immune cells, synaptic health, and stress hormones have received limited attention, and as demonstrated in this chapter, are warranted research targets moving forward in identifying efficacious nonpharmacological approaches for the prevention and treatment of dementia due to Alzheimer's disease.

Key facts of environmental enrichment

- Environmental enrichment refers to the addition of objects to an animal's environment and designed to enhance sensory stimulation.
- The first environmental enrichment studies were conducted in the 1960s, and since that time, research on enrichment for healthy rodents has robustly demonstrated a range of beneficial molecular, cellular, cognitive, and behavioral effects of the paradigm.
- The first experiments were conducted on healthy wild-type animals and then extended to a genetic model of Huntington's disease, and thereafter applied to models of Alzheimer's disease.
- Environmental enrichment has been demonstrated to enhance cognitive function in genetic models of Alzheimer's disease.
- Variable findings have been produced regarding the effect of environmental enrichment on amyloid pathology.

Summary points

- This chapter focuses on the effect of environmental enrichment on mouse models of familial Alzheimer's disease.
- Enrichment has consistently produced a range of beneficial brain changes in healthy wild-type rodents, including increased cortical thickness, increased synaptic density and plasticity, and enhanced learning and memory function.
- Most of the literature on Alzheimer's mouse models exposed to environmental enrichment show behavioral and cognitive benefits.
- The effects of enrichment on neuropathology have been more contentious.
- An array of extraneous factors putatively modulate the effect of environmental enrichment on the Alzheimer's disease brain.

References

Aliaga, E., Silhol, M., Bonneau, N., Maurice, T., Arancibia, S., & Tapia-Arancibia, L. (2010). Dual response of BDNF to sublethal concentrations of beta-amyloid peptides in cultured cortical neurons. *Neurobiology of Disease, 37*(1), 208–217. https://doi.org/10.1016/j.nbd.2009.10.004.
Arancibia, S., Silhol, M., Moulière, F., Meffre, J., Höllinger, I., Maurice, T., et al. (2008). Protective effect of BDNF against beta-amyloid induced neurotoxicity in vitro and in vivo in rats. *Neurobiology of Disease, 31*(3), 316–326. https://doi.org/10.1016/j.nbd.2008.05.012.

Arendash, G. W., Garcia, M. F., Costa, D. A., Cracchiolo, J. R., Wefes, I. M., & Potter, H. (2004). Environmental enrichment improves cognition in aged Alzheimer's transgenic mice despite stable beta-amyloid deposition. *NeuroReport, 15*(11), 1751—1754. https://doi.org/10.1097/01.wnr.0000137183.68847.

Arendt, T. (2009). Synaptic degeneration in Alzheimer's disease. *Acta Neuropathologica*. https://doi.org/10.1007/s00401-009-0536-x.

Cotel, M.-C., Jawhar, S., Christensen, D. Z., Bayer, T. A., & Wirths, O. (2012). Environmental enrichment fails to rescue working memory deficits, neuron loss, and neurogenesis in APP/PS1KI mice. *Neurobiology of Aging, 33*, 96—107.

Cracchiolo, J. R., Mori, T., Nazian, S. J., Tan, J., Potter, H., & Arendash, G. W. (2007). Enhanced cognitive activity-over and above social or physical activity-is required to protect Alzheimer's mice against cognitive impairment, reduce Aβ deposition, and increase synaptic immunoreactivity. *Neurobiology of Learning and Memory, 88*, 277—294.

Costa, D.a., Cracchiolo, J. R., Bachstetter, A. D., Hughes, T. F., Bales, K. R., Paul, S. M., et al. (2007). Enrichment improves cognition in AD mice by amyloid-related and unrelated mechanisms. *Neurobiology of Aging, 28*, 831—844. https://doi.org/10.1016/j.neurobiolaging.2006.04.009.

Csernansky, J. G., Dong, H., Ph, D., Fagan, A. M., Wang, L., Xiong, C., et al. (2006). Plasma cortisol and progression of dementia in DAT subjects. *American Journal of Psychiatry, 163*(December), 2164—2169. https://doi.org/10.1176/appi.ajp.163.12.2164.

DeKosky, S. T., & Scheff, S. W. (1990). Synapse loss in frontal cortex biopsies in Alzheimer's disease: Correlation with cognitive severity. *Annals of Neurology, 27*(5), 457—464. https://doi.org/10.1002/ana.410270502.

Derecki, N. C., Katzmarski, N., Kipnis, J., & Meyer-Luehmann, M. (2014). Microglia as a critical player in both developmental and late-life CNS pathologies. *Acta Neuropathologica*. https://doi.org/10.1007/s00401-014-1321-z.

Erickson, K. I., Miller, D. L., & Roecklein, K. A. (2012). The aging Hippocampus. *The Neuroscientist, 18*(1), 82—97. https://doi.org/10.1177/1073858410397054.

Garcia-Alloza, M., Robbins, E. M., Zhang-Nunes, S. X., Purcell, S. M., Betensky, R. A., Raju, S., et al. (2006). Characterization of amyloid deposition in the APPs we/PS1dE9 mouse model of Alzheimer disease. *Neurobiology of Disease, 24*(3), 516—524. https://doi.org/10.1016/j.nbd.2006.08.017. PubMed result.

Guo, Q., Zheng, H., & Justice, N. J. (2012). Central CRF system perturbation in an Alzheimer's disease knockin mouse model. *Neurobiology of Aging, 33*(11), 2678—2691. https://doi.org/10.1016/j.neurobiolaging.2012.01.002.

Hannan, a J. (2014). Environmental enrichment and brain repair: Harnessing the therapeutic effects of cognitive stimulation and physical activity to enhance experience-dependent plasticity. *Neuropathology and Applied Neurobiology, 40*(1), 13—25. https://doi.org/10.1111/nan.12102.

Heneka, M. T., Carson, M. J., Khoury, J. El, Landreth, G. E., Brosseron, F., Feinstein, D. L., et al. (2015). Neuroinflammation in Alzheimer's disease. *The Lancet Neurology*. https://doi.org/10.1016/S1474-4422(15)70016-5.

Herring, A., Lewejohann, L., Panzer, A.-L., Donath, A., Kröll, O., Sachser, N., et al. (2011). Preventive and therapeutic types of environmental enrichment counteract beta amyloid pathology by different molecular mechanisms. *Neurobiology of Disease, 42*(3), 530—538. https://doi.org/10.1016/j.nbd.2011.03.007.

Hock, C., Heese, K., Hulette, C., Rosenberg, C., & Otten, U. (2000). Region-specific neurotrophin imbalances in alzheimer disease: Decreased levels of brain-derived neurotrophic factor and increased levels of nerve growth factor in hippocampus and cortical areas. *Archives of Neurology, 57*(6), 846—851. https://doi.org/10.1001/archneur.57.6.846.

Hüttenrauch, M., Salinas, G., & Wirths, O. (2016). Effects of Long-Term Environmental Enrichment on Anxiety, Memory, Hippocampal Plasticity and Overall Brain Gene Expression in C57BL6 Mice. *Frontiers in molecular neuroscience, 9*(62). https://doi.org/10.3389/fnmol.2016.00062.

Ickes, B. R., Pham, T. M., Sanders, L. A., Albeck, D. S., Mohammed, A. H., & Granholm, A.-C. (2000). Long-term environmental enrichment leads to regional increases in neurotrophin levels in rat brain. *Experimental Neurology, 164*(1), 45—52. https://doi.org/10.1006/exnr.2000.7415.

Jankowsky, J. L., Melnikova, T., Fadale, D. J., Xu, G. M., Slunt, H. H., Gonzales, V., et al. (2005). Environmental enrichment mitigates cognitive deficits in a mouse model of Alzheimer's disease. *The Journal of Neuroscience: The Official Journal of the Society for Neuroscience, 25*(21), 5217–5224. https://doi.org/10.1523/JNEUROSCI.5080-04.2005.

Jankowsky, J. L., Xu, G., Fromholt, D., Gonzales, V., & Borchelt, D. R. (2003). Environmental enrichment exacerbates amyloid plaque formation in a transgenic mouse model of Alzheimer disease. *Journal of Neuropathology and Experimental Neurology, 62*(12), 1220–1227.

Jeong, Y. H., Kim, J. M., Yoo, J., Lee, S. H., Kim, H. S., & Suh, Y. H. (2011). Environmental enrichment compensates for the effects of stress on disease progression in Tg2576 mice, an Alzheimer's disease model. *Journal of Neurochemistry, 119*(6), 1282–1293. https://doi.org/10.1111/j.1471-4159.2011.07514.x.

Jung, C. K. E., & Herms, J. (2014). Structural dynamics of dendritic spines are influenced by an environmental enrichment: An in vivo imaging study. *Cerebral Cortex (New York, N.Y): 1991, 24*(2), 377–384. https://doi.org/10.1093/cercor/bhs317.

Kamenetz, F., Tomita, T., Hsieh, H., Seabrook, G., Borchelt, D., Iwatsubo, T., et al. (2003). APP processing and synaptic function. *Neuron, 37*(6), 925–937. https://doi.org/10.1016/S0896-6273(03)00124-7.

Lazarov, O., Robinson, J., Tang, Y. P., Hairston, I. S., Korade-Mirnics, Z., Lee, V. M. Y., et al. (2005). Environmental enrichment reduces Aβ levels and amyloid deposition in transgenic mice. *Cell, 120*(5), 701–713. https://doi.org/10.1016/j.cell.2005.01.015.

Levi, O., Jongen-Relo, A. L., Feldon, J., Roses, A. D., & Michaelson, D. M. (2003). ApoE4 impairs hippocampal plasticity isoform-specifically and blocks the environmental stimulation of synaptogenesis and memory. *Neurobiology of Disease, 13*(3), 273–282. https://doi.org/10.1016/S0969-9961(03)00045-7.

Lu, B., Nagappan, G., Guan, X., Nathan, P. J., & Wren, P. (2013). BDNF-based synaptic repair as a disease-modifying strategy for neurodegenerative diseases. *Nature Reviews Neuroscience, 14*(6), 1–16. https://doi.org/10.1038/nrn3505.

Mattson, M. P., Maudsley, S., & Martin, B. (2004). BDNF and 5-HT: A dynamic duo in age-related neuronal plasticity and neurodegenerative disorders. *Trends in Neurosciences*. https://doi.org/10.1016/j.tins.2004.08.001.

Mirochnic, S., Wolf, S., Staufenbiel, M., & Kempermann, G. (2009). Age effects on the regulation of adult hippocampal neurogenesis by physical activity and environmental enrichment in the APP23 mouse model of Alzheimer disease. *Hippocampus, 19*(10), 1008–1018. https://doi.org/10.1002/hipo.20560.

Morris, J. C., Storandt, M., Miller, J. P., McKeel, D. W., Price, J. L., Rubin, E. H., et al. (2001). Mild cognitive impairment represents early-stage Alzheimer disease. *Archives of Neurology, 58*(3), 397–405. https://doi.org/10.1001/archneur.58.3.397.

Nithianantharajah, J., & Hannan, A. J. (2006). Enriched environments, experience-dependent plasticity and disorders of the nervous system. *Nature Reviews Neuroscience, 7*(September), 697–709. https://doi.org/10.1038/nrn1970.

Parkhurst, C. N., Yang, G., Ninan, I., Savas, J. N., Yates, J. R., Lafaille, J. J., et al. (2013). Microglia promote learning-dependent synapse formation through brain-derived neurotrophic factor. *Cell, 155*(7), 1596–1609. https://doi.org/10.1016/j.cell.2013.11.030.

Van Praag, H., Kempermann, G., & Gage, F. H. (2000). Neural consequences of environmental enrichment. *Nature Reviews Neuroscience, 1*(December), 191–198. https://doi.org/10.1038/35044558.

Puzzo, D., Gulisano, W., Palmeri, A., & Arancio, O. (2015). Rodent models for Alzheimer's disease drug discovery. *Expert Opinion on Drug Discovery, 10*(7), 703–711. https://doi.org/10.1517/17460441.2015.1041913.

Ramírez-Rodríguez, G., Ocaña-Fernández, M. A., Vega-Rivera, N. M., Torres-Pérez, O. M., Gómez-Sánchez, A., Estrada-Camarena, E., et al. (2014). Environmental enrichment induces neuroplastic changes in middle age female BalbC mice and increases the hippocampal levels of BDNF, p-Akt and p-MAPK1/2. *Neuroscience, 260*, 158–170. https://doi.org/10.1016/j.neuroscience.2013.12.026.

Rosenzweig, M. R. (1996). Aspects of the search for neural mechanisms of memory. *Annual Review of Psychology, 47*, 1–32. https://doi.org/10.1146/annurev.psych.47.1.1.

Sorrentino, P., Iuliano, A., Polverino, A., Jacini, F., & Sorrentino, G. (2014). The dark sides of amyloid in Alzheimer's disease pathogenesis. *FEBS Letters*. https://doi.org/10.1016/j.febslet.2013.12.038.

Streit, W. J., Mrak, R. E., & Griffin, W. S. T. (2004). Microglia and neuroinflammation: A pathological perspective. *Journal of Neuroinflammation, 1*, 4. https://doi.org/10.1186/1742-2094-1-14.

Stuart, K. E., King, A. E., Fernandez-Martos, C. M., Summers, M. J., & Vickers, J. C. (2017). Environmental novelty exacerbates stress hormones and Aβ pathology in an Alzheimer's model. *Scientific Reports, 7*(1). https://doi.org/10.1038/s41598-017-03016-0.

Thornton, E., Vink, R., Blumbergs, P. C., & Van Den Heuvel, C. (2006). Soluble amyloid precursor protein reduces neuronal injury and improves functional outcome following diffuse traumatic brain injury in rats. *Brain Research, 1094*(1), 38—46. https://doi.org/10.1016/j.brainres.2006.03.107.

Tremblay, M.Ě., Lowery, R. L., & Majewska, A. K. (2010). Microglial interactions with synapses are modulated by visual experience. *PLoS Biology, 8*(11). https://doi.org/10.1371/journal.pbio.1000527.

Valero, J., España, J., Parra-Damas, A., Martín, E., Rodríguez-Álvarez, J., & Saura, C. A. (2011). Short-term environmental enrichment rescues adult neurogenesis and memory deficits in APPSw, Ind transgenic mice. *PLoS One, 6*(2), e16832. https://doi.org/10.1371/journal.pone.0016832.

Verghese, P. B., Castellano, J. M., & Holtzman, D. M. (2011). Apolipoprotein E in Alzheimer's disease and other neurological disorders. *The Lancet Neurology*. https://doi.org/10.1016/S1474-4422(10)70325-2.

Verret, L., Krezymon, A., Halley, H., Trouche, S., Zerwas, M., Lazouret, M., et al. (2013). Transient enriched housing before amyloidosis onset sustains cognitive improvement in Tg2576 mice. *Neurobiology of Aging, 34*(1), 211—225. https://doi.org/10.1016/j.neurobiolaging.2012.05.013.

Wang, J., Tanila, H., Puoliväli, J., Kadish, I., & Van Groen, T. (2003). Gender differences in the amount and deposition of amyloidβ in APPswe and PS1 double transgenic mice. *Neurobiology of Disease, 14*(3), 318—327. https://doi.org/10.1016/j.nbd.2003.08.009.

Wolf, S.a., Kronenberg, G., Lehmann, K., Blankenship, A., Overall, R., Staufenbiel, M., et al. (2006). Cognitive and physical activity differently modulate disease progression in the amyloid precursor protein (APP)-23 model of alzheimer's disease. *Biological Psychiatry, 60*(12), 1314—1323. https://doi.org/10.1016/j.biopsych.2006.04.004.

Xu, H., Gelyana, E., Rajsombath, M., Yang, T., Li, S., & Selkoe, D. (2016). Environmental enrichment potently prevents microglia-mediated neuroinflammation by human amyloid β-protein oligomers. *Journal Neuroscience, 36*, 9041—9056.

Ziv, Y., Ron, N., Butovsky, O., Landa, G., Sudai, E., Greenberg, N., et al. (2006). Immune cells contribute to the maintenance of neurogenesis and spatial learning abilities in adulthood. *Nature Neuroscience, 9*(2), 268—275. https://doi.org/10.1038/nn1629.

CHAPTER 44

Music therapy in dementia: the effects of music therapy and other musical interventions on behavior, emotion, and cognition

Alfredo Raglio[1], Lapo Attardo[2]

[1]Music Therapy Research Laboratory, Istituti Clinici Scientifici Maugeri, Pavia, Italy; [2]Music Therapist at ASP Istituti Milanesi Martinitt e Stelline e Pio Albergo Trivulzio, Milan, Italy

Mini-dictionary of terms

Neural plasticity is the sum of the experience-driven changes at behavioral, ensemble, cellular, and molecular levels in the brain.

Musical interventions are musical activities (e.g., listening to music, singing, rhythmical use of musical instruments) delivered in the absence of a music therapist, therapeutic setting, and specific intervention model. The aim is to temporally improve mood or socialization or provide motor or cognitive stimulation.

Music therapy is those musical interventions delivered in a clinical setting by a certified music therapist who acts in accordance with a reference model. Music therapy is distinguished by a relational component provided by the therapeutic relationship between the patient and therapist. The effects of the intervention yearn to become stable and long-lasting over time.

Therapeutic relationship According to clinical psychology process research, one of the most important therapeutic factors of psychotherapy resides in the therapeutic relationship between patient and therapist. This nonverbal factor is also an integral part of music therapy.

Behavioral and psychological symptoms in dementia (BPSDs) Besides cognitive deterioration, dementia is characterized by symptoms affecting both psychological well-being and behavior. These symptoms are very common and associated with a high level of distress and poor quality of life in patients and their caregivers. There is evidence that musical interventions are effective in reducing BPSDs.

Introduction

Music is a universal human aspect found in all cultures across ages (Nettl, 2000). It often assumes a central role in our lives being that music accompanies the most important social, religious, and cultural events, generating a strong impact on emotion and behavior. This can be explained by the fact that music represents one of the most powerful human experiences, engaging the person from different perspectives: perception, cognition, emotion, motion, and social interaction. Both music-making and listening trigger a wide range of cognitive, sensory, and motor functions and activate widespread neural networks involved in emotional processing and behavior regulation (Koelsch, 2014).

Diagnosis and Management in Dementia
ISBN 978-0-12-815854-8, https://doi.org/10.1016/B978-0-12-815854-8.00044-6

Musical practice dramatically impacts brain morphology and functions, and for this reason many neuroscientists consider musicians' brains a model of neural plasticity (Münte, Altenmüller, & Jäncke, 2002; Schlaug, 2015; Zatorre & McGill, 2005) and music-making a valuable rehabilitative tool (Altenmüller & Schlaug, 2015; François, Grau-Sánchez, Duarte, & Rodriguez-Fornells, 2015; Wan & Schlaug, 2010). Music can also provide a means of interaction that relies on primitive forms of effective communication and allows mutual understanding, intersubjective contact, and affect attunement (Di Cesare, Di Dio, Marchi, & Rizzolatti, 2015; Raglio et al., 2017; Stern, 2010). This is one of the most direct ways to express and comprehend our own and others' feelings and mental states in the nonverbal domain. Music-making thus represents a great chance to interact and express oneself when other forms of communication are lost and social interaction becomes too complex.

Music therapy and other musical interventions

Music therapy

In music therapy, the properties of both sound and music as well as those of nonverbal communication combine in defined clinical and methodological frameworks based on psychological and neuroscientific models. Previously a heuristic model has been proposed according to which music therapy might act through five main modulating factors affecting attention, emotion, cognition, behavior, and communication (Hillecke, Nickel, & Bolay, 2005; Koelsch, 2009). According to the attention modulation factor, sound and music can automatically capture attention more than is possible with other sensory stimuli due to the function of audition as an early warning system. This allows music to distract from negative experiences, such as pain or anxiety, or disruptive behaviors and draws a person's attention to the "here and now" of the music therapy session. The emotional regulation factor is sustained by evidence for the impact of music on emotions. A meta-analysis of functional neuroimaging studies revealed that music-evoked emotions modulate the activity of the major limbic and paralimbic structures involved in emotional processing such as the amygdala, nucleus accumbens, hypothalamus, hippocampus, insula, cingulate cortex, and orbitofrontal cortex (Koelsch, 2014). There is evidence that emotional responses to music also contribute to general health and well-being by positively modulating different neurochemical systems involved in reward, motivation, and pleasure; stress and arousal; immunity; and social interaction (Chanda & Levitin, 2013). The cognition modulation factor accounts for all cognitive processes involved in music cognition concerning melody, harmony, rhythm, syntax, musical meaning, and social cognition. Music therapy also involves memory abilities implied in processing musical information (e.g., melodies and lyrics) or music-associated memories that seem more resistant to cognitive decline than other kinds of episodic or semantic information (Cuddy et al., 2012; Vanstone et al., 2012). The behavior modulation factor refers the property of music to generate and condition movements and behaviors even without conscious will. The inner structures of music can organize perceptions and feelings and regulate behavior and autonomic responses when these are disrupted. Rhythm and tempo are powerful organizing factors that, for instance, allow music to be used to co-ordinate a multitude of persons acting at the same time in the same way. Some of the most striking evidence of this musical property resides in the auditory—motor entrain-ment, the neural connection between the auditory and motor systems that constitutes the foundation of music therapy treatment of movement disorders (Thaut, 2015). In addition, the strong impact of musical stimuli on the autonomic nervous system explains the calming effect of relaxing music (Chanda & Levitin, 2013). In conclusion, the communication modulation factor accounts for the nonverbal interpersonal elements involved in the musical interaction between patient and therapist. This is the case of active music therapy where the patient—therapist dyad interacts by playing a musical instrument or singing and thus developing a nonverbal relationship based on empathy and emotional sharing (Gold, Solli, Krüger, & Lie, 2009). This process promotes mutual understanding and allows for a regulation of emotions and mental states that takes place implicitly and thus bypasses cognitive control (Raglio et al., 2017).

Different music therapy approaches are available depending on the weight assumed by the just-mentioned factors, the clinical population, and the main aim of treatment. One might be tempted to consider music therapy any musical experience delivered in a pathological condition, but many experts would disagree with this. The World Feder-ation of Music Therapy defines music therapy as "the professional use of music and its elements as an intervention in medical, educational, and everyday environments with in-dividuals, groups, families, or communities who seek to optimize their quality of life and improve their physical, social, communicative, emotional, intellectual, and spiritual health and wellbeing" (World Federation of Music Therapy, 8th World Conference of Music Therapy Hamburg, 1996) (Wigram, Pedersen & Bonde, 2002). Some features are essential for musical intervention to be defined as music therapy; these are (1) the presence of a qualified professional (the music therapist); (2) a therapeutic rationale that specifies the theory and practice of the intervention; (3) the presence of a therapeutic setting; and (4) the possibility to define specific goals and provide relative evaluations (Raglio & Oasi, 2015). What distinguishes music therapy from other types of interven-tion is that musical experiences assume a central role in the therapy session. These musical experiences can include musical improvisation, music listening, singing, song reproduc-tion, songwriting, exercising on a musical instrument, etc.

Depending on the type of musical experience provided, different music therapy techniques are available. Two main classes of approach are predominant in proper music therapy practice: more relational approaches and more rehabilitative approaches. Rela-tional approaches refer mainly to clinical psychology models and aim primarily at reducing behavioral and psychological symptoms and promote communication and emotional expression. In this case, the most diffused techniques are those based on direct interaction between the patient and music therapist by means of musical instruments or singing (active techniques) or those that employ listening to music to verbally elaborate the emerging emotional and psychological content (receptive techniques). In relational approaches, the therapeutic factors reside both in the musical experience itself and in the therapeutic nonverbal relationship between patient and therapist. On the other hand, rehabilitative approaches refer mainly to neuroscientific models and actively engage the patient in musical exercises aimed at the rehabilitation of sensory, cognitive, and motor functions. In this case the main therapeutic factors consist of the activation and strengthening of impaired functions during the musical experience. However, the presence of a trained music therapist is an essential requirement for a musical intervention to be defined as music therapy (Gold et al., 2011; Raglio & Gianelli, 2009). The specific role of the music therapist is to accompany the patient through the more appropriate musical experience according to the objectives of the intervention. The same musical experience—e.g., listening to music, singing, or playing an instrument—can be present in different kinds of musical interventions as well as in everyday life. In music therapy the music therapist is a professional trained to guide the patient in a way that maximizes the beneficial effects that the musical experience and therapeutic relationship may have for that specific person. This implies delineation of a tailored intervention that accounts for the person's needs, residual skills, difficulties, strengths, and limits to reach specific therapeutic objectives, at the same time avoiding frustration due to impaired abilities or cognitive deterioration. Another important feature of music therapy is the definition of a therapeutic setting including the therapeutic contract with the patient, time and frequency of sessions, number and characteristics of patients, and any instruction or explicit/implicit request addressed to them. Only by considering all these features can the musical experience really represent a powerful means of stimulation and communication. The music therapist always acts in accordance with a clinical theoretical model of the functioning of mind and behavior, and this allows the musical experience to operate beyond simple recreational activity. The outcome of music therapy should determine both symptom attenuation and the prevention/stabilization of complications derived from the symptoms themselves. Music therapy actually goes through a systematization process according to "evidence-based practice" (Mace, Moorey, & Roberts, 2000; Vink & Bruinsma, 2003). Such a practice should involve reflections and applicative implications, which can be summarized as follows: clinical criteria for admission to treatment; music therapeutic assessment (finalized to either inclusion in or exclusion from treatment); definition of therapeutic goals; therapeutic contract; treatment; evaluation; and conclusion. Notwithstanding these factors, due to the generic beneficial potentialities of music, some interventions are improperly defined as "music therapy" even in absence of the just-mentioned characterizing criteria. A growing body of evidence-based literature concerns music therapy and its application to a variety of clinical conditions including severe mental illness (Geretsegger et al., 2017), depression (Aalbers et al., 2017), autism spectrum disorder (Geretsegger, Elefant, Mössler, & Gold, 2014), acquired brain injuries (Magee, Clark, Tamplin, & Bradt, 2017), cancer care (Bradt, Dileo, Magill, & Teague, 2016), and dementia (van der Steen et al., 2018). Music therapy and other musical interventions in neurological rehabilitation showed positive effects on cognition, motor function, and emotional well-being supported by processes of neural plasticity and psychological change (Raglio et al., 2015, pp. 1534—1539; Sihvonen et al., 2017).

Other musical interventions

Other musical interventions, different from proper music therapy treatment, are also extensively diffused and can be mainly categorized as listening to music and general musical activities. Listening to music includes both self-selected (individualized music listening) and experimenter-selected listening. Individualized music listening implies the administration of preferred music listening to attenuate behavioral and psychological symptoms and improve general well-being (Gerdner, 2012; Särkämö et al., 2008). On the other hand, experimenter-selected listening assumes specific musical pieces identified depending on the structural characteristics of music and sound parameters to regulate physical and biological processes or reduce behavioral and psychological symptoms (Haas & Brandes, 2009). In scientific literature, individualized music listening interventions are sometimes supported by a trained music therapist, whereas experimenter-selected listening is usually delivered by any member of the medical staff. General musical approaches are those delivered by a nontrained music therapy professional in the absence of a specific therapeutic setting and intervention model. This includes musical activities such as music-making, singing in a choir, dancing, caregiver singing, or listening to music with the aim to entertain or temporarily improve mood and socialization or provide motor or cognitive stimulation.

Music therapy in dementia

Scientific interest in using music in dementia care increased in the mid-19'80s, though humans have always used music to alleviate pain and suffering. The first clinical observations revealed that even in the latest stage of dementia when other kinds of stimulation were no longer effective, music could still elicit reactions (Norberg, Melin, & Asplund, 1986). Compared with a general deficit in semantic knowledge, preserved procedural memory and learning abilities in music initially elicited surprise and fascination (Crystal, Grober, & Masur, 1989). Persons with dementia still seek and engage with music that can be effectively used to evoke personal autobiographical memories (Cuddy, Sikka, & Vanstone, 2015). Music may also act as a memory enhancer in patients with dementia; the mechanisms that underlie this effect may reside in the activation of dopaminergic pathways, in increased arousal and attention, and in a more holistic coding that facilitates memory processes (Peck, Girard, Russo, & Fiocco, 2016; Simmons-Stern, Budson, & Ally, 2010). Familiar music listening can automatically stimulate emotion and memory at the same time, thus providing a useful tool to promote self-consciousness and well-being in people with dementia (Arroyo-Anlló, Poveda, & Gil, 2013; El Haj, Fasotti, & Allain, 2012; Raglio et al., 2015; Särkämö et al., 2014). Even in severe stages of dementia, familiar music can still produce pleasantness. This seems plausible because memory for familiar music seems to be processed by cerebral areas that remain relatively spared in terms of cortical atrophy as revealed by functional magnetic resonance (Jacobsen et al., 2015). In active music therapy, the nonverbal interpersonal processes that take place between the patient and the music therapist allow persons with dementia to organize and regulate their emotions and behavior through the nonverbal relationship with the music therapist (Raglio et al., 2010a, 2010b, 2008; Raglio & Gianelli, 2013). However, although musical skills may be relatively preserved compared with other cognitive functions, the precise profile of spared musical functions in people with dementia is likely to depend on the type and causes of the illness (Baird & Samson, 2015). For example, there might be an overlap between brain regions implicated in attributing mental states and coding emotional value, all of which are processes that are particularly impaired in frontotemporal degeneration (Downey et al., 2013; Omar et al., 2011). Nonetheless, even in such cases, music therapy seems to be helpful in reducing behavioral and psychological symptoms (Raglio et al., 2012).

There are now several reviews, systematic reviews, and meta-analyses of controlled clinical trials (CCTs) and randomized controlled trials (RCTs) assessing the effects of music therapy and other musical interventions in persons with dementia (Fusar-Poli, Bieleninik, Brondino, Chen, & Gold, 2017; Sihvonen et al., 2017; Steen et al., 2017; Zhang et al., 2016). Most included studies focus on neuropsychiatric and behavioral symptoms such as anxiety, agitation, aggression, and depression as well as cognitive deficits, emotional well-being, and quality of life. These meta-analyses suffer from the highly variable and generally low methodological quality of the included studies in terms of sample size, allocation concealment, active control group, blinding of outcome assessment, etc. Another limiting factor is the heterogeneity of musical interventions provided in the studies. However, albeit the methodological limitations restrict the generalizability of the results, the authors of these reviews conclude that music therapy is or may be effective in reducing symptoms and improving the quality of life in people with dementia. In the following subsections, some of the most relevant CCTs and RCTs concerning music therapy and other musical interventions in dementia are summarized. These are studies retrieved in PubMed and PsycINFO databases published in the English language during from approximately 2010 to 2020. Studies are presented in separate sections for different main outcomes (Fig. 44.1).

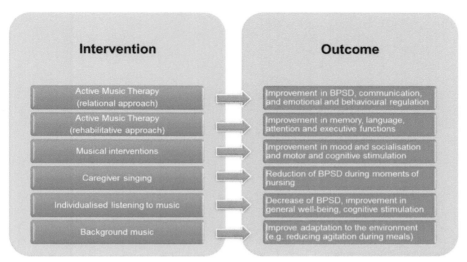

Figure 44.1 List of evidence-based musical interventions in dementia. The figure illustrates the different evidence-based musical interventions in dementia (left side) and their main outcomes (right side).

Effects on behavioral and psychological symptoms

Most studies on the effects of music therapy and other musical interventions in dementia have focused on agitation and other neuropsychiatric symptoms (Table 44.1). There is evidence that music therapy and musical interventions may be usefully employed to reduce behavioral and psychological symptoms in dementia (BPSDs).

Effects on cognitive functions

A few other studies have investigated the effect of musical interventions and music therapy on cognition in people with dementia (Table 44.2). This research suggests that music can be effectively used to boost memory, attention, orientation, and general cognition.

Effects on physiological functions

There is a scarcity of studies that evaluate the effects of music therapy and musical intervention on physiological parameters (Table 44.3). The available literature regards outcomes such as heart rate variability or plasma cytokine and catecholamine levels.

Despite differences in the music therapy approaches discussed in section 2.1, most interventions reported in the previously mentioned studies encompass multiple types of musical experiences, employing both active and receptive techniques.

Table 44.1 Effects of music therapy and other musical interventions on behavioral and psychological symptoms.

Author, year, journal	Stages of dementia	Outcomes	Intervention	Conductor	Results
Raglio et al., 2008, Alzheimer Dis Assoc Disord.	Moderate to severe	BPSDs (NPI)	Active music therapy	Music therapist	Decrease in BPSDs
Raglio et al., 2010, Aging Ment Health.	Severe	BPSDs (NPI)	Active music therapy	Music therapist	Reduction in BPSDs
Cook et al., 2010, J. Health Psychol.	Mild to moderate	Depression (GDS), quality of life (D–QoL)	Music listening, familiar song singing	Musician	Improvements in self-esteem and reduced depression
Cook et al., 2010, Aging Ment Health.	Mild to moderate	Agitation (CMAI-SF), anxiety (RAID)	Music listening, familiar song singing	Musician	No significant effects
Lin et al., 2011, Int J Geriatr Psychiatry.	Moderate to severe	Agitation (CMAI)	Musical activities (playing instruments, singing, listening to music)	Music therapist	Reduction in agitated behaviors
Janata, 2012, Music Med.	Moderate to severe	BPSDs (NPI), agitation (CMAI), depression (CSDD)	Individualized listening to music	Music therapist	Reduction in agitation, depression, and NPI
Sung et al., 2012, Int J Geriatr Psychiatry.	Mild to moderate	Anxiety (RAID), agitation (CMAI)	Musical activities (moving and playing along with familiar music)	Research assistant	Reduction in anxiety but not in agitation
Ridder et al., 2013, Aging Ment Health.	Moderate to severe	Agitation (CMAI), amount of medication	Active music therapy	Music therapist	Reduction in agitated behaviors and in medication

				Music therapist	Decrease in agitated behaviors
Vink et al., 2013, Int J Geriatr Psychiatry.	Moderate to severe	Agitation (CMAI)	Listening to music, singing, dancing, playing musical instruments	Music therapist	Decrease in agitated behaviors
Vink et al., 2014, J Am Geriatr Soc.	Moderate to severe	BPSDs (NPI-Q)	Music-making, singing, and listening to the music performed by the music therapist	Music therapist	Improvements in neuropsychiatric symptoms
Narme et al., 2014, J Alzheimers Dis.	Moderate to severe	BPSDs (NPI, CMAI), Caregiver distress (NPI distress score), cognition (SIB)	Music listening, singing, playing percussions, and music-prompted reminiscence	Supervisor with no education in music therapy	Improvements in emotional state, NPI, and caregiver distress; no effects on cognition
Raglio et al., 2015, J Am Geriatr Soc.	Moderate to severe	BPSDs (NPI), mood (CSDD), QoL (CBS–QoL)	(a) Active music therapy or (b) individualized listening to music	Music therapist	Improvements in NPI, mood, and QoL in all groups

The table summarizes the characteristics and the main results of clinical trials using musical intervention to reduce behavioral and psychological symptoms in dementia (BPSDs). List of abbreviations—see Table 44.3.

Table 44.2 Effects of music therapy and other musical intervention on cognitive functions.

Author, year, journal	Stages of dementia	Outcomes	Intervention	Conductor	Results
Guétin et al., 2009, Dement Geriatr Cogn Disord.	Mild to moderate	General cognition (MMSE), depression (GDS), anxiety (HRSD)	Individualized listening to music	Not specified therapist	No effects on cognition, improvements in anxiety and depression
Ceccato et al., 2012, Am J Alzheimers Dis Other Demen.	Mild to moderate	General cognition (MMSE), learning and memory (DPM, IPM) attention (DSF, DSB, attentional matrices)	Musical exercise to promote sensory and cognitive skills	Music therapist	Improvement in DPM, IPM, and attentional matrices
Särkämö et al., 2014, Gerontologist.	Mild to moderate	General cognition (MMSE), attention, executive functions (WAIS-III), memory (WMS-III)	(a) Group singing and rhythmic movements or (b) group music listening and discussion	Music teacher (a) and music therapist (b)	Improvements in general cognition, orientation, episodic and working memory, attention, and executive functions.
Chu et al., 2014, Biol Res Nurs.	Mild to moderate	General cognition, attention, memory and language (MMSE), salivary cortisol, depression (CSDD)	Singing, instrument playing, music listening, and music-prompted reminiscence	Music therapist	Improvements in depression and cognition (short-term recall function in particular); no effect on cortisol levels

The table summarizes the characteristics and the main results of clinical trials using musical intervention to boost cognitive functioning in dementia. List of abbreviations: see Table 44.3.

Table 44.3 Effects of music therapy and other musical intervention on physiological parameters.

Author, year, journal	Stages of dementia	Outcomes	Intervention	Conductor	Results
Okada et al., 2009, Int Heart J.	Advanced dementia	Plasma cytokine, catecholamine levels, heart rate variability (EGC Holter)	Not specified use of nursery rhymes, folk songs, hymns, and recent Japanese pop music	Music therapist	Improvement in cardiovascular status; reduction in plasma cytokines and catecholamine levels
Raglio et al., 2010, Curr Aging Sci.	Moderate–Severe	BPSDs (NPI), heart rate variability (EGC Holter)	Active music therapy	Music therapist	Improvement in depression and heart rate variability
Sakamoto et al., 2013, Int Psychogeriatr.	Severe	BPSDs (BEHAVE-AD), emotional state (face scale), heart rate variability	(a) Familiar music listening or (b) musical interactions	Music facilitator (music therapist and other professionals)	Improvements in BPSDs, emotional state, and heart rate variability

The table summarizes the characteristics and the main results of clinical trials using musical intervention to improve physiological parameters in dementia. *NPI*, neuropsychiatric inventory; *BPSDs*, behavioral and psychological symptoms in dementia; *GDS*, geriatric depression scale; *D-QoL*, dementia quality of lifequestionnaire; *CMAI-SF*, Cohen-Mansfield agitation inventory—short form; *RAID*, rating of anxiety in dementia scale; *CMAI*, Cohen-Mansfield agitation inventory; *CSDD*, Cornell scale for depression in dementia; *NPI-Q*, neuropsychiatric inventory—questionnaire; *SIB*, severe impairment battery; *CBS-QoL*, Cornell-Brown scale for quality of life in dementia; *MMSE*, mini-mental state examination; *HRSD*, Hamilton rating scale for depression; *DPM*, deferred prose memory test; *IPM*, immediate prose memory test; *DSF*, digit span forward; *DSB*, digit span backward; WAISS-III, Wechsler adult intelligence scale—III; *WMS-III*, Wechsler memory scale—III; *BEHAVE-AD*, behavioral pathology in Alzheimer's disease.

A global music approach to persons with dementia

Given the relevance of musical stimuli for persons with dementia, a structured intervention model named global music approach to persons with dementia (GMA-D) has been proposed (Raglio, Filippi, Bellandi, & Stramba-Badiale, 2014). GMA-D encompasses various evidence-based musical interventions that may normally take place in nursing homes and includes general music-based interventions, caregiver singing, individualized listening to music, background music, and proper music therapy sessions with patients and the patient—caregiver dyad. Each single intervention addresses the need of the person with dementia in different situations. Specifically, general music-based interventions comprehend activities like rhythmic use of instruments, singing, movements associated with music, or general listening to music. These activities are usually addressed to a patient group in the absence of a qualified therapist and thus do not have a therapeutic goal, a structured therapeutic setting, and any reference to a theoretical music therapy model. Caregiver singing consists of singing and/or vocalism performed by the caregiver during moments of nursing such as getting up in the morning, hygiene, getting dressed, escorting to the toilet, assistance in the bathing routine, and handling of patients. These are moments that can often generate stress, confusion, and disorientation to the person with dementia. When these feelings are the cause of aggressive, resisting, or defensive behavior that impedes nursing procedures, caregiver singing can be used as a distracting power that reduces behavioral problems and makes it much easier for caregivers to carry out their tasks. This also promotes contact in the assistance of the patient, inducing serenity while communicating the nonthreatening intentions of the caregiver. Individualized listening to music aims at reducing behavioral disturbances and promoting relaxation through listening to a personalized playlist based on individual preferences. This intervention should be preferred when there are difficulties in adaptation to the environment or in communicating with other persons. On the other hand, background music aims at reducing disturbed behaviors and improving the general emotional atmosphere during specific situations like lunchtime or moments of rest. This global music approach derives from a review of the scientific literature concerning musical interventions in dementia (Raglio et al., 2014). The general aim of the GMA-D is to provide a nonpharmacological treatment without side effects for reducing behavioral disturbances, providing cognitive and motor stimulation, maintaining social relationships, improving mood, and promoting the overall quality of life. A recent pilot testing of the GMA-D showed promising findings in terms of efficacy and implementation (Raglio, Filippi, Leonardelli, Trentini, & Bellandi, 2018).

Conclusions

There is evidence that music therapy can help people with dementia and their caregivers face problematic behaviors, provide sensory and cognitive stimulation, and improve general well-being and quality of life. The possible mechanisms of action of musical interventions include the specific impact of music on autonomic nervous system and neural plasticity as well as activation of systems and neural networks involved in reward, arousal, and emotion regulation. In addition, musical interaction represents a nonverbal form of communication that resists cognitive decline and allows social interaction ed emotional expression even in late stages of the illness.

Key facts of neuroscience of music

- Music is a complex human experience that engages different processes at the same time.
- From laboratory study of nonhuman animals, it is known that attention, repetition, and emotions are components known to produce plastic changes in the brain.
- Thanks to widespread neural activation, the repetition typical of constant musical practice, the high level of sustained attention, and the strong involvement of emotions, playing music produces remarkable anatomical and functional changes in the nervous system.
- Listening to music and music-making promote neural connectivity, induce gray and white matter changes, and favor processes of multisensory integration.
- For these reasons, many neuroscientists consider the musician's brain a model for the study of neural plasticity.
- The strong impact of musical experiences on brain areas and functions is the foundation of the rehabilitative potential of musical interventions.

Key facts of music therapy in dementia

- Persons with dementia show several behavioral and cognitive symptoms.
- Available drugs often produce modest symptom control and problematic side effects.
- Music represents an accessible and engaging stimulus to persons with dementia and can be organized in activities that boost cognitive functions and promote the regulation of emotions and behavior.
- Music therapy represents a promising, cost-effective nonpharmacological and intervention in dementia care absent of side effects.

Summary points

- Music is a complex stimulus that activates a wide range of neural, autonomic, cognitive, and motor functions.
- Musical processing resists cognitive deterioration and engages people with dementia.

- Music-making also represents a means of communication when other forms of social interaction are lost.
- Musical interventions exploit the properties of music to promote cognitive functioning, emotional and behavioral regulation, and quality of life.
- What distinguishes music therapy from other kinds of musical interventions is the presence of a therapeutic relationship with a qualified therapist who acts in accordance with a clinical model.
- Music therapy includes going through a systematized process according to evidence-based practice.
- There are now several CCTs and RCTs to assess the effects of music therapy and other musical interventions in BPSD, cognitive functioning, and physiological parameters in persons with dementia.
- A global music approach to persons with dementia that encompasses various evidence-based musical interventions that normally take place in nursing homes is proposed.

References

Aalbers, S., Fusar-Poli, L., Freeman, R. E., Spreen, M., Ket, J. C., Vink, A. C., et al. (2017). Music therapy for depression. *Cochrane Database of Systematic Reviews*. https://doi.org/10.1002/14651858.CD004517.pub3.

Altenmüller, E., & Schlaug, G. (2015). Apollo's gift: New aspects of neurologic music therapy. *Progress in Brain Research, 217*, 237–252. https://doi.org/10.1016/bs.pbr.2014.11.029.

Arroyo-Anlló, E. M., Poveda, J., & Gil, R. (2013). Familiar music as an enahancer of self-consciousness in patients with Alzheimer's disease. *BioMed Research International*, 1–11.

Baird, A., & Samson, S. (2015). Music and dementia. *Progress in Brain Research, 217*, 207–235. https://doi.org/10.1016/bs.pbr.2014.11.028.

Bradt, J., Dileo, C., Magill, L., & Teague, A. (2016). Music interventions for improving psychological and physical outcomes in cancer patients. Cochrane Database of Systematic Reviews. https://doi.org/10.1002/14651858.CD006911.pub3.

Chanda, M. L., & Levitin, D. J. (2013). The neurochemistry of music. *Trends in Cognitive Sciences, 17*(4), 179–191. https://doi.org/10.1016/j.tics.2013.02.007.

Crystal, H. A., Grober, E., & Masur, D. (1989). Preservation of musical memory in Alzheimer's disease. *Journal of Neurology Neurosurgery and Psychiatry, 52*(12), 1415–1416. Retrieved from http://www.ncbi.nlm.nih.gov/pubmed/2614438.

Cuddy, L. L., Duffin, J. M., Gill, S. S., Brown, C. L., Sikka, R., & Vanstone, A. D. (2012). Memory for melodies and lyrics in Alzheimer's disease. *Music Perception: An Interdisciplinary Journal, 29*(5), 479–491. https://doi.org/10.1525/mp.2012.29.5.479.

Cuddy, L. L., Sikka, R., & Vanstone, A. (2015). Preservation of musical memory and engagement in healthy aging and Alzheimer's disease. *Annals of the New York Academy of Sciences, 1337*, 223–231. https://doi.org/10.1111/nyas.12617.

Di Cesare, G., Di Dio, C., Marchi, M., & Rizzolatti, G. (2015). Expressing our internal states and understanding those of others. *Proceedings of the National Academy of Sciences, 112*(33), 10331–10335. https://doi.org/10.1073/pnas.1512133112.

Downey, L. E., Blezat, A., Nicholas, J., Omar, R., Golden, H. L., Mahoney, C. J., et al. (2013). Mentalising music in frontotemporal dementia. *Cortex, 49*(7), 1844–1855. https://doi.org/10.1016/j.cortex.2012.09.011.

El Haj, M., Fasotti, L., & Allain, P. (2012). The involuntary nature of music-evoked autobiographical memories in Alzheimer's disease. *Consciousness and Cognition, 21*(1), 238–246. https://doi.org/10.1016/j.concog.2011.12.005.

François, C., Grau-Sánchez, J., Duarte, E., & Rodriguez-Fornells, A. (2015). Musical training as an alternative and effective method for neuro-education and neuro-rehabilitation. *Frontiers in Psychology, 6*, 475. https://doi.org/10.3389/fpsyg.2015.00475.

Fusar-Poli, L., Bieleninik, L., Brondino, N., Chen, X.-J., & Gold, C. (2017). The effect of music therapy on cognitive functions in patients with dementia: A systematic review and meta-analysis. *Aging and Mental Health*, 1–10. https://doi.org/10.1080/13607863.2017.1348474.

Gerdner, L. A. (2012). Individualized music for dementia: Evolution and application of evidence-based protocol. *World Journal of Psychiatry, 2*(2), 26–32. https://doi.org/10.5498/wjp.v2.i2.26.

Geretsegger, M., Elefant, C., Mössler, K. A., & Gold, C. (2014). Music therapy for people with autism spectrum disorder. *Cochrane Database of Systematic Reviews, 6*, CD004381. https://doi.org/10.1002/14651858.CD004381.pub3.

Geretsegger, M., Mössler, K. A., Bieleninik, L., Chen, X.-J., Heldal, T. O., & Gold, C. (2017). Music therapy for people with schizophrenia and schizophrenia-like disorders. *Cochrane Database of Systematic Reviews, 5*, CD004025. https://doi.org/10.1002/14651858.CD004025.pub4.

Gold, C., Erkkilä, J., Bonde, L. O., Trondalen, G., Maratos, A., & Crawford, M. J. (2011). Music therapy or music medicine? *Psychotherapy and Psychosomatics, 80*(5), 304. https://doi.org/10.1159/000323166.

Gold, C., Solli, H. P., Krüger, V., & Lie, S. A. (2009). Dose–response relationship in music therapy for people with serious mental disorders: Systematic review and meta-analysis. *Clinical Psychology Review, 29*(3), 193–207. https://doi.org/10.1016/j.cpr.2009.01.001.

Haas, R., & Brandes, V. (Eds.). (2009). *Music that works*. Vienna: Springer Vienna. https://doi.org/10.1007/978-3-211-75121-3.

Hillecke, T., Nickel, A., & Bolay, H. V. (2005). Scientific perspectives on music therapy. *Annals of the New York Academy of Sciences, 1060*, 271–282. https://doi.org/10.1196/annals.1360.020.

Jacobsen, J. H., Stelzer, J., Fritz, T. H., Chételat, G., La Joie, R., & Turner, R. (2015). Why musical memory can Be preserved in advanced Alzheimer's disease. *Brain, 138*(December), 2438–2450. https://doi.org/10.1093/brain/awv135.

Koelsch, S. (2009). A neuroscientific perspective on music therapy. *Annals of the New York Academy of Sciences, 1169*, 374–384. https://doi.org/10.1111/j.1749-6632.2009.04592.x.

Koelsch, S. (2014). Brain correlates of music-evoked emotions. *Nature Reviews Neuroscience, 15*(3), 170–180. https://doi.org/10.1038/nrn3666.

Mace, C., Moorey, S., & Roberts, B. (2000). *Evidence in the psychological therapies: A critical guide for practitioners*. New York: Brunner-Routledge.

Magee, W. L., Clark, I., Tamplin, J., & Bradt, J. (2017). Music interventions for acquired brain injury. *Cochrane Database of Systematic Reviews, 1*, CD006787. https://doi.org/10.1002/14651858.CD006787.pub3.

Münte, T. F., Altenmüller, E., & Jäncke, L. (2002). The musician's brain as a model of neuroplasticity. *Nature Reviews Neuroscience, 3*(6), 473–478. https://doi.org/10.1038/nrn843.

Nettl, B. (2000). An enthnomusicologist contemplates universals in musical sound and musical culture. In N. L. Wallin, B. Merker, & S. Brown (Eds.), *The origins of music* (pp. 463–472). Cambridge, MA: MIT Press.

Norberg, A., Melin, E., & Asplund, K. (1986). Reactions to music, touch and object presentation in the final stage of dementia. An exploratory study. *International Journal of Nursing Studies, 23*(4), 315–323. Retrieved from http://www.ncbi.nlm.nih.gov/pubmed/3536774.

Omar, R., Henley, S. M. D., Bartlett, J. W., Hailstone, J. C., Gordon, E., Sauter, D. A., et al. (2011). The structural neuroanatomy of music emotion recognition: Evidence from frontotemporal lobar degeneration. *NeuroImage, 56*(3), 1814–1821. https://doi.org/10.1016/j.neuroimage.2011.03.002.

Peck, K. J., Girard, T. A., Russo, F. A., & Fiocco, A. J. (2016). Music and memory in Alzheimer's disease and the potential underlying mechanisms. *Journal of Alzheimer's Disease, 51*(4), 949–959. https://doi.org/10.3233/JAD-150998.

Raglio, A., Attardo, L., Gontero, G., Rollino, S., Groppo, E., & Granieri, E. (2015). Effects of music and music therapy on mood in neurological patients. *World Journal of Psychiatry, 5*(1), 68—78. https://doi.org/10.5498/wjp.v5.i1.68.

Raglio, A., Bellandi, D., Baiardi, P., Gianotti, M., Ubezio, M. C., & Granieri, E. (2012). Music therapy in frontal temporal dementia: A case report. *Journal of the American Geriatrics Society, 60*(8), 1578—1579. https://doi.org/10.1111/j.1532-5415.2012.04085.x.

Raglio, A., Bellandi, D., Baiardi, P., Gianotti, M., Ubezio, M. C., Zanacchi, E., et al. (2015). *Effect of active music therapy and individualized listening to music on Dementia : A multicenter randomized controlled.* https://doi.org/10.1111/jgs.13558.

Raglio, A., Bellelli, G., Traficante, D., Gianotti, M., Ubezio, M. C., Gentile, S., et al. (2010). Efficacy of music therapy treatment based on cycles of sessions: A randomised controlled trial. *Aging & Mental Health, 14*(8), 900—904. https://doi.org/10.1080/13607861003713158.

Raglio, A., Bellelli, G., Traficante, D., Gianotti, M., Ubezio, M. C., Villani, D., et al. (2008). Efficacy of music therapy in the treatment of behavioral and psychiatric symptoms of dementia. *Alzheimer Disease and Associated Disorders, 22*(2), 158—162. https://doi.org/10.1097/WAD.0b013e3181630b6f.

Raglio, A., Filippi, S., Bellandi, D., & Stramba-Badiale, M. (2014). Global music approach to persons with dementia: Evidence and practice. *Clinical Interventions in Aging, 9,* 1669—1676. https://doi.org/10.2147/CIA.S71388.

Raglio, A., Filippi, S., Leonardelli, L., Trentini, E., & Bellandi, D. (2018). The global music approach to dementia (GMA-D): Evidences from a case report. *Aging Clinical and Experimental Research, 0*(0), 0. https://doi.org/10.1007/s40520-018-0919-8.

Raglio, A., & Gianelli, M. V. (2009). Music therapy for individuals with dementia: Areas of interventions and research perspectives. *Current Alzheimer Research, 6*(3), 293—301. https://doi.org/10.2174/156720509788486617.

Raglio, A., & Gianelli, M. V. (2013). Music and music therapy in the management of behavioral disorders in dementia. *Neurodegenerative Disease Management, 3*(4), 295—298. https://doi.org/10.2217/nmt.13.27.

Raglio, A., Gnesi, M., Monti, M. C., Oasi, O., Gianotti, M., Attardo, L., et al. (2017). The Music Therapy Session Assessment Scale (MT-SAS): Validation of a new tool for music therapy process evaluation. *Clinical Psychology & Psychotherapy,* (March), 1—15. https://doi.org/10.1002/cpp.2115.

Raglio, A., & Oasi, O. (2015). Music and health: What interventions for what results? *Frontiers in Psychology, 6*(MAR), 1—3. https://doi.org/10.3389/fpsyg.2015.00230.

Raglio, A., Oasi, O., Gianotti, M., Manzoni, V., Bolis, S., Ubezio, M. C., et al. (2010). Effects of music therapy on psychological symptoms and heart rate variability in patients with dementia. A pilot study. *Current Aging Science, 3*(3), 242—246. Retrieved from http://www.ncbi.nlm.nih.gov/pubmed/20735342.

Särkämö, T., Tervaniemi, M., Laitinen, S., Forsblom, A., Soinila, S., Mikkonen, M., et al. (2008). Music listening enhances cognitive recovery and mood after middle cerebral artery stroke. *Brain: Journal of Neurology, 131*(Pt 3), 866—876. https://doi.org/10.1093/brain/awn013.

Särkämö, T., Tervaniemi, M., Laitinen, S., Numminen, A., Kurki, M., Johnson, J. K., et al. (2014). Cognitive, emotional, and social benefits of regular musical activities in early dementia: Randomized controlled study. *The Gerontologist, 54*(4), 634—650. https://doi.org/10.1093/geront/gnt100.

Schlaug, G. (2015). Musicians and music making as a model for the study of brain plasticity. *Progress in Brain Research, 217,* 37—55. https://doi.org/10.1016/bs.pbr.2014.11.020.

Sihvonen, A. J., Särkämö, T., Leo, V., Tervaniemi, M., Altenmüller, E., & Soinila, S. (2017). Music-based interventions in neurological rehabilitation. *The Lancet Neurology, 16*(8), 648—660. https://doi.org/10.1016/S1474-4422(17)30168-0.

Simmons-Stern, N. R., Budson, A. E., & Ally, B. A. (2010). Music as a memory enhancer in patients with Alzheimer's disease. *Neuropsychologia, 48*(10), 3164—3167. https://doi.org/10.1016/j.neuropsychologia.2010.04.033.

van der Steen, J. T., Smaling, H. J., van der Wouden, J. C., Bruinsma, M. S., Scholten, R. J., & Vink, A. C. (2018). Music-based therapeutic interventions for people with dementia. *Cochrane Database of Systematic Reviews, 7,* CD003477. https://doi.org/10.1002/14651858.CD003477.pub4.

Stern, D. (2010). *Forms of vitality: Exploring dynamic experience in psychology, the arts, psychotherapy, and development.* Oxford.

Thaut, M. H. (2015). The discovery of human auditory-motor entrainment and its role in the development of neurologic music therapy. *Progress in Brain Research, 217*, 253—266. https://doi.org/10.1016/bs.pbr.2014.11.030.

Vanstone, A. D., Sikka, R., Tangness, L., Sham, R., Garcia, A., & Cuddy, L. L. (2012). Episodic and semantic memory for melodies in Alzheimer's disease. *Music Perception: An Interdisciplinary Journal, 29*(5), 501—507. https://doi.org/10.1525/mp.2012.29.5.501.

Vink, A., & Bruinsma, M. (2003). Evidence based music therapy. *Music Therapy Today: A Quarterly Journal of Studies in Music and Music Therapy, 4*(5), 26. Retrieved from http://search.ebscohost.com/login.aspx?direct=true&db=rih&AN=2003-15197&site=ehost-live%5Cnhhttp://www.wfmt.info/Musictherapy-world/modules/mmmagazine/issues/20031103132043/20031103134548/Vink.pdf.

Wan, C. Y., & Schlaug, G. (2010). Music making as a tool for promoting brain plasticity across the life span. *The Neuroscientist: A Review Journal Bringing Neurobiology, Neurology and Psychiatry, 16*(5), 566—577. https://doi.org/10.1177/1073858410377805.

Wigram, A., Pedersen, I. N., & Bonde, L. O. (2002). *A comprehensive guide to music therapy: Theory, clinical practice, research and training.* London: Jessica Kingsley Publishers.

Zatorre, R., & McGill, J. (2005). Music, the food of neuroscience? *Nature, 434*(7031), 312—315. https://doi.org/10.1038/434312a.

Zhang, Y., Cai, J., An, L., Hui, F., Ren, T., Ma, H., et al. (2016). Does music therapy enhance behavioral and cognitive function in elderly dementia patients? A systematic review and meta-analysis. *Ageing Research Reviews, 35*, 1—11. https://doi.org/10.1016/j.arr.2016.12.003.

CHAPTER 45

Exploitation of aromatherapy in dementia—impact on pain and neuropsychiatric symptoms

Damiana Scuteri[1], Laura Rombolà[1], Luigi Antonio Morrone[1], Domenico Monteleone[2], Maria Tiziana Corasaniti[3], Tsukasa Sakurada[4], Shinobu Sakurada[5], Giacinto Bagetta[1]

[1]Section of Preclinical and Translational Pharmacology, Pharmacotechnological Documentation and Transfer Unit (PDTU), Department of Pharmacy, Health Science and Nutrition, University of Calabria, Rende, Italy; [2]DG Animal Health and Veterinary Drugs, Ministry of Health, Rome, Italy; [3]Department of Health Sciences, University "Magna Græcia" of Catanzaro, Catanzaro, Italy; [4]First Department of Pharmacology, Daiichi College of Pharmaceutical Sciences, Fukuoka, Japan; [5]Department of Physiology and Anatomy, Tohoku Pharmaceutical University, Sendai, Japan

List of abbreviations

AD Alzheimer's disease
Aβ β amyloid
BEO essential oil of bergamot
CMAI Cohen-Mansfield agitation inventory
i.pl. intraplantar
NPSs neuropsychiatric symptoms
NSAIDs nonsteroidal antiinflammatory drugs
PRECISION-ABPM Prospective Randomized Evaluation of Celecoxib Integrated Safety Versus Ibuprofen or Naproxen Ambulatory Blood Pressure Measurement
PSNL partial sciatic nerve ligation
5-HT serotonin

Mini-dictionary of terms

Aromatherapy essential oils administered for massage or inhalation to improve well-being.
Complementary medicine use of officinal plant products along with validated treatments.
Mitochondrial bioenergetics impairment reduced glucose metabolism and mitochondrial electron transport chain alterations involved in the formation of β amyloid plaques and neurofibrillary tangles.
Neuropsychiatric symptoms of dementia frequent multifactorial psychotic symptoms, including agitation and aggression in demented patients.
Unrelieved pain misdiagnosed and undertreated pain can cause agitation and aggression.

Introduction

Neurological diseases, characterized by memory loss and cognitive impairment, are gathered under the term "dementia," among which Alzheimer's disease (AD) represents

Diagnosis and Management in Dementia
ISBN 978-0-12-815854-8, https://doi.org/10.1016/B978-0-12-815854-8.00045-8

some 50%–70% of cases worldwide (Winblad et al., 2016). Its prevalence is subject to continuous increase that affects mainly people over 65 years of age, with an estimate of 131 million patients by 2050 (Prince et al., 2015). The pathophysiology of AD is very complex (Onyango, 2018): age and apolipoprotein E4 represent the main risk factors for sporadic AD, while gene mutations of amyloid precursor protein, presenilin 1, and presenilin 2 are thought to be responsible for familial AD. Other genes implicated are apolipoprotein J; gene encoding the complement component (3b/4b) receptor 1; gene encoding PI-binding clathrin assembly protein; gene encoding the bridging integrator 1; and disabled homolog 1 (Onyango, 2018). Some early alterations include impairment of mitochondrial bioenergetics and glucose metabolism (Gibson & Shi, 2010), as supported by positron emission tomography studies highlighting reduced glucose metabolism in the hippocampus and entorhinal cortex (Calsolaro & Edison, 2016) and by a lowered amount of mitochondria and the increased presence of free cytosolic mitochondrial DNA and proteins in the postmortem brains of AD patients (Arun, Liu, & Donmez, 2016). Such mitochondrial alterations may underlie β amyloid (Aβ) plaque formation and neurofibrillary tangles (Gibson & Shi, 2010) as well as the increased production of reactive oxygen species because of the derangement of the electron transport chain (Chen & Yan, 2010). Indeed, this alteration can induce hyperphosphorylation and polymerization of tau protein (Onyango, 2018; Simoncini et al., 2015). Moreover, there is strong evidence supporting involvement of the immune system and neuroinflammation in the pathophysiology of AD (Heneka et al., 2015), and the mitochondrion can induce activation of the inflammasome, involved in the pathway of proIL-1β and proIL-18 (Onyango, 2018). Due to the lack of a definite pathogenetic mechanism, a disease-modifying therapy is not yet available. Apart from memory loss and cognitive decline, underestimated pain (Sengstaken & King, 1993) and multifactorial neuropsychiatric symptoms (NPSs) of dementia (Husebo, Ballard, Sandvik, Nilsen, & Aarsland, 2011) remarkably reduce patient quality of life. There is promising evidence for the effectiveness of aromatherapy in the management of pain and NPSs (Ballard et al., 2009). The essential oil of bergamot (BEO), often used in aromatherapy, is endowed with properties that make it the most likely candidate for complementary therapy of pain and NPSs in persons with dementia (Scuteri, Morrone, et al., 2017).

Pain and neuropsychiatric symptoms

Pain signals are transmitted via the ascending pain pathway; peripheral nociceptors are responsible for the transduction and delivery of pain stimulus to the dorsal horn of the spinal cord where it is modulated by the descending pathway, whose main neurotransmitters are serotonin (5-HT) and norepinephrine (Brown & Boulay, 2013). Several neuropathological alterations occur in the impaired brain of AD patients:

neurodegeneration of the cholinergic nucleus basalis (Zarow, Lyness, Mortimer, & Chui, 2003) and the periaqueductal gray (Parvizi, Van Hoesen, & Damasio, 2000), as well as neuronal loss in the locus coeruleus (Zarow et al., 2003), have been reported. Furthermore, reduced expression of 5-HT 2A receptors has been shown in the frontal and temporal cortex of the severe AD postmortem brain (Lai et al., 2005). All these modifications could underlie an impairment of pain perception, and the T102C polymorphism of the 5HT2A receptor gene was identified as a possible risk factor for the development of NPSs (Holmes, Arranz, Powell, Collier, & Lovestone, 1998; Nacmias et al., 2001). At first glance, NPSs could solely result from cognitive decline, but with deeper insight they have been linked to misdiagnosed and consequently undertreated pain states (Sampson et al., 2015; Sengstaken & King, 1993). Indeed, due to aging, dementia patients are often subjected to chronic conditions that predispose them to pain: cancer pain; rheumatoid arthritis; knee osteoarthritis; postherpetic neuralgia; diabetic neuropathy; and stroke pain (Scherder et al., 2009). There is a dangerous bias in professionals' evaluation of pain owing to the misconception that dementia patients feel less pain than do healthy aged people based on their impaired communication skills (Scherder et al., 2009). Unrelieved pain can induce the development of NPSs (Husebo, Ballard, & Aarsland, 2011; Scherder et al., 2009), particularly agitation and aggression. Dementia patients who develop at least one NPS during the entire course of the disease amount to more than 90% and up to 85% of people within 5 years (Steinberg et al., 2008). Some 20% of community patients (Ballard & Corbett, 2013; Lyketsos et al., 2000) and about 40%—60% of those hosted in nursing homes (Ballard & Corbett, 2013; Margallo-Lana et al., 2001) develop the most uncontrollable of these symptoms—i.e., agitation and aggression. Unfortunately, the number of patients affected by dementia receive pharmacological pain treatment at a lower rate than do those in the general population of the elderly, thus demonstrating the undertreatment of pain states in these patients (Ballard, Smith, Husebo, Aarsland, & Corbett, 2011). In view of the lack of a reliable standardized method to distinguish pain-induced from not-pain-induced NPSs (Ballard et al., 2011), it is possible that patients receive unnecessary treatments. The pharmacological treatment of NPSs encounters the use of atypical antipsychotics, among which risperidone has proven to be the safest; however, this drug is indicated for no longer than 6—12 weeks because of increased risk of death and cerebrovascular accidents as well as poor handling of agitation (Ballard et al., 2009). Thus, careful evaluation and assessment of pain is needed for better management of NPSs (Husebo, Ballard, Sandvik, et al., 2011; Scuteri, Morrone, et al., 2017). Despite the lack of rigorous randomized, controlled parallel-group trials clarifying the relationship between pain and NPSs (Husebo, Ballard, & Aarsland, 2011), the importance of pain in their development has been corroborated by a multicenter cluster randomized controlled trial (RCT) (Husebo, Ballard, Sandvik, et al., 2011). This study compared the usual treatment of dementia patients of age ≥65 years suffering from NPSs with a standardized stepwise protocol of painkillers administered in the following order: oral paracetamol, oral morphine, buprenorphine transdermal patch, or oral pregabalin. According to the results of this study, average agitation was significantly reduced, by 17%, on the Cohen-Mansfield agitation inventory (CMAI), with an increase after withdrawal (Husebo, Ballard, Sandvik, et al., 2011). Previously, the effectiveness of analgesics in the management of NPSs was assessed by a double-blind trial conducted on nursing home residents affected by advanced dementia and serious agitation symptoms, for 4 weeks—over-85-year-old patients who completed the study showed a reduction in agitation on the CMAI after treatment with a long-acting opioid, in comparison with placebo (Manfredi et al., 2003). Based on evidence, some algorithms including the use of gabapentin to handle NPSs were proposed (Davies et al., 2018). Unfortunately, neither atypical antipsychotics nor analgesic drugs are devoid of serious side effects. Among the most used drugs for the control of pain are nonsteroidal antiinflammatory drugs (NSAIDs). The PRECISION-ABPM (Prospective Randomized Evaluation of Celecoxib Integrated Safety vs. Ibuprofen or Naproxen Ambulatory Blood Pressure Measurement) Trial, which is a prespecified substudy of PRECISION, a randomized, multicenter, double-blind noninferiority trial on 444 patients with rheumatoid arthritis or osteoarthritis for the evaluation of cardiovascular risk, highlighted the development of hypertension after treatment with ibuprofen, naproxen, and celecoxib in patients with normal baseline blood pressure (23.2% patients for ibuprofen, 19.0% for naproxen, and 10.3% for celecoxib) (Ruschitzka et al., 2017). Apart from such cardiovascular risk, another matter of importance is that even though NSAIDs and acetaminophen represent the first-line treatment for osteoarthritis-dependent pain conditions, the latter's results are often ineffective (Goodman & Brett, 2017). The most widespread analgesic drugs for severe pain are opioids, but although they are very effective in cancer pain, there is not strong evidence of effectiveness in chronic noncancer pain (Morrone, Scuteri, Rombola, Mizoguchi, & Bagetta, 2017). These drugs are associated with a high rate of mortality due to abuse (Evoy, Morrison, & Saklad, 2017), and the clinical response to opioid pharmacotherapy is influenced by the patient's pharmacogenetics (Morrone et al., 2017) with polymorphisms that affect their pharmacokinetics as well as pharmacodynamics (Solhaug & Molden, 2017). Finally, $\alpha2\delta$-1 ligands (e.g., gabapentin and pregabalin) are drugs of choice for neuropathic pain, though they are not devoid of the side effects (e.g., sedation, dizziness, and cognitive problems); in fact, the latter are often experienced when used in polytherapy with other psychotropic drugs in aged patients (Goodman & Brett, 2017), and their potential for abuse is underrated (Evoy et al., 2017). Therefore, a better education for correct diagnosis and assessment of pain in dementia patients is needed for pain control (Scuteri, Piro, et al., 2017) and, consequently, for the management of NPSs. Furthermore, the health risk posed by atypical antipsychotics and analgesics has prompted the search for alternative options for handling such symptoms.

Aromatherapy: clinical evidence for complementary use in dementia

Though limited and preliminary, evidence is emerging that supports the use of complementary therapies in the management of NPSs (Table 45.1).

Aromatherapy provides the best results in controlling agitation, the most resistant NPS (Ballard et al., 2009). The most studied essential oils belong to Melissa and Lavandula. The largest and most relevant RCT to demonstrate the effectiveness of aromatherapy in dementia was of 72 guests enrolled in National Health Service (UK) care

Table 45.1 Aromatherapy for the treatment of neuropsychiatric symptoms of dementia.

Ballard et al. (2002)	Placebo-controlled trial of 72 guests with dementia in UK care facilities. *Melissa officinalis* essential oil, applied topically on face and arms twice a day for 4 weeks, produced an improvement of agitation devoid of significant side effects.
Holmes et al. (2002)	Placebo-controlled trial of 15 dementia patients. Agitation was measured through the Pittsburgh agitation scale. Aromatherapy stream of 2% lavender essential oil provided modest efficacy.
Akhondzadeh et al. (2003)	Parallel group, double blind, randomized, placebo-controlled trial of 42 aged patients affected by mild-moderate AD. Administration of lemon balm extract. Effectiveness on both cognition and the agitation side effect at 4 months.
Lin et al. (2007)	Crossover randomized trial of 70 dementia patients. Inhalation of *Lavandula angustifolia* essential oil. Effectiveness as adjunctive therapy for agitation.
Jimbo et al. (2009)	Crossover study of 28 older patients with dementia, among whom 17 suffered from AD. Inhalation of lemon and rosemary essential oils in the morning and of lavender and orange essential oils in the evening. Effective also with concurrent use of more essential oils.
Yazdkhasti & Pirak (2016)	Single-blind, randomized clinical trial of 120 pregnant women. Inhalation of two drops of lavender corresponding with three different stages of cervical dilatation. Attenuation of pain.
Gok Metin et al. (2017)	Open-label randomized clinical trial of 46 affected by diabetic neuropathy. Massage with rosemary, geranium, lavender, eucalyptus, and chamomile essential oils blended in coconut oil. Effectiveness on pain.
Goepfert et al. (2017)	Randomized controlled pilot study. A control group of 10 people and an intervention group of 15 conscious and 5 unconscious palliative patients. Inhalation through a surgical mask of lemon and lavender. Opposite responses concerning physiological parameters.

Clinical evidence supports aromatherapy for the treatment of neuropsychiatric symptoms of dementia. The most relevant studies have been reported in the table.

facilities with severe dementia and agitation (Ballard, O'Brien, Reichelt, & Perry, 2002). The patients were divided into two equal groups who were treated with topical application on face and arms twice a day for 4 weeks: the intervention group was administered *M. officinalis* essential oil, while the placebo group was treated with sunflower oil (Ballard et al., 2002). Seventy-one patients completed the RCT. Agitation measured by CMAI was significantly reduced by aromatherapy, and the treatment was extremely well tolerated (Ballard et al., 2002). Another parallel group, double-blind, randomized, placebo-controlled trial was conducted for 4 months to study the effects of aromatherapy with Melissa essential oil on 42 AD patients, 65–85 years of age, divided into two equal groups (Akhondzadeh et al., 2003). The patients presented mild to moderate AD; probable AD; and a history of cognitive decline (Akhondzadeh et al., 2003). Administration of 60 drops/day of lemon balm extract prepared 1:1 in 45% alcohol from *M. officinalis* leaf and standardized to 500 µg citral/mL (Akhondzadeh et al., 2003) improved cognition (the main outcome of the study evaluated by the 11-item cognitive subscale of the AD assessment scale and the clinical dementia rating-sum of boxes) and agitation considered as a side effect (Akhondzadeh et al., 2003). A randomized multiple-dose, multiple time-point, double-blind, placebo-controlled, balanced crossover study demonstrated the dose- and time-dependent efficacy of lemon balm (*M. officinalis*) dried leaves to improve cognition and mood (Kennedy et al., 2003). Another quite large crossover randomized trial on 70 Chinese aged dementia patients studied the effectiveness of the inhalation of *Lavandula angustifolia* essential oil against agitation as assessed with the Chinese versions of the CMAI and the neuropsychiatric inventory (Lin, Chan, Ng, & Lam, 2007). The enrolled patients were divided into two equal groups alternatively administered both the intervention and the control. The results support the effectiveness of the inhalation of *L. angustifolia*, which may represent a useful adjunctive therapy to handle agitation (Lin et al., 2007). Previously, a small placebo-controlled trial on 15 patients affected with dementia and agitation, evaluated using the Pittsburgh agitation scale, had suggested modest effectiveness of an aromatherapeutic stream of 2% lavender essential oil (Holmes et al., 2002). The inhalation of combined essential oils was assessed for efficacy in dementia—this study was designed to evaluate the possibility to enhance concentration and mnemonic skills during the morning hours when these capabilities are mostly needed and to reduce agitation and improve calm during night hours (Jimbo, Kimura, Taniguchi, Inoue, & Urakami, 2009). To this aim, 28 aged patients suffering from dementia, and in particular 17 from AD, inhaled 0.04 mL of lemon and 0.08 mL of rosemary essential oils during the morning to foster action on the sympathetic nervous system and 0.08 mL of lavender and 0.04 mL of orange essential oils during the evening for the activation of the parasympathetic nervous system (Jimbo et al., 2009). These mixed essential oils strengthened the effectiveness of aromatherapy in AD (Jimbo et al., 2009). Some clinical studies have assessed the efficacy of aromatherapy on pain conditions. An open-label RCT on 46 diabetic patients with neuropathy highlighted the beneficial effects of massage using rosemary, geranium, lavender, eucalyptus, and chamomile essential oils blended in coconut oil as carrier 1:1:1:1:1 in 5% solution three times a week for 4 weeks on pain as measured through the Douleur neuropathique questionnaire, visual analog scale, and neuropathic pain impact on quality of life questionnaire (Gok Metin, Arikan Donmez, Izgu, Ozdemir, & Arslan, 2017). A single-blind, randomized clinical trial carried out on 120 pregnant patients underlined a reduction of perceived pain in the intervention group subjected to inhalation of two drops of lavender in correspondence with three different stages of cervical dilatation (Yazdkhasti & Pirak, 2016). Moreover, a randomized controlled pilot study was performed to evaluate the effects of aromatherapy on conscious and unconscious patients in palliative care compared with healthy subjects (Goepfert et al., 2017). The intervention consisted of administration through a surgical mask for inhalation of three to four drops of lemon (made up of citral, geraniol, limonene, and linalool contained in *Citrus medica limonum* essential oil) and lavender (constituted by cinnamal, eugenol, geraniol, limonene, and linalool from *L. angustifolia* essential oil) (Goepfert et al., 2017). The results pointed at the capability of these two different aromas to evoke opposite responses in terms of physiological parameters (Goepfert et al., 2017). Pharmacological activity of essential oils in persons with dementia, known to be often anosmic (Vance, 1999), is explained and exerted through their systemic absorption even when administered via the inhalatory route. The terpene constituents of the essential oils can cross the blood—brain barrier (Burns, Byrne, Ballard, & Holmes, 2002). The effects of *Melissa* and *Lavandula officinalis* essential oils seem to be due to their activities on the cholinergic system; moreover, *M. officinalis* was demonstrated to be endowed with cholinergic receptor binding properties (Kennedy et al., 2003), and lavender essential oil could inhibit glutamate and GABA binding to their receptors (Elisabetsky, Marschner, & Onofre Souza, 1995; Huang et al., 2008). Since neither the anticholinesterase drugs used in dementia nor these essential oils can definitely control agitation, the likely action on cholinergic neurotransmission is not the principal mechanism to exploit for this purpose (Scuteri, Rombola, et al., 2018). Aromatherapy with an essential oil effective in reducing pain, such as bergamot, may be very efficacious in the management of NPSs because of the tight link between unrelieved pain and former symptoms (Scuteri, Rombola, et al., 2018).

Rational basis for complementary use of bergamot in dementia

BEO, endowed with centuries-old traditional herbal medicine, has proven sound evidence for analgesic effects in *in vivo* studies in both neuropathic and inflammatory pain preclinical models. Intraplantar (i.pl.) administration of BEO to mice subjected to neuropathic pain models of spinal nerve ligation (Bagetta et al., 2010) or of partial sciatic nerve ligation (Kuwahata et al., 2013) produced an attenuation of tactile allodynia (Fig. 45.1A). Such an antiallodynic effect is enhanced when BEO is combined with morphine (Kuwahata et al., 2013) (Fig. 45.1B–C).

Also in the inflammatory pain models of the capsaicin (Sakurada et al., 2009) and formalin (Katsuyama et al., 2015) test the pretreatment with i.pl. BEO produced a dose-dependent antinociception (Fig. 45.2A and B) synergistic with intraperitoneal or intrathecal morphine and prevented by naloxone (Fig. 45.2C and D).

In both phases of the formalin test as well as in neuropathic experimental pain models, BEO was effective only when injected into the ipsilateral hind paw, thus suggesting the peripheral nature of the antinociceptive effect (Katsuyama et al., 2015), likely on the peripheral opioid system, being reduced by pretreatment with naloxone methiodide, an opioid receptor antagonist unable to cross the blood–brain barrier (Katsuyama et al., 2015; Sakurada et al., 2011). Also, inhalation of BEO showed antinociceptive activity in the formalin test (Scuteri, Crudo, et al., 2018). Microdialysis experiments along with studies on synaptosomes highlighted that BEO modulates hippocampal synaptic amino acid neurotransmitters and in low concentrations acts on presynaptic receptors of glutamate-releasing nerve endings, causing exocytosis of glutamate, while in high concentrations it may induce a Ca^{2+}-independent carrier-mediated process with the release of glutamate, fundamental in pain (Morrone et al., 2007). The enhancing activity of BEO on basal and induced autophagy (Russo et al., 2014), a highly conserved process deranged in neuropathic pain and aging (Berliocchi et al., 2011; Cuervo et al., 2005), may be involved in antinociception. Moreover, there is growing evidence to support the link between the AD risk factor apolipoprotein E4 and an impairment of the autophagic–endocytotic–lysosomal axis (Parcon et al., 2018; Schmukler, Michaelson, & Pinkas-Kramarski, 2018). Finally, for intraperitoneal administration, BEO exerted anxiolytic effects not superimposable to those of diazepam, thus documenting absence of sedative action, very relevant for the use of BEO in cognitively impaired patients (Rombola et al., 2017; Rombolà, Scuteri, et al., 2019). Altogether, the evidence gathered so far suggests the need for rigorous and large RCTs evaluating the effectiveness of aromatherapy using BEO in relieving the pain and NPSs of patients suffering from dementia.

Key facts of aromatherapy

- According to the World Health Organization, several countries still use traditional and complementary medicine as their main health care.
- The use of complementary medicine is increasing worldwide.
- Aromatherapy is a complementary therapy.
- A global strategy is essential for safe and effective access to traditional and complementary medicine.
- The World Health Organization developed a traditional medicine strategy for 2014–23.

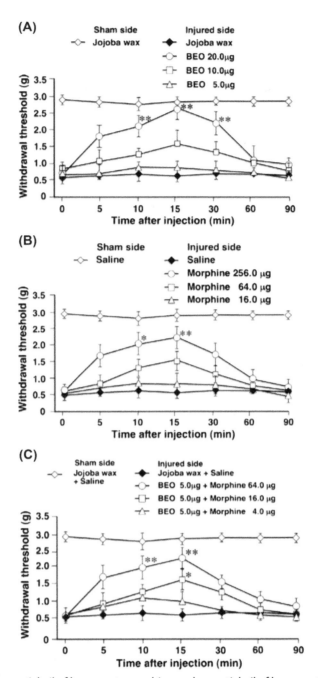

Figure 45.1 Effect of essential oil of bergamot, morphine and essential oil of bergamot +morphine on Partial Sciatic Nerve Ligation (PSNL)-induced mechanical allodynia of the ipsilateral hind paw. Baseline withdrawal thresholds were determined before intraplantar injection of essential oil of bergamot (BEO) (A), morphine (B), and essential oil of bergamot+morphine (C). The essential oil of bergamot produced an antiallodynic effect that is subjected to enhancement if it is administered in combination with morphine. Each value represents the mean ± standard error of the mean of 10 mice in each group. *$P < .05$ and **$P < .01$ compared with the jojoba wax-control group in PSNL mice. *(Taken with permission from Kuwahata, H., Komatsu, T., Katsuyama, S., Corasaniti, M. T., Bagetta, G., Sakurada, S., et al. (2013). Peripherally injected linalool and bergamot essential oil attenuate mechanical allodynia via inhibiting spinal ERK phosphorylation. Pharmacology Biochemistry and Behavior, 103(4), 735–741. https://doi.org/10.1016/j.pbb.2012.11.003.)*

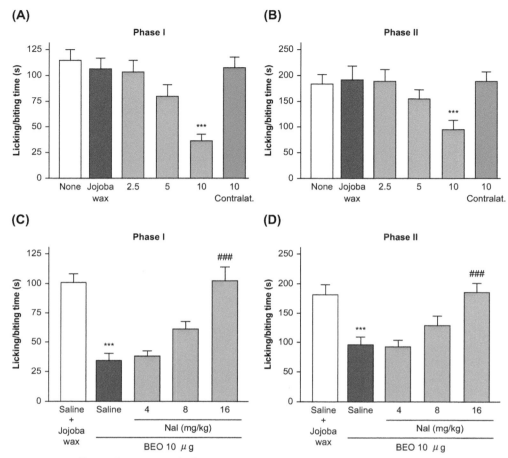

Figure 45.2 Effects of essential oil of bergamot injected into the hind paw, ipsilateral, or contralateral (Contralat.) to the site of formalin injection on the first (A) and second (B) phase of formalin test and antagonism by naloxone methiodide (C, D). Essential oil of bergamot (BEO) was injected 10 min before 2% formalin. Naloxone methiodide was preinjected into the hind paw 15 min before BEO. Jojoba wax was used as a control. Pretreatment with BEO into the ipsilateral hind paw produced a dose-dependent antinociception that was reduced by naloxone methiodide, an opioid receptor antagonist unable to cross the blood–brain barrier. Values represent the mean ± standard error of the mean for 10 mice in each group. Statistical differences between the groups were assessed by one-way ANOVA followed by Dunnett's test. (A, B): ***$P < .001$ compared with the jojoba wax control, ###$P < .001$ compared with saline intraperitoneal + BEO (10 μg). (C, D): ***$P < .001$ compared with the saline + jojoba wax control. #$P < .05$, ##$P < .01$, ###$P < .001$ compared with saline + BEO (10 μg). *(Taken with permission from Katsuyama, S., Otowa, A., Kamio, S., Sato, K., Yagi, T., Kishikawa, Y., et al., (2015). Effect of plantar subcutaneous administration of bergamot essential oil and linalool on formalin-induced nociceptive behavior in mice.* Biomedical Research, 36(1), 47–54. https://doi.org/10.2220/biomedres.36.47.)*

Summary points

- Alzheimer's disease is one of the most common causes of dementia.
- Mitochondrial derangement may be involved in the pathogenesis.
- A definite disease-modifying therapy is not available.
- Nearly the whole population of patients develops neuropsychiatric symptoms.
- Atypical neuroleptics are not safe and effective for the long term.
- Essential oils have proven efficacy against neuropsychiatric symptoms.
- Agitation is linked to unrelieved pain.
- BEO has antinociceptive properties.
- Aromatherapy with bergamot can be a novel strategy for neuropsychiatric symptoms management.

References

Akhondzadeh, S., Noroozian, M., Mohammadi, M., Ohadinia, S., Jamshidi, A. H., & Khani, M. (2003). Melissa officinalis extract in the treatment of patients with mild to moderate Alzheimer's disease: A double blind, randomised, placebo controlled trial. *Journal of Neurology Neurosurgery and Psychiatry*, 74(7), 863–866.

Arun, S., Liu, L., & Donmez, G. (2016). Mitochondrial biology and neurological diseases. *Current Neuropharmacology*, 14(2), 143–154.

Bagetta, G., Morrone, L. A., Rombola, L., Amantea, D., Russo, R., Berliocchi, L., et al. (2010). Neuropharmacology of the essential oil of bergamot. *Fitoterapia*, 81(6), 453–461. https://doi.org/10.1016/j.fitote.2010.01.013.

Ballard, C., & Corbett, A. (2013). Agitation and aggression in people with Alzheimer's disease. *Current Opinion in Psychiatry*, 26(3), 252–259. https://doi.org/10.1097/YCO.0b013e32835f414b.

Ballard, C. G., Gauthier, S., Cummings, J. L., Brodaty, H., Grossberg, G. T., Robert, P., et al. (2009). Management of agitation and aggression associated with Alzheimer disease. *Nature Reviews Neurology*, 5(5), 245–255. https://doi.org/10.1038/nrneurol.2009.39.

Ballard, C. G., O'Brien, J. T., Reichelt, K., & Perry, E. K. (2002). Aromatherapy as a safe and effective treatment for the management of agitation in severe dementia: The results of a double-blind, placebo-controlled trial with Melissa. *Journal of Clinical Psychiatry*, 63(7), 553–558.

Ballard, C., Smith, J., Husebo, B., Aarsland, D., & Corbett, A. (2011). The role of pain treatment in managing the behavioural and psychological symptoms of dementia (BPSD). *International Journal of Palliative Nursing*, 17(9), 420–422. https://doi.org/10.12968/ijpn.2011.17.9.420, 424.

Berliocchi, L., Russo, R., Maiaru, M., Levato, A., Bagetta, G., & Corasaniti, M. T. (2011). Autophagy impairment in a mouse model of neuropathic pain. *Molecular Pain*, 7, 83. https://doi.org/10.1186/1744-8069-7-83.

Brown, J. P., & Boulay, L. J. (2013). Clinical experience with duloxetine in the management of chronic musculoskeletal pain. A focus on osteoarthritis of the knee. *Therapeutic Advances in Musculoskeletal Disease*, 5(6), 291–304. https://doi.org/10.1177/1759720X13508508.

Burns, A., Byrne, J., Ballard, C., & Holmes, C. (2002). Sensory stimulation in dementia. *BMJ*, 325(7376), 1312–1313.

Calsolaro, V., & Edison, P. (2016). Alterations in glucose metabolism in Alzheimer's disease. *Recent Patents on Endocrine, Metabolic and Immune Drug Discovery*, 10(1), 31–39.

Chen, J. X., & Yan, S. S. (2010). Role of mitochondrial amyloid-beta in Alzheimer's disease. *Journal of Alzheimer's Disease*, 20(Suppl. 2), S569–S578. https://doi.org/10.3233/JAD-2010-100357.

Cuervo, A. M., Bergamini, E., Brunk, U. T., Droge, W., Ffrench, M., & Terman, A. (2005). Autophagy and aging: The importance of maintaining "clean" cells. *Autophagy*, 1(3), 131–140.

Davies, S. J., Burhan, A. M., Kim, D., Gerretsen, P., Graff-Guerrero, A., Woo, V. L., et al. (2018). Sequential drug treatment algorithm for agitation and aggression in Alzheimer's and mixed dementia. *Journal of Psychopharmacology.* https://doi.org/10.1177/0269881117744996, 269881117744996.

Elisabetsky, E., Marschner, J., & Onofre Souza, D. (1995). Effects of linalool on glutamatergic system in the rat cerebral cortex. *Neurochemical Research, 20*(4), 461–465.

Evoy, K. E., Morrison, M. D., & Saklad, S. R. (2017). Abuse and misuse of pregabalin and gabapentin. *Drugs, 77*(4), 403–426. https://doi.org/10.1007/s40265-017-0700-x.

Gibson, G. E., & Shi, Q. (2010). A mitocentric view of Alzheimer's disease suggests multi-faceted treatments. *Journal of Alzheimer's Disease, 20*(Suppl. 2), S591–S607. https://doi.org/10.3233/JAD-2010-100336.

Goepfert, M., Liebl, P., Herth, N., Ciarlo, G., Buentzel, J., & Huebner, J. (2017). Aroma oil therapy in palliative care: A pilot study with physiological parameters in conscious as well as unconscious patients. *Journal of Cancer Research and Clinical Oncology, 143*(10), 2123–2129. https://doi.org/10.1007/s00432-017-2460-0.

Gok Metin, Z., Arikan Donmez, A., Izgu, N., Ozdemir, L., & Arslan, I. E. (2017). Aromatherapy massage for neuropathic pain and quality of life in diabetic patients. *Journal of Nursing Scholarship, 49*(4), 379–388. https://doi.org/10.1111/jnu.12300.

Goodman, C. W., & Brett, A. S. (2017). Gabapentin and pregabalin for pain - is increased prescribing a cause for concern? *New England Journal of Medicine, 377*(5), 411–414. https://doi.org/10.1056/NEJMp1704633.

Heneka, M. T., Carson, M. J., El Khoury, J., Landreth, G. E., Brosseron, F., Feinstein, D. L., et al. (2015). Neuroinflammation in Alzheimer's disease. *The Lancet Neurology, 14*(4), 388–405. https://doi.org/10.1016/S1474-4422(15)70016-5.

Holmes, C., Arranz, M. J., Powell, J. F., Collier, D. A., & Lovestone, S. (1998). 5-HT2A and 5-HT2C receptor polymorphisms and psychopathology in late onset Alzheimer's disease. *Human Molecular Genetics, 7*(9), 1507–1509.

Holmes, C., Hopkins, V., Hensford, C., MacLaughlin, V., Wilkinson, D., & Rosenvinge, H. (2002). Lavender oil as a treatment for agitated behaviour in severe dementia: A placebo controlled study. *International Journal of Geriatric Psychiatry, 17*(4), 305–308. https://doi.org/10.1002/gps.593.

Huang, L., Abuhamdah, S., Howes, M. J., Dixon, C. L., Elliot, M. S., Ballard, C., et al. (2008). Pharmacological profile of essential oils derived from *Lavandula angustifolia* and *Melissa officinalis* with anti-agitation properties: Focus on ligand-gated channels. *Journal of Pharmacy and Pharmacology, 60*(11), 1515–1522. https://doi.org/10.1211/jpp/60.11.0013.

Husebo, B. S., Ballard, C., & Aarsland, D. (2011). Pain treatment of agitation in patients with dementia: A systematic review. *International Journal of Geriatric Psychiatry, 26*(10), 1012–1018. https://doi.org/10.1002/gps.2649.

Husebo, B. S., Ballard, C., Sandvik, R., Nilsen, O. B., & Aarsland, D. (2011). Efficacy of treating pain to reduce behavioural disturbances in residents of nursing homes with dementia: Cluster randomised clinical trial. *BMJ, 343*, d4065. https://doi.org/10.1136/bmj.d4065.

Jimbo, D., Kimura, Y., Taniguchi, M., Inoue, M., & Urakami, K. (2009). Effect of aromatherapy on patients with Alzheimer's disease. *Psychogeriatrics, 9*(4), 173–179. https://doi.org/10.1111/j.1479-8301.2009.00299.x.

Katsuyama, S., Otowa, A., Kamio, S., Sato, K., Yagi, T., Kishikawa, Y., et al. (2015). Effect of plantar subcutaneous administration of bergamot essential oil and linalool on formalin-induced nociceptive behavior in mice. *Biomedical Research, 36*(1), 47–54. https://doi.org/10.2220/biomedres.36.47.

Kennedy, D. O., Wake, G., Savelev, S., Tildesley, N. T., Perry, E. K., Wesnes, K. A., et al. (2003). Modulation of mood and cognitive performance following acute administration of single doses of *Melissa officinalis* (Lemon balm) with human CNS nicotinic and muscarinic receptor-binding properties. *Neuropsychopharmacology, 28*(10), 1871–1881. https://doi.org/10.1038/sj.npp.1300230.

Kuwahata, H., Komatsu, T., Katsuyama, S., Corasaniti, M. T., Bagetta, G., Sakurada, S., et al. (2013). Peripherally injected linalool and bergamot essential oil attenuate mechanical allodynia via inhibiting spinal ERK phosphorylation. *Pharmacology Biochemistry and Behavior, 103*(4), 735–741. https://doi.org/10.1016/j.pbb.2012.11.003.

Lai, M. K., Tsang, S. W., Alder, J. T., Keene, J., Hope, T., Esiri, M. M., et al. (2005). Loss of serotonin 5-HT2A receptors in the postmortem temporal cortex correlates with rate of cognitive decline in Alzheimer's disease. *Psychopharmacology, 179*(3), 673−677. https://doi.org/10.1007/s00213-004-2077-2.

Lin, P. W., Chan, W. C., Ng, B. F., & Lam, L. C. (2007). Efficacy of aromatherapy (*Lavandula angustifolia*) as an intervention for agitated behaviours in Chinese older persons with dementia: A cross-over randomized trial. *International Journal of Geriatric Psychiatry, 22*(5), 405−410. https://doi.org/10.1002/gps.1688.

Lyketsos, C. G., Steinberg, M., Tschanz, J. T., Norton, M. C., Steffens, D. C., & Breitner, J. C. (2000). Mental and behavioral disturbances in dementia: Findings from the cache county study on memory in aging. *American Journal of Psychiatry, 157*(5), 708−714. https://doi.org/10.1176/appi.ajp.157.5.708.

Manfredi, P. L., Breuer, B., Wallenstein, S., Stegmann, M., Bottomley, G., & Libow, L. (2003). Opioid treatment for agitation in patients with advanced dementia. *International Journal of Geriatric Psychiatry, 18*(8), 700−705. https://doi.org/10.1002/gps.906.

Margallo-Lana, M., Swann, A., O'Brien, J., Fairbairn, A., Reichelt, K., Potkins, D., et al. (2001). Prevalence and pharmacological management of behavioural and psychological symptoms amongst dementia sufferers living in care environments. *International Journal of Geriatric Psychiatry, 16*(1), 39−44.

Morrone, L. A., Rombola, L., Pelle, C., Corasaniti, M. T., Zappettini, S., Paudice, P., et al. (2007). The essential oil of bergamot enhances the levels of amino acid neurotransmitters in the hippocampus of rat: Implication of monoterpene hydrocarbons. *Pharmacological Research, 55*(4), 255−262. https://doi.org/10.1016/j.phrs.2006.11.010.

Morrone, L. A., Scuteri, D., Rombola, L., Mizoguchi, H., & Bagetta, G. (2017). Opioids resistance in chronic pain management. *Current Neuropharmacology, 15*(3), 444−456. https://doi.org/10.2174/1570159X14666161101092822.

Nacmias, B., Tedde, A., Forleo, P., Piacentini, S., Guarnieri, B. M., Bartoli, A., et al. (2001). Association between 5-HT(2A) receptor polymorphism and psychotic symptoms in Alzheimer's disease. *Biological Psychiatry, 50*(6), 472−475.

Onyango, I. G. (2018). Modulation of mitochondrial bioenergetics as a therapeutic strategy in Alzheimer's disease. *Neural Regeneration Research, 13*(1), 19−25. https://doi.org/10.4103/1673-5374.224362.

Parcon, P. A., Balasubramaniam, M., Ayyadevara, S., Jones, R. A., Liu, L., Shmookler Reis, R. J., et al. (2018). Apolipoprotein E4 inhibits autophagy gene products through direct, specific binding to CLEAR motifs. *Alzheimer's and Dementia, 14*(2), 230−242. https://doi.org/10.1016/j.jalz.2017.07.754.

Parvizi, J., Van Hoesen, G. W., & Damasio, A. (2000). Selective pathological changes of the periaqueductal gray matter in Alzheimer's disease. *Annals of Neurology, 48*(3), 344−353.

Prince, M., Wimo, A., Guerchet, M., Ali, G. C., Wu, Y., & Prina, A. M. (2015). *World Alzheimer Report 2015: The global impact of dementia. An analysis of prevalence, incidence, costs and trends*. London: Alzheimer's Disease International.

Rombolà, L., Scuteri, D., Adornetto, A., Straface, M., Sakurada, T., Mizoguchi, H., Straface, S., Corasaniti, M. T., Bagetta, G., Tonin, P., & Morrone, L. A. (2019). Anxiolytic-like effects of bergamot essential oil are insensitive to Flumazenil in Rats. *Evidence-Based Complementary and Alternative Medicine: eCAM. 2156873*. https://doi.org/10.1155/2019/2156873. eCollection 2019.

Rombola, L., Tridico, L., Scuteri, D., Sakurada, T., Sakurada, S., Mizoguchi, H., et al. (2017). Bergamot essential oil attenuates anxiety-like behaviour in rats. *Molecules, 22*(4). https://doi.org/10.3390/molecules22040614.

Ruschitzka, F., Borer, J. S., Krum, H., Flammer, A. J., Yeomans, N. D., Libby, P., et al. (2017). Differential blood pressure effects of ibuprofen, naproxen, and celecoxib in patients with arthritis: The PRECISION-ABPM (prospective randomized evaluation of celecoxib integrated safety versus ibuprofen or naproxen ambulatory blood pressure measurement) trial. *European Heart Journal, 38*(44), 3282−3292. https://doi.org/10.1093/eurheartj/ehx508.

Russo, R., Cassiano, M. G., Ciociaro, A., Adornetto, A., Varano, G. P., Chiappini, C., et al. (2014). Role of D-Limonene in autophagy induced by bergamot essential oil in SH-SY5Y neuroblastoma cells. *PLoS One, 9*(11), e113682. https://doi.org/10.1371/journal.pone.0113682.

Sakurada, T., Kuwahata, H., Katsuyama, S., Komatsu, T., Morrone, L. A., Corasaniti, M. T., et al. (2009). Intraplantar injection of bergamot essential oil into the mouse hindpaw: Effects on capsaicin-induced nociceptive behaviors. *International Review of Neurobiology, 85*, 237–248. https://doi.org/10.1016/S0074-7742(09)85018-6.

Sakurada, T., Mizoguchi, H., Kuwahata, H., Katsuyama, S., Komatsu, T., Morrone, L. A., et al. (2011). Intraplantar injection of bergamot essential oil induces peripheral antinociception mediated by opioid mechanism. *Pharmacology Biochemistry and Behavior, 97*(3), 436–443. https://doi.org/10.1016/j.pbb.2010.09.020.

Sampson, E. L., White, N., Lord, K., Leurent, B., Vickerstaff, V., Scott, S., et al. (2015). Pain, agitation, and behavioural problems in people with dementia admitted to general hospital wards: A longitudinal cohort study. *Pain, 156*(4), 675–683. https://doi.org/10.1097/j.pain.0000000000000095.

Scherder, E., Herr, K., Pickering, G., Gibson, S., Benedetti, F., & Lautenbacher, S. (2009). Pain in dementia. *Pain, 145*(3), 276–278. https://doi.org/10.1016/j.pain.2009.04.007.

Schmukler, E., Michaelson, D. M., & Pinkas-Kramarski, R. (2018). The interplay between apolipoprotein E4 and the autophagic-endocytic-lysosomal Axis. *Molecular Neurobiology*. https://doi.org/10.1007/s12035-018-0892-4.

Scuteri, D., Crudo, M., Rombola, L., Watanabe, C., Mizoguchi, H., Sakurada, S., et al. (2018). Antinociceptive effect of inhalation of the essential oil of bergamot in mice. *Fitoterapia, 129*, 20–24. https://doi.org/10.1016/j.fitote.2018.06.007.

Scuteri, D., Morrone, L. A., Rombola, L., Avato, P. R., Bilia, A. R., Corasaniti, M. T., et al. (2017). Aromatherapy and aromatic plants for the treatment of behavioural and psychological symptoms of dementia in patients with Alzheimer's disease: Clinical evidence and possible mechanisms. *Evidence Based Complementary and Alternative Medicine, 2017*, 9416305. https://doi.org/10.1155/2017/9416305.

Scuteri, D., Piro, B., Morrone, L. A., Corasaniti, M. T., Vulnera, M., & Bagetta, G. (2017). The need for better access to pain treatment: Learning from drug consumption trends in the USA. *Functional Neurology, 22*(4), 229–230.

Scuteri, D., Rombola, L., Tridico, L., Mizoguchi, H., Watanabe, C., Sakurada, T., et al. (2018). Neuropharmacological properties of the essential oil of bergamot for the clinical management of pain-related BPSDs. *Current Medicinal Chemistry*. https://doi.org/10.2174/0929867325666180307115546.

Sengstaken, E. A., & King, S. A. (1993). The problems of pain and its detection among geriatric nursing home residents. *Journal of the American Geriatrics Society, 41*(5), 541–544.

Simoncini, C., Orsucci, D., Caldarazzo Ienco, E., Siciliano, G., Bonuccelli, U., & Mancuso, M. (2015). Alzheimer's pathogenesis and its link to the mitochondrion. *Oxidative Medicine and Cellular Longevity, 2015*, 803942. https://doi.org/10.1155/2015/803942.

Solhaug, V., & Molden, E. (2017). Individual variability in clinical effect and tolerability of opioid analgesics - importance of drug interactions and pharmacogenetics. *Scandinavian Journal of Pain, 17*, 193–200. https://doi.org/10.1016/j.sjpain.2017.09.009.

Steinberg, M., Shao, H., Zandi, P., Lyketsos, C. G., Welsh-Bohmer, K. A., Norton, M. C., et al. (2008). Point and 5-year period prevalence of neuropsychiatric symptoms in dementia: The cache county study. *International Journal of Geriatric Psychiatry, 23*(2), 170–177. https://doi.org/10.1002/gps.1858.

Vance, D. (1999). Considering olfactory stimulation for adults with age-related dementia. *Perceptual and Motor Skills, 88*(2), 398–400. https://doi.org/10.2466/pms.1999.88.2.398.

Winblad, B., Amouyel, P., Andrieu, S., Ballard, C., Brayne, C., Brodaty, H., et al. (2016). Defeating Alzheimer's disease and other dementias: A priority for European science and society. *The Lancet Neurology, 15*(5), 455–532. https://doi.org/10.1016/S1474-4422(16)00062-4.

Yazdkhasti, M., & Pirak, A. (2016). The effect of aromatherapy with lavender essence on severity of labor pain and duration of labor in primiparous women. *Complementary Therapies in Clinical Practice, 25*, 81–86. https://doi.org/10.1016/j.ctcp.2016.08.008.

Zarow, C., Lyness, S. A., Mortimer, J. A., & Chui, H. C. (2003). Neuronal loss is greater in the locus coeruleus than nucleus basalis and substantia nigra in Alzheimer and Parkinson diseases. *Archives of Neurology, 60*(3), 337–341.

CHAPTER 46

Dancing in dementia

Lee-Fay Low, Helen Parker, Kathryn Dovey, Alycia Fong Yan

Faculty of Health Sciences, University of Sydney, Lidcombe, NSW, Australia

List of abbreviations

MMSE mini-mental state examination
SMD standardized mean difference

Dance can be defined as using specific movement sequences to move the body through space to music in a rhythmic and expressive way (Garfinkel, 2018). Dance may be performed solo, in pairs (e.g., Tango, waltz), or in a group (e.g., Arab dabke, line dancing) and is an art form that allows for cultural and personal expression (Garfinkel, 2018).

The history of dance

Dance is universal in human societies and commonly depicted in cave art around the world (Christensen, Cela-Conde, & Gomila, 2017). Historically, dance reflected courtship, rites of passage, religion, and ceremonies (Garfinkel, 2018). Early dance was a social activity performed for communities to participate in and watch, with rhythm set through instruments, singing, or clapping (Garfinkel, 2018).

Some forms of dance (e.g., ballet) have been professionalized and standardized (Abra, 2017). However, dance continues to be a leisure and cultural activity for communities (Buckland, 2006). Dance styles evolve and comingle. For instance, ballroom dance has European (e.g., waltz), American (e.g., East Coast Swing), and South American (e.g., Tango) roots (Centralhome Company, 2018). American hip-hop has influenced Korean K-pop (Oh, 2017) and the knife-wielding mahragan Egyptian street dance (Swedenburg, 2012).

Dance from an evolutionary perspective

It has been suggested that dance evolved as a method of demonstrating reproductive fitness and sexual attractiveness, increasing social cooperativeness through synchronizing group movements and social bonding, and telling stories and transmitting cultural knowledge (Christensen et al., 2017). Further, it has been theorized that dance and rhythm are inborn physiological reflexes, universal in humans, that evolved for extraverbal communication (Hagen & Bryant, 2003; Richter & Ostovar, 2016).

Diagnosis and Management in Dementia
ISBN 978-0-12-815854-8, https://doi.org/10.1016/B978-0-12-815854-8.00046-X

General health benefits of dance

Physical benefits

Cardiovascular. Dance can provide a full-body workout. Fast-paced dances like tap and jazz burn upwards of 500 calories per hour, and slower ballroom dances burn between 100 and 200 calories per hour (Alpert, 2011). Oxygen uptake and endurance improvements have been evident during high-intensity dance classes (Donath, Roth, Hohn, Zahner, & Faude, 2014), with heart rate and VO_2 max values improving similarly to other forms of physical activity (Fong Yan et al., 2018). Dancing can significantly decrease body mass index in children (Huang, Hogg, Zandieh, & Bostwick, 2012) and adults (Murrock & Gary, 2010).

Strength, flexibility, and balance. Muscle strength and power, particularly of the lower limbs and trunk, can be improved through dance (Fong Yan et al., 2018). Weight-transfer movements in dance increase lower-limb lean mass (Barene, Holtermann, Oseland, Brekke, & Krustrup, 2016), muscle strength, and endurance (Vordos et al., 2016). Muscular improvements from dance extend to lower limb flexibility, improved balance, and reduced postural sway (Barene, Holtermann, Oseland, Brekke, & Krustrup, 2016). Dance can improve proprioception, which can improve balance through increased muscle memory and strength (Cox & Herzog, 2013). Dance programs improve balance slightly more than home-based exercise programs do (Alpert, 2009) (Fig. 46.1).

Psychological benefits

Improved mood. Systematic reviews have shown that exercise programs reduce depression and anxiety (Arent, Landers, & Etnier, 2000; Long & Stavel, 1995). These improvements are thought to be caused by increases in blood circulation to the brain and changes to the stress response (Sharma, Madaan, & Petty, 2006) and may also be a result of increasing social connectedness (see below). A meta-analysis compared dance with no intervention in any sample (23 primary trials with a total of 1078 participants). Dance improved well-being (standardized mean difference [SMD]: 0.30, 95% CI 0.07—0.53), interpersonal outcomes (SMD: 0.45, 95% CI 0.07—0.83), and clinical outcomes (SMD: 0.44, 95% CI 0.22—0.66). In addition, dance decreased symptoms of depression (OR: 0.36, 95% CI 0.17—0.56) and anxiety (SMD: 0.44, 95%CI 0.15—0.72) (Koch, Kunz, Lykou, & Cruz, 2014).

Increased social connectedness. Group dance requires participants to synchronize movements to music and with each other and may also involve dancers having physical contact and making eye contact. Experimental studies have demonstrated that synchrony of movements to rhythm results in greater cooperation, group bonding, and elevated pain thresholds (Reddish, Fischer, & Bulbulia, 2013; Tarr, Launay, Cohen, & Dunbar, 2015; Tarr, Launay, & Dunbar, 2016). Interpersonal touch and eye contact also increase trust and cooperation as well as strengthen social bonds (Gallace & Spence, 2010).

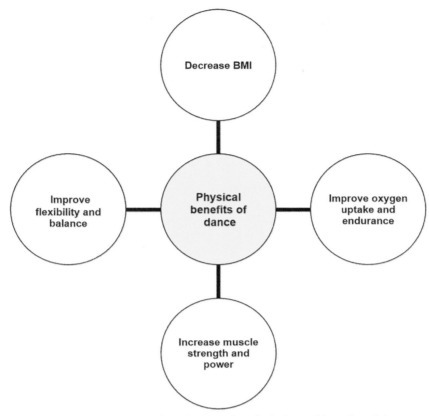

Figure 46.1 *Physical benefits of dance.* Identified physical benefits of dance.

Self-expression and self-identity. Dance is a medium through which people can express, define, and redefine their own and cultural identities (Nielsen & Koff, 2017). Exploration of identity through dance can contribute to self-confidence and well-being by enhancing connections with self and culture (Vincent, 2009; Wu et al., 2015). Dance therapy can assist people in processing their emotions and communicating their feelings through exploratory dance movements led by the dancer rather than an instructor (Devereaux, 2008; Koch & Weidinger-von der Recke, 2009) (Fig. 46.2).

Cognitive benefits

Learning dance requires motor imitation and encoding, and execution requires attention and recall (Bläsing et al., 2012). Imitation includes observation of the dance teacher and converting this auditory and visual information into motor actions (Laland, Wilkins, & Clayton, 2016). For dance routines to be practiced and performed, dancers must

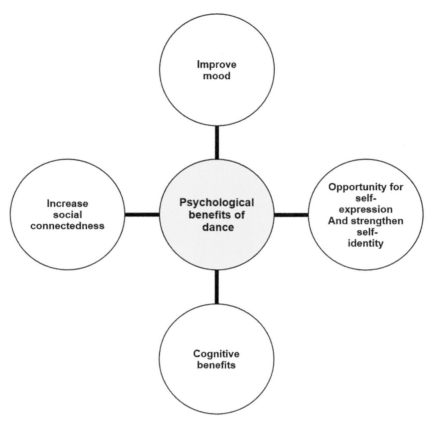

Figure 46.2 *Psychological benefits of dance.* Identified psychological benefits of dance.

memorize complex motor sequences, encoding information about movement, timing, and location in space (i.e., position relative to other dancers) as well as emotional valence (Stevens, Ginsborg, & Lester, 2011).

Rationale for dance to prevent dementia

Longitudinal cohort studies suggest that physical exercise reduces the risk of dementia (Blondell, Hammersley-Mather, & Veerman, 2014). Cognitive exercise also reduces the risk of cognitive decline, though the impact on incident dementia is not established (Valenzuela & Sachdev, 2009). Longitudinal cohort studies also suggest that low social participation, less frequent social contact, and more loneliness increase the risk of dementia (Kuiper et al., 2015). Hence dance might reduce the risk of dementia through physical activity, cognitive benefits, social channels, or a combination of these (Fig. 46.3).

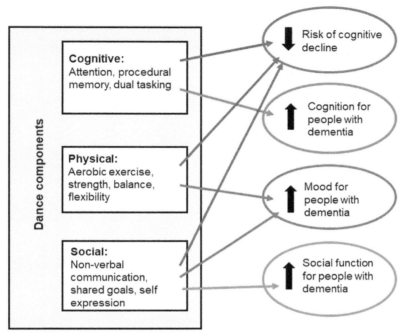

Figure 46.3 *Model of theoretical relationships between components of dance and outcomes relating to dementia.* Rationale for the potential influence of components of dance on cognitive and psychological outcomes in dementia.

Rationale for dance as therapy for people with dementia

Physical activity improves physical function, activities of daily living, cognition, and mood in people with dementia (Lee, Park, & Park, 2016). However, exercise trials for people with dementia have reported low adherence (Underwood et al., 2013). Physical activity in the form of dance, which people with dementia seem to naturally participate in and derive pleasure from, may be one method of harnessing the benefits of physical activity for people with dementia (Lapum & Bar, 2016).

The benefits of cognitive training on improving cognition for people with dementia have not been demonstrated (Bahar-Fuchs, Clare, & Woods, 2013), though a group cognitive-activity program called cognitive stimulation therapy has consistently shown benefits for cognition as well as quality of life (Spector, Orrell, & Woods, 2010). With regards to the capacity to engage in dance, procedural memory and motor learning are relatively preserved relative to declarative memory in Alzheimer's disease (Hirono et al., 1997), as is musical memory (Clark & Warren, 2015). The mirror neuron network, which is used during physical and emotional imitation, is also preserved in dementia (Farina et al., 2017).

Evidence is accumulating that simultaneous physical and cognitive activity (i.e., dual tasking) improves cognition in people with dementia (Tait, Duckham, Milte, Main, & Daly, 2017). Hence the combination of physical activity and cognitive challenge simultaneously during dance may have additional benefits on cognition.

The theory of unmet needs suggests that changes in mood and behavior in dementia arise because the person has needs that are not met, such as physiological (e.g., thirst, hunger, pain, excess energy) or psychological needs (e.g., boredom, loneliness) (Kolanowski, 1999). Consistent with this, psychosocial interventions that engage people with dementia reduce agitation behaviors (Cohen-Mansfield, 2016); furthermore, people with dementia have been observed to use dance as a method of self-expression (Nyström & Lauritzen, 2005). Dance might meet needs for meaningful activity, use of physical energy, and social contact.

Evidence for the relationship between dance and cognitive function in people with and without dementia

Dance movement therapy is the psychotherapeutic use of movement and dance to support intellectual, emotional, and motor functions of the body (Shim et al., 2017). A Cochrane review found no randomized controlled dance movement therapy trials for people with dementia (Karkou & Meekums, 2017).

In the studies below, dance was not used with psychotherapeutic intent.

What is the evidence that dance can reduce the risk of cognitive decline and dementia?

Cross-sectional observational studies

Three studies have reported that dancers have better memory and executive function than those of nondancers. One study compared 44 self-reported dancers of any experience level (mean 3 ± 7 years dance experience) with 43 participants without dance experience (mean age 70 ± 8 years) and found better auditory short- and long-term memory performance in dancers than in nondancers (Porat et al., 2016). Notably, this study also found that dancers had thinner cortical gray matter bilaterally (Porat et al., 2016), consistent with volumetric studies of younger dancers.

A study that compared 24 amateur ballroom dancers (17 ± 13 years experience) with nondancers (mean age 72 ± 1 years) found that selective attention and fluid intelligence were higher in dancers (Kattenstroth, Kolankowska, Kalisch, & Dinse, 2010). Better performance on selective attention, fluid intelligence, and reaction time was found when 11 competitive ballroom dancers (22 ± 3 years experience) were compared with 38 sedentary nondancers (mean age 71 ± 1 years) (Kattenstroth, Kalisch, Kolankowska, & Dinse, 2011).

However, two studies did not find cognitive differences between dancers and non-dancers. There were no differences between 28 nonprofessional dancers (mean 13 ± 8 years experience) and 29 matched nonsedentary nondancers (mean age $= 73 \pm 5$ years) on executive control, visual perceptual speed, and short- or long-term auditory memory (Niemann, Godde, & Voelcker-Rehage, 2016), nor in whole brain volume or hippocampal volume (Niemann et al., 2016). Similarly, no differences were found between 24 social dancers (mean 37 ± 27 years experience) and 84 matched nondancers (mean age 80 years) for auditory memory, executive function, or general mental status (Verghese, 2006).

These results suggest that dancing over several years might improve memory and executive function. However, diversity in the cognitive domains tested (areas such as attention and working memory might have greater impact from dance), average age of participants (older samples are more likely to show age-related cognitive decline), types of dance (e.g., only ballroom or social dancing), length and intensity of dance experience, and diversity in nondancer group (e.g., sedentary vs. physically active) make it difficult to generalize. Furthermore, self-selected dancers may have other characteristics (e.g., higher socioeconomic status) related to cognitive function.

Longitudinal studies

One epidemiological study compared dancers with nondancers over time, examining the relationship between leisure activities and risk of developing dementia. Over a median 5 years of follow-up, participants who danced at least several times a week (n = 155) were less likely to develop dementia than those who rarely danced (n = 438, HR: 0.24, 95%CI 0.06—0.99). Dancing was the only form of leisure physical activity associated with dementia risk in this study (Verghese et al., 2003).

Experimental studies—impact on cognition and brain volume or function compared with controls

Three studies compared dance with no-intervention controls and all reported that dance improved memory and/or executive function. Participants randomly assigned to Agilando ballroom dance (solo or partnered) (n = 25) 1 h a week for 24 weeks (1440 min total) showed improved attention, fluid intelligence, and reaction time compared with no-intervention controls (n = 10; mean age 69 ± 1 years) (Kattenstroth et al., 2010; Kattenstroth, Kalisch, Holt, Tegenthoff, & Dinse, 2013). A randomized controlled trial (Lazarou et al., 2017) compared 1 h twice a week for 10 months (4800 min total) of partnered ballroom dancing (n = 66) with no-intervention controls (n = 63) for persons with amnestic mild cognitive impairment (mean age $= 66 \pm 10$ years) (Lazarou et al., 2017). Improvements were found in the dance group relative to controls on global cognition, attention, auditory and visual memory, and executive function. A nonrandomized trial compared cha–cha 60 min twice a week for 6 months (2880 min total, n = 26) with usual activity controls (n = 12) in participants with metabolic syndrome (mean age 68 ± 4 years) (Kim et al., 2011). The dance group significantly improved in executive function, auditory memory, and global cognition.

However, four studies that compared dance with different types of exercise (aerobic and/or strength and balance) reported inconsistent results regarding impact on cognition. A nonrandomized trial compared 1 h a week for 24 weeks (1440 min total) of improvisational contemporary dance (n = 16) with fall-prevention exercise (n = 67) and tai chi chuan (n = 27; mean age 73 ± 6 years). The dance group improved on executive flexibility compared with the fall prevention and tai chi groups; there was no effect on attention setting or suppression in any group (Coubard, Duretz, Lefebvre, Lapalus, & Ferrufino, 2011). A randomized trial of ballroom dance 1 h twice a week for 8 months (n = 60, 4140 h total, 78% adherence) or home-based walking (n = 55, mean age 70 ± 6 years) found the dance group improved in short- and long-term visual memory. Executive function and auditory memory did not significantly change in either group (Merom et al., 2016). A randomized trial compared 90-min dance classes incorporating choreographic memorization, balance, and step challenges once to twice a week for 12 months (n = 26, 9180 min total) with fitness sport that incorporated endurance, strength-endurance, and flexibility training (n = 26; mean age 68 ± 4 years) (Müller et al., 2017; Rehfeld et al., 2017). The dance group showed increases in plasma brain-derived neurotrophic factor, left dentate gyrus and right subiculum, balance composite score, gray matter volume in the left precentral gyrus, and parahippocampal volume; both groups increased in hippocampal volume and attention and verbal memory. A randomized trial compared complex social dance (English country dancing, n = 49) with supervised brisk walking with (n = 42) or without (n = 40) a beta-alanine supplement, with balance and strength training controls (n = 43; mean age 65 ± 5 years) (Burzynska et al., 2017). All groups received three 1-h sessions a week for 6 months (4320 min total). The dance group showed stable white matter integrity in the fornix, whereas all other groups showed a reduction; there were no other differences between groups on diffusivity in other brain regions or on processing speed.

Two studies compared dance with a social attention control group. A nursing home study randomized cognitively intact participants (mean age 81 ± 6 years) to slow waltz (45 min a week for 10 weeks, 10 sessions total) or group discussions (Kosmat & Vranic, 2017). The dance group improved relative to controls on auditory memory and executive functioning; this was maintained after 5 months. A nonrandomized trial set in an independent living facility comparing 20 90-min tango classes over 12 weeks (n = 62, 1800 min total, n = 40 with full adherence) with health education (n = 12, mean age 84 ± 8 years) found no difference between groups over time on global cognitive function or executive function (Hackney et al., 2015).

What is the evidence that dance can improve outcomes for people with dementia?

Quantitative research—impact on cognition, mood, and behavior

One study randomized nursing home residents with dementia—baseline mini-mental state exam (MMSE) 12 ± 5—to nine 30—45 min weekly dance and movement sessions (n = 19, approximately 333 min total) or to usual treatment (n = 10) and reported improvements in clock drawing but not on global cognition or auditory memory, or behavioral symptoms (Hokkanen et al., 2008).

A randomized trial with permanent hospital patients with dementia (mean MMSE = 11 ± 5) compared daily 30-min dance lessons for 3 months (n = 15, 2700 min total) with daily conversation (n = 10, total mean age = 83 ± 8) and found that dance significantly improved MMSE but not behavior (Van de Winckel, Feys, De Weerdt, & Dom, 2004).

A nonrandomized trial in care homes compared residents (MMSE between 18 and 28) who participated in 2 h-long sessions of poco poco dance combined with 30 min each session of progressive muscle relaxation for 6 weeks (n = 44, 720 min total) with two 40-min sessions for 6 weeks of progressive muscle relaxation alone. The dance group showed significantly lower levels of depression and anxiety over time and significantly higher self-rated quality of life (Adam, Ramli, & Shahar, 2016).

A randomized controlled trial in a nursing home that included residents with cognitive impairment (MMSE 15 to 30, mean = 25 ± 4) compared adapted ballroom dance 1 h a week for 3 months (n = 79, 720 min total) with usual care (n = 83). Depression symptoms decreased in the dance group over time relative to controls (Vankova et al., 2014).

Qualitative research—impact on people with dementia

Qualitative research has used phenomenological and exploratory approaches including interviews with people with dementia, their family and staff, and observation. Studies consistently reported that people with dementia were engaged and enjoyed dancing and that communication increased during and following the dance sessions (Duignan, Hedley, & Milverton, 2009; Guzmán-García, Mukaetova-Ladinska, & James, 2012; Nyström & Lauritzen, 2005; Palo-Bengtsson & Ekman, 2002; Palo-Bengtsson, Winblad, & Ekman, 1998; Palo-Bengtsson, Winblad, & Ekman, 2003). A single uncontrolled qualitative evaluation of the impact of dance on community-dwelling people with mild to moderate dementia found positive impacts on general well-being, concentration, and communication after one 45-min session per week for 10 weeks (Hamill, Smith, & Röhricht, 2011).

Discussion

Observational and experimental literature have reported mixed results on the impact of dance and on cognition in older people. One longitudinal cohort study suggests that dance is protective against dementia. Cross-sectional studies that compared professional dancers with nondancers found differences in attention, executive function, auditory memory or reaction time, whereas studies of amateur dancers and nondancers reported no differences in cognition. This suggests that greater length and intensity of dance may be required to have cognitive benefits. Three intervention studies comparing partnered dance with no-intervention controls favored dance for different aspects of cognition. However, four studies that compared dance with exercise control groups, and two studies with social control groups, less consistently reported relative improvements in cognition, suggesting that the physical component of dance may be the most important for improving cognition.

There have only been a few controlled intervention studies of dance for people with dementia; these provide weak evidence that dance may improve mood and little evidence for improvements in cognition or behavior.

Most studies are small and probably underpowered for detecting medium to small effects. Observational studies have variable and somewhat arbitrary definitions of "dancers" vs. "nondancers" that might be subject to participant interpretation as well as recall bias. Intervention studies often did not include a comprehensive description of the dance intervention including the levels and types of physical, cognitive, and social components. Program adherence rates are rarely reported, so the level of adherence bias is not clear.

Future research

Researchers should develop a clear rationale for how the type of dance program will impact study outcomes and modify program components to maximize their impact on outcomes. This should be articulated in the study protocol. Choice of comparator or control group is also important—controlling for physical (e.g., walking program), cognitive (e.g., cognitive exercises), and/or social (e.g., socialization group) aspects would mean that the study was testing the impact of the uncontrolled components, not the dance program as a whole. A usual care intervention group might be the most appropriate for testing the impact of dance as a whole. Studies should be adequately powered, use intention-to-treat analyses, and fully account for missing data.

There were no trials of dance for people with dementia living in the community. Community-dwelling people with dementia tend to have milder levels of dementia than in nursing homes. Dance could be tested in this population (See Table 46.1 for considerations when designing future dance programs for people with dementia).

Table 46.1 Elements for consideration for future dance programs for people with dementia.

Component	Evidence	Suggestions
Dose	Research studies tested dance programs of sessions of 0.5–2 h at least once per week for between 1.5 and 3 months.	Consider the adequacy of the dose for the outcomes being targeted as well as participant ability and program logistics.
Content—balance between components	Some studies looked at mental health as well as cognitive factors.	Tailoring components of the dance program to address target outcomes may improve results. A dance program to improve depression in people with dementia might emphasize aerobic intensity and social contact rather than cognitive complexity.
Content—dance style and music	Two studies discussed choosing dance and music appropriate for participant age to increase adherence and enjoyment.	Choosing appropriate dance styles and music may facilitate participation. Dance is more effective when it is familiar and provides older participants with an opportunity to reminisce (Vankova et al. 2014).
Content—aerobic intensity	No studies tracked aerobic intensity. Two studies encouraged participants to increase intensity where appropriate.	Aerobic intensity is an important component for gaining the full benefits of an exercise program (Foulds, Bredin, Charlesworth, Ivey, & Warburton, 2014).
Content—cognitive complexity	Three studies; participants learned choreography with increasing complexity over the program.	Providing participants with appropriate levels of cognitive complexity may improve cognition. A study of dual-tasking activity with exercise found that dual tasking ability was significantly improved compared with exercise-only controls (Schwenk, Zieschang, Oster & Hauer, 2010).
Safety	One study described adapting movements for participants at different physical and cognitive levels. In some instances, staff were present to assist participants.	We showed that it is possible to deliver nonseated dance programs safely to participants with dementia in nursing homes (Low et al., 2016). Consider inclusion/exclusion criteria, dance instructor to resident ratio, and the dance movements to reduce risk of injury.

Continued

Table 46.1 Elements for consideration for future dance programs for people with dementia.—cont'd

Component	Evidence	Suggestions
Dance instructor	All dance groups were run by professional dance instructors.	It is not clear if dance programs are effective when delivered by nonprofessional dancers.
Staff or caregiver involvement	Some studies involved spouses, caregivers, and staff to assist participants or act as dance partners throughout the intervention.	Education and involvement of staff and family caregivers may facilitate participation (Palo-Bengtsson & Ekman, 2002).

Publishing a video of the dance program (e.g., as supplementary material) would allow other dance researchers to better understand the intervention and help with program replication. Describing strategies used during the dance program to ensure participant safety would also assist with future replication and implementation.

Further use of qualitative methods may allow a deeper and more nuanced understanding of how dance programs may work, the context, barriers, and facilitators to their operation, and impacts on persons with dementia.

Conclusions

A small number of trials of institutionalized persons with dementia suggest that dance may improve mood, but evidence is inconsistent for its impact on cognition or behavior. There currently is insufficient evidence to recommend dance as an intervention for prevention of or as a therapeutic intervention for dementia. Higher-quality methodological research is required.

Key facts of dance

- Dance is the act of moving through space in specific sequences in a rhythmic and expressive way, usually to music.
- Dance requires participants to synchronize movements with others and the music, recall movement sequences, and imitate movement.
- Dance is ubiquitous across cultures and thought to have evolved as a method of communication and social bonding.
- Dance has physical, cognitive, and social components and has been shown to be associated with physical, cognitive, and psychological benefits in healthy adults.
- A structured dance program can improve physical health outcomes equal to those from other forms of physical activity.

Summary points

- Dance is a physical activity intervention that people with dementia tend to derive pleasure from and naturally participate in.
- Dance has been shown in cross-sectional and longitudinal cohort studies to decrease the risk of cognitive decline and dementia.
- All controlled trials of dance for people with dementia have been in nursing home settings. These show that people with dementia are engaged and communicate better during dance sessions, and dance might improve mood; however, impact on cognition and behavior is unclear.
- There is insufficient evidence to recommend dance as an intervention for prevention or therapeutic intervention for dementia.
- Future research should strive for better methodological quality and to better describe the rationale for and content of dance intervention.

References

Abra, A. (2017). *Dancing in the English style: Consumption, Americanisation and national identity in Britain, 1918–50.* Manchester, United Kingdom: Manchester University Press.

Adam, D., Ramli, A., & Shahar, S. (2016). Effectiveness of a combined dance and relaxation intervention on reducing anxiety and depression and improving quality of life among the cognitively impaired elderly. *Sultan Qaboos University Medical Journal, 16*(1), e47–e53. https://doi.org/10.18295/squmj.2016.16.01.009.

Alpert, P. (2009). Exercise works. *Home Health Care Management and Practice, 21*(5), 371–374. https://doi.org/10.1177/1084822309334032.

Alpert, P. (2011). The health benefits of dance. *Home Health Care Management and Practice, 23*(2), 155–157. https://doi.org/10.1177/1084822310384689.

Arent, S., Landers, D., & Etnier, J. (2000). The effects of exercise on mood in older adults: A meta-analytic review. *Journal of Aging and Physical Activity, 8,* 407–430.

Bahar-Fuchs, A., Clare, L., & Woods, B. (2013). Cognitive training and cognitive rehabilitation for mild to moderate Alzheimer's disease and vascular dementia. *Cochrane Database of Systematic Reviews, 6,* Cd003260. https://doi.org/10.1002/14651858.CD003260.pub2.

Barene, S., Holtermann, A., Oseland, H., Brekke, O. L., & Krustrup, P. (2016). Effects on muscle strength, maximal jump height, flexibility and postural sway after soccer and zumba exercise among female hospital employees: A 9-month randomised controlled trial. *Journal of Sports Science, 34*(19), 1849–1858. https://doi.org/10.1080/02640414.2016.1140906.

Bläsing, B., Calvo-Merino, B., Cross, E. S., Jola, C., Honisch, J., & Stevens, C. J. (2012). Neurocognitive control in dance perception and performance. *Acta Psychologica, 139*(2), 300–308. https://doi.org/10.1016/j.actpsy.2011.12.005.

Blondell, S. J., Hammersley-Mather, R., & Veerman, J. L. (2014). Does physical activity prevent cognitive decline and dementia?: A systematic review and meta-analysis of longitudinal studies. *BMC Public Health, 14*(1), 510. https://doi.org/10.1186/1471-2458-14-510.

Buckland, T. (2006). *Dancing from past to present: Nation, culture, identities.* University of Wisconsin Press.

Burzynska, A. Z., Jiao, Y., Knecht, A. M., Fanning, J., Awick, E. A., Chen, T., et al. (2017). White matter integrity declined over 6-months, but dance intervention improved integrity of the fornix of older adults. *Frontiers in Aging Neuroscience, 9,* 59. https://doi.org/10.3389/fnagi.2017.00059.

Centralhome Company. (2018). *Dance history.* Retrieved from: https://www.centralhome.com/ballroomcountry/history.html.

Christensen, J. F., Cela-Conde, C. J., & Gomila, A. (2017). Not all about sex: Neural and biobehavioral functions of human dance. *Annals of the New York Academy of Sciences, 1400*(1), 8–32. https://doi.org/10.1111/nyas.13420.

Clark, C. N., & Warren, J. D. (2015). Music, memory and mechanisms in Alzheimer's disease. *Brain, 138*(8), 2122–2125. https://doi.org/10.1093/brain/awv148.

Cohen-Mansfield, J. (2016). Non-pharmacological interventions for agitation in dementia: Various strategies demonstrate effectiveness for care home residents; further research in home settings is needed. *Evidence-Based Nursing, 19*(1), 31. https://doi.org/10.1136/eb-2015-102059.

Coubard, O. A., Duretz, S., Lefebvre, V., Lapalus, P., & Ferrufino, L. (2011). Practice of contemporary dance improves cognitive flexibility in aging. *Frontiers in Aging Neuroscience, 3*, 13. https://doi.org/10.3389/fnagi.2011.00013.

Cox, R., & Herzog, V. (2013). The effect of pointe shoe toe box shape on proprioception in novice ballet dancers. *The Internet Journal of Allied Health Sciences and Practice, 11*(2).

Devereaux, C. (2008). Untying the knots: Dance/movement therapy with a family exposed to domestic violence. *American Journal of Dance Therapy, 30*(2), 58. https://doi.org/10.1007/s10465-008-9055-x.

Donath, L., Roth, R., Hohn, Y., Zahner, L., & Faude, O. (2014). The effects of Zumba training on cardiovascular and neuromuscular function in female college students. *European Journal of Sport Science, 14*(6), 569–577. https://doi.org/10.1080/17461391.2013.866168.

Duignan, D., Hedley, L., & Milverton, R. (2009). Exploring dance as a therapy for symptoms and social interaction in a dementia care unit. *Nursing Times, 105*(30), 19–22.

Farina, E., Baglio, F., Pomati, S., D'Amico, A., Campini, I. C., Di Tella, S., et al. (2017). The mirror neurons network in aging, mild cognitive impairment, and alzheimer disease: A functional MRI study. *Frontiers in Aging Neuroscience, 9*, 371. https://doi.org/10.3389/fnagi.2017.00371.

Fong Yan, A., Cobley, S., Chan, C., Pappas, E., Nicholson, L. L., Ward, R. E., et al. (2018). The effectiveness of dance interventions on physical health outcomes compared to other forms of physical activity: A systematic review and meta-analysis. *Sports Medicine, 48*(4), 933–951. https://doi.org/10.1007/s40279-017-0853-5.

Foulds, H. J., Bredin, S. S., Charlesworth, S. A., Ivey, A. C., & Warburton, D. E. (2014). Exercise volume and intensity: A dose-response relationship with health benefits. *European Journal of Applied Physiology, 114*(8), 1563–1571. https://doi.org/10.1007/s00421-014-2887-9.

Gallace, A., & Spence, C. (2010). The science of interpersonal touch: An overview. *Neuroscience and Biobehavioral Reviews, 34*(2), 246–259. https://doi.org/10.1016/j.neubiorev.2008.10.004.

Garfinkel, Y. (2018). The evolution of human dance: Courtship, rites of passage, trance, calendrical ceremonies and the professional dancer. *Cambridge Archaeological Journal, 28*(2), 283–298. https://doi.org/10.1017/S0959774317000865.

Guzmán-García, A., Mukaetova-Ladinska, E., & James, I. (2012). Introducing a Latin ballroom dance class to people with dementia living in care homes, benefits and concerns: A pilot study. *Dementia, 12*(5), 523–535. https://doi.org/10.1177/1471301211429753.

Hackney, M. E., Byers, C., Butler, G., Sweeney, M., Rossbach, L., & Bozzorg, A. (2015). Adapted tango improves mobility, motor—cognitive function, and gait but not cognition in older adults in independent living. *Journal of the American Geriatrics Society, 63*(10), 2105–2113. https://doi.org/10.1111/jgs.13650.

Hagen, E. H., & Bryant, G. A. (2003). Music and dance as a coalition signaling system. *Human Nature, 14*(1), 21–51. https://doi.org/10.1007/s12110-003-1015-z.

Hamill, M., Smith, L., & Röhricht, F. (2011). 'Dancing down memory lane': Circle dancing as a psychotherapeutic intervention in dementia—a pilot study. *Dementia, 11*(6), 709–724. https://doi.org/10.1177/1471301211420509.

Hirono, N., Mori, E., Ikejiri, Y., Imamura, T., Shimomura, T., Ikeda, M., et al. (1997). Procedural memory in patients with mild Alzheimer's disease. *Dementia and Geriatric Cognitive Disorders, 8*(4), 210–216. https://doi.org/10.1159/000106633.

Hokkanen, L., Rantala, L., Remes Anne, M., Härkönen, B., Viramo, P., & Winblad, I. (2008). Dance and movement therapeutic methods in management of dementia: A randomized, controlled study. *Journal of the American Geriatrics Society, 56*(4), 771–772. https://doi.org/10.1111/j.1532-5415.2008.01611.x.

Huang, S. Y., Hogg, J., Zandieh, S., & Bostwick, S. B. (2012). A ballroom dance classroom program promotes moderate to vigorous physical activity in elementary school children. *American Journal of Health Promotion, 26*(3), 160–165. https://doi.org/10.4278/ajhp.090625-QUAN-203.

Karkou, V., & Meekums, B. (2017). Dance movement therapy for dementia. *Cochrane Database of Systematic Reviews*, (2). https://doi.org/10.1002/14651858.CD011022.pub2.

Kattenstroth, J.-C., Kalisch, T., Holt, S., Tegenthoff, M., & Dinse, H. R. (2013). Six months of dance intervention enhances postural, sensorimotor, and cognitive performance in elderly without affecting cardio-respiratory functions. *Frontiers in Aging Neuroscience, 5*, 5. https://doi.org/10.3389/fnagi.2013.00005.

Kattenstroth, J.-C., Kalisch, T., Kolankowska, I., & Dinse, H. R. (2011). Balance, sensorimotor, and cognitive performance in long-year expert senior ballroom dancers. *Journal of Aging Research, 10*. https://doi.org/10.4061/2011/176709.

Kattenstroth, J.-C., Kolankowska, I., Kalisch, T., & Dinse, H. R. (2010). Superior sensory, motor, and cognitive performance in elderly individuals with multi-year dancing activities. *Frontiers in Aging Neuroscience, 2*, 31. https://doi.org/10.3389/fnagi.2010.00031.

Kim, S.-H., Kim, M., Ahn, Y.-B., Lim, H.-K., Kang, S.-G., Cho, J.-h., et al. (2011). Effect of dance exercise on cognitive function in elderly patients with metabolic syndrome: A pilot study. *Journal of Sports Science and Medicine, 10*(4), 671–678.

Koch, S., Kunz, T., Lykou, S., & Cruz, R. (2014). Effects of dance movement therapy and dance on health-related psychological outcomes: A meta-analysis. *The Arts in Psychotherapy, 41*(1), 46–64. https://doi.org/10.1016/j.aip.2013.10.004.

Koch, S., & Weidinger-von der Recke, B. (2009). Traumatised refugees: An integrated dance and verbal therapy approach. *The Arts in Psychotherapy, 36*(5), 289–296. https://doi.org/10.1016/j.aip.2009.07.002.

Kolanowski, A. M. (1999). An overview of the need-driven dementia-compromised behavior model. *Journal of Gerontological Nursing, 25*(9), 7–9.

Kosmat, H., & Vranic, A. (2017). The efficacy of a dance intervention as cognitive training for the old-old. *Journal of Aging and Physical Activity, 25*(1), 32–40.

Kuiper, J. S., Zuidersma, M., Oude Voshaar, R. C., Zuidema, S. U., van den Heuvel, E. R., Stolk, R. P., et al. (2015). Social relationships and risk of dementia: A systematic review and meta-analysis of longitudinal cohort studies. *Ageing Research Reviews, 22*, 39–57. https://doi.org/10.1016/j.arr.2015.04.006.

Laland, K., Wilkins, C., & Clayton, N. (2016). The evolution of dance. *Current Biology, 26*(1), R5–R9. https://doi.org/10.1016/j.cub.2015.11.031.

Lapum, J. L., & Bar, R. J. (2016). Dance for individuals with dementia. *Journal of Psychosocial Nursing and Mental Health Services, 54*(3), 31–34. https://doi.org/10.3928/02793695-20160219-05.

Lazarou, I., Parastatidis, T., Tsolaki, A., Gkioka, M., Karakostas, A., Douka, S., et al. (2017). International ballroom dancing against neurodegeneration: A randomized controlled trial in Greek community-dwelling elders with mild cognitive impairment. *American Journal of Alzheimer's Disease and Other Dementias, 32*(8), 489–499. https://doi.org/10.1177/1533317517725813.

Lee, H. S., Park, S. W., & Park, Y. J. (2016). Effects of physical activity programs on the improvement of dementia symptom: A meta-analysis. *BioMed Research International, 7*. https://doi.org/10.1155/2016/2920146.

Long, B. C., & Stavel, R.v. (1995). Effects of exercise training on anxiety: A meta-analysis. *Journal of Applied Sport Psychology, 7*(2), 167–189. https://doi.org/10.1080/10413209508406963.

Low, L. F., Carroll, S., Merom, D., Baker, J. R., Kochan, N., Moran, F., et al. (2016). We think you can dance! A pilot randomised controlled trial of dance for nursing home residents with moderate to severe dementia. *Complementary Therapies in Medicine, 29*, 42–44. https://doi.org/10.1016/j.ctim.2016.09.005.

Merom, D., Grunseit, A., Eramudugolla, R., Jefferis, B., McNeill, J., & Anstey, K. J. (2016). Cognitive benefits of social dancing and walking in old age: The dancing mind randomized controlled trial. *Frontiers in Aging Neuroscience, 8*, 26. https://doi.org/10.3389/fnagi.2016.00026.

Müller, P., Rehfeld, K., Schmicker, M., Hökelmann, A., Dordevic, M., Lessmann, V., et al. (2017). Evolution of neuroplasticity in response to physical activity in old age: The case for dancing. *Frontiers in Aging Neuroscience, 9*(56). https://doi.org/10.3389/fnagi.2017.00056.

Murrock, C. J., & Gary, F. A. (2010). Culturally specific dance to reduce obesity in african American women. *Health Promotion Practice, 11*(4), 465–473. https://doi.org/10.1177/1524839908323520.

Nielsen, C., & Koff, S. (2017). *Exploring identities in dance*. Retrieved from: https://ausdance.org.au/publications/details/exploring-identities-in-dance.

Niemann, C., Godde, B., & Voelcker-Rehage, C. (2016). Senior dance experience, cognitive performance, and brain volume in older women. *Neural Plasticity, 10*. https://doi.org/10.1155/2016/9837321.

Nyström, K., & Lauritzen, S. O. (2005). Expressive bodies: Demented persons' communication in a dance therapy context. *Health, 9*(3), 297–317. https://doi.org/10.1177/1363459305052902.

Oh, D. C. (2017). Black K-pop fan videos and polyculturalism. *Popular Communication, 15*(4), 269–282. https://doi.org/10.1080/15405702.2017.1371309.

Palo-Bengtsson, L., & Ekman, S.-L. (2002). Emotional response to social dencing and walks in persons with dementia. *American Journal of Alzheimer's Disease and Other Dementias, 17*(3), 149–153. https://doi.org/10.1177/153331750201700308.

Palo-Bengtsson, L., Winblad, B., & Ekman, S. L. (1998). Social dancing: A way to support intellectual, emotional and motor functions in persons with dementia. *Journal of Psychiatric and Mental Health Nursing, 5*(6), 545–554.

Palo-Bengtsson, L., Winblad, B., & Ekman, S. L. (2003). Social dancing: A way to support intellectual, emotional and motor functions in persons with dementia. *Journal of Psychiatric and Mental Health Nursing, 5*(6), 545–554. https://doi.org/10.1046/j.1365-2850.1998.560545.x.

Porat, S., Goukasian, N., Hwang, K. S., Zanto, T., Do, T., Pierce, J., et al. (2016). Dance experience and associations with cortical gray matter thickness in the aging population. *Dementia and Geriatric Cognitive Disorders Extra, 6*(3), 508–517. https://doi.org/10.1159/000449130.

Reddish, P., Fischer, R., & Bulbulia, J. (2013). Let's dance together: Synchrony, shared intentionality and cooperation. *PLoS One, 8*(8), e71182. https://doi.org/10.1371/journal.pone.0071182.

Rehfeld, K., Müller, P., Aye, N., Schmicker, M., Dordevic, M., Kaufmann, J., et al. (2017). Dancing or fitness sport? The effects of two training programs on hippocampal plasticity and balance abilities in healthy seniors. *Frontiers in Human Neuroscience, 11*, 305. https://doi.org/10.3389/fnhum.2017.00305.

Richter, J., & Ostovar, R. (2016). "It don't mean a thing if it ain't got that swing"– an alternative concept for understanding the evolution of dance and music in human beings. *Frontiers in Human Neuroscience, 10*(485). https://doi.org/10.3389/fnhum.2016.00485.

Schwenk, M., Zieschang, T., Oster, P., & Hauer, K. (2010). Dual-task performances can be improved in patients with dementia: A randomised controlled trial. *Neurology, 74*(24), 1961–1968. https://doi.org/10.1212/WNL.0b013e3181e39696.

Sharma, A., Madaan, V., & Petty, F. D. (2006). Exercise for mental health. *Primary Care Companion to the Journal of Clinical Psychiatry, 8*(2), 106.

Shim, M., Johnson, R. B., Gasson, S., Goodill, S., Jermyn, R., & Bradt, J. (2017). A model of dance/movement therapy for resilience-building in people living with chronic pain. *European Journal of Integrative Medicine, 9*, 27–40. https://doi.org/10.1016/j.eujim.2017.01.011.

Spector, A., Orrell, M., & Woods, B. (2010). Cognitive stimulation therapy (CST): Effects on different areas of cognitive function for people with dementia. *International Journal of Geriatric Psychiatry, 25*(12), 1253–1258. https://doi.org/10.1002/gps.2464.

Stevens, C., Ginsborg, J., & Lester, G. (2011). *Backwards and forwards in space and time: Recalling dance movement from long-term memory* (Vol. 4).

Swedenburg, T. (2012). Egypt's music of protest: From sayyid darwish to DJ haha. *Middle East Report*, (265), 39–43.

Tait, J. L., Duckham, R. L., Milte, C. M., Main, L. C., & Daly, R. M. (2017). Influence of sequential vs. Simultaneous dual-task exercise training on cognitive function in older adults. *Frontiers in Aging Neuroscience, 9*, 368. https://doi.org/10.3389/fnagi.2017.00368.

Tarr, B., Launay, J., Cohen, E., & Dunbar, R. (2015). Synchrony and exertion during dance independently raise pain threshold and encourage social bonding. *Biology Letters, 11*(10).

Tarr, B., Launay, J., & Dunbar, R. I. M. (2016). Silent disco: Dancing in synchrony leads to elevated pain thresholds and social closeness. *Evolution and Human Behavior, 37*(5), 343–349. https://doi.org/10.1016/j.evolhumbehav.2016.02.004.

Underwood, M., Lamb, S. E., Eldridge, S., Sheehan, B., Slowther, A.-M., Spencer, A., et al. (2013). Exercise for depression in elderly residents of care homes: A cluster-randomised controlled trial. *Lancet, 382*(9886), 41–49. https://doi.org/10.1016/S0140-6736(13)60649-2.

Valenzuela, M., & Sachdev, P. (2009). Can cognitive exercise prevent the onset of dementia? Systematic review of randomized clinical trials with longitudinal follow-up. *American Journal of Geriatric Psychiatry, 17*(3), 179–187. https://doi.org/10.1097/JGP.0b013e3181953b57.

Vankova, H., Holmerova, I., Machacova, K., Volicer, L., Veleta, P., & Celko, A. M. (2014). The effect of dance on depressive symptoms in nursing home residents. *Journal of the American Medical Directors Association, 15*(8), 582–587. https://doi.org/10.1016/j.jamda.2014.04.013.

Van de Winckel, A., Feys, H., De Weerdt, W., & Dom, R. (2004). Cognitive and behavioural effects of music-based exercises in patients with dementia. *Clinical Rehabilitation, 18*(3), 253–260. https://doi.org/10.1191/0269215504cr750oa.

Verghese, J. (2006). Cognitive and mobility profile of older social dancers. *Journal of the American Geriatrics Society, 54*(8), 1241–1244. https://doi.org/10.1111/j.1532-5415.2006.00808.x.

Verghese, J., Lipton, R. B., Katz, M. J., Hall, C. B., Derby, C. A., Kuslansky, G., et al. (2003). Leisure activities and the risk of dementia in the elderly. *New England Journal of Medicine, 348*, 2508–2516. https://doi.org/10.1056/NEJMoa022252.

Vincent, J. B. (2009). *In pursuit of a dancing 'body': Modernity, physicality and identity in Australia, 1919 to 1939.* The University of Melbourne.

Vordos, Z., Kouidi, E., Mavrovouniotis, F., Metaxas, T., Dimitros, E., Kaltsatou, A., et al. (2016). Impact of traditional Greek dancing on jumping ability, muscular strength and lower limb endurance in cardiac rehabilitation programmes. *European Journal of Cardiovascular Nursing, 16*(2), 150–156. https://doi.org/10.1177/1474515116636980.

Wu, E., Barnes, D. E., Ackerman, S. L., Lee, J., Chesney, M., & Mehling, W. E. (2015). Preventing loss of independence through exercise (PLIÉ): Qualitative analysis of a clinical trial in older adults with dementia. *Aging and Mental Health, 19*(4), 353–362. https://doi.org/10.1080/13607863.2014.935290.

CHAPTER 47

Hypoxic—hyperoxic conditioning and dementia

Robert T. Mallet[1,a], Johannes Burtscher[2], Eugenia B. Manukhina[1,3,4], H. Fred Downey[1,5], Oleg S. Glazachev[6], Tatiana V. Serebrovskaya[7], Martin Burtscher[8,9]

[1]Department of Physiology and Anatomy, University of North Texas Health Science Center, Fort Worth, TX, United States; [2]Laboratory of Molecular and Chemical Biology of Neurodegeneration, École Polytechnique Fédérale de Lausanne (EPFL), Lausanne, Switzerland; [3]Laboratory for Regulatory Mechanisms of Stress and Adaptation, Institute of General Pathology and Pathophysiology, Moscow, Russia; [4]Laboratory for Molecular Mechanisms of Stress, South Ural State University, Chelyabinsk, Russia; [5]Scientific Educational Center for Biomedical Technology, South Ural State University, Chelyabinsk, Russia; [6]Department of Normal Physiology, I.M.Sechenov First Moscow State Medical University (Sechenov University), Moscow, Russia; [7]Department of Hypoxic States, Bogomoletz Institute of Physiology, Kiev, Ukraine; [8]Department of Sport Science, University of Innsbruck, Innsbruck, Austria; [9]Austrian Society for Alpine and High-Altitude Medicine, Innsbruck, Austria

List of abbreviations

AD Alzheimer's disease
AP-1 Activator protein 1
Aβ Amyloid β
CAD Coronary artery disease
CBF Cerebral blood flow
cGMP Cyclic guanosine monophosphate
COPD Chronic obstructive pulmonary disease
CVRFs Cardiovascular risk factors
EPO Erythropoietin
GC Guanylyl cyclase
HBO Hyperbaric oxygen
HIF-1 Hypoxia inducible factor 1
HSP Heat shock protein
IH Intermittent hypoxia
IHC Intermittent hypoxia conditioning
IHHC Intermittent hypoxia—hyperoxia conditioning
mitoK_{ATP}-channel Mitochondrial ATP-sensitive potassium channel
mPTP Mitochondrial permeability transition pore
MTI Multimodal training intervention
NADPH Nicotinamide adenine dinucleotide phosphate
NF-κB Nuclear factor "kappa-light-chain-enhancer" of activated B cells
NO Nitric oxide
Nrf2 Nuclear factor erythroid 2-related factor 2
OSA Obstructive sleep apnea

[a] In absentia author

Diagnosis and Management in Dementia
ISBN 978-0-12-815854-8, https://doi.org/10.1016/B978-0-12-815854-8.00047-1

PIO$_2$ Partial pressure of inspired oxygen
PKC Protein kinase C
PKG Protein kinase G
ROS Reactive oxygen species
SOD Superoxide dismutase
UCP Uncoupling protein
VEGF Vascular endothelial growth factor

Mini-dictionary of terms

Hypoxia A condition of (1) low partial pressure of inspired oxygen and/or (2) insufficient oxygenation at the tissue level

Hyperoxia A higher than normal oxygen tension in the blood and tissues, typically produced by breathing a gas mixture with a high partial pressure of oxygen

Intermittent hypoxia conditioning Exposure of an organism to periods of hypoxia interspersed by periods of breathing normal air and eliciting adaptive physiological and molecular responses. Also known as intermittent hypoxia training

Intermittent hypoxia–hyperoxia conditioning Exposure of an organism to periods of hypoxia alternated with hyperoxic periods, eliciting adaptive physiological and molecular responses

Reactive oxygen species Partially reduced forms of oxygen, including free radicals, that modify the structure and function of proteins, phospholipids, and nucleic acids

Introduction

Hypoxia—i.e., low partial pressure of inspired oxygen (PIO$_2$)—is a dual-edged sword, a trigger of and treatment for chronic diseases. Paracelsus's adage (1537) regarding pharmaceuticals, *dosis sola facit venenum* (only the dose makes the poison), characterizes hypoxia, where "dose" represents the intensity and duration of hypoxia exposures. In hypobaric hypoxia, PIO$_2$ is decreased due to low barometric pressure—e.g., high altitudes; in normobaric hypoxia, the inspired air contains <21% O$_2$.

Paul Bert's classical study of the physiological effects of air pressure, *La Pression barométrique* (1878), described functional impairment or even death in aerobic species subjected to severe hypoxia. Surprisingly, within 50 years clinicians and scientists began considering hypoxia a potential *therapy* for chronic disease. A pioneer in clinical application of hypoxia, Nikolai Sirotinin asserted: "Hypoxia, even severe but *brief and intermittent,* causes beneficial effects on an organism" (authors' emphasis). Concordant with Paracelsus's maxim, Sirotinin's statement now is known to apply only to *moderate* hypoxia; more severe hypoxia is harmful. For example, worsening hypoxemia is associated with cognitive decline in patients with chronic obstructive pulmonary disease (COPD) or obstructive sleep apnea (OSA) syndrome (Daulatzai, 2013). Chronic continuous or intermittent hypoxia (IH) in those patients may contribute to neurodegeneration by altering the activity of ion channels or provoking amyloid β (Aβ) formation (Daulatzai, 2013). In contrast, calibrated IH programs proved beneficial for coronary artery disease (CAD) or COPD (Burtscher et al., 2004, 2009) and afforded neuroprotection in rats (Manukhina et al. 2016) and humans (Bayer et al., 2017; Schega, Peter, Törpel, Mutschler, & Isermann, 2013). These intermittent hypoxia conditioning (IHC) programs elicited neuroprotection by reducing cardiovascular and cerebrovascular risk factors such as hypertension (Lyamina et al., 2011) by upregulating neuroprotectants such as vascular endothelial growth factor (VEGF), erythropoietin (EPO), antioxidants, and nitric oxide (NO) as well as by suppressing apoptosis (Jung, Simpkins, Wilson, Downey, & Mallet, 2008).

A recent randomized controlled trial in elderly subjects with mild-moderate dementia (Bayer et al., 2017) demonstrated that alternating hypoxia and hyperoxia—i.e., intermittent hypoxia—hyperoxia conditioning (IHHC), produced robust neuroprotection. These preliminary results suggest exciting opportunities for clinical application of IHHC. This review presents and discusses IHC and IHHC for prophylaxis or therapy of neurodegenerative diseases, particularly dementia.

Methods to administer intermittent hypoxia and hypoxia—hyperoxia

Generally, IHC and IHHC programs utilize multiple 3—8 min exposures to hypoxia (10%—16% O_2) followed by 2—5 min exposures to normoxic (21% O_2) or hyperoxic (30%—40% O_2) gas totaling 30—40 min/session and applied at 1- or 2-day intervals over 2—8 weeks (Bayer et al., 2017; Burtscher et al., 2009). Gas mixtures with 10%—40% O_2 in N_2 (Bayer et al., 2017; Burtscher et al., 2004) may be delivered via face mask while peripheral O_2 saturation and heart rate are continuously monitored. Figs. 47.1 and 47.2 depict hypoxia—normoxia and hypoxia—hyperoxia cycles and associated data.

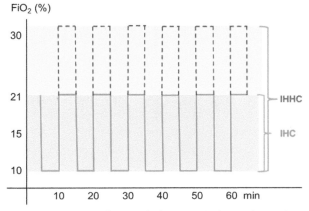

Figure 47.1 *Schematic presentation of typical therapeutic intermittent hypoxia and hypoxia—hyperoxia conditioning protocols.* Depicted are 10-min cycles of 5 min hypoxia and either 5 min normoxia (IHC) or hyperoxia (IHHC).

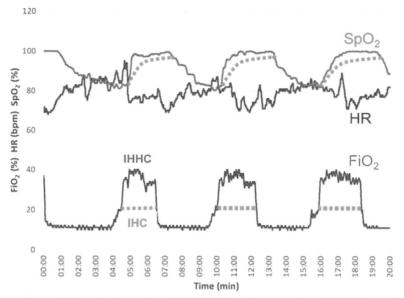

Figure 47.2 *Typical example of peripheral oxygen saturation (SpO₂) and heart rate (HR) responses to intermittent hypoxia conditioning and intermittent hypoxia–hyperoxia conditioning.* Gray trace: O_2 fraction (%) of inspired air (FIO₂); blue trace: HR (min⁻¹); brown trace: SpO₂ (%). Dotted red traces show FIO₂ and SpO₂ during normoxia periods of the IHC protocol.Data recording with ReOxy therapy device, AI Mediq S.A., Luxembourg.

Studies evaluating intermittent hypoxia and hypoxia–hyperoxia conditioning effects in dementia/Alzheimer's

Intermittent hypoxia conditioning and IHHC could potentially moderate the development and progression of dementia/Alzheimer's. In animal and human studies, IHC improved cardiovascular risk factors (CVRFs), augmented cerebral blood flow (CBF) and endothelial NO production, decreased oxidative stress, prevented neuronal degeneration, and stimulated neurogenesis and neuroregeneration (Burtscher et al., 2004; Haider et al., 2009; Manukhina et al., 2013, 2016). Despite these promising findings, few studies have evaluated IHC's effects on cognition in dementia patients. Schega et al. (2013) demonstrated that IHC augmented the cognitive benefits of strength-endurance training in a randomized controlled study in 60–70-year-old subjects. IHC was well tolerated and improved cardiovascular risk factors and exercise tolerance in patients with cardiovascular and respiratory diseases (Burtscher et al., 2004, 2009). At-risk alpine skiers who experienced hypobaric IHC during repeated downhill runs followed by ski lift ascents (Burtscher & Ruedl, 2015) or who lodged at moderate altitudes (≥1300 m) before skiing at higher elevations (Lo, Daniels, Levine, & Burtscher, 2013) had decreased incidence of sudden death.

IHHC might prove even more beneficial than IHC. Indeed, Sazontova, Bolotova, Bedareva, Kostina, & Arkhipenko (2012) demonstrated in rats that unlike IHC, IHHC augmented exercise training. Production of reactive oxygen species (ROS) and heat shock proteins (HSPs) during exercise were reduced to a greater extent after IHHC than with IHC. These findings prompted Bayer et al. (2017) to evaluate IHHC in patients ages 64—92 with mild-moderate dementia participating in a multimodal training intervention (MTI) program, the first study of IH treatment in dementia patients. The patients completed 15—20 sessions of physiotherapy, occupational therapy, and cycling over 5—7 weeks while receiving IHHC (four to eight cycles of 4—7 min hypoxia [10%—14% O_2] + 2—4 min hyperoxia [30%—40% O_2]) or normoxia. Cognitive function and functional exercise capacity were assessed before and after the program. IHHC was readily tolerated. Exercise tolerance improved in both MTI groups but more so in the IHHC group. However, cognition only improved after MTI plus IHHC. Thus, IHHC within an MTI program may improve cognitive and functional performance in geriatric patients with mild-moderate dementia. More well-controlled studies are necessary to confirm these findings and establish the most effective individualized hypoxia protocols.

Intermittent hypoxia conditioning preserves cerebrovascular function: Implications for Alzheimer's disease

Cardiovascular responses to IHC are potentially protective against Alzheimer's disease (AD) by augmenting CBF and vascularity and reducing CVRF for AD, such as systemic hypertension, atherogenic changes in the lipid profile, obesity and metabolic syndrome, ischemic heart disease, and psychological stress. In Western medicine, IH is commonly associated with the repeated, often severe arterial desaturations observed in chronic OSA, a recently proposed CVRF for AD (Andrade, Bubu, Varga, & Osorio, 2018). However, the intense IH employed in experimental OSA (e.g., ventilation with 100% N_2 for ~1 min, ~30 episodes/h, ~8 h/day, for many weeks) differs markedly from that of IHC, where inspired O_2 is reduced by ~50% for several minutes, <10 times per day, for 2—4 weeks (Liu et al., 2017; Zong et al., 2004). As Meerson, Beloshitskii, Vorontsova, Ustinova, & Rozhitskaia (1989) proposed, moderately stressful IHC activates powerful mechanisms that defend cells and tissues from other stresses, including those of chronic diseases.

Brain function demands adequate CBF, and not surprisingly, chronically reduced CBF is associated with neurodegeneration and Aβ deposition (Lourenço, Ledo, Barbosa, & Laranjinha, 2017; Toth, Tarantini, Csiszar, & Ungvari, 2017). Metabolic autoregulation and neurovascular coupling enable CBF to meet varying regional requirements for O_2 and metabolic fuels, and dysfunction of these mechanisms contributes to AD (Toth et al., 2017). NO's vasodilatory function, indispensable for cerebral autoregulation and neurovascular coupling, is compromised in AD (Lourenço et al., 2017), while NO over-production by microglia and astrocytes contributes to neurodegeneration (Brown & Neher 2010). IHC stimulates both vasoactive NO production (Goryacheva et al., 2010; Manukhina et al., 1999) and storage of excessive neurotoxic NO as nonreactive *S*-nitrosothiol and dinitrosyl iron complexes (Manukhina et al., 1999, 2011).

Manukhina et al. (2008) found that IHC adaptation stimulated systemic NO production and was associated with improved cognitive function in rats with experimental AD. In these rats, an NO-synthase (NOS) inhibitor potentiated the damaging effects of intra-cranial Aβ injection and abolished IHC's neuroprotection. Also, injections of an NO donor increased circulating NO and reduced memory disorders caused by Aβ injection to an extent similar to that afforded by IHC (Manukhina et al., 2008). Moreover, IHC improved memory, dampened hippocampal oxidative stress, suppressed brain NO over-production, and prevented cerebrocortical neuronal death or the pathomorphological hallmarks of experimental AD in rats (Manukhina et al. 2016). Thus, by improving NO-mediated cerebrovascular function, IHC suppressed oxido-nitrosative stress, mini-mized neurodegeneration, and maintained the cognitive function of rats with experi-mental AD.

Goryacheva et al. (2010) found that NOx concentration in brain tissue was elevated in experimental AD, and expression of all three NOS isoenzymes also increased. To what extent NOS III in astrocytes and glial cells contributed to elevated tissue NO is unknown. IHC blunted tissue increases in NOS and NOx (Goryacheva et al., 2010). In addition, the ability of IHC to elicit formation of NO stores may have blunted AD-related NO accumulation (Manukhina et al., 1999).

While there is convincing evidence that IHC improves cerebrovascular function in experimental AD, only Mashina et al. (2006) measured CBF, by positioning a laser Doppler flow probe on the parietal cortex. The decrease in CBF produced by Aβ injec-tion was blunted by IHC. Intracarotid injection of acetylcholine to activate NO release demonstrated that AD reduced endothelial function to 25% of normal, yet IHC completely preserved endothelial function while increasing NO storage. Future studies of IHC therapy for AD should include assessments of CBF.

Brain microvascular rarefaction is a hallmark of AD (Manukhina et al. 2016). Vascular density fell by 22%–25% in the hippocampus and cortex of rats with experimental AD but not in IHC-treated AD rats (Goryacheva, Barskov, Viktorov, Downey, & Manu-khina, 2011). Hypoxia-inducible factor-1 (HIF-1) initiates cerebral angiogenesis by acti-vating transcription of the VEGF gene (Chavez, Agani, Pichiule, & LaManna, 2000). Hypoxia-induced VEGF is concentrated primarily in astrocyte end feet surrounding the capillaries, where it can exert paracrine activation of angiogenesis (LaManna, Chavez, & Pichiule, 2004). By increasing cerebral vascularity (Manukhina et al. 2016), IHC improves CBF and its regional distribution. A NOS inhibitor blunted IHC-induced angiogenesis, underscoring NO's pivotal role in this adaptation (Barer et al., 2006).

In addition to its direct cerebrovascular effects, IHC alleviates several recognized CVRFs for AD. Aleshin et al. (1993) and Lyamina et al. (2011) reported antihypertensive responses of IHC-treated patients. In spontaneously hypertensive rats, IHC slowed hypertension development and averted vascular endothelial dysfunction (Manukhina, Jasti, Vanin, & Downey, 2011). With regard to obesity and metabolic syndrome, IHC reduced fat mass, body mass index, serum glucose, lactate and triglyceride concentrations, insulin activity, and blood pressure; it also augmented exercise-induced weight loss in overweight and obese human subjects (Hobbins, Hunter, Gaoua, & Girardm, 2017) and improved serum glucose homeostasis in prediabetic patients (Serebrovska et al., 2017). In ischemic heart disease patients, IHC lowered total serum cholesterol and low-density lipoprotein while increasing high-density lipoprotein (Tin'kov & Aksenov, 2002).

Pronounced antiarrhythmic effects of IHC are documented in patients (Ehrenbourg & Gorbachenkov, 1993) and in animal models of ischemic heart disease (Manukhina et al., 2013; Meerson et al., 1989; Zong et al., 2004). IHC reduced angina and improved cardiac function in patients with ischemic heart disease (Ehrenbourg & Gorbachenkov, 1993; Glazachev, Kopylov, Susta, Dudnik, & Zagaynaya, 2017) and reduced infarct size in animals with acute coronary artery obstruction (Manukhina et al., 2013; Zong et al., 2004).

Neuroprotective anti-stress effects of IHC were demonstrated in rats. IHC acted as an antidepressant and induced hippocampal neurogenesis in the face of a variety of mild stresses (Zhu et al., 2010). Following abrupt alcohol withdrawal, IHC attenuated oxidative damage to the brain, blunted mitochondrial permeability transition (mPTP) in the cerebral cortex and cerebellum, and mitigated behavioral abnormalities (Jung et al., 2008). IHC exerted similar beneficial effects in a rat model of posttraumatic stress disorder (Manukhina et al., 2018).

In summary, robust evidence from human and animal studies indicates that IHC improves cerebrovascular function in AD and reduces many CVRFs for AD.

Why might intermittent hypoxia—hyperoxia conditioning be more beneficial than intermittent hypoxia conditioning?

In recent decades, IHC has received increasing attention as a means of increasing the human organism's resistance to damaging factors (Burtscher et al., 2004). Reoxygenation generates ROS, which in excess inflicts damage, but in moderation triggers redox-signaling cascades that initiate adaptations that increase injury resistance (Arkhipenko, Sazontova, & Zhukova, 2005; Roy, Galano, Durand, Le Guennec, & Lee, 2017; Sazontova & Arkhipenko, 2005).

Apart from the well-known role of excess ROS in disease pathogenesis, more moderate ROS formation and ROS-initiated chemistry are physiological processes occurring continuously in all living organisms (Roy et al., 2017). Physiological functions of ROS include (1) oxidation of damaged molecules; (2) synthesis of messenger

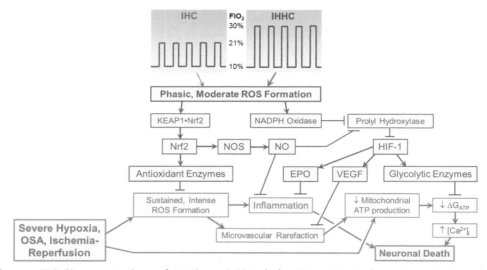

Figure 47.3 *Neuroprotective adaptations initiated by intermittent hypoxia and hypoxia—hyperoxia.* Moderate amounts of ROS generated during IHC, and to a greater extent IHHC, activate transcription factors Nrf2 and HIF-1, increasing production of antioxidant and glycolytic enzymes, nitric oxide synthase (NOS), erythropoietin (EPO), and vascular endothelial growth factor (VEGF). During severe hypoxia or ischemia-reperfusion of the IH-conditioned brain, these proteins suppress the intense, cytotoxic ROS and resultant inflammation, microvascular rarefaction, and ATP depletion that destroy neurons and thus cause dementia. ΔG_{ATP}: free energy of ATP hydrolysis.

molecules—e.g., eicosanoids from polyunsaturated fatty acyl moieties; and (3) redox transmission of external signals to regulate nuclear gene expression. ROS-initiated signaling mediates cell responses to hypoxia, oxidants, and reducing agents. These redox signaling processes constitute the basis for hormesis, where adaptation to one stress increases resistance to other stimuli.

Redox signaling activates transcription factors—e.g., NF-κB, AP-1, HIF-1α and nuclear factor erythroid 2-related factor 2 (Nrf2)—that induce expression of myriad protective molecules (Fig. 47.3)—e.g. antioxidant enzymes, HSPs, iron-regulating proteins, and repair enzymes (Sazontova & Arkhipenko, 2005). Thus, the moderate amounts of ROS generated by environmental stressors appear to be pivotal to the adaptive expression of multiple proteins affording antioxidant and antiinflammatory protection (Arkhipenko, Sazontova, & Zhukova, 2005; Roy et al., 2017; Sazontova & Arkhipenko, 2005). During IHC, tissue O_2 delivery and uptake are increased, especially in the brain, due to cerebral vasodilation combined with enhanced cerebral O_2 extraction (Liu et al., 2017).

Recent IH studies have demonstrated that replacing normoxic pauses with hyperoxic exposures (e.g. 30%—40% O_2) (Glazachev et al., 2017) augments brain protection. ROS are generated in a burst upon the abrupt increase in O_2 concentration. When hypoxia is followed by hyperoxia instead of normoxia, ROS-induced signaling may be reinforced without having to impose more severe hypoxia (Sazontova et al., 2012). Supporting this IHHC approach are reports of IHHC's enhancement of exercise tolerance in patients with stable CAD and metabolic syndrome (Glazachev et al., 2017). These findings indicate that alternating hypoxia and hyperoxia may provide an effective means of maintaining power output while enhancing blood lactate clearance during hyperoxic recovery. Moreover, regular exercise combined with interval hypoxic training improved cognitive function of elderly subjects (Schega et al., 2013), and the addition of hyperoxic intervals to IH can accelerate clearance of the metabolites that disrupt neuronal metabolism in dementia (Liguori et al., 2015).

In another preclinical study, 28 d continuous, moderate hypoxia or hyperoxia (10% or 30% O_2) both generated ROS but in different ways (Terraneo et al., 2017; Terraneo & Samaja, 2017). Subunit 4 of NADPH oxidase, the brain's major ROS source, increased in hypoxia but not hyperoxia, yet neither condition affected Nrf2, a regulator of NADPH oxidase expression. Regarding the underlying mechanism, hypoxia and hyperoxia produce imbalances between ROS generation and antioxidant defense, but hypoxia does so to a greater extent (Terraneo et al., 2017).

Few studies have directly compared the adaptive mechanisms of IHC and IHHC. IHHC in rats elicited greater upregulation of adaptive ROS signals than did IHC (Gonchar & Mankovska, 2012), and thus may elicit more robust protection. A 2-week program of daily, moderate IHHC (five cycles alternating 5 min 10% O_2 and 5 min 30% O_2) attenuated basal and Fe^{2+}/ascorbate-induced lipid peroxidation and formation of protein carbonyls and H_2O_2 in liver mitochondria during subsequent severe hypoxia (7% O_2 for 60 min). Induction of the Mn-dependent superoxide dismutase (SOD) isoform, MnSOD, and its coordinated antioxidant action with glutathione peroxidase might have contributed to the IHHC-related protection.

Other authors comparing IHC and IHHC efficacy and mechanisms (Arkhipenko et al., 2005; Sazontova et al., 2012) found improved exercise tolerance, increased rates of ROS detoxification, and antioxidant enzyme activities in the heart, liver, and brain of rats completing eight sessions of adaptation to swimming either alone or combined with IHC or IHHC. In rats completing a swimming training program and then acutely stressed by swimming to exhaustion, combined training and IHHC attenuated lipid peroxidation in brain and myocardium and also prevented overactivation of the antioxidant enzymes SOD and catalase compared with training alone or combined with IHC. The more robust defense afforded by training plus IHHC was accompanied by moderation of HSP expression. In rats subjected to training plus IHHC, acute stress increased constitutive HSC73 modestly with only minor changes in stress-induced HSP32 and HSP72, while in untrained and training-only rats, acute stress induced excessive expression of all three proteins. Thus IHHC, more powerfully than IHC, induced adaptations that limited oxidative damage of the heart and brain.

Mitochondria: targets for intermittent hypoxia–hyperoxia conditioning

As mitochondria are the primary O_2 consumers and ATP producers in cells, their pivotal role in adaptations to altered O_2 supply is not surprising. Indeed, molecular preconditioning mechanisms seem to converge on mitochondria (Correia et al., 2012; Jung et al., 2008). IHC improves cellular energy management when decreased O_2 supply limits oxidative phosphorylation. The adaptations include increased O_2-utilization efficiency of mitochondrial ATP production (Mela, 1979) and a shift from oxidative phosphorylation to glycolytic energy production with associated vascular and CBF adaptations (see above). Although the precise molecular mechanisms are not fully known, several mitochondrial processes may mediate pivotal IH-induced adaptations:

1. *Oxidative stress*: Mitochondria are major ROS producers, and mitochondrial ROS may mediate preconditioning by activating HIF-1 (Correia et al., 2012). Accordingly, antioxidants or ROS-scavengers can blunt preconditioning (Jung et al., 2008; Perez-Pinzon, Dave, & Raval, 2005). As discussed previously, IHC enhances antioxidant defenses including MnSOD (Sazontova et al., 2012).

2. *Mitochondrial ATP-sensitive potassium ($_{mito}K_{ATP}$)-channels*: $_{mito}K_{ATP}$ channels are activated via the protein kinase C ε isoform (PKC-ε) (Fig. 47.4) during preconditioning and seem to be particularly important for preconditioning in brain (Perez-Pinzon et al., 2005). Blocking $_{mito}K_{ATP}$ abolishes IHC neuroprotection, whereas $_{mito}K_{ATP}$ activation by diazoxide mimics IHC (Oldenburg, Cohen, & Downey, 2003).

3. *Mitochondrial permeability transition pore and apoptosis*: Triggered by ROS and Ca^{2+} overload, mPTP opening releases the proapoptotic factors, cytochrome c and apoptosis-inducing factor, into the cytosol, initiating the mitochondrial apoptotic cascade (Fig. 47.4). Inhibition of mPTP by various protein kinases may contribute to preconditioning cytoprotection (Hentia et al., 2018). NO and NO_2^- (Murillo, Kamga, Mo, & Shiva, 2011), redox-linked Ca^{2+} sensitivity (Halestrap, Clarke, & Khaliulin, 2007), and protein kinase pathways (Perez-Pinzon et al., 2005) have been implicated in preconditioning-induced mPTP regulation.

Reduced oxidative phosphorylation produces IHC-like protection. For example, blockers of mitochondrial respiratory complexes—e.g., 3-nitropropionic acid (complex II) and NS1619 (complex I)—can induce neuroprotection, in the latter case by reducing Ca^{2+} influx through glutamate receptors, upregulating SOD, and preserving ATP (Gáspár et al., 2009). Likewise, IHC diminishes oxidative phosphorylation in brain—e.g., by moderate uncoupling of electron transport from ATP-synthesis (Dirnagl & Meisel, 2008). This uncoupling might be mediated by UCP2, a mitochondrial uncoupling protein implicated in neuroprotection via cellular redox signaling or inhibition of apoptosis (Mattiasson et al., 2003). Specifically, UCP2 is upregulated during ischemic preconditioning in a manner prevented by ROS scavengers, thereby conferring tolerance to ischemia reperfusion.

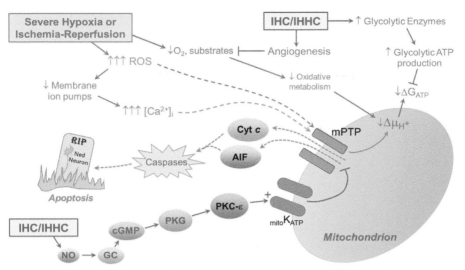

Figure 47.4 *Intermittent hypoxia and hypoxia—hyperoxia protection of neuronal energy-generating systems.* Severe hypoxia and ischemia-reperfusion compromise ATP production by interrupting delivery of O_2 and metabolic fuels and by generating toxic amounts of ROS that disable membrane Ca^{2+} transporters. The increased ROS and $(Ca^{2+})_i$ provoke opening of the mitochondrial permeability transition pores (mPTP), causing collapse of $\Delta\mu_{H^+}$, the driving force for ATP synthesis, and the mitochondrial release of cytochrome c and apoptosis-inducing factor (AIF) that activate proapoptotic caspases. IHC and IHHC protect neuronal ATP synthesis and free energy of ATP hydrolysis (ΔG_{ATP}) by (1) increasing NO production that activates $_{mito}K_{ATP}$ channels to suppress mPTP, (2) promoting angiogenesis to improve tissue perfusion, and (3) increasing glycolytic enzymes to increase O_2-independent ATP production. *GC*; guanylyl cyclase, *cGMP*; cyclic guanosine monophosphate, *PKG*; protein kinase G, *PKC*; protein kinase C.

Thus, IHC appears to induce substantial changes of mitochondrial function that mediate cellular adaptations via mitochondrial ROS. Overall, the mediators of IHC/IHHC neuroprotection may include decreased oxidative phosphorylation and associated enhancement of glycolysis, UCPs and $_{mito}K_{ATP}$-mediated modulation of mitochondrial membrane potential and ion homeostasis, improved antioxidant defense, and decreased apoptosis.

Precisely how IHHC impacts mitochondria is not yet clear. However, the results of hyperbaric (HBO) or normobaric oxygen therapy provide some hints. In HBO, application of 100% O_2 at 1—3 atm augments arterial PO$_2$ and O_2 content to maintain oxidative phosphorylation (Hentia et al., 2018). Repeated HBO enhances neuronal ischemic tolerance via augmented HSP-72 (Wada et al., 2006) and antioxidant defense (Li et al., 2009); these effects appear to be mainly attributable to hyperoxia, not hyperbaricity (Dong et al., 2002). Like IHC, HBO generates moderate ROS (Li et al., 2008) that may activate HIF1-driven transcription of the neuroprotective cytokine, erythropoietin (Gu et al., 2008), and also suppresses mitochondrial apoptosis (Li et al., 2009). The effects of mild hyperoxia may only partly overlap those of mild hypoxia (Terraneo & Samaja, 2017).

The robust benefits of IHHC (Bayer et al., 2017) may result from downregulation of oxidative phosphorylation and induction of antioxidant systems and neuroprotective pathways by hypoxia, followed by rapid reactivation of the highly efficient oxidative phosphorylation machinery by hyperoxia. Cellular systems could profit from both IHC's neuroprotection and hyperoxia-driven rapid repletion of ATP.

Conclusions

Recent studies in animal models and patients show IHC and IHHC to be powerfully neuroprotective and strongly support the application of this emerging intervention to improve neurocognitive function in dementia. IHC's and IHHC's mechanisms are not fully understood, and further work is essential to delineate the IHHC's molecular underpinnings for these interventions and to rigorously compare IHC vs. IHHC to determine whether IHHC is indeed superior to IHC.

Based on the limited mechanistic knowledge of IHHC and IHC, we speculate that the combination of hypoxic and hyperoxic intervals on one hand might synergistically improve antioxidant and antiinflammatory defenses, mitoprotection, redox signaling, and CBF (Fig. 47.5), and on the other hand, might activate complementary

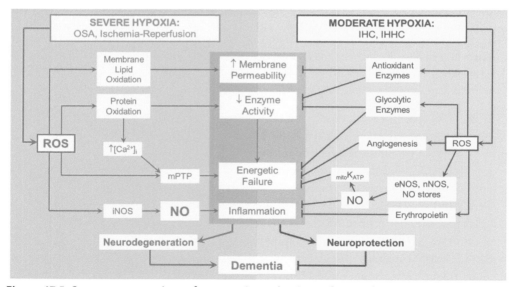

Figure 47.5 *Summary comparison of neurotoxic mechanisms of severe hypoxia (left) versus neuroprotective mechanisms of intermittent hypoxia and hypoxia–hyperoxia conditioning (right).* Reactive O_2 species (ROS) and NO play pivotal roles in both processes by damaging cellular components leading to Ca^{2+} mismanagement, collapse of mitochondrial ATP production and inflammation, or mediating adaptive mechanisms that protect membranes, enzymes, and mitochondria and suppress inflammation.

neuroprotective pathways where (1) hypoxia elicits cellular energy conservation by inducing HSP and downregulating oxidative phosphorylation, protein synthesis, and apoptosis, and then (2) during hyperoxia, waste products are cleared and oxidative phosphorylation efficiency is increased, thereby reinforcing energy balance for the ATP-consuming neuronal processes (e.g., membrane depolarization—repolarization, Ca^{2+} transport, synaptic signaling, and protein synthesis and trafficking) that are energetically costly but indispensable for neurocognitive function.

Key facts of hypoxia—hyperoxia

- The human brain requires adequate oxygen supply for its function and survival.
- Severe hypoxia is a well-known trigger for the development of chronic brain diseases including neurodegeneration and dementia.
- However, well-calibrated repeated hypoxia or hypoxia—hyperoxia exposures (hypoxia conditioning) activate adaptations that make the brain resistant to the causes of dementia.
- Hypoxia conditioning consists of moderately breathing hypoxic gas for a few minutes followed by a few minutes of breathing normal air or hyperoxic gas for 5—10 cycles per daily session.
- Hypoxia conditioning proved to favorably affect the development and progression of chronic diseases including neurodegeneration and dementia.
- Hypoxia conditioning does not use medications or surgery and can be modified to provide optimal treatment for an individual patient.

Summary points

- Intermittent exposures of the human organism to hypoxia or hypoxia—hyperoxia (IHC/IHHC) may potentially induce protective effects (conditioning) in the brain.
- IHC/IHHC have been suggested to beneficially moderate the development and progression of dementia.
- Cyclic hypoxia—normoxia and hypoxia—hyperoxia mobilize adaptive mechanisms that increase the brain's resistance to severe hypoxia, ischemia, and other insults that cause dementia.
- The mediators of IHC/IHHC neuroprotection may include ROS, nitric oxide, and transcription factors that activate antioxidant and hypoxia-responsive genes.
- The resultant adaptations may include complementary, neuroprotective improvements of antioxidant defenses, redox signaling, mitochondrial integrity, and CBF.

Acknowledgments

Portions of the work described in this chapter were supported by research grants from the Russian Science Foundation [to EBM] (grant #17-15-013418), the University of North Texas Health Science Center's Office of Research Development and Commercialization [to RTM], and the University of Innsbruck [to MB].

References

Aleshin, I. A., Kots Ia, I., Tverdokhlib, V. P., Galiautdinov, G. S., Vdovenko, L. G., Zabirov, M. R., et al. (1993). The nondrug treatment of hypertension patients by their adaptation to periodic hypoxia in a barochamber. *Terapevticheskii Arkhiv, 65,* 23–29.

Andrade, A. G., Bubu, O. M., Varga, A. W., & Osorio, R. S. (2018). The relationship between obstructive sleep apnea and Alzheimer's disease. *Journal of Alzheimer's Disease, 64.* S255–S270.

Arkhipenko, Y. V., Sazontova, T. G., & Zhukova, A. G. (2005). Adaptation to periodic hypoxia and hyperoxia improves resistance of membrane structures in heart, liver, and brain. *Bulletin of Experimental Biology and Medicine, 140,* 278–281.

Barer, G. R., Fairlie, J., Slade, J. Y., Ahmed, S., Laude, E. A., Emery, C. J., et al. (2006). Effects of NOS inhibition on the cardiopulmonary system and brain microvascular markers after intermittent hypoxia in rats. *Brain Research, 1098,* 196–203.

Bayer, U., Likar, R., Pinter, G., Stettner, H., Demschar, S., Trummer, B., et al. (2017). Intermittent hypoxic-hyperoxic training on cognitive performance in geriatric patients. *Alzheimer's Dement (N Y), 3,* 114–122.

Burtscher, M., Haider, T., Domej, W., Linser, T., Gatterer, H., Faulhaber, M., et al. (2009). Intermittent hypoxia increases exercise tolerance in patients at risk for or with mild COPD. *Respiratory Physiology & Neurobiology, 165,* 97–103.

Burtscher, M., Pachinger, O., Ehrenbourg, I., Mitterbauer, G., Faulhaber, M., Pühringer, R., et al. (2004). Intermittent hypoxia increases exercise tolerance in elderly men with and without coronary artery disease. *International Journal of Cardiology, 96,* 247–254.

Burtscher, M., & Ruedl, G. (2015). Favourable changes of the risk-benefit ratio in alpine skiing. *International Journal of Environmental Research and Public Health, 12,* 6092–6097.

Chavez, J. C., Agani, F., Pichiule, P., & LaManna, J. C. (2000). Expression of hypoxia-inducible factor-1α in the brain of rats during chronic hypoxia. *Journal of Applied Physiology, 89,* 1937–1942.

Correia, S. C., Santos, R. X., Cardoso, S. M., Santos, M. S., Oliveira, C. R., & Moreira, P. I. (2012). Cyanide preconditioning protects brain endothelial and NT2 neuron-like cells against glucotoxicity: Role of mitochondrial reactive oxygen species and HIF-1α. *Neurobiology of Disease, 45,* 206–218.

Daulatzai, M. A. (2013). Death by a thousand cuts in Alzheimer's disease: Hypoxia – the prodrome. *Neurotoxicity Research, 24,* 216–243.

Dirnagl, U., & Meisel, A. (2008). Endogenous neuroprotection: Mitochondria as gateways to cerebral preconditioning? *Neuropharmacology, 55,* 334–344.

Dong, H., Xiong, L., Zhu, Z., Chen, S., Hou, L., & Sakabe, T. (2002). Preconditioning with hyperbaric oxygen and hyperoxia induces tolerance against spinal cord ischemia in rabbits. *Anesthesiology, 96,* 907–912.

Ehrenbourg, I., & Gorbachenkov, A. (1993). Interval hypoxic training in ischemic heart disease. *Hypoxia Medical Journal, 1,* 13–16.

Gáspár, T., Domoki, F., Lenti, L., Katakam, P. V., Snipes, J. A., Bari, F., et al. (2009). Immediate neuronal preconditioning by NS1619. *Brain Research, 1285,* 196–207.

Glazachev, O., Kopylov, P., Susta, D., Dudnik, E., & Zagaynaya, E. (2017). Adaptations following an intermittent hypoxia-hyperoxia training in coronary artery disease patients: A controlled study. *Clinical Cardiology, 40,* 370–376.

Gonchar, O., & Mankovska, I. (2012). Moderate hypoxia/hyperoxia attenuates acute hypoxia-induced oxidative damage and improves antioxidant defense in lung mitochondria. *Acta Physiologica Hungarica, 99,* 436–446.

Goryacheva, A. V., Barskov, I. V., Viktorov, I. V., Downey, H. F., & Manukhina, E. B. (2011). Adaptation to intermittent hypoxia prevents rarefaction of the brain vascular net in rats with experimental Alzheimer's disease. *The FASEB Journal, 25*, 669, 3.

Goryacheva, A. V., Kruglov, S. V., Pshennikova, M. G., Smirin, B. V., Malyshev, I. Y., Barskov, I. V., et al. (2010). Adaptation to intermittent hypoxia restricts nitric oxide overproduction and prevents beta-amyloid toxicity in rat brain. *Nitric Oxide, 23*, 289—299.

Gu, G. J., Li, Y. P., Peng, Z. Y., Xu, J. J., Kang, Z. M., Xu, W. G., et al. (2008). Mechanism of ischemic tolerance induced by hyperbaric oxygen preconditioning involves upregulation of hypoxia-inducible fator-1α and erythropoietin in rats. *Journal of Applied Physiology, 104*, 1185—1191.

Haider, T., Casucci, G., Linser, T., Faulhaber, M., Gatterer, H., Ott, G., et al. (2009). Interval hypoxic training improves autonomic cardiovascular and respiratory control in patients with mild chronic obstructive pulmonary disease. *Journal of Hypertension, 27*, 1648—1654.

Halestrap, A. P., Clarke, S. J., & Khaliulin, I. (2007). The role of mitochondria in protection of the heart by preconditioning. *Biochimica et Biophysica Acta, 1767*, 1007—1031.

Hentia, C., Rizzato, A., Camporesi, E., Yang, Z., Muntean, D. M., Săndesc, D., et al. (2018). An overview of protective strategies against ischemia/reperfusion injury: The role of hyperbaric oxygen preconditioning. *Brain Behav, 8*. e00959.

Hobbins, L., Hunter, S., Gaoua, N., & Girard, O. (2017). Normobaric hypoxic conditioning to maximize weight loss and ameliorate cardio-metabolic health in obese populations: A systematic review. *American Journal of Physiology — Regulatory, Integrative and Comparative Physiology, 313*, R251—R264.

Jung, M. E., Simpkins, J. W., Wilson, A. M., Downey, H. F., & Mallet, R. T. (2008). Intermittent hypoxia conditioning prevents behavioral deficit and brain oxidative stress in ethanol-withdrawn rats. *Journal of Applied Physiology, 105*, 510—517.

LaManna, J. C., Chavez, J. C., & Pichiule, P. (2004). Structural and functional adaptation to hypoxia in the rat brain. *Journal of Experimental Biology, 207*, 3163—3169.

Liguori, C., Stefani, A., Sancesario, G., Sancesario, G. M., Marciani, M. G., & Pierantozzi, M. (2015). CSF lactate levels, τ proteins, cognitive decline: A dynamic relationship in Alzheimer's disease. *Journal of Neurology Neurosurgery and Psychiatry, 86*, 655—659.

Li, J., Liu, W., Ding, S., Xu, W., Guan, Y., Zhang, J. H., et al. (2008). Hyperbaric oxygen preconditioning induces tolerance against brain ischemia-reperfusion injury by upregulation of antioxidant enzymes in rats. *Brain Research, 1210*, 223—229.

Liu, X., Xu, D., Hall, J. R., Ross, S., Chen, S., Liu, H., et al. (2017). Enhanced cerebral perfusion during brief exposures to cyclic intermittent hypoxia. *Journal of Applied Physiology, 123*, 1689—1697.

Li, J. S., Zhang, W., Kang, Z. M., Ding, S. J., Liu, W. W., Zhang, J. H., et al. (2009). Hyperbaric oxygen preconditioning reduces ischemia-reperfusion injury by inhibition of apoptosis via mitochondrial pathway in rat brain. *Neuroscience, 159*, 1309—1315.

Lo, M. Y., Daniels, J. D., Levine, B. D., & Burtscher, M. (2013). Sleeping altitude and sudden cardiac death. *American Heart Journal, 166*, 71—75.

Lourenço, C. F., Ledo, A., Barbosa, R. M., & Laranjinha, J. (2017). Neurovascular-neuroenergetic coupling axis in the brain: Master regulation by nitric oxide and consequences in aging and neurodegeneration. *Free Radical Biology and Medicine, 108*, 668—682.

Lyamina, N. P., Lyamina, S. V., Senchiknin, V. N., Mallet, R. T., Downey, H. F., & Manukhina, E. B. (2011). Normobaric hypoxia conditioning reduces blood pressure and normalizes nitric oxide synthesis in patients with arterial hypertension. *Journal of Hypertension, 29*, 2265—2272.

Manukhina, E. B., Belkina, L. M., Terekhina, O. L., Abramochkin, D. V., Smirnova, E. A., Budanova, O. P., et al. (2013). Normobaric, intermittent hypoxia conditioning is cardio- and vasoprotective in rats. *Experimental Biology and Medicine (Maywood), 238*, 1413—1420.

Manukhina, E. B., Jasti, D., Vanin, A. F., & Downey, H. F. (2011). Intermittent hypoxia conditioning prevents endothelial dysfunction and improves nitric oxide storage in spontaneously hypertensive rats. *Experimental Biology and Medicine (Maywood), 236*, 867—873.

Manukhina, E. B., Malyshev, I. Y., Smirin, B. V., Mashina, S. Y., Saltykova, V. A., & Vanin, A. F. (1999). Production and storage of nitric oxide in adaptation to hypoxia. *Nitric Oxide, 3*, 393—401.

Manukhina, E. B., Pshennikova, M. G., Goryacheva, A. V., Khomenko, I. P., Mashina, S. Y., Pokidyshev, D. A., et al. (2008). Role of nitric oxide in prevention of cognitive disorders in neurodegenerative brain injuries in rats. *Bulletin of Experimental Biology and Medicine, 146*, 391—395.

Manukhina, E. B., Tseilikman, V. E., Tseilikman, O. B., Komelkova, M. V., Kondashevskaya, M. V., Goryacheva, A. V., et al. (2018). Intermittent hypoxia improves behavioral and adrenal gland dysfunction induced by post-traumatic stress disorder in rats. *Journal of Applied Physiology.*

Mashina, S. Y., Aleksandrin, V. V., Goryacheva, A. V., Vlasova, M. A., Vanin, A. F., Malyshev, I. Y., et al. (2006). Adaptation to hypoxia prevents disturbances in cerebral blood flow during neurodegenerative process. *Bulletin of Experimental Biology and Medicine, 142,* 169−172.

Mattiasson, G., Shamloo, M., Gido, G., Mathi, K., Tomasevic, G., Yi, S., et al. (2003). Uncoupling protein-2 prevents neuronal death and diminishes brain dysfunction after stroke and brain trauma. *Nature Medicine, 9,* 1062−1068.

Meerson, F. Z., Beloshitskii, P. V., Vorontsova, E., Ustinova, E. E., & Rozhitskaia, I. I. (1989). Effect of adaptation to continuous and intermittent hypoxia on heart resistance to ischemic and reperfusion arrhythmias. *Patologicheskaya Fiziologiya I Eksperimental'naya Terapiya,* 48−50.

Mela, L. (1979). Mitochondrial function in cerebral ischemia and hypoxia: Comparison of inhibitory and adaptive responses. *Neurological Research, 1,* 51−63.

Murillo, D., Kamga, C., Mo, L., & Shiva, S. (2011). Nitrite as a mediator of ischemic preconditioning and cytoprotection. *Nitric Oxide, 25,* 70−80.

Oldenburg, O., Cohen, M. V., & Downey, J. M. (2003). Mitochondrial K(ATP) channels in preconditioning. *Journal of Molecular and Cellular Cardiology, 35,* 569−575.

Perez-Pinzon, M. A., Dave, K. R., & Raval, A. P. (2005). Role of reactive oxygen species and protein kinase C in ischemic tolerance in the brain. *Antioxidants and Redox Signaling, 7,* 1150−1157.

Roy, J., Galano, J. M., Durand, T., Le Guennec, J. Y., & Lee, J. C. (2017). Physiological role of reactive oxygen species as promoters of natural defenses. *The FASEB Journal, 31,* 3729−3745.

Sazontova, T. G., & Arkhipenko, I. V. (2005). The role of free radical processes and redox-signalization in adaptation of the organism to changes in oxygen level. *Ross Fiziol Zh Im I M Sechenova, 91,* 636−655.

Sazontova, T., Bolotova, A., Bedareva, I., Kostina, N., & Arkhipenko, Y. (2012). Adaptation to intermittent hypoxia/hyperoxia enhances efficiency of exercise training. In L. Xi, & T. Serebrovskaya (Eds.), *Intermittent hypoxia and human diseases* (pp. 191−205). Heidelberg New York: London Springer.

Schega, L., Peter, B., Törpel, A., Mutschler, H., Isermann, B., & Hamacher, D. (2013). Effects of intermittent hypoxia on cognitive performance and quality of life in elderly adults: A pilot study. *Gerontology, 59,* 316−323.

Serebrovska, T. V., Portnychenko, A. G., Drevytska, T. I., Portnichenko, V. I., Xi, L., Egorov, E., et al. (2017). Intermittent hypoxia training in prediabetes patients: Beneficial effects on glucose homeostasis, hypoxia tolerance and gene expression. *Experimental Biology and Medicine (Maywood), 242,* 1542−1552.

Terraneo, L., Paroni, R., Bianciardi, P., Giallongo, T., Carelli, S., Gorio, A., et al. (2017). Brain adaptation to hypoxia and hyperoxia in mice. *Redox Biol, 11,* 12−20.

Terraneo, L., & Samaja, M. (2017). Comparative response of brain to chronic hypoxia and hyperoxia. *International Journal of Molecular Sciences, 18.*

Tin'kov, A. N., & Aksenov, V. A. (2002). Effects of intermittent hypobaric hypoxia on blood lipid concentrations in male coronary heart disease patients. *High Altitude Medicine & Biology, 3,* 277−282.

Toth, P., Tarantini, S., Csiszar, A., & Ungvari, Z. (2017). Functional vascular contributions to cognitive impairment and dementia: Mechanisms and consequences of cerebral autoregulatory dysfunction, endothelial impairment, and neurovascular uncoupling in aging. *American Journal of Physiology − Heart and Circulatory Physiology, 312,* H1−H20.

Wada, I., Otaka, M., Jin, M., Odashima, M., Komatsu, K., Konishi, N., et al. (2006). Expression of HSP72 in the gastric mucosa is regulated by gastric acid in rats-correlation of HSP72 expression with mucosal protection. *Biochemical and Biophysical Research Communications, 349,* 611−618.

Zhu, X. H., Yan, H. C., Zhang, J., Qu, H. D., Qiu, X. S., Chen, L., et al. (2010). Intermittent hypoxia promotes hippocampal neurogenesis and produces antidepressant-like effects in adult rats. *Journal of Neuroscience, 30,* 12653−12663.

Zong, P., Setty, S., Sun, W., Martinez, R., Tune, J. D., Ehrenburg, I. V., et al. (2004). Intermittent hypoxic training protects canine myocardium from infarction. *Experimental Biology and Medicine (Maywood), 229,* 806−812.

CHAPTER 48

Linking amyloid and depression in the development of Alzheimer's disease: effects of neuromodulatory interventions by brain stimulation

Akihiko Nunomura[1], Toshio Tamaoki[2], Kenji Tagai[1], Shinsuke Kito[1], Shunichiro Shinagawa[1], Masahiro Shigeta[1]

[1]Department of Psychiatry, Jikei University School of Medicine, Minato-ku, Tokyo, Japan; [2]Department of Neuropsychiatry, Graduate School of Medical Science, University of Yamanashi, Chuo, Yamanashi, Japan

List of abbreviations

AD Alzheimer's disease
ADAS-Cog Alzheimer's Disease Assessment Scale—Cognitive subscale
Aβ Amyloid-β
BPSD Behavioral and psychological symptoms of dementia
CSF Cerebrospinal fluid
DBS Deep brain stimulation
DLPFC Dorsolateral prefrontal cortex
ECT Electroconvulsive therapy
FDA Food and Drug Administration
LLD Late-life depression
MCI Mild cognitive impairment
RCT Randomized controlled trial
rTMS Repetitive transcranial magnetic stimulation
tDCS Transcranial direct current stimulation
VNS Vagus nerve stimulation

Mini-dictionary of terms

Brain stimulation Brain stimulation methods are available as neuromodulation techniques for the treatment of a variety of neuropsychiatric disorders. Brain stimulation methods are categorized into invasive and noninvasive methods. Invasive brain stimulation includes DBS and invasive VNS. Noninvasive brain stimulation includes ECT, transcranial magnetic stimulation, tDCS, transcranial alternating current stimulation, magnetic seizure therapy, cranial electrostimulation, and noninvasive VNS.

Deep brain stimulation DBS is a neurosurgical procedure for the treatment of movement disorders, including Parkinson's disease, essential tremor, and dystonia, as well as neuropsychiatric disorders such as epilepsy and obsessive—compulsive disorder. DBS involves implanting electrodes within certain areas of the brain. These electrodes produce electrical impulses that regulate abnormal impulses, or the electrical impulses can affect certain cells and chemicals within the brain. The amount of stimulation is controlled by a pacemaker-like device placed under the skin.

Diagnosis and Management in Dementia
ISBN 978-0-12-815854-8, https://doi.org/10.1016/B978-0-12-815854-8.00048-3

Electroconvulsive therapy ECT is a medical treatment most commonly used in patients with severe major depression or bipolar disorder that has not responded to other treatments. ECT is performed under general anesthesia, in which small electric currents are passed through the brain, intentionally triggering a brief seizure. ECT seems to cause changes in brain chemistry that can quickly reverse symptoms of certain mental health conditions.

Transcranial direct current stimulationt tDCS is a portable neuromodulatory technique that delivers a low electric current to the scalp. tDCS works by applying a positive (anodal) or negative (cathodal) current via electrodes to an area, facilitating the depolarization or hyperpolarization of neurons, respectively. tDCS is currently being explored for several neuropsychiatric conditions, including depression, schizophrenia, aphasia, addiction, epilepsy, chronic pain, and motor rehabilitation.

Transcranial magnetic stimulation TMS is a procedure that uses magnetic pulses to stimulate nerve cells in the brain. During a TMS procedure, a clinician applies an electromagnetic coil to the patient's forehead. The coil administers electromagnetic pulses that stimulate nerve cells in the brain. The use of TMS has been approved as a treatment for major depression, but is limited to those who have tried at least one antidepressant medication that has proven unsuccessful.

Introduction

Depression and dementia, particularly Alzheimer's disease (AD), are mental health problems that are commonly encountered in neuropsychiatric practice in the elderly. There is a close association between depression and AD, which is supported not only by high prevalence of depression in patients with AD (Chi, Yu, Tan, & Tan, 2014), but also by increased risk of developing AD in patients with depression. A history of depression approximately doubles the risk of developing AD later in life (Diniz, Butters, Albert, Dew, & Reynolds, 2013; Ownby, Crocco, Acevedo, John, & Loewenstein, 2006). Moreover, individuals with mild cognitive impairment (MCI) show a strikingly high conversion rate to AD when the MCI is accompanied by depression (Modrego & Ferrández, 2004). Indeed, several findings in the neurodegenerative process of AD have been observed also in patients with depression, including reduction in hippocampal volume (Campbell, Marriott, Nahmias, & MacQueen, 2004) as well as changes in amyloid-β (Aβ) protein metabolism (Namekawa et al., 2013). Of note, stress-induced increase in Aβ in brain interstitial fluid through corticotropin-releasing factor has been demonstrated in a transgenic mouse model of AD (Kang, Cirrito, Dong, Csernansky, & Holtzman, 2007), which indicates that behavioral stress can affect AD-related amyloid pathology. These findings suggest that a subgroup of late-life depression (LLD) is a prodromal or clinical high-risk state for AD that requires preventive intervention against development of dementia. A growing body of evidence is accumulating to show that brain stimulation interventions such as electroconvulsive therapy (ECT), a well-established treatment for LLD (Geduldig & Kellner, 2016; Hausner, Damian, Sartorius, & Frölich, 2011; van Diermen et al., 2018), and newer methods of noninvasive brain stimulation, such as repetitive transcranial magnetic stimulation (rTMS) and transcranial direct current stimulation (tDCS) (Birba et al., 2017; Iimori et al., 2019), may have "anti-dementia" mechanisms of action. In this chapter, we focus on possible neuromodulatory interventions using brain stimulation techniques for the prevention of dementia in association with an altered metabolism of Aβ in LLD at high risk for AD.

Amyloid-associated depression

The concept that LLD, MCI, and AD may represent a clinical continuum (Panza et al., 2010) is consistent with the hypothesis of "amyloid-associated depression" (Sun et al., 2008). This hypothesis has been originally proposed based on the findings that patients with LLD have a lower concentration of plasma Aβ peptide 42 (Aβ$_{42}$) combined with a higher concentration of plasma Aβ$_{40}$ compared with control elderly individuals (Sun et al., 2008), which is further supported by a recent metaanalysis (Nascimento, Silva, Malloy-Diniz, Butters, & Diniz, 2015). Indeed, an association between high plasma Aβ$_{40}$/Aβ$_{42}$ ratio and an increased risk of AD has been shown by large population studies (Graff-Radford et al., 2007; van Oijen, Hofman, Soares, Koudstaal, & Breteler, 2006). In 2016, a brain amyloid imaging study revealed that LLD patients with MCI are characterized by higher Aβ burden (Wu et al. 2016). Taken together, altered amyloid metabolism seems to be involved in the pathophysiology of LLD, in which antidepressants often demonstrate only modest efficacy (Mahgoub & Alexopoulos, 2016). Modulation of amyloid dysmetabolism may have a large impact on improving outcome in treatment-resistant LLD and preventing subsequent conversion to AD.

Electroconvulsive therapy for depression, mild cognitive impairment, and Alzheimer's disease

Electroconvulsive therapy for late-life depression

ECT has been used for the treatment of human diseases for more than 80 years, and is currently considered the most effective treatment for major depressive disorder (Liu, Sheng, Li, & Zhang, 2017). ECT is recommended for use by the American Psychiatric Association Task Force on ECT for severe depression (American Psychiatric Association, 2001; Hausner et al., 2011), and the safety of ECT for LLD has been reported regardless of a diagnosis of dementia, MCI, or no cognitive impairment (Hausner et al., 2011). Of particular interest, several studies have found that higher age of the patient with depression is associated with higher rate of response and remission by ECT (Geduldig & Kellner, 2016; van Diermen et al., 2018). Moreover, patients with LLD treated with ECT have a significantly shorter time to remission than those given pharmacological treatment (Gálvez et al., 2015; Spaans et al., 2015). In 2018, a large cohort study of patients with affective disorders investigated whether ECT alters the risk of dementia during a median follow-up of 4.9 years (Osler, Rozing, Christensen, Andersen, & Jørgensen, 2018). After correcting for the effects of patient selection or competing mortality, the authors concluded that ECT was not associated with risk of incidental dementia in patients with affective disorders. Surprisingly, in contrast to the general notion that elderly patients with recurrent ECT sessions might be vulnerable to cognitive decline, patients ages 70 years and older with affective disorders who received more than 10 ECT sessions had a decreased risk of dementia (hazard ratio 0.54, 95% confidence interval 0.36–0.80) (Osler et al., 2018).

Effects of electroconvulsive therapy on cognitive function

Although adverse cognitive effects may occur as short-term effects of ECT, they are usually limited and transient, even in the elderly patients (Geduldig & Kellner, 2016; Kumar et al., 2016). When right unilateral ultrabrief-pulse ECT is applied, fewer adverse cognitive effects are observed compared with brief-pulse ECT, with similar efficacy on depression (Gálvez et al., 2015). In contrast, several findings support the possibility that long-term effects of ECT reveal a procognitive function. A prospective study of patients with LLD has reported that there is a significant improvement in executive function in the patients 6 months after ECT compared with those examined within 1 week after ECT sessions (Verwijk et al., 2014). Similarly, in a systematic review of 39 publications describing effects of ECT on cognition, long-term (more than 6 months) cognitive outcomes with ECT were reported as either not changed or improved in patients with LLD (Kumar et al., 2016). A follow-up study of patients with depression has reported that a significant cognitive impairment demonstrated by 7 of 10 cognitive tests before ECT is improved 2 years after ECT. Importantly, this long-term cognitive improvement after ECT occurs regardless of remission status of depression, which suggests that ECT induces unique cognitive-boosting processes (Mohn & Rund, 2018). Indeed, an enhancement of neurogenesis and changes in gene expression of distinct neurotrophic signaling pathways have been demonstrated in the hippocampus of experimental animals subjected to electroconvulsive seizures (Altar et al., 2004; Madsen et al., 2000; Perera et al., 2007). Correspondingly, systematic reviews and metaanalyses have revealed that ECT for patients with depression increases brain-derived neurotrophic factor in blood (Rocha et al., 2016), as well as brain volume in certain areas, most consistently in the hippocampus (Gbyl and Videbech, 2018). In addition, ECT increases the antiaging hormone Klotho in cerebrospinal fluid (CSF) in patients with LLD (Hoyer et al., 2018).

Effects of electroconvulsive therapy on amyloid-β metabolism

ECT can affect the levels of Aβ in CSF and blood in patients with depression. In a study of CSF obtained from patients with depression before and after ECT, $A\beta_{42}$ in CSF was significantly elevated after ECT compared with baseline. Intriguingly, the increase in $A\beta_{42}$ after ECT was found in all patients with clinical response to the treatment, but not in those who did not respond (Kranaster et al., 2016). In 2019, the same research group reported that levels of CSF $A\beta_{40}$ at baseline can be identified as a prognostic biomarker of antidepressant efficacy of ECT in LLD (Kranaster et al., 2019). As for blood samples, it has been reported that plasma $A\beta_{40}$ and $A\beta_{42}$ show a transient increase immediately after ECT, followed by normalization 2 h after the ECT (Zimmermann et al., 2012), and show no significant change 1 week after ECT (Piccinni et al., 2013). We measured plasma $A\beta$ in patients with LLD within 2—4 weeks after ECT and found a tendency toward reduction of plasma $A\beta_{40}$ in patients who received ECT compared with those who received pharmacotherapy alone (Fig. 48.1). Indeed, a higher level of plasma $A\beta_{40}$ in the LLD patients at baseline is associated with more pronounced parahippocampal atrophy and poorer cognitive outcomes (Yamazaki et al., 2017). This is consistent with the findings of an 11-year follow-up study of an elderly cohort from the Rotterdam Study (Direk et al., 2013), in which a longitudinal association between depression and increased plasma $A\beta_{40}$ was demonstrated in the participants who developed dementia during the follow-up period. Taken together, these findings suggest beneficial effects of ECT against amyloid dysmetabolism. Clearly, well-designed longitudinal studies are required to elucidate the effects of ECT on cognition and $A\beta$ metabolism more rigorously (Sutton et al., 2015).

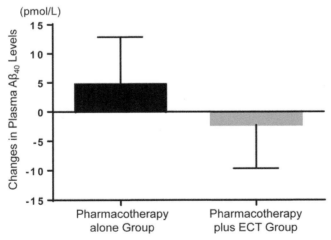

Figure 48.1 Levels of plasma $A\beta_{40}$ may be modified by electroconvulsive therapy (*ECT*). The patients with late-life depression who received pharmacotherapy plus ECT show a reduction in plasma $A\beta_{40}$ between pre- and posttreatment (mean -2.3 ± 7.4 [SD] pmol/L), while the patients who received pharmacotherapy alone show an increase in plasma $A\beta_{40}$ (mean $+4.7 \pm 8.2$ [SD] pmol/L). The difference in Δplasma $A\beta_{40}$ between the two groups is marginally significant by Mann—Whitney *U* test (*P* = .059). *(Reprinted from Yamazaki, C., Tamaoki, T., Nunomura, A., Tamai, K., Yasuda, K., & Motohashi, N. Plasma amyloid—β and Alzheimer's disease-related changes in late-life depression.* Journal of Alzheimer's Disease, 58, *349—354, Copyright (2017), with permission from IOS Press.)*

Electroconvulsive therapy for mild cognitive impairment and Alzheimer's disease

Previous studies have investigated the efficacy of ECT on several behavioral and psychological symptoms of dementia (BPSD), which suggests that ECT is a treatment option for depression and agitation in dementia (Sutor & Rasmussen, 2008; Oudman, 2012). Although several studies suggest that cognitive side effects of ECT are likely in later stages of dementia and in patients with vascular dementia (Oudman, 2012), controlled trials of ECT focusing on the cognitive effects in subjects with MCI and early-stage AD are scarce. Only one trial of ECT for AD patients has been registered at clinicaltrials.gov (Chang, Lane, & Lin, 2018). Further studies of ECT for subjects with MCI and AD are required to examine whether ECT can reduce cognitive impairment or slow the progression of cognitive decline.

rTMS for depression, mild cognitive impairment, and Alzheimer's disease

rTMS for late-life depression

Daily rTMS on prefrontal cortex for 4—6 weeks for treating depression was first proposed in 1994, and was approved by the US Food and Drug Administration (FDA) in 2008 (Iriate & George, 2018). rTMS has been shown to be a safe and effective treatment option for subjects with depression who are resistant to first-line pharmacotherapy and psychotherapy in the general adult population (Conelea et al., 2017). Moreover, several randomized controlled trials (RCTs) have suggested that rTMS has a synergistic effect with antidepressants in the treatment of major depressive disorder (Liu et al., 2017). The efficacy of rTMS among older patients has been also investigated in several open-label studies as well as RCTs (Gálvez et al., 2015). Sabesan et al. (2015) conducted a literature search of RCTs of rTMS for LLD and identified four RCTs, of which two reported no benefit from rTMS compared with sham, while two others reported a substantial benefit. Indeed, a lower intensity of stimulation and a smaller number of stimulations were used in the two negative studies compared with the two positive studies. Conelea et al. (2017) retrospectively collected the data of patients with treatment-resistant depression and found that response and remission rates of rTMS did not differ between younger and older patients, which indicated that the effectiveness of rTMS was not differentially modified by age.

Effects of rTMS on cognitive function

rTMS is consistently reported as a safe and well-tolerated treatment modality with no adverse cognitive side effects (Iriate and George, 2018). Iimori et al., (2019) conducted a systematic review of RCTs that examined the effectiveness of rTMS over the dorsolateral prefrontal cortex (DLPFC) and evaluated cognitive functions in patients with depression, and included 15 RCTs in which procognitive effects on executive function and attention were observed. Of note, performance improvement after rTMS in patients with LLD was not associated with mood, suggesting the possibility of a specific cumulative cognitive-enhancing effect with rTMS over the left DLPFC (Gálvez et al., 2015). Experimentally, rTMS can induce hippocampal neurogenesis (Ueyama et al., 2011) and synaptic plasticity (Lenz et al., 2016).

Effects of rTMS on Aβ-associated dysfunction

Experimentally, rTMS reverses $A\beta_{42}$-induced dysfunction in gamma oscillation during working memory in $A\beta_{42}$-injected rats (Bai et al., 2018). Moreover, rTMS ameliorates cognitive function and synaptic plasticity in a transgenic mouse model of AD expressing mutant human Aβ protein precursor and presenilin-1 genes (Huang et al., 2017). These findings suggest possible protective effects of rTMS against Aβ-associated brain dysfunction.

rTMS for mild cognitive impairment and Alzheimer's disease

Nardone et al. (2014) reviewed the studies that have employed rTMS in patients with MCI and AD and reported that rTMS showed considerable promise to reduce cognitive impairments, but results of the initial studies were considered as still preliminary. More recently, Dong et al. (2018) conducted a systematic review and metaanalysis in which five RCTs of rTMS for patients with AD were included. It was reported that high-frequency (10−20 Hz) rTMS over bilateral or left DLPFC significantly improved cognitive function as measured by the Alzheimer's Disease Assessment Scale−Cognitive subscale (ADAS-Cog) and Clinician's Global Impression of Change in AD patients compared with sham at 1−3 months after the treatment. There was no significant difference in Geriatric Depression Scale between the rTMS and the sham groups, suggesting that the cognitive improvement was independent of mood state in AD. Few mild adverse events, such as headache and fatigability, were observed in both the rTMS and the sham groups (Dong et al., 2018). Vacas et al. (2019) performed a systematic review of the studies of rTMS for the treatment of BPSD and found that two of the three RCTs using rTMS found statistically significant benefits, of which a metaanalysis established a tendency for efficacy of rTMS on BPSD. However, the accumulation of RCTs of rTMS for MCI subjects is still scarce. Drumond Marra et al. (2015) conducted an RCT of high-frequency rTMS on left DLPFC in nondepressed elderly with MCI and found a positive effect on everyday memory that was sustained at least 1 month after treatment. On the other hand, Turriziani et al. (2012) reported that inhibitory rTMS, i.e., low-frequency (1 Hz) rTMS, of the right DLPFC enhanced recognition memory in both healthy controls and subjects with MCI, while rTMS excitation of the same region in healthy controls deteriorated memory performance. Therefore, further studies are needed to investigate the most effective protocols and parameters of rTMS for the cognitive intervention to delay deterioration at a preclinical or early stage of AD.

tDCS for depression, mild cognitive impairment, and Alzheimer's disease

tDCS for late-life depression

tDCS has been used since the 1960s to generate modifications in cortical excitability in preclinical studies and as a therapeutic tool for major depressive disorder (Bennabi & Haffen, 2018). Anodal (or excitatory) tDCS has been typically applied to the left DLPFC to treat depression, with promising results (Gálvez et al., 2015). Of note, comparison among ECT, rTMS, and tDCS reveals that ECT is most effective, while rTMS and tDCS are the lower- and lowest-cost procedures, respectively (Liu et al., 2017). A synergistic effect of tDCS with traditional antidepressants in the treatment of major depressive disorder has also been shown by an RCT, which indicated that the combination therapy performs better than applying either treatment alone (Liu et al., 2017). However, as far as we are aware, there have been no RCTs examining the efficacy of tDCS specifically for the elderly population. Three studies of tDCS on LLD patients with small sample sizes (n = 6, 14, and 15) showed 25%–48% improvement, where the highest response rate was obtained with the strongest current intensity (2 mA) and greatest number of tDCS sessions (30 treatments over 6 weeks). Despite the differences in reported effectiveness, tDCS in elderly patients revealed no significant adverse events (Gálvez et al., 2015).

Effects of tDCS on cognitive function

It has been suggested that tDCS may have positive effects on cognition independent of its effects on mood (Gálvez et al., 2015). Metaanalyses of tDCS effects on healthy participants have focused largely on working memory tasks. A metaanalysis revealed that anodal tDCS produced a significant improvement in reaction time and a trend toward improvement in accuracy for working memory task following stimulation (Hill, Fitzgerald, & Hoy, 2016). Importantly, another metaanalysis of tDCS effects on healthy working memory indicated that positive results were limited to studies of anodal tDCS on left DLPFC combined with cognitive training (Mancuso, Ilieva, Hamilton, & Farah, 2016). No significant effects of tDCS have been found on episodic memory (Galli, Vadillo, Sirota, Feurra, & Medvedeva, 2019) or language function (picture naming and word reading) (Westwood & Romani, 2017) by recent metaanalyses.

Effects of tDCS on Aβ-associated dysfunction

An experimental study using a rat model of AD with bilateral injections of $A\beta_{40}$ peptide into the hippocampus showed that repetitive anodal tDCS (10 sessions in 2 weeks, 20 min per session) applied to the frontal cortex improved spatial learning and memory capability in the Morris water maze task. The memory improvement of the tDCS-treated animals was consistent with the histological findings of preservation of Nissl bodies and choline acetyltransferase density as well as lesser density of glial fibrillary acidic protein—positive cells, suggesting protective effects of tDCS against Aβ-induced brain lesion (Yu, Li, Wen, Zhang, & Tian, 2015).

tDCS for mild cognitive impairment and Alzheimer's disease

In 2018, Cruz Gonzalez et al (2018) conducted a systematic review of studies of tDCS in patients with MCI and dementia. They found 12 trials, including 5 RCTs and 3 crossover studies of tDCS for 195 patients with dementia (174 patients with AD and 21 patients with vascular dementia), as well 4 trials, including 1 RCT and 2 crossover studies, of tDCS for 53 patients with MCI. Among them, 11 studies applied tDCS alone and 5 studies paired tDCS with cognitive training. For metaanalysis, 4 RCTs with memory outcomes after tDCS on the left DLPFC were selected, which included 3 RCTs for AD and one RCT for vascular dementia. Indeed, Cruz Gonzalez et al. (2018) concluded that tDCS improved memory in patients with dementia in the short term but not in the long term. Also, tDCS seemed to have a mild positive effect on memory and language in patients with MCI. However, there was no conclusive advantage in coupling tDCS with cognitive training. Vacas et al. (2019) performed a systematic review of the studies of tDCS for the treatment of BPSD and found only two clinical trials, in which no evidence of efficacy on BPSD was reported. Clearly, more rigorous evidence is needed to establish whether tDCS can serve as an evidence-based intervention for MCI and AD.

Other brain stimulation methods

Not only noninvasive stimulation but also invasive stimulation methods, such as deep brain stimulation (DBS) and vagus nerve stimulation (VNS), can be used for the treatment of LLD and AD (Chang et al., 2018; Gálvez et al., 2015).

Deep brain stimulation

Although DBS is still an experimental technique for treatment of depression, a positive response to DBS with electrodes placed in the nucleus accumbens or subcallosal cingulate cortex has been reported in patients with treatment-resistant LLD (Gálvez et al., 2015). Moreover, effects of DBS targeted to the fornix or nucleus basalis of Meynert have been investigated in patients with AD (Chang et al., 2018). A randomized double-blind phase 2 trial of fornix DBS for mild AD (n = 42) with 2-year follow-up revealed that continuous DBS stimulation was well tolerated and made patients ages 65 years and over experience a slower progression of AD, but not those ages under 65 years (Leoutsakos et al., 2018). Of particular note, chronic fornix DBS significantly reduced Aβ deposition in the hippocampus and cortex of a transgenic rat model of AD expressing mutant human Aβ protein precursor and presenilin-1 genes, which was accompanied by decreases in astrogliosis, microglial activation, and neuronal loss (Leplus et al., 2019).

Vagus nerve stimulation

Invasive and noninvasive (transcutaneous) VNS has been approved by the US FDA as an adjunctive treatment for treatment-resistant depression, while there are no specific data on LLD (Gálvez et al., 2015). However, the use of VNS to treat symptoms of AD is still experimental. A pilot study of VNS for patients with AD (n = 17) showed that VNS was well tolerated, and 41.2% and 70.6% of the patients improved or did not decline from baseline after 1 year on the ADAS-Cog and Mini-Mental State Examination, respectively (Merrill et al., 2006).

Conclusion

LLD, MCI, and AD may represent a clinical continuum that is associated with Aβ pathology, which suggests that a subgroup of LLD requires preventive interventions against development of dementia. Neuromodulation using noninvasive brain stimulation techniques such as ECT, rTMS, and tDCS (Table 48.1) and invasive brain stimulation techniques such as DBS and VNS have the potential to treat LLD, MCI, and AD. Given the high prevalence of LLD and AD in the aging society, noninvasive methods are generally preferable for developing a strategy to reduce the risk of incidental dementia. Further investigations are needed to confirm whether these neuromodulatory interventions are able to delay the progression of the pathological pathway from LLD to dementia.

Key facts of depression in association with dementia

- A history of depression approximately doubles the risk of developing AD.
- Depression significantly accelerates the conversion rate from MCI to AD.
- Patients with depression have lower hippocampal volume.
- Patients with LLD have an altered concentration of plasma Aβ.
- LLD with MCI is characterized by higher Aβ burden in neuroimaging.

Table 48.1 Clinical effects of noninvasive brain stimulation methods.

	ECT	rTMS	tDCS
Depression	A well-established treatment for severe depression	An FDA-approved treatment option for treatment-resistant depression	Possible use suggested by a positive RCT showing a synergistic effect with antidepressants
MCI	No studies on cognitive effects	Possible use suggested by a few positive RCTs on cognitive improvement	Possible use suggested by a positive metaanalysis on cognitive improvement
AD	No studies on cognitive effects; possible use suggested by a positive case series study for BPSD treatment	Possible use suggested by a positive metaanalysis on cognitive improvement and another positive metaanalysis on BPSD treatment	Possible use suggested by a positive metaanalysis on cognitive improvement, but a negative systematic review for BPSD treatment

AD, Alzheimer's disease; *BPSD*, behavioral and psychological symptoms of dementia; *ECT*, electroconvulsive therapy; *FDA*, US Food and Drug Administration; *MCI*, mild cognitive impairment; *RCTs*, randomized controlled trials; *rTMS*, repetitive transcranial magnetic stimulation; *tDCS*, transcranial direct current stimulation.

Summary points

- ECT is recommended as a treatment of severe depression, and higher age of the patient is associated with higher rates of response and remission.
- ECT increases hippocampal volumes and CSF $A\beta_{42}$.
- Experimentally, ECT increases hippocampal neurogenesis.
- Patients of 70 years and older who receive more than 10 ECT sessions have a lower risk of dementia.
- rTMS is a treatment option for a treatment-resistant depression in either young or elderly patients.
- A systematic review of patients with depression suggests procognitive effects of rTMS on executive function and attention.
- Experimentally, rTMS increases hippocampal neurogenesis and synaptic plasticity and is protective against $A\beta$-induced dysfunction.
- An RCT suggests a synergistic effect of tDCS with antidepressants in the treatment of depression.
- A metaanalysis of healthy individuals suggests procognitive effects of tDCS on working memory.
- Experimentally, tDCS is protective against $A\beta$-induced behavioral and histopathological changes.

References

Altar, C. A., Laeng, P., Jurata, L. W., Brockman, J. A., Lemire, A., Bullard, J., et al. (2004). Electroconvulsive seizures regulate gene expression of distinct neurotrophic signaling pathways. *Journal of Neuroscience, 24*, 2667–2677.

American Psychiatric Association. (2001). *The practice of electroconvulsive therapy: Recommendations for treatment, training and privileging* (2nd ed.). Washington, DC: American Psychiatric Press (Chapter 2).

Bai, W., Liu, T., Dou, M., Xia, M., Lu, J., & Tian, X. (2018). Repetitive transcranial magnetic stimulation reverses Aβ1-42-induced dysfunction in gamma oscillation during working memory. *Current Alzheimer Research, 15*, 570–577.

Bennabi, D., & Haffen, E. (2018). Transcranial direct current stimulation (tDCS): A promising treatment for major depressive disorder? *Brain Sciences, 8*. pii: E81.

Birba, A., Ibáñez, A., Sedeño, L., Ferrari, J., García, A. M., & Zimerman, M. (2017). Non-invasive brain stimulation: A new strategy in mild cognitive impairment? *Frontiers in Aging Neuroscience, 9*, 16.

Campbell, S., Marriott, M., Nahmias, C., & MacQueen, G. M. (2004). Lower hippocampal volume in patients suffering from depression: A meta-analysis. *American Journal of Psychiatry, 161*, 598–607.

Chang, C. H., Lane, H. Y., & Lin, C. H. (2018). Brain stimulation in Alzheimer's disease. *Frontiers in Psychiatry, 9*, 201.

Chi, S., Yu, J. T., Tan, M. S., & Tan, L. (2014). Depression in Alzheimer's disease: Epidemiology, mechanisms, and management. *Journal of Alzheimer's Disease, 42*, 739–755.

Conelea, C. A., Philip, N. S., Yip, A. G., Barnes, J. L., Niedzwiecki, M. J., Greenberg, B. D., et al. (2017). Transcranial magnetic stimulation for treatment-resistant depression: Naturalistic treatment outcomes for younger versus older patients. *Journal of Affective Disorders, 217*, 42–47.

Cruz Gonzalez, P., Fong, K. N. K., Chung, R. C. K., Ting, K. H., Law, L. L. F., & Brown, T. (2018). Can transcranial direct-current stimulation alone or combined with cognitive training be used as a clinical intervention to improve cognitive functioning in persons with mild cognitive impairment and dementia? A systematic review and meta-analysis. *Frontiers in Human Neuroscience, 12*, 416.

van Diermen, L., van den Ameele, S., Kamperman, A. M., Sabbe, B. C. G., Vermeulen, T., Schrijvers, D., et al. (2018). Prediction of electroconvulsive therapy response and remission in major depression: Meta-analysis. *British Journal of Psychiatry, 212*, 71–80.

Diniz, B. S., Butters, M. A., Albert, S. M., Dew, M. A., & Reynolds, C. F., 3rd (2013). Late-life depression and risk of vascular dementia and Alzheimer's disease: Systematic review and meta-analysis of community-based cohort studies. *British Journal of Psychiatry, 202*, 329–335.

Direk, N., Schrijvers, E. M., de Bruijn, R. F., Mirza, S., Hofman, A., Ikram, M. A., et al. (2013). Plasma amyloid β, depression, and dementia in community-dwelling elderly. *Journal of Psychiatric Research, 47*, 479–485.

Dong, X., Yan, L., Huang, L., Guan, X., Dong, C., Tao, H., et al. (2018). Repetitive transcranial magnetic stimulation for the treatment of Alzheimer's disease: A systematic review and meta-analysis of randomized controlled trials. *PloS One, 13*, e0205704.

Drumond Marra, H. L., Myczkowski, M. L., Maia Memória, C., Arnaut, D., Leite Ribeiro, P., Sardinha Mansur, C. G., et al. (2015). Transcranial magnetic stimulation to address mild cognitive impairment in the elderly: A randomized controlled study. *Behavioural Neurology, 2015*, 287843.

Galli, G., Vadillo, M. A., Sirota, M., Feurra, M., & Medvedeva, A. (2019). A systematic review and meta-analysis of the effects of transcranial direct current stimulation (tDCS) on episodic memory. *Brain Stimul, 12*, 231–241.

Gálvez, V., Ho, K. A., Alonzo, A., Martin, D., George, D., & Loo, C. K. (2015). Neuromodulation therapies for geriatric depression. *Current Psychiatry Reports, 17*, 59.

Gbyl, K., & Videbech, P. (2018). Electroconvulsive therapy increases brain volume in major depression: A systematic review and meta-analysis. *Acta Psychiatrica Scandinavica, 138*, 180–195.

Geduldig, E. T., & Kellner, C. H. (2016). Electroconvulsive therapy in the elderly: New findings in geriatric depression. *Current Psychiatry Reports, 18*, 40.

Graff-Radford, N. R., Crook, J. E., Lucas, J., Boeve, B. F., Knopman, D. S., Ivnik, R. J., et al. (2007). Association of low plasma Aβ42/Aβ40 ratios with increased imminent risk for mild cognitive impairment and Alzheimer disease. *Archives of Neurology, 64*, 354–362.

Hausner, L., Damian, M., Sartorius, A., & Frölich, L. (2011). Efficacy and cognitive side effects of electroconvulsive therapy (ECT) in depressed elderly inpatients with coexisting mild cognitive impairment or dementia. *Journal of Clinical Psychiatry, 72*, 91–97.

Hill, A. T., Fitzgerald, P. B., & Hoy, K. E. (2016). Effects of anodal transcranial direct current stimulation on working memory: A systematic review and meta-analysis of findings from healthy and neuropsychiatric populations. *Brain Stimul, 9*, 197–208.

Hoyer, C., Sartorius, A., Aksay, S. S., Bumb, J. M., Janke, C., Thiel, M., et al. (2018). Electroconvulsive therapy enhances the anti-ageing hormone Klotho in the cerebrospinal fluid of geriatric patients with major depression. *European Neuropsychopharmacology, 28*, 428–435.

Huang, Z., Tan, T., Du, Y., Chen, L., Fu, M., Yu, Y., et al. (2017). Low-frequency repetitive transcranial magnetic stimulation ameliorates cognitive function and synaptic plasticity in APP23/PS45 mouse model of Alzheimer's disease. *Frontiers in Aging Neuroscience, 9*, 292.

Iimori, T., Nakajima, S., Miyazaki, T., Tarumi, R., Ogyu, K., Wada, M., et al. (2019). Effectiveness of the prefrontal repetitive transcranial magnetic stimulation on cognitive profiles in depression, schizophrenia, and Alzheimer's disease: A systematic review. *Progress In Neuro-Psychopharmacology & Biological Psychiatry, 88*, 31–40.

Iriarte, I. G., & George, M. S. (2018). Transcranial magnetic stimulation (TMS) in the elderly. *Current Psychiatry Reports, 20*, 6.

Kang, J. E., Cirrito, J. R., Dong, H., Csernansky, J. G., & Holtzman, D. M. (2007). Acute stress increases interstitial fluid amyloid-β via corticotropin-releasing factor and neuronal activity. *Proceedings of the National Academy of Sciences of the United States of America, 104*, 10673–10678.

Kranaster, L., Aksay, S. S., Bumb, J. M., Janke, C., Alonso, A., Hoyer, C., et al. (2016). Electroconvulsive therapy selectively enhances amyloid β 1-42 in the cerebrospinal fluid of patients with major depression: A prospective pilot study. *European Neuropsychopharmacology, 26*, 1877–1884.

Kranaster, L., Hoyer, C., Aksay, S. S., Bumb, J. M., Müller, N., Zill, P., et al. (2019). Biomarkers for antidepressant efficacy of electroconvulsive therapy: An exploratory cerebrospinal fluid study. *Neuropsychobiology, 77*, 13–22.

Kumar, S., Mulsant, B. H., Liu, A. Y., Blumberger, D. M., Daskalakis, Z. J., & Rajji, T. K. (2016). Systematic review of cognitive effects of electroconvulsive therapy in late-life depression. *American Journal of Geriatric Psychiatry, 24*, 547–565.

Lenz, M., Galanis, C., Müller-Dahlhaus, F., Opitz, A., Wierenga, C. J., Szabó, G., et al. (2016). Repetitive magnetic stimulation induces plasticity of inhibitory synapses. *Nature Communications, 7*, 10020.

Leoutsakos, J. S., Yan, H., Anderson, W. S., Asaad, W. F., Baltuch, G., Burke, A., et al. (2018). Deep brain stimulation targeting the fornix for mild Alzheimer dementia (the ADvance trial): A two year follow-up including results of delayed activation. *Journal of Alzheimer's Disease, 64*, 597–606.

Leplus, A., Lauritzen, I., Melon, C., Kerkerian-Le Goff, L., Fontaine, D., & Checler, F. (2019). Chronic fornix deep brain stimulation in a transgenic Alzheimer's rat model reduces amyloid burden, inflammation, and neuronal loss. *Brain Structure and Function, 224*, 363–372.

Liu, S., Sheng, J., Li, B., & Zhang, X. (2017). Recent advances in non-invasive brain stimulation for major depressive disorder. *Frontiers in Human Neuroscience, 11*, 526.

Madsen, T. M., Treschow, A., Bengzon, J., Bolwig, T. G., Lindvall, O., & Tingström, A. (2000). Increased neurogenesis in a model of electroconvulsive therapy. *Biological Psychiatry, 47*, 1043–1049.

Mahgoub, N., & Alexopoulos, G. S. (2016). Amyloid hypothesis: Is there a role for antiamyloid treatment in late-life depression? *American Journal of Geriatric Psychiatry, 24*, 239–247.

Mancuso, L. E., Ilieva, I. P., Hamilton, R. H., & Farah, M. J. (2016). Does transcranial direct current stimulation improve healthy working memory?: A meta-analytic review. *Journal of Cognitive Neuroscience, 28*, 1063–1089.

Merrill, C. A., Jonsson, M. A., Minthon, L., Ejnell, H., C-son Silander, H., Blennow, K., et al. (2006). Vagus nerve stimulation in patients with Alzheimer's disease: Additional follow-up results of a pilot study through 1 year. *Journal of Clinical Psychiatry, 67*, 1171–1178.

Modrego, P. J., & Ferrández, J. (2004). Depression in patients with mild cognitive impairment increases the risk of developing dementia of Alzheimer type: A prospective cohort study. *Archives of Neurology, 61,* 1290–1293.

Mohn, C., & Rund, B. R. (2018). Neurognitive function and symptom remission 2 years after ECT in major depressive disorders. *Journal of Affective Disorders, 246,* 368–375.

Namekawa, Y., Baba, H., Maeshima, H., Nakano, Y., Satomura, E., Takebayashi, N., et al. (2013). Heterogeneity of elderly depression: Increased risk of Alzheimer's disease and Aβ protein metabolism. *Progress In Neuro-Psychopharmacology & Biological Psychiatry, 43,* 203–208.

Nardone, R., Tezzon, F., Höller, Y., Golaszewski, S., Trinka, E., & Brigo, F. (2014). Transcranial magnetic stimulation (TMS)/repetitive TMS in mild cognitive impairment and Alzheimer's disease. *Acta Neurologica Scandinavica, 129,* 351–366.

Nascimento, K. K., Silva, K. P., Malloy-Diniz, L. F., Butters, M. A., & Diniz, B. S. (2015). Plasma and cerebrospinal fluid amyloid-β levels in late-life depression: A systematic review and meta-analysis. *Journal of Psychiatric Research, 69,* 35–41.

van Oijen, M., Hofman, A., Soares, H. D., Koudstaal, P. J., & Breteler, M. M. (2006). Plasma $A\beta_{1-40}$ and $A\beta_{1-42}$ and the risk of dementia: A prospective case-cohort study. *The Lancet Neurology, 5,* 655–660.

Osler, M., Rozing, M. P., Christensen, G. T., Andersen, P. K., & Jørgensen, M. B. (2018). Electroconvulsive therapy and risk of dementia in patients with affective disorders: A cohort study. *Lancet Psychiatry, 5,* 348–356.

Oudman, E. (2012). Is electroconvulsive therapy (ECT) effective and safe for treatment of depression in dementia? A short review. *The Journal of ECT, 28,* 34–38.

Ownby, R. L., Crocco, E., Acevedo, A., John, V., & Loewenstein, D. (2006). Depression and risk for Alzheimer disease: Systematic review, meta-analysis, and metaregression analysis. *Archives of General Psychiatry, 63,* 530–538.

Panza, F., Frisardi, V., Capurso, C., D'Introno, A., Colacicco, A. M., Imbimbo, B. P., et al. (2010). Late-life depression, mild cognitive impairment, and dementia: Possible continuum? *American Journal of Geriatric Psychiatry, 18,* 98–116.

Perera, T. D., Coplan, J. D., Lisanby, S. H., Lipira, C. M., Arif, M., Carpio, C., et al. (2007). Antidepressant-induced neurogenesis in the hippocampus of adult nonhuman primates. *Journal of Neuroscience, 27,* 4894–4901.

Piccinni, A., Veltri, A., Vizzaccaro, C., Catena Dell'Osso, M., Medda, P., Domenici, L., et al. (2013). Plasma amyloid-β levels in drug-resistant bipolar depressed patients receiving electroconvulsive therapy. *Neuropsychobiology, 67,* 185–191.

Rocha, R. B., Dondossola, E. R., Grande, A. J., Colonetti, T., Ceretta, L. B., Passos, I. C., et al. (2016). Increased BDNF levels after electroconvulsive therapy in patients with major depressive disorder: A meta-analysis study. *Journal of Psychiatric Research, 83,* 47–53.

Sabesan, P., Lankappa, S., Khalifa, N., Krishnan, V., Gandhi, R., & Palaniyappan, L. (2015). Transcranial magnetic stimulation for geriatric depression: Promises and pitfalls. *World Journal of Psychiatry, 5,* 170–181.

Spaans, H. P., Sienaert, P., Bouckaert, F., van den Berg, J. F., Verwijk, E., Kho, K. H., et al. (2015). Speed of remission in elderly patients with depression: Electroconvulsive therapy v. medication. *British Journal of Psychiatry, 206,* 67–71.

Sun, X., Steffens, D. C., Au, R., Folstein, M., Summergrad, P., Yee, J., et al. (2008). Amyloid-associated depression: A prodromal depression of Alzheimer disease? *Archives of General Psychiatry, 65,* 542–550.

Sutor, B., & Rasmussen, K. G. (2008). Electroconvulsive therapy for agitation in Alzheimer disease: A case series. *The Journal of ECT, 24,* 239–241.

Sutton, T. A., Sohrabi, H. R., Rainey-Smith, S. R., Bird, S. M., Weinborn, M., & Martins, R. N. (2015). The role of APOE-ε4 and beta amyloid in the differential rate of recovery from ECT: A review. *Translational Psychiatry, 5,* e539.

Turriziani, P., Smirni, D., Zappalà, G., Mangano, G. R., Oliveri, M., & Cipolotti, L. (2012). Enhancing memory performance with rTMS in healthy subjects and individuals with mild cognitive impairment: The role of the right dorsolateral prefrontal cortex. *Frontiers in Human Neuroscience, 6,* 62.

Ueyama, E., Ukai, S., Ogawa, A., Yamamoto, M., Kawaguchi, S., Ishii, R., et al. (2011). Chronic repetitive transcranial magnetic stimulation increases hippocampal neurogenesis in rats. *Psychiatry and Clinical Neurosciences, 65,* 77–81.

Vacas, S. M., Stella, F., Loureiro, J. C., do Couto, F. S., Oliveira-Maia, A. J., & Forlenza, O. V. (2019). Noninvasive brain stimulation for behavioural and psychological symptoms of dementia: A systematic review and meta-analysis. *International Journal of Geriatric Psychiatry.* https://doi.org/10.1002/gps.5003 (Epub ahead of print).

Verwijk, E., Comijs, H. C., Kok, R. M., Spaans, H. P., Tielkes, C. E., Scherder, E. J., et al. (2014). Short- and long-term neurocognitive functioning after electroconvulsive therapy in depressed elderly: A prospective naturalistic study. *International Psychogeriatrics, 26,* 315–324.

Westwood, S. J., & Romani, C. (2017). Transcranial direct current stimulation (tDCS) modulation of picture naming and word reading: A meta-analysis of single session tDCS applied to healthy participants. *Neuropsychologia, 104,* 234–249.

Wu, K. Y., Liu, C. Y., Chen, C. S., Chen, C. H., Hsiao, I. T., Hsieh, C. J., et al. (2016). Beta-amyloid deposition and cognitive function in patients with major depressive disorder with different subtypes of mild cognitive impairment: (18)F-florbetapir (AV-45/Amyvid) PET study. *European Journal of Nuclear Medicine and Molecular Imaging, 43,* 1067–1076.

Yamazaki, C., Tamaoki, T., Nunomura, A., Tamai, K., Yasuda, K., & Motohashi, N. (2017). Plasma amyloid-β and Alzheimer's disease-related changes in late-life depression. *Journal of Alzheimer's Disease, 58,* 349–354.

Yu, X., Li, Y., Wen, H., Zhang, Y., & Tian, X. (2015). Intensity-dependent effects of repetitive anodal transcranial direct current stimulation on learning and memory in a rat model of Alzheimer's disease. *Neurobiology of Learning and Memory, 123,* 168–178.

Zimmermann, R., Schmitt, H., Rotter, A., Sperling, W., Kornhuber, J., & Lewczuk, P. (2012). Transient increase of plasma concentrations of amyloid β peptides after electroconvulsive therapy. *Brain Stimul, 5,* 25–29.

CHAPTER 49

Regional and local dementia care networks

Franziska Laporte Uribe[1], Karin Wolf-Ostermann[2], Jochen René Thyrian[3], Bernhard Holle[1]

[1]German Center for Neurodegenerative Diseases e.V. (DZNE), DZNE site Witten, Witten, Germany; [2]University of Bremen, Department 11, Human and Health Sciences, Bremen, Germany; [3]German Center for Neurodegenerative Diseases e.V. (DZNE), DZNE site Rostock/Greifswald, Greifswald, Germany

List of abbreviation

DemNet-D Multicentered, interdisciplinary evaluation study of dementia care networks in Germany

Mini-dictionary of terms

Care arrangements Have been defined as "the domestic setting plus the combination of any type of informal and formal carers and services that are necessary to maintain the care situation and enable the person with dementia to remain at home" (von Kutzleben, Reuther, Dortmann, & Holle, 2015).

Case management may be defined as "A process of care planning and co-ordination of the services and resources used by people with disabilities and their families. Primary functions of case management include assessment, development of a care plan, securing access to services, and monitoring to ensure service timeliness, comprehensiveness, and quality over time" (Dill, 2006) as cited in Iliffe et al. (2017). However, in dementia, a wide range of case management models exist across Europe (Iliffe et al., 2017).

Dementia care networks May be considered models of integrated care that link different support services and stakeholder, such as general practitioners, clinics, Alzheimer societies, charity organizations, or local authorities.

Integrated care Is defined by the World Health Organization as "(t)he management and delivery of health services so that clients receive a continuum of preventive and curative services, according to their needs over time and across different levels of the health system" (2008).

Knowledge management "Is the process of creating, sharing, using and managing the knowledge and information of an organization" (Girard & Girard, 2015).

Introduction

Dementia is a major cause of care dependency among elderly people worldwide (World Health Organization & Alzheimer's Disease International, 2012). Currently, there are an estimated 1.6 million people living with dementia in Germany, and this number is expected to double within the next 30 years (Bickel, 2016). Most people prefer to live at home if they become care dependent (Tarricone & Tsouros, 2008). In Germany, approximately half (1.76 million) of persons in need of care, including persons with

Diagnosis and Management in Dementia
ISBN 978-0-12-815854-8, https://doi.org/10.1016/B978-0-12-815854-8.00049-5

dementia, are exclusively supported by an informal caregiver, such as a family member, and approximately one-quarter (830,000) receive support from a professional home care provider (Statistisches Bundesamt, 2018). Informal caregivers are crucial for enabling the person with dementia to live in their own home for as long as possible (Brodaty, Green, & Koschera, 2003); however, caregivers may experience adverse physical, psychological, social, and financial consequences (Brodaty et al., 2003; Clyburn, Stones, Hadjistavropoulos, & Tuokko, 2000; Schulz & Sherwood, 2008), which increase the likelihood that the person with dementia is placed in a long-term care facility (Gaugler, Yu, Krichbaum, & Wyman, 2009). Therefore, adequate care for people with dementia must acknowledge their preferences and reduce caregiver burden by involving a mix of formal and informal support, medical and nonmedical specialists, and professionals. However, it remains a challenge to provide timely and appropriate care despite the numerous community-based social, medical, and psychological support services available. "In Europe people with dementia syndrome and their carers all too often encounter services that are limited in resources, poorly coordinated, variable in quality and quantity, protocol-driven, inequitable, sometimes stigmatising and only weakly tailored to individual and family needs" (Iliffe et al., 2017). The German health care system is characterized by fragmentation. For caregivers, this indicates that the complex care situation is linked to an untransparent bureaucracy that might prevent them from accessing available support options. Moreover, services and care providers are often clustered in urban areas, whereas rural areas are underserviced. In addition, a stigma remains surrounding dementia that translates into a perception of persons affected by dementia that available services are not suitable for them. To overcome these challenges, the German government has made networking and the establishment of regional dementia alliances one of its goals (Federal Ministry for Family Affairs Senior Citizens Women & Youth & Federal Ministry of Health, 2014), and in recent years, regional and local dementia care networks have been established.

What are dementia care networks?

In general, dementia care networks are cooperation structures created to provide timely and appropriate support for people living with dementia, thereby increasing their quality of life and reducing caregiver burden. They are intended to further the development and coordination of support structures and promote collaboration between stakeholders. They may be considered models of integrated care (Thyrian, Wübbeler, & Hoffmann, 2013). Some networks have a more health-oriented approach and link services often related to the diagnosis, treatment, and care of a person with dementia, whereas other networks take a broader approach that focuses on person-centered coordinated care and support (Laporte Uribe, Gräske et al., 2017).

Dementia care networks bring together various primary and secondary health care providers, outpatient and inpatient services, community and private sector organizations, doctors, therapists, specialists, and volunteers. Formal contracts or agreements typically specify their cooperation (Gräske et al., 2016). People with dementia and their families are often not explicitly part of a dementia care network or are aware of their involvement because the point of entry is typically through one of the stakeholders, key persons, or core institutions, such as a general practitioner or a memory clinic. Dementia care networks vary widely in size, number of stakeholders, and staff, as well as funding and cooperation structures (Michalowsky et al., 2017). Therefore, there are different ways to characterize them. One potential distinction is related to their founding organization/institution and their stakeholders: medically associated networks versus community oriented networks (Wubbeler et al., 2017). Medically associated networks (Fig. 49.1) are

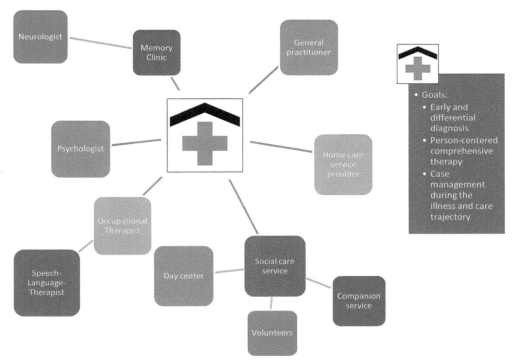

Figure 49.1 Example of a medically associated dementia care network, its stakeholders and goals. This figure illustrates the exemplary structure of a medically associated dementia care network. Typical stakeholders (e.g., memory clinics, day centers, general practitioners, specialist physicians) are linked with the founding institution or organization, such as, a clinic. Medically associated dementia care networks usually aim at providing an early and differential diagnosis as well as person-centered comprehensive care often using case management processes to achieve these goals. *(This figure was developed by F. Laporte Uribe for this book chapter based on a definition of dementia care networks by Wubbeler, M., Thyrian, J. R., Michalowsky, B., Erdmann, P., Hertel, J., Holle, B. et al. (2017). How do people with dementia utilise primary care physicians and specialists within dementia networks? Results of the dementia networks in Germany (DemNet-D) study.* Health and Social Care in the Community, *25(1), 285-294. https://doi:10.1111/hsc.12315.)*

often established around a clinic or a clinic associated organization that involves several medical care disciplines, such as general practitioners, specialist physicians, and (psycho-) social services. Their primary aim is an early and formal differential diagnosis, as well as providing person-centered comprehensive therapy to the person with dementia. These networks will typically establish a dementia care pathway to achieve their aims.

A community-oriented network is typically founded by a local government or government organization that involves different stakeholders, such as doctors, clinics, private sector organizations, or health fund representatives (Wubbeler et al., 2017). However, nongovernment welfare organizations (e.g., the church-affiliated Diakonie, the Red Cross, or the Welfare Association) also often initiate and/or finance community-oriented networks. In general, community-oriented networks (Fig. 49.2) aim to improve public awareness and decrease stigmatization.

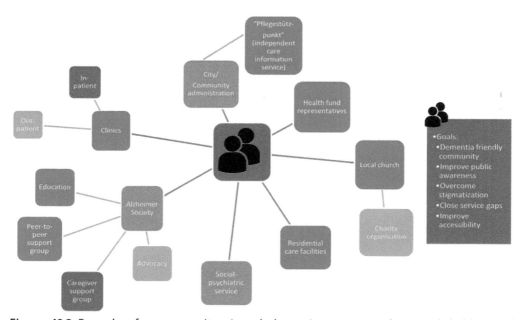

Figure 49.2 Example of a community-oriented dementia care network, its stakeholders and goals. This figure illustrates the exemplary structure of a community-oriented dementia care network. Typical stakeholders (e.g., social services, Alzheimer societies, general practitioners, community administrations) are linked with the founding institution or organization. Community-oriented dementia care networks might aim at addressing stigmatization, becoming a dementia-friendly community, or improving accessibility to support services. *(This figure was developed by F. Laporte Uribe for this book chapter based on a definition of dementia care networks by Wubbeler, M., Thyrian, J. R., Michalowsky, B., Erdmann, P., Hertel, J., Holle, B. et al. (2017). How do people with dementia utilise primary care physicians and specialists within dementia networks? Results of the dementia networks in Germany (DemNet-D) study. Health and Social Care in the Community, 25(1), 285–294. https://doi:10.1111/hsc. 12315.)*

Dementia care networks may also be characterized by their geographical service area (rural vs. urban) or their degree of formalization (formal organization vs. less formal structures and processes). Based on their governance structure, four different dementia care network types have been described: "Stakeholder network" (focused on providing support through information and education by linking regional care providers and actively involving stakeholders); "Organization network" (characterized by a high level of internal formal governance and a defined control center); "Hybrid network" (characterized by their ability to rapidly adapt strategies to changing requirements); and "Mission network" (characterized by user-focused, specific care-related aims that are followed through consistently) (Laporte Uribe, Heinrich et al., 2017; Schäfer-Walkmann, Traub, & Peitz, 2017). However, these definitions are intended to provide the primary characteristics of a network rather than an exclusive description; therefore, they overlap and complement each other.

Dementia care networks are often initiated if a service gap is identified. The network subsequently aims to address this gap, for example, by setting up a home-care support service or a dementia café. Similarly, network goals might include the alignment of existing services, the avoidance of double structures, or the reduction of provider competition. However, the goals of networks are often less specific, such as the development of a dementia-friendly community. Many networks aim to improve the accessibility of services and the visibility or knowledge with regard to existing support structures. To achieve their goals, networks might develop their own public relations strategy, conduct meetings, and provide training for professionals and family caregivers, develop guidelines for residential care facilities, initiate mobile consultant services such as a "dementia bus", or develop case management concepts (Heinrich, Laporte Uribe, Wubbeler, Hoffmann, & Roes, 2016; Landesinitiative Demenz-Service Nordrhein-Westfalen, 2012). In general, the motivation for interested parties to initiate a dementia care network is diverse; therefore, the aims of dementia care networks and their structure are equally diverse.

Research on dementia care networks

Dementia care networks have previously been established in Germany; however, research on this topic remains scarce. Internationally, there are some exceptions with studies on integrated dementia care models in Canada (Lemieux-Charles et al., 2005), the United States (Bass et al., 2014; Kally, Cherry, Howland, & Villarruel, 2014), the United Kingdom (Banerjee et al., 2007), and the Netherlands (Van Mierlo, Meiland, Van Hout, & Droes, 2014). A comprehensive evaluation of case management approaches was recently published in the Cochrane Database (Reilly et al., 2015). In Germany, there is sound scientific evidence from a controlled intervention trial that the provision of dementia-specific medication and the utilization of medical treatments were higher for persons with dementia affiliated with a rural dementia care network compared to individuals who received care as usual (Köhler et al., 2014). In another study, six dementia care networks were evaluated, and it was determined that the following factors contributed to the successful setup and functioning of these networks: regional parameters, the systematic development of case management, and the inclusion of facilitators, e.g., general practitioners and other medical specialists (Reichert, Koehler, Leve, & Zimmer, 2011).

Building on this preliminary work, the DemNet-D study was conducted (Heinrich, Laporte Uribe, Wubbeler et al., 2016; Laporte Uribe, Gräske et al., 2017; Wubbeler et al., 2017). In this multicentered evaluation study, barriers and facilitators of successful dementia care networks in Germany were comprehensively described for the first time. Moreover, the authors gained a deeper understanding of network users and the services that they utilize in this setting. Data were collected from 560 persons with dementia and their caregivers in 13 dementia care networks at baseline and 1-year follow-up. Furthermore, data were collected from interviews with key persons, stakeholders, and network documents. Several results and conclusions are highlighted in the following sections.

First, it was found that people with dementia had a moderate quality of life, with better ratings for those persons who lived together with their caregivers and were less impaired in their daily living abilities (Gräske et al., 2016). No regional differences (urban vs. rural networks) were identified regarding network users' quality of life. At baseline, participants reported low feelings of loneliness and high levels of perceived social inclusion (Wolf-Ostermann et al., 2017). At follow-up, the authors identified sustained levels of quality of life and social participation for network users (Forschungsverbund DemNet-D (DemNet-D Research Group), 2016). Compared to other studies in which people living with dementia received care and support as usual rather than through a dementia care network, these findings provided the first indications that dementia care networks provide adequate support for persons living with dementia and contribute to their social integration and their feelings of being part of a community. However, this research comprised an initial, comprehensive evaluation study rather than a controlled intervention trial. A comparison with other studies must be treated with caution.

Second, analyses of caregiver burden and the positive aspects of care have shown that family caregivers reported a low to moderate burden that was associated with the ability of the person with dementia to perform activities of daily living and symptoms of challenging behaviors (Laporte Uribe, Gräske et al., 2017; Laporte Uribe, Heinrich et al., 2017). Moreover, nonspousal caregivers felt less burdened than spouses or partners. Interestingly, women felt more burdened than men; however, women's caregiving role simultaneously provided them with more room for personal development. At follow-up, caregivers reported a reduced burden for numerous aspects, for example, the burden related to practical care tasks and role conflicts. Women and caregivers who had a better health status at the beginning of the study were more likely to feel a decrease in burden over the course of 1 year than men. It was concluded that, independent of the network governance type, all evaluated networks were similarly well suited to support people with dementia and their families and were able to reduce caregiver burden.

Third, care arrangements differed between men and women, i.e., men were supported by twice as many informal caregivers as women (Laporte Uribe, Wolf-Ostermann, Wübbeler, & Holle, 2017). The care situation also differed between urban and rural networks, with some formal services utilized less often in rural areas. Most people with dementia used a mix of formal (71%) and informal (88%) support, with numbers slightly increasing over the course of 1 year. Home care nursing services and day centers were the most frequently utilized services, whereas low-threshold services were less often used (Fig. 49.3).

Numerous reasons were identified to explain why services that had been used in the past were discontinued by some network users (Table 49.1).

Care arrangements were considered as stable by approximately 88% of caregivers at baseline and 92% at follow-up (Laporte Uribe, Wolf-Ostermann et al., 2017). The authors concluded that dementia care networks contribute to a high degree of perceived stability in the care situation. Furthermore, they suggested that gender differences, the person with dementia and caregiver relationship, and the availability and accessibility of low-threshold services should be considered when developing support structures that are available through dementia care networks. Moreover, it was concluded that indicators of positive caregiving aspects could be used by dementia care networks to advance support structures for informal caregivers.

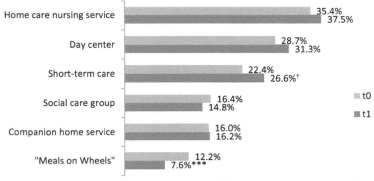

Figure 49.3 DemNet-D study results on the utilization of formal support services at baseline (t_0) and follow-up (t_1). This figure shows results from the DemNet-D study (multicentered, interdisciplinary evaluation study of dementia care networks in Germany). It illustrates the percentage of persons with dementia utilizing a formal support service (i.e., home care nursing service, day center, social care group, companion home service, lunch service) at the time of the assessment, at baseline (t_0) and follow-up (t_1). The information was provided by the same caregivers at both assessments. Differences in utilization between baseline and follow-up are statistically significant for ***$P = .001$, approaching statistical significance for †$P = .052$ (McNemar Test). *(Data for this figure were taken from the final DemNet-D study report (original German version: F. Laporte Uribe (Forschungsverbund DemNet-D (DemNet-D Research Group). (2016). Final report of the DemNet-D study [Multizentrische, interdisziplinäre Evaluationsstudie von Demenznetzwerken in Deutschland (DemNet-D). Sachbericht des Forschungsverbundes]. Retrieved from Greifswald, Witten, Berlin, Bremen, Stuttgart (10.04.2018) https:// demenznetzwerke.de/wp-content/uploads/2018/02/A3_1_1a_2_3_Abschlussbericht_DZNE-Greifswald__ Witten_Berlin_IfaS.pdf), English translation: F. Laporte Uribe) with permission.)*

Table 49.1 DemNet-D study results on reasons for nonutilization of formal support services at baseline (t0).

Home-care nursing service	
The two most frequently cited reasons for discontinuation (more than one answer possible, valid percentage, n = 33)	
Support service was no longer suitable	30.3% (10)
The person with dementia no longer wanted the support	18.2% (6)
Day center	
The three most frequently cited reasons for discontinuation (more than one answer possible, valid percentage, n = 24)	
Support service was no longer suitable	41.6% (10)
The person with dementia no longer wanted the support	29.2% (7)
Difficulties to fit the service in with daily routine	20.8% (5)
Social care group	
The three most frequently cited reasons for discontinuation (more than one answer possible, valid percentage, n = 16)	
Support service was no longer suitable	43.8% (7)
The person with dementia no longer wanted the support	31.3% (5)
Companion home service	
The two most frequently cited reasons for discontinuation (more than one answer possible, valid percentage, n = 11)	
The person with dementia no longer wanted the support	45.5% (5)
Financial reasons	18.2% (2)
"Meals on wheels"	
The two most frequently cited reasons for discontinuation (more than one answer possible, valid percentage, n = 23)	
The person with dementia no longer wanted the support	34.8% (8)
Support service was no longer suitable	26.1% (6)

This table lists results from the DemNet-D study (multicentered, interdisciplinary evaluation study of dementia care networks in Germany) on the two most frequently cited reasons why a formal support service (i.e., home care nursing service, day center, social care group, companion home service, lunch service) that had been utilized in the past was discontinued at the time of the assessment (i.e., study baseline). The information was provided by the caregiver on services that had been utilized in the past by the person with dementia.

Data for this table were taken from the final DemNet-D study report (original German version: F. Laporte Uribe (Forschungsverbund DemNet-D (DemNet-D Research Group). (2016). Final report of the DemNet-D study [Multizentrische, interdisziplinäre Evaluationsstudie von Demenznetzwerken in Deutschland (DemNet-D). Sachbericht des Forschungsverbundes]. Retrieved from Greifswald, Witten, Berlin, Bremen, Stuttgart (10.04.2018). https://demenznetzwerke.de/wp-content/uploads/2018/02/A3_1_1a_2_3_Abschlussbericht_DZNE-Greifswald__Witten_Berlin_IfaS.pdf), English translation: F. Laporte Uribe) with permission.

Health services were consistently more often utilized by persons with dementia supported by one of the 13 dementia care networks in DemNet-D than by the general population (Thyrian et al., 2018). The authors found that persons with dementia were more likely to use a service if they had previously used it at baseline and if they had a higher capacity of daily functioning (more frequent contact with a specialist physician or occupational therapy). Being a woman decreased the likelihood to contact a specialist physician and increased the probability to utilize no service at all. More precisely, users of dementia care networks more regularly received consultations with neurologists or psychiatrists (Wubbeler et al., 2017), were more frequently treated with antidementia drugs (Wubbeler, Wucherer, et al., 2015), and more frequently received nonpharmacological therapies and aids (Wubbeler, Thyrian, et al., 2015). Overall, the authors concluded that people with dementia in this study benefited from the support of a dementia care network. The findings also raised attention to a gender bias in this setting.

Furthermore, financing structures were assessed (Michalowsky et al., 2017). The authors determined that dementia care networks were primarily financed by membership fees, earnings from services provided, public funds, and payments by municipalities or health care providers. Most (8/13) networks considered themselves as being financially sustainable. However, it must be considered that only networks considered eligible to participate by the Federal Ministry of Health based on a previous governmental evaluation were included in the study (Laporte Uribe, Heinrich, et al., 2017). These networks had previously been set up some time ago and were well established in the community. Interestingly, rural networks experienced more difficulties to acquire sustainable funding resources than networks servicing an urban area with a clustering of service providers. Another barrier for network sustainability included services that had been funded through provisional subsidies, decreasing the likelihood of these services to become permanent support structures. Furthermore, a minimum funding of 50,000€ per year for human resources seemed to be a threshold for a sustainable dementia care network.

Finally, knowledge management processes were evaluated in the DemNet-D study (Heinrich, Laporte Uribe, Roes et al., 2016; Heinrich, Laporte Uribe, Wubbeler et al., 2016). The authors identified a multitude of different knowledge evaluation and knowledge management processes and tools with different degrees of formalization. There was a mix of structures with less specific knowledge management tools utilized in the majority of networks. Competition between stakeholders was a difficult barrier to overcome in many dementia care networks. Networks also reported difficulties to reach potential network users. Limited time resources and stigmatization were described as main reasons for persons with dementia and their family caregivers not to access available support services, such as information meetings. However, in dementia care networks with less formalized knowledge management structures, only 5% of caregivers indicated that they had a need for dementia-specific information but did not know where to obtain this information. In networks with highly formalized structures, this percentage was lower (2%). The authors concluded that dementia care networks, particularly networks with highly formalized structures, in this study were very successful in disseminating information on support services to family caregivers.

It is noteworthy that the results of DemNet-D had an impact on German legislation making funding options for dementia care networks part of the new §45c of the Second Act to Strengthening Long Term Care (Pflegestärkungsgesetz II), which entered into force on January 1, 2017 (German Center for Neurodegenerative Diseases, 2016). As a result of this paragraph, nursing health care insurances can financially support dementia care networks with an annual funding of up to 20,000€ per network and district. The funding source is the national health fund. Therefore, more than 8 million euros are available for dementia care networks every year. A financial participation of the regional municipality is not required. Furthermore, the funding is available for all health care networks. The funding will support the establishment and development of dementia care networks and therefore improve treatment and care in dementia diseases throughout Germany.

What makes dementia care networks successful?

Research has shown that numerous factors contribute to a sustainable funding structure of dementia care networks: a minimum funding of 50,000€ per annum for funding human resources that coordinate the dementia care network, financial support from local governments/municipalities, a network of 40—50 stakeholders, and a mix of different funding sources (Michalowsky et al., 2017). As a result of this research, a general financing model for dementia care networks was developed (Fig. 49.4).

Furthermore, research has shown that agency neutral facilitators/network leaders and common network goals clearly defined, for example, in formalized mission statements and position papers, were important to overcome potential stakeholder competition (Heinrich, Laporte Uribe, Wubbeler et al., 2016).

The use of information dissemination strategies that are based on active contacts has been shown to be more successful than passive strategies when reaching and involving potential network users (Heinrich, Laporte Uribe, Roes et al., 2016). Thus, general practitioners or members of medical assessment teams as external gatekeepers may be considered key partners when disseminating information, for example, on services available through a network. Furthermore, more formalized knowledge management processes, such as working groups, quality management systems, IT-system-based-feedback acquisition, or formal evaluation processes, may increase the level of knowledge with regard to dementia-specific support options among family caregivers. Nevertheless, to react flexibly to clients' needs, and given limited time and financial resources, dementia care networks might choose a mix of formal and less formal structures, including knowledge management processes, depending on their goals and network size.

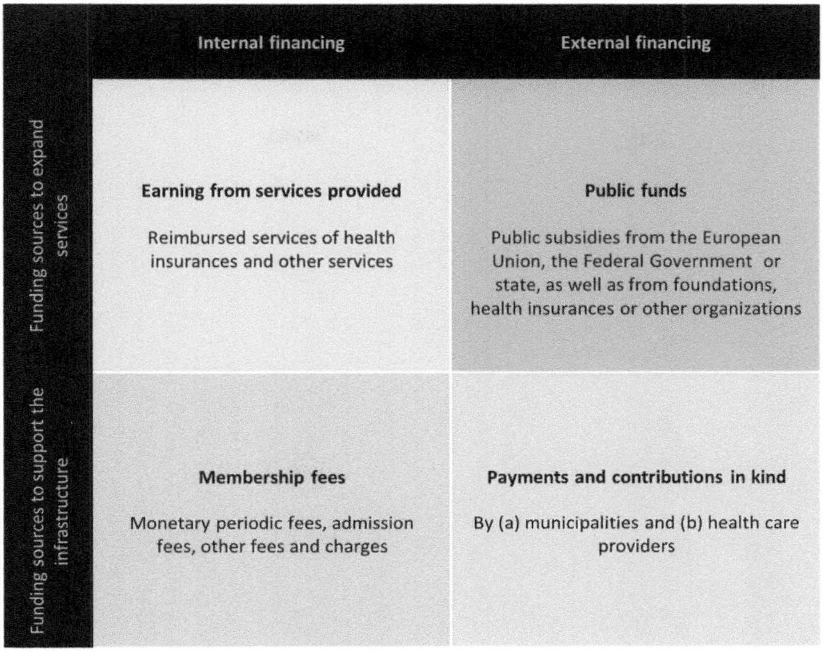

Figure 49.4 Generalizable financing model of dementia care networks. This figure provides a generalizable financing model of dementia care networks. It illustrates the different sources of funding in the inner matrix distinguished by the types of funding (internal, external) and the purpose of funding (to expand services, to support the infrastructure). *(The model was developed based on data from the DemNet-D study (multicentered, interdisciplinary evaluation study of dementia care networks in Germany). This figure was taken from Michalowsky, B., Wübbeler, M., Thyrian, J. R., Holle, B., Gräske, J., Schäfer-Walkmann, S., et al. (2017). Finanzierung regionaler Demenznetzwerke: Determinanten einer nachhaltigen Finanzierung am Beispiel spezialisierter Gesundheitsnetzwerke. Das Gesundheitswesen, 79(12), 1031–1035. https://doi.org/10.1055/s-0042-102344. Final report of the DemNet-D study [Multizentrische, interdisziplinäre Evaluationsstudie von Demenznetzwerken in Deutschland (DemNet-D). Sachbericht des Forschungsverbundes]. Retrieved from Greifswald, Witten, Berlin, Bremen, Stuttgart (10.04.2018). https://demenznetzwerke.de/wp-content/uploads/2018/02/A3_1_1a_2_3_Abschlussbericht_DZNE-Greifswald__Witten_Berlin_IfaS.pdf), English translation: B. Michalowsky) with permission.)*

Although dementia care networks have been created in Germany in recent years, they have not been implemented systematically or nationwide, and their concepts and structures are often not known to other stakeholders within the health care system or other dementia care networks (Heinrich et al., 2017). As a result of the DemNet-D study, an internet portal (www.demenznetzwerke.de) was developed and implemented to transfer research findings into practice (Fig. 49.5). The portal provides information for other interested dementia care networks and health care stakeholders who want to optimize existing networks or build new dementia care networks (Heinrich et al., 2017). In addition to several findings from the study and a link to the study report that is mostly intended for researchers and interested persons from policy and clinical practice, the portal provides practice-oriented explanations and definitions (often as videos), as well

Figure 49.5 Dementia Care Network internet portal. This figure shows a couple of screenshots merged into one picture of the Dementia Care Network internet portal (www.demenznetzwerke.de). This internet portal was developed and implemented as a result of the DemNet-D study (multicentered, interdisciplinary evaluation study of dementia care networks in Germany) to transfer research findings into practice. The portal provides information for interested dementia care networks and health care stakeholders who want to optimize existing networks or build new dementia care networks. *(This figure was developed by S. Heinrich for the final DemNet-D study report (Forschungsverbund DemNet-D (DemNet-D Research Group). (2016). Final report of the DemNet-D study [Multizentrische, interdisziplinäre Evaluationsstudie von Demenznetzwerken in Deutschland (DemNet-D). Sachbericht des Forschungsverbundes]. Retrieved from Greifswald, Witten, Berlin, Bremen, Stuttgart (10.04.2018) https://demenznetzwerke.de/wp-content/uploads/2018/02/A3_1_1a_2_3_Abschlussbericht_DZNE-Greifswald_Witten_Berlin_IfaS.pdf) and used with permission/Copyright 2015 demenznetzwerke.de.)*

as contact details of "best practice partners", i.e., networks and stakeholders with expertise on a specific aspect. Moreover, the portal is intended as a "toolbox" that provides tools based on real-life dementia care network materials, such as organization charts, legal and financing structures, or evaluation forms that website users can collect in a "shopping basket" to develop their individualized dementia care network toolkit. Since the internet portal went online in October 2015, a total of 10,536 internet users visited the portal and the tools and materials provided were downloaded 19,983 times (as of April 18, 2018).

Key facts of family caregivers

- Most people with dementia prefer to live at home and are cared for by family members or friends.
- Family caregivers (those who provide informal unpaid care) are crucial for enabling a person with dementia to stay at home for as long as possible since the likelihood to be admitted to residential care rises sharply for persons with dementia living alone.
- Family caregivers often experience not only physical and psychological adverse events but also negative social and economic consequences.
- Usually it is the wife or a daughter of a person living with dementia who becomes a family caregiver. Similarly, daughters-in-law are more likely to assume the role of caregivers than sons or sons-in-law.
- Female caregivers may experience higher levels of burden but also show higher levels of personal development.
- Caregiving has been the focus of research for the past 40 years; however, worldwide there is still a need to translate research findings more effectively into individually tailored social and economic support structures for families affected by dementia.
- Furthermore, there is a lack of longitudinal studies considering the complex care situation and care process often evolving over the course of several years.

Summary points

- As models of integrated care, dementia care networks in Germany have shown the potential to overcome a fragmented health care system.
- Linking different support services and stakeholders, including general practitioners, clinics, Alzheimer societies, charity organizations, local authorities, and other groups, dementia care networks are often initiated to close service gaps, align existing services, avoid double structures, and reduce provider competition.
- Depending on the type of dementia care network, the network goals may be more health oriented (e.g., early and differential diagnosis) or community focused (e.g., overcoming stigmatization and a dementia-friendly community).
- Research has shown that dementia care networks, independent of their governance type, may be successful in sustaining the quality of life of persons living with

dementia, contributing to their feeling of social participation, reducing caregiver burden, and increasing service utilization. Gender differences in care arrangements and service utilization must be considered in this setting.

- A minimum funding of 50,000€ per year seems to be necessary to achieve sustainability for a dementia care network in Germany. Networks should reflect early into their development process on their aims and move toward formal structures and management processes over time because these networks seem to have better chances of becoming permanent structures.
- The dementia care internet portal www.demenznetzwerke.de provides information and tools for interested networks and stakeholders who want to optimize existing networks or build new dementia care networks.

References

Banerjee, S., Willis, R., Matthews, D., Contell, F., Chan, J., & Murray, J. (2007). Improving the quality of care for mild to moderate dementia: An evaluation of the Croydon memory service model. *International Journal of Geriatric Psychiatry, 22*(8), 782–788. https://doi.org/10.1002/gps.1741.

Bass, D. M., Judge, K. S., Snow, A. L., Wilson, N. L., Morgan, R. O., Maslow, K., et al. (2014). A controlled trial of partners in dementia care: Veteran outcomes after six and twelve months. *Alzheimer's Research & Therapy, 6*(1), 9. https://doi.org/10.1186/alzrt242.

Bickel, H. (2016). *Incidence and prevalence of dementia (key aspects, 1) [die Häufigkeit von Demenzerkrankungen (Das Wichtigste 1)].* Retrieved from Berlin https://www.deutsche-alzheimer.de/fileadmin/alz/pdf/fact-sheets/infoblatt1_haeufigkeit_demenzerkrankungen_dalzg.pdf (27.03.2018).

Brodaty, H., Green, A., & Koschera, A. (2003). Meta-Analysis of psychosocial interventions for caregivers of people with dementia. *Journal of the American Geriatrics Society, 51*(5), 657–664. https://doi.org/10.1034/j.1600-0579.2003.00210.x.

Clyburn, L. D., Stones, M. J., Hadjistavropoulos, T., & Tuokko, H. (2000). Predicting caregiver burden and depression in Alzheimer's disease. *Journals of Gerontology Series B: Psychological Sciences and Social Sciences, 55*(1), S2–S13.

Dill, A. (2006). Case management. In G. L. Albrecht, D. T. Mitchell, & S. L. Snyder (Eds.), *Encyclopedia of disability* (pp. 228–231). Thousand Oaks, CA: Sage Publication.

Federal Ministry for Family Affairs Senior Citizens Women and Youth, & Federal Ministry of Health. (2014). *Alliance for people with dementia. The fields of action.* Retrieved from http://www.allianz-fuer-demenz.de/fileadmin/de.allianz-fuer-demenz/content.de/downloads/BF1501_001_Allianz-Nationale_Demenz-strategie_EN_RZ.pdf (15.04.2018).

Forschungsverbund DemNet-D (DemNet-D Research Group). (2016). *Final report of the DemNet-D study [Multizentrische, interdisziplinäre Evaluationsstudie von Demenznetzwerken in Deutschland (DemNet-D). Sach-bericht des Forschungsverbundes].* Retrieved from Greifswald, Witten, Berlin, Bremen, Stuttgart (10.04.2018) https://demenznetzwerke.de/wp-content/uploads/2018/02/A3_1_1a_2_3_Abschluss-bericht_DZNE-Greifswald__Witten_Berlin_IfaS.pdf.

Gaugler, J. E., Yu, F., Krichbaum, K., & Wyman, J. F. (2009). Predictors of nursing home admission for persons with dementia. *Medical Care, 47*(2), 191–198. https://doi.org/10.1097/MLR.0b013e31818457ce, 00005650-200902000-00009 [pii].

German Center for Neurodegenerative Diseases. (2016). *Results of a study on dementia care networks impacts German legislation [Ergebnisse einer Studie zu Demenznetzwerken fließen in Gesetzesänderung ein]* (Press release). Retrieved from www.dzne.de/aktuelles/presse-und-oeffentlichkeitsarbeit/pressemitteilungen/presse/detail/ergebnisse-einer-studie-zu-demenznetzwerken-fliessen-in-gesetzesaenderung-ein/ (18.04.2018).

Girard, J., & Girard, J. (2015). Defining knowledge management: Toward an applied compendium. *Online Journal of Applied Knowledge Management, 3*(1), 1—20.

Gräske, J., Meyer, S., Schmidt, A., Schmidt, S., Laporte Uribe, F., Thyrian, J. R., et al. (2016). Regionale Demenznetzwerke in Deutschland — Ergebnisse der DemNet-D-Studie zur Lebensqualität der Nutzer/innen. *Pflege — Die wissenschaftliche Zeitschrift für Pflegeberufe, 29*(2), 93—101. https://doi.org/10.1024/1012-5302/a000xxx.

Heinrich, S., Laporte Uribe, F., Roes, M., Hoffmann, W., Thyrian, J. R., Wolf-Ostermann, K., et al. (2016). Knowledge management in dementia care networks: A qualitative analysis of successful information and support strategies for people with dementia living at home and their family caregivers. *Public Health, 131*, 40—48. https://doi.org/10.1016/j.puhe.2015.10.021.

Heinrich, S., Laporte Uribe, F., Wubbeler, M., Hoffmann, W., & Roes, M. (2016). Knowledge evaluation in dementia care networks: A mixed-methods analysis of knowledge evaluation strategies and the success of informing family caregivers about dementia support services. *International Journal of Mental Health Systems, 10*, 69. https://doi.org/10.1186/s13033-016-0100-8.

Heinrich, S., Sommerfeld, U., Michalowsky, B., Hoffmann, W., Thyrian, J. R., Wolf-Ostermann, K., et al. (2017). How to initiate dementia care networks? Processes, barriers, and facilitators during the development process of a practice-oriented website toolkit out of research results. *The International Quarterly of Community Health Education, 37*(3—4), 151—160. https://doi.org/10.1177/0272684X17736245.

Iliffe, S., Wilcock, J., Synek, M., Carboch, R., Hradcova, D., & Holmerova, I. (2017). Case management for people with dementia and its translations: A discussion paper. *Dementia.* https://doi.org/10.1177/1471301217697802, 1471301217697802.

Kally, Z., Cherry, D. L., Howland, S., & Villarruel, M. (2014). Asian Pacific Islander dementia care network: A model of care for underserved communities. *Journal of Gerontological Social Work, 57*(6—7), 710—727. https://doi.org/10.1080/01634372.2013.854856.

Köhler, L., Meinke-Franze, C., Hein, J., Fendrich, K., Heymann, R., Thyrian, J. R., et al. (2014). Does an interdisciplinary network improve dementia care? Results from the IDemUck-study. *Current Alzheimer Research, 11*, 538—548.

von Kutzleben, M., Reuther, S., Dortmann, O., & Holle, B. (2015). Care arrangements for community-dwelling people with dementia in Germany as perceived by informal carers — a cross-sectional pilot survey in a provincial—rural setting. *Health and Social Care in the Community, 24*(3), 283—296. https://doi.org/10.1111/hsc.12202.

Landesinitiative Demenz-Service Nordrhein-Westfalen. (2012). *Leitfaden für den Aufbau und die Umsetzung von regionalen Demenznetzwerken.* Retrieved from https://www.demenz-service-nrw.de/tl_files/Landesinitiative/Die%20Landesinitiative/Arbeitsgruppen/Netzwerke-Leitfaden_PDF.pdf (19.04.2018).

Laporte Uribe, F., Gräske, J., Grill, S., Heinrich, S., Schäfer-Walkmann, S., Thyrian, J. R., et al. (2017). Regional dementia care networks in Germany: Changes in caregiver burden at one-year follow-up and associated factors. *International Psychogeriatrics, 29*(6), 991—1004. https://doi.org/10.1017/S1041610217000126.

Laporte Uribe, F., Heinrich, S., Wolf-Ostermann, K., Schmidt, S., Thyrian, J. R., Schafer-Walkmann, S., et al. (2017). Caregiver burden assessed in dementia care networks in Germany: Findings from the DemNet-D study baseline. *Aging and Mental Health, 21*(9), 926—937. https://doi.org/10.1080/13607863.2016.1181713.

Laporte Uribe, F., Wolf-Ostermann, K., Wübbeler, M., & Holle, B. (2017). Care arrangements in dementia care networks: Findings from the DemNet-D study baseline and 1-year follow-up. *Journal of Aging and Health.* https://doi.org/10.1177/0898264317696778, 0898264317696778.

Lemieux-Charles, L., Chambers, L. W., Cockerill, R., Jaglal, S., Brazil, K., Cohen, C., et al. (2005). Evaluating the effectiveness of community-based dementia care networks: The dementia care networks' study. *The Gerontologist, 45*(4), 456—464. https://doi.org/10.1093/geront/45.4.456.

Michalowsky, B., Wübbeler, M., Thyrian, J. R., Holle, B., Gräske, J., Schäfer-Walkmann, S., et al. (2017). Finanzierung regionaler Demenznetzwerke: Determinanten einer nachhaltigen Finanzierung am Beispiel spezialisierter Gesundheitsnetzwerke. *Das Gesundheitswesen, 79*(12), 1031—1035. https://doi.org/10.1055/s-0042-102344.

Reichert, M., Koehler, K., Leve, V., & Zimmer, B. (2011). Evaluating service networks for people with dementia and their informal caregivers: The "EVIDENT"-project. *The Gerontologist, 51*(Suppl. 1_2), 356. https://doi.org/10.1093/geront/gns068.

Reilly, S., Miranda-Castillo, C., Malouf, R., Hoe, J., Toot, S., Challis, D., et al. (2015). Case management approaches to home support for people with dementia. *Cochrane Database of Systematic Reviews, 1,* CD008345. https://doi.org/10.1002/14651858.CD008345.pub2.

Schäfer-Walkmann, S., Traub, F., & Peitz, A. (2017). The art of governance of dementia care networks in Germany — results from the DemNet-D study [Die hohe Kunst der Steuerung von Demenznetzwerken in Deutschland — Ergebnisse der DemNet-D-Studie]. In S. Schäfer-Walkmann, & F. Traub (Eds.), *Evolution durch Vernetzung: Beiträge zur interdisziplinären Versorgungsforschung. Edition Centaurus: Perspektiven Sozialer Arbeit in Theorie und Praxis* (pp. 47—58). https://doi.org/10.1007/978-3-658-14809-6.

Schulz, R., & Sherwood, P. R. (2008). Physical and mental health effects of family caregiving. *American Journal of Nursing, 108*(9 Suppl. 1), 23—27. https://doi.org/10.1097/01.NAJ.0000336406.45248.4c. quiz 27.

Statistisches Bundesamt (Destatis). (2018). *Pflegestatistik 2017 — Pflege im Rahmen der Pflegeversicherung. Deutschlandergebnisse* Retrieved from Wiesbaden. https://www.destatis.de/DE/Publikationen/Thematisch/Gesundheit/Pflege/PflegeDeutschlandergebnisse5224001139004.pdf?__blob=publicationFile (14.02.2020)

Tarricone, R., & Tsouros, A. D. (2008). *Home care in Europe — the solid facts. WHO Regional Office for Europe.* Kopenhagen. Retrieved from http://www.euro.who.int/__data/assets/pdf_file/0005/96467/E91884.pdf?ua=1 (20.04.2018).

Thyrian, J. R., Michalowsky, B., Hertel, J., Wübbeler, M., Gräske, J., Holle, B., et al. (2018). How does utilization of health care services change in people with dementia served by dementia care networks? Results of the longitudinal, observational DemNet-D-study. *J Alzheimers Dis., 66*(4), 1609—1617.

Thyrian, J. R., Wübbeler, M., & Hoffmann, W. (2013). Interventions into the care system for dementia. *Geriatric Mental Health Care, 1*(3), 67—71. https://doi.org/10.1016/j.gmhc.2013.04.001.

Van Mierlo, L. D., Meiland, F. J., Van Hout, H. P., & Droes, R. M. (2014). Towards personalized integrated dementia care: A qualitative study into the implementation of different models of case management. *BMC Geriatrics, 14,* 84. https://doi.org/10.1186/1471-2318-14-84.

Wolf-Ostermann, K., Meyer, S., Schmidt, A., Schritz, A., Holle, B., Wübbeler, M., et al. (2017). [Users of regional dementia care networks in Germany: First results of the evaluation study DemNet-D]Nutzer und Nutzerinnen regionaler Demenznetzwerke in Deutschland: Erste Ergebnisse der Evaluationsstudie DemNet-D. *Zeitschrift für Gerontologie und Geriatrie, 50*(1), 21—27. https://doi.org/10.1007/s00391-015-1006-9.

World Health Organization. (2008). *Integrated health services - what and why?* Geneva. Retrieved from http://www.who.int/healthsystems/service_delivery_techbrief1.pdf (20.04.2018).

World Health Organization, & Alzheimer's Disease International. (2012). *Dementia: A public health priority.* Geneva. Retrieved from http://apps.who.int/iris/bitstream/10665/75263/1/9789241564458_eng.pdf (20.04.2018).

Wubbeler, M., Thyrian, J. R., Michalowsky, B., Erdmann, P., Hertel, J., Holle, B., et al. (2017). How do people with dementia utilise primary care physicians and specialists within dementia networks? Results of the dementia networks in Germany (DemNet-D) study. *Health and Social Care in the Community, 25*(1), 285—294. https://doi.org/10.1111/hsc.12315.

Wubbeler, M., Thyrian, J. R., Michalowsky, B., Hertel, J., Laporte Uribe, F., Wolf-Ostermann, K., et al. (2015). Nonpharmacological therapies and provision of aids in outpatient dementia networks in Germany: Utilization rates and associated factors. *Journal of Multidisciplinary Healthcare, 8,* 229—236. https://doi.org/10.2147/jmdh.s80560.

Wubbeler, M., Wucherer, D., Hertel, J., Michalowsky, B., Heinrich, S., Meyer, S., et al. (2015). Antidementia drug treatment in dementia networks in Germany: Use rates and factors associated with treatment use. *BMC Health Services Research, 15,* 205. https://doi.org/10.1186/s12913-015-0855-7.

CHAPTER 50

Cognitive behavioral therapy use in Alzheimer's disease: mitigating risks associated with the olfactory brain

Larry D. Reid, Dylan Z. Taylor, Alicia A. Walf
Department of Cognitive Science, Rensselaer Polytechnic Institute, Troy, NY, United States

List of abbreviations

Ab Amyloid beta
AD Alzheimer' disease
CBT Cognitive behavioral therapy
IIS Innate immune system
LOAD Late onset Alzheimer' disease
OTC Over-the-counter

Mini-dictionary of terms

Anosmia complete loss of ability to detect odors.
Brain nasal cavity interface the complex interplay of bones and tissues in between the brain and the nasal cavity.
Hyposmia reduced ability to detect odors.
Innate immune system (nasal cavity) nonspecific subsection of the immune system that is largely responsible for cleaning the contents of breathed air.
Mucociliary clearance pathogen self-clearing mechanism involving an interplay between mucus and cilia.
Olfactory mucosa upper region of the nasal cavity composed of epithelium lamina propria, connective tissue, and olfactory neurons.
Respiratory mucosa (of the nasal cavity) lower region of the nasal cavity consisting of epithelium, lamina propria, and connective tissue, but lacking olfactory neurons.
Salient risk factors (for Alzheimer's disease) The more prominent risk factors for Alzheimer's disease consist of cognitive decline, poor diet, and lack of exercise(sedentary life style), inadequate sleep, high levels of stress, frequent exposure to poor air quality, and lost olfactory perception.

Introduction

Late-onset Alzheimer's disease (LOAD), the common form of Alzheimer' disease (AD), is a slowly developing, insidious disease usually taking a decade or more from its earliest manifestations to the full development of the disease, i.e., dementia and death (Alzheimer's Association, 2018). LOAD's beginning symptoms (mild loss of autobiographical and procedural memory) occur during or just before the 6th decade of life. Autopsies, and now

Diagnosis and Management in Dementia
ISBN 978-0-12-815854-8, https://doi.org/10.1016/B978-0-12-815854-8.00050-1

brain scans, confirm what is manifest in daily behavior—initially there is little brain damage, but LOAD's progression eventually culminates in massive losses of brain tissue.

Less than half of individuals reaching the middle of the 8th decade of life develop LOAD (Alzheimer's Association, 2018). Many living to their 9th decade sustain adequate cognition and this happens even among those with a known genetic risk of developing LOAD (Tindale, Leach, Spinelli, & Brooks-Wilson, 2017). Unfortunately, that favorable outcome is matched by those who eventually suffer extensive brain damage. There is no conceivable way of restoring the lost brain tissue of advanced AD; and even if a way would be implemented, there would be a considerable loss of mind.

The mild onset of LOAD and its slow development opens the possibility that an intervention in the early stages of the disease might postpone or even halt the progression of LOAD. However, that possibility by itself is without much substance unless efficacious interventions are specified.

Research has focused on why an individual's well-functioning brain in the 6th decade of life gradually deteriorates, throughout a decade or more, eventually to dementia. Research has led to the modern theory of AD (Selkoe & Hardy, 2016). The theory posits that amyloid beta (Ab), a polypeptide cleaved from a larger precursor, is usually produced nearly continually by neurons (Saido & Iwata, 2006) and might accumulate in the brain's interstitial fluid. If Ab is not nearly continuously removed from interstitial fluid, one molecule tends to clump with others to form an amyloid plaque. Amyloid plaques are setting conditions leading to neuronal death. And, focal neuronal death can steadily progress from a place of neuronal death and step by step march throughout the brain (Franks, Chuah, King, & Vickers, 2015) (the mechanism for this might be an excessive immune process leading to inflammation).

Although Ab is a well-known molecular target of AD, we and others suggest that instead of focusing on the molecular level to focus at a different level of analysis, that is, on behavior (Fig. 50.1). This is especially true given what we know now about the propensity of the brain to not only underlie behaviors but also be shaped by them. Importantly, given the nature of AD as an incurable, progressive disease with some genetic risks, it is of benefit to individuals to know that there can be cognitive behavioral therapy (CBT) approaches that they can take themselves to minimally improve brain (and body) health and potentially prevent AD. Thus, we propose the use of broader cognitive behavioral therapies to improve brain health and prevent AD.

Our review (Reid, Avens, & Walf, 2017) concluded that prevention was possible if adequate interventions were applied to what epidemiological research had identified as salient risk factors associated with the development of AD. Also, we noted that the current technology of CBT, if applied, would likely reduce problematic features of the identified risks. We also noted that behavioral interventions are more apt to succeed if programmed early in the development of AD.

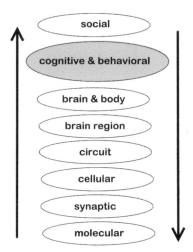

Figure 50.1 Levels of analysis to consider in the neurobiology of dementia and Alzheimer's disease. These levels have bidirectional effects (e.g., brain and body can affect behavior; behavior can affect brain and body function). This chapter primarily focuses on the cognitive and behavioral levels and interactions at organ systems such as brain and olfactory systems.

Behaviorally focused interventions are likely to be beneficial because they engage the ordinary physiology of the brain. Addressing multiple risk factors can, for example, improve the fluid flow throughout the brain and, therefore, prevent an accumulation of amyloid plaques and other metabolic debris. In our review of CBT for the treatment of AD, we (Reid et al., 2017) provided considerably more detail about how CBT might be a way (and perhaps, the only way) of preventing advanced AD. The focus of our group in this regard has expanded to seven general categories of risks that can be mitigated by cognitive behavioral approaches (see Fig. 50.2): olfaction, sleep, diet, physical exercise, cognitive exercise, air quality, and stress. Here, we consider prevention from a somewhat different perspective.

The nasal cavity—brain interface is the periphery of the olfactory brain

Loss of olfactory perception, i.e., hyposmia and anosmia, is an early warning sign of impending cognitive decline and, potentially, a prodromal sign of LOAD (Devanand et al., 2015). Further, there is evidence that LOAD may begin in the periphery of the olfactory brain (Daulatzai, 2015; Ethell, 2014; Franks et al., 2015). Seemingly, these conclusions would immediately focus considerable attention on the nasal cavity—brain interface. Despite the possible importance of the nasal cavity—brain interface to the development of LOAD, those studying risk factors and the development of LOAD have largely ignored risks associated with olfactory loss.

Figure 50.2 Seven deadly risks of Alzheimer' disease. Olfaction is depicted in the middle of this figure to show the focus of this chapter and relate how olfaction may interact with these other risks. Another consideration is that these varied risks are sensitive to cognitive-behavioral approaches.

Considerable attention has been paid to the risks of insufficient physical exercise and unhealthy diets. Consequently, programs have been developed to encourage retirees to develop habits of regular physical exercise and healthy eating and drinking to postpone, or even halt, LOAD's development. Recently, there has been considerable emphasis on poor sleeping habits as being a significant risk (Nedergaard, 2013). Each and all of these are salient to general health, as well as being related to adequate fluid flow throughout the brain. Adequate fluid flow throughout the brain is the mechanism that usually clears excess proteins, including Ab, from the interstitial fluid of the brain, hence preventing LOAD's development.

Perhaps, the salient, well-studied risks are a focus of attention because they are amenable to being corrected by CBT, whereas, until recently there was no effective treatment for loss of olfactory perception not due to obvious blockage of the nasal passages. Fortunately, however, a behavioral treatment for hyposmia and anosmia has been developed that does improve olfactory perception in many, but not all, patients (Sorokowska, Drechsler, Karwowski, & Hummel, 2017).

Unfortunately, behavioral treatment for olfactory loss due to lingering effects of infections is only effective with some individuals (about 60%). All of this leads to these questions: What causes loss of olfactory abilities and can the circumstances for lost olfaction be prevented, hence reducing the prevalence of AD? Toward getting answers to those questions, we first review the anatomy and physiology of the nasal cavity. That review then provides information about how the cavity's innate immune system (IIS) protects the brain from harm. When the system is compromised, there is a marked increase in the risk of serious diseases of the lungs and brain. Much of our review comes from information provided by Beule (2010) and Hariri and Cohen (2016).

Some anatomy and physiology of the nasal cavity

Daily about 12,000 L of air flow through the openings of the nose and mouth and into and out of the lungs, ultimately providing the vital functions of delivering oxygen and exhaling carbon dioxide. Unfortunately, among the vapors we take in, there can be a wide array of potentially dangerous contents, including toxic chemicals (e.g., volatile solvents, various kinds of smoke) and organic matter (e.g., pollens, viruses, bacteria, and fungi), none of which should reach the interiors of the lungs or the brain. Fortunately, day in and day out, the IIS, embodied in the tissues of the respiratory airway, manage these threats without discomforting us as we breathe. Unfortunately, sometimes the system fails, and when it does the consequences can lead to fatal diseases including chronic obstructive pulmonary disease and, we posit here, begins processes that lead to LOAD. To explicate how disease of the sinonasal cavity can be a causal event in LOAD's development, we review the anatomy and physiology of the cavity and relate some interesting new findings in so doing.

The configuration of the bones of the nasal cavity (Fig. 50.3) distributes incoming air throughout the cavity so that large proportions (actually nearly all) of the air come into contact with the tissue, i.e., the mucosa, lining the bones of the skull (Keck & Lindemann, 2010). However, the narrow dome of the cavity, the area sensing odorants, ordinarily receives only 10%–15% of inhaling air, thereby limiting the area's exposure to potential contaminants. Sniffing brings more air to the apex of the cavity, facilitating olfactory perception. The configuration of the sinonasal cavity is such that air eventually reaching the delicate tissues of the lungs' numerous alveoli is at 100% humidity and warmed to body temperature.

A thin layer of tissue lines the bones of the sinonasal cavity as well as other tissues of the airway. The cavity is lined by two similar tissues: the olfactory mucosa and the respiratory mucosa. Their names are derived from the layer of mucus that coats the tissues. The mucosa has two distinct layers, the epithelial layer and the lamina propria, separated by a basal membrane. Facing the airway is a layer of mucus that floats on a layer of serous fluid.

The olfactory mucosa is the tissue lining the superior part of the nasal cavity. Its surface is only a few square centimeters. The olfactory mucosa is the place for sensation due to odorants—the initial activity eventually yielding olfactory perception. The olfactory mucosa contains the bipolar neurons whose dendritic processes have receptors for odorants.

Extending through and on the mucus layer of the olfactory mucosa are the olfactory neurons' dendrites (cilia) ready to be stimulated by odorants in the air. The fine, unmyelinated axons of mature olfactory neurons come together in the lamina propria to form bundles (olfactory fila) and as such ascend through the foramina of the cribriform plate where they synapse with neurons of the olfactory bulb. The supporting cells surrounding

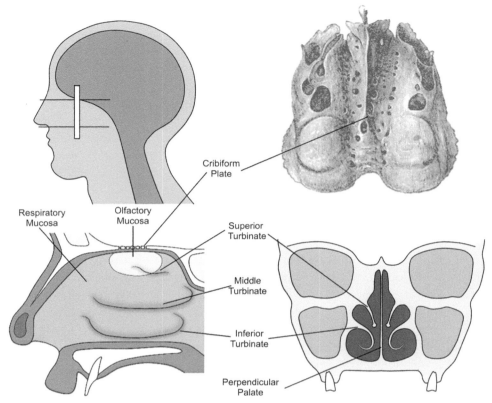

Figure 50.3 The bones of the nasal cavity. The sagittal section is of the bones lateral to the midline. Notice the three bones that protrude into the cavity. The coronal section shows that the nasal cavity is separated into its halves by a bone running down the middle of the cavity. The top of the apex is the cribriform plate. The view of the cribriform is looking at the floor of skull. The narrow long grooves is where the olfactory bulbs would lie. The picture of the brain side of the cribriform plate is by Vandyke, C. H. (1858). Ethmoid bone from above. *(Figure by Caroline R. Mann.)*

the neurons have numerous, short, densely packed villi whose function is to ensure that the upper portion of the nasal cavity is sufficiently moisturized (Fig. 50.4).

Branches of the maxillary artery supply the densely vascularized lamina propria. The artery's branches eventually become arterioles and venules (capillary beds) whose contents drain into the facial vein. Within the lamina propria, there are lymph vessels, macrophages, and mast cells. The lymph vessels, as they do elsewhere, collect and dispose of metabolic debris and any contamination that might breach the protection provided by the olfactory mucosa's epithelial layer.

The bipolar neurons of the olfactory nerve have a limited lifetime (approximately 90 days) and they are replaced by new ones. The olfactory mucosa has basal cells that can develop into bipolar neurons. Attendant to the turnover of neurons in the olfactory

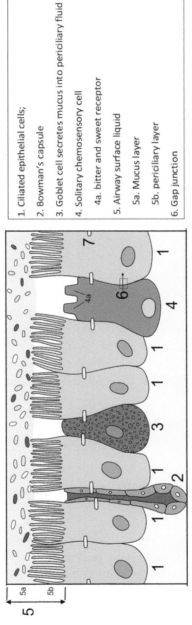

1. Ciliated epithelial cells;
2. Bowman's capsule
3. Goblet cell secretes mucus into periciliary fluid
4. Solitary chemosensory cell
4a. bitter and sweet receptor
5. Airway surface liquid
   5a. Mucus layer
   5b. periciliary layer
6. Gap junction

Figure 50.4 A depiction of a small segment of the olfactory mucosa and olfactory bulb. *(Figure by Caroline R. Mann.)*

mucosa, there is replacement of neurons in the olfactory bulb. Neuroblasts generated in the subventricular zone along the walls of the lateral ventricles migrate to the olfactory bulb to eventually function as interneurons in the bulb (Liu & Guthrie, 2011). This unique and complicated process (for which germane knowledge is limited) is probably dependent, to a considerable degree, on the usual activities of a healthy olfactory mucosa and probably would be disturbed by insults to the mucosa. Along the same lines, there are neuronal inputs to the bulb by neurons from other regions of the brain. For example, cholinergic axons innervate all layers of the bulb, thereby being involved with the processing of incoming signals from the olfactory nerve (Liu & Guthrie, 2011). Many variables must be critical to the functional integration of this complicated turnover of neurons necessary for olfactory perception. Therefore, caution should be used when prescribing drugs that might disturb the neurogenesis necessary to replace dying neurons of the olfactory brain's periphery.

In combination with the physical attributes of the boney structure of the nasal cavity, the respiratory mucosa (the lining most of nasal cavity and the frontal sinuses) protects against a variety of deleterious constituents that might be, and usually are, in the incoming air—a remarkable feat for 120 cm^2 of tissue. This feat is accomplished by the IIS of the respiratory mucosa.

Unlike the olfactory mucosa, the respiratory mucosa has no neurons populating the olfactory nerve and the epithelial cells facing the airway have much longer cilia. Hariri and Cohen (2016) pointed out that the tissues of the sinonasal cavity have resident, nonpathogenic bacteria. Consequently, the IIS of the cavity must be sensitive to harmful exposure by microbes and other contaminants. Hariri and Cohen (2016) described the recent developments in the understanding of how the IIS usually responds to increases in microbes (say a large influx of rhinoviruses) and other contaminants in the air.

Mechanism of mucociliary clearance

The basic action preventing microbes and dirt from doing harm to the sinuses, lungs, and brain is called mucociliary clearance. The combined actions of epithelial, goblet, and glandular cells (Bowman's capsule) of the respiratory mucosa produce two layers of watery substances that coat the entire surface of the tissues of the sinonasal cavity (Fig. 50.5). The layer surrounding the cilia of the epithelial cells is called, appropriately, the periciliary fluid layer. This layer is serous fluid and is produced by the epithelial cells themselves and glandular cells of the lamina propria. This layer, by its fluidity, allows for the movement of the cilia, a critical event in mucociliary clearance. Floating on top of the periciliary fluid is a layer of mucus.

Mucus is a viscous substance of considerable water in which resides mucin proteins produced by goblet cells and to some extent by glandular cells whose bodies are in the lamina propria. Mucus contains various minerals, antiseptic enzymes (such as lysozymes),

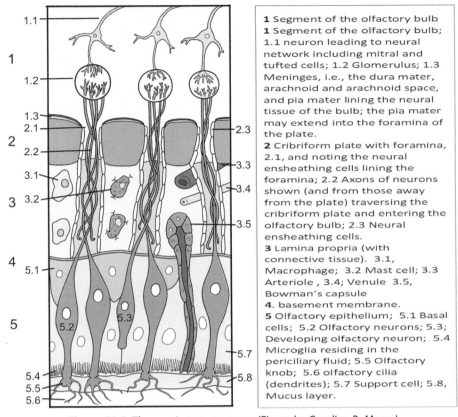

1 Segment of the olfactory bulb
1 Segment of the olfactory bulb; 1.1 neuron leading to neural network including mitral and tufted cells; 1.2 Glomerulus; 1.3 Meninges, i.e., the dura mater, arachnoid and arachnoid space, and pia mater lining the neural tissue of the bulb; the pia mater may extend into the foramina of the plate.
2 Cribriform plate with foramina, 2.1, and noting the neural ensheathing cells lining the foramina; 2.2 Axons of neurons shown (and from those away from the plate) traversing the cribriform plate and entering the olfactory bulb; 2.3 Neural ensheathing cells.
3 Lamina propria (with connective tissue). 3.1, Macrophage; 3.2 Mast cell; 3.3 Arteriole , 3.4; Venule 3.5, Bowman's capsule
4. basement membrane.
5 Olfactory epithelium; 5.1 Basal cells; 5.2 Olfactory neurons; 5.3; Developing olfactory neuron; 5.4 Microglia residing in the periciliary fluid; 5.5 Olfactory knob; 5.6 olfactory cilia (dendrites); 5.7 Support cell; 5.8, Mucus layer.

Figure 50.5 The respiratory mucosa. *(Figure by Caroline R. Mann.)*

immunoglobulins, and glycoproteins (e.g., lactoferrin, mucin). The contents of mucus function to bind to both inorganic and organic contents in air as a first step toward keeping contamination from reaching the sinuses, lungs, and brain.

The beating of the epithelial cells' cilia moves the mucus layer at a velocity of 2—25 mm/min (mucociliary clearance). Faster beating is a response generated by increased contaminants in the airway. Captured microbes and particulate matter are moved along with mucus out of the airways by sneezing, coughing, spitting, and runny noses. However, the usual exit of mucus and its captured contaminants is to move to the throat, be swallowed, and dumped into the acid bath of the stomach, an adaptive thing to do with pathogens. In response to pathogens, the epithelial cells can, by way of a series of biochemical steps, induce the release of nitric oxide. Increments of nitric oxide increase ciliary beating and can directly kill bacteria when diffused into the surface liquids of the respiratory mucosa (Hariri & Cohen, 2016).

Notably, from the perspective of this chapter, the lamina propria has mast cells that release histamine, which, in turn, produces a dilation of blood vessels in the lamina propria, which has the effect of providing more fluids to the fluid layer of the outer surface of the mucosa, hence replacing that which is moved away from the nasal cavity by more rapid mucociliary clearance. The dilation also allows for the release of proteins and immune cells useful in containing infections. The trigger for the release of histamine is an increment in deleterious contents in the air.

Recently, Hariri and Cohen (2016) have described several actions, in addition to the release of histamine, necessary for an effective response to an increase in contaminants in air. Basically, mucociliary clearance cleanses the air swirling in the cavity, hence protecting the tissue itself as well as protecting the brain and lungs from becoming diseased. The effectiveness of the IIS is dependent upon sufficient fluid, production of mucus, and increased mucociliary clearance. Recent findings detail how the dilation of blood vessels, which sustain the necessary moisture of the surface area allowing for faster beating of cilia, and the release of chemicals that literally disable and kill pathogens is coordinated. Hariri and Cohen (2016) reported that the tissue of the sinonasal cavity has sensors for increased contaminates in the cavity, which, in turn, produce responses to deal with threats.

One means of sensing a threat is by way of receptors associated with epithelial cells, i.e., receptors similar to the bitter taste receptor T2R38. These receptors are sensitive to a secretion of gram-negative bacteria, i.e., acyl-homo-serine lactone quorum sensing molecules. This secretion of bacteria is an adaptive response of bacteria regulating their timing of reproduction for optimal survival. A response to secretions of the bacteria is, by way of a series of chemical reactions, the release of nitric oxide into the outer fluid layers of the tissue. Nitric oxide stimulates faster ciliary beating and notably kills bacteria (Hariri & Cohen, 2016).

Hariri and Cohen (2016) also describe another regulatory process that is responsive to increments in contamination of the airways. This involves the discovery of a cell type populating the epithelial layer of the respiratory tissue, i.e., the solitary chemosensory cell. This cell has receptors activated by an array of "bitter" substances and another class of receptors activated by glucose. The net effect of the actions of these two kinds of receptors is to regulate the production of calcium ion, which, in turn, regulates the secretion of antimicrobial peptides. The glucose receptor activity tends to suppress the release of calcium ions, however, when there are multiple microbes using glucose there is less stimulation of the glucose receptors, which, in turn, produces less inhibition of calcium. Consequently, there is sophisticated regulation of the many responses in the service of killing microbes when they are abundant and conserving resources when there are only the resident nonpathological microbes. That is, the tissues of the airway present a dynamic system with a functionality that is regulated by a complex of activities with multiple dependencies. It seems prudent not to disturb this complexity.

The periphery of the olfactory brain is vulnerable

The brain is at risk because some of the neural tissue of olfaction resides outside of the skull in a thin layer of tissue (i.e., the olfactory mucosa), the exterior of which is exposed to breathed air. Given that disease of the olfactory mucosa and olfactory bulb may be the beginnings of LOAD, we asked: What steps could be taken to prevent disease of the periphery of the olfactory brain?

A common cause of olfactory problems is lingering effects of infections (Doty, Philip, Reddy, & Kerr, 2003; Hummel et al., 2009). The most common kind of infection of the nasal cavity is the common cold, i.e., the infections of the 150 or so viruses with similar effects. The common cold is indeed common. In children, there are usually 4 to 6 incidences a year. Adults usually have a couple of colds a year (De Sutter et al., 2012, 2015). Year after year, the IIS of the sinonasal cavity cures a common cold, albeit slowly, taking about 8 days (2 days of mild prodromal signs; 3 days of intense symptoms; followed by days of decreasing symptoms) (Dasaraju & Liu, 1996). There are no vaccines or medicines to prevent or cure a common cold, only some drugs to relieve symptoms.

The periphery of the olfactory brain is embedded in the mucosal tissue lining the bones of the nasal cavity. The IIS is inherent to the tissue of the airways. The IIS protects against disease of the airway by constantly adapting to increases in contaminants in the air and neutralizing their ability to induce disease while at the same time arranging for the moment-to-moment necessity to get air in and out of the lungs. IIS usually prevents an accumulation of infectious material and other contaminants in the air from reaching the depths of the lungs. That same IIS protects the olfactory bulbs from deleterious contents (particularly viruses) in the air. That same IIS also prevents allergens from accumulating sufficiently so that they evoke the sensitization leading to an adaptive immune process (e.g., allergies, chronic rhinitis). In brief, the IIS is a dynamic set of processes whose net effect usually protects us from a variety of diseases in the service of vital functions.

For the IIS to function well, the periciliary fluids of the cilia must be plentiful and the entire mucosa needs to be surrounded by a warm, moist environment. Basically, therefore, fluids that sustain this liquid-essential environment must be nearly continuously supplied via blood vessels and most likely by some drainage of cerebral spinal fluid (which might also serve to carry metabolic debris from the interstitial fluids of the brain) (Nagra, Koh, Zakharov, Armstrong, & Johnston, 2006).

Because the IIS ordinarily protects against chronic diseases of the lungs and brain and deters allergenic diseases, it follows that any event that might compromise the system's functionality is problematic.

The effects of over-the-counter drugs

Here we focus on the widespread use of over-the-counter (OTC) drugs sold to treat the symptoms of the common cold. The active ingredients of many of these drugs are also used to reduce insomnia. The most common active ingredients of the most common medicines used to treat the symptoms of the common cold and insomnia are diphenhydramine, phenylephrine, and acetaminophen. Diphenhydramine and phenylephrine, albeit via different means, are drugs that hinder vasodilation and hence have the supposed therapeutic effect of relieving some symptoms of the common cold and allergies. When in combination, as in some OTC drugs, they are doubly problematic. Acetaminophen is an analgesic given to lessen any pain with a cold. Acetaminophen also reduces fever. However, generally, colds do not produce a fever (Dasaraju & Liu, 1996). When fever occurs, it usually indicates a bacterial infection of the bronchi and lungs. So, acetaminophen might mute the warning that a cold has led to a bacterial infection.

These OTC drugs only temporarily reduce some symptoms of colds. Further, and critically, the ordinary activities of the IIS of the respiratory mucosa usually cure a common cold within days of an infection. Despite the fact that OTC drugs, marketed to help manage symptoms of a cold, have limited utility, individuals often buy and take OTC drugs to control some of the symptoms of a common cold (Johnsen, 2016).

The Cochrane Database of Systematic Reviews (De Sutter, Saraswat, & van Driel, 2015) reviewed 17 publications, involving 4130 adults, assessing the effectiveness of antihistamines (as monotherapies) as medicines for the common cold. The review concluded that antihistamines have a limited effect on obvious symptoms during the first 1 or 2 days of administration, but no significant effect 3—10 days post administration. An earlier review (De Sutter et al., 2012) assessed the effectiveness of antihistamines-decongestant-analgesic combinations for colds across 27 trials and indicated more positive results, but with more severe side effects. There are clearly studies with OTC medicines showing no significant effects with respect to severity of colds and some showing modest beneficial effects at recommended doses and for short periods of use. In other words, the best that can be expected from taking an OTC medicine for a common cold is some temporary relief from some unpleasant symptoms (e.g., reduction in subjective measures of congestion and dry mouth).

Knowledge of the physiology of the IIS leads to the conclusion that for the system to function optimally the tissues need to be continuously moist and warm. It is critical that there is an adequate supply of fluid surrounding the cilia thereby allowing for mucociliary actions to clear the tissues of the sinonasal cavity from contaminants in breathed air. In response to a surge in contaminants, there is a dilation of blood vessels in the lamina propria, which allows more fluid and contents of blood into the mucosa. The dilation of blood vessels and the consequent increases in the size of the tissue covering the nasal bones often blocks easy breathing through the nose and produces feelings of congestion.

These outward signs are manifestations of the IIS's response to a surge in microbes or particulate matter being lodged in the mucosa.

People, however, may not be aware of the possibility that many of these OTC cough-cold and allergy drugs might hinder the IIS's ability to cure the cold. And more seriously, these drugs might be a setting condition for bacteria to reach the lungs (George et al., 2014) and for contaminants in the air (e.g., viruses) to breach the olfactory mucosa (Dasaraju & Liu, 1996) and possibly invade the olfactory bulb. The potential for serious risk is greatly enhanced when these drugs are taken in larger than recommended doses and taken for many days in a row. A common ingredient of these products is one of several antihistamines. These drugs are risky because, by way of preventing vasodilation, they dry the mucosa, which leads to poor mucociliary clearance of contaminants in the airway. A dry mucosal layer will not capture contaminants (e.g., bacteria) and hence they can descend into the depths of the lungs where they multiply. It is not rare that a cold is followed by a bacterial infection of the bronchi and lungs (George et al., 2014). With a dried olfactory mucosa, there is a greater chance of a viral infection merely because the contamination is not moved swiftly away from the cells of the mucosa. Viruses might invade an epithelial cell where they multiply sufficiently to kill the cell, thereby, shedding many more viruses to contaminate all cells in the surrounding area (Dasaraju & Liu, 1996), including breaching the protection of the olfactory bulb. Once a viral infection is in the olfactory bulb, such infection induces the adaptive immune system. If the adaptive immune system does not succeed, the condition of toxic inflammation can develop and that can kill neurons.

Methods by which pathogens reach the brain

Microbial pathogens might get into the brain in a number of ways, including viral infection of olfactory neurons, eventually traveling to the bulb by axonal transport, by travel within the olfactory ensheathing cells, and via perineural spaces allowing fluids of the lamina propria to mix with the fluid of the subarachnoid space (Dando et al., 2014). The IIS of the olfactory mucosa is the protection from pathogens invading the olfactory bulb, and subsequently inducing disease in adjacent parts of the olfactory brain.

The outward manifestations of a common cold are products of enhanced fluids being delivered to the mucosa by dilation of the blood vessels of the lamina propria thereby "flooding" the tissue with blood plasma, proteins, and immune cells in blood. This adaptive response along with the secretions of the mucosa's cells support greater frequency of beating of cilia and thereby faster and more efficient movement of mucus, which has captured contaminants. The chain of events that lead to vasodilation are (a) sensing contamination, (b) mast cells releasing histamine, and (c) histamine inducing vasodilation

(Jackson & Burson, 1977). The vasodilation provides the necessary increment in fluids, facilitating faster mucociliary clearance of disease-producing contaminants (those in air and any viruses being released from infected but dying epithelial cells). The increment in fluids also increases the flow of proteins and chemicals such as nitric oxide into the fluids to kill biological pathogens. Notice, *the active ingredients in most medicines for the common cold produce the opposite of the adaptive response of vasodilation, which is necessary for the enhanced removal of a surge in infectious material.*

Other drug-induced disruptions of the functions of the respiratory mucosa

Beule (2010) has listed a number of drugs likely to disrupt the respiratory mucosa's ordinary functioning. These potentially disruptive therapeutic drugs include benzodiazepines, tricyclic antidepressants as well as some selective-serotonin-reuptake inhibitors, antipsychotics, and the α1-adrenergic receptor agonist phenylephrine (a drug that induces vasoconstriction and is used as a medicine for the common cold, even in combination of diphenhydramine). Habitually inhaling through the nose or lungs via smoking other drugs, such as heroin, cocaine, or nicotine, produces diseases of the nasal cavity and adjacent tissues (Choi, Krantz, Smith, & Trick, 2015). Thus, many drugs used therapeutically or recreationally can impact the nasal mucosa.

There are serious concerns about the widespread use of drugs with anticholinergic activity. Taking drugs inducing anticholinergic effects, as the desired effect or as a side effect, for prolonged periods continues to disrupt memory and concomitantly induces cognitive decline (Carrière et al., 2009; Risacher et al., 2016; Ruxton, Woodman, & Mangoni, 2015), hence hinders regularly engaging in utilitarian behaviors and increases the risk of accidents, e.g., falls (Campbell et al., 2009). Naples et al. (2015) determined that roughly 24%—48% of adults 65 years old or older in English-speaking nations took one or more drugs with anticholinergic effects.

Because the IIS of the mucosa is the critical, first-line defense against serious disease, any drug-induced disruption of the IIS of the respiratory mucosa increases the risk of diseases of the brain, lung, and sinuses. A well-functioning IIS is all that protects against the development of several serious diseases, *except for behaviors (acts) that, in turn, protect the IIS of the airway's mucosa.*

The degree of risk for serious diseases due to drug-induced disruptions of the functionality of the respiratory mucosa may be small. However, it is highly likely that the risk might dramatically increase with the addition of activities such as smoking tobacco and marijuana products and living in an air-polluted environment while, at the same time, having a sedentary lifestyle, and while taking several drugs that might otherwise be medicines if used alone. Conversely, the risk for serious disease, including LOAD, is reduced when *behaviors are engaged* that have the net effect of avoiding the disruption of the IIS of the mucosa of the airway.

Conclusion

Here, we focused on one potential insult to IIS that we posit repeatedly hinders the capabilities of the IIS to effectively protect the brain from infections and dirt, thereby, increasing the risk of LOAD. The act of taking diphenhydramine or phenylephrine, or drugs producing similar effects, dries up the necessary fluids for effective mucociliary clearance, the essential feature of the IIS's way of protecting the airway from chronic diseases. Perhaps these drugs produce some small relief from some unpleasantness, but their potentially serious consequences need to be considered.

An extensive scientific literature (e.g., De Sutter et al., 2012) supports the conclusion that the widespread use of "medicines" for the common cold and insomnia only temporarily relieves some unpleasantness. At their worst, they disrupt the ability of the IIS to protect the brain and lungs from serious diseases, diseases that are among the top 10 causes of death.

There are cognitions and behaviors to engage to aid and abet the IIS's ability to cure common colds, and, further, they may be more effective in controlling the unpleasantness of a cold than the available medicines. The construal that the symptoms are helping to cure a cold is better than catastrophizing. In brief, catastrophizing about the symptoms of a common cold and abruptly taking easily available medications may be a catastrophic risk because such acts can prevent the IIS from functioning optimally.

Given that LOAD is a slowly developing disease, it is difficult to verify that a given instance of a common cold, or other disease affecting the olfactory mucosa, can be a causal event in the development of LOAD. However, it may be that any event that might often disrupt the functions of the IIS of the sinonasal cavity might eventually lead to a breach in the olfactory mucosa.

Runny noses, coughing, sneezing, spitting, hacking up debris from the throat and swallowing postnasal drip associated with a common cold are adaptive processes that probably should not be stopped. In rare circumstances, these adaptive responses may be prevented from becoming excessive, but ordinarily, they should be valued. Calling these symptoms bad, needing to be stopped, is like calling firemen bad because they are trying to put out a fire. What firemen are doing is controlling the spread of fire and preventing further damage, and that is analogous to the role played by the IIS of the mucosa of the airway. The idea that these adaptive responses are problematic (and should somehow be stopped no matter the cost) does not make coherent sense. These adaptive responses should probably be construed to be signs that the IIS is winning a battle against infections and dirty air. In brief, snot's not bad.

Key facts of mucociliary clearance

- Prevents pathogens from gaining access to the brain and lungs, thereby preventing disease of these organs.
- A major component of the innate immune system.
- Subject to interference by over-the-counter drugs.

Summary points

- There is evidence that reducing the risk of Alzheimer's disease may halt the disease's progression.
- The innate immune system employs multiple mechanisms to combat invading microbes.
- A disrupted innate immune system jeopardizes the olfactory bulb (a hallmark of Alzheimer's disease).
- Over-the-counter medicines for the common cold can hinder the innate immune system.
- Behaviors benefiting the innate immune system may reduce Alzheimer's disease risk.

References

Alzheimer's Association. (2018). 2018 Alzheimer's disease facts and figures. *Alzheimer's & Dementia, 14*(3), 367—429.

Beule, A. G. (2010). *Physiology and pathophysiology of respiratory mucosa of the nose and the paranasal sinuses.* German Medical Science GMS Publishing House. https://doi.org/10.3205/cto000071.

Campbell, N., Boustani, M., Limbil, T., et al. (2009). The cognitive impact of anticholinergics: a clinical review. *Clinical Interventions in Aging, 4*, 225—233.

Carrière, I., Fourrier-Reglat, A., Dartigues, J. F., et al. (2009). Drugs with anticholinergic properties, cognitive decline, and dementia in an elderly general population: the 3-city study. *Archives of Internal Medicine, 169*(14), 1317—1324. https://doi.org/10.1001/archinternmed.2009.229.

Choi, H., Krantz, A., Smith, J., & Trick, W. (2015). Medical diagnoses associated with substance dependence among inpatients at a large urban hospital. *PLoS ONE, 10*(6), e0131324. https://doi.org/10.1371/journal.pone.0131324.

Dando, S. J., Mackay-Sim, A., Norton, R., et al. (2014). Pathogens penetrating the central nervous system: infection pathways and the cellular and molecular mechanisms of invasion. *Clinical Microbiology Reviews, 27*(4), 691—726. https://doi.org/10.1128/CMR.00118-13.

Dasaraju, P. V., & Liu, C. (1996). Infections of the respiratory system. In S. Baron (Ed.), *Medical Microbiology* (4th ed.) Galveston (TX): University of Texas Medical Branch at Galveston. Retrieved from http://www.ncbi.nlm.nih.gov/books/NBK8142/.

Daulatzai, M. A. (2015). Olfactory dysfunction: its early temporal relationship and neural correlates in the pathogenesis of Alzheimer's disease. *Journal of Neural Transmission, 122*(10), 1475—1497. https://doi.org/10.1007/s00702-015-1404-6 (Vienna, Austria: 1996).

De Sutter, A. I., Saraswat, A., & van Driel, M. L. (2015). Antihistamines for the common cold. *Cochrane Database of Systematic Reviews.* https://doi.org/10.1002/14651858.CD009345.pub2.

De Sutter, A. I., van Driel, M. L., Kumar, A. A., Lesslar, O., & Skrt, A. (2012). The Cochrane Collaboration Oral antihistamine-decongestant-analgesic combinations for the common cold. In *Cochrane Database of Systematic Reviews.* Chichester, UK: John Wiley & Sons, Ltd. https://doi.org/10.1002/14651858.CD004976.pub3.

Devanand, D. P., Lee, S., Manly, J., et al. (2015). Olfactory deficits predict cognitive decline and Alzheimer dementia in an urban community. *Neurology*, *84*(2), 182–189. https://doi.org/10.1212/WNL.0000000000001132.

Doty, R. L., Philip, S., Reddy, K., & Kerr, K. L. (2003). Influences of antihypertensive and antihyperlipidemic drugs on the senses of taste and smell: a review. *Journal of Hypertension*, *21*(10), 1805–1813. https://doi.org/10.1097/01.hjh.0000084769.37215.16.

Ethell, D. W. (2014). Disruption of cerebrospinal fluid flow through the olfactory system may contribute to Alzheimer's disease pathogenesis. *J Alzheimers Dis*, *41*(4), 1021–1030. https://doi.org/10.3233/jad-130659.

Franks, K. H., Chuah, M. I., King, A. E., & Vickers, J. C. (2015). Connectivity of pathology: the olfactory system as a model for network-driven mechanisms of Alzheimer's disease pathogenesis. *Frontiers in Aging Neuroscience*, *7*, 234. https://doi.org/10.3389/fnagi.2015.00234.

George, S. N., Garcha, D. S., Mackay, A. J., et al. (2014). Human rhinovirus infection during naturally occurring COPD exacerbations. *The European Respiratory Journal*, *44*(1), 87–96. https://doi.org/10.1183/09031936.00223113.

Hariri, B. M., & Cohen, N. A. (2016). New insights into upper airway innate immunity. *American Journal of Rhinology & Allergy*, *30*(5), 319–323. https://doi.org/10.2500/ajra.2016.30.4360.

Hummel, T., Rissom, K., Reden, J., et al. (2009). Effects of olfactory training in patients with olfactory loss. *The Laryngoscope*, *119*(3), 496–499. https://doi.org/10.1002/lary.20101.

Jackson, R. T., & Burson, J. H. (1977). Effect of inflammatory mediators on nasal mucosa. *Archives of Otolaryngology (Chicago, Ill.: 1960)*, *103*(8), 441–444.

Johnsen, M. (February 2016). *Cough-Cold Report 2016*. Drug Store News.

Keck, T., & Lindemann, J. (2010). Numerical simulation and nasal air-conditioning. *GMS Current Topics in Otorhinolaryngology, Head and Neck Surgery*, *9*, Doc08. https://doi.org/10.3205/cto000072.

Liu, H., & Guthrie, K. M. (2011). Neuronal replacement in the injured olfactory bulb. *Experimental Neurology*, *228*(2), 270–282. https://doi.org/10.1016/j.expneurol.2011.01.021.

Nagra, G., Koh, L., Zakharov, A., Armstrong, D., & Johnston, M. (2006). Quantification of cerebrospinal fluid transport across the cribriform plate into lymphatics in rats. *American Journal of Physiology. Regulatory, Integrative and Comparative Physiology*, *291*(5), R1383–1389. https://doi.org/10.1152/ajpregu.00235.2006.

Naples, J. G., Marcum, Z. A., Perera, S., Health, aging and body composition study., et al. (2015). Concordance between anticholinergic burden scales. *Journal of the American Geriatrics Society*, *63*(10), 2120–2124. https://doi.org/10.1111/jgs.13647.

Nedergaard, M. (2013). Neuroscience. Garbage truck of the brain. *Science (New York, N.Y.)*, *340*(6140), 1529–1530. https://doi.org/10.1126/science.1240514.

Reid, L. D., Avens, F. E., & Walf, A. A. (2017). Cognitive behavioral therapy (CBT) for preventing Alzheimer's disease. *Behavioural Brain Research*, *334*, 163–177. https://doi.org/10.1016/j.bbr.2017.07.024.

Risacher, S. L., McDonald, B. C., Tallman, E. F., Alzheimer's disease neuroimaging initiative., et al. (2016). Association between anticholinergic medication use and cognition, brain metabolism, and brain atrophy in cognitively normal older adults. *JAMA Neurology*, *73*(6), 721–732. https://doi.org/10.1001/jamaneurol.2016.0580.

Ruxton, K., Woodman, R. J., & Mangoni, A. A. (2015). Drugs with anticholinergic effects and cognitive impairment, falls and all-cause mortality in older adults: A systematic review and meta-analysis. *British Journal of Clinical Pharmacology*, *80*(2), 209–220. https://doi.org/10.1111/bcp.12617.

Saido, T. C., & Iwata, N. (2006). Metabolism of amyloid beta peptide and pathogenesis of Alzheimer's disease. Towards presymptomatic diagnosis, prevention and therapy. *Neuroscience Research*, *54*(4), 235–253. https://doi.org/10.1016/j.neures.2005.12.015.

Selkoe, D. J., & Hardy, J. (2016). The amyloid hypothesis of Alzheimer's disease at 25 years. *EMBO Molecular Medicine*, *8*(6), 595–608. https://doi.org/10.15252/emmm.201606210.

Sorokowska, A., Drechsler, E., Karwowski, M., & Hummel, T. (2017). Effects of olfactory training: a meta-analysis. *Rhinology*, *55*(1), 17–26. https://doi.org/10.4193/Rhin16.195.

Tindale, L. C., Leach, S., Spinelli, J. J., & Brooks-Wilson, A. R. (2017). Lipid and Alzheimer's disease genes associated with healthy aging and longevity in healthy oldest-old. *Oncotarget*, *8*(13), 20612–20621. https://doi.org/10.18632/oncotarget.15296.

Vandyke, C. H. (1858). Ethmoid bone from above. Anatomy of the Human Body. *Gray, H. Plate*, 149.

CHAPTER 51

Maximizing cognition in mild cognitive impairment and early stage dementia

Bridget Regan[1], Yvonne Wells[1], Paul O'Halloran[2]

[1]Lincoln Centre for Research on Ageing, Australian Institute for Primary Care & Ageing, School of Nursing and Midwifery, La Trobe University, Melbourne, VIC, Australia; [2]School of Psychology and Public Health, La Trobe University, Melbourne, VIC, Australia

List of abbreviations

AD Alzheimer dementia
CR Cognitive rehabilitation
MCI Mild cognitive impairment
RCT Randomized controlled trial
VD Vascular dementia

Mini-dictionary of terms

Cognitive rehabilitation utilizes a problem-solving approach individually tailored to each person's goals in day-to-day life.
Cognitive strategy training teaching a range of strategies to combat cognitive problems, typically in a group context.
Cognitive training repeated practice of a cognitive activity, typically via computer (e.g., brain training).
Effortful learning techniques that maximize the effort involved in learning new information or activities.
Errorless learning techniques that minimize the errors made when learning new information or activities.
Mild cognitive impairment (MCI) a condition characterized by a cognitive concern, cognitive impairment on psychometric testing, largely intact activities of daily living (ADLs), and not meeting criteria for dementia (Albert et al., 2011).
Rehearsal-based approaches techniques that rely on rehearsal to learn new information.
Single-component training standardized teaching or training one particular technique (e.g., diary use) to facilitate cognitive function.

Background

Although it might be possible to reduce the risk of cognitive decline through a healthy lifestyle, increased exercise, and an active social life, there is no way to avoid cognitive decline or dementia altogether in later life (see Table 51.1). In the absence of curative pharmaceutical options, it is essential that effective psychosocial interventions that support people to live well with cognitive impairment and assist in tackling the challenges

Diagnosis and Management in Dementia
ISBN 978-0-12-815854-8, https://doi.org/10.1016/B978-0-12-815854-8.00051-3

Table 51.1 Features of dementia.

Domain	Issues
Cognitive	Memory loss
	Difficulty communicating or finding words
	Difficulty reasoning or problem-solving
	Difficulty with planning and organizing
	Confusion and disorientation
Psychological	Personality changes
	Depression
	Anxiety
	Inappropriate behavior
	Paranoia, agitation, and hallucinations
Functional	Difficulty handling complex tasks
Muscular	Difficulty with coordination and motor functions

This table provides an overview of the features of dementia that may occur as the disease progresses.
From Mayo Clinic. (1998–2018). *Dementia: Symptoms and causes.* Retrieved from https://www.mayoclinic.org/diseases-conditions/dementia/symptoms-causes/syc-20352013.

that people face in day-to-day life are identified. Interventions delivered at an early stage in the dementia process, if effective, have the potential to increase the duration of independence, decrease symptoms of depression or behavioral difficulties, improve quality of life, and, ultimately, delay institutionalization. Economic modeling has demonstrated that even small improvements in function, or delays in decline, create substantial cost savings (Access Economics, 2004).

MCI and dementia were once seen as conditions that were not amenable for rehabilitation because of their associated cognitive impairment and progressive course. Nevertheless, research has demonstrated that people with MCI and early stage dementia do retain some of the necessary cognitive capabilities to learn new strategies, although extra support may be required (Fernández-Ballesteros, Zamarrón, Tárraga, Moya, & Iñiguez, 2003). Evidence also suggests that some people with MCI and early dementia experience excess disability, where functional disability is greater than would be predicted by the degree of impairment. This can occur because a range of factors, such as an unsupportive environment or personal factors including stigma. Therefore it should be possible to assist an individual with MCI or early dementia to function optimally by teaching new strategies and finding ways to tackle excess disability via positive and supportive environments (Clare et al., 2019).

Given these factors, there has been growing interest in whether people with MCI and early dementia may benefit from cognitive intervention and rehabilitation techniques (Chandler, Parks, Marsiske, Rotblatt, & Smith, 2016).

Different approaches to cognitive interventions

All cognitive interventions aim to impact cognitive function positively and, ideally, to maximize everyday function. They can be contrasted with interventions that focus primarily on behavioral symptoms (e.g., wandering) or emotional states (e.g., depression).

A range of different cognitive approaches have been developed. Specifically, the focus of this chapter is mainly on individualized cognitive rehabilitation (CR). This type of approach shows the most promise in relation to facilitating real everyday outcomes for people with MCI and dementia, whereas the evidence for the impact of cognitive training and standardized programs on everyday life is more equivocal. However, a brief description of the other three common techniques is provided below. The chapter then goes on to describe the CR approach in detail and to outline research on the topic to date. Finally, implications for future research and clinical practice will be discussed.

CR is compared with cognitive training, cognitive strategy training, and single component training in Table 51.2. Cognitive training relies largely on rehearsal. The other

Table 51.2 Different approaches to cognitive interventions.

	Cognitive rehabilitation (CR)	Cognitive training	Cognitive strategy training	Single-component training
Description	Teaching a range of techniques tailored to individual goals in day-to-day life	Use of cognitive exercises to improve cognitive function	Teaching a range of techniques to improve function in day-to-day life	Utilizing a single technique (e.g., diary use) to improve function
Target	Function and participation restriction	Impairment	Function	Function
Format	Individualized	Individual or group	Typically group	Individual or group
Techniques taught/ utilized	All types of cognitive techniques	Mainly rehearsal	All types of cognitive techniques	A single technique, typically either rehearsal or external or internal strategies
Goals	Performance and function in relation to collaboratively set goals	Improved or maintained ability in specific cognitive domains	To learn a new range of strategies that can be used in daily life	To learn to apply a single technique

This table provides an overview and description of the four main approaches to cognitive interventions.

interventions use a variety of techniques, including compensatory strategies (e.g., external aids) and internal strategies (see below). CR is unique in tailoring the intervention to individuals' goals.

Common techniques utilized in cognitive interventions

A range of common techniques utilized in cognitive interventions are outlined in Table 51.3. Rehearsal-based approaches involve repeated exposure to a stimulus. External compensatory aids include the use of memory aids to support the individual to function better in a task. Such aids may include diaries, calendars, or smart phones. Internal compensatory aids are cognitive "tools" used to assist with new learning and organization of information and help to process the information at a deeper level (Hampstead, Gillis, & Stringer, 2014). These can include semantic organization, semantic elaboration, and mental imagery.

A range of environmental and psychosocial approaches can also be utilized to assist an individual to manage their cognitive problems. For instance, finding ways to self-soothe and manage anxiety may assist an individual to stay more alert and better process what is happening.

During learning sessions, instructional strategies can be applied such as using an errorless-learning paradigm and/or strategies that require effortful processing (e.g., trial

Table 51.3 Examples of techniques utilized in cognitive interventions.

Rehearsal-based approaches	External aids	Internal approaches (Mnemonics)	Environmental/ practical approaches	Psychosocial approaches
Repeated exposure	Diary/ Calendar/ Lists/ Notes	Semantic organization	Reduce distractions in environment	Anxiety management
Spaced retrieval/ Vanishing Cues	Smart phones/ Alarms	Semantic elaboration	Plan activities at the best time of day	Ask people to repeat what they have said
Computer training programs/ games	Dosette box	Mental imagery	Establish regular weekly routine	Take someone along to help with recall

This table provides an overview of different techniques used in cognitive interventions.
Adapted from Fig. 2 in Hampstead, B. M., Gillis, M. M., & Stringer, A. Y. (2014). Cognitive rehabilitation of memory for mild cognitive impairment: A metholodological review and model for future research. *Journal of the International Neuropsychological Society, 20*(2), 135−151, and from the MAXCOG handouts; see: Regan, B., & Wells, Y. (2018). Effectiveness of the maximising cognition (MAXCOG) information resource for clients with mild cognitive impairment and their families. *Australasian Journal on Ageing, 37*(1), E29−E32.

and error) (Clare & Jones, 2008). However, Clare and Jones (2018) in their review of the literature found that people with early stage dementia appear to learn equally well regardless of the method of learning that is applied.

Cognitive training

Cognitive training (CT), also known as "brain training" (Bahar-Fuchs, Martyr, Goh, Sabates, & Clare, 2018), aims to provide specific manualized and usually standardized training in a particular domain of cognition, such as speed of information processing, memory, attention or problem-solving. It is geared toward clients who have sufficient cognitive resources to engage in supported practice of tasks either with a therapist or computer. The aim is either to target cognitively impaired domains or to practice relatively intact cognitive skills to support more-impaired cognitive skills. The technique is based on the concept of *neuroplasticity*—the idea that repeated practice in a domain may help to improve or at least maintain performance in that domain.

There is, however, controversy as to the benefits of cognitive training approaches (Owen et al., 2010). Although research has often demonstrated an improvement on cognitive testing in the domain in which the training occurred, this approach may not generalize to improved function in day-to-day life (Bahar-Fuchs et al., 2018).

The evidence for the benefits of cognitive training in dementia remains relatively poor. A Cochrane systematic review found no evidence for significant benefits in early stage dementia on any outcome (Bahar-Fuchs, Claire, & Woods, 2013) and a more recent systematic review also showed no improvements (Hill et al., 2016).

Findings have been more positive for MCI groups. For instance, both Chandler et al. (2016) and Hill et al. (2016) concluded that despite a lack of benefit in everyday activities, cognitive training does improve mood.

Cognitive strategy training

Group programs involving cognitive strategy training have been attempted, particularly for those with relatively mild cognitive difficulties. Typically, a range of more generic strategies are discussed and practiced in a group setting, rather than as a subset of strategies targeted to individual goals. Sessions are not conducted in the home environment but in an external location. Consequently, in a group program there is often limited time to explore and practice individual functional goals and to ensure they will impact everyday life. Nevertheless, group programs offer advantages, including opportunities to make friends and share experiences with other participants (Kinsella et al., 2016).

The extent to which cognitive strategy training results in improvements remains equivocal, with some research suggesting positive outcomes (e.g., Kinsella et al., 2009; Rojas et al., 2013), but other research reporting few or no effects (Kinsella et al., 2016; Troyer, Murphy, Anderson, Moscovitch, & Craik, 2008; Unverzagt et al., 2007). Further, the extent to which cognitive strategy training translates from performance on cognitive tests to everyday function remains unclear.

Single component training

A small group of studies have utilized a single standardized technique or strategy taught to all participants (e.g., Finn & McDonald, 2015; Greenaway, Duncan, & Smith, 2013; Jean et al., 2010). The technique or strategy learned in a single component intervention may be relevant to the individual's day-to-day life, depending on their particular difficulties. So far, all single-component interventions have focused on improving memory in groups of clients with amnestic MCI. For instance, Greenaway et al. (2013) trained participants in the use of a diary, Finn and McDonald (2015) utilized repetition-lag training, and Jean et al. (2010) compared errorless learning and spaced retrieval with effortful learning. Such techniques differ from the tailored approach taken in CR, where a more diverse group of individuals with a range of cognitive difficulties are included and strategies are targeted to individual goals. To date, all these single component type studies demonstrate an improvement in the specific task participants have been trained to do. However, studies vary in the extent to which improvements generalize to day-to-day function, mood, and self-efficacy.

Cognitive rehabilitation

Cognitive rehabilitation (CR) refers to a more individualized approach involving goal setting with clients and often a family member. It is based on a problem-solving approach (Clare, 2017; Wilson, 2002) and uses rehabilitation principles to address the impacts of cognitive impairment. The aim is to enable an individual to function at their best possible level, given the nature and extent of their cognitive impairments. It targets daily functions and engagement in worthwhile and meaningful activities to sustain as much independence as possible. Cognitive rehabilitation utilizes a person-centered approach, in that each person's unique life experience, motivations, values, preferences, skills, and needs are taken into account. It is also holistic, in that relationships and the environment are also considered.

CR involves firstly the setting of realistic personal goals. The process of goal setting itself is a powerful behavioral strategy (Locke & Latham, 2002) and is widely used in rehabilitation interventions in a variety of populations, including brain injury (Rockwood, Joyce, & Stolee, 1997; Trombly, Radomski, Trexel, & Burnett-Smith, 2002) and stroke

(Pan, Chung, & Hsin-Hwei, 2003). A therapist assists the individual to develop SMART goals (i.e., specific, measurable, achievable, and realistic within a defined timescale). Examples of goals used in the MAXCOG intervention include developing a simpler system to manage paperwork or learning all the names of the people at the bowls club (Regan, Wells, Farrow, O'Halloran, & Workman, 2017).

After goals are collaboratively identified and realistic targets selected, the therapist selects from an array of strategies that can be taught to assist an individual to reach their goal (Clare, 2007). The choice of strategies depends on the therapist's assessment of barriers facing an individual undertaking a task. Barriers can include cognitive (e.g., not remembering what to do or struggling to concentrate), emotional (e.g., feeling anxious or fearful), environmental (e.g., being in a place that is not conducive to carrying out the activity), social (e.g., not having someone to undertake the task with), or behavioral (e.g., lacking some of the necessary skills) factors, or a combination of these. Understanding barriers provides a starting point for the problem-solving process to generate strategies that might be useful in overcoming such issues. For instance, if the problem is related to difficulty remembering, a memory aid such as an alarm might be useful. However, if the problem is emotional, the solution might be to find ways of regulating emotions. Once possible solutions are developed, a plan for goal attainment is devised.

How effective is cognitive rehabilitation in dementia?

Several randomized controlled trials (RCTs) have now been conducted using an individualized cognitive rehabilitation approach in clients with mild-to-moderate dementia (see Table 51.4).

A pioneering randomized controlled trial was conducted by Clare et al. (2010) in the United Kingdom. This trial, with 69 participants with mild-to-moderate dementia (MMSE 18 or above), was a single-site, single-blind study that compared an eight-session cognitive rehabilitation intervention with relaxation therapy (with equivalent therapist time) and a no-treatment control. Intervention clients reported improved performance and higher satisfaction with goal achievement, while there was no change in either of the comparison groups. An innovative aspect of this study was their client-centered approach; the client's role in collaborative goal-setting was central, and perception of change was the primary outcome measure.

Since then, several randomized control trials of cognitive rehabilitation have been completed. The GREAT trial involved 426 participants with a diagnosis of dementia (Alzheimer, vascular, or mixed dementia) and was conducted in multiple centers in the United Kingdom by occupational therapists (Clare et al., 2019). Clare and colleagues compared CR with treatment as usual. Participants underwent baseline assessment and goal setting to identify areas of everyday function that could be improved or managed better prior to randomization. All participants had mild-to-moderate cognitive

Table 51.4 RCT intervention studies in MCI and dementia.

Reference/ location	Intervention group	Control group	Intervention	Sessions/ Duration	Follow-up	Outcomes	Effect sizes
Amieva and Dartigues (2013), Ameiva et al. (2016), France	AD (n = 157)	AD reminiscence (n = 172) AD cognitive training (n = 170) AD no treatment (n = 154)	Individualized cognitive rehabilitation	Weekly 90 min sessions for 3 months followed by 90 min maintenance sessions for the next 21 months	3 months 6 months 12 months 18 months 24 months	Lower functional disability and 6 month delay in institutionalizations	Not reported
Clare et al. (2010); UK	AD (n = 20)	AD relaxation (n = 23) AD no treatment (n = 22)	Individualized cognitive rehabilitation	Weekly 60 min session for 8 weeks	Postintervention 6 months	Significant improvement in goal performance and satisfaction	Large-effect sizes CR vs relaxation = 1.18 CR vs NT = 0.91
Clare et al. (2019); UK	AD, VD, and mixed D (n = 208)	AD, VD, and mixed D no treatment (n = 218)	Individualized cognitive rehabilitation	Ten 60 min sessions over 3 months followed by four maintenance sessions over 6 months	3 months 9 months	Significant improvement in goal performance and satisfaction	Large-effect sizes for 3 and 9 months with 0.81 and 0.8, respectively.
Regan et al. (2017); Australia	AD and MCI (n = 25)	AD and MCI no treatment (n = 15)	Individualized cognitive rehabilitation	Weekly 60 min sessions held over 4 weeks	At conclusion	Significant improvement in goal performance and satisfaction	Moderate effect sizes of 0.11 for the first performance goal

This table provides a summary of recent research studies with cognitive rehabilitation interventions, including details about sample size, duration, and outcomes. *AD*, Alzheimer dementia; *MCI*, mild cognitive impairment; *mixed D*, mixed dementia; *VD*, vascular dementia.

impairment (MMSE score 18 or more) and a family member to participate, and were stable on their medications (if prescribed). A similar approach to their initial pilot study was used, in that self-reported goal attainment was the primary outcome measure, and carers also provided independent ratings of goal attainment at both points. Secondary outcomes included participant quality of life, mood, self-efficacy and cognition, and carer stress, health status, and quality of life. Statistically, significant large positive effects were reported for participant-rated goal attainment (at both 3 and 9 months), which were consistent with carers' ratings. However, no significant effects on any of the secondary outcomes were detected. Limitations associated with the study included the absence of any functional measure or any long-term follow-up to assess whether the intervention had an impact on rates of institutionalization.

The ETNA-3 trial is another large-scale trial (n = 653) of individualized cognitive rehabilitation recently conducted in France (Amieva & Dartigues, 2013; Amieva et al., 2016). This trial had four separate arms—individualized cognitive rehabilitation, group-based cognitive training, group-based reminiscence therapy, and usual care—undertaken with individuals with mild-to-moderate Alzheimer disease (MMSE 16–26). Each of the three therapy arms comprised weekly sessions (duration 1.5 h) for 3 months followed by maintenance sessions held every 6 weeks for the following 21 months. The primary outcome was the rate of survival without progression to moderately severe-to-severe dementia at 2 years. Secondary outcomes included cognitive impairment, functional disability, behavioral disturbance, quality of life, and carer burden and resource utilization. None of the therapies impacted on rate of dementia progression. However, individualized cognitive rehabilitation resulted in lower functional disability and a 6-month delay in institutionalization at 2 years.

The ETNA-3 study employed experienced psychologists to carry out the intervention. The first two sessions were devoted to selecting meaningful activities with the person with dementia and their carer. Goals could be amended at any time. Unfortunately, other details of the methodology have not been published. The study did not appear to have used a structured interview schedule to facilitate identification of concerns and goals. Therefore, although the authors indicate that activities to be trained were consistent with personally relevant goals, it is not clear how this was achieved. Also, the "training" approach was not clearly specified. While the psychologist was required to adapt the program depending on the cognitive difficulties of the participant and to use an "errorless learning procedure" when appropriate, it remains unclear whether the approach to CR in the ETNA-3 trial resembled the flexible and multifaceted problem-solving approach targeting everyday tasks used in the GREAT study or if it was more similar to a "cognitive training" approach (i.e., with repeated rehearsal) teaching functional skills, as has been utilized by other researchers (see Thivierge, Jean, & Simard, 2014; Voigt-Radloff et al., 2017).

How effective is cognitive rehabilitation in MCI and mixed populations?

To date, the bulk of studies including MCI participants have involved group programs or single-component interventions. As far as we are aware there are currently no studies of individualized cognitive rehabilitation utilizing only MCI clients.

One study has been conducted utilizing a mixed sample of participants, most with MCI (n = 34), but including a small number with early stage dementia (n = 6) (Regan et al., 2017). Regan and colleagues from Australia conducted the MAXCOG RCT (n = 40), which compared a relatively short four-session CR intervention with treatment as usual. This study was more translational than many of the other studies, as the intervention coopted an existing early intervention team from the local Alzheimer Association. A manual was developed to encourage consistency between counsellors in implementing the approach. A set of information handouts provided easy access to a range of ideas and strategies that could be utilized as part of a face-to-face intervention (Regan & Wells, 2018). The primary outcomes were goal performance and satisfaction. The main finding was that participants in the intervention group reported higher posttest levels of performance and satisfaction. This was the case for the first goal and partially the case for the second goal (where satisfaction but not performance increased for the intervention group). These improvements had moderate effect sizes. Findings suggest that the MAXCOG intervention was effective at assisting clients to reach at least one goal in comparison with the control group. Impacts of the intervention on secondary measures, including mood, quality of life, and functional status, were largely nonsignificant. Further qualitative analysis was conducted on the intervention participants in this study using the method of "most significant change." This research, which analyzed narrative accounts provided by intervention participants, identified some broader outcomes such as participants' improved acceptance of their cognitive difficulties (Regan & Wells, 2017).

Summary of results

A growing body of research suggests that individualized CR is successful at helping individuals with early stage dementia achieve their goals in everyday life. Certainly, most studies involving individualized CR demonstrate improvements in goal attainment. The ETNA-3 study also provides evidence to suggest that individualized CR may also result in key real-world outcomes such as improved functional status and delayed institutionalization. The evidence for the success of individualized CR for individuals with MCI is currently less robust, with only one small study involving a short four-session intervention showing promise to date.

Of note is that, except for the ETNA-3 trial and its impact on functional status, no studies have succeeded in demonstrating any impact on broader outcomes such as mood, quality of life, or carer burden as measured by questionnaires. Reasons for this are unclear. One possibility is that the changes in individuals' abilities to carry out specific activities that occur in response to CR, whilst very important in their own right, simply do not impact on broader appraisals of carer burden or quality of life. Alternatively, it is possible that functional change does result in changes to quality of life or carer burden, but the available measures are not sufficiently sensitive to detect these. Results from qualitative studies that do show evidence of broader impacts for individuals, such as improved confidence and insight into their memory difficulties, support this conclusion (Clare et al., 2019; Regan & Wells, 2017).

Implications for future research

The most obvious gap in research currently is the lack of individualized CR studies with MCI clients. Arguably, the earlier in the disease process that one intervenes, the more likely it is that an individual will retain sufficient cognitive capabilities to take on board new strategies and approaches to solving problems. Therefore, it may be that implementing individualized CR in clients with MCI, as opposed to those with dementia, will result in stronger, longer-lasting improvements in day-to-day function, which in turn may increase the duration of independence and potentially further delay institutionalization. Further longitudinal research is needed to test this hypothesis.

As mentioned earlier, most research in this area has utilized group-based or standardized approaches to clients with MCI that may be less expensive to run than individualized programs, which are time-intensive. Whilst group programs may confer some advantages, such as the capacity to share strategy use, typically there is a lack of time in such programs to develop and practice individualized approaches to solving the unique goals of each individual. Such programs may also be too ambitious, in that attempting to teach a range of strategies to people with memory impairment in a group setting may be too taxing on the memory capabilities of such individuals. A one-on-one targeted approach that minimizes the amount of material to be learned may be more appropriate and effective.

Similarly, whilst there may be benefits to the standardized approaches used in single-component studies with homogeneous groups such as those with amnestic MCI, such approaches may not effectively meet all the individual needs of participants, particularly those in more diverse groups. For instance, teaching diary use may be ineffective if the person is struggling more with their language than memory. Likewise, it may not be helpful to focus on a particular strategy or technique for learning if anxiety is the main factor that needs to be addressed. Research is needed to explore the questions of which approach is more effective at improving everyday outcomes for MCI clients and whether the value add of offering a time-intensive individualized program justifies the cost.

Two large-scale clinical trials of CR in dementia have now been conducted: the UK GREAT study and the French ETNA-3 study. These larger-scale trials add to the body of evidence suggesting that individualized CR interventions, although time-consuming and costly, result in important positive outcomes for people with early stage dementia, including goal attainment and delay of institutionalization. Further research is needed, however, to replicate the ETNA-3 finding that individualized CR may delay institutionalization. A challenge in undertaking this task is insufficient detail on the interventions utilized in the ETNA-3 trial. Researchers need to describe carefully the methods utilized to enable replication and avoid confusion. For instance, it is important to distinguish between a nuanced problem-solving approach in which a range of techniques (e.g., internal, external, rehearsal, environmental, or social) can be utilized and a "cognitive training"—type approach in which repeated rehearsal of a particular functional task is undertaken.

The literature in this area is plagued by the vague and imprecise use of terms. For instance, some authors have utilized the term "cognitive rehabilitation" as an overarching term (e.g., Huckans et al., 2013). In their Cochrane review, Bahar-Fuchs et al. (2013) emphasized the specific meaning of CR and its differences from cognitive training. In this chapter we have emphasized the need to differentiate further between different intervention approaches, such as cognitive strategy training and single component training. Grouping disparate intervention approaches in meta-analyses makes it difficult to compare the impacts of different approaches.

Implications for clinical practice

Research suggests that CR can be administered by a range of allied health professions. Amieva et al. (2016) utilized psychologists in the ETNA-3 study. In contrast, Clare's research group have exclusively utilized occupational therapists to implement CR (Clare et al., 2019), but are hoping to adapt their program so that it can be applied within the UK National Health System. They have undertaken some feasibility testing that suggests that improvements in goal attainment can be achieved, utilizing a more pragmatic approach involving fewer sessions and less-qualified staff under supervision. Certainly, Regan et al. (2017) were able to demonstrate improvements in goal attainment for their mixed sample of participants (including early stage dementia and MCI) utilizing counsellors from the local Alzheimer Association, most of whom had completed only basic studies in psychology.

The necessary duration of a CR intervention is less clear and may vary depending on the issues for the individual involved. The GREAT study implemented a total of 10 sessions with a further four maintenance sessions over 6 months, and the ETNA-3 study comprised a total of 12 sessions with a further 21 maintenance sessions over 21 months. In contrast, Regan et al. (2017) were able to show some benefits with a total of four sessions, albeit with predominantly higher-functioning MCI clients. Indeed, as Clare et al. (2019) have pointed out, the exact duration of the program may vary depending on the individual's needs and stage of dementia progression.

One of the issues with the MAXCOG study (Regan et al., 2017) was that a subset of individuals who initially expressed interest in being part of the research program subsequently struggled to identify any areas of need in which they could formulate goals. Clare et al. (2019) have also acknowledged a subgroup of individuals who did not proceed to randomization for this reason. Therefore, it may be that CR is not appropriate for some individuals, as it requires active engagement. Such individuals may decide later that they do wish to engage. Alternatively, particularly in the case of clients with limited insight, it may be possible to work more directly with the carers to try to manage any difficulties and work on goals.

Conclusion

Individualized CR stands out as a type of cognitive intervention that shows promise in its capacity to assist people with MCI and early dementia to improve their day-to-day function and, potentially, to delay institutionalization. Further research is needed, particularly to confirm its efficacy for people with MCI and to compare it with other approaches, such as cognitive strategy training and single component training.

Key facts of cognitive interventions

- Intervening at an early stage in the dementia process potentially increases the duration of independence, decreases symptoms of depression or behavioral difficulties, improves quality of life, and delays institutionalization.
- Small delays in decline in function for people with dementia create substantial cost savings.
- The four main types of cognitive interventions are cognitive training, cognitive strategy training, single-component training, and cognitive rehabilitation.
- All cognitive interventions involve one or more techniques, including rehearsal, external aids, internal approaches, and environmental and psychosocial approaches.
- Cognitive rehabilitation refers to an individualized approach involving goal setting with clients and often a family member.

Summary points

- Research findings into the efficacy of cognitive training, cognitive strategy training, and single-component training are equivocal. In particular, the extent to which improvements on trained tasks generalize to day-to-day function is uncertain.

- At least five randomized controlled trials have investigated an individualized cognitive rehabilitation approach in early stage dementia. The bulk of these trials demonstrate goal attainment with moderate-to-large effect sizes.
- Most studies have looked at the impact of cognitive rehabilitation for people with dementia; there is a gap in research for people with mild cognitive impairment.
- Future research should focus on the efficacy of cognitive rehabilitation versus other approaches, such as cognitive strategy training.
- Further research is also needed to identify the optimal duration of cognitive rehabilitation for different groups.

References

Access Economics. (2004). *Delaying the onset of Alzheimer's disease: Projections and issues.* Retrieved from: http://www.fightdementia.org.au/common/files/NAT/20040820_Nat_AE_DelayOnsetADProjIssues.pdf.

Albert, M. S., DeKosky, S. T., Dickson, D., Dubois, B., Feldman, H. H., Fox, N. C., et al. (2011). The diagnosis of mild cognitive impairment due to Alzheimer's disease: Recommendations from the National Institute on Aging-Alzheimer's Association workgroups on diagnostic guidelines for Alzheimer's disease. *Alzheimer's and Dementia, 7*(3), 270–279.

Amieva, H., & Dartigues, J.-F. (2013). ETNA3, A clinical randomized study assessing three cognitive-oriented therapies in dementia: Rationale and general design. *Revue Neurologique, 169*(10), 752–756.

Amieva, H., Robert, P. H., Grandoulier, A.-S., Meillon, C., De Rotrou, J., Andrieu, S., et al. (2016). Group and individual cognitive therapies in Alzheimer's disease: The ETNA3 randomized trial. *International Psychogeriatrics, 28*(5), 707–717.

Bahar-Fuchs, A., Claire, L., & Woods, B. (2013). Cognitive rehabilitation and cognitive training for early-stage Alzheimer's disease and vascular dementia. *Cochrane Database of Systematic Reviews.* https://doi.org/10.1002/14651858.CD003260.pub2.

Bahar-Fuchs, A., Martyr, A., Goh, A. M. Y., Sabates, J., & Clare, L. (2018). Cognitive training for people with mild to moderate dementia. *Cochrane Database of Systematic Reviews,* (7).https://doi.org/10.1002/14651858.CD013069.

Chandler, M. J., Parks, A. C., Marsiske, M., Rotblatt, L. J., & Smith, G. E. (2016). Everyday impact of cognitive interventions in mild cognitive impairment: A systematic review and meta-analysis. *Neuropsychology Review, 26*(3), 225–251. https://doi.org/10.1007/s11065-016-9330-4.

Clare, L. (2007). *Neuropsychological rehabilitation and people with dementia.* Psychology Press.

Clare, L. (2017). Rehabilitation for people living with dementia: A practical framework of positive support. *PLoS Medicine, 14*(3). e1002245.

Clare, L., & Jones, R. S. (2008). Errorless learning in the rehabilitation of memory impairment: A critical review. *Neuropsychology Review, 18*(1), 1–23.

Clare, L., Kudlicka, A., Oyebode, J. R., Jones, R. W., Bayer, A., Leroi, I., et al. (2019). Individual goal-oriented cognitive rehabilitation to improve everyday functioning for people with early-stage dementia, A multicentre randomised controlled trial (the GREAT trial). *International Journal of Geriatric Psychiatry, 34*(5), 709–721.

Clare, L., Linden, D. E., Woods, R. T., Whitaker, R., Evans, S. J., Parkinson, C. H., et al. (2010). Goal-oriented cognitive rehabilitation for people with early-stage Alzheimer disease: A single-blind randomized controlled trial of clinical efficacy. *American Journal of Geriatric Psychiatry, 18*(10), 928–939. https://doi.org/10.1097/JGP.0b013e3181d5792a.

Fernández-Ballesteros, R., Zamarrón, M. D., Tárraga, L., Moya, R., & Iñiguez, J. (2003). Cognitive plasticity in healthy, mild cognitive impairment (MCI) subjects and Alzheimer's disease patients: A research project in Spain. *European Psychologist, 8*(3), 148.

Finn, M., & McDonald, S. (2015). Repetition-lag training to improve recollection memory in older people with amnestic mild cognitive impairment. A randomized controlled trial. *Aging, Neuropsychology, and Cognition, 22*(2), 244–258.

Greenaway, M., Duncan, N., & Smith, G. (2013). The memory support system for mild cognitive impairment: Randomized trial of a cognitive rehabilitation intervention. *International Journal of Geriatric Psychiatry, 28*(4), 402–409.

Hampstead, B. M., Gillis, M. M., & Stringer, A. Y. (2014). Cognitive rehabilitation of memory for mild cognitive impairment: A metholodological review and model for future research. *Journal of the International Neuropsychological Society, 20*(2), 135–151.

Hill, N. T., Mowszowski, L., Naismith, S. L., Chadwick, V. L., Valenzuela, M., & Lampit, A. (2016). Computerized cognitive training in older adults with mild cognitive impairment or dementia: A systematic review and meta-analysis. *American Journal of Psychiatry, 174*(4), 329–340.

Huckans, M., Hutson, L., Twamley, E., Jak, A., Kaye, J., & Storzbach, D. (2013). Efficacy of cognitive rehabilitation therapies for mild cognitive impairment (MCI) in older adults: Working toward a theroretical model and evidence-based interventions. *Neuropsychology Review, 23*(1), 63–80.

Jean, L., Simard, M., Wiederkehr, S., Bergeron, M.-È., Turgeon, Y., Hudon, C., et al. (2010). Efficacy of a cognitive training programme for mild cognitive impairment: Results of a randomised controlled study. *Neuropsychological Rehabilitation, 20*(3), 377–405.

Kinsella, G. J., Ames, D., Storey, E., Ong, B., Pike, K. E., Saling, M. M., et al. (2016). Strategies for improving memory: A randomized trial of memory groups for older people, including those with mild cognitive impairment. *Journal of Alzheimer's Disease, 49*(1), 31–43.

Kinsella, G. J., Mullaly, E., Rand, E., Ong, B., Burton, C., Price, S., et al. (2009). Early intervention for mild cognitive impairment: A randomised controlled trial. *Journal of Neurology, Neurosurgery and Psychiatry, 80*(7), 730–736.

Locke, E. A., & Latham, G. P. (2002). Building a practically useful theory of goal setting and task motivation: A 35-year odyssey. *American Psychologist, 57*(9), 705.

Mayo Clinic. (1998–2018). *Dementia: Symptoms and causes.* Retrieved from https://www.mayoclinic.org/diseases-conditions/dementia/symptoms-causes/syc-20352013.

Owen, A. M., Hampshire, A., Grahn, J. A., Stenton, R., Dajani, S., Burns, A. S., et al. (2010). Putting brain training to the test. *Nature, 465*(7299), 775.

Pan, A. W., Chung, L., & Hsin-Hwei, G. (2003). Reliability and validity of the Canadian occupational performance measure for clients with psychiatric disorders in Taiwan. *Occupational Therapy International, 10*(4), 269–277.

Regan, B., & Wells, Y. (2017). 'Most significant change' and the maximising cognition (MAXCOG) intervention: The views of clients, supporters and counsellors. *Australasian Journal on Ageing, 36*(4), 324–326.

Regan, B., & Wells, Y. (2018). Effectiveness of the maximising cognition (MAXCOG) information resource for clients with mild cognitive impairment and their families. *Australasian Journal on Ageing, 37*(1), E29–E32.

Regan, B., Wells, Y., Farrow, M., O'Halloran, P., & Workman, B. (2017). MAXCOG—maximizing cognition: A randomized controlled trial of the efficacy of goal-oriented cognitive rehabilitation for people with mild cognitive impairment and early Alzheimer disease. *American Journal of Geriatric Psychiatry, 25*(3), 258–269.

Rockwood, K., Joyce, B., & Stolee, P. (1997). Use of goal attainment scaling in measuring clinically important change in cognitive rehabilitation patients. *Journal of Clinical Epidemiology, 50*(5), 581–588.

Rojas, G. J., Villar, V., Iturry, M., Harris, P., Serrano, C. M., Herrera, J. A., et al. (2013). Efficacy of a cognitive intervention program in patients with mild cognitive impairment. *International Psychogeriatrics, 25*(5), 825–831.

Thivierge, S., Jean, L., & Simard, M. (2014). A randomized cross-over controlled study on cognitive rehabilitation of instrumental activities of daily living in Alzheimer disease. *The American Journal of Geriatric Psychiatry, 22*(11), 1188–1199.

Trombly, C. A., Radomski, M. V., Trexel, C., & Burnett-Smith, S. E. (2002). Occupational therapy and achievement of self-identified goals by adults with acquired brain injury: Phase II. *American Journal of Occupational Therapy, 56*(5), 489–498.

Troyer, A. K., Murphy, K. J., Anderson, N. D., Moscovitch, M., & Craik, F. I. (2008). Changing everyday memory behaviour in amnestic mild cognitive impairment: A randomised controlled trial. *Neuropsychological Rehabilitation, 18*(1), 65—88.

Unverzagt, F. W., Kasten, L., Johnson, K. E., Rebok, G. W., Marsiske, M., Koepke, K. M., et al. (2007). Effect of memory impairment on training outcomes in ACTIVE. *Journal of the International Neuropsychological Society, 13*(6), 953—960.

Voigt-Radloff, S., de Werd, M. M., Leonhart, R., Boelen, D. H., Rikkert, M. G. O., Fliessbach, K., et al. (2017). Structured relearning of activities of daily living in dementia: The randomized controlled REDALI-DEM trial on errorless learning. *Alzheimer's Research and Therapy, 9*(1), 22.

Wilson, B. A. (2002). Towards a comprehensive model of cognitive rehabilitation. *Neuropsychological Rehabilitation, 12*(2), 97—110.

CHAPTER 52

Exercise, cognitive creativity, and dementia

Emily Frith[1], Paul D. Loprinzi[2]

[1]Physical Activity Epidemiology Laboratory, Exercise & Memory Laboratory, Department of Health, Exercise Science and Recreation Management, The University of Mississippi, University, MS, United States; [2]The University of Mississippi, Physical Activity Epidemiology Laboratory, Exercise & Memory Laboratory, Department of Health, Exercise Science, and Recreation Management, 229 Turner Center, University, MS, United States

Mini-dictionary of terms

Brain—derived neurotrophic factor Integral protein that supports the production and survival of neurons, cognitive function, and synaptic plasticity, which contribute to learning and memory

Convergent thinking Characterizes the ability to generate a single, unifying solution to a problem consisting of multiple, distinct elements

Creativity Broad term encompassing convergent thinking, divergent thinking, and insight creativity as well as any idea, imagination, or production that is both original and of value

Divergent thinking Characterizes the ability to generate multiple ideas from a single prompt or cue

Executive function Higher-order, complex cognitions that largely depend on prefrontal cortex activity and include goal-oriented behaviors, such as attention, planning, inhibition, and flexible thinking

Hippocampus Medial temporal lobe brain structure known to play an integral role in learning and memory function

Insight problem-solving Characterizes the evolution of specialized mental search strategies from problem presentation to solution, often denoted by an "Aha!" moment of enlightenment upon connecting remote concepts to arrive at a correct solution

Population attributable risk (relative to dementia) An estimation of the incidence of Alzheimer's disease or dementia that may be expected if specific corollary risk factors, such as physical inactivity, were eliminated

Prefrontal cortex Frontal lobe brain structure important for the execution of many complex cognitions, such as decision-making and executive function

Unfortunately, dementia prevalence is estimated to affect 10.5% of the approximately 48 million individuals living in the Unites States above the age of 65 (Hudomiet, Hurd, & Rohwedder, 2018). Symptoms of mild cognitive impairment occur intermediately between normal aging and the onset of dementia. Dementia is characterized by a progression of moderate-to-severe, clinically observable impairments in cognitive, emotional, social, and behavioral function with underlying causes attributable to structural and morphological changes in brain tissue and neural architecture (Gustafson, 1996). No medical cure exists for this diagnosis, which tends to confine those suffering to a debilitating prognosis of 20 years or less of remaining life as the disease progresses (Heyn, Abreu, & Ottenbacher, 2004). As individuals begin to experience worsening neurological deficits, their ability to maintain social relationships (APA, 2000), functional independence, and physical and psychological vitality are severely limited. To this end,

Diagnosis and Management in Dementia
ISBN 978-0-12-815854-8, https://doi.org/10.1016/B978-0-12-815854-8.00052-5

it is of critical importance to consider the efficacy of alternative preventative and therapeutic strategies, such as physical activity participation, involvement in cognitively stimulating activities that encourage creative expression, or perhaps a combination of both regular physical activity and creative engagement to reduce the risk of late-life cognitive impairment or to attenuate the severity and rapid progression of physical and psychological complications typified in dementia-related pathologies. This chapter will discuss how dementia may influence creativity and whether exercise and creativity may each exert individual effects on cognitive health and prevention of dementia or attenuate debilitating mental impairments as dementia symptoms worsen over time. We will also explore these isolated effects in the context of biological and theoretical mechanisms of action and will provide recommendations for future empirical work combining physical exercise and creative activities into tailored interventions designed to better comprehend this disease and perhaps counteract the devastating implications a dementia prognosis presents to quality of life and optimal physical and mental functioning across one's life span.

Does dementia influence creativity?

Creativity is an umbrella term for the variegated mental processes necessary for both generating many diverse ideas and solving problems that require detection of a singular, unifying solution. Importantly, creative works must be both original and appropriate to satisfy task demands of the creativity problem or prompt (Sternberg & Lubart, 1999). Complex executive cognitions allow individuals to readily update information, remain focused when confronted with interfering stimuli, and employ working memory for strategic manipulation of problem-associated components (Nusbaum & Silvia, 2011). As creativity involves "meta-cognitive" processes, creative thought is heavily reliant on executive cognitions and prefrontal cortex function. Deficits in creativity may be expected to progress in line with the severity of cognitive decline among those diagnosed with dementia. Interestingly, however, some researchers have posited that too much reliance on executive processes may be a disadvantage to certain types of creative thinking, namely divergent thinking (Barr, Pennycook, Stolz, & Fugelsang, 2015). Divergent thought requires mental search strategies inclusive of seemingly unrelated concepts. Focused attention coupled with reliance on working memory systems may lead to fixation upon only those cues that are relevant and obviously connected to a given prompt (Wiley & Jarosz, 2012) while a piece of the mental puzzle remains obscured. Thus, divergence from conventional ideas is achieved when an individual has the capacity to exert their creative potential as well as the motivation to do so given the influence of multiple factors including the social and physical environment in addition to individual personality. An ability to overcome fixation may be largely dispositional and is perhaps better understood as nonanalytic and non-goal-oriented reasoning that may partially explain why even individuals with lower cognitive capacity for creativity may still be capable of displaying magnificent creative potential (Allen & Thomas, 2011; Sternberg & Lubart, 1996).

Research suggests that specific dementia diagnoses differentially impact creative output, as lateralized frontotemporal dementia (FTD) has been associated with increased talent in artistic pursuits, such as music production, painting, sculpture, and photography (Miller et al., 1998; Miller, Ponton, Benson, Cummings, & Mena, 1996; Palmiero, Di Giacomo, & Passafiume, 2012). For some patients, semantic dementia (SD) symptomology may elicit motivations for individuals to participate in artistic activities despite never having previously engaged in those behaviors (Palmiero, Di Giacomo, & Passafiume, 2012). Mechanistically, this may be explained by a maintenance of visuospatial functionality with a concomitant increase in behavioral compulsions, repetitive fixations, and related changes in personality initiated by frontal damage (Miller et al., 1998; Thomas Ante'rion, Honore-Masson, Dirson, & Laurent, 2002). In addition, frontal lesions have been demonstrated to favorably influence insight creativity, potentially by promoting access to a broader mental search space (Reverberi, Toraldo, D'Agostini, & Skrap, 2005), which would also theoretically subserve creativity tasks requiring original connections between distant concepts and overcome threats of inhibition. Lessened neural control driven by morphological brain changes across the temporal and parietal cortices (Drago et al., 2006) may allow inclusive mental search strategies to override cognitive inhibition. Inclusive thinking may offer a boon to creative ideation among certain dementia subpopulations, as disinhibition coupled with a resultant willingness to search unconventional responses is a cornerstone of creative insight problem-solving (Reverberi et al., 2005).

Semantic variant primary aggressive aphasia, or SD, is a subtype of FTD and typically initiates progressive neural degeneration of anterior temporal areas of the brain. Consequently, deficits in language, verbalization, and memory function evolve over time, and SD patients have been shown to display associated behavioral symptoms, such as mental rigidity and reduced compassion for others (Rosen et al., 2006) as well as insatiable compulsions to write (clinically termed hypergraphia) (Wu et al., 2015). Similar to other FTD variants that contribute to behavioral compulsions to produce art, dementia patients diagnosed with the SD variant have been shown to write excessively despite dysfunction in anterior temporal lobe dysfunction. Case studies have demonstrated that three SD patients with unaffected dorsolateral temporal language centers experienced impulses to write a novel and poems (Wu et al., 2015). It appears that deteriorating function of the medial temporal lobes, amygdala, insula, and parahippocampal gyri coupled with maintenance of dorsolateral temporal cortex function may be a plausible mechanism of explanation for this abnormal behavioral manifestation (Wu et al., 2015). Conversely, among other FTD patients motivated to create artwork, dominant temporal lobe degeneration appears to be coupled with maintenance of dorsolateral frontal and parietal function (Miller et al., 1996), perhaps providing additional support for this paradoxical emergence of intensified functions of the nondominant hemisphere favorably influencing creative output in the face of a devastating disease (Seeley et al., 2008).

Notably, however, a compulsive drive to write or produce art does not appear to be sufficient to yield creativity. When dementia complications compromise frontotemporal regions, visuospatial and organizational skills may be undermined, resulting in artwork that fails to indicate proficient or original creative ability (Khaskie & Storandt, 1995; Rankin et al., 2007; Thomas Ante'rion, 2002). When scientists (i.e., nonprofessional artists) were asked to judge the creativity of FTD patients who had drawn landscapes, foliage, and animal and human figures, the artworks were rated not as original or appropriate but rather disordered, oversimplistic, and juvenile, which theoretically may be a byproduct of dementia patients' reduced capacity for goal-directed planning and abstraction related to prefrontal cortex (PFC) decrements in executive function (Rankin et al., 2007). Another study utilized art specialists as judges asked to rate artwork created by either professional artists or individuals who were not well established or formally trained as artists but had participated in an art-education program (i.e., dementia patients) (Belver & Ullán, 2017). The professional artist judges were blinded to the identity and health status of all participants in the study; nonetheless, over 50% of the art specialists attributed the pictures of four patients to professional artists, while 19.1%–48.4% of specialists tended to rate the remaining 23 inexperienced artworks as such (Belver & Ullán, 2017).

To this end, there appears to be weak evidence for widespread facilitation of artistic talent among individuals suffering from neurological degeneration. Even so, it should be recognized that aging individuals with lower cognitive functioning and/or dementia may be capable of achieving nontrivial creative performance enhancements or motivation to engage in de novo creative pursuits. It appears that in some cases dementia may improve creativity at an individual, relative level. That is, perhaps compared with healthy individuals, the creativity of dementia sufferers would not be noteworthy. Of critical importance, however, is that the exceptional case studies and clinical observations reporting facilitation of artistic expression despite evolving dementia symptomology often fail to assess creativity with valid and reliable measures designed to answer the question of whether dementia may incite latent creative abilities (De Souza et al., 2014). Although related, creative and artistic outputs are not ubiquitously synonymous; thus, more experimental work employing controlled laboratory assessments of creative performance and neuroimaging procedures is warranted (De Souza et al., 2014).

Regarding performance on laboratory assessments of divergent thinking, individuals with dementia tend to experience difficulty generating original ideas (Cruz de Souza et al., 2010; Hart & Wade, 2006). The memory decrements that progress in dementia impair the ability to access previously stored semantic information. As semantic knowledge may be preserved in some individuals, another potential challenge to divergent creativity is impaired cognitive flexibility or deployment of mental search schemas couched within frontal lobe dependent functionality (Palmiero et al., 2012). A reduced capability for retrieving conceptual and contextual features related to a divergent thinking cue, such as thinking of many novel uses for a box (Guilford, 1967), may limit performance on open-ended problems (Palmiero et al., 2012). Furthermore, progressive decline in prefrontal lobe integrity may negatively affect divergent thinking and frontal-dependent cognitive and behavioral functions similarly. Specifically, evidence for poor performance on frontal executive tasks requiring set-shifting, inhibition, abstract thinking, and initiation of appropriate mental search strategies has been reported among patients diagnosed with FTD (Cruz de Souza et al., 2010).

In addition, due to the complex role the PFC plays in human cognition, this brain region is thought to play a pivotal role in directing biological processes crucial to creativity (Carlsson, Wendt, & Risberg, 2000; De Souza et al., 2014; Dietrich, 2004). Abnormal functioning of the PFC introduces a host of risks to effective learning and memory, and by default, creativity. Prefrontal lesions prevent accuracy and precision in environmental perception, combinatory thought processes, such as integrating information to evaluate against problem-solving requirements, and motivational and affectual responses to task stimuli, which are governed by intimate connections between the PFC and limbic system (De Souza, et al., 2014; Fellows, 2013). The PFC also shares neural connections with sensorimotor systems, which allow humans to adapt to changing environments and to physically perform planned actions (Cole et al., 2013).

Exercise as dementia prevention

Biology of the effects of exercise on dementia symptomology

There are various metabolic pathways by which physical activities and exercise may influence biomarkers of Alzheimer's disease and dementia. Alzheimer's disease and vascular dementia are the two most commonly diagnosed subtypes of dementia. Other predominant subtypes include Lewy body dementia, FTD, and SD (previously mentioned) (Brunnström, Gustafson, Passant, & Englund, 2009). Although an explanation of all plausible physiological targets of exercise on dementia pathogenesis is beyond the scope of this chapter, several plausible mechanisms will be discussed to provide an overview of the evidence for exercise to exert an impact on dementia risk and prognosis.

The pathophysiology of dementia may be potentially altered through exercise participation via disruptions in the formation of beta-amyloid peptides, which are a primary constituent of senile plaques and neurofibrillary tangles that form in the brains of dementia patients over time (Kim et al., 2014). Beta-amyloid peptides are produced extracellularly, as well as inside neurons, contributing to structural changes in both synapses and neurons (Capetillo-Zarate, Tampellini, Gracia, & Gouras, 2012) This results in disrupted protein conformation and leads to disease presentation of the aforementioned aggregation of plaques and tangles within the brain. Specifically, a rat model indicated that treadmill running may counteract short-term memory impairments associated with Alzheimer's disease in rats receiving an intravenous beta-amyloid injection. The injection was initially confirmed to shorten dendritic spine length in the dentate gyrus of the principal brain region serving memory storage and formation, the hippocampus. Following 30 min a day of progressive-intensity treadmill running, dendritic spine length was increased, suggesting a combative effect of exercise on the physiological evolution of dementia (Loprinzi, Frith, & Ponce, 2018). In addition, although amyloid injections concomitantly impaired the expression of brain-derived neurotrophic factor (BDNF) in the hippocampus, BDNF proliferation was augmented in this brain structure following daily running (Kim et al., 2014). Beta-amyloid peptides may also bind to acetylcholine receptors, which may then elevate phosphatase release and consequently induce long-term synaptic depression (LTD). As long-term potentiation is a suggested cellular correlate of memory, LTD would reciprocally impair memory. Additionally, beta-amyloid levels can decrease levels of the NR2B subunit of N-methyl-D-aspartate (NMDA), with the NMDA receptor being a critical receptor in memory function (Loprinzi et al., 2018).

While the hippocampus is the primary brain region controlling autobiographical or episodic human memory storage of personally salient events, norepinephrine also plays a critical role in the regulation of learning and emotional memory (Loprinzi, Frith, & Edwards, 2018). Among healthy patients and those with mild cognitive impairments as seen in early-stage dementia, high-intensity exercise has been shown to favorably influence recall accuracy and increase norepinephrine concentration (Segal & Cahill, 2009). This effect was not observed in matched, nonexercise control groups. It remains unclear whether exercise effectuates similar outcomes in more advanced stages of dementia or Alzheimer's disease; nevertheless, research suggests that exercise may be beneficial to neutralize initial memory deficits associated with mild cognitive impairment (Segal & Cahill, 2009).

Although not described in exhaustive detail herein, exercise may influence various pathways in the brain, including counteracting the formation of senile beta-amyloid plaques in the brain, regulating the proliferation of neurotrophic growth factors, improving cerebrovascular functioning, attenuating inflammation in the brain (Nascimento et al., 2014; Souza et al., 2013), and increasing synaptic plasticity (Kim et al., 2014; McAuley, Kramer, & Colombe, 2004), all of which may impact learning and memory.

Preventative and therapeutic considerations

A healthy brain begins to atrophy in the third decade of life, with significant atrophy in frontal, parietal, and temporal cortices over time (Jernigan et al., 2001). The hippocampus is a medial temporal lobe structure and appears to be a primary target for dementia-associated atrophy (Dickerson & Sperling, 2008). Hippocampal dysfunction manifests early in the pathogenesis of dementia and may even precede morphological alterations observable through neuroimaging techniques (Smith et al., 2007). The prefrontal cortex is also a structure that tends to substantially atrophy in dementia, resulting in compromised abilities to reason and engage in high-level cognitive functions including failure to inhibit prepotent responses and related deficits in analogical and associative thinking (Waltz et al., 1999). Therefore, it is important to provide an overview of how exercise may influence brain regions and structures targeted in dementia as well as associated secondary biomarkers of cognitive disorder and impairment. Although not exhaustive, we discuss several relationships between physical activity and cognition within this population and provide recommendations for future research focused on dementia prevention and therapy in the context of exercise interventions.

Dementia is linked with metabolic risk factors for a host of chronic diseases including but certainly not limited to an excess production of proinflammatory cytokines, elevated low-density lipoproteins driving abnormal cholesterol metabolism, and reduced glucose sensitivity. Exercise is proposed to impact these biomarkers similarly for both healthy aging individuals and those clinically diagnosed with degenerative neurological disorders (Nascimento et al., 2014). Interestingly, exercise demonstrated similar effects in reducing inflammation in a sample of healthy and cognitively impaired elders, but the effects of exercise on cognitive performance were significantly pronounced among seniors with a lower baseline level of cognitive function (Nascimento et al., 2014). Relatedly, findings from our laboratory have shown that various durations of acute, moderate-intensity ambulatory exercise benefit prefrontal executive inhibitory capacity and memory-based cognitive performance among healthy, college-aged participants. We speculate that elderly individuals suffering from mild cognitive impairments may experience marked exercise-induced improvements in baseline cognition, as a ceiling effect may exist for the extent of cognitive facilitation that may be derived from acute, ambulatory exercise (Crush & Loprinzi, 2017). Additionally, other research has yielded similar outcomes, demonstrating that the greatest exercise-associated cognitive benefits may be detected within samples exhibiting lower cognitive function at baseline (Sibley & Beilock, 2007).

Reduced BDNF is thought to drive hippocampal atrophy in the brains of Alzheimer's disease patients as well as among cognitively healthy older individuals and is thought to heavily contribute to the prevalence of memory deficits and depression throughout the aging process (Erickson et al., 2010; Yeh et al., 2015). Exercise, however, is documented to mitigate the degree of pathology by augmenting plasma BDNF circulation in human interventions (Erickson, Miller et al., 2012) as well as increase concentrations of BDNF in the neural tissue of mice models (Loprinzi & Frith, 2018). One transgenic mouse model provided evidence for exercise as a plausible inhibitory mechanism against underlying pathogenic dementia phenotypes (Um et al., 2008). Exercise exerted a robust influence on BDNF enhancement and cognitive-behavioral irregularities in mice having a neuron-specific enolase mutation of amyloid precursor protein. This effect was suggested to occur via reductions in amyloid-beta load and an associated proliferation of growth factors crucial to both neural preservation and plasticity of neural structures essential to memory (Kouznetsova et al., 2006; Trejo, Carro, & Torres-Aleman, 2001; Um et al., 2008).

Exercise may mitigate secondary risk factors for early disability. Hypertension is suggested to serve as a potent risk factor for cognitive impairment and dementia. Hypertension has been shown to inhibit healthy blood flow to the temporal and frontal lobes (Fujishima, Ibayashi, Fujii, & Mori, 1995), may contribute to reduced white matter volume (Jennings & Ryan, 1998; Waldstein, 2003), and may also stimulate the production of neurofibrillary tangles in the brain (Sparks et al., 1995). Cross-sectional research suggests that, compared with their sedentary counterparts, meeting physical activity guidelines is associated with higher cognitive functioning among a nationally representative sample of hypertensive older adults (Frith & Loprinzi, 2017). Physical exercise may also be capable of attenuating naturally occurring, age-related reductions of vital central nervous system metabolites (Erickson, Weinstein et al., 2012), such as acetylcholinesterase and N-acetylaspartate, which have been implicated in Alzheimer's diagnoses, and has been shown to contribute to healthy memory function (Fibiger, 1991; Kwo-On-Yuen et al., 1994). Higher physical activity participation is associated with higher neural activation and survival even among individuals genetically susceptible to dementia (Smith et al., 2011).

One meta-analysis identified 30 randomized control trials that investigated the effects of exercise training interventions on the preservation of health status among patients with a clinical diagnosis of Alzheimer's disease or dementia (Heyn et al., 2004). Physical, behavioral, functional, and cognitive parameters were assessed in this synthesis of experimental work. Moderate to large effects were reported for the effects of physical fitness on physical health and performance, with moderate treatment effects determined for physical, behavioral, functional, and cognitive measures when considered together. Physical exercise interventions commonly employed in the selected interventions included aerobic exercise (often walking, strength training, and flexibility programs, all of which produced moderate effects relative to physical performance among patients with mild to severe cognitive impairment) (Heyn et al., 2004).

Both cross-sectional and longitudinal evidence supports the claim that higher levels of physical activity support healthy cognitive function in later life (Smith, 1998; Colcombe et al., 2003; Heyn et al., 2004; Laurin, Verreault, Lindsay, MacPherson, & Rockwood, 2001; Li et al., 1991). In fact, physical inactivity is proposed as the highest population-attributable risk for Alzheimer's disease (Norton, Matthews, Barnes, Yaffe, & Brayne, 2014). It has been estimated that increasing physical activity may prevent up to 3% of all-cause dementia cases with a decrease in physical activity often preceding dementia diagnoses (Kivimäki & Singh-Manoux, 2018). Although this may seem marginal, physical activities and/or exercise may be important preventative and therapeutic strategies for offsetting several of the psychological complications linked to dementia.

Brief recommendations for integrating exercise, creativity, and dementia

In our previous narrative review introducing the putative implications of "memorcise" in the context of sustaining health among those facing Alzheimer's disease or dementia, we described various metabolic pathways by which exercise may serve as a promising therapeutic strategy to ameliorate some of the ill effects on memory and cognition that emerge as these diseases advance. Much research thus far has focused on the effects of exercise in diseased mice models with considerable evidence demonstrating that memory function may be preserved or increase despite a clinically controlled induction of dementia status. This research has identified numerous modulatory mechanisms, specifically including beneficial changes in the strength of synaptic connections reinforced and created via long-term potentiation, and reduced beta-amyloid concentrations, which prevent cell death of neurons essential for particularly frontal and temporal-dependent memory processes. For a more in-depth picture of neuromolecular alterations, please refer to our recent review (Loprinzi et al., 2018).

Research thus far has shown promising evidence for moderate-intensity and high-intensity exercise regimens to influence the greatest degree of cognitive protection and improvements in mental performance (Frith & Loprinzi, 2018; Jensen et al., 2015). Notably, a recent synthesis of the research in this arena cautioned future investigators to consider the potentially differential influences of varying exercise prescriptions on pathological indices of dementia risk as well as general upregulations of neural functionality that are suggested to occur in line with habitual exercise (Jensen et al., 2015). These distinctions may prove useful when developing interventions designed to answer specific research questions concerning prevention or therapeutic utility of exercise for dementia-related outcomes versus the broader application of exercise as a cognitive preservation strategy to supplement healthy aging directives.

No research to our knowledge has examined exercise and cognitive creativity interventions as a dualistic approach to consider for dementia prevention and/or therapy. Although previous work has examined the influences of exercise on creativity and exercise on dementia, as well as evaluating the associations of dementia with select creative outputs, and whether certain higher-order cognitions, including creativity, may be negatively impacted following a dementia diagnosis, no empirical venture has aimed to employ both exercise and creativity measures as tools to effectuate modifications in dementia risk or existing symptomology. To this end, we will next detail relevant research exploring creativity and dementia, providing a framework to reinforce our recommendations for future research needed in this area.

Embodied cognitions, creativity, and dementia

Exercise affects many of the brain functions dementia diagnoses threaten to debilitate via widespread effects on the entirety of human neural infrastructure. Exercise also influences higher-order cognitions, learning, and memory, specifically with morphological and operational alterations to the medial temporal lobe (e.g., specifically the hippocampus within this region), frontal lobe, PFC, parietal lobes, etc. The mental processes that contribute to creative problem-solving and idea generation also rely on synchrony and preservation of these aforementioned brain systems. Thus, incorporating both physical activity and creativity in a unified intervention approach may be a worthwhile endeavor for researchers interested in developing new methods for understanding the modifiable trajectories of dementia prognoses and perhaps preventing its inception. Importantly, individuals suffering from dementia complications may also benefit emotionally from therapeutic engagement in creative activities (Palmiero et al., 2012), with creative pursuits promoting a sense of competence, purpose, and growth in aging populations (Fisher & Specht, 1999). Taken together, we propose that creative movement is a likely avenue of interest when implementing physical activity and creativity programs to influence cognition among those at risk for or suffering from dementia.

Embodied cognition is an emerging field, with previous work suggesting that memory, learning, creativity, and brain activity are facilitated when actions coordinate with thought (Cook, Yip, & Goldin-Meadow, 2010; Cotterill, 2001; Madan & Singhal, 2012). As aforementioned, physical inactivity is anticipated to pose the highest population-attributable risk for Alzheimer's disease (Norton et al., 2014); thus, the body appears to profoundly impact the mind. Relatedly, the mind may equally as profoundly impact the body (Madan & Singhal, 2012). Less research has been conducted in this area, especially among aging individuals, but the preventative and therapeutic prospects for elderly dementia sufferers with increasing cognitive deficits must not be overlooked. For example, the concept of mental time travel, or the argument that mental representations of both past and future events are optimally activated in an embodied state, has been demonstrated in individuals asked to either recall experiences from the past or imagine what their lives would be like in the future. Prospective cognitions were behaviorally correlated to overt forward movements, while retrospective cognitions were related to observations of backward movement (Miles, Nind, & Macrae, 2010). Performance in the video game Tetris, which requires spatial manipulations to appropriately align different configurations of blocks on a screen, was also suggested to be augmented with supplementary epistemic actions or extra preliminary rotations of potential game configurations in order to offload perceptual processing and computation. The more frequent the initial rotations, the better individuals played, which may be indicative of interactive domain actions replicating the often internally deployed visuospatial sketchpad often utilized during working memory processing (Maglio & Kirsh, 1996).

Studies specifically examining embodied creativity have provided evidence for various physical actions to favorably influence creative output. Interaction with problem elements have been shown to facilitate insight problem solving (Vallée-Tourangeau, Steffensen, Vallée-Tourangeau, & Sirota, 2016; Weller, Villejoubert, & Vallée-Tourangeau, 2011), perhaps as physically manipulating objects directly serving the problem would arguably counter the magnitude of cognitive resources necessary for mentally interacting, updating, and retrieving imagined constituents of the problem. To this end, the efficacy of imagery in the mental workspace may be accentuated as sufficient motor activations consolidate sensory and motor processes (Madan & Singhal, 2012) The type of movements employed may also differentially affect cognitive processes, as fluid hand motion (e.g., drawing curved figures) has been associated with higher scores on divergent and convergent thinking assessments with no influence observed with both fluid and rigid motions (e.g., drawing straight lines) on analytic problem-solving (Slepian & Ambady, 2012). Similarly, squeezing a soft ball was related to higher divergent and convergent creative performances, while squeezing a hard ball did not correlate with creativity scores (Kim, 2015). Roaming or freely walking is also suggested to promote divergent thought in comparison with predetermined walking along a fixed route (Kuo & Yeh, 2016; Zhou, Zhang, Hommel, & Zhang, 2017). Taken together, these findings offer preliminary indications for fluid movements to serve as a potential conduit to actualizing creative potential across a variety of tasks, while rigid, constrained movements may not portend equivalent outcomes. Notably, data from these experiments were collected in healthy adult populations and are not without their methodological limitations; however, future research should consider implementing comparative protocols within dementia populations to discern whether embodied cognitions, and particularly embodied creativity, may benefit mental and physical health among these individuals.

Conclusion

This chapter introduced the proposed individual effects of exercise and creativity on cognitive function, particularly as these parameters may relate to the development and progression of dementia. A comprehensive, yet not exhaustive, discussion of several biological and theoretical mechanisms of action for these suggested influential pathways allows us to provide informed recommendations for future empirical work in this arena. We contend that interactive physical activity and creative thinking training programs are a worthwhile endeavor to consider for those facing dementia or cognitive impairments. Tailoring interventions to accommodate cognitive deficits by encouraging embodied cognition may reduce the likelihood of cognitive overload and/or facilitate learning, memory, and creativity. Mental and physical interactions should be assessed in combination by designing rigorous experiments to optimize scientific comprehension of the appropriate physical modality, temporal trajectory of mind-body interactions, and targeted cognitions that embodiment may subserve. Such aims will bring researchers closer to understanding viable courses of action for the treatment, and idealistically eventual prevention, of this insidious disease.

Summary points

- This chapter focuses on the interrelationship between exercise, cognitive creativity, and dementia
- We explore these interrelationships in the context of biological and theoretical contributions
- Research demonstrates that dementia may influence creative behavior and performance
- Dementia may also influence engagement in exercise behavior
- There is some evidence to suggest that exercise participation may influence cognitive creativity as well as reduce the risk of dementia
- We discuss the implications of these interrelationships as well as highlight areas in need of future research

Key facts

- Dementia prevalence is estimated to affect 10.5% of the 48 million individuals living in the Unites States above the age of 65.
- Both cross-sectional and longitudinal evidence supports the claim that higher levels of physical activity support healthy cognitive function in later life.
- Research thus far has shown promising evidence for moderate-intensity and high-intensity exercise regimens to influence the greatest degree of cognitive protection and improvements in mental performance.
- Aging individuals with lower cognitive functioning and/or dementia may be capable of demonstrating moderate-to-high levels of creative performance.
- Embodied cognition and creativity are emerging disciplines, with previous work suggesting that memory, learning, creativity, and brain activity are facilitated when bodily actions coordinate with thought.

References

Allen, A. P., & Thomas, K. E. (2011). A dual process account of creative thinking. *Creativity Research Journal,* 23(109–118).

APA. (2000). *DSM-V: Diagnostic and statistical manual of mental disorders* (Vol. 5). Washington (DC): APA.

Barr, N., Pennycook, G., Stolz, J. A., & Fugelsang, J. A. (2015). Reasoned connections: A dual-process perspective on creative thought. *Thinking and Reasoning, 21*(1), 61–75.

Belver, M. H., & Ullán, A. M. (2017). Artistic creativity and dementia. A study of assessment by experts. *Arte, Individuo Y Sociedad, 29*(3), 127–138.

Brunnström, H., Gustafson, L., Passant, U., & Englund, E. (2009). Prevalence of dementia subtypes: A 30-year retrospective survey of neuropathological reports. *Archives of Gerontology and Geriatrics, 49*(1), 146–149.

Capetillo-Zarate, E., Gracia, L., Tampellini, D., & Gouras, G. K. (2012). Intraneuronal Abeta accumulation, amyloid plaques, and synapse pathology in Alzheimer's disease. *Neurodegenerative Diseases, 10*(1–4), 56–59.

Carlsson, I., Wendt, P. E., & Risberg, J. (2000). On the neurobiology of creativity. Differences in frontal activity between high and low creative subjects. *Neuropsychologia, 38*, 873–885.

Colcombe, S. J., Erickson, K., Raz, N., et al. (2003). Aerobic fitness reduces brain tissue loss in aging humans. *The Journals of Gerontology. Series A, Biological Sciences and Medical Sciences, 58*, 176–180.

Cole, M. W., Reynolds, J. R., Power, J. D., Repovs, G., Anticevic, A., & Braver, T. S. (2013). Multi-task connectivity reveals flexible hubs for adaptive task control. *Nature Neuroscience, 16*(16), 1348–1355.

Cook, S. W., Yip, T. K., & Goldin-Meadow, S. (2010). Gesturing makes memories that last. *Journal of Memory and Language, 63*, 465–475.

Cotterill, R. M. (2001). Cooperation of the basal ganglia, cerebellum, sensory cerebrum and hippocampus: Possible implications for cognition, consciousness, intelligence and creativity. *Progress in Neurobiology, 64*(1), 1–33.

Crush, E. A., & Loprinzi, P. D. (2017). Dose-response effects of exercise duration and recovery on cognitive functioning. *Perceptual and Motor Skills, 124*(6), 1164–1193.

Cruz de Souza, L., Volle., E., Bertoux, M., Czernecki, V., Funkiewiez, A., Allali, G., et al. (2010). Poor creativity in frontotemporal dementia: A window into the neural bases of the creative mind. *Neuropsychologia, 48*, 3733–3742.

De Souza, L. C., Guimarães, H. C., Teixeira, A. L., Caramelli, P., Levy, R., Dubois, B., et al. (2014). Frontal lobe neurology and the creative mind. *Frontiers in Psychology, 5*, 761.

Dickerson, B. C., & Sperling, R. (2008). Functional abnormalities of the medial temporal lobe memory system in mild cognitive impairment and Alzheimer's disease: Insights from functional MRI studies. *Neuropsychologia, 46*(6), 1624–1635.

Dietrich, A. (2004). The cognitive neuroscience of creativity. *Psychonomic Bulletin and Review, 11*, 1011–1026.

Drago, V., Foster, P. S., Trifiletti, D., FitzGerald, D. B., Kluger, B. M., Crucian, G. P., et al. (2006). What's inside the art? The influence of frontotemporal dementia in art production. *Neurology, 67*, 1285–1287.

Erickson, K. I., Prakash, R., Voss, M. W., Chaddock, L., Heo, S., McLaren, M., et al. (2010). Brain-derived neurotrophic factor is associated with age-related decline in hippocampal volume. *Journal of Neuroscience, 30*, 5368–5375.

Erickson, K. I., Weinstein, A., Sutton, B. P., et al. (2012a). Beyond vascularization: Aerobic fitness is associated with N-acetylaspartate and working memory. *Brain and Behavior, 2*(1), 32–41.

Erickson, K., Miller, D., & Roecklein, K. (2012b). The aging hippocampus:interactions between exercise, depression, and BDNF. *The Neuroscientist, 18*, 82–97.

Fellows, L. K. (2013). Decision-making: Executive functions meet motivation. In D. T. Stuss, & R. T. Knight (Eds.), *Principles of frontal lobe function* (Vol. 2, pp. 490–500). New York, NY: Oxford University Press.

Fibiger, H. C. (1991). Cholinergic mechanisms in learning, memory and dementia: A review of recent evidence. *Trends in Neurosciences, 14*(6), 220–223.

Fisher, B. J., & Specht, D. K. (1999). Successful aging and creativity in later life. *Journal of Aging Studies, 13*(4), 457–472.

Frith, E., & Loprinzi, P. D. (2017). Physical activity and cognitive function among older adults with hypertension. *Journal of Hypertension, 35*(6), 1271–1275.

Fujishima, M., Ibayashi, S., Fujii, K., & Mori, S. (1995). Cerebral blood flow and brain function in hypertension. *Hypertension Research, 18*(2), 111–117.

Guilford, J. P. (1967). *The nature of human intelligence*. New York: McGrawHill.

Gustafson, L. (1996). What is dementia? *Acta Neurologica Scandinavica, 94*, 22–24.

Hart, R. P., & Wade, J. (2006). Divergent thinking in Alzheimer's and frontotemporal dementia. *Aging, Neuropsychology, and Cognition, 13*, 281–290.

Heyn, P., Abreu, B. C., & Ottenbacher, K. J. (2004). The effects of exercise training on elderly persons with cognitive impairment and dementia: A meta-analysis. *Archives of Physical Medicine and Rehabilitation, 85*(10), 1694–1704.

Hudomiet, P., Hurd, M. D., & Rohwedder, S. (2018). Dementia prevalence in the United States in 2000 and 2012. Estimates based on a nationally representative study. *Journal of Gerontology: Serie Bibliographique, 73*(1), S10–S19.

Jennings, M. M., Jr., Ryan, C. M., et al. (1998). Cerebral blood flow in hypertensive patients: An initial report of reduced and compensatory blood flow responses during performance of two cognitive tasks. *Hypertension, 31*(6), 1216–1222.

Jensen, C. S., Hasselbalch, S. G., Waldemar, G., & Simonsen, A. H. (2015). Biochemical markers of physical exercise on mild cognitive impairment and dementia: Systematic review and perspectives. *Frontiers in Neurology, (6)*, 187.

Jernigan, T. L., Archibald, S., Fennema-Notestine, C., et al. (2001). Effects of age on tissues and regions of the cerebrum and cerebellum. *Neurobiology of Aging, 22*(4), 581–594.

Khaskie, M., & Storandt, B. (1995). Visuospatial deficit in dementia of the Alzheimer type. *Archives of Neurology, 52*(4), 422–425.

Kim, J. (2015). Physical activity benefits creativity: Squeezing a ball for enhancing creativity. *Creativity Research Journal, 27*(4), 328–333.

Kim, B. K., Shin, M. S., Kim, C. J., Baek, S. B., Ko, Y. C., & Kim, Y. P. (2014). Treadmill exercise improves short-term memory by enhancing neurogenesis in amyloid beta-induced Alzheimer disease rats. *Journal of Exercise Rehabilitation, 10*(1), 2.

Kivimäki, M., & Singh-Manoux, A. (2018). Prevention of dementia by targeting risk factors. *The Lancet, 391*(10130), 1574–1575.

Kouznetsova, E., Klingner, M., Sorger, D., Sabri, O., Grossmann, U., Steinbach, J., et al. (2006). Developmental and amyloid plaque-related changes in cerebral cortical capillaries in transgenic Tg2576 Alzheimer mice. *International Journal of Developmental Neuroscience, 24*, 187–193.

Kuo, C. Y., & Yeh, Y. Y. (2016). Sensorimotor-conceptual integration in free walking enhances divergent thinking for young and older adults. *Frontiers in Psychology, 7*(1580).

Kwo-On-Yuen, P. F., Newmark, R., Budinger, T. F., Kaye, J. A., Ball, M. J., & Jagust, W. J. (1994). Brain N-acetyl-L- aspartic acid in Alzheimer's disease: A proton magnetic resonance spectroscopy study. *Brain Research, 667*(2), 167–174.

Laurin, D., Verreault, R., Lindsay, J., MacPherson, K., & Rockwood, K. (2001). Physical activity and risk of cognitive impairment and dementia in elderly persons. *Archives of Neurology, 58*, 498–504.

Li, G., Shen, Y., Chen, C. H., Zhau, Y. W., Li, S. R., & Lu, M. (1991). A three-year follow-up study of age-related dementia in an urban area of Beijing. *Acta Psychiatrica Scandinavica, 83*, 99–104.

Loprinzi, P. D., & Frith, E. (2018). A brief primer on the mediational role of BDNF in the exercise-memory link. *Clinical Physiology and Functional Imaging, 39*(1), 9–14.

Loprinzi, P. D., Frith, E., & Edwards, M. K. (2018a). Exercise and emotional memory: A systematic review. *Journal of Cognitive Enhancement*, 1–10.

Loprinzi, P. D., Frith, E., & Ponce, P. (2018b). Memorcise and Alzheimer's disease. *The Physician and Sportsmedicine, 46*(2), 145–154.

Madan, C. R., & Singhal, A. (2012). Using actions to enhance memory: Effects of enactment, gestures, and exercise on human memory. *Frontiers in Psychology, 3*(507).

Maglio, P. P., & Kirsh, D. (1996). Epistemic action increases with skill. *Paper Presented at the Proceedings of the Eighteenth Annual Conference of the Cognitive Science Society, 16*, 391–396.

McAuley, E., Kramer, A., & Colombe, S. (2004). Cardiovascular fitness and neurocognitive function in older adults: A brief review. *Brain, Behavior, and Immunity, 18*, 214–220.

Miles, L., Nind, L., & Macrae, C. (2010). Moving through time. *Psychological Science, 21*(2), 222.

Miller, B. L., Cummings, J., Mishkin, F., et al. (1998). Emergence of artistic talent in frontotemporal dementia. *Neurology, 51*, 978–982.

Miller, B. L., Ponton, M., Benson, D. F., Cummings, J. L., & Mena, I. (1996). Enhanced artistic creativity with temporal lobe degeneration [letter]. *Lancet, 348*(9043), 1744–1745.

Nascimento, C. M. C., Pereira, J., de Andrade, L. P., Garuffi, M., Talib, L. L., Forlenza, O. V., et al. (2014). Physical exercise in MCI elderly promotes reduction of pro-inflammatory cytokines and improvements on cognition and BDNF peripheral levels. *Current Alzheimer Research, 11*, 799—805.

Norton, S., Matthews, F., Barnes, D. E., Yaffe, K., & Brayne, C. (2014). Potential for primary prevention of Alzheimer's disease: An analysis of population-based data. *The Lancet Neurology, 13*(8), 788-779.

Nusbaum, E. C., & Silvia, P. J. (2011). Are intelligence and creativity really so different? Fluid intelligence, executive processes, and strategy use in divergent thinking. *Intelligence, 39*, 36—45.

Palmiero, M., Di Giacomo, D., & Passafiume, D. (2012). Creativity and dementia: A review. *Cognitive Processing, 13*(3), 193—209.

Rankin, K. P., Liu, A. A., Howard, S., Slama, H., Hou, C. E., Shuster, K., et al. (2007). A case-controlled study of altered visual art production in Alzheimer's and FTLD. *Cognitive and Behavioral Neurology, 20*, 48—61.

Reverberi, C., Toraldo, A., D'Agostini, S., & Skrap, M. (2005). Better without (lateral) frontal cortex? Insight problems solved by frontal patients. *Brain, 128*, 2882—2890.

Rosen, H. J., Allison, S., Ogar, J. M., Amici, S., Rose, K., Dronkers, N., et al. (2006). Behavioral features in semantic dementia vs other forms of progressive aphasias. *Neurology, 67*(10), 1752—1756.

Seeley, W. W., Matthews, B., Crawford, R. K., Gorno-Tempini, M. L., Foti, D., Mackenzie, I. R., & Miller, B. L. (2008). Unravelling Boléro: Progressive aphasia, transmodal creativity and the right posterior neocortex. *Brain, 131*(Pt 1), 39—49.

Segal, S. K., & Cahill, L. (2009). Endogenous noradrenergic activation and memory for emotional material in men and women. *Psychoneuroendocrinology, 34*, 1263—71.

Sibley, B. A., & Beilock, S. L. (2007). Exercise and working memory: An individual differences investigation. *Journal of Sport and Exercise Psychology, 29*(6), 783—791.

Slepian, M. L., & Ambady, N. (2012). Fluid movement and creativity. *Journal of Experimental Psychology: General, 141*(4), 625.

Smith, A. (1998). Lifelong exercise may help ward off Alzheimer's disease. In *Paper presented at the American Academy of Neurology's Annual Meeting, Minneapolis, MN.*

Smith, C. D., Chebrolu, H., Wekstein, D. R., et al. (2007). Brain structural alterations before mild cognitive impairment. *Neurology, 68*(16), 1268—1273.

Smith, J. C., Nielson, K., Woodard, J. L., et al. (2011). Interactive effects of physical activity and APOE-epsilon4 on BOLD semantic memory activation in healthy elders. *NeuroImage, 54*(1), 635—644.

Souza, L. C., Filho, C., Goes, A. T. R., Fabbro, L., Del, de Gomes, M. G., Savegnago, L., et al. (2013). Neuroprotective effect of physical exercise in a mouse model of Alzheimer's disease induced by β-amyloid1—40 peptide. *Neurotoxicity Research, 24*, 148—163.

Sparks, D. L., Scheff, S., Liu, H., Landers, T. M., Coyne, C. M., & Hunsaker, J. C. (1995). Increased incidence of neurofibrillary tangles (NFT) in non-demented individuals with hypertension. *Journal of Neurological Sciences, 131*(2), 162—169.

Sternberg, R. J., & Lubart, T. I. (1996). Investing in creativity. *American Psychologist, 51*(7), 677.

Sternberg, R. J. L., & Lubart, T. (1999). The concept of creativity: Prospects and paradigms. In R. J. L. Sternberg (Ed.), *Handbook of creativity* (pp. 3—15). Cambridge: Cambridge University Press.

Thomas Anterion, C., Honore-Masson, S., Dirson, S., & Laurent, B. (2002). Lonely cowboy's thoughts. *Neurology, 59*, 1812—1813.

Trejo, J. L., Carro, E., & Torres-Aleman, I. (2001). Circulating insulin-like growth factor I mediates exercise-induced increases in the number of new neurons in the adult hippocampus. *Journal of Neuroscience, 21*, 1628—1634.

Um, H. S., Kang, E. B., Leem, Y. H., Cho, I. H., Yang, C. H., Chae, K. R., et al. (2008). Exercise training acts as a therapeutic strategy for reduction of the pathogenic phenotypes for Alzheimer's disease in an NSE/APPsw-transgenic model. *International Journal of Molecular Medicine, 22*(4), 529—539.

Vallée-Tourangeau, F., Steffensen, S. V., Vallée-Tourangeau, G., & Sirota, M. (2016). Insight with hands and things. *Acta Psychologica, 170*, 195—205.

Waldstein, S. R. (2003). The relation of hypertension to cognitive function. *Current Directions in Psychological Science, 12*(1), 9—12.

Waltz, J. A., Knowlton, B. J., Holyoak, K. J., Boone, K. B., Mishkin, F. S., de Menezes Santos, M., et al. (1999). A system for relational reasoning in human prefrontal cortex. *Psychological Science, 10*(2), 119—125.

Weller, A., Villejoubert, G., & Vallée-Tourangeau, F. (2011). Interactive insight problem solving. *Thinking and Reasoning, 17*(4), 424—439.

Wiley, J., & Jarosz, A. (2012). Working memory capacity, attentional focus, and problem solving. *Current Directions in Psychological Science, 21*, 258—262.

Wu, T. Q., Miller, Z. A., Adhimoolam, B., Zackey, D. D., Khan, B. K., Ketelle, R., et al. (2015). Verbal creativity in semantic variant primary progressive aphasia. *Neurocase, 21*(1), 73—78.

Yeh, S., Li, L., Chuang, Y., Liu, C., Tsai, F., Lee, M., et al. (2015). Effects of music aerobic exercise on depression and brain-derived neurotrophic factor levels in community dwelling women. *BioMed Research International*, 135893.

Zhou, Y., Zhang, Y., Hommel, B., & Zhang, H. (2017). The impact of bodily states on divergent thinking: Evidence for a control-depletion account. *Frontiers in Psychology, 8*, 1546.

CHAPTER 53

Person-centered communication among formal caregivers of persons with dementia

Marie Y. Savundranayagam[1], Pabiththa Kamalraj[2]
[1]School of Health Studies, Western University, London, ON, Canada; [2]Health and Rehabilitation Sciences, Western University, London, ON, Canada

List of abbreviations

CEM communication enhancement model
CPA communication predicament of aging
LTC long-term care
NDB need-driven dementia-compromised behavior
PwD persons with dementia

Mini-dictionary of terms

Formal caregiver trained health care professional or paraprofessional who provides paid care for individuals requiring assistance.
Home care various caregiving services provided in one's own home.
Long-term care home a care facility in which various caregiving services are provided to persons requiring frequent support. These facilities are also commonly referred to as nursing homes, assisted living facilities, convalescent care homes, etc.
Pragmatics the arm of linguistics focused on the way in which language is used in differing contexts.
Semantics the arm of linguistics concerned with language content (i.e., the meanings of words).

Introduction

Persons with dementia (PwD) experience changes in their ability to communicate, which hinders the opportunity for social interaction. However, PwD are capable of communicating when caregivers account for their unique needs. This chapter examines the use of person-centered communication among formal caregivers of PwD, particularly within the long-term care (LTC) context. It begins with an overview of dementia, with emphasis on the consequences of communication impairment for PwD. The chapter shifts to the concept of personhood and the ways in which person-centered care can preserve the personhood of PwD. Next, person-centered communication in dementia care is highlighted, with emphasis from theoretical contributions and an overview of methods

Diagnosis and Management in Dementia
ISBN 978-0-12-815854-8, https://doi.org/10.1016/B978-0-12-815854-8.00053-7

to measure person-centered communication. We end with a discussion of the effects of person-centered communication on PwD and their formal caregivers.

Dementia is a syndrome characterized by progressive changes in memory, cognition, and behavior (American Psychiatric Association [APA], 2013). The diagnostic features of the condition, as outlined by the *Diagnostic and Statistical Manual of Mental Disorders*, fifth edition, involve a significant decline in one or more cognitive domains, including attention, executive functioning, learning and memory, language, perceptual—motor skills, and social cognition (APA, 2013). Cognitive deficits stemming from this decline interfere with independence in everyday activities, do not occur exclusively in the context of delirium, and are not better explained by the presence of another mental disorder, such as schizophrenia or depression. The clinical manifestations of dementia, which may include impaired memory, diminished language functioning, and alterations in personality, emphasize the need for care among PwD (Mitchell & Agnelli, 2015), as the condition ultimately inhibits one's ability to communicate effectively and satisfy essential needs (Edvardsson, Winblad, & Sandman, 2008). A gradual decline in communication capabilities is one of the major characteristics of PwD and can be one of the first noticeable symptoms exhibited (Downs & Collins, 2015).

Consequences of communication impairment on persons with dementia

Communication is an essential aspect of human life. It is used to express feelings, wishes, and needs, and thus plays a significant role in facilitating a sense of security and belonging, maintaining quality of life, and preserving identity (Jootun & McGhee, 2011). PwD, however, experience changes in their ability to communicate (Rousseaux, Seve, Vallet, Pasquier, & Mackowiak-Cordoliani, 2010). Changes in semantic and pragmatic levels of language processing result in communicative deficits that hinder the opportunity for engagement in meaningful convers ations (Ryan, Byrne, Spykerman, Orange, & 2005). Semantic impairments in PwD include word-finding difficulties, paraphasia (unintended utterances), issues with word and sentence comprehension, and an overall loss of verbal fluency, whereas pragmatic problems involve difficulties associated with prosody, logical organization of discourse, the use of gestures, conversational turn-taking, adapting to the knowledge of the social partner, and displaying feedback (Rousseaux et al., 2010). Semantic and pragmatic deficits impede the ability of the social partner to follow the thought processes of PwD and sustain conversation (Ryan et al., 2005).

As dementia progresses, and vocalizations become increasingly difficult for PwD, their attempts at communication may be misconstrued as meaningless and confused by social partners (Acton, Yauk, Hopkins, & Mayhew, 2007). When communication is viewed in this manner, attempts aimed at establishing a social connection with PwD diminish, resulting in social isolation and, ultimately, denial of the basic human need to belong (Acton et al., 2007; Stein-Parbury et al., 2012). When coupled with reduced opportunities for communication, social isolation stemming from dementia-related communication changes results in negative consequences for PwD (Downs & Collins, 2015). PwD can be subjected to interactions that result in depersonalization, through processes such as disempowerment and objectification (Kitwood, 1990).

The communication predicament of aging (CPA) model, depicted as a negative feedback loop, shows that social partners modify their communication based on inaccurate stereotypes rather than on actual needs and deficits (Hummert, Garstka, Ryan, & Bonnesen, 2004; Ryan, Hummert, & Boich, 1995). The CPA model begins with recognition of cues that indicate age-related changes, including those related to physiological, psychological, and sociocultural factors, which are then interpreted by others as indicators of incompetence and dependence. These negative stereotypes lead to speech modifications characterized by overaccommodation (secondary baby talk, oversimplified speech, patronizing communication, elderspeak, or ignoring) and result in reinforcement of age-stereotyped behaviors and constraining conditions that limit the opportunities available for satisfying and effective communication. Constraints stemming from consistent exposure to overaccommodation create adverse outcomes for older adults, including negative effects on psychological well-being and self-esteem.

Barriers related to aging, cognitive impairments, and institutionalization make PwD, particularly those living in LTC homes, vulnerable to communication predicaments (Savundranayagam, Ryan, Anas, & Orange, 2007). Although LTC homes present multiple opportunities for encounters to occur between caregivers and residents, PwD typically experience extremely limited amounts of social interaction with formal caregivers (Ward, Vass, Aggarwal, Garfield, & Cybyk, 2008), due to communicative deficits (Norbergh, Asplund, Rassmussen, Nordahl, & Sandman, 2001). Interactions between PwD and formal caregivers are often brief, are task oriented, contain a series of directives, and are primarily controlled by formal staff (Ward et al., 2008).

Elderspeak is prevalent among formal caregivers who interact with PwD in LTC settings (Williams, 2006; Williams, Herman, Gajewski, & Wilson, 2009; Williams, Kemper, & Hummert, 2003). Elderspeak is characterized by simplistic grammar and vocabulary, slowed speech, shortened sentences, elevated volume and pitch, and inappropriate use of terms of endearment (Kemper & Harden, 1999). Some aspects of elderspeak, such as repetition and elaboration, may help to improve performance in older adults. Other features, like slowed speech, are not helpful, and can cause older adults to report lower levels of communication-related competence (Kemper & Harden, 1999). Elderspeak poses a significant threat to the maintenance of personhood for PwD (Kitwood, 1997; Williams et al., 2009).

Preserving personhood using person-centered care

Personhood, as defined by Kitwood (1997), refers to "a standing or status that is bestowed upon one human being, by others, in the context of relationship and social being.

It implies recognition, respect and trust" (p. 8). The notion of personhood is significant in dementia care because the condition can be viewed through the lens of loss, with a focus on the diminishing cognitive abilities of the individual (Savundranayagam, Sibalija, & Scotchmer, 2016). Normative brain functions, especially memory, are highly valued in society (Post, 2006), given that cognition enables functions such as informed choice, decision-making, rationality, and responsible action (Dewing, 2008).

When formal caregivers perceive the personhood of PwD as diminished, PwD are at risk of being viewed as meaningless, making care provision and the role of caregiving appear similarly insignificant (Edvardsson et al., 2008). Negative attitudes toward PwD are prevalent among many formal caregivers, particularly in the LTC context (Brodaty, Draper, & Low, 2003), where the relationship between formal care staff and PwD is arguably the most critical (Savundranayagam, 2012). PwD are viewed by formal caregivers as anxious, lonely, frightened/vulnerable, unpredictable, and exercising little control over their behavior (Brodaty et al., 2003), and as being more difficult and requiring greater workload than caring for those without the condition (Lee, Hui, Kng, & Auyeung, 2013).

Personhood of PwD may change or be concealed over time, but it is never lost (Edvardsson et al., 2008). The persistence of personhood was evidenced by persons with severe dementia who demonstrated episodes of lucidity (Normann, Asplund, & Norberg, 1998). Episodes of lucidity, however, occurred when PwD acted closely with formal care staff who viewed them as valuable, perceived their behavior as meaningful manifestations of experience, and refrained from the use of excessive demands. This demonstrates that personhood is upheld within the context of a mutually respectful and trusting relationship (Fazio, Pace, Flinner, & Kallmyer, 2018). The focus on preserving personhood among PwD highlights a significant shift from a problem-centered focus to one that is person-centered (Savundranayagam et al., 2007). It seeks to stress the importance of health and well-being, rather than decline.

Person-centered care is a strength-based approach that emphasizes supporting PwD with retaining and utilizing their abilities and skills (Edvardsson et al., 2008). Person-centered care incorporates one's life history, personal preferences, and values and seeks to contribute to the development and maintenance of meaningful, rewarding relationships, characterized by supportive social interaction (Kitwood, 1997). In contrast with the medical model of care, in which processes, schedules, staff, and organizational needs are prioritized, the person-centered approach focuses on developing interpersonal relationships to learn more about the person (Fazio et al., 2018). Person-centered care offers an effective way to restore personhood of PwD (Edvardsson et al., 2008; Ryan et al., 2005).

Although literature regarding person-centered care is abundant, the concept often lacks a single, agreed-upon definition (White, Newton-Curtis, & Lyons, 2008). The seminal work by Kitwood (1997) outlines key features that define person-centered dementia care, including acknowledgment of the individual as a person who is capable of experiencing life and relationships, despite being affected by a progressive disease; incorporation of personal history into care; offering and respecting choices; and attention to retained abilities. Brooker (2004), a mentee of Kitwood, further described person-centered dementia care, in which four major elements integral to the concept, collectively forming the acronym VIPS, are outlined: *valuing* PwD and those who care for them (V), treating PwD as *individuals* with unique needs (I), viewing the world from the *perspective* of PwD (P), and a positive *social* environment in which the PwD is able to experience relative well-being (S). Although there are various conceptualizations of person-centered care, this approach ultimately seeks to demonstrate a shift in focus from traditional biomedical models of care to one that upholds autonomy and personal choice (Fazio et al., 2018).

Person-centered communication in dementia care

Communication is essential in the provision of person-centered care and maintaining personhood (Passalacqua & Harwood, 2012; Savundranayagam, 2014). Within caregiving relationships, formal caregivers are an essential source of physical and psychological support for PwD (Fazio et al., 2018). There is a responsibility for caregivers to promote personhood, well-being, and respect through communication (Savundranayagam, 2014).

Theoretical frameworks that support person-centered approaches to care

Person-centered approaches to communication can aid in enhancing retained communication abilities and compensating for communication-related deficits (Downs & Collins, 2015). Person-centered communication does not need to be cumbersome (Kitwood, 1997). Communication is more likely to be enhanced when care is focused on the PwD instead of the disease (Ryan, Meredith, Maclean, & Orange, 1995). The communication enhancement model (CEM) is a useful framework by which health care providers can maximize positive and appropriate interactions with older adults (Ryan, Meredith, et al., 1995). In contrast to the CPA model (Hummert et al., 2004; Ryan, Hummert et al., 1995), this model is presented as a positive feedback loop, based on three primary areas: (1) appropriate speech modification that is based on the actual abilities of older adults, (2) supportive physical environment, and (3) positive social environment in which caregivers demonstrate genuine interest, respect, and sensitivity. The model posits that appropriate changes made to these areas will reinforce a sense of autonomy and competence among older adults. Personhood complements the model, with its emphasis on positive social environment, life history, and existing abilities.

The need-driven dementia-compromised behavior (NDB) model also takes a person-centered approach, with its focus on exploring the unique source of unmet needs in PwD (Algase et al., 1996). The NDB model was developed in response to the growing concern associated with the "disruptive" behaviors of dementia. This model views behaviors such as aggression and resisting care as responses to unmet needs stemming from background and proximal factors. Background factors include relatively stable cognitive, neurological, health status, and psychosocial factors. Proximal factors are the changing aspects of the immediate social and physical environment and the dynamic states and needs associated with PwD. Communication by formal caregivers serves as a proximal factor because it reflects the social environment. Unmet needs stemming from ineffective communication by formal caregivers can result in responsive or need-driven behaviors by PwD (Acton et al., 2007; Medvene & Lann-Wolcott, 2010; Williams et al., 2009). Need-driven behaviors tend to occur when formal caregivers use elderspeak over neutral communication (Williams et al., 2009). Communication barriers, therefore, have negative effects on the quality of life of PwD, the quality of care provided, and the relationship between formal caregivers and PwD (Eggenberger, Heimerl, & Bennett, 2013). Consequently, it is imperative that formal caregivers possess appropriate and suitable communication skills to interact effectively with PwD (Savundranayagam et al., 2016).

Methods for measuring person-centered communication

Person-centered communication has been measured in the following ways: (1) Kitwood's communication-focused indicators of person-centered care, (2) language-based communication strategies that support person-centered care, (3) reduction in the use of elderspeak, (4) person-centered topic analysis, and (5) emotional tone.

Kitwood's communication-focused indicators of person-centered care

Four indicators of person-centered care, as outlined by Kitwood (1997), are communication focused: recognition, negotiation, facilitation, and validation (Ryan et al., 2005; Savundranayagam, 2014; Savundranayagam & Moore-Nielsen, 2015). Recognition occurs when PwD are acknowledged as unique persons by name, profiles, or accomplishments. It includes the use of nonverbal behaviors, such as eye contact. Negotiation involves inquiring about one's needs and preferences, whereas facilitation is concerned with enabling PwD to carry out actions that would not be possible otherwise, through filling in the missing parts of the intended action. Validation involves acknowledging the reality of an individual's feelings and emotions and accepting that their reality can differ from those who are cognitively intact.

Research by Ryan et al. (2005) and Savundranayagam (2014) provides examples of Kitwood's four indicators of person-centered care. Ryan et al. (2005) focused solely on conversations between experienced communicators and one LTC resident with dementia. Moreover, the purpose of each recorded encounter in their study was to have a conversation. In contrast, most interactions in LTC homes tend to involve tasks. This is especially true of conversations between LTC residents and their formal caregivers. Savundranayagam (2014) examined the occurrences of person-centered communication during routine care interactions. Over a third of recorded staff utterances used one of Kitwood's four indicators. This was the first study to demonstrate that person-centered communication was possible during routine care by formal caregivers who did not receive training in person-centered communication.

Language-based strategies that support person-centered care

Another approach to measuring person-centered communication is derived from evidence on language strategies that promote comprehension and production in PwD. Although language strategies may be effective in promoting comprehension and production, they may not necessarily be person centered. Thus, Savundranayagam and Moore-Nielsen (2015) examined the relationship between language-based and person-centered strategies that support communication in PwD. They coded conversations between formal caregivers and PwD living in an LTC home. Transcripts were coded separately using Kitwood's four indicators of person-centered communication and using the current knowledge on effective language-based strategies. Examples of effective language-based strategies included open-ended questions (Tappen, Williams-Burgess, Edelstein, Touhy, & Fishman, 1997), verbatim repetition (Small, Kemper, & Lyons, 1997), paraphrasing, confirming understanding through seeking clarification (Watson, Chenery, & Carter, 1999), affirmations (Ramanathan, 1997), prompting repetition (Savundranayagam & Orange, 2014), and using politeness to address resistive behaviors (Medvene & Lann-Wolcott, 2010).

Analyses of the overlap between the two sets of coded transcripts revealed specific language-based strategies that support person-centered communication. For example, recognition was used by formal caregivers in dementia care through the use of greetings, affirmations, questions (yes/no and open-ended), and rephrasing (Savundranayagam & Moore-Nielsen, 2015). The use of the individual's name is viewed as recognition (Ryan et al., 2005), and this occurs with the use of such strategies as greetings, affirmations, and rephrasing (Savundranayagam & Moore-Nielsen, 2015). Furthermore, the frequent use of open-ended questions overlapped with recognition; open-ended questions were used to inquire about and acknowledge the life history of PwD. Relatedly, negotiation was commonly demonstrated by formal caregivers through the use of questions (Ryan et al., 2005), predominantly yes/no style questions, and rephrasing (Savundranayagam & Moore-Nielsen, 2015). Similarly, facilitation involved the use of questions (Ryan et al., 2005), both yes/no and open-ended (Savundranayagam & Moore-Nielsen, 2015), and included, albeit to a lesser extent, the use of affirmations and rephrasing. Validation was observed through the use of affirmations, that is, statements that demonstrate agreement or acknowledge emotion, which are frequently utilized alongside requests or instructions (Ramanathan, 1997; Santo Pietro & Ostuni, 2003). Taken together, these findings show that specific language-based strategies can facilitate person-centered communication.

Reduction of elderspeak

One way to assess whether a conversation is person centered is to examine the absence of person-centered communication, namely elderspeak (Williams et al., 2018). Elderspeak has been analyzed using computer-assisted behavioral analysis (Williams, 2009; Williams, Perkhounkova, Herman, & Bossen, 2016) and psycholinguistic analysis (Williams, 2006; Williams et al., 2003). Behavioral analysis calculates the percentage of *time* that a formal caregiver used elderspeak in a video-recorded sample. Elderspeak was operationalized using verbal definitions (e.g., diminutives and collective pronoun substitutions), nonverbal prosody (e.g., high pitch, exaggerated intonation), and nonverbal actions (e.g., hands on hips, looking away) (Williams et al., 2009). In contrast, psycholinguistic analysis of elderspeak focuses on the *number* of diminutives and collective pronoun substitutions. Although it is time consuming and costly, the computer-assisted behavioral analysis of elderspeak captures verbal and nonverbal features of communication and provides a complete view of communicative strategies and interactional context. Video observation and computer-assisted analysis aid in facilitating a more dynamic and accurate analysis and further permit measurement of the duration of communication behaviors.

Person-centered topic analysis

Content analysis of transcribed conversations between formal caregivers and LTC residents has been used to assesses whether topics are task oriented, person centered, or superficial (Williams et al., 2018; Williams, Ilten, & Bower, 2005). Task-related interactions focus on routine care for PwD, whereas person-centered communication involves topics that are tailored to the individual's past, family, or overall life. Superficial encounters focus on general topics, such as the weather, and do not relate to the individual with dementia.

Emotional tone

Emotional tone analysis gauges the emotional components of communication for tones deemed as controlling and person-centered. The Emotional Tone Rating Scale (Williams, Boyle, Herman, Coleman, & Hummert, 2012) has been used to measure emotional tone of audio-recordings (Williams et al., 2018). Using audio-recordings alleviates concerns regarding privacy and confidentiality of research participants. The scale comprises eight items that identify both controlling and person-centered tones.

Descriptors of the control dimension include directive, bossy, domineering, and controlling, whereas items in the person-centered dimension involve respectful, supportive, polite, and caring. Each descriptor is assessed on a 1 to 5 scale, with higher scores signaling a stronger message. Emotional tone analysis examines both verbal and certain nonverbal aspects of contextually diverse communication, including voice pitch and volume.

Effects of person-centered communication on formal caregivers and persons with dementia

Both the CEM and the NDB model posit that modifying communication to address unique life history and unmet needs will result in more satisfying interactions between PwD and their caregivers and in fewer need-driven behaviors (Algase et al., 1996; Ryan, Hummert et al.,1995; Ryan, Meredith et al.,1995). In support of these two models, there is a growing literature on the effects of person-centered communication for both PwD and their formal caregivers. Formal caregivers who use person-centered communication strategies in their interactions with PwD were perceived as being more affirming, respectful, helpful, competent, and satisfied with interactions than those who used directive language (Savundranayagam et al., 2007). Person-centered communication can aid staff in feeling more content with their encounters with PwD (Savundranayagam et al., 2007).

Person-centered communication by formal caregivers results in positive outcomes for PwD, including cooperation, partaking in conversion, self-disclosure, politeness, and asking for clarification (Savundranayagam et al., 2016). This highlights the ability of PwD to engage with others and demonstrates that person-centered communication promotes reciprocity between PwD and formal caregivers (Savundranayagam et al., 2016). PwD can increase their contributions to communicative encounters when the social partner uses person-centered communication (Ryan et al., 2005). In such situations, the PwD takes control of the interaction and the formal caregiver aids in affirming the encounter (Kitwood, 1997; Ryan et al., 2005). The processes of *creation* and *giving* were identified by Kitwood (1997) as reflecting this form of interaction (Ryan et al., 2005). Creation is marked by the PwD contributing something distinct to the communicative encounter (Kitwood, 1997). Giving involves the expression of affection or concern or offering to provide assistance (Kitwood, 1997). Conversely, missed opportunities for person-centered communication increase the potential for negative reactions by PwD, including resistiveness to care and distress (Savundranayagam et al., 2016; Williams, 2006; Williams & Herman, 2011; Williams et al., 2003, 2009). There are two types of missed opportunities for person-centered communication: alternatives and omissions (Savundranayagam, 2014). Alternatives are utterances utilized by formal caregivers that could have been otherwise stated in a person-centered way. Omissions are instances in which formal caregivers could have made a person-centered utterance but said nothing. Residents with dementia were more likely to respond to missed opportunities by formal caregivers with expressions of distress or resisting care.

Person-centered interventions with a communication component have demonstrated benefits to PwD and their formal caregivers. For example, there were significant reductions in agitation, discomfort, and aggression among PwD when they were showered and bathed by formal caregivers who used person-centered approaches, compared with a control group (Sloane et al., 2004). Their findings support the NDB model, which states that adjusting the social environment to fit the unmet needs of PwD will result in fewer need-driven behaviors. That same intervention showed greater use of verbal support and a more positive work experience among formal caregivers (Hoeffer et al. 2006), lending support to the CEM's prediction that enhanced communication influences the relationship between PwD and their caregivers. Similarly, the VIPS communication skills intervention showed positive results for formal caregivers (Passalacqua & Harwood, 2012). It was based on the four elements of person-centered dementia care, as outlined by Brooker (2004): *valuing* people, *individualized* care, *personal* perspectives, and *social* environment. Formal caregivers reported increases in the use of gestures, yes/no questions, offering choice, and humor. They also reported reduced depersonalization of residents, increased hope for PwD, and more empathy toward them.

Despite the benefits of person-centered communication in dementia care, formal caregivers engage in ineffective communication behaviors characterized by patronizing and task-oriented speech, which stem from a lack of training in person-centered communication (Ward et al., 2008; Williams et al., 2009). To address this issue, we developed a person-centered communication intervention called Be EPIC (Savundranayagam, Basque, & Johnson, 2018), which was derived from Kitwood's four person-centered communication indicators and existing work on language-based strategies that support person-centered communication (Savundranayagam & Moore-Nielsen, 2015). It has four foci: assessing the environment (E), using person-centered communication (P), focusing on client relationships (I matter, too), and incorporating the client's abilities, life history, and preferences during routine care (C). Be EPIC is interactive and includes practice with simulated and actual PwD. Our preliminary findings show that formal caregivers perceive the training to be relevant and increased their confidence in communicating with their clients with dementia. Equally important are lessons learned about structural barriers that disempower formal caregivers from implementing their newly learned knowledge and skills. For example, formal caregivers received little knowledge on the social history of their clients with dementia and had heavy workloads stemming from a shortage of formal caregivers. The workload can make it challenging for care staff to avoid reverting to the use of inadequate and inappropriate communication behaviors (Savundranayagam et al., 2007; Stanyon, Griffiths, Thomas, & Gordon, 2016). Thus, the success of person-centered care approaches is contingent on leadership and managerial support

(Chenoweth et al., 2009; Stein-Parbury et al., 2012). Introducing person-centered care into dementia care settings requires the support of leaders and managers to alter care practices and systems.

Conclusion

The literature on person-centered communication has made great strides, especially in its theoretical grounding and measurement approaches. These advances in theory and measurement offer fruitful opportunities for developing and assessing the impact of person-centered communication interventions on PwD and their formal caregivers. The research discussed in this chapter has focused on the LTC home context. There is very little research on person-centered communication approaches in home care (Riachi, 2017). Keeping PwD at home is a global priority (Wimo & Prince, 2010). Thus, future research should focus on the communication experiences of both PwD and their formal caregivers within a home care context. Formal caregivers working in home care are arguably in more need of training in person-centered communication compared with their LTC counterparts because they often work alone. Given the profound effects of person-centered communication on PwD and their formal caregivers, it is imperative that this significant gap in literature is addressed further.

Key facts related to dementia caregiving

- Informal caregivers, defined as family members or friends who provide unpaid care, deliver a substantial amount of dementia care.
- There will be a decrease in the future availability of informal caregivers, due to decreasing fertility and an increasing number of women in the workforce.
- This decreasing availability of informal caregivers and an increased rate of dementia are expected to contribute to the need for formal caregivers.
- Examples of formal caregivers include nurses, personal support workers, and home health aides.
- Front-line formal caregivers provide assistance with the most intimate care, including activities of daily living, such as bathing, and instrumental activities of daily living, such as meal preparation.

Summary points

- PwD living in LTC homes experience limited amounts of social interaction due to barriers related to aging, their cognitive and communicative impairments, and their social context.

- Formal caregivers use ineffective communication characterized by patronizing and task-oriented speech.
- Person-centered communication involves recognizing the individual as unique, validating feelings and emotions, facilitating actions and behaviors, and negotiating needs and preferences.
- Four methods were identified for measuring person-centered communication: Kitwood's communication focused indicators of person-centered care, language-based strategies that support person-centered care, reduction of patronizing communication, person-centered topic analysis, and emotional tone.

References

Acton, G. J., Yauk, S., Hopkins, B. A., & Mayhew, P. A. (2007). Increasing social communication in persons with dementia. *Research and Theory for Nursing Practice, 21*(1), 32–44.

Algase, D. L., Beck, C., Kolanowski, A., Whall, A., Berent, S., Richards, K., et al. (1996). Need-driven dementia-compromised behavior: An alternative view of disruptive behavior. *American Journal of Alzheimer's Disease, 11*(6), 10–19.

American Psychiatric Association. (2013). *Diagnostic and statistical manual of mental disorders* (5th ed.).

Brodaty, H., Draper, B., & Low, L. (2003). Nursing home staff attitudes towards residents with dementia: Strain and satisfaction with work. *Journal of Advanced Nursing, 44*(6), 583–590.

Brooker, D. (2004). What is person-centred care in dementia? *Reviews in Clinical Gerontology, 13*(3), 215–222.

Chenoweth, L., King, M. T., Jeon, Y. H., Brodaty, H., Stein-Parbury, J., Norman, R., et al. (2009). Caring for aged dementia care resident study (CADRES) of person-centred care, dementia-care mapping, and usual care in dementia: A cluster-randomised trial. *The Lancet Neurology, 8*(4), 317–325.

Dewing, J. (2008). Personhood and dementia: Revisiting Tom Kitwood's ideas. *International Journal of Older People Nursing, 3*(1), 3–13.

Downs, M., & Collins, L. (2015). Person-centred communication in dementia care. *Nursing Standard, 30*(11), 37–41.

Edvardsson, D., Winblad, B., & Sandman, P. O. (2008). Person-centred care of people with severe Alzheimer's disease: Current status and ways forward. *The Lancet Neurology, 7*(4), 362–367.

Eggenberger, E., Heimerl, K., & Bennett, M. I. (2013). Communication skills training in dementia care: A systematic review of effectiveness, training content, and didactic methods in different care settings. *International Psychogeriatrics, 25*(3), 345–358.

Fazio, S., Pace, D., Flinner, J., & Kallmyer, B. (2018). The fundamentals of person-centered care for individuals with dementia. *The Gerontologist, 58*(S1), S10–S19.

Hoeffer, B., Talerico, K. A., Rasin, J., Mitchell, C. M., Stewart, B. J., McKenzie, D., et al. (2006). Assisting cognitively impaired nursing home residents with bathing: Effects of two interventions on caregiving. *The Gerontologist, 46*(4), 524–532.

Hummert, M. L., Garstka, T. A., Ryan, E. B., & Bonnensen, J. L. (2004). The role of age stereotypes in interpersonal communication. In J. F. Nussbaum, & J. Coupland (Eds.), *Handbook of communication and aging research* (pp. 91–114). Mahwah, NJ: Lawrence Erlbaum Associates Inc.

Jootun, D., & McGhee, G. (2011). Effective communication with people who have dementia. *Nursing Standard, 25*(25), 40–46.

Kemper, S., & Harden, T. (1999). Disentangling what is beneficial about elderspeak from what is not. *Psychology and Aging, 14*(4), 656–670.

Kitwood, T. (1990). The dialectics of dementia: With particular reference to Alzheimer's Disease. *Ageing and Society, 10*(02), 177–196.

Kitwood, T. (1997). *Dementia reconsidered: The person comes first*. Berkshire, UK: Open University Press.

Lee, J., Hui, E., Kng, C., & Auyeung, T. W. (2013). Attitudes of long-term care staff toward dementia and their related factors. *International Psychogeriatrics, 25*(1), 140–147.

Medvene, L. J., & Lann-Wolcott, H. (2010). An exploratory study of nurse aides' communication behaviors: Giving "positive regard" as a strategy. *International Journal of Older People Nursing, 5*(1), 41–50.

Mitchell, G., & Agnelli, J. (2015). Person-centred care for people with dementia: Kitwood reconsidered. *Nursing Standard, 30*(7), 46–50.

Norbergh, K. G., Asplund, K., Rassmussen, B. H., Nordahl, G., & Sandman, P. O. (2001). How patients with dementia spend their time in a psycho-geriatric unit. *Scandinavian Journal of Caring Sciences, 15*(3), 215–221.

Normann, H. K., Asplund, K., & Norberg, A. (1998). Episodes of lucidity in people with severe dementia as narrated by formal carers. *Journal of Advanced Nursing, 28*(6), 1295–1300.

Passalacqua, S. A., & Harwood, J. (2012). VIPS communication skills training for paraprofessional dementia caregivers: An intervention to increase person-centered dementia care. *Clinical Gerontologist, 35*(5), 425–445.

Post, S. G. (2006). Respectare: Moral respect for the lives of the deeply forgetful. In J. C. Hughes, S. J. Louw, & S. R. Sabat (Eds.), *Dementia: Mind meaning and the person* (pp. 223–234). Oxford, UK: Oxford University Press.

Ramanathan, V. (1997). *Alzheimer discourse: Some sociolinguistic dimensions.* Mahwah, NJ: Lawrence Erlbaum Associates Inc.

Riachi, R. (2017). Person-centred communication in dementia care: A qualitative study of the use of the SPECAL® method by care workers in the UK. *Journal of Social Work Practice, 32*(3), 303–321.

Rousseaux, M., Sève, A., Vallet, M., Pasquier, F., & Mackowiak-Cordoliani, M. A. (2010). An analysis of communication in conversation in patients with dementia. *Neuropsychologia, 48*(13), 3884–3890.

Ryan, E. B., Byrne, K., Spykerman, H., & Orange, J. B. (2005). Evidencing Kitwood's personhood strategies conversation as care in dementia. In B. H. Davis (Ed.), *Alzheimer talk, text and context: Enhancing communication* (pp. 18–36). New York: Palgrave Macmillan.

Ryan, E. B., Hummert, M. L., & Boich, L. H. (1995). Communication predicaments of aging: Patronizing behavior toward older adults. *Journal of Language and Social Psychology, 14*(1–2), 144–166.

Ryan, E. B., Meredith, S. D., Maclean, M. J., & Orange, J. B. (1995). Changing the way we talk with elders: Promoting health using the communication enhancement model. *The International Journal of Aging and Human Development, 41*(2), 89–107.

Santo Pietro, M. J., & Ostuni, E. (2003). *Successful communication with persons with Alzheimer's disease: An inservice manual.* St. Louis, MO: Butterworth Heinemann.

Savundranayagam, M. Y. (2012). Person-centered care: Measurement, implementation, and outcomes. *Clinical Gerontologist, 35*(5), 357–359.

Savundranayagam, M. Y. (2014). Missed opportunities for person-centered communication: Implications for staff-resident interactions in long-term care. *International Psychogeriatrics, 26*(4), 645–654.

Savundranayagam, M. Y., Basque, S., & Johnson, K. (2018). *Feasibility of Be EPIC: A dementia-focused person-centered communication intervention for home care workers.* Boston, MA: To be presented at the annual meeting of the Gerontological Society of America.

Savundranayagam, M. Y., & Moore-Nielsen, K. (2015). Language-based communication strategies that support person-centered communication with persons with dementia. *International Psychogeriatrics, 27*(10), 1707–1718.

Savundranayagam, M. Y., & Orange, J. B. (2014). Matched and mismatched appraisals of the effectiveness of communication strategies by family caregivers of persons with Alzheimer's disease. *International Journal of Language and Communication Disorders, 49*(1), 49–59.

Savundranayagam, M. Y., Ryan, E. B., Anas, A. P., & Orange, J. B. (2007). Communication and dementia: Staff perceptions of conversational strategies. *Clinical Gerontologist, 31*(2), 47–63.

Savundranayagam, M. Y., Sibalija, J., & Scotchmer, E. (2016). Resident reactions to person-centered communication by long-term care staff. *American Journal of Alzheimer's Disease and Other Dementias, 31*(6), 530–537.

Sloane, P. D., Hoeffer, B., Mitchell, M., Mckenzie, D., Barrick, A. L., Rader, J., et al. (2004). Effect of person-centered showering and the towel bath on bathing-associated aggression, agitation, and discomfort in nursing home residents with dementia: A randomized controlled trial. *Journal of the American Geriatrics Society, 52*(11), 1795—1804.

Small, J. A., Kemper, S., & Lyons, K. (1997). Sentence comprehension in Alzheimer's disease: Effects of grammatical complexity, speech rate, and repetition. *Psychology and Aging, 12*(1), 3—11.

Stanyon, M. R., Griffiths, A., Thomas, S. A., & Gordon, A. L. (2016). The facilitators of communication with people with dementia in a care setting: An interview study with healthcare workers. *Age and Ageing, 45*(1), 164—170.

Stein-Parbury, J., Chenoweth, L., Jeon, Y. H., Brodaty, H., Haas, M., & Norman, R. (2012). Implementing person-centered care in residential dementia care. *Clinical Gerontologist, 35*(5), 404—424.

Tappen, R. M., Williams-Burgess, C., Edelstein, J., Touhy, T., & Fishman, S. (1997). Communicating with individuals with Alzheimer's disease: Examination of recommended strategies. *Archives of Psychiatric Nursing, 11*(5), 249—256.

Ward, R., Vass, A. A., Aggarwal, N., Garfield, C., & Cybyk, B. (2008). A different story: Exploring patterns of communication in residential dementia care. *Ageing and Society, 28*(5), 629—651.

Watson, C., Chenery, H., & Carter, M. (1999). An analysis of trouble and repair in the natural conversations of people with dementia of the Alzheimer's type. *Aphasiology, 13*(3), 195—218.

White, D. L., Newton-Curtis, L., & Lyons, K. S. (2008). Development and initial testing of a measure of person-directed care. *The Gerontologist, 48*(Suppl. 1), 114—123.

Williams, K. N. (2006). Improving outcomes of nursing home interactions. *Research in Nursing and Health, 29*(2), 121—133.

Williams, K. N., Boyle, D. K., Herman, R. E., Coleman, C. K., & Hummert, M. L. (2012). Psychometric analysis of the emotional tone rating scale: A measure of person-centered communication. *Clinical Gerontologist, 35*(5), 376—389.

Williams, K. N., & Herman, R. E. (2011). Linking resident behavior to dementia care communication: Effects of emotional tone. *Behavior Therapy, 42*(1), 42—46.

Williams, K. N., Herman, R., Gajweski, B., & Wilson, K. (2009). Elderspeak communication: Impact on dementia care. *American Journal of Alzheimer's Disease and Other Dementias, 24*(1), 11—20.

Williams, K. N., Ilten, T. B., & Bower, H. (2005). Meeting communication needs: Topics of talk in the nursing home. *Journal of Psychosocial Nursing and Mental Health Services, 43*(7), 38—45.

Williams, K., Kemper, S., & Hummert, M. L. (2003). Improving nursing home communication: An intervention to reduce elderspeak. *The Gerontologist, 43*(2), 242—247.

Williams, K. N., Perkhounkova, Y., Herman, R., & Bossen, A. (2016). A communication intervention to reduce resistiveness in dementia care: A cluster randomized controlled trial. *The Gerontologist, 57*(4), 707—718.

Williams, K. N., Perkhounkova, Y., Jao, Y.-L., Bossen, A., Hein, M., Chung, S., et al. (2018). Person-centered communication for nursing home residents with dementia. *Western Journal of Nursing Research, 40*(7), 1012—1031.

Wimo, A., & Prince, M. (2010). World Alzheimer report 2010: The global economic impact of dementia. *Alzheimer's Disease International.*

Index

Note: 'Page numbers followed by "*f*" indicate figures and "*t*" indicate tables.'